Special Volume:
Numerical Methods for Non-Newtonian Fluids
Guest Editors: R. Glowinski and Jinchao Xu

Handbook of Numerical Analysis

General Editor:

P.G. Ciarlet

Laboratoire Jacques-Louis Lions
Université Pierre et Marie Curie
4 Place Jussieu
75005 PARIS, France

and

Department of Mathematics
City University of Hong Kong
Tat Chee Avenue
KOWLOON, Hong Kong

ELSEVIER

AMSTERDAM • BOSTON • HEIDELBERG • LONDON
NEW YORK • OXFORD • PARIS • SAN DIEGO
SAN FRANCISCO • SINGAPORE • SYDNEY • TOKYO
North-Holland is an imprint of Elsevier

N·H

Volume XVI

Special Volume: Numerical Methods for Non-Newtonian Fluids

Guest Editors:

R. Glowinski

Department of Mathematics,
University of Houston,
Houston, TX USA

and

Laboratoire Jacques-Louis Lions,
Université Pierre et Marie Curie,
4 Place Jussieu,
75005 PARIS, France

J. Xu

Department of Mathematics,
Pennsylvania State University,
University Park, PA 16802

ELSEVIER

AMSTERDAM • BOSTON • HEIDELBERG • LONDON
NEW YORK • OXFORD • PARIS • SAN DIEGO
SAN FRANCISCO • SINGAPORE • SYDNEY • TOKYO
North-Holland is an imprint of Elsevier

N·H

North-Holland is an imprint of Elsevier
The Boulevard, Langford Lane, Kidlington, Oxford OX5 1GB, UK
Radarweg 29, PO Box 211, 1000 AE Amsterdam, The Netherlands

British Library Cataloguing in Publication Data
A catalogue record for this book is available from the British Library

Library of Congress Cataloging-in-Publication Data
A catalog record for this book is available from the Library of Congress

ISBN: 978-0-444-53047-9

For information on all North-Holland publications
visit our website at *elsevierdirect.com*

Printed and bound by CPI Group (UK) Ltd, Croydon, CR0 4YY

Transferred to Digital Print 2012

General Preface

In the early eighties, when Jacques-Louis Lions and I considered the idea of a *Handbook of Numerical Analysis*, we carefully laid out specific objectives, outlined in the following excerpts from the "General Preface" which has appeared at the beginning of each of the volumes published so far:

> During the past decades, giant needs for ever more sophisticated mathematical models and increasingly complex and extensive computer simulations have arisen. In this fashion, two indissociable activities, *mathematical modeling* and *computer simulation*, have gained a major status in all aspects of science, technology and industry.
>
> In order that these two sciences be established on the safest possible grounds, mathematical rigor is indispensable. For this reason, two companion sciences, *Numerical Analysis* and *Scientific Software*, have emerged as essential steps for validating the mathematical models and the computer simulations that are based on them.
>
> *Numerical Analysis* is here understood as the part of *Mathematics* that describes and analyzes all the numerical schemes that are used on computers; its objective consists in obtaining a clear, precise, and faithful, representation of all the "information" contained in a mathematical model; as such, it is the natural extension of more classical tools, such as analytic solutions, special transforms, functional analysis, as well as stability and asymptotic analysis.
>
> The various volumes comprising the *Handbook of Numerical Analysis* will thoroughly cover all the major aspects of Numerical Analysis, by presenting accessible and in-depth surveys, which include the most recent trends.
>
> More precisely, the Handbook will cover the *basic methods of Numerical Analysis*, gathered under the following general headings:
>
> – Solution of Equations in $\mathbb{R}^n$,
> – Finite Difference Methods,
> – Finite Element Methods,
> – Techniques of Scientific Computing.

It will also cover the *numerical solution of actual problems of contemporary interest in Applied Mathematics*, gathered under the following general headings:

- Numerical Methods for Fluids,
- Numerical Methods for Solids.

In retrospect, it can be safely asserted that Volumes I to IX, which were edited by both of us, fulfilled most of these objectives, thanks to the eminence of the authors and the quality of their contributions.

After Jacques-Louis Lions' tragic loss in 2001, it became clear that Volume IX would be the last one of the type published so far, i.e., edited by both of us and devoted to some of the general headings defined above. It was then decided, in consultation with the publisher, that each future volume will instead be devoted to a single "specific application" and called for this reason a Special Volume. "Specific applications" will include mathematical finance, meteorology, celestial mechanics, computational chemistry, living systems, electromagnetism, computational mathematics, etc. It is worth noting that the inclusion of such "specific applications" in the *Handbook of Numerical Analysis* was part of our initial project.

To ensure the continuity of this enterprise, I will continue to act as the Editor of each Special Volume, whose conception will be jointly coordinated and supervised by a Guest Editor.

P.G. CIARLET
July 2002

Contents of Volume XVI

Contents of the Handbook

SOLUTION OF EQUATIONS IN $\mathbb{R}^n$ (PART 2)

VOLUME IV

FINITE ELEMENT METHODS (PART 2)

NUMERICAL METHODS FOR SOLIDS (PART 2)

VOLUME V

TECHNIQUES OF SCIENTIFIC COMPUTING (PART 2)

VOLUME VI

NUMERICAL METHODS FOR SOLIDS (PART 3)

NUMERICAL METHODS FOR FLUIDS (PART 1)

VOLUME VII

SOLUTION OF EQUATIONS IN $\mathbb{R}^n$ (PART 3)

TECHNIQUES OF SCIENTIFIC COMPUTING (PART 3)

VOLUME VIII

SOLUTION OF EQUATIONS IN $\mathbb{R}^n$ (PART 4)

TECHNIQUES OF SCIENTIFIC COMPUTING (PART 4)

NUMERICAL METHODS FOR FLUIDS (PART 2)

VOLUME IX

NUMERICAL METHODS FOR FLUIDS (PART 3)

VOLUME X

SPECIAL VOLUME: COMPUTATIONAL CHEMISTRY

VOLUME XII

SPECIAL VOLUME: COMPUTATIONAL MODELS FOR THE HUMAN BODY

VOLUME XIII

SPECIAL VOLUME: NUMERICAL METHODS IN ELECTROMAGNETICS

VOLUME XIV

SPECIAL VOLUME: COMPUTATIONAL METHODS FOR THE ATMOSPHERE AND THE OCEANS

VOLUME XV

SPECIAL VOLUME: MATHEMATICAL MODELING AND NUMERICAL METHODS IN FINANCE

MATHEMATICAL MODELS (PART I)

Numerical Methods for Non-Newtonian Fluids

"I learned that the stems are built up of several dozen smaller tubes, each containing a magnetic slurry: iron powder in a viscous liquid."

Jack Vance
The Killing Machine
Book Two of
The Demon Princes, Volume One
Tom Doherty Associates Inc., New York, 1997

"Il est, il est, en lieu d'écumes et d'eaux vertes, comme aux clairières en feu de la Mathématique, des vérités plus ombrageuses à notre approche que l'encolure des bêtes fabuleuses." (*)

Saint-John Perse
Amers
Editions Gallimard, Paris, 1957

(*) Approximate translation:

"There are, there are, in places of foams and green waters, as in the burning clearings of Mathematics, some truths more prickly to our nearness than the neck of the fantastic beasts."

Foreword

Few years ago, after the completion of Volume IX of the *Handbook of Numerical Analysis*, one of the guest editors of the present volume wondered which topics deserve a dedicated volume. Among the topics he considered, two in particular stood out: a methodology-oriented topic, *Operator-Splitting*, and a thematic topic, *Computational Non-Newtonian Fluid Mechanics*. As operator-splitting methods already had a strong presence in several volumes of the *Handbook of Numerical Analysis* (starting with a 266-page article by G.I. Marchuk in Volume 1), he focused on the second topic. And, although the *Handbook* had already covered some problems from non-Newtonian fluid mechanics, analytically and computationally – problems from *Viscoelasticity* in FERNÁNDEZ-CARA, GUILLÉN and ORTEGA [2002] and from *Viscoelasticity* and *Viscoplasticity* in GLOWINSKI [2003] – more work remained to be done. Given that the first of these two articles is essentially analytical and the second is mostly dedicated to Newtonian flow, there is a strong rationale for a volume that concentrates on the numerical simulation of a variety of non-Newtonian fluid flows.

There is no doubt that non-Newtonian flows and their numerical simulation have generated abundant literature, including the *Journal of Non-Newtonian Fluid Mechanics* (another Elsevier publication) and books such as those by BINGHAM [1922], LODGE [1964], DUVAUT and LIONS [1972a,b, 1976], JOSEPH [1990], HUILGOL and PHANTIEN [1997], and OWENS and PHILLIPS [2002], as well as additional publications, references to which can be found in the articles of this volume. This abundance of publications can be explained by the fact that non-Newtonian fluids occur in many real-life situations, such as the food industry, the oil and gas industry, chemical, civil and mechanical engineering, and the biosciences, to name just a few. Moreover, the mathematical and numerical analyses of non-Newtonian fluid flow models provide very challenging problems to partial differential equations specialists and applied and computational mathematicians alike.

Finite elements and *finite volumes* have been the methods of choice for the numerical simulation of non-Newtonian fluid flows (see e.g., MARCHAL and CROCHET [1986, 1987], FORTIN and FORTIN [1989], FORTIN and PIERRE [1989], EL HADJ and PA TANGUY [1990], GUENETTE and FORTIN [1995], FORTIN and ESSELAOUI [1987], SINGH and LEAL [1993], BAAIJENS [1994, 1998], VAN KEMENADE [1994a]; VAN KEMENADE and DEVILLE [1994b], FIÉTIER and DEVILLE [2003], XUE et al. [1998], SINGH, JOSEPH, HESLA, GLOWINSKI and PAN [2000], PATANKAR et al. [2000], PILLAPAKKAM and SINGH [2001], CHAUVIERES and OWENS [2001], BEHR, ARORA, CORONADO and PASQUALI [2005], CORONADO, ARORA, BEHR and PASQUALI [2007], DEAN, GLOWINSKI and GUIDOBONI [2007]; see also the many references within these articles as well as in the articles in this volume).

The purpose of this volume is twofold:

(1) Provide a review of well-known computational methods for the simulation of non-Newtonian fluid flows, particularly of the viscoelastic and viscoplastic types.
(2) Discuss new numerical methods that have proven to be more efficient and more accurate than traditional methods.

Even though the articles in this volume investigate a significant range of applications, we strongly believe that the methods discussed herein will find applications in many more areas.

This volume is divided into three parts, each of which presents one or more articles relevant to a key problem inherent to non-Newtonian flows:

- *Part I* is dedicated to the numerical analysis and simulation of grade-two fluids. *V. Girault* and *F. Hecht's* article addresses the mathematical and computational difficulties associated with the grade-two model, thereby providing a good introduction to the analysis of flows with more complicated constitutive laws.
- *Part II* has four articles dedicated to the modeling and mathematical and numerical analysis of *viscoelastic flows*. The article by *A. Lozinski, R.G. Owens* and *T.N. Phillips* follows the stochastic approach advocated by LASO and ÖTTINGER [1993] for deriving constitutive laws for polymeric flows. The article takes these laws, which connect microscopic stochastic models with macroscopic ones, as the basis for its approach because they are expected to be more accurate than the more phenomenological ones encountered in the classical literature. The article by *A. Bonito, Ph. Clement* and *M. Picasso* addresses the modeling, numerical analysis, and simulation of viscoelastic flows, using models obtained via a two-scale analysis operating at mesoscopic and macroscopic levels. In addition, this article discusses the simulation of viscoelastic flow with free surface, a highly nontrivial problem. The article by *Y.J. Lee, J. Xu*, and *C.S. Zhang* is mostly methodological and investigates the difficult problem (at a large Weissenberg number) associated with the advection of the viscoelastic extra-stress tensor. This article also shows that multilevel and parallelization methods can significantly speed up viscoelastic calculations. Part II concludes with a article by *T.W. Pan, J. Hao*, and *R. Glowinski*, which investigates several methods that can be used to guarantee the definite positiveness of the viscoelastic extra-stress tensor. The article also discusses the numerical simulation of particulate flow for viscoelastic fluids.
- *Part III* has two articles, both of which discuss the simulation of *viscoplastic fluid flows* where the viscoplastic properties are possibly coupled with additional physical properties such as temperature dependence, compressibility, thixotropy, interaction with solid particles, and an electric field. The first article, by *R. Glowinski* and *A. Wachs*, investigates a variety of viscoplastic flows encountered in the oil and gas industry, such as waxy crude oil flow in pipelines at low temperatures. The second article, by *R.H.W. Hoppe* and *W.G. Litvinov*, is dedicated to the modeling and simulation of electrorheological fluid flows and to the optimal design of devices that use these fluids.

This volume offers investigations, results, and conclusions that will no doubt be useful to engineers and computational and applied mathematicians who are concerned with the various aspects of non-Newtonian fluid mechanics. Special thanks are due to Gavin Becker, Philippe G. Ciarlet, Arjen Sevenster, Lauren Schultz, and Mageswaran Babusivakumar, all of whom played major roles in bringing this volume into existence.

ROLAND GLOWINSKI
JINCHAO XU

Bibliography

BAAIJENS, F.P.T. (1994). Application of low-order Discontinuous Galerkin methods to the analysis of viscoelastic flows. *J. Non-Newton. Fluid Mech.* **52** (1), 37–57.

BAAIJENS, F.P.T. (1998). Mixed finite element methods for viscoelastic flow analysis: a review. *J. Non-Newton. Fluid Mech.* **79** (2–3), 361–385.

BINGHAM, E.C. (1922). *Fluidity and Plasticity* (McGraw-Hill, New York, NY).

DUVAUT, G., LIONS, J.L. (1972). *Les Inéquations en Mécanique et en Physique* (Dunod, Paris).

DUVAUT, G., Lions, J.L. (1976). *Inequalities in Mechanics and Physics* (Springer, Berlin).

EL HADJ, M., TANGUY, P.A. (1990). A finite element procedure coupled with the method of characteristics for simulation of viscoelastic fluid flow. *J. Non-Newton. Fluid Mech.* **36**, 333–349.

FIÉTIER, N., DEVILLE, M.O. (2003). Linear stability analysis of time-dependent algorithms with spectral element methods for the simulation of viscoelastic flows. *J. Non-Newton. Fluid Mech.* **115** (2–3), 157–190.

FORTIN, M., ESSELAOUI, D. (1987). A finite element procedure for viscoelastic flows. *Int. J. Numer. Methods Fluids* **7** (10), 989–1145.

FORTIN, M., FORTIN, A. (1989). A new approach for the FEM simulation of viscoelastic flows. *J. Non-Newton. Fluid Mech.* **32** (3), 295–310.

FORTIN, M., PIERRE, R. (1989). On the convergence of the mixed method of Crochet and Marchal for viscoelastic flows. *Comput. Methods Appl. Mech. Eng.* **73** (3), 341–350.

GUENETTE, R., FORTIN, M. (1995). A new mixed finite element method for computing viscoelastic flows. *J. Non-Newton. Fluid Mech.* **60**, 27–52.

HUILGOL, R.R., PHAN-THIEN, N. (1997). *Fluid Mechanics of Viscoelasticity* (Elsevier, Amsterdam).

JOSEPH, D.D. (1990). *Fluid Dynamics of Viscoelastic Liquids* (Springer, Berlin).

LODGE, A.S. (1964). *Elastic Liquids* (Academic Press, New York, NY).

MARCHAL, J.M., CROCHET, M.J. (1986). Hermitian finite elements for calculating viscoelastic flow. *J. Non-Newton. Fluid Mech.* **20**, 187–207.

MARCHAL, J.M., CROCHET, M.J. (1987). A new mixed finite element for calculating viscoelastic flow. *J. Non-Newton. Fluid Mech.* **26** (1), 77–114.

OWENS, R.G., PHILLIPS, T.N. (2002). *Computational Rheology* (Imperial College Press, London, UK).

PATANKAR, N.A., SINGH, P., JOSEPH, D.D., GLOWINSKI, R., PAN, T.W. (2000). A new formulation of the distributed Lagrange multiplier/fictitious domain method for particulate flows. *J. Non-Newton. Fluid Mech.* **26** (9), 1509–1524.

SINGH, P., JOSEPH, D.D., HESLA, T.I., GLOWINSKI, R., PAN, T.W. (2000). A distributed Lagrange multiplier/fictitious domain method for viscoelastic particulate flows. *J. Non-Newton. Fluid Mech.* **91** (2–3), 165–188.

SINGH, P., LEAL, L.G. (1993). Finite-element simulation of the start-up problem for a viscoelastic fluid in an eccentric rotating cylinder geometry using a third-order upwind scheme. *Theor. Comput. Fluid Dyn.* **5** (2–3), 107–137.

VAN KEMENADE, V., DEVILLE, M.O. (1994a). Application of spectral elements to viscoelastic creeping flows. *J. Non-Newton. Fluid Mech.* **51** (3), 277–308.

VAN KEMENADE, V., DEVILLE, M.O. (1994b). Spectral elements for viscoelastic flows with change of type. *J. Rheol.* **38** (2), 291–307.

XUE, S.C., PHAN-THIEN, N., TANNER, R.I. (1998). Three dimensional numerical simulations of viscoelastic flows through planar contractions. *J. Non-Newton. Fluid Mech.* **74** (1–3), 195–245.

Numerical Methods for Grade-Two Fluid Models: Finite-Element Discretizations and Algorithms

Vivette Girault

UPMC, Univ. Paris 06, UMR 7598, F-75005 Paris, France and
Department of Mathematics, TAMU, College Station TX 77843, USA
E-mail: girault@ann.jussieu.fr

Frédéric Hecht

UPMC, Univ. Paris 06, UMR 7598, F-75005 Paris, France
E-mail: hecht@ann.jussieu.fr

Numerical Methods for Non-Newtonian Fluids
Special Volume (R. Glowinski and Jinchao Xu, Guest Editors) of
HANDBOOK OF NUMERICAL ANALYSIS, VOL. XVI
P.G. Ciarlet (Editor)

ISSN 1570-8659
DOI 10.1016/B978-0-444-53047-9.00001-0

Contents

CHAPTER 1

Theoretical Results

1.0. Foreword

The numerical analysis of schemes and algorithms used in discretizing non-Newtonian fluid models is a challenging task. To this date, there are only very few models for which a complete numerical analysis, namely stability, error estimates, and convergence of algorithms, is known. The two-dimensional grade-two fluid model with tangential Dirichlet boundary conditions studied in this work is one of them. This is made possible by the fact that, owing to the dimension, this model has a formulation that yields good discrete a priori estimates. In three dimensions, discrete a priori estimates for the same formulation are not yet known. Tangential boundary conditions alone, i.e., with no inflow or outflow, are studied here because the problem may be ill-posed if complete Dirichlet boundary conditions are prescribed.

The material in this work is fairly well self-contained and all prerequisite notions are recalled. It is accessible to advanced graduate students and part of this work was taught by the first author in an advanced graduate course at the Mathematics Department of the University of Pittsburgh.

This work is divided into six chapters. In order to present clearly the main ideas, without obscuring the discussion by too many technical details, the first four chapters are devoted to the problem with homogeneous Dirichlet boundary conditions. The first chapter presents a short survey of theoretical results with particular emphasis on the two-dimensional problem. Chapter 2 is devoted to the discretization of the steady-state problem, and Chapter 3 is devoted to the discretization of the time-dependent problem. Chapter 4 presents an interesting heuristic least-squares scheme and gradient algorithm for the steady and unsteady problems. The steady model with tangential Dirichlet boundary conditions is treated in Chapter 5. Numerical experiments are presented in Chapter 6.

1.1. Introduction and preliminaries

A grade-two fluid belongs to the class of non-Newtonian fluids of differential type. Non-Newtonian fluid models are used to describe the behavior of liquids frequently encountered in nature and industry, such as many polymeric liquids, biological fluids, foams, and slurries. Unlike water, these liquids are characterized by the fact that they exhibit at least one behavior such as shear-thinning or shear-thickenning, stress-relaxation, nonlinear creep, normal stress differences or yielding. Grade-two fluids cannot exhibit stress-relaxation, but they can develop normal stress differences and they can experience creep.

In a fluid of differential type, be it Newtonian or non-Newtonian, the Cauchy stress tensor is determined explicitly by the symmetric part of the velocity gradient and possibly its various higher time derivatives. But in contrast to Newtonian fluid models where the constitutive relation for the Cauchy stress tensor is a linear function of the symmetric part of the velocity gradient, in a non-Newtonian fluid model, this constitutive relation is nonlinear.

A grade-two fluid is considered an appropriate model for the motion of a water solution of polymers, cf. DUNN and RAJAGOPAL [1995]. Interestingly, its equations can also be interpreted as a model of turbulence; we refer to the work of HOLM, MARSDEN and RATIU (cf. for instance [1998a, 1998b]). In the simplest case, its equations of motion have the form

$$\frac{\partial}{\partial t}(\boldsymbol{u} - \alpha \Delta \boldsymbol{u}) - \nu \Delta \boldsymbol{u} + \mathbf{curl}(\boldsymbol{u} - \alpha \Delta \boldsymbol{u}) \times \boldsymbol{u} + \nabla p = \boldsymbol{f}. \tag{1.1.1}$$

As the fluid is incompressible, it satisfies the constraint

$$\operatorname{div} \boldsymbol{u} = 0, \tag{1.1.2}$$

and (1.1.1) and (1.1.2) are complemented by a Dirichlet boundary condition and an initial condition.

In some sense, the theoretical results that have been proven up to date for this model are fairly satisfactory, but there still remain important open questions such as the problem posed by nonhomogeneous Dirichlet boundary conditions or that posed by a rough exterior force, such as an L^2 force, to mention just these two "simple" questions. At least for the steady two-dimensional problem, we can handle tangential Dirichlet boundary conditions, i.e., with no ingoing or outgoing flow. But if there is an ingoing or outgoing flow, the problem is ill-posed and we still do not know what additional boundary condition must be added to make the problem well-posed.

In contrast, numerical results obtained so far are very scanty. We now know how to do the numerical analysis of some carefully chosen schemes for the steady and time-dependent problems in dimension $d = 2$. But up to now, the numerical analysis of schemes that approximate this problem in dimension $d = 3$, be it steady or unsteady, is not resolved. The explanation is simple: we lack some discrete a priori estimates, estimates that appear plausible, but for which we have yet no proof, except perhaps for very crude schemes. These estimates are a crucial ingredient in the numerical analysis of several models of non-Newtonian fluids, and this analysis will remain an open question as long as such estimates are not established.

For this reason, the present work is dedicated only to numerical methods for the model in two dimensions.

1.1.1. Notation

The following notation will be used in the sequel. We state them in dimension $d = 3$ because the theoretical problem is, of course, three-dimensional, but the numerical study will be done mainly in dimension $d = 2$. Unless otherwise specified, the domains of interest Ω will all be *bounded, connected,* and have a boundary $\partial\Omega$ that is at least $\mathcal{C}^{0,1}$, i.e., Lipschitz-continuous (cf. GRISVARD [1985]), and we shall call them *Lipschitz-continuous domains*. We denote by $\mathcal{D}(\Omega)$ the subspace of functions of $\mathcal{C}^{\infty}(\overline{\Omega})$ with compact support in Ω. Let $k = (k_1, k_2, k_3)$ be a triple of non-negative integers and set $|k| = k_1 + k_2 + k_3$; we define the partial derivative

∂^k of order $|k|$ by:

$$\partial^k v = \frac{\partial^{|k|} v}{\partial x_1^{k_1} \partial x_2^{k_2} \partial x_3^{k_3}}.$$

Recall the standard Sobolev spaces, for a non-negative integer m and a number $r \geq 1$ (cf. ADAMS [1975] or NEČAS [1967])

$$W^{m,r}(\Omega) = \{v \in L^r(\Omega);\ \partial^k v \in L^r(\Omega)\ \forall |k| \leq m\},$$

equipped with the seminorm

$$|v|_{W^{m,r}(\Omega)} = \left[\sum_{|k|=m} \int_\Omega |\partial^k v|^r \, \mathrm{d}\boldsymbol{x} \right]^{1/r},$$

and the norm (for which it is a Banach space)

$$\|v\|_{W^{m,r}(\Omega)} = \left[\sum_{0 \leq k \leq m} |v|^r_{W^{k,r}(\Omega)} \right]^{1/r},$$

with the usual modification when $r = \infty$; we refer to GRISVARD [1985], LIONS and MAGENES [1968] or ADAMS [1975] for extending this definition to fractional Sobolev spaces. When $r = 2$, this space is the Hilbert space $H^m(\Omega)$. In particular, the scalar product of $L^2(\Omega)$ is denoted by $(\cdot, \cdot)$. These definitions are extended straightforwardly to vector-valued functions, with the same notation, except for non-Hilbert norms. In the case of a vector or tensor $\boldsymbol{u}$, we set

$$\|\boldsymbol{u}\|_{L^r(\Omega)} = \left[\int_\Omega |\boldsymbol{u}(\boldsymbol{x})|^r \, \mathrm{d}\boldsymbol{x} \right]^{1/r},$$

where $|\cdot|$ denotes the Euclidian norm when $\boldsymbol{u}$ is a vector: $|\boldsymbol{u}|^2 = \boldsymbol{u} \cdot \boldsymbol{u}$, or the Frobenius norm when $\boldsymbol{u}$ is a tensor: $|\boldsymbol{u}|^2 = \boldsymbol{u} : \boldsymbol{u}$.

For imposing vanishing boundary values on $\partial\Omega$, we define

$$H_0^1(\Omega) = \{v \in H^1(\Omega);\ v|_{\partial\Omega} = 0\},$$

and more generally, for a number $r \geq 1$, we define

$$W_0^{1,r}(\Omega) = \{v \in W^{1,r}(\Omega);\ v|_{\partial\Omega} = 0\}.$$

We shall frequently use Sobolev imbeddings: for a real number $p \in \mathbb{R}$, $p \geq 1$ in dimension $d = 2$ or $1 \leq p \leq 6$ in dimension $d = 3$, the space $H^1(\Omega)$ is imbedded into $L^p(\Omega)$. In particular, there exists a constant S_p (that depends only on p, the dimension and the domain) such that

$$\forall v \in H_0^1(\Omega),\ \|v\|_{L^p(\Omega)} \leq S_p |v|_{H^1(\Omega)}. \tag{1.1.3}$$

When $p = 2$, this is Poincaré's inequality and S_2 is Poincaré's constant. In the case of the maximum norm, the following imbedding holds:

$$\forall r > d,\ W^{1,r}(\Omega) \subset L^\infty(\Omega). \tag{1.1.4}$$

In particular, for each $r > d$, there exists a constant $S_{\infty,r}$, such that

$$\forall v \in W^{1,r}(\Omega) \cap H_0^1(\Omega),\ \|v\|_{L^\infty(\Omega)} \le S_{\infty,r}\|\nabla v\|_{L^r(\Omega)}. \tag{1.1.5}$$

Owing to Poincaré's inequality, the seminorm $|\cdot|$ is a norm on $H_0^1(\Omega)$, equivalent to the full norm. As it is directly related to the gradient operator, we choose this seminorm as norm on $H_0^1(\Omega)$, and in particular, we use it to define the dual norm on its dual space $H^{-1}(\Omega)$:

$$\forall f \in H^{-1}(\Omega),\ \|f\|_{H^{-1}(\Omega)} = \sup_{v\neq 0, v\in H_0^1(\Omega)} \frac{\langle f, v\rangle}{|v|_{H^1(\Omega)}}, \tag{1.1.6}$$

where $\langle\cdot,\cdot\rangle$ is the duality pairing between $H^{-1}(\Omega)$ and $H_0^1(\Omega)$.

For imposing tangential boundary conditions, we define

$$H_\tau^1(\Omega) = \{\boldsymbol{v} \in H^1(\Omega)^3;\ \boldsymbol{v}\cdot\boldsymbol{n} = 0 \text{ on } \partial\Omega\}, \tag{1.1.7}$$

where $\boldsymbol{n} = (n_1, n_2, n_3)$ is the unit normal vector to $\partial\Omega$, directed outside Ω, and $\boldsymbol{v} = (v_1, v_2, v_3)$. An easy application of Peetre–Tartar's Theorem (cf. PEETRE [1966], and TARTAR [1978], or GIRAULT and RAVIART [1986]) proves the analog of Sobolev's imbeddings in $H_\tau^1(\Omega)$ for any real number $p \ge 1$ if $d = 2$ or $1 \le p \le 6$ if $d = 3$:

$$\forall \boldsymbol{v} \in H_\tau^1(\Omega),\ \|\boldsymbol{v}\|_{L^p(\Omega)} \le \tilde{S}_p|\boldsymbol{v}|_{H^1(\Omega)}. \tag{1.1.8}$$

In particular, for $p = 2$, the mapping $\boldsymbol{v} \mapsto |\boldsymbol{v}|_{H^1(\Omega)}$ is a norm on $H_\tau^1(\Omega)$, equivalent to the H^1 norm and $\tilde{S}_2$ is the analog of Poincaré's constant. Moreover, the analog of (1.1.5) holds: for each $r > d$, there exists a constant $\widetilde{S}_{\infty,r}$, such that

$$\forall \boldsymbol{v} \in W^{1,r}(\Omega)^3 \cap H_\tau^1(\Omega),\ \|\boldsymbol{v}\|_{L^\infty(\Omega)} \le \widetilde{S}_{\infty,r}\|\nabla \boldsymbol{v}\|_{L^r(\Omega)}, \tag{1.1.9}$$

where $\nabla \boldsymbol{v}$ denotes the gradient tensor: $(\nabla \boldsymbol{v})_{ij} = \partial v_i/\partial x_j$. We shall also use the classical spaces for Navier–Stokes equations:

$$V = \{\boldsymbol{v} \in H_0^1(\Omega)^3;\ \operatorname{div}\boldsymbol{v} = 0 \text{ in } \Omega\}, \tag{1.1.10}$$

where $\operatorname{div}\boldsymbol{v} = \sum_{i=1}^3 \partial v_i/\partial x_i$,

$$V^\perp = \{\boldsymbol{v} \in H_0^1(\Omega)^3;\ \forall \boldsymbol{w} \in V,\ (\nabla\boldsymbol{v}, \nabla\boldsymbol{w}) = 0\}, \tag{1.1.11}$$

$$W = \{\boldsymbol{v} \in H_\tau^1(\Omega);\ \operatorname{div}\boldsymbol{v} = 0 \text{ in } \Omega\}, \tag{1.1.12}$$

$$L_0^2(\Omega) = \{q \in L^2(\Omega);\ \int_\Omega q\,\mathrm{d}\boldsymbol{x} = 0\},$$

more generally,

$$L_0^r(\Omega) = \{q \in L^r(\Omega);\ \int_\Omega q\,\mathrm{d}\boldsymbol{x} = 0\},$$

$$H(\mathrm{div}, \Omega) = \{\boldsymbol{v} \in L^2(\Omega)^3;\ \mathrm{div}\,\boldsymbol{v} \in L^2(\Omega)\},$$

$$H_0(\mathrm{div}, \Omega) = \{\boldsymbol{v} \in H(\mathrm{div}, \Omega);\ \boldsymbol{v} \cdot \boldsymbol{n} = 0 \text{ on } \partial\Omega\},$$

$$H(\mathbf{curl}, \Omega) = \{\boldsymbol{v} \in L^2(\Omega)^3;\ \mathbf{curl}\,\boldsymbol{v} \in L^2(\Omega)^3\},$$

where

$$\mathbf{curl}\,\boldsymbol{v} = \left(\frac{\partial v_3}{\partial x_2} - \frac{\partial v_2}{\partial x_3}, \frac{\partial v_1}{\partial x_3} - \frac{\partial v_3}{\partial x_1}, \frac{\partial v_2}{\partial x_1} - \frac{\partial v_1}{\partial x_2}\right). \tag{1.1.13}$$

These definitions carry over to $d = 2$ with one exception: when $d = 2$, the **curl** operator is considered a scalar because it has only one component:

$$\forall \boldsymbol{v} = (v_1, v_2),\ \mathrm{curl}\,\boldsymbol{v} = \frac{\partial v_2}{\partial x_1} - \frac{\partial v_1}{\partial x_2}, \tag{1.1.14}$$

and we define

$$H(\mathrm{curl}, \Omega) = \{\boldsymbol{v} \in L^2(\Omega)^2;\ \mathrm{curl}\,\boldsymbol{v} \in L^2(\Omega)\}. \tag{1.1.15}$$

We also recall the following identity, valid in a Lipschitz domain of $I\!R^d$, $d = 2, 3$:

$$\forall \boldsymbol{v} \in H_0^1(\Omega)^d,\ |\boldsymbol{v}|^2_{H^1(\Omega)} = \|\mathrm{div}\,\boldsymbol{v}\|^2_{L^2(\Omega)} + \|\mathbf{curl}\,\boldsymbol{v}\|^2_{L^2(\Omega)}. \tag{1.1.16}$$

As usual, for handling time-dependent problems, it is convenient to consider functions defined on a time interval $]a, b[$ with values in a functional space, say X (cf. LIONS and MAGENES [1968]). More precisely, let $\|\cdot\|_X$ denote the norm of X; then for any number r, $1 \le r \le \infty$, we define

$$L^r(a, b; X) = \{f \text{ measurable in }]a, b[;\ \int_a^b \|f(t)\|_X^r \mathrm{d}t < \infty\}$$

equipped with the norm

$$\|f\|_{L^r(a,b;X)} = \left(\int_a^b \|f(t)\|_X^r \mathrm{d}t\right)^{1/r},$$

with the usual modification if $r = \infty$. It is a Banach space if X is a Banach space and, when $r = 2$, it is a Hilbert space if X is a Hilbert space. For example, $L^2(a, b; H^m(\Omega))$ is a Hilbert space and, in particular, $L^2(a, b; L^2(\Omega))$ coincides with $L^2(\Omega\times]a, b[)$. In addition, we shall also use spaces with derivatives in time, such as

$$H^1(a, b; X) = \{f \in L^2(]a, b[; X);\ \frac{\partial f}{\partial t} \in L^2(]a, b[; X)\},$$

equipped with the graph norm

$$\|f\|_{H^1(a,b;X)} = \left(\|f\|^2_{L^2(a,b;X)} + \|\frac{\partial f}{\partial t}\|^2_{L^2(a,b;X)} \right)^{1/2},$$

for which it is a Hilbert space.

1.1.2. Properties of the Laplace and Stokes operators

We close this introduction by recalling useful properties of the Laplace and Stokes equations in dimension $d = 2$ or $d = 3$. The presentation is restricted to homogeneous Dirichlet boundary conditions.

Let us start with the Laplace equation with a homogeneous Dirichlet boundary condition in a bounded Lipschitz domain: For f given in $H^{-1}(\Omega)$, find u in $H^1_0(\Omega)$ such that

$$-\Delta u = f \quad \text{in } \Omega. \tag{1.1.17}$$

It can be set into the following equivalent variational formulation: Find u in $H^1_0(\Omega)$ solving

$$\forall v \in H^1_0(\Omega), \ (\nabla u, \nabla v) = \langle f, v \rangle.$$

By Lax–Milgram's Lemma (cf. LAX and MILGRAM [1954]), this problem has one and only one solution that depends continuously on f. Furthermore, increasing the regularity of f, increases up to a certain extent, the regularity of u. This is stated in the following theorems; the first one is proved by GRISVARD [1985] and the second one by DAUGE [1992].

THEOREM 1.1.1. *Let Ω be a polygon in $\mathbb{R}^2$. If f belongs to $L^r(\Omega)$ for some r with $1 < r \leq 4/3$, then the solution u of (1.1.17) belongs to $W^{2,r}(\Omega)$ with continuous dependence on f.*

THEOREM 1.1.2. *Let Ω be a polyhedron in $\mathbb{R}^3$ with a Lipschitz-continuous boundary. If f belongs to $H^{s-1}(\Omega)$ for some s with $0 \leq s < 1/2$, then the solution u of (1.1.17) belongs to $H^{s+1}(\Omega)$ with continuous dependence on f. If f belongs to $L^{3/2}(\Omega)$, then u belongs to $H^{3/2}(\Omega)$ with continuous dependence on f.*

When f is smoother than in the above statements, the solution is also smoother provided the inner angles of $\partial\Omega$ are suitably restricted. For instance, it is well known that the next regularity holds in a convex domain (cf. GRISVARD [1985]).

THEOREM 1.1.3. *If f belongs to $L^2(\Omega)$ and the domain is a convex polygon or polyhedron, then the solution u of (1.1.17) belongs to $H^2(\Omega)$, with continuous dependence on f.*

None of the results listed above address the major question: When is the solution in $W^{1,\infty}$? This property has no clear-cut answer (cf. DAUGE [1992], KOZLOV, MAZ'YA and ROSSMANN [2000]), but a sufficient condition can be given in view of the Sobolev imbedding (1.1.4) applied to gradients: for each $r > d$, there exists a constant $C_{\infty,r}$ such that

$$\forall v \in W^{2,r}(\Omega), \ \|\nabla v\|_{L^\infty(\Omega)} \leq C_{\infty,r} \|v\|_{W^{2,r}(\Omega)}. \tag{1.1.18}$$

Thus, the question can be reformulated as follows : When does a right-hand side f in $L^r(\Omega)$ for some real number $r > d$ imply that u belongs to $W^{2,r}(\Omega)$? The answer is given by GRISVARD [1985] when $d = 2$ and by DAUGE [1992] when $d = 3$.

THEOREM 1.1.4. *(1) Let Ω be a convex polygon in $\mathbb{R}^2$. Then there exists a real number $r_\Omega > 2$ depending on the largest inner angle of $\partial\Omega$ such that for all r with $2 \le r \le r_\Omega$, f in $L^r(\Omega)$ implies that the solution u of (1.1.17) belongs to $W^{2,r}(\Omega)$ with continuous dependence on f.*
(2) In $\mathbb{R}^3$, let Ω be a polyhedron with its largest inner dihedral angle strictly smaller than $2\pi/3$. Then there exists a real number $r_\Omega > 3$ depending on the largest inner dihedral angle of $\partial\Omega$ such that for all r with $2 \le r \le r_\Omega$, f in $L^r(\Omega)$ implies that the solution u of (1.1.17) belongs to $W^{2,r}(\Omega)$ with continuous dependence on f.

Now, we turn to the Stokes problem with homogeneous Dirichlet boundary conditions in a bounded, connected Lipschitz domain. It reads: For $\boldsymbol{f}$ given in $H^{-1}(\Omega)^d$ and constant $\nu > 0$, find $\boldsymbol{u}$ in $H_0^1(\Omega)^d$ and p in $L_0^2(\Omega)$, solution of

$$-\nu\,\Delta\,\boldsymbol{u} + \nabla\,p = \boldsymbol{f} \quad \text{in } \Omega, \tag{1.1.19}$$

$$\operatorname{div}\boldsymbol{u} = 0 \quad \text{in } \Omega. \tag{1.1.20}$$

It is well known (see for instance GIRAULT and RAVIART [1986]) that this problem has the two equivalent variational formulations:
1. Find $(\boldsymbol{u}, p) \in H_0^1(\Omega)^d \times L_0^2(\Omega)$, such that

$$\forall \boldsymbol{v} \in H_0^1(\Omega)^d, \;\; \nu\,(\nabla\,\boldsymbol{u}, \nabla\,\boldsymbol{v}) - (p, \operatorname{div}\boldsymbol{v}) = \langle \boldsymbol{f}, \boldsymbol{v}\rangle, \tag{1.1.21}$$

$$\forall q \in L_0^2(\Omega), \;\; (q, \operatorname{div}\boldsymbol{u}) = 0. \tag{1.1.22}$$

2. Find $\boldsymbol{u} \in V$ such that

$$\forall \boldsymbol{v} \in V, \;\; \nu\,(\nabla\,\boldsymbol{u}, \nabla\,\boldsymbol{v}) = \langle \boldsymbol{f}, \boldsymbol{v}\rangle. \tag{1.1.23}$$

Problem (1.1.19)–(1.1.20) has a unique solution $(\boldsymbol{u}, p)$ that depends continuously on $\boldsymbol{f}$:

$$|\boldsymbol{u}|_{H^1(\Omega)} \le \frac{1}{\nu}\|\boldsymbol{f}\|_{H^{-1}(\Omega)}, \;\; \|p\|_{L^2(\Omega)} \le \frac{1}{\beta}\|\boldsymbol{f}\|_{H^{-1}(\Omega)}, \tag{1.1.24}$$

where $\frac{1}{\beta} > 0$ is the constant of the divergence isomorphism of $V^\perp$ onto $L_0^2(\Omega)$:

$$\forall \boldsymbol{v} \in V^\perp, \;\; |\boldsymbol{v}|_{H^1(\Omega)} \le \frac{1}{\beta}\|\operatorname{div}\boldsymbol{v}\|_{L^2(\Omega)}. \tag{1.1.25}$$

This is equivalent to the inf-sup condition (cf. BABUŠKA [1973], BRENNER and SCOTT [1994], BREZZI [1974], BREZZI and FORTIN [1991], DURÁN and MUSCHIETTI [2001], and GIRAULT and RAVIART [1986], or ERN and GUERMOND [2004]):

$$\forall q \in L_0^2(\Omega), \quad \sup_{\boldsymbol{v}\in H_0^1(\Omega)^d} \frac{1}{|\boldsymbol{v}|_{H^1(\Omega)}} \int_\Omega q \operatorname{div}\boldsymbol{v}\,\mathrm{d}\boldsymbol{x} \ge \beta\|q\|_{L^2(\Omega)}. \tag{1.1.26}$$

The regularity properties of the solution of the Stokes problem are fairly similar to those of the Laplace equation. The following result is now well known (cf. KELLOG and OSBORN [1976], or GRISVARD [1985], if $d = 2$, and DAUGE [1989], if $d = 3$).

THEOREM 1.1.5. *If f belongs to $L^2(\Omega)^d$ and the domain is a convex polygon or polyhedron, then the solution $(\boldsymbol{u}, p)$ of (1.1.19)–(1.1.20) belongs to $H^2(\Omega)^d \times H^1(\Omega)$, with continuous dependence on $\boldsymbol{f}$.*

Of course when Ω is convex, we obtain by interpolation for $0 \leq s \leq 1$, that $(\boldsymbol{u}, p)$ belongs to $H^{s+1}(\Omega)^d \times H^s(\Omega)$, with continuous dependence on $\boldsymbol{f}$, whenever $\boldsymbol{f}$ belongs to $H^{s-1}(\Omega)^d$. But for small s, the restrictions on the angles of the domain can be substantially relaxed. Indeed, without restriction on the angles of $\partial\Omega$, the following theorems hold, analogous to Theorems 1.1.1 and 1.1.2; the first one can be found in GRISVARD [1985] and the second one in DAUGE [1989].

THEOREM 1.1.6. *Let Ω be a polygon in $\mathbb{R}^2$. If $\boldsymbol{f}$ belongs to $L^r(\Omega)^2$ for some r with $1 < r \leq 4/3$, then the solution $(\boldsymbol{u}, p)$ of (1.1.19)–(1.1.20) belongs to $W^{2,r}(\Omega)^2 \times W^{1,r}(\Omega)$ with continuous dependence on $\boldsymbol{f}$.*

THEOREM 1.1.7. *Let Ω be a polyhedron in $\mathbb{R}^3$ with a Lipschitz-continuous boundary. If $\boldsymbol{f}$ belongs to $H^{s-1}(\Omega)^3$ for some s with $0 \leq s < 1/2$, then the solution $(\boldsymbol{u}, p)$ of (1.1.19)–(1.1.20) belongs to $H^{s+1}(\Omega)^3 \times H^s(\Omega)$ with continuous dependence on $\boldsymbol{f}$.*

The result for the borderline case $s = 1/2$, which extends a result of FABES, KENIG and VERCHOTTA [1988], is due to DAUGE and COSTABEL [2000] and can be found in GIRAULT and LIONS [2001a]:

THEOREM 1.1.8. *Let Ω be a polyhedron in $\mathbb{R}^3$ with a Lipschitz-continuous boundary. If $\boldsymbol{f}$ belongs to $L^{3/2}(\Omega)^3$, then the solution $(\boldsymbol{u}, p)$ of (1.1.19)–(1.1.20) belongs to $H^{3/2}(\Omega)^3 \times H^{1/2}(\Omega)$ with continuous dependence on $\boldsymbol{f}$.*

The case when the velocity is in $W^{1,\infty}$ will play an important part in the sequel. Again, we formulate it as follows: When does a right-hand side $\boldsymbol{f}$ in $L^r(\Omega)^d$ for some real number $r > d$ imply that $\boldsymbol{u}$ belongs to $W^{2,r}(\Omega)^d$? The answer is given by GRISVARD [1985] when $d = 2$ and by DAUGE [1989], KOZLOV, MAZ'YA and ROSSMANN [2000] when $d = 3$.

THEOREM 1.1.9. *(1) Let Ω be a convex polygon in $\mathbb{R}^2$. Then there exists a real number $r_\Omega > 2$ depending on the largest inner angle of $\partial\Omega$ such that for all r with $2 \leq r \leq r_\Omega$, $\boldsymbol{f}$ in $L^r(\Omega)^2$ implies that the solution $(\boldsymbol{u}, p)$ of (1.1.19)–(1.1.20) belongs to $W^{2,r}(\Omega)^2 \times W^{1,r}(\Omega)$ with continuous dependence on $\boldsymbol{f}$.*
(2) In $\mathbb{R}^3$, let Ω be a polyhedron with its largest inner dihedral angle strictly smaller than $2\pi/3$. Then there exists a real number $r_\Omega > 3$ depending on the largest inner dihedral angle of $\partial\Omega$ such that for all r with $2 \leq r \leq r_\Omega$, $\boldsymbol{f}$ in $L^r(\Omega)^3$ implies that the solution $(\boldsymbol{u}, p)$ of (1.1.19)–(1.1.20) belongs to $W^{2,r}(\Omega)^3 \times W^{1,r}(\Omega)$ with continuous dependence on $\boldsymbol{f}$.

1.2. Constitutive and momentum equations

There are several references on the mechanics of grade-two fluid models; for example, the reader can refer to TRUESDELL and RAJAGOPAL [2000], DUNN and FOSDICK [1974], or TRUESDELL and NOLL [1975]. Before writing the constitutive equation of a grade-two fluid, let us recall the Rivlin–Ericksen tensors. They are defined recursively by (cf. RIVLIN and ERICKSEN [1955]):

$$\boldsymbol{A}_1 = \boldsymbol{L} + \boldsymbol{L}^T,$$

and for $n \geq 2$:

$$\boldsymbol{A}_n = \frac{\mathrm{d}}{\mathrm{d}t}\boldsymbol{A}_{n-1} + \boldsymbol{A}_{n-1}\boldsymbol{L} + \boldsymbol{L}^T\boldsymbol{A}_{n-1}, \tag{1.2.1}$$

where $\boldsymbol{L} = \boldsymbol{L}(\boldsymbol{u})$ denotes the velocity gradient

$$\boldsymbol{L} = \boldsymbol{L}(\boldsymbol{u}) = \nabla\,\boldsymbol{u}, \tag{1.2.2}$$

i.e., denoting the symmetric part of the velocity gradient by $\boldsymbol{D}$, we have

$$\boldsymbol{A}_1 = 2\boldsymbol{D}.$$

As usual $\frac{\mathrm{d}}{\mathrm{d}t}$ denotes the material time derivative: for any tensor $\boldsymbol{A}$,

$$\frac{\mathrm{d}\boldsymbol{A}}{\mathrm{d}t} = \frac{\partial \boldsymbol{A}}{\partial t} + \boldsymbol{u}\cdot\nabla\boldsymbol{A}, \tag{1.2.3}$$

where $\boldsymbol{u}\cdot\nabla\boldsymbol{A}$ denotes the product:

$$\boldsymbol{u}\cdot\nabla\boldsymbol{A} = \sum_{i=1}^{d} u_i \frac{\partial \boldsymbol{A}}{\partial x_i}.$$

Note that all Rivlin–Ericksen tensors of order n, defined by the recursive relation (1.2.1) are frame-indifferent.

The constitutive equation for the Cauchy stress tensor of a grade-two fluid is given by

$$\boldsymbol{T} = \boldsymbol{T}(\boldsymbol{u}, \pi) = -\pi\boldsymbol{I} + \mu\boldsymbol{A}_1 + \alpha_1\boldsymbol{A}_2 + \alpha_2\boldsymbol{A}_1^2, \tag{1.2.4}$$

where π denotes the pressure and $\boldsymbol{I}$ is the identity tensor. The parameter μ is the viscosity of the fluid and the parameters α_1 and α_2 are normal stress moduli. Formula (1.2.4) is indeed the equation of a differential fluid because $\boldsymbol{T}$ is defined explicitly in terms of $\boldsymbol{A}_1$ and $\boldsymbol{A}_2$. Furthermore, the presence of $\boldsymbol{A}_1^2$ and of the products in the definition of $\boldsymbol{A}_2$ makes this relation nonlinear. To compare, the constitutive relation for the Navier–Stokes fluid model is the linear relation

$$\boldsymbol{T} = \boldsymbol{T}(\boldsymbol{u}, \pi) = -\pi\boldsymbol{I} + \mu\boldsymbol{A}_1. \tag{1.2.5}$$

We observe that when the normal stress moduli vanish, (1.2.4) and (1.2.5) coincide.

When substituting (1.2.4) into the balance of linear momentum:

$$\operatorname{div} \boldsymbol{T}(\boldsymbol{u}, \pi) + \varrho \boldsymbol{f} = \varrho \frac{\mathrm{d}\boldsymbol{u}}{\mathrm{d}t}, \tag{1.2.6}$$

where $\boldsymbol{f}$ is the specific body force, ϱ is the density, and $(\operatorname{div} \boldsymbol{T})_i = \operatorname{div}(\boldsymbol{T}_i)$, we obtain the equation of motion of a grade-two fluid. Dividing by the density ϱ, setting

$$\nu = \frac{\mu}{\varrho},$$

and without changing the symbols for the normal stress moduli and the pressure divided by the density, the equation of motion reads:

$$\begin{aligned}\frac{\partial}{\partial t}(\boldsymbol{u} - \alpha_1 \Delta \boldsymbol{u}) - \nu \Delta \boldsymbol{u} + \mathbf{curl}\,(\boldsymbol{u} - (2\alpha_1 + \alpha_2)\Delta \boldsymbol{u}) \times \boldsymbol{u} \\ - (\alpha_1 + \alpha_2)\Delta(\boldsymbol{u} \cdot \nabla \boldsymbol{u}) + 2(\alpha_1 + \alpha_2)\,(\boldsymbol{u} \cdot \nabla(\Delta \boldsymbol{u})) \\ + \nabla\left(\pi - (2\alpha_1 + \alpha_2)\left(\boldsymbol{u} \cdot \Delta \boldsymbol{u} + \frac{1}{4}|\boldsymbol{A}_1|^2\right) + \frac{1}{2}|\boldsymbol{u}|^2\right) = \boldsymbol{f}.\end{aligned} \tag{1.2.7}$$

It is shown by FOSDICK and RAJAGOPAL [1978a,b] that in order for the fluid to be thermodynamically compatible, the parameters must satisfy

$$\mu \geq 0, \ \alpha_1 \geq 0, \ \alpha_1 + \alpha_2 = 0. \tag{1.2.8}$$

In this case, setting $\alpha = \alpha_1 = -\alpha_2$, (1.2.7) simplifies to

$$\begin{aligned}\frac{\partial}{\partial t}(\boldsymbol{u} - \alpha \Delta \boldsymbol{u}) - \nu \Delta \boldsymbol{u} + \mathbf{curl}(\boldsymbol{u} - \alpha \Delta \boldsymbol{u}) \times \boldsymbol{u} \\ + \nabla\left(\pi - \alpha\left(\boldsymbol{u} \cdot \Delta \boldsymbol{u} + \frac{1}{4}|\boldsymbol{A}_1|^2\right) + \frac{1}{2}|\boldsymbol{u}|^2\right) = \boldsymbol{f}.\end{aligned}$$

Finally, denoting by p the term involving the gradient in the second line:

$$p = \pi - \alpha\left(\boldsymbol{u} \cdot \Delta \boldsymbol{u} + \frac{1}{4}|\boldsymbol{A}_1|^2\right) + \frac{1}{2}|\boldsymbol{u}|^2,$$

the equation of motion of a grade-two fluid becomes

$$\frac{\partial}{\partial t}(\boldsymbol{u} - \alpha \Delta \boldsymbol{u}) - \nu \Delta \boldsymbol{u} + \mathbf{curl}(\boldsymbol{u} - \alpha \Delta \boldsymbol{u}) \times \boldsymbol{u} + \nabla p = \boldsymbol{f}.$$

As the fluid is incompressible, it must satisfy the constraint

$$\operatorname{div} \boldsymbol{u} = 0.$$

To close the system, we complement these two equations with adequate boundary conditions and an initial condition.

REMARK 1.2.1. The condition $\alpha \geq 0$ has been (and is still) a source of rough controversy; we refer to DUNN and RAJAGOPAL [1995] for an interesting discussion on this subject. Apart from mechanical considerations, mathematically speaking, the term $-\frac{\partial}{\partial t}\alpha\Delta \boldsymbol{u}$ in the left-hand side of the momentum equation makes the model unstable when α is negative (see Remark 1.3.3), and therefore, we shall not study this case here. □

1.3. A brief survey of theoretical results

The results presented here are for homogeneous boundary conditions. The theory of the steady two-dimensional problem with tangential boundary conditions is discussed in Chapter 5.

1.3.1. The no-slip three-dimensional problem

Let $[0, T]$ be an interval of time, with $T > 0$, and let Ω be a bounded, connected domain of $I\!R^3$, with a Lipschitz-continuous boundary $\partial\Omega$. Consider the problem: Find a velocity vector $\boldsymbol{u}$ and a scalar pressure p, solution of

$$\frac{\partial}{\partial t}(\boldsymbol{u} - \alpha\Delta \boldsymbol{u}) - \nu\Delta \boldsymbol{u} + \mathbf{curl}(\boldsymbol{u} - \alpha\Delta \boldsymbol{u}) \times \boldsymbol{u} + \nabla p = \boldsymbol{f} \quad \text{in } \Omega\times]0, T[, \tag{1.3.1}$$

with the incompressibility condition:

$$\operatorname{div} \boldsymbol{u} = 0 \quad \text{in } \Omega\times]0, T[; \tag{1.3.2}$$

to simplify, we only impose here a homogeneous Dirichlet boundary condition, i.e., a no-slip condition:

$$\boldsymbol{u} = \boldsymbol{0} \quad \text{on } \partial\Omega\times]0, T[, \tag{1.3.3}$$

and the initial condition:

$$\boldsymbol{u}(0) = \boldsymbol{u}_0 \text{ in } \Omega \text{ with } \operatorname{div} \boldsymbol{u}_0 = 0 \text{ in } \Omega \text{ and } \boldsymbol{u}_0 = \boldsymbol{0} \text{ on } \partial\Omega. \tag{1.3.4}$$

REMARK 1.3.1. Considering that (1.3.1) involves a third derivative, we can ask the question: does (1.3.3) impose enough boundary conditions to determine the solution of (1.3.1)–(1.3.4)? We shall see further on that the answer is "yes." More generally, GIRAULT and SCOTT [1999] prove that in dimension $d = 2$, the answer is also "yes" for the steady-state problem in the case when (1.3.3) is replaced by a tangential Dirichlet condition:

$$\boldsymbol{u} = \boldsymbol{g} \quad \text{on } \partial\Omega\times]0, T[\quad \text{with} \quad \boldsymbol{g}\cdot\boldsymbol{n} = 0, \tag{1.3.5}$$

see Section 5.1.2. It is likely that, with adequate conditions on $\boldsymbol{g}$, this result extends to the evolution equation (1.3.1)–(1.3.4). But when the boundary values are not tangential, there are examples where the problem is ill-posed, cf. RAJAGOPAL [1995], RAJAGOPAL and KALONI [1989], and Remarks 1.3.4, 6.2.1, parts (2) and (3). □

Problem (1.3.1)–(1.3.4) is difficult because its nonlinear term involves a third-order derivative, whereas its elliptic part only comes from a Laplace operator; for this reason, it behaves mostly as a hyperbolic problem. From 1993 onward, many publications have been devoted to this problem, but by far the best proof of existence, due to Cioranescu and Ouazar, goes back to more than 25 years ago (1981) and is found in the thesis of OUAZAR [1981]; it was published later by CIORANESCU and OUAZAR [1984a, 1984b]. The reader can also refer to CIORANESCU, GIRAULT, GLOWINSKI and SCOTT [1999] and to CIORANESCU and GIRAULT [1997].

Here is a brief description of the construction of solutions by Cioranescu and Ouazar. Some of its ideas will be very helpful for discretizing the problem. First, we make precise assumptions on the data and the domain: Ω simply-connected with boundary of class $\mathcal{C}^{3,1}$, $\boldsymbol{f}$ in $L^2(0,T;H^1(\Omega)^3)$ and $\boldsymbol{u}_0$ in $H^3(\Omega)^3$. Formally, observe first that (1.3.1) yields the energy equality:

$$\frac{1}{2}\frac{\mathrm{d}}{\mathrm{d}t}\|\boldsymbol{u}(t)\|^2_{L^2(\Omega)} + \frac{\alpha}{2}\frac{\mathrm{d}}{\mathrm{d}t}|\boldsymbol{u}(t)|^2_{H^1(\Omega)} + \nu|\boldsymbol{u}(t)|^2_{H^1(\Omega)} = (\boldsymbol{f}(t), \boldsymbol{u}(t)). \tag{1.3.6}$$

This equality shows in particular that, if a solution $\boldsymbol{u}$ exists, then it is unconditionally bounded in $L^\infty(0,T;H^1(\Omega)^3)$ by the data $\boldsymbol{f}$. Now, set

$$\boldsymbol{z} = \mathbf{curl}(\boldsymbol{u} - \alpha\,\Delta\,\boldsymbol{u}). \tag{1.3.7}$$

This choice is crucial because Cioranescu and Ouazar prove that if a function $\boldsymbol{u} \in V$ satisfies $\mathbf{curl}(\boldsymbol{u} - \alpha\,\Delta\,\boldsymbol{u}) \in L^2(\Omega)^3$ and Ω is simply connected, then $\boldsymbol{u} \in H^3(\Omega)^3 \cap V$ and there exists a constant C such that

$$\|\boldsymbol{u}\|_{H^3(\Omega)} \le C\,\|\mathbf{curl}(\boldsymbol{u} - \alpha\,\Delta\,\boldsymbol{u})\|_{L^2(\Omega)}. \tag{1.3.8}$$

Next, take formally the **curl** of (1.3.1); this gives a transport equation, (that we multiply here by α):

$$\alpha\frac{\partial}{\partial t}\boldsymbol{z} + \nu\boldsymbol{z} + \alpha\,\boldsymbol{u}\cdot\nabla\,\boldsymbol{z} - \alpha\,\boldsymbol{z}\cdot\nabla\,\boldsymbol{u} = \nu\,\mathbf{curl}\,\boldsymbol{u} + \alpha\,\mathbf{curl}\,\boldsymbol{f} \quad \text{in } \Omega\times]0,T[, \tag{1.3.9}$$

and formally multiply (1.3.9) by $\boldsymbol{z}$. As $\boldsymbol{u} \in V$, the following Green's formula holds formally:

$$\int_\Omega (\boldsymbol{u}\cdot\nabla\,\boldsymbol{z})\cdot\boldsymbol{z}\,\mathrm{d}\boldsymbol{x} = 0, \tag{1.3.10}$$

and yields the inequality:

$$\begin{aligned}\frac{\alpha}{2}\frac{\mathrm{d}}{\mathrm{d}t}\|\boldsymbol{z}(t)\|^2_{L^2(\Omega)} + (\nu - \alpha\|\nabla\,\boldsymbol{u}(t)\|_{L^\infty(\Omega)})\|\boldsymbol{z}(t)\|^2_{L^2(\Omega)} &\le (\nu\,\|\mathbf{curl}\,\boldsymbol{u}(t)\|_{L^2(\Omega)} \\ &\quad + \alpha\,\|\mathbf{curl}\,\boldsymbol{f}(t)\|_{L^2(\Omega)})\|\boldsymbol{z}(t)\|_{L^2(\Omega)}.\end{aligned} \tag{1.3.11}$$

By applying the Sobolev bound (1.1.18) to $\|\nabla\,\boldsymbol{u}(t)\|_{L^\infty(\Omega)}$ and by using (1.3.8) and (1.3.7), we obtain with another constant C

$$\|\nabla\,\boldsymbol{u}(t)\|_{L^\infty(\Omega)} \le C\,\|\boldsymbol{z}(t)\|_{L^2(\Omega)}.$$

Then by substituting this bound into the left-hand side of (1.3.11), and by substituting the estimate deduced from (1.3.6) to bound $\|\mathbf{curl}\,\boldsymbol{u}(t)\|_{L^2(\Omega)}$ in its right-hand side, we find that $\|z(t)\|^2_{L^2(\Omega)}$ is bounded by the solution of a Riccati differential equation on the time interval $[0, T^*]$, for some $T^* > 0$, $T^* \leq T$. This shows that, if a solution $\boldsymbol{u}$ exists, then it is bounded in $L^\infty(0, T^*; H^3(\Omega)^3)$, see CODDINGTON and LEVINSON [1955]. Finally, on multiplying formally (1.3.1) by $\partial \boldsymbol{u}/\partial t$ and using the previous bound for $\boldsymbol{u}$, we infer that $\partial \boldsymbol{u}/\partial t$ is also bounded in $L^2(0, T^*; H^1(\Omega)^3)$.

These bounds only hold provided a solution exists, but constructing a solution by making use of (1.3.1), (1.3.7), and (1.3.9) is very difficult because these three equations are redundant and no fixed-point can use all three at the same time. The originality and power of construction by Cioranescu and Ouazar lie in that they did use all three equations. Their idea consists in discretizing (1.3.1) by a Galerkin method with the basis of eigenfunctions of the operator $\mathbf{curl}\,\mathbf{curl}(\boldsymbol{u} - \alpha\,\Delta\,\boldsymbol{u})$. This special basis has the effect that, on multiplying the ith equation that discretizes (1.3.1) by the eigenvalue λ_i and on summing over i, we derive a discrete version of the transport equation (1.3.9). This allows to recover (1.3.11) in the discrete case. Thus, we construct a discrete solution $\boldsymbol{u}_m$ that is bounded uniformly in $L^\infty(0, T^*; H^3(\Omega)^3)$, with $\partial \boldsymbol{u}_m/\partial t$ also bounded uniformly in $L^2(0, T^*; H^1(\Omega)^3)$. Note that all the above steps (which were hitherto formal), and in particular the delicate Green's formula (1.3.10), are justified because the basis functions are sufficiently smooth. Furthermore, passing to the limit is standard because this limit is only taken in the discrete version of (1.3.1). The above bounds allow us to pass to the limit in the discrete equations and prove local existence in time of a solution. Global existence in time for suitably restricted data can also be established, by taking better advantage of the small damping effect of the viscous term $-\nu\Delta\,\boldsymbol{u}$. The precise conditions are somewhat technical, and we refer the reader to CIORANESCU and GIRAULT [1997]. The next theorem summarizes the local existence result that was obtained by CIORANESCU and OUAZAR [1984a, 1984b].

THEOREM 1.3.2. *Let Ω be simply connected with boundary of class $\mathcal{C}^{3,1}$. Then, for any force $\boldsymbol{f}$ in $L^2(0, T; H^1(\Omega)^3)$, any initial velocity $\boldsymbol{u}_0$ in $H^3(\Omega)^3$ and any parameters $\nu > 0$ and $\alpha > 0$, there exists a time $T^* > 0$, such that problem (1.3.1)–(1.3.4) has a unique solution $(\boldsymbol{u}, p)$ in $L^\infty(0, T^*; H^3(\Omega)^3) \times L^2(0, T^*; L^2_0(\Omega))$ with $\partial \boldsymbol{u}/\partial t$ in $L^2(0, T^*; H^1(\Omega)^3)$.*

Regarding the regularity hypotheses on the data, it follows from (1.3.11) that $\mathbf{curl}\boldsymbol{f} \in L^2(\Omega)^3$ is sufficient (instead of $\boldsymbol{f}$ in $H^1(\Omega)^3$). Furthermore, finding $\boldsymbol{u}$ in $H^3(\Omega)^3$ is not necessary; if we accept solutions that are less smooth, we can lower the regularity of $\partial\Omega$. Indeed, (1.3.11) only requires $\boldsymbol{u}$ in $W^{1,\infty}(\Omega)^3$. Thus applying Sobolev's imbedding (1.1.18), it suffices that $\boldsymbol{u} \in W^{2,r}(\Omega)^3$ for some $r > 3$. This is also sufficient for estimating $\|\partial \boldsymbol{u}/\partial t\|_{L^2(\Omega)}$. As (1.3.8) is based on the regularity of a Stokes problem with data in $H^1(\Omega)^3$, it can be replaced by a weaker statement with data in $L^r(\Omega)^3$, and Theorem 1.1.9 in the case $d = 3$ implies that it suffices that the largest inner dihedral angle of $\partial\Omega$ be strictly smaller than $2\pi/3$. Finally, BERNARD [1998] and BERNARD [1999] prove that Ω can be multiply-connected, if $\partial\Omega$ is of class $\mathcal{C}^{2,1}$. This makes use of the material in AMROUCHE, BERNARDI, DAUGE and GIRAULT [1998].

REMARK 1.3.3. The importance of the positivity of α is made clear by the energy equality (1.3.6). □

REMARK 1.3.4. The derivation of (1.3.11) requires eliminating the term $\alpha(\boldsymbol{u}\cdot\nabla z, z)$. In view of (1.3.2), and assuming that Green's formula is valid, we have:

$$\alpha\int_{\Omega}(\boldsymbol{u}\cdot\nabla z)z\,\mathrm{d}\boldsymbol{x}=\frac{\alpha}{2}\int_{\partial\Omega}(\boldsymbol{u}\cdot\boldsymbol{n})|z|^2\mathrm{d}s. \tag{1.3.12}$$

Hence this term vanishes either if $\boldsymbol{u}\cdot\boldsymbol{n}=0$, that is the case of a tangential boundary condition, or if $z=\mathbf{0}$ where $\boldsymbol{u}\cdot\boldsymbol{n}\neq 0$. In the second case, what is the physical meaning of this condition on z ? And what is the mathematical meaning of this condition on z, when z is only in $L^2(\Omega)^3$, as it is here? □

REMARK 1.3.5. At first sight, the energy equality (1.3.6) seems minor because it gives a bound in $H^1(\Omega)^3$, whereas (1.3.8) gives a bound in $H^3(\Omega)^3$. But in fact, (1.3.6) is crucial in estimating the term $\|\mathbf{curl}\,\boldsymbol{u}(t)\|_{L^2(\Omega)}$ in the right-hand side of (1.3.9) *in terms of the data* $\boldsymbol{f}$. If we replace it by (1.3.8), then $\boldsymbol{f}$ is replaced by z, and the resulting loss of optimality is devastating. This loss of optimality will be clearly apparent when studying the problem in two dimensions. □

REMARK 1.3.6. The above construction does not apply when the **curl** of $\boldsymbol{f}$ is not in $L^2(\Omega)^3$. This case is rarely met in practice because $\boldsymbol{f}$ is usually the gravitational force and is very smooth. Nonetheless, the case of rough data is interesting from the mathematical point of view. We refer to the work of BRESCH and LEMOINE in [1998], where $\boldsymbol{f}$ belongs to $L^r(\Omega)^3$ with $r>3$. This work complements the results presented in this text, but does not extend them because Bresch and Lemoine lose (1.3.6) and thus cannot recover our results even when $\mathbf{curl}\,\boldsymbol{f}\in L^2(\Omega)^3$. Finally, existence of solutions when $\boldsymbol{f}\in L^r(\Omega)^3$ with $r\leq 3$ (for instance, $r=2$) is an open problem. □

1.3.2. The two-dimensional problem

In two dimensions, the analysis of problem (1.3.1)–(1.3.4) simplifies substantially by virtue of the following identity, valid for all vectors $z=(0,0,z)$ and $\boldsymbol{u}=(u_1,u_2,0)$ in two variables $\boldsymbol{x}=(x_1,x_2,0)$:

$$\mathbf{curl}(z\times\boldsymbol{u})=\boldsymbol{u}\cdot\nabla z. \tag{1.3.13}$$

As a consequence, assuming that f belongs to $H(\text{curl},\Omega)$ defined in (1.1.15), the energy inequality (1.3.11) reduces to:

$$\begin{aligned}\frac{\alpha}{2}\frac{\mathrm{d}}{\mathrm{d}t}\|z(t)\|^2_{L^2(\Omega)}+\nu\,\|z(t)\|^2_{L^2(\Omega)}&\leq\big(\nu\,\|\text{curl}\,\boldsymbol{u}(t)\|_{L^2(\Omega)}\\&\quad+\alpha\,\|\text{curl}\,\boldsymbol{f}(t)\|_{L^2(\Omega)}\big)\|z(t)\|_{L^2(\Omega)},\end{aligned} \tag{1.3.14}$$

an inequality that no longer involves the gradient of $\boldsymbol{u}$ in $L^\infty(\Omega)^{2\times 2}$. This fact enabled Ouazar to prove in OUAZAR [1981] that problem (1.3.1)–(1.3.4) has always at least one global in time solution, in a simply-connected plane domain with sufficiently smooth boundary, for all positive parameters ν and α and all forces $\boldsymbol{f}$ in $H^1(\Omega)^2$. Published in 1981, this result is remarkable considering that none of the many publications devoted to grade-two fluids that appeared from 1993 onward and used other techniques, are able to prove existence of solutions without heavy restrictions on the size of the data and parameters, see for instance VIDEMANN [1997] and references therein.

The fact that the gradient of $\boldsymbol{u}$ does not need to be in $L^\infty(\Omega)^{2\times 2}$ will enable us to perform the numerical analysis of discrete schemes that solve (1.3.1)–(1.3.4) in two dimensions, and in particular these discrete schemes will not require a CFL condition (cf. COURANT, FRIEDRICHS and LEWY [1928]). However, discretizing the special basis of eigenfunctions of the operator $\mathbf{curl}\,\mathbf{curl}(\boldsymbol{u} - \alpha\,\Delta\,\boldsymbol{u})$ does not appear realistic, and hence we shall use a less sophisticated approach. It consists in splitting the grade-two problem into a Stokes-like system and a transport equation. This transport equation is discussed in the next subsection.

1.3.3. The steady transport equation in arbitrary dimension

The analysis of the two-dimensional problem (1.3.1)–(1.3.4), essentially relies on the well-posedness of a steady scalar transport equation in a Lipschitz domain Ω. Transport equations, steady or transient, have been addressed by a large number of mathematicians, and we can only quote a small number of them: AMBROSIO [2004], BARDOS [1970], BEIRÃO DA VEIGA [1987], COLOMBINI and LERNER [2002], DESJARDINS [1996], DIPERNA and LIONS [1989], FERNÁNDEZ CARA, GUILLÉN GONZÁLEZ and ROBLES ORTEGA [2002], GIRAULT and SCOTT [1999], HÖRMANDER [1983], PUEL and ROPTIN [1967], WALKINGTON [2005]. Here we present the work of GIRAULT and SCOTT [1999] because it is adapted to the situation of grade-two fluids.

The analysis of the equation studied here is independent of the dimension and therefore, we consider Ω in $\mathbb{R}^d$. This problem reads: For f given in $L^2(\Omega)$ and $\boldsymbol{u}$ given in W (see (1.1.12)), find z in $L^2(\Omega)$ satisfying

$$\nu z + \gamma\,\boldsymbol{u}\cdot\nabla z = f \text{ in } \Omega, \tag{1.3.15}$$

where $\nu > 0$ and $\gamma \neq 0$ are given parameters and $d \geq 2$ is an arbitrary integer. Albeit linear, this problem is difficult because of the poor regularity of the domain and the driving velocity $\boldsymbol{u}$. Note that, for $\boldsymbol{u} \in W$, the product $\boldsymbol{u}\cdot\nabla z$ is well-defined in the sense of distributions, by virtue of the following identity that holds for all divergence-free vectors $\boldsymbol{u}$ and scalars z:

$$\boldsymbol{u}\cdot\nabla z = \operatorname{div}(z\boldsymbol{u}).$$

Furthermore, as z and f belong to $L^2(\Omega)$, (1.3.15) implies that z is slightly more regular and belongs to:

$$X_{\boldsymbol{u}} = \{z \in L^2(\Omega);\ \boldsymbol{u}\cdot\nabla z \in L^2(\Omega)\}; \tag{1.3.16}$$

$X_{\boldsymbol{u}}$ is a Hilbert space for the norm

$$\|z\|_{\boldsymbol{u}} = \left(\|z\|^2_{L^2(\Omega)} + \|\boldsymbol{u}\cdot\nabla z\|^2_{L^2(\Omega)}\right)^{1/2}. \tag{1.3.17}$$

Constructing a solution of (1.3.15) by Galerkin's method is an easy exercise.

PROPOSITION 1.3.7. *Let $\Omega \subset \mathbb{R}^d$ be bounded and Lipschitz-continuous. For all $\boldsymbol{u}$ in W, all f in $L^2(\Omega)$ and all real numbers $\gamma \neq 0$ and $\nu > 0$, the transport equation (1.3.15) has at least one solution z in $X_{\boldsymbol{u}}$ and z satisfies*

$$\|z\|_{L^2(\Omega)} \leq \frac{1}{\nu}\|f\|_{L^2(\Omega)}, \quad \|\boldsymbol{u} \cdot \nabla z\|_{L^2(\Omega)} \leq \frac{1}{|\gamma|}\|f\|_{L^2(\Omega)}. \tag{1.3.18}$$

But proving uniqueness of the solution is hard because $\boldsymbol{u}$ is not smooth, it does not vanish on the boundary and this boundary is not smooth. Ideally, uniqueness relies on the validity of the following Green's formula:

$$\forall \boldsymbol{u} \in W,\ \forall z \in X_{\boldsymbol{u}},\ \sum_{i=1}^{d} \int_\Omega u_i \frac{\partial z}{\partial x_i} z \,\mathrm{d}\boldsymbol{x} = 0. \tag{1.3.19}$$

This formula holds for z in $H^1(\Omega)$, and if it were known that $H^1(\Omega)$ is dense in $X_{\boldsymbol{u}}$, then (1.3.19) would stem trivially by density. Unfortunately, when $\boldsymbol{u}$ has only H^1 regularity, this density must be established, and this is just as difficult as Green's formula itself; in fact, it is shown in GIRAULT and SCOTT [1999] that these two properties are equivalent. This density is established by proving the following results. The first one is based on regularization by convolution with a special mollifier (a variant of an idea of PUEL and ROPTIN [1967]), and the second one relies on the renormalization technique of DIPERNA and LIONS [1989]. The details can be found in GIRAULT and SCOTT [1999].

THEOREM 1.3.8. *Let Ω be a bounded Lipschitz-continuous domain of $\mathbb{R}^d$ and let $\boldsymbol{u}$ be given in $H^1(\Omega)^d$. Then for each z in $L^2(\Omega)$ such that $\boldsymbol{u} \cdot \nabla z$ belongs to $L^1(\Omega)$ (e.g., if $z \in X_{\boldsymbol{u}}$), there exists a sequence $(z_k)_{k \geq 1}$ of functions $z_k \in \mathcal{C}^\infty(\overline{\Omega})$ such that*

$$\lim_{k\to\infty} z_k = z \text{ in } L^2(\Omega), \quad \lim_{k\to\infty} \boldsymbol{u} \cdot \nabla z_k = \boldsymbol{u} \cdot \nabla z \text{ in } L^1(\Omega).$$

PROPOSITION 1.3.9. *Let $\Omega \subset \mathbb{R}^d$ be bounded and Lipschitz-continuous. For all $\boldsymbol{u}$ in W, all f in $L^2(\Omega)$ and all real numbers $\gamma \neq 0$ and $\nu > 0$, the transport equation (1.3.15) has one and only one solution z in $X_{\boldsymbol{u}}$.*

These two results have important consequences.

PROPOSITION 1.3.10. *Let $\Omega \subset \mathbb{R}^d$ be bounded and Lipschitz-continuous and let $\boldsymbol{u}$ be given in W. Then (1.3.19) holds for all z in $X_{\boldsymbol{u}}$.*

COROLLARY 1.3.11. *Let $\Omega \subset \mathbb{R}^d$ be bounded and Lipschitz-continuous and let $\boldsymbol{u}$ be given in W. Then*

$$\forall v \in X_{\boldsymbol{u}},\ \forall w \in X_{\boldsymbol{u}},\ \int_\Omega (\boldsymbol{u} \cdot \nabla v) w \,\mathrm{d}\boldsymbol{x} + \int_\Omega (\boldsymbol{u} \cdot \nabla w) v \,\mathrm{d}\boldsymbol{x} = 0. \tag{1.3.20}$$

COROLLARY 1.3.12. *Let $\Omega \subset \mathbb{R}^d$ be bounded and Lipschitz-continuous and let $\boldsymbol{u}$ be given in W. Then any h in $L^2(\Omega)$ has the orthogonal decomposition:*

$$h = z + \boldsymbol{u} \cdot \nabla z \quad in\ \Omega,$$

where z belongs to $X_{\boldsymbol{u}}$, and

$$\|z\|_{\boldsymbol{u}} = \|h\|_{L^2(\Omega)}. \tag{1.3.21}$$

THEOREM 1.3.13. *Let $\Omega \subset \mathbb{R}^d$ be Lipschitz-continuous and let $\boldsymbol{u}$ be given in W. Then $\mathcal{D}(\Omega)$ is dense in $X_{\boldsymbol{u}}$.*

REMARK 1.3.14. The density statement of Theorem 1.3.8 holds without restriction on $\boldsymbol{u}$. But what do we know of the density of $\mathcal{D}(\overline{\Omega})$ in $X_{\boldsymbol{u}}$ when $\boldsymbol{u}$ is arbitrary in $H^1(\Omega)^d$? If this density were true, we could give meaning to the left-hand side of (1.3.12), and we could solve the steady grade-two problem with any Dirichlet boundary condition, by prescribing z where $\boldsymbol{u} \cdot \boldsymbol{n} \neq 0$. □

The unique solvability of (1.3.15) in $L^2(\Omega)$ extends immediately to the equation

$$c\,z + \gamma\, \boldsymbol{u} \cdot \nabla z = f \text{ in } \Omega,$$

where $c \in L^\infty(\Omega)$ is uniformly bounded below: There exists $c_0 > 0$ such that

$$c(\boldsymbol{x}) \geq c_0 \text{ a.e. } \boldsymbol{x} \in \Omega.$$

More generally, it extends straightforwardly to the following transport system: Find $\boldsymbol{z} \in L^2(\Omega)^d$ solution of

$$\boldsymbol{C}\boldsymbol{z} + \gamma\, \boldsymbol{u} \cdot \nabla \boldsymbol{z} = \boldsymbol{f} \text{ in } \Omega, \tag{1.3.22}$$

where $\boldsymbol{f}$ is given in $L^2(\Omega)^d$, $\gamma \neq 0$ in $\mathbb{R}$, $\boldsymbol{u}$ in W, and $\boldsymbol{C} \in L^\infty(\Omega)^{d\times d}$ is a uniformly positive definite matrix, i.e., satisfying: There exists a constant $c_0 > 0$ such that

$$\forall \boldsymbol{\chi} \in \mathbb{R}^d,\ (\boldsymbol{C}(\boldsymbol{x})\boldsymbol{\chi}, \boldsymbol{\chi}) \geq c_0 |\boldsymbol{\chi}|^2 \text{ a.e. } \boldsymbol{x} \in \Omega. \tag{1.3.23}$$

PROPOSITION 1.3.15. *Let $\Omega \subset \mathbb{R}^d$ be bounded and Lipschitz-continuous. For all $\boldsymbol{u}$ in W, all $\boldsymbol{f}$ in $L^2(\Omega)^d$, all matrix-valued functions $\boldsymbol{C} \in L^\infty(\Omega)^{d\times d}$ satisfying (1.3.23), and all real numbers $\gamma \neq 0$, the transport system (1.3.22) has one and only one solution $\boldsymbol{z}$ in $L^2(\Omega)^d$, and $\boldsymbol{z}$ satisfies the bound*

$$\|\boldsymbol{z}\|_{L^2(\Omega)} \leq \frac{1}{c_0}\|\boldsymbol{f}\|_{L^2(\Omega)}. \tag{1.3.24}$$

There are several generalizations of Proposition 1.3.9 to L^p, $p > 2$, cf. FERNÁNDEZ CARA, GUILLÉN GONZÁLEZ and ROBLES ORTEGA [2002], ROBLES ORTEGA [1995], or GIRAULT and SCOTT [2002a]. The first two references construct the solution by a characteristic method on a smooth domain with driving velocity in $W^{2,r}$ for $r > d$. The third reference adapts this proof to a Lipschitz domain and a driving velocity $\boldsymbol{u}$ in W by extension to a smooth ball and regularization of $\boldsymbol{u}$. It yields the following result.

PROPOSITION 1.3.16. *Let $p > 2$ be a real number, Ω a bounded Lipschitz domain in $\mathbb{R}^d$, $f \in L^p(\Omega)$, $\gamma \neq 0$, $\nu > 0$ in $\mathbb{R}$, and $\boldsymbol{u}$ in W. Then the unique solution z of the transport equation (1.3.15) belongs to $L^p(\Omega)$ and*

$$\|z\|_{L^p(\Omega)} \leq \frac{1}{\nu}\|f\|_{L^p(\Omega)}. \tag{1.3.25}$$

Unfortunately, the approach in FERNÁNDEZ CARA, GUILLÉN GONZÁLEZ and ROBLES ORTEGA [2002] does not seem to extend to the system (1.3.22). Nevertheless, L^p results are derived for (1.3.22) by GIRAULT and TARTAR [2010] when $d \leq 4$. The proof is based on an elliptic regularization of (1.3.22), whence the restriction on the dimension, and a Yosida approximation of the elliptic regularization. Thus, we have the next result.

THEOREM 1.3.17. *Let $p > 2$ be a real number, Ω a bounded Lipschitz domain in $\mathbb{R}^d$, $d = 2, 3, 4$, $\boldsymbol{f} \in L^p(\Omega)^d$, $\gamma \neq 0$ in $\mathbb{R}$, $\boldsymbol{u}$ in W, and $\boldsymbol{C} \in L^\infty(\Omega)^{d\times d}$ satisfying (1.3.23). Then, the unique solution $\boldsymbol{z}$ of (1.3.22) belongs to $L^p(\Omega)^d$ and*

$$\|\boldsymbol{z}\|_{L^p(\Omega)} \leq \frac{1}{c_0}\|\boldsymbol{f}\|_{L^p(\Omega)}. \tag{1.3.26}$$

If $\frac{2d}{d+2} \leq p < 2$ for $d = 3, 4$ or $1 < p < 2$ when $d = 2$, a proof by duality and transposition (cf. LIONS and MAGENES [1968]), shows that the transposed formulation of the transport system (1.3.22) has one and only one solution $\boldsymbol{z} \in L^p(\Omega)^d$ that satisfies (1.3.25) and that solves (1.3.22).

Regarding the H^1 regularity of the solution of (1.3.15), several authors (cf. for instance BEIRÃO DA VEIGA [1987] and references therein) have established in a smooth domain that if $f \in H^1(\Omega)$ and if $\boldsymbol{u}$ is in $W^{1,\infty}(\Omega)^d \cap W$ small enough, then z belongs to $H^1(\Omega)$ and is suitably bounded by the data. There are several proofs of this result, but all either require a smooth boundary or rely on the H^2 regularity of the Laplace equation with homogeneous Dirichlet or Neumann boundary conditions. This regularity holds either if the boundary is smooth or if the domain is a convex polygon or polyhedron. For instance, BEIRÃO DA VEIGA [1987] discretizes (1.3.15) in the basis of eigenfunctions of the Laplace operator, with a Neumann boundary condition:

$$-\Delta v_k = \lambda_k v_k \text{ in } \Omega, \quad \frac{\partial v_k}{\partial \boldsymbol{n}} = 0 \text{ on } \partial\Omega, \quad \int_\Omega v_k \, \mathrm{d}\boldsymbol{x} = 0.$$

The convexity of Ω, or the regularity of its boundary, guarantees that v_k belongs to $H^2(\Omega)$ because Theorem 1.1.3 applies also to homogeneous Neumann boundary conditions, cf. GRISVARD [1985]. This approach leads to the following result.

THEOREM 1.3.18. *Let Ω be convex or have a boundary of class $\mathcal{C}^{1,1}$; assume that f belongs to $H^1(\Omega)$ and $\boldsymbol{u}$ belongs to $W^{1,\infty}(\Omega)^d \cap W$ with*

$$\frac{|\gamma|}{\nu}\|\nabla \boldsymbol{u}\|_{L^\infty(\Omega)} := \delta < 1. \tag{1.3.27}$$

Then the solution z of the transport equation (1.3.15) belongs to $H^1(\Omega)$ and is bounded as follows:

$$|z|_{H^1(\Omega)} \leq \frac{1}{1-\delta}|f|_{H^1(\Omega)}. \tag{1.3.28}$$

The L^p estimates of Theorem 1.3.17 can be used to derive the $W^{1,p}$ regularity of the solution of (1.3.15). Indeed, under the assumptions of Theorem 1.3.18, since z belongs to $H^1(\Omega)$, the gradient of each term in (1.3.15) is well defined in the sense of distributions and ∇z solves: Find $\boldsymbol{w}$ in $L^2(\Omega)^d$ such that

$$\nu \boldsymbol{w} + \gamma \boldsymbol{u} \cdot \nabla \boldsymbol{w} + \gamma\ (\nabla \boldsymbol{u})^T \boldsymbol{w} = \nabla f. \tag{1.3.29}$$

It is a particular case of (1.3.22) with $\boldsymbol{C} = \nu \boldsymbol{I} + \gamma\ (\nabla \boldsymbol{u})^T$. The fact that $\boldsymbol{u}$ belongs to $W^{1,\infty}(\Omega)^d$ implies that $\boldsymbol{C}$ is uniformly bounded in Ω and owing to (1.3.27), $\boldsymbol{C}$ satisfies (1.3.23) with $c_0 = \nu - |\gamma|\,|\boldsymbol{u}|_{W^{1,\infty}(\Omega)}$. Hence Theorem 1.3.17 implies immediately the next result.

THEOREM 1.3.19. *Let $p > 2$ be a real number. Let Ω be a bounded convex or $\mathcal{C}^{1,1}$ domain in $\mathbb{R}^d$, $d = 2, 3, 4$, $f \in W^{1,p}(\Omega)$, $\nu > 0$, $\gamma \neq 0$ in $\mathbb{R}$, and $\boldsymbol{u} \in W^{1,\infty}(\Omega)^d \cap W$ satisfying (1.3.27). Then the unique solution z of (1.3.15) belongs to $W^{1,p}(\Omega)$ and*

$$|z|_{W^{1,p}(\Omega)} \leq \frac{1}{1-\delta}|f|_{W^{1,p}(\Omega)}. \tag{1.3.30}$$

REMARK 1.3.20. Finally, as observed in GIRAULT and TARTAR [2010], the statement of Theorem 1.3.19 is valid in a bounded Lipschitz domain in the case when the full trace of $\boldsymbol{u}$ vanishes on $\partial\Omega$. □

Analogous results for a time-dependent transport equation are derived in Section 3.1.

1.4. Splitting the two-dimensional problem

In this section, we propose to extend the two-dimensional results of OUAZAR [1981] to domains with rough boundaries. More precisely, we solve the two-dimensional problem (1.3.1)–(1.3.4) in an arbitrary bounded, connected domain Ω with a Lipschitz-continuous boundary $\partial\Omega$, by putting it into what is known to numerical analysts as a mixed formulation. The reader can refer to BREZZI and FORTIN [1991], GIRAULT and RAVIART [1986], or ERN and GUERMOND [2004] for current examples of mixed formulations. We follow the approach of GIRAULT and SCOTT [1999], but we treat here the simpler case of a steady fluid with a no-slip boundary condition. The transient problem is postponed to Chapter 3, and the case of nonhomogeneous boundary conditions to Chapter 5.

1.4.1. The steady no-slip two-dimensional problem

Considering the material of Section 1.3.1, we assume that Ω is a bounded, connected domain in $\mathbb{R}^2$, with a Lipschitz-continuous boundary $\partial\Omega$, $\boldsymbol{f}$ is a given function in $H(\text{curl}, \Omega)$ and $\nu > 0$ and $\alpha > 0$ are two given real constants. Following OUAZAR [1981] and GIRAULT and

SCOTT [1999], we shall look for the pressure p in $L_0^2(\Omega)$ and the velocity $\boldsymbol{u}$ in the space of functions $\boldsymbol{v}$ in V such that $\mathrm{curl}(\boldsymbol{v} - \alpha\Delta\boldsymbol{v})$ is in $L^2(\Omega)$; this may be written more concisely as $\boldsymbol{u} \in V^\alpha$, where

$$V^\alpha = \{\boldsymbol{v} \in V;\ \alpha\,\mathrm{curl}\,\Delta\boldsymbol{v} \in L^2(\Omega)\}. \tag{1.4.1}$$

It is a Hilbert space equipped with the norm

$$\|\boldsymbol{v}\|_{V^\alpha} = \left(|\boldsymbol{v}|^2_{H^1(\Omega)} + \|\alpha\,\mathrm{curl}\,\Delta\boldsymbol{v}\|^2_{L^2(\Omega)}\right)^{1/2}. \tag{1.4.2}$$

Of course, when $\alpha = 0$, V^α reduces to V. Then, the steady version of problem (1.3.1)–(1.3.4) reads: Find a pair $(\boldsymbol{u}, p) \in V^\alpha \times L_0^2(\Omega)$, solution of

$$-\nu\Delta\boldsymbol{u} + \mathbf{curl}(\boldsymbol{u} - \alpha\Delta\boldsymbol{u}) \times \boldsymbol{u} + \nabla p = \boldsymbol{f} \quad \text{in } \Omega. \tag{1.4.3}$$

Strictly speaking, the sign of α does not influence the mathematical analysis of this problem, but we choose it positive to be consistent with thermodynamics.

Now, let $(\boldsymbol{u} = (u_1, u_2), p) \in V^\alpha \times L_0^2(\Omega)$ be a solution of (1.4.3), and introduce the auxiliary variable

$$z = \mathrm{curl}(\boldsymbol{u} - \alpha\Delta\boldsymbol{u}), \qquad \boldsymbol{z} = (0, 0, z). \tag{1.4.4}$$

Note that $z \in L^2(\Omega)$, $\boldsymbol{z} \in L^2(\Omega)^3$,

$$\mathrm{div}\,\boldsymbol{z} = 0,\ \boldsymbol{z} \times \boldsymbol{u} = (-zu_2, zu_1). \tag{1.4.5}$$

Then (1.4.3) becomes a generalized Stokes equation

$$-\nu\,\Delta\boldsymbol{u} + \boldsymbol{z} \times \boldsymbol{u} + \nabla p = \boldsymbol{f} \quad \text{in } \Omega. \tag{1.4.6}$$

Next, let us take the curl of (1.4.6) in the sense of distributions and apply (1.3.13); we obtain

$$-\nu\,\Delta(\mathrm{curl}\,\boldsymbol{u}) + \boldsymbol{u} \cdot \nabla z = \mathrm{curl}\,\boldsymbol{f}. \tag{1.4.7}$$

Then, we can write

$$-\nu\,\alpha\,\Delta(\mathrm{curl}\,\boldsymbol{u}) = \nu\,(\mathrm{curl}(\boldsymbol{u} - \alpha\,\Delta\boldsymbol{u}) - \mathrm{curl}\,\boldsymbol{u}) = \nu(z - \mathrm{curl}\,\boldsymbol{u}),$$

and, passing $\nu\,\mathrm{curl}\,\boldsymbol{u}$ to the right-hand side, we see that z satisfies the transport equation

$$\nu\,z + \alpha\,\boldsymbol{u} \cdot \nabla z = \nu\,\mathrm{curl}\,\boldsymbol{u} + \alpha\,\mathrm{curl}\,\boldsymbol{f}. \tag{1.4.8}$$

Finally, we observe that the only regularity that is explicitly used by (1.4.6), (1.4.8) is: $\boldsymbol{u} \in V$, $p \in L_0^2(\Omega)$ and $z \in L^2(\Omega)$.

Conversely, let $\boldsymbol{u} \in V$, $p \in L_0^2(\Omega)$ and $\boldsymbol{z} = (0, 0, z)$ with $z \in L^2(\Omega)$ be a solution of (1.4.6), (1.4.8). Then $\boldsymbol{z}$ satisfies (1.4.5) and taking the curl of (1.4.6) in the sense of distributions yields again (1.4.7). Next multiplying by α and comparing with (1.4.8), we obtain

$$-\nu\,\alpha\,\Delta(\operatorname{curl}\boldsymbol{u}) = \nu\,z - \nu\operatorname{curl}\boldsymbol{u},$$

i.e., $z = \operatorname{curl}(\boldsymbol{u} - \alpha\Delta\,\boldsymbol{u})$. Therefore, $\boldsymbol{u}$ belongs to V^α and substituting the expression of z into (1.4.6) shows that $(\boldsymbol{u}, p)$ is a solution of the original equation (1.4.3). This is summarized in the following lemma.

LEMMA 1.4.1. *Problem (1.4.3) with* $(\boldsymbol{u}, p)$ *in* $V^\alpha \times L_0^2(\Omega)$ *is equivalent to: Find* $(\boldsymbol{u}, p, z)$ *in* $V \times L_0^2(\Omega) \times L^2(\Omega)$ *solution of the generalized Stokes problem (1.4.6) and the transport equation (1.4.8), namely*

$$\begin{aligned} -\nu\,\Delta\,\boldsymbol{u} + \boldsymbol{z}\times\boldsymbol{u} + \nabla\,p = \boldsymbol{f} \quad & \text{in } \Omega,\\ \operatorname{div}\boldsymbol{u} = 0 \quad & \text{in } \Omega,\\ \boldsymbol{u} = \boldsymbol{0} \quad & \text{on } \partial\Omega,\\ \nu\,z + \alpha\,\boldsymbol{u}\cdot\nabla\,z = \nu\operatorname{curl}\boldsymbol{u} + \alpha\operatorname{curl}\boldsymbol{f}. & \end{aligned} \tag{1.4.9}$$

REMARK 1.4.2. When $\alpha = 0$, equation (1.4.6) is unchanged and equation (1.4.8) reduces to $z = \operatorname{curl}\boldsymbol{u}$. In this case, z is simply the vorticity of $\boldsymbol{u}$ and problem (1.4.9) can be interpreted as a velocity-vorticity-pressure formulation of the Navier–Stokes equations. □

REMARK 1.4.3. As mentioned in Section 1.3.1, (1.4.4), (1.4.6), and (1.4.8) are redundant. For a standard discretization, one of them must be discarded because discretizing the special basis of eigenfunctions of the operator $\mathbf{curl}\,\mathbf{curl}(\boldsymbol{u} - \alpha\,\Delta\,\boldsymbol{u})$ appears unrealistic. In view of the importance of (1.4.6) and (1.4.8), we choose to discard the relation (1.4.4) between z and $\boldsymbol{u}$. The price to pay is that the regularity of a solution $(\boldsymbol{u}, p)$ of (1.4.9) can only be deduced from (1.4.6). □

Existence of a solution of (1.4.9) is easily derived by Galerkin's method. We sketch the construction and refer to GIRAULT and SCOTT [1999] for details and proofs. First, we note that for a given z in $L^2(\Omega)$, with the notation of (1.4.5), the generalized Stokes problem: Find $(\boldsymbol{v}(z), q(z))$ in $V \times L_0^2(\Omega)$, such that

$$-\nu\,\Delta\,\boldsymbol{v} + \boldsymbol{z}\times\boldsymbol{v} + \nabla\,q = \boldsymbol{f} \quad \text{in } \Omega, \tag{1.4.10}$$

has a unique solution. Indeed, this problem has the following equivalent variational formulation: Find $(\boldsymbol{v}(z), q(z))$ in $H_0^1(\Omega)^2 \times L_0^2(\Omega)$, such that

$$\forall \boldsymbol{w} \in H_0^1(\Omega)^2,\ a_z(\boldsymbol{v}(z), \boldsymbol{w}) + b(\boldsymbol{w}, q(z)) = (\boldsymbol{f}, \boldsymbol{w}), \tag{1.4.11}$$

$$\forall r \in L_0^2(\Omega),\ b(\boldsymbol{v}(z), r) = 0, \tag{1.4.12}$$

where

$$\begin{aligned} a_z(\boldsymbol{v}, \boldsymbol{w}) &= \nu(\nabla\,\boldsymbol{v}, \nabla\,\boldsymbol{w}) + (\boldsymbol{z}\times\boldsymbol{v}, \boldsymbol{w}),\\ b(\boldsymbol{w}, r) &= -(r, \operatorname{div}\boldsymbol{w}). \end{aligned}$$

The following proposition states that it is well posed.

PROPOSITION 1.4.4. *Let Ω be bounded, connected, and Lipschitz-continuous, $\nu > 0$ and $\boldsymbol{f} \in L^2(\Omega)^2$. For any z in $L^2(\Omega)$, the generalized Stokes problem (1.4.11)–(1.4.12) has a unique solution $(\boldsymbol{v}(z), q(z))$ in $V \times L_0^2(\Omega)$. This solution satisfies the following bounds in $H^1(\Omega)^2 \times L^2(\Omega)$:*

$$|\boldsymbol{v}(z)|_{H^1(\Omega)} \le \frac{S_2}{\nu}\|\boldsymbol{f}\|_{L^2(\Omega)}, \tag{1.4.13}$$

$$\|q(z)\|_{L^2(\Omega)} \le \frac{1}{\beta}\Big(S_2\|\boldsymbol{f}\|_{L^2(\Omega)} + S_4^2|\boldsymbol{v}(z)|_{H^1(\Omega)}\|z\|_{L^2(\Omega)}\Big), \tag{1.4.14}$$

where $\beta > 0$ is the constant of the inf-sup condition (1.1.26) and S_p the constant of Sobolev's imbedding (1.1.3).

Observe that the bound (1.4.13) is independent of z.

Next, let c be the standard trilinear form associated with the Navier–Stokes equations

$$c(\boldsymbol{u}; z, \theta) = \sum_{i=1}^{2}\int_\Omega u_i\left(\frac{\partial z}{\partial x_i}\right)\theta\,\mathrm{d}\boldsymbol{x}. \tag{1.4.15}$$

It stems from the results of Section 1.3.3, that c satisfies (1.3.19); but for existence, we only require the much simpler statement:

$$\forall \boldsymbol{u} \in W,\ \forall z \in H^1(\Omega),\ \ c(\boldsymbol{u}; z, z) = 0. \tag{1.4.16}$$

Let $\{w_i\}_{i\ge 1}$ be a basis of $H^2(\Omega)$, let Z_m be the vector space spanned by $(w_i)_{i=1}^m$, and let us discretize z by Galerkin's method in this basis. For each z_m in Z_m, we set $\boldsymbol{z}_m = (0, 0, z_m)$, we denote by $\boldsymbol{u}(z_m)$ the unique solution of the generalized Stokes problem (1.4.11)–(1.4.12), and we discretize the transport equation (1.4.8) by: Find z_m in Z_m such that, for $1 \le i \le m$,

$$\nu(z_m, w_i) + \alpha c(\boldsymbol{u}(z_m); z_m, w_i) = \nu(\operatorname{curl}\boldsymbol{u}(z_m), w_i) + \alpha(\operatorname{curl}\boldsymbol{f}, w_i). \tag{1.4.17}$$

As $\boldsymbol{u}(z_m)$ is determined by z_m, the only unknown in (1.4.17) is z_m, and thus (1.4.17) is a system of m nonlinear equations in m unknowns, the components of z_m in Z_m. Hence, it can be solved by Brouwer's fixed point theorem, see for instance GIRAULT and RAVIART [1986]. The result is stated in the next proposition.

PROPOSITION 1.4.5. *Let Ω be bounded, connected, and Lipschitz-continuous. For all integers $m \ge 1$, all $\nu > 0$, all $\alpha > 0$, and all $\boldsymbol{f} \in H(\text{curl}, \Omega)$, the discrete problem (1.4.17) has at least one solution z_m in Z_m and each solution z_m satisfies the uniform estimate with respect to m:*

$$\|z_m\|_{L^2(\Omega)} \le \frac{S_2}{\nu}\|\boldsymbol{f}\|_{L^2(\Omega)} + \frac{\alpha}{\nu}\|\operatorname{curl}\boldsymbol{f}\|_{L^2(\Omega)}. \tag{1.4.18}$$

The last estimate is derived by using (1.1.16).

The uniform estimate (1.4.18) allows to pass to the limit in (1.4.17) and leads to the following existence result.

THEOREM 1.4.6. *Let Ω be bounded, connected, and Lipschitz-continuous. For all $\nu > 0$, all $\alpha > 0$, and all $\boldsymbol{f} \in H(\text{curl}, \Omega)$, problem (1.4.9) has at least one solution $(\boldsymbol{u}, p, z)$ and each solution $(\boldsymbol{u}, p, z)$ of (1.4.9) satisfies the following estimates:*

$$\|z\|_{L^2(\Omega)} \leq \frac{S_2}{\nu}\|\boldsymbol{f}\|_{L^2(\Omega)} + \frac{\alpha}{\nu}\|\text{curl}\,\boldsymbol{f}\|_{L^2(\Omega)}, \tag{1.4.19}$$

$$|\boldsymbol{u}|_{H^1(\Omega)} \leq \frac{S_2}{\nu}\|\boldsymbol{f}\|_{L^2(\Omega)}, \tag{1.4.20}$$

$$\|p\|_{L^2(\Omega)} \leq \frac{1}{\beta}\left(S_2\|\boldsymbol{f}\|_{L^2(\Omega)} + S_4^2|\boldsymbol{u}|_{H^1(\Omega)}\|z\|_{L^2(\Omega)}\right), \tag{1.4.21}$$

where β is the constant of the inf-sup condition (1.1.26) and S_p the constant of Sobolev's imbedding (1.1.3).

REMARK 1.4.7. The estimates (1.4.19)–(1.4.21) hold on a bounded, connected, Lipschitz-continuous domain, without restriction on the size of the data. But their derivation, and in particular deriving an unconditional estimate for z depends drastically on the choice of this auxiliary variable and the space to which it should belong. With our choice, that dates back to OUAZAR [1981], the transport equation (1.4.8) for z has only one nonlinear term and the regularity of the Galerkin solution is such that $\|z_m\|_{L^2(\Omega)}$ can be bounded unconditionally by $\text{curl}\,\boldsymbol{u}_m$ (hence by $\boldsymbol{f}$) and $\text{curl}\,\boldsymbol{f}$. There are, of course, several possibilities for splitting the original equation, but no other choice seems to produce this happy result. In this respect, the splitting achieved by problem (1.4.9) is optimal. □

Additional regularity and uniqueness

The material of this paragraph can be found in GIRAULT and SCOTT [1999]. When Ω is a polygon, any solution $(\boldsymbol{u}, p)$ of (1.4.9) has additional regularity because, for $\boldsymbol{f}$ sufficiently smooth, the homogeneous generalized Stokes operator in (1.4.10) has a regularizing effect. In contrast, the transport operator in (1.4.8) brings no regularization. Therefore, Theorem 1.1.6 has the following consequence.

THEOREM 1.4.8. *Let Ω be a connected polygon and assume that all the inner angles of $\partial\Omega$ belong to $]0, 2\pi[$. Let $\nu > 0$, $\alpha > 0$, and $\boldsymbol{f} \in L^{4/3}(\Omega)^2$ be given. Then all solutions $(\boldsymbol{u}, p, z)$ of the first three equations of problem (1.4.9) satisfy*

$$\boldsymbol{u} \in W^{2,4/3}(\Omega)^2, \quad p \in W^{1,4/3}(\Omega),$$

with continuous dependence on the data

$$\|\boldsymbol{u}\|_{W^{2,4/3}(\Omega)} + \|p\|_{W^{1,4/3}(\Omega)} \leq C_1\|\boldsymbol{f}\|_{L^{4/3}(\Omega)}\left(1 + \frac{S_4^2}{\nu}\|z\|_{L^2(\Omega)}\right), \tag{1.4.22}$$

where C_1 is the continuity constant of Theorem 1.1.6. For $\boldsymbol{f} \in H(\text{curl}, \Omega)$, the regularity of z is unchanged.

PROOF. This is a very simple bootstrap argument. Sobolev's imbedding gives immediately the following analog of (1.4.13):

$$|\boldsymbol{u}|_{H^1(\Omega)} \leq \frac{S_4}{\nu} \|\boldsymbol{f}\|_{L^{4/3}(\Omega)}.$$

Then $z \times \boldsymbol{u}$ belongs to $L^{4/3}(\Omega)^2$, with

$$\|z \times \boldsymbol{u}\|_{L^{4/3}(\Omega)} \leq S_4 \|z\|_{L^2(\Omega)} |\boldsymbol{u}|_{H^1(\Omega)},$$

and (1.4.22) follows by applying Theorem 1.1.6 to the Stokes problem (1.1.19)–(1.1.20) with right-hand side $\boldsymbol{f} - z \times \boldsymbol{u}$. □

To simplify, we introduce the notation for any z in $L^2(\Omega)$:

$$K_1(z) = 1 + \frac{S_4^2}{\nu} \|z\|_{L^2(\Omega)}. \tag{1.4.23}$$

For $\boldsymbol{f} \in L^2(\Omega)^2$, higher regularity can be derived by restricting adequately the inner angles of $\partial\Omega$, the best result being achieved when Ω is convex. We skip the proof, which is also based on a bootstrap argument.

THEOREM 1.4.9. *Let Ω be a convex polygon. Let $\nu > 0$, $\alpha > 0$, and $\boldsymbol{f} \in L^2(\Omega)^2$ be given. Then all solutions $(\boldsymbol{u}, p, z)$ of the first three equations in problem (1.4.9) satisfy*

$$\boldsymbol{u} \in H^2(\Omega)^2, \quad p \in H^1(\Omega),$$

with continuous dependence on the data

$$\|\boldsymbol{u}\|_{H^2(\Omega)} + \|p\|_{H^1(\Omega)} \leq C_2 \left(\|\boldsymbol{f}\|_{L^2(\Omega)} + C_\infty C_1 K_1(z) \, \|z\|_{L^2(\Omega)} \|\boldsymbol{f}\|_{L^{4/3}(\Omega)}\right), \tag{1.4.24}$$

where C_1 and C_2 are respectively the continuity constants of Theorems 1.1.6 and 1.1.5, and C_∞ is the constant of the imbedding:

$$\forall v \in W^{2,4/3}(\Omega), \ \|v\|_{L^\infty(\Omega)} \leq C_\infty \|v\|_{W^{2,4/3}(\Omega)}. \tag{1.4.25}$$

For $\boldsymbol{f} \in H(\text{curl}, \Omega)$, the regularity of z is unchanged.

These two regularity theorems are not sufficient for establishing uniqueness because the proof requires that the solution $\boldsymbol{u}$ belong to $W^{1,\infty}(\Omega)^2$. By Sobolev's imbedding theorem, this holds if $\boldsymbol{u}$ is in $W^{2,r}(\Omega)^2$ for some $r > 2$. But if the regularity of z is restricted to $L^2(\Omega)$, we cannot expect the solution $\boldsymbol{v}$ of the generalized Stokes problem (1.4.10) to have higher regularity than $H^2(\Omega)^2$. And, if $\boldsymbol{f}$ belongs only to $H(\text{curl}, \Omega)$, the solution z of the transport equation (1.4.8) has no higher regularity than $L^2(\Omega)$. However, problems (1.4.9) and (1.4.3) are equivalent and by using (1.4.4), which we have not used so far, we can improve somewhat the statement of Theorem 1.4.9, without additional assumption on Ω and $\boldsymbol{f}$. More precisely, we have the following results.

LEMMA 1.4.10. *Let Ω be a convex polygon; let $\boldsymbol{v} \in V$ and $y \in L^2(\Omega)$ be related by*

$$y = \text{curl}\, \Delta \boldsymbol{v}.$$

Then there exists $r_0 > 2$ depending on the inner angles of $\partial\Omega$, such that for all $r \in [2, r_0]$, $\boldsymbol{v} \in W^{2,r}(\Omega)^2$, and there exists a constant C, depending only on r and Ω, such that

$$\|\boldsymbol{v}\|_{W^{2,r}(\Omega)} \leq C\,\|y\|_{L^2(\Omega)}. \tag{1.4.26}$$

PROPOSITION 1.4.11. *Under the assumptions of Theorem 1.4.9, there exists a real number $r_0 > 2$, depending on the inner angles of $\partial\Omega$, such that for all $r \in [2, r_0]$, any solution $\boldsymbol{u} \in V^\alpha$ of (1.4.3) belongs to $W^{2,r}(\Omega)^2$, and there exists a constant C_r, depending only on r and Ω, such that*

$$\|\boldsymbol{u}\|_{W^{2,r}(\Omega)} \leq C_r\left(\frac{1}{\alpha}\|\operatorname{curl}\boldsymbol{u}\|_{L^2(\Omega)} + \frac{1}{\nu}\|\operatorname{curl}\boldsymbol{f}\|_{L^2(\Omega)}\right). \tag{1.4.27}$$

This proposition implies uniqueness.

THEOREM 1.4.12. *With the assumptions of Theorem 1.4.9 and notation of Proposition 1.4.11, the solution of problem (1.4.3) is unique if the data satisfies, for some r with $2 < r \leq r_0$,*

$$\nu^2 > S_2\left(S_4^2 + C_{\infty,r}\frac{C_r}{\alpha}\right)\|\boldsymbol{f}\|_{L^2(\Omega)} + C_{\infty,r}C_r\|\operatorname{curl}\boldsymbol{f}\|_{L^2(\Omega)}, \tag{1.4.28}$$

where S_p is the constant of (1.1.3), C_r is the constant of (1.4.27), and $C_{\infty,r}$ is the constant of (1.1.18).

We shall see in the next chapter another sufficient condition for uniqueness, less sharp, but better adapted to the discrete form of problem (1.4.9).

Regarding the regularity of z, by applying to (1.4.8) the material of Section 1.3.3, we immediately derive the next result from (1.3.25):

PROPOSITION 1.4.13. *Let $(\boldsymbol{u}, p, z)$ be a solution of Problem (1.4.9), let $2 < r < \infty$ and $\operatorname{curl}\boldsymbol{f} \in L^r(\Omega)$. Assume that $\operatorname{curl}\boldsymbol{u} \in L^r(\Omega)$. Then $z \in L^r(\Omega)$ and*

$$\|z\|_{L^r(\Omega)} \leq \|\operatorname{curl}\boldsymbol{u}\|_{L^r(\Omega)} + \frac{\alpha}{\nu}\|\operatorname{curl}\boldsymbol{f}\|_{L^r(\Omega)}. \tag{1.4.29}$$

Likewise, we deduce the next theorem from Proposition 1.4.11, Theorems 1.3.18 and 1.3.19, and Remark 1.3.20.

THEOREM 1.4.14. *Let $(\boldsymbol{u}, p, z)$ be a solution of Problem (1.4.9), such that $\boldsymbol{u} \in W^{1,\infty}(\Omega)^2$ satisfies (1.3.27) with $|\gamma| = \alpha$:*

$$\frac{\alpha}{\nu}\|\nabla\boldsymbol{u}\|_{L^\infty(\Omega)} := \delta < 1. \tag{1.4.30}$$

If for some $r \geq 2$, $\operatorname{curl}\boldsymbol{f}$ belongs to $W^{1,r}(\Omega)$ and $\operatorname{curl}\boldsymbol{u}$ belongs to $W^{1,r}(\Omega)$, then z belongs to $W^{1,r}(\Omega)$ and is bounded by

$$|z|_{W^{1,r}(\Omega)} \leq \frac{1}{1-\delta}\left(|\operatorname{curl}\boldsymbol{u}|_{W^{1,r}(\Omega)} + \frac{\alpha}{\nu}|\operatorname{curl}\boldsymbol{f}|_{W^{1,r}(\Omega)}\right). \tag{1.4.31}$$

CHAPTER 2

Discretizing the Steady Split No-Slip Problem

In this chapter, we present several simple finite-element schemes for discretizing Problem (1.4.9) in a *bounded, connected, polygonal* domain Ω of $\mathbb{R}^2$: Find $(\boldsymbol{u}, p, z)$ in $H^1(\Omega)^2 \times L^2_0(\Omega) \times L^2(\Omega)$ solution of

$$\begin{aligned} -\nu\,\Delta\boldsymbol{u} + z \times \boldsymbol{u} + \nabla p = \boldsymbol{f} \quad &\text{in } \Omega,\\ \operatorname{div}\boldsymbol{u} = 0 \quad &\text{in } \Omega,\\ \boldsymbol{u} = \boldsymbol{0} \quad &\text{on } \partial\Omega,\\ \nu\, z + \alpha\,\boldsymbol{u}\cdot\nabla z = \nu\,\operatorname{curl}\boldsymbol{u} + \alpha\,\operatorname{curl}\boldsymbol{f} \quad &\text{in } \Omega. \end{aligned}$$

A large choice of finite elements are available for discretizing this problem, and we shall first present their numerical analysis in general finite-element spaces. But considering its complexity, only examples of finite-elements of order one or two will be proposed. Nevertheless, even within this narrow range, there are several possibilities. Indeed, approximating the generalized Stokes problem (1.4.6) by means of a good Stokes solver is fairly straightforward, but devising a good scheme for approximating the transport equation (1.4.8) is more delicate. Here, we shall present centered and upwind schemes for (1.4.8). As all these schemes are nonlinear, they must be implemented with suitable numerical algorithms, and we shall discuss a simple successive approximations algorithm. For the sake of brevity, it is presented for centered schemes, but it adapts easily to upwind schemes. Again, first, we study the problem with a no-slip boundary condition, because its numerical analysis is much simpler, and postpone the case of nonhomogeneous boundary conditions to Chapter 5.

2.1. General centered schemes

The material presented in this section is mainly taken from GIRAULT and SCOTT [2002a]. Let Ω be a bounded connected polygon. We discretize the auxiliary variable z in a finite-dimensional space $Z_h \subset H^1(\Omega)$ and the velocity and pressure in a pair of finite-dimensional spaces, $X_h \subset H_0^1(\Omega)^2$ and $M_h \subset L_0^2(\Omega)$, satisfying a uniform discrete inf-sup condition: There exists a constant $\beta^* > 0$, independent of h, such that

$$\forall q_h \in M_h, \ \sup_{\boldsymbol{v}_h \in X_h} \frac{\int_\Omega q_h \operatorname{div} \boldsymbol{v}_h \, \mathrm{d}\boldsymbol{x}}{|\boldsymbol{v}_h|_{H^1(\Omega)}} \geq \beta^* \|q_h\|_{L^2(\Omega)}. \tag{2.1.1}$$

Then we define the discrete analogs of V and $V^\perp$ by

$$V_h = \left\{ \boldsymbol{v}_h \in X_h;\ \forall q_h \in M_h,\ \int_\Omega q_h \operatorname{div} \boldsymbol{v}_h \, \mathrm{d}\boldsymbol{x} = 0 \right\}, \tag{2.1.2}$$

$$V_h^\perp = \{\boldsymbol{v}_h \in X_h;\ \forall \boldsymbol{w}_h \in V_h,\ (\nabla \boldsymbol{v}_h, \nabla \boldsymbol{w}_h) = 0\}. \tag{2.1.3}$$

For the transport term $\boldsymbol{u} \cdot \nabla z$, we propose the consistent trilinear form:

$$\forall \boldsymbol{v} \in H^1(\Omega)^2,\ \forall \varphi, \theta \in H^1(\Omega),\ \tilde{c}(\boldsymbol{v}; \varphi, \theta) = (\boldsymbol{v} \cdot \nabla \varphi, \theta) + \frac{1}{2}((\operatorname{div} \boldsymbol{v})\varphi, \theta). \tag{2.1.4}$$

It is consistent with $c(\cdot; \cdot, \cdot)$ in the sense that

$$\forall \boldsymbol{v} \in W,\ \forall \varphi, \theta \in H^1(\Omega),\ \tilde{c}(\boldsymbol{v}; \varphi, \theta) = c(\boldsymbol{v}; \varphi, \theta).$$

Furthermore, $\tilde{c}$ is antisymmetric because by Green's formula

$$\forall \boldsymbol{v} \in H_\tau^1(\Omega),\ \forall \varphi, \theta \in H^1(\Omega),\ \tilde{c}(\boldsymbol{v}; \varphi, \theta) = -\tilde{c}(\boldsymbol{v}; \theta, \varphi). \tag{2.1.5}$$

With these spaces and trilinear form, we choose the following general centered scheme for approximating problem (1.4.9): Find $(\boldsymbol{u}_h, p_h)$ in $X_h \times M_h$ and $\boldsymbol{z}_h = (0, 0, z_h)$ with z_h in Z_h, such that

$$\forall \boldsymbol{v}_h \in X_h,\ \nu(\nabla \boldsymbol{u}_h, \nabla \boldsymbol{v}_h) + (\boldsymbol{z}_h \times \boldsymbol{u}_h, \boldsymbol{v}_h) - (p_h, \operatorname{div} \boldsymbol{v}_h) = (\boldsymbol{f}, \boldsymbol{v}_h), \tag{2.1.6}$$

$$\forall q_h \in M_h,\ (q_h, \operatorname{div} \boldsymbol{u}_h) = 0, \tag{2.1.7}$$

$$\forall \theta_h \in Z_h,\ \nu\,(z_h, \theta_h) + \alpha\, \tilde{c}(\boldsymbol{u}_h; z_h, \theta_h) = \nu\,(\operatorname{curl} \boldsymbol{u}_h, \theta_h) + \alpha\,(\operatorname{curl} \boldsymbol{f}, \theta_h). \tag{2.1.8}$$

The system (2.1.6), (2.1.7) is a generalized version of the discrete Stokes problem: For $\nu > 0$ and $\boldsymbol{f}$ given in $L^2(\Omega)^2$, find $(\boldsymbol{v}_h, q_h)$ in $X_h \times M_h$, solution of

$$\forall \boldsymbol{w}_h \in X_h,\ \nu(\nabla \boldsymbol{v}_h, \nabla \boldsymbol{w}_h) - (q_h, \operatorname{div} \boldsymbol{w}_h) = (\boldsymbol{f}, \boldsymbol{w}_h), \tag{2.1.9}$$

$$\forall r_h \in M_h,\ (r_h, \operatorname{div} \boldsymbol{v}_h) = 0. \tag{2.1.10}$$

The next two lemmas recall the properties of (2.1.9)–(2.1.10). The proofs are an easy consequence of (2.1.1) and the Babuška–Brezzi theory (cf. for instance BABUŠKA [1973], BRENNER and SCOTT [1994], BREZZI [1974], BREZZI and FORTIN [1991], GIRAULT and RAVIART [1986], or ERN and GUERMOND [2004]).

LEMMA 2.1.1. *Assume that (2.1.1) holds. Then for each $q_h \in M_h$, there exists a unique $\boldsymbol{v}_h \in V_h^{\perp}$ such that:*

$$\forall s_h \in M_h,\ (s_h, \operatorname{div} \boldsymbol{v}_h) = (q_h, s_h), \quad |\boldsymbol{v}_h|_{H^1(\Omega)} \leq \frac{1}{\beta^*}\|q_h\|_{L^2(\Omega)}. \tag{2.1.11}$$

LEMMA 2.1.2. *Assume that (2.1.1) holds. Then the discrete Stokes problem (2.1.9)–(2.1.10) has a unique solution $(\boldsymbol{v}_h, q_h) \in X_h \times M_h$, and this solution satisfies the uniform bound:*

$$|\boldsymbol{v}_h|_{H^1(\Omega)} \leq \frac{S_2}{\nu}\|\boldsymbol{f}\|_{L^2(\Omega)}, \quad \|q_h\|_{L^2(\Omega)} \leq \frac{S_2}{\beta^*}\|\boldsymbol{f}\|_{L^2(\Omega)}, \tag{2.1.12}$$

where β^* and S_2 are the constants of respectively (2.1.1) and (1.1.3).

Hence the discrete generalized Stokes problem: For a given z_h in Z_h, find $(\boldsymbol{v}_h(z_h), q_h(z_h))$ in $V_h \times M_h$, solution of

$$\forall \boldsymbol{w}_h \in X_h,\ \nu(\nabla \boldsymbol{v}_h, \nabla \boldsymbol{w}_h) + (z_h \times \boldsymbol{v}_h, \boldsymbol{w}_h) - (q_h, \operatorname{div} \boldsymbol{v}_h) = (\boldsymbol{f}, \boldsymbol{v}_h), \tag{2.1.13}$$

has a unique solution with uniform a priori estimates given in the following proposition. The proof is skipped because it is trivial.

PROPOSITION 2.1.3. *Assume that (2.1.1) holds. Let $\nu > 0$ and $\boldsymbol{f} \in L^2(\Omega)^2$. For any z_h in Z_h, the generalized Stokes problem (2.1.13) has a unique solution $(\boldsymbol{v}_h(z_h), q_h(z_h))$ in $V_h \times M_h$. This solution satisfies the following bounds in $H^1(\Omega)^2 \times L^2(\Omega)$:*

$$|\boldsymbol{v}_h(z_h)|_{H^1(\Omega)} \leq \frac{S_2}{\nu}\|\boldsymbol{f}\|_{L^2(\Omega)}, \tag{2.1.14}$$

$$\|q_h(z_h)\|_{L^2(\Omega)} \leq \frac{1}{\beta^*}\big(S_2\|\boldsymbol{f}\|_{L^2(\Omega)} + S_4^2|\boldsymbol{v}_h(z_h)|_{H^1(\Omega)}\|z_h\|_{L^2(\Omega)}\big). \tag{2.1.15}$$

The next theorem gives existence of at least one solution of (2.1.6)–(2.1.8).

THEOREM 2.1.4. *Assume that (2.1.1) holds. Then for all $\nu > 0$, $\alpha > 0$, and for all $\boldsymbol{f}$ in $H(curl, \Omega)$, the discrete problem (2.1.6)–(2.1.8) has at least one solution $(\boldsymbol{u}_h, p_h, z_h) \in V_h \times M_h \times Z_h$, and each solution satisfies the a priori estimates (2.1.14), (2.1.15) and*

$$\|z_h\|_{L^2(\Omega)} \leq \frac{S_2}{\nu}\|\boldsymbol{f}\|_{L^2(\Omega)} + \frac{\alpha}{\nu}\|\operatorname{curl} \boldsymbol{f}\|_{L^2(\Omega)}. \tag{2.1.16}$$

PROOF. It follows from Proposition 2.1.3 that problem (2.1.6)–(2.1.8) is equivalent to: Find z_h in Z_h such that

$$\forall \theta_h \in Z_h,\ \nu\,(z_h, \theta_h) + \alpha\,\tilde{c}(\boldsymbol{u}_h(z_h); z_h, \theta_h) = \nu\,(\operatorname{curl} \boldsymbol{u}_h(z_h), \theta_h) + \alpha\,(\operatorname{curl} \boldsymbol{f}, \theta_h), \tag{2.1.17}$$

where $(\boldsymbol{u}_h(z_h), p_h(z_h)) \in V_h \times M_h$ is the solution of (2.1.13). Let us solve (2.1.17) by Brouwer's Fixed Point Theorem. To this end, for fixed λ_h in Z_h, we define $H(\lambda_h)$ in Z_h by

$$\forall \mu_h \in Z_h,\ (H(\lambda_h), \mu_h) = \nu(\lambda_h, \mu_h) + \alpha\, \tilde{c}(\boldsymbol{u}_h(\lambda_h); \lambda_h, \mu_h) \\ - \nu(\operatorname{curl} \boldsymbol{u}_h(\lambda_h), \mu_h) - \alpha\, (\operatorname{curl} \boldsymbol{f}, \mu_h).$$

This finite-dimensional, square system of linear equations defines a continuous mapping $H : Z_h \mapsto Z_h$. Moreover, the H^1 regularity of λ_h, the antisymmetry (2.1.5) of $\tilde{c}(\cdot; \cdot, \cdot)$, and (2.1.14) imply that, for all $\lambda_h \in Z_h$,

$$\begin{aligned}(H(\lambda_h), \lambda_h) &= \nu\, \|\lambda_h\|^2_{L^2(\Omega)} - \nu(\operatorname{curl} \boldsymbol{u}_h(\lambda_h), \lambda_h) - \alpha\, (\operatorname{curl} \boldsymbol{f}, \lambda_h) \\ &\geq \nu\, \|\lambda_h\|^2_{L^2(\Omega)} - \big(\nu\, |\boldsymbol{u}_h(\lambda_h)|_{H^1(\Omega)} + \alpha\, \|\operatorname{curl} \boldsymbol{f}\|_{L^2(\Omega)}\big)\, \|\lambda_h\|_{L^2(\Omega)} \\ &\geq \nu\, \|\lambda_h\|^2_{L^2(\Omega)} - \big(S_2 \|\boldsymbol{f}\|_{L^2(\Omega)} + \alpha\, \|\operatorname{curl} \boldsymbol{f}\|_{L^2(\Omega)}\big)\, \|\lambda_h\|_{L^2(\Omega)}.\end{aligned}$$

Hence $(H(\lambda_h), \lambda_h) \geq 0$ for all λ_h in Z_h satisfying

$$\|\lambda_h\|_{L^2(\Omega)} = \frac{S_2}{\nu} \|\boldsymbol{f}\|_{L^2(\Omega)} + \frac{\alpha}{\nu} \|\operatorname{curl} \boldsymbol{f}\|_{L^2(\Omega)}.$$

By Brouwer's Fixed Point Theorem, this proves existence of at least one solution z_h in Z_h of (2.1.17).

Finally, by choosing $\boldsymbol{v}_h = \boldsymbol{u}_h$ in (2.1.6) and $\theta_h = z_h$ in (2.1.8), we immediately derive that every solution of (2.1.6)–(2.1.8) satisfies (2.1.14) and (2.1.16). Then the estimate (2.1.15) for p_h follows from (2.1.1). □

2.1.1. *Convergence*

It stems from the uniform bounds (2.1.14)–(2.1.16), that there exists a subsequence of h (still denoted by h) and functions $\boldsymbol{u} \in H^1_0(\Omega)^2$, $p \in L^2(\Omega)$, $z \in L^2(\Omega)$ such that

$$\begin{aligned}&\lim_{h\to 0} \boldsymbol{u}_h = \boldsymbol{u} \ \text{ weakly in } H^1_0(\Omega)^2, \\ &\lim_{h\to 0} p_h = p \ \text{ weakly in } L^2(\Omega), \\ &\lim_{h\to 0} z_h = z \ \text{ weakly in } L^2(\Omega).\end{aligned} \tag{2.1.18}$$

The compactness of the imbedding of $H^1(\Omega)$ into $L^p(\Omega)$ for any real number $p \geq 2$ implies also

$$\forall p \in [2, \infty[,\ \lim_{h\to 0} \boldsymbol{u}_h = \boldsymbol{u} \ \text{ strongly in } L^p(\Omega).$$

Now, for passing to the limit in (2.1.6)–(2.1.8), we need the following approximation properties of the discrete spaces.

HYPOTHESIS 2.1.5. *(1) There exists an operator $P_h \in \mathcal{L}(H_0^1(\Omega)^2; X_h)$ that preserves the discrete divergence:*

$$\forall \boldsymbol{v} \in H_0^1(\Omega)^2,\ \forall q_h \in M_h,\ \int_\Omega q_h \operatorname{div} P_h(\boldsymbol{v})\, d\boldsymbol{x} = \int_\Omega q_h \operatorname{div} \boldsymbol{v}\, d\boldsymbol{x}, \tag{2.1.19}$$

and such that

$$\forall \boldsymbol{v} \in H_0^1(\Omega)^2,\ \lim_{h\to 0} \|P_h(\boldsymbol{v}) - \boldsymbol{v}\|_{H^1(\Omega)} = 0.$$

(2) There exists an operator $r_h \in \mathcal{L}(L_0^2(\Omega); M_h)$ such that

$$\forall q \in L_0^2(\Omega),\ \lim_{h\to 0} \|r_h(q) - q\|_{L^2(\Omega)} = 0.$$

(3) There exists an operator $R_h \in \mathcal{L}(L^2(\Omega); Z_h)$ such that

$$\forall \theta \in L^2(\Omega),\ \lim_{h\to 0} \|R_h(\theta) - \theta\|_{L^2(\Omega)} = 0,$$

$$\forall p \in [2, \infty],\ \forall \theta \in W^{1,p}(\Omega),\ \lim_{h\to 0} \|R_h(\theta) - \theta\|_{W^{1,p}(\Omega)} = 0.$$

These assumptions will be sharpened in the next sections, but for the moment, this is all we require.
Let us first pass to the limit in (2.1.6) and (2.1.7).

PROPOSITION 2.1.6. *Under the first two assumptions of Hypothesis 2.1.5, the limit functions $(\boldsymbol{u}, p)$ in (2.1.18) belong to $V \times L_0^2(\Omega)$ and the triple $(\boldsymbol{u}, p, z)$ satisfies the first two equations of (1.4.9).*

PROOF. (1) First we prove that $(\boldsymbol{u}, p)$ is in $V \times L_0^2(\Omega)$. Let q be any function in $L_0^2(\Omega)$ and choose $q_h = r_h(q) \in M_h$ in (2.1.7). The weak convergence of $\operatorname{div} \boldsymbol{u}_h$ and the strong convergence of $r_h(q)$ imply that $(q, \operatorname{div} \boldsymbol{u}) = 0$; then the fact that $\operatorname{div} \boldsymbol{u}$ is in $L_0^2(\Omega)$ shows that $\boldsymbol{u}$ belongs to V. Similarly, the fact that p_h belongs to $L_0^2(\Omega)$ and its weak convergence show that p is in $L_0^2(\Omega)$.

(2) Next, we pass to the limit in (2.1.6). Let $\boldsymbol{v}$ be any function in $H_0^1(\Omega)^2$ and choose $\boldsymbol{v}_h = P_h(\boldsymbol{v}) \in X_h$ in (2.1.6). The convergence of the bilinear and linear terms stem from the weak convergence of $\boldsymbol{u}_h$ and the strong convergence of $P_h(\boldsymbol{v})$, both in $H^1(\Omega)^2$. For the nonlinear term $(z_h \times \boldsymbol{u}_h, P_h(\boldsymbol{v}))$, we use the weak convergence of z_h in $L^2(\Omega)$ and the strong convergence of $\boldsymbol{u}_h$ and $P_h(\boldsymbol{v})$, both in $L^4(\Omega)^2$. Thus $(\boldsymbol{u}, p)$ satisfies

$$\forall \boldsymbol{v} \in H_0^1(\Omega)^2,\ \nu(\nabla \boldsymbol{u}, \nabla \boldsymbol{v}) + (z \times \boldsymbol{u}, \boldsymbol{v}) - (p, \operatorname{div} \boldsymbol{v}) = (\boldsymbol{f}, \boldsymbol{v}), \tag{2.1.20}$$

that is equivalent to the first equation of (1.4.9). □

At first sight, one can think that passing to the limit in (2.1.8) follows the same lines by choosing $\theta_h = R_h(\theta)$, for a sufficiently smooth function θ. But this process is not conclusive because the stabilizing term $((\operatorname{div} \boldsymbol{u}_h) z_h, R_h(\theta))$ in the trilinear form involves the product of two weakly convergent sequences. As was pointed out by Chacón-Rebollo in CHACÓN-REBOLLO [2001], this difficulty can be bypassed by establishing first the strong convergence of $\boldsymbol{u}_h$ in $H^1(\Omega)^2$.

PROPOSITION 2.1.7. *Under the first two assumptions of Hypothesis 2.1.5, the first convergence in (2.1.18) holds strongly:*

$$\lim_{h \to 0} \|\boldsymbol{u}_h - \boldsymbol{u}\|_{H^1(\Omega)} = 0. \tag{2.1.21}$$

PROOF. By taking the difference between (2.1.6) and (2.1.20) with test function $\boldsymbol{v}_h$ in V_h and by inserting $P_h(\boldsymbol{u})$, we derive the preliminary error equation, for any $q_h \in M_h$:

$$\begin{aligned} \nu(\nabla(\boldsymbol{u}_h - P_h(\boldsymbol{u})), \nabla \boldsymbol{v}_h) &= \nu\,(\nabla(\boldsymbol{u} - P_h(\boldsymbol{u})), \nabla \boldsymbol{v}_h) + (z \times \boldsymbol{u} - z_h \times \boldsymbol{u}_h, \boldsymbol{v}_h) \\ &\quad - (p - q_h, \operatorname{div} \boldsymbol{v}_h). \end{aligned}$$

Let us choose $q_h = r_h(p)$ and $\boldsymbol{v}_h = \boldsymbol{u}_h - P_h(\boldsymbol{u})$ that belongs to V_h by virtue of (2.1.19). Thus,

$$\begin{aligned} \nu\,|\boldsymbol{u}_h - P_h(\boldsymbol{u})|^2_{H^1(\Omega)} &= \nu\,(\nabla(\boldsymbol{u} - P_h(\boldsymbol{u})), \nabla(\boldsymbol{u}_h - P_h(\boldsymbol{u}))) \\ &\quad + (z \times \boldsymbol{u} - z_h \times \boldsymbol{u}_h, \boldsymbol{u}_h - P_h(\boldsymbol{u})) - (p - r_h(p), \operatorname{div}(\boldsymbol{u}_h - P_h(\boldsymbol{u}))). \end{aligned} \tag{2.1.22}$$

Owing to the weak convergence of $\boldsymbol{u}_h$ and the strong convergence of $P_h(\boldsymbol{u})$, both in $H^1(\Omega)^2$, the first term in the right-hand side of (2.1.22) tends to zero. Similarly, the strong convergence of $r_h(p)$ and the weak convergence of $\operatorname{div} \boldsymbol{u}_h$, both in $L^2(\Omega)$, imply that the last term in the right-hand side of (2.1.22) tends to zero. Finally, the weak convergence of z_h in $L^2(\Omega)$ and the strong convergence of $\boldsymbol{u}_h$ and $P_h(\boldsymbol{u})$, both in $L^4(\Omega)^2$ show that the nonlinear term in the right-hand side of (2.1.22) also tends to zero. Consequently,

$$\lim_{h \to 0} \nu\,|\boldsymbol{u}_h - P_h(\boldsymbol{u})|^2_{H^1(\Omega)} = 0,$$

thus yielding the strong convergence of $\boldsymbol{u}_h$ to $\boldsymbol{u}$ in $H^1_0(\Omega)^2$. □

Now, we are in a position to pass to the limit in (2.1.8).

PROPOSITION 2.1.8. *Assume that Hypothesis 2.1.5 holds. Then the limit functions $(\boldsymbol{u}, z)$ in (2.1.18) satisfy (1.4.8).*

PROOF. Let θ be any function in $W^{1,4}(\Omega)$ and take $\theta_h = R_h(\theta)$ in (2.1.8). Passing to the limit in the bilinear and linear forms of (2.1.8) is routine and there remains the limit of the transport term. As all functions here are sufficiently smooth, we can apply (2.1.5):

$$\tilde{c}(\boldsymbol{u}_h; z_h, R_h(\theta)) = -\tilde{c}(\boldsymbol{u}_h; R_h(\theta), z_h).$$

On one hand, $\operatorname{div} \boldsymbol{u}_h$ converges strongly in $L^2(\Omega)$ and $R_h(\theta)$ converges strongly in $L^\infty(\Omega)$. On the other hand, $\boldsymbol{u}_h$ and $\nabla R_h(\theta)$ both converge strongly in $L^4(\Omega)^2$. Therefore,

$$\lim_{h \to 0} \tilde{c}(\boldsymbol{u}_h; z_h, R_h(\theta)) = -c(\boldsymbol{u}; \theta, z)$$

because $\operatorname{div} \boldsymbol{u} = 0$. Hence, for all θ in $W^{1,4}(\Omega)$, we obtain

$$\nu(z, \theta) - \alpha\, c(\boldsymbol{u}; \theta, z) = \nu(\operatorname{curl} \boldsymbol{u}, \theta) + \alpha\, (\operatorname{curl} \boldsymbol{f}, \theta).$$

In the sense of distributions, this is equivalent to (1.4.8). □

It remains to establish the strong convergence of z_h and p_h and derive the main convergence result of this section.

THEOREM 2.1.9. *Under Hypothesis 2.1.5, there exists a subsequence of h (still denoted by h) and a solution $(\boldsymbol{u}, p, z) \in V \times L^2_0(\Omega) \times L^2(\Omega)$ of problem (1.4.9) such that*

$$\begin{aligned}
&\lim_{h\to 0} |\boldsymbol{u}_h - \boldsymbol{u}|_{H^1(\Omega)} = 0,\\
&\lim_{h\to 0} \|z_h - z\|_{L^2(\Omega)} = 0,\\
&\lim_{h\to 0} \|p_h - p\|_{L^2(\Omega)} = 0.
\end{aligned} \tag{2.1.23}$$

PROOF. It remains to prove the last two strong convergences.

(1) First, we consider the limit of z_h: we write

$$\|z_h - z\|^2_{L^2(\Omega)} = (z_h - z, z_h) - (z_h - z, z),$$

and it suffices to study the first term. By taking the difference between (2.1.8) and (1.4.8) multiplied by the test function z_h, we obtain the following equation:

$$\nu(z_h - z, z_h) + \alpha\tilde{c}(\boldsymbol{u}_h; z_h, z_h) - \alpha c(\boldsymbol{u}; z, z_h) = \nu(\operatorname{curl}(\boldsymbol{u}_h - \boldsymbol{u}), z_h).$$

Applying (2.1.5), this reduces to

$$(z_h - z, z_h) = \frac{\alpha}{\nu} c(\boldsymbol{u}; z, z_h) + (\operatorname{curl}(\boldsymbol{u}_h - \boldsymbol{u}), z_h).$$

On one hand, the fact that z belongs to $X_{\boldsymbol{u}}$, the weak convergence of z_h in $L^2(\Omega)$, and Corollary 1.3.11 imply that

$$\lim_{h\to 0} c(\boldsymbol{u}; z, z_h) = c(\boldsymbol{u}; z, z) = 0.$$

On the other hand, the strong convergence of $\operatorname{curl} \boldsymbol{u}_h$ in $L^2(\Omega)$ and the weak convergence of z_h imply that

$$\lim_{h\to 0} (\operatorname{curl}(\boldsymbol{u}_h - \boldsymbol{u}), z_h) = 0.$$

Hence

$$\lim_{h\to 0} (z_h - z, z_h) = 0,$$

thus proving the strong convergence of z_h.

(2) Now, we turn to p_h. By subtracting (2.1.20) from (2.1.6) with test function $\boldsymbol{v}_h \in X_h$ and by inserting $r_h(p)$, we obtain

$$\forall \boldsymbol{v}_h \in X_h,\ (p_h - r_h(p), \operatorname{div} \boldsymbol{v}_h) = (p - r_h(p), \operatorname{div} \boldsymbol{v}_h) + \nu(\nabla(\boldsymbol{u}_h - \boldsymbol{u}), \nabla \boldsymbol{v}_h) + (z_h \times \boldsymbol{u}_h - z \times \boldsymbol{u}, \boldsymbol{v}_h).$$

Let us choose the function $\boldsymbol{v}_h$ associated by Lemma 2.1.1 with the function $q_h = p_h - r_h(p)$; as $\boldsymbol{v}_h$ belongs to $V_h^{\perp}$, this gives

$$\|p_h - r_h(p)\|^2_{L^2(\Omega)} = (p - r_h(p), \operatorname{div} \boldsymbol{v}_h) + \nu(\nabla(P_h(\boldsymbol{u}) - \boldsymbol{u}), \nabla \boldsymbol{v}_h) + (z_h \times \boldsymbol{u}_h - z \times \boldsymbol{u}, \boldsymbol{v}_h), \tag{2.1.24}$$

and (2.1.11) yields

$$|\boldsymbol{v}_h|_{H^1(\Omega)} \leq \frac{1}{\beta^*} \|p_h - r_h(p)\|_{L^2(\Omega)}.$$

This last relation implies the weak convergence of $\boldsymbol{v}_h$ in $H_0^1(\Omega)^2$. Then the strong convergence of p_h follows by taking the limit of the right-hand side of (2.1.24) and using the weak convergence of $\boldsymbol{v}_h$ and the strong convergence of $\boldsymbol{u}_h$ and z_h. □

2.1.2. *Further estimates for the discrete velocity*

Here, we need to sharpen the approximation properties in the statement of Hypothesis 2.1.5. As Ω is assumed to be a polygon, it can be entirely triangulated. For an arbitrary triangle T, we denote by h_T the diameter of T and by ρ_T the radius of the ball inscribed in T. Let $h > 0$ be a discretization parameter and let $\mathcal{T}_h$ be a family of triangulations of $\overline{\Omega}$, consisting of triangles with maximum mesh size h

$$h := \max_{T \in \mathcal{T}_h} h_T,$$

that is *regular* (also called *nondegenerate*):

$$\max_{T \in \mathcal{T}_h} \frac{h_T}{\rho_T} \leq \sigma_0, \tag{2.1.25}$$

with the constant σ_0 independent of h (cf. CIARLET [1991], and BRENNER and SCOTT [1994]). Here we assume that the triangulation is conforming, i.e., it is such that any two triangles are either disjoint or share a vertex or a complete side. Moreover, we suppose that in each triangle T, the finite-element functions of X_h, M_h and Z_h are all *polynomials*, but for the moment, the degrees of these polynomials are not specified. Then, we complement 2.1.5 by the following assumptions:

HYPOTHESIS 2.1.10. *The operators P_h and r_h satisfy, for each real number $s \in [0, 1]$ and for each number $r \geq 2$:*

(1) There exists a constant C, independent of h, such that

$$\forall \boldsymbol{v} \in \left(W^{s+1,r}(\Omega) \cap H_0^1(\Omega)\right)^2, \ |P_h(\boldsymbol{v}) - \boldsymbol{v}|_{W^{1,r}(\Omega)} \leq C h^s |\boldsymbol{v}|_{W^{s+1,r}(\Omega)}. \tag{2.1.26}$$

(2) There exists a constant C, independent of h, such that

$$\forall q \in W^{s,r}(\Omega) \cap L_0^2(\Omega), \ \|r_h(q) - q\|_{L^r(\Omega)} \leq C h^s |q|_{W^{s,r}(\Omega)}. \tag{2.1.27}$$

Now, observe that for fixed z_h, the pair $(\boldsymbol{u}_h, p_h)$ approximates the solution of a generalized Stokes problem of the form (1.4.10) with z_h instead of z: Find $(\boldsymbol{v}(z_h), q(z_h)) \in H_0^1(\Omega)^2 \times L_0^2(\Omega)$ solution of

$$-\nu \, \Delta \, \boldsymbol{v}(z_h) + z_h \times \boldsymbol{v}(z_h) + \nabla \, q(z_h) = \boldsymbol{f} \quad \text{in } \Omega, \tag{2.1.28}$$

$$\operatorname{div} \boldsymbol{v}(z_h) = 0 \quad \text{in } \Omega. \tag{2.1.29}$$

It is interesting to compare $\boldsymbol{u}_h$ and $\boldsymbol{v}(z_h)$, when $(\boldsymbol{v}(z_h), q(z_h))$ has sufficient regularity. This regularity is a direct consequence of (1.4.22) and (1.4.24): Without restriction on the angles of $\partial\Omega$, $(\boldsymbol{v}(z_h), q(z_h))$ belong to $W^{2,4/3}(\Omega)^2 \times W^{1,4/3}(\Omega)$ and satisfy

$$\|\boldsymbol{v}(z_h)\|_{W^{2,4/3}(\Omega)} + \|q(z_h)\|_{W^{1,4/3}(\Omega)} \leq C_1 K_1(z_h) \, \|\boldsymbol{f}\|_{L^{4/3}(\Omega)}, \tag{2.1.30}$$

where $K_1(z)$ is defined by (1.4.23). If Ω is a convex polygon, $(\boldsymbol{v}(z_h), q(z_h))$ belong to $H^2(\Omega)^2 \times H^1(\Omega)$ and are bounded by

$$\begin{aligned} &\|\boldsymbol{v}(z_h)\|_{H^2(\Omega)} + \|q(z_h)\|_{H^1(\Omega)} \\ &\qquad \leq C_2 \left(\|\boldsymbol{f}\|_{L^2(\Omega)} + C_\infty C_1 K_1(z_h) \|z_h\|_{L^2(\Omega)} \|\boldsymbol{f}\|_{L^{4/3}(\Omega)}\right). \end{aligned} \tag{2.1.31}$$

Furthermore, the following lemma compares $(\boldsymbol{u}, p)$ and $(\boldsymbol{v}(z_h), q(z_h))$. Note that its statement is independent of the particular functions z and z_h. It is valid for any pair of solutions of the generalized Stokes problem (2.1.28)–(2.1.29) associated with any pair of functions z and z_h in $L^2(\Omega)$.

LEMMA 2.1.11. *Let Ω be a connected polygon with all inner angles in $]0, 2\pi[$; then*

$$\begin{aligned} &\|\boldsymbol{u} - \boldsymbol{v}(z_h)\|_{W^{2,4/3}(\Omega)} + \|p - q(z_h)\|_{W^{1,4/3}(\Omega)} \leq C_1 K_1(z) \|\boldsymbol{v}(z_h)\|_{L^4(\Omega)} \|z - z_h\|_{L^2(\Omega)}, \\ &\|\boldsymbol{u} - \boldsymbol{v}(z_h)\|_{L^\infty(\Omega)} \leq C_\infty C_1 K_1(z) \|\boldsymbol{v}(z_h)\|_{L^4(\Omega)} \|z - z_h\|_{L^2(\Omega)}. \end{aligned} \tag{2.1.32}$$

If in addition, Ω is convex, we have

$$\begin{aligned} &\|\boldsymbol{u} - \boldsymbol{v}(z_h)\|_{H^2(\Omega)} + \|p - q(z_h)\|_{H^1(\Omega)} \leq C_2 \|z - z_h\|_{L^2(\Omega)} \\ &\qquad \times \left(\|\boldsymbol{v}(z_h)\|_{L^\infty(\Omega)} + C_\infty C_1 K_1(z) \|\boldsymbol{v}(z_h)\|_{L^4(\Omega)} \|z\|_{L^2(\Omega)}\right). \end{aligned} \tag{2.1.33}$$

PROOF. Subtracting (2.1.28) from the first equation in (1.4.9), we find that $(\boldsymbol{u} - \boldsymbol{v}(z_h), p - q(z_h)) \in H_0^1(\Omega)^2 \times L_0^2(\Omega)$ solve the first three equations of problem (1.4.9) with right-hand side $-(z - z_h) \times \boldsymbol{v}(z_h)$:

$$-\nu\Delta(\boldsymbol{u} - \boldsymbol{v}(z_h)) + z \times (\boldsymbol{u} - \boldsymbol{v}(z_h)) + \nabla(p - q(z_h)) = -(z - z_h) \times \boldsymbol{v}(z_h),$$
$$\operatorname{div}(\boldsymbol{u} - \boldsymbol{v}(z_h)) = 0 \text{ in } \Omega.$$

Therefore (2.1.32) and (2.1.33) follow from (1.4.22), (1.4.25), and (1.4.24). □

The next lemma presents a bound for $\boldsymbol{u}_h - P_h(\boldsymbol{v}(z_h))$.

LEMMA 2.1.12. *Let Ω be a polygon with all inner angles in $]0, 2\pi[$; then, under Hypotheses 2.1.5 and 2.1.10, there exists a constant $C > 0$, independent of h, such that*

$$|\boldsymbol{u}_h - P_h(\boldsymbol{v}(z_h))|_{H^1(\Omega)} \leq C h^{1/2} \left(K_1(z_h)|\boldsymbol{v}(z_h)|_{H^{3/2}(\Omega)} + \frac{1}{\nu}|q(z_h)|_{H^{1/2}(\Omega)} \right). \tag{2.1.34}$$

If, in addition, Ω is convex, then there exists another constant $C > 0$ independent of h, such that

$$|\boldsymbol{u}_h - P_h(\boldsymbol{v}(z_h))|_{H^1(\Omega)} \leq C h \left(K_1(z_h)|\boldsymbol{v}(z_h)|_{H^2(\Omega)} + \frac{1}{\nu}|q(z_h)|_{H^1(\Omega)} \right). \tag{2.1.35}$$

PROOF. To shorten the text, we momentarily drop the dependence of $\boldsymbol{v}$ and q on z_h. As in the proof of Proposition 2.1.7, we derive from (2.1.6) and (2.1.28), for all $\boldsymbol{w}_h \in V_h$,

$$\nu(\nabla(\boldsymbol{u}_h - P_h(\boldsymbol{v})), \nabla \boldsymbol{w}_h) + (z_h \times (\boldsymbol{u}_h - P_h(\boldsymbol{v})), \boldsymbol{w}_h) - (r_h(q) - q, \operatorname{div} \boldsymbol{w}_h)$$
$$= \nu(\nabla(\boldsymbol{v} - P_h(\boldsymbol{v})), \nabla \boldsymbol{w}_h) + (z_h \times (\boldsymbol{v} - P_h(\boldsymbol{v})), \boldsymbol{w}_h).$$

Then, choosing $\boldsymbol{w}_h = \boldsymbol{u}_h - P_h(\boldsymbol{v}) \in V_h$, we obtain

$$|\boldsymbol{u}_h - P_h(\boldsymbol{v})|_{H^1(\Omega)} \leq K_1(z_h)|\boldsymbol{v} - P_h(\boldsymbol{v})|_{H^1(\Omega)} + \frac{1}{\nu}\|r_h(q) - q\|_{L^2(\Omega)}, \tag{2.1.36}$$

and (2.1.34) follows from Theorem 1.1.6, the imbedding of $W^{2,4/3}(\Omega)$ into $H^{3/2}(\Omega)$, and Hypothesis 2.1.10 with $s = 1/2$ and $r = 2$.

Similarly, we derive (2.1.35) from (2.1.36) by applying Theorem 1.1.5 and Hypothesis 2.1.10 with $s = 1$. □

At this stage, we can derive a variety of bounds for $\boldsymbol{u}_h - P_h(\boldsymbol{v}(z_h))$, depending on different assumptions on the domain and triangulation. They are based on the inverse inequality of the next lemma (cf. CIARLET [1991]), which is valid in arbitrary dimensions d, and generally rely on the following assumption on the triangulation: the family of triangulations $\mathcal{T}_h$ is *uniformly regular* (also called *quasi-uniform*) if there exist two constants $\tau > 0$ and $\sigma_0 > 0$, independent of h, such that

$$\forall T \in \mathcal{T}_h, \ \tau h \leq h_T \leq \sigma_0 \rho_T. \tag{2.1.37}$$

LEMMA 2.1.13. *Let the triangulation $\mathcal{T}_h$ satisfy (2.1.25). For any finite-element space Θ_h constructed on $\mathcal{T}_h$, and for any number $r \geq 2$, there exists a constant C, independent of h,* *such that*

$$\forall v_h \in \Theta_h, \ \|v_h\|_{L^r(\Omega)} \le C \frac{1}{\rho_{\min}^{d\left(\frac{1}{2}-\frac{1}{r}\right)}} \|v_h\|_{L^2(\Omega)}, \tag{2.1.38}$$

where

$$\varrho_{\min} = \inf_{T \in \mathfrak{T}_h} \rho_T.$$

If in addition, $\mathfrak{T}_h$ satisfies (2.1.37), then there exists a constant C, independent of h, such that

$$\forall v_h \in \Theta_h, \ \|v_h\|_{L^r(\Omega)} \le C h^{d\left(\frac{1}{r}-\frac{1}{2}\right)} \|v_h\|_{L^2(\Omega)}. \tag{2.1.39}$$

With this material, we can establish $W^{1,r}$ bounds for $\boldsymbol{u}_h - P_h(\boldsymbol{v}(z_h))$; they stem directly from Lemma 2.1.12 and (2.1.39).

THEOREM 2.1.14. *Let Ω be a connected polygon with all inner angles in $]0, 2\pi[$, and let Hypotheses 2.1.5 and 2.1.10 hold. If $\mathfrak{T}_h$ satisfies (2.1.37), then for any real number $r \in [2, 4]$, there exists a constant C, depending on r but not on h, such that*

$$|\boldsymbol{u}_h - P_h(\boldsymbol{v}(z_h))|_{W^{1,r}(\Omega)} \le C h^{2/r-1/2} \left(K_1(z_h) |\boldsymbol{v}(z_h)|_{H^{3/2}(\Omega)} + \frac{1}{\nu} |q(z_h)|_{H^{1/2}(\Omega)} \right). \tag{2.1.40}$$

If in addition, Ω is a convex polygon, then for any number r in $[2, \infty]$, there exists another constant C, depending on r but not on h, such that

$$|\boldsymbol{u}_h - P_h(\boldsymbol{v}(z_h))|_{W^{1,r}(\Omega)} \le C h^{2/r} \left(K_1(z_h) |\boldsymbol{v}(z_h)|_{H^2(\Omega)} + \frac{1}{\nu} |q(z_h)|_{H^1(\Omega)} \right). \tag{2.1.41}$$

Considering the stability of P_h given by (2.1.26) with $s = 0$, the bounds for $\boldsymbol{v}(z_h)$ and $q(z_h)$ given by (2.1.30) and (2.1.31), and the uniform bound for z_h given by (2.1.16), we have the following corollary.

COROLLARY 2.1.15. *Under the assumptions of the first part of Theorem 2.1.14, there exists a constant C, independent of h, such that any velocity solution $\boldsymbol{u}_h$ of (2.1.6)–(2.1.8) satisfies the uniform bound:*

$$|\boldsymbol{u}_h|_{W^{1,4}(\Omega)} \le C. \tag{2.1.42}$$

Moreover under the assumptions of the second part of Theorem 2.1.14, for each real number $r \ge 2$, there exists another constant C, depending on r but not on h, such that

$$|\boldsymbol{u}_h|_{W^{1,r}(\Omega)} \le C. \tag{2.1.43}$$

REMARK 2.1.16. We cannot extend (2.1.43) to $r = \infty$ because we have no bound for $\boldsymbol{v}(z_h)$ in $W^{1,\infty}(\Omega)^2$. Such a bound would require a uniform estimate for z_h in $L^q(\Omega)$ for some $q > 2$, and so far, this appears to be an open problem. □

REMARK 2.1.17. The bound (2.1.42) implies that $\boldsymbol{u}_h$ is uniformly bounded in maximum norm:

$$\|\boldsymbol{u}_h\|_{L^\infty(\Omega)} \le C. \tag{2.1.44}$$

This property will be used in a convex domain in Section 2.4. But in this case, the restriction (2.1.37) on the mesh can be substantially relaxed. Indeed, for proving (2.1.44), we only need that $\boldsymbol{u}_h$ be bounded in $W^{1,r}$ for some $r > 2$. Owing to the stability of P_h and to (2.1.30), we write

$$\begin{aligned}|\boldsymbol{u}_h|_{W^{1,r}(\Omega)} &\le |\boldsymbol{u}_h - P_h(\boldsymbol{v}(z_h))|_{W^{1,r}(\Omega)} + c_1|\boldsymbol{v}(z_h)|_{W^{1,r}(\Omega)}\\ &\le |\boldsymbol{u}_h - P_h(\boldsymbol{v}(z_h))|_{W^{1,r}(\Omega)} + c_2K_1(z_h)\|\boldsymbol{f}\|_{L^{4/3}(\Omega)},\end{aligned}$$

where all constants c_i are independent of h. Then applying (2.1.38), (2.1.35), and (2.1.31) to the first term in the right-hand side, we obtain

$$\begin{aligned}|\boldsymbol{u}_h - P_h(\boldsymbol{v}(z_h))|_{W^{1,r}(\Omega)} &\le \frac{c_3}{\varrho_{\min}^{1-2/r}}|\boldsymbol{u}_h - P_h(\boldsymbol{v}(z_h))|_{H^1(\Omega)}\\ &\le \frac{c_4 h}{\varrho_{\min}^{1-2/r}}\left(K_1(z_h)|\boldsymbol{v}(z_h)|_{H^2(\Omega)} + \frac{1}{\nu}|q(z_h)|_{H^1(\Omega)}\right) \le \frac{c_5 h}{\varrho_{\min}^{1-2/r}}.\end{aligned}$$

Thus, the condition on the mesh is

$$h \le C\,\varrho_{\min}^{1-2/r}, \tag{2.1.45}$$

for some $r > 2$. For example, if we choose $r = 2.1$, then the exponent of $\varrho_{\min}$ is $1/21$, and (2.1.45) hardly restricts the mesh. □

2.1.3. *Another view on uniqueness*

The statement of the uniqueness Theorem 1.4.12 does not extend to the discrete problem because a discrete analog of (1.4.4) is not available. Hence, it is useful to derive a sufficient condition for uniqueness (albeit less sharp), directly from the equations of Problem (1.4.9). To this end, let $(\boldsymbol{u}_1, z_1)$ and $(\boldsymbol{u}_2, z_2)$ be any two solutions of (1.4.9) (we eliminate the pressure because it is determined by the other variables). Then arguing as in Lemma 2.1.11, under its assumptions, we easily derive

$$|\boldsymbol{u}_1 - \boldsymbol{u}_2|_{H^1(\Omega)} \le \frac{S_4}{\nu}\|\boldsymbol{u}_2\|_{L^4(\Omega)}\|z_1 - z_2\|_{L^2(\Omega)}, \tag{2.1.46}$$

$$\|\boldsymbol{u}_1 - \boldsymbol{u}_2\|_{L^\infty(\Omega)} \le C_\infty C_1 K_1(z_1)\|\boldsymbol{u}_2\|_{L^4(\Omega)}\|z_1 - z_2\|_{L^2(\Omega)}. \tag{2.1.47}$$

Similarly, by writing

$$\nu(z_1 - z_2) + \alpha\boldsymbol{u}_2\cdot\nabla(z_1 - z_2) = \nu\,\mathrm{curl}(\boldsymbol{u}_1 - \boldsymbol{u}_2) - \alpha(\boldsymbol{u}_1 - \boldsymbol{u}_2)\cdot\nabla z_1,$$

we infer, assuming that z_1 belongs to $H^1(\Omega)$,

$$\|z_1 - z_2\|_{L^2(\Omega)} \le \|\operatorname{curl}(\boldsymbol{u}_1 - \boldsymbol{u}_2)\|_{L^2(\Omega)} + \frac{\alpha}{\nu}\|\boldsymbol{u}_1 - \boldsymbol{u}_2\|_{L^\infty(\Omega)}|z_1|_{H^1(\Omega)}.$$

By substituting (2.1.46) and (2.1.47) into this inequality, we obtain

$$\|z_1 - z_2\|_{L^2(\Omega)} \le \frac{1}{\nu}\|\boldsymbol{u}_2\|_{L^4(\Omega)}\left(S_4 + \alpha C_\infty C_1 K_1(z_1)|z_1|_{H^1(\Omega)}\right)\|z_1 - z_2\|_{L^2(\Omega)},$$

whence the following variant of Theorem 1.4.12. The proof stems directly from this inequality and estimate (1.4.20).

PROPOSITION 2.1.18. *In addition to the assumptions of Lemma 2.1.11 Part 1, suppose that Problem (1.4.9) has a solution $(\boldsymbol{u}, p, z)$ in $V \times L_0^2(\Omega) \times L^2(\Omega)$ such that $z \in H^1(\Omega)$ and*

$$\frac{S_4 S_2}{\nu^2}\|\boldsymbol{f}\|_{L^2(\Omega)}\left(S_4 + \alpha C_\infty C_1 K_1(z)|z|_{H^1(\Omega)}\right) < 1. \tag{2.1.48}$$

Then Problem (1.4.9) has no other solution in $V \times L_0^2(\Omega) \times L^2(\Omega)$.

Note that (2.1.48) holds, for instance, when the force $\boldsymbol{f}$ is small or the viscosity is large.

2.1.4. A priori error bounds

From the exact Problem (1.4.9) and the discrete problem (2.1.6)–(2.1.8), we readily obtain, for all $\boldsymbol{v}_h$ in V_h, all q_h in M_h, and all θ_h in Z_h:

$$\nu(\nabla(\boldsymbol{u}_h - \boldsymbol{u}), \nabla\boldsymbol{v}_h) + ((z_h - z) \times \boldsymbol{u}, \boldsymbol{v}_h) + (z_h \times (\boldsymbol{u}_h - \boldsymbol{u}), \boldsymbol{v}_h) - (q_h - p, \operatorname{div}\boldsymbol{v}_h) = 0, \tag{2.1.49}$$

$$\nu(z_h - z, \theta_h) + \alpha\tilde{c}(\boldsymbol{u}_h - \boldsymbol{u}; z_h, \theta_h) + \alpha\tilde{c}(\boldsymbol{u}; z_h - z, \theta_h) = \nu(\operatorname{curl}(\boldsymbol{u}_h - \boldsymbol{u}), \theta_h). \tag{2.1.50}$$

Then (2.1.49), Lemma 2.1.1, and (2.1.24), imply the following lemma.

LEMMA 2.1.19. *Let $(\boldsymbol{u}_h, p_h, z_h)$ be a solution of (2.1.6)–(2.1.8) and let $(\boldsymbol{u}, p, z)$ be a solution of Problem (1.4.9). Under the first two assumptions of Hypothesis 2.1.5, we have:*

$$|\boldsymbol{u} - \boldsymbol{u}_h|_{H^1(\Omega)} \le 2|\boldsymbol{u} - P_h(\boldsymbol{u})|_{H^1(\Omega)} + \frac{S_4}{\nu}\|\boldsymbol{u}\|_{L^4(\Omega)}\|z - z_h\|_{L^2(\Omega)} + \frac{S_4}{\nu}\|z_h\|_{L^2(\Omega)}\|\boldsymbol{u} - P_h(\boldsymbol{u})\|_{L^4(\Omega)} + \frac{1}{\nu}\|p - r_h(p)\|_{L^2(\Omega)}, \tag{2.1.51}$$

$$\|p - p_h\|_{L^2(\Omega)} \le \left(1 + \frac{1}{\beta^*}\right)\|p - r_h(p)\|_{L^2(\Omega)} + \frac{1}{\beta^*}\Big(\nu\,|\boldsymbol{u} - P_h(\boldsymbol{u})|_{H^1(\Omega)} + S_4\left(\|\boldsymbol{u}\|_{L^4(\Omega)}\|z - z_h\|_{L^2(\Omega)} + \|z_h\|_{L^2(\Omega)}\|\boldsymbol{u} - \boldsymbol{u}_h\|_{L^4(\Omega)}\right)\Big), \tag{2.1.52}$$

where β^* is the constant of (2.1.1).

Now, let us examine (2.1.50).

LEMMA 2.1.20. *Let* $(\boldsymbol{u}_h, p_h, z_h)$ *be a solution of (2.1.6)–(2.1.8) and let* $(\boldsymbol{u}, p, z)$ *be a solution of Problem (1.4.9). For any* λ_h *in* Z_h*, we have*

$$\begin{aligned}\|z - z_h\|_{L^2(\Omega)} \le 2\,\|z - \lambda_h\|_{L^2(\Omega)} &+ \|\mathrm{curl}(\boldsymbol{u} - \boldsymbol{u}_h)\|_{L^2(\Omega)} \\ &+ \frac{\alpha}{\nu}\big(\|(\boldsymbol{u} - \boldsymbol{u}_h)\cdot\nabla\,\lambda_h\|_{L^2(\Omega)} + \|\boldsymbol{u}\cdot\nabla(z - \lambda_h)\|_{L^2(\Omega)} \\ &+ \frac{1}{2}\,\|\lambda_h\,\mathrm{div}(\boldsymbol{u} - \boldsymbol{u}_h)\|_{L^2(\Omega)}\big).\end{aligned} \tag{2.1.53}$$

PROOF. Inserting any $\lambda_h \in Z_h$ into (2.1.50), we derive for all $\theta_h \in Z_h$

$$\begin{aligned}\nu(z_h - \lambda_h, \theta_h) + \alpha\tilde{c}(\boldsymbol{u}_h; z_h - \lambda_h, \theta_h) &= \nu(\mathrm{curl}(\boldsymbol{u}_h - \boldsymbol{u}), \theta_h) \\ &+ \nu(z - \lambda_h, \theta_h) + \alpha\tilde{c}(\boldsymbol{u}; z - \lambda_h, \theta_h) + \alpha\tilde{c}(\boldsymbol{u} - \boldsymbol{u}_h; \lambda_h, \theta_h).\end{aligned}$$

Then (2.1.53) follows by choosing $\theta_h = z_h - \lambda_h$ and applying (2.1.5). □

Note that the statement of this lemma requires no particular regularity assumption on the data and the domain. However, if we want to deduce from it a useful error inequality, we must assume that z belongs to $W^{1,r}(\Omega)$, for some $r > 2$. This is caused by the hyperbolic character of the transport equation. Then we obtain the following corollary.

COROLLARY 2.1.21. *Let* Ω *be convex,* $(\boldsymbol{u}, p, z)$ *a solution of Problem (1.4.9),* r_0 *the number of Proposition 1.4.11, and let the assumptions of Theorem 1.4.14 hold, so that* $z \in W^{1,r}(\Omega)$*, for some real number* r *in* $]2, r_0[$*. Let* $(\boldsymbol{u}_h, p_h, z_h)$ *be a solution of (2.1.6)–(2.1.8) and let* R_h *be the operator of Hypothesis 2.1.5; we have*

$$\begin{aligned}\|z - z_h\|_{L^2(\Omega)} \le 2\,\|z - R_h(z)\|_{L^2(\Omega)} &+ |\boldsymbol{u} - \boldsymbol{u}_h|_{H^1(\Omega)} \\ &+ \frac{\alpha}{\nu}\Big(\|\boldsymbol{u} - \boldsymbol{u}_h\|_{L^{r^*}(\Omega)}|R_h(z)|_{W^{1,r}(\Omega)} + \|\boldsymbol{u}\|_{L^\infty(\Omega)}|z - R_h(z)|_{H^1(\Omega)} \\ &+ \frac{1}{2}|\boldsymbol{u} - \boldsymbol{u}_h|_{H^1(\Omega)}\|R_h(z)\|_{L^\infty(\Omega)}\Big),\end{aligned} \tag{2.1.54}$$

where $\frac{1}{r^*} = \frac{1}{2} - \frac{1}{r}$.

By substituting (2.1.51) into (2.1.54), we immediately derive the next theorem. Its statement makes use of the following notation and bounds:

$$\begin{aligned}&\|R_h(z)\|_{W^{1,r}(\Omega)} \le E_r\|z\|_{W^{1,r}(\Omega)}, \quad \|R_h(z)\|_{L^\infty(\Omega)} \le C_{\infty,r}E_r\|z\|_{W^{1,r}(\Omega)}, \\ &K_2(r, z) = \Big(S_{r^*} + \frac{1}{2}C_{\infty,r}\Big)E_r\|z\|_{W^{1,r}(\Omega)},\end{aligned} \tag{2.1.55}$$

where $C_{\infty,r}$ is the constant of (1.1.18).

THEOREM 2.1.22. *We retain the assumptions of Corollary 2.1.21. Assume that $\mathcal{T}_h$ satisfies (2.1.25) and that Hypothesis 2.1.5 holds. Then, if the data are small enough so that*

$$\left(1 + \frac{\alpha}{\nu} K_2(r, z)\right) S_4 \|\boldsymbol{u}\|_{L^4(\Omega)} \leq \frac{\nu}{2}, \tag{2.1.56}$$

we have the following error estimate:

$$\begin{aligned}\|z - z_h\|_{L^2(\Omega)} \leq\ & 4\|z - R_h(z)\|_{L^2(\Omega)} + 2\frac{\alpha}{\nu}\|\boldsymbol{u}\|_{L^\infty(\Omega)}|z - R_h(z)|_{H^1(\Omega)} \\ & + 2\left(1 + \frac{\alpha}{\nu} K_2(r, z)\right)\left(\frac{S_4}{\nu}\|z_h\|_{L^2(\Omega)}\|\boldsymbol{u} - P_h(\boldsymbol{u})\|_{L^4(\Omega)}\right. \\ & \left. + 2\,|\boldsymbol{u} - P_h(\boldsymbol{u})|_{H^1(\Omega)} + \frac{1}{\nu}\|p - r_h(p)\|_{L^2(\Omega)}\right).\end{aligned} \tag{2.1.57}$$

The above statement calls for the following comments.

1. In order to recover an error estimate of the same order for the three unknowns, the auxiliary variable z must have more regularity than expected, compared with that of the velocity and pressure. This well known imbalance results from the hyperbolic nature of the transport equation. It can be partially remedied by the use of suitable upwind schemes, see Section 2.4.
2. The assumptions of Theorem 2.1.22 can all be checked on the data and the domain. The factors in (2.1.56) are bounded independently of h and can be expressed in terms of the data.
3. The left-hand sides of (2.1.56) and (2.1.48) have related structures.
4. The statement of Theorem 2.1.22 remains valid when α tends to zero.

2.1.5. *Remarks on uniqueness of the discrete solution*

The proof of uniqueness of the discrete solution is still an open problem (even assuming uniqueness of the exact solution), if we want to keep the regularity of the exact solution compatible with a polygonal domain. Indeed, any pair of solutions $(\boldsymbol{u}_h, p_h, z_h)$, $(\boldsymbol{u}'_h, p'_h, z'_h)$ of (2.1.6)–(2.1.8) in $V_h \times M_h \times Z_h$ satisfies: $\boldsymbol{u}_h - \boldsymbol{u}'_h \in V_h$, $p_h - p'_h \in M_h$ $z_h - z'_h \in Z_h$,

$$\begin{aligned}&\forall \boldsymbol{v}_h \in V_h,\ \nu(\nabla(\boldsymbol{u}_h - \boldsymbol{u}'_h), \nabla \boldsymbol{v}_h) + (z'_h \times (\boldsymbol{u}_h - \boldsymbol{u}'_h), \boldsymbol{v}_h) \\ &\qquad = -((z_h - z'_h) \times \boldsymbol{u}_h, \boldsymbol{v}_h),\end{aligned} \tag{2.1.58}$$

$$\begin{aligned}&\forall \theta_h \in Z_h,\ \nu(z_h - z'_h, \theta_h) + \alpha\tilde{c}(\boldsymbol{u}_h; z_h - z'_h, \theta_h) \\ &\qquad + \alpha\tilde{c}(\boldsymbol{u}_h - \boldsymbol{u}'_h; z'_h, \theta_h) = \nu(\operatorname{curl}(\boldsymbol{u}_h - \boldsymbol{u}'_h), \theta_h).\end{aligned} \tag{2.1.59}$$

Therefore,

$$\begin{aligned}\|z_h - z'_h\|_{L^2(\Omega)} \leq\ & \frac{\alpha}{\nu}\left(\|(\boldsymbol{u}_h - \boldsymbol{u}'_h)\cdot\nabla z'_h\|_{L^2(\Omega)} + \frac{1}{2}\|z'_h \operatorname{div}(\boldsymbol{u}_h - \boldsymbol{u}'_h)\|_{L^2(\Omega)}\right) \\ & + |\boldsymbol{u}_h - \boldsymbol{u}'_h|_{H^1(\Omega)}.\end{aligned}$$

The difficulty comes from the first two terms in the right-hand side of this inequality: for instance, we can derive a bound for $\|\boldsymbol{u}_h - \boldsymbol{u}'_h\|_{L^\infty(\Omega)}$ and $|\boldsymbol{u}_h - \boldsymbol{u}'_h|_{W^{1,r}(\Omega)}$, but we have no bound for $|z'_h|_{H^1(\Omega)}$, unless we assume that $z \in H^2(\Omega)$. Indeed, the only way in which we can estimate this term is by writing that

$$|z'_h|_{H^1(\Omega)} \leq |z'_h - R_h(z)|_{H^1(\Omega)} + |R_h(z)|_{H^1(\Omega)}$$
$$\leq \frac{C}{h}\|z'_h - R_h(z)\|_{L^2(\Omega)} + |R_h(z)|_{H^1(\Omega)}.$$

In view of (2.1.57), this gives a bound for $|z'_h|_{H^1(\Omega)}$ in the best of cases when R_h is suitably accurate, if we assume that $z \in H^2(\Omega)$, but we cannot check this assumption on a polygonal domain.

2.2. Centered schemes: Examples

Recall that Ω is a connected polygon. The three examples described here are chosen in order to satisfy the uniform inf-sup condition (2.1.1). They are presented for the homogeneous problem, but they will be easily adapted to nonhomogeneous boundary conditions in Chapter 5. Their study can be found in several texts (for instance BREZZI and FORTIN [1991], GIRAULT and RAVIART [1986], and ERN and GUERMOND [2004]), but we shall mostly use the material in GIRAULT and SCOTT [2003], because this reference emphasizes the local character of the approximation operator P_h, which is crucial in the numerical analysis of nonhomogeneous boundary conditions. The simplest examples are the "mini-element" and the Bernardi–Raugel element; both are of order one, and the Bernardi–Raugel element has the advantage of being locally mass-conservative. The Taylor–Hood element is of order two. The Crouzeix–Raviart element of order one (cf. CROUZEIX and RAVIART [1973]), also locally mass-conservative, is a simple interesting variant, but it is nonconforming and its theory requires a slightly different treatment. The same applies to the second order nonconforming Crouzeix–Raviart element (cf. FORTIN and SOULIÉ [1983]) or the third-order nonconforming Crouzeix–Raviart element (cf. CROUZEIX and FALK [1989]). As written at the beginning of this chapter, we have concentrated on elements of low degree, but of course, we could have used higher degree elements.

2.2.1. *The mini-element*

The mini-element, introduced by ARNOLD, BREZZI and FORTIN [1984], is of order one for the velocity and order two for the pressure. Let $\mathbb{P}_k$ denote the space of polynomials in two variables with total degree less than or equal to k. In each triangle T, the pressure p is a polynomial of $\mathbb{P}_1$ and each component of the velocity is the sum of a polynomial of $\mathbb{P}_1$ and a "bubble" function. Denoting the vertices of T by $\boldsymbol{a}_i$, $1 \leq i \leq 3$, and its corresponding barycentric coordinate by λ_i, the basic bubble function b_T is the polynomial of degree three

$$b_T(\boldsymbol{x}) = \lambda_1(\boldsymbol{x})\lambda_2(\boldsymbol{x})\lambda_3(\boldsymbol{x}),$$

that vanishes on the boundary of T. Thus, we take

$$X_h = \Big\{ \boldsymbol{v}_h \in H_0^1(\Omega)^2;\ \forall T \in \mathcal{T}_h,\ \boldsymbol{v}_h|_T \in (\mathbb{P}_1 \oplus \mathrm{Vect}(b_T))^2 \Big\}, \tag{2.2.1}$$

$$M_h = \Big\{ q_h \in H^1(\Omega) \cap L_0^2(\Omega);\ \forall T \in \mathcal{T}_h,\ q_h|_T \in \mathbb{P}_1 \Big\}. \tag{2.2.2}$$

Considering the velocity's approximation order, it is reasonable to choose also an approximation of order one for z:

$$Z_h = \{ \theta_h \in H^1(\Omega);\ \forall T \in \mathcal{T}_h,\ \theta_h|_T \in \mathbb{P}_1 \}. \tag{2.2.3}$$

Observe that $M_h = Z_h \cap L_0^2(\Omega)$. The next lemma constructs P_h for the mini-element. It uses the following notation: a *macroelement* Δ_T is the union of elements of $\mathcal{T}_h$ sharing at least a vertex with T. When the family of triangulations $\mathcal{T}_h$ satisfies (2.1.25), the number of elements of Δ_T is bounded by a constant, say L_1, independent of h and T; and a given element T can belong to at most a fixed number of macroelements Δ_S, say L_2, also independent of h and T.

LEMMA 2.2.1. *If the family of triangulations $\mathcal{T}_h$ satisfies (2.1.25), there exists an operator $P_h \in \mathcal{L}(H_0^1(\Omega)^2; X_h)$ satisfying (2.1.19):*

$$\forall \boldsymbol{v} \in H_0^1(\Omega)^2,\ \forall q_h \in M_h,\ \int_\Omega q_h \mathrm{div}\, P_h(\boldsymbol{v})\, \mathrm{d}\boldsymbol{x} = \int_\Omega q_h \mathrm{div}\, \boldsymbol{v}\, \mathrm{d}\boldsymbol{x},$$

and the following approximation properties

$$\forall \boldsymbol{v} \in W^{s,r}(\Omega)^2,\ \forall T \in \mathcal{T}_h,$$
$$|P_h(\boldsymbol{v}) - \boldsymbol{v}|_{W^{m,q}(T)} \le C_1 h_T^{\,s-m+2\left(\frac{1}{q}-\frac{1}{r}\right)} |\boldsymbol{v}|_{W^{s,r}(\Delta_T)}, \tag{2.2.4}$$

for integers $m = 0$ or 1, for all real numbers $1 \le s \le 2$, and all numbers $1 \le r, q \le \infty$, such that

$$W^{s,r}(\Omega) \subset W^{m,q}(\Omega),$$

with a constant C_1 independent of h and T.

PROOF. Take $\boldsymbol{v}$ in $H_0^1(\Omega)^2$ and let Π_h be a regularization operator, such as the SCOTT and ZHANG [1990] operator that is a polynomial of $\mathbb{P}_1$ in each triangle, is globally continuous, and preserves the polynomials of $\mathbb{P}_1$, so that it preserves in particular the zero boundary value. We choose

$$P_h(\boldsymbol{v}) = \Pi_h(\boldsymbol{v}) - \sum_{T \in \mathcal{T}_h} \boldsymbol{c}_T b_T, \tag{2.2.5}$$

where the constant vectors $\boldsymbol{c}_T$ are adjusted so that P_h satisfies (2.1.19). But q_h belongs to M_h and by construction, $P_h(\boldsymbol{v}) - \boldsymbol{v}$ vanishes on the boundary of Ω, therefore, (2.1.19) amounts to

$$\forall q_h \in M_h, \int_{\Omega} (P_h(\boldsymbol{v}) - \boldsymbol{v}) \cdot \nabla q_h \, \mathrm{d}\boldsymbol{x} = 0. \tag{2.2.6}$$

Now, ∇q_h is a constant vector in each triangle T. Therefore, (2.2.6) holds provided that

$$\forall T \in \mathcal{T}_h, \int_T (P_h(\boldsymbol{v}) - \boldsymbol{v}) \, \mathrm{d}\boldsymbol{x} = \boldsymbol{0}.$$

From the definition (2.2.5) of P_h and the disjoint supports of the bubble functions, this last equation determines the constant vectors $\boldsymbol{c}_T$:

$$\forall T \in \mathcal{T}_h, \ \boldsymbol{c}_T = \frac{1}{\int_T b_T \, \mathrm{d}\boldsymbol{x}} \int_T (\Pi_h(\boldsymbol{v}) - \boldsymbol{v}) \, \mathrm{d}\boldsymbol{x}. \tag{2.2.7}$$

Let us estimate first $|\boldsymbol{c}_T|$ and next $\|b_T\|_{L^2(\Omega)}$ and $|b_T|_{W^{1,q}(T)}$. Let $\hat{T}$ be the reference unit triangle and B_T the matrix of the affine transformation that maps $\hat{T}$ onto T. On one hand, for any $q \geq 2$,

$$|\boldsymbol{c}_T| \leq \hat{c}|T|^{-1/q} \|\Pi_h(\boldsymbol{v}) - \boldsymbol{v}\|_{L^q(T)},$$

where $\hat{c}$ denote various constants that depend only on $\hat{T}$ and the exponent q. On the other hand,

$$\|b_T\|_{L^2(T)} \leq \hat{c}|T|^{1/2}, \ \|b_T\|_{W^{1,q}(T)} \leq \hat{c}|T|^{1/q}|B_T^{-1}|.$$

Therefore,

$$|\boldsymbol{c}_T| \|b_T\|_{L^2(T)} \leq \hat{c}\|\Pi_h(\boldsymbol{v}) - \boldsymbol{v}\|_{L^2(T)},$$

$$|\boldsymbol{c}_T| |b_T|_{W^{1,q}(T)} \leq \hat{c}|B_T^{-1}| \|\Pi_h(\boldsymbol{v}) - \boldsymbol{v}\|_{L^q(T)}.$$

From the disjoint support of the bubble functions b_T, we infer that

$$\|P_h(\boldsymbol{v}) - \boldsymbol{v}\|_{L^2(T)} \leq (1 + \hat{c})\|\Pi_h(\boldsymbol{v}) - \boldsymbol{v}\|_{L^2(T)},$$

$$|P_h(\boldsymbol{v}) - \boldsymbol{v}|_{W^{1,q}(T)} \leq |\Pi_h(\boldsymbol{v}) - \boldsymbol{v}|_{W^{1,q}(T)} + \hat{c}|B_T^{-1}| \|\Pi_h(\boldsymbol{v}) - \boldsymbol{v}\|_{L^q(T)}. \tag{2.2.8}$$

Then (2.2.4) follows from (2.2.8), the regularity (2.1.25) of the triangulation and the local approximation properties of Π_h, see SCOTT and ZHANG [1990]. □

It can be easily checked that (2.1.19) and (2.2.4) with $r = q = 2$ and $m = s = 1$ imply the uniform inf-sup condition (2.1.1) between X_h and M_h. As far as M_h and Z_h are concerned, either the variant of the Scott & Zhang regularization operator defined by GIRAULT and LIONS [2001b], or the regularization operators defined by CLÉMENT [1975] or BERNARDI and GIRAULT [1998] (still denoted by Π_h) are good candidates for r_h and R_h. Then, R_h satisfies the analog of (2.2.4) with the same notation

$$\forall z \in W^{s,r}(\Omega),\ \forall T \in \mathcal{T}_h,\ |R_h(z) - z|_{W^{m,q}(T)} \leq C_2 h_T^{s-m+2\left(\frac{1}{q}-\frac{1}{r}\right)} |z|_{W^{s,r}(\Delta_T)}, \tag{2.2.9}$$

with another constant C_2 independent of h, T, and Δ_T. Moreover, Π_h can be easily adjusted to the zero mean-value for r_h by setting

$$\forall q \in H^1(\Omega),\ r_h(q) = \Pi_h(q) - \frac{1}{|\Omega|} \int_\Omega (\Pi_h(q) - q)\, \mathrm{d}\boldsymbol{x}. \tag{2.2.10}$$

Considering this global zero mean-value constraint, instead of the local estimate (2.2.9), r_h satisfies (2.1.27):

$$\forall q \in W^{s,r}(\Omega) \cap L_0^2(\Omega),\ \|r_h(q) - q\|_{L^r(\Omega)} \leq C_3 h^s |q|_{W^{s,r}(\Omega)},$$

for all numbers $r \geq 2$ and all real numbers $s \in [0, 2]$, with a constant C_3 independent of h.

Hence all assumptions of Hypotheses 2.1.5 and 2.1.10 are satisfied by the spaces X_h, M_h, and Z_h. As (2.2.4) only yields an error of order h for the velocity, the error estimate of Theorem 2.1.22 gives the same order. More precisely, we have the following theorem:

THEOREM 2.2.2. *Let the family of triangulations $\mathcal{T}_h$ satisfy (2.1.25), let $(\boldsymbol{u}, p, z)$ be a solution of Problem (1.4.9), with $z \in H^2(\Omega)$, $\boldsymbol{u} \in H^2(\Omega)^2$ and $p \in H^1(\Omega)$, satisfying (2.1.56), and let $(\boldsymbol{u}_h, p_h, z_h)$ be a solution of (2.1.6)–(2.1.8) with the finite-element spaces (2.2.1)–(2.2.3). Then, there exists a constant C, independent of h, such that*

$$\|z - z_h\|_{L^2(\Omega)} + |\boldsymbol{u} - \boldsymbol{u}_h|_{H^1(\Omega)} + \|p - p_h\|_{L^2(\Omega)} \leq Ch.$$

2.2.2. *The Bernardi–Raugel element*

Now we turn to the Bernardi–Raugel finite-element cf. BERNARDI and RAUGEL [1985]. Let f_i denote the side of T opposite $\boldsymbol{a}_i$ and let $\boldsymbol{n}_i$ be the unit normal vector to f_i pointing outside T. We define the three edge "bubble functions"

$$\boldsymbol{p}_{1,T} = \boldsymbol{n}_1 \lambda_2 \lambda_3,\ \boldsymbol{p}_{2,T} = \boldsymbol{n}_2 \lambda_1 \lambda_3,\ \boldsymbol{p}_{3,T} = \boldsymbol{n}_3 \lambda_1 \lambda_2,$$

and we set

$$\mathcal{P}_1(T) = \mathbb{P}_1^2 \oplus \mathrm{Vect}\{\boldsymbol{p}_{1,T}, \boldsymbol{p}_{2,T}, \boldsymbol{p}_{3,T}\}.$$

The finite-element spaces for the Bernardi–Raugel element are

$$X_h = \{\boldsymbol{v}_h \in H_0^1(\Omega)^2;\ \forall T \in \mathcal{T}_h,\ \boldsymbol{v}_h|_T \in \mathcal{P}_1(T)\}, \tag{2.2.11}$$

$$M_h = \{q_h \in L_0^2(\Omega);\ \forall T \in \mathcal{T}_h,\ q_h|_T \in \mathbb{P}_0\}. \tag{2.2.12}$$

The local mass conservativity of this pair of spaces (i.e., in each element T) follows from the fact that we can now choose a test pressure that takes the value one in T and zero elsewhere. As the approximation error of these two spaces are of order one, we take for Z_h the space defined by (2.2.3), i.e., the same as for the mini-element. Because the discrete pressures have no continuity requirements across elements, we can take for r_h the L^2 projection on the constant functions in each T, corrected so that it globally has a mean-value of zero. The following lemma constructs a suitable operator P_h.

LEMMA 2.2.3. *If the family of triangulations $\mathcal{T}_h$ verifies (2.1.25), there exists an operator $P_h \in \mathcal{L}(H_0^1(\Omega)^2; X_h)$ satisfying (2.1.19) and (2.2.4) with the same values of s, m, r, and q as in Lemma 2.2.1.*

PROOF. We only sketch the proof; it is similar to that of Lemma 2.2.1. For $\boldsymbol{v}$ in $H_0^1(\Omega)^2$, we choose

$$P_h(\boldsymbol{v}) = \Pi_h(\boldsymbol{v}) - \sum_{T\in\mathcal{T}_h}\sum_{i=1}^{3}\alpha_{i,T}\boldsymbol{p}_{i,T}. \tag{2.2.13}$$

It can be easily checked that, for satisfying (2.1.19), it suffices to take

$$\alpha_{i,T} = \frac{1}{\int_{f_i}\lambda_j\lambda_k \mathrm{d}s}\int_{f_i}(\Pi_h(\boldsymbol{v}) - \boldsymbol{v})\cdot\boldsymbol{n}_i \mathrm{d}s,\ j \neq k \neq i.$$

On one hand, passing to $\hat{T}$, applying a trace theorem on $\partial\hat{T}$ and reverting to T, we find

$$|\alpha_{i,T}| \leq \hat{c}|T|^{-1/q}\left(\|\Pi_h(\boldsymbol{v}) - \boldsymbol{v}\|_{L^q(T)} + |B_T||\Pi_h(\boldsymbol{v}) - \boldsymbol{v}|_{W^{1,q}(T)}\right).$$

On the other hand,

$$|\boldsymbol{p}_{i,T}|_{W^{1,q}(T)} \leq \hat{c}|T|^{1/q}|B_T^{-1}|.$$

Therefore,

$$|\alpha_{i,T}|\|\boldsymbol{p}_{i,T}\|_{L^2(T)} \leq \hat{c}\left(\|\Pi_h(\boldsymbol{v}) - \boldsymbol{v}\|_{L^2(T)} + |B_T||\Pi_h(\boldsymbol{v}) - \boldsymbol{v}|_{H^1(T)}\right),$$

$$|\alpha_{i,T}||\boldsymbol{p}_{i,T}|_{W^{1,q}(T)} \leq \hat{c}\left(|\Pi_h(\boldsymbol{v}) - \boldsymbol{v}|_{W^{1,q}(T)} + |B_T^{-1}|\|\Pi_h(\boldsymbol{v}) - \boldsymbol{v}\|_{L^q(T)}\right).$$

The proof finishes as in Lemma 2.2.1. □

The conclusion is the same: all assumptions of Hypotheses 2.1.5 and 2.1.10 are satisfied by the spaces X_h, M_h, and Z_h, and the resulting scheme has order one.

THEOREM 2.2.4. *Let the family of triangulations $\mathcal{T}_h$ satisfy (2.1.25), let $(\boldsymbol{u}, p, z)$ be a solution of Problem (1.4.9), with $z \in H^2(\Omega)$, $\boldsymbol{u} \in H^2(\Omega)^2$ and $p \in H^1(\Omega)$, satisfying (2.1.56), and let $(\boldsymbol{u}_h, p_h, z_h)$ be a solution of (2.1.6)–(2.1.8) with the finite-element spaces (2.2.11), (2.2.12), and (2.2.3). Then, there exists a constant C, independent of h, such that*

$$\|z - z_h\|_{L^2(\Omega)} + |\boldsymbol{u} - \boldsymbol{u}_h|_{H^1(\Omega)} + \|p - p_h\|_{L^2(\Omega)} \leq Ch.$$

2.2.3. *The Taylor–Hood element*

Finally, consider the classical conforming Taylor–Hood element of degree two with continuous pressures (cf. HOOD and TAYLOR [1973]):

$$X_h = \{v_h \in H_0^1(\Omega)^2;\ \forall T \in \mathcal{T}_h,\ v_h|_T \in \mathbb{P}_2^2\}, \tag{2.2.14}$$

$$M_h = \{q_h \in H^1(\Omega) \cap L_0^2(\Omega);\ \forall T \in \mathcal{T}_h,\ q_h|_T \in \mathbb{P}_1\}. \tag{2.2.15}$$

Note that M_h is the pressure space of the mini-element. Because the approximation error of these spaces is of the order of h^2, we choose

$$Z_h = \{\theta_h \in H^1(\Omega);\ \forall T \in \mathcal{T}_h,\ \theta_h|_T \in \mathbb{P}_2\}. \tag{2.2.16}$$

The inf-sup condition for this element was established by BERCOVIER and PIRONNEAU [1979], then by VERFÜRTH [1984], and by GIRAULT and RAVIART [1986]; this last reference gives a proof with a semilocal argument based on the approach of BOLAND and NICOLAIDES [1983] and STENBERG [1984], under the assumption that the family $\mathcal{T}_h$ is nondegenerate and each triangle T has at most one edge on $\partial\Omega$. But none of these references propose an approximation operator satisfying (2.1.19) and (2.2.4). For a long time, it was an open problem, the only remedy being the introduction of additional degrees of freedom as in the work of DURÁN, NOCHETTO and WANG [1988]. The construction presented here fills this gap. It is due to GIRAULT and SCOTT [2003], and it proceeds along the following lines:

1. Define a preliminary operator Π_h that preserves the mean-value of the divergence in each T;
2. Decompose the domain into a union of suitable macroelements;
3. Correct Π_h in each macroelement, so that a local inf-sup condition is satisfied there.

If this is adequately done, the corrected operator satisfies a global inf-sup condition, and its support is a neighborhood of the support of the function to which it is applied.

First, let us construct a preliminary operator $\Pi_h \in \mathcal{L}(H_0^1(\Omega)^2; X_h)$ that satisfies

$$\forall v \in H_0^1(\Omega)^2,\ \forall T \in \mathcal{T}_h,\ \int_T \operatorname{div}(\Pi_h(v) - v)\,\mathrm{d}x = 0, \tag{2.2.17}$$

and the approximation property analogous to (2.2.4):

$$\begin{aligned} &\forall v \in W^{s,r}(\Omega)^2,\ \forall T \in \mathcal{T}_h,\\ &|\Pi_h(v) - v|_{W^{m,q}(T)} \le C_3\, h_T^{\,s-m+2\left(\frac{1}{q}-\frac{1}{r}\right)} |v|_{W^{s,r}(\Delta_T)}, \end{aligned} \tag{2.2.18}$$

for integers $m = 0$ or 1, for all real numbers $1 \le s \le 3$, and all numbers $1 \le r, q \le \infty$, such that

$$W^{s,r}(\Omega) \subset W^{m,q}(\Omega),$$

with a constant C_3 independent of h and T. This operator can be derived from the reference SCOTT and ZHANG [1990] as follows. Let T be a triangle with vertices $\boldsymbol{a}_i$, and opposite sides f_i, $1 \le i \le 3$. A polynomial p of degree two is uniquely determined in T by the six values:

$$p(\boldsymbol{a}_i), \int_{f_i} p(s)\,\mathrm{d}s,\ 1 \le i \le 3.$$

For $1 \le i \le 3$, let $\varphi_{\boldsymbol{a}_i} \in \mathbb{P}_2$ and $\varphi_{f_i} \in \mathbb{P}_2$ be the Lagrange basis functions associated with these values, i.e.,

$$\varphi_{\boldsymbol{a}_i}(\boldsymbol{a}_j) = \delta_{i,j}, \int_{f_k} \varphi_{\boldsymbol{a}_i}(s)\,\mathrm{d}s = 0,\ 1 \le j,k \le 3,$$

$$\varphi_{f_i}(\boldsymbol{a}_k) = 0, \int_{f_j} \varphi_{f_i}(s)\,\mathrm{d}s = \delta_{i,j},\ 1 \le j,k \le 3.$$

For defining Π_h on $H^1(\Omega)$, we regularize the above point values as follows. With each vertex $\boldsymbol{a}_i$, we choose once and for all a segment κ_i of $\mathcal{T}_h$ with end point $\boldsymbol{a}_i$. This choice is arbitrary, with one exception: for preserving vanishing boundary values, we impose in addition that κ_i be contained in $\partial\Omega$, whenever $\boldsymbol{a}_i$ lies on $\partial\Omega$. Let $\psi_{\boldsymbol{a}_i} \in \mathbb{P}_2(\kappa_i)$ be the dual basis function on κ_i, i.e.,

$$\int_{\kappa_i} \psi_{\boldsymbol{a}_i}(s)\varphi_b(s)\,\mathrm{d}s = \delta_{\boldsymbol{a}_i,b}, \tag{2.2.19}$$

where b denotes the segment κ_i itself or its two end points. Then, we replace the point-value $p(\boldsymbol{a}_i)$ by the degree of freedom

$$\int_{\kappa_i} p(s)\psi_{\boldsymbol{a}_i}(s)\,\mathrm{d}s.$$

Thus, we define Π_h by

$$\Pi_h(v)(\boldsymbol{x}) = \sum_{\boldsymbol{a}_i \in \mathcal{S}_h} \left(\int_{\kappa_i} v(s)\psi_{\boldsymbol{a}_i}(s)\,\mathrm{d}s \right) \varphi_{\boldsymbol{a}_i}(\boldsymbol{x}) + \sum_{f \in \Gamma_h} \left(\int_f v(s)\,\mathrm{d}s \right) \varphi_f(\boldsymbol{x}), \tag{2.2.20}$$

where $\mathcal{S}_h$ denotes the set of all vertices $\boldsymbol{a}_i$ of $\mathcal{T}_h$ and Γ_h denotes the set of all segments f of $\mathcal{T}_h$. It stems from the above choice of degrees of freedom on the segments f and the corresponding choice of basis functions that

$$\forall f \in \Gamma_h, \int_f (\Pi_h(v) - v)\,\mathrm{d}s = 0. \tag{2.2.21}$$

Of course, this implies (2.2.17). Furthermore, it is easy to check that Π_h is a projection, by virtue of (2.2.19); i.e., if v_h is globally continuous in Ω and a polynomial of $\mathbb{P}_2$ in each triangle T, then

$$\Pi_h(v_h) = v_h.$$

This property allows one to apply the argument of SCOTT and ZHANG [1990] and show that Π_h satisfies the optimal approximation property (2.2.18) for $s \in [1,3]$.

Next, let us define the following spaces: for each function q_h in M_h we define in each T,

$$\tilde{q}_h = q_h - \frac{1}{|T|}\int_T q_h \, \mathrm{d}\boldsymbol{x},$$

$$\tilde{M}_h = \{\tilde{q}_h;\ q_h \in M_h\},$$

$$\tilde{X}_h = \{\boldsymbol{v}_h \in X_h;\ \forall T \in \mathcal{T}_h,\ \int_T \operatorname{div} \boldsymbol{v}_h \, \mathrm{d}\boldsymbol{x} = 0\}.$$

Note that, in contrast to the functions of M_h, the functions of $\tilde{M}_h$ are not continuous. On the other hand, they have zero mean-value in each T. This will enable us to eliminate the piecewise constant pressures. And to begin with, let us state a variant of the inf-sup condition on a triangle T. On each side f_i of T, we choose once and for all a tangent vector $\boldsymbol{t}_i$ with length $|f_i|$, we denote by $\boldsymbol{n}_i$ the unit normal to f_i exterior to T, and we denote by $\boldsymbol{b}_i$ the midpoint of this side. Let $q_h \in \mathbb{P}_1 \cap L_0^2(T)$; following BERCOVIER and PIRONNEAU [1979], we define $\boldsymbol{v}_h \in \mathbb{P}_2^2$ as follows:

$$\boldsymbol{v}_h(\boldsymbol{a}_i) = \boldsymbol{0},\ 1 \le i \le 3,$$

and on any side f_i of T that is not on $\partial\Omega$:

$$(\boldsymbol{v}_h \cdot \boldsymbol{t}_i)(\boldsymbol{b}_i) = -(\nabla q_h \cdot \boldsymbol{t}_i)(\boldsymbol{b}_i),\ (\boldsymbol{v}_h \cdot \boldsymbol{n}_i)(\boldsymbol{b}_i) = 0; \tag{2.2.22}$$

if f_i lies on $\partial\Omega$, we set $\boldsymbol{v}_h(\boldsymbol{b}_i) = \boldsymbol{0}$.

LEMMA 2.2.5. *Assume that the family $\mathcal{T}_h$ satisfies (2.1.25) and each triangle T has at most one edge on $\partial\Omega$. Then for any triangle T and for all $q_h \in \mathbb{P}_1 \cap L_0^2(T)$, the function $\boldsymbol{v}_h$ defined above satisfies:*

$$\forall r \ge 2,\ \int_T q_h \operatorname{div} \boldsymbol{v}_h \, \mathrm{d}\boldsymbol{x} \ge \hat{c}\|q_h\|_{L^r(T)}\|q_h\|_{L^{r'}(T)},\ \frac{1}{r} + \frac{1}{r'} = 1, \tag{2.2.23}$$

$$\forall r \ge 2,\ \|\boldsymbol{v}_h\|_{L^r(T)} \le \hat{c} h_T^{2/r}\|q_h\|_{L^2(T)}, \tag{2.2.24}$$

$$\forall r \ge 2,\ |\boldsymbol{v}_h|_{W^{1,r}(T)} \le \hat{c}\|q_h\|_{L^r(T)}, \tag{2.2.25}$$

where $\hat{c}$ denote several constants, depending possibly on r, but independent of h, T, q_h, and $\boldsymbol{v}_h$. In these three inequalities, the exponents r are independent of each other and can be infinite.

We skip the proof because it is based on a straightforward extension of the arguments of Theorem II.4.2, pp. 178,179 in GIRAULT and RAVIART [1986].

Let q_h be an arbitrary element of M_h and let $\tilde{q}_h$ be its associated function in $\tilde{M}_h$. The function $\boldsymbol{v}_h$ defined earlier, associated with the restriction of $\tilde{q}_h$ in each T belongs globally to $\tilde{X}_h$ because, on one hand, $\nabla \tilde{q}_h \cdot \boldsymbol{t} = \nabla q_h \cdot \boldsymbol{t}$ is continuous at the interfaces f between adjacent elements, and on the other hand, by construction, $\boldsymbol{v}_h = \boldsymbol{0}$ on f if $f \subset \partial\Omega$. A global inf-sup condition can be derived by summing (2.2.23) over all triangles of $\mathcal{T}_h$, but this does not serve our purpose because this process yields a global approximation operator. However, if this is performed on a suitable macroelement, then the inf-sup condition is local and its corresponding approximation operator is quasi-local. Whence the idea of proceeding by macroelements.

To this end, $\overline{\Omega}$ is decomposed into a finite union of macroelements $\{\mathcal{O}_i\}_{i=1}^R$, mutually distinct, but with possible overlaps:

$$\overline{\Omega} = \cup_{i=1}^R \mathcal{O}_i. \tag{2.2.26}$$

They are obtained by choosing an adequate set of internal vertices $\{\boldsymbol{a}_i\}_{i=1}^R$ of $\mathcal{T}_h$ and by taking for $\mathcal{O}_i$ the union of all triangles of $\mathcal{T}_h$ that share the vertex $\boldsymbol{a}_i$. For instance, (2.2.26) is satisfied by choosing the set of *all internal* vertices of $\mathcal{T}_h$. Of course, this choice is not unique and (2.2.26) still holds while many vertices are deleted. The important features of this decomposition are as follows:

1. The choice of internal vertices implies that each triangle T of $\mathcal{O}_i$ has at most one side on $\partial\mathcal{O}_i$;
2. Each $\mathcal{O}_i$ is connected because Ω is a connected polygon;
3. The regularity of the family $\mathcal{T}_h$ implies that the maximum number of triangles T in $\mathcal{O}_i$ is bounded by a constant L_1, independent of h;
4. Each triangle T belongs to at most three macroelements; therefore, the maximum number of macroelements intersecting a given macroelement is bounded by another constant L_2, independent of h.

In order to derive an inf-sup condition on each macroelement, we define spaces analogous to $\tilde{M}_h$ and $\tilde{X}_h$ as follows:

$$\tilde{M}_h(\mathcal{O}_i) = \{\tilde{q}_h|_{\mathcal{O}_i};\ \tilde{q}_h \in \tilde{M}_h\},$$
$$\tilde{X}_h(\mathcal{O}_i) = \{\boldsymbol{v}_h \in \tilde{X}_h;\ \text{supp}(\boldsymbol{v}_h) \subset \mathcal{O}_i\},$$

and spaces analogous to V_h and $V_h{}^{\perp}$:

$$\tilde{V}_h(\mathcal{O}_i) = \{\boldsymbol{v}_h \in \tilde{X}_h(\mathcal{O}_i);\ \forall q_h \in \tilde{M}_h(\mathcal{O}_i),\ \int_{\mathcal{O}_i} q_h \operatorname{div} \boldsymbol{v}_h \mathrm{d}\boldsymbol{x} = 0\},$$

$$\tilde{V}_h(\mathcal{O}_i)^{\perp} = \{\boldsymbol{v}_h \in \tilde{X}_h(\mathcal{O}_i);\ \forall \boldsymbol{w}_h \in \tilde{V}_h(\mathcal{O}_i),\ \int_{\mathcal{O}_i} \nabla \boldsymbol{v}_h \cdot \nabla \boldsymbol{w}_h \,\mathrm{d}\boldsymbol{x} = 0\}.$$

Then the statement of Lemma 2.2.5 is easily extended to a macroelement.

LEMMA 2.2.6. *Assume that the family $\mathcal{T}_h$ satisfies (2.1.25) and each triangle T has at most one edge on $\partial\Omega$. Let $\boldsymbol{a}_i$ be an interior vertex of $\mathcal{T}_h$, L the number of triangles in $\mathcal{O}_i$, and let $2 \le r \le \infty$ be a number. For each $q_h \in \tilde{M}_h(\mathcal{O}_i)$, there exists $\boldsymbol{v}_h \in \tilde{X}_h(\mathcal{O}_i)$ such that*

$$\int_{\mathcal{O}_i} q_h \mathrm{div}\boldsymbol{v}_h \mathrm{d}\boldsymbol{x} \ge \hat{c}(r, L)\|q_h\|_{L^r(\mathcal{O}_i)}\|q_h\|_{L^{r'}(\mathcal{O}_i)}, \quad \frac{1}{r} + \frac{1}{r'} = 1, \tag{2.2.27}$$

$$|\boldsymbol{v}_h|_{W^{1,r}(\mathcal{O}_i)} \le \|q_h\|_{L^r(\mathcal{O}_i)}, \tag{2.2.28}$$

where the constant $\hat{c}(r, L)$ depends on r and L, but is independent of h, i, q_h, or $\boldsymbol{v}_h$.

The choice $r = 2$ in (2.2.27) and (2.2.28), and the fact that $L \le L_1$ imply the following local inf-sup condition, with a constant λ^* independent of h and i:

$$\forall q_h \in \tilde{M}_h(\mathcal{O}_i), \quad \sup_{\boldsymbol{v}_h \in \tilde{X}_h(\mathcal{O}_i)} \frac{\int_{\mathcal{O}_i} q_h \,\mathrm{div}\, \boldsymbol{v}_h \,\mathrm{d}\boldsymbol{x}}{|\boldsymbol{v}_h|_{H^1(\mathcal{O}_i)}} \ge \lambda^* \|q_h\|_{L^2(\mathcal{O}_i)}. \tag{2.2.29}$$

Owing to (2.2.29), we can construct an approximation operator P_h by correcting Π_h in each $\mathcal{O}_i$. Roughly speaking, P_h is defined through

$$P_h(\boldsymbol{v}) = \Pi_h(\boldsymbol{v}) + \boldsymbol{c}_h(\boldsymbol{v}), \tag{2.2.30}$$

where $\boldsymbol{c}_h(\boldsymbol{v}) \in \tilde{X}_h$ is constructed so that

$$\forall q_h \in \tilde{M}_h, \int_{\Omega} q_h \,\mathrm{div}\, \boldsymbol{c}_h(\boldsymbol{v}) \,\mathrm{d}\boldsymbol{x} = \int_{\Omega} q_h \,\mathrm{div}(\boldsymbol{v} - \Pi_h(\boldsymbol{v})) \,\mathrm{d}\boldsymbol{x}. \tag{2.2.31}$$

We shall see below that because (2.2.17) holds, then (2.2.31) and the constraint on $\tilde{X}_h$ imply that P_h satisfies (2.1.19). More precisely, we can prove the main result of this paragraph.

THEOREM 2.2.7. *Assume that the family $\mathcal{T}_h$ satisfies (2.1.25) and each triangle T has at most one edge on $\partial\Omega$. Then there exists an operator $P_h \in \mathcal{L}(H_0^1(\Omega)^2; X_h)$ of the form (2.2.30) satisfying (2.1.19) and*

$$\begin{aligned} &\forall \boldsymbol{v} \in W^{s,r}(\Omega)^2, \ \forall 1 \le i \le R, \\ &|P_h(\boldsymbol{v}) - \boldsymbol{v}|_{W^{m,q}(\mathcal{O}_i)} \le C_1 h_i^{s-m+2\left(\frac{1}{q}-\frac{1}{r}\right)} |\boldsymbol{v}|_{W^{s,r}(\tilde{\Delta}_i)}, \end{aligned} \tag{2.2.32}$$

for all real numbers $1 \le s \le 3$, $1 \le r, q \le \infty$, and integers $m = 0$ or 1, such that

$$W^{s,r}(\Omega) \subset W^{m,q}(\Omega),$$

where $\tilde{\Delta}_i$ is a suitable macroelement with

$$\mathrm{diam}(\tilde{\Delta}_i) \le C_2 h_i; \tag{2.2.33}$$

the constants C_1 and C_2 are independent of h and R and $h_i = \max_{T \subset \mathcal{O}_i} h_T$.

PROOF. (1) In order to deal with possible macroelements overlaps, we associate a partition of $\overline{\Omega}$ to the set $\{\mathcal{O}_i\}$. To this end, we define $\Delta_1 = \mathcal{O}_1$, then we take for Δ_2 the union of all elements T that belong to $\mathcal{O}_2$, but not to Δ_1 and by induction, we choose for Δ_i the set (possibly empty) of all T that belong to $\mathcal{O}_i$ but not to $\cup_{j=1}^{i-1}\Delta_j$. By construction, the Δ_i are mutually disjoint,

$$\Omega = \cup_{i=1}^{R}\Delta_i,\ \ \Delta_i \subset \mathcal{O}_i,\ 1 \le i \le R.$$

The uniform local inf-sup condition (2.2.29) implies that, for each i, there exists a unique function $\boldsymbol{c}_{h,i} \in \tilde{V}_h(\mathcal{O}_i)^{\perp}$ solution of

$$\forall q_h \in \tilde{M}_h(\mathcal{O}_i),\ \int_{\mathcal{O}_i} q_h \operatorname{div} \boldsymbol{c}_{h,i}\, \mathrm{d}\boldsymbol{x} = \int_{\Delta_i} q_h \operatorname{div}(\boldsymbol{v} - \Pi_h(\boldsymbol{v}))\, \mathrm{d}\boldsymbol{x}. \tag{2.2.34}$$

(Note that $\boldsymbol{c}_{h,i} = \boldsymbol{0}$ when Δ_i is empty). Then we extend each $\boldsymbol{c}_{h,i}$ by zero outside $\mathcal{O}_i$, and we set

$$\boldsymbol{c}_h(\boldsymbol{v}) = \sum_{i=1}^{R} \boldsymbol{c}_{h,i}.$$

By construction, $\boldsymbol{c}_h(\boldsymbol{v}) \in \tilde{X}_h$; moreover, the support of $\boldsymbol{c}_{h,i}$ and the partitioning of Ω into $\{\Delta_i\}_{i=1}^{R}$ imply that $\boldsymbol{c}_h(\boldsymbol{v})$ satisfies (2.2.31). Indeed, we have

$$\begin{aligned}
\int_{\Omega} q_h \operatorname{div} \boldsymbol{c}_h(\boldsymbol{v})\, \mathrm{d}\boldsymbol{x} &= \int_{\Omega} q_h \operatorname{div}\left(\sum_{i=1}^{R} \boldsymbol{c}_{h,i}\right) \mathrm{d}\boldsymbol{x} = \sum_{i=1}^{R} \int_{\Omega} q_h \operatorname{div} \boldsymbol{c}_{h,i}\, \mathrm{d}\boldsymbol{x}\\
&= \sum_{i=1}^{R} \int_{\mathcal{O}_i} q_h \operatorname{div} \boldsymbol{c}_{h,i}\, \mathrm{d}\boldsymbol{x} = \sum_{i=1}^{R} \int_{\Delta_i} q_h \operatorname{div}(\boldsymbol{v} - \Pi_h(\boldsymbol{v}))\, \mathrm{d}\boldsymbol{x}\\
&= \int_{\Omega} q_h \operatorname{div}(\boldsymbol{v} - \Pi_h(\boldsymbol{v}))\, \mathrm{d}\boldsymbol{x}.
\end{aligned}$$

(2) The local inf-sup condition (2.2.29) implies that

$$|\boldsymbol{c}_{h,i}|_{H^1(\mathcal{O}_i)} \le \frac{1}{\lambda^*} \|\operatorname{div}(\boldsymbol{v} - \Pi_h(\boldsymbol{v}))\|_{L^2(\Delta_i)}. \tag{2.2.35}$$

Let $\hat{T}$ denote the unit reference element and $\hat{\boldsymbol{c}}_i$ the composition of $\boldsymbol{c}_{h,i}|_T$ with the affine transformation that maps $\hat{T}$ onto T. Because each $\hat{\boldsymbol{c}}_i$ belongs to a finite-dimensional space, of dimension bounded by a fixed constant, on which all norms are equivalent, we can write for any $q \ge 2$:

$$\begin{aligned}
\|\boldsymbol{c}_{h,i}\|_{L^q(\mathcal{O}_i)} &\le \hat{C}\left(\sum_{T\subset\mathcal{O}_i} |T|\,\|\hat{\boldsymbol{c}}_i\|_{L^2(\hat{T})}^q\right)^{1/q} \le \hat{C}\, h_i^{2/q}\left(\sum_{T\subset\mathcal{O}_i} \|\hat{\boldsymbol{c}}_i\|_{L^2(\hat{T})}^q\right)^{1/q}\\
&\le \hat{C}\, h_i^{2/q}\left(\sum_{T\subset\mathcal{O}_i} \|\hat{\boldsymbol{c}}_i\|_{L^2(\hat{T})}^2\right)^{1/2} \le \hat{C}\,\frac{h_i^{2/q}}{\rho_i}\|\boldsymbol{c}_{h,i}\|_{L^2(\mathcal{O}_i)},
\end{aligned} \tag{2.2.36}$$

where $\rho_i = \min_{T \subset \mathcal{O}_i} \rho_T$ and $\hat{C}$ denotes contants that are independent of h and i. The third inequality follows from Jensen's inequality. If $1 \le q < 2$, Hölder's inequality and the fact that $\mathcal{O}_i$ contains at most L_1 elements give directly

$$\|\boldsymbol{c}_{h,i}\|_{L^q(\mathcal{O}_i)} \le \hat{C}\, h_i^{2(1/q-1/2)} \|\boldsymbol{c}_{h,i}\|_{L^2(\mathcal{O}_i)}. \tag{2.2.37}$$

Because $\boldsymbol{c}_{h,i} \in H_0^1(\mathcal{O}_i)^2$, Poincaré's inequality (1.1.3) gives

$$\|\boldsymbol{c}_{h,i}\|_{L^2(\mathcal{O}_i)} \le \hat{C}\mathrm{diam}(\mathcal{O}_i)|\boldsymbol{c}_{h,i}|_{H^1(\mathcal{O}_i)} \le \hat{C}\, h_i |\boldsymbol{c}_{h,i}|_{H^1(\mathcal{O}_i)}. \tag{2.2.38}$$

When substituting (2.2.38) and (2.2.35) into (2.2.36), we derive for any $q \ge 2$:

$$\|\boldsymbol{c}_{h,i}\|_{L^q(\mathcal{O}_i)} \le \hat{C}\,\frac{h_i^{1+2/q}}{\rho_i}|\boldsymbol{c}_{h,i}|_{H^1(\mathcal{O}_i)} \le \frac{\hat{C}}{\lambda^*}\frac{h_i^{1+2/q}}{\rho_i}\|\mathrm{div}(\boldsymbol{v} - \Pi_h(\boldsymbol{v}))\|_{L^2(\Delta_i)}, \tag{2.2.39}$$

and if $1 \le q < 2$,

$$\|\boldsymbol{c}_{h,i}\|_{L^q(\mathcal{O}_i)} \le \frac{\hat{C}}{\lambda^*} h_i^{2/q}\|\mathrm{div}(\boldsymbol{v} - \Pi_h(\boldsymbol{v}))\|_{L^2(\Delta_i)}.$$

A similar argument, somewhat simpler because there is no need for Poincaré's inequality, yields for $q \ge 2$:

$$|\boldsymbol{c}_{h,i}|_{W^{1,q}(\mathcal{O}_i)} \le \frac{\hat{C}}{\lambda^*}\frac{h_i^{2/q}}{\rho_i}\|\mathrm{div}(\boldsymbol{v} - \Pi_h(\boldsymbol{v}))\|_{L^2(\Delta_i)}, \tag{2.2.40}$$

and for $1 \le q < 2$,

$$|\boldsymbol{c}_{h,i}|_{W^{1,q}(\mathcal{O}_i)} \le \frac{\hat{C}}{\lambda^*} h_i^{2/q-1}\|\mathrm{div}(\boldsymbol{v} - \Pi_h(\boldsymbol{v}))\|_{L^2(\Delta_i)}. \tag{2.2.41}$$

(3) The expression of $\boldsymbol{c}_h$ gives

$$\|\boldsymbol{c}_h\|_{L^q(\mathcal{O}_i)} = \left(\int_{\mathcal{O}_i} |\sum_{j=1}^{R} \boldsymbol{c}_{h,j}|^q \,\mathrm{d}\boldsymbol{x}\right)^{1/q}.$$

But because $\boldsymbol{c}_{h,j}$ vanishes outside $\mathcal{O}_j$, the above sum runs over all indices j, such that $\mathcal{O}_j$ intersects $\mathcal{O}_i$. Let us number these indices from 1 to $R_i \le L_3$. Thus, the sum on j has at most L_3 terms. Hence,

$$\|\boldsymbol{c}_h\|_{L^q(\mathcal{O}_i)} \le L_3^{\alpha}\left(\int_{\mathcal{O}_i} \sum_{j=1}^{R_i} |\boldsymbol{c}_{h,j}|^q \,\mathrm{d}\boldsymbol{x}\right)^{1/q} \le L_3^{\alpha}\left(\sum_{j=1}^{R_i} \|\boldsymbol{c}_{h,j}\|^q_{L^q(\mathcal{O}_i \cap \mathcal{O}_j)}\right)^{1/q},$$

where $\alpha = 1/2$ if $1 \le q < 2$ and $\alpha = 1 - 1/q$, if $q \ge 2$. Hence (2.2.39) implies, if $q \ge 2$

$$\|\boldsymbol{c}_h\|_{L^q(\mathcal{O}_i)} \le \frac{\hat{C}}{\lambda^*}\left(\sum_{j=1}^{R_i} \frac{h_j^{q+2}}{\rho_j^q}\|\mathrm{div}(\boldsymbol{v} - \Pi_h(\boldsymbol{v}))\|^q_{L^2(\Delta_j)}\right)^{1/q},$$

where the index j is such that $\mathcal{O}_j$ intersects $\mathcal{O}_i$. If $1 \le q < 2$, we have instead

$$\|\boldsymbol{c}_h\|_{L^q(\mathcal{O}_i)} \le \frac{\hat{C}}{\lambda^*}\left(\sum_{j=1}^{R_i} h_j^2 \|\operatorname{div}(\boldsymbol{v} - \Pi_h(\boldsymbol{v}))\|^q_{L^2(\Delta_j)}\right)^{1/q}.$$

Therefore, the local quasi-uniformity of $\mathcal{T}_h$ (cf. for instance BERNARDI [1989]) and Jensen's inequality if $q \ge 2$ or Hölder's inequality if $q < 2$ yield the following bound:

$$\|\boldsymbol{c}_h\|_{L^q(\mathcal{O}_i)} \le \hat{C} h_i^{2/q} \|\operatorname{div}(\boldsymbol{v} - \Pi_h(\boldsymbol{v}))\|_{L^2(D_i)}, \tag{2.2.42}$$

where D_i is the union of Δ_j for all j such that $\mathcal{O}_j$ intersect $\mathcal{O}_i$. Similarly, we derive from (2.2.40):

$$|\boldsymbol{c}_h|_{W^{1,q}(\mathcal{O}_i)} \le \hat{C} h_i^{2/q-1} \|\operatorname{div}(\boldsymbol{v} - \Pi_h(\boldsymbol{v}))\|_{L^2(D_i)}. \tag{2.2.43}$$

Then (2.2.32) follows from (2.2.42) or (2.2.43) and (2.2.18) with $m = 1$ and $q = 2$.

Of course, if we integrate over Ω instead of $\mathcal{O}_i$, we obtain for $m = 0$ or $m = 1$:

$$|\boldsymbol{c}_h|_{W^{m,q}(\Omega)} \le \hat{C} h^{1-m+\min(0,2/q-1)} \|\operatorname{div}(\boldsymbol{v} - \Pi_h(\boldsymbol{v}))\|_{L^2(\Omega)}.$$

□

REMARK 2.2.8. Observe that (2.2.18) implies that

$$\operatorname{dist}(\operatorname{supp}(\Pi_h(\boldsymbol{v})), \operatorname{supp}(\boldsymbol{v})) \le C_3 h, \tag{2.2.44}$$

with a constant C_3 that is independent of h. □

REMARK 2.2.9. Furthermore,

$$\operatorname{dist}(\operatorname{supp}(P_h(\boldsymbol{v})), \operatorname{supp}(\boldsymbol{v})) \le C_4 h, \tag{2.2.45}$$

where the constant C_4 is independent of h. Indeed, if $\Delta_{i_1} \cup \cdots \cup \Delta_{i_k}$ is the union of all sets where $\operatorname{div}(\boldsymbol{v} - \Pi_h(\boldsymbol{v}))$ is not identically zero, then the support of $\boldsymbol{c}_h$ is contained in $\Delta_{i_1} \cup \cdots \cup \Delta_{i_k}$. Because each macroelement in this union contains at least one element where $\operatorname{div}(\boldsymbol{v} - \Pi_h(\boldsymbol{v}))$ does not vanish, the distance between the supports of $\boldsymbol{c}_h$ and $\operatorname{div}(\boldsymbol{v} - \Pi_h(\boldsymbol{v}))$ is smaller than the largest diameter of the macroelements. Then (2.2.44) and the assumptions on the macroelements imply (2.2.45). □

The conclusion of this section is analogous to that of the preceding ones: all assumptions of Hypotheses 2.1.5 and 2.1.10 are satisfied by the spaces X_h, M_h, and Z_h, and the resulting scheme has order two.

THEOREM 2.2.10. *Let the family of triangulations $\mathcal{T}_h$ satisfy (2.1.25) and be such that each triangle T has at most one edge on $\partial\Omega$. Let $(\boldsymbol{u}, p, z)$ be a solution of Problem (1.4.9), with $z \in H^3(\Omega)$, $\boldsymbol{u} \in H^3(\Omega)^2$ and $p \in H^2(\Omega)$, satisfying (2.1.56), and let $(\boldsymbol{u}_h, p_h, z_h)$ be a solution of (2.1.6)–(2.1.8) with the finite-element spaces (2.2.14)–(2.2.16). Then, there exists a constant C, independent of h, such that*

$$\|z - z_h\|_{L^2(\Omega)} + |\boldsymbol{u} - \boldsymbol{u}_h|_{H^1(\Omega)} + \|p - p_h\|_{L^2(\Omega)} \le C h^2.$$

2.3. Centered schemes: Successive approximations

Let us revert to the general situation of Section 2.1, with the same assumptions. The nonlinear discrete scheme (2.1.6)–(2.1.8) cannot be implemented as such, but is easily linearized by successive approximations. In this section, we present one of these algorithms studied in GIRAULT and SCOTT [2002a]. Starting from an arbitrary z_h^0 in Z_h, we define the sequence $(\boldsymbol{u}_h^n, p_h^n, z_h^n) \in X_h \times M_h \times Z_h$ for $n \geq 1$, knowing z_h^{n-1}, by

$$\forall \boldsymbol{v}_h \in X_h,\ \nu(\nabla \boldsymbol{u}_h^n, \nabla \boldsymbol{v}_h) + (z_h^{n-1} \times \boldsymbol{u}_h^n, \boldsymbol{v}_h) - (p_h^n, \operatorname{div} \boldsymbol{v}_h) = (\boldsymbol{f}, \boldsymbol{v}_h), \tag{2.3.1}$$

$$\forall q_h \in M_h,\ (q_h, \operatorname{div} \boldsymbol{u}_h^n) = 0, \tag{2.3.2}$$

$$\forall \theta_h \in Z_h,\ \nu(z_h^n, \theta_h) + \alpha\, \tilde{c}(\boldsymbol{u}_h^n; z_h^n, \theta_h) = \nu(\operatorname{curl} \boldsymbol{u}_h^n, \theta_h) + \alpha\, (\operatorname{curl} \boldsymbol{f}, \theta_h). \tag{2.3.3}$$

Clearly, given z_h^{n-1}, (2.3.1)–(2.3.2) has a unique solution $(\boldsymbol{u}_h^n, p_h^n)$; in fact, with the notation of Proposition 2.1.3, $\boldsymbol{u}_h^n = \boldsymbol{v}_h(z_h^{n-1})$ and $p_h^n = q_h(z_h^{n-1})$. Similarly, knowing $\boldsymbol{u}_h^n$, (2.3.3) has a unique solution z_h^n. In both cases, this is valid without restriction on the data. The next lemma shows that this sequence satisfies the same bounds as each solution of (2.1.6)–(2.1.8). The proof is the same as that of Proposition 2.1.3 and the beginning of Theorem 2.1.4.

LEMMA 2.3.1. *Assume that (2.1.1) holds. Then for all $\nu > 0$ and $\alpha > 0$, for all $\boldsymbol{f} \in H(\operatorname{curl}, \Omega)$ and all starting functions $z_h^0 \in Z_h$, the solution $(\boldsymbol{u}_h^n, p_h^n, z_h^n)$ of (2.3.1)–(2.3.3) is bounded as follows:*

$$|\boldsymbol{u}_h^n|_{H^1(\Omega)} \leq \frac{S_2}{\nu} \|\boldsymbol{f}\|_{L^2(\Omega)}, \tag{2.3.4}$$

$$\|z_h^n\|_{L^2(\Omega)} \leq \frac{S_2}{\nu} \|\boldsymbol{f}\|_{L^2(\Omega)} + \frac{\alpha}{\nu} \|\operatorname{curl} \boldsymbol{f}\|_{L^2(\Omega)}, \tag{2.3.5}$$

$$\|p_h^n\|_{L^2(\Omega)} \leq \frac{1}{\beta^*} \Big(S_2 \|\boldsymbol{f}\|_{L^2(\Omega)} + S_4^2 |\boldsymbol{u}_h^n|_{H^1(\Omega)} \|z_h^n\|_{L^2(\Omega)} \Big). \tag{2.3.6}$$

Without restriction on the data, these bounds imply convergence, but only up to subsequences, and hence not necessarily to a solution of (2.1.6)–(2.1.8). Convergence to a solution can be obtained by restricting the data and the solution, so that problem (1.4.9) has a unique solution, but it is more easily derived by introducing the “fixed point algorithm” of the next subsection.

2.3.1. *A “fixed point algorithm”*

Let us adapt (2.3.1)–(2.3.3) to Problem (1.4.9): Starting from an arbitrary smooth enough z^0 and knowing z^{n-1}, find $(\boldsymbol{u}^n, p^n, z^n)$ in $V \times L_0^2(\Omega) \times L^2(\Omega)$ for $n \geq 1$, solution of

$$\begin{aligned} -\nu\, \Delta \boldsymbol{u}^n + z^{n-1} \times \boldsymbol{u}^n + \nabla p^n &= \boldsymbol{f} \quad \text{in } \Omega, \\ \operatorname{div} \boldsymbol{u}^n &= 0 \quad \text{in } \Omega, \\ \boldsymbol{u}^n &= \boldsymbol{0} \quad \text{on } \partial\Omega, \\ \nu\, z^n + \alpha\, \boldsymbol{u}^n \cdot \nabla z^n &= \nu \operatorname{curl} \boldsymbol{u}^n + \alpha \operatorname{curl} \boldsymbol{f} \quad \text{in } \Omega. \end{aligned} \tag{2.3.7}$$

Clearly, $(\boldsymbol{u}^n, p^n, z^n)$ satisfies the same uniform bounds as any solution $(\boldsymbol{u}, p, z)$ of Problem (1.4.9):

$$
\begin{aligned}
|\boldsymbol{u}^n|_{H^1(\Omega)} &\leq \frac{S_2}{\nu}\|\boldsymbol{f}\|_{L^2(\Omega)},\\
\|z^n\|_{L^2(\Omega)} &\leq \frac{S_2}{\nu}\|\boldsymbol{f}\|_{L^2(\Omega)} + \frac{\alpha}{\nu}\|\operatorname{curl}\boldsymbol{f}\|_{L^2(\Omega)},\\
\|p^n\|_{L^2(\Omega)} &\leq \frac{1}{\beta}\Big(S_2\|\boldsymbol{f}\|_{L^2(\Omega)} + S_4^2|\boldsymbol{u}^n|_{H^1(\Omega)}\|z^n\|_{L^2(\Omega)}\Big).
\end{aligned}
\tag{2.3.8}
$$

Moreover, by assuming that the solution of (1.4.9) is sufficiently smooth and the data sufficiently small, so that condition (2.1.48) that guarantees uniqueness holds, we can prove that the fixed point algorithm (2.3.7) is contracting.

THEOREM 2.3.2. *We retain the assumptions and notation of Theorem 1.4.14, and we suppose in addition that the data are sufficiently small so that the following variant of (2.1.48) holds:*

$$
\begin{aligned}
\theta :&= \frac{1}{\nu}\|\boldsymbol{u}\|_{L^4(\Omega)}\\
&\times\left(S_4 + \alpha C_\infty C_1|z|_{H^1(\Omega)}\left(1 + \frac{S_4^2}{\nu^2}\big(S_2\|\boldsymbol{f}\|_{L^2(\Omega)} + \alpha\|\operatorname{curl}\boldsymbol{f}\|_{L^2(\Omega)}\big)\right)\right) < 1,
\end{aligned}
\tag{2.3.9}
$$

where C_1 is the continuity constant of Theorem 1.1.6 and C_∞ the constant of (1.4.25). Then, for any $n \geq 1$,

$$
\|z^n - z\|_{L^2(\Omega)} \leq \theta\|z^{n-1} - z\|_{L^2(\Omega)}. \tag{2.3.10}
$$

PROOF. By observing that $(\boldsymbol{u}^n - \boldsymbol{u}, p^n - p) = (\boldsymbol{v}(z^{n-1}), q(z^{n-1}))$ with $\boldsymbol{f} = -(z^{n-1} - z)\times\boldsymbol{u}$, i.e.,

$$
-\nu\,\Delta\,\boldsymbol{v}(z^{n-1}) + z^{n-1}\times\boldsymbol{v}(z^{n-1}) + \nabla\,q(z^{n-1}) = -(z^{n-1} - z)\times\boldsymbol{u}, \tag{2.3.11}
$$

we obtain first

$$
|\boldsymbol{u}^n - \boldsymbol{u}|_{H^1(\Omega)} \leq \frac{S_4}{\nu}\|\boldsymbol{u}\|_{L^4(\Omega)}\|z^{n-1} - z\|_{L^2(\Omega)}, \tag{2.3.12}
$$

and next, by applying (2.1.30):

$$
\|\boldsymbol{u}^n - \boldsymbol{u}\|_{L^\infty(\Omega)} \leq C_\infty C_1 K_1(z^{n-1})\|\boldsymbol{u}\|_{L^4(\Omega)}\|z^{n-1} - z\|_{L^2(\Omega)}. \tag{2.3.13}
$$

Let $\zeta^n = z^n - z$; then ζ^n solves the transport equation:

$$
\nu\,\zeta^n + \alpha\,\boldsymbol{u}^n\cdot\nabla\,\zeta^n = \nu\operatorname{curl}\boldsymbol{v}(z^{n-1}) - \alpha\,\boldsymbol{v}(z^{n-1})\cdot\nabla\,z, \tag{2.3.14}
$$

and the above assumptions are such that $\boldsymbol{v}(z^{n-1}) \cdot \nabla z$ belongs to $L^2(\Omega)$. Therefore,

$$\|\zeta^n\|_{L^2(\Omega)} \le |\boldsymbol{v}(z^{n-1})|_{H^1(\Omega)} + \frac{\alpha}{\nu}\|\boldsymbol{v}(z^{n-1})\|_{L^\infty(\Omega)}|z|_{H^1(\Omega)},$$

and (2.3.10) follows by substituting (2.3.12), (2.3.13), (1.4.23), and (2.3.8) into this inequality. □

As far as the pressure is concerned, $p^n - p$ satisfies

$$\|p^n - p\|_{L^2(\Omega)} \le \frac{S_4}{\beta}\Big(\|z^{n-1}\|_{L^2(\Omega)}\|\boldsymbol{u}^n - \boldsymbol{u}\|_{L^4(\Omega)} + \|z^{n-1} - z\|_{L^2(\Omega)}\|\boldsymbol{u}\|_{L^4(\Omega)}\Big). \tag{2.3.15}$$

Therefore, Theorem 2.3.2 yields the following strong convergences for the whole sequences to the unique solution $(\boldsymbol{u}, p, z)$ of (1.4.9):

$$\begin{aligned} &\lim_{n\to\infty} \boldsymbol{u}^n = \boldsymbol{u} \text{ strongly in } H_0^1(\Omega)^2,\\ &\lim_{n\to\infty} p^n = p \text{ strongly in } L^2(\Omega),\\ &\lim_{n\to\infty} z^n = z \text{ strongly in } L^2(\Omega). \end{aligned} \tag{2.3.16}$$

It also leads to the next two corollaries.

COROLLARY 2.3.3. *Under the assumptions of Theorem 2.3.2, the whole sequences $(\boldsymbol{u}^n, p^n)$ converge strongly to $(\boldsymbol{u}, p)$ in $H^2(\Omega)^2 \times H^1(\Omega)$.*

We skip the proof because it is a straightforward application of (2.3.11), (2.1.30), and (2.1.31).

COROLLARY 2.3.4. *In addition to the hypotheses of Theorem 2.3.2, we suppose that z^0 is chosen in $H^1(\Omega)$ and $z_h^0 = R_h(z^0)$. Then for each $n \ge 1$, the whole sequences in h, $(\boldsymbol{u}_h^n, p_h^n, z_h^n)$ converge strongly to $(\boldsymbol{u}^n, p^n, z^n)$:*

$$\begin{aligned} &\lim_{h\to 0} \boldsymbol{u}_h^n = \boldsymbol{u}^n \textit{ strongly in } H_0^1(\Omega)^2,\\ &\lim_{h\to 0} p_h^n = p^n \textit{ strongly in } L^2(\Omega),\\ &\lim_{h\to 0} z_h^n = z^n \textit{ strongly in } L^2(\Omega). \end{aligned}$$

PROOF. We argue by induction. Assume that for some $n \ge 1$, the whole sequence in h, z_h^{n-1} satisfies

$$\lim_{h\to 0} z_h^{n-1} = z^{n-1} \text{ strongly in } L^2(\Omega).$$

By assumption, this is true for $n = 1$. Proceeding as in Section 2.1.1, the uniform estimates (2.3.8) imply that there exist functions $\boldsymbol{u}^n \in V$, $p^n \in L_0^2(\Omega)$, and $z^n \in L^2(\Omega)$ such that, up to subsequences, as h tends to zero, $(\boldsymbol{u}_h^n, p_h^n, z_h^n)$ tend to $(\boldsymbol{u}^n, p^n, z^n)$ weakly in $H^1(\Omega)^2 \times L^2(\Omega) \times L^2(\Omega)$. Passing to the limit in (2.3.1)–(2.3.2), and using the induction hypothesis, we see that the limit functions $(\boldsymbol{u}^n, p^n, z^{n-1})$ solve the first two equations in (2.3.7). Moreover, the convergence of $\boldsymbol{u}_h^n$ is strong in $H^1(\Omega)^2$, and that of p_h^n is strong in $L^2(\Omega)$. The strong convergence of $\boldsymbol{u}_h^n$ permits to pass to the limit in (2.3.3) and show that $(\boldsymbol{u}^n, z^n)$ solve the last row of (2.3.7). In addition, the convergence of z_h^n is strong in $L^2(\Omega)$. As (2.3.7) has a unique solution (given z^{n-1}), the whole sequences $(\boldsymbol{u}_h^n, p_h^n, z_h^n)$ converge. Hence, z_h^n satisfies the induction hypothesis. □

Corollary 2.3.4 neither addresses uniformity of the convergence with respect to n, nor gives a rate of convergence. Therefore, when combined with Theorem 2.3.2, it implies convergence for h depending on n. For instance, it guarantees that for each $\varepsilon > 0$, there exists an integer $n_0 \geq 1$ and a positive number h_0 depending on n_0, such that for all $h \leq h_0$

$$|\boldsymbol{u}_h^{n_0} - \boldsymbol{u}|_{H^1(\Omega)} + \|p_h^{n_0} - p\|_{L^2(\Omega)} + \|z_h^{n_0} - z\|_{L^2(\Omega)} \leq \varepsilon.$$

A rate of convergence can be derived, but requires additional uniform bounds. They are obtained when the solution of (1.4.9) is sufficiently smooth, by applying to $\boldsymbol{v}(z^{n-1})$ and ζ^n, (2.1.30)–(2.1.31), and the results of Proposition 1.4.13, and Theorem 1.4.14. However, the assumptions below on z cannot be checked on the data in a domain with corners.

COROLLARY 2.3.5. *Let Ω be a convex polygon, let r_0 be the constant of Proposition 1.4.11, let r belong to $]2, r_0]$, and let $z^0 \in W^{2,r}(\Omega)$, $z \in W^{2,r}(\Omega)$, and suppose that $\boldsymbol{u}$ satisfies (1.4.30). Then, under the assumptions of Theorem 2.3.2, there exists an integer $n_0 \geq 2$ such that for all $n \geq n_0$, the function $\zeta^n := z^n - z$ belongs to $W^{1,r}(\Omega)$ and there exists a constant C, independent of n, such that*

$$\forall n \geq n_0,\ |\zeta^n|_{W^{1,r}(\Omega)} \leq C\left(\|\zeta^{n-2}\|_{L^2(\Omega)} + \|\zeta^{n-1}\|_{L^2(\Omega)}\right). \tag{2.3.17}$$

PROOF. Recall that $\boldsymbol{u}^n - \boldsymbol{u} = \boldsymbol{v}(z^{n-1})$ defined by (2.3.11). First, let us derive a bound for $\boldsymbol{v}(z^{n-1})$ in $W^{2,r}(\Omega)^2$ for some $r > 2$. This requires an L^r bound for both $z^{n-1} \times \boldsymbol{v}(z^{n-1})$ and $\zeta^{n-1} \times \boldsymbol{u}$. As we are only interested in r slightly larger than two, we can assume that $2 < r_0 \leq 4$. For $2 < r \leq r_0$, (2.1.30) and (2.3.8) imply the following bound for $\boldsymbol{v}(z^{n-1})$:

$$|\boldsymbol{v}(z^{n-1})|_{W^{1,r}(\Omega)} \leq c_1 K_1(z^{n-1})\|\boldsymbol{u}\|_{L^4(\Omega)}\|\zeta^{n-1}\|_{L^2(\Omega)} \leq c_2\|\zeta^{n-1}\|_{L^2(\Omega)}, \tag{2.3.18}$$

where all constants c_i are independent of n. Then owing to (2.3.18) and (2.3.13), the right-hand side of (2.3.14) is bounded in L^r:

$$\|\nu \operatorname{curl} \boldsymbol{v}(z^{n-1}) - \alpha\, \boldsymbol{v}(z^{n-1}) \cdot \nabla z\|_{L^r(\Omega)} \leq \nu c_2\|\zeta^{n-1}\|_{L^2(\Omega)}$$
$$+ \alpha c_3 |z|_{W^{1,r}(\Omega)}\|\zeta^{n-1}\|_{L^2(\Omega)} \leq c_4\|\zeta^{n-1}\|_{L^2(\Omega)}.$$

As a consequence, (1.3.25) yields that ζ^n belongs to $L^r(\Omega)$ and

$$\|\zeta^n\|_{L^r(\Omega)} \leq \frac{c_4}{\nu}\|\zeta^{n-1}\|_{L^2(\Omega)}. \tag{2.3.19}$$

We also have

$$\|z^n\|_{L^r(\Omega)} \leq c_5.$$

This implies that $\zeta^{n-1} \times \boldsymbol{u}$ and $z^{n-1} \times \boldsymbol{v}(z^{n-1})$ are bounded in L^r:

$$\|\zeta^{n-1} \times \boldsymbol{u}\|_{L^r(\Omega)} \leq \frac{c_4}{\nu}\|\boldsymbol{u}\|_{L^\infty(\Omega)}\|\zeta^{n-2}\|_{L^2(\Omega)} \leq c_6\|\zeta^{n-2}\|_{L^2(\Omega)},$$

$$\|z^{n-1} \times \boldsymbol{v}(z^{n-1})\|_{L^r(\Omega)} \leq c_5\|\boldsymbol{v}(z^{n-1})\|_{L^\infty(\Omega)} \leq c_7\|\zeta^{n-1}\|_{L^2(\Omega)}.$$

Therefore, the convexity of Ω and Theorem 1.1.9 imply that $\boldsymbol{v}(z^{n-1})$ belongs to $W^{2,r}(\Omega)^2$, $q(z^{n-1}) = p^n - p$ belongs to $W^{1,r}(\Omega)$, and

$$|\boldsymbol{v}(z^{n-1})|_{W^{2,r}(\Omega)} + |q(z^{n-1})|_{W^{1,r}(\Omega)} \leq c_8\left(\|\zeta^{n-2}\|_{L^2(\Omega)} + \|\zeta^{n-1}\|_{L^2(\Omega)}\right). \tag{2.3.20}$$

Next, (2.3.18) and (2.3.20) imply

$$|\nu \operatorname{curl} \boldsymbol{v}(z^{n-1}) - \alpha\, \boldsymbol{v}(z^{n-1}) \cdot \nabla z|_{W^{1,r}(\Omega)} \leq \left(\|\zeta^{n-2}\|_{L^2(\Omega)} + \|\zeta^{n-1}\|_{L^2(\Omega)}\right) \times \left(\nu c_8 + c_9\|\nabla z\|_{W^{1,r}(\Omega)}\right). \tag{2.3.21}$$

Finally, since the assumptions of Theorem 2.3.2 are satisfied, the right-hand side of (2.3.20) tends to zero and hence, for each $\varepsilon > 0$, there exists an integer $n_0 \geq 2$ such that for all $n \geq n_0$,

$$\frac{\alpha}{\nu}\|\nabla \boldsymbol{v}(z^{n-1})\|_{W^{1,r}(\Omega)} \leq \varepsilon.$$

The desired result follows by choosing $\varepsilon = \frac{1-\delta}{2}$, where δ is the constant of (1.4.30), and by applying Theorem 1.4.14 to (2.3.14). □

2.3.2. *Successive approximations: Rate of convergence*

For fixed n, (2.3.1)–(2.3.3) is a straightforward discretization of (2.3.7), and by virtue of Theorem 2.3.2, it suffices to estimate $\boldsymbol{u}_h^n - \boldsymbol{u}^n$, $p_h^n - p^n$ and $z_h^n - z^n$. This is derived through the approach of Section 2.1.4. First, we have the analogs of (2.1.51) and (2.1.52):

$$|\boldsymbol{u}^n - \boldsymbol{u}_h^n|_{H^1(\Omega)} \leq 2|\boldsymbol{u}^n - P_h(\boldsymbol{u}^n)|_{H^1(\Omega)} + \frac{S_4}{\nu}\|\boldsymbol{u}^n\|_{L^4(\Omega)}\|z^{n-1} - z_h^{n-1}\|_{L^2(\Omega)} + \frac{S_4}{\nu}\|z_h^{n-1}\|_{L^2(\Omega)}\|\boldsymbol{u}^n - P_h(\boldsymbol{u}^n)\|_{L^4(\Omega)} + \frac{1}{\nu}\|p^n - r_h(p^n)\|_{L^2(\Omega)}, \tag{2.3.22}$$

$$\|p^n - p_h^n\|_{L^2(\Omega)} \leq \left(1 + \frac{1}{\beta^*}\right)\|p^n - r_h(p^n)\|_{L^2(\Omega)} + \frac{1}{\beta^*}\Big(\nu\, |\boldsymbol{u}^n - P_h(\boldsymbol{u}^n)|_{H^1(\Omega)} + S_4\left(\|\boldsymbol{u}_h^n\|_{L^4(\Omega)}\|z^{n-1} - z_h^{n-1}\|_{L^2(\Omega)} + \|z^{n-1}\|_{L^2(\Omega)}\|\boldsymbol{u}^n - \boldsymbol{u}_h^n\|_{L^4(\Omega)}\right)\Big). \tag{2.3.23}$$

Next, we have the analog of (2.1.54)

$$\begin{aligned}\|z^n - z_h^n\|_{L^2(\Omega)} &\le 2\,\|z^n - R_h(z^n)\|_{L^2(\Omega)} + |\boldsymbol{u}^n - \boldsymbol{u}_h^n|_{H^1(\Omega)}\\ &+ \frac{\alpha}{\nu}\Big(\|\boldsymbol{u}^n - \boldsymbol{u}_h^n\|_{L^{r^*}(\Omega)}|R_h(z^n)|_{W^{1,r}(\Omega)} + \|\boldsymbol{u}^n\|_{L^\infty(\Omega)}|z^n - R_h(z^n)|_{H^1(\Omega)}\\ &+ \frac{1}{2}|\boldsymbol{u}^n - \boldsymbol{u}_h^n|_{H^1(\Omega)}\|R_h(z^n)\|_{L^\infty(\Omega)}\Big),\end{aligned} \tag{2.3.24}$$

where $\frac{1}{r^*} = \frac{1}{2} - \frac{1}{r}$. Then, we have the following error theorem.

THEOREM 2.3.6. *We retain the assumptions and notation of Corollary 2.3.5 and we suppose that $n \ge n_0$, with n_0 defined in Corollary 2.3.5. If the data are restricted so that*

$$\frac{S_4}{\nu}\|\boldsymbol{u}^n\|_{L^4(\Omega)}\left(1 + \frac{\alpha}{\nu}\left(S_{r^*}|R_h(z^n)|_{W^{1,r}(\Omega)} + \frac{1}{2}\|R_h(z^n)\|_{L^\infty(\Omega)}\right)\right) \le \delta, \tag{2.3.25}$$

where $0 < \delta < 1$ is independent of h and n, then for all integer $k \ge 0$,

$$\begin{aligned}&\|z^{n_0+k} - z_h^{n_0+k}\|_{L^2(\Omega)}\\ &\le C_1\left(\|z - R_h(z)\|_{H^1(\Omega)} + |\boldsymbol{u} - P_h(\boldsymbol{u})|_{H^1(\Omega)} + \|p - r_h(p)\|_{L^2(\Omega)}\right)\\ &\quad + \delta^{k+1}\|z^{n_0-1} - z_h^{n_0-1}\|_{L^2(\Omega)} + C_2\theta^{n_0-2}\max(\theta^k, \delta^k)\|z^0 - z\|_{L^2(\Omega)},\end{aligned} \tag{2.3.26}$$

with constants C_1 and C_2 independent of h and k, and where θ is defined by (2.3.9).

PROOF. By substituting (2.3.22) into (2.3.24) and using (2.3.25), we easily derive

$$\begin{aligned}\|z^n - z_h^n\|_{L^2(\Omega)} &\le \delta\|z^{n-1} - z_h^{n-1}\|_{L^2(\Omega)} + 2\|z^n - R_h(z^n)\|_{L^2(\Omega)}\\ &+ \frac{\alpha}{\nu}\|\boldsymbol{u}^n\|_{L^\infty(\Omega)}|z^n - R_h(z^n)|_{H^1(\Omega)}\\ &+ \left(1 + \frac{\alpha}{\nu}\left(S_{r^*}|R_h(z^n)|_{W^{1,r}(\Omega)} + \frac{1}{2}\|R_h(z^n)\|_{L^\infty(\Omega)}\right)\right)\\ &\times\Big(2\,|\boldsymbol{u}^n - P_h(\boldsymbol{u}^n)|_{H^1(\Omega)} + \frac{S_4}{\nu}\|z^{n-1}\|_{L^2(\Omega)}\|\boldsymbol{u}^n - P_h(\boldsymbol{u}^n)\|_{L^4(\Omega)}\\ &+ \frac{1}{\nu}\|p^n - r_h(p^n)\|_{L^2(\Omega)}\Big).\end{aligned}$$

First, we have

$$|R_h(z^n)|_{W^{1,r}(\Omega)} \le c_1|z^n|_{W^{1,r}(\Omega)},\quad \|R_h(z^n)\|_{L^\infty(\Omega)} \le c_2\|z^n\|_{W^{1,r}(\Omega)},$$

where all constants c_i are independent of n. Next,

$$\|\boldsymbol{u}^n\|_{L^\infty(\Omega)} \le c_3,\quad \|z^{n-1}\|_{L^2(\Omega)} \le c_4.$$

Therefore,

$$
\begin{aligned}
\|z^n - z_h^n\|_{L^2(\Omega)} &\le \delta \|z^{n-1} - z_h^{n-1}\|_{L^2(\Omega)} \\
&\quad + c_5 \left(\|z^n - R_h(z^n)\|_{H^1(\Omega)} + |\boldsymbol{u}^n - P_h(\boldsymbol{u}^n)|_{H^1(\Omega)} + \|p^n - r_h(p^n)\|_{L^2(\Omega)} \right) \\
&\le \delta \|z^{n-1} - z_h^{n-1}\|_{L^2(\Omega)} \\
&\quad + c_5 \left(\|z - R_h(z)\|_{H^1(\Omega)} + |\boldsymbol{u} - P_h(\boldsymbol{u})|_{H^1(\Omega)} + \|p - r_h(p)\|_{L^2(\Omega)} \right) \\
&\quad + c_6 |z - z^n|_{H^1(\Omega)} + c_7 h \left(|\boldsymbol{u} - \boldsymbol{u}^n|_{H^2(\Omega)} + |p - p^n|_{H^1(\Omega)} \right).
\end{aligned}
$$

With (2.3.10), (2.3.17), and (2.3.20), this becomes for all $n \ge n_0$:

$$
\begin{aligned}
\|z^n - z_h^n\|_{L^2(\Omega)} &\le \delta \|z^{n-1} - z_h^{n-1}\|_{L^2(\Omega)} \\
&\quad + c_5 \left(\|z - R_h(z)\|_{H^1(\Omega)} + |\boldsymbol{u} - P_h(\boldsymbol{u})|_{H^1(\Omega)} + \|p - r_h(p)\|_{L^2(\Omega)} \right) \\
&\quad + c_8 \theta^{n-2} \|z - z^0\|_{L^2(\Omega)}.
\end{aligned}
\tag{2.3.27}
$$

An easy induction yields for all integers $k \ge 0$:

$$
\begin{aligned}
&\|z^{n_0+k} - z_h^{n_0+k}\|_{L^2(\Omega)} \\
&\le c_5 \left(\sum_{i=0}^{k} \delta^i \right) \left(\|z - R_h(z)\|_{H^1(\Omega)} + |\boldsymbol{u} - P_h(\boldsymbol{u})|_{H^1(\Omega)} + \|p - r_h(p)\|_{L^2(\Omega)} \right) \\
&\quad + \delta^{k+1} \|z^{n_0-1} - z_h^{n_0-1}\|_{L^2(\Omega)} + c_8 \theta^{n_0-2} \left(\sum_{i=0}^{k} \delta^i \theta^{k-i} \right) \|z^0 - z\|_{L^2(\Omega)}.
\end{aligned}
$$

Considering that both θ and δ belong to $]0, 1[$, this implies (2.3.26). □

REMARK 2.3.7. Owing to the stability of R_h and Corollary 2.3.5, the left-hand side of (2.3.25) is bounded uniformly with respect to h and n:

$$
\begin{aligned}
&\frac{S_4}{\nu} \|\boldsymbol{u}^n\|_{L^4(\Omega)} \left(1 + \frac{\alpha}{\nu} \left(S_{r^*} |R_h(z^n)|_{W^{1,r}(\Omega)} + \frac{1}{2} \|R_h(z^n)\|_{L^\infty(\Omega)} \right) \right) \\
&\le \frac{S_4^2 S_2}{\nu^2} \|\boldsymbol{f}\|_{L^2(\Omega)} \left(1 + \frac{\alpha}{\nu} \left(S_{r^*} c_1 |z^n|_{W^{1,r}(\Omega)} + \frac{1}{2} c_2 \|z^n\|_{W^{1,r}(\Omega)} \right) \right) \\
&\le \frac{S_4^2 S_2}{\nu^2} \|\boldsymbol{f}\|_{L^2(\Omega)} \left(1 + \frac{\alpha}{\nu} \left(S_{r^*} c_1 |z|_{W^{1,r}(\Omega)} + \frac{1}{2} c_2 \|z\|_{W^{1,r}(\Omega)} + c_3 \theta^{n-2} \|z^0 - z\|_{L^2(\Omega)} \right) \right),
\end{aligned}
$$

with constants independent of h and n. All quantities appearing in the right-hand side of this last relation are bounded in terms of the data. □

REMARK 2.3.8. Because n_0 is fixed and z_h^n is bounded in L^2, the terms in the right-hand side of (2.3.26), second row, tend uniformly and geometrically to zero as k tends to infinity. The terms in the first row represent the standard approximation error of the discrete spaces of (2.1.6)–(2.1.8). □

2.4. Upwind schemes

Knowing z_h, computing the solution of (2.1.6)–(2.1.7) is time consuming, because it is a system coupled by a constraint, and its matrix is not definite. Many techniques have been devised to deal with this difficulty, which is inherent to the Stokes problem, and is now well-documented. There is no space here to list all the references on the subject; for instance, the reader can refer to BREZZI and FORTIN [1991], ERN and GUERMOND [2004], GIRAULT and RAVIART [1986], or the very complete work of GLOWINSKI [2003]. As it involves no constraint, computing the solution of (2.1.8) is comparatively faster, but experiments show that the accuracy of this solution may be disappointing. This is because of the hyperbolic character of the transport equation (1.4.8); we have already pointed out the great imbalance between the regularity assumption for z and that for $\boldsymbol{u}$, when deriving error estimates. Upwinding techniques have been introduced many years ago in order to reduce this imbalance and enhance convergence in approximating transport equations. There are several upwinding methods; we cannot describe them all and we present two methods: upwinding by streamline diffusion and upwinding by Lesaint–Raviart's discontinuous Galerkin method.

2.4.1. Upwinding by streamline diffusion

The technique of streamline diffusion was first introduced by HUGUES [1978] and studied by JOHNSON, NAVERT, and PITKARANTA [1985] (cf. also JOHNSON [1987], and PIRONNEAU [1989]). It consists in adding a suitable transport term to the test function. On one hand, it allows to derive an estimate for $\sqrt{h}\boldsymbol{u}_h \cdot \nabla z_h$, which cannot be obtained with a centered scheme, and on the other hand, it enhances convergence. As (2.1.6)–(2.1.7) are unchanged, the analysis of this upwind scheme uses several results established in the preceding sections, and therefore, we shall only sketch most of the proofs. Details can be found in GIRAULT and SCOTT [2002a].

As previously, Ω is a connected polygon. We retain the notation and assumptions of Section 2.1, and we discretize Problem (1.4.9) by: Find $(\boldsymbol{u}_h, p_h, z_h)$ in $X_h \times M_h \times Z_h$, solution of

$$\begin{aligned}
&\forall \boldsymbol{v}_h \in X_h,\ \nu(\nabla \boldsymbol{u}_h, \nabla \boldsymbol{v}_h) + (\boldsymbol{z}_h \times \boldsymbol{u}_h, \boldsymbol{v}_h) - (p_h, \operatorname{div} \boldsymbol{v}_h) = (\boldsymbol{f}, \boldsymbol{v}_h),\\
&\forall q_h \in M_h,\ (q_h, \operatorname{div} \boldsymbol{u}_h) = 0,\\
&\forall \theta_h \in Z_h,\ \nu(z_h, \theta_h + h\boldsymbol{u}_h \cdot \nabla \theta_h) + \alpha\,(\boldsymbol{u}_h \cdot \nabla z_h, \theta_h + h\,\boldsymbol{u}_h \cdot \nabla \theta_h)\\
&\qquad + \frac{1}{2}(\alpha + h\,\nu)\,((\operatorname{div} \boldsymbol{u}_h) z_h, \theta_h) \qquad\qquad (2.4.1)\\
&\qquad = \nu(\operatorname{curl} \boldsymbol{u}_h, \theta_h + h\,\boldsymbol{u}_h \cdot \nabla \theta_h) + \alpha\,(\operatorname{curl} \boldsymbol{f}, \theta_h + h\,\boldsymbol{u}_h \cdot \nabla \theta_h),
\end{aligned}$$

where $\boldsymbol{z}_h = (0, 0, z_h)$. The first two equations coincide with (2.1.6)–(2.1.7) and the last equation is obtained by testing a discrete version of (1.4.8) with $\theta_h + h\,\boldsymbol{u}_h \cdot \nabla \theta_h$ and stabilizing it with the consistent term in the second row. The factor h multiplying $\boldsymbol{u}_h \cdot \nabla \theta_h$ can first be chosen arbitrarily (positive), but the value h is required to establish satisfactory error bounds.

Streamline diffusion: Convergence

The discrete problem (2.1.6), (2.1.7), (2.4.1) satisfies Proposition 2.1.3 and the analog of Theorem 2.1.4. More precisely, we have the following result.

THEOREM 2.4.1. *Assume that (2.1.1) holds. Then for all $\nu > 0$, $\alpha > 0$, and for all f in $H(curl, \Omega)$, the discrete problem (2.1.6), (2.1.7), (2.4.1) has at least one solution $(\boldsymbol{u}_h, p_h) \in V_h \times M_h$, $z_h \in Z_h$, and each solution satisfies the a priori estimates (2.1.14), (2.1.15):*

$$|\boldsymbol{u}_h|_{H^1(\Omega)} \leq \frac{S_2}{\nu} \|\boldsymbol{f}\|_{L^2(\Omega)},$$
$$\|p_h\|_{L^2(\Omega)} \leq \frac{1}{\beta^*} \left(S_2 \|\boldsymbol{f}\|_{L^2(\Omega)} + S_4^2 |\boldsymbol{u}_h|_{H^1(\Omega)} \|z_h\|_{L^2(\Omega)} \right),$$

and

$$\nu \, \|z_h\|^2_{L^2(\Omega)} + \alpha \, h \, \|\boldsymbol{u}_h \cdot \nabla \, z_h\|^2_{L^2(\Omega)} \leq 2 \, S_h \left(\frac{S_2^2}{\alpha \, \nu} \|\boldsymbol{f}\|^2_{L^2(\Omega)} + \frac{\alpha}{\nu} \|\text{curl}\, \boldsymbol{f}\|^2_{L^2(\Omega)} \right), \quad (2.4.2)$$

where $S_h = \alpha + \nu \, h$.

PROOF. The a priori estimates (2.1.14)–(2.1.15) are the same as in Proposition 2.1.3. The estimate (2.4.2) is an easy consequence of (2.1.4)–(2.1.5), and repeated applications of Young's inequality. Existence of a solution then follows from Brouwer's Fixed Point Theorem as in the proof of Theorem 2.1.4. □

REMARK 2.4.2. Note that, in contrast to (2.1.16), (2.4.2) does not allow α to tend to zero. □

Then, we have the following analog of Theorem 2.1.9.

THEOREM 2.4.3. *Under the assumptions of Theorem 2.1.9, there exists a subsequence of h (still denoted by h) and a solution $(\boldsymbol{u}, p, z) \in V \times L_0^2(\Omega) \times L^2(\Omega)$ of Problem (1.4.9) such that*

$$\begin{aligned} \lim_{h \to 0} |\boldsymbol{u}_h - \boldsymbol{u}|_{H^1(\Omega)} &= 0, \\ \lim_{h \to 0} \|z_h - z\|_{L^2(\Omega)} &= 0, \\ \lim_{h \to 0} \|p_h - p\|_{L^2(\Omega)} &= 0, \\ \lim_{h \to 0} \sqrt{h} \|\boldsymbol{u}_h \cdot \nabla \, z_h\|_{L^2(\Omega)} &= 0. \end{aligned} \quad (2.4.3)$$

PROOF. The uniform bounds of Theorem 2.4.1 allow us to prove that (a subsequence of) the sequences $\boldsymbol{u}_h, p_h, z_h$, and $\sqrt{h}\boldsymbol{u}_h \cdot \nabla \, z_h$ converge weakly to $\boldsymbol{u}$ in V, to p in $L_0^2(\Omega)$, to z in $L^2(\Omega)$, and to some function w in $L^2(\Omega)$, respectively, as h tends to zero. As in Proposition 2.1.6, the triple $(\boldsymbol{u}, p, z)$ satisfies (1.4.6). Similarly, as in Proposition 2.1.7, the convergence of $\boldsymbol{u}_h$ holds strongly:

$$\lim_{h \to 0} \|\boldsymbol{u}_h - \boldsymbol{u}\|_{H^1(\Omega)} = 0.$$

Furthermore, (2.4.2) shows that

$$\lim_{h \to 0} h \, \|\boldsymbol{u}_h \cdot \nabla \, z_h\|_{L^2(\Omega)} = 0.$$

With these two strong convergences, the argument of Proposition 2.1.8 allows us to pass to the limit in (2.4.1) and prove that $(\boldsymbol{u}, p, z)$ is indeed a solution of Problem (1.4.9).

To establish the strong convergence of z_h and $\sqrt{h}(\boldsymbol{u}_h \cdot \nabla z_h)$, we take the difference between (2.4.1) with test function z_h and (1.4.8) tested against $z_h + h\,\boldsymbol{u}_h \cdot \nabla z_h$:

$$\nu \|z_h\|^2_{L^2(\Omega)} + \alpha\, h\, \|\boldsymbol{u}_h \cdot \nabla z_h\|^2_{L^2(\Omega)} = \nu(z, z_h + h\,\boldsymbol{u}_h \cdot \nabla z_h)$$
$$+ \alpha(\boldsymbol{u} \cdot \nabla z, z_h + h\,\boldsymbol{u}_h \cdot \nabla z_h) + \nu(\operatorname{curl}(\boldsymbol{u}_h - \boldsymbol{u}), z_h + h\,\boldsymbol{u}_h \cdot \nabla z_h).$$

By passing to the limit, this gives

$$\lim_{h\to 0}\left(\nu \|z_h\|^2_{L^2(\Omega)} + \alpha\, h\, \|\boldsymbol{u}_h \cdot \nabla z_h\|^2_{L^2(\Omega)}\right) = \nu \|z\|^2_{L^2(\Omega)}.$$

Hence $\lim_{h\to 0} \nu \|z_h\|^2_{L^2(\Omega)} \le \nu \|z\|^2_{L^2(\Omega)}$, thus implying first that

$$\lim_{h\to 0} \|z_h\|_{L^2(\Omega)} = \|z\|_{L^2(\Omega)},$$

owing to the lower semicontinuity of the norm for the weak topology, and next

$$\lim_{h\to 0} \sqrt{\alpha\, h}\|\boldsymbol{u}_h \cdot \nabla z_h\|_{L^2(\Omega)} = 0.$$

Finally, the strong convergence of p_h is established as in Theorem 2.1.9. □

Streamline diffusion: Error estimates

As far as $\boldsymbol{u}_h$ and p_h are concerned, all results of Sections 2.1.1 and 2.1.2, carry over here, as well as the statement of Lemma 2.1.19, namely (2.1.51) and (2.1.52)

$$|\boldsymbol{u} - \boldsymbol{u}_h|_{H^1(\Omega)} \le 2|\boldsymbol{u} - P_h(\boldsymbol{u})|_{H^1(\Omega)} + \frac{S_4}{\nu}\|\boldsymbol{u}\|_{L^4(\Omega)}\|z - z_h\|_{L^2(\Omega)}$$
$$+ \frac{S_4}{\nu}\|z_h\|_{L^2(\Omega)}\|\boldsymbol{u} - P_h(\boldsymbol{u})\|_{L^4(\Omega)} + \frac{1}{\nu}\|p - r_h(p)\|_{L^2(\Omega)},$$
$$\|p - p_h\|_{L^2(\Omega)} \le \left(1 + \frac{1}{\beta^*}\right)\|p - r_h(p)\|_{L^2(\Omega)} + \frac{1}{\beta^*}\big(\nu\,|\boldsymbol{u} - P_h(\boldsymbol{u})|_{H^1(\Omega)}$$
$$+ S_4\left(\|\boldsymbol{u}\|_{L^4(\Omega)}\|z - z_h\|_{L^2(\Omega)} + \|z_h\|_{L^2(\Omega)}\|\boldsymbol{u} - \boldsymbol{u}_h\|_{L^4(\Omega)}\right)\big),$$

where β^* is the constant of (2.1.1).

Of course, the treatment of $z - z_h$ differs, but an error inequality is more easily derived from the upwinded transport equation (2.4.1) than from (2.1.8), because its structure yields directly an upper bound for $\sqrt{\alpha\, h}\boldsymbol{u}_h \cdot \nabla(z_h - \lambda_h)$, with any choice of λ_h. Furthermore, the factor h is chosen so as to enhance convergence. The next result is established under the hypotheses of Corollary 2.1.21.

THEOREM 2.4.4. *Let Ω be convex, $(\boldsymbol{u}, p, z)$ a solution of Problem (1.4.9), r_0 the number of Proposition 1.4.11, and let the assumptions of Theorem 1.4.14 hold, so that $z \in W^{1,r}(\Omega)$, for some real number r in $]2, r_0[$. Let $(\boldsymbol{u}_h, p_h, z_h)$ be any solution of (2.1.6), (2.1.7), (2.4.1).*

Then, we have the following estimate for $z_h - \lambda_h$, for any λ_h in Z_h:

$$\begin{aligned}
&\frac{\nu}{2}\|z_h - \lambda_h\|^2_{L^2(\Omega)} + \frac{\alpha\, h}{2}\|\boldsymbol{u}_h \cdot \nabla(z_h - \lambda_h)\|^2_{L^2(\Omega)} \leq 2\alpha\, h\, \|\boldsymbol{u}_h\|^2_{L^\infty(\Omega)}|z - \lambda_h|^2_{H^1(\Omega)} \\
&\quad + 2\left(3\,\nu + 2\,h\frac{\nu^2}{\alpha} + \frac{\alpha}{h}\right)\|z - \lambda_h\|^2_{L^2(\Omega)} + \alpha\left(3\frac{\alpha}{\nu} + 2\,h\right)\|(\boldsymbol{u} - \boldsymbol{u}_h)\cdot\nabla\, z\|^2_{L^2(\Omega)} \\
&\quad + \frac{3}{4}\left(\frac{\alpha^2}{\nu} + \nu\, h^2\right)\|\mathrm{div}(\boldsymbol{u} - \boldsymbol{u}_h)\lambda_h\|^2_{L^2(\Omega)} + 6\,\frac{\alpha^2}{\nu}\|\mathrm{div}(\boldsymbol{u} - \boldsymbol{u}_h)(\lambda_h - z)\|^2_{L^2(\Omega)} \\
&\quad + \frac{\nu}{2}\left(3 + 8\,h\frac{\nu}{\alpha}\right)|\boldsymbol{u} - \boldsymbol{u}_h|^2_{H^1(\Omega)}.
\end{aligned} \tag{2.4.4}$$

PROOF. By taking the difference between (2.4.1) and (1.4.8) tested against $\theta_h + h\,\boldsymbol{u}_h \cdot \nabla\,\theta_h$, inserting λ_h and choosing $\theta_h = z_h - \lambda_h$, we obtain

$$\begin{aligned}
&\nu\|z_h - \lambda_h\|^2_{L^2(\Omega)} + \alpha\, h\, \|\boldsymbol{u}_h \cdot \nabla(z_h - \lambda_h)\|^2_{L^2(\Omega)} \\
&\quad = -\nu(\lambda_h - z, z_h - \lambda_h + h\,\boldsymbol{u}_h \cdot \nabla(z_h - \lambda_h)) \\
&\qquad - \alpha(\boldsymbol{u}_h \cdot \nabla(\lambda_h - z), z_h - \lambda_h + h\,\boldsymbol{u}_h \cdot \nabla(z_h - \lambda_h)) \\
&\qquad - \alpha((\boldsymbol{u}_h - \boldsymbol{u})\cdot\nabla\, z, z_h - \lambda_h + h\,\boldsymbol{u}_h \cdot \nabla(z_h - \lambda_h)) \\
&\qquad - \frac{1}{2}(\alpha + \nu\, h)(\mathrm{div}(\boldsymbol{u}_h - \boldsymbol{u})\lambda_h, z_h - \lambda_h) \\
&\qquad + \nu(\mathrm{curl}(\boldsymbol{u}_h - \boldsymbol{u}), z_h - \lambda_h + h\,\boldsymbol{u}_h \cdot \nabla(z_h - \lambda_h)).
\end{aligned} \tag{2.4.5}$$

The estimates for all terms in the right-hand side of (2.4.5) are standard except for the second term because it involves the gradient of $\lambda_h - z$, whose upper bound requires more regularity. Applying Green's formula, we have

$$\begin{aligned}
-\alpha(\boldsymbol{u}_h \cdot \nabla(\lambda_h - z), z_h - \lambda_h) &= \alpha(\boldsymbol{u}_h \cdot \nabla(z_h - \lambda_h), \lambda_h - z) \\
&\quad + \alpha(\mathrm{div}(\boldsymbol{u}_h - \boldsymbol{u})(z_h - \lambda_h), \lambda_h - z).
\end{aligned}$$

Thus, for any $\gamma > 0$ and $\varepsilon > 0$,

$$\begin{aligned}
|\alpha(\boldsymbol{u}_h \cdot \nabla(\lambda_h - z), z_h - \lambda_h)| &\leq \frac{\alpha}{2}\left[h\,\gamma\,\|\boldsymbol{u}_h \cdot \nabla(z_h - \lambda_h)\|^2_{L^2(\Omega)} + \frac{1}{h\,\gamma}\|\lambda_h - z\|^2_{L^2(\Omega)}\right] \\
&\quad + \frac{1}{2}\left[\nu\,\varepsilon\|z_h - \lambda_h\|^2_{L^2(\Omega)} + \frac{\alpha^2}{\nu\,\varepsilon}\|\mathrm{div}(\boldsymbol{u}_h - \boldsymbol{u})(\lambda_h - z)\|^2_{L^2(\Omega)}\right].
\end{aligned}$$

Therefore, for any $\zeta > 0$, $\gamma > 0$, and $\varepsilon > 0$,

$$\begin{aligned}
&|\alpha(\boldsymbol{u}_h\cdot\nabla(\lambda_h - z), z_h - \lambda_h + h\,\boldsymbol{u}_h \cdot \nabla(z_h - \lambda_h))| \\
&\quad \leq \frac{\alpha}{2}\left[h\,\gamma\,\|\boldsymbol{u}_h \cdot \nabla(z_h - \lambda_h)\|^2_{L^2(\Omega)} + \frac{1}{h\,\gamma}\|\lambda_h - z\|^2_{L^2(\Omega)}\right] \\
&\qquad + \frac{1}{2}\left[\nu\,\varepsilon\|z_h - \lambda_h\|^2_{L^2(\Omega)} + \frac{\alpha^2}{\nu\,\varepsilon}\|\mathrm{div}(\boldsymbol{u}_h - \boldsymbol{u})(\lambda_h - z)\|^2_{L^2(\Omega)}\right] \\
&\qquad + \frac{h\,\alpha}{2}\left[\zeta\,\|\boldsymbol{u}_h \cdot \nabla(z_h - \lambda_h)\|^2_{L^2(\Omega)} + \frac{1}{\zeta}\|\boldsymbol{u}_h\|^2_{L^\infty(\Omega)}|\lambda_h - z|^2_{H^1(\Omega)}\right].
\end{aligned} \tag{2.4.6}$$

Then (2.4.4) follows readily by substituting (2.4.6) into (2.4.5), and by repeated applications of Young's inequality. □

The choice $\lambda_h = R_h(z)$ in (2.4.4), Hölder's inequality in the nonlinear products, and the stability properties of R_h give the next result.

COROLLARY 2.4.5. *With the assumptions and notation of Theorem 2.4.4, and if Hypotheses 2.1.5 hold, z_h satisfies the following error estimate*

$$\begin{aligned}\frac{\nu}{2}\|z_h - z\|^2_{L^2(\Omega)} &+ \frac{\alpha\, h}{2}\|\boldsymbol{u}_h\cdot\nabla(z_h - R_h(z))\|^2_{L^2(\Omega)} \le 2\alpha\, h\,\|\boldsymbol{u}_h\|^2_{L^\infty(\Omega)}|z - R_h(z)|^2_{H^1(\Omega)}\\ &+ 2(\frac{13}{4}\,\nu + 2\,h\,\frac{\nu^2}{\alpha} + \frac{\alpha}{h})\|z - R_h(z)\|^2_{L^2(\Omega)}\\ &+ |\boldsymbol{u}-\boldsymbol{u}_h|^2_{H^1(\Omega)}\left(|z|^2_{W^{1,r}(\Omega)}\left(\alpha\, S^2_{r^*}\left(\frac{3\,\alpha}{\nu} + 2\,h\right)\right.\right.\\ &+\left.\left. (1+C_r)^2\frac{3}{4}\left(9\frac{\alpha^2}{\nu} + \nu\, h^2\right)\right) + \frac{\nu}{2}\left(3 + 8\,h\frac{\nu}{\alpha}\right)\right), \qquad (2.4.7)\end{aligned}$$

where C_r is the stability constant of R_h in $W^{1,r}(\Omega)$.

Because Ω is convex, by slightly restricting the mesh as in Remark 2.1.17, (see (2.1.45)), we have $\boldsymbol{u}_h$ uniformly bounded in L^∞. Then by substituting (2.1.51):

$$\begin{aligned}|\boldsymbol{u}-\boldsymbol{u}_h|_{H^1(\Omega)} \le 2|\boldsymbol{u} - P_h(\boldsymbol{u})|_{H^1(\Omega)} &+ \frac{S_4}{\nu}\|\boldsymbol{u}\|_{L^4(\Omega)}\|z - z_h\|_{L^2(\Omega)}\\ &+ \frac{S_4}{\nu}\|z_h\|_{L^2(\Omega)}\|\boldsymbol{u} - P_h(\boldsymbol{u})\|_{L^4(\Omega)} + \frac{1}{\nu}\|p - r_h(p)\|_{L^2(\Omega)},\end{aligned}$$

into (2.4.7), we derive the following error bound for small enough data and smooth enough solutions, if the hypotheses of Theorem 2.4.4 and (2.1.45) hold (for simplicity, we do not detail the constants):

$$\begin{aligned}\nu\|z_h - z\|^2_{L^2(\Omega)} + \alpha\, h\,\|\boldsymbol{u}_h\cdot\nabla(z_h - z)\|^2_{L^2(\Omega)} &\le C\left(\frac{1}{h}\|z - R_h(z)\|^2_{L^2(\Omega)}\right.\\ &+\left. h\,|z - R_h(z)|^2_{H^1(\Omega)} + |\boldsymbol{u} - P_h(\boldsymbol{u})|^2_{H^1(\Omega)} + \|p - r_h(p)\|^2_{L^2(\Omega)}\right).\end{aligned} \qquad (2.4.8)$$

This upwind scheme can be used with the three examples of Section 2.2. It has the same order as the centered scheme of Section 2.1, but its error estimates are of particular interest when the regularity of z is small. For example, if $z \in W^{1,r}(\Omega)$, then the mini-element produces an error of the order of $\sqrt{h}$, whereas we cannot establish any order of convergence for the centered scheme. Finally, the discrete solution can also be computed with the successive approximation algorithm discussed in Section 2.3.

2.4.2. *Upwinding by Lesaint–Raviart's Discontinuous Galerkin*

The upwind Discontinuous Galerkin scheme analyzed in this section works in each element with polynomial functions for z_h that are totally discontinuous across element interfaces.

The upwinding in the term $\boldsymbol{u}_h \cdot \nabla z_h$ is achieved by using in each element only the values of $\boldsymbol{u}_h$ entering the element. This scheme was first introduced in a 1973 Los Alamos Report on neutron transport by REED and HILL (cf. [1973]) and it was first analyzed in this context by LESAINT and RAVIART [1974]. Since then, it has been widely used, adapted to a variety of situations, and generalized. The relevant list of references is far too long to be included here. As a few examples, the reader can refer to GIRAULT and RAVIART [1982, 1986], PIRONNEAU [1989], or GIRAULT, RIVIÈRE and WHEELER [2004] for an application of this scheme to steady, incompressible Navier–Stokes equations, to DAWSON, SUN and WHEELER [2004] for applying it to coupled flow and transport, or to BREZZI, MARINI and SÜLI [2004] for an interesting generalization of this scheme. The material presented here is taken from the work of GIRAULT and SCOTT [2002c]; see also the reference by AMARA, BERNARDI, GIRAULT and HECHT [2005], where a variant of this Discontinuous Galerkin method is applied to a regularized version of (2.1.6)–(2.4.13) below.

The consequence of the above-mentioned discontinuity is that the discrete space Z_h must consist of globally L^2 functions, whereas the discrete spaces for the velocity and pressure, X_h and M_h, can be chosen as in the previous sections. Thus, we discretize z in a finite-dimensional space $Z_h \subset L^2(\Omega)$, such as

$$Z_h = \{\theta_h \in L^2(\Omega);\ \forall T \in \mathcal{T}_h,\ \theta_h|_T \in I\!P_k\}, \tag{2.4.9}$$

where $k \geq 1$ is an integer. Although the functions of Z_h are discontinuous, we shall use (except in one instance) a continuous approximation operator of z. Then, the third assumption of Hypothesis 2.1.5 is satisfied if we choose for R_h the GIRAULT and LIONS [2001b] variant of the SCOTT and ZHANG [1990] regularization operator, the CLÉMENT [1975] operator, or the BERNARDI and GIRAULT [1998] operator. This choice of R_h satisfies for any number $r \geq 1$, for $m = 0, 1$, and $0 \leq s \leq k$,

$$\forall \theta \in W^{s+1,r}(\Omega),\ |R_h(\theta) - \theta|_{W^{m,r}(\Omega)} \leq C\,h^{s+1-m}|\theta|_{W^{s+1,r}(\Omega)}. \tag{2.4.10}$$

First, let us recall how upwinding can be achieved by this Discontinuous Galerkin approximation. Let $\boldsymbol{u}_h$ be a discrete velocity in $H_0^1(\Omega)^2$, and for each triangle T, let

$$\partial T_- = \{\boldsymbol{x} \in \partial T;\ \boldsymbol{u}_h(\boldsymbol{x}) \cdot \boldsymbol{n}_T(\boldsymbol{x}) < 0\}, \tag{2.4.11}$$

where $\boldsymbol{n}_T$ denotes the unit normal to ∂T, exterior to T. This is the portion of ∂T, where the flux driven by $\boldsymbol{u}_h$ enters T. Note that, when running over all triangles T of $\mathcal{T}_h$, ∂T_- only involves interior segments of $\mathcal{T}_h$ because $\boldsymbol{u}_h \cdot \boldsymbol{n}_T = 0$ on $\partial\Omega$. Then, we approximate the nonlinear term $(\boldsymbol{u} \cdot \nabla z, \theta)$ by

$$\tilde{c}^{\mathrm{DG}}(\boldsymbol{u}_h; z_h, \theta_h) = \sum_{T \in \mathcal{T}_h} \left(\int_T (\boldsymbol{u}_h \cdot \nabla z_h)\theta_h \,\mathrm{d}\boldsymbol{x} + \int_{\partial T_-} |\boldsymbol{u}_h \cdot \boldsymbol{n}_T|(z_h^{\mathrm{int}} - z_h^{\mathrm{ext}})\theta_h^{\mathrm{int}} \mathrm{d}s \right) + \frac{1}{2}\int_\Omega (\operatorname{div} \boldsymbol{u}_h) z_h \theta_h \,\mathrm{d}\boldsymbol{x}, \tag{2.4.12}$$

where the superscript int (resp. ext) refers to the trace on the segments of ∂T of the function taken inside (resp. outside) T. When $\boldsymbol{u}_h$ is replaced by $\boldsymbol{u} \in V$ and z_h by $z \in H^1(\Omega)$, this form is a consistent approximation of $(\boldsymbol{u} \cdot \nabla z, \theta_h)$. Note also that when summing over all triangles, ∂T_- is counted exactly once because $\boldsymbol{u}_h \cdot \boldsymbol{n}_T$ changes sign across adjacent elements. Rather, in the above sum, the boundary integrations are taken once over complete interior segments. With this form, the upwind scheme reads: Find $\boldsymbol{u}_h$ in X_h, p_h in M_h, and z_h in Z_h satisfying (2.1.6)–(2.1.7):

$$\forall \boldsymbol{v}_h \in X_h,\ \nu(\nabla \boldsymbol{u}_h, \nabla \boldsymbol{v}_h) + (z_h \times \boldsymbol{u}_h, \boldsymbol{v}_h) - (p_h, \operatorname{div} \boldsymbol{v}_h) = (\boldsymbol{f}, \boldsymbol{v}_h),$$
$$\forall q_h \in M_h,\ (q_h, \operatorname{div} \boldsymbol{u}_h) = 0,$$

and

$$\forall \theta_h \in Z_h,\ \nu(z_h, \theta_h) + \alpha\, \widetilde{c}^{\mathrm{DG}}(\boldsymbol{u}_h; z_h, \theta_h) = \nu(\operatorname{curl} \boldsymbol{u}_h, \theta_h) + \alpha\, (\operatorname{curl} \boldsymbol{f}, \theta_h). \tag{2.4.13}$$

REMARK 2.4.6. The possibility that z_h be constant in each element of the triangulation is not considered here, but the subsequent analysis can readily be adapted to this space coupled with the mini-element or the Bernardi–Raugel element. The error of the resulting scheme is of the order of $h^{1/2}$. □

Some properties of form $\widetilde{c}^{\mathrm{DG}}$

The upwinding effect of $\widetilde{c}^{\mathrm{DG}}$ is made clearer by expressing it in the following form. Let Γ_h^i denote the set of interior segments of $\mathcal{T}_h$. A unit normal vector $\boldsymbol{n}_e$ can be assigned to each segment e of Γ_h^i by numbering the triangles of $\mathcal{T}_h$, say from 1 to N_h, and by setting $\boldsymbol{n}_e = \boldsymbol{n}_{T_k}$, the unit normal to e directed outside T_k if e is adjacent to T_k and T_ℓ with $k < \ell$. Then, we define formally the jump of a function φ through e in the direction of $\boldsymbol{n}_e$ by

$$[\varphi]_e = \left(\varphi|_{T_k} - \varphi|_{T_\ell}\right)|_e. \tag{2.4.14}$$

Next, let

$$e_- = \{\boldsymbol{x} \in e;\ \boldsymbol{u}_h(\boldsymbol{x}) \cdot \boldsymbol{n}_e(\boldsymbol{x}) < 0\},\ e_+ = \{\boldsymbol{x} \in e;\ \boldsymbol{u}_h(\boldsymbol{x}) \cdot \boldsymbol{n}_e(\boldsymbol{x}) > 0\},$$

and note that by reversing the orientation of the normal, e_+ is changed into e_-. Then, with the above notation,

$$\sum_{T \in \mathcal{T}_h} \int_{\partial T_-} |\boldsymbol{u}_h \cdot \boldsymbol{n}_T| (z_h^{\mathrm{int}} - z_h^{\mathrm{ext}}) \theta_h^{\mathrm{int}} \mathrm{d}s$$
$$= - \sum_{e \in \Gamma_h^i} \left(\int_{e_-} (\boldsymbol{u}_h \cdot \boldsymbol{n}_e)[z_h] \theta_h|_{T_k} \mathrm{d}s + \int_{e_+} (\boldsymbol{u}_h \cdot \boldsymbol{n}_e)[z_h] \theta_h|_{T_\ell} \mathrm{d}s \right).$$

Set

$$\theta_{h,d} = \theta_h|_{T_k} \text{ if } \boldsymbol{u}_h \cdot \boldsymbol{n}_e < 0,\ \theta_{h,d} = 0 \text{ if } \boldsymbol{u}_h \cdot \boldsymbol{n}_e = 0,\ \theta_{h,d} = \theta_h|_{T_\ell} \text{ if } \boldsymbol{u}_h \cdot \boldsymbol{n}_e > 0.$$

Then

$$\sum_{T\in\mathfrak{T}_h}\int_{\partial T_-} |\boldsymbol{u}_h\cdot\boldsymbol{n}_T|(z_h^{\text{int}}-z_h^{\text{ext}})\theta_h^{\text{int}}\mathrm{d}s = -\sum_{e\in\Gamma_h^i}\int_e (\boldsymbol{u}_h\cdot\boldsymbol{n}_e)[z_h]_e\theta_{h,d}\mathrm{d}s. \tag{2.4.15}$$

In spite of its representation, form $\tilde{c}^{\text{DG}}$ defined by (2.4.12) is not trilinear because its dependence on the first argument $\boldsymbol{u}_h$ is highly nonlinear. Nevertheless, it satisfies the following valuable identity established by LESAINT and RAVIART [1974]. We reproduce its proof for the reader's convenience.

LEMMA 2.4.7. *For all $\boldsymbol{u}_h$ in X_h, for all z_h and θ_h in Z_h, we have*

$$\tilde{c}^{\text{DG}}(\boldsymbol{u}_h; z_h, \theta_h) = \sum_{T\in\mathfrak{T}_h}\Bigg(-\int_T (\boldsymbol{u}_h\cdot\nabla\theta_h)z_h\,\mathrm{d}\boldsymbol{x} + \int_{\partial T_-}|\boldsymbol{u}_h\cdot\boldsymbol{n}_T|(\theta_h^{\text{ext}}-\theta_h^{\text{int}})z_h^{\text{ext}}\mathrm{d}s\Bigg) - \frac{1}{2}\int_\Omega (\operatorname{div}\boldsymbol{u}_h)\theta_h z_h\,\mathrm{d}\boldsymbol{x}. \tag{2.4.16}$$

PROOF. An application of Green's formula in each T gives the following equation:

$$\sum_{T\in\mathfrak{T}_h}\int_T (\boldsymbol{u}_h\cdot\nabla z_h)\theta_h\mathrm{d}\boldsymbol{x} = -\sum_{T\in\mathfrak{T}_h}\int_T ((\boldsymbol{u}_h\cdot\nabla\theta_h)+(\operatorname{div}\boldsymbol{u}_h)\theta_h)\,z_h\,\mathrm{d}\boldsymbol{x} + \sum_{T\in\mathfrak{T}_h}\int_{\partial T}(\boldsymbol{u}_h\cdot\boldsymbol{n}_T)(z_h\theta_h)|_T\,\mathrm{d}s.$$

When substituted into (2.4.12), we obtain

$$\begin{aligned}\tilde{c}^{\text{DG}}(\boldsymbol{u}_h; z_h, \theta_h) = &-\sum_{T\in\mathfrak{T}_h}\int_T\left(\boldsymbol{u}_h\cdot\nabla\theta_h+\frac{1}{2}(\operatorname{div}\boldsymbol{u}_h)\theta_h\right)z_h\,\mathrm{d}\boldsymbol{x}\\ &-\sum_{T\in\mathfrak{T}_h}\int_{(\partial T)_-}(\boldsymbol{u}_h\cdot\boldsymbol{n}_T)(z_h^{\text{int}}-z_h^{\text{ext}})\theta_h^{\text{int}}\,\mathrm{d}s\\ &+\sum_{T\in\mathfrak{T}_h}\int_{\partial T}(\boldsymbol{u}_h\cdot\boldsymbol{n}_T)(z_h\theta_h)|_T\,\mathrm{d}s.\end{aligned} \tag{2.4.17}$$

Let e belong to Γ_h^i, let T_1 and T_2 denote the two elements of $\mathfrak{T}_h$ adjacent to e, and set $\boldsymbol{n}_e = \boldsymbol{n}_{T_1}$. The last term in (2.4.17) reads

$$\sum_{T\in\mathfrak{T}_h}\int_{\partial T}(\boldsymbol{u}_h\cdot\boldsymbol{n}_T)(z_h\theta_h)|_T\,\mathrm{d}s = \sum_{e\in\Gamma_h^i}\int_e(\boldsymbol{u}_h\cdot\boldsymbol{n}_e)[z_h\theta_h]_e\mathrm{d}s,$$

and by substituting into (2.4.17), we derive

$$\widetilde{c}^{\mathrm{DG}}(\boldsymbol{u}_h; z_h, \theta_h) = -\sum_{T\in\mathfrak{T}_h}\int_T \left(\boldsymbol{u}_h\cdot\nabla\theta_h + \frac{1}{2}(\operatorname{div}\boldsymbol{u}_h)\theta_h\right) z_h\,\mathrm{d}\boldsymbol{x}$$
$$-\sum_{T\in\mathfrak{T}_h}\int_{(\partial T)_-}(\boldsymbol{u}_h\cdot\boldsymbol{n}_T)(z_h^{\mathrm{int}} - z_h^{\mathrm{ext}})\theta_h^{\mathrm{int}}\,\mathrm{d}s + \sum_{e\in\Gamma_h^i}\int_e(\boldsymbol{u}_h\cdot\boldsymbol{n}_e)[z_h\theta_h]_e\mathrm{d}s. \tag{2.4.18}$$

Let us compare the two terms in the second row of (2.4.18). With the above notation, the contribution of e to the first term of this row is

$$A = -\int_{e_-}(\boldsymbol{u}_h\cdot\boldsymbol{n}_e)[z_h]_e\theta_h|_{T_1}\mathrm{d}s - \int_{e_+}(\boldsymbol{u}_h\cdot\boldsymbol{n}_e)[z_h]_e\theta_h|_{T_2}\mathrm{d}s.$$

The contribution of e to the second term is

$$B = \int_{e_-}(\boldsymbol{u}_h\cdot\boldsymbol{n}_e)\left((z_h\theta_h)|_{T_1} - (z_h\theta_h)|_{T_2}\right)\mathrm{d}s + \int_{e_+}(\boldsymbol{u}_h\cdot\boldsymbol{n}_e)\left((z_h\theta_h)|_{T_1} - (z_h\theta_h)|_{T_2}\right)\mathrm{d}s.$$

Thus $A+B$ has the expression

$$A+B = \int_{e_-}(\boldsymbol{u}_h\cdot\boldsymbol{n}_e)[\theta_h]_e z_h|_{T_2}\mathrm{d}s + \int_{e_+}(\boldsymbol{u}_h\cdot\boldsymbol{n}_e)[\theta_h]_e z_h|_{T_1}\mathrm{d}s.$$

But $\boldsymbol{n}_e = \boldsymbol{n}_{T_1} = -\boldsymbol{n}_{T_2}$ and by reversing the orientation of the normal, e_+ is changed into e_-. Hence,

$$-\sum_{T\in\mathfrak{T}_h}\int_{(\partial T)_-}(\boldsymbol{u}_h\cdot\boldsymbol{n}_T)(z_h^{\mathrm{int}} - z_h^{\mathrm{ext}})\theta_h^{\mathrm{int}}\,\mathrm{d}s + \sum_{e\in\Gamma_h^i}\int_e(\boldsymbol{u}_h\cdot\boldsymbol{n}_e)[z_h\theta_h]_e\mathrm{d}s$$
$$= \sum_{T\in\mathfrak{T}_h}\int_{(\partial T)_-}(\boldsymbol{u}_h\cdot\boldsymbol{n}_T)(\theta_h^{\mathrm{int}} - \theta_h^{\mathrm{ext}})z_h^{\mathrm{ext}}\,\mathrm{d}s. \tag{2.4.19}$$

Then, (2.4.16) follows by substituting (2.4.19) into (2.4.18). □

Note that when θ_h is in $H^1(\Omega)$, (2.4.16) reduces to

$$\widetilde{c}^{\mathrm{DG}}(\boldsymbol{u}_h; z_h, \theta_h) = -\int_\Omega(\boldsymbol{u}_h\cdot\nabla\theta_h)z_h\,\mathrm{d}\boldsymbol{x} - \frac{1}{2}\int_\Omega(\operatorname{div}\boldsymbol{u}_h)\theta_h z_h\,\mathrm{d}\boldsymbol{x}. \tag{2.4.20}$$

By summing (2.4.12) and (2.4.16) with $\theta_h = z_h \in Z_h$, we derive the positivity of $\widetilde{c}^{\mathrm{DG}}$ for all $\boldsymbol{u}_h \in X_h$, and all $z_h \in Z_h$:

$$\widetilde{c}^{\mathrm{DG}}(\boldsymbol{u}_h; z_h, z_h) = \frac{1}{2}\sum_{T\in\mathfrak{T}_h}\int_{\partial T_-}|\boldsymbol{u}_h\cdot\boldsymbol{n}_T|(z_h^{\mathrm{ext}} - z_h^{\mathrm{int}})^2\mathrm{d}s. \tag{2.4.21}$$

Existence of solutions and convergence
Again, as (2.1.6) and (2.1.7) are unchanged, problems (2.1.6), (2.1.7), (2.4.13) satisfy Proposition 2.1.3 and the analog of Theorem 2.1.4. The proof is skipped because it is a straightforward consequence of (2.4.21) and Brouwer's Fixed Point Theorem.

PROPOSITION 2.4.8. *Assume that (2.1.1) holds. Then for all $\nu > 0$, $\alpha > 0$, and for all f in $H(curl, \Omega)$, the discrete problem (2.1.6), (2.1.7), (2.4.13) has at least one solution $(\boldsymbol{u}_h, p_h) \in V_h \times M_h$, $z_h \in Z_h$, and each solution satisfies the a priori estimates (2.1.14)–(2.1.15):*

$$|\boldsymbol{u}_h|_{H^1(\Omega)} \le \frac{S_2}{\nu}\|\boldsymbol{f}\|_{L^2(\Omega)},$$

$$\|p_h\|_{L^2(\Omega)} \le \frac{1}{\beta^*}\left(S_2\|\boldsymbol{f}\|_{L^2(\Omega)} + S_4^2|\boldsymbol{u}_h|_{H^1(\Omega)}\|z_h\|_{L^2(\Omega)}\right),$$

and

$$\begin{aligned}\nu\,\|z_h\|^2_{L^2(\Omega)} + \alpha \sum_{T\in\mathfrak{T}_h}\int_{\partial T_-} |\boldsymbol{u}_h\cdot\boldsymbol{n}_T|(z_h^{\text{ext}} - z_h^{\text{int}})^2 ds \\ \le \frac{2}{\nu}\left(S_2^2\|\boldsymbol{f}\|^2_{L^2(\Omega)} + \alpha^2\|\text{curl}\,\boldsymbol{f}\|^2_{L^2(\Omega)}\right).\end{aligned} \tag{2.4.22}$$

REMARK 2.4.9. By comparing with Remark 2.4.2, we see that the upwinding in (2.4.13) does not have the drawback of (2.4.1), in the sense that (2.4.22) allows α to tend to zero. On the other hand, discontinuous functions involve more degrees of freedom. □

Extracting subsequences (that we still denote by the index h), the uniform a priori estimates of Proposition 2.4.8 show that, on one hand, $(\boldsymbol{u}_h, p_h, z_h)$ converge weakly to functions $(\boldsymbol{u}, p, z)$ in $V \times L_0^2(\Omega) \times L^2(\Omega)$, and on the other hand, the quantity defined by $\sum_{T\in\mathfrak{T}_h}\int_{\partial T_-} |\boldsymbol{u}_h\cdot\boldsymbol{n}_T|(z_h^{\text{ext}} - z_h^{\text{int}})^2 ds$ converges to a non-negative number, say S. The next theorem proves that this convergence is strong and the limit functions solve Problem (1.4.9).

THEOREM 2.4.10. *Under the assumptions of Theorem 2.1.9, there exists a subsequence of h (still denoted by h) and a solution $(\boldsymbol{u}, p, z) \in V \times L_0^2(\Omega) \times L^2(\Omega)$ of Problem (1.4.9) such that*

$$\begin{aligned}\lim_{h\to 0} |\boldsymbol{u}_h - \boldsymbol{u}|_{H^1(\Omega)} &= 0,\\ \lim_{h\to 0} \|z_h - z\|_{L^2(\Omega)} &= 0,\\ \lim_{h\to 0} \|p_h - p\|_{L^2(\Omega)} &= 0,\\ \lim_{h\to 0}\sum_{T\in\mathfrak{T}_h}\int_{\partial T_-} |\boldsymbol{u}_h\cdot\boldsymbol{n}_T|(z_h^{\text{ext}} - z_h^{\text{int}})^2 ds &= 0 \text{ in } \mathbb{R}.\end{aligned} \tag{2.4.23}$$

PROOF. As the discontinuity of z_h plays no part in (2.1.6), the argument used for the centered schemes shows that $(\boldsymbol{u}, p, z)$ solves (1.4.6) and the convergence of $\boldsymbol{u}_h$ to $\boldsymbol{u}$ in $H^1(\Omega)^2$

is strong

$$\lim_{h\to 0} \|\boldsymbol{u}_h - \boldsymbol{u}\|_{H^1(\Omega)} = 0.$$

In order to pass to the limit in (2.4.13), we take an arbitrary function $\theta \in H^2(\Omega)$ and we choose $\theta_h = R_h(\theta) \in H^1(\Omega)$. Then, in view of (2.4.20),

$$\tilde{c}^{\mathrm{DG}}(\boldsymbol{u}_h; z_h, R_h(\theta)) = -\int_\Omega (\boldsymbol{u}_h \cdot \nabla R_h(\theta)) z_h \, \mathrm{d}\boldsymbol{x} - \frac{1}{2}\int_\Omega (\operatorname{div} \boldsymbol{u}_h) R_h(\theta) z_h \, \mathrm{d}\boldsymbol{x}.$$

Hence, we are back to the situation of Proposition 2.1.8, and with the above strong convergence, its argument allows us to pass to the limit in (2.4.13) and prove that $(\boldsymbol{u}, p, z)$ is indeed a solution of Problem (1.4.9).

To establish the strong convergence of z_h, we take $\theta_h = z_h$ in (2.4.13), apply (2.4.21), and compare with (1.4.8) tested against z_h:

$$\nu\|z_h\|^2_{L^2(\Omega)} + \frac{\alpha}{2}\sum_{T\in\mathcal{T}_h}\int_{\partial T_-} |\boldsymbol{u}_h \cdot \boldsymbol{n}_T|(z_h^{\mathrm{ext}} - z_h^{\mathrm{int}})^2 \mathrm{d}s = \nu(z, z_h) + \alpha(\boldsymbol{u} \cdot \nabla z, z_h) + \nu(\operatorname{curl}(\boldsymbol{u}_h - \boldsymbol{u}), z_h).$$

By passing to the limit, the strong convergence of $\boldsymbol{u}_h$ gives

$$\lim_{h\to 0}\left(\nu\|z_h\|^2_{L^2(\Omega)}\right) + \frac{\alpha}{2}S = \nu\|z\|^2_{L^2(\Omega)}.$$

Hence $\lim_{h\to 0}\|z_h\|^2_{L^2(\Omega)} \le \|z\|^2_{L^2(\Omega)}$. This yields, on one hand, the strong convergence of z_h and, on the other hand, the fact that $S = 0$. Finally, the strong convergence of p_h is established as in Theorem 2.1.9. □

Discontinuous Galerkin: Error estimates

Here also, all the results of Sections 2.1.1 and 2.1.2 concerning $\boldsymbol{u}_h$ and p_h carry over here, as well as the statement of Lemma 2.1.19, namely (2.1.51) and (2.1.52):

$$|\boldsymbol{u} - \boldsymbol{u}_h|_{H^1(\Omega)} \le 2|\boldsymbol{u} - P_h(\boldsymbol{u})|_{H^1(\Omega)} + \frac{S_4}{\nu}\|\boldsymbol{u}\|_{L^4(\Omega)}\|z - z_h\|_{L^2(\Omega)} + \frac{S_4}{\nu}\|z_h\|_{L^2(\Omega)}\|\boldsymbol{u} - P_h(\boldsymbol{u})\|_{L^4(\Omega)} + \frac{1}{\nu}\|p - r_h(p)\|_{L^2(\Omega)},$$

$$\|p - p_h\|_{L^2(\Omega)} \le \left(1 + \frac{1}{\beta^*}\right)\|p - r_h(p)\|_{L^2(\Omega)} + \frac{1}{\beta^*}\big(\nu\,|\boldsymbol{u} - P_h(\boldsymbol{u})|_{H^1(\Omega)} + S_4\big(\|\boldsymbol{u}\|_{L^4(\Omega)}\|z - z_h\|_{L^2(\Omega)} + \|z_h\|_{L^2(\Omega)}\|\boldsymbol{u} - \boldsymbol{u}_h\|_{L^4(\Omega)}\big)\big),$$

where β^* is the constant of (2.1.1).

Clearly, the treatment of $z - z_h$ differs, both from that of the centered scheme and that of the upwind scheme by streamline diffusion. First, assuming that z belongs to $H^1(\Omega)$, by taking the difference between (2.4.13) and (1.4.8) tested against θ_h, inserting any function

$\lambda_h \in Z_h$, choosing $\theta_h = z_h - \lambda_h$, and applying (2.4.16) and (2.4.21), we obtain the error equality:

$$\begin{aligned}
&\nu\|z_h - \lambda_h\|^2_{L^2(\Omega)} + \alpha \sum_{T\in\mathcal{T}_h} \frac{1}{2} \int_{\partial T_-} |\boldsymbol{u}_h \cdot \boldsymbol{n}_T|((z_h - \lambda_h)^{\text{ext}} - (z_h - \lambda_h)^{\text{int}})^2 \mathrm{d}s \\
&+ \alpha \sum_{T\in\mathcal{T}_h} \Bigg(- \int_T \boldsymbol{u}_h \cdot \nabla (z_h - \lambda_h)(\lambda_h - z)\,\mathrm{d}\boldsymbol{x} \\
&+ \alpha \int_{\partial T_-} |\boldsymbol{u}_h \cdot \boldsymbol{n}_T|((z_h - \lambda_h)^{\text{ext}} - (z_h - \lambda_h)^{\text{int}})(\lambda_h - z)^{\text{ext}} \mathrm{d}s \Bigg) \qquad (2.4.24) \\
&- \frac{\alpha}{2} \int_\Omega \operatorname{div}(\boldsymbol{u}_h - \boldsymbol{u})(\lambda_h - z)(z_h - \lambda_h)\,\mathrm{d}\boldsymbol{x} \\
&+ \frac{\alpha}{2} \int_\Omega \operatorname{div}(\boldsymbol{u}_h - \boldsymbol{u}) z (z_h - \lambda_h) \mathrm{d}\boldsymbol{x} + \alpha \int_\Omega (\boldsymbol{u}_h - \boldsymbol{u}) \cdot \nabla z (z_h - \lambda_h)\,\mathrm{d}\boldsymbol{x} \\
&= \nu(z - \lambda_h, z_h - \lambda_h) + \nu(\operatorname{curl}(\boldsymbol{u}_h - \boldsymbol{u}), z_h - \lambda_h).
\end{aligned}$$

Let us look more closely at the third and fourth terms in the left-hand side of (2.4.24). They require a special treatment because they mean that the unknown $z_h - \lambda_h$ should be measured in a norm that is finer than the L^2 norm. In each element T, the third term can be split into:

$$\begin{aligned}
\int_T \boldsymbol{u}_h \cdot \nabla (z_h - \lambda_h)(\lambda_h - z)\,\mathrm{d}\boldsymbol{x} &= \int_T (\boldsymbol{u}_h - \boldsymbol{u}) \cdot \nabla (z_h - \lambda_h)(\lambda_h - z)\,\mathrm{d}\boldsymbol{x} \\
&+ \int_T \boldsymbol{u} \cdot \nabla (z_h - \lambda_h)(\lambda_h - z)\,\mathrm{d}\boldsymbol{x}. \qquad (2.4.25)
\end{aligned}$$

If we choose $\lambda_h = R_h(z)$, all benefit from using (2.4.16) is lost because the error of $\lambda_h - z$ in the L^2 norm must balance the gradient of $z_h - \lambda_h$. Instead, let us take advantage of the discontinuity of the space Z_h and choose $\lambda_h = \varrho_h(z)$, the L^2 projection of z on $\mathbb{P}_k$ in each triangle T: $\varrho_h(z) \in \mathbb{P}_k$ is defined by

$$\forall q \in \mathbb{P}_k, \quad \int_T (\varrho_h(z) - z) q\,\mathrm{d}\boldsymbol{x} = 0.$$

This operator has locally the same accuracy as R_h. Moreover, we have for any constant vector $\boldsymbol{c}$:

$$\int_T \boldsymbol{u} \cdot \nabla (z_h - \varrho_h(z))(\varrho_h(z) - z)\,\mathrm{d}\boldsymbol{x} = \int_T (\boldsymbol{u} - \boldsymbol{c}) \cdot \nabla (z_h - \varrho_h(z))(\varrho_h(z) - z)\,\mathrm{d}\boldsymbol{x}, \qquad (2.4.26)$$

because the components of $\nabla(z_h - \varrho_h(z))$ belong to $\mathbb{P}_{k-1}$. With this choice, we have the following error inequality.

THEOREM 2.4.11. *Let Ω be convex, $(\boldsymbol{u}, p, z)$ a solution of Problem (1.4.9), r_0 the number of Proposition 1.4.11, and let the assumptions of Theorem 1.4.14 hold, so that $z \in W^{1,r}(\Omega)$, for some real number r in $]2, r_0[$. Let the family of triangulations satisfy (2.1.25) and let $(\boldsymbol{u}_h, p_h, z_h)$ be any solution of (2.1.6), (2.1.7), (2.4.13). Then, we have the following inequality for $z_h - \varrho_h(z)$:*

$$\begin{aligned}\nu \|z_h - \varrho_h(z)\|^2_{L^2(\Omega)} &\le \frac{7}{\nu}\Big(\nu^2\Big(\|z - \varrho_h(z)\|^2_{L^2(\Omega)} + |\boldsymbol{u}_h - \boldsymbol{u}|^2_{H^1(\Omega)}\Big)\\ &+ \alpha^2\sigma_0^2 C_1\Big(|\boldsymbol{u}|^2_{W^{1,\infty}(\Omega)}\|z - \varrho_h(z)\|^2_{L^2(\Omega)} + S^2_{r^*}|z|^2_{W^{1,r}(\Omega)}|\boldsymbol{u}_h - \boldsymbol{u}|^2_{H^1(\Omega)}\Big)\\ &+ \frac{\alpha^2}{4}\Big(\|\operatorname{div}(\boldsymbol{u}_h - \boldsymbol{u})(z - \varrho_h(z))\|^2_{L^2(\Omega)} + \|\operatorname{div}(\boldsymbol{u}_h - \boldsymbol{u})z\|^2_{L^2(\Omega)}\\ &+ 4\,\|(\boldsymbol{u}_h - \boldsymbol{u})\cdot\nabla z\|^2_{L^2(\Omega)}\Big)\Big) + \alpha\, C_2\sigma_0^2\|\boldsymbol{u}_h\|_{L^\infty(\Omega)}\sum_{T\in\mathfrak{T}_h} h_T|z - \varrho_h(z)|^2_{H^1(T)},\end{aligned} \tag{2.4.27}$$

where σ_0 is the constant of (2.1.25), C_1 and C_2 are constants independent of h, and

$$\frac{1}{r^*} = \frac{1}{2} - \frac{1}{r}.$$

PROOF. We bound the terms in the left-hand side of (2.4.24), starting with the third term. In view of (2.4.25), we begin with (2.4.26). In each T, a *local* inverse inequality, a suitable choice for $\boldsymbol{c}$, such as the mean value of $\boldsymbol{u}$ in T, and (2.1.25) yield:

$$\begin{aligned}&\left|\int_T (\boldsymbol{u} - \boldsymbol{c})\cdot\nabla(z_h - \varrho_h(z))(\varrho_h(z) - z)\,\mathrm{d}\boldsymbol{x}\right|\\ &\qquad \le \frac{c_1}{\rho_T}\|\boldsymbol{u} - \boldsymbol{c}\|_{L^\infty(T)}\|\varrho_h(z) - z\|_{L^2(T)}\|z_h - \varrho_h(z)\|_{L^2(T)}\\ &\qquad \le c_2\sigma_0|\boldsymbol{u}|_{W^{1,\infty}(T)}\|\varrho_h(z) - z\|_{L^2(T)}\|z_h - \varrho_h(z)\|_{L^2(T)},\end{aligned} \tag{2.4.28}$$

where c_i denote various constants independent of h and T. Similarly, by applying the same local inverse inequality and the approximation property of ϱ_h, we obtain in each T

$$\begin{aligned}&\left|\int_T (\boldsymbol{u}_h - \boldsymbol{u})\cdot\nabla(z_h - \varrho_h(z))(\varrho_h(z) - z)\,\mathrm{d}\boldsymbol{x}\right|\\ &\qquad \le c_3\sigma_0|z|_{W^{1,r}(T)}\|\boldsymbol{u}_h - \boldsymbol{u}\|_{L^{r^*}(T)}\|z_h - \varrho_h(z)\|_{L^2(T)}.\end{aligned} \tag{2.4.29}$$

The fourth term in (2.4.24) can be split into

$$\left|\sum_{T\in\mathfrak{T}_h}\int_{\partial T_-} |\boldsymbol{u}_h\cdot\boldsymbol{n}_T|((z_h - \varrho_h(z))^{\mathrm{ext}} - (z_h - \varrho_h(z))^{\mathrm{int}})(\varrho_h(z) - z)^{\mathrm{ext}}\mathrm{d}s\right|$$

$$\leq \frac{1}{2} \sum_{T \in \mathcal{T}_h} \int_{\partial T_-} |\boldsymbol{u}_h \cdot \boldsymbol{n}_T| ((z_h - \varrho_h(z))^{\text{ext}} - (z_h - \varrho_h(z))^{\text{int}})^2 \mathrm{d}s$$

$$+ \frac{1}{2} \sum_{T \in \mathcal{T}_h} \int_{\partial T_-} |\boldsymbol{u}_h \cdot \boldsymbol{n}_T| \left((\varrho_h(z) - z)^{\text{ext}}\right)^2 \mathrm{d}s.$$

The first term cancels with the first boundary term in the left-hand side of (2.4.24), but the second term requires more work. Let $e \in \Gamma_h^i$ denote a side of T (recall that the above sum only involves interior segments of $\mathcal{T}_h$), and let $\tilde{T}$ be the other element sharing e. Then by passing to the reference segment $\hat{e}$, and by denoting with a hat the composition with the affine transformation that maps $\hat{e}$ onto e, we can write

$$\int_e |\boldsymbol{u}_h \cdot \boldsymbol{n}_T| \left((\varrho_h(z) - z)^{\text{ext}}\right)^2 \mathrm{d}s \leq \|\boldsymbol{u}_h\|_{L^\infty(e)} \|(\varrho_h(z) - z)^{\text{ext}}\|^2_{L^2(e)}$$
$$\leq |e| \|\hat{\boldsymbol{u}}_h\|_{L^\infty(\hat{e})} \|(\hat{\varrho}_h(\hat{z}) - \hat{z})^{\text{ext}}\|^2_{L^2(\hat{e})}.$$

For the second factor, we use a trace theorem on $\hat{T}$, the reference element corresponding to $\tilde{T}$, and for $\hat{\boldsymbol{u}}_h$, we use an equivalence of norms on $\hat{T}$. This gives

$$\int_e |\boldsymbol{u}_h \cdot \boldsymbol{n}_T| \left((\varrho_h(z) - z)^{\text{ext}}\right)^2 \mathrm{d}s \leq c_4 |e| \|\hat{\boldsymbol{u}}_h\|_{L^\infty(\hat{T})} \|\hat{\varrho}_h(\hat{z}) - \hat{z}\|^2_{H^1(\hat{T})}.$$

But by definition of the L^2 projection (that is invariant under affine transformations),

$$\int_{\hat{T}} (\hat{\varrho}_h(\hat{z}) - \hat{z}) \mathrm{d}\hat{x} = 0,$$

and therefore there exists a constant c_5 independent of h, such that

$$\|\hat{\varrho}_h(\hat{z}) - \hat{z}\|^2_{H^1(\hat{T})} \leq c_5 |\hat{\varrho}_h(\hat{z}) - \hat{z}|^2_{H^1(\hat{T})}.$$

Collecting these two inequalities, reverting to $\tilde{T}$, and using the regularity of $\mathcal{T}_h$, we derive

$$\int_e |\boldsymbol{u}_h \cdot \boldsymbol{n}_T| \left((\varrho_h(z) - z)^{\text{ext}}\right)^2 \mathrm{d}s \leq c_6 \frac{|e|}{|\tilde{T}|} h_{\tilde{T}}^2 \|\boldsymbol{u}_h\|_{L^\infty(\tilde{T})} |\varrho_h(z) - z|^2_{H^1(\tilde{T})}$$
$$\leq c_7 \sigma_0^2 h_{\tilde{T}} \|\boldsymbol{u}_h\|_{L^\infty(\tilde{T})} |\varrho_h(z) - z|^2_{H^1(\tilde{T})}. \tag{2.4.30}$$

All the other terms are easily bounded and (2.4.27) follows by summing the above inequalities over all T in $\mathcal{T}_h$ and applying Young's inequality. □

As in Section 2.4.1, because Ω is convex, by slightly restricting the mesh as in Remark 2.1.17, we have $\boldsymbol{u}_h$ uniformly bounded in L^∞. Then by substituting (2.1.51) into

(2.4.27), we derive the following error bound for small enough data and smooth enough solutions, if the hypotheses of Theorem 2.4.4 and (2.1.45) hold (again, for simplicity we do not detail the constants):

$$\nu\|z_h - z\|^2_{L^2(\Omega)} \le C\Big(\|\varrho_h(z) - z\|^2_{L^2(\Omega)} + |\boldsymbol{u} - P_h(\boldsymbol{u})|^2_{H^1(\Omega)} + \|p - r_h(p)\|^2_{L^2(\Omega)} + \sum_{T\in\mathcal{T}_h} h_T|z - \varrho_h(z)|^2_{H^1(T)}\Big). \tag{2.4.31}$$

Of course, by suitably changing the constants, the boundary term

$$\alpha \sum_{T\in\mathcal{T}_h} \frac{1}{2}\int_{\partial T_-} |\boldsymbol{u}_h\cdot\boldsymbol{n}_T|((z_h - z)^{\text{ext}} - (z_h - z)^{\text{int}})^2 \mathrm{d}s$$

is also bounded by the right-hand side of (2.4.31). When the mini-element or the Bernardi–Raugel element are used in this scheme, the choice $k = 1$ in the definition of Z_h yields an error of order h if z is in $H^{3/2}(\Omega)$, in which case it belongs automatically to $W^{1,r}(\Omega)$ because r is supposed to be close to 2. Of course, the convexity assumption on the domain implies that $\boldsymbol{u}$ is in $H^2(\Omega)^2$ and p in $H^1(\Omega)$. Finally, we infer from (2.4.31) that approximating z by piecewise constant functions (i.e., $k = 0$) yields an error of the order of $h^{1/2}$ when z belongs to $W^{1,r}(\Omega)$.

CHAPTER 3

Discretizing the Time-Dependent No-Slip Problem

3.1. Introduction

In this chapter, we first split the time-dependent problem (1.1.1)–(1.1.2) in a bounded, connected Lipschitz domain Ω of $\mathbb{R}^2$, with a homogeneous Dirichlet boundary condition on $\partial\Omega$. Considering the a priori estimates of Section 1.3.1, we formulate the problem as follows. For given real numbers $T > 0$, $\nu > 0$, and $\alpha > 0$, $\boldsymbol{f}$ given in $L^2(0, T; H(\text{curl}, \Omega))$, and $\boldsymbol{u}_0$ given in V^α, find $\boldsymbol{u} \in L^\infty(0, T; V^\alpha) \cap H^1(0, T; V)$ and $p \in L^2(0, T, L_0^2(\Omega))$, solution of

$$\begin{aligned}\frac{\partial}{\partial t}(\boldsymbol{u} - \alpha\Delta\,\boldsymbol{u}) - \nu\,\Delta\,\boldsymbol{u} + \mathbf{curl}(\boldsymbol{u} - \alpha\Delta\,\boldsymbol{u}) \times \boldsymbol{u} + \nabla p = \boldsymbol{f} \quad &\text{in } \Omega\times]0, T[,\\ \boldsymbol{u}(0) = \boldsymbol{u}_0 \quad &\text{in } \Omega,\end{aligned} \tag{3.1.1}$$

the divergence-free condition and the homogeneous Dirichlet boundary condition on $\partial\Omega$ being prescribed on the functions of V^α. Recall that $H(\text{curl}, \Omega)$ is defined in (1.1.15):

$$H(\text{curl}, \Omega) = \{\boldsymbol{v} \in L^2(\Omega)^2;\ \text{curl}\,\boldsymbol{v} \in L^2(\Omega)\},$$

and V^α is defined in (1.4.1):

$$V^\alpha = \{\boldsymbol{v} \in V;\ \alpha\,\text{curl}\,\Delta\,\boldsymbol{v} \in L^2(\Omega)\},$$

where, in this case,

$$V = \{\boldsymbol{v} \in H_0^1(\Omega)^2;\ \text{div}\,\boldsymbol{v} = 0 \text{ in } \Omega\}.$$

This problem is split into a linearized time-dependent system and a time-dependent transport equation, equivalent to (3.1.1). Next, we semi-discretize these two problems in time with a backward Euler scheme. This scheme is used in SAADOUNI [2007] and in GIRAULT and SAADOUNI [2007] to establish global existence in time of a solution of the split problem, for all data, in a Lipschitz domain. Finally, we discretize in space this semi-discrete scheme with the finite elements methods studied in Chapter II and establish convergence and error estimates under *no CFL condition.* We discretize the transport equation with centered schemes, but to save space we only discuss Discontinuous Galerkin upwind schemes.

Part of the material presented in this chapter is taken from SAADOUNI [2007] and ABBOUD and SAYAH [2009].

The theoretical and numerical analysis of time-dependent problems usually rely on the following Gronwall's Lemma and its discrete counterparts.

LEMMA 3.1.1. *Let $T > 0$ and let κ be a non-negative function in $L^1(0,T)$. Let $C \geq 0$ be a constant and $\varphi \in \mathcal{C}^0([0,T])$ a function satisfying*

$$\forall t \in [0,T],\ 0 \leq \varphi(t) \leq C + \int_0^t \kappa(s)\varphi(s)\,\mathrm{d}s. \tag{3.1.2}$$

Then φ verifies the bound

$$\forall t \in [0,T],\ \varphi(t) \leq C\exp\left(\int_0^t \kappa(s)\,\mathrm{d}s\right). \tag{3.1.3}$$

LEMMA 3.1.2. *Let $(a_n)_{n\geq 0}$, $(b_n)_{n\geq 0}$, and $(c_n)_{n\geq 0}$ be three sequences of non-negative real numbers, such that $(c_n)_{n\geq 0}$ is monotonic increasing,*

$$a_0 + b_0 \leq c_0,$$

and there exists a real number $\lambda > 0$ such that

$$\forall n \geq 1,\ a_n + b_n \leq c_n + \lambda \sum_{m=0}^{n-1} a_m. \tag{3.1.4}$$

Then these sequences are bounded by

$$\forall n \geq 0,\ a_n + b_n \leq c_n \mathrm{e}^{n\lambda}. \tag{3.1.5}$$

LEMMA 3.1.3. *Let $A > 0$ and let $(\zeta_n)_{n\geq 0}$ and $(b_n)_{n\geq 1}$ be two sequences of non-negative real numbers satisfying*

$$\forall n \geq 1,\ \zeta_n \leq (1+A)\zeta_{n-1} + b_n. \tag{3.1.6}$$

Then for all $n \geq 1$, ζ_n satisfies the bound

$$\zeta_n \leq (1+A)^n \zeta_0 + \sum_{i=1}^{n} b_i (1+A)^{n-i}. \tag{3.1.7}$$

Because we are dealing with a nonlinear time-dependent problem, we require compactness in time in order to pass to the limit in the semi-discrete problem. For this, we shall use the following theorem established by SIMON [1990]. It generalizes the Aubin–Lions Lemma, see LIONS [1969] or SHOWALTER [1997].

THEOREM 3.1.4. *Let X, E, Y be three Banach spaces with continuous imbeddings: $X \subset E \subset Y$, the imbedding of X into E being compact. Then for any number $q \in [1, \infty]$, the space*

$$\{v \in L^q(0, T; X); \frac{\partial v}{\partial t} \in L^1(0, T; Y)\} \tag{3.1.8}$$

is compactly imbedded into $L^q(0, T; E)$.

It will also be convenient to use the norm $\|\cdot\|_\alpha$ defined by

$$\|v\|_\alpha = \left(\|v\|^2_{L^2(\Omega)} + \alpha|v|^2_{H^1(\Omega)}\right)^{1/2}, \tag{3.1.9}$$

and the space

$$\mathcal{W} = \{(\boldsymbol{v}, q) \in L^\infty(0, T; V^\alpha) \times L^2(0, T; L^2_0(\Omega)); \frac{\partial \boldsymbol{v}}{\partial t} \in L^2(0, T; V)\}. \tag{3.1.10}$$

3.1.1. The transient tranport equation in arbitrary dimension

By analogy with the steady case, the subsequent analysis relies on sharp estimates for the solution of a time-dependent transport equation: Find $z \in L^\infty(0, T; L^2(\Omega))$ satisfying

$$\text{a.e. in } \Omega\times]0, T[, \ \frac{\partial z}{\partial t} + \gamma\, \boldsymbol{u}\cdot\nabla z = f, \tag{3.1.11}$$

$$\text{a.e. in } \Omega, \ z(0) = z_0, \tag{3.1.12}$$

where

$$\boldsymbol{u}\cdot\nabla z = u_1\frac{\partial z}{\partial x_1} + u_2\frac{\partial z}{\partial x_2},$$

with the data: $\gamma \neq 0$, f given in $L^2(\Omega\times]0, T[)$, z_0 given in $L^2(\Omega)$. Because any solution $z \in L^\infty(0, T; L^2(\Omega))$ of (3.1.11)–(3.1.12) belongs to $H^1(0, T; W^{-1,q}(\Omega))$, where

$$q = \frac{d}{d-1} \text{ if } d \geq 3, \ q < 2 \text{ if } d = 2,$$

the initial condition (3.1.12) makes sense. Establishing existence of a solution of (3.1.11)–(3.1.12) is straightforward. For instance, semi-discretization in time (which regularizes the effect of the time derivative) gives readily the following result.

PROPOSITION 3.1.5. *Let $\Omega \subset \mathbb{R}^d$ be bounded and Lipschitz-continuous. For any real numbers $T > 0$, $\gamma \neq 0$ and any functions $\boldsymbol{u}$ in $L^2(0, T; W)$, f in $L^2(\Omega\times]0, T[)$, and z_0 in $L^2(\Omega)$, problem (3.1.11)–(3.1.12) has at least one solution z in $L^\infty(0, T; L^2(\Omega))$, and this solution satisfies the bound*

$$\|z\|^2_{L^\infty(0,T;L^2(\Omega))} \leq \left(\|z_0\|^2_{L^2(\Omega)} + \|f\|^2_{L^2(\Omega\times]0,T[)}\right)\exp(T). \tag{3.1.13}$$

But again, proving uniqueness is not straightforward, considering the low regularity of the domain and of the driving velocity. By adapting the regularization technique of Theorem 1.3.8, GIRAULT and SCOTT [2010] establish the following uniqueness result.

THEOREM 3.1.6. *Let $\Omega \subset \mathbb{R}^d$ be bounded and Lipschitz-continuous. For any real numbers $T > 0$, $\gamma \neq 0$ and any functions $\boldsymbol{u}$ in $L^2(0, T; W)$, f in $L^2(\Omega \times]0, T[)$, and z_0 in $L^2(\Omega)$, problem (3.1.11)–(3.1.12) has exactly one solution z in $L^\infty(0, T; L^2(\Omega))$.*

When (3.1.11) has the additional term νz in the left-hand side, as is the case subsequently:

$$\text{a.e. in } \Omega \times]0, T[, \quad \frac{\partial z}{\partial t} + \nu z + \gamma \boldsymbol{u} \cdot \nabla z = f, \tag{3.1.14}$$

an easy variant of Theorem 3.1.6 gives:

PROPOSITION 3.1.7. *Let $\Omega \subset \mathbb{R}^d$ be bounded and Lipschitz-continuous. For any real numbers $T > 0$, $\gamma \neq 0$, and $\nu > 0$, and any functions $\boldsymbol{u}$ in $L^2(0, T; W)$, f in $L^2(\Omega \times]0, T[)$, and z_0 in $L^2(\Omega)$, problem (3.1.14), (3.1.12) has exactly one solution z in $L^\infty(0, T; L^2(\Omega))$ and this solution is bounded as follows*

$$\begin{aligned} \|z\|^2_{L^\infty(0,T;L^2(\Omega))} &\leq \|z_0\|^2_{L^2(\Omega)} + \frac{1}{2\nu}\|f\|^2_{L^2(\Omega\times]0,T[)}, \\ \|z\|^2_{L^2(0,T;L^2(\Omega))} &\leq \frac{1}{\nu}\|z_0\|^2_{L^2(\Omega)} + \frac{1}{\nu^2}\|f\|^2_{L^2(\Omega\times]0,T[)}. \end{aligned} \tag{3.1.15}$$

For adequate data, the unique solution of (3.1.14), (3.1.12) belongs to L^p or $W^{1,p}$ in space. This can be derived from a semi-discrete approximation of (3.1.14), (3.1.12) and is postponed to the end of Section 3.2.2.

3.2. Splitting the problem

In this section, we consider again a bounded, connected, Lipschitz domain in $\mathbb{R}^2$, and we retain the above assumptions on the data: $T > 0$, $\alpha > 0$, and $\nu > 0$ are given real numbers, f is given in $L^2(0, T; H(\text{curl}, \Omega))$ and $\boldsymbol{u}_0$ is given in V^α.

Let us revert to Section 1.4.1. Recall (1.4.4) and (1.4.5) defining the auxiliary variable z and its vector product with $\boldsymbol{u}$:

$$\begin{aligned} &z = \text{curl}(\boldsymbol{u} - \alpha \Delta \boldsymbol{u}), \quad \boldsymbol{z} = (0, 0, z), \\ &\text{div}\, \boldsymbol{z} = 0, \ \boldsymbol{z} \times \boldsymbol{u} = (-z u_2, z u_1). \end{aligned}$$

Then by substituting the expression for $\boldsymbol{z}$ into the first row of (3.1.1), we obtain the following linearized system: Find $\boldsymbol{u} \in H^1(0, T; V)$ and $p \in L^2(0, T, L^2_0(\Omega))$ such that

$$\begin{aligned} &\frac{\partial}{\partial t}(\boldsymbol{u} - \alpha \Delta \boldsymbol{u}) - \nu \Delta \boldsymbol{u} + \boldsymbol{z} \times \boldsymbol{u} + \nabla p = \boldsymbol{f} \quad \text{a.e. in } \Omega \times]0, T[, \\ &\qquad \boldsymbol{u}(0) = \boldsymbol{u}_0 \quad \text{a.e. in } \Omega. \end{aligned} \tag{3.2.1}$$

As for the steady problem, we note that the variational formulation of (3.2.1) only requires that $\boldsymbol{u}$ take its values in V. Then, we take formally the curl of both sides of this equation, and taking advantage of (1.3.13),

$$\mathrm{curl}(z \times \boldsymbol{u}) = \boldsymbol{u} \cdot \nabla z,$$

we obtain a transport equation, after multiplying both sides by α: Find $z \in L^\infty(0, T; L^2(\Omega))$, such that

$$\alpha \frac{\partial z}{\partial t} + \nu z + \alpha \boldsymbol{u} \cdot \nabla z = \nu \,\mathrm{curl}\, \boldsymbol{u} + \alpha \,\mathrm{curl}\, \boldsymbol{f} \quad \text{a.e. in } \Omega \times]0, T[,$$
$$z(0) = z_0 = \mathrm{curl}(\boldsymbol{u}_0 - \alpha \Delta \boldsymbol{u}_0) \quad \text{a.e. in } \Omega. \tag{3.2.2}$$

It is established in GIRAULT and SAADOUNI [2007] that the coupled problem (3.2.1)–(3.2.2) is equivalent to the original system (3.1.1). More precisely, we have:

PROPOSITION 3.2.1. *For all real numbers $T > 0$, $\alpha > 0$, and $\nu > 0$, all functions $\boldsymbol{f} \in L^2(0, T; H(\mathrm{curl}, \Omega))$ and $\boldsymbol{u}_0 \in V^\alpha$, problems (3.1.1) and (3.2.1)–(3.2.2) are equivalent.*

Moreover, all solutions of (3.2.1)–(3.2.2) satisfy the following unconditional a priori estimates. The estimates for $\boldsymbol{u}$ are straightforward and those for z follow from Proposition 3.1.7.

PROPOSITION 3.2.2. *For all real numbers $T > 0$, $\alpha > 0$, and $\nu > 0$, all functions $\boldsymbol{f} \in L^2(0, T; H(\mathrm{curl}, \Omega))$ and $\boldsymbol{u}_0 \in V^\alpha$, any solution $(\boldsymbol{u}, p, z)$ in $\mathcal{W} \times L^2(\Omega)$ of (3.2.1)–(3.2.2) satisfies almost everywhere in $]0, T[$:*

$$\|\boldsymbol{u}(t)\|_\alpha^2 \le \|\boldsymbol{u}_0\|_\alpha^2 + \frac{S_2^2}{2\nu} \|\boldsymbol{f}\|^2_{L^2(\Omega \times]0,t[)},$$
$$\nu \|\nabla \boldsymbol{u}\|^2_{L^2(\Omega \times]0,t[)} \le \|\boldsymbol{u}_0\|_\alpha^2 + \frac{S_2^2}{\nu} \|\boldsymbol{f}\|^2_{L^2(\Omega \times]0,t[)}, \tag{3.2.3}$$

$$\|z(t)\|^2_{L^2(\Omega)} \le \frac{1}{\alpha} \|\boldsymbol{u}_0\|_\alpha^2 + \|z_0\|^2_{L^2(\Omega)} + \frac{S_2^2}{\alpha\nu} \|\boldsymbol{f}\|^2_{L^2(\Omega \times]0,t[)} + \frac{\alpha}{\nu} \|\mathrm{curl}\, \boldsymbol{f}\|^2_{L^2(\Omega \times]0,t[)}, \tag{3.2.4}$$

$$\|z\|^2_{L^2(\Omega \times]0,t[)} \le \frac{2}{\nu} \|\boldsymbol{u}_0\|_\alpha^2 + \frac{\alpha}{\nu} \|z_0\|^2_{L^2(\Omega)} + \frac{2}{\nu^2} \left(S_2^2 \|\boldsymbol{f}\|^2_{L^2(\Omega \times]0,t[)} + \alpha^2 \|\mathrm{curl}\, \boldsymbol{f}\|^2_{L^2(\Omega \times]0,t[)} \right), \tag{3.2.5}$$

$$\int_0^t \left\| \frac{\partial \boldsymbol{u}}{\partial t} \right\|_\alpha^2 dt \le \nu |\boldsymbol{u}_0|^2_{H^1(\Omega)} + \|\boldsymbol{f}\|^2_{L^2(\Omega \times]0,t[)} + \frac{S_4^4}{\alpha\nu} \|z\|^2_{L^\infty(]0,t[; L^2(\Omega))} \left(\|\boldsymbol{u}_0\|_\alpha^2 + \frac{{S_2}^2}{\nu} \|\boldsymbol{f}\|^2_{L^2(\Omega \times]0,t[)} \right), \tag{3.2.6}$$

$$\|p\|^2_{L^2(\Omega\times]0,t[)} \leq \frac{3}{\beta^2}\left(S_2^2\left(\nu|\boldsymbol{u}_0|^2_{H^1(\Omega)} + 2\|\boldsymbol{f}\|^2_{L^2(\Omega\times]0,t[)}\right)\right.$$
$$\left. + \frac{S_4^4}{\nu}\|z\|^2_{L^\infty(]0,t[;L^2(\Omega))}\left(\|\boldsymbol{u}_0\|^2_\alpha + \frac{S_2^2}{\nu}\|\boldsymbol{f}\|^2_{L^2(\Omega\times]0,t[)}\right)(1+\frac{{S_2}^2}{\alpha})\right). \tag{3.2.7}$$

3.2.1. *Further a priori estimates for the velocity and pressure*

When Ω is a polygon, we can sharpen the a priori estimates of the velocity and pressure part of the solutions of (3.2.1)–(3.2.2).

THEOREM 3.2.3. *Let Ω be a connected polygon. For all real numbers $T > 0$, $\alpha > 0$ and $\nu > 0$, all functions $\boldsymbol{f} \in L^2(0, T; H(\text{curl}, \Omega))$ and $\boldsymbol{u}_0 \in V^\alpha$, the velocity and pressure part of any solution $(\boldsymbol{u}, p, z)$ in $\mathcal{W} \times L^2(\Omega)$ of (3.2.1)–(3.2.2) has the following regularity*

$$\boldsymbol{u} \in H^1(0, T; W^{2,4/3}(\Omega)^2),\ p \in L^2(0, T; W^{1,4/3}(\Omega)), \tag{3.2.8}$$

and satisfies the following a priori estimates a.e. in $]0, T[$ (for simplicity, we do not detail the constants):

$$\|\boldsymbol{u}(t)\|_{W^{2,4/3}(\Omega)} \leq \|\boldsymbol{u}_0\|_{W^{2,4/3}(\Omega)} + C\left(\left\|\frac{\partial \boldsymbol{u}}{\partial t}\right\|_{L^2(\Omega\times]0,t[)}\right.$$
$$\left. + \|z\|_{L^\infty(]0,t[;L^2(\Omega))}\|\boldsymbol{u}\|_{L^\infty(0,t;H^1(\Omega)^2)} + \|\boldsymbol{f}\|_{L^2(\Omega\times]0,t[)}\right), \tag{3.2.9}$$

$$\|p\|_{L^2(0,t;W^{1,4/3}(\Omega))} \leq C\left(\left\|\frac{\partial \boldsymbol{u}}{\partial t}\right\|_{L^2(\Omega\times]0,t[)}\right.$$
$$\left. + \|z\|_{L^\infty(]0,t[;L^2(\Omega))}\|\boldsymbol{u}\|_{L^2(0,t;H^1(\Omega)^2)} + \|\boldsymbol{f}\|_{L^2(\Omega\times]0,t[)}\right), \tag{3.2.10}$$

$$\left\|\frac{\partial \boldsymbol{u}}{\partial t}\right\|_{L^2(0,t;W^{2,4/3}(\Omega))} \leq C\left(\left\|\frac{\partial \boldsymbol{u}}{\partial t}\right\|_{L^2(\Omega\times]0,t[)}\right.$$
$$\left. + \|\Delta\,\boldsymbol{u}\|_{L^2(0,t;L^{4/3}(\Omega)^2)} + \|z\|_{L^\infty(]0,t[;L^2(\Omega))}\|\boldsymbol{u}\|_{L^2(0,t;H^1(\Omega)^2)} + \|\boldsymbol{f}\|_{L^2(\Omega\times]0,t[)}\right). \tag{3.2.11}$$

If in addition, Ω is convex, then

$$\boldsymbol{u} \in H^1(0, T; H^2(\Omega)^2),\ p \in L^2(0, T; H^1(\Omega)), \tag{3.2.12}$$

and

$$\|\boldsymbol{u}(t)\|_{H^2(\Omega)} \leq C\left(\left\|\frac{\partial \boldsymbol{u}}{\partial t}\right\|_{L^2(\Omega\times]0,t[)} + \|z\|_{L^\infty(]0,t[;L^2(\Omega))} + \|\boldsymbol{f}\|_{L^2(\Omega\times]0,t[)}\right)$$
$$+ \|\boldsymbol{u}_0\|_{H^2(\Omega)}, \tag{3.2.13}$$

$$\|p\|_{L^2(0,t;H^1(\Omega))} \le C\left(\left\|\frac{\partial \boldsymbol{u}}{\partial t}\right\|_{L^2(\Omega\times]0,t[)} + \|z\|_{L^\infty(]0,t[;L^2(\Omega))} + \|\boldsymbol{f}\|_{L^2(\Omega\times]0,t[)}\right), \tag{3.2.14}$$

$$\left\|\frac{\partial \boldsymbol{u}}{\partial t}\right\|_{L^2(0,T;H^2(\Omega)^2)} \le C\left(\left\|\frac{\partial \boldsymbol{u}}{\partial t}\right\|_{L^2(\Omega\times]0,t[)} + \|\Delta\,\boldsymbol{u}\|_{L^2(0,t;L^2(\Omega)^2)} + \|z\|_{L^\infty(]0,t[;L^2(\Omega))} + \|\boldsymbol{f}\|_{L^2(\Omega\times]0,t[)}\right). \tag{3.2.15}$$

PROOF. A detailed proof can be found in SAADOUNI [2007], but it is sketched here for the reader's convenience because its argument will be used in the sequel. We set

$$\boldsymbol{w} = \alpha\frac{\partial \boldsymbol{u}}{\partial t} + \nu\,\boldsymbol{u},\ \boldsymbol{g} = \boldsymbol{f} - \frac{\partial \boldsymbol{u}}{\partial t} - z\times\boldsymbol{u}.$$

Then the pair $(\boldsymbol{w}, p)$ is the solution of the steady Stokes system, almost everywhere in $]0, T[$:

$$-\Delta\,\boldsymbol{w} + \nabla p = \boldsymbol{g},\ \operatorname{div}\boldsymbol{w} = 0 \text{ in } \Omega,\ \boldsymbol{w} = \boldsymbol{0} \text{ on } \partial\Omega. \tag{3.2.16}$$

But $\boldsymbol{g} \in L^2(0, T; L^{4/3}(\Omega)^2)$, as $\boldsymbol{f} \in L^2(\Omega\times]0, T[)^2$, $z \in L^\infty(0, T; L^2(\Omega))$, and $(\boldsymbol{u}, p) \in \mathcal{W}$. Hence Theorem 1.1.6 implies that $\boldsymbol{w}$ is in $L^2(0, T; W^{2,4/3}(\Omega)^2)$ and p in $L^2(0, T; W^{1,4/3}(\Omega))$. In turn, because $\Delta\,\boldsymbol{w}$ is in $L^2(0, T; L^{4/3}(\Omega)^2)$, recalling the definition of $\boldsymbol{w}$, we infer that

$$\frac{\partial}{\partial t}(e^{\frac{\nu}{\alpha}t}\Delta\,\boldsymbol{u}) \in L^2(0, T; L^{4/3}(\Omega)^2). \tag{3.2.17}$$

Assuming for the moment that $\boldsymbol{u}_0$ belongs to $W^{2,4/3}(\Omega)^2$, (3.2.17) implies that $\Delta\,\boldsymbol{u}$ belongs to $L^\infty(0, T; L^{4/3}(\Omega)^2)$. Then (3.2.8) follows from Theorem 1.1.1 and the definition of $\boldsymbol{w}$. It remains to prove that $\boldsymbol{u}_0$ is in $W^{2,4/3}(\Omega)^2$. This is established by an easy variant of Lemma 1.4.10. Indeed, for proving this regularity, convexity of the domain is not necessary and it can be shown that in any polygon, $\boldsymbol{u}_0 \in V^\alpha$ implies $\boldsymbol{u}_0 \in W^{2,4/3}(\Omega)^2$. In fact, this last argument proves directly that $\boldsymbol{u}$ belongs to $L^\infty(0, T; W^{2,4/3}(\Omega)^2)$, but it does not yield the regularity of p. Moreover, the above proof will be used further on because it only relies on (3.2.1) but not on (1.4.4).

Finally (3.2.12) follows easily by the same argument, because (3.2.8) implies that $z\times\boldsymbol{u}$ belongs to $L^\infty(0, T; L^2(\Omega)^2)$. All estimates are straightforward consequences of the above arguments and the continuous dependence of the solution of the Laplace and Stokes equations on their data. □

We can also establish improved a priori estimates first for z and next for $\boldsymbol{u}$, but these will be more conveniently derived from the semi-discrete scheme below.

3.2.2. A semi-discrete scheme

A solution of (3.2.1)–(3.2.2) can be constructed as the limit of the solutions of the following semi-discrete scheme. Let $N > 1$ be an integer, define the time step k by

$$k = \frac{T}{N},$$

and the subdivision points by $t_n = nk$. For each $n \geq 1$, we approximate $\boldsymbol{f}(t_n)$ by its average defined almost everywhere in Ω:

$$\boldsymbol{f}^n(\boldsymbol{x}) = \frac{1}{k}\int_{t_{n-1}}^{t_n} \boldsymbol{f}(\boldsymbol{x}, s)\,\mathrm{d}s. \tag{3.2.18}$$

We set

$$\boldsymbol{u}^0 = \boldsymbol{u}_0, \quad z^0 = z_0 = \operatorname{curl}(\boldsymbol{u}_0 - \alpha\Delta\,\boldsymbol{u}_0). \tag{3.2.19}$$

Clearly, for each n, $\boldsymbol{f}^n$ belongs to $H(\mathrm{curl}, \Omega)$ and by assumption, z^0 is in $L^2(\Omega)$. Let us solve (3.2.1)–(3.2.2) by semi-discretization in time (i.e., exact in space and discrete in time): Knowing $\boldsymbol{u}^0 \in V^\alpha$ and $z^0 \in L^2(\Omega)$, find sequences $(\boldsymbol{u}^n)_{n\geq 1}$, $(z^n)_{n\geq 1}$, and $(p^n)_{n\geq 1}$ such that $\boldsymbol{u}^n \in V$, $z^n \in L^2(\Omega)$, and $p^n \in L_0^2(\Omega)$ solve for $1 \leq n \leq N$,

$$\frac{1}{k}(\boldsymbol{u}^n - \boldsymbol{u}^{n-1}) - \alpha\frac{1}{k}\Delta(\boldsymbol{u}^n - \boldsymbol{u}^{n-1}) - \nu\Delta\,\boldsymbol{u}^n + z^{n-1} \times \boldsymbol{u}^n + \nabla p^n = \boldsymbol{f}^n \quad \text{in } \Omega, \tag{3.2.20}$$

$$\alpha\frac{1}{k}(z^n - z^{n-1}) + \nu z^n + \alpha\,\boldsymbol{u}^n \cdot \nabla z^n = \nu\operatorname{curl}\boldsymbol{u}^n + \alpha\operatorname{curl}\boldsymbol{f}^n \quad \text{in } \Omega. \tag{3.2.21}$$

Given z^{n-1} and $\boldsymbol{u}^{n-1}$, (3.2.20) is essentially a steady Stokes problem, and it is easy to check that it has a unique solution $(\boldsymbol{u}^n, p^n)$. In turn, given $\boldsymbol{u}^n$ and z^{n-1}, (3.2.21) is a steady transport equation, and owing to Proposition 1.3.9, it has a unique solution because $\operatorname{curl}\boldsymbol{f}^n$ belongs to $L^2(\Omega)$.

A priori estimates and convergence

The following proposition gives basic uniform a priori estimates for $(\boldsymbol{u}^n)_{n\geq 1}$ and $(z^n)_{n\geq 1}$. Its proof is a straightforward variant of that of (1.4.13) and (1.4.18).

PROPOSITION 3.2.4. *The sequences $(\boldsymbol{u}^n)_{n\geq 1}$ and $(z^n)_{n\geq 1}$ satisfy the following uniform a priori estimates for $1 \leq n \leq N$:*

$$\|\boldsymbol{u}^n\|_\alpha^2 + \sum_{i=1}^n \|\boldsymbol{u}^i - \boldsymbol{u}^{i-1}\|_\alpha^2 \leq \frac{S_2^2}{2\nu}\|\boldsymbol{f}\|_{L^2(\Omega\times]0,t_n[)}^2 + \|\boldsymbol{u}^0\|_\alpha^2, \tag{3.2.22}$$

$$\begin{aligned}\|z^n\|_{L^2(\Omega)}^2 + \sum_{i=1}^n \|z^i - z^{i-1}\|_{L^2(\Omega)}^2 \leq{}& \frac{S_2^2}{\alpha\nu}\|\boldsymbol{f}\|_{L^2(\Omega\times]0,t_n[)}^2 + \frac{\alpha}{\nu}\|\operatorname{curl}\boldsymbol{f}\|_{L^2(\Omega\times]0,t_n[)}^2 \\ &+ \frac{1}{\alpha}\|\boldsymbol{u}^0\|_\alpha^2 + \|z^0\|_{L^2(\Omega)}^2.\end{aligned} \tag{3.2.23}$$

As is usual for transient incompressible flow problems, an estimate for the pressure can only be obtained by deriving first an estimate for the derivative of the velocity; here this corresponds to the difference quotient of the velocity. This is the object of the next proposition. We skip the proof, which is straightforward.

PROPOSITION 3.2.5. *Let*

$$C_z = \sup_{0 \le n \le N-1} \|z^n\|_{L^2(\Omega)}.$$

The sequences $((\boldsymbol{u}^n - \boldsymbol{u}^{n-1})/k)_{n\ge1}$ and $(p^n)_{n\ge1}$ satisfy the following uniform a priori estimates for $1 \le n \le N$:

$$\sum_{i=1}^{n} \frac{1}{k}\|\boldsymbol{u}^i - \boldsymbol{u}^{i-1}\|_\alpha^2 \le \nu|\boldsymbol{u}^0|_{H^1(\Omega)}^2 + \|\boldsymbol{f}\|_{L^2(\Omega\times]0,t_n[)}^2 + \frac{1}{\alpha\nu}S_4^4 C_z^2\left(\|\boldsymbol{u}^0\|_\alpha^2 + \frac{S_2{}^2}{\nu}\|\boldsymbol{f}\|_{L^2(\Omega\times]0,t_n[)}^2\right), \tag{3.2.24}$$

$$\sum_{i=1}^{n} k\|p^i\|_{L^2(\Omega)}^2 \le \frac{3}{\beta^2}\left(S_2^2\left(\nu|\boldsymbol{u}^0|_{H^1(\Omega)}^2 + 2\|\boldsymbol{f}\|_{L^2(\Omega\times]0,t_n[)}^2\right) + C_z^2\frac{S_4^4}{\nu}\left(\|\boldsymbol{u}^0\|_\alpha^2 + \frac{S_2^2}{\nu}\|\boldsymbol{f}\|_{L^2(\Omega\times]0,t_n[)}^2\right)(1 + \frac{S_2{}^2}{\alpha})\right). \tag{3.2.25}$$

For passing to the limit in (3.2.20)–(3.2.21), it is convenient to transform the sequences $(\boldsymbol{u}^n)$, (p^n), and (z^n) into functions. Because both $(\boldsymbol{u}^n)$ and (z^n) need to be "differentiated," we define the piecewise linear functions in time as follows:

$$\forall t \in [t_n, t_{n+1}],\ \boldsymbol{u}_k(t) = \boldsymbol{u}^n + \frac{t - t_n}{k}(\boldsymbol{u}^{n+1} - \boldsymbol{u}^n),\ 0 \le n \le N-1,$$
$$\forall t \in [t_n, t_{n+1}],\ z_k(t) = z^n + \frac{t - t_n}{k}(z^{n+1} - z^n),\ 0 \le n \le N-1.$$

Next, in view of the other terms in (3.2.20)–(3.2.21), we define the step functions:

$$\forall t \in]t_n, t_{n+1}],\ \boldsymbol{f}_k(t) = \boldsymbol{f}^{n+1},\ 0 \le n \le N-1,$$
$$\forall t \in]t_n, t_{n+1}],\ \boldsymbol{w}_k(t) = \boldsymbol{u}^{n+1},\ 0 \le n \le N-1,$$
$$\forall t \in]t_n, t_{n+1}],\ p_k(t) = p^{n+1},\ 0 \le n \le N-1,$$
$$\forall t \in]t_n, t_{n+1}],\ \zeta_k(t) = z^{n+1},\ 0 \le n \le N-1,$$
$$\forall t \in [t_n, t_{n+1}[,\ \lambda_k(t) = z^n,\ 0 \le n \le N-1.$$

With this notation, (3.2.20)–(3.2.21) read

$$\frac{\partial \boldsymbol{u}_k}{\partial t} - \alpha\frac{\partial \Delta\boldsymbol{u}_k}{\partial t} - \nu\Delta\boldsymbol{w}_k + \lambda_k \times \boldsymbol{w}_k + \nabla p_k = \boldsymbol{f}_k \quad \text{a.e. in } \Omega\times]0,T[,$$
$$\alpha\frac{\partial z_k}{\partial t} + \nu\,\zeta_k + \alpha\,\boldsymbol{w}_k\cdot\nabla\zeta_k = \nu\,\mathrm{curl}\,\boldsymbol{w}_k + \alpha\,\mathrm{curl}\,\boldsymbol{f}_k \quad \text{a.e. in } \Omega\times]0,T[.$$

The uniform bounds of Propositions 3.2.4 and 3.2.5 imply that uniformly $\boldsymbol{u}_k$ and $\boldsymbol{w}_k$ are bounded in $L^\infty(0,T;H^1(\Omega)^2)$, p_k is bounded in $L^2(\Omega\times]0,T[)$, and z_k, ζ_k, and λ_k are bounded in $L^\infty(0, T; L^2(\Omega))$. Moreover, an easy calculation shows that on one hand

$$\lim_{k\to 0}(\boldsymbol{f}_k - \boldsymbol{f}) = \boldsymbol{0} \text{ strongly in } L^2(\Omega\times]0, T[)^2,$$

and on the other hand,

$$\begin{aligned}
\|\boldsymbol{w}_k - \boldsymbol{u}_k\|^2_{L^2(0,T;H^1(\Omega)^2)} &\le \frac{1}{3}k\sum_{n=1}^{N}|\boldsymbol{u}^n - \boldsymbol{u}^{n-1}|^2_{H^1(\Omega)},\\
\|\zeta_k - z_k\|^2_{L^2(\Omega\times]0,T[)} &\le \frac{1}{3}k\sum_{n=1}^{N}\|z^n - z^{n-1}\|^2_{L^2(\Omega)},\\
\|\lambda_k - z_k\|^2_{L^2(\Omega\times]0,T[)} &\le \frac{1}{3}k\sum_{n=1}^{N}\|z^n - z^{n-1}\|^2_{L^2(\Omega)}.
\end{aligned} \tag{3.2.26}$$

The following convergence results are established in GIRAULT and SAADOUNI [2007].

PROPOSITION 3.2.6. *There exist functions $\boldsymbol{u} \in H^1(0, T; V)$, $p \in L^2(0, T; L^2_0(\Omega))$ and $z \in L^\infty(0, T; L^2(\Omega))$ such that a subsequence of k, still denoted by k, satisfies:*

$$\begin{aligned}
&\lim_{k\to 0}\boldsymbol{u}_k = \lim_{k\to 0}\boldsymbol{w}_k = \boldsymbol{u} \text{ weakly * in } L^\infty(0, T; V),\\
&\lim_{k\to 0} z_k = \lim_{k\to 0}\zeta_k = \lim_{k\to 0}\lambda_k = z \text{ weakly * in } L^\infty(0, T; L^2(\Omega)),\\
&\lim_{k\to 0} p_k = p \text{ weakly in } L^2(0, T; L^2_0(\Omega)),\\
&\lim_{k\to 0}\frac{\partial}{\partial t}\boldsymbol{u}_k = \frac{\partial}{\partial t}\boldsymbol{u} \text{ weakly in } L^2(0, T; V).
\end{aligned}$$

Furthermore,

$$\begin{aligned}
&\lim_{k\to 0}(\boldsymbol{w}_k - \boldsymbol{u}_k) = \boldsymbol{0} \text{ strongly in } L^2(0, T; H^1(\Omega)^2),\\
&\lim_{k\to 0}(\zeta_k - z_k) = \lim_{k\to 0}(\lambda_k - z_k) = 0 \text{ strongly in } L^2(\Omega\times]0, T[),\\
&\lim_{k\to 0}\boldsymbol{u}_k = \boldsymbol{u} \text{ strongly in } L^2(0, T; L^4(\Omega)^2).
\end{aligned} \tag{3.2.27}$$

Again, we skip the proof, which is straightforward. In particular, the strong convergence of $\boldsymbol{u}_k$ in (3.2.27) follows from the fact that $(\boldsymbol{u}_k)$ is bounded uniformly in $H^1(0, T; H^1_0(\Omega)^2)$ and because the imbedding of $H^1(\Omega)$ into $L^4(\Omega)$ is compact, Theorem 3.1.4 implies that $\boldsymbol{u}_k$ converges strongly to $\boldsymbol{u}$ in $L^2(0, T; L^4(\Omega)^2)$.

Proposition 3.2.6 is sufficient to establish existence of a solution of problem (3.1.1), but beforehand we prove sharper estimates for the semi-discrete solution.

Further a priori estimates for the semi-discrete solution

First, we show that the solution $(\boldsymbol{u}^n, p^n)$ of (3.2.20) satisfies the analogs of (3.2.9)–(3.2.11), and (3.2.13)–(3.2.15).

THEOREM 3.2.7. *If Ω is a connected polygon, then all $\boldsymbol{u}^n$ belong to $W^{2,4/3}(\Omega)^2$, all p^n belong to $W^{1,4/3}(\Omega)$, and there exists a constant C_1, independent of n and k, such that*

$$\sup_{1\le n\le N} \|\boldsymbol{u}^n\|_{W^{2,4/3}(\Omega)} + \left(\sum_{n=1}^{N} k|p^n|^2_{W^{1,4/3}(\Omega)}\right)^{1/2} + \left(\sum_{n=1}^{N} \frac{1}{k}\|\boldsymbol{u}^n - \boldsymbol{u}^{n-1}\|^2_{W^{2,4/3}(\Omega)}\right)^{1/2} \le C_1. \tag{3.2.28}$$

Hence the following weak limits hold up to subsequences:

$$\lim_{k\to 0} \boldsymbol{u}_k = \boldsymbol{u} \text{ weakly in } H^1(0,T;\, W^{2,4/3}(\Omega)^2),$$

$$\lim_{k\to 0} p_k = p \text{ weakly in } L^2(0,T;\, W^{1,4/3}(\Omega)).$$

If in addition, Ω is convex, then all $\boldsymbol{u}^n$ belong to $H^2(\Omega)^2$, all p^n belong $H^1(\Omega)$, and there exists a constant C_2, independent of n and k, such that

$$\sup_{1\le n\le N} \|\boldsymbol{u}^n\|_{H^2(\Omega)} + \left(\sum_{n=1}^{N} k|p^n|^2_{H^1(\Omega)}\right)^{1/2} + \left(\sum_{n=1}^{N} \frac{1}{k}\|\boldsymbol{u}^n - \boldsymbol{u}^{n-1}\|^2_{H^2(\Omega)}\right)^{1/2} \le C_2. \tag{3.2.29}$$

Thus, up to subsequences,

$$\lim_{k\to 0} \boldsymbol{u}_k = \boldsymbol{u} \text{ weakly in } H^1(0,T;\, H^2(\Omega)^2),$$

$$\lim_{k\to 0} p_k = p \text{ weakly in } L^2(0,T;\, H^1(\Omega)).$$

PROOF. The proof is a semi-discrete analog of that of Theorem 3.2.3. For $1 \le n \le N$, set

$$\boldsymbol{\varphi}^n = \frac{\alpha}{k}(\boldsymbol{u}^n - \boldsymbol{u}^{n-1}) + \nu\, \boldsymbol{u}^n.$$

By (3.2.20), for each n the pair $(\boldsymbol{\varphi}^n, p^n)$ solves a Stokes problem with data $\boldsymbol{g}^n$ defined by

$$\boldsymbol{g}^n = -\frac{1}{k}(\boldsymbol{u}^n - \boldsymbol{u}^{n-1}) - z^{n-1} \times \boldsymbol{u}^n + \boldsymbol{f}^n,$$

to be specific, the pair $(\boldsymbol{\varphi}^n, p^n) \in V \times L^2_0(\Omega)$ satisfies

$$-\Delta\, \boldsymbol{\varphi}^n + \nabla\, p^n = \boldsymbol{g}^n \ \text{ in } \Omega. \tag{3.2.30}$$

But for each n, $\boldsymbol{g}^n$ belongs to $L^{4/3}(\Omega)^2$:

$$\|\boldsymbol{g}^n\|_{L^{4/3}(\Omega)} \le |\Omega|^{1/4}\frac{1}{k}\|\boldsymbol{u}^n - \boldsymbol{u}^{n-1}\|_{L^2(\Omega)} + S_4 C_z |\boldsymbol{u}^n|_{H^1(\Omega)} + |\Omega|^{1/4}\|\boldsymbol{f}^n\|_{L^2(\Omega)}. \tag{3.2.31}$$

Hence Theorem 1.1.6 implies that $\boldsymbol{\varphi}^n$ is in $W^{2,4/3}(\Omega)^2$, and there exists a constant c_1, such that

$$\|\boldsymbol{\varphi}^n\|_{W^{2,4/3}(\Omega)} + |p^n|_{W^{1,4/3}(\Omega)} \leq c_1 \|\boldsymbol{g}^n\|_{L^{4/3}(\Omega)}, \tag{3.2.32}$$

where all constants c_i below are independent of n and k. Therefore, combining the expression of $\boldsymbol{\varphi}^n$ with (3.2.31) substituted into (3.2.32), we obtain

$$\left(\frac{\alpha}{k} + \nu\right) \|\boldsymbol{u}^n\|_{W^{2,4/3}(\Omega)} \leq \frac{\alpha}{k} \|\boldsymbol{u}^{n-1}\|_{W^{2,4/3}(\Omega)} + c_1 \left(|\Omega|^{1/4} \frac{1}{k} \|\boldsymbol{u}^n - \boldsymbol{u}^{n-1}\|_{L^2(\Omega)} \right.$$
$$\left. + S_4 C_z |\boldsymbol{u}^n|_{H^1(\Omega)} + |\Omega|^{1/4} \|\boldsymbol{f}^n\|_{L^2(\Omega)} \right).$$

By summing over n and multiplying by k, this yields for $1 \leq n \leq N$,

$$\|\boldsymbol{u}^n\|_{W^{2,4/3}(\Omega)} \leq \|\boldsymbol{u}_0\|_{W^{2,4/3}(\Omega)} + \frac{c_1}{\alpha}\sqrt{T} \left(|\Omega|^{1/4} \left(\sum_{i=1}^{n} \frac{1}{k} \|\boldsymbol{u}^i - \boldsymbol{u}^{i-1}\|^2_{L^2(\Omega)} \right)^{1/2} \right.$$
$$\left. + \sqrt{T} S_4 C_z \sup_{1 \leq i \leq n} |\boldsymbol{u}^i|_{H^1(\Omega)} + |\Omega|^{1/4} \|\boldsymbol{f}\|_{L^2(\Omega \times]0,t_n[)} \right). \tag{3.2.33}$$

Then the first part of (3.2.28) follows by substituting (3.2.22) and (3.2.24) into (3.2.33). Similarly, we easily derive for $1 \leq n \leq N$,

$$\sum_{i=1}^{n} k |p^i|^2_{W^{1,4/3}(\Omega)} \leq 3c_1^2 \left(|\Omega|^{1/2} \sum_{i=1}^{n} \frac{1}{k} \|\boldsymbol{u}^i - \boldsymbol{u}^{i-1}\|^2_{L^2(\Omega)} \right.$$
$$\left. + S_4^2 C_z^2 \sum_{i=1}^{n} k |\boldsymbol{u}^i|^2_{H^1(\Omega)} + |\Omega|^{1/2} \|\boldsymbol{f}\|^2_{L^2(\Omega \times]0,t_n[)} \right). \tag{3.2.34}$$

The bound for the difference quotient stems from (3.2.32)–(3.2.34):

$$\sum_{i=1}^{n} \frac{1}{k} \|\boldsymbol{u}^i - \boldsymbol{u}^{i-1}\|^2_{W^{2,4/3}(\Omega)} \leq \frac{4}{\alpha^2} c_1^2 \left(|\Omega|^{1/2} \sum_{i=1}^{n} \frac{1}{k} \|\boldsymbol{u}^i - \boldsymbol{u}^{i-1}\|^2_{L^2(\Omega)} \right.$$
$$\left. + \nu^2 \sum_{i=1}^{n} k \|\Delta \boldsymbol{u}^i\|^2_{L^{4/3}(\Omega)} + S_4^2 C_z^2 \sum_{i=1}^{n} k |\boldsymbol{u}^i|^2_{H^1(\Omega)} + |\Omega|^{1/2} \|\boldsymbol{f}\|^2_{L^2(\Omega \times]0,t_n[)} \right).$$

Next, (3.2.28) implies that $\boldsymbol{u}^n$ is bounded in L^∞, for $1 \leq n \leq N$,

$$\|\boldsymbol{u}^n\|_{L^\infty(\Omega)} \leq C_{\boldsymbol{u}}. \tag{3.2.35}$$

Hence for each n, $\boldsymbol{g}^n$ belongs to $L^2(\Omega)^2$:

$$\|\boldsymbol{g}^n\|_{L^2(\Omega)} \leq \frac{1}{k} \|\boldsymbol{u}^n - \boldsymbol{u}^{n-1}\|_{L^2(\Omega)} + C_{\boldsymbol{u}} C_z + \|\boldsymbol{f}^n\|_{L^2(\Omega)},$$

and if Ω is convex, Theorem 1.1.5 implies that $\boldsymbol{\varphi}^n$ belongs to $H^2(\Omega)^2$, and there exists a constant c_2, such that

$$\|\boldsymbol{\varphi}^n\|_{H^2(\Omega)} + |p^n|_{H^1(\Omega)} \leq c_2 \|\boldsymbol{g}^n\|_{L^2(\Omega)}. \tag{3.2.36}$$

Thus, we conclude as above, for $1 \leq n \leq N$,

$$\|\boldsymbol{u}^n\|_{H^2(\Omega)} \leq \frac{c_2}{\alpha}\sqrt{T}\left(\left(\sum_{i=1}^{n}\frac{1}{k}\|\boldsymbol{u}^i - \boldsymbol{u}^{i-1}\|^2_{L^2(\Omega)}\right)^{1/2} + C_{\boldsymbol{u}} C_z \sqrt{T} + \|\boldsymbol{f}\|_{L^2(\Omega\times]0,t_n[)}\right) + \|\boldsymbol{u}_0\|_{H^2(\Omega)}.$$

Likewise, we have for $1 \leq n \leq N$,

$$\sum_{i=1}^{n} k|p^i|^2_{H^1(\Omega)} \leq 3c_2^2 \left(\sum_{i=1}^{n}\frac{1}{k}\|\boldsymbol{u}^i - \boldsymbol{u}^{i-1}\|^2_{L^2(\Omega)} + C_{\boldsymbol{u}}^2 C_z^2 T + \|\boldsymbol{f}\|^2_{L^2(\Omega\times]0,t_n[)}\right).$$

Finally,

$$\sum_{i=1}^{n}\frac{1}{k}\|\boldsymbol{u}^i - \boldsymbol{u}^{i-1}\|^2_{H^2(\Omega)} \leq \frac{4}{\alpha^2}c_2^2\left(\sum_{i=1}^{n}\frac{1}{k}\|\boldsymbol{u}^i - \boldsymbol{u}^{i-1}\|^2_{L^2(\Omega)} + \nu^2\sum_{i=1}^{n} k\|\Delta\,\boldsymbol{u}^i\|^2_{L^2(\Omega)} + C_{\boldsymbol{u}}^2 C_z^2 T + \|\boldsymbol{f}\|^2_{L^2(\Omega\times]0,t_n[)}\right).$$

Then (3.2.29) follows from these three bounds. □

Without further information on z^n, (3.2.29) cannot be improved, because it reduces to a Stokes system with data in L^2. Therefore, we now prove that z^n satisfies the analog of (1.3.25).

PROPOSITION 3.2.8. *Let Ω be a connected polygon and $r \in [2,4]$. If $z_0 \in L^r(\Omega)$ and $\operatorname{curl} \boldsymbol{f} \in L^2(0,T;L^r(\Omega))$, then the solution z^n of (3.2.21) belongs to $L^r(\Omega)$ and there exists a constant C_3, independent of n and k, such that*

$$\sup_{1\leq n\leq N} \|z^n\|_{L^r(\Omega)} \leq C_3. \tag{3.2.37}$$

Therefore,

$$\lim_{k\to 0} z_k = z \textit{ weakly } * \textit{ in } L^\infty(0,T;L^r(\Omega)).$$

PROOF. The proof is a simpler version of that of the preceding theorem. Let $n \geq 1$; with the notation

$$\ell^n = \frac{\alpha}{k}z^{n-1} + \nu \operatorname{curl} \boldsymbol{u}^n + \alpha \operatorname{curl} \boldsymbol{f}^n,$$

(3.2.21) reduces to the transport equation

$$\left(\frac{\alpha}{k}+\nu\right) z^n + \alpha\, \boldsymbol{u}^n \cdot \nabla z^n = \ell^n. \tag{3.2.38}$$

Now, we proceed by induction. Assuming that z^{n-1} is in $L^r(\Omega)$, which is true for z^0, it follows from (3.2.28) and the imbedding of $W^{2,4/3}$ into $W^{1,4}$, that ℓ^n is in $L^r(\Omega)$, and

$$\|\ell^n\|_{L^r(\Omega)} \leq \frac{\alpha}{k}\|z^{n-1}\|_{L^r(\Omega)} + \sqrt{2}\nu\, |\boldsymbol{u}^n|_{W^{1,r}(\Omega)} + \alpha\, \|\operatorname{curl} \boldsymbol{f}^n\|_{L^r(\Omega)}. \tag{3.2.39}$$

Therefore, (1.3.25) yields in particular

$$\alpha\|z^n\|_{L^r(\Omega)} \leq \alpha\|z^{n-1}\|_{L^r(\Omega)} + \sqrt{2}\nu k|\boldsymbol{u}^n|_{W^{1,r}(\Omega)} + \alpha k\|\operatorname{curl} \boldsymbol{f}^n\|_{L^r(\Omega)}.$$

Then summing over n, we obtain, for $1 \leq n \leq N$,

$$\|z^n\|_{L^r(\Omega)} \leq \|z^0\|_{L^r(\Omega)} + \sqrt{2}\frac{\nu}{\alpha}T \sup_{1\leq i\leq n} |\boldsymbol{u}^i|_{W^{1,r}(\Omega)} + \sqrt{T}\|\operatorname{curl} \boldsymbol{f}\|_{L^2(0,t_n;L^r(\Omega)^2)}, \tag{3.2.40}$$

whence (3.2.37). □

This result permits to improve the statement of Theorem 3.2.7 and derive in particular a bound for $\boldsymbol{u}^n$ in $W^{1,\infty}$.

COROLLARY 3.2.9. *Let Ω be a convex polygon, and let $r_\Omega > 2$ be the number defined in Theorem 1.1.9. If, for some $r \in [2, r_\Omega]$, $\boldsymbol{f} \in L^2(0, T; L^r(\Omega)^2)$,* curl $\boldsymbol{f} \in L^2(0, T; L^r(\Omega))$, *and $z^0 \in L^r(\Omega)$, then all pairs $(\boldsymbol{u}^n, p^n)$ belong to $W^{2,r}(\Omega)^2 \times W^{1,r}(\Omega)$ for $1 \leq n \leq N$, and there exists a constant C_4, independent of n and k, such that*

$$\sup_{1\leq i\leq N} \|\boldsymbol{u}^n\|_{W^{2,r}(\Omega)} + \left(\sum_{n=1}^{N} k|p^n|^2_{W^{1,r}(\Omega)}\right)^{1/2} + \left(\sum_{n=1}^{N} \frac{1}{k}\|\boldsymbol{u}^n - \boldsymbol{u}^{n-1}\|^2_{W^{2,r}(\Omega)}\right)^{1/2} \leq C_4. \tag{3.2.41}$$

Thus,

$$\lim_{k\to 0} \boldsymbol{u}_k = \boldsymbol{u} \text{ weakly in } H^1(0, T; W^{2,r}(\Omega)^2),$$

$$\lim_{k\to 0} p_k = p \text{ weakly in } L^2(0, T; W^{1,r}(\Omega)).$$

PROOF. The proof follows the lines of Theorem 3.2.7. With the same notation, we have $\boldsymbol{g}^n$ in $L^r(\Omega)^2$ for $1 \leq n \leq N$:

$$\|\boldsymbol{g}^n\|_{L^r(\Omega)} \leq \frac{S_r}{k}|\boldsymbol{u}^n - \boldsymbol{u}^{n-1}|_{H^1(\Omega)} + \|z^{n-1}\|_{L^r(\Omega)}\|\boldsymbol{u}^n\|_{L^\infty(\Omega)} + \|\boldsymbol{f}^n\|_{L^r(\Omega)}.$$

Thus, Theorem 1.1.9 implies that there exists a constant c, such that

$$\|\boldsymbol{u}^n\|_{W^{2,r}(\Omega)} \leq \|\boldsymbol{u}^{n-1}\|_{W^{2,r}(\Omega)} + c\frac{k}{\alpha}\left(\frac{S_r}{k}|\boldsymbol{u}^n - \boldsymbol{u}^{n-1}|_{H^1(\Omega)} + C_{\boldsymbol{u}}\|z^{n-1}\|_{L^r(\Omega)} + \|\boldsymbol{f}^n\|_{L^r(\Omega)}\right),$$

where C_u is the constant of (3.2.35). By summing, we readily obtain for $1 \leq n \leq N$,

$$\|u^n\|_{W^{2,r}(\Omega)} \leq \|u^0\|_{W^{2,r}(\Omega)} + \frac{c}{\alpha}\sqrt{T}\left(S_r\left(\sum_{i=1}^{n}\frac{1}{k}|u^i - u^{i-1}|^2_{H^1(\Omega)}\right)^{1/2}\right. \\ \left. + C_u\sqrt{T}\Big(\sup_{0\leq i\leq n-1}\|z^i\|_{L^r(\Omega)}\Big) + \|f\|_{L^2(0,t_n;L^r(\Omega)^2)}\right). \tag{3.2.42}$$

Similarly, we easily derive that

$$\sum_{i=1}^{n} k|p^i|^2_{W^{1,r}(\Omega)} \leq 3c^2\left(S_r^2\sum_{i=1}^{n}\frac{1}{k}|u^i - u^{i-1}|^2_{H^1(\Omega)}\right. \\ \left. + C_u^2 T\Big(\sup_{0\leq i\leq n-1}\|z^i\|^2_{L^r(\Omega)}\Big) + \|f\|^2_{L^2(0,t_n;L^r(\Omega)^2)}\right). \tag{3.2.43}$$

Finally,

$$\sum_{i=1}^{n}\frac{1}{k}\|u^i - u^{i-1}\|^2_{W^{2,r}(\Omega)} \leq \frac{4}{\alpha^2}c^2\left(S_r^2\sum_{i=1}^{n}\frac{1}{k}|u^i - u^{i-1}|^2_{H^1(\Omega)} + \nu^2\sum_{i=1}^{n}k\|\Delta u^i\|^2_{L^r(\Omega)}\right. \\ \left. + C_u^2 T\left(\sup_{0\leq i\leq n-1}\|z^i\|^2_{L^r(\Omega)}\right) + \|f\|^2_{L^2(0,t_n;L^r(\Omega)^2)}\right). \tag{3.2.44}$$

Then (3.2.41) follows from (3.2.42)–(3.2.44), (3.2.37), and (3.2.24). □

As a consequence, there exists a constant $C_{\nabla u}$, independent of n and k, such that

$$\sup_{1\leq n\leq N}\|\nabla u^n\|_{L^\infty(\Omega)} \leq C_{\nabla u}. \tag{3.2.45}$$

For sufficiently smooth data, this uniform bound enables to derive bounds for ∇z^n and the difference quotient of z^n.

THEOREM 3.2.10. *In addition to the assumptions of Corollary 3.2.9, suppose that z^0 is in $H^1(\Omega)$ and* curl f *is in $L^2(0, T; H^1(\Omega))$.*

(1) Then there exists a real number $k_0 > 0$ such that for all $k \leq k_0$ and all n with $1 \leq n \leq N$, z^n belongs to $H^1(\Omega)$ and

$$\sup_{1\leq n\leq N}|z^n|_{H^1(\Omega)} + \left(\sum_{n=1}^{N}\frac{1}{k}\|z^n - z^{n-1}\|^2_{L^2(\Omega)}\right)^{1/2} \leq C_5, \tag{3.2.46}$$

with a constant C_5, independent of n and k. Therefore,

$$\lim_{k\to 0} z_k = z \text{ weakly in } H^1(0, T; L^2(\Omega)) \text{ and weakly * in } L^\infty(0, T; H^1(\Omega)).$$

(2) *If in addition,* $z^0 \in W^{1,r}(\Omega)$ *and* $\mathbf{curl} \boldsymbol{f} \in W^{1,r}(\Omega)$ *for some* $r \in]2, r_\Omega[$, *then for all* $k \le k_0$, *we have* $z^n \in W^{1,r}(\Omega)$ *for* $1 \le n \le N$ *and*

$$\sup_{1\le n\le N} |z^n|_{W^{1,r}(\Omega)} + \left(\sum_{n=1}^{N} \frac{1}{k}\|z^n - z^{n-1}\|^2_{L^r(\Omega)}\right)^{1/2} \le C_6, \tag{3.2.47}$$

with a constant C_6, *independent of n and k. Similarly,*

$$\lim_{k\to 0} z_k = z \text{ weakly in } H^1(0,T; L^r(\Omega)) \text{ and weakly * in } L^\infty(0,T; W^{1,r}(\Omega)).$$

PROOF. (1) We argue by induction. For $n \ge 1$, assume that $z^{n-1} \in H^1(\Omega)$, which is true for z^0. Then, in view of (3.2.38) and Theorem 1.3.18, it follows from the regularity of $\boldsymbol{f}^n$, (3.2.45), and the positivity of ν that $z^n \in H^1(\Omega)$ provided

$$k \le \frac{1}{C_{\nabla \boldsymbol{u}}}. \tag{3.2.48}$$

Let us make this assumption for the moment. As (3.2.48) is independent of n, it follows by induction that indeed $z^n \in H^1(\Omega)$ for $0 \le n \le N$ and ∇z^n is the unique solution of the transport equation

$$\left(\frac{\alpha}{k} + \nu\right) \boldsymbol{w}^n + \alpha\, \boldsymbol{u}^n \cdot \nabla \boldsymbol{w}^n + \alpha\, (\nabla \boldsymbol{u}^n)^T \boldsymbol{w}^n = \boldsymbol{\ell}^n, \tag{3.2.49}$$

where

$$\boldsymbol{\ell}^n = \frac{\alpha}{k} \nabla z^{n-1} + \nabla \left(\mathrm{curl}(\nu\, \boldsymbol{u}^n + \alpha \boldsymbol{f}^n)\right). \tag{3.2.50}$$

Then taking the scalar product of (3.2.49) with $\boldsymbol{w}^n$, and applying Young's inequality, we derive

$$|z^n|^2_{H^1(\Omega)} - |z^{n-1}|^2_{H^1(\Omega)} + |z^n - z^{n-1}|^2_{H^1(\Omega)} \le 2k\|\nabla \boldsymbol{u}^n\|_{L^\infty(\Omega)} |z^n|^2_{H^1(\Omega)}$$
$$+ \frac{\nu}{\alpha} k |\mathrm{curl}\, \boldsymbol{u}^n|^2_{H^1(\Omega)} + \frac{\alpha}{\nu} k |\mathrm{curl}\, \boldsymbol{f}^n|^2_{H^1(\Omega)}.$$

Now, we write

$$|z^n|^2_{H^1(\Omega)} \le 2|z^n - z^{n-1}|^2_{H^1(\Omega)} + 2|z^{n-1}|^2_{H^1(\Omega)},$$

and we sharpen (3.2.48) by assuming that

$$k \le \frac{1}{4C_{\nabla \boldsymbol{u}}}. \tag{3.2.51}$$

By summing over n, this yields

$$|z^n|^2_{H^1(\Omega)} + \sum_{i=1}^{n-1} |z^i - z^{i-1}|^2_{H^1(\Omega)} \le |z^0|^2_{H^1(\Omega)} + 4\, C_{\nabla \boldsymbol{u}} \sum_{i=0}^{n-1} k |z^i|^2_{H^1(\Omega)}$$
$$+ \frac{2\nu}{\alpha} \sum_{i=1}^{n} k |\boldsymbol{u}^i|^2_{H^2(\Omega)} + \frac{\alpha}{\nu} \|\mathbf{curl}\, \boldsymbol{f}\|^2_{L^2(0,t_n; H^1(\Omega))}.$$

Hence the discrete Gronwall's Lemma 3.1.2 gives for $1 \leq n \leq N$

$$\sup_{1\leq i\leq n} |z^i|^2_{H^1(\Omega)} \leq \left(|z^0|^2_{H^1(\Omega)} + \frac{2\nu}{\alpha} \sum_{i=1}^{n} k|\boldsymbol{u}^i|^2_{H^2(\Omega)} + \frac{\alpha}{\nu} \|\text{curl}\, \boldsymbol{f}\|^2_{L^2(0,t_n;H^1(\Omega))} \right) \times \exp(4C_{\nabla \boldsymbol{u}} t_n). \tag{3.2.52}$$

In turn, this uniform bound for ∇z^n yields a bound for the difference quotient:

$$\sum_{i=1}^{n} \frac{1}{k} \|z^i - z^{i-1}\|^2_{L^2(\Omega)} \leq \frac{\nu}{\alpha} \|z^0\|^2_{L^2(\Omega)} + 3 \left(\frac{\nu^2}{\alpha^2} \sum_{i=1}^{n} k|\boldsymbol{u}^i|^2_{H^1(\Omega)} + C^2_{\boldsymbol{u}} \sum_{i=1}^{n} k|z^i|^2_{H^1(\Omega)} + \|\text{curl}\, \boldsymbol{f}\|^2_{L^2(\Omega\times]0,t_n[)} \right).$$

(2) Similarly, we infer by induction from Theorem 1.3.19, Corollary 3.2.9, and (3.2.48) that z^n belongs to $W^{1,r}(\Omega)$ for $1 \leq n \leq N$. Hence with the above notation, the function $|\boldsymbol{w}^n|^{r-2}\boldsymbol{w}^n$ belongs to $L^{r'}(\Omega)$ where $1/r + 1/r' = 1$. Therefore, by taking the scalar product of (3.2.49) and (3.2.50) with this function, and applying (3.2.45), we derive

$$\left(1 - kC_{\nabla \boldsymbol{u}} + \frac{\nu}{\alpha} k\right) \|\boldsymbol{w}^n\|_{L^r(\Omega)} \leq \|\boldsymbol{w}^{n-1}\|_{L^r(\Omega)} + k\left(\frac{\nu}{\alpha} |\text{curl}\, \boldsymbol{u}^n|_{W^{1,r}(\Omega)} + |\text{curl}\, \boldsymbol{f}^n|_{W^{1,r}(\Omega)}\right).$$

Now, assuming (3.2.51), we have

$$1 - kC_{\nabla \boldsymbol{u}} \geq \frac{3}{4},$$

and the above inequality reduces to

$$\|\boldsymbol{w}^n\|_{L^r(\Omega)} \leq \left(1 + \frac{4}{3} kC_{\nabla \boldsymbol{u}}\right) \|\boldsymbol{w}^{n-1}\|_{L^r(\Omega)} + \frac{4}{3} k \left(\frac{\nu}{\alpha} |\text{curl}\, \boldsymbol{u}^n|_{W^{1,r}(\Omega)} + |\text{curl}\, \boldsymbol{f}^n|_{W^{1,r}(\Omega)}\right).$$

Then, it suffices to apply Lemma 3.1.3 with

$$A = \frac{4}{3} kC_{\nabla \boldsymbol{u}},\ b_n = \frac{4}{3} k \left(\frac{\nu}{\alpha} |\text{curl}\, \boldsymbol{u}^n|_{W^{1,r}(\Omega)} + |\text{curl}\, \boldsymbol{f}^n|_{W^{1,r}(\Omega)}\right).$$

More precisely, we easily derive that

$$\sum_{i=1}^{n} b_i (1 + A)^{n-i} \leq \exp\left(\frac{4}{3} C_{\nabla \boldsymbol{u}} t_n\right) \left(\frac{\nu}{\alpha C_{\nabla \boldsymbol{u}}} \sup_{1\leq i\leq n} |\text{curl}\, \boldsymbol{u}^i|_{W^{1,r}(\Omega)} + \frac{2}{\sqrt{6C_{\nabla \boldsymbol{u}}}} \|\text{curl}\, \boldsymbol{f}\|_{L^2(0,t_n;W^{1,r}(\Omega))} \right),$$

and Lemma 3.1.3 yields for $1 \leq n \leq N$,

$$|z^n|_{W^{1,r}(\Omega)} \leq \exp\left(\frac{4}{3} C_{\nabla \boldsymbol{u}} t_n\right) \left(|z^0|_{W^{1,r}(\Omega)} + \frac{2}{\sqrt{6C_{\nabla \boldsymbol{u}}}} \|\text{curl}\, \boldsymbol{f}\|_{L^2(0,t_n;W^{1,r}(\Omega))} + \frac{\nu}{\alpha C_{\nabla \boldsymbol{u}}} \sup_{1\leq i\leq n} |\text{curl}\, \boldsymbol{u}^i|_{W^{1,r}(\Omega)} \right). \tag{3.2.53}$$

This gives a bound for the difference quotient in L^r:

$$\sum_{i=1}^{n} \frac{1}{k} \|z^i - z^{i-1}\|^2_{L^r(\Omega)} \leq 4 \left(\frac{\nu^2}{\alpha^2} \sum_{i=1}^{n} k \|z^i\|^2_{L^r(\Omega)} + C_{\boldsymbol{u}}^2 \sum_{i=1}^{n} k |z^i|^2_{W^{1,r}(\Omega)} \right.$$
$$\left. + \frac{\nu^2}{\alpha^2} \sum_{i=1}^{n} k \|\mathrm{curl}\, \boldsymbol{u}^i\|^2_{L^r(\Omega)} + \|\mathrm{curl}\, \boldsymbol{f}\|^2_{L^2(0,t_n;L^r(\Omega))} \right). \qquad \square$$

Another a priori estimate from an application to the transport equation
Let us revert to the time-dependent transport equations (3.1.14) and (3.1.12). In the statement of Proposition 3.2.8, the value of the exponent r is restricted in order to guarantee that the right-hand side of (3.2.21) is in $L^r(\Omega)$. But when the data f of (3.1.14) belongs to $L^r(\Omega)$, this restriction is no longer necessary. Then, applying the above argument to a semi-discrete scheme for (3.1.14) and (3.1.12), and passing to the limit, the next result follows easily from Proposition 3.2.8 and Theorem 3.1.6.

PROPOSITION 3.2.11. *Let $\Omega \subset \mathbb{R}^d$ be bounded and Lipschitz-continuous and let $2 < r < \infty$. For any real numbers $\gamma \neq 0$ and $\nu > 0$, and any functions $\boldsymbol{u}$ in $L^2(0,T;W)$, f in $L^2(0,T;L^r(\Omega))$, and z_0 in $L^r(\Omega)$, the solution z of problems (3.1.14), (3.1.12) belongs to $L^\infty(0,T;L^r(\Omega))$ and it satisfies*

$$\|z\|_{L^\infty(0,T;L^r(\Omega))} \leq \|z_0\|_{L^r(\Omega)} + \sqrt{T}\|f\|_{L^2(0,T;L^r(\Omega))}. \tag{3.2.54}$$

Similarly, Theorems 3.2.10 and 3.1.6 yield the following proposition.

PROPOSITION 3.2.12. *Let $d = 2, 3$ and let $\Omega \subset \mathbb{R}^d$ be a bounded convex polygon or polyhedron. For any real numbers $T > 0$, $\gamma \neq 0$, and $\nu > 0$, and any functions $\boldsymbol{u} \in L^2(0,T;W \cap W^{1,\infty}(\Omega)^d)$, $f \in L^2(0,T;H^1(\Omega))$, and $z_0 \in H^1(\Omega)$, the unique solution z of (3.1.14)–(3.1.12) belongs to $L^\infty(0,T;H^1(\Omega)) \cap H^1(0,T;L^2(\Omega))$ and*

$$\|z\|_{L^\infty(0,T;H^1(\Omega))} \leq \left(|z_0|^2_{H^1(\Omega)} + \frac{1}{2\nu}\|f\|^2_{L^2(0,T;H^1(\Omega))} \right)^{1/2} \exp(c_1|\gamma|T),$$
$$\left\|\frac{\partial z}{\partial t}\right\|_{L^2(\Omega\times]0,T[)} \leq \left(\nu\|z_0\|^2_{L^2(\Omega)} + 2\|f\|^2_{L^2(\Omega\times]0,T[)} + 2\gamma^2 c_2^2 \|z\|^2_{L^\infty(0,T;H^1(\Omega))} \right)^{1/2}, \tag{3.2.55}$$

where

$$c_1 = \|\nabla \boldsymbol{u}\|_{L^\infty(\Omega\times]0,T[)},\quad c_2 = \|\boldsymbol{u}\|_{L^2(0,T;L^\infty(\Omega)^d)}.$$

Let $2 < r < \infty$. If in addition, $f \in L^2(0,T;W^{1,r}(\Omega))$, and $z_0 \in W^{1,r}(\Omega)$, the unique solution z of (3.1.14), (3.1.12) belongs to $L^\infty(0,T;W^{1,r}(\Omega)) \cap H^1(0,T;L^r(\Omega))$ and

$$\|z\|_{L^\infty(0,T;W^{1,r}(\Omega))} \leq \left(|z_0|_{W^{1,r}(\Omega)} + \frac{1}{\nu}\|f\|_{L^r(0,T;W^{1,r}(\Omega))} \right) \exp(c_1|\gamma|T), \tag{3.2.56}$$

$$\left\| \frac{\partial z}{\partial t} \right\|_{L^2(0,T;L^r(\Omega))} \leq \sqrt{3} \left(\nu^2 \|z\|^2_{L^2(0,T;L^r(\Omega))} + c_2^2 \gamma^2 \|z\|^2_{L^\infty(0,T;W^{1,r}(\Omega))} + \|f\|^2_{L^2(0,T;L^r(\Omega))} \right)^{1/2}. \quad (3.2.57)$$

When the first part of Theorem 3.2.7 is applied to (3.2.1) and Proposition 3.2.11 is applied to (3.2.2), we obtain immediately the following a priori bound for the solution of (3.2.1)–(3.2.2).

PROPOSITION 3.2.13. *Let $r \in [2,4]$, and let Ω be a connected polygon. If $z_0 \in L^r(\Omega)$ and $\operatorname{curl} \boldsymbol{f} \in L^2(0,T;L^r(\Omega))$, then all solutions $(\boldsymbol{u}, p, z)$ of (3.2.1)–(3.2.2) satisfy $z \in L^\infty(0,T;L^r(\Omega))$ and a.e. in $]0,t[$,*

$$\|z(t)\|_{L^r(\Omega)} \leq \|z_0\|_{L^r(\Omega)} + \sqrt{2}\frac{\nu}{\alpha} t \|\boldsymbol{u}\|_{L^\infty(0,t;W^{1,r}(\Omega)^2)} + \sqrt{t} \|\operatorname{curl} \boldsymbol{f}\|_{L^2(0,t;L^r(\Omega))}. \quad (3.2.58)$$

With Proposition 3.2.13, we can prove that $\nabla \boldsymbol{u}$ is bounded if Ω is a convex polygon or a smooth domain. However, we shall see in the next section that convexity or regularity of Ω imply uniqueness of the solution. Therefore, the boundedness of $\nabla \boldsymbol{u}$ and its consequences will easily be deduced from the convergence of the semi-discrete solution.

3.2.3. Existence, regularity, and uniqueness

The convergences of Proposition 3.2.6 allow us to pass to the limit in (3.2.20)–(3.2.21) and hence show unconditional existence of a solution of the coupled problem (3.2.1)–(3.2.2). As this problem is equivalent to the original system (3.1.1), it also yields existence of a solution to (3.1.1). The following theorem collects the results established so far.

THEOREM 3.2.14. *Let Ω be a bounded connected Lipschitz-continuous domain in two dimensions.*

(1) For any $\alpha > 0$, $\nu > 0$, $\boldsymbol{f}$ in $L^2(0,T;H(\operatorname{curl};\Omega))$, and $\boldsymbol{u}_0 \in V^\alpha$, problem (3.1.1) has at least one solution $(\boldsymbol{u}, p)$ in $\mathcal{W}$. All solutions satisfy (3.2.3)–(3.2.7).

(2) If in addition, Ω is a polygon, all solutions satisfy also (3.2.9)–(3.2.11).

(3) If moreover Ω is a convex polygon or has a smooth boundary, all solutions satisfy also (3.2.13)–(3.2.15).

(4) Finally, if Ω is a connected polygon, if $r \in [2,4]$, $z_0 \in L^r(\Omega)$, and $\operatorname{curl} \boldsymbol{f} \in L^2(0,T;L^r(\Omega))$, then all solutions satisfy (3.2.58).

The arguments of the proof are standard, see GIRAULT and SAADOUNI [2007]. The regularity of the solution follows from Theorem 3.2.3. Additional regularity stems from Proposition 3.2.8, Corollary 3.2.9, Theorem 3.2.10, and Proposition 3.2.13.

Now, we can sharpen (3.2.27) and establish the strong convergence of $\boldsymbol{u}_k$ in $\mathcal{C}^0(0,T;H^1(\Omega)^2)$.

THEOREM 3.2.15. *Under the assumptions of Part 1 of Theorem 3.2.14, we have, up to subsequences,*

$$\lim_{k \to 0} \sup_{t \in [0,T]} \|\boldsymbol{u}(t) - \boldsymbol{u}_k(t)\|_\alpha = 0. \quad (3.2.59)$$

PROOF. We can write for a.e. t in $[0, T]$:

$$\frac{1}{2}\|\boldsymbol{u}(t) - \boldsymbol{u}_k(t)\|_\alpha^2 + \nu\|\nabla(\boldsymbol{u} - \boldsymbol{u}_k)\|_{L^2(\Omega\times]0,t[)}^2 = A - B - C, \tag{3.2.60}$$

where

$$A = \int_0^t \left(\left(\frac{\partial}{\partial s}(\boldsymbol{u} - \boldsymbol{u}_k), \boldsymbol{u}\right) + \alpha\left(\frac{\partial}{\partial s}\nabla(\boldsymbol{u} - \boldsymbol{u}_k), \nabla \boldsymbol{u}\right) + \nu\,(\nabla(\boldsymbol{u} - \boldsymbol{u}_k), \nabla \boldsymbol{u})\right) \mathrm{d}s,$$

$$B = \int_0^t \left(\left(\frac{\partial}{\partial s}(\boldsymbol{u} - \boldsymbol{u}_k), \boldsymbol{u}_k - \boldsymbol{w}_k\right) + \alpha\left(\frac{\partial}{\partial s}\nabla(\boldsymbol{u} - \boldsymbol{u}_k), \nabla(\boldsymbol{u}_k - \boldsymbol{w}_k)\right)\right)\mathrm{d}s$$

$$+ \nu\int_0^t (\nabla(\boldsymbol{u} - \boldsymbol{u}_k), \nabla(\boldsymbol{u}_k - \boldsymbol{w}_k))\mathrm{d}s,$$

$$C = \int_0^t \left(\left(\frac{\partial}{\partial s}(\boldsymbol{u} - \boldsymbol{u}_k), \boldsymbol{w}_k\right) + \alpha\left(\frac{\partial}{\partial s}\nabla(\boldsymbol{u} - \boldsymbol{u}_k), \nabla \boldsymbol{w}_k\right)\right)\mathrm{d}s$$

$$+ \nu\int_0^t \big(\nabla(\boldsymbol{u} - \boldsymbol{u}_k), \nabla \boldsymbol{w}_k)\big)\mathrm{d}s.$$

The weak convergences of Proposition 3.2.6 and strong convergences of (3.2.26) imply that both A and B tend to zero with k. To establish the convergence of C, we subtract the semi-discrete equation (3.2.20) from the exact equation (3.2.1), both tested against $\boldsymbol{w}_k$ and we compare with C. This gives

$$C = \int_0^t \big(-\nu\,(\nabla(\boldsymbol{u}_k - \boldsymbol{w}_k), \nabla \boldsymbol{w}_k)) - (z \times \boldsymbol{u}, \boldsymbol{w}_k) + (\boldsymbol{f} - \boldsymbol{f}_k, \boldsymbol{w}_k)\big)\mathrm{d}s.$$

Then C tends to zero, owing to the weak convergences of Proposition 3.2.6, strong convergences of (3.2.26), and the strong convergence of $\boldsymbol{f} - \boldsymbol{f}_k$. □

Next we investigate uniqueness. But, whereas existence holds without restriction, we shall see that we can only prove uniqueness in a convex polygon. This is due to the possible lack of regularity of the solutions, as will be made clear in the proposition below. Exceptionally, let $c(\cdot;\cdot,\cdot)$ denote the extension to vectors of the trilinear form c defined in (1.4.15):

$$c(\boldsymbol{u}; \boldsymbol{v}, \boldsymbol{w}) = \sum_{i=1}^{2}\sum_{j=1}^{2}\int_\Omega u_i \frac{\partial v_j}{\partial x_i} w_j \,\mathrm{d}\boldsymbol{x}. \tag{3.2.61}$$

It satisfies the analog of (1.4.16):

$$\forall \boldsymbol{u} \in V,\ \forall \boldsymbol{v} \in H^1(\Omega)^2,\ c(\boldsymbol{u}; \boldsymbol{v}, \boldsymbol{v}) = 0. \tag{3.2.62}$$

With this notation, we have the following proposition. Its proof is a straightforward adaptation of that of a similar result established in GIRAULT and SCOTT [1999] for the steady problem. See also OUAZAR [1981].

PROPOSITION 3.2.16. *Assume that Ω is a connected polygon and $\boldsymbol{u}_0$ belongs to V^α. Let $(\boldsymbol{u}^1, p^1)$ and $(\boldsymbol{u}^2, p^2)$ be two solutions of (3.1.1) in $\mathcal{W}$ and let $\boldsymbol{w} = \boldsymbol{u}^1 - \boldsymbol{u}^2$, $q = p^1 - p^2$. Then $(\boldsymbol{w}, q)$ satisfies almost everywhere in $]0, T[$,*

$$\frac{1}{2}\frac{\mathrm{d}}{\mathrm{d}t}\|\boldsymbol{w}\|_\alpha^2 + \nu|\boldsymbol{w}|_{H^1(\Omega)}^2 + c(\boldsymbol{w}; \boldsymbol{u}^1, \boldsymbol{w}) + \alpha\, c(\boldsymbol{w}; \operatorname{curl}\boldsymbol{u}^1, \operatorname{curl}\boldsymbol{w}) - 2\alpha\int_\Omega \operatorname{curl}\boldsymbol{w}(\nabla u_1^1 \cdot \nabla w_2 - \nabla u_2^1 \cdot \nabla w_1)\,\mathrm{d}\boldsymbol{x} = 0. \tag{3.2.63}$$

All terms in (3.2.63) make sense, but unfortunately, without additional regularity, (3.2.63) does not seem to imply that $\boldsymbol{w} = \boldsymbol{0}$. Indeed, the last two terms in (3.2.63) have no particular sign and in order to be controlled by the first two terms, they must be bounded in terms of the H^1 norm of $\boldsymbol{w}$. This is the case if we assume that $\boldsymbol{u}^1$ belongs to $W^{2,r}(\Omega)^2$ for some $r > 2$. Since by Sobolev's imbedding, $\boldsymbol{w}$ belongs to $L^q(\Omega)^2$ for any finite $q > 2$, then we can choose q so that the product $(\operatorname{curl}\boldsymbol{w})\boldsymbol{w}$ belongs to $L^{r'}(\Omega)^2$, the dual exponent of r, i.e., $q = 2r/(r-2)$. Hence, with the notation of (1.1.3),

$$\left|c(\boldsymbol{w}; \operatorname{curl}\boldsymbol{u}^1, \operatorname{curl}\boldsymbol{w})\right| \le S_{\frac{2r}{r-2}}\, |\boldsymbol{w}|_{H^1(\Omega)}^2\, \|\Delta\,\boldsymbol{u}^1\|_{L^r(\Omega)}, \tag{3.2.64}$$

$$\left|\int_\Omega \operatorname{curl}\boldsymbol{w}(\nabla u_1^1 \cdot \nabla w_2 - \nabla u_2^1 \cdot \nabla w_1)\,\mathrm{d}\boldsymbol{x}\right| \le |\boldsymbol{w}|_{H^1(\Omega)}^2 \left|\boldsymbol{u}^1\right|_{W^{1,\infty}(\Omega)}. \tag{3.2.65}$$

With these remarks, it is easy to see that if problem (3.1.1) has one solution $\boldsymbol{u} \in W^{2,r}(\Omega)^2$ for some $r > 2$, then it has no other solution $(\boldsymbol{u}, p) \in \mathcal{W}$. More precisely, we have the following theorem. Existence stems from Corollary 3.2.9, and uniqueness follows readily by applying the regularity result of Lemma 1.4.10, because Ω is convex, substituting (3.2.64) and (3.2.65) into (3.2.63) and concluding that $\boldsymbol{w} = \boldsymbol{0}$ by Gronwall's Lemma 3.1.1.

THEOREM 3.2.17. *Assume that Ω is a convex polygon. Then for any $\alpha > 0$, $\nu > 0$, $z_0 \in L^r(\Omega)$, and $\boldsymbol{f}$ in $L^2(0, T; L^r(\Omega)^2)$ with $\operatorname{curl}\boldsymbol{f}$ in $L^2(0, T; L^r(\Omega))$, for some $r > 2$, problem (3.1.1) has exactly one solution $(\boldsymbol{u}, p) \in \mathcal{W}$.*

Finally, Theorems 3.2.17 and 3.2.10 imply the next result.

COROLLARY 3.2.18. *If the assumptions of the first part of Theorem 3.2.10 hold, then $z \in L^\infty(0, T; H^1(\Omega)) \cap H^1(0, T; L^2(\Omega))$, and if the assumptions of the second part hold, then $z \in L^\infty(0, T; W^{1,r}(\Omega)) \cap H^1(0, T; L^r(\Omega))$.*

This corollary yields the strong convergence of z^n in $\mathcal{C}^0(0, T; L^2(\Omega))$.

COROLLARY 3.2.19. *Under the assumptions of Corollary 3.2.18, we have, up to subsequences,*

$$\lim_{k\to 0} \sup_{t\in[0,T]} \|z(t) - z_k(t)\|_{L^2(\Omega)} = 0. \tag{3.2.66}$$

PROOF. The proof follows the lines of that of Theorem 3.2.15. We have for a.e. t in $[0, T]$:

$$\frac{\alpha}{2}\|z(t) - z_k(t)\|^2_{L^2(\Omega)} + \nu\|z(t) - z_k(t)\|^2_{L^2(\Omega)} = A - B - C, \tag{3.2.67}$$

where

$$A = \int_0^t \left(\alpha\left(\frac{\partial}{\partial s}(z - z_k), z\right) + \nu\,(z - z_k, z)\right)\mathrm{d}s,$$

$$B = \int_0^t \left(\alpha\left(\frac{\partial}{\partial s}(z - z_k), z_k - \zeta_k\right) + \nu\,(z - z_k, z_k - \zeta_k)\right)\mathrm{d}s,$$

$$C = \int_0^t \left(\alpha\left(\frac{\partial}{\partial s}(z - z_k), \zeta_k\right) + \nu\,(z - z_k, \zeta_k)\right)\mathrm{d}s.$$

The weak convergence of z_k and its derivatives, and the strong convergences of (3.2.26) imply that both A and B tend to zero with k. Also C tends to zero considering that it can be written as

$$C = \int_0^t (\nu\,(\zeta_k - z_k, \zeta_k) - \alpha\,(\boldsymbol{u}\cdot\nabla z, \zeta_k) + \nu\,(\operatorname{curl}(\boldsymbol{u} - \boldsymbol{w}_k), \zeta_k))\mathrm{d}s$$
$$+\,\alpha\int_0^t \big(\operatorname{curl}(\boldsymbol{f} - \boldsymbol{f}_k, \zeta_k\big)\,\mathrm{d}s.$$

□

In turn, this result gives the strong convergences of $\boldsymbol{u}_k$ in $H^1(0, T; H^1(\Omega)^2)$ and of p_k in $L^2(\Omega\times]0, T[)$.

COROLLARY 3.2.20. *Under the assumptions of Corollary 3.2.18, we have, up to subsequences,*

$$\lim_{k\to 0} \|\frac{\partial}{\partial t}(\boldsymbol{u}_k - \boldsymbol{u})\|_{L^2(0,T;H^1(\Omega)^2)} = 0, \quad \lim_{k\to 0} \|p_k - p\|_{L^2(\Omega\times]0,T[)} = 0. \tag{3.2.68}$$

PROOF. We write for a.e. t in $[0, T]$:

$$\int_0^t \|\frac{\partial}{\partial t}(\boldsymbol{u}_k - \boldsymbol{u})\|^2_\alpha \mathrm{d}s + \frac{\nu}{2}|\boldsymbol{u}_k - \boldsymbol{u}|^2_{H^1(\Omega)} = A - B - C,$$

where

$$A = \int_0^t \left(\left(\frac{\partial}{\partial s}(\boldsymbol{u} - \boldsymbol{u}_k), \frac{\partial \boldsymbol{u}}{\partial s} \right) + \alpha \left(\nabla \frac{\partial}{\partial s}(\boldsymbol{u} - \boldsymbol{u}_k), \nabla \frac{\partial \boldsymbol{u}}{\partial s} \right) \right) \mathrm{d}s$$

$$+ \nu \int_0^t \left(\nabla(\boldsymbol{u} - \boldsymbol{u}_k), \nabla \frac{\partial \boldsymbol{u}}{\partial s} \right) \mathrm{d}s,$$

$$B = \int_0^t \left(\left(\frac{\partial}{\partial s}(\boldsymbol{u} - \boldsymbol{u}_k), \frac{\partial \boldsymbol{u}_k}{\partial s} \right) + \alpha \left(\nabla \frac{\partial}{\partial s}(\boldsymbol{u} - \boldsymbol{u}_k), \nabla \frac{\partial \boldsymbol{u}_k}{\partial s} \right) \right) \mathrm{d}s$$

$$+ \nu \int_0^t \left(\nabla(\boldsymbol{u} - \boldsymbol{w}_k), \nabla \frac{\partial \boldsymbol{u}_k}{\partial s} \right) \mathrm{d}s,$$

$$C = \nu \int_0^t \left(\nabla(\boldsymbol{w}_k - \boldsymbol{u}_k), \nabla \frac{\partial \boldsymbol{u}_k}{\partial s} \right) \mathrm{d}s.$$

Then we observe that B has the form

$$B = \int_0^t \left(\left(\boldsymbol{f} - \boldsymbol{f}_k, \frac{\partial \boldsymbol{u}_k}{\partial s} \right) - \left(z \times \boldsymbol{u}, \frac{\partial \boldsymbol{u}_k}{\partial s} \right) + \left(\lambda_k \times \boldsymbol{w}_k, \frac{\partial \boldsymbol{u}_k}{\partial s} \right) \right) \mathrm{d}s,$$

and the strong convergence of the derivative of $\boldsymbol{u}_k$ follows from the previous results.

Finally, we write

$$\int_0^t (p_k - p, \operatorname{div} \boldsymbol{v})\, \mathrm{d}s = \int_0^t \left(\left(\frac{\partial}{\partial s}(\boldsymbol{u}_k - \boldsymbol{u}), \boldsymbol{v} \right) + \alpha \left(\nabla \frac{\partial}{\partial s}(\boldsymbol{u}_k - \boldsymbol{u}), \nabla \boldsymbol{v} \right) \right) \mathrm{d}s$$

$$+ \int_0^t (\nu\, (\nabla(\boldsymbol{u}_k - \boldsymbol{u}), \nabla \boldsymbol{v}) + (\lambda_k \times \boldsymbol{w}_k - z \times \boldsymbol{u}, \boldsymbol{v}))\mathrm{d}s$$

$$- \int_0^t (\boldsymbol{f}_k - \boldsymbol{f}, \boldsymbol{v})\, \mathrm{d}s,$$

and we choose the function $\boldsymbol{v}$ associated by (1.1.25) to $p_k - p$; whence the strong convergence of p_k. □

3.3. Fully discrete centered schemes

Let us revert to the material of Section 2.1, namely, we discretize the auxiliary variable z in a finite-dimensional space $Z_h \subset H^1(\Omega)$ and the velocity and pressure in a pair of finite-dimensional spaces, $X_h \subset H_0^1(\Omega)^2$ and $M_h \subset L_0^2(\Omega)$, satisfying the uniform discrete inf-sup condition (2.1.1): There exists a constant $\beta^* > 0$, independent of h, such that

$$\forall q_h \in M_h, \sup_{v_h \in X_h} \frac{\int_\Omega q_h \operatorname{div} v_h \, dx}{|v_h|_{H^1(\Omega)}} \geq \beta^* \|q_h\|_{L^2(\Omega)}.$$

Moreover, we make the assumptions of Hypothesis 2.1.5. The transport term, $u \cdot \nabla z$, is discretized by the consistent antisymmetric form defined in (2.1.4):

$$\forall v \in H^1(\Omega)^2, \ \forall \varphi, \theta \in H^1(\Omega), \ \tilde{c}(v; \varphi, \theta) = (v \cdot \nabla \varphi, \theta) + \frac{1}{2}((\operatorname{div} v)\varphi, \theta).$$

The interval $[0, T]$ is divided into N equal segments of length k, with end points $t_i = ik$, $0 \leq i \leq N$. Then, by discretizing in space the semi-discrete scheme (3.2.20)–(3.2.21), we approximate problem (3.2.1)–(3.2.2) with the following backward Euler, fully discrete scheme:

- Set

$$u_h^0 = P_h(u_0), \ z_h^0 = R_h(z_0), \ \mathbf{z}_h^0 = (0, 0, z_h^0), \tag{3.3.1}$$

  where P_h and R_h satisfy parts 1 and 3 of Hypothesis 2.1.5.
- Knowing $u_h^0 \in X_h$ and $z_h^0 \in Z_h$, find sequences $(u_h^n)_{n\geq 1}$, $(z_h^n)_{n\geq 1}$, and $(p_h^n)_{n\geq 1}$ such that $u_h^n \in X_h$, $z_h^n \in Z_h$, and $p_h^n \in M_h$ solve for $1 \leq n \leq N$,

$$\forall v_h \in X_h, \ \frac{1}{k}(u_h^n - u_h^{n-1}, v_h) + \frac{\alpha}{k}(\nabla(u_h^n - u_h^{n-1}), \nabla v_h) + \nu(\nabla u_h^n, \nabla v_h)$$
$$+ (\mathbf{z}_h^{n-1} \times u_h^n, v_h) - (p_h^n, \operatorname{div} v_h) = (f^n, v_h), \tag{3.3.2}$$
$$\forall q_h \in M_h, \ (q_h, \operatorname{div} u_h^n) = 0, \tag{3.3.3}$$
$$\forall \theta_h \in Z_h, \ \frac{\alpha}{k}(z_h^n - z_h^{n-1}, \theta_h) + \nu(z_h^n, \theta_h) + \alpha\tilde{c}(u_h^n; z_h^n, \theta_h)$$
$$= \nu(\operatorname{curl} u_h^n, \theta_h) + \alpha(\operatorname{curl} f^n, \theta_h), \tag{3.3.4}$$

  where f^n is defined by (3.2.18):

$$f^n(x) = \frac{1}{k}\int_{t_{n-1}}^{t_n} f(x, s) \, ds.$$

At each step n, the system (3.3.2)–(3.3.3) is a linear discrete problem of Stokes type, and owing to the discrete inf-sup condition (2.1.1), it has a unique solution. Likewise, (3.3.4) is a discrete transport equation and owing to the antisymmetry of $\tilde{c}$, it also has a unique solution. Thus, with the starting values (3.3.1), the equations (3.3.2)–(3.3.4) determine unique sequences $(u_h^n)_{n\geq 1}$, $(z_h^n)_{n\geq 1}$, and $(p_h^n)_{n\geq 1}$. These sequences satisfy the following a priori estimates, for $1 \leq n \leq N$ (compare with Propositions 3.2.4 and 3.2.5):

$$\|u_h^n\|_\alpha^2 + \sum_{i=1}^n \|u_h^i - u_h^{i-1}\|_\alpha^2 \leq \frac{S_2^2}{2\nu}\|f\|_{L^2(\Omega\times]0,t_n[)}^2 + \|u_h^0\|_\alpha^2, \tag{3.3.5}$$

$$\nu \sum_{i=1}^{n} k|\boldsymbol{u}_h^i|^2_{H^1(\Omega)} \leq \frac{S_2^2}{\nu}\|\boldsymbol{f}\|^2_{L^2(\Omega\times]0,t_n[)} + \|\boldsymbol{u}_h^0\|^2_\alpha, \tag{3.3.6}$$

$$\|z_h^n\|^2_{L^2(\Omega)} + \sum_{i=1}^{n}\|z_h^i - z_h^{i-1}\|^2_{L^2(\Omega)} \leq \frac{S_2^2}{\alpha\nu}\|\boldsymbol{f}\|^2_{L^2(\Omega\times]0,t_n[)} + \frac{\alpha}{\nu}\|\operatorname{curl}\boldsymbol{f}\|^2_{L^2(\Omega\times]0,t_n[)} + \frac{1}{\alpha}\|\boldsymbol{u}_h^0\|^2_\alpha + \|z_h^0\|^2_{L^2(\Omega)}, \tag{3.3.7}$$

$$\sum_{i=1}^{n}\frac{1}{k}\|\boldsymbol{u}_h^i - \boldsymbol{u}_h^{i-1}\|^2_\alpha \leq \nu|\boldsymbol{u}_h^0|^2_{H^1(\Omega)} + \|\boldsymbol{f}\|^2_{L^2(\Omega\times]0,t_n[)} + \frac{1}{\alpha\nu}S_4^4 C_{hz}^2\left(\|\boldsymbol{u}_h^0\|^2_\alpha + \frac{{S_2}^2}{\nu}\|\boldsymbol{f}\|^2_{L^2(\Omega\times]0,t_n[)}\right), \tag{3.3.8}$$

$$\sum_{i=1}^{n} k\|p_h^i\|^2_{L^2(\Omega)} \leq \frac{3}{\beta^2}\left(S_2^2\left(\nu|\boldsymbol{u}_h^0|^2_{H^1(\Omega)} + 2\|\boldsymbol{f}\|^2_{L^2(\Omega\times]0,t_n[)}\right) + C_{hz}^2\frac{S_4^4}{\nu}\left(\|\boldsymbol{u}_h^0\|^2_\alpha + \frac{S_2^2}{\nu}\|\boldsymbol{f}\|^2_{L^2(\Omega\times]0,t_n[)}\right)\left(1 + \frac{{S_2}^2}{\alpha}\right)\right), \tag{3.3.9}$$

where

$$C_{hz} = \sup_{0\leq n\leq N-1}\|z_h^n\|_{L^2(\Omega)}. \tag{3.3.10}$$

An L^∞ estimate

Let us prove that the sequence $(\boldsymbol{u}_h^n)$ is also uniformly bounded in L^∞ in space and time. This property will be useful for deriving error estimates further on. As in the steady problem, an L^∞ bound for $(\boldsymbol{u}_h^n)$ is conveniently derived by comparing it with the solution of a similar exact system, where z^n is replaced by z_h^n. More precisely, starting with $\boldsymbol{v}^0 = \boldsymbol{u}_0$, we consider for each $1 \leq n \leq N$, the solution $(\boldsymbol{v}^n, q^n)$ in $V \times L_0^2(\Omega)$ of

$$\frac{1}{k}(\boldsymbol{v}^n - \boldsymbol{v}^{n-1}) - \alpha\frac{1}{k}\Delta(\boldsymbol{v}^n - \boldsymbol{v}^{n-1}) - \nu\Delta\boldsymbol{v}^n + z_h^{n-1}\times\boldsymbol{v}^n + \nabla q^n = \boldsymbol{f}^n \quad \text{in } \Omega. \tag{3.3.11}$$

The initial data and the sequence (z_h^n) determine uniquely the sequences $(\boldsymbol{v}^n)$ and (q^n). Note that $(\boldsymbol{u}_h^n)$ and (p_h^n) are a standard finite-element approximation of $(\boldsymbol{v}^n)$ and (q^n). Clearly, $(\boldsymbol{v}^n)$ and (q^n) satisfy the same uniform bounds as $(\boldsymbol{u}^n)$ and (p^n) with z_h^{n-1} instead of z^{n-1}. In addition, we have the following result for the difference $\boldsymbol{u}_h^n - P_h(\boldsymbol{v}^n)$.

PROPOSITION 3.3.1. *Under the first two assumptions of Hypothesis 2.1.5, we have for each n, $1 \leq n \leq N$,*

$$\|\boldsymbol{u}_h^n - P_h(\boldsymbol{v}^n)\|^2_\alpha + \sum_{i=1}^{n}\|\boldsymbol{u}_h^i - \boldsymbol{u}_h^{i-1} - P_h(\boldsymbol{v}^i - \boldsymbol{v}^{i-1})\|^2_\alpha \leq \frac{2}{\nu}(S_2^2 + \alpha)\sum_{i=1}^{n}\frac{1}{k}\|P_h(\boldsymbol{v}^i - \boldsymbol{v}^{i-1}) - (\boldsymbol{v}^i - \boldsymbol{v}^{i-1})\|^2_\alpha$$

$$+ 4\nu \sum_{i=1}^{n} k|P_h(\boldsymbol{v}^i) - \boldsymbol{v}^i|^2_{H^1(\Omega)} + \frac{4}{\nu} C_{hz}^2 S_4^2 \sum_{i=1}^{n} k\|P_h(\boldsymbol{v}^i) - \boldsymbol{v}^i\|^2_{L^4(\Omega)}$$

$$+ \frac{4}{\nu} \sum_{i=1}^{n} k\|r_h(q^i) - q^i\|^2_{L^2(\Omega)}. \tag{3.3.12}$$

Moreover, if Ω is convex, if the triangulation satisfies (2.1.25), and if the assumptions of Hypothesis 2.1.10 hold, we have for each n, $1 \leq n \leq N$,

$$\|\boldsymbol{u}_h^n - P_h(\boldsymbol{v}^n)\|^2_\alpha + \sum_{i=1}^{n} \|\boldsymbol{u}_h^i - \boldsymbol{u}_h^{i-1} - P_h(\boldsymbol{v}^i - \boldsymbol{v}^{i-1})\|^2_\alpha$$

$$\leq C^2 h^2 \left(\frac{2}{\nu}(S_2^2 + \alpha) \sum_{i=1}^{n} \frac{1}{k} \|\boldsymbol{v}^i - \boldsymbol{v}^{i-1}\|^2_{H^2(\Omega)} + 4\nu \sum_{i=1}^{n} k|\boldsymbol{v}^i|^2_{H^2(\Omega)} \right. \tag{3.3.13}$$

$$\left. + h \frac{4}{\nu} C_{hz}^2 S_4^2 \sum_{i=1}^{n} k|\boldsymbol{v}^i|^2_{H^2(\Omega)} + \frac{4}{\nu} \sum_{i=1}^{n} k|q^i|^2_{H^1(\Omega)} \right).$$

Considering that z_h^n is uniformly bounded in L^2 with respect to h and n, the estimate (3.2.29) in a convex domain is valid for $(\boldsymbol{v}^n, q^n)$. We can apply it to evaluate the right-hand side of (3.3.13), because an L^∞ bound for $\boldsymbol{u}_h^n$ will be used in a convex domain. Thus, we derive the following corollary. Its proof is skipped because it proceeds as in Remark 2.1.17.

COROLLARY 3.3.2. *Under the assumptions of the second part of Proposition 3.3.1, there exists a constant C_1, independent of h and k, such that for all n, $1 \leq n \leq N$,*

$$\|\boldsymbol{u}_h^n - P_h(\boldsymbol{v}^n)\|^2_\alpha + \sum_{i=1}^{n} \|\boldsymbol{u}_h^i - \boldsymbol{u}_h^{i-1} - P_h(\boldsymbol{v}^i - \boldsymbol{v}^{i-1})\|^2_\alpha \leq C_1 h^2. \tag{3.3.14}$$

In addition, if the mesh satisfies (2.1.45) for some $r > 2$:

$$h \leq C_2 \,\varrho_{\min}^{1-2/r},$$

then there exists a constant C_3, independent of h and k, such that for all n, $1 \leq n \leq N$,

$$\|\boldsymbol{u}_h^n\|_{W^{1,r}(\Omega)} \leq C_3. \tag{3.3.15}$$

This implies that $\boldsymbol{u}_h^n$ is uniformly bounded:

$$\|\boldsymbol{u}_h^n\|_{L^\infty(\Omega)} \leq C_4. \tag{3.3.16}$$

Convergence

PROPOSITION 3.3.3. *Under the assumptions of Hypothesis 2.1.5, we have the following strong convergences for each n, $1 \leq n \leq N$, as h tends to zero, without restriction to* *subsequences:*

$$\lim_{h\to 0} \|\boldsymbol{u}_h^n - \boldsymbol{u}^n\|_\alpha = 0,$$
$$\lim_{h\to 0} \|p_h^n - p^n\|_{L^2(\Omega)} = 0, \tag{3.3.17}$$
$$\lim_{h\to 0} \|z_h^n - z^n\|_{L^2(\Omega)} = 0,$$

where $(\boldsymbol{u}^n, p^n, z^n)$ *solves (3.2.20)–(3.2.21).*

PROOF. The proof proceeds by induction.
(1) For $n = 0$, the approximation properties of the operators P_h and R_h in Hypothesis 2.1.5 imply

$$\lim_{h\to 0} \boldsymbol{u}_h^0 = \boldsymbol{u}^0 \text{ strongly in } H^1(\Omega)^2, \ \lim_{h\to 0} z_h^0 = z^0 \text{ strongly in } L^2(\Omega).$$

(2) For $n \geq 0$, on one hand, we assume that the following strong convergences hold as h tends to zero, without restriction to subsequences:

$$\lim_{h\to 0} \boldsymbol{u}_h^n = \boldsymbol{u}^n \text{ strongly in } H^1(\Omega)^2, \ \lim_{h\to 0} z_h^n = z^n \text{ strongly in } L^2(\Omega).$$

On the other hand, it follows from (3.3.5), (3.3.7), and (3.3.9), that the following weak convergences hold, as h tends to zero up to subsequences, to some functions $\boldsymbol{u}^{n+1} \in H_0^1(\Omega)^2$, $z^{n+1} \in L^2(\Omega)$, and $p^{n+1} \in L_0^2(\Omega)$:

$$\lim_{h\to 0} \boldsymbol{u}_h^{n+1} = \boldsymbol{u}^{n+1} \text{ weakly in } H^1(\Omega)^2,$$
$$\lim_{h\to 0} z_h^{n+1} = z^{n+1} \text{ weakly in } L^2(\Omega),$$
$$\lim_{h\to 0} p_h^{n+1} = p^{n+1} \text{ weakly in } L^2(\Omega).$$

By passing to the limit in (3.3.2) and (3.3.3) and arguing as in Proposition 2.1.6, we easily derive that the limit functions $(\boldsymbol{u}^{n+1}, p^{n+1})$ satisfy (3.2.20) at time t_{n+1}:

$$\frac{1}{k}(\boldsymbol{u}^{n+1} - \boldsymbol{u}^n) - \alpha\frac{1}{k}\Delta(\boldsymbol{u}^{n+1} - \boldsymbol{u}^n) - \nu\Delta\,\boldsymbol{u}^{n+1} + z^n \times \boldsymbol{u}^{n+1} + \nabla\, p^{n+1} = \boldsymbol{f}^{n+1} \quad \text{in } \Omega,$$
$$\operatorname{div} \boldsymbol{u}^{n+1} = 0 \quad \text{in } \Omega.$$

But given $(\boldsymbol{u}^n, z^n) \in H_0^1(\Omega)^2 \times L^2(\Omega)$, this system has a unique solution $(\boldsymbol{u}^{n+1}, p^{n+1}) \in H_0^1(\Omega)^2 \times L_0^2(\Omega)$. The uniqueness of the limit functions implies the convergence of the whole sequences $(\boldsymbol{u}_h^{n+1}, p_h^{n+1})$. Furthermore, using the induction hypothesis, the strong convergence of $(\boldsymbol{u}_h^{n+1}, p_h^{n+1})$ is derived as in Proposition 2.1.7 and the last part of Theorem 2.1.9. Then, we can pass to the limit in (3.3.4) at time t_{n+1} and we recover (3.2.21). Again, given z^n, the equation has a unique solution and therefore, the whole sequence z_h^{n+1} tends to z^{n+1}.

Finally, owing to the induction hypothesis, the strong convergence of z_h^{n+1} is established as in the first part of Theorem 2.1.9. □

Here again, it is convenient to transform the sequences $(\boldsymbol{u}_h^n)$, (p_h^n), and (z_h^n) into functions: first, we define the piecewise linear functions in time:

$$\forall t \in [t_n, t_{n+1}],\ \boldsymbol{u}_{hk}(t) = \boldsymbol{u}_h^n + \frac{t - t_n}{k}\left(\boldsymbol{u}_h^{n+1} - \boldsymbol{u}_h^n\right),\ 0 \le n \le N-1,$$

$$\forall t \in [t_n, t_{n+1}],\ z_{hk}(t) = z_h^n + \frac{t - t_n}{k}\left(z_h^{n+1} - z_h^n\right),\ 0 \le n \le N-1,$$

and next we define the step functions:

$$\forall t \in]t_n, t_{n+1}],\ \boldsymbol{w}_{hk}(t) = \boldsymbol{u}_h^{n+1},\ 0 \le n \le N-1,$$
$$\forall t \in]t_n, t_{n+1}],\ p_{hk}(t) = p_h^{n+1},\ 0 \le n \le N-1,$$
$$\forall t \in]t_n, t_{n+1}],\ \zeta_{hk}(t) = z_h^{n+1},\ 0 \le n \le N-1,$$
$$\forall t \in [t_n, t_{n+1}[,\ \lambda_{hk}(t) = z_h^n,\ 0 \le n \le N-1.$$

With this notation, (3.3.2)–(3.3.4) read

$$\forall \boldsymbol{v}_h \in X_h,\ \left(\frac{\partial \boldsymbol{u}_{hk}}{\partial t}, \boldsymbol{v}_h\right) + \alpha\left(\nabla \frac{\partial \boldsymbol{u}_{hk}}{\partial t}, \nabla \boldsymbol{v}_h\right) + \nu(\nabla \boldsymbol{w}_{hk}, \nabla \boldsymbol{v}_h) + (\lambda_{hk} \times \boldsymbol{w}_{hk}, \boldsymbol{v}_h)$$
$$- (p_{hk}, \operatorname{div} \boldsymbol{v}_h) = (\boldsymbol{f}_k, \boldsymbol{v}_h) \quad \text{a.e. in }]0, T[,$$
$$\forall q_h \in M_h,\ (q_h, \operatorname{div} \boldsymbol{u}_{hk}) = 0 \quad \text{a.e. in }]0, T[,$$
$$\forall \theta_h \in Z_h,\ \alpha\left(\frac{\partial z_{hk}}{\partial t}, \theta_h\right) + \nu(\zeta_{hk}, \theta_h) + \alpha\tilde{c}(\boldsymbol{w}_{hk}; \zeta_{hk}, \theta_h) = \nu(\operatorname{curl} \boldsymbol{w}_{hk}, \theta_h)$$
$$+ \alpha(\operatorname{curl} \boldsymbol{f}_k, \theta_h) \quad \text{a.e. in }]0, T[.$$

It follows from (3.3.5) to (3.3.9) that the following sequences of functions are uniformly bounded with respect to h and k: $\boldsymbol{u}_{hk}$ in $H^1(0, T; H^1(\Omega)^2)$, $\boldsymbol{w}_{hk}$ in $L^\infty(0, T; H^1(\Omega)^2)$, p_{hk} in $L^2(\Omega\times]0, T[)$, and z_{hk}, ζ_{hk}, and λ_{hk} in $L^\infty(0, T; L^2(\Omega))$. Moreover, for fixed k, Proposition 3.3.3 implies the following convergences, without restriction to subsequences,

$$\lim_{h\to 0} \|\boldsymbol{u}_{hk} - \boldsymbol{u}_k\|_{H^1(0,T;H^1(\Omega)^2)} = 0,\ \lim_{h\to 0} \|\boldsymbol{w}_{hk} - \boldsymbol{w}_k\|_{L^\infty(0,T;H^1(\Omega)^2)} = 0,$$
$$\lim_{h\to 0} \|p_{hk} - p_k\|_{L^2(\Omega\times]0,T[)} = 0,$$
$$\lim_{h\to 0} \|z_{hk} - z_k\|_{L^\infty(0,T;L^2(\Omega))} = \lim_{h\to 0} \|\zeta_{hk} - \zeta_k\|_{L^\infty(0,T;L^2(\Omega))}$$
$$= \lim_{h\to 0} \|\lambda_{hk} - \lambda_k\|_{L^\infty(0,T;L^2(\Omega))} = 0,$$

where the functions $\boldsymbol{u}_k$, $\boldsymbol{w}_k$, p_k, z_k, ζ_k, and λ_k are defined in Section 3.2.2. Up to subsequence of k, they satisfy (see Proposition 3.2.6):

$$\lim_{k\to 0} \boldsymbol{u}_k = \lim_{k\to 0} \boldsymbol{w}_k = \boldsymbol{u} \text{ weakly * in } L^\infty(0, T; V),$$

$$\lim_{k\to 0} z_k = \lim_{k\to 0} \zeta_k = \lim_{k\to 0} \lambda_k = z \text{ weakly } * \text{ in } L^\infty(0,T;L^2(\Omega)),$$

$$\lim_{k\to 0} p_k = p \text{ weakly in } L^2(0,T;L^2_0(\Omega)),$$

$$\lim_{k\to 0} \frac{\partial}{\partial t}\boldsymbol{u}_k = \frac{\partial}{\partial t}\boldsymbol{u} \text{ weakly in } L^2(0,T;V).$$

A priori error estimates

Let $(\boldsymbol{u}_h^n, p_h^n, z_h^n)$ be the solution of (3.3.2)–(3.3.4), and $(\boldsymbol{u}, p, z) \in \mathcal{W} \times L^\infty(0,T;L^2(\Omega))$ a solution of (3.2.1)–(3.2.2). We assume that $z \in \mathcal{C}^0(0,T;L^2(\Omega))$. For each $n \geq 1$, we set

$$\boldsymbol{e}_{\boldsymbol{u}}^n = \boldsymbol{u}_h^n - P_h(\boldsymbol{u}(t_n)),\ e_p^n = p_h^n - \frac{1}{k}\int_{t_{n-1}}^{t_n} r_h(p(s))\,\mathrm{d}s,\ e_z^n = z_h^n - R_h(z(t_n)),$$

$$\boldsymbol{e}_{\boldsymbol{u}}^0 = \boldsymbol{0},\ e_z^0 = 0,$$

and we easily derive the next error equation for $\boldsymbol{u}_h^n$. To simplify the formulas below, we denote with a prime the derivative with respect to time.

LEMMA 3.3.4. *We have for each n with $1 \leq n \leq N$ and for all $\boldsymbol{v}_h \in X_h$,*

$$\begin{aligned}
&\left(\boldsymbol{e}_{\boldsymbol{u}}^n - \boldsymbol{e}_{\boldsymbol{u}}^{n-1}, \boldsymbol{v}_h\right) + \alpha\left(\nabla(\boldsymbol{e}_{\boldsymbol{u}}^n - \boldsymbol{e}_{\boldsymbol{u}}^{n-1}), \nabla \boldsymbol{v}_h\right) + \nu k\left(\nabla \boldsymbol{e}_{\boldsymbol{u}}^n, \nabla \boldsymbol{v}_h\right)\\
&+ k\left(z_h^{n-1} \times \boldsymbol{e}_{\boldsymbol{u}}^n, \boldsymbol{v}_h\right) - k\left(e_p^n, \operatorname{div} \boldsymbol{v}_h\right) = -\int_{t_{n-1}}^{t_n}\left(e_z^{n-1} \times \boldsymbol{u}(t_n), \boldsymbol{v}_h\right)\mathrm{d}s\\
&+ E_{\boldsymbol{u}'}^n + E_{\boldsymbol{u}}^n + E_p^n + N_{\boldsymbol{u}}^n + N_z^n,
\end{aligned} \tag{3.3.18}$$

where

$$E_{\boldsymbol{u}'}^n = \int_{t_{n-1}}^{t_n}\left(\left(\boldsymbol{u}'(s) - P_h(\boldsymbol{u}'(s)), \boldsymbol{v}_h\right) + \alpha\left(\nabla(\boldsymbol{u}'(s) - P_h(\boldsymbol{u}'(s))), \nabla \boldsymbol{v}_h\right)\right)\mathrm{d}s,$$

$$E_{\boldsymbol{u}}^n = \nu\int_{t_{n-1}}^{t_n}(\nabla(\boldsymbol{u}(s) - P_h(\boldsymbol{u}(s))) + \nabla P_h(\boldsymbol{u}(s) - \boldsymbol{u}(t_n)), \nabla \boldsymbol{v}_h)\mathrm{d}s,$$

$$E_p^n = -\int_{t_{n-1}}^{t_n}(p(s) - r_h(p(s)), \operatorname{div} \boldsymbol{v}_h)\mathrm{d}s,$$

$$N_{\boldsymbol{u}}^n = -\int_{t_{n-1}}^{t_n}\left(z_h^{n-1} \times (P_h(\boldsymbol{u}(t_n)) - \boldsymbol{u}(t_n)) + z(s) \times (\boldsymbol{u}(t_n) - \boldsymbol{u}(s))\,, \boldsymbol{v}_h\right)\mathrm{d}s,$$

$$N_z^n = -\int_{t_{n-1}}^{t_n}(R_h\,(z(t_{n-1}) - z(s)) \times \boldsymbol{u}(t_n) + (R_h(z(s)) - z(s)) \times \boldsymbol{u}(s), \boldsymbol{v}_h)\mathrm{d}s.$$

By choosing $\boldsymbol{v}_h = \boldsymbol{e}_{\boldsymbol{u}}^n \in V_h$ in (3.3.18), by using the equality

$$\int_{t_{n-1}}^{t_n} (g(s) - g(t_n))\,\mathrm{d}s = -\int_{t_{n-1}}^{t_n} (s - t_{n-1})g'(s)\mathrm{d}s,$$

that holds for all functions $g \in W^{1,1}(0, T)$, by suitably applying Young's inequality so that the term $\nu k|\boldsymbol{e}_{\boldsymbol{u}}^n|^2_{H^1(\Omega)}$ in the left-hand side absorbs all contributions of $|\boldsymbol{e}_{\boldsymbol{u}}^n|^2_{H^1(\Omega)}$ in the right-hand side, and by summing over n, we obtain the following error estimate for the velocity.

THEOREM 3.3.5. *If z belongs to $H^1(0, T; L^2(\Omega))$, we have for each n with $1 \le n \le N$,*

$$\begin{aligned}\|\boldsymbol{e}_{\boldsymbol{u}}^n\|_\alpha^2 &+ \sum_{i=1}^n \|\boldsymbol{e}_{\boldsymbol{u}}^i - \boldsymbol{e}_{\boldsymbol{u}}^{i-1}\|_\alpha^2 + \nu \sum_{i=1}^n k|\boldsymbol{e}_{\boldsymbol{u}}^i|^2_{H^1(\Omega)}\\ &\le 9\frac{S_4^2}{\nu}\|\boldsymbol{u}\|^2_{L^\infty(0,t_n;L^4(\Omega)^2)} \sum_{i=0}^{n-1} k\|e_z^i\|^2_{L^2(\Omega)}\\ &+ \mathcal{A}_{1,\boldsymbol{u}'}^n + \mathcal{A}_{1,\boldsymbol{u}}^n + \mathcal{A}_{1,z}^n + \mathcal{A}_{1,p}^n + \mathcal{A}_{1,k}^n,\end{aligned} \tag{3.3.19}$$

where the $\mathcal{A}_{1,\cdot}$ refer to the following approximation errors:

$$\begin{aligned}\mathcal{A}_{1,\boldsymbol{u}'}^n &= \frac{9}{\nu}\left(S_2^2 + \alpha^2\right)\|\boldsymbol{u}' - P_h(\boldsymbol{u}')\|^2_{L^2(0,t_n;H^1(\Omega)^2)},\\ \mathcal{A}_{1,\boldsymbol{u}}^n &= 9\left(\nu\|\boldsymbol{u} - P_h(\boldsymbol{u})\|^2_{L^2(0,t_n;H^1(\Omega)^2)} + t_n\frac{S_4^2}{\nu}C_{hz}^2\|\boldsymbol{u} - P_h(\boldsymbol{u})\|^2_{L^\infty(0,t_n;L^4(\Omega)^2)}\right),\\ \mathcal{A}_{1,z}^n &= \frac{9}{\nu}S_4^2\|\boldsymbol{u}\|^2_{L^\infty(0,t_n;L^4(\Omega)^2)}\|z - R_h(z)\|^2_{L^2(\Omega\times]0,t_n[)},\\ \mathcal{A}_{1,p}^n &= \frac{9}{\nu}\|p - r_h(p)\|^2_{L^2(\Omega\times]0,t_n[)},\\ \mathcal{A}_{1,k}^n &= 3k^2\left(\frac{S_4^2}{\nu}\left(\|\boldsymbol{u}\|^2_{L^\infty(0,t_n;L^4(\Omega))}\|R_h(z')\|^2_{L^2(\Omega\times]0,t_n[)}\right.\right.\\ &\quad\left.\left. + \|\boldsymbol{u}'\|^2_{L^2(0,t_n;L^4(\Omega))}\|z\|^2_{L^\infty(0,t_n;L^2(\Omega))}\right) + \nu\|P_h(\boldsymbol{u}')\|^2_{L^2(0,t_n;H^1(\Omega)^2)}\right).\end{aligned}$$

For z_h^n, we have the next error equation. Its proof is easy.

LEMMA 3.3.6. *Assume that z belongs to $L^2(0, T; H^1(\Omega)) \cap H^1(0, T; L^2(\Omega))$. We have for each n with $1 \le n \le N$, and for all $\theta_h \in Z_h$,*

$$\begin{aligned}\alpha\left(e_z^n - e_z^{n-1}, \theta_h\right) &+ k\left(\nu\left(e_z^n, \theta_h\right) + \alpha\left(\boldsymbol{u}_h^n \cdot \nabla e_z^n, \theta_h\right) + \frac{\alpha}{2}\left((\operatorname{div}\boldsymbol{u}_h^n)e_z^n, \theta_h\right)\right)\\ &= \int_{t_{n-1}}^{t_n}\left(\nu \operatorname{curl}\boldsymbol{e}_{\boldsymbol{u}}^n - \alpha\boldsymbol{e}_{\boldsymbol{u}}^n \cdot \nabla z(s) - \frac{\alpha}{2}(\operatorname{div}\boldsymbol{e}_{\boldsymbol{u}}^n)R_h(z(t_n)), \theta_h\right)\mathrm{d}s\\ &+ \left(E_z^n + N_z^n + N_{\mathrm{div}}^n + E_{\mathrm{curl}}^n, \theta_h\right),\end{aligned} \tag{3.3.20}$$

where

$$E_z^n = \nu \int_{t_{n-1}}^{t_n} (z(s) - R_h(z(s)) + R_h(z(s) - z(t_n)))\mathrm{d}s$$
$$+\alpha \int_{t_{n-1}}^{t_n} \big(z'(s) - R_h(z'(s))\big)\mathrm{d}s,$$

$$N_z^n = -\alpha \int_{t_{n-1}}^{t_n} \big(\boldsymbol{u}_h^n \cdot \nabla (R_h(z(t_n)) - z(s)) + R_h(z(s)) - z(s)\big)$$
$$+ P_h(\boldsymbol{u}(t_n) - \boldsymbol{u}(s)) \cdot \nabla z(s) + (P_h(\boldsymbol{u}(s)) - \boldsymbol{u}(s)) \cdot \nabla z(s))\,\mathrm{d}s,$$

$$N_{\mathrm{div}}^n = -\frac{\alpha}{2} \int_{t_{n-1}}^{t_n} \mathrm{div}\,(P_h(\boldsymbol{u}(t_n)) - \boldsymbol{u}(t_n))\, R_h(z(t_n))\mathrm{d}s,$$

$$E_{\mathrm{curl}}^n = \nu \int_{t_{n-1}}^{t_n} (\mathrm{curl}\, P_h(\boldsymbol{u}(t_n) - \boldsymbol{u}(s)) + \mathrm{curl}(P_h(\boldsymbol{u}(s)) - \boldsymbol{u}(s)))\mathrm{d}s.$$

By analogy with (3.3.10), we set

$$C_{h\boldsymbol{u}} = \sup_{1 \le n \le N} \|\boldsymbol{u}_h^n\|_{L^\infty(\Omega)}. \tag{3.3.21}$$

Under the assumptions of the second part of Proposition 3.3.1 and if the mesh satisfies (2.1.45) for $r > 2$, then $C_{h\boldsymbol{u}}$ is bounded independently of h and k, see (3.3.16).

Next, by choosing $\theta_h = z_h^n$ in (3.3.20), we derive a first error inequality for z_h^n. The proof is straightforward, but the regularity assumption on z' cannot be checked on the data in a domain with corners.

LEMMA 3.3.7. *Assume that z is in $L^\infty(0, T; W^{1,r}(\Omega))$ for some $r > 2$, and z' is in $L^2(0, T; H^1(\Omega))$. We have for each n with $1 \le n \le N$,*

$$\alpha \left(\|e_z^n\|_{L^2(\Omega)}^2 - \|e_z^{n-1}\|_{L^2(\Omega)}^2 + \|e_z^n - e_z^{n-1}\|_{L^2(\Omega)}^2\right) + 2\nu k \|e_z^n\|_{L^2(\Omega)}^2$$
$$\le 2\sqrt{k}\|e_z^n\|_{L^2(\Omega)} \left(K_3(r, z)\sqrt{k}|e_{\boldsymbol{u}}^n|_{H^1(\Omega)} + \mathcal{A}_{2,k}^n + \mathcal{A}_{2,\boldsymbol{u}}^n + \mathcal{A}_{2,z}^n\right), \tag{3.3.22}$$

where

$$K_3(r, z) = \nu + \alpha \left(S_{r^\star} + \frac{1}{2} C_{\infty,r} E_r\right) \|z\|_{L^\infty(0,t_n; W^{1,r}(\Omega))},$$

$C_{\infty,r}$ is the constant of (1.1.18), E_r is defined in (2.1.55), $S_{r^\star}$ is the Sobolev imbedding constant of (1.1.3), and the $\mathcal{A}_{2,\cdot}$ refer to the following approximation errors:

$$\mathcal{A}_{2,\boldsymbol{u}}^n = \nu \|\boldsymbol{u} - P_h(\boldsymbol{u})\|_{L^2(t_{n-1},t_n; H^1(\Omega)^2)} + \alpha \|z\|_{L^\infty(0,t_n; H^1(\Omega))} \|\boldsymbol{u} - P_h(\boldsymbol{u})\|_{L^2(t_{n-1},t_n; L^\infty(\Omega)^2)}$$
$$+ \frac{\alpha}{2} C_{\infty,r} E_r \sqrt{k} \|z\|_{L^\infty(0,t_n; W^{1,r}(\Omega))} \|\boldsymbol{u} - P_h(\boldsymbol{u})\|_{L^\infty(0,t_n; H^1(\Omega)^2)},$$

$$\mathcal{A}^n_{2,z} = \alpha\Big(C_{hu}\|z - R_h(z)\|_{L^2(t_{n-1},t_n;H^1(\Omega))} + \|z' - R_h(z')\|_{L^2(\Omega\times]t_{n-1},t_n[)}\Big) + \nu\|z - R_h(z)\|_{L^2(\Omega\times]t_{n-1},t_n[)},$$

$$\mathcal{A}^n_{2,k} = \frac{k}{\sqrt{3}}\Big(\alpha\|z\|_{L^\infty(0,t_n;H^1(\Omega))}\|P_h(\boldsymbol{u}')\|_{L^2(t_{n-1},t_n;L^\infty(\Omega)^2)} + \nu\|P_h(\boldsymbol{u}')\|_{L^2(t_{n-1},t_n;H^1(\Omega)^2)} + \nu\|R_h(z')\|_{L^2(\Omega\times]t_{n-1},t_n[)} + \alpha\, C_{hu}\|R_h(z')\|_{L^2(t_{n-1},t_n;H^1(\Omega))}\Big).$$

Theorem 3.3.5 gives the following error inequality for z^n_h.

THEOREM 3.3.8. *Suppose that z belongs to $L^\infty(0,T;W^{1,r}(\Omega))$ for some $r > 2$ and z' belongs to $L^2(0,T;H^1(\Omega))$. Then under the assumptions of the second part of Proposition 3.3.1 and if the mesh satisfies (2.1.45) for this r, we have for each n with $1 \le n \le N$,*

$$\|e^n_z\|^2_{L^2(\Omega)} + \sum_{i=1}^n \|e^i_z - e^{i-1}_z\|^2_{L^2(\Omega)} \le \exp\Big(\frac{18}{\alpha\nu^3}t_n\,\big(S_4K_3(r,z)\|\boldsymbol{u}\|_{L^\infty(0,t_n;L^4(\Omega)^2)}\big)^2\Big) \times \frac{2}{\nu}\Big(\mathcal{A}^n_{3,\boldsymbol{u}} + \mathcal{A}^n_{3,\boldsymbol{u}'} + \mathcal{A}^n_{3,z} + \mathcal{A}^n_{3,z'} + \mathcal{A}^n_{3,p} + \mathcal{A}^n_{3,k}\Big), \tag{3.3.23}$$

where the $\mathcal{A}_{3,\cdot}$ refer to the following approximation errors:

$$\mathcal{A}^n_{3,\boldsymbol{u}'} = \frac{9}{\alpha\nu^2}K_3^2(r,z)\left(S_2^2 + \alpha^2\right)\|\boldsymbol{u}' - P_h(\boldsymbol{u}')\|^2_{L^2(0,t_n;H^1(\Omega)^2)}, \tag{3.3.24}$$

$$\mathcal{A}^n_{3,\boldsymbol{u}} = \frac{3}{\alpha}\left(\nu^2 + 3K_3^2(r,z)\right)\|\boldsymbol{u} - P_h(\boldsymbol{u})\|^2_{L^2(0,t_n;H^1(\Omega)^2)} + \frac{9}{\alpha\nu^2}t_nS_4^2C_{hz}^2\|\boldsymbol{u} - P_h(\boldsymbol{u})\|^2_{L^\infty(0,t_n;L^4(\Omega)^2)} + 3\alpha\Big(\|z\|^2_{L^\infty(0,t_n;H^1(\Omega))}\|\boldsymbol{u} - P_h(\boldsymbol{u})\|^2_{L^2(0,t_n;L^\infty(\Omega)^2)} + \frac{1}{4}\,t_nC^2_{\infty,r}E_r^2\|\boldsymbol{u} - P_h(\boldsymbol{u})\|^2_{L^\infty(0,t_n;H^1(\Omega)^2)}\Big), \tag{3.3.25}$$

$$\mathcal{A}^n_{3,z'} = 3\,\alpha\|z' - R_h(z')\|^2_{L^2(\Omega\times]0,t_n[)}, \tag{3.3.26}$$

$$\mathcal{A}^n_{3,z} = 3\,\alpha C_{hu}^2\|z - R_h(z)\|^2_{L^2(0,t_n;H^1(\Omega))} + 3\left(\alpha\nu^2 + \frac{3}{\alpha\nu^2}K_3^2(r,z)S_4^2\|\boldsymbol{u}\|^2_{L^\infty(0,t_n;L^4(\Omega)^2)}\right)\|z - R_h(z)\|^2_{L^2(\Omega\times]0,t_n[)}, \tag{3.3.27}$$

$$\mathcal{A}^n_{3,p} = \frac{9}{\alpha\nu^2}K_3^2(r,z)\|p - r_h(p)\|^2_{L^2(\Omega\times]0,t_n[)}, \tag{3.3.28}$$

$$\mathcal{A}^n_{3,k} = \frac{k^2}{\alpha}\Big(\frac{3}{\nu^2}K_3^2(r,z)S_4^2\Big(\|\boldsymbol{u}\|^2_{L^\infty(0,t_n;L^4(\Omega)^2)}\|R_h(z')\|^2_{L^2(\Omega\times]0,t_n[)} + \|\boldsymbol{u}'\|^2_{L^2(0,t_n;L^4(\Omega)^2)}\|z\|^2_{L^\infty(0,t_n;L^2(\Omega))}\Big) + 3\,K_3^2(r,z)\|P_h(\boldsymbol{u}')\|^2_{L^2(0,t_n;H^1(\Omega)^2)} + \frac{4}{3}\Big(\alpha^2\|z\|^2_{L^\infty(0,t_n;H^1(\Omega))}\|P_h(\boldsymbol{u}')\|^2_{L^2(0,t_n;L^\infty(\Omega)^2)}$$

$$+ \nu^2 \left(\|P_h(\boldsymbol{u}')\|^2_{L^2(0,t_n;H^1(\Omega)^2)} + \|R_h(z')\|^2_{L^2(\Omega\times]0,t_n[)} \right)$$

$$+ \alpha^2 C_{hu}^2 \|R_h(z')\|^2_{L^2(0,t_n;H^1(\Omega))} \Bigg)\Bigg). \tag{3.3.29}$$

PROOF. Let us sketch the proof, it is technical but involves no difficulty. First, we make repeated applications of Young's inequality in order to absorb the factor $k\|e_z^n\|^2_{L^2(\Omega)}$ by the corresponding term in the left-hand side of (3.3.22). Then we sum the resulting inequality over n. This eliminates entirely the unknowns in the right-hand side of (3.3.22) with the exception of

$$\frac{2}{\alpha\nu} K_3^2(r,z) \sum_{i=1}^{n} k|\boldsymbol{e}_{\boldsymbol{u}}^i|^2_{H^1(\Omega)},$$

that is bounded in (3.3.19) by terms that only involve interpolation or approximation errors and

$$\frac{18}{\alpha\nu^3} K_3^2(r,z) S_4^2 \|\boldsymbol{u}\|^2_{L^\infty(0,t_n;L^4(\Omega)^2)} \sum_{i=0}^{n-1} k\|e_z^i\|^2_{L^2(\Omega)}.$$

Then (3.3.23) follows from Lemma 3.1.2. □

By combining this error estimate for z with the error estimate for $\boldsymbol{u}$ of Theorem 3.3.5, we obtain the following order of convergence for $\boldsymbol{u}$ and z. Of course, the order of convergence in space is determined by the regularity of the solution and the accuracy of P_h, r_h, and R_h. If the solution is smoother and higher-order elements are used, such as the Taylor–Hood element, then the order of convergence in space increases.

THEOREM 3.3.9. *Let Ω be convex and suppose that the solution $(\boldsymbol{u}, p, z)$ of (3.2.1)–(3.2.2) satisfies $z \in L^\infty(0,T;W^{1,r}(\Omega)) \cap L^2(0,T;H^2(\Omega))$ for some $r > 2$, $z' \in L^2(0,T;H^1(\Omega))$, $\boldsymbol{u} \in H^1(0,T;H^2(\Omega)^2)$, and $p \in L^2(0,T;H^1(\Omega))$. Then if the triangulation satisfies (2.1.45) for this r, the assumptions of Hypothesis 2.1.10 hold, and R_h satisfies (2.4.10), the scheme (3.3.2)–(3.3.4) is of order one in time and space:*

$$\sup_{1\le n\le N} \left(\|\boldsymbol{e}_{\boldsymbol{u}}^n\|_\alpha + \|e_z^n\|_{L^2(\Omega)} \right) \le C(h+k),$$

with a constant C independent of h and k.

As usual, an error estimate for the pressure must be preceded by an error estimate for the difference quotient. By testing (3.3.18) with $\boldsymbol{v}_h = \boldsymbol{e}_{\boldsymbol{u}}^n - \boldsymbol{e}_{\boldsymbol{u}}^{n-1}$, dividing by k, summing over n and substituting the bounds of Theorems 3.3.5 and 3.3.8, we readily derive the next extension of Theorem 3.3.9 under the same assumptions:

$$\left(\sum_{n=1}^{N} \frac{1}{k} \|\boldsymbol{e}_{\boldsymbol{u}}^n - \boldsymbol{e}_{\boldsymbol{u}}^{n-1}\|^2_\alpha \right)^{1/2} \le C(h+k),$$

with another constant C independent of h and k. Then the error estimate for the pressure:

$$\left(\sum_{n=1}^{N} k\|e_p^n\|_{L^2(\Omega)}^2\right)^{1/2} \leq C(h+k),$$

with a constant C independent of h and k, follows from the discrete inf-sup condition (2.1.1) and the above estimates. Of course, the order h in space is dictated by the above regularity of the solution and the choice of finite-elements. If the Taylor–Hood element is used and the solution is correspondingly smoother then an order of h^2 in space can be obtained. It is important to point out that all the above results are derived without requiring a CFL condition.

3.4. Fully discrete upwind scheme with discontinuous Galerkin

We follow the approach of ABBOUD and SAYAH [2009]. Reverting to the setting of Section 2.4.2, let $\mathcal{T}_h$ be a family of triangulations satisfying (2.1.25), let X_h and M_h be chosen as in Section 2.1.2, with the same assumptions, and for each t, let z be discretized in a finite-dimensional space $Z_h \subset L^2(\Omega)$, such as (2.4.9):

$$Z_h = \{\theta_h \in L^2(\Omega);\ \forall T \in \mathcal{T}_h,\ \theta_h|_T \in \mathbb{P}_k\},$$

with degree $k \geq 1$. The operators P_h and r_h are unchanged, and R_h is chosen as in Section 2.4.2 satisfying (2.4.10):

$$\forall \theta \in W^{s+1,r}(\Omega),\ |R_h(\theta) - \theta|_{W^{m,r}(\Omega)} \leq C\, h^{s+1-m}|\theta|_{W^{s+1,r}(\Omega)},$$

for any number $r \geq 1$, for $m = 0, 1$, and $0 \leq s \leq k$. In addition, we introduce right away the local projection $\varrho_h(z) \in \mathbb{P}_k$ defined by

$$\forall q \in \mathbb{P}_k,\ \int_T (\varrho_h(z) - z) q\, \mathrm{d}\boldsymbol{x} = 0.$$

We have seen in Section 2.4.2 that this local projection permits to recover sharper error estimates. In particular, it is a convenient tool for starting the algorithm.

The nonlinear term $(\boldsymbol{u} \cdot \nabla z, \theta)$ is approximated by (2.4.12)

$$\tilde{c}^{\mathrm{DG}}(\boldsymbol{u}_h; z_h, \theta_h) = \sum_{T \in \mathcal{T}_h} \left(\int_T (\boldsymbol{u}_h \cdot \nabla z_h)\theta_h\, \mathrm{d}\boldsymbol{x} + \int_{\partial T_-} |\boldsymbol{u}_h \cdot \boldsymbol{n}|(z_h^{\mathrm{int}} - z_h^{\mathrm{ext}})\theta_h^{\mathrm{int}} \mathrm{d}\sigma \right) + \frac{1}{2} \int_\Omega (\operatorname{div} \boldsymbol{u}_h) z_h \theta_h\, \mathrm{d}\boldsymbol{x},$$

where the element of arc length is denoted by σ to avoid confusion with the element of time, and as previously, the superscript int (resp. ext) denotes the trace on the segments of ∂T of the function taken inside (resp. outside) T. It coincides with $(\boldsymbol{u} \cdot \nabla z, \theta_h)$ when $\boldsymbol{u}$ belongs to V and z is in $H^1(\Omega)$.

The time stepping is the same as in Section 3.3, namely, the interval $[0, T]$ is divided into N equal segments of length k, with end points $t_i = i\,k$, $0 \le i \le N$, and we approximate problem (3.2.1)–(3.2.2) with:

• Set

$$\boldsymbol{u}_h^0 = P_h(\boldsymbol{u}_0),\ z_h^0 = \varrho_h(z_0),\ \boldsymbol{z}_h^0 = (0, 0, z_h^0). \tag{3.4.1}$$

• Knowing $\boldsymbol{u}_h^0 \in X_h$ and $z_h^0 \in Z_h$, find sequences $(\boldsymbol{u}_h^n)_{n\ge1}$, $(z_h^n)_{n\ge1}$, and $(p_h^n)_{n\ge1}$ such that $\boldsymbol{u}_h^n \in X_h$, $z_h^n \in Z_h$, and $p_h^n \in M_h$ solve for $1 \le n \le N$,

$$\forall \boldsymbol{v}_h \in X_h,\ \frac{1}{k}(\boldsymbol{u}_h^n - \boldsymbol{u}_h^{n-1}, \boldsymbol{v}_h) + \frac{\alpha}{k}(\nabla(\boldsymbol{u}_h^n - \boldsymbol{u}_h^{n-1}), \nabla \boldsymbol{v}_h) + \nu(\nabla \boldsymbol{u}_h^n, \nabla \boldsymbol{v}_h)$$

$$+ (\boldsymbol{z}_h^{n-1} \times \boldsymbol{u}_h^n, \boldsymbol{v}_h) - (p_h^n, \operatorname{div} \boldsymbol{v}_h) = (\boldsymbol{f}^n, \boldsymbol{v}_h), \tag{3.4.2}$$

$$\forall q_h \in M_h,\ (q_h, \operatorname{div} \boldsymbol{u}_h^n) = 0, \tag{3.4.3}$$

$$\forall \theta_h \in Z_h,\ \frac{\alpha}{k}(z_h^n - z_h^{n-1}, \theta_h) + \nu(z_h^n, \theta_h) + \alpha \tilde{c}^{\mathrm{DG}}(\boldsymbol{u}_h^n; z_h^n, \theta_h) = \nu(\operatorname{curl} \boldsymbol{u}_h^n, \theta_h)$$

$$+ \alpha(\operatorname{curl} \boldsymbol{f}^n, \theta_h), \tag{3.4.4}$$

where $\boldsymbol{f}^n$ is defined by (3.2.18):

$$\boldsymbol{f}^n(\boldsymbol{x}) = \frac{1}{k} \int_{t_{n-1}}^{t_n} \boldsymbol{f}(\boldsymbol{x}, s)\, ds.$$

We already know that at each step, (3.4.2)–(3.4.3) has a unique solution $(\boldsymbol{u}_h^n, p_h^n)$. Consequently, recalling that $\tilde{c}^{\mathrm{DG}}$ satisfies (2.4.16) for all $\boldsymbol{u}_h \in X_h$, θ_h and $z_h \in Z_h$:

$$\tilde{c}^{\mathrm{DG}}(\boldsymbol{u}_h; z_h, \theta_h) = \sum_{T \in \mathcal{T}_h} \left(-\int_T (\boldsymbol{u}_h \cdot \nabla \theta_h) z_h \, d\boldsymbol{x} + \int_{\partial T_-} |\boldsymbol{u}_h \cdot \boldsymbol{n}_T| (\theta_h^{\mathrm{ext}} - \theta_h^{\mathrm{int}}) z_h^{\mathrm{ext}} d\sigma \right) - \frac{1}{2} \int_\Omega (\operatorname{div} \boldsymbol{u}_h) \theta_h z_h \, d\boldsymbol{x},$$

that implies (2.4.21) for all $\boldsymbol{u}_h \in X_h$, and all $z_h \in Z_h$:

$$\tilde{c}^{\mathrm{DG}}(\boldsymbol{u}_h; z_h, z_h) = \frac{1}{2} \sum_{T \in \mathcal{T}_h} \int_{\partial T_-} |\boldsymbol{u}_h \cdot \boldsymbol{n}_T| (z_h^{\mathrm{ext}} - z_h^{\mathrm{int}})^2 d\sigma,$$

then at each step, (3.4.4) also has a unique solution.

Regarding a priori estimates, as (3.4.2), (3.4.3) are unchanged, clearly the sequence $(\boldsymbol{u}_h^n)_{n\geq 1}$, satisfies the a priori estimates (3.3.5) and (3.3.6):

$$\|\boldsymbol{u}_h^n\|_\alpha^2 + \sum_{i=1}^n \|\boldsymbol{u}_h^i - \boldsymbol{u}_h^{i-1}\|_\alpha^2 \leq \frac{S_2^2}{2\nu}\|\boldsymbol{f}\|^2_{L^2(\Omega\times]0,t_n[)} + \|\boldsymbol{u}_h^0\|_\alpha^2,$$

$$\nu\sum_{i=1}^n k|\boldsymbol{u}_h^i|^2_{H^1(\Omega)} \leq \frac{S_2^2}{\nu}\|\boldsymbol{f}\|^2_{L^2(\Omega\times]0,t_n[)} + \|\boldsymbol{u}_h^0\|_\alpha^2,$$

while $(z_h^n)_{n\geq 1}$ satisfies

$$\|z_h^n\|^2_{L^2(\Omega)} + \sum_{i=1}^n \|z_h^i - z_h^{i-1}\|^2_{L^2(\Omega)} + \alpha\sum_{i=1}^n \sum_{T\in\mathcal{T}_h}\int_{\partial T_-} |\boldsymbol{u}_h^i\cdot\boldsymbol{n}_T|((z_h^i)^{\text{ext}} - (z_h^i)^{\text{int}})^2 d\sigma$$

$$\leq \frac{S_2^2}{\alpha\nu}\|\boldsymbol{f}\|^2_{L^2(\Omega\times]0,t_n[)} + \frac{\alpha}{\nu}\|\text{curl}\,\boldsymbol{f}\|^2_{L^2(\Omega\times]0,t_n[)} + \frac{1}{\alpha}\|\boldsymbol{u}_h^0\|_\alpha^2 + \|z_h^0\|^2_{L^2(\Omega)}. \tag{3.4.5}$$

With this bound, the divided differences and pressures satisfy (3.3.8) and (3.3.9), respectively,

$$\sum_{i=1}^n \frac{1}{k}\|\boldsymbol{u}_h^i - \boldsymbol{u}_h^{i-1}\|_\alpha^2 \leq \nu|\boldsymbol{u}_h^0|^2_{H^1(\Omega)} + \|\boldsymbol{f}\|^2_{L^2(\Omega\times]0,t_n[)}$$
$$+ \frac{1}{\alpha\nu}S_4^4 C_{hz}^2\left(\|\boldsymbol{u}_h^0\|_\alpha^2 + \frac{S_2^2}{\nu}\|\boldsymbol{f}\|^2_{L^2(\Omega\times]0,t_n[)}\right),$$

$$\sum_{i=1}^n k\|p_h^i\|^2_{L^2(\Omega)} \leq \frac{3}{\beta^2}\left(S_2^2\left(\nu|\boldsymbol{u}_h^0|^2_{H^1(\Omega)} + 2\|\boldsymbol{f}\|^2_{L^2(\Omega\times]0,t_n[)}\right)\right.$$
$$\left. + C_{hz}^2\frac{S_4^4}{\nu}\left(\|\boldsymbol{u}_h^0\|_\alpha^2 + \frac{S_2^2}{\nu}\|\boldsymbol{f}\|^2_{L^2(\Omega\times]0,t_n[)}\right)\left(1 + \frac{{S_2}^2}{\alpha}\right)\right),$$

where

$$C_{hz} = \sup_{0\leq n\leq N-1}\|z_h^n\|_{L^2(\Omega)}, \tag{3.4.6}$$

z_h^n being the sequence generated by (3.4.2)–(3.4.4). In view of (3.4.5), C_{hz} is uniformly bounded with respect to h, and hence so are the sequences of divided differences and pressures. Similarly, considering that the L^∞ estimate for $\boldsymbol{u}_h^n$ derived in Section 3.3 only depends on a uniform upper bound for z_h^n, we can easily check that the statement of Corollary 3.3.2 applies to (3.4.2)–(3.4.4).

COROLLARY 3.4.1. *Under the assumptions of the second part of Proposition 3.3.1, there exists a constant C_1, independent of h and k, such that for all n, $1 \leq n \leq N$, the sequence*

$(\boldsymbol{u}_h^n)_{n=1}^N$ *constructed by (3.4.2)–(3.4.4) satisfies*

$$\|\boldsymbol{u}_h^n - P_h(\boldsymbol{v}^n)\|_\alpha^2 + \sum_{i=1}^n \|\boldsymbol{u}_h^i - \boldsymbol{u}_h^{i-1} - P_h(\boldsymbol{v}^i - \boldsymbol{v}^{i-1})\|_\alpha^2 \le C_1 h^2. \tag{3.4.7}$$

In addition, if the mesh satisfies (2.1.45) for some $r > 2$:

$$h \le C_2\, \varrho_{\min}^{1-2/r},$$

then there exists a constant C_3, independent of h and k, such that for all n, $1 \le n \le N$,

$$\|\boldsymbol{u}_h^n\|_{W^{1,r}(\Omega)} \le C_3. \tag{3.4.8}$$

This implies that $\boldsymbol{u}_h^n$ is uniformly bounded: There exists another constant C_{hu}, independent of h and k, such that for all n, $1 \le n \le N$,

$$\|\boldsymbol{u}_h^n\|_{L^\infty(\Omega)} \le C_{hu}. \tag{3.4.9}$$

Convergence

For convergence as h tends to zero, we have the analog of Proposition 3.3.3.

PROPOSITION 3.4.2. *We keep the assumptions of Hypothesis 2.1.5 for P_h and r_h, and we take (2.4.10) for R_h. Then the following strong convergences hold for each n, $1 \le n \le N$, as h tends to zero, without restriction to subsequences:*

$$\begin{aligned} \lim_{h\to 0} \|\boldsymbol{u}_h^n - \boldsymbol{u}^n\|_\alpha &= 0,\\ \lim_{h\to 0} \|p_h^n - p^n\|_{L^2(\Omega)} &= 0,\\ \lim_{h\to 0} \|z_h^n - z^n\|_{L^2(\Omega)} &= 0,\\ \lim_{h\to 0}\left(\sum_{T\in\mathcal{T}_h}\int_{\partial T_-} |\boldsymbol{u}_h^n\cdot\boldsymbol{n}_T|((z_h^n)^{\text{ext}} - (z_h^n)^{\text{int}})^2 d\sigma\right) &= 0, \end{aligned} \tag{3.4.10}$$

where $(\boldsymbol{u}^n, p^n, z^n)$ solve the semi-discrete scheme (3.2.20)–(3.2.21).

PROOF. Here also, we argue by induction.

1. For $n = 0$, the approximation properties of the operators P_h and ϱ_h imply

$$\lim_{h\to 0} \boldsymbol{u}_h^0 = \boldsymbol{u}^0 \text{ strongly in } H^1(\Omega)^2,\ \lim_{h\to 0} z_h^0 = z^0 \text{ strongly in } L^2(\Omega).$$

2. For $n \ge 0$, on one hand, we assume that the following strong convergences hold as h tends to zero, without restriction to subsequences:

$$\lim_{h\to 0} \boldsymbol{u}_h^n = \boldsymbol{u}^n \text{ strongly in } H^1(\Omega)^2,\ \lim_{h\to 0} z_h^n = z^n \text{ strongly in } L^2(\Omega).$$

On the other hand, it follows from (3.3.5), (3.4.5), and (3.3.9), that the following weak convergences hold, as h tends to zero up to subsequences, to some functions $\boldsymbol{u}^{n+1} \in H_0^1(\Omega)^2$, $z^{n+1} \in L^2(\Omega)$, and $p^{n+1} \in L_0^2(\Omega)$:

$$\lim_{h\to 0} \boldsymbol{u}_h^{n+1} = \boldsymbol{u}^{n+1} \text{ weakly in } H^1(\Omega)^2,$$
$$\lim_{h\to 0} z_h^{n+1} = z^{n+1} \text{ weakly in } L^2(\Omega),$$
$$\lim_{h\to 0} p_h^{n+1} = p^{n+1} \text{ weakly in } L^2(\Omega).$$

By passing to the limit in (3.4.2) and (3.4.3) and arguing as in Proposition 2.1.6, we easily derive that the limit functions $(\boldsymbol{u}^{n+1}, p^{n+1})$ satisfy (3.2.20) at time t_{n+1}:

$$\frac{1}{k}(\boldsymbol{u}^{n+1} - \boldsymbol{u}^n) - \alpha\frac{1}{k}\Delta(\boldsymbol{u}^{n+1} - \boldsymbol{u}^n) - \nu\Delta\,\boldsymbol{u}^{n+1} + z^n \times \boldsymbol{u}^{n+1} + \nabla p^{n+1} = \boldsymbol{f}^{n+1} \quad \text{in } \Omega,$$
$$\operatorname{div} \boldsymbol{u}^{n+1} = 0 \quad \text{in } \Omega,$$

that the whole sequences converge, and that the convergence is strong. Next, we pass to the limit in (3.4.4) at time t_{n+1}. For this, we proceed as in Theorem 2.4.10, and in particular, we use (2.4.16) and the continuity of $R_h(\theta)$ to pass to the limit in $\tilde{c}^{\mathrm{DG}}(\boldsymbol{u}_h^{n+1}; z_h^{n+1}, R_h(\theta))$ for smooth enough θ. The strong convergence of z_h^{n+1} is also established as in Theorem 2.4.10. To be specific, (3.4.4) is tested with z_h^{n+1} and compared with (3.2.21) at time t_{n+1}:

$$\begin{aligned}\Big(\frac{\alpha}{k} + \nu\Big)\|z_h^{n+1}\|^2_{L^2(\Omega)} - \frac{\alpha}{k}\Big(z_h^n, z_h^{n+1}\Big) + \sum_{T\in\mathcal{T}_h}\int_{\partial T_-} |\boldsymbol{u}_h^{n+1}\cdot\boldsymbol{n}_T|\Big(\big(z_h^{n+1}\big)^{\mathrm{ext}} - \big(z_h^{n+1}\big)^{\mathrm{int}}\Big)^2 d\sigma \\ = \nu\Big(\operatorname{curl}\Big(\boldsymbol{u}_h^{n+1} - \boldsymbol{u}^{n+1}\Big), z_h^{n+1}\Big) + \Big(\frac{\alpha}{k} + \nu\Big)\Big(z^{n+1}, z_h^{n+1}\Big) - \frac{\alpha}{k}\Big(z^n, z_h^{n+1}\Big) \\ + \alpha\Big(\boldsymbol{u}^{n+1}\cdot\nabla z^{n+1}, z_h^{n+1}\Big).\end{aligned} \tag{3.4.11}$$

Therefore,

$$\begin{aligned}\Big(\frac{\alpha}{k} + \nu\Big)\|z_h^{n+1}\|^2_{L^2(\Omega)} - \frac{\alpha}{k}\Big(z_h^n, z_h^{n+1}\Big) \le \nu\Big(\operatorname{curl}\Big(\boldsymbol{u}_h^{n+1} - \boldsymbol{u}^{n+1}\Big), z_h^{n+1}\Big) \\ + \Big(\frac{\alpha}{k} + \nu\Big)\Big(z^{n+1}, z_h^{n+1}\Big) - \frac{\alpha}{k}\Big(z^n, z_h^{n+1}\Big) \\ + \alpha\Big(\boldsymbol{u}^{n+1}\cdot\nabla z^{n+1}, z_h^{n+1}\Big).\end{aligned}$$

Owing to the induction hypothesis, the strong convergence of $\boldsymbol{u}_h^{n+1}$, and the weak convergence of z_h^{n+1}, this gives

$$\lim_{h\to 0}\|z_h^{n+1}\|^2_{L^2(\Omega)} \le \|z^{n+1}\|^2_{L^2(\Omega)},$$

whence the strong convergence of z_h^{n+1}. The last convergence in (3.4.10) follows immediately by using this strong convergence in (3.4.11). Finally, uniqueness of the solution of (3.2.21) yields convergence of the whole sequence. □

A priori error estimates

Let $(\boldsymbol{u}_h^n, p_h^n, z_h^n)$ be the solution of (3.4.2)–(3.4.4), and $(\boldsymbol{u}, p, z) \in \mathcal{W} \times L^\infty(0, T; L^2(\Omega))$ a solution of (3.2.1)–(3.2.2). We suppose that z belongs to $\mathcal{C}^0(0, T; L^2(\Omega))$. For each $n \ge 1$, we set

$$\boldsymbol{e}_{\boldsymbol{u}}^n = \boldsymbol{u}_h^n - P_h(\boldsymbol{u}(t_n)),\ e_p^n = p_h^n - \frac{1}{k}\int_{t_{n-1}}^{t_n} r_h(p(s))\,\mathrm{d}s,\ e_z^n = z_h^n - \varrho_h(z(t_n)).$$

We start with

$$\boldsymbol{e}_{\boldsymbol{u}}^0 = \boldsymbol{0},\ e_z^0 = 0,$$

and since the equations for $(\boldsymbol{u}_h^n, p_h^n)$ are the same as in Section 3.3, the error equation for $\boldsymbol{u}_h^n$ given in the statement of Lemma 3.3.4 also applies here with R_h replaced by ϱ_h. This is made possible by the fact that the continuity of $R_h(z)$ is not used in this lemma.

LEMMA 3.4.3. *We have for each n with $1 \leq n \leq N$ and for all $\boldsymbol{v}_h \in X_h$,*

$$\left(\boldsymbol{e}_{\boldsymbol{u}}^n - \boldsymbol{e}_{\boldsymbol{u}}^{n-1}, \boldsymbol{v}_h\right) + \alpha\left(\nabla(\boldsymbol{e}_{\boldsymbol{u}}^n - \boldsymbol{e}_{\boldsymbol{u}}^{n-1}), \nabla \boldsymbol{v}_h\right) + \nu k(\nabla \boldsymbol{e}_{\boldsymbol{u}}^n, \nabla \boldsymbol{v}_h)$$
$$+ k\left(z_h^{n-1} \times \boldsymbol{e}_{\boldsymbol{u}}^n, \boldsymbol{v}_h\right) - k\left(e_p^n, \operatorname{div} \boldsymbol{v}_h\right) = -\int_{t_{n-1}}^{t_n} \left(e_z^{n-1} \times \boldsymbol{u}(t_n), \boldsymbol{v}_h\right)\mathrm{d}s$$
$$+ E_{\boldsymbol{u}'}^n + E_{\boldsymbol{u}}^n + E_p^n + N_{\boldsymbol{u}}^n + N_z^n, \tag{3.4.12}$$

where

$$E_{\boldsymbol{u}'}^n = \int_{t_{n-1}}^{t_n} \left(\left(\boldsymbol{u}'(s) - P_h(\boldsymbol{u}'(s)), \boldsymbol{v}_h\right) + \alpha\left(\nabla(\boldsymbol{u}'(s) - P_h(\boldsymbol{u}'(s))), \nabla \boldsymbol{v}_h\right)\right)\mathrm{d}s,$$

$$E_{\boldsymbol{u}}^n = \nu \int_{t_{n-1}}^{t_n} (\nabla(\boldsymbol{u}(s) - P_h(\boldsymbol{u}(s))) + \nabla P_h(\boldsymbol{u}(s) - \boldsymbol{u}(t_n)), \nabla \boldsymbol{v}_h)\mathrm{d}s,$$

$$E_p^n = -\int_{t_{n-1}}^{t_n} (p(s) - r_h(p(s)), \operatorname{div} \boldsymbol{v}_h)\mathrm{d}s,$$

$$N_{\boldsymbol{u}}^n = -\int_{t_{n-1}}^{t_n} \left(z_h^{n-1} \times (P_h(\boldsymbol{u}(t_n)) - \boldsymbol{u}(t_n)) + z(s) \times (\boldsymbol{u}(t_n) - \boldsymbol{u}(s)), \boldsymbol{v}_h\right)\mathrm{d}s,$$

$$N_z^n = -\int_{t_{n-1}}^{t_n} (\varrho_h\,(z(t_{n-1}) - z(s)) \times \boldsymbol{u}(t_n) + (\varrho_h(z(s)) - z(s)) \times \boldsymbol{u}(s), \boldsymbol{v}_h)\,\mathrm{d}s.$$

Proceeding as in Section 3.3, the choice $\boldsymbol{v}_h = \boldsymbol{e}_{\boldsymbol{u}}^n \in V_h$ in (3.4.12), and the fact that

$$\int_{t_{n-1}}^{t_n} (g(s) - g(t_n))\mathrm{d}s = -\int_{t_{n-1}}^{t_n} (s - t_{n-1})g'(s)\mathrm{d}s,$$

yields the following error estimate for the velocity.

THEOREM 3.4.4. *If z belongs to $H^1(0,T;L^2(\Omega))$, we have for each n with $1 \le n \le N$,*

$$\|\boldsymbol{e}_{\boldsymbol{u}}^n\|_\alpha^2 + \sum_{i=1}^n \|\boldsymbol{e}_{\boldsymbol{u}}^i - \boldsymbol{e}_{\boldsymbol{u}}^{i-1}\|_\alpha^2 + \nu \sum_{i=1}^n k |\boldsymbol{e}_{\boldsymbol{u}}^i|_{H^1(\Omega)}^2 \le 9\frac{S_4^2}{\nu}\|\boldsymbol{u}\|_{L^\infty(0,t_n;L^4(\Omega)^2)}^2 \sum_{i=0}^{n-1} k\|e_z^i\|_{L^2(\Omega)}^2 + \mathcal{A}_{1,\boldsymbol{u}'}^n + \mathcal{A}_{1,\boldsymbol{u}}^n + \mathcal{A}_{1,z}^n + \mathcal{A}_{1,p}^n + \mathcal{A}_{1,k}^n, \tag{3.4.13}$$

where the $\mathcal{A}_{1,\cdot}$ refer to the following approximation errors:

$$\mathcal{A}_{1,\boldsymbol{u}'}^n = \frac{9}{\nu}\left(S_2^2 + \alpha^2\right)\|\boldsymbol{u}' - P_h(\boldsymbol{u}')\|_{L^2(0,t_n;H^1(\Omega)^2)}^2,$$

$$\mathcal{A}_{1,\boldsymbol{u}}^n = 9\left(\nu\|\boldsymbol{u} - P_h(\boldsymbol{u})\|_{L^2(0,t_n;H^1(\Omega)^2)}^2 + t_n\frac{S_4^2}{\nu}C_{hz}^2\|\boldsymbol{u} - P_h(\boldsymbol{u})\|_{L^\infty(0,t_n;L^4(\Omega)^2)}^2\right),$$

$$\mathcal{A}_{1,z}^n = \frac{9}{\nu}S_4^2\|\boldsymbol{u}\|_{L^\infty(0,t_n;L^4(\Omega)^2)}^2\|z - \varrho_h(z)\|_{L^2(\Omega\times]0,t_n[)}^2,$$

$$\mathcal{A}_{1,p}^n = \frac{9}{\nu}\|p - r_h(p)\|_{L^2(\Omega\times]0,t_n[)}^2,$$

$$\mathcal{A}_{1,k}^n = 3k^2\left(\frac{S_4^2}{\nu}\left(\|\boldsymbol{u}\|_{L^\infty(0,t_n;L^4(\Omega))}^2\|\varrho_h(z')\|_{L^2(\Omega\times]0,t_n[)}^2 + \|\boldsymbol{u}'\|_{L^2(0,t_n;L^4(\Omega))}^2\|z\|_{L^\infty(0,t_n;L^2(\Omega))}^2\right) + \nu\|P_h(\boldsymbol{u}')\|_{L^2(0,t_n;H^1(\Omega)^2)}^2\right).$$

The next lemma gives an error equation for z_h. As expected, it differs slightly in its treatment of the nonlinear term from the statement of Lemma 3.3.6.

LEMMA 3.4.5. *Assume that z belongs to $\mathcal{C}^0(0,T;H^1(\Omega)) \cap H^1(0,T;L^2(\Omega))$. We have for each n with $1 \le n \le N$, and for all $\theta_h \in Z_h$,*

$$\alpha(e_z^n - e_z^{n-1}, \theta_h) + k\left(\nu(e_z^n,\theta_h) + \alpha\tilde{c}^{\mathrm{DG}}(\boldsymbol{u}_h^n; e_z^n, \theta_h)\right) = \alpha\int_{t_{n-1}}^{t_n}\left(\tilde{c}^{\mathrm{DG}}(\boldsymbol{u}_h^n; z(t_n) - \varrho_h(z(t_n)), \theta_h) - \tilde{c}(\boldsymbol{e}_{\boldsymbol{u}}^n; z(t_n), \theta_h)\right)\mathrm{d}s + \nu\int_{t_{n-1}}^{t_n}(\operatorname{curl}\boldsymbol{e}_{\boldsymbol{u}}^n, \theta_h)\,\mathrm{d}s + \left(E_z^n + N_z^n + N_{\mathrm{div}}^n + E_{\mathrm{curl}}^n, \theta_h\right), \tag{3.4.14}$$

where $\tilde{c}$ is defined in (2.1.4),

$$E_z^n = \nu\int_{t_{n-1}}^{t_n}\left(z(s) - \varrho_h(z(s)) + (s - t_{n-1})\varrho_h(z'(s))\right)\mathrm{d}s + \alpha\int_{t_{n-1}}^{t_n}\left(z'(s) - \varrho_h(z'(s))\right)\mathrm{d}s,$$

$$N_z^n = \alpha \int_{t_{n-1}}^{t_n} \left((s - t_{n-1})\left(\boldsymbol{u}'(s) \cdot \nabla z(s) + \boldsymbol{u}(t_n) \cdot \nabla z'(s)\right)\right.$$
$$\left. + (\boldsymbol{u}(t_n) - P_h(\boldsymbol{u}(t_n)) \cdot \nabla z(t_n)\right) \mathrm{d}s,$$

$$N_{\mathrm{div}}^n = \frac{\alpha}{2} \int_{t_{n-1}}^{t_n} \operatorname{div}(\boldsymbol{u}(t_n) - P_h(\boldsymbol{u}(t_n))z(t_n) \, \mathrm{d}s,$$

$$E_{\mathrm{curl}}^n = \nu \int_{t_{n-1}}^{t_n} (\operatorname{curl} P_h(\boldsymbol{u}(t_n) - \boldsymbol{u}(s)) + \operatorname{curl}(P_h(\boldsymbol{u}(s)) - \boldsymbol{u}(s))) \, \mathrm{d}s.$$

By choosing $\theta_h = e_z^n$ and applying (2.4.16) and (2.4.21), (3.4.14) gives

$$\frac{\alpha}{2}\left(\|e_z^n\|_{L^2(\Omega)}^2 - \|e_z^{n-1}\|_{L^2(\Omega)}^2 + \|e_z^n - e_z^{n-1}\|_{L^2(\Omega)}^2\right) + k\nu\|e_z^n\|_{L^2(\Omega)}^2$$
$$+ \alpha k \sum_{T \in \mathfrak{T}_h} \frac{1}{2} \int_{\partial T_-} |\boldsymbol{u}_h^n \cdot \boldsymbol{n}_T|((e_z^n)^{\mathrm{ext}} - (e_z^n)^{\mathrm{int}})^2 \mathrm{d}\sigma$$
$$= \alpha k \left(\sum_{T \in \mathfrak{T}_h}\left(-\int_T (\boldsymbol{u}_h^n \cdot \nabla e_z^n + \frac{1}{2}\operatorname{div}(\boldsymbol{e}_{\boldsymbol{u}}^n) e_z^n)(z(t_n) - \varrho_h(z(t_n)) \, \mathrm{d}\boldsymbol{x}\right.\right.$$
$$\left.\left. + \int_{\partial T_-} |\boldsymbol{u}_h^n \cdot \boldsymbol{n}_T|((e_z^n)^{\mathrm{ext}} - (e_z^n)^{\mathrm{int}})\left(z(t_n) - \varrho_h(z(t_n))^{\mathrm{ext}} \mathrm{d}\sigma\right)\right) - \tilde{c}(\boldsymbol{e}_{\boldsymbol{u}}^n; z(t_n), e_z^n)\right)$$
$$+ \nu \int_{t_{n-1}}^{t_n} (\operatorname{curl} \boldsymbol{e}_{\boldsymbol{u}}^n, e_z^n) \, \mathrm{d}s + \left(E_z^n + N_z^n + N_{\mathrm{div}}^n + E_{\mathrm{curl}}^n, e_z^n\right). \tag{3.4.15}$$

The first and third terms in the right-hand side of (3.4.15) are handled as in Section 2.4.2:

$$\int_T (\boldsymbol{u}_h^n \cdot \nabla e_z^n)(z(t_n) - \varrho_h(z(t_n))) \, \mathrm{d}\boldsymbol{x} = \int_T (\boldsymbol{e}_{\boldsymbol{u}}^n \cdot \nabla e_z^n)(z(t_n) - \varrho_h(z(t_n))) \, \mathrm{d}\boldsymbol{x}$$
$$+ \int_T (P_h(\boldsymbol{u}(t_n) - \boldsymbol{c}) \cdot \nabla e_z^n (z(t_n) - \varrho_h(z(t_n))) \, \mathrm{d}\boldsymbol{x}, \tag{3.4.16}$$

where (2.4.26) is used for inserting an arbitrary constant $\boldsymbol{c}$ in each T. In view of (2.4.29), we have

$$\left|\int_T (\boldsymbol{e}_{\boldsymbol{u}}^n \cdot \nabla e_z^n)(z(t_n) - \varrho_h(z(t_n))) \, \mathrm{d}\boldsymbol{x}\right| \le c\sigma_0 |z(t_n)|_{W^{1,r}(T)} \|\boldsymbol{e}_{\boldsymbol{u}}^n\|_{L^{r^*}(T)} \|e_z^n\|_{L^2(T)}, \tag{3.4.17}$$

and in view of (2.4.28) and the stability of P_h, we have

$$\left|\int_T (P_h(\boldsymbol{u}(t_n)) - \boldsymbol{c}) \cdot \nabla\, e_z^n(z(t_n) - \varrho_h(z(t_n)))\, \mathrm{d}\boldsymbol{x}\right| \le c\sigma_0 |\boldsymbol{u}(t_n)|_{W^{1,\infty}(T)}$$
$$\|z(t_n) - \varrho_h(z(t_n))\|_{L^2(T)} \|e_z^n\|_{L^2(T)}, \tag{3.4.18}$$

where σ_0 is the constant of (2.1.25). To simplify, here we denote by c a generic constant independent of h and k. We split the third term as in Section 2.4.2:

$$\left|\sum_{T\in\mathfrak{T}_h} \int_{\partial T_-} |\boldsymbol{u}_h^n \cdot \boldsymbol{n}_T|((e_z^n)^{\text{ext}} - (e_z^n)^{\text{int}})(z(t_n) - \varrho_h(z(t_n))^{\text{ext}} \mathrm{d}\sigma\right|$$
$$\le \frac{1}{2} \sum_{T\in\mathfrak{T}_h} \int_{\partial T_-} |\boldsymbol{u}_h^n \cdot \boldsymbol{n}_T|((e_z^n)^{\text{ext}} - (e_z^n)^{\text{int}})^2 \mathrm{d}\sigma$$
$$+ \frac{1}{2} \sum_{T\in\mathfrak{T}_h} \int_{\partial T_-} |\boldsymbol{u}_h^n \cdot \boldsymbol{n}| \left(z(t_n) - \varrho_h(z(t_n))^{\text{ext}}\right)^2 \mathrm{d}\sigma,$$

and the first part cancels with the same one in the left-hand side of (3.4.15). The second part is bounded according to (2.4.30):

$$\int_e |\boldsymbol{u}_h^n \cdot \boldsymbol{n}_T| \left((z(t_n) - \varrho_h(z(t_n)))^{\text{ext}}\right)^2 \mathrm{d}\sigma \le c\sigma_0^2 h_{\tilde{T}} \|\boldsymbol{u}_h^n\|_{L^\infty(\tilde{T})} |z(t_n) - \varrho_h(z(t_n))|_{H^1(\tilde{T})}^2. \tag{3.4.19}$$

With this, we have the analog of Lemma 3.3.7.

LEMMA 3.4.6. *Assume that z is in $L^\infty(0,T; W^{1,r}(\Omega))$ for some $r > 2$, and z' is in $L^2(0,T; H^1(\Omega))$. We have for each n with $1 \le n \le N$,*

$$\alpha\Big(\|e_z^n\|_{L^2(\Omega)}^2 - \|e_z^{n-1}\|_{L^2(\Omega)}^2 + \|e_z^n - e_z^{n-1}\|_{L^2(\Omega)}^2\Big) + 2\nu k \|e_z^n\|_{L^2(\Omega)}^2$$
$$\le 2\sqrt{k}\|e_z^n\|_{L^2(\Omega)} \Big(\big(\nu + C_1\alpha\|z\|_{L^\infty(0,T;W^{1,r}(\Omega))}\big)\sqrt{k}|e_{\boldsymbol{u}}^n|_{H^1(\Omega)} \tag{3.4.20}$$
$$+ \mathcal{A}_{4,k}^n + \mathcal{A}_{4,\boldsymbol{u}}^n + \mathcal{A}_{4,z}^n\Big) + C_2\alpha hk C_{h\boldsymbol{u}} \sum_{T\in\mathfrak{T}_h} \|z - \varrho_h(z)\|_{L^\infty(0,t_n;H^1(T))}^2,$$

where C_i denote constants independent of h, k, ν, and α, $C_{h\boldsymbol{u}}$ is defined in (3.4.9), and the $\mathcal{A}_{4,\cdot}$ refer to the following approximation errors:

$$\mathcal{A}_{4,\boldsymbol{u}}^n = \nu\|\boldsymbol{u} - P_h(\boldsymbol{u})\|_{L^2(t_{n-1},t_n;H^1(\Omega)^2)} + \frac{\alpha}{2}\sqrt{k}\|\operatorname{div}(\boldsymbol{u} - P_h(\boldsymbol{u}))z\|_{L^\infty(0,t_n;L^2(\Omega))}$$
$$+ \alpha\sqrt{k}\|z\|_{L^\infty(0,t_n;H^1(\Omega))}\|\boldsymbol{u} - P_h(\boldsymbol{u})\|_{L^\infty(\Omega\times]t_{n-1},t_n[)},$$
$$\mathcal{A}_{4,z}^n = \alpha\Big(\|z' - \varrho_h(z')\|_{L^2(\Omega\times]t_{n-1},t_n[)} + \frac{\nu}{\alpha}\|z - \varrho_h(z)\|_{L^2(\Omega\times]t_{n-1},t_n[)}$$
$$+ C_3\sqrt{k}\|\nabla\,\boldsymbol{u}\|_{L^\infty(\Omega\times]0,t_n[)}\|z - \varrho_h(z)\|_{L^\infty(0,t_n;L^2(\Omega))}\Big),$$

$$\mathcal{A}_{4,k}^n = \frac{k}{\sqrt{3}}\Big(\nu\|\varrho_h(z')\|_{L^2(\Omega\times]t_{n-1},t_n[)} + \alpha\|\boldsymbol{u}' \cdot \nabla z\|_{L^2(\Omega\times]t_{n-1},t_n[)}$$

$$+ \nu\|P_h(\boldsymbol{u}')\|_{L^2(t_{n-1},t_n;H^1(\Omega)^2)} + \alpha\|\boldsymbol{u}\|_{L^\infty(\Omega\times]0,t_n[)}\|z'\|_{L^\infty(0,t_n;H^1(\Omega))}\Big).$$

By summing (3.4.20) over n and substituting into the resulting inequality, the estimate (3.4.13) for $\nu\sum_{i=1}^n k|\boldsymbol{e}_{\boldsymbol{u}}^i|^2_{H^1(\Omega)}$, we derive the main error estimate of this section. To simplify the presentation, we do not detail the constant factors apart from the exponential factor.

THEOREM 3.4.7. *Suppose that z belongs to $L^\infty(0,T;W^{1,r}(\Omega))$ for some $r>2$ and z' belongs to $L^2(0,T;H^1(\Omega))$. Then under the assumptions of the second part of Proposition 3.3.1 and if the mesh satisfies (2.1.45) for this r, we have for each n with $1 \le n \le N$,*

$$\|e_z^n\|^2_{L^2(\Omega)} + \sum_{i=1}^n \|e_z^i - e_z^{i-1}\|^2_{L^2(\Omega)}$$

$$\le \exp\left(\frac{27}{2\alpha\nu^3} t_n \left(\nu + C_1\alpha\|z\|_{L^\infty(0,T;W^{1,r}(\Omega))}\right)^2 S_4^2\|\boldsymbol{u}\|^2_{L^\infty(0,t_n;L^4(\Omega))}\right)$$

$$\times\Big(c_1\|\boldsymbol{u} - P_h(\boldsymbol{u})\|^2_{H^1(0,t_n;H^1(\Omega)^2)} + c_2\|\boldsymbol{u} - P_h(\boldsymbol{u})\|^2_{L^\infty(\Omega\times]0,t_n[)}$$

$$+ c_3\|p - r_h(p)\|^2_{L^2(\Omega\times]0,t_n[)} + c_4\|z - \varrho_h(z)\|^2_{H^1(0,t_n;L^2(\Omega))}$$

$$+ c_5 h\sum_{T\in\mathcal{T}_h}\|z - \varrho_h(z)\|^2_{L^\infty(0,t_n;H^1(T))} + c_6k^2\Big), \qquad (3.4.21)$$

where the c_i denote constants independent of h and k.

Clearly, the factor h in the next to last term of (3.4.21) permits to reduce the regularity of z, from $L^\infty(0,T;H^2(\Omega))$ to $L^\infty(0,T;H^{3/2}(\Omega))$, while maintaining an error of the order of $h+k$. Hence, we can somewhat improve the statement of Theorem 3.3.9.

THEOREM 3.4.8. *Let Ω be convex and suppose that the solution $(\boldsymbol{u},p,z)$ of (3.2.1)–(3.2.2) satisfies $z\in L^\infty(0,T;H^{3/2}(\Omega))$, $z'\in L^2(0,T;H^1(\Omega))$, $\boldsymbol{u}\in H^1(0,T;H^2(\Omega)^2)$, and $p\in L^2(0,T;H^1(\Omega))$. Then if the triangulation satisfies (2.1.45) for some $r>2$, the assumptions of Hypothesis 2.1.10 hold, and R_h satisfies (2.4.10), the scheme (3.4.2)–(3.4.4) is of order one in time and space:*

$$\sup_{1\le n\le N}\left(\|\boldsymbol{e}_{\boldsymbol{u}}^n\|_\alpha + \|e_z^n\|_{L^2(\Omega)}\right) \le C(h+k),$$

with a constant C independent of h and k.

We have the same conclusion for the pressure:

$$\left(\sum_{n=1}^N k\|e_p^n\|^2_{L^2(\Omega)}\right)^{1/2} \le C(h+k),$$

with another constant C independent of h and k. Note again that none of these results require a CFL condition.

CHAPTER 4

A Least-Squares Approach for the No-Slip Problem

The fairly straightforward schemes for discretizing (1.4.3) and (3.1.1) presented in the preceding chapters are obtained by eliminating the redundant relation (1.4.4) between the velocity $\boldsymbol{u}$ and the auxiliary variable z:

$$z = \operatorname{curl}(\boldsymbol{u} - \alpha \Delta \boldsymbol{u}).$$

In contrast, least-squares schemes can take advantage of this redundancy. In this chapter, we examine two closely related least-squares schemes and gradient algorithms based on a slightly different splitting. The first scheme has been studied by PARK [1998]. We also briefly describe a least-square scheme and gradient algorithm taken from PARK [1998] for approximating a similar semi-discrete variant of (3.1.1); see also CIORANESCU, GIRAULT, GLOWINSKI and SCOTT [1999]. All schemes and algorithms presented in this chapter are heuristic.

4.1. Least-squares schemes for the steady no-slip problem

Recall the steady version of problem (1.3.1)–(1.3.4) in a bounded, connected Lipschitz domain Ω of $\mathbb{R}^2$: Find a pair $(\boldsymbol{u}, p) \in V^\alpha \times L_0^2(\Omega)$, solution of (1.4.3)

$$-\nu \Delta \boldsymbol{u} + \mathbf{curl}(\boldsymbol{u} - \alpha \Delta \boldsymbol{u}) \times \boldsymbol{u} + \nabla p = \boldsymbol{f} \quad \text{in } \Omega,$$

with $\nu > 0$, $\alpha > 0$, and $\boldsymbol{f} \in L^2(\Omega)^2$. Instead of choosing $\operatorname{curl}(\boldsymbol{u} - \alpha \Delta \boldsymbol{u})$ for auxiliary variable, let us take

$$\boldsymbol{\zeta} = \boldsymbol{u} - \alpha \Delta \boldsymbol{u}.$$

As $\boldsymbol{u}$ belongs to V^α, see (1.4.1), $\boldsymbol{\zeta}$ is in the space

$$Y = \{\boldsymbol{\zeta} \in H^{-1}(\Omega)^2;\ \operatorname{curl} \boldsymbol{\zeta} \in L^2(\Omega)\}, \tag{4.1.1}$$

equipped with the graph norm

$$\|\boldsymbol{\zeta}\|_Y = \left(\|\boldsymbol{\zeta}\|^2_{H^{-1}(\Omega)} + \|\operatorname{curl}\boldsymbol{\zeta}\|^2_{L^2(\Omega)}\right)^{1/2}. \tag{4.1.2}$$

Thus, the relation between $\boldsymbol{\zeta}$ and $\boldsymbol{u}$ reads: Given $\boldsymbol{\zeta} \in Y$, find $\boldsymbol{u} \in H^1_0(\Omega)^2$ such that a.e. in Ω,

$$\boldsymbol{u} - \alpha\Delta\boldsymbol{u} = \boldsymbol{\zeta}. \tag{4.1.3}$$

Similarly, the relation between $\boldsymbol{\zeta}$ and $\boldsymbol{u}$, expressed by (1.4.3), reads: Given $\nu > 0$, $\alpha > 0$, $\boldsymbol{f} \in L^2(\Omega)^2$, and $\boldsymbol{\zeta} \in Y$, find a pair $(\boldsymbol{u}, p)$ in $H^1_0(\Omega)^2 \times L^2_0(\Omega)$ such that a.e. in Ω,

$$-\nu\Delta\boldsymbol{u} + \operatorname{curl}\boldsymbol{\zeta} \times \boldsymbol{u} + \nabla p = \boldsymbol{f}, \tag{4.1.4}$$

$$\operatorname{div}\boldsymbol{u} = 0. \tag{4.1.5}$$

Since $\boldsymbol{\zeta}$ is given, each problem (4.1.3) and (4.1.4)–(4.1.5) has a unique solution. Let us denote by $\boldsymbol{u}_1 = \boldsymbol{u}_1(\boldsymbol{\zeta})$ the solution of (4.1.3) and by $(\boldsymbol{u}_2, p) = (\boldsymbol{u}_2(\boldsymbol{\zeta}), p(\boldsymbol{\zeta}))$ the solution of (4.1.4)–(4.1.5). There is no reason why they should define the same function $\boldsymbol{u}$, but a least-squares constraint minimizing the norm of their difference can hopefully "force" them to coincide. For example, by defining the functional

$$\forall\boldsymbol{\zeta} \in Y,\ J(\boldsymbol{\zeta}) = \frac{1}{2}\|\nabla(\boldsymbol{u}_1(\boldsymbol{\zeta}) - \boldsymbol{u}_2(\boldsymbol{\zeta}))\|^2_{L^2(\Omega)}, \tag{4.1.6}$$

we can solve the minimum equation: Find $\boldsymbol{\zeta} \in Y$ such that

$$J(\boldsymbol{\zeta}) = \inf_{\boldsymbol{\lambda}\in Y} J(\boldsymbol{\lambda}). \tag{4.1.7}$$

Then $(\boldsymbol{u}_2, p)$ solves (1.4.3) only when the minimum in (4.1.7) is zero and $\boldsymbol{\zeta}$ realizes this minimum. The next result specifies the relation between (4.1.3)–(4.1.7) and (1.4.3).

PROPOSITION 4.1.1. *For all $\nu > 0$, all $\alpha > 0$, and all $\boldsymbol{f} \in H(\operatorname{curl}, \Omega)$, problems (4.1.3)–(4.1.7) and (1.4.3) are equivalent in a bounded, connected Lipschitz domain Ω of $\mathbb{R}^2$.*

PROOF. Let $(\boldsymbol{u}, p) \in V^\alpha \times L^2_0(\Omega)$ be any solution of (1.4.3) and set $\boldsymbol{\zeta} = \boldsymbol{u} - \alpha\Delta\boldsymbol{u}$. Then for this $\boldsymbol{\zeta}$, $\boldsymbol{u}$ solves (4.1.3), $(\boldsymbol{u}, p)$ solves (4.1.4)–(4.1.5), and the minimum in (4.1.7) is realized by $\boldsymbol{\zeta}$ and is zero. Therefore, if (1.4.3) has a solution, then (4.1.3)–(4.1.7) has also a solution and the minimum in (4.1.7) is zero. Because this minimum is zero, all solutions of (4.1.3)–(4.1.7) must satisfy (1.4.3). Hence, (4.1.3)–(4.1.7) and (1.4.3) are equivalent as soon as (1.4.3) has a solution. But we know from Theorem 1.4.6 that (1.4.3) has at least one solution, whence the proposition. □

REMARK 4.1.2. The incompressibility constraint has only been prescribed on the solution of (4.1.4). From a theoretical point of view, the statement of Proposition 4.1.1 would have been unchanged if instead we had only prescribed incompressibility on the solution of (4.1.3). As far as discretization is concerned, there appears to be some advantage in (4.1.4)–(4.1.5) because the divergence constraint on u_2 permits to relax somewhat the regularity of ζ; see Section 4.1.2. □

4.1.1. *Variational formulations and properties of J*

Both (4.1.3) and (4.1.4)–(4.1.5) have straightforward variational formulations that lead to the following scheme. Given $\boldsymbol{\zeta} \in Y$, find $\boldsymbol{u}_1 \in H_0^1(\Omega)^2$ solution of

$$\forall \boldsymbol{v} \in H_0^1(\Omega)^2,\ (\boldsymbol{u}_1, \boldsymbol{v}) + \alpha(\nabla \boldsymbol{u}_1, \nabla \boldsymbol{v}) = \langle \boldsymbol{\zeta}, \boldsymbol{v}\rangle. \tag{4.1.8}$$

For the same function $\boldsymbol{\zeta} \in Y$, find $(\boldsymbol{u}_2, p) \in H_0^1(\Omega)^2 \times L_0^2(\Omega)$ solution of

$$\forall \boldsymbol{v} \in H_0^1(\Omega)^2,\ \nu(\nabla \boldsymbol{u}_2, \nabla \boldsymbol{v}) + (\operatorname{curl} \boldsymbol{\zeta} \times \boldsymbol{u}_2, \boldsymbol{v}) - (p, \operatorname{div} \boldsymbol{v}) = (\boldsymbol{f}, \boldsymbol{v}), \tag{4.1.9}$$

$$\forall q \in L_0^2(\Omega),\ (q, \operatorname{div} \boldsymbol{u}_2) = 0. \tag{4.1.10}$$

The system is closed by solving the minimum equation:

$$J(\boldsymbol{\zeta}) = \inf_{\boldsymbol{\lambda} \in Y} J(\boldsymbol{\lambda}). \tag{4.1.11}$$

Of course, $\boldsymbol{u}_1(\boldsymbol{\zeta})$ is well-defined as soon as $\boldsymbol{\zeta}$ belongs to $H^{-1}(\Omega)^2$ and $(\boldsymbol{u}_2(\boldsymbol{\zeta}), p(\boldsymbol{\zeta}))$ is well defined as soon as $\operatorname{curl} \boldsymbol{\zeta}$ belongs to $L^2(\Omega)$. We shall discretize this formulation in the next section by means of a gradient algorithm. To this end, it is useful to study the properties of the functional J. First, we have

$$\forall \boldsymbol{\zeta} \in H^{-1}(\Omega)^2,\ |\boldsymbol{u}_1(\boldsymbol{\zeta})|_{H^1(\Omega)} \leq \frac{1}{\alpha}\|\boldsymbol{\zeta}\|_{H^{-1}(\Omega)}, \tag{4.1.12}$$

$$\forall \boldsymbol{\zeta} \in Y,\ |\boldsymbol{u}_2(\boldsymbol{\zeta})|_{H^1(\Omega)} \leq \frac{S_2}{\nu}\|\boldsymbol{f}\|_{L^2(\Omega)}, \tag{4.1.13}$$

$$\forall \boldsymbol{\zeta} \in Y,\ \|p(\boldsymbol{\zeta})\|_{L^2(\Omega)} \leq \frac{S_2}{\beta}\|\boldsymbol{f}\|_{L^2(\Omega)}\left(1 + \frac{S_4^2}{\nu}\|\operatorname{curl} \boldsymbol{\zeta}\|_{L^2(\Omega)}\right), \tag{4.1.14}$$

where β is the inf-sup constant of (1.1.25), and the S are constants of Sobolev's imbedding (1.1.3). As the bound (4.1.13) is independent of $\boldsymbol{\zeta}$, the behavior at infinity of the difference $(\boldsymbol{u}_1 - \boldsymbol{u}_2)(\boldsymbol{\zeta})$ is determined by that of $\boldsymbol{u}_1(\boldsymbol{\zeta})$, when $\boldsymbol{\zeta}$ tends to infinity.

LEMMA 4.1.3. *If Ω is bounded and Lipschitz, we have*

$$\lim_{\|\boldsymbol{\zeta}\|_{H^{-1}(\Omega)} \to \infty} |(\boldsymbol{u}_1 - \boldsymbol{u}_2)(\boldsymbol{\zeta})|_{H^1(\Omega)} = \infty. \tag{4.1.15}$$

PROOF. As mentioned above, it suffices to study the limit of $\boldsymbol{u}_1(\boldsymbol{\zeta})$ as $\boldsymbol{\zeta}$ tends to infinity. Now, (4.1.8) implies

$$\forall \boldsymbol{v} \in H_0^1(\Omega)^2,\ \langle \boldsymbol{\zeta}, \boldsymbol{v}\rangle \leq \left(\alpha + S_2^2\right) |\boldsymbol{u}_1(\boldsymbol{\zeta})|_{H^1(\Omega)} |\boldsymbol{v}|_{H^1(\Omega)}.$$

Therefore,

$$\|\boldsymbol{\zeta}\|_{H^{-1}(\Omega)} \le \left(\alpha + S_2^2\right) |\boldsymbol{u}_1(\boldsymbol{\zeta})|_{H^1(\Omega)},$$

and

$$|\boldsymbol{u}_1(\boldsymbol{\zeta})|_{H^1(\Omega)} \ge \frac{1}{\alpha + S_2^2} \|\boldsymbol{\zeta}\|_{H^{-1}(\Omega)},$$

whence the limit; note that this result is independent of the dimension. □

REMARK 4.1.4. It would be interesting to prove (4.1.15) when $\boldsymbol{\zeta}$ tends to infinity in Y, but this result does not seem to be true because it is possible for $\boldsymbol{\zeta}$ to stay bounded in $H^{-1}(\Omega)^2$ while $\operatorname{curl} \boldsymbol{\zeta}$ has an infinite limit in $L^2(\Omega)$. □

Next, we show that $\boldsymbol{u}_1$, $\boldsymbol{u}_2$, and p are Lipschitz-continuous in $\boldsymbol{\zeta}$. We skip the proof as it is straightforward.

LEMMA 4.1.5. *The functions $\boldsymbol{u}_1(\boldsymbol{\zeta})$ and $\boldsymbol{u}_2(\boldsymbol{\zeta})$ are uniformly Lipschitz-continuous with respect to $\boldsymbol{\zeta}$:*

$$\forall \boldsymbol{\zeta}, \boldsymbol{\lambda} \in H^{-1}(\Omega)^2,\ |\boldsymbol{u}_1(\boldsymbol{\zeta}) - \boldsymbol{u}_1(\boldsymbol{\lambda})|_{H^1(\Omega)} \le \frac{1}{\alpha} \|\boldsymbol{\zeta} - \boldsymbol{\lambda}\|_{H^{-1}(\Omega)}, \tag{4.1.16}$$

$$\forall \boldsymbol{\zeta}, \boldsymbol{\lambda} \in Y,\ |\boldsymbol{u}_2(\boldsymbol{\zeta}) - \boldsymbol{u}_2(\boldsymbol{\lambda})|_{H^1(\Omega)} \le \frac{S_2}{\nu^2} S_4^2 \|\boldsymbol{f}\|_{L^2(\Omega)} \|\operatorname{curl}(\boldsymbol{\zeta} - \boldsymbol{\lambda})\|_{L^2(\Omega)}. \tag{4.1.17}$$

For all bounded $\boldsymbol{\zeta}$ in Y, the function $\boldsymbol{\zeta} \mapsto p(\boldsymbol{\zeta})$ is Lipschitz-continuous:

$$\begin{aligned}\forall \boldsymbol{\zeta}, \boldsymbol{\lambda} \in Y,\ \|p(\boldsymbol{\zeta}) - p(\boldsymbol{\lambda})\|_{L^2(\Omega)} \le {} & \frac{S_2}{\nu} \frac{S_4^2}{\beta} \|\boldsymbol{f}\|_{L^2(\Omega)} \|\operatorname{curl}(\boldsymbol{\zeta} - \boldsymbol{\lambda})\|_{L^2(\Omega)} \\ & \left(1 + \frac{S_4^2}{\nu} \|\operatorname{curl} \boldsymbol{\lambda}\|_{L^2(\Omega)}\right).\end{aligned} \tag{4.1.18}$$

To alleviate notation, we denote derivatives with respect to the variable $\boldsymbol{\zeta}$ with a prime. From the statement of Lemma 4.1.5, it is easy to prove that $\boldsymbol{u}_1$, $\boldsymbol{u}_2$, and p are differentiable with respect to $\boldsymbol{\zeta}$. Indeed, for any $\boldsymbol{\zeta}$ and $\boldsymbol{\lambda}$ in $H^{-1}(\Omega)^2$, setting $\boldsymbol{w}_1 = \boldsymbol{u}_1'(\boldsymbol{\zeta}) \cdot \boldsymbol{\lambda}$, we have that $\boldsymbol{w}_1 \in H_0^1(\Omega)^2$ is the only solution of

$$\forall \boldsymbol{v} \in H_0^1(\Omega)^2,\ (\boldsymbol{w}_1, \boldsymbol{v}) + \alpha(\nabla \boldsymbol{w}_1, \nabla \boldsymbol{v}) = \langle \boldsymbol{\lambda}, \boldsymbol{v} \rangle. \tag{4.1.19}$$

In terms of operators, we can write that $\boldsymbol{u}_1'(\boldsymbol{\zeta}) \cdot \boldsymbol{\lambda} = D_\alpha^{-1}(\boldsymbol{\lambda})$, where

$$D_\alpha = I - \alpha \Delta : H_0^1(\Omega)^2 \mapsto H^{-1}(\Omega)^2, \tag{4.1.20}$$

is an isomorphism from $H_0^1(\Omega)^2$ onto $H^{-1}(\Omega)^2$, and is self-adjoint:

$$\forall \boldsymbol{u}, \boldsymbol{v} \in H_0^1(\Omega)^2,\ \langle \boldsymbol{u}, D_\alpha(\boldsymbol{v}) \rangle = \langle D_\alpha(\boldsymbol{u}), \boldsymbol{v} \rangle. \tag{4.1.21}$$

Similarly, setting $w_2 = u_2'(\zeta) \cdot \lambda$ and $p_2 = p'(\zeta) \cdot \lambda$ for any ζ and λ in Y, and recalling that

$$V = \{v \in H_0^1(\Omega)^2;\ \operatorname{div} v = 0\},$$

we have that $(w_2, p_2) \in V \times L_0^2(\Omega)$ is the only solution of

$$\forall v \in H_0^1(\Omega)^2,\ \nu(\nabla w_2, \nabla v) + (\operatorname{curl} \zeta \times w_2, v) - (p_2, \operatorname{div} v) = -(\operatorname{curl} \lambda \times u_2, v). \tag{4.1.22}$$

In terms of operators, (4.1.22) reads

$$(u_2'(\zeta) \cdot \lambda, p'(\zeta) \cdot \lambda) = N_\zeta^{-1}(-\operatorname{curl} \lambda \times u_2),$$

where for each ζ in Y, $N_\zeta : V \times L_0^2(\Omega) \mapsto H^{-1}(\Omega)^2$ is the Stokes-like operator

$$(u, p) \mapsto -\nu \Delta u + \operatorname{curl} \zeta \times u + \nabla p. \tag{4.1.23}$$

Then we can easily prove the next result.

LEMMA 4.1.6. *The functions u_1, u_2, and p are differentiable with respect to ζ and we have, with the notation of Lemma 4.1.5,*

$$\forall \zeta, \lambda \in H^{-1}(\Omega)^2,\ |u_1'(\zeta) \cdot \lambda|_{H^1(\Omega)} \le \frac{1}{\alpha} \|\lambda\|_{H^{-1}(\Omega)}, \tag{4.1.24}$$

$$\forall \zeta, \lambda \in Y,\ |u_2'(\zeta) \cdot \lambda|_{H^1(\Omega)} \le \frac{S_2}{\nu^2} S_4^2 \|f\|_{L^2(\Omega)} \|\operatorname{curl} \lambda\|_{L^2(\Omega)}, \tag{4.1.25}$$

$$\forall \zeta, \lambda \in Y,\ \|p'(\zeta) \cdot \lambda\|_{L^2(\Omega)} \le \frac{S_4^2}{\beta} \frac{S_2}{\nu} \|f\|_{L^2(\Omega)} \|\operatorname{curl} \lambda\|_{L^2(\Omega)} \left(1 + \frac{S_4^2}{\nu} \|\operatorname{curl} \zeta\|_{L^2(\Omega)}\right). \tag{4.1.26}$$

Regarding second derivatives, it is clear from (4.1.19) that the second derivative of $u_1(\zeta)$ is always zero. From (4.1.22), it stems that $u_2(\zeta)$ and $p(\zeta)$ are twice differentiable and denoting $u_2''(\zeta) \cdot (\lambda, \mu)$ and $p''(\zeta) \cdot (\lambda, \mu)$ by (w_3, p_3), we have that $(w_3, p_3) \in V \times L_0^2(\Omega)$ is the only solution of

$$\begin{aligned} \forall \zeta, \lambda, \mu \in Y,\ \nu(\nabla w_3, \nabla v) + (\operatorname{curl} \zeta \times w_3, v) - (p_3, \operatorname{div} v) \\ = -(\operatorname{curl} \lambda \times u_2'(\zeta) \cdot \mu, v) - (\operatorname{curl} \mu \times w_2, v), \end{aligned} \tag{4.1.27}$$

i.e.,

$$(w_3, p_3) = N_\zeta^{-1}(-\operatorname{curl} \lambda \times u_2'(\zeta) \cdot \mu - \operatorname{curl} \mu \times w_2).$$

This implies that $u_2''(\zeta) \cdot (\lambda, \mu)$ and $p''(\zeta) \cdot (\lambda, \mu)$ satisfy the bounds for all ζ, λ, and μ in Y

$$\begin{aligned} &|u_2''(\zeta) \cdot (\lambda, \mu)|_{H^1(\Omega)} \\ &\quad \le \frac{S_4^2}{\nu} \left(\|\operatorname{curl} \lambda\|_{L^2(\Omega)} |u_2'(\zeta) \cdot \mu|_{H^1(\Omega)} + \|\operatorname{curl} \mu\|_{L^2(\Omega)} |u_2'(\zeta) \cdot \lambda|_{H^1(\Omega)}\right), \end{aligned} \tag{4.1.28}$$

$$\|p''(\boldsymbol{\zeta})\cdot(\boldsymbol{\lambda},\boldsymbol{\mu})\|_{L^2(\Omega)} \le \frac{S_4^2}{\beta}\big(\|\operatorname{curl}\boldsymbol{\lambda}\|_{L^2(\Omega)}|\boldsymbol{u}_2'(\boldsymbol{\zeta})\cdot\boldsymbol{\mu}|_{H^1(\Omega)} + \|\operatorname{curl}\boldsymbol{\mu}\|_{L^2(\Omega)}|\boldsymbol{u}_2'(\boldsymbol{\zeta})\cdot\boldsymbol{\lambda}|_{H^1(\Omega)} + \|\operatorname{curl}\boldsymbol{\zeta}\|_{L^2(\Omega)}|\boldsymbol{u}_2''(\boldsymbol{\zeta})\cdot(\boldsymbol{\lambda},\boldsymbol{\mu})|_{H^1(\Omega)}\big). \tag{4.1.29}$$

In particular, when $\boldsymbol{\lambda}=\boldsymbol{\mu}$, (4.1.27)–(4.1.29) reduce to, for all $\boldsymbol{\zeta}$ and $\boldsymbol{\lambda}$ in Y:

$$\nu(\nabla\boldsymbol{w}_3,\nabla\boldsymbol{v}) + (\operatorname{curl}\boldsymbol{\zeta}\times\boldsymbol{w}_3,\boldsymbol{v}) - (p_3,\operatorname{div}\boldsymbol{v}) = -2(\operatorname{curl}\boldsymbol{\lambda}\times\boldsymbol{w}_2,\boldsymbol{v}), \tag{4.1.30}$$

$$|\boldsymbol{u}_2''(\boldsymbol{\zeta})\cdot(\boldsymbol{\lambda},\boldsymbol{\lambda})|_{H^1(\Omega)} \le 2\frac{S_4^2}{\nu}\|\operatorname{curl}\boldsymbol{\lambda}\|_{L^2(\Omega)}|\boldsymbol{u}_2'(\boldsymbol{\zeta})\cdot\boldsymbol{\lambda}|_{H^1(\Omega)}, \tag{4.1.31}$$

$$\|p''(\boldsymbol{\zeta})\cdot(\boldsymbol{\lambda},\boldsymbol{\lambda})\|_{L^2(\Omega)} \le \frac{S_4^2}{\beta}\big(\|\operatorname{curl}\boldsymbol{\zeta}\|_{L^2(\Omega)}|\boldsymbol{u}_2''(\boldsymbol{\zeta})\cdot(\boldsymbol{\lambda},\boldsymbol{\lambda})|_{H^1(\Omega)} + 2\|\operatorname{curl}\boldsymbol{\lambda}\|_{L^2(\Omega)}|\boldsymbol{u}_2'(\boldsymbol{\zeta})\cdot\boldsymbol{\lambda}|_{H^1(\Omega)}\big). \tag{4.1.32}$$

Although we shall not need higher-order derivatives, it follows readily by induction from (4.1.27) that $\boldsymbol{u}_2(\boldsymbol{\zeta})$ and $p(\boldsymbol{\zeta})$ are infinitely differentiable.

Now, to simplify, we set

$$\boldsymbol{H}(\boldsymbol{\zeta}) = (\boldsymbol{u}_1-\boldsymbol{u}_2)(\boldsymbol{\zeta}).$$

Then (4.1.12), (4.1.13) imply

$$\forall\boldsymbol{\zeta}\in Y,\ |\boldsymbol{H}(\boldsymbol{\zeta})|_{H^1(\Omega)} \le \frac{1}{\alpha}\|\boldsymbol{\zeta}\|_{H^{-1}(\Omega)} + \frac{S_2}{\nu}\|\boldsymbol{f}\|_{L^2(\Omega)}, \tag{4.1.33}$$

and it stems from Lemma 4.1.5 that $\boldsymbol{H}$ is uniformly Lipschitz continuous with respect to $\boldsymbol{\zeta}$:

$$\forall\boldsymbol{\zeta},\boldsymbol{\lambda}\in Y,\ |\boldsymbol{H}(\boldsymbol{\zeta})-\boldsymbol{H}(\boldsymbol{\lambda})|_{H^1(\Omega)} \le \left(\frac{1}{\alpha^2}+\frac{S_2^2}{\nu^4}S_4^4\|\boldsymbol{f}\|^2_{L^2(\Omega)}\right)^{1/2}\|\boldsymbol{\zeta}-\boldsymbol{\lambda}\|_Y. \tag{4.1.34}$$

Similarly, it follows from Lemma 4.1.6 that $\boldsymbol{H}$ has derivatives of all orders, and in particular, we have

$$\forall\boldsymbol{\zeta},\boldsymbol{\lambda}\in Y,\ |\boldsymbol{H}'(\boldsymbol{\zeta})\cdot\boldsymbol{\lambda}|_{H^1(\Omega)} \le \left(\frac{1}{\alpha^2}+\frac{S_2^2}{\nu^4}S_4^4\|\boldsymbol{f}\|^2_{L^2(\Omega)}\right)^{1/2}\|\boldsymbol{\lambda}\|_Y. \tag{4.1.35}$$

$$\forall\boldsymbol{\zeta},\boldsymbol{\lambda}\in Y,\ |\boldsymbol{H}''(\boldsymbol{\zeta})\cdot(\boldsymbol{\lambda},\boldsymbol{\lambda})|_{H^1(\Omega)} \le 2\frac{S_4^4}{\nu^3}S_2\|\boldsymbol{f}\|_{L^2(\Omega)}\|\operatorname{curl}\boldsymbol{\lambda}\|^2_{L^2(\Omega)}. \tag{4.1.36}$$

Thus, we have the following propositions.

PROPOSITION 4.1.7. *The functional $\boldsymbol{\zeta}\mapsto J(\boldsymbol{\zeta})$ is Lipschitz continuous provided $\boldsymbol{\zeta}$ is bounded in $H^{-1}(\Omega)^2$:*

$$\forall\boldsymbol{\zeta},\boldsymbol{\lambda}\in Y,\ |J(\boldsymbol{\zeta})-J(\boldsymbol{\lambda})| \le \frac{1}{2}\left(\frac{1}{\alpha}\big(\|\boldsymbol{\zeta}\|_{H^{-1}(\Omega)}+\|\boldsymbol{\lambda}\|_{H^{-1}(\Omega)}\big)+2\frac{S_2}{\nu}\|\boldsymbol{f}\|_{L^2(\Omega)}\right) \times\left(\frac{1}{\alpha^2}+\frac{S_2^2}{\nu^4}S_4^4\|\boldsymbol{f}\|^2_{L^2(\Omega)}\right)^{1/2}\|\boldsymbol{\zeta}-\boldsymbol{\lambda}\|_Y. \tag{4.1.37}$$

PROPOSITION 4.1.8. *The functional $\boldsymbol{\zeta} \mapsto J(\boldsymbol{\zeta})$ has derivatives of all orders, and its derivatives are bounded on all bounded sets of Y. In particular,*

$$\forall \boldsymbol{\zeta}, \boldsymbol{\lambda} \in Y,\ J'(\boldsymbol{\zeta}) \cdot \boldsymbol{\lambda} = (\nabla \boldsymbol{H}'(\boldsymbol{\zeta}) \cdot \boldsymbol{\lambda}, \nabla \boldsymbol{H}(\boldsymbol{\zeta})),$$

$$\forall \boldsymbol{\zeta}, \boldsymbol{\lambda} \in Y,\ |J'(\boldsymbol{\zeta}) \cdot \boldsymbol{\lambda}| \le \left(\frac{1}{\alpha}\|\boldsymbol{\zeta}\|_{H^{-1}(\Omega)} + \frac{S_2}{\nu}\|\boldsymbol{f}\|_{L^2(\Omega)}\right) \times \left(\frac{1}{\alpha^2} + \frac{S_2^2}{\nu^4} S_4^4 \|\boldsymbol{f}\|^2_{L^2(\Omega)}\right)^{1/2} \|\boldsymbol{\lambda}\|_Y, \tag{4.1.38}$$

$$\forall \boldsymbol{\zeta}, \boldsymbol{\lambda} \in Y,\ J''(\boldsymbol{\zeta}) \cdot (\boldsymbol{\lambda}, \boldsymbol{\lambda}) = (\nabla \boldsymbol{H}''(\boldsymbol{\zeta}) \cdot (\boldsymbol{\lambda}, \boldsymbol{\lambda}), \nabla \boldsymbol{H}(\boldsymbol{\zeta})) + |\boldsymbol{H}'(\boldsymbol{\zeta}) \cdot \boldsymbol{\lambda}|^2_{H^1(\Omega)},$$

$$\forall \boldsymbol{\zeta}, \boldsymbol{\lambda} \in Y,\ |J''(\boldsymbol{\zeta}) \cdot (\boldsymbol{\lambda}, \boldsymbol{\lambda})| \le 2\frac{S_4^4}{\nu^3} S_2 \|\boldsymbol{f}\|_{L^2(\Omega)} \|\mathrm{curl}\, \boldsymbol{\lambda}\|^2_{L^2(\Omega)} |\boldsymbol{u}_1(\boldsymbol{\zeta}) - \boldsymbol{u}_2(\boldsymbol{\zeta})|_{H^1(\Omega)} + \left(\frac{1}{\alpha^2} + \frac{S_2^2}{\nu^4} S_4^4 \|\boldsymbol{f}\|^2_{L^2(\Omega)}\right) \|\boldsymbol{\lambda}\|^2_Y. \tag{4.1.39}$$

The above expression for $J''(\boldsymbol{\zeta}) \cdot (\boldsymbol{\lambda}, \boldsymbol{\lambda})$ does not allow to conclude that J is convex. In addition, considering that

$$J(\boldsymbol{\zeta}) \ge \frac{1}{2}\left(|\boldsymbol{u}_1(\boldsymbol{\zeta})| - |\boldsymbol{u}_2(\boldsymbol{\zeta})|\right)^2,$$

Lemma 4.1.3 implies that $J(\boldsymbol{\zeta})$ tends to infinity as $\boldsymbol{\zeta}$ tends to infinity in $H^{-1}(\Omega)^2$, but as stated in Remark 4.1.4, this does not give coercivity of J when $\boldsymbol{\zeta}$ tends to infinity in Y.

4.1.2. *A related least-squares scheme*

Here we retain the variational formulation (4.1.8) of (4.1.3), but we formulate differently (4.1.4)–(4.1.5) by taking into account the incompressibility constraint. Indeed, owing to the zero divergence of $\boldsymbol{u}_2$, the term $(\mathrm{curl}\, \boldsymbol{\zeta} \times \boldsymbol{u}_2, \boldsymbol{v})$ has more than one expression. We use the following identity, valid when $d = 2$ or 3.

LEMMA 4.1.9. *Let $d = 2$ or $d = 3$. For any $\boldsymbol{u}$ and $\boldsymbol{v} \in H^1(\Omega)^d$ satisfying $\boldsymbol{u} \cdot \boldsymbol{n} = \boldsymbol{v} \cdot \boldsymbol{n} = 0$ on $\partial\Omega$ and $\mathrm{div}\, \boldsymbol{u} = 0$ in Ω, and for any $\boldsymbol{\zeta} \in Y$ satisfying $\boldsymbol{\zeta} \in L^3(\Omega)^d$ if $d = 3$ and $\boldsymbol{\zeta} \in L^r(\Omega)^d$ for some $r > 2$ if $d = 2$, we have*

$$(\mathbf{curl}\, \boldsymbol{\zeta} \times \boldsymbol{u}, \boldsymbol{v}) = (\boldsymbol{v} \cdot \nabla \boldsymbol{u} - \boldsymbol{u} \cdot \nabla \boldsymbol{v}, \boldsymbol{\zeta}) + (\boldsymbol{u} \cdot \boldsymbol{\zeta}, \mathrm{div}\, \boldsymbol{v}). \tag{4.1.40}$$

PROOF. The proof of (4.1.40) is straightforward when $\boldsymbol{\zeta}$ is smooth enough, e.g., $\boldsymbol{\zeta} \in H^2(\Omega)^d$. When $d = 3$ and $\boldsymbol{\zeta} \in Y \cap L^3(\Omega)^3$, then $\boldsymbol{\zeta} \in H(\mathrm{curl}; \Omega) \cap L^3(\Omega)^3$, $H^2(\Omega)^3$ is dense in this space, and the proof of (4.1.40) proceeds by density. The same argument is valid when $d = 2$. □

Clearly, in view of the right-hand side of (4.1.40), we can set (4.1.4)–(4.1.5) into an equivalent variational formulation that is meaningful for all $\boldsymbol{\zeta}$ in $L^r(\Omega)^2$ as soon as $r > 2$. Note that if Ω has a smooth boundary or is a convex polygon, then each solution $\boldsymbol{\zeta}$ belongs indeed to $L^r(\Omega)^2$ for some number $r > 2$; see Proposition 1.4.11. Based on these remarks, we propose another scheme; we denote its solutions with a tilde to distinguish them from the solutions of the previous scheme. Given $\boldsymbol{\zeta} \in Y \cap L^r(\Omega)^2$, find $\tilde{\boldsymbol{u}}_1 \in H_0^1(\Omega)^2$ solution of

$$\forall \boldsymbol{v} \in H_0^1(\Omega)^2,\ (\tilde{\boldsymbol{u}}_1, \boldsymbol{v}) + \alpha(\nabla \tilde{\boldsymbol{u}}_1, \nabla \boldsymbol{v}) = (\boldsymbol{\zeta}, \boldsymbol{v}). \tag{4.1.41}$$

For the same function $\boldsymbol{\zeta} \in Y \cap L^r(\Omega)^2$, find $(\tilde{\boldsymbol{u}}_2, \tilde{p}) \in H_0^1(\Omega)^2 \times L_0^2(\Omega)$ solution of

$$\forall \boldsymbol{v} \in H_0^1(\Omega)^2,\ \nu(\nabla \tilde{\boldsymbol{u}}_2, \nabla \boldsymbol{v}) + (\boldsymbol{v} \cdot \nabla \tilde{\boldsymbol{u}}_2 - \tilde{\boldsymbol{u}}_2 \cdot \nabla \boldsymbol{v}, \boldsymbol{\zeta}) - (\tilde{p}, \operatorname{div} \boldsymbol{v}) = (\boldsymbol{f}, \boldsymbol{v}), \tag{4.1.42}$$

$$\forall q \in L_0^2(\Omega),\ (q, \operatorname{div} \tilde{\boldsymbol{u}}_2) = 0. \tag{4.1.43}$$

The system is closed by solving the minimum equation:

$$\tilde{J}(\tilde{\xi}) = \inf_{\boldsymbol{\lambda} \in Y \cap L^r(\Omega)^2} \tilde{J}(\boldsymbol{\lambda}) = \inf_{\boldsymbol{\lambda} \in Y \cap L^r(\Omega)^2} \frac{1}{2} \|\nabla(\tilde{\boldsymbol{u}}_1(\boldsymbol{\zeta}) - \tilde{\boldsymbol{u}}_2(\boldsymbol{\zeta}))\|_{L^2(\Omega)}^2. \tag{4.1.44}$$

The pressure p is related to $\tilde{p}$ by

$$\tilde{p} = p - \tilde{\boldsymbol{u}}_2 \cdot \boldsymbol{\zeta} + \gamma, \tag{4.1.45}$$

where γ is a constant adjusted so that $\tilde{p}$ belongs to $L_0^2(\Omega)$:

$$\gamma = \frac{1}{|\Omega|} \int_\Omega \tilde{\boldsymbol{u}}_2 \cdot \boldsymbol{\zeta} \, d\boldsymbol{x}.$$

Of course, $\tilde{\boldsymbol{u}}_1(\boldsymbol{\zeta})$ is well-defined as soon as $\boldsymbol{\zeta}$ belongs to $H^{-1}(\Omega)^2$ so that (4.1.41) coincides with (4.1.8). In contrast, $(\tilde{\boldsymbol{u}}_2(\boldsymbol{\zeta}), \tilde{p}(\boldsymbol{\zeta}))$ is well-defined as soon as $\boldsymbol{\zeta}$ belongs to $L^r(\Omega)^2$ for some $r > 2$. Thus, the resulting scheme differs from the previous one, but according to Proposition 1.4.11, (4.1.41)–(4.1.45) is equivalent to (1.4.3) when Ω is smooth or is a convex polygon. The present scheme has the advantage of somewhat reducing the regularity of $\boldsymbol{\zeta}$, at the expense of adding another term in the computation of the second velocity.

Let us go quickly over the properties of this scheme. Clearly $\tilde{\boldsymbol{u}}_1(\boldsymbol{\zeta})$ and $\tilde{\boldsymbol{u}}_2(\boldsymbol{\zeta})$ satisfy the analogs of (4.1.12) and (4.1.13), respectively:

$$\forall \boldsymbol{\zeta} \in H^{-1}(\Omega)^2,\ |\tilde{\boldsymbol{u}}_1(\boldsymbol{\zeta})|_{H^1(\Omega)} \le \frac{1}{\alpha} \|\boldsymbol{\zeta}\|_{H^{-1}(\Omega)},$$

$$\forall \boldsymbol{\zeta} \in L^r(\Omega)^2,\ |\tilde{\boldsymbol{u}}_2(\boldsymbol{\zeta})|_{H^1(\Omega)} \le \frac{S_2}{\nu} \|\boldsymbol{f}\|_{L^2(\Omega)}.$$

But the bound for $\tilde{p}(\boldsymbol{\zeta})$ is slightly different:

$$\forall \boldsymbol{\zeta} \in L^r(\Omega)^2,\ \|\tilde{p}(\boldsymbol{\zeta})\|_{L^2(\Omega)} \le \frac{1}{\beta} \left(S_2 \|\boldsymbol{f}\|_{L^2(\Omega)} + 2 S_{r^\star} |\tilde{\boldsymbol{u}}_2(\boldsymbol{\zeta})|_{H^1(\Omega)} \|\boldsymbol{\zeta}\|_{L^r(\Omega)}\right), \tag{4.1.46}$$

for $r > 2$ and $\frac{1}{r^\star} = \frac{1}{2} - \frac{1}{r}$.

Again, $\tilde{\boldsymbol{u}}_1 - \tilde{\boldsymbol{u}}_2$ satisfies the same weak coercivity as $\boldsymbol{u}_1 - \boldsymbol{u}_2$:

$$\lim_{\|\boldsymbol{\zeta}\|_{H^{-1}(\Omega)}\to\infty} |(\tilde{\boldsymbol{u}}_1 - \tilde{\boldsymbol{u}}_2)(\boldsymbol{\zeta})|_{H^1(\Omega)} = \infty,$$

but not strong coercivity because it is possible for $\boldsymbol{\zeta}$ to stay bounded in $H^{-1}(\Omega)^2$ while it has an infinite limit in $L^r(\Omega)^2$.

Here also $\tilde{\boldsymbol{u}}_1$ and $\tilde{\boldsymbol{u}}_2$ are Lipschitz-continuous with respect to $\boldsymbol{\zeta}$:

$$\forall \boldsymbol{\zeta}, \boldsymbol{\lambda} \in H^{-1}(\Omega)^2,\ |\tilde{\boldsymbol{u}}_1(\boldsymbol{\zeta}) - \tilde{\boldsymbol{u}}_1(\boldsymbol{\lambda})|_{H^1(\Omega)} \le \frac{1}{\alpha}\|\boldsymbol{\zeta} - \boldsymbol{\lambda}\|_{H^{-1}(\Omega)},$$

$$\forall \boldsymbol{\zeta}, \boldsymbol{\lambda} \in L^r(\Omega)^2,\ |\tilde{\boldsymbol{u}}_2(\boldsymbol{\zeta}) - \tilde{\boldsymbol{u}}_2(\boldsymbol{\lambda})|_{H^1(\Omega)} \le \frac{2}{\nu^2} S_2 S_{r^\star}\|\boldsymbol{f}\|_{L^2(\Omega)}\|\boldsymbol{\zeta} - \boldsymbol{\lambda}\|_{L^r(\Omega)}.$$

For all bounded $\boldsymbol{\zeta}$ in $L^r(\Omega)^2$, the function $\tilde{p}(\boldsymbol{\zeta})$ is Lipschitz-continuous with respect to $\boldsymbol{\zeta}$:

$$\forall \boldsymbol{\zeta}, \boldsymbol{\lambda} \in L^r(\Omega)^2,\ \|\tilde{p}(\boldsymbol{\zeta}) - \tilde{p}(\boldsymbol{\lambda})\|_{L^2(\Omega)} \le \frac{2S_{r^\star}}{\beta}\Bigg(\|\boldsymbol{\zeta}\|_{L^r(\Omega)}|\tilde{\boldsymbol{u}}_2(\boldsymbol{\zeta}) - \tilde{\boldsymbol{u}}_2(\boldsymbol{\lambda})|_{H^1(\Omega)} + \frac{S_2}{\nu}\|\boldsymbol{f}\|_{L^2(\Omega)}\|\boldsymbol{\zeta} - \boldsymbol{\lambda}\|_{L^r(\Omega)}\Bigg). \tag{4.1.47}$$

Similarly, $\tilde{\boldsymbol{u}}_1$, $\tilde{\boldsymbol{u}}_2$, and $\tilde{p}$ are differentiable with respect to $\boldsymbol{\zeta}$. Indeed, for any $\boldsymbol{\zeta}$ and $\boldsymbol{\lambda}$ in $H^{-1}(\Omega)^2$, $\tilde{\boldsymbol{u}}_1'(\boldsymbol{\zeta})\cdot\boldsymbol{\lambda}$, is the only solution $\tilde{\boldsymbol{w}}_1 \in H_0^1(\Omega)^2$ of

$$\forall \boldsymbol{v} \in H_0^1(\Omega)^2,\ (\tilde{\boldsymbol{w}}_1, \boldsymbol{v}) + \alpha(\nabla\tilde{\boldsymbol{w}}_1, \nabla\boldsymbol{v}) = \langle\tilde{\boldsymbol{\lambda}}, \boldsymbol{v}\rangle. \tag{4.1.48}$$

Next, recalling the trilinear form c defined by (3.2.61):

$$c(\boldsymbol{u}; \boldsymbol{v}, \boldsymbol{w}) = \sum_{i=1}^{2}\sum_{j=1}^{2}\int_\Omega u_i\frac{\partial v_j}{\partial x_i}w_j\,d\boldsymbol{x},$$

and defining the bilinear form $\mathcal{B}_{\boldsymbol{\zeta}}$ by

$$\mathcal{B}_{\boldsymbol{\zeta}}(\boldsymbol{u}, \boldsymbol{v}) = c(\boldsymbol{v}; \boldsymbol{u}, \boldsymbol{\zeta}) - c(\boldsymbol{u}; \boldsymbol{v}, \boldsymbol{\zeta}), \tag{4.1.49}$$

we have that for any $\boldsymbol{\zeta}$ and $\boldsymbol{\lambda}$ in $L^r(\Omega)^2$, $r > 2$, $\tilde{\boldsymbol{u}}_2'(\boldsymbol{\zeta})\cdot\boldsymbol{\lambda}$ and $\tilde{p}'(\boldsymbol{\zeta})\cdot\boldsymbol{\lambda}$ are the only solution $(\tilde{\boldsymbol{w}}_2, \tilde{p}_2) \in V \times L_0^2(\Omega)$ of

$$\forall \boldsymbol{v} \in H_0^1(\Omega)^2,\ \nu(\nabla\tilde{\boldsymbol{w}}_2, \nabla\boldsymbol{v}) + \mathcal{B}_{\boldsymbol{\zeta}}(\tilde{\boldsymbol{w}}_2, \boldsymbol{v}) - (\tilde{p}_2, \operatorname{div}\boldsymbol{v}) = -\mathcal{B}_{\boldsymbol{\lambda}}(\tilde{\boldsymbol{u}}_2, \boldsymbol{v}). \tag{4.1.50}$$

The form $\mathcal{B}_{\boldsymbol{\zeta}}$ is antisymmetric:

$$\forall \boldsymbol{v} \in H_0^1(\Omega)^2,\ \mathcal{B}_{\boldsymbol{\zeta}}(\boldsymbol{v}, \boldsymbol{v}) = 0,$$

and continuous

$$\forall \boldsymbol{u}, \boldsymbol{v} \in H_0^1(\Omega)^2, \forall\boldsymbol{\zeta} \in L^r(\Omega)^2,\ |\mathcal{B}_{\boldsymbol{\zeta}}(\boldsymbol{u}, \boldsymbol{v})| \le 2S_{r^\star}\|\boldsymbol{\zeta}\|_{L^r(\Omega)}|\boldsymbol{u}|_{H^1(\Omega)}|\boldsymbol{v}|_{H^1(\Omega)}. \tag{4.1.51}$$

Moreover, the derivative of $\tilde{\boldsymbol{u}}_1$, $\tilde{\boldsymbol{u}}_2$, and $\tilde{p}_2$ satisfy the bounds

$$\begin{aligned}
&\forall \boldsymbol{\zeta}, \boldsymbol{\lambda} \in H^{-1}(\Omega)^2,\ |\tilde{\boldsymbol{u}}_1'(\boldsymbol{\zeta}) \cdot \boldsymbol{\lambda}|_{H^1(\Omega)} \le \frac{1}{\alpha} \|\boldsymbol{\lambda}\|_{H^{-1}(\Omega)},\\
&\forall \boldsymbol{\zeta}, \boldsymbol{\lambda} \in L^r(\Omega)^2,\ |\tilde{\boldsymbol{u}}_2'(\boldsymbol{\zeta}) \cdot \boldsymbol{\lambda}|_{H^1(\Omega)} \le \frac{2}{\nu^2} S_2 S_{r^\star} \|\boldsymbol{f}\|_{L^2(\Omega)} \|\boldsymbol{\lambda}\|_{L^r(\Omega)},\\
&\forall \boldsymbol{\zeta}, \boldsymbol{\lambda} \in L^r(\Omega)^2,\ \|\tilde{p}'(\boldsymbol{\zeta}) \cdot \boldsymbol{\lambda}\|_{L^2(\Omega)} \le \frac{2S_{r^\star}}{\beta} \Big(\|\boldsymbol{\zeta}\|_{L^r(\Omega)} |(\tilde{\boldsymbol{u}}_2'(\boldsymbol{\zeta}) \cdot \boldsymbol{\lambda}|_{H^1(\Omega)}\\
&\qquad + \frac{S_2}{\nu} \|\boldsymbol{f}\|_{L^2(\Omega)} \|\boldsymbol{\lambda}\|_{L^r(\Omega)} \Big).
\end{aligned} \tag{4.1.52}$$

The second derivative of $\tilde{\boldsymbol{u}}_1(\boldsymbol{\zeta})$ is always zero. Formula (4.1.50) implies that $\tilde{\boldsymbol{u}}_2(\boldsymbol{\zeta})$ and $\tilde{p}(\boldsymbol{\zeta})$ are twice differentiable and denoting $\tilde{\boldsymbol{u}}_2''(\boldsymbol{\zeta}) \cdot (\boldsymbol{\lambda}, \boldsymbol{\mu}), \tilde{p}''(\boldsymbol{\zeta}) \cdot (\boldsymbol{\lambda}, \boldsymbol{\mu}))$ by $(\tilde{\boldsymbol{w}}_3, \tilde{p}_3)$, we have that $(\tilde{\boldsymbol{w}}_3, \tilde{p}_3) \in V \times L_0^2(\Omega)$ is the only solution of

$$\begin{aligned}
\forall \boldsymbol{\zeta}, \boldsymbol{\lambda}, \boldsymbol{\mu} \in L^r(\Omega)^2,\ &\nu(\nabla \tilde{\boldsymbol{w}}_3, \nabla \boldsymbol{v}) + \mathcal{B}_{\boldsymbol{\zeta}}(\tilde{\boldsymbol{w}}_3, \boldsymbol{v}) - (\tilde{p}_3, \operatorname{div} \boldsymbol{v})\\
&= -\mathcal{B}_{\boldsymbol{\lambda}}(\tilde{\boldsymbol{u}}_2'(\boldsymbol{\zeta}) \cdot \boldsymbol{\mu}, \boldsymbol{v}) - \mathcal{B}_{\boldsymbol{\mu}}(\tilde{\boldsymbol{w}}_2, \boldsymbol{v}).
\end{aligned} \tag{4.1.53}$$

In view of (4.1.51), $\tilde{\boldsymbol{u}}_2''(\boldsymbol{\zeta}) \cdot (\boldsymbol{\lambda}, \boldsymbol{\mu})$ and $\tilde{p}''(\boldsymbol{\zeta}) \cdot (\boldsymbol{\lambda}, \boldsymbol{\mu})$ are bounded as follows

$$\begin{aligned}
&\forall \boldsymbol{\zeta}, \boldsymbol{\lambda}, \boldsymbol{\mu} \in L^r(\Omega)^2,\ |\tilde{\boldsymbol{u}}_2''(\boldsymbol{\zeta}) \cdot (\boldsymbol{\lambda}, \boldsymbol{\mu})|_{H^1(\Omega)}\\
&\quad \le \frac{2S_{r^\star}}{\nu} \big(\|\boldsymbol{\lambda}\|_{L^r(\Omega)} |\tilde{\boldsymbol{u}}_2'(\boldsymbol{\zeta}) \cdot \boldsymbol{\mu}|_{H^1(\Omega)} + \|\boldsymbol{\mu}\|_{L^r(\Omega)} |\tilde{\boldsymbol{u}}_2'(\boldsymbol{\zeta}) \cdot \boldsymbol{\lambda}|_{H^1(\Omega)} \big),
\end{aligned} \tag{4.1.54}$$

$$\begin{aligned}
&\forall \boldsymbol{\zeta}, \boldsymbol{\lambda}, \boldsymbol{\mu} \in L^r(\Omega)^2,\ \|\tilde{p}''(\boldsymbol{\zeta}) \cdot (\boldsymbol{\lambda}, \boldsymbol{\mu})\|_{L^2(\Omega)}\\
&\quad \le \frac{2S_{r^\star}}{\beta} \big(\|\boldsymbol{\lambda}\|_{L^r(\Omega)} |\tilde{\boldsymbol{u}}_2'(\boldsymbol{\zeta}) \cdot \boldsymbol{\mu}|_{H^1(\Omega)} + \|\boldsymbol{\mu}\|_{L^r(\Omega)} |\tilde{\boldsymbol{u}}_2'(\boldsymbol{\zeta}) \cdot \boldsymbol{\lambda}|_{H^1(\Omega)}\\
&\quad + \|\boldsymbol{\zeta}\|_{L^r(\Omega)} |\tilde{\boldsymbol{u}}_2''(\boldsymbol{\zeta}) \cdot (\boldsymbol{\lambda}, \boldsymbol{\mu})|_{H^1(\Omega)} \big).
\end{aligned} \tag{4.1.55}$$

When $\boldsymbol{\lambda} = \boldsymbol{\mu}$, (4.1.53)–(4.1.55) reduce to, for all $\boldsymbol{\zeta}$ and $\boldsymbol{\lambda}$ in $L^r(\Omega)^2$,

$$\nu(\nabla \tilde{\boldsymbol{w}}_3, \nabla \boldsymbol{v}) + \mathcal{B}_{\boldsymbol{\zeta}}(\tilde{\boldsymbol{w}}_3, \boldsymbol{v}) - (\tilde{p}_3, \operatorname{div} \boldsymbol{v}) = -2\mathcal{B}_{\boldsymbol{\lambda}}(\tilde{\boldsymbol{u}}_2'(\boldsymbol{\zeta}) \cdot \boldsymbol{\lambda}, \boldsymbol{v}). \tag{4.1.56}$$

$$|\tilde{\boldsymbol{u}}_2''(\boldsymbol{\zeta}) \cdot (\boldsymbol{\lambda}, \boldsymbol{\lambda})|_{H^1(\Omega)} \le \frac{4S_{r^\star}}{\nu} \|\boldsymbol{\lambda}\|_{L^r(\Omega)} |\tilde{\boldsymbol{u}}_2'(\boldsymbol{\zeta}) \cdot \boldsymbol{\lambda}|_{H^1(\Omega)}, \tag{4.1.57}$$

$$\|\tilde{p}''(\boldsymbol{\zeta}) \cdot (\boldsymbol{\lambda}, \boldsymbol{\lambda})\|_{L^2(\Omega)} \le \frac{4S_{r^\star}}{\beta} \|\boldsymbol{\lambda}\|_{L^r(\Omega)} |\tilde{\boldsymbol{u}}_2'(\boldsymbol{\zeta}) \cdot \boldsymbol{\lambda}|_{H^1(\Omega)} \left(1 + \frac{2S_{r^\star}}{\nu} \|\boldsymbol{\zeta}\|_{L^r(\Omega)} \right). \tag{4.1.58}$$

By induction, $\tilde{\boldsymbol{u}}_2$ and $\tilde{p}$ are infinitely differentiable.

Let

$$\tilde{\boldsymbol{H}}(\boldsymbol{\zeta}) = (\tilde{\boldsymbol{u}}_1 - \tilde{\boldsymbol{u}}_2)(\boldsymbol{\zeta}).$$

It satisfies the following bound

$$\forall \boldsymbol{\zeta} \in L^r(\Omega)^2,\ |\tilde{\boldsymbol{H}}(\boldsymbol{\zeta})|_{H^1(\Omega)} \le \frac{1}{\alpha} \|\boldsymbol{\zeta}\|_{H^{-1}(\Omega)} + \frac{S_2}{\nu} \|\boldsymbol{f}\|_{L^2(\Omega)},$$

and is uniformly Lipschitz continuous with respect to $\boldsymbol{\zeta}$:

$$\forall \boldsymbol{\zeta}, \boldsymbol{\lambda} \in L^r(\Omega)^2, \ |\tilde{\boldsymbol{H}}(\boldsymbol{\zeta}) - \tilde{\boldsymbol{H}}(\boldsymbol{\lambda})|_{H^1(\Omega)} \leq \frac{1}{\alpha}\|\boldsymbol{\zeta} - \boldsymbol{\lambda}\|_{H^{-1}(\Omega)} + \frac{2}{\nu^2} S_2 S_{r^\star} \|\boldsymbol{f}\|_{L^2(\Omega)} \|\boldsymbol{\zeta} - \boldsymbol{\lambda}\|_{L^r(\Omega)}.$$

Similarly, $\tilde{\boldsymbol{H}}$ has derivatives of all orders, and in particular, we have for all $\boldsymbol{\zeta}$ and $\boldsymbol{\lambda}$ in $L^r(\Omega)^2$,

$$|\tilde{\boldsymbol{H}}'(\boldsymbol{\zeta}) \cdot \boldsymbol{\lambda}|_{H^1(\Omega)} \leq \frac{1}{\alpha}\|\boldsymbol{\lambda}\|_{H^{-1}(\Omega)} + \frac{2}{\nu^2} S_2 S_{r^\star} \|\boldsymbol{f}\|_{L^2(\Omega)} \|\boldsymbol{\lambda}\|_{L^r(\Omega)},$$

$$|\tilde{\boldsymbol{H}}''(\boldsymbol{\zeta}) \cdot (\boldsymbol{\lambda}, \boldsymbol{\lambda})|_{H^1(\Omega)} \leq \frac{8S_2}{\nu^3} S_{r^\star}^2 \|\boldsymbol{f}\|_{L^2(\Omega)} \|\boldsymbol{\lambda}\|_{L^r(\Omega)}^2.$$

Therefore, $\tilde{J}(\boldsymbol{\zeta})$ is Lipschitz continuous for all bounded $\boldsymbol{\zeta}$ in $H^{-1}(\Omega)^2$:

$$\forall \boldsymbol{\zeta}, \boldsymbol{\lambda} \in L^r(\Omega)^2, \ |\tilde{J}(\boldsymbol{\zeta}) - \tilde{J}(\boldsymbol{\lambda})| \leq \frac{1}{2}\left(\frac{1}{\alpha}\left(\|\boldsymbol{\zeta}\|_{H^{-1}(\Omega)} + \|\boldsymbol{\lambda}\|_{H^{-1}(\Omega)}\right) + 2\frac{S_2}{\nu}\|\boldsymbol{f}\|_{L^2(\Omega)}\right) \times \left(\frac{1}{\alpha}\|\boldsymbol{\zeta} - \boldsymbol{\lambda}\|_{H^{-1}(\Omega)} + \frac{2}{\nu^2} S_2 S_{r^\star} \|\boldsymbol{f}\|_{L^2(\Omega)} \|\boldsymbol{\zeta} - \boldsymbol{\lambda}\|_{L^r(\Omega)}\right). \tag{4.1.59}$$

The functional $\tilde{J}(\boldsymbol{\zeta})$ has derivatives of all orders, and its derivatives are bounded on all bounded sets of $L^r(\Omega)^2$. In particular, we have for all $\boldsymbol{\zeta}$ and $\boldsymbol{\lambda}$ in $L^r(\Omega)^2$,

$$\tilde{J}'(\boldsymbol{\zeta}) \cdot \boldsymbol{\lambda} = (\nabla \tilde{\boldsymbol{H}}'(\boldsymbol{\zeta}) \cdot \boldsymbol{\lambda}, \nabla \tilde{\boldsymbol{H}}(\boldsymbol{\zeta})),$$

$$\tilde{J}''(\boldsymbol{\zeta}) \cdot (\boldsymbol{\lambda}, \boldsymbol{\lambda}) = (\nabla \tilde{\boldsymbol{H}}''(\boldsymbol{\zeta}) \cdot (\boldsymbol{\lambda}, \boldsymbol{\lambda}), \nabla \tilde{\boldsymbol{H}}(\boldsymbol{\zeta})) + \left|\tilde{\boldsymbol{H}}'(\boldsymbol{\zeta}) \cdot \boldsymbol{\lambda}\right|_{H^1(\Omega)}^2,$$

$$|\tilde{J}'(\boldsymbol{\zeta}) \cdot \boldsymbol{\lambda}| \leq \left(\frac{1}{\alpha}\|\boldsymbol{\lambda}\|_{H^{-1}(\Omega)} + \frac{2}{\nu^2} S_2 S_{r^\star} \|\boldsymbol{f}\|_{L^2(\Omega)} \|\boldsymbol{\lambda}\|_{L^r(\Omega)}\right) \times |\tilde{\boldsymbol{u}}_1(\boldsymbol{\zeta}) - \tilde{\boldsymbol{u}}_2(\boldsymbol{\zeta})|_{H^1(\Omega)}, \tag{4.1.60}$$

$$|\tilde{J}''(\boldsymbol{\zeta}) \cdot (\boldsymbol{\lambda}, \boldsymbol{\lambda})| \leq \frac{8S_2}{\nu^3} S_{r^\star}^2 \|\boldsymbol{f}\|_{L^2(\Omega)} \|\boldsymbol{\lambda}\|_{L^r(\Omega)}^2 |\tilde{\boldsymbol{u}}_1(\boldsymbol{\zeta}) - \tilde{\boldsymbol{u}}_2(\boldsymbol{\zeta})|_{H^1(\Omega)} + \left(\frac{1}{\alpha}\|\boldsymbol{\lambda}\|_{H^{-1}(\Omega)} + \frac{2}{\nu^2} S_2 S_{r^\star} \|\boldsymbol{f}\|_{L^2(\Omega)} \|\boldsymbol{\lambda}\|_{L^r(\Omega)}\right)^2. \tag{4.1.61}$$

Again, the above considerations guarantee neither the convexity nor the coercivity of $\tilde{J}$.

4.2. An approximate gradient algorithm

Gradient algorithms for approximating the minimum equations (4.1.7) or (4.1.44) are bound to be heuristic because neither the convexity of J or $\tilde{J}$, nor their behavior at infinity are guaranteed. Besides, the minimum of J or $\tilde{J}$ is not necessarily unique. Moreover, even if an approximate minimum is searched along a "line," the fact that J or $\tilde{J}$ are not quadratic implies that the equations for solving this approximate minimum must themselves be suitably approximated. Therefore, the gradient algorithm we present here is derived heuristically;

but nonetheless, it gives interesting numerical results, see PARK [1998]. To simplify the discussion, the algorithm is presented for approximately minimizing J, but it can readily be adapted to $\tilde{J}$ by interchanging $(\operatorname{curl} \boldsymbol{\zeta} \times \boldsymbol{u}, \boldsymbol{v})$ and $\mathcal{B}_{\boldsymbol{\zeta}}(\boldsymbol{u}, \boldsymbol{v})$. Furthermore, the algorithm is described at the continuous level, but we must keep in mind that it will be applied to a finite-element discretization of (4.1.3)–(4.1.7).

It is convenient to define $\boldsymbol{g}(\boldsymbol{\zeta})$ the gradient of $J(\boldsymbol{\zeta})$ by: $\boldsymbol{g}(\boldsymbol{\zeta}) \in H^1(\Omega)^2$ is the solution of

$$\forall \boldsymbol{\lambda} \in H^1(\Omega)^2, \ (\boldsymbol{g}(\boldsymbol{\zeta}), \boldsymbol{\lambda}) + (\nabla \boldsymbol{g}(\boldsymbol{\zeta}), \nabla \boldsymbol{\lambda}) = J'(\boldsymbol{\zeta}) \cdot \boldsymbol{\lambda}. \tag{4.2.1}$$

This is not the gradient in the usual sense; indeed, it should be defined for all $\boldsymbol{\lambda}$ in Y, and the left-hand side should be the scalar product of Y. However, the practical computation of the scalar product of Y is expensive, and because all norms are equivalent in a finite-dimensional space, (4.2.1) is adequate at the discrete level, as long as Y is approximated by a finite-element subspace of $H^1(\Omega)^2$.

Let $\boldsymbol{\zeta}^0 \in Y$ be a first approximation of $\boldsymbol{\zeta}$ and let $\boldsymbol{g}^0 = \boldsymbol{g}(\boldsymbol{\zeta}^0)$. Then for any integer $m \geq 0$, knowing $\boldsymbol{\zeta}^m$ and $\boldsymbol{g}^m = \boldsymbol{g}(\boldsymbol{\zeta}^m)$, we want to find a positive real number, ρ^m, that solves

$$J(\boldsymbol{\zeta}^m - \rho^m \boldsymbol{g}^m) = \inf_{\rho \in \mathbb{R}_+} J(\boldsymbol{\zeta}^m - \rho \boldsymbol{g}^m). \tag{4.2.2}$$

The results of the preceeding section show that the mapping $\rho \mapsto J(\boldsymbol{\zeta}^m - \rho \boldsymbol{g}^m)$ is very smooth, but give no information neither on its coercivity nor on its convexity. Nevertheless, let us assume that the solution ρ^m of (4.2.2) can be found among the solutions of

$$\frac{\mathrm{d}}{\mathrm{d}\rho} J(\boldsymbol{\zeta}^m - \rho \boldsymbol{g}^m) = 0,$$

i.e.,

$$J'(\boldsymbol{\zeta}^m - \rho \boldsymbol{g}^m) \cdot \boldsymbol{g}^m = 0, \quad \text{i.e.,} \quad (\nabla \boldsymbol{H}'(\boldsymbol{\zeta}^m - \rho \boldsymbol{g}^m) \cdot \boldsymbol{g}^m, \nabla \boldsymbol{H}(\boldsymbol{\zeta}^m - \rho \boldsymbol{g}^m)) = 0. \tag{4.2.3}$$

As J is not quadratic, solving (4.2.3) for ρ is not obvious, and instead, we propose to expand $\boldsymbol{H}(\boldsymbol{\zeta}^m - \rho \boldsymbol{g}^m)$ up to second order in terms of ρ. We write:

$$\boldsymbol{H}(\boldsymbol{\zeta}^m - \rho \boldsymbol{g}^m) \simeq \boldsymbol{H}(\boldsymbol{\zeta}^m) - \rho \boldsymbol{H}'(\boldsymbol{\zeta}^m) \cdot \boldsymbol{g}^m + \frac{\rho^2}{2} \boldsymbol{H}''(\boldsymbol{\zeta}^m) \cdot (\boldsymbol{g}^m, \boldsymbol{g}^m),$$
$$\boldsymbol{H}'(\boldsymbol{\zeta}^m - \rho \boldsymbol{g}^m) \cdot \boldsymbol{g}^m \simeq \boldsymbol{H}'(\boldsymbol{\zeta}^m) \cdot \boldsymbol{g}^m - \rho \boldsymbol{H}''(\boldsymbol{\zeta}^m) \cdot (\boldsymbol{g}^m, \boldsymbol{g}^m).$$

By substituting the right-hand side of each expression into (4.2.3), we replace (4.2.3) by a polynomial equation of degree three in ρ:

$$\begin{aligned} -\frac{1}{2}\rho^3 |\boldsymbol{H}''(\boldsymbol{\zeta}^m) \cdot (\boldsymbol{g}^m, \boldsymbol{g}^m)|^2_{H^1(\Omega)} &+ \frac{3}{2}\rho^2 \left(\nabla \boldsymbol{H}''(\boldsymbol{\zeta}^m) \cdot (\boldsymbol{g}^m, \boldsymbol{g}^m), \nabla \boldsymbol{H}'(\boldsymbol{\zeta}^m) \cdot \boldsymbol{g}^m\right) \\ &- \rho\Big(\big(\nabla \boldsymbol{H}''(\boldsymbol{\zeta}^m) \cdot (\boldsymbol{g}^m, \boldsymbol{g}^m), \nabla \boldsymbol{H}(\boldsymbol{\zeta}^m)\big) + |\boldsymbol{H}'(\boldsymbol{\zeta}^m) \cdot \boldsymbol{g}^m|^2_{H^1(\Omega)}\Big) \\ &+ \|\boldsymbol{g}^m\|^2_{H^1(\Omega)} = 0. \end{aligned} \tag{4.2.4}$$

In view of (4.2.1), the last term coincides with $(\nabla \boldsymbol{H}'(\boldsymbol{\zeta}^m) \cdot \boldsymbol{g}^m, \nabla \boldsymbol{H}(\boldsymbol{\zeta}^m))$. The relation (4.2.4) is an equation of the form $h(\rho) = 0$, where h is a polynomial of degree three that satisfies $h(0) \geq 0$. Therefore, it has at least one positive root, but there is no simple criterion that guarantees uniqueness of this root. If there are several positive roots, we set

$$\rho^m = \rho_{\min},$$

the smallest of the positive roots.

REMARK 4.2.1. The roots of the equation $h'(\rho) = 0$:

$$\begin{aligned} -\frac{3}{2}\rho^2 |\boldsymbol{H}''(\boldsymbol{\zeta}^m) \cdot (\boldsymbol{g}^m, \boldsymbol{g}^m)|^2_{H^1(\Omega)} + 3\rho\big(\nabla \boldsymbol{H}''(\boldsymbol{\zeta}^m) \cdot (\boldsymbol{g}^m, \boldsymbol{g}^m), \nabla \boldsymbol{H}'(\boldsymbol{\zeta}^m) \cdot \boldsymbol{g}^m\big) \\ - \Big(\big(\nabla \boldsymbol{H}''(\boldsymbol{\zeta}^m) \cdot (\boldsymbol{g}^m, \boldsymbol{g}^m), \nabla \boldsymbol{H}(\boldsymbol{\zeta}^m)\big) + |\boldsymbol{H}'(\boldsymbol{\zeta}^m) \cdot \boldsymbol{g}^m|^2_{H^1(\Omega)}\Big) = 0, \end{aligned} \tag{4.2.5}$$

can give some insight on $\rho_{\min}$. If the roots of (4.2.5) are complex or both negative, then (4.2.4) has a unique positive solution, but this condition is not necessary. If the roots of (4.2.5) are real and have an opposite sign, then $\rho_{\min}$ is located between these roots. But if the roots are both positive, then they give no immediate information on $\rho_{\min}$. □

4.2.1. Computational details

For the moment, we drop the superscript m. The right-hand side of (4.2.1) defining the gradient $\boldsymbol{g}$ is the difference of two terms:

$$\forall \boldsymbol{v} \in H^1(\Omega)^2,\ J'(\boldsymbol{\zeta}) \cdot \boldsymbol{v} = \big(\nabla \boldsymbol{u}_1'(\boldsymbol{\zeta}) \cdot \boldsymbol{v}, \nabla(\boldsymbol{u}_1 - \boldsymbol{u}_2)\big) - \big(\nabla \boldsymbol{u}_2'(\boldsymbol{\zeta}) \cdot \boldsymbol{v}, \nabla(\boldsymbol{u}_1 - \boldsymbol{u}_2)\big). \tag{4.2.6}$$

Let us derive for each of them an expression that is better adapted to computation. For the first one, we consider the unique solution of the "dual" problem: Find $\boldsymbol{\omega}_1 \in H_0^1(\Omega)^2$, solution of

$$\forall \boldsymbol{\theta} \in H_0^1(\Omega)^2,\ (\boldsymbol{\omega}_1, \boldsymbol{\theta}) + \alpha(\nabla \boldsymbol{\omega}_1, \nabla \boldsymbol{\theta}) = (\boldsymbol{u}_1 - \boldsymbol{u}_2, \boldsymbol{\theta}). \tag{4.2.7}$$

With the notation of (4.1.20), we have $\boldsymbol{\omega}_1 = D_\alpha^{-1}(\boldsymbol{u}_1 - \boldsymbol{u}_2)$. Recalling (4.1.19), we also have

$$\boldsymbol{u}_1'(\boldsymbol{\zeta}) \cdot \boldsymbol{v} = D_\alpha^{-1}(\boldsymbol{v}). \tag{4.2.8}$$

The relationship between the derivative of $\boldsymbol{u}_1$ and $\boldsymbol{\omega}_1$ is given by the next lemma.

LEMMA 4.2.2. *For all $\boldsymbol{v} \in H^{-1}(\Omega)^2$, we have*

$$\big(\nabla \boldsymbol{u}_1'(\boldsymbol{\zeta}) \cdot \boldsymbol{v}, \nabla(\boldsymbol{u}_1 - \boldsymbol{u}_2)\big) = \frac{1}{\alpha}\langle \boldsymbol{u}_1 - \boldsymbol{u}_2 - \boldsymbol{\omega}_1, \boldsymbol{v}\rangle. \tag{4.2.9}$$

PROOF. Let $\boldsymbol{v} \in H^{-1}(\Omega)^2$; from (4.1.19), we have

$$(\boldsymbol{u}_1'(\boldsymbol{\zeta}) \cdot \boldsymbol{v}, \boldsymbol{u}_1 - \boldsymbol{u}_2) + \alpha(\nabla \boldsymbol{u}_1'(\boldsymbol{\zeta}) \cdot \boldsymbol{v}, \nabla(\boldsymbol{u}_1 - \boldsymbol{u}_2)) = \langle \boldsymbol{v}, \boldsymbol{u}_1 - \boldsymbol{u}_2\rangle.$$

In view of (4.2.8), this gives

$$(\nabla \boldsymbol{u}_1'(\boldsymbol{\zeta}) \cdot \boldsymbol{v}, \nabla(\boldsymbol{u}_1 - \boldsymbol{u}_2)) = \frac{1}{\alpha}\langle \boldsymbol{v}, \boldsymbol{u}_1 - \boldsymbol{u}_2\rangle - \frac{1}{\alpha}(D_\alpha^{-1}(\boldsymbol{v}), \boldsymbol{u}_1 - \boldsymbol{u}_2).$$

But $\boldsymbol{u}_1 - \boldsymbol{u}_2 = D_\alpha(\boldsymbol{\omega}_1)$, and because $\boldsymbol{\omega}_1$ belongs to $H_0^1(\Omega)^2$, (4.1.21) implies:

$$(D_\alpha^{-1}(\boldsymbol{v}), \boldsymbol{u}_1 - \boldsymbol{u}_2) = (D_\alpha^{-1}(\boldsymbol{v}), D_\alpha(\boldsymbol{\omega}_1)) = (D_\alpha\, D_\alpha^{-1}(\boldsymbol{v}), \boldsymbol{\omega}_1) = \langle \boldsymbol{v}, \boldsymbol{\omega}_1\rangle,$$

whence (4.2.9). □

To handle the second term in (4.2.6), we consider the unique solution $(\boldsymbol{\omega}_2, q_2)$ of the "dual" problem: Given $\boldsymbol{\zeta}$ in Y, find $(\boldsymbol{\omega}_2, q_2) \in V \times L_0^2(\Omega)$ solution of

$$\forall \boldsymbol{\theta} \in H_0^1(\Omega)^2,\ \nu(\nabla \boldsymbol{\omega}_2, \nabla \boldsymbol{\theta}) - (\operatorname{curl} \boldsymbol{\zeta} \times \boldsymbol{\omega}_2, \boldsymbol{\theta}) - (q_2, \operatorname{div} \boldsymbol{\theta}) = \nu(\nabla(\boldsymbol{u}_1 - \boldsymbol{u}_2), \nabla \boldsymbol{\theta}). \tag{4.2.10}$$

With the notation of (4.1.23), (4.2.10) is equivalent to

$$(\boldsymbol{\omega}_2, q_2) = N_{-\zeta}^{-1}(-\nu\Delta(\boldsymbol{u}_1 - \boldsymbol{u}_2)).$$

An easy calculation yields the following relation between $\boldsymbol{\omega}_2$ and the derivative of $\boldsymbol{u}_2$.

LEMMA 4.2.3. *For all $\boldsymbol{\zeta}$ and $\boldsymbol{v}$ in Y, we have*

$$\big(\nabla \boldsymbol{u}_2'(\boldsymbol{\zeta}) \cdot \boldsymbol{v}, \nabla(\boldsymbol{u}_1 - \boldsymbol{u}_2)\big) = -\frac{1}{\nu}(\operatorname{curl} \boldsymbol{v} \times \boldsymbol{u}_2, \boldsymbol{\omega}_2). \tag{4.2.11}$$

Note that both $\boldsymbol{\omega}_1$ and $\boldsymbol{\omega}_2$ depend on $\boldsymbol{\zeta}$; the dependence of $\boldsymbol{\omega}_2$ is obvious and that of $\boldsymbol{\omega}_1$ follows from the dependence of $\boldsymbol{u}_1$ and $\boldsymbol{u}_2$ on $\boldsymbol{\zeta}$. By collecting (4.2.9) and (4.2.11), we obtain the following expression for $\boldsymbol{g}$, for all $\boldsymbol{\zeta}$ in Y:

$$\forall \boldsymbol{v} \in H^1(\Omega)^2,\ (\boldsymbol{g}, \boldsymbol{v}) + (\nabla \boldsymbol{g}, \boldsymbol{v}) = \frac{1}{\alpha}(\boldsymbol{u}_1 - \boldsymbol{u}_2 - \boldsymbol{\omega}_1, \boldsymbol{v}) + \frac{1}{\nu}(\operatorname{curl} \boldsymbol{v} \times \boldsymbol{u}_2, \boldsymbol{\omega}_2). \tag{4.2.12}$$

Now, we are in a position to list the steps of the approximate gradient algorithm. Choose a starting function $\boldsymbol{\zeta}^0 \in Y$, e.g., $\boldsymbol{\zeta}^0 = \boldsymbol{0}$. Choose a threshold ε for the stopping criterion. If $\boldsymbol{\zeta}^0 = \boldsymbol{0}$, then $\boldsymbol{u}_1^0 = \boldsymbol{0}$. If $\boldsymbol{\zeta}^0 \neq \boldsymbol{0}$, then compute $\boldsymbol{u}_1^0 = D_\alpha^{-1}(\boldsymbol{\zeta}^0) \in H_0^1(\Omega)^2$ solution of (4.1.8):

$$\forall \boldsymbol{v} \in H_0^1(\Omega)^2,\ (\boldsymbol{u}_1^0, \boldsymbol{v}) + \alpha(\nabla \boldsymbol{u}_1^0, \nabla \boldsymbol{v}) = \langle \boldsymbol{\zeta}^0, \boldsymbol{v}\rangle.$$

Then for $m \geq 0$, knowing $\boldsymbol{\zeta}^m \in Y$ and $\boldsymbol{u}_1^m \in H_0^1(\Omega)^2$:

1. Compute $(\boldsymbol{u}_2^m, p^m) = N_{\zeta^m}^{-1}(\boldsymbol{f}) \in V \times L_0^2(\Omega)$ solution of (4.1.9):

$$\forall \boldsymbol{v} \in H_0^1(\Omega)^2,\ \nu(\nabla \boldsymbol{u}_2^m, \nabla \boldsymbol{v}) + (\operatorname{curl} \boldsymbol{\zeta}^m \times \boldsymbol{u}_2^m, \boldsymbol{v}) - (p^m, \operatorname{div} \boldsymbol{v}) = (\boldsymbol{f}, \boldsymbol{v}),$$

2. Compute $\boldsymbol{\omega}_1^m = D_\alpha^{-1}(\boldsymbol{u}_1^m - \boldsymbol{u}_2^m) \in H_0^1(\Omega)^2$:

$$\forall \boldsymbol{\theta} \in H_0^1(\Omega)^2,\ (\boldsymbol{\omega}_1^m, \boldsymbol{\theta}) + \alpha(\nabla \boldsymbol{\omega}_1^m, \nabla \boldsymbol{\theta}) = (\boldsymbol{u}_1^m - \boldsymbol{u}_2^m, \boldsymbol{\theta}).$$

3. Compute $(\boldsymbol{\omega}_2^m, p_2^m) = N_{-\boldsymbol{\zeta}^m}^{-1}(-\nu\Delta\left(\boldsymbol{u}_1^m - \boldsymbol{u}_2^m\right)) \in V \times L_0^2(\Omega)$, i.e., solution of (4.2.10):

$$\forall \boldsymbol{\theta} \in H_0^1(\Omega)^2,\ \nu(\nabla \boldsymbol{\omega}_2^m, \nabla \boldsymbol{\theta}) - (\operatorname{curl} \boldsymbol{\zeta}^m \times \boldsymbol{\omega}_2^m, \boldsymbol{\theta}) - (q_2^m, \operatorname{div} \boldsymbol{\theta}) = \nu(\nabla(\boldsymbol{u}_1^m - \boldsymbol{u}_2^m), \nabla \boldsymbol{\theta}).$$

4. Compute $\boldsymbol{g}^m \in H^1(\Omega)^2$ solution of (4.2.12):

$$\forall \boldsymbol{v} \in H^1(\Omega)^2,\ (\boldsymbol{g}^m, \boldsymbol{v}) + (\nabla \boldsymbol{g}^m, \boldsymbol{v}) = \frac{1}{\alpha}(\boldsymbol{u}_1^m - \boldsymbol{u}_2^m - \boldsymbol{\omega}_1^m, \boldsymbol{v}) + \frac{1}{\nu}(\operatorname{curl} \boldsymbol{v} \times \boldsymbol{u}_2^m, \boldsymbol{\omega}_2^m).$$

5. Compute $A = \|\boldsymbol{g}^m\|_{H^1(\Omega)}$. If $A \le \varepsilon$, the loop ends and $(\boldsymbol{u}_2^m, p^m)$ is the approximate solution.
   Otherwise,
6. Compute

$$(\boldsymbol{u}_1^m)' \cdot \boldsymbol{g}^m = D_\alpha^{-1}(\boldsymbol{g}^m).$$

7. Compute the pair

$$\left((\boldsymbol{u}_2^m)' \cdot \boldsymbol{g}^m, q_1^m\right) = N_{\boldsymbol{\zeta}^m}^{-1}(-\operatorname{curl} \boldsymbol{g}^m \times \boldsymbol{u}_2^m).$$

8. Compute

$$\boldsymbol{H}'(\boldsymbol{\zeta}^m) \cdot \boldsymbol{g}^m = (\boldsymbol{u}_1^m)' \cdot \boldsymbol{g}^m - (\boldsymbol{u}_2^m)' \cdot \boldsymbol{g}^m,\ B = |\boldsymbol{H}'(\boldsymbol{\zeta}^m) \cdot \boldsymbol{g}^m|_{H^1(\Omega)}^2.$$

9. Compute the pair

$$(\boldsymbol{H}''(\boldsymbol{\zeta}^m) \cdot (\boldsymbol{g}^m, \boldsymbol{g}^m), q_2^m) = N_{\boldsymbol{\zeta}^m}^{-1}\left(-2\operatorname{curl} \boldsymbol{g}^m \times (\boldsymbol{u}_2^m)' \cdot \boldsymbol{g}^m\right).$$

10. Compute

$$C = -\left(\nabla \boldsymbol{H}''(\boldsymbol{\zeta}^m) \cdot (\boldsymbol{g}^m, \boldsymbol{g}^m), \nabla \boldsymbol{H}(\boldsymbol{\zeta}^m)\right) - B,$$
$$D = \frac{3}{2}\left(\nabla \boldsymbol{H}''(\boldsymbol{\zeta}^m) \cdot (\boldsymbol{g}^m, \boldsymbol{g}^m), \nabla \boldsymbol{H}'(\boldsymbol{\zeta}^m) \cdot \boldsymbol{g}^m\right),$$
$$E = -\frac{1}{2}|\boldsymbol{H}''(\boldsymbol{\zeta}^m) \cdot (\boldsymbol{g}^m, \boldsymbol{g}^m)|_{H^1(\Omega)}^2.$$

11. Compute an approximation of the smallest positive root ρ^m of

$$A^2 + C\rho + D\rho^2 + E\rho^3,$$

Update:

$$\zeta^{m+1} = \zeta^m - \rho^m g^m,$$
$$u_1^{m+1} = u_1^m - \rho^m (u_1^m)' \cdot g^m,$$

replace m by $m+1$ and go to Step 1.

4.3. Application to the time-dependent problem

From a computational point of view, a least-squares scheme and gradient algorithm is less interesting for solving a time-dependent problem, because a linearized time-stepping scheme seems more economical. However, for the sake of completeness, we briefly describe one such scheme.

Let us revert to the time-dependent problem (3.1.1) in a bounded, connected Lipschitz domain Ω of $\mathbb{R}^2$: Find $\boldsymbol{u} \in L^\infty(0, T; V^\alpha)$ and $p \in L^2(0, T, L_0^2(\Omega))$ such that

$$\frac{\partial}{\partial t}(\boldsymbol{u} - \alpha\Delta \boldsymbol{u}) - \nu\,\Delta \boldsymbol{u} + \mathbf{curl}(\boldsymbol{u} - \alpha\Delta \boldsymbol{u}) \times \boldsymbol{u} + \nabla p = \boldsymbol{f} \quad \text{in } \Omega\times]0, T[,$$
$$\boldsymbol{u}(0) = \boldsymbol{u}_0 \quad \text{in } \Omega,$$

where $T > 0$, $\nu > 0$, and $\alpha > 0$ are given real numbers, V^α is defined in (1.4.1), $\boldsymbol{f}$ is given in $L^2(0, T; H(\text{curl}, \Omega))$, and $\boldsymbol{u}_0$ is given in V^α.

We propose to adapt the first least-squares scheme and gradient algorithm of the preceding section to the following semi-discrete version of (3.1.1). Let $N > 1$ be an integer, define the time step k by

$$k = \frac{T}{N},$$

the subdivision points by $t_n = n\,k$, and the approximate value $\boldsymbol{f}^n$ of $\boldsymbol{f}(t_n)$ by (3.2.18)

$$\boldsymbol{f}^n(\boldsymbol{x}) = \frac{1}{k}\int_{t_{n-1}}^{t_n} \boldsymbol{f}(\boldsymbol{x}, s)\,\mathrm{d}s.$$

Starting from

$$\boldsymbol{u}^0 = \boldsymbol{u}_0, \tag{4.3.1}$$

find sequences $(\boldsymbol{u}^n)_{n\geq 1}$, $(\boldsymbol{\zeta}^n)_{n\geq 1}$, and $(p^n)_{n\geq 1}$ such that $\boldsymbol{u}^n \in V$, $\boldsymbol{\zeta}^n \in Y$, and $p^n \in L_0^2(\Omega)$ solve for $1 \leq n \leq N$,

$$\frac{1}{k}(\boldsymbol{u}^n - \boldsymbol{u}^{n-1}) - \alpha\frac{1}{k}\Delta(\boldsymbol{u}^n - \boldsymbol{u}^{n-1}) - \nu\Delta \boldsymbol{u}^n + \text{curl}\,\boldsymbol{\zeta}^n \times \boldsymbol{u}^n + \nabla p^n = \boldsymbol{f}^n \quad \text{in } \Omega, \tag{4.3.2}$$

$$\boldsymbol{\zeta}^n = \boldsymbol{u}_n - \alpha\Delta\,\boldsymbol{u}_n. \tag{4.3.3}$$

For each n, (4.3.2)–(4.3.3) has the same form as (4.1.3)–(4.1.5), with ν replaced by $\frac{\alpha}{k} + \nu$, with the additional elliptic term $\frac{1}{k}\boldsymbol{u}^n$ in the left-hand side of (4.1.4) and $\boldsymbol{f}$ replaced by

$$\boldsymbol{f}^n + \frac{1}{k}\left(\boldsymbol{u}^{n-1} - \alpha\Delta\,\boldsymbol{u}^{n-1}\right)$$

in its right-hand side. Thus, the operator D_α is unchanged, while the operator N_ζ is replaced by $N_{\zeta,k} : V \times L_0^2(\Omega) \mapsto H^{-1}(\Omega)^2$ defined by

$$(u, p) \mapsto \frac{1}{k}\boldsymbol{u} - \left(\frac{\alpha}{k} + \nu\right)\Delta\,\boldsymbol{u} + \operatorname{curl}\boldsymbol{\zeta} \times \boldsymbol{u} + \nabla p. \tag{4.3.4}$$

With these modifications, we can readily adapt to this situation, the least-squares scheme and approximate gradient algorithm of Section 4.2.1.

More precisely, for each n, at the ith step of the algorithm, denoting the iterate by the superscript i, knowing $(\boldsymbol{u}_1^n)^i$, $\boldsymbol{u}^{n-1}$ and $(\boldsymbol{\zeta}^n)^i$, $\left((\boldsymbol{u}_2^n)^i, (p^n)^i\right)$ is computed by

$$\left((\boldsymbol{u}_2^n)^i, (p^n)^i\right) = N^{-1}_{(\boldsymbol{\zeta}^n)^i,k}(\boldsymbol{F}^n),$$

where

$$\boldsymbol{F}^n = \boldsymbol{f}^n + \frac{1}{k}\boldsymbol{u}^{n-1} - \frac{\alpha}{k}\Delta\,\boldsymbol{u}^{n-1}.$$

The function $(\boldsymbol{\omega}_1^n)^i$ is computed as in Step 2, the pair $\left((\boldsymbol{\omega}_2^n)^i, (p_2^n)^i\right)$ is computed by

$$\left((\boldsymbol{\omega}_2^n)^i, (p_2^n)^i\right) = N^{-1}_{-(\boldsymbol{\zeta}^n)^i,k}\left(-\nu\Delta((\boldsymbol{u}_1^n)^i - (\boldsymbol{u}_2^n)^i)\right),$$

the gradient $(\boldsymbol{g}^n)^i$ is computed as in Step 4, the derivative $\left((\boldsymbol{u}_1^n)^i\right)' \cdot (\boldsymbol{g}^n)^i$ is computed as in Step 6, the first derivative pair $\left(((\boldsymbol{u}_2^n)^i)' \cdot (\boldsymbol{g}^n)^i, (q_1^n)^i\right)$ is computed by

$$\left(((\boldsymbol{u}_2^n)^i)' \cdot (\boldsymbol{g}^n)^i, (q_1^n)^i\right) = N^{-1}_{(\boldsymbol{\zeta}^n)^i,k}\left(-\operatorname{curl}(\boldsymbol{g}^n)^i \times (\boldsymbol{u}_2^n)^i\right),$$

the derivative $\boldsymbol{H}'((\boldsymbol{\zeta}^n)^i) \cdot (\boldsymbol{g}^n)^i$ is computed as in Step 8, the second derivative pair $\left(\boldsymbol{H}''((\boldsymbol{\zeta}^n)^i) \cdot ((\boldsymbol{g}^n)^i, (\boldsymbol{g}^n)^i), (q_2^n)^i\right)$ is computed by

$$\left(\boldsymbol{H}''((\boldsymbol{\zeta}^n)^i) \cdot ((\boldsymbol{g}^n)^i, (\boldsymbol{g}^n)^i), (q_2^n)^i\right) = N^{-1}_{(\boldsymbol{\zeta}^n)^i,k}\left(-2\operatorname{curl}(\boldsymbol{g}^n)^i \times ((\boldsymbol{u}_2^n)^i)' \cdot (\boldsymbol{g}^n)^i\right).$$

The remaining steps are unchanged. For each n, the algorithm can be started with $(\boldsymbol{u}_1^n)^1 = \boldsymbol{u}^{n-1}$ and $(\boldsymbol{\zeta}^n)^1 = \boldsymbol{\zeta}^{n-1}$.

CHAPTER 5

The Steady Problem with Tangential Boundary Conditions

As stated in Remark 1.3.1, the analysis of a grade-two fluid model with fully nonhomogeneous Dirichlet boundary conditions is not clear. This difficulty is caused by the behavior of the transport equation, when the normal component of the driving velocity does not vanish on the boundary. This is commented in Remark 1.3.14. Consequently, we restrict our study to the case of tangential boundary conditions.

As for the Navier–Stokes equations, nonhomogeneous Dirichlet boundary conditions for the grade-two fluid model are handled by means of a suitable lifting of the type introduced by LERAY [1933], HOPF [1951], see also LIONS [1969], TEMAM [1979], or GIRAULT and RAVIART [1986]. In the case of smooth tangential boundary conditions on a smooth boundary, this lifting is simplified by TEMAM [1997]. Here we shall use a variant of the lifting in TEMAM [1997] that applies to a Lipschitz boundary. It is developed by GIRAULT and SCOTT [2002a].

5.1. Some theoretical results

The material presented here is taken from GIRAULT and SCOTT [1999]. Let Ω be a bounded connected domain in $\mathbb{R}^2$, with a Lipschitz-continuous boundary $\partial\Omega$ and exterior normal $\boldsymbol{n}$, $\boldsymbol{f}$ a given function in $H(\text{curl}, \Omega)$, $\boldsymbol{g}$ a given tangential vector field in $H^{1/2}(\partial\Omega)^2$, and $\nu > 0$ and $\alpha > 0$ two given real constants. Recall the spaces $H^1_\tau(\Omega)$ and W, respectively, defined by (1.1.7) and (1.1.12):

$$H^1_\tau(\Omega) = \{\boldsymbol{v} \in H^1(\Omega)^2;\ \boldsymbol{v}\cdot\boldsymbol{n} = 0 \text{ on } \partial\Omega\},$$
$$W = \{\boldsymbol{v} \in H^1_\tau(\Omega);\ \operatorname{div}\boldsymbol{v} = 0 \text{ in } \Omega\},$$

and define the analog of V^α:

$$W^\alpha = \{\boldsymbol{v} \in W;\ \text{curl}(\boldsymbol{v} - \alpha\,\Delta\,\boldsymbol{v}) \in L^2(\Omega)\}. \tag{5.1.1}$$

Consider the steady grade-two fluid model in Ω: Find a pair $(\boldsymbol{u} = (u_1, u_2), p) \in W^\alpha \times L^2_0(\Omega)$ such that

$$-\nu\,\Delta\,\boldsymbol{u} + z \times \boldsymbol{u} + \nabla p = \boldsymbol{f} \quad \text{in } \Omega,$$
$$\operatorname{div}\boldsymbol{u} = 0 \quad \text{in } \Omega,$$

$$\boldsymbol{u} = \boldsymbol{g} \quad \text{on } \partial\Omega \quad \text{with} \quad \boldsymbol{g}\cdot\boldsymbol{n} = 0,$$
$$\boldsymbol{z} = (0,0,z),\; z = \operatorname{curl}(\boldsymbol{u} - \alpha\,\Delta\,\boldsymbol{u}),$$

where

$$\boldsymbol{z}\times\boldsymbol{u} = (-z\,u_2, z\,u_1).$$

This problem is equivalent to: Find $(\boldsymbol{u}, p, z)$ in $W \times L^2_0(\Omega) \times L^2(\Omega)$ solution of

$$\begin{aligned} -\nu\,\Delta\,\boldsymbol{u} + \boldsymbol{z}\times\boldsymbol{u} + \nabla\, p = \boldsymbol{f} \quad &\text{in } \Omega,\\ \operatorname{div}\boldsymbol{u} = 0 \quad &\text{in } \Omega,\\ \boldsymbol{u} = \boldsymbol{g} \quad &\text{on } \partial\Omega,\\ \nu\, z + \alpha\,\boldsymbol{u}\cdot\nabla\, z = \nu \operatorname{curl}\boldsymbol{u} + \alpha \operatorname{curl}\boldsymbol{f} \quad &\text{in } \Omega. \end{aligned} \tag{5.1.2}$$

We have seen in Proposition 1.3.9 that, given $\boldsymbol{u}$ in $H^1_\tau(\Omega)$, the last equation in (5.1.2) has a unique solution because it is a transport equation. Furthermore,

$$\|z\|_{L^2(\Omega)} \le \|\operatorname{curl}\boldsymbol{u}\|_{L^2(\Omega)} + \frac{\alpha}{\nu}\|\operatorname{curl}\boldsymbol{f}\|_{L^2(\Omega)}. \tag{5.1.3}$$

Therefore, the analysis of (5.1.2) reduces essentially to that of its first system, which is a Stokes-like problem. In variational form, for given z in $L^2(\Omega)$, it reads: Find a pair $(\boldsymbol{u}, p)$ in $H^1(\Omega)^2 \times L^2(\Omega)$, such that

$$\forall \boldsymbol{v} \in H^1_0(\Omega)^2,\; \nu(\nabla\,\boldsymbol{u}, \nabla\,\boldsymbol{v}) + (\boldsymbol{z}\times\boldsymbol{u}, \boldsymbol{v}) - (p, \operatorname{div}\boldsymbol{v}) = (\boldsymbol{f}, \boldsymbol{v}), \tag{5.1.4}$$
$$\forall q \in L^2_0(\Omega),\; (q, \operatorname{div}\boldsymbol{u}) = 0, \tag{5.1.5}$$
$$\boldsymbol{u} = \boldsymbol{g} \quad \text{on } \partial\Omega. \tag{5.1.6}$$

5.1.1. A lifting

It is well known that the analysis of problem (5.1.4)–(5.1.6) requires a lifting function $\boldsymbol{w} \in W$ that has the same trace as $\boldsymbol{u}$ on $\partial\Omega$. The obvious (but not necessarily best) candidate is given by the following consequence of the inf-sup condition (1.1.26), see for instance GIRAULT and RAVIART [1986]. It is valid in arbitrary dimensions.

PROPOSITION 5.1.1. *Let Ω be a bounded, connected Lipschitz domain of $\mathbb{R}^d$. For each $\boldsymbol{g} \in H^{1/2}(\partial\Omega)^d$ satisfying the compatibility condition*

$$\int_{\partial\Omega} \boldsymbol{g}\cdot\boldsymbol{n}\,\mathrm{d}s = 0, \tag{5.1.7}$$

there exists a unique function $\boldsymbol{w} \in H^1(\Omega)^d$ depending linearly on $\boldsymbol{g}$, and a constant L independent of $\boldsymbol{g}$ satisfying

$$\begin{aligned} \forall \boldsymbol{v} \in V;\; (\nabla\,\boldsymbol{w}, \nabla\,\boldsymbol{v}) &= 0,\\ \operatorname{div}\boldsymbol{w} &= 0,\\ \boldsymbol{w}|_{\partial\Omega} &= \boldsymbol{g},\\ \|\boldsymbol{w}\|_{H^1(\Omega)} &\le L\|\boldsymbol{g}\|_{H^{1/2}(\partial\Omega)}. \end{aligned} \tag{5.1.8}$$

But a straightforward computation shows that $\boldsymbol{w}$ is not always an appropriate lifting because it does not yield existence of a solution of (5.1.2) for all data. Indeed, when introduced into (5.1.4), it generates a term of the form $(\boldsymbol{z} \times \boldsymbol{w}, \boldsymbol{v})$ that cannot be dominated by the elliptic part of (5.1.4) if ν is small while the other data are large. This observation dates back to the work of LERAY [1933]. Based on the fact that this additional term involves no derivative of $\boldsymbol{w}$, Leray proposed to truncate $\boldsymbol{w}$ in order to make this term arbitrarily small. In order to preserve the zero divergence of $\boldsymbol{w}$, this truncation was achieved by truncating the stream function of $\boldsymbol{w}$. Of course, this process changes the constant L in (5.1.8) and, unfortunately, this new constant grows exponentially with $\frac{1}{\nu}$, cf. HOPF [1951], LIONS [1969], TEMAM [1979], or GIRAULT and RAVIART [1986]. However, at least in two dimensions, when the normal component of $\boldsymbol{g}$ vanishes on $\partial\Omega$, a lifting can be constructed so that the new constant L only has a polynomial growth. We refer to TEMAM [1997] when both $\boldsymbol{g}$ and $\partial\Omega$ are smooth and to GIRAULT and SCOTT [2002a], when $\partial\Omega$ is Lipschitz and $\boldsymbol{g}$ is in a slightly smaller space that $H^{1/2}(\partial\Omega)^2$.

In what follows, we suppose that ν is small and the other data are large, otherwise, the results stated below are not necessary because we can use the lifting $\boldsymbol{w}$ of Proposition 5.1.1.

The following result is established in GIRAULT and SCOTT [2002a]. Its statement requires the set Ω_ε, defined for all ε sufficiently small by

$$\Omega_\varepsilon = \{\boldsymbol{x} \in \Omega;\ d(\boldsymbol{x}) \le C\varepsilon\}, \tag{5.1.9}$$

where $d(\boldsymbol{x})$ denotes the distance from $\boldsymbol{x}$ to the boundary $\partial\Omega$ and $C > 0$ is a suitable constant. The parameter ε is sufficiently small so that Ω_ε is a tubular neighborhood of $\partial\Omega$, and in particular, if $\partial\Omega$ is not connected, then Ω_ε consists of mutually disjoint neighborhoods of each connected component of $\partial\Omega$.

THEOREM 5.1.2. *Let Ω be a bounded, connected Lipschitz domain of $\mathbb{R}^2$, and let $\boldsymbol{g}$, satisfying $\boldsymbol{g} \cdot \boldsymbol{n} = 0$, be given in $W^{1-1/r,r}(\partial\Omega)^2$ for some real number $r > 2$. There exists a real number $\varepsilon_0 > 0$, only depending on Ω, such that for each ε with $0 < \varepsilon \le \varepsilon_0$, there exists a function $\boldsymbol{u_g}$ in $W \cap W^{1,r}(\Omega)^2$, supported by Ω_ε, such that:*

$$\boldsymbol{u_g}|_{\partial\Omega} = \boldsymbol{g}.$$

For any number s with $1 \le s \le \infty$, we have

$$\|\boldsymbol{u_g}\|_{L^s(\Omega_\varepsilon)} \le C\varepsilon^{1/s}\|\boldsymbol{g}\|_{W^{1-1/r,r}(\partial\Omega)}. \tag{5.1.10}$$

In addition, $\boldsymbol{u_g}$ satisfies

$$\forall \boldsymbol{v} \in H_0^1(\Omega)^2,\ \big\||\boldsymbol{u_g}|\,|\boldsymbol{v}|\big\|_{L^2(\Omega_\varepsilon)} \le C\varepsilon|\boldsymbol{v}|_{H^1(\Omega_\varepsilon)}\|\boldsymbol{g}\|_{W^{1-1/r,r}(\partial\Omega)}, \tag{5.1.11}$$

$$\|\nabla\,\boldsymbol{u_g}\|_{L^2(\Omega_\varepsilon)} \le C\frac{1}{\sqrt{\varepsilon}}\|\boldsymbol{g}\|_{W^{1-1/r,r}(\partial\Omega)}, \tag{5.1.12}$$

$$\|\nabla\,\boldsymbol{u_g}\|_{L^r(\Omega_\varepsilon)} \le C\varepsilon^{1/r-1}\|\boldsymbol{g}\|_{W^{1-1/r,r}(\partial\Omega)}. \tag{5.1.13}$$

REMARK 5.1.3. The assumption $\boldsymbol{g} \in W^{1-1/r,r}(\partial\Omega)^2$ is used in proving (5.1.11), where $\boldsymbol{u_g}$ is bounded in $L^\infty(\Omega_\varepsilon)^2$ and (5.1.10) is applied with $s = \infty$. The L^∞ bound for $\boldsymbol{u_g}$ is obtained by Sobolev's imbedding, when $\boldsymbol{u_g}$ belongs to $W^{1,r}(\Omega)^2$ for some r slightly larger than 2, and $W^{1-1/r,r}(\partial\Omega)$ is precisely the trace space of $W^{1,r}(\Omega)$. □

5.1.2. *Existence, regularity, and uniqueness*

For the sake of simplicity, we choose from now on to make the assumption that $\boldsymbol{g} \in W^{1-1/r,r}(\partial\Omega)^2$. Indeed, when $\boldsymbol{g}$ is only in $H^{1/2}(\partial\Omega)^2$, Theorem 5.1.2 is replaced by a similar result with a slightly deteriorated exponent of ε in (5.1.12), but this complicates the subsequent a priori bounds.

We have the following existence result; cf. GIRAULT and SCOTT [1999].

THEOREM 5.1.4. *Let Ω be a bounded, connected Lipschitz domain of $\mathbb{R}^2$. For all data $\nu > 0$, $\alpha > 0$, $\boldsymbol{f} \in H(\text{curl}, \Omega)$, and $\boldsymbol{g} \in W^{1-1/r,r}(\partial\Omega)^2$ for some real number $r > 2$, satisfying $\boldsymbol{g} \cdot \boldsymbol{n} = 0$, problem (5.1.2) has at least one solution $(\boldsymbol{u}, p, z)$ in $W \times L^2_0(\Omega) \times L^2(\Omega)$, and all solutions of (5.1.2) satisfy the a priori estimates:*

$$|\boldsymbol{u}|_{H^1(\Omega)} \le 2\frac{S_2}{\nu}\|\boldsymbol{f}\|_{L^2(\Omega)} + \frac{\alpha}{\nu}\|\text{curl}\,\boldsymbol{f}\|_{L^2(\Omega)} + \frac{C}{\sqrt{\nu}}\|\boldsymbol{g}\|^{3/2}_{W^{1-1/r,r}(\partial\Omega)}, \tag{5.1.14}$$

$$\|z\|_{L^2(\Omega)} \le \|\text{curl}\,\boldsymbol{u}\|_{L^2(\Omega)} + \frac{\alpha}{\nu}\|\text{curl}\,\boldsymbol{f}\|_{L^2(\Omega)}, \tag{5.1.15}$$

$$\|p\|_{L^2(\Omega)} \le \frac{1}{\beta}\Big(S_2\|\boldsymbol{f}\|_{L^2(\Omega)} + |\boldsymbol{u}|_{H^1(\Omega)}(\nu + S_4\tilde{S}_4\|z\|_{L^2(\Omega)})\Big), \tag{5.1.16}$$

where β is the inf-sup constant of (1.1.25), S_i are the constants of Sobolev's imbedding (1.1.3), $\tilde{S}_i$ are those of (1.1.8), and C depends on the constants of (5.1.11) and (5.1.12).

REMARK 5.1.5. The contribution of the nonhomogeneous boundary data $\boldsymbol{g}$ to the a priori estimate (5.1.14) is of the order of $\nu^{-1/2}$. When ν is small, it is negligible compared with the contribution of the interior force $\boldsymbol{f}$. Hence the above approach gives a much sharper estimate than the classical Leray–Hopf's Lemma. □

When Ω is a polygon, considering that $\boldsymbol{f} \in H(\text{curl}, \Omega)$ is sufficiently smooth, the velocity, and pressure solutions of (5.1.2) are more regular provided $\boldsymbol{g}$ is smoother. The two theorems below are analogs of Theorems 1.4.8 and 1.4.9, and are derived from results of GRISVARD [1985]. Their statements require a more precise description of the geometry of Ω. We do not assume that the boundary $\partial\Omega$ is connected, and for each connected component γ_j, $0 \le j \le k$, of $\partial\Omega$, we denote by Γ_i, for $1 \le i \le N$, the straight line segments of γ_j, with the convention that Γ_i is adjacent to Γ_{i+1} and Γ_{N+1} coincides with Γ_1. Also, we denote by $\boldsymbol{n}_i$ the unit normal to Γ_i pointing outside Ω, by $\boldsymbol{t}_i$ the unit tangent vector along Γ_i pointing in the clockwise direction, by $\boldsymbol{x}_i$ the common vertex of Γ_i and Γ_{i+1}, and by ω_i the inner angle between them. Strictly speaking, we should use the notation Γ_i^j and N_j to specify the dependence on j, but we drop it to alleviate notation. By definition, a polygon has no cracks and, therefore, all the inner angles of $\partial\Omega$ must satisfy $0 < \omega_i < 2\pi$.

THEOREM 5.1.6. *Assume that Ω is a bounded, connected polygon. Let $\nu > 0$, $\alpha > 0$, and $\boldsymbol{f}$ be given in $H(\mathrm{curl}, \Omega)$. If the boundary data $\boldsymbol{g}$ satisfies on each γ_j, $0 \le j \le k$,*

$$\boldsymbol{g} \in W^{5/4,4/3}(\Gamma_i)^2 \ \textit{for}\ 1 \le i \le N, \quad \boldsymbol{g}\cdot\boldsymbol{n} = 0 \ \textit{on}\ \partial\Omega, \tag{5.1.17}$$

then all solutions of problem (5.1.2) satisfy

$$\boldsymbol{u} \in W^{2,4/3}(\Omega)^2, \quad p \in W^{1,4/3}(\Omega),$$

with continuous dependence on the data

$$\begin{aligned}\|\boldsymbol{u}\|_{W^{2,4/3}(\Omega)} + \|p\|_{W^{1,4/3}(\Omega)} \le C_1 \Bigg(& \|\boldsymbol{f}\|_{L^2(\Omega)} + \sum_{j=0}^{k} [\boldsymbol{g}]_{W^{5/4,4/3}(\gamma_j)} \\ & + C_2 \left(\|\boldsymbol{f}\|_{L^2(\Omega)} + \|\boldsymbol{g}\|_{H^{1/2}(\partial\Omega)}\right) \|z\|_{L^2(\Omega)} + C_3 \|\boldsymbol{g}\|_{H^{1/2}(\partial\Omega)} \|z\|^2_{L^2(\Omega)} \Bigg),\end{aligned} \tag{5.1.18}$$

where

$$[\boldsymbol{g}]_{W^{5/4,4/3}(\gamma_j)} = \sum_{i=1}^{N} \|\boldsymbol{g}\|_{W^{5/4,4/3}(\Gamma_i)}.$$

The regularity of z is unchanged.

THEOREM 5.1.7. *We retain the hypotheses of Theorem 5.1.6 and in addition, we suppose Ω is a convex polygon and the boundary data $\boldsymbol{g}$ satisfies on each γ_j, $0 \le j \le k$,*

$$\boldsymbol{g} \in H^{3/2}(\Gamma_i)^2 \ \textit{for}\ 1 \le i \le N, \quad \boldsymbol{g}\cdot\boldsymbol{n} = 0 \ \textit{on}\ \partial\Omega, \tag{5.1.19}$$

$$\int_0^{\varepsilon} \frac{1}{s} \left| \frac{\partial \boldsymbol{g}_{i+1}\cdot\boldsymbol{n}_i}{\partial \boldsymbol{t}_{i+1}}(\boldsymbol{x}_i + s\boldsymbol{t}_{i+1}) - \frac{\partial \boldsymbol{g}_i\cdot\boldsymbol{n}_{i+1}}{\partial \boldsymbol{t}_i}(\boldsymbol{x}_i - s\boldsymbol{t}_i)\right|^2 ds < \infty, \tag{5.1.20}$$

where $\varepsilon = \min_{1\le i\le N}|\Gamma_i|$. Then all solutions of problem (5.1.2) satisfy

$$\boldsymbol{u} \in H^2(\Omega)^2, \quad p \in H^1(\Omega),$$

with continuous dependence on the data

$$\begin{aligned}\|\boldsymbol{u}\|_{H^2(\Omega)} + \|p\|_{H^1(\Omega)} \le C_1 \Bigg(& \|\boldsymbol{f}\|_{L^2(\Omega)} + \sum_{j=0}^{k} [\boldsymbol{g}]_{H^{3/2}(\gamma_j)} \\ & + C_2 \left(\|\boldsymbol{f}\|_{L^2(\Omega)} + \sum_{j=0}^{k} [\boldsymbol{g}]_{W^{5/4,4/3}(\gamma_j)} \right) \|z\|_{L^2(\Omega)} \\ & + C_3 (\|\boldsymbol{f}\|_{L^2(\Omega)} + \|\boldsymbol{g}\|_{H^{1/2}(\partial\Omega)}) \|z\|^2_{L^2(\Omega)} + C_4 \|\boldsymbol{g}\|_{H^{1/2}(\partial\Omega)} \|z\|^3_{L^2(\Omega)} \Bigg),\end{aligned} \tag{5.1.21}$$

where

$$[\boldsymbol{g}]^2_{H^{3/2}(\gamma_j)} = \sum_{i=1}^{N} \|\boldsymbol{g}\|^2_{H^{3/2}(\Gamma_i)} + \sum_{i=1}^{N} \int_0^{\varepsilon} \frac{1}{s} \left| \frac{\partial \boldsymbol{g}_{i+1} \cdot \boldsymbol{n}_i}{\partial \boldsymbol{t}_{i+1}}(\boldsymbol{x}_i + s\boldsymbol{t}_{i+1}) - \frac{\partial \boldsymbol{g}_i \cdot \boldsymbol{n}_{i+1}}{\partial \boldsymbol{t}_i}(\boldsymbol{x}_i - s\boldsymbol{t}_i) \right|^2 \mathrm{d}s.$$

The regularity of z is unchanged.

As in the homogeneous case, the proof of uniqueness requires that the solution $\boldsymbol{u}$ be in $W^{2,r}(\Omega)^2$ for some $r > 2$, thus implying that it is in $W^{1,\infty}(\Omega)^2$. The next proposition gives a sufficient condition on the boundary data $\boldsymbol{g}$ for this regularity. This analog of Proposition 1.4.11 is essentially based on regularity results of GRISVARD [1985].

PROPOSITION 5.1.8. *In addition to the hypotheses of Theorem 5.1.6, we suppose that Ω is a convex polygon. There exists a real number $r_0 > 2$, depending on the inner angles of $\partial\Omega$, such that: if for some real r with $2 < r < r_0$, and on each γ_j, $0 \le j \le k$,*

$$\boldsymbol{g} \in W^{2-1/r,r}(\Gamma_i)^2 \ \text{for}\ 1 \le i \le N, \quad \boldsymbol{g} \cdot \boldsymbol{n} = 0 \ \text{on}\ \partial\Omega, \tag{5.1.22}$$

$$\left(\frac{\partial \boldsymbol{g}_{i+1} \cdot \boldsymbol{n}_i}{\partial \boldsymbol{t}_{i+1}} - \frac{\partial \boldsymbol{g}_i \cdot \boldsymbol{n}_{i+1}}{\partial \boldsymbol{t}_i} \right)(\boldsymbol{x}_i) = 0 \ \text{for}\ 1 \le i \le N, \tag{5.1.23}$$

then any solution $\boldsymbol{u} \in W^{\alpha}$ of (5.1.2) belongs to $W^{2,r}(\Omega)^2$ and

$$\|\boldsymbol{u}\|_{W^{2,r}(\Omega)} \le C_r \left(\frac{1}{\alpha} \|\operatorname{curl} \boldsymbol{u}\|_{L^2(\Omega)} + \frac{1}{\nu} \|\operatorname{curl} \boldsymbol{f}\|_{L^2(\Omega)} + \sum_{j=0}^{k} [\boldsymbol{g}]_{W^{2-1/r,r}(\gamma_j)} \right), \tag{5.1.24}$$

where C_r is a constant independent of α and ν and

$$[\boldsymbol{g}]_{W^{2-1/r,r}(\gamma_j)} = \sum_{i=1}^{N} \|\boldsymbol{g}\|_{W^{2-1/r,r}(\Gamma_i)}.$$

This gives the following uniqueness result. It can be stated in terms of the data, but for the sake of simplicity, we state it in terms of one solution.

THEOREM 5.1.9. *Under the assumptions of Proposition 5.1.8, problem (5.1.2) has a unique solution as soon as one of its solutions satisfy, for some $r > 2$,*

$$\nu > 2\alpha \|\nabla \boldsymbol{u}\|_{L^\infty(\Omega)} + S_2 \tilde{S}_4 |\boldsymbol{u}|_{H^1(\Omega)} + \alpha S_{r^\star} |\boldsymbol{u}|_{W^{2,r}(\Omega)}, \tag{5.1.25}$$

where $\frac{1}{r^\star} = \frac{1}{2} - \frac{1}{r}$.

5.2. Centered schemes for the nonhomogeneous problem

From a numerical point of view, the nonhomogeneous problem (5.1.2) is discretized with the same test functions as the homogeneous problem, by passing the contribution of a suitable discretization of the boundary data $\boldsymbol{g}$ on the right-hand side. Its numerical analysis relies on a good approximation operator that preserves the tangential character of $\boldsymbol{g}$ and approximates the local support of the lifting $\boldsymbol{u_g}$ described in Theorem 5.1.2. As for the homogeneous problem (1.4.9), we first present and analyze a general centered scheme and afterward apply it to simple finite-element spaces. Part of the material presented here is taken from GIRAULT and SCOTT [2002a].

5.2.1. *A general finite-element scheme*

From now on, we assume that Ω is a connected polygon. As previously, we suppose that the boundary data $\boldsymbol{g}$, satisfying $\boldsymbol{g} \cdot \boldsymbol{n} = 0$, is given in $W^{1-1/r,r}(\partial\Omega)^2$ for some real number $r > 2$. Reverting to the material of Section 2.1, we discretize z in the same finite-dimensional space $Z_h \subset H^1(\Omega)$, and we discretize the transport term with the trilinear form $\tilde{c}$ defined in (2.1.4):

$$\forall \boldsymbol{v} \in H^1(\Omega)^2,\ \forall \varphi, \theta \in H^1(\Omega),\ \tilde{c}(\boldsymbol{v}; \varphi, \theta) = (\boldsymbol{v} \cdot \nabla \varphi, \theta) + \frac{1}{2}((\operatorname{div} \boldsymbol{v})\varphi, \theta).$$

As far as the velocity is concerned, let $X_{h,\tau}$ be a finite-dimensional subspace of $H^1_\tau(\Omega)^2$ and set $X_h = X_{h,\tau} \cap H^1_0(\Omega)^2$. Let M_h be a finite-dimensional subspace of $L^2_0(\Omega)$ and assume that the pair (X_h, M_h) satisfies the uniform discrete inf-sup condition (2.1.1): There exists a constant $\beta^* > 0$, independent of h, such that

$$\forall q_h \in M_h,\ \sup_{\boldsymbol{v}_h \in X_h} \frac{\int_\Omega q_h \operatorname{div} \boldsymbol{v}_h \, d\boldsymbol{x}}{|\boldsymbol{v}_h|_{H^1(\Omega)}} \geq \beta^* \|q_h\|_{L^2(\Omega)}.$$

Then we discretize the velocity and pressure in $(X_{h,\tau}, M_h)$. Let W_h denote the discrete analog of W:

$$W_h = \{\boldsymbol{v}_h \in X_{h,\tau};\ \forall q_h \in M_h,\ \int_\Omega q_h \operatorname{div} \boldsymbol{v}_h \, d\boldsymbol{x} = 0\}, \tag{5.2.1}$$

and recall the discrete analogs of V, and $V^\perp$ defined, respectively, by (2.1.2) and (2.1.3):

$$V_h = \{\boldsymbol{v}_h \in X_h;\ \forall q_h \in M_h,\ \int_\Omega q_h \operatorname{div} \boldsymbol{v}_h \, d\boldsymbol{x} = 0\},$$

$$V_h^\perp = \{\boldsymbol{v}_h \in X_h;\ \forall \boldsymbol{w}_h \in V_h,\ (\nabla \boldsymbol{v}_h, \nabla \boldsymbol{w}_h) = 0\}.$$

Regarding the nonhomogeneous boundary condition, let G_h denote the trace space of $X_{h,\tau}$, and let $\boldsymbol{g}_h$ be a suitable approximation of $\boldsymbol{g}$ in G_h; it satisfies in particular $\boldsymbol{g}_h \cdot \boldsymbol{n} = 0$. We make the following assumptions on $\boldsymbol{g}_h$ that mimick the results of Theorem 5.1.2.

HYPOTHESIS 5.2.1. *There exists $\varepsilon_0 > 0$, only depending on Ω, such that for each ε with $0 < \varepsilon < \varepsilon_0$ and for all h with $0 < h < C_b\varepsilon$, for a constant C_b independent of h and ε, there exists a function $\boldsymbol{u}_{h,\boldsymbol{g}} \in W_h$ satisfying*

$$\boldsymbol{u}_{h,\boldsymbol{g}}|_{\partial\Omega} = \boldsymbol{g}_h, \tag{5.2.2}$$

$$|\boldsymbol{u}_{h,\boldsymbol{g}}|_{H^1(\Omega)} \leq \frac{C}{\sqrt{\varepsilon}} \|\boldsymbol{g}\|_{W^{1-1/r,r}(\partial\Omega)}, \tag{5.2.3}$$

$$\forall \boldsymbol{v}_h \in X_h,\ \big\| |\boldsymbol{u}_{h,\boldsymbol{g}}|\, |\boldsymbol{v}_h| \big\|_{L^2(\Omega)} \leq C\sqrt{\varepsilon} |\boldsymbol{v}|_{H^1(\Omega_\varepsilon)} \|\boldsymbol{g}\|_{W^{1-1/r,r}(\partial\Omega)}. \tag{5.2.4}$$

REMARK 5.2.2. These assumptions are verified in the next section by approximating $\boldsymbol{u}_{\boldsymbol{g}}$ in familiar spaces used for the Stokes problem. As a result of this approximation, the factor $\sqrt{\varepsilon}$ in (5.2.4) is not as favorable as in (5.1.11). □

With the lifting $\boldsymbol{u}_{h,\boldsymbol{g}}$, we approximate problem (5.1.2) by the following general centered scheme: Find $\boldsymbol{u}_h$ in $X_h + \boldsymbol{u}_{h,\boldsymbol{g}}$, p_h in M_h and $\boldsymbol{z}_h = (0, 0, z_h)$ with z_h in Z_h, such that

$$\forall \boldsymbol{v}_h \in X_h,\ \nu(\nabla \boldsymbol{u}_h, \nabla \boldsymbol{v}_h) + (\boldsymbol{z}_h \times \boldsymbol{u}_h, \boldsymbol{v}_h) - (p_h, \operatorname{div} \boldsymbol{v}_h) = (\boldsymbol{f}, \boldsymbol{v}_h), \tag{5.2.5}$$

$$\forall q_h \in M_h,\ (q_h, \operatorname{div} \boldsymbol{u}_h) = 0, \tag{5.2.6}$$

$$\forall \theta_h \in Z_h,\ \nu(z_h, \theta_h) + \alpha\, \tilde{c}(\boldsymbol{u}_h; z_h, \theta_h) = \nu(\operatorname{curl} \boldsymbol{u}_h, \theta_h) + \alpha\, (\operatorname{curl} \boldsymbol{f}, \theta_h). \tag{5.2.7}$$

Given $z_h \in Z_h$ and $\boldsymbol{u}_{h,\boldsymbol{g}} \in W_h$, problem (5.2.5)–(5.2.6) is a discrete nonhomogeneous generalized Stokes problem of the form: Find $(\boldsymbol{v}_h(z_h), q_h(z_h))$ in $W_h \times M_h$, solution of

$$\forall \boldsymbol{w}_h \in X_h,\ \nu(\nabla \boldsymbol{v}_h, \nabla \boldsymbol{w}_h) + (\boldsymbol{z}_h \times \boldsymbol{v}_h, \boldsymbol{w}_h) - (q_h, \operatorname{div} \boldsymbol{v}_h) = (\boldsymbol{f}, \boldsymbol{v}_h), \tag{5.2.8}$$

$$(\boldsymbol{u}_h - \boldsymbol{u}_{h,\boldsymbol{g}})|_{\partial\Omega} = \boldsymbol{0}. \tag{5.2.9}$$

Owing to (2.1.1), this problem has a unique solution.

The next theorem gives existence of at least one solution of (5.2.5)–(5.2.7).

THEOREM 5.2.3. *Let $\boldsymbol{g}$, satisfying $\boldsymbol{g} \cdot \boldsymbol{n} = 0$, be given in $W^{1-1/r,r}(\partial\Omega)^2$ for some real number $r > 2$. Assume that (2.1.1) and Hypothesis 5.2.1 hold. Then there exists $h_b > 0$ such that for all $h \leq h_b$, all $\nu > 0$, all $\alpha > 0$, and all $\boldsymbol{f}$ in $H(\text{curl}, \Omega)$, the discrete problem (5.2.5)–(5.2.7) has at least one solution $(\boldsymbol{u}_h, p_h) \in W_h \times M_h$, $z_h \in Z_h$, and each solution satisfies a priori estimates similar to (5.1.14)–(5.1.16):*

$$|\boldsymbol{u}_h|_{H^1(\Omega)} \leq 2\frac{S_2}{\nu} \|\boldsymbol{f}\|_{L^2(\Omega)} + \frac{\alpha}{\nu} \|\operatorname{curl} \boldsymbol{f}\|_{L^2(\Omega)} + \frac{C}{\nu} \|\boldsymbol{g}\|^2_{W^{1-1/r,r}(\partial\Omega)}, \tag{5.2.10}$$

$$\|z_h\|_{L^2(\Omega)} \leq \|\operatorname{curl} \boldsymbol{u}_h\|_{L^2(\Omega)} + \frac{\alpha}{\nu} \|\operatorname{curl} \boldsymbol{f}\|_{L^2(\Omega)}, \tag{5.2.11}$$

$$\|p_h\|_{L^2(\Omega)} \leq \frac{1}{\beta^\star} \Big(S_2 \|\boldsymbol{f}\|_{L^2(\Omega)} + |\boldsymbol{u}_h|_{H^1(\Omega)} \big(\nu + S_4 \tilde{S}_4 \|z_h\|_{L^2(\Omega)} \big) \Big), \tag{5.2.12}$$

where C depends on the constants of (5.2.3) and (5.2.4).

PROOF. The proof is much the same as in the homogeneous case. First, problem (5.2.5)–(5.2.7) is equivalent to (2.1.17): Find z_h in Z_h such that

$$\forall \theta_h \in Z_h,\ \nu(z_h, \theta_h) + \alpha\, \tilde{c}(\boldsymbol{u}_h(z_h); z_h, \theta_h) = \nu(\operatorname{curl} \boldsymbol{u}_h(z_h), \theta_h) + \alpha(\operatorname{curl} \boldsymbol{f}, \theta_h),$$

where $(\boldsymbol{u}_h(z_h), p_h(z_h)) \in W_h \times M_h$ is the solution of (5.2.8)–(5.2.9). Next, we solve it by Brouwer's Fixed Point Theorem. For fixed λ_h in Z_h, we define $H(\lambda_h)$ in Z_h by

$$\begin{aligned}\forall \mu_h \in Z_h,\ (H(\lambda_h), \mu_h) &= \nu(\lambda_h, \mu_h) + \alpha\, \tilde{c}(\boldsymbol{u}_h(\lambda_h); \lambda_h, \mu_h)\\ &\quad - \nu(\operatorname{curl} \boldsymbol{u}_h(\lambda_h), \mu_h) - \alpha(\operatorname{curl} \boldsymbol{f}, \mu_h).\end{aligned}$$

In order to derive a lower bound for $(H(\lambda_h), \lambda_h)$, we set $\boldsymbol{u}_{h,\mathbf{0}} = \boldsymbol{u}_h - \boldsymbol{u}_{h,\boldsymbol{g}}$ and test (5.2.8) with $\boldsymbol{u}_{h,\mathbf{0}}$, using (5.2.4) with arbitrary ε in the trilinear term. This gives, for all $h \leq C_b\varepsilon$,

$$\begin{aligned}|\boldsymbol{u}_{h,\mathbf{0}}|_{H^1(\Omega)} &\leq \frac{S_2}{\nu}\|\boldsymbol{f}\|_{L^2(\Omega)} + |\boldsymbol{u}_{h,\boldsymbol{g}}|_{H^1(\Omega)}\\ &\quad + \frac{C}{\nu}\sqrt{\varepsilon}\|\boldsymbol{g}\|_{W^{1-1/r,r}(\partial\Omega)}\left(|\boldsymbol{u}_{h,\mathbf{0}}|_{H^1(\Omega)} + \sqrt{2}|\boldsymbol{u}_{h,\boldsymbol{g}}|_{H^1(\Omega)} + \frac{\alpha}{\nu}\|\operatorname{curl} \boldsymbol{f}\|_{L^2(\Omega)}\right).\end{aligned}$$

With the choice

$$\varepsilon = \left(\frac{\nu}{2C\|\boldsymbol{g}\|_{W^{1-1/r,r}(\partial\Omega)}}\right)^2, \tag{5.2.13}$$

we obtain

$$|\boldsymbol{u}_{h,\mathbf{0}}|_{H^1(\Omega)} \leq 2\frac{S_2}{\nu}\|\boldsymbol{f}\|_{L^2(\Omega)} + \frac{\alpha}{\nu}\|\operatorname{curl} \boldsymbol{f}\|_{L^2(\Omega)} + (2+\sqrt{2})|\boldsymbol{u}_{h,\boldsymbol{g}}|_{H^1(\Omega)}. \tag{5.2.14}$$

Therefore

$$\begin{aligned}(H(\lambda_h), \lambda_h) \geq \|\lambda_h\|_{L^2(\Omega)}\nu\Bigg(&\|\lambda_n\|_{L^2(\Omega)} - 2(1+\sqrt{2})|\boldsymbol{u}_{h,\boldsymbol{g}}|_{H^1(\Omega)} - 2\frac{S_2}{\nu}\|\boldsymbol{f}\|_{L^2(\Omega)}\\ &- 2\frac{\alpha}{\nu}\|\operatorname{curl} \boldsymbol{f}\|_{L^2(\Omega)}\Bigg) \geq 0,\end{aligned}$$

for all $\lambda_h \in Z_h$ with $\|\lambda_h\|_{L^2(\Omega)}$ large enough. By Brouwer's Fixed Point Theorem this proves existence of at least one solution z_h in Z_h of (5.2.5)–(5.2.7). From (5.2.13), the condition $h_b \leq C_b\varepsilon$ implies that we must choose

$$h_b \leq C_b\left(\frac{\nu}{2C\|\boldsymbol{g}\|_{W^{1-1/r,r}(\partial\Omega)}}\right)^2. \tag{5.2.15}$$

Finally, by testing (5.2.8) with $\boldsymbol{v}_h = \boldsymbol{u}_{h,\mathbf{0}}$, choosing ε according to (5.2.13), and applying (5.2.3), we easily derive the bound (5.2.10). The other bounds are straightforward. □

REMARK 5.2.4. We shall see in the examples of the next section that the condition $h \le h_b$ in the statement of Theorem 5.2.3 need only be prescribed on layers of elements that intersect a neighborhood of $\partial\Omega$. When the viscosity ν is small, it is restrictive, but if we were to use the original Leray–Hopf lifting, the bound (5.2.15) for h_b would be replaced by an exponential function of $-1/\nu$, which is far more restrictive. □

Convergence

Here and in the next paragraph, we assume that (2.1.1) and Hypothesis 5.2.1 hold, as well as $h \le h_b$, with h_b defined by (5.2.15), so as to guarantee the existence of solutions.

Owing to the uniform bounds in Theorem 5.2.3, convergence of solutions of (5.2.5)–(5.2.7) is established as in Section 2.1.1. For passing to the limit in the discrete scheme, we retain the second and third assumptions of Hypothesis 2.1.5:
There exists an operator $r_h \in \mathcal{L}(L^2_0(\Omega); M_h)$ such that

$$\forall q \in L^2_0(\Omega),\ \lim_{h\to 0} \|r_h(q) - q\|_{L^2(\Omega)} = 0.$$

There exists an operator $R_h \in \mathcal{L}(L^2(\Omega); Z_h)$ such that

$$\forall \theta \in L^2(\Omega),\ \lim_{h\to 0} \|R_h(\theta) - \theta\|_{L^2(\Omega)} = 0,$$
$$\forall p \in [2, \infty],\ \forall \theta \in W^{1,p}(\Omega),\ \lim_{h\to 0} \|R_h(\theta) - \theta\|_{W^{1,p}(\Omega)} = 0,$$

and we combine assumption (1) of 2.1.5 and Hypothesis 5.2.1 as follows:

HYPOTHESIS 5.2.5. *There exists an operator $\tilde{P}_h$ in $\mathcal{L}(H^1_\tau(\Omega); X_{h,\tau})$ and in $\mathcal{L}(H^1_0(\Omega)^2; X_h)$ that preserves the discrete divergence: For all $\boldsymbol{v} \in H^1_\tau(\Omega)$,*

$$\forall q_h \in M_h,\ \int_\Omega q_h \operatorname{div} \tilde{P}_h(\boldsymbol{v})\, d\boldsymbol{x} = \int_\Omega q_h \operatorname{div} \boldsymbol{v}\, d\boldsymbol{x}, \tag{5.2.16}$$

that is convergent in $H^1_\tau(\Omega)$,

$$\forall \boldsymbol{v} \in H^1_\tau(\Omega),\ \lim_{h\to 0} \|\tilde{P}_h(\boldsymbol{v}) - \boldsymbol{v}\|_{H^1(\Omega)} = 0,$$

is such that the trace of $\tilde{P}_h(\boldsymbol{v})$ on $\partial\Omega$ only depends on the trace of $\boldsymbol{v}$, and is such that $\tilde{P}_h(\boldsymbol{u_g})$ satisfies (5.2.3) and (5.2.4).

In view of these last two properties, we take

$$\boldsymbol{g}_h = \tilde{P}_h(\boldsymbol{u})|_{\partial\Omega} = \tilde{P}_h(\boldsymbol{u_g})|_{\partial\Omega}, \tag{5.2.17}$$

and we construct a solution $(\boldsymbol{u}_h, p_h, z_h)$ of (5.2.5)–(5.2.7). With the above assumptions, we readily derive that there exists a subsequence of h (still denoted by h) such that

$$\lim_{h\to 0} \boldsymbol{u}_h = \boldsymbol{u} \ \text{ weakly in } H^1_\tau(\Omega),$$

$$\lim_{h\to 0} p_h = p \ \text{ weakly in } L^2(\Omega),$$

$$\lim_{h\to 0} z_h = z \ \text{ weakly in } L^2(\Omega),$$

where $(\boldsymbol{u}, p, z)$ solves (5.1.4)–(5.1.6). Next, the proof of strong convergence in Proposition 2.1.7 carries over here because it is based on the equation satisfied by $\boldsymbol{u}_h - \tilde{P}_h(\boldsymbol{u})$. Hence,

$$\lim_{h\to 0} \|\boldsymbol{u}_h - \boldsymbol{u}\|_{H^1(\Omega)} = 0. \tag{5.2.18}$$

Consequently, we can pass to the limit in (5.2.7), and conclude that $(\boldsymbol{u}, p, z)$ solves (5.1.2). Finally, the strong convergence of z_h and p_h is proved as in Theorem 2.1.9, thereby yielding the following convergence result.

THEOREM 5.2.6. *We retain the second and third assumptions of Hypothesis 2.1.5, and we suppose that Hypothesis 5.2.5 holds. Then there exists a subsequence of h (still denoted by h) and a solution* $(\boldsymbol{u}, p, z) \in W \times L_0^2(\Omega) \times L^2(\Omega)$ *of problem (5.1.2) such that*

$$\lim_{h\to 0} \|\boldsymbol{u}_h - \boldsymbol{u}\|_{H^1(\Omega)} = 0,$$

$$\lim_{h\to 0} \|z_h - z\|_{L^2(\Omega)} = 0,$$

$$\lim_{h\to 0} \|p_h - p\|_{L^2(\Omega)} = 0.$$

A priori error bounds

Error bounds for (5.2.5)–(5.2.7) are proved in much the same way as for the homogeneous problem. Indeed, they rely on the error formula (2.1.50), that is also valid here for $z - z_h$:

$$\begin{aligned}\forall \theta_h \in Z_h,\ & \nu(z_h - z, \theta_h) + \alpha\, \tilde{c}(\boldsymbol{u}_h - \boldsymbol{u}; z_h, \theta_h) + \alpha\, \tilde{c}(\boldsymbol{u}; z_h - z, \theta_h)\\ &= \nu(\operatorname{curl}(\boldsymbol{u}_h - \boldsymbol{u}), \theta_h),\end{aligned}$$

and on the error formula for the velocity, that holds for all $\boldsymbol{v}_h$ in V_h, all q_h in M_h:

$$\begin{aligned}&\nu(\nabla(\boldsymbol{u}_h - \tilde{P}_h(\boldsymbol{u})), \nabla \boldsymbol{v}_h) + (z_h \times (\boldsymbol{u}_h - \tilde{P}_h(\boldsymbol{u})), \boldsymbol{v}_h) + ((z_h - z) \times \boldsymbol{u}, \boldsymbol{v}_h)\\ &\quad - (q_h - p, \operatorname{div} \boldsymbol{v}_h) = \nu(\nabla(\boldsymbol{u} - \tilde{P}_h(\boldsymbol{u})), \nabla \boldsymbol{v}_h) + (z_h \times (\boldsymbol{u} - \tilde{P}_h(\boldsymbol{u})), \boldsymbol{v}_h).\end{aligned} \tag{5.2.19}$$

By virtue of Hypothesis 5.2.5, $\boldsymbol{u}_h - \tilde{P}_h(\boldsymbol{u})$ belongs to V_h and can be used as test function in (5.2.19). Then the statement of Lemma 2.1.19 carries over here by changing P_h into $\tilde{P}_h$.

LEMMA 5.2.7. *Let* $(\boldsymbol{u}_h, p_h, z_h)$ *be a solution of (5.2.5)–(5.2.7) and let* $(\boldsymbol{u}, p, z)$ *be a solution of Problem (5.1.2). Under Hypothesis 5.2.5 and the second assumption of Hypothesis 2.1.5, we have:*

$$\begin{aligned}|\boldsymbol{u} - \boldsymbol{u}_h|_{H^1(\Omega)} &\le 2|\boldsymbol{u} - \tilde{P}_h(\boldsymbol{u})|_{H^1(\Omega)} + \frac{S_4}{\nu}\|\boldsymbol{u}\|_{L^4(\Omega)}\|z - z_h\|_{L^2(\Omega)}\\ &\quad + \frac{S_4}{\nu}\|z_h\|_{L^2(\Omega)}\|\boldsymbol{u} - \tilde{P}_h(\boldsymbol{u})\|_{L^4(\Omega)} + \frac{1}{\nu}\|p - r_h(p)\|_{L^2(\Omega)},\end{aligned} \tag{5.2.20}$$

$$\|p - p_h\|_{L^2(\Omega)} \leq \left(1 + \frac{1}{\beta^*}\right) \|p - r_h(p)\|_{L^2(\Omega)} + \frac{1}{\beta^*} \left(\nu \, |\boldsymbol{u} - \tilde{P}_h(\boldsymbol{u})|_{H^1(\Omega)}\right.$$
$$\left. + S_4 \left(\|\boldsymbol{u}\|_{L^4(\Omega)} \|z - z_h\|_{L^2(\Omega)} + \|z_h\|_{L^2(\Omega)} \|\boldsymbol{u} - \boldsymbol{u}_h\|_{L^4(\Omega)}\right)\right), \quad (5.2.21)$$

where β^* is the constant of (2.1.1).

Similarly, the statement of Lemma 2.1.20 is also valid here.

LEMMA 5.2.8. *Let $(\boldsymbol{u}_h, p_h, z_h)$ be a solution of (5.2.5)–(5.2.7) and let $(\boldsymbol{u}, p, z)$ be a solution of Problem (5.1.2). For any λ_h in Z_h, we have*

$$\|z - z_h\|_{L^2(\Omega)} \leq 2 \, \|z - \lambda_h\|_{L^2(\Omega)} + \|\text{curl}(\boldsymbol{u} - \boldsymbol{u}_h)\|_{L^2(\Omega)}$$
$$+ \frac{\alpha}{\nu} \left(\|(\boldsymbol{u} - \boldsymbol{u}_h) \cdot \nabla \lambda_h\|_{L^2(\Omega)} + \|\boldsymbol{u} \cdot \nabla(z - \lambda_h)\|_{L^2(\Omega)}\right.$$
$$\left. + \frac{1}{2} \|\lambda_h \operatorname{div}(\boldsymbol{u} - \boldsymbol{u}_h)\|_{L^2(\Omega)}\right).$$

Clearly, exploiting Lemma 5.2.8 requires z in $W^{1,r}(\Omega)$, for some $r > 2$. In view of Theorem 1.4.14, this holds when Ω is convex, $\text{curl}\, \boldsymbol{f}$ belongs to $W^{1,r}(\Omega)$, $\text{curl}\, \boldsymbol{u}$ belongs to $W^{1,r}(\Omega)$, and $\boldsymbol{u} \in W^{1,\infty}(\Omega)^2$ satisfies (1.4.30):

$$\frac{\alpha}{\nu} \|\nabla \boldsymbol{u}\|_{L^\infty(\Omega)} := \delta < 1.$$

The next proposition sharpens the statement of Lemma 5.2.8.

PROPOSITION 5.2.9. *Let Ω be convex, let $(\boldsymbol{u}_h, p_h, z_h)$ be a solution of (5.2.5)–(5.2.7), let $(\boldsymbol{u}, p, z)$ be a solution of Problem (5.1.2), let r_0 be the number of Proposition 1.4.11, and let the previous assumptions hold, so that $z \in W^{1,r}(\Omega)$, for some real number r in $]2, r_0[$. Under the third assumption of Hypothesis 2.1.5, we have*

$$\|z - z_h\|_{L^2(\Omega)} \leq 2 \, \|z - R_h(z)\|_{L^2(\Omega)} + \sqrt{2} |\boldsymbol{u} - \boldsymbol{u}_h|_{H^1(\Omega)}$$
$$+ \frac{\alpha}{\nu} \left(\|\boldsymbol{u} - \boldsymbol{u}_h\|_{L^{r^*}(\Omega)} |R_h(z)|_{W^{1,r}(\Omega)} + \|\boldsymbol{u}\|_{L^\infty(\Omega)} |z - R_h(z)|_{H^1(\Omega)}\right.$$
$$\left. + \frac{1}{\sqrt{2}} |\boldsymbol{u} - \boldsymbol{u}_h|_{H^1(\Omega)} \|R_h(z)\|_{L^\infty(\Omega)}\right), \quad (5.2.22)$$

where $\frac{1}{r^*} = \frac{1}{2} - \frac{1}{r}$.

The error on z is obtained by substituting (5.2.20) into (5.2.22). It uses similar notation as in the homogeneous case:

$$\|R_h(z)\|_{W^{1,r}(\Omega)} \leq E_r \|z\|_{W^{1,r}(\Omega)}, \quad \|R_h(z)\|_{L^\infty(\Omega)} \leq C_{\infty,r} E_r \|z\|_{W^{1,r}(\Omega)},$$
$$\tilde{K}_2(r, z) = \left(\tilde{S}_{r^*} + \frac{1}{\sqrt{2}} C_{\infty,r}\right) E_r \|z\|_{W^{1,r}(\Omega)}, \quad (5.2.23)$$

where $C_{\infty,r}$ is the constant of (1.1.18).

THEOREM 5.2.10. *Suppose that the second and third assumptions of Hypothesis 2.1.5 hold, as well as Hypothesis 5.2.5. Then under the assumptions of Proposition 5.2.9, and if the data are small enough so that*

$$\frac{1}{\nu}\Big(\sqrt{2}+\frac{\alpha}{\nu}\tilde{K}_2(r,z)\Big)S_4\|\boldsymbol{u}\|_{L^4(\Omega)} \le \frac{1}{2}, \tag{5.2.24}$$

we have the following error estimate:

$$\begin{aligned}\|z-z_h\|_{L^2(\Omega)} \le{}& 4\|z-R_h(z)\|_{L^2(\Omega)} + 2\frac{\alpha}{\nu}\|\boldsymbol{u}\|_{L^\infty(\Omega)}|z-R_h(z)|_{H^1(\Omega)}\\ &+2\Big(\sqrt{2}+\frac{\alpha}{\nu}\tilde{K}_2(r,z)\Big)\Big(\frac{S_4}{\nu}\|z_h\|_{L^2(\Omega)}\|\boldsymbol{u}-\tilde{P}_h(\boldsymbol{u})\|_{L^4(\Omega)}\\ &+2\,|\boldsymbol{u}-\tilde{P}_h(\boldsymbol{u})|_{H^1(\Omega)}+\frac{1}{\nu}\|p-r_h(p)\|_{L^2(\Omega)}\Big).\end{aligned} \tag{5.2.25}$$

The factor $\sqrt{2}$ comes from the bound of the rotation and divergence of functions that do not vanish on the boundary. As in the homogeneous case, the assumptions of Theorem 5.2.10 can all be checked on the data and the domain. All factors in (5.2.24) are bounded independently of h and can be expressed in terms of the data. The statement of Theorem 5.2.10 remains valid when α tends to zero.

Finally, it is worthwhile comparing (5.2.24) with sufficient conditions for uniqueness adapted to the splitting (5.1.4)–(5.1.6). Because uniqueness only involves difference of solutions, we immediately derive from Section 2.1.3 the next estimate for any pair of solutions $(\boldsymbol{u}_1,p_1,z_1)$ and $(\boldsymbol{u}_2,p_2,z_2)$:

$$\|z_1-z_2\|_{L^2(\Omega)} \le \frac{1}{\nu}\|\boldsymbol{u}_2\|_{L^4(\Omega)}\big(S_4+\alpha C_\infty C_1K_1(z_1)|z_1|_{H^1(\Omega)}\big)\|z_1-z_2\|_{L^2(\Omega)},$$

where C_1 is the continuity constant of Theorem 1.1.6, $K_1(z)$ is defined in (1.4.23), and C_∞ is the constant of (1.4.25). Then (5.1.4)–(5.1.6) has a unique solution provided

$$\frac{1}{\nu}\|\boldsymbol{u}_2\|_{L^4(\Omega)}\big(S_4+\alpha C_\infty C_1K_1(z_1)|z_1|_{H^1(\Omega)}\big) < 1,$$

a condition that is somewhat similar to (5.2.24).

5.2.2. *Examples of centered schemes*

In this section, $h>0$ is a discretization parameter and $\mathcal{T}_h$ is a regular family of conforming triangulations of $\overline{\Omega}$, consisting of triangles with maximum mesh size h satisfying (2.1.25): There exists a constant $\sigma_0>0$ independent of h, such that

$$\max_{T\in\mathcal{T}_h}\frac{h_T}{\rho_T}\le\sigma_0,$$

where h_T is the diameter of T and ρ_T the radius of the ball inscribed in T. For each element T, Δ_T denotes the union of elements of $\mathcal{T}_h$ sharing at least a vertex with T.

The three examples studied in Section 2.2 satisfy all the assumptions introduced in Section 5.2.1. In particular, owing to the local character of their approximation operator, they satisfy Hypothesis 5.2.5 that guarantees a good approximation of the lifting $\boldsymbol{u_g}$. As expected, the main tool in the numerical analysis of the forthcoming examples is the approximate lifting function $\boldsymbol{u}_{h,\boldsymbol{g}}$. Let ε_0 be the parameter of Theorem 5.1.2; recall that for each ε with $0 < \varepsilon \le \varepsilon_0$, $\boldsymbol{u_g}$ is the lifting constructed in Theorem 5.1.2, supported by the tubular neighborhood Ω_ε of $\partial\Omega$ defined by (5.1.9):

$$\Omega_\varepsilon = \{\boldsymbol{x} \in \Omega;\ d(\boldsymbol{x}) \le C\varepsilon\}.$$

Let $\Omega_{h,\varepsilon}$ denote the union of elements of $\mathcal{T}_h$ for which Δ_T intersects Ω_ε

$$\Omega_{h,\varepsilon} = \{\boldsymbol{x} \in T;\ |\Delta_T \cap \Omega_\varepsilon| > 0\}, \tag{5.2.26}$$

and let h_b be the maximum diameter of all elements T in $\Omega_{h,\varepsilon}$

$$h_b = \sup_{T \in \Omega_{h,\varepsilon}} h_T. \tag{5.2.27}$$

Therefore, the condition (5.2.15)

$$h_b \le C_b \left(\frac{\nu}{2C\|\boldsymbol{g}\|_{W^{1-1/r,r}(\partial\Omega)}} \right)^2$$

restricts the meshsize of $\mathcal{T}_h$ in the neighborhood of $\partial\Omega$. This condition amounts to

$$h_b \le C_b \varepsilon \tag{5.2.28}$$

because the factor of C_b in the right-hand side of (5.2.15) is the value of ε used in proving existence of solutions of the discrete problem.

The mini-element

In the nonhomogeneous case that we are considering, the finite-element spaces for the mini-element are

$$X_{h,\tau} = \left\{ \boldsymbol{v}_h \in H^1_\tau(\Omega);\ \forall T \in \mathcal{T}_h,\ \boldsymbol{v}_h|_T \in (\mathbb{P}_1 \oplus \mathrm{Vect}(b_T))^2 \right\}, \tag{5.2.29}$$

where b_T is the bubble function

$$b_T(\boldsymbol{x}) = \lambda_1(\boldsymbol{x})\lambda_2(\boldsymbol{x})\lambda_3(\boldsymbol{x}),$$

that vanishes on the boundary of T. The spaces M_h and Z_h are, respectively, defined in (2.2.2) and (2.2.3):

$$M_h = \{q_h \in H^1(\Omega) \cap L^2_0(\Omega);\ \forall T \in \mathcal{T}_h,\ q_h|_T \in \mathbb{P}_1\},$$
$$Z_h = \{\theta_h \in H^1(\Omega);\ \forall T \in \mathcal{T}_h,\ \theta_h|_T \in \mathbb{P}_1\}.$$

REMARK 5.2.11. The functions of $X_{h,\tau}$ reduce to polynomials of degree one on each side e of the elements T, and hence also on each boundary side. As a result, the condition $\boldsymbol{v}_h \cdot \boldsymbol{n} = 0$ on the boundary implies that $\boldsymbol{v}_h$ vanishes at each corner of $\partial\Omega$. On a fixed polygon, with a small number of corners, this extra condition is acceptable. But it would not be suitable if $\partial\Omega$ were a polygonal approximation of a curved boundary. □

With this pair of spaces, we have the analog of Lemma 2.2.1.

LEMMA 5.2.12. *If the family of triangulations $\mathcal{T}_h$ satisfies (2.1.25), there exists an operator $\tilde{P}_h \in \mathcal{L}(H^1_\tau(\Omega); X_{h,\tau}) \cap \mathcal{L}(H^1_0(\Omega)^2; X_h)$ preserving the discrete divergence:*

$$\forall \boldsymbol{v} \in H^1_\tau(\Omega),\ \forall q_h \in M_h,\ \int_\Omega q_h \operatorname{div} \tilde{P}_h(\boldsymbol{v})\,\mathrm{d}\boldsymbol{x} = \int_\Omega q_h \operatorname{div} \boldsymbol{v}\,\mathrm{d}\boldsymbol{x}. \tag{5.2.30}$$

It has the following local approximation properties

$$\begin{aligned}&\forall \boldsymbol{v} \in W^{s,r}(\Omega)^2,\ \forall T \in \mathcal{T}_h,\\ &\qquad |\tilde{P}_h(\boldsymbol{v}) - \boldsymbol{v}|_{W^{m,q}(T)} \le C_1 h_T^{s-m+2\left(\frac{1}{q}-\frac{1}{r}\right)} |\boldsymbol{v}|_{W^{s,r}(\Delta_T)},\end{aligned} \tag{5.2.31}$$

for integers $m = 0$ or 1, for all real numbers $1 \le s \le 2$, and all numbers $1 \le r, q \le \infty$, such that

$$W^{s,r}(\Omega) \subset W^{m,q}(\Omega),$$

with a constant C_1 independent of h and T. And it approximates the support of $\boldsymbol{v}$:

$$\operatorname{dist}\big(\operatorname{supp}(\tilde{P}_h(\boldsymbol{v})), \operatorname{supp}(\boldsymbol{v})\big) \le C_2 h, \tag{5.2.32}$$

where the constant C_2 is independent of h.

PROOF. The operator $\tilde{P}_h$ is defined as P_h in Lemma 2.2.1, except that the underlying operator Π_h must preserve the tangential character of the boundary data. To this end, we choose Π_h to be the following SCOTT and ZHANG operator [1990]. In a triangle T, for each v in $W^{1,1}(T)$, $\Pi_h(v)$ is a polynomial of $\mathbb{P}_1$ whose degrees of freedom are defined at the vertices of T as follows. For each vertex $\boldsymbol{a}$, we choose a side $f_{\boldsymbol{a}}$ with $\boldsymbol{a}$ as one end point and such that $f_{\boldsymbol{a}}$ lies on the boundary $\partial\Omega$ if $\boldsymbol{a}$ belongs to $\partial\Omega$. On this side $f_{\boldsymbol{a}}$, we define the dual basis function $\psi_{\boldsymbol{a}}$ of the Lagrange basis functions on $f_{\boldsymbol{a}}$ associated with these degrees of freedom (there are two of them in this case), and we set:

$$\Pi_h(v)(\boldsymbol{a}) = \int_{f_{\boldsymbol{a}}} v(s)\psi_{\boldsymbol{a}}(s)\mathrm{d}s.$$

By construction, Π_h is a projection, and in particular, if v is continuous on $\partial\Omega$, and is a polynomial of $\mathbb{P}_1$ on each boundary side of $\partial\Omega$, then $\Pi_h(v) = v$ on $\partial\Omega$. Therefore Π_h preserves the zero trace, and when applied to a vector function, it preserves the zero normal component (because the normal vector $\boldsymbol{n}$ is constant on each boundary side of $\partial\Omega$), i.e., $\Pi_h \in \mathcal{L}(H^1_\tau(\Omega); X_{h,\tau})$ and $\Pi_h \in \mathcal{L}(H^1_0(\Omega)^2; X_h)$. In addition, it is established in SCOTT and ZHANG [1990] that Π_h satisfies the approximation estimate of (5.2.31). Now, in view of the correction introduced in (2.2.5), we define $\tilde{P}_h$ by

$$\tilde{P}_h(\boldsymbol{v}) = \Pi_h(\boldsymbol{v}) - \sum_{T\in\mathcal{T}_h} \boldsymbol{c}_T b_T, \tag{5.2.33}$$

where $\boldsymbol{c}_T$ is defined in (2.2.7)

$$\forall T \in \mathcal{T}_h,\ \boldsymbol{c}_T = \frac{1}{\int_T b_T \,\mathrm{d}\boldsymbol{x}} \int_T (\Pi_h(\boldsymbol{v}) - \boldsymbol{v}) \,\mathrm{d}\boldsymbol{x}.$$

The estimates for this correction established in Lemma 2.2.1 show that $\tilde{P}_h$ has the same local approximation error as Π_h, whence (5.2.31). Because this correction only involves bubble functions that vanish on the boundary of each T, the operator $\tilde{P}_h$ defined by (5.2.33) has the same trace on $\partial\Omega$ as $\Pi_h(\boldsymbol{v})$; hence $\tilde{P}_h \in \mathcal{L}(H^1_\tau(\Omega); X_{h,\tau}) \cap \mathcal{L}(H^1_0(\Omega)^2; X_h)$. The conservation of these boundary values and the above construction imply that $\tilde{P}_h$ preserves the discrete divergence of functions of $H^1_\tau(\Omega)$. Finally, the support of $\Pi_h(\boldsymbol{v})$ is close to the support of $\boldsymbol{v}$ because the degrees of freedom in T are restricted to the macroelement Δ_T. Therefore, it verifies (2.2.44):

$$\mathrm{dist}(\mathrm{supp}(\Pi_h(\boldsymbol{v})), \mathrm{supp}(\boldsymbol{v})) \le C_2 h,$$

with a constant C_2 that is independent of h. As each correction is supported by a single element, it follows that $\tilde{P}_h(\boldsymbol{v})$ has the same support as $\Pi_h(\boldsymbol{v})$. This gives (5.2.32). □

Let $\boldsymbol{u}_{h,g} = \tilde{P}_h(\boldsymbol{u}_g)$ and set $\boldsymbol{g}_h = \boldsymbol{u}_{h,g}|_{\partial\Omega}$. The next proposition checks the estimates of Hypothesis 5.2.1.

PROPOSITION 5.2.13. *Fix ε with $0 < \varepsilon < \varepsilon_0$. Let $\mathcal{T}_h$ satisfy (2.1.25); let the meshsize in the neighborhood of $\partial\Omega$ satisfy $h_b \le C_b\varepsilon$. Then, if $\boldsymbol{g} \in W^{1-1/r,r}(\partial\Omega)^2$ for some $r > 2$, $\boldsymbol{u}_{h,g}$ satisfies (5.2.3):*

$$|\boldsymbol{u}_{h,g}|_{H^1(\Omega)} \le \frac{C}{\sqrt{\varepsilon}} \|\boldsymbol{g}\|_{W^{1-1/r,r}(\partial\Omega)},$$

more generally,

$$\|\boldsymbol{u}_{h,g}\|_{L^\infty(\Omega)} \le \frac{C}{\varepsilon^{1/r}} \|\boldsymbol{g}\|_{W^{1-1/r,r}(\partial\Omega)}, \tag{5.2.34}$$

and $\boldsymbol{u}_{h,g}$ satisfies a slightly sharper estimate than (5.2.4)

$$\forall \boldsymbol{v}_h \in X_h,\ \big\| |\boldsymbol{u}_{h,g}|\, |\boldsymbol{v}_h| \big\|_{L^2(\Omega)} \le C\varepsilon^{1-1/r} |\boldsymbol{v}|_{H^1(\Omega_\varepsilon)} \|\boldsymbol{g}\|_{W^{1-1/r,r}(\partial\Omega)}. \tag{5.2.35}$$

PROOF. To establish (5.2.3), we write

$$|\boldsymbol{u}_{h,g}|_{H^1(\Omega)} \le |\tilde{P}_h(\boldsymbol{u_g}) - \boldsymbol{u_g}|_{H^1(\Omega)} + |\boldsymbol{u_g}|_{H^1(\Omega)},$$

and we apply (5.2.31) in any T of $\Omega_{h,\varepsilon}$ with $m = s = 1$ and $q = r = 2$,

$$|\tilde{P}_h(\boldsymbol{u_g}) - \boldsymbol{u_g}|_{H^1(T)} \le c_1 |\boldsymbol{u_g}|_{H^1(\Delta_T)},$$

where all constants are independent of h and ε. Then (5.2.3) follows from (5.1.12).

We cannot argue as above for proving (5.2.34) because the definition of Π_h does not guarantee its stability in $L^\infty(\Omega)$. Instead, we deduce from (5.2.31) with $m = 0$, $q = \infty$, and $s = 1$ that for any T in $\Omega_{h,\varepsilon}$,

$$\|\tilde{P}_h(\boldsymbol{u_g}) - \boldsymbol{u_g}\|_{L^\infty(T)} \le c_2 h_T^{1-2/r} |\boldsymbol{u_g}|_{W^{1,r}(\Delta_T)}.$$

Therefore,

$$\|\tilde{P}_h(\boldsymbol{u_g}) - \boldsymbol{u_g}\|_{L^\infty(\Omega_{h,\varepsilon})} \le c_3 h_b^{1-2/r} |\boldsymbol{u_g}|_{W^{1,r}(\Omega_\varepsilon)},$$

and (5.1.13) implies

$$\|\tilde{P}_h(\boldsymbol{u_g}) - \boldsymbol{u_g}\|_{L^\infty(\Omega_{h,\varepsilon})} \le c_4 h_b^{1-2/r} \varepsilon^{1/r-1} \|\boldsymbol{g}\|_{W^{1-1/r,r}(\partial\Omega)}.$$

As $r > 2$, the exponent of h_b is positive, and hence the assumption (5.2.28) gives

$$\|\tilde{P}_h(\boldsymbol{u_g}) - \boldsymbol{u_g}\|_{L^\infty(\Omega_{h,\varepsilon})} \le c_5 \varepsilon^{-1/r} \|\boldsymbol{g}\|_{W^{1-1/r,r}(\partial\Omega)}.$$

Then (5.2.34) follows from this and (5.1.10) applied with $s = \infty$.

Finally, we turn to (5.2.35). For any $\boldsymbol{v}_h$ in X_h, we have

$$\big\| |\boldsymbol{u}_{h,g}|\, |\boldsymbol{v}_h| \big\|_{L^2(\Omega)} = \big\| |\boldsymbol{u}_{h,g}|\, |\boldsymbol{v}_h| \big\|_{L^2(\Omega_{h,\varepsilon})} \le \|\boldsymbol{u}_{h,g}\|_{L^\infty(\Omega_{h,\varepsilon})} \|\boldsymbol{v}_h\|_{L^2(\Omega_{h,\varepsilon})}.$$

As $\boldsymbol{v}_h$ vanishes on $\partial\Omega$, it satisfies an extension of Poincaré's inequality in $\Omega_{h,\varepsilon}$:

$$\|\boldsymbol{v}_h\|_{L^2(\Omega_{h,\varepsilon})} \le c_6 \varepsilon |\boldsymbol{v}_h|_{H^1(\Omega)}, \tag{5.2.36}$$

and (5.2.35) is a consequence of (5.2.36) and (5.2.34). □

Because r is taken slightly larger than 2, (5.2.35) somewhat improves (5.2.4). Of course, it becomes better as r increases.

Because R_h and r_h are the same as in the homogeneous case, we can conclude with the same error estimate as in Theorem 2.2.2.

THEOREM 5.2.14. *Let the family of triangulations $\mathcal{T}_h$ satisfy (2.1.25), and let the maximum meshsize h_b of $\mathcal{T}_h$ in the neighborhood of $\partial\Omega$ satisfy (5.2.15). Let $(\boldsymbol{u}, p, z)$ be a solution of Problem (5.1.2), with $z \in H^2(\Omega)$, $\boldsymbol{u} \in H^2(\Omega)^2$, and $p \in H^1(\Omega)$ satisfying (5.2.24), and let $(\boldsymbol{u}_h, p_h, z_h)$ be a solution of (5.2.5)–(5.2.7) with the finite-element spaces (5.2.29), (2.2.2),* *and (2.2.3). Then, there exists a constant C, independent of h, such that*

$$\|z - z_h\|_{L^2(\Omega)} + |\boldsymbol{u} - \boldsymbol{u}_h|_{H^1(\Omega)} + \|p - p_h\|_{L^2(\Omega)} \le C h.$$

Note that the condition $\boldsymbol{u} \in H^2(\Omega)^2$ implies necessarily that $\boldsymbol{g}$ is sufficiently smooth, and therefore the condition $\boldsymbol{g} \in W^{1-1/r,r}(\partial\Omega)^2$ for r slightly larger than 2 holds.

The Bernardi–Raugel element
Here, the finite-element spaces of the Bernardi–Raugel element are

$$X_{h,\tau} = \{\boldsymbol{v}_h \in H^1_\tau(\Omega);\ \forall T \in \mathcal{T}_h,\ \boldsymbol{v}_h|_T \in \mathcal{P}_1(T)\}, \tag{5.2.37}$$

where

$$\mathcal{P}_1(T) = I\!P_1^2 \oplus \text{Vect}\{\boldsymbol{p}_{1,T}, \boldsymbol{p}_{2,T}, \boldsymbol{p}_{3,T}\},$$

and $\boldsymbol{p}_{i,T}$ are the three edge "bubble functions"

$$\boldsymbol{p}_{1,T} = \boldsymbol{n}_1\lambda_2\lambda_3,\ \boldsymbol{p}_{2,T} = \boldsymbol{n}_2\lambda_1\lambda_3,\ \boldsymbol{p}_{3,T} = \boldsymbol{n}_3\lambda_1\lambda_2.$$

The spaces M_h and Z_h are defined, respectively, by (2.2.12) and (2.2.3)

$$M_h = \{q_h \in L^2_0(\Omega);\ \forall T \in \mathcal{T}_h,\ q_h|_T \in I\!P_0\},$$
$$Z_h = \{\theta_h \in H^1(\Omega);\ \forall T \in \mathcal{T}_h,\ \theta_h|_T \in I\!P_1\}.$$

Reverting to the proof of Lemma 2.2.3, we see that the obvious interior degree of freedom of a function $\boldsymbol{v}_h$ of $\mathcal{P}_1(T)$ on a side f_i of T is

$$\int_{f_i} \boldsymbol{v}_h \cdot \boldsymbol{n}_i \mathrm{d}s.$$

Therefore, if $\boldsymbol{v}_h$ belongs to $X_{h,\tau}$, it reduces to a polynomial of $I\!P_1^2$ on each segment that lies on $\partial\Omega$. Thus, Remark 5.2.11 on the validity of the mini-element for approximating functions of $H^1_\tau(\Omega)$ also applies to the Bernardi–Raugel element. Furthermore, the statement of Lemma 5.2.12 is also valid here without modification.

LEMMA 5.2.15. *If the family of triangulations $\mathcal{T}_h$ verifies (2.1.25), there exists an operator $\tilde{P}_h \in \mathcal{L}(H^1_\tau(\Omega); X_{h,\tau}) \cap \mathcal{L}(H^1_0(\Omega)^2; X_h)$ satisfying the local conservation of divergence (5.2.30), the local support (5.2.32), and the local approximation properties (5.2.31) with the same values of s, m, r, and q as in Lemma 5.2.12.*

PROOF. For $\boldsymbol{v}$ in $H^1_\tau(\Omega)$, we choose the same Scott & Zhang operator Π_h as in the proof of Lemma 5.2.12, and we correct it as in Lemma 2.2.3:

$$\tilde{P}_h(\boldsymbol{v}) = \Pi_h(\boldsymbol{v}) - \sum_{T\in\mathcal{T}_h}\sum_{i=1}^{3} \alpha_{i,T}\boldsymbol{p}_{i,T}, \tag{5.2.38}$$

where

$$\alpha_{i,T} = \frac{1}{\int\limits_{f_i} \lambda_j \lambda_k \mathrm{d}s} \int\limits_{f_i} (\Pi_h(\boldsymbol{v}) - \boldsymbol{v}) \cdot \boldsymbol{n}_i \mathrm{d}s \quad j \neq k \neq i.$$

Hence, for $\boldsymbol{v} \in H^1_\tau(\Omega)$, if the element T is adjacent to $\partial\Omega$ and f_i is a boundary side of T, then

$$\int\limits_{f_i} \tilde{P}_h(\boldsymbol{v}) \cdot \boldsymbol{n}_i \mathrm{d}s = 0.$$

Therefore $\tilde{P}_h(\boldsymbol{v}) = \Pi_h(\boldsymbol{v})$ on this side. In other words, for all $\boldsymbol{v} \in H^1_\tau(\Omega)$, we have $\tilde{P}_h(\boldsymbol{v}) = \Pi_h(\boldsymbol{v})$ on $\partial\Omega$. As Π_h preserves the zero trace and the zero normal component, the same is true for $\tilde{P}_h$. The remainder of the proof proceeds as in Lemmas 5.2.12 and 2.2.3. □

Again, we approximate $\boldsymbol{u_g}$ with $\tilde{P}_h$. Then the situation is exactly the same as for the mini-element, the statement of Proposition 5.2.13 is valid here and we derive the same conclusion, namely that the resulting scheme has order one:

THEOREM 5.2.16. *Let the family of triangulations $\mathcal{T}_h$ satisfy (2.1.25), and let the maximum meshsize h_b of $\mathcal{T}_h$ in the neighborhood of $\partial\Omega$ satisfy (5.2.15). Let $(\boldsymbol{u}, p, z)$ be a solution of Problem (5.1.2), with $z \in H^2(\Omega)$, $\boldsymbol{u} \in H^2(\Omega)^2$ and $p \in H^1(\Omega)$ satisfying (5.2.24), and let $(\boldsymbol{u}_h, p_h, z_h)$ be a solution of (5.2.5)–(5.2.7) with the finite-element spaces (5.2.37), (2.2.12), and (2.2.3). Then, there exists a constant C, independent of h, such that*

$$\|z - z_h\|_{L^2(\Omega)} + |\boldsymbol{u} - \boldsymbol{u}_h|_{H^1(\Omega)} + \|p - p_h\|_{L^2(\Omega)} \leq C h.$$

The Taylor–Hood element

For the nonhomogeneous problem, the Taylor–Hood finite-element method of degree two uses the velocity space

$$X_{h,\tau} = \{\boldsymbol{v}_h \in H^1_\tau(\Omega);\ \forall T \in \mathcal{T}_h,\ \boldsymbol{v}_h|_T \in \mathbb{P}_2^2\}, \tag{5.2.39}$$

and the same spaces (2.2.2) for the pressure and (2.2.16) for the auxiliary variable as in the homogeneous case

$$M_h = \{q_h \in H^1(\Omega) \cap L^2_0(\Omega);\ \forall T \in \mathcal{T}_h,\ q_h|_T \in \mathbb{P}_1\},$$
$$Z_h = \{\theta_h \in H^1(\Omega);\ \forall T \in \mathcal{T}_h,\ \theta_h|_T \in \mathbb{P}_2\}.$$

When $\mathcal{T}_h$ is a regular family of triangulations that consist of triangles with at most one edge on $\partial\Omega$, a suitable approximation operator is constructed in Section 2.2.3, see (2.2.30), that preserves the discrete divergence (5.2.30), has local support (5.2.32), and has local approximation properties analogous to (5.2.31). These properties are listed in Theorem 2.2.7. We take the same operator as in (2.2.30), i.e.,

$$\tilde{P}_h(\boldsymbol{v}) = \Pi_h(\boldsymbol{v}) + \boldsymbol{c}_h(\boldsymbol{v}), \tag{5.2.40}$$

and it remains to check that it maps $H^1_\tau(\Omega)$ into $X_{h,\tau}$. This follows immediately from the construction of the Scott & Zhang operator Π_h because by construction, Π_h preserves the zero trace and the zero normal component. Because the correction $\boldsymbol{c}_h(\boldsymbol{v})$ belongs to $H^1_0(\Omega)^2$, then $\tilde{P}_h(\boldsymbol{v})$ belongs to $\mathcal{L}(H^1_\tau(\Omega); X_{h,\tau})$ and to $\mathcal{L}(H^1_0(\Omega)^2; X_h)$. Hence we have the following analog of Theorem 2.2.10.

THEOREM 5.2.17. *Let the family of triangulations $\mathcal{T}_h$ satisfy (2.1.25) and be such that each triangle T has at most one edge on $\partial\Omega$. In addition, let the maximum meshsize h_b of $\mathcal{T}_h$ in the neighborhood of $\partial\Omega$ satisfy (5.2.15). Let $(\boldsymbol{u}, p, z)$ be a solution of Problem (5.1.2), with $z \in H^3(\Omega)$, $\boldsymbol{u} \in H^3(\Omega)^2$, and $p \in H^2(\Omega)$, satisfying (5.2.24), and let $(\boldsymbol{u}_h, p_h, z_h)$ be a solution of (5.2.5)–(5.2.7) with the finite-element spaces (5.2.39), (2.2.2), (2.2.16). Then, there exists a constant C, independent of h, such that*

$$\|z - z_h\|_{L^2(\Omega)} + |\boldsymbol{u} - \boldsymbol{u}_h|_{H^1(\Omega)} + \|p - p_h\|_{L^2(\Omega)} \le C h^2.$$

5.3. Upwind schemes for the nonhomogeneous problem

The upwinding proposed in Sections 2.4.1 and 2.4.2 is only applied to the transport equation. Although the properties of this equation are not affected by a nonhomogeneous tangential boundary condition, its solution depends on the fluid's velocity, and hence on this boundary condition.

5.3.1. Streamline diffusion

We retain the notation and assumptions of Section 5.2.1, in particular, we suppose that the boundary data $\boldsymbol{g}$, satisfying $\boldsymbol{g} \cdot \boldsymbol{n} = 0$, is given in $W^{1-1/r,r}(\partial\Omega)^2$ for some real number $r > 2$. Let $\mathcal{T}_h$ be a family of triangulations satisfying (2.1.25). The pressure is discretized in a finite-dimensional subspace M_h of $L^2_0(\Omega)$ and the velocity in a finite-dimensional subspace $X_{h,\tau}$ of $H^1_\tau(\Omega)$, chosen so that the pair (X_h, M_h) satisfies (2.1.1), where $X_h = X_{h,\tau} \cap H^1_0(\Omega)^2$. We approximate $\boldsymbol{g}$ with a function $\boldsymbol{g}_h$ in the trace space G_h of $X_{h,\tau}$, satisfying Hypothesis 5.2.1. The space Z_h is a finite-dimensional subspace of $H^1(\Omega)$, the same as in Section 2.4.1.

With the lifting $\boldsymbol{u}_{h,g}$ of Hypothesis 5.2.1, we approximate problem (5.1.2) by the following general streamline diffusion scheme: Find $\boldsymbol{u}_h$ in $X_h + \boldsymbol{u}_{h,g}$, p_h in M_h and $\boldsymbol{z}_h = (0, 0, z_h)$ with z_h in Z_h, solution of (5.2.5), (5.2.6):

$$\forall \boldsymbol{v}_h \in X_h,\ \nu(\nabla \boldsymbol{u}_h, \nabla \boldsymbol{v}_h) + (\boldsymbol{z}_h \times \boldsymbol{u}_h, \boldsymbol{v}_h) - (p_h, \operatorname{div} \boldsymbol{v}_h) = (\boldsymbol{f}, \boldsymbol{v}_h),$$
$$\forall q_h \in M_h,\ (q_h, \operatorname{div} \boldsymbol{u}_h) = 0,$$

and

$$\begin{aligned}\forall \theta_h \in Z_h,\ &\nu\,(z_h, \theta_h + h\,\boldsymbol{u}_h \cdot \nabla \theta_h) + \alpha\,(\boldsymbol{u}_h \cdot \nabla z_h, \theta_h + h\,\boldsymbol{u}_h \cdot \nabla \theta_h)\\ &+ \frac{1}{2}(\alpha + h\,\nu)\,((\operatorname{div} \boldsymbol{u}_h) z_h, \theta_h)\\ &= \nu\,(\operatorname{curl} \boldsymbol{u}_h, \theta_h + h\,\boldsymbol{u}_h \cdot \nabla \theta_h) + \alpha\,(\operatorname{curl} \boldsymbol{f}, \theta_h + h\,\boldsymbol{u}_h \cdot \nabla \theta_h).\end{aligned} \tag{5.3.1}$$

Streamline diffusion: Convergence
As previously, by Brouwer's Fixed Point Theorem, existence of a solution follows from uniform a priori estimates. On one hand, arguing as in the homogeneous case, we readily

derive that any solution satisfies

$$\|z_h\|^2_{L^2(\Omega)} \le (2\alpha + \nu h)\left(\frac{1}{\alpha}\|\text{curl}\,\boldsymbol{u}_h\|^2_{L^2(\Omega)} + \frac{\alpha}{\nu^2}\|\text{curl}\,\boldsymbol{f}\|^2_{L^2(\Omega)}\right). \tag{5.3.2}$$

On the other hand, owing to Hypothesis 5.2.1, we have the same estimate as in the centered scheme for $\boldsymbol{u}_{h,\mathbf{0}} = \boldsymbol{u}_h - \boldsymbol{u}_{h,\boldsymbol{g}}$:

$$|\boldsymbol{u}_{h,\mathbf{0}}|_{H^1(\Omega)} \le \frac{S_2}{\nu}\|\boldsymbol{f}\|_{L^2(\Omega)} + |\boldsymbol{u}_{h,\boldsymbol{g}}|_{H^1(\Omega)} + \frac{C}{\nu}\sqrt{\varepsilon}\|\boldsymbol{g}\|_{W^{1-1/r,r}(\partial\Omega)}\|z_h\|_{L^2(\Omega)}. \tag{5.3.3}$$

By substituting (5.3.2) into (5.3.3), and choosing, for instance,

$$\varepsilon = \frac{\alpha}{2\alpha+\nu}\left(\frac{\nu}{2C\|\boldsymbol{g}\|_{W^{1-1/r,r}(\partial\Omega)}}\right)^2 \le \frac{\alpha}{2\alpha+\nu h}\left(\frac{\nu}{2C\|\boldsymbol{g}\|_{W^{1-1/r,r}(\partial\Omega)}}\right)^2, \tag{5.3.4}$$

we obtain the same upper bound as (5.2.14):

$$|\boldsymbol{u}_{h,\mathbf{0}}|_{H^1(\Omega)} \le 2\frac{S_2}{\nu}\|\boldsymbol{f}\|_{L^2(\Omega)} + \frac{\alpha}{\nu}\|\text{curl}\,\boldsymbol{f}\|_{L^2(\Omega)} + (2+\sqrt{2})|\boldsymbol{u}_{h,\boldsymbol{g}}|_{H^1(\Omega)}. \tag{5.3.5}$$

Then the next existence theorem follows easily from the two bounds (5.3.2) and (5.3.5).

THEOREM 5.3.1. *Let $\boldsymbol{g}$, satisfying $\boldsymbol{g}\cdot\boldsymbol{n} = 0$, be given in $W^{1-1/r,r}(\partial\Omega)^2$ for some real number $r > 2$. Assume that (2.1.1) and Hypothesis 5.2.1 hold. Then there exists $h_b > 0$ such that for all $h \le h_b$, all $\nu > 0$, $\alpha > 0$, and for all $\boldsymbol{f}$ in $H(curl, \Omega)$, the discrete problem (5.2.5), (5.2.6), (5.3.1) has at least one solution $(\boldsymbol{u}_h, p_h, z_h) \in W_h \times M_h \times Z_h$, and each solution satisfies the uniform a priori estimate*

$$|\boldsymbol{u}_h|_{H^1(\Omega)} \le 2\frac{S_2}{\nu}\|\boldsymbol{f}\|_{L^2(\Omega)} + \frac{\alpha}{\nu}\|\text{curl}\,\boldsymbol{f}\|_{L^2(\Omega)} + \frac{C}{\nu}\left(\alpha + \frac{\nu}{\alpha}\right)^{1/2}\|\boldsymbol{g}\|^2_{W^{1-1/r,r}(\partial\Omega)}, \tag{5.3.6}$$

together with (5.3.2) and (5.2.12)

$$\|z_h\|^2_{L^2(\Omega)} \le (2\alpha + \nu h)\left(\frac{1}{\alpha}\|\text{curl}\,\boldsymbol{u}_h\|^2_{L^2(\Omega)} + \frac{\alpha}{\nu^2}\|\text{curl}\,\boldsymbol{f}\|^2_{L^2(\Omega)}\right),$$

$$\|p_h\|_{L^2(\Omega)} \le \frac{1}{\beta^\star}\left(S_2\|\boldsymbol{f}\|_{L^2(\Omega)} + |\boldsymbol{u}_h|_{H^1(\Omega)}\left(\nu + S_4\tilde{S}_4\|z_h\|_{L^2(\Omega)}\right)\right),$$

where C depends on the constants of (5.2.3) and (5.2.4). The value of h_b is determined by ε in (5.3.4):

$$h_b \le C_b\frac{\alpha}{2\alpha+\nu}\left(\frac{\nu}{2C\|\boldsymbol{g}\|_{W^{1-1/r,r}(\partial\Omega)}}\right)^2. \tag{5.3.7}$$

By the same argument as in Section 2.4.1, convergence, first weak and next strong, of a solution $(\boldsymbol{u}_h, p_h, z_h)$ of (5.2.5), (5.2.6), (5.3.1) to a solution $(\boldsymbol{u}, p, z)$ of (5.1.2) follows easily from the uniform estimates of Theorem 5.3.1.

THEOREM 5.3.2. *We retain the second and third assumptions of Hypothesis 2.1.5, and we suppose that Hypothesis 5.2.5 holds. Let* $(\boldsymbol{u}_h, p_h, z_h)$ *be a solution of (5.2.5), (5.2.6), (5.3.1). Then there exists a subsequence of h (still denoted by h) and a solution* $(\boldsymbol{u}, p, z) \in W \times L_0^2(\Omega) \times L^2(\Omega)$ *of problem (5.1.2) such that*

$$\lim_{h\to 0} \|\boldsymbol{u}_h - \boldsymbol{u}\|_{H^1(\Omega)} = 0,$$

$$\lim_{h\to 0} \|z_h - z\|_{L^2(\Omega)} = 0,$$

$$\lim_{h\to 0} \|p_h - p\|_{L^2(\Omega)} = 0,$$

$$\lim_{h\to 0} \sqrt{h}\|\boldsymbol{u}_h \cdot \nabla z_h\|_{L^2(\Omega)} = 0.$$

A maximum norm estimate for the discrete velocity

The a priori estimates of Theorem 5.2.3 are sufficient for deriving error estimates for centered schemes, but as is observed in the homogeneous case, an L^∞ estimate for $\boldsymbol{u}_h$ is useful for upwind schemes. The situation is the same here.

By analogy with the material of Section 2.1.2, we suppose that $\mathcal{T}_h$ is a regular family of conforming triangulations of $\overline{\Omega}$, consisting of triangles with maximum mesh size h, and we suppose that in each triangle T, the finite-element functions of X_h, M_h, and Z_h are all polynomials.

Given $z_h \in Z_h$, let $\boldsymbol{v}(z_h) \in H_\tau^1(\Omega)$ and $q(z_h) \in L_0^2(\Omega)$ be the unique solution of

$$\forall \boldsymbol{w} \in H_0^1(\Omega)^2,\ \nu(\nabla \boldsymbol{v}(z_h), \nabla \boldsymbol{w}) + (z_h \times \boldsymbol{v}(z_h), \boldsymbol{w}) - (q(z_h), \operatorname{div} \boldsymbol{w}) = (\boldsymbol{f}, \boldsymbol{w}), \quad (5.3.8)$$

$$\forall r \in L_0^2(\Omega),\ (r, \operatorname{div} \boldsymbol{v}(z_h)) = 0, \quad (5.3.9)$$

$$\boldsymbol{v}(z_h) = \boldsymbol{g} \quad \text{on } \partial\Omega. \quad (5.3.10)$$

Thus the pair $(\boldsymbol{u}_h, p_h)$ is a finite-element discretization of (5.3.8)–(5.3.10), and higher-order estimates for $\boldsymbol{u}_h$ can be established if $\boldsymbol{v}(z_h)$ is sufficiently smooth. Considering that the proof of error estimates requires a convex domain, we can use directly Theorem 5.1.7. Therefore, we also assume that $\boldsymbol{g}$ satisfies (5.1.19) and (5.1.20)

$$\boldsymbol{g} \in H^{3/2}(\Gamma_i)^2 \text{ for } 1 \le i \le N, \quad \boldsymbol{g}\cdot\boldsymbol{n} = 0 \text{ on } \partial\Omega,$$

$$\int_0^\varepsilon \frac{1}{s}\left|\frac{\partial \boldsymbol{g}_{i+1}\cdot\boldsymbol{n}_i}{\partial \boldsymbol{t}_{i+1}}(\boldsymbol{x}_i + s\boldsymbol{t}_{i+1}) - \frac{\partial \boldsymbol{g}_i\cdot\boldsymbol{n}_{i+1}}{\partial \boldsymbol{t}_i}(\boldsymbol{x}_i - s\boldsymbol{t}_i)\right|^2 ds < \infty,$$

where $\varepsilon = \min_{1\le i\le N}|\Gamma_i|$. Then $\boldsymbol{v}(z_h) \in H^2(\Omega)^2$, $q(z_h) \in H^1(\Omega)$ and (to simplify, we do not detail the dependence on the data $\boldsymbol{f}$ and $\boldsymbol{g}$)

$$\|\boldsymbol{v}(z_h)\|_{H^2(\Omega)} + \|q(z_h)\|_{H^1(\Omega)} \le C_1 + C_2\|z_h\|_{L^2(\Omega)} + C_3\|z_h\|^2_{L^2(\Omega)} + C_4\|z_h\|^3_{L^2(\Omega)}. \quad (5.3.11)$$

The following proposition states the analog of Remark 2.1.17.

PROPOSITION 5.3.3. *Let Ω be convex and let $\boldsymbol{g}$ satisfy (5.1.19) and (5.1.20). Let $(\boldsymbol{u}_h, p_h, z_h)$ be any solution of (5.2.5), (5.2.6), (5.3.1), and suppose that the operator r_h and the operator $\tilde{P}_h$ of Hypothesis 5.2.5 satisfy Hypothesis 2.1.10 for each real number $s \in [0, 1]$ and for each number $r \geq 2$:*

$$\forall \boldsymbol{v} \in \left(W^{s+1,r}(\Omega) \cap H_0^1(\Omega)\right)^2, \ |\tilde{P}_h(\boldsymbol{v}) - \boldsymbol{v}|_{W^{1,r}(\Omega)} \leq C h^s |\boldsymbol{v}|_{W^{s+1,r}(\Omega)},$$

$$\forall q \in W^{s,r}(\Omega) \cap L_0^2(\Omega), \ \|r_h(q) - q\|_{L^r(\Omega)} \leq C h^s |q|_{W^{s,r}(\Omega)}.$$

If the triangulation is chosen so that, in addition to (2.1.25), (2.1.45) holds for some $r > 2$, close to 2:

$$h \leq C \varrho_{\min}^{1-2/r},$$

then there exists a constant C independent of h such that

$$\|\boldsymbol{u}_h\|_{L^\infty(\Omega)} \leq C. \tag{5.3.12}$$

PROOF. Let $r > 2$, close to 2, and assume that $\mathcal{T}_h$ satisfies (2.1.45) with this value of r. Arguing as in Remark 2.1.17 and using the stability of $\tilde{P}_h$, we write

$$|\boldsymbol{u}_h|_{W^{1,r}(\Omega)} \leq |\boldsymbol{u}_h - \tilde{P}_h(\boldsymbol{v}(z_h))|_{W^{1,r}(\Omega)} + c_1 |\boldsymbol{v}(z_h)|_{W^{1,r}(\Omega)}.$$

On one hand, the hypotheses on Ω and $\boldsymbol{g}$ imply that $\boldsymbol{v}(z_h)$ is uniformly bounded in $W^{1,r}(\Omega)^2$. On the other hand, considering that $\boldsymbol{u}_h - \tilde{P}_h(\boldsymbol{v}(z_h))$ belongs to V_h, the bound (2.1.36) derived in proving Lemma 2.1.12 is valid here and the regularity of $\boldsymbol{v}(z_h)$ and $q(z_h)$ give the analog of (2.1.35)

$$|\boldsymbol{u}_h - \tilde{P}_h(\boldsymbol{v}(z_h))|_{H^1(\Omega)} \leq C h \left(K_1(z_h) |\boldsymbol{v}(z_h)|_{H^2(\Omega)} + \frac{1}{\nu} |q(z_h)|_{H^1(\Omega)} \right). \tag{5.3.13}$$

Then (5.3.12) follows from an inverse inequality, (2.1.45), (5.3.11), and the uniform estimate (5.3.2) for z_h. □

Streamline diffusion: Error estimates

As in the homogeneous case, the error inequalities for the velocity and pressure are the same as in centered schemes, and the statement of Lemma 5.2.7 is valid here for any solution $(\boldsymbol{u}_h, p_h, z_h)$ of (5.2.5), (5.2.6), (5.3.1), and any solution $(\boldsymbol{u}, p, z)$ of Problem (5.1.2):

$$\begin{aligned} |\boldsymbol{u} - \boldsymbol{u}_h|_{H^1(\Omega)} &\leq 2|\boldsymbol{u} - \tilde{P}_h(\boldsymbol{u})|_{H^1(\Omega)} + \frac{S_4}{\nu} \|\boldsymbol{u}\|_{L^4(\Omega)} \|z - z_h\|_{L^2(\Omega)} \\ &\quad + \frac{S_4}{\nu} \|z_h\|_{L^2(\Omega)} \|\boldsymbol{u} - \tilde{P}_h(\boldsymbol{u})\|_{L^4(\Omega)} + \frac{1}{\nu} \|p - r_h(p)\|_{L^2(\Omega)}, \end{aligned} \tag{5.3.14}$$

$$\begin{aligned} \|p - p_h\|_{L^2(\Omega)} &\leq \left(1 + \frac{1}{\beta^*}\right) \|p - r_h(p)\|_{L^2(\Omega)} + \frac{1}{\beta^*} \big(\nu \, |\boldsymbol{u} - \tilde{P}_h(\boldsymbol{u})|_{H^1(\Omega)} \\ &\quad + S_4 \left(\|\boldsymbol{u}\|_{L^4(\Omega)} \|z - z_h\|_{L^2(\Omega)} + \|z_h\|_{L^2(\Omega)} \|\boldsymbol{u} - \boldsymbol{u}_h\|_{L^4(\Omega)}\right)\big), \end{aligned} \tag{5.3.15}$$

where β^* is the constant of (2.1.1).

Similarly, the error inequality for $z - z_h$ is the same as in the homogeneous upwind scheme. Therefore, if z is sufficiently smooth to give meaning to all terms in the right-hand side of the inequality below, the analog of (2.4.4) holds for any λ_h in Z_h:

$$\frac{\nu}{2}\|z_h - \lambda_h\|^2_{L^2(\Omega)} + \frac{\alpha h}{2}\|\boldsymbol{u}_h \cdot \nabla(z_h - \lambda_h)\|^2_{L^2(\Omega)} \le 2\alpha h \|\boldsymbol{u}_h\|^2_{L^\infty(\Omega)}|z - \lambda_h|^2_{H^1(\Omega)}$$
$$+ 2\left(3\nu + 2h\frac{\nu^2}{\alpha} + \frac{\alpha}{h}\right)\|z - \lambda_h\|^2_{L^2(\Omega)} + \alpha\left(3\frac{\alpha}{\nu} + 2h\right)\|(\boldsymbol{u} - \boldsymbol{u}_h)\cdot\nabla z\|^2_{L^2(\Omega)}$$
$$+ \frac{3}{4}\left(\frac{\alpha^2}{\nu} + \nu h^2\right)\|\operatorname{div}(\boldsymbol{u} - \boldsymbol{u}_h)\lambda_h\|^2_{L^2(\Omega)} + 6\frac{\alpha^2}{\nu}\|\operatorname{div}(\boldsymbol{u} - \boldsymbol{u}_h)(\lambda_h - z)\|^2_{L^2(\Omega)}$$
$$+ \frac{\nu}{2}\left(3 + 8h\frac{\nu}{\alpha}\right)|\boldsymbol{u} - \boldsymbol{u}_h|^2_{H^1(\Omega)}.$$

Again, the choice $\lambda_h = R_h(z)$ in this inequality, Hölder's inequality in the nonlinear products, and the stability properties of R_h give the next upper bound, provided that z belongs to $W^{1,r}(\Omega)$:

$$\frac{\nu}{2}\|z_h - z\|^2_{L^2(\Omega)} + \frac{\alpha h}{2}\|\boldsymbol{u}_h \cdot \nabla(z_h - R_h(z))\|^2_{L^2(\Omega)} \le 2\alpha h \|\boldsymbol{u}_h\|^2_{L^\infty(\Omega)}|z - R_h(z)|^2_{H^1(\Omega)}$$
$$+ 2\left(\frac{13}{4}\nu + 2h\frac{\nu^2}{\alpha} + \frac{\alpha}{h}\right)\|z - R_h(z)\|^2_{L^2(\Omega)}$$
$$+ |\boldsymbol{u} - \boldsymbol{u}_h|^2_{H^1(\Omega)}\left(|z|^2_{W^{1,r}(\Omega)}\left(\alpha S^2_{r^*}\left(\frac{3\alpha}{\nu} + 2h\right)\right.\right.$$
$$\left.\left. + (1 + C_r)^2\frac{3}{4}\left(9\frac{\alpha^2}{\nu} + \nu h^2\right)\right) + \frac{\nu}{2}\left(3 + 8h\frac{\nu}{\alpha}\right)\right), \tag{5.3.16}$$

where C_r is the stability constant of R_h in $W^{1,r}(\Omega)$.

The assumptions of Proposition 5.1.8 guarantee that $\boldsymbol{u}$ belongs to $W^{2,r}(\Omega)^2$, which in turn implies that z belongs to $W^{1,r}(\Omega)$. As a by-product, Proposition 5.3.3 yields that $\boldsymbol{u}_h$ is uniformly bounded in $L^\infty(\Omega)^2$. Therefore, we can conclude that, under the assumptions of Proposition 5.1.8 for some $r > 2$, close to 2, if $\mathcal{T}_h$ satisfies (2.1.45) and (5.3.7), and if the data are sufficiently small, then

$$\nu\|z_h - z\|^2_{L^2(\Omega)} + \alpha h \|\boldsymbol{u}_h \cdot \nabla(z_h - z)\|^2_{L^2(\Omega)} \le C\left(\frac{1}{h}\|z - R_h(z)\|^2_{L^2(\Omega)}\right.$$
$$\left. + h|z - R_h(z)|^2_{H^1(\Omega)} + |\boldsymbol{u} - \tilde{P}_h(\boldsymbol{u})|^2_{H^1(\Omega)} + \|p - r_h(p)\|^2_{L^2(\Omega)}\right). \tag{5.3.17}$$

This upwind scheme can be used with the three examples of Section 5.2.2. It yields the same asymptotic error as if it were applied to a homogeneous problem, and it is recommended in the same circumstances, especially when z has little regularity.

5.3.2. *Lesaint–Raviart's Discontinuous Galerkin Method*

Let $\mathcal{T}_h$ be a family of triangulations satisfying (2.1.25). We keep the above approximation spaces $X_{h,\tau}$ for the velocity and M_h for the pressure, with the same notation and assumptions, and we discretize z in a finite-dimensional subspace Z_h of $H^1(\Omega)$ such as (2.4.9)

$$Z_h = \{\theta_h \in L^2(\Omega);\ \forall T \in \mathcal{T}_h,\ \theta_h|_T \in \mathbb{P}_k\},$$

where $k \geq 1$ is an integer. It is also possible to approximate z by piecewise constant functions, but this option is not studied here. We choose for R_h a suitable continuous approximation operator such as the Girault and Lions variant of the Scott and Zhang regularization operator, or the Bernardi and Girault operator. Recall that it has the approximation error (2.4.10)

$$\forall \theta \in W^{s+1,r}(\Omega),\ |R_h(\theta) - \theta|_{W^{m,r}(\Omega)} \leq C\, h^{s+1-m} |\theta|_{W^{s+1,s}(\Omega)},$$

for any number $r \geq 1$, for $m = 0,\, 1$, and $0 \leq s \leq k$.

Now observe that the definition (2.4.11) of the inflow boundary of a triangle T

$$\partial T_- = \{\boldsymbol{x} \in \partial T;\ \boldsymbol{u}_h(\boldsymbol{x}) \cdot \boldsymbol{n}_T(\boldsymbol{x}) < 0\},$$

can be applied without modification to a function $\boldsymbol{u}_h$ of $X_{h,\tau}$ because it only involves the normal trace of $\boldsymbol{u}_h$, and hence only involves interior segments of the triangulation. Therefore, we keep the same consistent approximation (2.4.12) of the nonlinear term $(\boldsymbol{u} \cdot \nabla z, \theta)$

$$\tilde{c}^{\mathrm{DG}}(\boldsymbol{u}_h; z_h, \theta_h) = \sum_{T \in \mathcal{T}_h} \left(\int_T (\boldsymbol{u}_h \cdot \nabla z_h)\theta_h \,\mathrm{d}\boldsymbol{x} + \int_{\partial T_-} |\boldsymbol{u}_h \cdot \boldsymbol{n}_T|(z_h^{\mathrm{int}} - z_h^{\mathrm{ext}})\theta_h^{\mathrm{int}} \mathrm{d}s \right) + \frac{1}{2} \int_\Omega (\operatorname{div} \boldsymbol{u}_h) z_h \theta_h \,\mathrm{d}\boldsymbol{x},$$

where the superscript int (resp. ext) refers to the trace on the segment of ∂T of the function taken inside (resp. outside) T. When $\boldsymbol{u}_h$ is replaced by $\boldsymbol{u} \in W$ and z_h by $z \in H^1(\Omega)$, this form is a consistent approximation of $(\boldsymbol{u} \cdot \nabla z, \theta_h)$. Furthermore, formula (2.4.16)

$$\tilde{c}^{\mathrm{DG}}(\boldsymbol{u}_h; z_h, \theta_h) = \sum_{T \in \mathcal{T}_h} \left(-\int_T (\boldsymbol{u}_h \cdot \nabla \theta_h) z_h \,\mathrm{d}\boldsymbol{x} + \int_{\partial T_-} |\boldsymbol{u}_h \cdot \boldsymbol{n}_T|(\theta_h^{\mathrm{ext}} - \theta_h^{\mathrm{int}}) z_h^{\mathrm{ext}} \mathrm{d}s \right) - \frac{1}{2} \int_\Omega (\operatorname{div} \boldsymbol{u}_h)\theta_h z_h \,\mathrm{d}\boldsymbol{x},$$

and its consequence (2.4.21)

$$\tilde{c}^{\mathrm{DG}}(\boldsymbol{u}_h; z_h, z_h) = \frac{1}{2} \sum_{T \in \mathcal{T}_h} \int_{\partial T_-} |\boldsymbol{u}_h \cdot \boldsymbol{n}_T|(z_h^{\mathrm{ext}} - z_h^{\mathrm{int}})^2 \mathrm{d}s,$$

established in Section 2.4.2, are valid here because they only rely on the fact that $\boldsymbol{u}_h \cdot \boldsymbol{n} = 0$ on $\partial\Omega$.

With this form and the lifting $\boldsymbol{u}_{h,g}$ of Hypothesis 5.2.1, we approximate problem (5.1.2) by the upwind scheme: Find $\boldsymbol{u}_h$ in $X_h + \boldsymbol{u}_{h,g}$, p_h in M_h and $z_h = (0, 0, z_h)$ with z_h in Z_h, solution of (5.2.5), (5.2.6):

$$\forall \boldsymbol{v}_h \in X_h,\ \nu(\nabla \boldsymbol{u}_h, \nabla \boldsymbol{v}_h) + (z_h \times \boldsymbol{u}_h, \boldsymbol{v}_h) - (p_h, \operatorname{div} \boldsymbol{v}_h) = (\boldsymbol{f}, \boldsymbol{v}_h),$$

$$\forall q_h \in M_h,\ (q_h, \operatorname{div} \boldsymbol{u}_h) = 0,$$

and

$$\forall \theta_h \in Z_h,\ \nu\,(z_h, \theta_h) + \alpha\,\tilde{c}^{\mathrm{DG}}(\boldsymbol{u}_h; z_h, \theta_h) = \nu\,(\operatorname{curl} \boldsymbol{u}_h, \theta_h) + \alpha\,(\operatorname{curl} \boldsymbol{f}, \theta_h). \tag{5.3.18}$$

Discontinuous Galerkin: convergence
Existence of a solution follows from Brouwer's Fixed Point Theorem and uniform a priori estimates. By arguing as in Section 2.4.2, we immediately derive that the third component of any solution $(\boldsymbol{u}_h, p_h, z_h)$ of (5.2.5), (5.2.6), (5.3.18) is bounded as follows:

$$\|z_h\|_{L^2(\Omega)}^2 + \frac{\alpha}{\nu} \sum_{T \in \mathcal{T}_h} \int_{\partial T_-} |\boldsymbol{u}_h \cdot \boldsymbol{n}_T| (z_h^{\mathrm{ext}} - z_h^{\mathrm{int}})^2 \mathrm{d}s \\ \leq 2 \left(\|\operatorname{curl} \boldsymbol{u}_h\|_{L^2(\Omega)}^2 + \frac{\alpha^2}{\nu^2} \|\operatorname{curl} \boldsymbol{f}\|_{L^2(\Omega)}^2 \right). \tag{5.3.19}$$

Next, $\boldsymbol{u}_{h,\mathbf{0}} = \boldsymbol{u}_h - \boldsymbol{u}_{h,g}$ satisfies (5.3.3)

$$|\boldsymbol{u}_{h,\mathbf{0}}|_{H^1(\Omega)} \leq \frac{S_2}{\nu} \|\boldsymbol{f}\|_{L^2(\Omega)} + |\boldsymbol{u}_{h,g}|_{H^1(\Omega)} + \frac{C}{\nu} \sqrt{\varepsilon} \|\boldsymbol{g}\|_{W^{1-1/r,r}(\partial\Omega)} \|z_h\|_{L^2(\Omega)}.$$

By combining these two bounds, and choosing for instance

$$\varepsilon = \frac{1}{2} \left(\frac{\nu}{2C \|\boldsymbol{g}\|_{W^{1-1/r,r}(\partial\Omega)}} \right)^2, \tag{5.3.20}$$

we obtain the same upper bound as (5.2.14) and (5.3.5):

$$|\boldsymbol{u}_{h,\mathbf{0}}|_{H^1(\Omega)} \leq 2 \frac{S_2}{\nu} \|\boldsymbol{f}\|_{L^2(\Omega)} + \frac{\alpha}{\nu} \|\operatorname{curl} \boldsymbol{f}\|_{L^2(\Omega)} + (2 + \sqrt{2}) |\boldsymbol{u}_{h,g}|_{H^1(\Omega)}. \tag{5.3.21}$$

The bounds (5.3.19) and (5.3.21) readily yield the following existence theorem.

THEOREM 5.3.4. *Let $\boldsymbol{g}$, satisfying $\boldsymbol{g} \cdot \boldsymbol{n} = 0$, be given in $W^{1-1/r,r}(\partial\Omega)^2$ for some real number $r > 2$. Assume that (2.1.1) and Hypothesis 5.2.1 hold. Then there exists $h_b > 0$ such that for all $h \leq h_b$, all $\nu > 0$, and all $\boldsymbol{f}$ in $H(curl, \Omega)$, the discrete problem (5.2.5), (5.2.6), (5.3.18) has at least one solution $(\boldsymbol{u}_h, p_h, z_h) \in W_h \times M_h \times Z_h$, and each solution satisfies the uniform a priori estimate*

$$|\boldsymbol{u}_h|_{H^1(\Omega)} \leq 2 \frac{S_2}{\nu} \|\boldsymbol{f}\|_{L^2(\Omega)} + \frac{\alpha}{\nu} \|\operatorname{curl} \boldsymbol{f}\|_{L^2(\Omega)} + \frac{C}{\nu} \|\boldsymbol{g}\|_{W^{1-1/r,r}(\partial\Omega)}^2, \tag{5.3.22}$$

together with (5.3.19) and (5.2.12)

$$\|z_h\|^2_{L^2(\Omega)} + \frac{\alpha}{\nu}\sum_{T\in\mathcal{T}_h}\int_{\partial T_-}|\boldsymbol{u}_h\cdot\boldsymbol{n}_T|(z_h^{\text{ext}} - z_h^{\text{int}})^2 ds$$

$$\leq 2\left(\|\operatorname{curl}\boldsymbol{u}_h\|^2_{L^2(\Omega)} + \frac{\alpha^2}{\nu^2}\|\operatorname{curl}\boldsymbol{f}\|^2_{L^2(\Omega)}\right),$$

$$\|p_h\|_{L^2(\Omega)} \leq \frac{1}{\beta^\star}\left(S_2\|\boldsymbol{f}\|_{L^2(\Omega)} + |\boldsymbol{u}_h|_{H^1(\Omega)}\left(\nu + S_4\tilde{S}_4\|z_h\|_{L^2(\Omega)}\right)\right),$$

where C depends on the constants of (5.2.3) and (5.2.4). The value of h_b is determined by ε in (5.3.20):

$$h_b \leq \frac{1}{2}C_b\left(\frac{\nu}{2C\|\boldsymbol{g}\|_{W^{1-1/r,r}(\partial\Omega)}}\right)^2. \tag{5.3.23}$$

By extracting subsequences (that we still denote by the index h), and proceeding as in Section 2.4.2, these uniform a priori estimates allow us to prove convergence, first weak and next strong, of a solution $(\boldsymbol{u}_h, p_h, z_h)$ of (5.2.5), (5.2.6), (5.3.18) to a solution $(\boldsymbol{u}, p, z)$ of (5.1.2).

THEOREM 5.3.5. *We retain the second and third assumptions of Hypothesis 2.1.5, and we suppose that Hypothesis 5.2.5 holds. Let $(\boldsymbol{u}_h, p_h, z_h)$ be a solution of (5.2.5), (5.2.6), (5.3.18). Then there exists a subsequence of h (still denoted by h) and a solution $(\boldsymbol{u}, p, z) \in W \times L^2_0(\Omega) \times L^2(\Omega)$ of problem (5.1.2) such that*

$$\lim_{h\to 0}\|\boldsymbol{u}_h - \boldsymbol{u}\|_{H^1(\Omega)} = 0,$$

$$\lim_{h\to 0}\|z_h - z\|_{L^2(\Omega)} = 0,$$

$$\lim_{h\to 0}\|p_h - p\|_{L^2(\Omega)} = 0,$$

$$\lim_{h\to 0}\sum_{T\in\mathcal{T}_h}\int_{\partial T_-}|\boldsymbol{u}_h\cdot\boldsymbol{n}_T|(z_h^{\text{ext}} - z_h^{\text{int}})^2 ds = 0 \ \text{ in } \mathbb{R}.$$

Discontinuous Galerkin: Error estimates

As in the homogeneous case, the error inequalities for the velocity and pressure are the same as in centered schemes, and the statement of Lemma 5.2.7 is valid here for any solution $(\boldsymbol{u}_h, p_h, z_h)$ of (5.2.5), (5.2.6), (5.3.18) and a solution $(\boldsymbol{u}, p, z)$ of (5.1.2):

$$|\boldsymbol{u} - \boldsymbol{u}_h|_{H^1(\Omega)} \leq 2|\boldsymbol{u} - \tilde{P}_h(\boldsymbol{u})|_{H^1(\Omega)} + \frac{S_4}{\nu}\|\boldsymbol{u}\|_{L^4(\Omega)}\|z - z_h\|_{L^2(\Omega)}$$
$$+ \frac{S_4}{\nu}\|z_h\|_{L^2(\Omega)}\|\boldsymbol{u} - \tilde{P}_h(\boldsymbol{u})\|_{L^4(\Omega)} + \frac{1}{\nu}\|p - r_h(p)\|_{L^2(\Omega)}, \tag{5.3.24}$$

$$\|p - p_h\|_{L^2(\Omega)} \leq \left(1 + \frac{1}{\beta^*}\right)\|p - r_h(p)\|_{L^2(\Omega)} + \frac{1}{\beta^*}\left(\nu\,|\boldsymbol{u} - \tilde{P}_h(\boldsymbol{u})|_{H^1(\Omega)}\right.$$
$$\left. + S_4\left(\|\boldsymbol{u}\|_{L^4(\Omega)}\|z - z_h\|_{L^2(\Omega)} + \|z_h\|_{L^2(\Omega)}\|\boldsymbol{u} - \boldsymbol{u}_h\|_{L^4(\Omega)}\right)\right), \tag{5.3.25}$$

where β^* is the constant of (2.1.1).

Similarly, the error inequality for $z - z_h$ is the same as in the homogeneous Lesaint–Raviart Discontinuous Galerkin scheme. Assuming that $z \in H^1(\Omega)$, we begin with (2.4.24)

$$
\begin{aligned}
&\nu\|z_h - \lambda_h\|^2_{L^2(\Omega)} + \alpha \sum_{T\in\mathcal{T}_h} \frac{1}{2} \int\limits_{\partial T_-} |\boldsymbol{u}_h \cdot \boldsymbol{n}_T|((z_h - \lambda_h)^{\text{ext}} - (z_h - \lambda_h)^{\text{int}})^2 \mathrm{d}s \\
&\quad + \alpha \sum_{T\in\mathcal{T}_h} \Bigg(- \int\limits_T \boldsymbol{u}_h \cdot \nabla(z_h - \lambda_h)(\lambda_h - z)\,\mathrm{d}\boldsymbol{x} \\
&\quad + \alpha \int\limits_{\partial T_-} |\boldsymbol{u}_h \cdot \boldsymbol{n}_T|((z_h - \lambda_h)^{\text{ext}} - (z_h - \lambda_h)^{\text{int}})(\lambda_h - z)^{\text{ext}} \mathrm{d}s \Bigg) \\
&\quad - \frac{\alpha}{2} \int\limits_\Omega \operatorname{div}(\boldsymbol{u}_h - \boldsymbol{u})(\lambda_h - z)(z_h - \lambda_h)\,\mathrm{d}\boldsymbol{x} \\
&\quad + \frac{\alpha}{2} \int\limits_\Omega \operatorname{div}(\boldsymbol{u}_h - \boldsymbol{u}) z (z_h - \lambda_h)\mathrm{d}\boldsymbol{x} + \alpha \int\limits_\Omega (\boldsymbol{u}_h - \boldsymbol{u}) \cdot \nabla z (z_h - \lambda_h)\,\mathrm{d}\boldsymbol{x} \\
&= \nu(z - \lambda_h, z_h - \lambda_h) + \nu(\operatorname{curl}(\boldsymbol{u}_h - \boldsymbol{u}), z_h - \lambda_h).
\end{aligned}
$$

Then the proof of Theorem 2.4.11 extends directly to the nonhomogeneous problem.

THEOREM 5.3.6. *Let Ω be convex, $(\boldsymbol{u}, p, z)$ a solution of Problem (5.1.2), r_0 the number of Proposition 5.1.8, let* **g** *be as in Proposition 5.1.8, and let the assumptions of Theorem 1.4.14 hold, so that $z \in W^{1,r}(\Omega)$, for some real number r in $]2, r_0[$. Let $\mathcal{T}_h$ be a family of triangulations satisfying (2.1.25) and let $(\boldsymbol{u}_h, p_h, z_h)$ be any solution of (5.2.5), (5.2.6), (5.3.18). Then we have the following inequality for $z_h - \varrho_h(z)$, where ϱ_h is the local L^2 projection on $\mathbb{P}_k$:*

$$
\begin{aligned}
\nu\,\|z_h - \varrho_h(z)\|^2_{L^2(\Omega)} &\le \frac{7}{\nu}\Big(\nu^2\Big(\|z - \varrho_h(z)\|^2_{L^2(\Omega)} + |\boldsymbol{u}_h - \boldsymbol{u}|^2_{H^1(\Omega)}\Big) \\
&\quad + \alpha^2\sigma_0^2 C_1\Big(|\boldsymbol{u}|^2_{W^{1,\infty}(\Omega)}\|z - \varrho_h(z)\|^2_{L^2(\Omega)} + S^2_{r^*}|z|^2_{W^{1,r}(\Omega)}|\boldsymbol{u}_h - \boldsymbol{u}|^2_{H^1(\Omega)}\Big) \\
&\quad + \frac{\alpha^2}{4}\Big(\|\operatorname{div}(\boldsymbol{u}_h - \boldsymbol{u})(z - \varrho_h(z))\|^2_{L^2(\Omega)} + \|(\operatorname{div}(\boldsymbol{u}_h - \boldsymbol{u}))z\|^2_{L^2(\Omega)} \\
&\quad + 4\,\|(\boldsymbol{u}_h - \boldsymbol{u}) \cdot \nabla z\|^2_{L^2(\Omega)}\Big)\Big) + \alpha\, C_2\sigma_0^2\|\boldsymbol{u}_h\|_{L^\infty(\Omega)} \sum_{T\in\mathcal{T}_h} h_T|z - \varrho_h(z)|^2_{H^1(T)},
\end{aligned}
$$

where σ_0 is the constant of (2.1.25), C_1 and C_2 are constants independent of h, and

$$\frac{1}{r^*} = \frac{1}{2} - \frac{1}{r}.$$

Because all solutions of (5.2.5), (5.2.6), (5.3.18) are such that z_h is uniformly bounded in $L^2(\Omega)$, the statement of Proposition 5.3.3 extends immediately to the Discontinuous

Galerkin scheme: By slightly restricting the mesh, there exists a constant C independent of h such that

$$\|\boldsymbol{u}_h\|_{L^\infty(\Omega)} \le C.$$

Under all the above assumptions, by combining (5.3.24) with the bounds of Theorem 5.3.6 and this bound for $\boldsymbol{u}_h$, we derive the following error estimate for small enough data and smooth enough solutions:

$$\begin{aligned}
&\nu\|z_h - z\|^2_{L^2(\Omega)} + \alpha \sum_{T\in\mathfrak{T}_h} \frac{1}{2} \int_{\partial T_-} |\boldsymbol{u}_h \cdot \boldsymbol{n}_T|((z_h - z)^{\text{ext}} - (z_h - z)^{\text{int}})^2 \mathrm{d}s \\
&\le C\Bigg(\|\varrho_h(z) - z\|^2_{L^2(\Omega)} + |\boldsymbol{u} - \tilde{P}_h(\boldsymbol{u})|^2_{H^1(\Omega)} + \|p - r_h(p)\|^2_{L^2(\Omega)} \\
&\quad + \sum_{T\in\mathfrak{T}_h} h_T |z - \varrho_h(z)|^2_{H^1(T)} \Bigg).
\end{aligned} \tag{5.3.26}$$

Thus, the asymptotic error of this scheme is the same as for the homogeneous problem.

CHAPTER 6

Numerical Experiments

In this chapter, we present several numerical experiments both in the steady and time-dependent cases. Some examples are academic and have a known explicit solution. They allow to check convergence rates and confirm the theoretical analysis of the preceding chapters. Others are benchmark problems for which no explicit solution is known, but their qualitative results agree with the behavior expected from grade-two fluids. All results are obtained with the package FreeFem++, see HECHT, LE HYARIC, PIRONNEAU and OHTSUKA [2008]. Numerical experiments with the methods developed in this work have also been performed by CHAMMAI [2006], KANAAN [2007], SAYAH [2007]. As stated in Chapter 4, the least-squares method has been implemented by PARK [1998].

6.1. The steady problem

We consider first examples in an academic situation where an explicit solution is known, and next in the more realistic benchmark cases of the step and driven cavity domains. In each case, the nonlinear scheme is solved by Newton's algorithm. Let us recall it for the reader's convenience. The scheme has the form

$$\forall \boldsymbol{v}_h \in X_h,\ \nu(\nabla \boldsymbol{u}_h, \nabla \boldsymbol{v}_h) + (z_h \times \boldsymbol{u}_h, \boldsymbol{v}_h) - (p_h, \operatorname{div} \boldsymbol{v}_h) = (\boldsymbol{f}, \boldsymbol{v}_h), \tag{6.1.1}$$

$$\forall q_h \in M_h,\ (q_h, \operatorname{div} \boldsymbol{u}_h) = 0, \tag{6.1.2}$$

$$\forall \theta_h \in Z_h,\ \nu\,(z_h, \theta_h) + \alpha\,\tilde{c}(\boldsymbol{u}_h; z_h, \theta_h) = \nu\,(\operatorname{curl} \boldsymbol{u}_h, \theta_h) + \alpha\,(\operatorname{curl} \boldsymbol{f}, \theta_h). \tag{6.1.3}$$

For given z_h, (6.1.1)–(6.1.2) is a discrete Stokes system with right-hand side $(\boldsymbol{f} - z_h \times \boldsymbol{u}_h, \boldsymbol{v}_h)$. To be specific, let $\mathcal{S}_h\boldsymbol{g}$ denote the solution $(\boldsymbol{v}_h, q_h) \in X_h \times M_h$ of the discrete Stokes system

$$\forall \boldsymbol{w}_h \in X_h,\ \nu(\nabla \boldsymbol{v}_h, \nabla \boldsymbol{w}_h) - (q_h, \operatorname{div} \boldsymbol{w}_h) = (\boldsymbol{g}, \boldsymbol{w}_h), \tag{6.1.4}$$

$$\forall r_h \in M_h,\ (r_h, \operatorname{div} \boldsymbol{v}_h) = 0. \tag{6.1.5}$$

Then $(\boldsymbol{u}_h, p_h) = \mathcal{S}_h(\boldsymbol{f} - z_h \times \boldsymbol{u}_h)$. Next, for given $\boldsymbol{u}_h \in X_h$, (6.1.3) is a discrete transport equation with right-hand side $(\operatorname{curl}(\nu \boldsymbol{u}_h + \alpha \boldsymbol{f}), \theta_h)$. More precisely, let $\mathcal{T}_h(\boldsymbol{g}, \boldsymbol{v}_h)$ be the solution $\zeta_h \in Z_h$ of

$$\forall \theta_h \in Z_h,\ \nu\,(\zeta_h, \theta_h) + \alpha\,\tilde{c}(\boldsymbol{v}_h; \zeta_h, \theta_h) = (\operatorname{curl}(\nu \boldsymbol{v}_h + \alpha \boldsymbol{g}), \theta_h). \tag{6.1.6}$$

With this notation, (6.1.1)–(6.1.3) can be expressed implicitly as

$$\mathcal{F}_h(\boldsymbol{u}_h, p_h) := (\boldsymbol{u}_h, p_h) - \mathcal{S}_h\,(\boldsymbol{f} - \mathcal{T}_h(\boldsymbol{f}, \boldsymbol{u}_h) \times \boldsymbol{u}_h) = 0. \tag{6.1.7}$$

Hence Newton's algorithm applied to (6.1.7) reads: Compute $(\boldsymbol{u}_h^{n+1}, p_h^{n+1}) \in X_h \times M_h$ such that

$$\mathcal{F}_h'(\boldsymbol{u}_h^n, p_h^n) \cdot (\boldsymbol{u}_h^{n+1} - \boldsymbol{u}_h^n, p_h^{n+1} - p_h^n) = -\mathcal{F}_h(\boldsymbol{u}_h^n, p_h^n), \tag{6.1.8}$$

starting from a given pair $(\boldsymbol{u}_h^0, p_h^0) \in X_h \times M_h$. Although the auxiliary function $z_h^n \in Z_h$ is not explicitly mentioned in (6.1.7), its computation is included above. Here is the general step, knowing $(\boldsymbol{u}_h^n, p_h^n, z_h^n) \in X_h \times M_h \times Z_h$; let $(\boldsymbol{u}_h^{n+1}, p_h^{n+1}, z_h^{n+1}) \in X_h \times M_h \times Z_h$ be the next iterate and set

$$\boldsymbol{w}_{\boldsymbol{u}_h} = \boldsymbol{u}_h^{n+1} - \boldsymbol{u}_h^n, \; w_{p_h} = p_h^{n+1} - p_h^n, \; w_{z_h} = z_h^{n+1} - z_h^n.$$

Then $(\boldsymbol{w}_{\boldsymbol{u}_h}, w_{p_h}, w_{z_h})$ satisfies

$$\begin{aligned}
&\forall \boldsymbol{v}_h \in X_h, \nu(\nabla \boldsymbol{w}_{\boldsymbol{u}_h}, \nabla \boldsymbol{v}_h) + (\boldsymbol{w}_{z_h} \times \boldsymbol{u}_h^n + \boldsymbol{z}_h^n \times \boldsymbol{w}_{\boldsymbol{u}_h}, \boldsymbol{v}_h) - (w_{p_h}, \operatorname{div} \boldsymbol{v}_h)\\
&\qquad = -\nu(\nabla \boldsymbol{u}_h^n, \nabla \boldsymbol{v}_h) - (z_h^n \times \boldsymbol{u}_h^n, \boldsymbol{v}_h) + (p_h^n, \operatorname{div} \boldsymbol{v}_h) + (\boldsymbol{f}, \boldsymbol{v}_h),\\
&\forall q_h \in M_h, (q_h, \operatorname{div} \boldsymbol{w}_{\boldsymbol{u}_h}) = 0,\\
&\forall \theta_h \in Z_h, \nu\,(w_{z_h}, \theta_h) + \alpha\, \tilde{c}(\boldsymbol{w}_{\boldsymbol{u}_h}; z_h^n, \theta_h) + \alpha\, \tilde{c}(\boldsymbol{u}_h^n; w_{z_h}, \theta_h) - \nu(\operatorname{curl} \boldsymbol{w}_{\boldsymbol{u}_h}, \theta_h)\\
&\qquad = -\nu\,(z_h^n, \theta_h) - \alpha\, \tilde{c}(\boldsymbol{u}_h^n; z_h^n, \theta_h) + (\operatorname{curl}(\nu \boldsymbol{u}_h^n + \alpha \boldsymbol{f}), \theta_h),
\end{aligned} \tag{6.1.9}$$

provided this linear system has a solution. As is usual in Newton's method, the solvability of (6.1.9) is not always guaranteed. In order to reduce the bandwidth of the matrix, the zero mean-value constraint on the test functions of M_h are ignored (the value of $(q_h, \operatorname{div} \boldsymbol{w}_{\boldsymbol{u}_h})$ is unchanged), and the same constraint on p_h is relaxed by adding to the second equation in (6.1.9) the term $\varepsilon(p_h, q_h)$ for a small parameter $\varepsilon > 0$. The starting functions $(\boldsymbol{u}_h^0, p_h^0)$ can be computed, for instance, by solving an adequate Stokes problem and the starting auxiliary function z_h^0 can be taken to be the L^2 projection of $\operatorname{curl} \boldsymbol{u}_h^0$ onto Z_h. The algorithm stops either when the norm of $(\boldsymbol{w}_{\boldsymbol{u}_h}, w_{p_h}, w_{z_h})$ is adequately small or the number of iterations exceeds a chosen threshold (or (6.1.9) is not solvable!).

6.1.1. Explicit case

In these series of tests, the domain Ω is the unit square $]0, 1[\times]0, 1[$, the solution is $\boldsymbol{u} = (u_1, u_2)$ and p with

$$\begin{aligned}
u_1 &= 2\left(y^2(1-y) - y(1-y)^2\right)\sin(\pi x)^2,\\
u_2 &= 4\pi y^2(1-y)^2 \sin(\pi x)\cos(\pi x),\\
p &= \cos(x)\cos(y) - C,
\end{aligned}$$

where the constant C is adjusted so that p has zero mean-value in Ω, i.e.,

$$C = (\sin(1))^2.$$

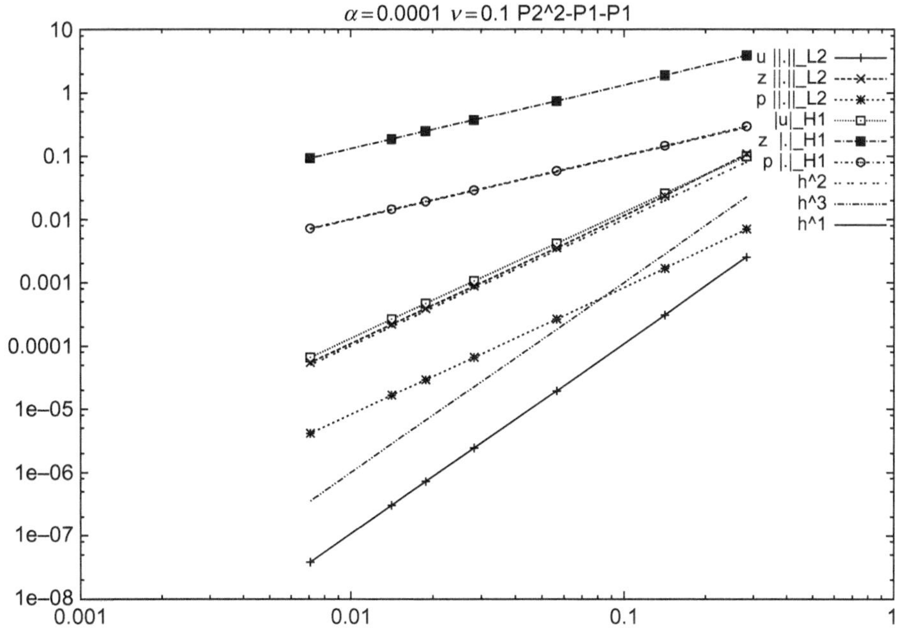

FIG. 6.1 Convergence curves: Taylor–Hood, $\alpha = 10^{-4}$, $\nu = 10^{-1}$, close to Navier–Stokes.

The auxiliary variable z is computed through the two-dimensional version of (1.3.7):

$$z = \operatorname{curl}(\boldsymbol{u} - \alpha\, \Delta\, \boldsymbol{u}).$$

The associated right hand-side f is computed by Maple so that the generalized Stokes problem (1.4.6) is satisfied, namely: $(\boldsymbol{u}, p, z)$ in $V \times L_0^2(\Omega) \times L^2(\Omega)$ solves

$$\begin{aligned} -\nu\, \Delta\, \boldsymbol{u} + \boldsymbol{z} \times \boldsymbol{u} + \nabla\, p &= \boldsymbol{f} \quad \text{in } \Omega, \\ \operatorname{div} \boldsymbol{u} &= 0 \quad \text{in } \Omega, \\ \boldsymbol{u} &= \boldsymbol{0} \quad \text{on } \partial\Omega, \end{aligned}$$

and consequently, the transport equation (1.4.8) is also satisfied:

$$\nu\, z + \alpha\, \boldsymbol{u} \cdot \nabla\, z = \nu \operatorname{curl} \boldsymbol{u} + \alpha \operatorname{curl} \boldsymbol{f}.$$

The nonlinear scheme (6.1.1)–(6.1.3) is applied with the Taylor–Hood finite-element (2.2.14), (2.2.15), with the mini-element (2.2.1), (2.2.2), and with the Bernardi–Raugel finite element (2.2.11), (2.2.12). It is solved by Newton's method with continuation in α and ν, more precisely, we take

$$\alpha^{n+1} = \sqrt{3}\alpha^n, \; \nu^{n+1} = \frac{1}{2}\nu^n,$$

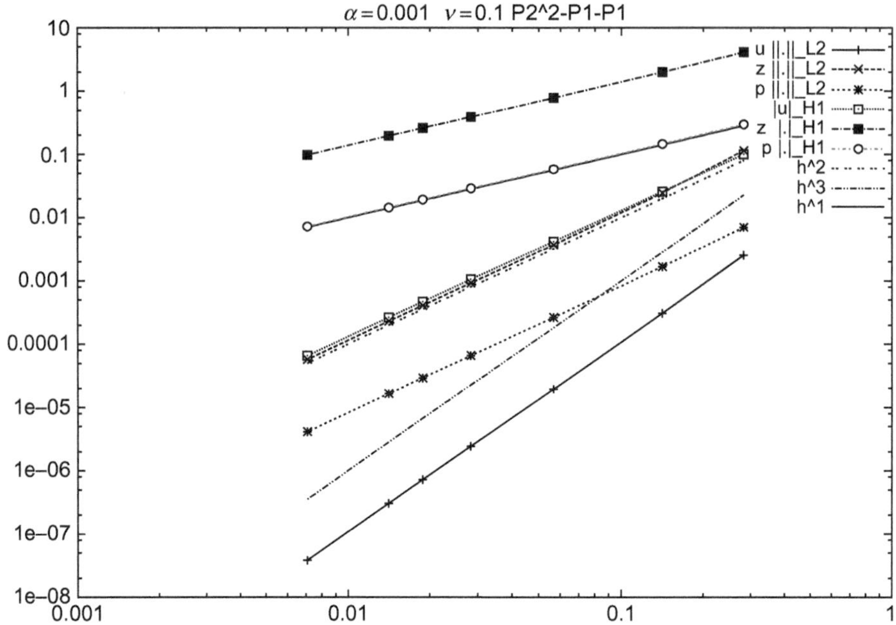

FIG. 6.2 Convergence curves: Taylor–Hood, $\alpha = 10^{-3}$, $\nu = 10^{-1}$.

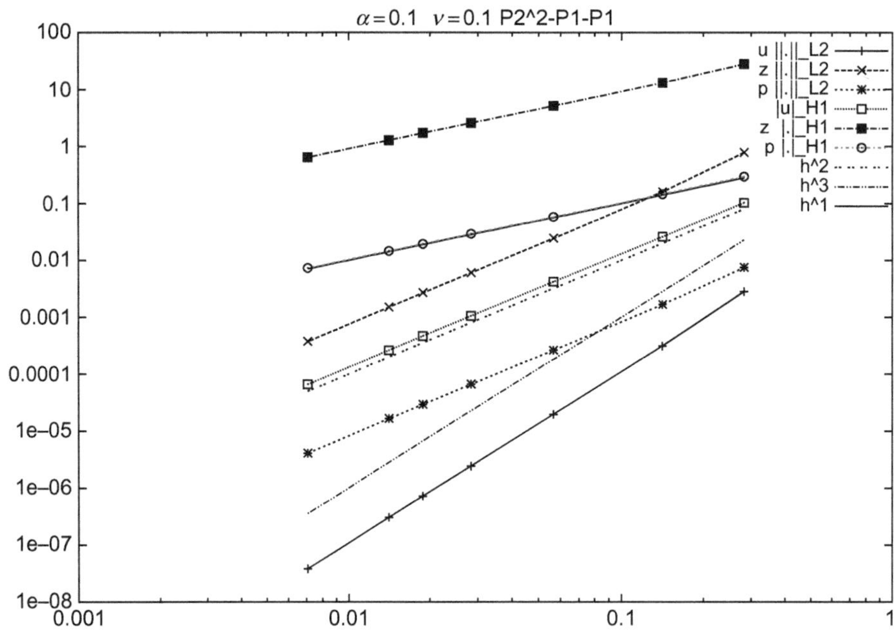

FIG. 6.3 Convergence curves: Taylor–Hood, $\alpha = 10^{-1}$, $\nu = 10^{-1}$.

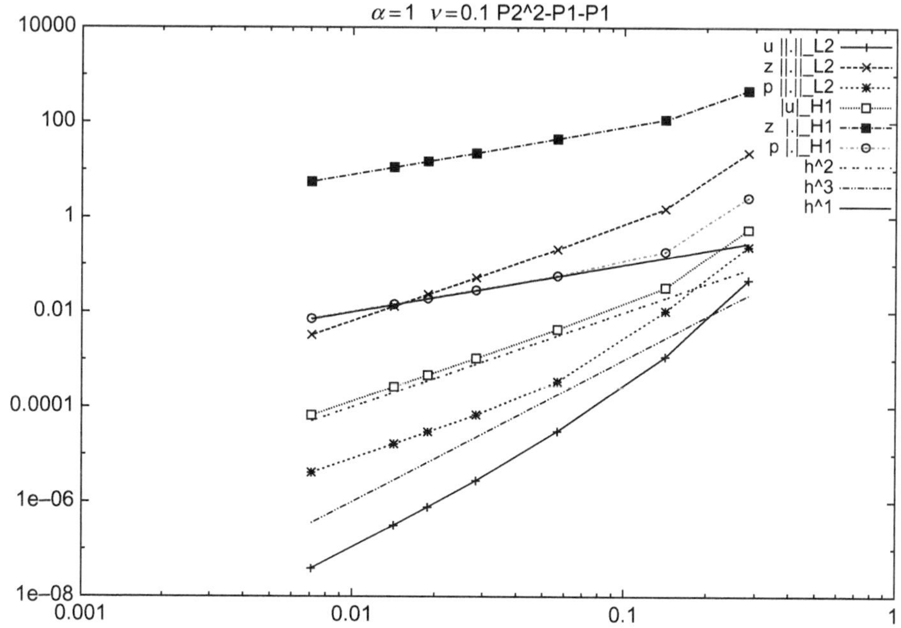

FIG. 6.4 Convergence curves: Taylor–Hood, $\alpha = 1$, $\nu = 10^{-1}$.

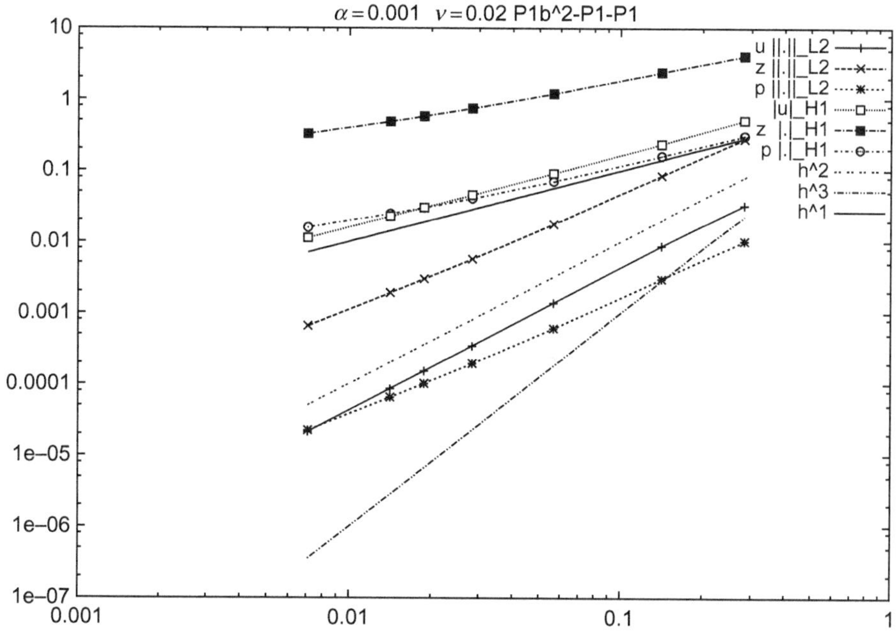

FIG. 6.5 Convergence curves: Mini-element, $\alpha = 10^{-3}$, $\nu = 10^{-1}$.

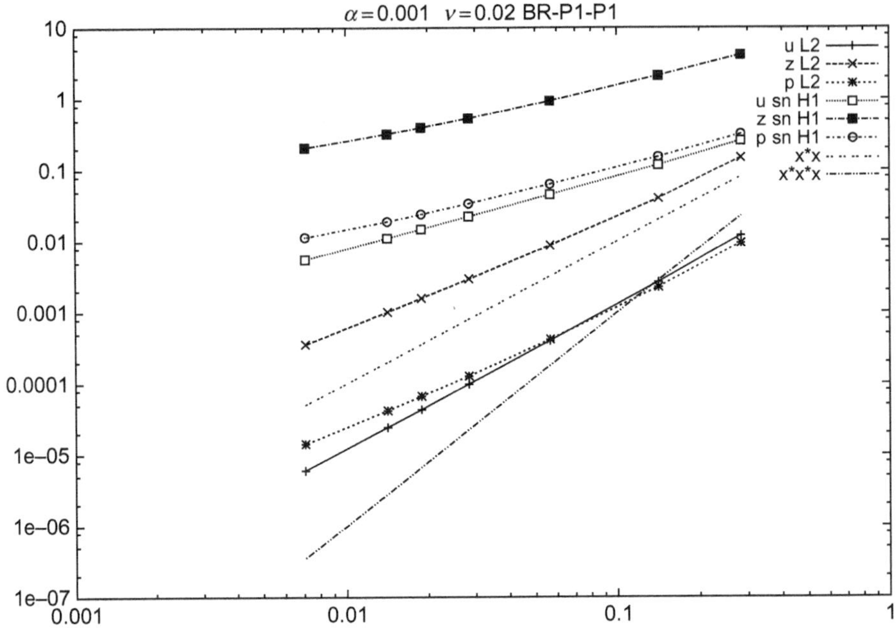

FIG. 6.6 Convergence curves: Bernardi–Raugel, $\alpha = 10^{-3}$, $\nu = 10^{-1}$.

starting with

$$\alpha^0 = \min(\alpha, 0.1),\ \nu^0 = \max(\nu, 0.01).$$

The absolute errors of the three finite-element schemes are depicted in Figures 6.1–6.6, for typical values of α and ν.

The following tables show the progression with decreasing meshsize of the absolute velocity error in the H^1 norm and the velocity, pressure, and auxiliary variable errors in the L^2 norm. The meshsize is h and n is the number of meshpoints on each side of the square. The CPU time gives an indication of the complexity of each algorithm, but note that in these experiments, the implementation of the mini-element is not optimal because the interior bubbles are not eliminated.

$\nu = 0.02,\ \alpha = 0$, Taylor–Hood

h	n	err $\|\boldsymbol{u}\|_{L^2}$	err $\|p\|_{L^2}$	err $\|z\|_{L^2}$	err $\|\boldsymbol{u}\|_{H^1}$	Cpu s
0.2828	5	0.1075	0.1414	3.889	0.2938	0.5
0.1414	10	0.02321	0.02911	1.887	0.1453	1.0
0.0566	25	0.00352	0.0043	0.741	0.05795	6.1
0.0283	50	0.0008726	0.001061	0.3693	0.02896	28.1
0.0189	75	0.0003872	0.0004702	0.2461	0.0193	69.2
0.0141	100	0.0002177	0.0002642	0.1845	0.01448	124.3
0.0071	200	5.44e-05	6.6e-05	0.09224	0.007239	593.2

$\nu = 0.02,\ \alpha = 0.01$, Taylor–Hood

h	n	err $\|\boldsymbol{u}\|_{L^2}$	err $\|p\|_{L^2}$	err $\|z\|_{L^2}$	err $\|\boldsymbol{u}\|_{H^1}$	Cpu s
0.2828	5	0.1732	0.1428	6.289	0.2937	0.4
0.1414	10	0.03665	0.02917	3.001	0.1453	1.2
0.0566	25	0.005562	0.004301	1.178	0.05795	7.9
0.0283	50	0.001379	0.001061	0.587	0.02896	34.9
0.0189	75	0.0006121	0.0004702	0.3911	0.0193	86.2
0.0141	100	0.0003441	0.0002642	0.2933	0.01448	167.2
0.0071	200	8.598e-05	6.6e-05	0.1466	0.007239	758.1

$\nu = 0.02,\ \alpha = 0.1$, Taylor–Hood

h	n	err $\|\boldsymbol{u}\|_{L^2}$	err $\|p\|_{L^2}$	err $\|z\|_{L^2}$	err $\|\boldsymbol{u}\|_{H^1}$	Cpu s
0.2828	5	1.249	0.2611	34.86	0.318	1.0
0.1414	10	0.1637	0.03118	13.21	0.1454	3.5
0.0566	25	0.02428	0.004344	5.162	0.05795	20.8
0.0283	50	0.006018	0.001063	2.573	0.02896	95.3
0.0189	75	0.00267	0.0004707	1.714	0.0193	232.8
0.0141	100	0.001501	0.0002644	1.285	0.01448	441.7
0.0071	200	0.0003751	6.601e-05	0.6425	0.007239	1857.8

$\nu = 0.1,\ \alpha = 0$, Taylor–Hood

h	n	err $\|\boldsymbol{u}\|_{L^2}$	err $\|p\|_{L^2}$	err $\|z\|_{L^2}$	err $\|\boldsymbol{u}\|_{H^1}$	Cpu s
0.2828	5	0.1072	0.09946	3.878	0.295	0.3
0.1414	10	0.02321	0.02591	1.887	0.1454	0.9
0.0566	25	0.00352	0.004209	0.741	0.05795	6.0
0.0283	50	0.0008726	0.001055	0.3693	0.02896	26.7
0.0189	75	0.0003872	0.000469	0.2461	0.0193	62.8
0.0141	100	0.0002177	0.0002639	0.1845	0.01448	117.5
0.0071	200	5.44e-05	6.598e-05	0.09224	0.007239	561.3

$\nu = 0.1,\ \alpha = 0.01$, Taylor–Hood

h	n	err $\|\boldsymbol{u}\|_{L^2}$	err $\|p\|_{L^2}$	err $\|z\|_{L^2}$	err $\|\boldsymbol{u}\|_{H^1}$	Cpu s
0.2828	5	0.1687	0.09954	6.179	0.295	0.3
0.1414	10	0.03663	0.02591	3	0.1454	0.9
0.0566	25	0.005562	0.004209	1.178	0.05795	6.0
0.0283	50	0.001379	0.001055	0.587	0.02896	26.9
0.0189	75	0.0006121	0.000469	0.3911	0.0193	63.2
0.0141	100	0.0003441	0.0002639	0.2933	0.01448	118.7
0.0071	200	8.598e-05	6.598e-05	0.1466	0.007239	560.9

$\nu = 0.1,\ \alpha = 0.1$, Taylor–Hood

h	n	err $\|\boldsymbol{u}\|_{L^2}$	err $\|p\|_{L^2}$	err $\|z\|_{L^2}$	err $\|\boldsymbol{u}\|_{H^1}$	Cpu s
0.2828	5	0.7768	0.103	28.09	0.2955	0.9
0.1414	10	0.1597	0.026	13.15	0.1454	2.8
0.0566	25	0.02426	0.004211	5.161	0.05795	17.4
0.0283	50	0.006017	0.001055	2.572	0.02896	76.2
0.0189	75	0.00267	0.000469	1.714	0.0193	184.6
0.0141	100	0.001501	0.0002639	1.285	0.01448	349.7
0.0071	200	0.0003751	6.598e-05	0.6425	0.007239	1587.8

$\nu = 0.1,\ \alpha = 1$, Taylor–Hood

h	n	err $\|\boldsymbol{u}\|_{L^2}$	err $\|p\|_{L^2}$	err $\|z\|_{L^2}$	err $\|\boldsymbol{u}\|_{H^1}$	Cpu s
0.2828	5	22.86	0.5571	478.5	2.625	3.1
0.1414	10	1.527	0.0331	116.1	0.1906	9.8
0.0566	25	0.2125	0.004362	45.09	0.058	55.9
0.0283	50	0.05255	0.001064	22.47	0.02896	261.5
0.0189	75	0.02331	0.0004708	14.97	0.0193	616.3
0.0141	100	0.01311	0.0002644	11.23	0.01448	1170.7
0.0071	200	0.003275	6.601e-05	5.612	0.007239	4896.2

$\nu = 1,\ \alpha = 0$, Taylor–Hood

h	n	err $\|\boldsymbol{u}\|_{L^2}$	err $\|p\|_{L^2}$	err $\|z\|_{L^2}$	err $\|\boldsymbol{u}\|_{H^1}$	Cpu s
0.2828	5	0.1085	0.09734	3.877	0.3942	0.2
0.1414	10	0.02326	0.02577	1.887	0.1509	0.7
0.0566	25	0.00352	0.004205	0.741	0.05806	4.4
0.0283	50	0.0008726	0.001055	0.3693	0.02897	19.3
0.0189	75	0.0003872	0.000469	0.2461	0.01931	45.9
0.0141	100	0.0002177	0.0002638	0.1845	0.01448	86.2
0.0071	200	5.44e-05	6.597e-05	0.09224	0.007239	419.9

$\nu = 1,\ \alpha = 0.01$, Taylor–Hood

h	n	err $\|\boldsymbol{u}\|_{L^2}$	err $\|p\|_{L^2}$	err $\|z\|_{L^2}$	err $\|\boldsymbol{u}\|_{H^1}$	Cpu s
0.2828	5	0.1694	0.09734	6.171	0.3936	0.3
0.1414	10	0.03666	0.02577	3	0.1509	0.7
0.0566	25	0.005562	0.004205	1.178	0.05806	6.0
0.0283	50	0.001379	0.001055	0.587	0.02897	26.3
0.0189	75	0.0006121	0.000469	0.3911	0.01931	63.3
0.0141	100	0.0003441	0.0002638	0.2933	0.01448	118.0
0.0071	200	8.598e-05	6.597e-05	0.1466	0.007239	559.5

$\nu = 1,\ \alpha = 0.1$, Taylor–Hood						
h	n	err $\|\boldsymbol{u}\|_{L^2}$	err $\|p\|_{L^2}$	err $\|z\|_{L^2}$	err $\|\boldsymbol{u}\|_{H^1}$	Cpu s
0.2828	5	0.7299	0.09737	27.09	0.3889	0.7
0.1414	10	0.1594	0.02577	13.14	0.1509	2.5
0.0566	25	0.02426	0.004205	5.161	0.05806	12.5
0.0283	50	0.006017	0.001055	2.572	0.02897	54.8
0.0189	75	0.00267	0.000469	1.714	0.01931	130.1
0.0141	100	0.001501	0.0002638	1.285	0.01448	243.4
0.0071	200	0.0003751	6.597e-05	0.6425	0.007239	1303.8

$\nu = 1,\ \alpha = 1$, Taylor–Hood						
h	n	err $\|\boldsymbol{u}\|_{L^2}$	err $\|p\|_{L^2}$	err $\|z\|_{L^2}$	err $\|\boldsymbol{u}\|_{H^1}$	Cpu s
0.2828	5	6.752	0.1003	245.1	0.4396	2.4
0.1414	10	1.393	0.02585	114.9	0.1511	8.2
0.0566	25	0.2117	0.004206	45.08	0.05805	48.3
0.0283	50	0.05253	0.001055	22.47	0.02897	196.5
0.0189	75	0.02331	0.000469	14.97	0.01931	450.1
0.0141	100	0.01311	0.0002638	11.23	0.01448	807.6
0.0071	200	0.003275	6.597e-05	5.612	0.007239	3370.3

$\nu = 0.02,\ \alpha = 0$, Mini-element						
h	n	err $\|\boldsymbol{u}\|_{L^2}$	err $\|p\|_{L^2}$	err $\|z\|_{L^2}$	err $\|\boldsymbol{u}\|_{H^1}$	Cpu s
0.2828	5	0.2774	0.5072	3.926	0.3059	0.3
0.1414	10	0.08504	0.2353	2.336	0.1601	0.7
0.0566	25	0.01769	0.09105	1.166	0.07032	4.5
0.0283	50	0.005687	0.04518	0.7287	0.04022	18.9
0.0189	75	0.002985	0.03005	0.5665	0.02986	45.1
0.0141	100	0.001902	0.02252	0.4777	0.02448	81.1
0.0071	200	0.0006521	0.01124	0.3237	0.01573	359.8

$\nu = 0.02,\ \alpha = 0.01$, Mini-element						
h	n	err $\|\boldsymbol{u}\|_{L^2}$	err $\|p\|_{L^2}$	err $\|z\|_{L^2}$	err $\|\boldsymbol{u}\|_{H^1}$	Cpu s
0.2828	5	0.304	0.5077	5.74	0.3075	0.3
0.1414	10	0.08833	0.2353	3.208	0.1602	1.0
0.0566	25	0.01801	0.09106	1.474	0.07031	5.7
0.0283	50	0.005745	0.04518	0.8587	0.04022	24.0
0.0189	75	0.003006	0.03005	0.6427	0.02986	57.5
0.0141	100	0.001912	0.02252	0.5294	0.02448	103.8
0.0071	200	0.000654	0.01124	0.3433	0.01573	458.6

$\nu = 0.02,\ \alpha = 0.1$, Mini-element						
h	n	err $\|\boldsymbol{u}\|_{L^2}$	err $\|p\|_{L^2}$	err $\|z\|_{L^2}$	err $\|\boldsymbol{u}\|_{H^1}$	Cpu s
0.2828	5	2.014	0.545	40.21	0.4039	0.8
0.1414	10	0.4518	0.2372	15.17	0.1785	2.9
0.0566	25	0.06967	0.09116	5.371	0.07113	16.5
0.0283	50	0.01758	0.04519	2.672	0.04027	65.0
0.0189	75	0.007961	0.03006	1.797	0.02987	153.6
0.0141	100	0.004562	0.02252	1.362	0.02448	283.2
0.0071	200	0.001223	0.01124	0.7122	0.01573	1221.0

$\nu = 0.1,\ \alpha = 0$, Mini-element						
h	n	err $\|\boldsymbol{u}\|_{L^2}$	err $\|p\|_{L^2}$	err $\|z\|_{L^2}$	err $\|\boldsymbol{u}\|_{H^1}$	Cpu s
0.2828	5	0.2794	0.4409	3.931	0.5203	0.3
0.1414	10	0.08646	0.227	2.339	0.3625	0.7
0.0566	25	0.01796	0.09052	1.165	0.2071	4.5
0.0283	50	0.00577	0.04511	0.7286	0.1425	18.6
0.0189	75	0.003028	0.03003	0.5664	0.1155	43.9
0.0141	100	0.001929	0.02251	0.4777	0.09973	79.1
0.0071	200	0.0006613	0.01124	0.3237	0.07018	350.7

$\nu = 0.1,\ \alpha = 0.01$, Mini-element						
h	n	err $\|\boldsymbol{u}\|_{L^2}$	err $\|p\|_{L^2}$	err $\|z\|_{L^2}$	err $\|\boldsymbol{u}\|_{H^1}$	Cpu s
0.2828	5	0.3074	0.4409	5.739	0.521	0.2
0.1414	10	0.09052	0.227	3.217	0.3623	0.7
0.0566	25	0.0184	0.09052	1.474	0.2071	4.5
0.0283	50	0.005853	0.04511	0.8582	0.1425	18.8
0.0189	75	0.003059	0.03003	0.6423	0.1155	44.2
0.0141	100	0.001944	0.02251	0.5291	0.09972	79.8
0.0071	200	0.0006641	0.01124	0.3432	0.07018	350.5

$\nu = 0.1,\ \alpha = 0.1$, Mini-element						
h	n	err $\|\boldsymbol{u}\|_{L^2}$	err $\|p\|_{L^2}$	err $\|z\|_{L^2}$	err $\|\boldsymbol{u}\|_{H^1}$	Cpu s
0.2828	5	0.8983	0.4422	27.16	0.5374	0.7
0.1414	10	0.2132	0.2272	13.1	0.3623	2.1
0.0566	25	0.03495	0.09053	5.218	0.2069	13.0
0.0283	50	0.009405	0.04511	2.641	0.1424	53.9
0.0189	75	0.004472	0.03003	1.785	0.1155	138.2
0.0141	100	0.002672	0.02251	1.356	0.09971	252.1
0.0071	200	0.0008062	0.01124	0.7125	0.07018	1099.5

$\nu = 0.1,\ \alpha = 1$, Mini-element						
h	n	err $\|\boldsymbol{u}\|_{L^2}$	err $\|p\|_{L^2}$	err $\|z\|_{L^2}$	err $\|\boldsymbol{u}\|_{H^1}$	Cpu s
0.2828	5	43.18	0.6873	709.5	3.234	2.7
0.1414	10	6.188	0.2352	173.7	1.041	8.3
0.0566	25	1.01	0.09093	54.77	0.2561	46.8
0.0283	50	0.2488	0.04516	24.29	0.1467	190.9
0.0189	75	0.1102	0.03005	15.69	0.1164	414.7
0.0141	100	0.06189	0.02252	11.61	0.1	770.7
0.0071	200	0.01544	0.01124	5.693	0.07019	3400.1

$\nu = 1,\ \alpha = 0$, Mini-element						
h	n	err $\|\boldsymbol{u}\|_{L^2}$	err $\|p\|_{L^2}$	err $\|z\|_{L^2}$	err $\|\boldsymbol{u}\|_{H^1}$	Cpu s
0.2828	5	0.4039	0.4379	3.934	4.287	0.2
0.1414	10	0.1391	0.2267	2.34	3.32	0.5
0.0566	25	0.03111	0.0905	1.165	1.989	3.3
0.0283	50	0.01049	0.04511	0.7286	1.395	13.7
0.0189	75	0.005623	0.03003	0.5664	1.139	31.6
0.0141	100	0.003624	0.02251	0.4777	0.9868	58.0
0.0071	200	0.001267	0.01124	0.3237	0.6981	259.6

$\nu = 1,\ \alpha = 0.01$, Mini-element						
h	n	err $\|\boldsymbol{u}\|_{L^2}$	err $\|p\|_{L^2}$	err $\|z\|_{L^2}$	err $\|\boldsymbol{u}\|_{H^1}$	Cpu s
0.2828	5	0.4242	0.4379	5.755	4.287	0.2
0.1414	10	0.1419	0.2267	3.22	3.32	0.5
0.0566	25	0.0314	0.0905	1.473	1.989	4.5
0.0283	50	0.01054	0.04511	0.858	1.395	18.6
0.0189	75	0.005643	0.03003	0.6422	1.139	43.7
0.0141	100	0.003634	0.02251	0.529	0.9868	79.6
0.0071	200	0.001268	0.01124	0.3432	0.6981	350.4

$\nu = 1,\ \alpha = 0.1$, Mini-element						
h	n	err $\|\boldsymbol{u}\|_{L^2}$	err $\|p\|_{L^2}$	err $\|z\|_{L^2}$	err $\|\boldsymbol{u}\|_{H^1}$	Cpu s
0.2828	5	0.8252	0.4379	26.07	4.282	0.6
0.1414	10	0.2096	0.2267	13	3.317	1.9
0.0566	25	0.03918	0.0905	5.213	1.989	11.7
0.0283	50	0.01204	0.04511	2.642	1.395	49.0
0.0189	75	0.006204	0.03003	1.786	1.139	114.0
0.0141	100	0.003912	0.02251	1.357	0.9868	206.8
0.0071	200	0.001319	0.01124	0.7131	0.6981	1002.0

$\nu = 1,\ \alpha = 1$, Mini-element						
h	n	err $\Vert\boldsymbol{u}\Vert_{L^2}$	err $\Vert p\Vert_{L^2}$	err $\Vert z\Vert_{L^2}$	err $\vert\boldsymbol{u}\vert_{H^1}$	Cpu s
0.2828	5	7.719	0.4391	246.7	4.363	1.9
0.1414	10	1.812	0.2268	116.1	3.304	6.5
0.0566	25	0.2822	0.0905	45.16	1.986	38.1
0.0283	50	0.07047	0.04511	22.48	1.395	159.4
0.0189	75	0.03142	0.03003	14.98	1.139	371.3
0.0141	100	0.01775	0.02251	11.23	0.9867	674.1
0.0071	200	0.004519	0.01124	5.619	0.6981	2954.5

$\nu = 0.02,\ \alpha = 0$, Bernardi–Raugel						
h	n	err $\Vert\boldsymbol{u}\Vert_{L^2}$	err $\Vert p\Vert_{L^2}$	err $\Vert z\Vert_{L^2}$	err $\vert\boldsymbol{u}\vert_{H^1}$	Cpu s
0.2828	5	0.1442	0.2576	3.913	0.3177	0.4
0.1414	10	0.03932	0.1179	2.028	0.1533	0.8
0.0566	25	0.008685	0.04538	0.9029	0.06338	5.0
0.0283	50	0.002943	0.02245	0.5191	0.03396	21.8
0.0189	75	0.001579	0.01492	0.3855	0.0241	51.3
0.0141	100	0.001018	0.01117	0.3159	0.01911	95.3
0.0071	200	0.0003562	0.005573	0.2027	0.01139	434.6

$\nu = 0.02,\ \alpha = 0.01$, Bernardi–Raugel						
h	n	err $\Vert\boldsymbol{u}\Vert_{L^2}$	err $\Vert p\Vert_{L^2}$	err $\Vert z\Vert_{L^2}$	err $\vert\boldsymbol{u}\vert_{H^1}$	Cpu s
0.2828	5	0.2014	0.2582	6.251	0.318	0.3
0.1414	10	0.04785	0.118	3.061	0.1533	1.0
0.0566	25	0.00957	0.04538	1.279	0.06336	6.4
0.0283	50	0.003106	0.02245	0.6889	0.03396	28.8
0.0189	75	0.001639	0.01492	0.4898	0.02409	69.5
0.0141	100	0.001048	0.01117	0.3888	0.0191	126.1
0.0071	200	0.0003616	0.005573	0.2322	0.01139	575.5

$\nu = 0.02,\ \alpha = 0.1$, Bernardi–Raugel						
h	n	err $\Vert\boldsymbol{u}\Vert_{L^2}$	err $\Vert p\Vert_{L^2}$	err $\Vert z\Vert_{L^2}$	err $\vert\boldsymbol{u}\vert_{H^1}$	Cpu s
0.2828	5	1.417	0.3108	35	0.3813	0.9
0.1414	10	0.2055	0.1185	13.49	0.1537	3.0
0.0566	25	0.03163	0.04539	5.204	0.06331	18.0
0.0283	50	0.008131	0.02245	2.598	0.03393	77.3
0.0189	75	0.003723	0.01492	1.738	0.02408	188.9
0.0141	100	0.002152	0.01117	1.309	0.0191	338.0
0.0071	200	0.0005927	0.005573	0.6658	0.01139	1544.5

$\nu = 0.1,\ \alpha = 0$, Bernardi–Raugel						
h	n	err $\|\boldsymbol{u}\|_{L^2}$	err $\|p\|_{L^2}$	err $\|z\|_{L^2}$	err $\|\boldsymbol{u}\|_{H^1}$	Cpu s
0.2828	5	0.1461	0.253	3.866	0.6368	0.3
0.1414	10	0.04013	0.1178	2.025	0.2845	0.8
0.0566	25	0.008835	0.04538	0.9022	0.1407	5.1
0.0283	50	0.002994	0.02245	0.519	0.09329	22.3
0.0189	75	0.001607	0.01492	0.3855	0.07463	53.1
0.0141	100	0.001036	0.01117	0.3159	0.06398	99.9
0.0071	200	0.0003624	0.005573	0.2027	0.04457	448.5

$\nu = 0.1,\ \alpha = 0.01$, Bernardi–Raugel						
h	n	err $\|\boldsymbol{u}\|_{L^2}$	err $\|p\|_{L^2}$	err $\|z\|_{L^2}$	err $\|\boldsymbol{u}\|_{H^1}$	Cpu s
0.2828	5	0.1966	0.253	6.075	0.636	0.2
0.1414	10	0.04901	0.1178	3.061	0.2843	0.8
0.0566	25	0.009804	0.04538	1.281	0.1407	5.0
0.0283	50	0.003174	0.02245	0.6899	0.09329	22.0
0.0189	75	0.001673	0.01492	0.4904	0.07463	51.6
0.0141	100	0.001069	0.01117	0.3892	0.06398	95.4
0.0071	200	0.0003683	0.005573	0.2324	0.04457	433.2

$\nu = 0.1,\ \alpha = 0.1$, Bernardi–Raugel						
h	n	err $\|\boldsymbol{u}\|_{L^2}$	err $\|p\|_{L^2}$	err $\|z\|_{L^2}$	err $\|\boldsymbol{u}\|_{H^1}$	Cpu s
0.2828	5	0.792	0.2537	27.82	0.6274	0.7
0.1414	10	0.1654	0.1178	13.13	0.2829	2.3
0.0566	25	0.0258	0.04538	5.174	0.1405	14.5
0.0283	50	0.006708	0.02245	2.594	0.09325	63.2
0.0189	75	0.00311	0.01492	1.737	0.07461	162.1
0.0141	100	0.00182	0.01117	1.309	0.06398	300.8
0.0071	200	0.0005207	0.005573	0.6667	0.04457	1403.8

$\nu = 0.1,\ \alpha = 1$, Bernardi–Raugel						
h	n	err $\|\boldsymbol{u}\|_{L^2}$	err $\|p\|_{L^2}$	err $\|z\|_{L^2}$	err $\|\boldsymbol{u}\|_{H^1}$	Cpu s
0.2828	5	30.69	0.5207	537.9	3.387	2.8
0.1414	10	2.353	0.12	124.7	0.3687	8.6
0.0566	25	0.3905	0.04541	47.22	0.1414	48.5
0.0283	50	0.09911	0.02246	22.91	0.09304	204.5
0.0189	75	0.04436	0.01492	15.14	0.07449	481.4
0.0141	100	0.02504	0.01117	11.32	0.0639	884.6
0.0071	200	0.006292	0.005573	5.631	0.04456	4033.8

$\nu = 1,\ \alpha = 0$, Bernardi–Raugel						
h	n	err $\|\boldsymbol{u}\|_{L^2}$	err $\|p\|_{L^2}$	err $\|z\|_{L^2}$	err $\|\boldsymbol{u}\|_{H^1}$	Cpu s
0.2828	5	0.3387	0.2528	3.857	5.647	0.2
0.1414	10	0.08436	0.1177	2.024	2.455	0.6
0.0566	25	0.01819	0.04538	0.9021	1.284	3.7
0.0283	50	0.006236	0.02245	0.5189	0.8873	16.5
0.0189	75	0.003365	0.01492	0.3855	0.7211	38.7
0.0141	100	0.002177	0.01117	0.3158	0.6234	72.9
0.0071	200	0.0007653	0.005573	0.2027	0.4399	345.4

$\nu = 1,\ \alpha = 0.01$, Bernardi–Raugel						
h	n	err $\|\boldsymbol{u}\|_{L^2}$	err $\|p\|_{L^2}$	err $\|z\|_{L^2}$	err $\|\boldsymbol{u}\|_{H^1}$	Cpu s
0.2828	5	0.3628	0.2528	6.047	5.646	0.2
0.1414	10	0.08898	0.1177	3.06	2.455	0.6
0.0566	25	0.01869	0.04538	1.281	1.284	5.2
0.0283	50	0.006326	0.02245	0.69	0.8873	22.8
0.0189	75	0.003399	0.01492	0.4905	0.7211	55.3
0.0141	100	0.002193	0.01117	0.3893	0.6234	101.6
0.0071	200	0.0007682	0.005573	0.2325	0.4399	478.1

$\nu = 1,\ \alpha = 0.1$, Bernardi–Raugel						
h	n	err $\|\boldsymbol{u}\|_{L^2}$	err $\|p\|_{L^2}$	err $\|z\|_{L^2}$	err $\|\boldsymbol{u}\|_{H^1}$	Cpu s
0.2828	5	0.7977	0.2528	26.84	5.63	0.6
0.1414	10	0.1787	0.1177	13.09	2.453	2.2
0.0566	25	0.03008	0.04538	5.176	1.284	13.5
0.0283	50	0.008614	0.02245	2.595	0.8872	60.3
0.0189	75	0.004276	0.01492	1.738	0.7211	142.6
0.0141	100	0.002634	0.01117	1.31	0.6234	264.7
0.0071	200	0.0008502	0.005573	0.6671	0.4399	1165.9

$\nu = 1,\ \alpha = 1$, Bernardi–Raugel						
h	n	err $\|\boldsymbol{u}\|_{L^2}$	err $\|p\|_{L^2}$	err $\|z\|_{L^2}$	err $\|\boldsymbol{u}\|_{H^1}$	Cpu s
0.2828	5	6.797	0.2533	243.8	5.495	2.1
0.1414	10	1.433	0.1178	115	2.435	7.0
0.0566	25	0.2186	0.04538	45.09	1.282	43.3
0.0283	50	0.05441	0.02245	22.47	0.8869	190.5
0.0189	75	0.02422	0.01492	14.97	0.721	439.6
0.0141	100	0.01366	0.01117	11.23	0.6233	807.6
0.0071	200	0.003455	0.005573	5.614	0.4399	3565.2

These experiments call for the following comments.

1. From a computational point of view, the convergence of Newton's algorithm deteriorates sharply with increasing values of α with respect to ν. We have omitted tables corresponding to $\nu = 0.02$ and $\alpha = 1$ for Taylor–Hood's method, $\nu = 0.005$ and $\alpha = 0.1$, $\nu = 0.005$ and $\alpha = 1$, $\nu = 0.01$ and $\alpha = 1$, $\nu = 0.02$ and $\alpha = 1$ for the mini-element method, and $\nu = 0.02$ and $\alpha = 1$ for the Bernardi–Raugel method because Newton's method clearly does not converge for these values.
2. Even if the error is large, what can be observed from the asymptotic behavior of the error is consistent with the order predicted by the theory, and sometimes it is better. In particular, the velocity error in the L^2 norm seems to be systematically one order larger than the order of the error in the H^1 norm. But proving this fact is an open problem because a duality argument on the auxiliary variable z is difficult. Similarly, the pressure error for the mini-element is sometimes better than $O(h)$ and so is the error for the auxiliary variable z in the L^2 norm. This can also be due to some superconvergence behavior.
3. The auxiliary variable's error in L^2 is usually larger than the velocity's error in H^1. This is not surprizing when α is large. When α is small, this discrepancy is due to the ill-conditioning of the matrix.
4. The Bernardi–Raugel element is harder to implement than the mini-element, and its CPU time is slightly larger, but, at least in these examples, its results are also more accurate.

6.1.2. Benchmark problems

The driven cavity

The reader can compare the tests below with those performed in [GLOWINSKI, 2003, §IX.44] for the numerical simulation of an incompressible Navier–Stokes flow in a square cavity.

In the following tests, the domain Ω is the unit square $]0,1[\times]0,1[$. The boundary conditions for $\boldsymbol{u}$ are: $\boldsymbol{u} = (0,0)$ on the vertical sides and bottom horizontal side of Ω, i.e., $x_1 = 0$, $x_1 = 1$, $x_2 = 0$, and

$$\boldsymbol{u} = (4x_1(1-x_1), 0) \text{ on the top horizontal side } x_2 = 1.$$

Note that on one hand this is a regularized driven cavity because $\boldsymbol{u}$ is continuous on the boundary of Ω, and on the other hand $\boldsymbol{u}\cdot\boldsymbol{n} = 0$. The nonlinear scheme (6.1.1)–(6.1.3), applied with Taylor–Hood finite-elements (2.2.14), (2.2.15), is solved by Newton's method with continuation on ν: α varies from 0 to 1000 and ν varies from 0.001 to 1000. The following Figures 6.7–6.16 depict the isovalues of the auxiliary variable z_h, the pressure p_h, and the Euclidean norm of the velocity vector $\boldsymbol{u}_h$ at different Reynolds numbers and for different values of α. We observe some oscillations in z_h, as expected considering that the equation for z_h has no upwinding. We also observe that p_h does not behave like a pressure. The reason is that the variable p_h approximates the modified pressure p, which is related to the fluid's pressure π by (see Section 1.2)

$$p = \pi - \alpha\left(\boldsymbol{u}\cdot\Delta\boldsymbol{u} + \frac{1}{4}|\boldsymbol{A}_1|^2\right) + \frac{1}{2}|\boldsymbol{u}|^2.$$

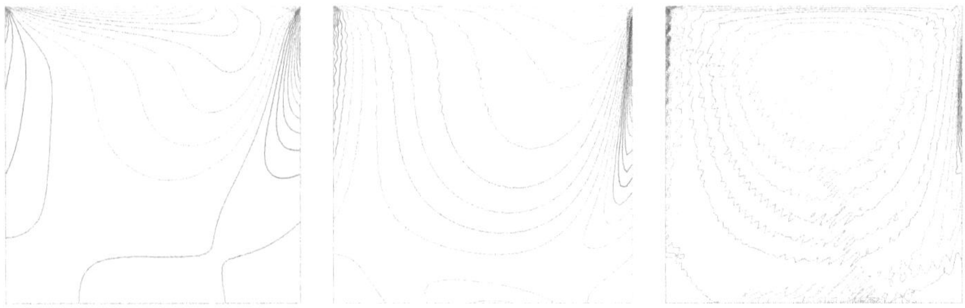

FIG. 6.7 Isovalues of z_h at Reynolds number 100, for $\alpha = 0.001, 0.01, 0.1$.

FIG. 6.8 Isovalues of p_h at Reynolds number 100, for $\alpha = 0.001, 0.01, 0.1$.

FIG. 6.9 Isovalues of $|\boldsymbol{u}_h|$ at Reynolds number 100, for $\alpha = 0.001, 0.01, 0.1$.

The backward step

The reader can compare the tests below see Figures 6.17–6.24 with those performed in [GLOWINSKI, 2003, §IX.44] for the numerical simulation of an incompressible Navier–Stokes flow in a two-dimensional channel with a backward facing step.

In the following tests, the domain Ω is obtained by deleting the rectangle $[-4, 0] \times [-1/2, 0]$ from the rectangular region $]-4, 13[\times]-1/2, 1[$. The boundary Γ is split into four parts: Γ_i the left part (inlet), Γ_o the right part (outlet), Γ_u the upper part, and Γ_b the bottom part. In addition to the velocity and pressure, we compute an approximation ψ_h of the stream function ψ defined by $\boldsymbol{u} = \mathbf{curl}\, \psi$ (see [GIRAULT and RAVIART, 1986, § 5.2]). When the flow is stationary, the streamlines for the flow are the isovalues of the stream function ψ.

FIG. 6.10 Isovalues of z_h at Reynolds number 400, for $\alpha = 0.001, 0.01, 0.1$.

FIG. 6.11 Isovalues of p_h at Reynolds number 400, for $\alpha = 0.001, 0.01, 0.1$.

FIG. 6.12 Isovalues of $|\boldsymbol{u}_h|$ at Reynolds number 400, for $\alpha = 0.001, 0.01, 0.1$.

The discrete stream function ψ_h belongs to the finite-element space:

$$\Phi_h = \{\varphi \in H^1(\Omega);\ \forall T \in \mathcal{T}_h,\ \varphi_{|T} \in \mathbb{P}_2\},$$

and solves the problem: Find $\psi_h \in \Phi_h$ such that

$$\begin{aligned}
(\operatorname{curl}\psi_h, \operatorname{curl}\varphi_h) &= (\operatorname{curl}\boldsymbol{u}_h, \varphi_h) && \forall \varphi_h \in \Phi_{0h}\,,\\
\psi_h &= 0 && \text{on} \quad \Gamma_b,\\
\psi_h &= -\int_{\Gamma_i} \boldsymbol{u}_h \cdot \boldsymbol{n}\, ds && \text{on} \quad \Gamma_u\,,
\end{aligned} \tag{6.1.10}$$

where $\Phi_{0h} = \{\varphi_h \in \Phi_h;\ \varphi_{h|\Gamma_b \cup \Gamma_u} = 0\}$.

FIG. 6.13 Isovalues of z_h at Reynolds number 1000, for $\alpha = 0.001, 0.01, 0.1$.

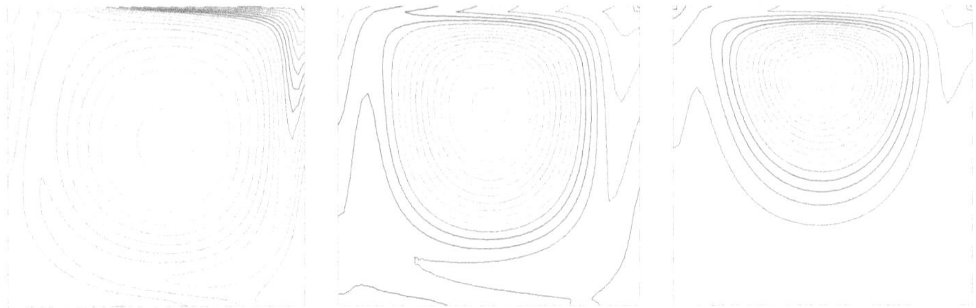

FIG. 6.14 Isovalues of p_h at Reynolds number 1000, for $\alpha = 0.001, 0.01, 0.1$.

FIG. 6.15 Isovalues of $|\boldsymbol{u}_h|$ at Reynolds number 1000, for $\alpha = 0.001, 0.01, 0.1$.

6.2. The time-dependent case

In these tests, see Figures 6.25–6.36 (compare with the numerical simulation of an incompressible Navier–Stokes flow past a cylinder in [GLOWINSKI, 2003, §X.53.7]) we take $\boldsymbol{f} = \boldsymbol{0}$, and the domain Ω is a rectangle with a circular hole: $\Omega =]-4, 17[\times]-4, 4[\setminus\{\boldsymbol{x} \in \mathbb{R}^2;\ |\boldsymbol{x}| \le \frac{1}{2}\}$. In the sequel, the boundary of the rectangle is denoted by Γ_r, the hole is named cylinder and its boundary is denoted by Γ_c. The boundary conditions for $\boldsymbol{u}$ are as follows:

- $\boldsymbol{u} = (1, 0)$ on the segments of Γ_r, i.e., $\boldsymbol{x} = (x, y)$ with $-4 \le x \le 17$ and $y = \pm 4$, or $-4 \le y \le 4$ and $x = -4$, $x = 17$,

FIG. 6.16 Isovalues of $z_h, p_h, |\boldsymbol{u}_h|$ at Reynolds number 1000, for $\alpha = 0.001$ (very close to the solution of the classical Navier–Stokes equation).

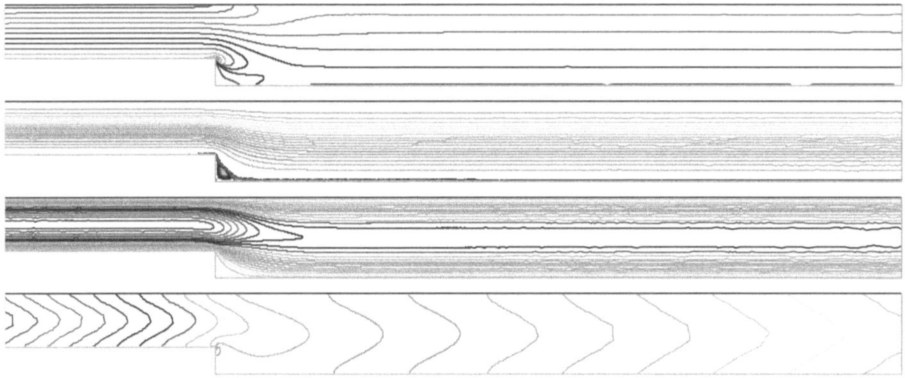

FIG. 6.17 Isovalues of $z_h, \psi_h, p_h, |\boldsymbol{u}_h|$ at Reynolds number 10, for $\alpha = 0.1$.

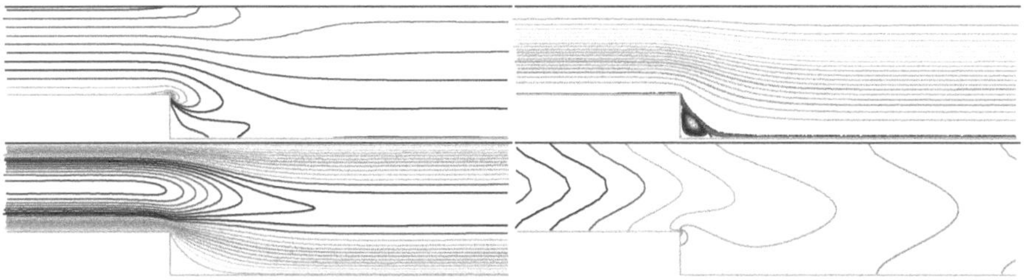

FIG. 6.18 Zoom around the step of the isovalues of $z_h, \psi_h, p_h, |\boldsymbol{u}_h|$ at Reynolds number 10, for $\alpha = 0.1$.

- $\boldsymbol{u} = u_{c(t)}(-y, x)$ on the circle Γ_c, i.e., on $|\boldsymbol{x}| = \frac{1}{2}$, where $u_{c(t)}$ is the angular velocity of the cylinder at time t. This angular velocity is used to initialize the Von Kármán vortex street.

We approximate problem (3.2.1), (3.2.2) with the linear backward Euler scheme defined by (3.3.2), (3.3.3), (3.3.4), with the initial condition (3.3.1). Note, however, that since the boundary data are not tangential on the inflow and outflow part of Γ_r, the space for the unknown velocity $\boldsymbol{u}_h^n$ is a variant, say $\bar{X}_h$, of X_h, and similarly, the operator approximating

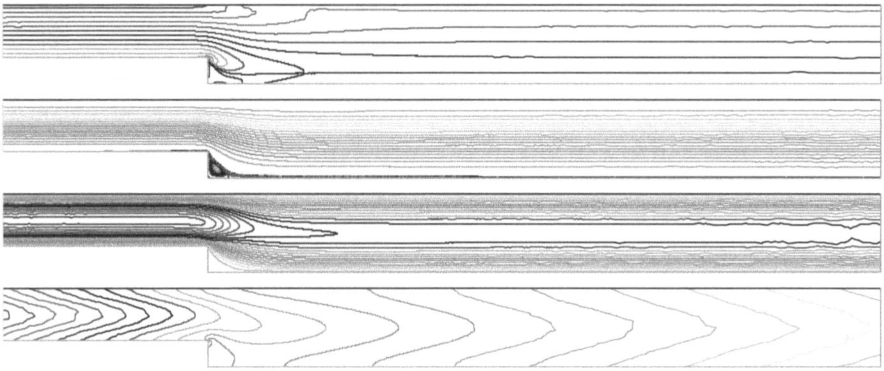

FIG. 6.19 Isovalues of z_h, ψ_h, p_h, $|\boldsymbol{u}_h|$ at Reynolds number 50, for $\alpha = 0.1$.

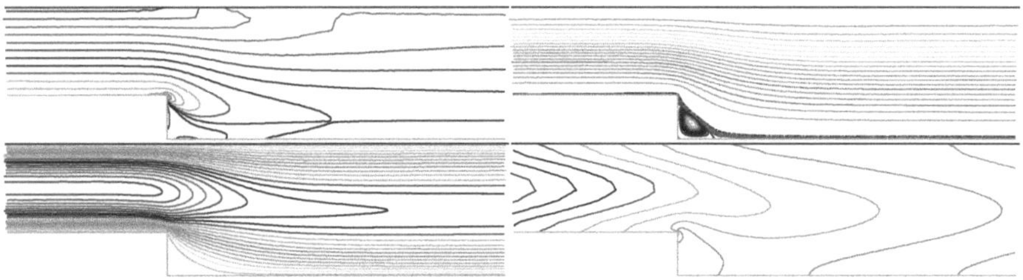

FIG. 6.20 Zoom around the step of the isovalues of z_h, ψ_h, p_h, $|\boldsymbol{u}_h|$ at Reynolds number 50, for $\alpha = 0.1$.

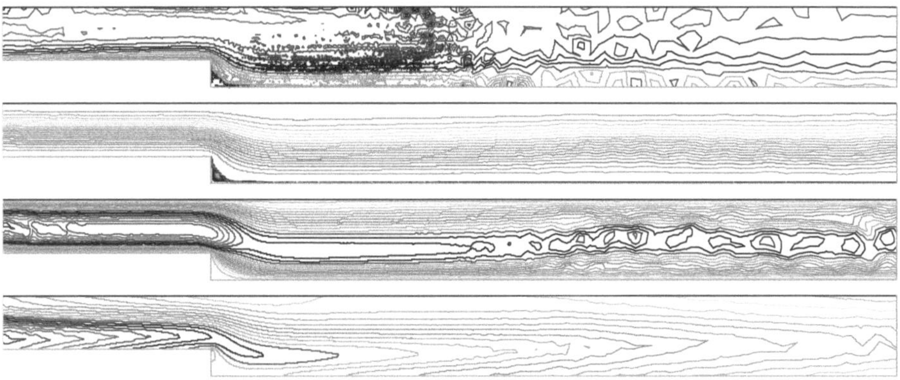

FIG. 6.21 Isovalues of z_h, ψ_h, p_h, $|\boldsymbol{u}_h|$ at Reynolds number 113, for $\alpha = 0.1$.

the initial velocity $\boldsymbol{u}_h^0$ is a variant $\bar{P}_h$ of P_h. Of course, the test functions $\boldsymbol{v}_h$ must belong to X_h. With these straightforward modifications, we solve:

• Initial step

$$\boldsymbol{u}_h^0 = \bar{P}_h(\boldsymbol{u}_0),\ z_h^0 = R_h(z_0),\ \boldsymbol{z}_h^0 = (0, 0, z_h^0),$$

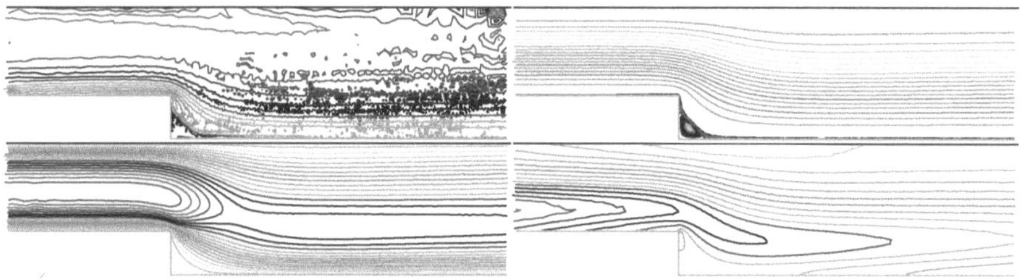

FIG. 6.22 Zoom around the step of the isovalues of z_h, ψ_h, p_h, $|\boldsymbol{u}_h|$ at Reynolds number 113, for $\alpha = 0.1$.

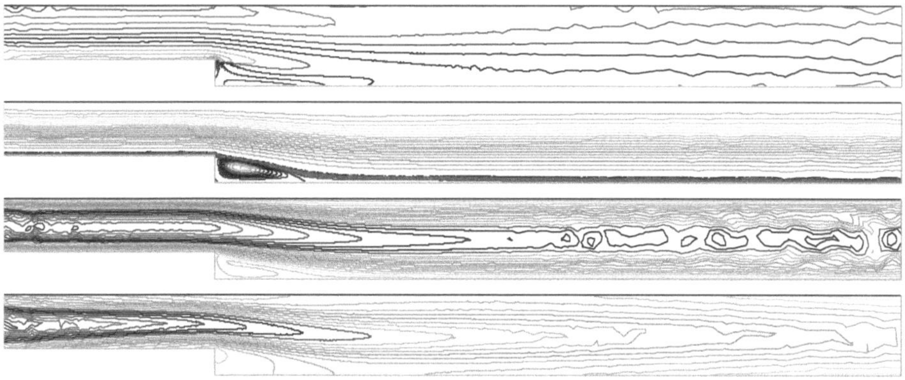

FIG. 6.23 Isovalues of z_h, ψ_h, p_h, $|\boldsymbol{u}_h|$ at Reynolds number 150, for $\alpha = 0.1$.

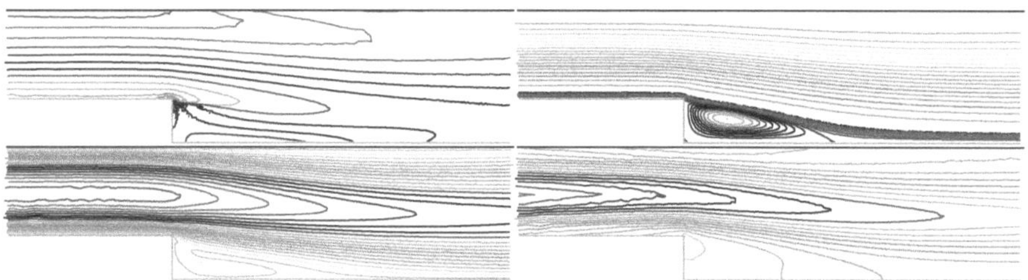

FIG. 6.24 Zoom around the step of the isovalues of z_h, ψ_h, p_h, $|\boldsymbol{u}_h|$ at Reynolds number 150, for $\alpha = 0.1$.

- Knowing $\boldsymbol{u}_h^0 \in \bar{X}_h$ and $z_h^0 \in Z_h$, find sequences $(\boldsymbol{u}_h^n)_{n\geq 1}$, $(z_h^n)_{n\geq 1}$, and $(p_h^n)_{n\geq 1}$ such that $\boldsymbol{u}_h^n \in \bar{X}_h$, $z_h^n \in Z_h$, and $p_h^n \in M_h$ solve for $1 \leq n \leq N$,

$$\forall \boldsymbol{v}_h \in X_h,\ \frac{1}{k}(\boldsymbol{u}_h^n - \boldsymbol{u}_h^{n-1}, \boldsymbol{v}_h) + \frac{\alpha}{k}(\nabla(\boldsymbol{u}_h^n - \boldsymbol{u}_h^{n-1}), \nabla \boldsymbol{v}_h) + \nu(\nabla \boldsymbol{u}_h^n, \nabla \boldsymbol{v}_h)$$
$$+ (\boldsymbol{z}_h^n \times \boldsymbol{u}_h^n, \boldsymbol{v}_h) - (p_h^n, \operatorname{div} \boldsymbol{v}_h) = (\boldsymbol{f}^n, \boldsymbol{v}_h),$$
$$\forall q_h \in M_h,\ (q_h, \operatorname{div} \boldsymbol{u}_h^n) = 0,$$
$$\forall \theta_h \in Z_h,\ \frac{\alpha}{k}(z_h^n - z_h^{n-1}, \theta_h) + \nu(z_h^n, \theta_h) + \alpha\tilde{c}(\boldsymbol{u}_h^n; z_h^n, \theta_h) = \nu(\operatorname{curl} \boldsymbol{u}_h^n, \theta_h)$$
$$+ \alpha(\operatorname{curl} \boldsymbol{f}^n, \theta_h),$$

where $\boldsymbol{f}^n$ is defined by (3.2.18):

$$\boldsymbol{f}^n(\boldsymbol{x}) = \frac{1}{k}\int_{t_{n-1}}^{t_n} \boldsymbol{f}(\boldsymbol{x}, s)\,\mathrm{d}s,$$

so that in our experiments, $\boldsymbol{f}^n = \mathbf{0}$.

In order to generate an unsteady solution, we start with a solution of the steady grade-two fluid that has been computed with the methods of Section 6.1 for $\nu = 0.01$ and $\alpha = 0.005$. We prescribe on the cylinder the following oscillatory rotation

$$u_c(t) = \begin{cases} \cos(\frac{2\pi}{5.2}t) & \text{if } 0 \le t \le \frac{3\pi}{10.4} \\ 0 & \text{if } t > \frac{3\pi}{10.4} \end{cases}$$

during three-fourth of the 5.2 period (corresponding to a classical Strouhal number for this flow), with 100 time steps per period so that $k = 0.052$, and we stop the computation just after time $t = 100$ (that is 1998 time steps), when the flow becomes periodic in time.

In Figures 6.25–6.30, we plot the drag and lift on the cylinder, and the velocity $\boldsymbol{u}^n(\boldsymbol{a}) = (u_1^n(\boldsymbol{a}), u_2^n(\boldsymbol{a}))$ at time step n and point $\boldsymbol{a} = (0, 2)$. The drag and lift are the two components of the resulting force on the cylinder:

$$\boldsymbol{F}^n = \int_{\Gamma_c} \boldsymbol{T}(\boldsymbol{u}^n, \pi^n)\cdot \boldsymbol{n}\mathrm{d}\sigma.$$

By applying Green's formula to the discrete analog of the normal stress on the body, these components are given, for $i = 1, 2$, by

$$F_i^n = -\nu(\nabla \boldsymbol{u}_h^n, \nabla \boldsymbol{v}_i) - (z_h^n \times \boldsymbol{u}_h^n, \boldsymbol{v}_i) - (p_h^n, \operatorname{div} \boldsymbol{v}_i), \tag{6.2.1}$$

where the two functions $\boldsymbol{v}_1$ and $\boldsymbol{v}_2$ belong to $\bar{X}_h$, vanish on Γ_r and are unit vectors on Γ_c, $\boldsymbol{v}_1 = (1, 0)$, $\boldsymbol{v}_2 = (0, 1)$.

We also compute an approximation of the stream function ψ, i.e., such that $\boldsymbol{u} = \mathbf{curl}\,\psi$. In the present case, the streamlines of the flow are not the isovalues of the stream function ψ, but these isovalues give some information on the flow. The discrete stream function ψ_h is the solution of problem (6.1.10), with other constant Dirichlet boundary conditions on Γ_c and on the top and bottom parts of Γ_r, adjusted so that the flux between these three boundaries is compatible with the flow.

Because the solutions at time steps 1914, 1939, and 1964 correspond respectively to extreme values and to the point of inflection of the drag F_2, we present the solution at these time steps in Figs 6.34, 6.35, 6.36, and an enlarged view in Figs 6.31, 6.32, 6.33.

REMARK 6.2.1. (1) A large number of tests had to be performed in order to obtain a really unsteady solution because the parameters had to be carefully tuned. For example, if $\alpha = 0.01$ and $\nu = 0.01$, the steady state occurs at time 50. Another example, if the time step is too large, for instance $k \ge 0.1$, we observe a continuous decay of the lift and drag variation in time.

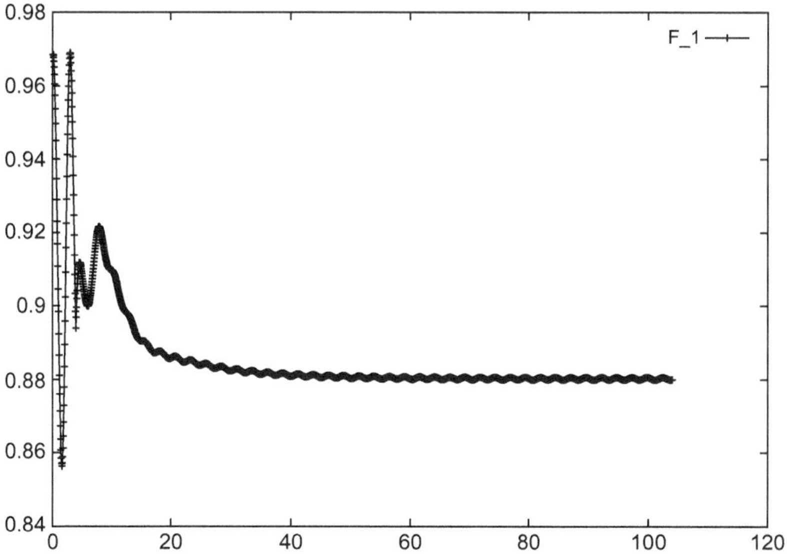

FIG. 6.25 Drag F_1 versus time, $\nu = 0.01$, $\alpha = 0.005$, $k = 0.052$.

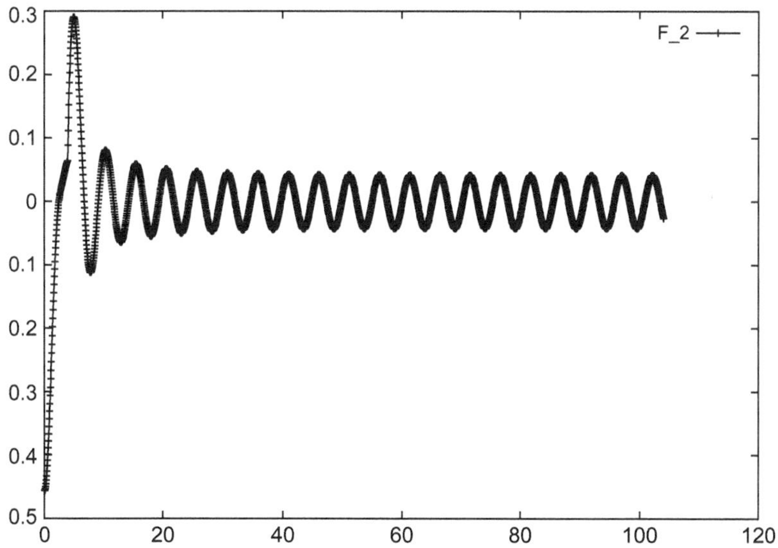

FIG. 6.26 Lift F_2 versus time, $\nu = 0.01$, $\alpha = 0.005$, $k = 0.052$.

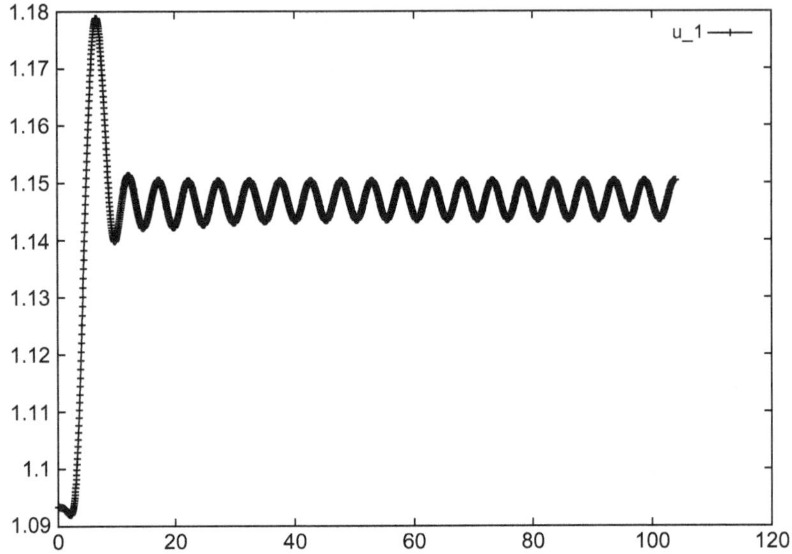

FIG. 6.27 $u_1^n(\boldsymbol{a})$ versus time, $\nu = 0.01$, $\alpha = 0.005$, $k = 0.052$.

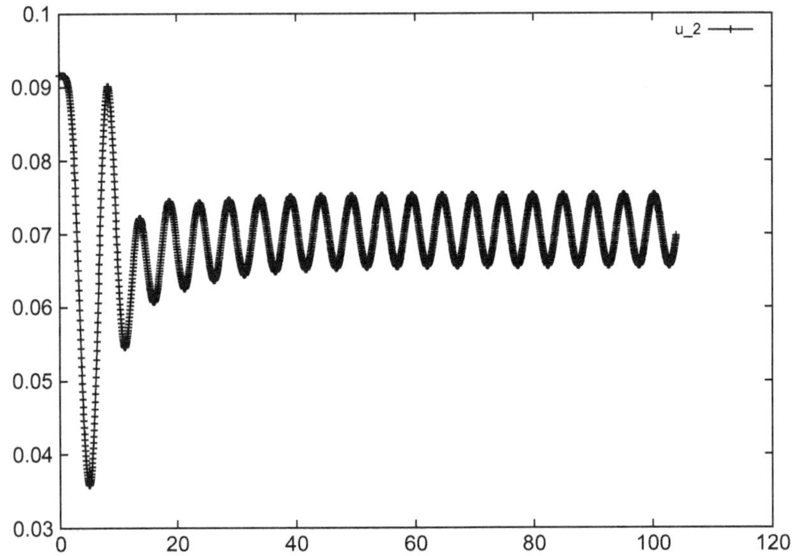

FIG. 6.28 $u_2^n(\boldsymbol{a})$ versus time, $\nu = 0.01$, $\alpha = 0.005$, $k = 0.052$.

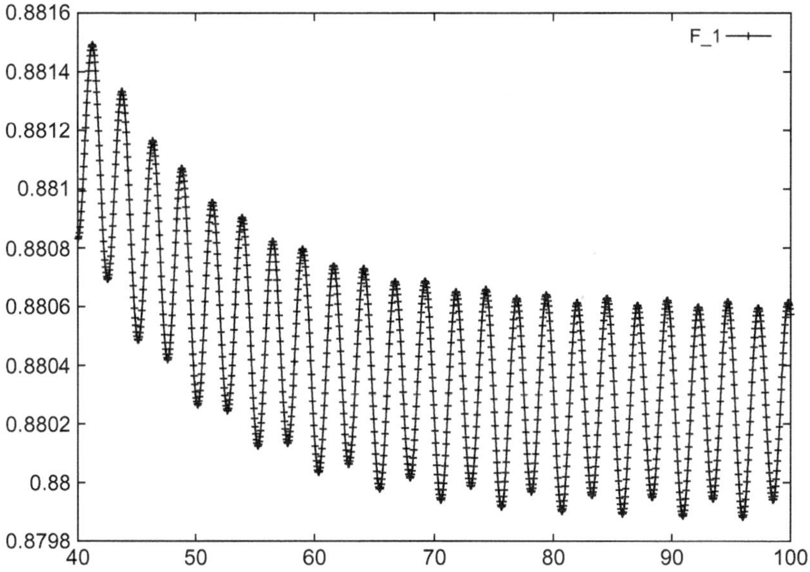

FIG. 6.29 Zoom of the drag F_1 versus time, $\nu = 0.01$, $\alpha = 0.005$, $k = 0.052$.

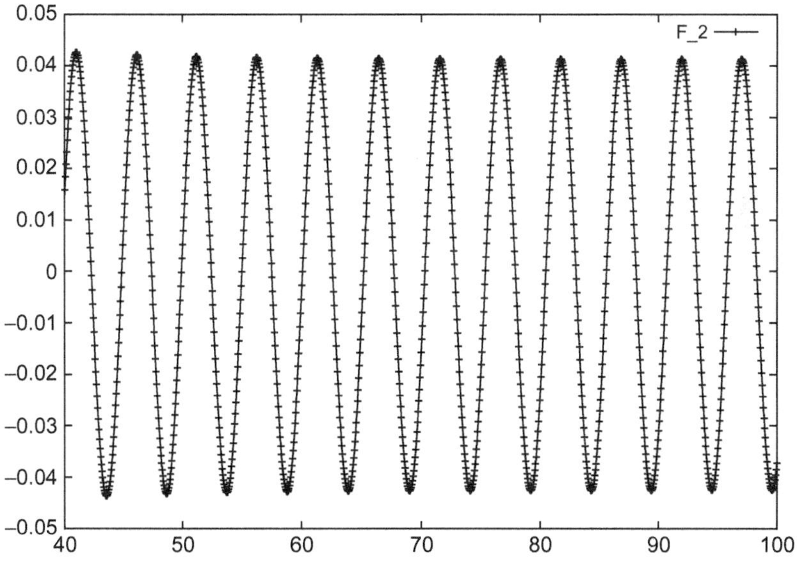

FIG. 6.30 Zoom of the lift F_2 versus time, $\nu = 0.01$, $\alpha = 0.005$, $k = 0.052$.

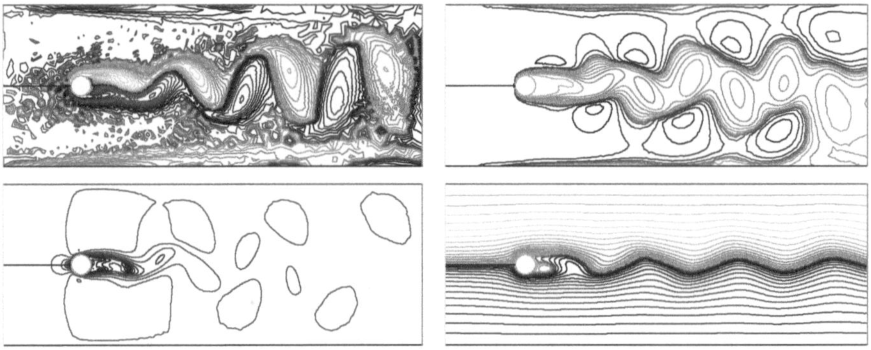

FIG. 6.31 Isovalues of z_h, p_h, $|\boldsymbol{u}_h|$, ψ_h at time step 1914.

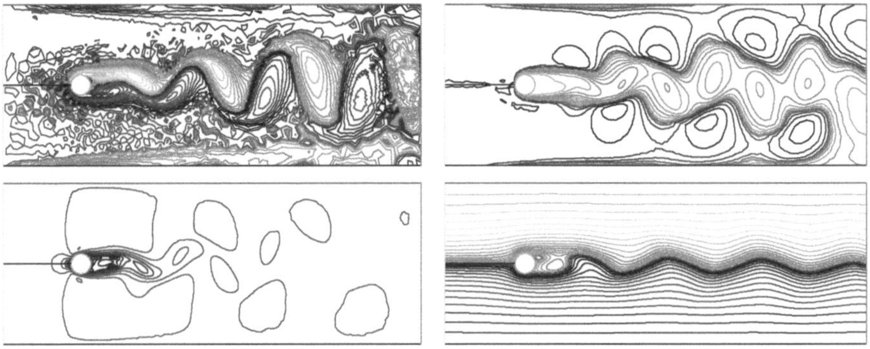

FIG. 6.32 Isovalues of z_h, p_h, $|\boldsymbol{u}_h|$, ψ_h at time step 1939.

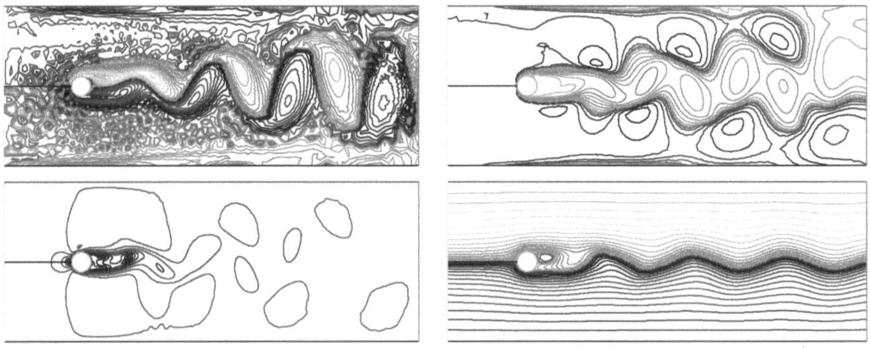

FIG. 6.33 Isovalues of z_h, p_h, $|\boldsymbol{u}_h|$, ψ_h at time step 1964.

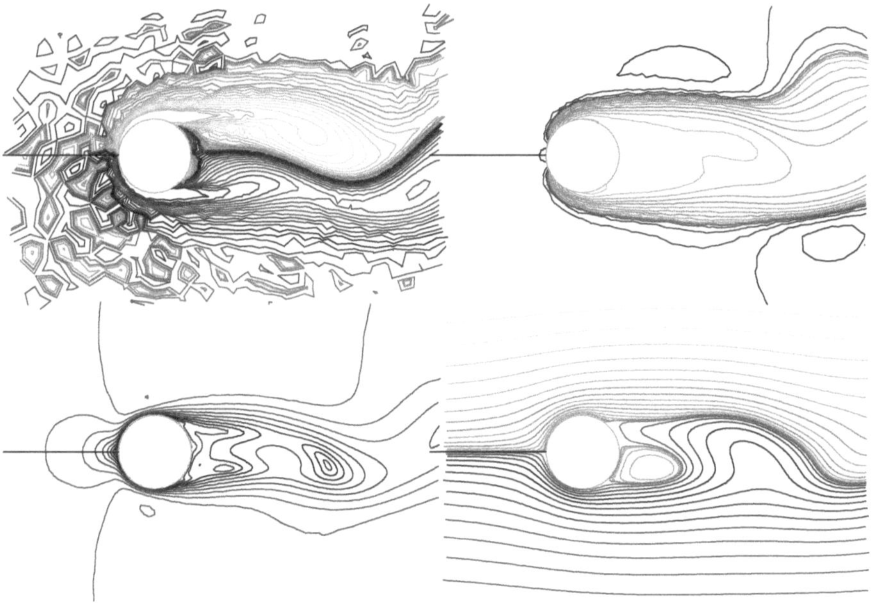

FIG. 6.34 Zoom of isovalues of z_h, p_h, $|\boldsymbol{u}_h|$, ψ_h at time step 1914.

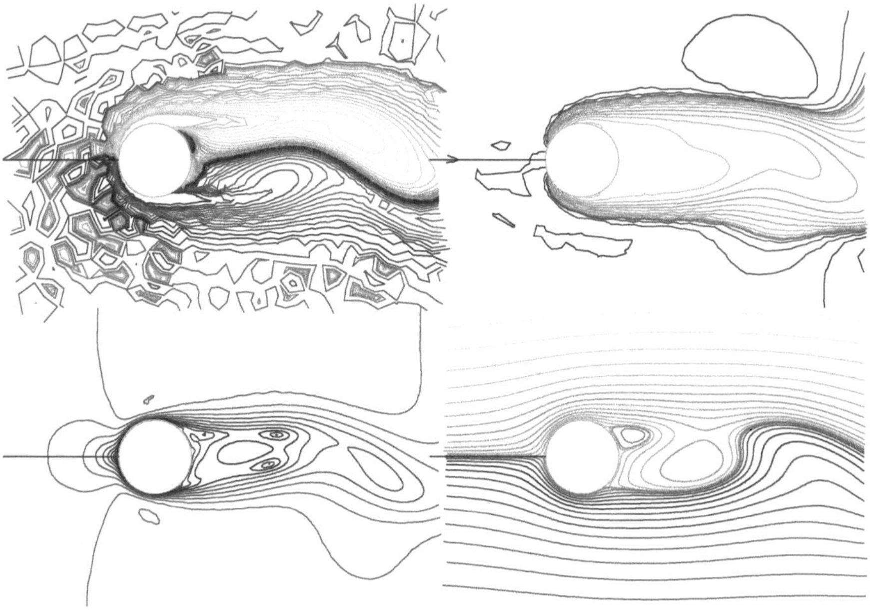

FIG. 6.35 Zoom of isovalues of z_h, p_h, $|\boldsymbol{u}_h|$, ψ_h at time step 1939.

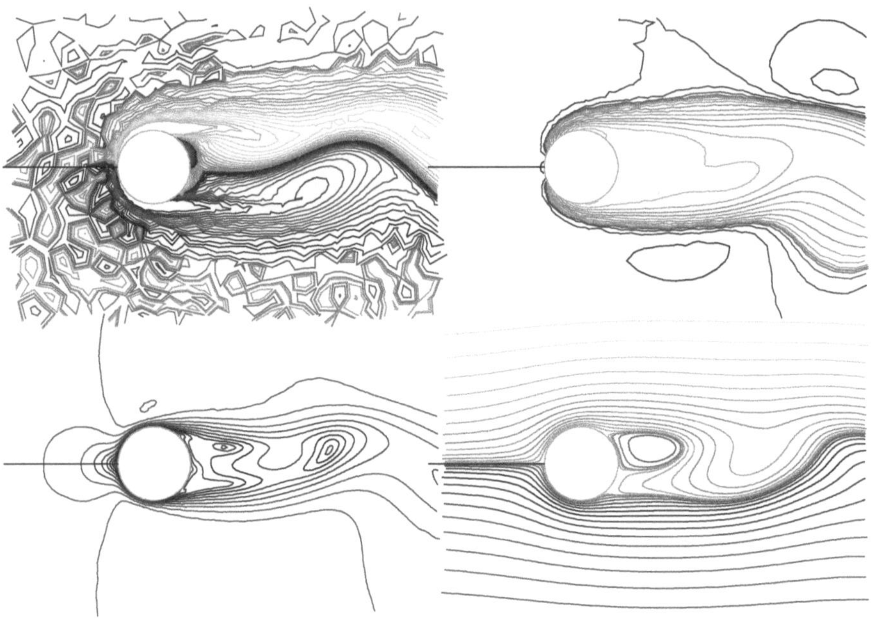

FIG. 6.36 Zoom of isovalues of z_h, p_h, $|\boldsymbol{u}_h|$, ψ_h at time step 1964.

(2) In these experiments, the velocity is not tangential on the inflow and outflow parts of the boundary. As a result, we have to prescribe a boundary condition on the auxiliary variable z where the flow is entering Γ_r because $\boldsymbol{u} \cdot \boldsymbol{n} < 0$ there. This is made clear by formula (1.3.12)

$$\alpha \int_{\Omega} (\boldsymbol{u} \cdot \nabla z) z \, \mathrm{d}\boldsymbol{x} = \frac{\alpha}{2} \int_{\partial\Omega} (\boldsymbol{u} \cdot \boldsymbol{n}) |z|^2 \mathrm{d}s.$$

The theory developed here does not address this situation (see Remark 1.3.14), but in the present experiments, we observed a stable flow with a suitable boundary condition on z, whereas the flow became unstable without this boundary condition.

(3) We were not able to compute a solution satisfying natural boundary conditions on the outflow part of Γ_r. The difficulties arose from the transport equation, but we do not know what boundary conditions should complement this equation.

Bibliography

ABBOUD, H., SAYAH, T. (2009). Upwind discretization of a time-dependent two-dimensional grade-two fluid model. *Comp. Math. Appl.* **57** (8), 1249–1264.

ADAMS, R.A. (1975). *Sobolev Spaces* (Academic Press, New York, NY).

AMARA, M., BERNARDI, C., GIRAULT, V., HECHT, F. (2005). Regularized finite element discretizations of a grade-two fluid model. *Int. J. Numer. Methods Fluids* **48**, 1375–1414.

AMBROSIO, L. (2004). Transport equations and Cauchy problems for BV vector fields. *Invent. Math.* **158**, 227–260.

AMROUCHE, C., BERNARDI, C., DAUGE, M., GIRAULT, V. (1998). Vector potentials in three-dimensional non-smooth domains. *Math. Methods Appl. Sci.* **21**, 823–864.

ARNOLD, D., BREZZI, F., FORTIN, M. (1984). A stable finite element for the Stokes equations. *Calcolo* **21** (4), 337–344.

BABUŠKA, I. (1973). The finite element method with Lagrangian multipliers. *Numer. Math.* **20**, 179–192.

BARDOS, C. (1970). Problèmes aux limites pour les équations aux dérivées partielles du premier ordre à coefficients réels; Théorèmes d'approximation; Application à l'équation de transport. *Ann. Scient. Éc. Norm. Sup., Série* 4 **3**, 185–233.

BEIRÃO DA VEIGA, H. (1987). Existence results in Sobolev spaces for a stationary transport equation. *Ricerche di Matematica* Suppl. **36**, 173–184.

BERCOVIER, M., PIRONNEAU, O. (1979). Error estimates for finite element method solution of the Stokes problem in the primitive variables. *Numer. Math.* **33**, 211–224.

BERNARD, J.M. (1998). Fluides de Second et Troisième Grade en Dimension trois: Solution Globale et Régularité. Thèse de l'Université Pierre et Marie Curie, Paris VI.

BERNARD, J.M. (1999). Stationary problem of second-grade fluids in three dimensions: existence, uniqueness and regularity. *Math. Methods Appl. Sci.* **22**, 655–687.

BERNARDI, C. (1989). Optimal finite element interpolation on curved domains. *SIAM J. Numer. Anal.* **26**, 1212–1240.

BERNARDI, C., GIRAULT, V. (1998). A local regularization operator for triangular and quadrilateral finite elements. *SIAM J. Numer. Anal.* **35** (5), 1893–1916.

BERNARDI, C., RAUGEL, G. (1985). Analysis of some finite elements for the Stokes problem. *Math. Comp.* **44** (169), 71–79.

BOLAND, J., NICOLAIDES, R. (1983). Stablility of finite elements under divergence constraints. *SIAM J. Numer. Anal.* **20** (4), 722–731.

BRENNER, S., SCOTT, L.R. (1994). *The Mathematical Theory of Finite Element Methods*, TAM 15 (Springer-Verlag, Berlin).

BRESCH, D., LEMOINE, J. (1998). On the existence of solutions for nonstationary second-grade fluids. In: Ammann, H., et al. (eds.), *Navier-Stokes Equations and Related Nonlinear Problems* (VSP/TEV, Vilnius, Lithuania), pp. 15–30.

BREZZI, F. (1974). On the existence, uniqueness and approximation of saddlepoint problems arising from Lagrange multipliers. *RAIRO, Anal. Num.* **R2**, 129–151.

BREZZI, F., FORTIN, M. (1991). *Mixed and Hybrid Finite Element Methods* (Springer-Verlag, New York).

BREZZI, F., MARINI, L.D., SÜLI, E. (2004). Discontinuous Galerkin methods for first-order hyperbolic problems. *Math. Mod. Methods Appl. Sci.* **14**, 1893–1903.

CHACÓN-REBOLLO, T. (2001). Private oral communication.

CHAMMAI, F. (2006). Discrétisation par éléments finis d'un modèle à deux dimensions d'un fluide de grade deux, Master Thesis (Fac. Sciences Université Saint-Joseph, Beirut).

CIARLET, P.G. (1991). Basic error estimates for elliptic problems. In: Ciarlet, P.G., Lions, J.L. (eds.), *Handbook of Numerical Analysis*, Finite Element Methods, Part 1 (Elsevier, North-Holland, Amsterdam).

CIORANESCU, D., GIRAULT, V. (1997). Weak and classical solutions of a family of second grade fluids. *Int. J. Non Linear Mech.* **32**, 317–335.

CIORANESCU, D., GIRAULT, V., GLOWINSKI, R., SCOTT, L.R. (1999). Some theoretical and numerical aspects of grade-two fluid models. In: Jaeger, W., Nečas, J., John, O., Najzar, K., Stara, J. (eds.), *Partial Differential Equations - Theory and Numerical Solution*, Research Notes in Mathematics 406 (Chapman & Hall/CRC, New York, NY), pp. 99–110.

CIORANESCU, D., OUAZAR, E.H. (1984a). Existence et unicité pour les fluids de second grade. *C. R. Acad. Sci. Paris* **298**, Série I, 285–287.

CIORANESCU, D., OUAZAR, E.H. (1984b). Existence and uniqueness for fluids of second grade, In: *Nonlinear Partial Differential Equations* (Collège de France Seminar, Pitman 109, Boston, MA), pp. 178–197.

CLÉMENT, P. (1975). Approximation by finite element functions using local regularization. *RAIRO, Anal. Num.* **R-2**, 77–84.

CODDINGTON, E.A., LEVINSON, N. (1955). *Theory of Ordinary Differential Equations* (Mc Graw Hill, New York, NY).

COLOMBINI, F., LERNER, N. (2002). Uniqueness of continuous solutions for BV vector fields. *Duke Math. J.* **111** (2), 357–384.

COSTABEL, M., DAUGE, M. (2000). Private email communication.

COURANT, R., FRIEDRICHS, K., LEWY, H. (1928). ber die partiellen Differenzengleichungen der mathematischen Physik. *Mathematische Annalen* **100** (1), 3274. English translation: On the partial difference equations of mathematical physics. *IBM J.*, (1967), 215–234.

CROUZEIX, M., FALK, R.S. (1989). Nonconforming finite elements for the Stokes problem. *Math. Comp.* **52** (186), 437–456.

CROUZEIX, M., RAVIART, P.A. (1973). Conforming and non-conforming finite element methods for solving the stationary Stokes problem. *RAIRO Anal. Numér.* **8**, 33–76.

DAUGE, M. (1989). Stationary Stokes and Navier-Stokes systems on two or three-dimensional domains with corners. *SIAM J. Math. Anal.* **20** (1), 74–97.

DAUGE, M. (1992). Neumann and mixed problems on curvilinear polyhedra. *Integr. Equat. Oper. Th.* **15**, 227–261.

DAWSON, C., SUN, S., WHEELER, M.F. (2004). Compatible algorithms for coupled flow and transport. *Comput. Methods Appl. Mech. Eng.* **194**, 2565–2580.

DESJARDINS, B. (1996). A few remarks on ordinary differential equations. *Comm. Partial Diff. Equ.* **11–12** (21), 1667–1703.

DIPERNA, R.J., LIONS, P.L. (1989). Ordinary differential equations, transport theory and Sobolev spaces. *Invent. Math.* **98**, 511–547.

DUNN, J.E., FOSDICK, R.L. (1974). Thermodynamics, stability, and boundedness of fluids of complexity two and fluids of second grade. *Arch. Ration. Mech. Anal.* **56**, 191–252.

DUNN, J.E., RAJAGOPAL, K.R. (1995). Fluids of differential type: Critical review and thermodynamic analysis. *Int. J. Eng. Sci.* **33**, 689–729.

DURÁN, R., MUSCHIETTI, A. (2001). An explicit right inverse of the divergence operator which is continuous in weighted norms. *Studia Mathematica* **148**, 207–219.

DURÁN, R., NOCHETTO, R.H., WANG, J. (1988). Sharp maximum norm error estimates for finite element approximations of the Stokes problem in 2 – D. *Math. Comp.* **51** (184), 1177–1192.

ERN, A., GUERMOND, J.-L. (2004). *Theory and Practice of Finite Elements*. Applied Mathematical Sciences Volume 159 (Springer-Verlag, New York).

FABES, E., KENIG, C., VERCHOTTA, G. (1988). The Dirichlet problem for the Stokes system on Lipschitz domains. *Duke Math. J.* **57** (3), 769–793.

FERNÁNDEZ CARA, E., GUILLÉN GONZÁLEZ, F., ROBLES ORTEGA, R. (2002). Mathematical modelling and analysis of viscoelastic fluids of the Oldroyd kind. In: (Ciarlet, P.G., Lions, J.L. (eds.)), *Handbook of Numerical Analysis*, VIII, Elsevier, North-Holland, pp. 543–661.

FORTIN, M., SOULIÉ, M. (1983). A non-conforming piecewise quadratic finite element on triangles. *Int. J. Numer. Methods Eng.* **19**, 505–520.

FOSDICK, R.L., RAJAGOPAL, K.R. (1978a). Anomalous features in the model of "Second order fluids". *Arch. Ration. Mech. Anal.* **70**, 145–152.

FOSDICK, R.L., RAJAGOPAL, K.R. (1978b). Uniqueness and drag for fluids of second grade in steady motion. *Int. J. Non Linear Mech.* **13**, 131–137.

GIRAULT, V., LIONS, J.-L. (2001a). Two-grid finite-element schemes for the steady Navier-Stokes problem in polyhedra. *Portug. Math.* **58** (1), 25–57.

GIRAULT, V., LIONS, J.-L. (2001b). Two-grid finite-element schemes for the transient Navier-Stokes equations. *M2AN* **35**, 945–980.

GIRAULT, V., RAVIART, P.A. (1982). An analysis of upwind schemes for the Navier-Stokes Equations. *SIAM J. Numer. Anal.* **19** (2), 312–333.

GIRAULT, V., RAVIART, P.A. (1986). *Finite Element Methods for Navier-Stokes Equations. Theory and Algorithms*, SCM 5 (Springer-Verlag, Berlin).

GIRAULT, V., RIVIÈRE, B., WHEELER, M.F. (2004). A discontinuous Galerkin method with non-overlapping domain decomposition for the Stokes and Navier-Stokes problems. *Math. Comp.* **74**, 53–84.

GIRAULT, V., SAADOUNI, M. (2007). On a time-dependent grade-two fluid model in two dimensions. *Comput. Math.* **53**, 347–360.

GIRAULT, V., SCOTT, L.R. (1999). Analysis of a two-dimensional grade-two fluid model with a tangential boundary condition. *J. Math. Pures Appl.* **78**, 981–1011.

GIRAULT, V., SCOTT, L.R. (2002a). Finite-element discretizations of a two-dimensional grade-two fluid model. M2AN **35**, 1007–1053.

GIRAULT, V., SCOTT, L.R. (2002b). Hermite interpolation of non-smooth functions preserving boundary conditions. *Math. Comp.* **71**, 1043–1074.

GIRAULT, V., SCOTT, L.R. (2002c). Upwind discretizations of a steady grade-two fluid model in two dimensions. In: Cioranescu, D., Lions, J.-L. (eds.), *Nonlinear Partial Differential Equations and their Applications* (College de France Seminar, Volume XIV) **31**, pp. 393–414.

GIRAULT, V., SCOTT, L.R. (2003). A quasi-local interpolation operator preserving the discrete divergence. *Calcolo* **40**, 1–19.

GIRAULT, V., SCOTT, L.R. (2010). On a time-dependent transport equation in a Lipschitz domain. *SIAM J. Math. Anal.* **42**, 1721–1731.

GIRAULT, V., TARTAR, L. (2010). L^p and $W^{1,p}$ regularity of the solution of a steady transport equation, *C. R. Acad. Sci. Paris, Sir I* **348**, 885–890.

GLOWINSKI, R. (2003). Finite element methods for incompressible viscous flow. In: Ciarlet, P.G., Lions, J.L. (eds.), *Handbook of Numerical Analysis*, IX, Elsevier, North-Holland, pp. 9–1176.

GRISVARD, P. (1985). *Elliptic Problems in Nonsmooth Domains*. Pitman Monographs and Studies in Mathematics, Volume 24 (Pitman, Boston, MA).

HECHT, F., LE HYARIC, A., PIRONNEAU, O., OHTSUKA, K. (2008). Freefem++, Second Edition, Version 2.24-2-2 (Laboratoire J.-L. Lions, UPMC, Paris).

HOLM, D.D., MARSDEN, J.E., RATIU, T.S. (1998a). Euler-Poincaré models of ideal fluids with nonlinear dispersion. *Phys. Rev. Lett.* **349**, 4173–4177.

HOLM, D.D., MARSDEN, J.E., RATIU, T.S. (1998b). The Euler-Poincaré equations and semidirect products with applications to continuum theories. *Adv. Math.* **137**, 1–81.

HOOD, P., TAYLOR, C. (1973). A numerical solution of the Navier-Stokes equations using the finite element technique. *Comp. Fluids* **1**, 73–100.

HOPF, E. (1951). Uber die Aufangswertaufgabe fur die hydrodynamischen Grundleichungen. *Math. Nachr.* **4**, 213–231.

HÖRMANDER, L. (1983). *The Analysis of Linear Partial Differential Operators*, Part III (Springer-Verlag, Berlin).

HUGUES, T.J.R. (1978). A simple finite element scheme for developping upwind finite elements. *Int. J. Numer. Methods Eng.* **12**, 1359–1365.

JOHNSON, C. (1987). *Numerical Solution of PDE by the Finite Element Method* (Cambridge University Press, Cambridge).

JOHNSON, C., NAVERT, U., PITKARANTA, J. (1985). Finite element methods for linear hyperbolic problems. *Comput. Methods Appl. Mech. Eng.* **45**, 285–312.

KANAAN, M. (2007). Méthode de deux niveaux pour les fluides de grade deux stationnaires, Master Thesis (Fac. Sciences, Université Saint-Joseph, Beirut).

KELLOG, R.B., OSBORN, J.E. (1976). A regularity for the Stokes problem in a convex polygon. *J. Funct. Anal.* **21**, 397–431.

KOZLOV, V.A., MAZ'YA, V.G., ROSSMANN, J. (2000). *Spectral Problems Associated with Corner Singularities of Solutions to Elliptic Equations*. Mathematical Surveys and Monographs, Volume 85 (AMS, Providence).

LAX, P., MILGRAM, N. (1954). *Parabolic Equations*. Contributions to the Theory of Partial Differential Equations (Princeton).

LERAY, J. (1933). Etude de diverses équations intégrales nonlinéaires et de quelques problèmes que pose l'hydrodynamique. *J. Math. Pures Appl.* **12**, 1–82.

LESAINT, P., RAVIART, P.A. (1974). On a finite element method for solving the neutron transport equation. In: *Mathematical Aspects of finite Elements in Partial Differential Equations* (Academic Press, New York, NY), pp. 89–122.

LIONS, J.L. (1969). *Quelques Méthodes de Résolution des Problèmes aux Limites Non Linéaires* (Dunod, Paris).

LIONS, J.L., MAGENES, E. (1968). *Problèmes aux Limites non Homogènes et Applications, I* (Dunod, Paris).

NEČAS, J. (1967). *Les Méthodes Directes en Théorie des Équations Elliptiques* (Masson, Paris).

NITSCHE, J. (1970). Über ein Variationsprinzip zur Lösung von Dirichlet-Problemen bei Verwendung von Teilräumen, die keinen Randbedingungen unterworfen sind. *Abh. Math. Hamburg.* **36**, 9–15.

OUAZAR, E.H. (1981). Sur les Fluides de Second Grade, Thèse de 3ème Cycle de l'Université Pierre et Marie Curie, Paris VI.

PARK, K.H. (1998). Least-squares methods for the simulation of the flow of an incompressible fluid of second grade, Ph.D. Thesis, The University of Houston, TX, USA.

PEETRE, J. (1966). Espaces d'interpolation et théorème de Soboleff. *Ann. Inst. Fourier* **16**, 279–317.

PIRONNEAU, O. (1989). *Finite Element Methods for Fluids* (Wiley, New-York).

PUEL, J.P., ROPTIN, M.C. (1967). Lemme de Friedrichs. Théorème de densité résultant du lemme de Friedrichs, Report, Diplôme d'Etudes Approfondies, Advisor C. Goulaouic, Université de Rennes.

RAJAGOPAL, K.R. (1995). On boundary conditions for fluids of the differential type. In: Sequeira, A. (ed.), *Navier-Stokes Equations and Related Problems* (Plenum Press).

RAJAGOPAL, K.R., KALONI, P.N. (1989). Some remarks on boundary conditions for the flow of fluids of the differential type. In: *Continuum Mechanics and its Applications* (Hemisphere Press).

REED, W. H., HILL, T.R. (1973). Triangular mesh methods for the neutron transport equation, Los Alamos Scientific Laboratory Report, LA-UR-73–479.

RIVLIN, R.S., ERICKSEN, J.L. (1955). Stress-deformation relations for isotropic materials. *Arch. Ration. Mech. Anal.* **4**, 323–425.

ROBLES ORTEGA, R. (1995). Contribución al estudio teórico de algunas E.D.P. no lineales relacionadas con fluidos no Newtonianos, Thesis of University of Sevilla, Spain.

SAADOUNI, M. (2007). Un modèle instationnaire bidimensionnel de fluide de grade deux, Thèse de Doctorat de l'Université Pierre et Marie Curie, Paris VI.

SAYAH, T. (2007). Numerical solution of a time-dependent two-dimensional Grade-Two fluid model, Internal Report, Fac. Sciences, Université Saint-Joseph, Beirut.

SCOTT, L.R., ZHANG, S. (1990). Finite element interpolation of non-smooth functions satisfying boundary conditions. *Math. Comp.* **54**, 483–493.

SHOWALTER, R.E. (1997). *Monotone operators in Banach space and nonlinear partial differential equations*. Mathematical Surveys and Monographs, Volume 49 (American Mathematical Society, Providence, RI).

Simon, J. (1990). Compact sets in the space $L^p(0, T; B)$. *Ann. Math. Pures Appl.* **146**, 1093–1117.

Stein, E. (1970). *Singular Integrals and Differentiability Properties of Functions* (Princeton University Press, Princeton, NJ).

Stenberg, R. (1984). Analysis of finite element methods for the Stokes problem: a unified approach. *Math. Comp.* **42**, 9–23.

Tartar, L. (1978). *Topics in Nonlinear Analysis*, Publications Mathématiques d'Orsay, Université Paris-Sud, Orsay.

Temam, R. (1979). *Navier-Stokes Equations, Theory and Numerical Analysis* (North-Holland, Amsterdam).

Temam, R. (1997). *Infinite-Dimensional Dynamical Systems in Mechanics and Physics*. Applied Mathematical Sciences, Volume 68 (Springer-Verlag, Berlin).

Truesdell, C., Noll, W. (1975). *The Nonlinear Field Theory of Mechanics in Handbuch der Physik*, Vol. III, (Springer-Verlag, Berlin).

Truesdell, C., Rajagopal, K.R. (2000). An Introduction to the Mechanics of Fluids, Modeling and Simulation. In: *Science, Engineering and Technology*, (Birkhauser, Basel).

Verfürth, R. (1984). Error estimates for a mixed finite element approximation of the Stokes equations. *RAIRO Anal. Numér.* **18** (2), 175–182.

Videmann, J.H. (1997). Mathematical analysis of visco-elastic non Newtonian fluids, Thesis, University of Lisbon, Portugal.

Walkington, N.J. (2005). Convergence of the discontinuous Galerkin method for discontinuous solutions. *SIAM J. Numer. Anal.* **42**, 1801–1817.

List of Notation

τ	quasi-uniformity parameter of $\mathcal{T}_h$, (2.1.37)	40
$\boldsymbol{u} \cdot \nabla \boldsymbol{A}$	$\sum_{i=1}^{d} u_i \frac{\partial \boldsymbol{A}}{\partial x_i}$	13
$\boldsymbol{u_g}$	lifting of $\boldsymbol{g}$ supported by Ω_ε, Theorem 5.1.2	145
$\boldsymbol{u}_{h,\boldsymbol{g}}$	lifting of $\boldsymbol{g}_h \in G_h$ in Ω_ε, Hypothesis 5.2.1	150
V	H_0^1 with zero divergence, (1.1.10)	8
$V^\perp$	orthogonal of V in H_0^1, (1.1.11)	9
V^α	V with α curl $\Delta \boldsymbol{v}$ in L^2, (1.4.1)	24
$\\|\boldsymbol{v}\\|_{V^\alpha}$	graph norm of V^α, (1.4.2)	24
$\\|\boldsymbol{v}\\|_\alpha$	$\left(\\|v\\|_{L^2(\Omega)}^2 + \alpha \|v\|_{H^1(\Omega)}^2\right)^{1/2}$, (3.1.9)	83
V_h	approximately divergence-free functions of X_h, (2.1.2)	32
$V_h^\perp$	orthogonal of V_h in X_h, (2.1.3)	32
W	H_τ^1 with zero divergence, (1.1.12)	9
$W^{m,r}(\Omega)$	Sobolev space	7
$\mathcal{W}$	(3.1.10)	83
W^α	W with curl$(\boldsymbol{v} - \alpha\, \Delta \boldsymbol{v})$ in L^2, (5.1.1)	143
W_h	approximately divergence-free functions of $X_{h,\tau}$, (5.2.1)	149
$\|v\|_{W^{m,r}(\Omega)}$	seminorm of Sobolev space	7
$\\|v\\|_{W^{m,r}(\Omega)}$	norm of Sobolev space	7
$\\|z\\|_{\boldsymbol{u}}$	graph norm of $X_{\boldsymbol{u}}$, (1.3.17)	19
$\\|\boldsymbol{\zeta}\\|_Y$	graph norm of Y, (4.1.2)	126
$\|\cdot\|$	Euclidian or Frobenius norm	7
X_h, M_h, Z_h	discrete spaces for velocity, pressure, auxiliary variable	32
$X_{\boldsymbol{u}}$	L^2 with $\boldsymbol{u} \cdot \nabla z$ in L^2, (1.3.16)	19
$X_{h,\tau}$	discrete subspace for velocity in $H_\tau^1(\Omega)^2$	149
$\boldsymbol{x}_i$	common vertex of Γ_i and Γ_{i+1}	146
Y	$\boldsymbol{\zeta}$ in H^{-1} with curl $\boldsymbol{\zeta}$ in L^2, (4.1.1)	125

The Langevin and Fokker–Planck Equations in Polymer Rheology

Alexei Lozinski
Université Paul Sabatier, Institut de Mathématiques de Toulouse, 118 route de Narbonne, F-31062 Toulouse Cedex 9, France
E-mail: alexei.lozinski@math.univ-toulouse.fr

Robert G. Owens
Département de mathématiques et de statistique, Université de Montréal, CP 6128 succ. Centre-Ville, Montréal QC H3C 3J7, Canada
E-mail: owens@dms.umontreal.ca

Timothy N. Phillips
School of Mathematics, Cardiff University, Cardiff CF24 4AG, United Kingdom
E-mail: PhillipsTN@cardiff.ac.uk

Numerical Methods for Non-Newtonian Fluids
Special Volume (R. Glowinski and Jinchao Xu, Guest Editors) of
HANDBOOK OF NUMERICAL ANALYSIS, VOL. XVI
P.G. Ciarlet (Editor)

ISSN 1570-8659
DOI 10.1016/B978-0-444-53047-9.00002-2

Contents

CHAPTER 1

Introduction

Traditionally, and as exemplified, for example, in the classic treatise on the subject by CROCHET, DAVIES and WALTERS [1984], the mathematical description and numerical simulation of the flow of complex (polymeric) fluids have involved the coupling of the macroscopic equations for the conservation of linear momentum and of mass with the determination of the polymeric contribution to the Cauchy stress tensor through some differential or integral constitutive equation. The system has to be completed with the addition of appropriate boundary and initial conditions, of course, and the exact solution being unavailable to all but the simplest of problems, a perturbation method or grid-based numerical method has usually been employed for its solution. So stood the state of the art in 1984, at least.

Progress since then, and in particular since the ground-breaking paper of LASO and ÖTTINGER [1993] with its introduction of the CONNFFESSIT (*Calculation of Non-Newtonian Flow: Finite Elements and Stochastic Simulation Technique*) approach to the stochastic simulation of the polymer dynamics as an alternative to solving a constitutive equation for the polymeric stress, has been rapid. A whole new class of numerical methods, commonly referred to as "micro–macro" or multiscale methods, has been spawned, these methods having as their common theme that the polymeric stress is calculated from a kinetic theory model that takes account of the configurations $\boldsymbol{X}_t$ (say) of some coarse-grained microstructure in the fluid. A detailed and thorough survey of micro–macro methods up to the start of the twenty-first century has been prepared by KEUNINGS [2004] so that we will mainly be concerned in this Introduction with highlighting major research contributions over the past 6 years or so.

The starting point for micro–macro methods in the context of incompressible isothermal complex fluids is the system of equations describing the conservation of linear momentum and of mass. This may be written as

$$\rho \frac{D\boldsymbol{v}}{Dt} - \eta_N \nabla^2 \boldsymbol{v} + \nabla p = \nabla \cdot \boldsymbol{\tau}, \tag{1.1}$$

$$\nabla \cdot \boldsymbol{v} = 0, \tag{1.2}$$

where $\boldsymbol{v}$ denotes the fluid velocity, ρ is the fluid density, η_N is the viscosity of the (Newtonian) solvent, and p is the pressure. $\boldsymbol{\tau}$ is the polymeric contribution to the Cauchy stress tensor, and although closure approximations allow an integral or differential constitutive equation to be solved for $\boldsymbol{\tau}$, it is now widely recognized (see, for example, SHAQFEH and JAGADEESHAN [2008]) that this approach has not been promising and, indeed, often leads to models that are unsympathetic to the underlying fluid physics. The coupling of stochastic microscale equations for the microstructure configurations with the macroscopic momentum-continuity equations is viewed much more favorably by the rheological community as a consequence. If the first two elements in the multiscale modeling of complex fluids are the macroscale equations of motion and a suitable mathematical description of the evolving microstructure, the third is the coupling of the first two elements via a stress calculator. That is, we have to be able to compute the macroscopic polymeric stress field $\boldsymbol{\tau}$ by taking suitable ensemble averages of the microstructural configuration $\boldsymbol{X}_t$.

1.1. The Langevin and Fokker–Planck equations

In general, the probability density function (pdf) $\psi(x, t)$ of a time-dependent stochastic process X_t satisfying some stochastic differential equation

$$\mathrm{d}X_t = A(x, t)\mathrm{d}t + B(x, t)\mathrm{d}W_t, \qquad \text{(Langevin)} \tag{1.3}$$

also evolves in time and satisfies a *Fokker–Planck equation*. In (1.3), the function A is a drift term and represents the deterministic part of this equation. W_t denotes a Wiener process, which is Gaussian process having zero mean and covariance $\langle W_t W_{t'}\rangle = \min(t, t')$. The use of Itô's formula and Kolmogorov's forward equation (see ÖTTINGER [1996] or GARDINER [2003] for the details) allows one to conclude that the Fokker–Planck equation associated with the Itô stochastic differential equation (1.3) is

$$\frac{\partial \psi}{\partial t} = -\frac{\partial}{\partial x}\left(A(x, t)\psi(x, t)\right) + \frac{1}{2}\frac{\partial^2}{\partial x^2}\left(B^2(x, t)\psi(x, t)\right). \tag{1.4}$$

The function B^2 in (1.4) is a diffusion coefficient. For the multivariate Itô stochastic differential equation,

$$\mathrm{d}\boldsymbol{X}_t = \boldsymbol{A}(\boldsymbol{X}_t, t)\mathrm{d}t + \boldsymbol{B}(\boldsymbol{X}_t, t)\mathrm{d}\boldsymbol{W}_t, \tag{1.5}$$

the equivalent Fokker–Planck equation reads

$$\frac{\partial \psi}{\partial t} = -\frac{\partial}{\partial \boldsymbol{x}}\cdot\left[\boldsymbol{A}(\boldsymbol{x}, t)\psi(\boldsymbol{x}, t)\right] + \frac{1}{2}\frac{\partial}{\partial \boldsymbol{x}}\frac{\partial}{\partial \boldsymbol{x}} : \left[\boldsymbol{B}(\boldsymbol{x}, t)\boldsymbol{B}^T(\boldsymbol{x}, t)\psi(\boldsymbol{x}, t)\right], \tag{1.6}$$

where now $\boldsymbol{A}$ is a vector-valued function, $\boldsymbol{B}$ is a matrix function, and $\boldsymbol{W}_t$ is a multidimensional Wiener process. We use the notation $\partial/\partial\boldsymbol{x}\partial/\partial\boldsymbol{x}$: when applied to a Cartesian tensor to mean the divergence of its divergence, i.e., given a twice differentiable Cartesian tensor $\boldsymbol{C}$,

$$\frac{\partial}{\partial \boldsymbol{x}}\frac{\partial}{\partial \boldsymbol{x}} : \boldsymbol{C} := \frac{\partial}{\partial \boldsymbol{x}}\cdot\left(\frac{\partial}{\partial \boldsymbol{x}}\cdot \boldsymbol{C}\right).$$

We now proceed to provide several examples drawn from the modeling of melts and both concentrated and dilute polymer solutions where both Langevin and the equivalent Fokker–Planck equations may be written down.

1.1.1. Concentrated and dilute polymer solutions modeled by dumbbells or chains of dumbbells

We suppose that a polymer molecule may be adequately modeled using a single dumbbell or a chain of such dumbbells joined end-to-end. Figure 1.1 shows a typical dumbbell that consists of two point beads joined by a massless spring. The end-to-end vector is denoted by $\boldsymbol{q}$, the position vector of the ith bead is denoted by $\boldsymbol{r}_i$, and the velocity of the fluid at the point having position vector $\boldsymbol{r}_i$ is denoted by $\boldsymbol{v}_i$ $(i = 1, 2)$. In the subsequent pages, let $\boldsymbol{r}_c$ denote the position vector of the center of mass of a dumbbell or chain of dumbbells and $\boldsymbol{v}_c := \boldsymbol{v}(\boldsymbol{r}_c)$ the fluid velocity at the center of mass. The spring force $\boldsymbol{F}_1$ acting on bead "1" will be that of the finitely extensible nonlinear elastic (FENE) model (see WARNER [1972] or BIRD, CURTISS, ARMSTRONG and HASSAGER [1987b], for example):

$$\boldsymbol{F}_1 := \frac{H\boldsymbol{q}}{1 - q^2/q_{\max}^2}, \tag{1.7}$$

where H is a spring constant, $q = \|\boldsymbol{q}\|_2$ is the inter-bead distance, and $q_{\max}$ is the maximum extensibility of the spring. Obviously, $\boldsymbol{F}_2 = -\boldsymbol{F}_1$.

Concentrated polymer solutions: the encapsulated dumbbell model of BIRD *and* DEAGUIAR *[1983]* For concentrated solutions, one of the simplest models is the encapsulated FENE dumbbell model of BIRD and DEAGUIAR [1983]. In a concentrated polymer solution or melt, the motion of a molecule is restricted by the presence of other molecules, and it is appropriate under these circumstances that the hydrodynamic drag on the beads of the dumbbell should be anisotropic.

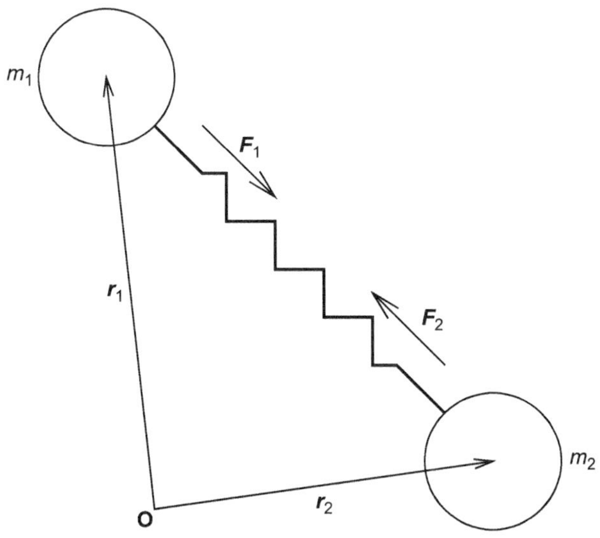

FIG. 1.1 Simple dumbbell consisting of two point masses having position vectors $\boldsymbol{r}_1$ and $\boldsymbol{r}_2$ joined by a spring. The end-to-end vector $\boldsymbol{q} := \boldsymbol{r}_2 - \boldsymbol{r}_1$ and the spring force $\boldsymbol{F} := \boldsymbol{F}_1 = -\boldsymbol{F}_2$.

Let $\boldsymbol{u} = \boldsymbol{q}/q$ denote a unit vector in the direction of $\boldsymbol{q}$. Then, BIRD and DEAGUIAR [1983] defined a drag force $\boldsymbol{F}_i^{\mathrm{d}}$ on bead i of the dumbbell as

$$\boldsymbol{F}_i^{\mathrm{d}} = \boldsymbol{\zeta} \cdot \left(\boldsymbol{v}_i - \frac{\mathrm{d}\boldsymbol{r}_i}{\mathrm{d}t} \right), \tag{1.8}$$

where the friction tensor $\boldsymbol{\zeta}$ is defined as

$$\boldsymbol{\zeta} := \zeta(\boldsymbol{uu} + \sigma^{-1}(\boldsymbol{\delta} - \boldsymbol{uu})), \tag{1.9}$$

and ζ and σ are parameters. We note that when $\sigma = 1$, the friction tensor is isotropic and that when $\sigma < 1$, the molecule experiences greater resistance to movement in a direction normal to the connector vector $\boldsymbol{q}$.

We denote the masses of the two beads by m_1 and m_2 and may write down their equations of motion as follows:

$$m_1 \frac{\mathrm{d}^2 \boldsymbol{r}_1}{\mathrm{d}t^2} = \boldsymbol{\zeta} \left(\boldsymbol{v}_1 - \frac{\mathrm{d}\boldsymbol{r}_1}{\mathrm{d}t} \right) + \boldsymbol{F}_1 + \boldsymbol{B}_1, \tag{1.10}$$

$$m_2 \frac{\mathrm{d}^2 \boldsymbol{r}_2}{\mathrm{d}t^2} = \boldsymbol{\zeta} \left(\boldsymbol{v}_2 - \frac{\mathrm{d}\boldsymbol{r}_2}{\mathrm{d}t} \right) + \boldsymbol{F}_2 + \boldsymbol{B}_2, \tag{1.11}$$

where $\boldsymbol{B}_i$ denotes a Brownian force on the ith bead due to bombardment by the surrounding solvent molecules and is defined in terms of the infinitesimal increment in a Wiener process $\boldsymbol{W}_i$ as

$$\boldsymbol{B}_i \mathrm{d}t = \sqrt{2k_B T \zeta} \left(\boldsymbol{uu} + \frac{1}{\sqrt{\sigma}} (\boldsymbol{\delta} - \boldsymbol{uu}) \right) \mathrm{d}\boldsymbol{W}_i. \tag{1.12}$$

In (1.12), k_B denotes Boltzmann's constant, and T is the absolute temperature. In the developments that follow, we will assume, for simplicity, that the bead masses $m_1 = m_2 = m$ (some common value) and that the flow is homogeneous, this last assumption meaning that the velocity difference $\boldsymbol{v}_2 - \boldsymbol{v}_1$ may be written as

$$\boldsymbol{v}_2 - \boldsymbol{v}_1 = \nabla \boldsymbol{v}_{\mathrm{c}} \cdot \boldsymbol{q}.$$

We note here that SCHIEBER and ÖTTINGER [1988], SCHIEBER [1992] considered a generalization of (1.10) and (1.11) in the context of Rouse chains whereby the rate of change of the *relative* velocity $\dot{\boldsymbol{r}} - \boldsymbol{v}_i$ of the ith bead (rather than its acceleration $\ddot{\boldsymbol{r}}_i$) appeared on the left-hand sides of these equations. The validity of doing this rests on the assumption that the fluid appears to be in equilibrium locally in the frame of reference that stays concomitant with the macroscopic streaming velocity of the fluid. However, in order to follow their subsequent ideas for the derivation of a Fokker–Planck equation, it is not necessary to make this assumption. Accordingly, let us subtract (1.10) from (1.11) and introduce the relative velocity

$$\boldsymbol{V} := \frac{\mathrm{d}\boldsymbol{q}}{\mathrm{d}t} - \nabla \boldsymbol{v}_{\mathrm{c}} \cdot \boldsymbol{q}. \tag{1.13}$$

Then, we may write down the following first-order system of equations, which is equivalent to (1.10) and (1.11):

$$m\mathrm{d}\boldsymbol{V} = -(\boldsymbol{\zeta}\boldsymbol{V} + 2\boldsymbol{F}_1 + m(\nabla\dot{\boldsymbol{v}}_c \cdot \boldsymbol{q} + \nabla\boldsymbol{v}_c(\boldsymbol{V} + \nabla\boldsymbol{v}_c \cdot \boldsymbol{q})))\,\mathrm{d}t + 2\sqrt{k_B T\zeta}\left(\boldsymbol{uu} + \frac{1}{\sqrt{\sigma}}(\boldsymbol{\delta} - \boldsymbol{uu})\right)\mathrm{d}\boldsymbol{W}, \tag{1.14}$$

$$\mathrm{d}\boldsymbol{q} = (\boldsymbol{V} + \nabla\boldsymbol{v}_c \cdot \boldsymbol{q})\,\mathrm{d}t, \tag{1.15}$$

where $\nabla\dot{\boldsymbol{v}}_c$ denotes the time derivative of the fluid velocity gradient. The Wiener process $\boldsymbol{W}$ appearing in (1.14) is defined by

$$\boldsymbol{W} := \frac{(\boldsymbol{W}_2 - \boldsymbol{W}_1)}{\sqrt{2}}.$$

Let $\Psi = \Psi(\boldsymbol{q}, \boldsymbol{V}, t)$ denote the pdf of the stochastic process $\boldsymbol{X}_t = (\boldsymbol{q}, \boldsymbol{V})^T$. Then, from (1.5) and (1.6), we see that the Fokker–Planck equation corresponding to the stochastic system (1.14) and (1.15) is

$$\frac{\partial\Psi}{\partial t} = -\frac{\partial}{\partial\boldsymbol{q}} \cdot ((\boldsymbol{V} + \nabla\boldsymbol{v}_c \cdot \boldsymbol{q})\Psi) + \frac{\partial}{\partial\boldsymbol{V}} \cdot \left(\frac{1}{\zeta\lambda_B}(\boldsymbol{\zeta}\boldsymbol{V} + 2\boldsymbol{F}_1)\Psi + (\nabla\dot{\boldsymbol{v}}_c \cdot \boldsymbol{q} + \nabla\boldsymbol{v}_c(\boldsymbol{V} + \nabla\boldsymbol{v}_c \cdot \boldsymbol{q}))\Psi\right) + \frac{2k_B T}{m\lambda_B}\frac{\partial}{\partial\boldsymbol{V}}\frac{\partial}{\partial\boldsymbol{V}} : \left(\left(\boldsymbol{uu} + \frac{1}{\sigma}(\boldsymbol{\delta} - \boldsymbol{uu})\right)\Psi\right), \tag{1.16}$$

where $\lambda_B := m/\zeta$ is a characteristic timescale for velocity fluctuations of the beads due to the action of the Brownian forces. In general, $\lambda_B \ll \lambda_H$, where $\lambda_H := \zeta/4H$ is a characteristic relaxation time for the dumbbell configuration. Let us now define the marginal pdf $\psi = \psi(\boldsymbol{q}, t)$ by integrating Ψ over the entire velocity space $\mathcal{V}$ (say):

$$\psi(\boldsymbol{q}, t) := \int_{\mathcal{V}} \Psi(\boldsymbol{q}, \boldsymbol{V}, t)\,\mathrm{d}\boldsymbol{V}. \tag{1.17}$$

Then, integrating the Fokker–Planck equation (1.16) throughout with respect to $\boldsymbol{V}$ over all of $\mathcal{V}$ and using the divergence theorem leads to the following continuity equation:

$$\frac{\partial\psi}{\partial t} = -\frac{\partial}{\partial\boldsymbol{q}} \cdot ((\ll \boldsymbol{V} \gg + \nabla\boldsymbol{v}_c \cdot \boldsymbol{q})\psi), \tag{1.18}$$

where the velocity space average $\ll \cdot \gg$ is defined by

$$\ll \cdot \gg := \frac{1}{\psi}\int_{\mathcal{V}} \cdot\, \Psi(\boldsymbol{q}, \boldsymbol{V}, t)\,\mathrm{d}\boldsymbol{V}. \tag{1.19}$$

Let us now multiply (1.16) throughout by $\lambda_B \boldsymbol{V}$. Integrating with respect to $\boldsymbol{V}$ over all of $\mathcal{V}$, using the divergence theorem, taking the limit $\lambda_B \to 0$, and retaining only terms of order $(\lambda_B \boldsymbol{V})^0$ and $\sqrt{\lambda_B}\boldsymbol{V}$ then leads to

$$\lambda_B \frac{\partial}{\partial \boldsymbol{q}} \cdot (\ll \boldsymbol{VV} \gg \psi) + \frac{\boldsymbol{\zeta}}{\zeta} \ll \boldsymbol{V} \gg \psi + \frac{2}{\zeta} \boldsymbol{F}_1 \psi = \boldsymbol{0}. \tag{1.20}$$

Finally, we multiply (1.16) throughout by $\lambda_B^2 \boldsymbol{VV}$, integrate throughout with respect to $\boldsymbol{V}$ over all of $\mathcal{V}$, use the divergence theorem several times, allow $\lambda_B \to 0$, and again keep only terms of order $(\lambda_B \boldsymbol{V})^0$ or $\sqrt{\lambda_B}\boldsymbol{V}$, to arrive at the Maxwell–Boltzmann relation for the kinetic energy of the dumbbell in equilibrium:

$$\lambda_B \ll \boldsymbol{VV} \gg = \frac{2k_BT}{\zeta} \boldsymbol{\delta} \Rightarrow \frac{1}{2} m \ll \boldsymbol{VV} \gg = k_BT\boldsymbol{\delta}. \tag{1.21}$$

Combining Eqns (1.18), (1.20), and (1.21), we deduce that a Fokker–Planck equation may be written for the pdf $\psi(\boldsymbol{q}, t)$ in the form

$$\frac{\partial \psi}{\partial t} = -\frac{\partial}{\partial \boldsymbol{q}} \cdot \left(\nabla \boldsymbol{v}_{\mathrm{c}} \cdot \boldsymbol{q} \psi - 2\boldsymbol{\zeta}^{-1} \boldsymbol{F}_1 \psi - 2k_BT\boldsymbol{\zeta}^{-1} \frac{\partial \psi}{\partial \boldsymbol{q}} \right), \tag{1.22}$$

where $\boldsymbol{\zeta}^{-1}$, the inverse of the friction tensor $\boldsymbol{\zeta}$, is given by

$$\boldsymbol{\zeta}^{-1} = \zeta^{-1}(\boldsymbol{uu} + \sigma(\boldsymbol{\delta} - \boldsymbol{uu})).$$

We now introduce a dimensionless end-to-end vector

$$\boldsymbol{q}^* = \boldsymbol{q}/\ell_0, \tag{1.23}$$

where $\ell_0 := \sqrt{k_BT/H}$ and define $\tau_s = 2\lambda_H$ and $D = \sigma/(2\lambda_H)$. We henceforth drop the asterisk on $\boldsymbol{q}$. The Fokker–Planck equation (1.22) now simplifies to

$$\frac{\partial \psi}{\partial t} = -\frac{\partial}{\partial \boldsymbol{q}} \cdot \left[\nabla \boldsymbol{v}_{\mathrm{c}} \cdot \boldsymbol{q} \psi - \frac{1}{2\lambda_H} \boldsymbol{F} \psi - \frac{1}{2\lambda_H} [\boldsymbol{uu} + \sigma(\boldsymbol{\delta} - \boldsymbol{uu})] \cdot \frac{\partial \psi}{\partial \boldsymbol{q}} \right], \tag{1.24}$$

and a corresponding Itô stochastic differential equation is easily shown to be

$$\mathrm{d}\boldsymbol{q} = \left[\nabla \boldsymbol{v}_{\mathrm{c}} \cdot \boldsymbol{q} - \frac{1}{\tau_s} \boldsymbol{F} + \left(\frac{2}{\tau_s} - 2D \right) \frac{\boldsymbol{u}}{q} \right] \mathrm{d}t + \left[\sqrt{\frac{2}{\tau_s}} \boldsymbol{uu} + \sqrt{2D}(\boldsymbol{\delta} - \boldsymbol{uu}) \right] \mathrm{d}\boldsymbol{W}. \tag{1.25}$$

In (1.24) and (1.25), the nondimensional force law

$$\boldsymbol{F} = \boldsymbol{F}(\boldsymbol{q}) = \boldsymbol{q} \Big/ \left(1 - \frac{q^2}{b} \right), \tag{1.26}$$

and b is a dimensionless maximum spring extensibility defined by

$$b = q_{\max}^2. \tag{1.27}$$

Dilute polymer solutions: the FENE dumbbell and FENE chain models In the isotropic case ($\sigma = 1$), (1.24) becomes

$$\frac{\partial \psi}{\partial t} = -\frac{\partial}{\partial \boldsymbol{q}} \cdot \left(\nabla \boldsymbol{v}_{\mathrm{c}} \cdot \boldsymbol{q} \psi - \frac{1}{2\lambda_H} \boldsymbol{F} \psi - \frac{1}{2\lambda_H} \frac{\partial \psi}{\partial \boldsymbol{q}} \right), \tag{1.28}$$

and this is just the Fokker–Planck equation for a dilute solution of dumbbells. A more general Fokker–Planck equation for dilute polymer solutions may be derived by starting with the system of stochastic differential equations for a bead-spring FENE chain consisting of $d+1$ identical beads joined by d massless FENE springs. Suppose that $\boldsymbol{r}_j$ ($j = 1, \ldots, d+1$) denotes the position vector of the jth bead in the chain. Introducing the nondimensionalized jth connector vector

$$\boldsymbol{q}_j = \frac{1}{\ell_0}(\boldsymbol{r}_{j+1} - \boldsymbol{r}_j), \qquad j = 1, \ldots, d, \tag{1.29}$$

and the position vector of the center of mass

$$\boldsymbol{r}_{\mathrm{c}} = \frac{1}{d+1} \sum_{j=1}^{d+1} \boldsymbol{r}_j, \tag{1.30}$$

it may be shown (see DELAUNAY, LOZINSKI and OWENS [2007], for example) that these satisfy the stochastic differential equations

$$\mathrm{d}\boldsymbol{q}_j(t) = \left[\frac{1}{\ell_0}(\boldsymbol{v}(\boldsymbol{r}_{j+1}) - \boldsymbol{v}(\boldsymbol{r}_j)) - \frac{1}{4\lambda_H} \sum_{k=1}^{d} A_{jk} \boldsymbol{F}(\boldsymbol{q}_k) \right] \mathrm{d}t + \sqrt{\frac{1}{\lambda_H}} \mathrm{d}\boldsymbol{W}_j^q(t), \tag{1.31}$$

$$\mathrm{d}\boldsymbol{r}_{\mathrm{c}}(t) = \frac{1}{d+1} \sum_{j=1}^{d+1} \boldsymbol{v}(\boldsymbol{r}_j)\, \mathrm{d}t + \sqrt{\frac{\ell_0^2}{2(d+1)\lambda_H}}\, \mathrm{d}\boldsymbol{W}^c(t), \tag{1.32}$$

where both $\boldsymbol{W}_j^q(t) = \frac{\boldsymbol{W}_{j+1} - \boldsymbol{W}_j}{\sqrt{2}}$ and $\boldsymbol{W}^c(t) = \frac{1}{\sqrt{d+1}} \sum_{j=1}^{d+1} \boldsymbol{W}_j(t)$ have the same distribution as $\boldsymbol{W}_j$, where $\boldsymbol{W}_j$ is a multidimensional Wiener process. The matrix

$$\boldsymbol{A} := \begin{pmatrix} 2 & -1 & & (0) \\ -1 & \ddots & \ddots & \\ & \ddots & \ddots & -1 \\ (0) & & -1 & 2 \end{pmatrix}$$

is tridiagonal and is known as the Rouse matrix. Suppose, first, that the flow is homogeneous so that we are permitted to write $\boldsymbol{v}_{j+1} - \boldsymbol{v}_j = \ell_0 \nabla \boldsymbol{v} \cdot \boldsymbol{q}_j$, where $\boldsymbol{v}$ is evaluated at the center of mass of the jth connector vector. Then, again using the equivalence noted above between (1.5) and (1.6), the Fokker–Planck equation corresponding to (1.31) and (1.32) and describing the evolution of the configuration pdf ψ is

$$\begin{aligned}\frac{\partial\psi}{\partial t} &= \sum_{k=1}^{d}\frac{\partial}{\partial \boldsymbol{q}_k}\cdot\left(-\nabla\boldsymbol{v}\cdot\boldsymbol{q}_k\psi+\frac{1}{2\lambda_H}\boldsymbol{F}(\boldsymbol{q}_k)\psi+\frac{1}{2\lambda_H}\frac{\partial\psi}{\partial\boldsymbol{q}_k}\right)\\ &\quad-\frac{1}{4\lambda_H}\sum_{k=1}^{d-1}\frac{\partial}{\partial \boldsymbol{q}_k}\cdot\left(\boldsymbol{F}(\boldsymbol{q}_{k+1})\psi+\frac{\partial\psi}{\partial\boldsymbol{q}_{k+1}}\right)\\ &\quad-\frac{1}{4\lambda_H}\sum_{k=2}^{d}\frac{\partial}{\partial \boldsymbol{q}_k}\cdot\left(\boldsymbol{F}(\boldsymbol{q}_{k-1})\psi+\frac{\partial\psi}{\partial\boldsymbol{q}_{k-1}}\right).\end{aligned} \tag{1.33}$$

It may be seen that Eq (1.28) is a particular case of (1.33) when $d = 1$.

A second case of interest of a Fokker–Planck equation that may be inferred from the system of stochastic differential equations is when $d = 1$, but we do not make the homogeneous flow assumption. Then, it follows in a fully nonhomogeneous flow, and we would have a Fokker–Planck equation

$$\frac{D\psi}{Dt} = -\frac{\partial}{\partial\boldsymbol{q}}\cdot\left((\boldsymbol{v}_2-\boldsymbol{v}_1)\psi-\frac{1}{2\lambda_H}\boldsymbol{F}(\boldsymbol{q})\psi-\frac{1}{2\lambda_H}\frac{\partial\psi}{\partial\boldsymbol{q}}\right)+\frac{\ell_0^2}{8\lambda_H}\frac{\partial^2\psi}{\partial\boldsymbol{r}_c^2}, \tag{1.34}$$

where the material derivative D/Dt is defined by

$$\frac{D}{Dt} = \frac{\partial}{\partial t}+\frac{(\boldsymbol{v}_1+\boldsymbol{v}_2)}{2}\cdot\frac{\partial}{\partial\boldsymbol{r}_c}.$$

Let L be some macroscopic length scale. Suppose that the velocity field $\boldsymbol{v}$ varies approximately linearly over the length of a dumbbell so that l_0/L is sufficiently small that terms which are quadratic (or higher) in l_0/L may be neglected. This will be referred to throughout this chapter as *locally homogeneous flow*. Thus, there is no diffusion term in real space and $\boldsymbol{v}_1 - \boldsymbol{v}_2 \approx -\nabla\boldsymbol{v}_c\cdot\boldsymbol{q}$ and $\boldsymbol{v}_1+\boldsymbol{v}_2 \approx 2\boldsymbol{v}_c$. The Fokker–Planck equation now becomes

$$\frac{D\psi}{Dt} = -\frac{\partial}{\partial\boldsymbol{q}}\cdot\left(\nabla\boldsymbol{v}_c\cdot\boldsymbol{q}\psi-\frac{1}{2\lambda_H}\boldsymbol{F}\psi-\frac{1}{2\lambda_H}\frac{\partial\psi}{\partial\boldsymbol{q}}\right). \tag{1.35}$$

In an interesting recent short communication, SCHIEBER [2006] has identified the Fokker–Planck equation (1.35) with the stochastic differential equation

$$\mathrm{d}\boldsymbol{q} = \left[-\boldsymbol{v}_c\cdot\nabla\boldsymbol{q}+\nabla\boldsymbol{v}_c\cdot\boldsymbol{q}-\frac{1}{2\lambda_H}\boldsymbol{F}\right]\mathrm{d}t+\sqrt{\frac{1}{\lambda_H}}\mathrm{d}\boldsymbol{W}, \tag{1.36}$$

which is satisfied by a so-called Brownian configuration field $\boldsymbol{q}(\boldsymbol{x}, t)$ (HULSEN, VAN HEEL and VAN DEN BRULE [1997], and ÖTTINGER, VAN DEN BRULE and HULSEN [1997]). Equation (1.36) is the same as (1.25) in the case $\sigma = 1$ except for the presence of a convective derivative acting on $\boldsymbol{q}$, which is now an Eulerian vector field defined over the entire flow domain. We note that the Wiener process in (1.36) is a function only of time so that the physical interpretation of the Brownian configuration field is as a continuous-in-space ensemble of dumbbells having the same initial configuration and subject to the same sequence of random forces throughout the domain (ÖTTINGER, VAN DEN BRULE and HULSEN [1997]). SCHIEBER [2006] also proposed a stochastic differential equation for a Brownian configuration field that corresponds to a fully nonhomogeneous Fokker–Planck equation, an example of which is that given in (1.34). Note that in an entirely homogeneous flow, the material derivative in (1.35) is replaced by a partial derivative, and we once again have Eq (1.28).

It is worth adding here and for future reference that a useful approximation to the FENE spring law (1.26), leading to a closed-form differential equation for the elastic stress $\boldsymbol{\tau}$, is that of the FENE-P model, given by

$$\boldsymbol{F} = \boldsymbol{q} \Big/ \left(1 - \frac{\langle q^2 \rangle}{b}\right). \tag{1.37}$$

In the definition above, $\langle \cdot \rangle$ is an ensemble average, defined as

$$\langle \cdot \rangle := \int_{\mathcal{Q}} \cdot \, \psi(\boldsymbol{x}, \boldsymbol{q}, t) \, \mathrm{d}\boldsymbol{q}, \tag{1.38}$$

where ψ is the appropriate pdf. The "P" in the name of the model stands for PETERLIN [1966], the author who originally proposed the closure approximation.

The elastic stress for a simple dumbbell model or for a single segment of a bead-spring chain with force law $\boldsymbol{F}$ is given by the KRAMERS [1944] expression (see page 69 of BIRD, ARMSTRONG and HASSAGER [1987a]):

$$\boldsymbol{\tau}(\boldsymbol{x}, t) \doteq \frac{\eta_p}{\lambda_H} \alpha_{b,d} \left(-\boldsymbol{\delta} + \langle \boldsymbol{qF} \rangle\right), \tag{1.39}$$

where η_p denotes the zero shear-rate polymeric viscosity, and, if d now denotes the number of space dimensions,

$$\alpha_{b,d} := \begin{cases} 1 & \text{Hookean } (b \to \infty) \text{ dumbbell,} \\ \dfrac{b+d+2}{b} & d - \text{dimensional FENE dumbbell,} \\ \dfrac{b+d}{b} & d - \text{dimensional FENE-P dumbbell.} \end{cases} \tag{1.40}$$

In the case of a stochastic description of the process $\boldsymbol{q}$, the ensemble average $\langle \boldsymbol{qF} \rangle$ would be computed as an average of the diadic product $\boldsymbol{qF}$ over a large number of realizations $\boldsymbol{q}$ of the solution to the stochastic equations (1.25), (1.31), or (1.36), the limit as the number of realizations tends to infinity being $\langle \boldsymbol{qF} \rangle$ by the strong law of large numbers. In the (Lagrangian) CONNFFESSIT approach of LASO and ÖTTINGER [1993] calculating the stress in a finite element thus implies tracking huge numbers of model polymers and performing the averages in (1.39) over the dumbbells that are present at time t in that finite element. Difficulties may be encountered however in maintaining an adequate number of dumbbells in a given finite element. The Brownian configuration field method of HULSEN, VAN HEEL and VAN DEN BRULE [1997] circumvents the difficulty of tracking a large number of individual polymers since the fields are defined throughout the domain at all times. One major issue of concern with the Lagrangian stochastic approach to the evaluation of the elastic stress is that a great number of dumbbells may have to be employed in order to reduce noise in the solution. A variant of the CONNFFESSIT method that is more efficient in this respect is the Lagrangian particle method of HALIN, LIELENS, KEUNINGS and LEGAT [1998], which tracks ensembles of polymer molecules rather than individual dumbbells. Depletion of particles in some finite elements may be overcomed either by the direct insertion, where needed, of more particles (GALLEZ, HALIN, LIELENS, KEUNINGS and LEGAT [1999]) or by specifying particle positions at the current time a priori in each element of the mesh and calculating the particle trajectories leading to these locations at the specified time (WAPPEROM, KEUNINGS and LEGAT [2000]). Variance reduction methods make it possible to reduce the noise levels in Lagrangian simulations (BONVIN and PICASSO [1999] and JOURDAIN, LE BRIS and LELIÈVRE [2004a]), but these may become less efficient as the relaxation time increases and the fluid becomes more elastic (BONVIN [2000]). It is pertinent to remark here that a further advantage of the use of Brownian configuration fields over the tracking of individual configuration vectors is that the strong spatial correlations that exist in the configuration fields formulation leads to natural variance reduction. Giving rise to a deterministic set of equations, the Fokker–Planck formulation completely obviates the problem of variance in the calculated stress, however.

*1.1.2. Reptation models: The Doi–Edwards model (*DOI and EDWARDS [1978a,b,c]*)*

Arguably the most popular coarse-grained models for describing the behavior of concentrated solutions of polymers and of polymer melts are those of reptation type, originally introduced by DE GENNES [1971] and later extended in scope by DOI and EDWARDS [1978a,b,c]. In the Doi–Edwards model, a molecule moves through a tube formed from the surrounding polymers except for the chain ends which are free to move in any direction. Doi and Edwards assumed that after deformation the polymer molecules retract immediately back to their equilibrium lengths. They also assumed affine tube deformation by the flow and that the tube segments are independently aligned. We denote by $\boldsymbol{u}$ the unit orientation vector of a tube segment and by $s \in [0, 1]$ the normalized arc length of a polymer chain. A configuration pdf $\psi(\boldsymbol{u}, s, \boldsymbol{x}, t)$ may be introduced so that $\psi(\boldsymbol{u}, s, \boldsymbol{x}, t)\mathrm{d}\boldsymbol{u}\mathrm{d}s$ is the joint probability that at time t a tube segment with position vector $\boldsymbol{x}$ has an orientation vector in the interval $[\boldsymbol{u}, \boldsymbol{u} + \mathrm{d}\boldsymbol{u}]$ and contains the part of the polymer chain labeled in the interval $[s, s + \mathrm{d}s]$. In the Doi–Edwards model, the configuration pdf is the solution to the Fokker–Planck equation

$$\frac{D\psi}{Dt} = -\frac{\partial}{\partial \boldsymbol{u}} \cdot [(\boldsymbol{\delta} - \boldsymbol{uu}) \cdot \nabla \boldsymbol{v} \boldsymbol{u} \psi] + \frac{1}{\pi^2 \tau_d} \frac{\partial^2 \psi}{\partial s^2}, \tag{1.41}$$

where τ_d denotes the reptation time, defined as the characteristic time for a polymer to escape its original tube. The boundary conditions at $s = 0$ and $s = 1$ for the Fokker–Planck equation (1.41) are

$$\psi(\boldsymbol{u}, s, \boldsymbol{x}, t) = \frac{1}{4\pi} \delta(|\boldsymbol{u}| - 1), \quad s = 0, 1. \tag{1.42}$$

Equivalent to (1.41) is the following system of stochastic differential equations for the processes $\boldsymbol{U}_t$ and S_t:

$$\mathrm{d}\boldsymbol{U}_t = ((\boldsymbol{\delta} - \boldsymbol{U}_t\boldsymbol{U}_t) \cdot \nabla \boldsymbol{v} \boldsymbol{U}_t)\,\mathrm{d}t, \tag{1.43}$$

$$\mathrm{d}S_t = \frac{1}{\pi}\sqrt{\frac{1}{\tau_d}}\mathrm{d}W_t, \tag{1.44}$$

where W_t is the Wiener process. As pointed out by ÖTTINGER [1996], the boundary conditions (1.42) may be shown to be reflecting boundary conditions for S_t, and when S_t reaches one of these boundaries, $\boldsymbol{U}_t$ is chosen as a random unit vector. In this sense, then, $\boldsymbol{U}_t$ and S_t are coupled through the boundary conditions (1.42).

The polymeric stress $\boldsymbol{\tau}$ is now calculated from

$$\boldsymbol{\tau} = 5G_N^0\langle \boldsymbol{uu}\rangle, \tag{1.45}$$

where G_N^0 is an elastic modulus and the orientation tensor $\langle \boldsymbol{uu}\rangle$ is defined by

$$\langle \boldsymbol{uu}\rangle = \int_{s=0}^{1}\int_{B(0,1)} \boldsymbol{uu}\psi(\boldsymbol{u}, s, \boldsymbol{x}, t)\,\mathrm{d}\boldsymbol{u}\mathrm{d}s, \tag{1.46}$$

$B(0, 1)$ denoting the surface of the unit sphere centered at the origin. Again, where a system of stochastic equations (1.44) is solved for the processes $(\boldsymbol{U}_t, S_t)$, an average of $\boldsymbol{U}_t\boldsymbol{U}_t$ would be computed over a large number of realizations.

One important difference between modern reptation theory and that of Doi and Edwards is the incorporation of a release of constraints by motion of the members of the matrix that forms the tube around a polymer molecule. Some modern reptation models have also been described in the literature in terms of both stochastic processes and an evolution equation for a pdf. For example, the so-called simplified uniform model (ÖTTINGER [1999]) was evaluated in various transient and steady shear and extensional flows by FANG, KRÖGER and ÖTTINGER [2000] using a stochastic method and comparing the results with experimental data. Then, FANG, LOZINSKI and OWENS [2004] used a spectral method to solve the equivalent Fokker–Planck equation satisfied by the configurational distribution function.

1.2. Recent progress in the mathematical analysis and numerical simulation of flows of polymeric fluids

Mathematical analyses Despite the insurmountable difficulty of solving exactly the field equations for fluids with microstructure in all but the most basic of cases, recent developments have led to what now constitutes a considerable body of literature on the mathematical analysis of the macroscopic equations (1.1) and (1.2) coupled with either a Fokker–Planck or stochastic equation for the microstructural configuration. Studies of the well posedness of the equations for dilute solutions of Hookean dumbbells (leading to the Oldroyd-B model) exist (see GUILLOPÉ and SAUT [1990], and LIN, LIU and ZHANG [2005] for the Oldroyd-B model and LIONS and MASMOUDI [2000] for a corotational variant, for example) but our interest here is to briefly review some published results on analyses when no closed-form constitutive equation exists.

In one of the earliest analyses of the model equations resulting from the kinetic theory of polymer solutions, RENARDY [1991] proved a local existence and uniqueness theorem in the case of both infinitely extensible and finitely extensible dumbbells in the absence of a solvent. Technical assumptions were made about the spring force, which excluded the FENE model, however. Local existence results for solutions of the nonlinear dumbbell equations have also been derived by WEINAN, LI and ZHANG [2004], LI, ZHANG and ZHANG [2004], ZHANG and ZHANG [2006]. WEINAN, LI and ZHANG [2004] proved well-posedness for general nonlinear spring laws with smooth potential by directly deriving a priori estimates on the stochastic equation satisfied by a Brownian configuration field, whereas LI, ZHANG and ZHANG [2004], ZHANG and ZHANG [2006] chose to couple the momentum-continuity pair with a Fokker–Planck equation. The analysis of LI, ZHANG and ZHANG [2004] extended that of RENARDY [1991] but again excluded the case of a FENE spring. All three groups of authors required high regularity of the initial data. ZHANG and ZHANG [2006] showed additionally that, subject to the technical limitations outlined in the paper, a preassigned boundary condition of the FENE-type Fokker–Planck equation was unnecessary as a result of the singularity on the boundary of the configuration domain. A similar conclusion was drawn by LIU and LIU [2008] in the case of FENE models under a steady flow field when $b \geq 2$.

JOURDAIN, LELIÈVRE and LE BRIS [2004b] have studied the FENE dumbbell model in the case of a simple shear flow and b sufficiently large. Existence of a unique solution to the FENE Langevin equation was proved, and a local-in-time existence and uniqueness result for the system coupling the stochastic differential equation and the linear momentum equation was deduced. The long-time behavior of some micro–macro models for polymeric fluids was investigated by JOURDAIN, LE BRIS, LELIÈVRE and OTTO [2006], and entropy inequalities were used to prove exponential convergence to equilibrium. LIONS and MASMOUDI [2007] had previously based their proof of global-in-time weak solutions to the corotational Oldroyd-B model on propagation of compactness (see LIONS and MASMOUDI [2000]) and now adopted a similar approach in order to prove global existence of weak solutions for the corotational FENE dumbbell model (where $\nabla \boldsymbol{v}_c$ in (1.35) is replaced by its antisymmetric part) and the Doi model for rigid rods (see DOI and EDWARDS [1988]). Use of the antisymmetric part of the velocity gradient in (1.35) enabled better estimates to be obtained for the pdf ψ. MASMOUDI [2008] proved global existence for the FENE dumbbell model if the initial state was close to equilibrium and for the corotational FENE dumbbell model in two dimensions. Global existence for smooth solutions for the coupled microscopic–macroscopic two-dimensional corotational FENE model has also been established by LIN, ZHANG and ZHANG [2008]. Continuity of velocity gradients in dilute suspensions of rod-like molecules described by the Doi model was investigated recently by OTTO and TZAVARAS [2008], and the authors considered perturbations of the stationary steady state. It was proved that discontinuities could not occur in finite time.

Although almost all mathematical studies of the well posedness of the micro–macro equations for polymeric fluids are limited to homogeneous or locally homogeneous flows an exception is the study of BARRETT and SÜLI [2007]. Here, the authors worked with the coupled macroscopic, nonhomogeneous Fokker–Planck equation system for the bead-spring model and established the existence of global-in-time weak solutions for a general class of spring-force potentials including that for the FENE spring. The directional Friedrichs mollifiers in the Kramers expression for the stress (see their Eq (1.11)) were replaced, however, by their isotropic counterparts to simplify the analysis. Earlier, BARRETT, SCHWAB and SÜLI [2005] had proved the existence of global-in-time weak solutions to the coupled locally homogeneous system, with an $\boldsymbol{x}$-mollified velocity gradient in the Fokker–Planck equation and an $\boldsymbol{x}$-mollified pdf ψ in the Kramers expression for the stress.

Stochastic techniques for the solution of a Langevin equation As intimated in Section 1.1.1 above, a major advance in the numerical multiscale modeling of viscoelastic fluids was taken with the introduction by HULSEN, VAN HEEL and VAN DEN BRULE [1997] of the Brownian configuration field method as an alternative to the Lagrangian frameworks of the original CONNFFESSIT method of LASO and ÖTTINGER [1993] or the Lagrangian particle method (HALIN, LIELENS, KEUNINGS and LEGAT [1998], GALLEZ, HALIN, LIELENS, KEUNINGS and LEGAT [1999], and WAPPEROM, KEUNINGS and LEGAT [2000]) and avoiding the costly particle tracking of these methods. The original Brownian configuration field method has undergone continuous improvement since its inception. Noting that micro–macro methods have been put at a disadvantage through the common use of explicit time marching schemes, LASO, RAMÍREZ and PICASSO [2004], RAMÍREZ and LASO [2005] have developed fully implicit methods for the Brownian configuration field method and CONNFFESSIT. LASO, RAMÍREZ and PICASSO [2004] showed how the size of the linear system resulting from the discretization of start-up planar Couette flow for both Brownian configuration fields and CONNFFESSIT could be reduced using a Schur complement approach. These ideas were extended for Brownian configuration fields simulations for complex flows in their 2005 paper. The earlier Brownian configuration field-based semi-implicit method of SOMASI and KHOMAMI [2000] was combined to good effect with a predictor-corrector scheme coming from SOMASI, KHOMAMI, WOO, HUR and SHAQFEH [2002] by KOPPOL, SURESHKUMAR and KHOMAMI [2007] to solve benchmark problems with bead-spring chain descriptions of the polymer molecules. Their algorithm was approximately 50 times faster than the method of RAMÍREZ and LASO [2005] and scaled linearly with the number of processors and with the number of chain segments. Free surface slot coating flows were solved with the Brownian configuration field method coupled with an arbitrary Lagrange–Eulerian method by BAJAJ, BHAT, PRAKASH and PASQUALI [2006]. A fully implicit time integration scheme was used for simulations using Hookean and FENE-P dumbbells, and in this case, the algorithm was found to be stable at much higher numbers than purely macroscopic calculations. The authors also extended the semi-implicit scheme of SOMASI and KHOMAMI [2000] to non-homogeneous flows of nonlinear dumbbells with hydrodynamic interactions and coupled this with a predictor-corrector method due to SOMASI, KHOMAMI, WOO, HUR and SHAQFEH [2002].

For integral type models, another Eulerian multiscale simulation technique, the deformation field method of PETERS, HULSEN and VAN DEN BRULE [2000a], does away completely with any need for particle tracking and monitoring of the deformation history along particle paths. An improved deformation field method based on a change of the reference time relative from an absolute time to one relative to the current time was proposed by HULSEN, PETERS and VAN DEN BRULE [2001]. Both the Brownian configuration field method and the deformation field method have been used for simulations involving single segment reptation models in complex geometries (see, for example, VAN HEEL, HULSEN and VAN DEN BRULE [1999], PETERS, VAN HEEL, HULSEN and VAN DE BRULE [2000b], and WAPPEROM and KEUNINGS [2000]). However, GIGRAS and KHOMAMI [2002] showed that these methods cannot be used for all reptation-type models and, in particular, proposed a new hybrid

Eulerian multiscale simulation technique for simulations with Öttinger's simplified uniform model (ÖTTINGER [1999]).

Numerical methods for the solution of the Fokker–Planck equation (1.6) Until comparatively recently, little had been done to advocate the use of numerical methods for the solution of an equivalent Fokker–Planck equation. Some exceptions include the earlier work of WARNER [1972], FAN [1985a,b,c] on steady-state shearing flows and small amplitude oscillatory shear flows of solutions and melts described with dumbbell models. Computations in the paper of FAN [1989b] were, to the best of our knowledge, the first to be published of complex flows using the Fokker–Planck equation. Significant contributions to the corpus of literature on Fokker–Planck-based numerical techniques in the 1990s and early twenty-first century include those of the Armstrong group at MIT (ARMSTRONG, NAYAK, GHOSH and BROWN [1996], NAYAK [1998], and SUEN, JOO and ARMSTRONG [2002]), who used discontinuous Galerkin methods to spatially discretize the Fokker–Planck equation for dumbbell and Doi–Edwards models and a Daubechies wavelet basis for representations in configurational space.

Operator splitting methods have proved to be important for the efficient solution of the Fokker–Planck equation. Although Brownian configuration fields are continuous across the whole flow domain, JENDREJACK, DE PABLO and GRAHAM [2002] observed that they may encounter a loss of smoothness in physical space for strong flows. In an attempt to overcome this loss of smoothness, the authors used an operator splitting method for the Fokker–Planck equation for a bead-spring chain model by solving the configuration (diffusion) and convection parts separately. A Brownian configuration field method was used for the first part to give a delta function representation of the intermediate configuration distribution function, and the convective update was performed in an orthogonal (Legendre-type) polynomial representation. An important factor in the affordability of some recent Fokker–Planck methods proposed by CHAUVIÈRE, FANG, LOZINSKI and OWENS [2003], CHAUVIÈRE and LOZINSKI [2004a,b], LOZINSKI and CHAUVIÈRE [2003], LOZINSKI, CHAUVIÈRE, FANG and OWENS [2003] was the time-splitting employed for the solution of a configuration distribution function ψ whereby two half-time steps were performed: one corresponding to a solution in configuration space and the other in physical space (convective step). CHAUVIÈRE, FANG, LOZINSKI and OWENS [2003], CHAUVIÈRE and LOZINSKI [2004b] used a mixed finite difference-spectral method to solve for the distribution function for a dilute solution of FENE dumbbells and showed that considerable time savings could be gained over an equally accurate Brownian configuration field method. In LOZINSKI and CHAUVIÈRE [2003], the authors introduced a fast solver based on a rotation operator applied to all terms in the configuration step, and this led to yet greater savings in CPU time over stochastic techniques. Significant computational savings over stochastic methods were also made by using a Fokker–Planck solver for single segment reptation models in the paper by LOZINSKI, CHAUVIÈRE, FANG and OWENS [2003]. The fast solver of LOZINSKI and CHAUVIÈRE [2003] was employed by LOZINSKI, OWENS and FANG [2004] to solve for nonhomogeneous tube flow of a FENE-type fluid. More recently, the splitting of the Fokker–Planck operator into a transport (convective) part and a diffusion part allowed HELZEL and OTTO [2006] to study the Doi model for suspensions of rod-like molecules and to investigate the spurt phenomenon both in the dilute and concentrated regimes. A finite volume method was used for the convective transport step, and a second-order accurate finite difference method was employed for the solution of a heat equation.

The papers described above all treat problems where the total number of configuration space dimensions is low. Unfortunately, the cost advantage of Fokker–Planck-based methods when compared to stochastic techniques such as CONNFFESSIT or the Brownian configuration field method is rapidly lost once simulations in strong flows (with highly localized distribution functions) or involving high-dimensional spaces are attempted. In the former case, this is because the accurate representation of the configuration distribution function necessarily increases the number of basis functions employed, even if an adaptive strategy is employed. In the latter case, loss of competitiveness is due to the so-called curse of dimension: the size of standard tensor-product bases for the representation of the configuration distribution function can grow exponentially with total dimension. DELAUNAY, LOZINSKI and OWENS [2007] attempted to use sparse tensor product spaces to represent the distribution function, and the authors performed calculations with bead-spring chains having up to 5 spring links. Results for steady simple shear flow were demonstrably superior in terms of cost and accuracy to a traditional full tensor product representation of the polymer configuration distribution function.

The number of degrees of freedom involved in the solution of the Fokker–Planck equation in high dimensions was significantly reduced by AMMAR, RYCKELYNCK, CHINESTA and KEUNINGS [2006b] by using an a priori reduction method based on the Karhunen-Love decomposition (also known as the "method of snapshots" in the proper orthogonal decomposition literature). The basis was then enriched using Krylov subspaces related to the solution residual at an earlier time. Although the number of degrees of freedom was reduced compared to techniques based on standard tensor product basis representations, extension of the basis reduction technique to multi-bead-spring models is not direct. Despite lacking the generality of solvers such as the sparse grid technique for the resolution of multidimensional partial differential equations, the new generation of Fokker–Planck solvers proposed by AMMAR, MOKDAD, CHINESTA and KEUNINGS [2006a], AMMAR, MOKDAD, CHINESTA and KEUNINGS [2007] is nonetheless suitable for highly multidimensional parabolic partial differential equations with homogeneous boundary conditions. However, this is not seen as a serious limitation since the distribution function of some of the commonest microstructures is defined in some bounded domain and vanishes on its boundary. The new basis reduction technique is based on a finite separated representation of the distribution function and the use of an iterative method whereby the basis is progressively enriched until a calculated residual drops below a prescribed tolerance. The viability of the method was demonstrated in the steady case for a bead-spring chain of up to 10 FENE connector vectors in a two-dimensional flow. In the unsteady case, calculations in one space dimension of dumbbell chains having up to 7 springs and in two space dimensions of a simple FENE model demonstrated the feasibility of the proposed technique. Regrettably, however, no calculations were performed for complex flows or comparisons made with stochastic techniques in order to compare the relative costs. Further improvements in Fokker–Planck-based methods as a viable alternative to stochastic techniques may be anticipated in the future.

1.3. Article summary

The outline of the remainder of this article is as follows:

Chapter 2 is devoted to stochastic simulation techniques for solving the Langevin equation. Section 2.1 provides an introduction to stochastic differential equations for dilute polymer solutions modeled by dumbbells. The basic concepts and theoretical results related to stochastic differential equations are presented in this section. Sections 2.2 and 2.3 describe micro–macro techniques for simulating flows of polymeric fluids. These methods are based on coupling macroscopic techniques for solving the conservation equations with microscopic methods for determining the polymeric stress in the fluid. Section 2.2 charts the early developments in this field based on the original CONNFFESSIT method of LASO and ÖTTINGER [1993]. Some of the early attempts to reduce the statistical error in the stochastic simulations without increasing the number of realizations are described. Section 2.3 presents some of the major advances in the development and implementation of micro–macro techniques. Of particular note here is the method of Brownian configuration fields of HULSEN, VAN HEEL and VAN DEN BRULE [1997]. This method avoids the need to track a large number of particles, which was an inherent component of the first-generation techniques.

Section 2.4 is devoted to a description of efficient implicit schemes for micro–macro simulations developed by LASO, RAMÍREZ and PICASSO [2004]. These schemes give rise to a large nonlinear system of algebraic equations for both the macroscopic and microscopic degrees of freedom at each time step with efficiency being achieved using size reduction techniques. Section 2.5 provides a brief account of the solution of stochastic differential equations for linear polymer melts based on the Doi–Edwards model. Section 3 is devoted to the deterministic numerical methods based on the Fokker–Planck equation for several kinetic theory models of polymer fluids. We begin in Section 3.1 with a detailed presentation of such methods for dilute solutions of polymer molecules modeled by FENE dumbbells. We deal there with globally or locally homogeneous flows. A much more complicated situation of a fully nonhomogeneous flow is treated in Section 3.2. Our methods are based on spectral representation of the pdf using higher order polynomials and Fourier modes. This enables us to construct numerical methods, which compete favorably with the stochastic simulations. Essentially the same techniques can be employed to simulate flows of concentrated solutions and melts. This is illustrated in Section 3.3 on the example of the Doi–Edwards reptation model. The number of configuration space dimensions in all the models mentioned above is relatively low (≤ 3). A more detailed description of the polymer molecules, however, often gives rise to models with significantly more degrees of freedom. Developing deterministic numerical methods for such models is a challenging problem, which is a topic of active research. In Section 3.4, we review two promising approaches on the example of the model of FENE chains: sparse tensor product representation and the low-rank separation algorithms. While the former method is already well studied (albeit usually for the problems that are simpler than the Fokker–Planck equation of interest here), the latter approach is in its very beginnings. We attempt therefore to put it in a broader context of greedy algorithms extensively studied by Temlyakov et al. (see, for example, DEVORE and TEMLYAKOV [1996]).

Chapter 4 reviews some of the simulations that have been performed using these techniques and presents some results and comparisons between the approaches in terms of computational cost, accuracy, and stability. In particular, we compare the performance of stochastic methods based on the Langevin equation with deterministic methods based on the Fokker–Planck equation for both simple and complex flows. Section 4.1 is concerned with one of the major benchmark problems in computational rheology, viz. flow past a cylinder in a channel. Numerical results are presented comparing macroscopic simulations with equivalent micro–macro simulations for Hookean dumbbells. Numerical predictions using the FENE dumbbell model, for which there is no closed-form constitutive equation, are also presented. In Sections 4.2.1 and 4.2.2, the problems of start-up of plane Couette flow and of steady Poiseuille flow are solved using both a Fokker–Planck-based spectral method and Brownian dynamics, and comparisons are made. The capacity of spectral methods to solve the nonhomogeneous Fokker–Planck equation (1.34) is demonstrated in Section 4.3 in the case of steady Poiseuille flow in a narrow channel. Stochastic and deterministic discretization methods are used for solving start-up Couette flow of a Doi–Edwards fluid in Section 4.4. Finally, a comparative study of sparse tensor product spectral methods and low-rank separation algorithms is made for some simple problems involving bead-spring chains in Section 4.5.

CHAPTER 2

Stochastic Simulation Techniques

2.1. Introduction to stochastic differential equations

In Chapter 1, the Fokker–Planck equation

$$\frac{\partial \psi}{\partial t} = -\frac{\partial}{\partial \boldsymbol{q}} \cdot \left[\nabla \boldsymbol{v}_{\mathrm{c}} \cdot \boldsymbol{q}\psi - \frac{1}{2\lambda_H}\boldsymbol{F}\psi - \frac{1}{2\lambda_H}\frac{\partial \psi}{\partial \boldsymbol{q}} \right], \tag{2.1}$$

for the configuration pdf, ψ, was derived (cf. Eqn (1.24) with $\sigma = 1$) starting from a description of the polymer dynamics of a dilute polymer solution using kinetic theory. In terms of the general Fokker–Planck equation

$$\frac{\partial \psi}{\partial t} = -\frac{\partial}{\partial \boldsymbol{q}} \cdot [\boldsymbol{A}(\boldsymbol{q}, t)\psi(\boldsymbol{q}, t)] + \frac{1}{2}\frac{\partial}{\partial \boldsymbol{q}}\frac{\partial}{\partial \boldsymbol{q}} : [\boldsymbol{D}(\boldsymbol{q}, t)\psi(\boldsymbol{q}, t)], \tag{2.2}$$

the drift term, $\boldsymbol{A}$, and the diffusion tensor, $\boldsymbol{D}$, are given by

$$\boldsymbol{A}(\boldsymbol{q}, t) = \nabla \boldsymbol{v}_{\mathrm{c}} \cdot \boldsymbol{q} - \frac{1}{2\lambda_H}\boldsymbol{F}, \tag{2.3}$$

and

$$\boldsymbol{D}(\boldsymbol{q}, t) = \boldsymbol{B}(\boldsymbol{q}, t)\boldsymbol{B}^T(\boldsymbol{q}, t) = \frac{1}{2\lambda_H}\boldsymbol{\delta}, \tag{2.4}$$

respectively.

The solution of the Fokker–Planck equation enables the stress in a polymeric liquid to be determined. This is achieved by performing an ensemble average over configuration space using the Kramers expression (1.39). In a general macroscopic flow, the polymer chains will experience a viscous drag force, in addition to the spring force and Brownian force on the beads, which influences their configuration. Thus, in general, the configuration pdf will vary in space and time. Nonequilibrium conditions will induce a polymeric contribution to the extra-stress tensor as a result of the anisotropic orientation and stretch of the polymer chains. Therefore, in a general macroscopic flow, one must solve the Fokker–Planck equation at each discretization point in space and time. Clearly, this is a computationally expensive process, particularly if the dimension of configuration space is large, and is only feasible if computationally efficient techniques can be utilized. This will form the subject of the next chapter. For the remainder of this chapter, we will consider numerical techniques for solving the equivalent stochastic differential equation

$$\mathrm{d}\boldsymbol{q}_t = \boldsymbol{A}(\boldsymbol{q}_t, t)\mathrm{d}t + \boldsymbol{B}(\boldsymbol{q}_t, t)\mathrm{d}\boldsymbol{W}_t, \tag{2.5}$$

associated with the Fokker–Planck equation (2.2) where $\boldsymbol{q}_t$ is a d-dimensional stochastic process. The components of the multidimensional Wiener process $\boldsymbol{W}_t$ are independent Wiener processes, i.e., Gaussian processes with zero mean and covariance $\langle \boldsymbol{W}_t \boldsymbol{W}_{t'} \rangle = \min(t, t')\boldsymbol{\delta}$. The increments $\boldsymbol{W}_t - \boldsymbol{W}_{t'}$ of the Wiener process are themselves Gaussian processes for which

$$\langle \boldsymbol{W}_t - \boldsymbol{W}_{t'} \rangle = 0, \quad \langle (\boldsymbol{W}_t - \boldsymbol{W}_{t'})^2 \rangle = |t - t'|\boldsymbol{\delta}. \tag{2.6}$$

For $t_1 \le t_2 \le t_3 \le t_4$, we can show using these properties that

$$\begin{aligned}
\langle (\boldsymbol{W}_{t_4} - \boldsymbol{W}_{t_3})(\boldsymbol{W}_{t_2} - \boldsymbol{W}_{t_1}) \rangle &= \langle \boldsymbol{W}_{t_4} \boldsymbol{W}_{t_2} \rangle - \langle \boldsymbol{W}_{t_3} \boldsymbol{W}_{t_2} \rangle \\
&\quad + \langle \boldsymbol{W}_{t_3} \boldsymbol{W}_{t_1} \rangle - \langle \boldsymbol{W}_{t_4} \boldsymbol{W}_{t_1} \rangle \\
&= (t_2 - t_2 + t_1 - t_1)\boldsymbol{\delta} \\
&= \boldsymbol{0}.
\end{aligned}$$

Thus, increments of the Wiener process for disjoint time intervals are uncorrelated. Therefore, since the increments are Gaussian random variables, they are also independent. It follows, from the expression for the mean-square of the increment of the Wiener process given in (2.6), that typical increments of the Wiener process are of the order of $\sqrt{\Delta t}$ in a time interval of size Δt.

Let $\mathrm{T} = [0, T]$. We have the following existence and uniqueness result (ÖTTINGER [1996]).

THEOREM 1. *If the functions* $\boldsymbol{A}$ *and* $\boldsymbol{B}$ *on* $\mathbb{R}^d \times \mathrm{T}$ *satisfy the Lipschitz conditions*

$$\begin{aligned}
|\boldsymbol{A}(\overline{\boldsymbol{Q}}, t) - \boldsymbol{A}(\widetilde{\boldsymbol{Q}}, t)| &\le c|\overline{\boldsymbol{Q}} - \widetilde{\boldsymbol{Q}}|, \\
|\boldsymbol{B}(\overline{\boldsymbol{Q}}, t) - \boldsymbol{B}(\widetilde{\boldsymbol{Q}}, t)| &\le c|\overline{\boldsymbol{Q}} - \widetilde{\boldsymbol{Q}}|,
\end{aligned} \tag{2.7}$$

and the linear growth conditions

$$\begin{aligned}
|\boldsymbol{A}(\boldsymbol{Q}, t)| &\le c(1 + |\boldsymbol{Q}|), \\
|\boldsymbol{B}(\boldsymbol{Q}, t)| &\le c(1 + |\boldsymbol{Q}|),
\end{aligned} \tag{2.8}$$

for all $\overline{\boldsymbol{Q}}, \widetilde{\boldsymbol{Q}}, \boldsymbol{Q} \in \mathbb{R}^d$, $t \in \mathrm{T}$, *for some constant* c, *then the unique solution of the stochastic differential (2.5) is a Markov process. The infinitesimal generator of this Markov process is given by the second-order differential operator,* $\mathcal{L}_t$, *defining the right-hand side of (2.2), i.e.,*

$$\mathcal{L}_t = \boldsymbol{A}(\boldsymbol{q}, t) \cdot \frac{\partial}{\partial \boldsymbol{q}} + \frac{1}{2}\boldsymbol{D}(\boldsymbol{q}, t) : \frac{\partial}{\partial \boldsymbol{q}} \frac{\partial}{\partial \boldsymbol{q}}. \tag{2.9}$$

Note that for this theorem to hold, the force $\boldsymbol{F}$ must be Lipschitz continuous. Although this is the case for the Hookean force law, it is not for the FENE force law. As far as we are aware there is no corresponding existence and uniqueness result for FENE dumbbells. However, some partial results have been derived by JOURDAIN, LELIÈVRE and LE BRIS [2004b] for simple Couette flow provided b is sufficiently large.

Consider the numerical integration of the stochastic differential equation (2.5) with initial condition $\boldsymbol{Q} = \boldsymbol{Q}_0$ at time $t = 0$. The time interval T is partitioned into subintervals $[t_i, t_{i+1})$, $i = 0, \ldots, n-1$, where $0 = t_0 < t_1 < \cdots < t_n = T$. Let $\boldsymbol{Q}_i$ denote the approximation to $\boldsymbol{Q}_t$ at time $t = t_i$. To solve the stochastic differential equation (2.5), an ensemble of trajectories is generated using a suitable integration scheme. To introduce some of the basic concepts associated with the solution of stochastic differential equations, the Euler–Maruyama scheme, which is perhaps the simplest scheme for integrating (2.5), is considered. The Euler–Maruyama scheme applied to (2.5) is

$$\boldsymbol{Q}_{i+1} = \boldsymbol{Q}_i + \boldsymbol{A}(\boldsymbol{Q}_i, t_i)\Delta t_i + \boldsymbol{B}(\boldsymbol{Q}_i, t_i)\Delta \boldsymbol{W}_i, \tag{2.10}$$

where the components of the d-dimensional vector of increments $\Delta \boldsymbol{W}_i = \boldsymbol{W}_{t_{i+1}} - \boldsymbol{W}_{t_i}$ are Gaussian random variables with zero mean and variance $\Delta t_i = t_{i+1} - t_i$.

The concept of the order of strong and weak convergence is used to obtain a quantitative description of the accuracy of an approximation scheme for solving the stochastic differential equation (2.5). Let $\Delta t = \max_{1 \le i \le n} \Delta t_i$, then a given approximation scheme is said to converge strongly of order $\nu > 0$ at time $T = t_n$ if there exists a constant C independent of Δt such that

$$\langle |\boldsymbol{Q}(T) - \boldsymbol{Q}_n|^2 \rangle \le C(\Delta t)^{2\nu}.$$

We have the following result concerning the strong convergence of the Euler–Maruyama scheme.

THEOREM 2. *Consider the stochastic differential equation (2.5), where the ensemble average* $\langle \boldsymbol{Q}_0^2 \rangle$ *is finite, the functions* $\boldsymbol{A}$ *and* $\boldsymbol{B}$ *satisfy (2.7) and (2.8), and*

$$\begin{aligned} |\boldsymbol{A}(\boldsymbol{Q}, t) - \boldsymbol{A}(\boldsymbol{Q}, t')| &\le c(1 + |\boldsymbol{Q}|)|t - t'|^{1/2}, \\ |\boldsymbol{B}(\boldsymbol{Q}, t) - \boldsymbol{B}(\boldsymbol{Q}, t')| &\le c(1 + |\boldsymbol{Q}|)|t - t'|^{1/2}, \end{aligned} \tag{2.11}$$

for all $\boldsymbol{Q} \in \mathbb{R}^d$ *and* $t, t' \in \mathrm{T}$, *where* c *is a constant. Then, the Euler scheme (2.10) for solving (2.5) converges strongly with order* $\nu = 1/2$.

PROOF. See ÖTTINGER [1996], for example. □

From this result, we see that the order of convergence of the Euler scheme applied to a stochastic differential equation is one half lower than when the scheme is applied to a corresponding deterministic equation. To proffer an explanation for this, consider the integrated form of the stochastic differential equation (2.5) over the time interval $[t_i, t_{i+1}]$:

$$\boldsymbol{Q}_{t_{i+1}} = \boldsymbol{Q}_{t_i} + \int_{t_i}^{t_{i+1}} \boldsymbol{A}(\boldsymbol{Q}_t, t)\,\mathrm{d}t + \int_{t_i}^{t_{i+1}} \boldsymbol{B}(\boldsymbol{Q}_t, t) \cdot \mathrm{d}\boldsymbol{W}_t, \tag{2.12}$$

The Euler–Maruyama scheme (2.10) is obtained from (2.12) by approximating the integrands at the initial time t_i in the interval of integration.

Although the error introduced in this approximation by replacing t by t_i is $O(\Delta t_i)$, the deviation between $\boldsymbol{Q}_t$ and $\boldsymbol{Q}_{t_i}$ is of order $(t - t_i)^{1/2}$. The corresponding leading order corrections to $\boldsymbol{B}$ are obtained from the Itô formula

$$\boldsymbol{B}(\boldsymbol{Q}_t, t) \approx \boldsymbol{B}(\boldsymbol{Q}_{t_i}, t_i) + (\boldsymbol{W}_t - \boldsymbol{W}_{t_i}) \cdot \boldsymbol{B}^T(\boldsymbol{Q}_{t_i}, t_i) \cdot \frac{\partial}{\partial \boldsymbol{Q}} \boldsymbol{B}(\boldsymbol{Q}_{t_i}, t_i). \tag{2.13}$$

These corrections, which are of order $O(\Delta t_i^{\frac{1}{2}})$, should be accommodated into an approximation of the stochastic integral in (2.12) in order to obtain a method with strong order of convergence $\nu = 1$.

In the Euler–Maruyama scheme, $\boldsymbol{B}$ is evaluated at the initial configuration $(\boldsymbol{Q}_{t_i}, t_i)$ throughout the interval $[t_i, t_{i+1}]$ instead of a time-dependent configuration. As we have shown in (2.13), the time-dependent configurations differ from the initial ones by stochastic terms of $O(\Delta t^{1/2})$, which is responsible for the Euler–Maruyama scheme having strong order of convergence $\nu = 1/2$. Note, however, that when the diffusion coefficient functions $\boldsymbol{B}$ are independent of $\boldsymbol{Q}$ (additive noise), the scheme has strong order of convergence $\nu = 1$.

MIL'SHTEIN [1974] used the correction (2.13) to obtain the following scheme with strong order of convergence $\nu = 1$:

$$\begin{aligned} \boldsymbol{Q}_{i+1} &= \boldsymbol{Q}_i + \boldsymbol{A}(\boldsymbol{Q}_i, t_i)\Delta t_i + \boldsymbol{B}(\boldsymbol{Q}_i, t_i)\boldsymbol{W}_i\sqrt{\Delta t_i} \\ &\quad + \left\{ \int_{t_i}^{t_{i+1}} (\boldsymbol{W}_t - \boldsymbol{W}_{t_i}) \mathrm{d}\boldsymbol{W}_t \right\}^T : \boldsymbol{B}^T(\boldsymbol{Q}_{t_i}, t_i) \cdot \frac{\partial}{\partial \boldsymbol{Q}} \boldsymbol{B}(\boldsymbol{Q}_{t_i}, t_i). \end{aligned}$$

Although, the symmetric part of the stochastic integral in this correction term can be evaluated in terms of increments of the Wiener process, the full integral is expensive to simulate since it requires the time evolution of the Wiener process throughout the interval $[t_i, t_{i+1}]$. Under the conditions of Theorem 2 supplemented with similar Lipschitz and growth conditions on certain first- and second-order spatial derivatives of the coefficient functions $\boldsymbol{A}$ and $\boldsymbol{B}$, KLOEDEN and PLATEN [1992] have established rigorously the order of strong convergence for this scheme.

The concept of strong convergence provides information on the accuracy of individual trajectories. However, if the main interest is in the accuracy of the averages of certain quantities, then a more meaningful measure of accuracy is weak convergence. An approximation scheme is said to converge weakly with order $\nu > 0$ at time T if, for all sufficiently smooth functions $g : \mathbb{R}^d \to \mathbb{R}$ with polynomial growth, there exists a constant $C_g > 0$, independent of Δt, such that for sufficiently small Δt

$$|\langle g(\boldsymbol{Q}(T)) \rangle - \langle g(\boldsymbol{Q}_n) \rangle| \le C_g(\Delta t)^\nu.$$

If the drift and diffusion coefficients, $\boldsymbol{A}$ and $\boldsymbol{B}$, satisfy suitable smoothness and growth conditions, the Euler–Maruyama scheme (2.10) can be shown to be weakly convergent of order $\nu = 1$ (see theorem 14.5.1 of KLOEDEN and PLATEN [1992]).

Since the whole point of computing the solution to the stochastic differential equation (2.5) is to generate an approximation to the polymeric contribution to the extra stress rather than individual trajectories, the concept of weak convergence is more relevant in the context of polymer dynamics. In particular, to determine the polymeric contribution to the extra stress at a particular point in space, the stochastic differential equation (2.5) is solved for a large number of realizations, N_R, of the stochastic process $\boldsymbol{Q}$. Suppose that $\boldsymbol{Q}^{(k)}$, $k = 1, \ldots, N_R$, are the realizations generated in this way. The configuration pdf is not known since it has not been computed explicitly. Therefore, the polymeric extra stress given by the Kramers expression (1.39) is approximated by the ensemble average

$$\boldsymbol{\tau} = \frac{\eta_p}{\lambda_H}\alpha_{b,d}\left(-\boldsymbol{\delta} + \frac{1}{N_R}\sum_{k=1}^{N_R}\boldsymbol{Q}^{(k)}\boldsymbol{F}(\boldsymbol{Q}^{(k)})\right), \tag{2.14}$$

where $\alpha_{b,d}$ is a model-dependent parameter, defined in (1.40).

2.2. First-generation micro–macro techniques

2.2.1. CONNFFESSIT

Until fairly recently, the established approach to determining the stress in a viscoelastic flow calculation was to solve a closed form constitutive equation. As mentioned in Chapter 1, it was not until the early 1990s that an alternative approach to determining the polymer stress in a viscoelastic fluid, which bypassed the need to solve a constitutive equation with its incumbent problems, was advocated by LASO and ÖTTINGER [1993]. The new approach developed by LASO and ÖTTINGER [1993] is based on the solution of a stochastic differential equation to determine the polymer stress. In their paper, LASO and ÖTTINGER [1993] described the first micro–macro technique for solving viscoelastic flow problems based on kinetic theory models. The essence of the micro–macro approach is that it couples the coarse-grained molecular scale of kinetic theory to the macroscopic scale of continuum mechanics. In their particular implementation of the micro–macro technique, LASO and ÖTTINGER [1993] combined a finite element solution of the conservation equations with a stochastic simulation technique for computing the polymer stress. The polymeric contribution to the extra stress is computed from the configuration of a large ensemble of model polymers. The time evolution of this ensemble is calculated using Brownian dynamics, in which, for each model polymer, the trajectory of the center of mass is calculated from the local velocity field, and the evolution of the configuration of the molecules is determined by integrating the stochastic differential equation corresponding to the internal degrees of freedom of the model being considered. This hybrid method was named CONNFFESSIT (see Chapter 1), and micro–macro methods for polymer dynamics were born.

A major advantage of the micro–macro approach is that it allows for greater flexibility in the kinetic theory models that can be studied and simulated numerically since it does not require the existence of an equivalent or approximate closed-form constitutive equation. This means that models based on kinetic theory considerations that do not possess a closed-form equivalent, such as the FENE model, can be simulated directly without resorting to closure approximations, such as the FENE-P approximation, that are not universally accurate. Another advantage of the micro–macro approach is that effects such as polydispersity and hydrodynamic interactions can be fairly easily incorporated into the numerical procedure since the motion of individual polymer molecules is simulated (FEIGL, LASO and ÖTTINGER [1995]). The function of the model polymer "molecules" is to permit the computation of the polymer stress. This is achieved through the configurations of the molecules which contain information about the strain history.

Most numerical methods that are based on the micro–macro approach, with the notable exception of SOMASI and KHOMAMI [2000], LASO, RAMÍREZ and PICASSO [2004], decouple the solution of the conservation laws from the integration of the stochastic differential equation for the polymer configurations. The latter serves to determine the polymer contribution to the extra stress. At each time step (for transient flows) or iteration (for steady flows), the micro–macro algorithm proceeds as follows:

(1) Using the current approximation to the polymer stress as a source term in the momentum equation, the conservation equations are solved using standard discretization techniques (finite element, finite volume, or spectral element, for example) to obtain updated approximations to the velocity and pressure fields.
(2) The new velocity field is then used to convect a sufficiently large number of model polymer molecules through the flow domain. This is achieved by integrating the stochastic differential equation associated with the kinetic theory model along particle trajectories.
(3) The polymer stress within an element is determined using Kramers expression in which the ensemble average is computed by taking an average over a large number of realizations of the configurations of the polymer molecules in that element.

These steps are repeated until convergence is obtained.

The nonlinear coupling between the conservation laws and the Langevin equation has been investigated theoretically by JOURDAIN, LELIÈVRE and LE BRIS [2002, 2004b], LELIÈVRE [2004] for start-up of 2D planar shear flow of Hookean and FENE dumbbells. In this flow problem, all convection terms vanish in the governing equations. The Cauchy problem for Hookean dumbbells is shown to be well posed. Furthermore, the authors establish convergence of a finite element approximation to the exact solution to the problem. An optimal error estimate was derived by LELIÈVRE [2004].

The extension of this analysis for FENE dumbbells is more complicated due to the nonlinear and singular nature of the drift term in the Langevin equation (2.5) and the fact that the model is not closed and both components of the dumbbell connector vector depend on the spatial variable. For b sufficiently large, JOURDAIN, LELIÈVRE and LE BRIS [2004b] have proved the existence of a unique solution to the Langevin equation (2.5) for simple Couette flow. In the same paper, the authors established a local-in-time existence and uniqueness result for the coupled micro/macro problem for data that is sufficiently regular. The extension of this existence result for arbitrary large time or for a class of data that is less regular remains elusive.

Let us look at the stochastic part of the calculation for the FENE dumbbell model in more detail. The Euler–Maruyama scheme for solving the stochastic differential equation (2.5) with $\boldsymbol{A}$ and $\boldsymbol{B}$ defined by (2.3) and (2.4), respectively, for the FENE dumbbell model is

$$\boldsymbol{Q}_{i+1} = \boldsymbol{Q}_i + \left(\nabla \boldsymbol{v}_c \cdot \boldsymbol{Q}_i - \frac{1}{2\lambda_H}\boldsymbol{F}(\boldsymbol{Q}_i)\right)\Delta t_i + \frac{1}{\sqrt{\lambda_H}}\Delta \boldsymbol{W}_i. \tag{2.15}$$

The scheme (2.15) may be used to generate a set of N_R independent trajectories, $\boldsymbol{Q}^{(j)}$, $j = 1, \dots, N_R$, by selecting realizations of the random vectors $\boldsymbol{W}_i$. These trajectories may be characterized in time by $(\boldsymbol{Q}_0^{(j)}, \ldots, \boldsymbol{Q}_n^{(j)})$, $j = 1, \ldots, N_R$, where n is the current value of the time counter. At each time $t = t_i$, the polymeric stress is computed by taking an ensemble average over individual realizations of the stochastic process $\boldsymbol{Q}$, i.e.,

$$\boldsymbol{\tau}(t_i) = \frac{\eta_p}{\lambda_H}\left(\frac{b+d+2}{b}\right)\left(-\boldsymbol{\delta} + \frac{1}{N_R}\sum_{k=1}^{N_R} \boldsymbol{Q}_i^{(k)} \boldsymbol{F}(\boldsymbol{Q}_i^{(k)})\right). \tag{2.16}$$

Since $\boldsymbol{Q}^{(j)}, j = 1, \ldots, N_R$, are independent, the strong law of large numbers guarantees that the arithmetic mean in (2.16) converges to the ensemble average as $N_R \to \infty$.

Higher order time integration schemes based on the predictor-corrector approach may also be used for stochastic differential equations. For example, the Euler-trapezoidal predictor-corrector pair

$$\begin{aligned}
\overline{\boldsymbol{Q}}_{i+1} &= \boldsymbol{Q}_i + \boldsymbol{A}(\boldsymbol{Q}_i, t_i)\Delta t_i + \boldsymbol{B}(\boldsymbol{Q}_i, t_i)\Delta \boldsymbol{W}_i, \\
\boldsymbol{Q}_{i+1} &= \mathbf{D}\boldsymbol{Q}_i + \frac{1}{2}\left[\boldsymbol{A}(\overline{\boldsymbol{Q}}_{i+1}, t_{i+1}) + \boldsymbol{A}(\boldsymbol{Q}_i, t_i)\right]\Delta t_i \\
&\quad + \frac{1}{2}\left[\boldsymbol{B}(\overline{\boldsymbol{Q}}_{i+1}, t_{i+1}) + \boldsymbol{B}(\boldsymbol{Q}_i, t_i)\right]\Delta \boldsymbol{W}_i.
\end{aligned} \tag{2.17}$$

for solving the general stochastic differential equation (2.5) has weak order of convergence $\nu = 2$ in the case of additive noise, i.e., when the diffusion coefficient $\boldsymbol{B}$ depends only on t. In the case of multiplicative noise, the order of weak convergence of this scheme is reduced to $\nu = 1$. The diffusion coefficient defined in (2.4) is independent of $\boldsymbol{q}$, and therefore, one can expect weak order of convergence $\nu = 2$ when this predictor-corrector scheme is applied to the numerical solution of the stochastic differential equation (2.5).

Stability problems may be experienced when the Euler–Maruyama scheme is applied to the stochastic differential equation (2.5) for the FENE model. In particular, if the time step Δt used in the stochastic part of the calculation is too large, then it is possible for the dimensionless length of a dumbbell to exceed $\sqrt{b}$, the dimensionless finite extensibility parameter, which is clearly physically unrealistic. It is possible to circumvent this problem in several ways. For example, one can reset the dimensionless length of offending dumbbells to a value less than $\sqrt{b}$ by "reflecting" the excess length back through the maximum value $\sqrt{b}$ (LASO and ÖTTINGER [1993]). An alternative means of avoiding dumbbell configurations that violate the maximum extensibility constraint is to use an implicit scheme such as the predictor-corrector scheme (2.17) of weak order 2 in which the spring force law is treated implicitly:

$$\begin{aligned}
\overline{\boldsymbol{Q}}_{i+1} &= \boldsymbol{Q}_i + \left[\nabla \boldsymbol{v}_c \cdot \mathbf{Q}_i - \frac{1}{2\lambda_H}\boldsymbol{F}(\boldsymbol{Q}_i)\right]\Delta t_i \\
&\quad + \Delta \mathbf{W}_i,
\end{aligned} \tag{2.18}$$

$$\begin{aligned}
\left[1 + \frac{\Delta t_i}{4\lambda_H}\boldsymbol{F}(\boldsymbol{Q}_{i+1})\right]\boldsymbol{Q}_{i+1} &= \boldsymbol{Q}_i + \frac{1}{2}\left[\nabla \boldsymbol{v}_c \cdot \overline{\boldsymbol{Q}}_{i+1} + \nabla \boldsymbol{v}_c \cdot \mathbf{Q}_i\right] \\
&\quad - \frac{\Delta t_i}{4\lambda_H}\boldsymbol{F}(\boldsymbol{Q}_i) + \Delta \mathbf{W}_i.
\end{aligned} \tag{2.19}$$

However, note that the corrector stage of the scheme (2.19) is a cubic equation for $\boldsymbol{Q}_{i+1}$. The direction of $\boldsymbol{Q}_{i+1}$ is given by the direction of the right-hand side of (2.19). The length of $\boldsymbol{Q}_{i+1}$ is determined by solving a cubic equation of the form

$$q^3 - Lq^2 - \left(1 + \frac{\Delta t}{4\lambda}\right) bq + Lb = 0, \tag{2.20}$$

where L is the length of the vector on the right-hand side of (2.19). This equation possesses exactly one real root in the interval $[0, \sqrt{b}]$ for an arbitrary value of L (see ÖTTINGER [1996], for example).

LASO and ÖTTINGER [1993] described the basic components of the CONNFFESSIT approach by simulating the time development of plane Couette flow using several model polymers such as Hookean and FENE dumbbells. In this one-dimensional problem, the gap between the two plates was divided into M finite elements and an ensemble of model polymers assigned to each element. During the transient development of this flow, the dumbbells do not migrate across finite element boundaries. Instead they reside in their initial element for all time. In this problem, therefore, there is no need to keep track of the spatial locations of the dumbbells in order to allocate them to particular finite elements as a precursor to the evaluation of the polymer stress. In their computations, LASO and ÖTTINGER [1993] chose $M = 40$ with $O(10^6)$ dumbbells per finite element. The time step for the stochastic part of the calculation was chosen to be $\Delta t_{\text{stoch}} = \lambda_H/1000$, where λ_H is a characteristic relaxation time of the dumbbell.

An initial distribution of dumbbell configuration vectors, $\boldsymbol{Q}_0^{(j)}, j = 1, \ldots, N_R$, is required to begin the numerical simulations. It is natural to choose these from the equilibrium distribution, ψ_{eq}, found by solving the Fokker–Planck equation (2.1) with $\boldsymbol{u} \equiv \boldsymbol{0}$. For 3D Hookean dumbbells, with dimensionless force law

$$\boldsymbol{F} = \boldsymbol{q},$$

the equilibrium distribution is

$$\psi_{\text{eq}} = \frac{1}{(2\pi)^{3/2}} \exp\left(-\frac{q^2}{2}\right), \tag{2.21}$$

(see BIRD, CURTISS, ARMSTRONG and HASSAGER [1987b]), a three-dimensional Gaussian distribution with zero mean and unit covariance matrix.

Comparisons of stochastic simulations using Hookean dumbbells were made with conventional finite element computations using the equivalent Oldroyd-B model. Excellent agreement was found between the two approaches within statistical error. These initial studies using the CONNFFESSIT methodology also revealed important differences in the dynamic behavior of the FENE and FENE-P models. However, the early implementations of CONNFFESSIT proved to be considerably more expensive in terms of computer memory and CPU time than the corresponding macroscopic computations. The expense is attributable to the very large number of dumbbells that need to be simulated to obtain convergence and also the generation of MN_Rd random numbers at each time step in the simulation of the stochastic differential equation in d dimensions, where M is the number of finite elements.

A feature of stochastic simulations is the presence of temporal and spatial fluctuations in the computed stress and velocity fields. The temporal fluctuations arise from the statistical error incurred in approximating ensemble averages such as (2.16), for example, using a finite number, N_R, of trajectories. These fluctuations may be controlled by increasing the number of trajectories that are simulated. However, as we have already remarked, this approach to statistical error reduction is undesirable because of the computational cost. Spatial fluctuations arise through the computation of the divergence of a nonsmooth stress field in the momentum equation. As we shall see these fluctuations may be reduced using correlated local ensembles of model polymers to approximate (2.16).

Despite the advantages of the CONNFFESSIT approach in terms of the kinetic theory models that can be simulated, there were a number of computational shortcomings in the original implementations of the idea. First, the trajectories of a large number of molecules have to be determined. Second, to evaluate the local polymer stress, the model polymer molecules must be sorted according to elements every time step. Naturally, problems will also arise if elements become deficient of particles. Third, the computed stress may be nonsmooth, and this may cause problems when it is differentiated to form the source term in the momentum equation. In subsequent developments of the technique, these shortcomings have been overcome to a certain extent. Some of the new ideas will be described in the remainder of this chapter.

2.2.2. *Variance reduction*

Variance reduction is a means of reducing the statistical error in a stochastic simulation without increasing the number of trajectories that are simulated. To compute the polymeric stress at some time $\bar{t}$ in the case of the Hookean or FENE dumbbell models, we need to determine

$$\langle \boldsymbol{Q}(\bar{t})\boldsymbol{F}(\boldsymbol{Q}(\bar{t}))\rangle \approx \frac{1}{N_R}\sum_{k=1}^{N_R} \boldsymbol{Q}^{(k)}(\bar{t})\boldsymbol{F}(\boldsymbol{Q}^{(k)}(\bar{t})). \tag{2.22}$$

Let $\Theta = \mathrm{Var}[\boldsymbol{Q}(\bar{t})\boldsymbol{F}(\boldsymbol{Q}(\bar{t}))]$ denote the variance of $\boldsymbol{QF}(\boldsymbol{Q})$. Then, the statistical error incurred in approximating the ensemble (2.22) is $\sqrt{\Theta/N}$. In a variance-reduced simulation, the idea is to construct a random variable Y such that $\langle Y\rangle = \langle \boldsymbol{Q}(\bar{t})\boldsymbol{F}(\boldsymbol{Q}(\bar{t}))\rangle$ and $\mathrm{Var}(Y) = \Theta_Y < \Theta$.

MELCHIOR and ÖTTINGER [1995, 1996] proposed a number of variance reduction methods in the context of the CONNFFESSIT methodology based on importance sampling strategies and the idea of control variables. Large variances in stochastic simulations invariably occur when very few realizations make a significant contribution to the mean value (2.22) giving rise to a large variance. Since the equilibrium distributions for Hookean and FENE dumbbells are peaked at the origin, the majority of the configurations that are generated are centered there. These configurations do not make a significant contribution to the approximation of the ensemble average. The idea in importance sampling is to introduce a bias that gives greater weight to the realizations that make a substantial contribution to the average. The bias is constructed from an approximate solution of a stochastic differential equation for a modified stochastic process $\tilde{\boldsymbol{Q}}$ that gives greater weight to the important realizations (MELCHIOR and ÖTTINGER [1995]). In the second approach, based on control variables

(MELCHIOR and ÖTTINGER [1996]), the idea is to find a random variable that possesses the same fluctuations as the random variable of interest, $\boldsymbol{Q}$, but with a zero mean. When the control variable is subtracted from the original variable, then the mean remains unchanged while the fluctuations are reduced. MELCHIOR and ÖTTINGER [1995, 1996] described two techniques for constructing control variables: direct control and parallel process simulations.

ÖTTINGER, VAN DEN BRULE and HULSEN [1997] applied the control variable method to the problem of start-up of weak homogeneous shear flow in order to demonstrate the power of variance reduction techniques within the CONNFFESSIT framework for micro–macro simulations. Two stochastic simulations were performed for an ensemble of FENE dumbbells, one assuming equilibrium (no flow) and the other under flowing conditions. Using the same initial ensemble and with the same sequence of random numbers in the simulations, the fluctuations in the transient development of the shear stress were virtually indistinguishable. In this application of the control variable technique, the equilibrium simulation is suited to the role of parallel process simulation since it has virtually the same fluctuations as the original simulation at small shear-rates, and the average of the shear stress vanishes at equilibrium due to symmetry considerations. Consequently, the subtraction of the equilibrium control variable from the original variable results in substantial variance reduction, and reliable results were obtained with far fewer dumbbells in the ensemble than would otherwise have been necessary.

An alternative approach to variance reduction is to use local correlated ensembles of model polymers. The idea is that corresponding model polymers in each material element feel the same Brownian force. More precisely, the same initial ensemble of model polymers is defined in each material element, and corresponding model polymers in each material element are allowed to evolve using the same sequence of random numbers. This approach leads to strong correlations in the stress fluctuations in neighboring material elements. The evaluation of the divergence of the stress in the momentum equation involves the difference between stresses and leads to a cancellation of fluctuations and dramatic variance reduction. The Brownian configuration field method of HULSEN, VAN HEEL and VAN DEN BRULE [1997], ÖTTINGER, VAN DEN BRULE and HULSEN [1997] and the Lagrangian particle method of HALIN, LIELENS, KEUNINGS and LEGAT [1998] are examples of variance-reduced stochastic simulation methods based on the idea of correlated local ensembles of model polymers. Not only do these techniques reduce the spatial fluctuations in the computed velocity and stress fields but they also require the generation of fewer random numbers. This greatly reduces the computational cost associated with these stochastic simulation techniques. The cost of achieving variance reduction is that unphysical correlations in the random forces are introduced into the simulations. For problems in which physical fluctuations are important, one must revert to calculations based on uncorrelated Brownian forces even though this is likely to be more expensive.

In the continuous formulation of the problem, the velocity and polymeric contribution to the extra stress are deterministic, while the microstructural variables are random. In the corresponding discrete problem, all the variables are random. Therefore, the accuracy of a given simulation depends on the variance of the discrete variables. A large variance would imply that independent simulations would yield vastly different solutions for the same problem. Clearly, this would not be a satisfactory state of affairs.

Numerical results performed using Brownian configuration fields (ÖTTINGER, VAN DEN BRULE and HULSEN [1997]) showed that the variance of the discrete velocity is reduced significantly. This led the authors to regard the method to be a powerful variance reduction technique. However, this only tells part of the story. It has been shown by HALIN, LIELENS,

KEUNINGS and LEGAT [1998], BONVIN and PICASSO [1999] that although the use of correlated ensembles reduces the variance in the velocity approximation, the variance in the polymeric stress is increased (for nonlinear FENE dumbbells). This must be due to the nonlinear interaction between the macroscopic and microscopic parts of the calculation. JOURDAIN, LE BRIS and LELIÈVRE [2004a] have confirmed this experimental finding in a theoretical analysis in the case of start-up of shear flow of Hookean dumbbells.

2.3. Second-generation micro–macro techniques

2.3.1. *Brownian configuration fields*

To overcome the problem of having to track particle trajectories, HULSEN, VAN HEEL and VAN DEN BRULE [1997] developed a method based on the evolution of Brownian configuration fields (see Chapter 1). The collection of discrete particles that is used in the CONNFFESSIT approach is replaced by an ensemble of configuration fields. Rather than computing the configuration of discrete particles along flow trajectories, as in the CONNFFESSIT approach, the method determines the evolution of a finite number of continuous configuration fields. The configuration fields are convected by the flow, and their deformation is governed by the kinematics, elastic retraction, and Brownian motion. As far as the determination of the stress is concerned, the evolution of configuration fields is equivalent to the tracking of individual particles. At each point $(\boldsymbol{x}, t)$, an ensemble of configuration vectors $\boldsymbol{Q}_i$, $i = 1, \ldots, N_f$, is generated that experienced the same history in terms of the kinematics but which underwent different stochastic processes.

The method of Brownian configuration fields may be interpreted as an Eulerian implementation of the concept of correlated local ensembles. The idea behind the use of correlated local ensembles of model polymer molecules is to use the same local ensemble of model polymer molecules initially in each Lagrangian particle and then to subject corresponding model polymer molecules in each particle to the same sequence of random forces. In the method of Brownian configuration fields, the configurations of corresponding model polymer molecules within each local ensemble are combined to form a configuration field. Therefore, at each instant in time, each configuration field experiences a random kick that is uniform in space.

The method of Brownian configuration fields is based on the evolution of an ensemble of N_f configuration fields $\boldsymbol{Q}_i(\boldsymbol{x}, t)$, $i = 1, \ldots, N_f$, defined throughout the flow domain. One can identify the configuration field $\boldsymbol{Q}_i(\boldsymbol{x}, t)$ with the configuration of the ith model polymer molecule in the local ensemble at the point having position vector $\boldsymbol{x}$ at time t. Initially these fields are chosen to be spatially uniform, and their values are independently sampled from the equilibrium distribution function of the dumbbell model so that $\boldsymbol{Q}_i(\boldsymbol{x}, 0) = \boldsymbol{Q}_i^0(\boldsymbol{x})$. The evolution of the ith configuration field, $\boldsymbol{Q}_i(\boldsymbol{x}, t)$, is governed by the stochastic differential equation

$$\mathrm{d}\boldsymbol{Q}_i(\boldsymbol{x}, t) = (-\boldsymbol{u}(\boldsymbol{x}, t) \cdot \nabla \boldsymbol{Q}_i(\boldsymbol{x}, t) + \boldsymbol{A}(\boldsymbol{Q}_i(\boldsymbol{x}, t)))\mathrm{d}t + \boldsymbol{B}(\boldsymbol{Q}_i(\boldsymbol{x}, t))\,\mathrm{d}\boldsymbol{W}_i(t). \tag{2.23}$$

This equation is similar to (2.5) except that an additional term has been included to account for the convection of the configuration field by the flow. It is the Eulerian formulation of the stochastic differential equation (2.5). Note that the spatial gradients of the configuration fields are well-defined and smooth since $\mathrm{d}\boldsymbol{W}_i(t)$ only depends on time. Since these fields are spatially smooth, they can be approximated by finite elements (HULSEN, VAN HEEL and VAN DEN BRULE [1997]) or spectral elements (PHILLIPS and SMITH [2006]), for example. Therefore, the evaluation of the divergence of the polymeric contribution to the extra-stress tensor, which forms the source term in the momentum equation, contributes to a reduction in the stress fluctuations due to the strong spatial correlations in these quantities that develop at neighboring points in the flow domain.

The integration of the stochastic differential equation (2.23) only requires dN_f random variables in d dimensions since $d\boldsymbol{W}(t)$ only depends on time and not on position. This means that the same sequence of random numbers is used to determine the configuration of the ith model polymer molecule in the local ensemble throughout the flow domain.

HULSEN, VAN HEEL and VAN DEN BRULE [1997] used a discontinuous Galerkin method for solving the stochastic differential equation (2.23) for each configuration field. Let $\mathcal{Q}$ denote the appropriate function space for the configuration field. Then, the weak formulation of (2.23) is as follows: for each finite element e, find $\boldsymbol{Q}_i \in \mathcal{Q}$ such that

$$\begin{aligned}&\left(d\boldsymbol{Q}_i + (\boldsymbol{u}\cdot\nabla\boldsymbol{Q}_i + \boldsymbol{A}(\boldsymbol{Q}_i))dt - \boldsymbol{B}(\boldsymbol{Q}_i)d\boldsymbol{W}_i, \boldsymbol{R}\right)_e \\ &\qquad + (\boldsymbol{n}\cdot\boldsymbol{u}(\boldsymbol{Q}_i^+ - \boldsymbol{Q}_i)dt, \boldsymbol{R})_{\gamma^{\text{in}}} = 0, \qquad (2.24)\end{aligned}$$

$\forall\, \boldsymbol{R} \in \mathcal{Q}$, for $i = 1, \ldots, N_f$, where $\boldsymbol{Q}_i^+$ is the value of $\boldsymbol{Q}_i$ in the neighboring upstream element or the value imposed at inflow and γ^{in} is the part of the boundary of e for which $\boldsymbol{u}\cdot\boldsymbol{n} < 0$ ($\boldsymbol{n}$ is the unit outward normal to the boundary of e). Discontinuous bilinear polynomials were used to approximate $\boldsymbol{Q}_i$ in each element.

The polymeric contribution to the extra-stress tensor is determined by projecting the ensemble average

$$\frac{1}{N_f}\sum_{i=1}^{N_f} \boldsymbol{Q}_i \boldsymbol{F}(\boldsymbol{Q}_i)$$

onto $\boldsymbol{\mathcal{Q}}$ to obtain a tensor $\boldsymbol{C}$ and then evaluating $\boldsymbol{\tau}$ using

$$\boldsymbol{\tau} = \frac{\eta_p}{\lambda_H}\alpha_{b,\text{d}}\,(-\boldsymbol{\delta} + \boldsymbol{C}),$$

where $\alpha_{b,\text{d}}$ is the model-dependent parameter defined in (1.40).

HULSEN, VAN HEEL and VAN DEN BRULE [1997] have used this technique to simulate the start-up of planar flow of an Oldroyd-B fluid past a cylinder between two parallel plates. The Brownian configuration field method overcomes the weaknesses of the original methods based on Brownian simulation techniques in that it does not require individual particles to be tracked and sorted according to residency in a particular finite element at each time step. Another advantage is that the statistical error at any point in the flow domain is governed by the number, N_f, of configuration fields, which is independent of the mesh. In contrast, for particle-based methods, mesh refinement requires an increase in the number of particles in order to maintain a given statistical error. In fact, since the smallest elements typically contain the fewest number of particles, the statistical error will be largest in the regions of the flow where the greatest accuracy is required.

BONVIN and PICASSO [1999] have applied a variance reduction method, viz. the control variable method, in order to reduce the level of noise in their simulations based on Brownian configuration fields. However, BONVIN [2000] points out that this variance reduction technique does not always reduce the noise. In particular, for flows with large Deborah numbers, the scheme without control variates can produce less noisy solutions than the variance reduced method based on control variates.

CHAUVIÈRE [2002] has derived an efficient method for solving the stochastic differential equation (2.23) that removes the noise from the micro–macro simulations. It should noted, however, that the original analysis, performed for the Oldroyd-B and FENE-P models, is only applicable to models where a closed form constitutive equation exists. A more efficient version of this method that is competitive with macroscopic computations was developed by CHAUVIÈRE and LOZINSKI [2003]. CHAUVIÈRE and LOZINSKI [2003] express the backward Euler scheme for solving the stochastic differential equation (2.5) in the case of the Hookean dumbbell model in the form

$$\boldsymbol{E}_u\boldsymbol{Q}(\boldsymbol{x}, t_i) = \boldsymbol{Q}(\boldsymbol{x}, t_{i-1}) + \sqrt{\frac{1}{\lambda_H}}(\boldsymbol{W}(t_i) - \boldsymbol{W}(t_{i-1})) \tag{2.25}$$

where the linear operator $\boldsymbol{E}_u$ is defined by

$$\begin{aligned}\boldsymbol{E}_u(\boldsymbol{x}, t_i) = \boldsymbol{Q}(\boldsymbol{x}, t_i) + \Bigg(&\boldsymbol{u}(\boldsymbol{x}, t_i)\cdot\nabla\boldsymbol{Q}(\boldsymbol{x}, t_i) \\ &-\boldsymbol{\kappa}(\boldsymbol{x}, t_i)\boldsymbol{Q}(\boldsymbol{x}, t_i) + \frac{1}{2\lambda_H}\boldsymbol{Q}(\boldsymbol{x}, t_i)\Bigg)\Delta t,\end{aligned} \tag{2.26}$$

where $\boldsymbol{\kappa}$ is the velocity gradient. Now each of the random vectors $\boldsymbol{W}(t_i) - \boldsymbol{W}(t_{i-1})$ contains d independent Gaussian random scalar variables each having zero mean and variance Δt. Therefore, at time t_i, the Gaussian random vector $\boldsymbol{\Phi}(t_i)$ can be introduced such that $\langle\boldsymbol{\Phi}(t_i)\rangle = \boldsymbol{0}$, $\langle\boldsymbol{\Phi}(t_i)\boldsymbol{\Phi}^T(t_i)\rangle = \boldsymbol{\delta}$, and $\boldsymbol{W}(t_i) - \boldsymbol{W}(t_{i-1}) = \sqrt{\Delta t}\boldsymbol{\Phi}(t_i)$. Note that $\boldsymbol{\Phi}(t_i)$ are mutually independent, i.e., $\langle\boldsymbol{\Phi}(t_i)\boldsymbol{\Phi}^T(t_j)\rangle = \boldsymbol{0}$, if $i \neq j$. Therefore, in terms of $\boldsymbol{\Phi}$, we can write (2.25) in the form

$$\boldsymbol{E}_u\boldsymbol{Q}(\boldsymbol{x}, t_i) = \boldsymbol{Q}(\boldsymbol{x}, t_{i-1}) + \sqrt{\frac{\Delta t}{\lambda_H}}\boldsymbol{\Phi}(t_i). \tag{2.27}$$

Initially, the fluid is in an equilibrium state so that the initial condition for (2.27) is of the form

$$\boldsymbol{Q}(\boldsymbol{x}, t_0) = \boldsymbol{\Phi}(t_0). \tag{2.28}$$

This ensures that the extra stress defined by (2.16) is zero at time $t = t_0$.

In the method of Brownian configuration fields, we have N_f configuration fields $\{\boldsymbol{Q}_m(\boldsymbol{x}, t)\}$, $m = 1, \ldots, N_f$, which model the random quantities $\boldsymbol{Q}(\boldsymbol{x}, t)$. Each configuration field, $\boldsymbol{Q}_m$, satisfies an equation of the form (2.27) with $\boldsymbol{\Phi}$ replaced by a pseudo-random vector $\boldsymbol{\Phi}_m$ having approximately Gaussian distribution. An approximation to the extra-stress tensor is then determined through

$$\boldsymbol{\tau}(\boldsymbol{x}, t_i) \approx \frac{\eta_p}{\lambda_H}\left(-\boldsymbol{\delta} + \frac{1}{N_f}\sum_{m=1}^{Nf}\boldsymbol{Q}(\boldsymbol{x}, t_i)\boldsymbol{Q}^T(\boldsymbol{x}, t_i)\right), \tag{2.29}$$

in the case of Hookean dumbbells.

THEOREM 3. *Let $\boldsymbol{Q}(\boldsymbol{x}, t_i)$ be the solution to the stochastic differential equation (2.5) with initial condition (2.28) at time $t_i = i\Delta t$, where $\boldsymbol{E}_{\boldsymbol{u}}$ is defined by (2.27).*
Let $\boldsymbol{A}(\boldsymbol{x}, t_i)$ for $i \geq 1$ denote symmetric positive definite $d \times d$ matrices defined by

$$\boldsymbol{A}^2(\boldsymbol{x}, t_i) = \widehat{\boldsymbol{A}}(\boldsymbol{x}, t_i)\widehat{\boldsymbol{A}}^T(\boldsymbol{x}, t_i) + \widetilde{\boldsymbol{A}}(\boldsymbol{x}, t_i)\widetilde{\boldsymbol{A}}^T(\boldsymbol{x}, t_i), \tag{2.30}$$

where $\widetilde{\boldsymbol{A}}(\boldsymbol{x}, t_i)$ and $\widehat{\boldsymbol{A}}(\boldsymbol{x}, t_i)$ are $d \times d$ matrices satisfying the following equations

$$\boldsymbol{E}_{\boldsymbol{u}}\widetilde{\boldsymbol{A}}(\boldsymbol{x}, t_i) = \boldsymbol{A}(\boldsymbol{x}, t_{i-1}) \tag{2.31}$$

$$\boldsymbol{E}_{\boldsymbol{u}}\widehat{\boldsymbol{A}}(\boldsymbol{x}, t_i) = \sqrt{\frac{\Delta t}{\lambda_H}}\boldsymbol{\delta}, \tag{2.32}$$

and

$$\boldsymbol{A}(\boldsymbol{x}, t_0) = \boldsymbol{\delta}. \tag{2.33}$$

Then, we have

$$\boldsymbol{Q}(\boldsymbol{x}, t_i) \sim \boldsymbol{A}(\boldsymbol{x}, t_i)\widetilde{\boldsymbol{\Phi}}(t_i), \tag{2.34}$$

where $\widetilde{\boldsymbol{\Phi}}(t_i)$ is a d-dimensional Gaussian random vector.

PROOF. See CHAUVIÈRE and LOZINSKI [2003]. □

This theorem can be used to derive an expression for the extra-stress tensor.

THEOREM 4. *The extra-stress tensor $\boldsymbol{\tau}(\boldsymbol{x}, t_i)$ at time $t = i\Delta t$ is given by*

$$\boldsymbol{\tau}(\boldsymbol{x}, t_i) = \frac{\eta_p}{\lambda_H}\Big(-\boldsymbol{\delta} + \boldsymbol{A}^2(\boldsymbol{x}, t_i)\Big), \tag{2.35}$$

where $\boldsymbol{A}(\boldsymbol{x}, t_i)$ is the $d \times d$ matrix defined by eqn (2.30).

CHAUVIÈRE and LOZINSKI [2003] proceed to show that the extra-stress tensor computed using (2.27), (2.28), (2.30)–(2.33), and (2.35) satisfies the Oldroyd-B constitutive equation in the limit $\Delta t \to 0$.

CHAUVIÈRE and LOZINSKI [2003] embed this method for determining a noise-free extra stress within a micro–macro technique as follows. Suppose that a generalized Stokes problem has been solved using the extra stress at time $t_{i-1} = (i-1)\Delta t$ to obtain the velocity field $\boldsymbol{u}(\boldsymbol{x}, t_{i-1})$. Let the computational grid consist of points $\boldsymbol{x}_j$, $j = 1, \ldots, N_c$, and let $\overline{\boldsymbol{E}}_{\boldsymbol{u}}(t_i)$ be the matrix arising from the spatial discretization of the linear operator $\boldsymbol{E}_{\boldsymbol{u}}$ at time t_i for a given velocity $\boldsymbol{u}$. The N_c $d \times d$ matrices $\boldsymbol{A}(\boldsymbol{x}_j, t_{i-1}), j = 1, \ldots, N_c$ are formed into a single $(dN_c) \times d$ matrix $\boldsymbol{\mathcal{A}}(t_{i-1})$ by stacking them on top of each other. The $(dN_c) \times d$ matrix $\boldsymbol{\mathcal{I}}$ is formed from N_c identity $d \times d$ matrices in the same way. New $(dN_c) \times d$ matrices are calculated by solving the equations

$$\overline{\boldsymbol{E}}_u(t_i)\widetilde{\boldsymbol{\mathcal{A}}}(t_i) = \boldsymbol{\mathcal{A}}(t_{i-1})$$

$$\overline{\boldsymbol{E}}_u(t_i)\widehat{\boldsymbol{\mathcal{A}}}(t_i) = \sqrt{\frac{\Delta t}{\lambda_H}}\boldsymbol{\mathcal{I}}.$$

The $d \times d$ symmetric positive definite matrices $\boldsymbol{A}(x_j, t_i)$ are computed by solving the algebraic equation

$$\boldsymbol{A}^2(\boldsymbol{x}_j, t_i) = \widehat{\boldsymbol{A}}(\boldsymbol{x}_j, t_i)\widehat{\boldsymbol{A}}^T(\boldsymbol{x}_j, t_i) + \widetilde{\boldsymbol{A}}(\boldsymbol{x}_j, t_i)\widetilde{\boldsymbol{A}}^T(\boldsymbol{x}_j, t_i), \tag{2.36}$$

at every grid point $\boldsymbol{x}_j, j = 1, \ldots, N_c$. Then finally, the extra-stress tensor is computed using (2.35) at each grid point.

This process can be extended quite simply to the FENE-P model, in which case the expression for the extra-stress tensor is

$$\boldsymbol{\tau}(\boldsymbol{x}, t_i) = \frac{\eta_p}{\lambda_H}\left(\frac{b + \mathrm{d}}{b}\right)\left(-\boldsymbol{\delta} + \frac{\boldsymbol{A}^2(\boldsymbol{x}, t_i)}{1 - \frac{1}{b}\mathrm{tr}(\boldsymbol{A}^2(\boldsymbol{x}, t_i))}\right). \tag{2.37}$$

In fact, this procedure is able to deliver noise-free expressions for the stress for the Hookean and FENE-P dumbbell models precisely because they have Gaussian statistics and an equivalent closed-form constitutive equation. This approach can be applied to any model for which this is the case.

2.3.2. Lagrangian particle method

The Lagrangian particle method (LPM) (see Chapter 1) introduced by HALIN, LIELENS, KEUNINGS and LEGAT [1998] provides a refinement of the CONNFFESSIT approach. In common with CONNFFESSIT, LPM is based on a decoupled micro/macro scheme in which the solution of the conservation laws using finite element techniques is combined with the solution of the evolution equation

$$\mathrm{d}\boldsymbol{Q}(\boldsymbol{x}, t) = (-\boldsymbol{u}(\boldsymbol{x}, t) \cdot \nabla\boldsymbol{Q}(\boldsymbol{x}, t) + \boldsymbol{A}(\boldsymbol{Q}(\boldsymbol{x}, t)))\,\mathrm{d}t + \boldsymbol{B}(\boldsymbol{Q}(\boldsymbol{x}, t))\,\mathrm{d}\boldsymbol{W}(t), \tag{2.38}$$

using a method of characteristics. The polymeric contribution to the extra stress is computed at a number of Lagrangian particles, N_p, that are convected by the flow.

Let $\{\boldsymbol{x}_i^n : 1 \le i \le N_p\}$ denote the positions of a set of Lagrangian particles at time $t_n = n\Delta t$. The trajectory, $\boldsymbol{x}_i(t)$, $t_n \le t \le t_{n+1}$, of each Lagrangian particle is determined using the velocity field, $\boldsymbol{u}(t_n)$, at time $t = t_n$ by solving the initial value problem

$$\frac{\mathrm{d}\boldsymbol{x}_i}{\mathrm{d}t} = \boldsymbol{u}^n(\boldsymbol{x}_i), \ \boldsymbol{x}_i(t_n) = \boldsymbol{x}_i^n. \tag{2.39}$$

This problem can be solved using high-order methods such as the fourth-order Runge- Kutta method (GALLEZ, HALIN, LIELENS, KEUNINGS and LEGAT [1999]), for example, using an intermediate time step $\Delta t_{\mathrm{int}} = \Delta t / K$, where $K \ge 1$ is an integer.

Although the particle tracking in the microscopic part of LPM is similar to that used in CONNFFESSIT, it differs from it in a computationally significant way since each Lagrangian particle carries information about a number, N_d, of model polymer molecules. Therefore, LPM reduces the number of particles that need to be tracked from $N_p \times N_d$ in a CONNFFESSIT calculation to N_p.

Over the time interval $[t_n, t_{n+1}]$, the microscopic part of LPM involves the following calculations: along the trajectory of each of the Lagrangian particles, the stochastic differential equation (2.38) is solved for each of the N_d model polymer molecules using $\boldsymbol{Q} = \boldsymbol{Q}(\boldsymbol{x}, t_n)$ as the initial condition. The polymer stress associated with each Lagrangian particle is then approximated by taking an ensemble average over the N_d realizations of the stochastic process $\boldsymbol{Q}$

$$\boldsymbol{\tau}(\boldsymbol{x}(t_{n+1})) \approx \frac{\eta_p}{\lambda_H}\alpha_{b,d}\left(-\boldsymbol{\delta} + \frac{1}{N_d}\sum_{i=1}^{N_d}\boldsymbol{Q}^{(i)}\boldsymbol{F}\left(\boldsymbol{Q}^{(i)}\right)\right), \tag{2.40}$$

where $\boldsymbol{Q}^{(i)}$ is an individual realization of the stochastic process and $\alpha_{b,d}$ is defined in (1.40). The polymeric stress associated with the Lagrangian particles is then known at time $t = t_{n+1}$. The stress approximation within each finite element is determined by computing the best linear least squares polynomial that fits the data associated with the Lagrangian particles within that element. Clearly, this requires that there are at least three particles within each element at all times. The least-squares approximation to the stress in each element is then used to form the source term in the momentum equation as a precursor to the macroscopic part of the calculation.

The LPM may be implemented in one of two modes which correspond to using either uncorrelated or correlated local ensembles of model polymer molecules. In the first mode, a total of $N_p \times N_d$ independent Wiener processes are required for the stochastic part of the calculation. This mode corresponds to the original CONNFFESSIT method in the case when N_d particles are located at each of N_p positions at each time step rather than all $N_p \times N_d$ particles being dispersed throughout the flow. The second mode is very close in philosophy to the method of Brownian configuration fields in that it uses correlated local ensembles of model polymer molecules. It is implemented by using the same initial ensemble of model polymer molecules in each Lagrangian particle and the same N_d independent Wiener processes to compute the configurations of corresponding polymer molecules within each particle. Thus, LPM may be viewed as a discrete version of the method of Brownian configuration fields.

The adaptive Lagrangian particle method (ALPM) was developed by GALLEZ, HALIN, LIELENS, KEUNINGS and LEGAT [1999] to ensure that there are sufficiently many Lagrangian particles within each element for all time. More precisely, the ALPM ensures that the number of Lagrangian particles in each element lies within a specified interval by creating or destroying particles. Although GALLEZ, HALIN, LIELENS, KEUNINGS and LEGAT [1999] found that ALPM was superior to LPM in terms of numerical accuracy and computational cost, they found that it was more elaborate to implement due to the work involved when new particles are created. When a new Lagrangian particle is created within a given element, to ensure that there are sufficiently many particles, the configuration of the local N_d model polymer molecules must be initialized appropriately before the stochastic differential equation (2.38) can be integrated along the trajectory of the particle. GALLEZ, HALIN, LIELENS, KEUNINGS and LEGAT [1999] perform the initialization of each $\boldsymbol{Q}^{(i)}$, $i = 1, \ldots, N_d$, associated with the new particle using a linear least-squares approximation based on the corresponding configurations of the particles present in the element. Obviously, this procedure can only be implemented in conjunction with correlated local ensembles of model polymer molecules.

A further development of the LPM methodology was the backward-tracking Lagrangian particle method (BLPM) (WAPPEROM, KEUNINGS and LEGAT [2000]). Rather than tracing the trajectories of the Lagrangian particles through the flow in time, the approach adopted in BLPM for determining the polymeric contribution to the extra stress is to fix the positions of the particles on the background finite element mesh and to track them backwards in time over a single time step. Once the location of the particles has been determined at the previous time step, suitable initial values of the configurations are evaluated at those points using the procedure outlined above for the ALPM, and then the stochastic differential equation (2.38) is integrated forwards in time over a single time step along the trajectories that have been computed. The polymeric contribution to the extra stress is then evaluated at the fixed points on the mesh. WAPPEROM, KEUNINGS and LEGAT [2000] choose the fixed positions to be the nine nodal points associated with a quadrilateral finite element mesh. Therefore, instead of solving the initial value problem (2.39) to determine the trajectories of the particles, in the BLPM, they are found by integrating the initial value problem

$$\frac{\mathrm{d}\boldsymbol{x}_i}{\mathrm{d}t} = \boldsymbol{u}^{n+1}(\boldsymbol{x}_i), \quad \boldsymbol{x}_i(t_{n+1}) = \boldsymbol{x}_i^{n+1}, \tag{2.41}$$

backwards in time over the interval $[t_n, t_{n+1}]$. Note that since the velocity field, in general, changes from one time step to the next, the particle trajectory that terminates at the fixed particle locations on the mesh will also change in time so that different Lagrangian particles arrive at these points as time proceeds.

Of the different versions of the LPM that have been developed, the BLPM, which uses correlated ensembles of model polymer molecules, possesses the more attractive features in terms of the numerical properties of the method and the computational overhead.

2.4. Implicit micro–macro schemes

Most micro–macro schemes for time-dependent problems are based on schemes in which the solution of the microscopic and macroscopic equations is decoupled within each time step. This is because the utilization of fully implicit methods in which the macroscopic and microscopic unknowns are updated simultaneously is expensive in terms of computer memory. Decoupled micro–macro schemes are explicit with respect to the degrees of freedom associated with the other stage within the time step. More specifically, in the macroscopic stage, the polymeric stress is treated explicitly in the momentum equation, and in the microscopic stage, the velocity gradient is treated explicitly in the Langevin equation. The restriction on the time step dictated by stability requirements in decoupled micro–macro schemes is generally much more severe than required for accuracy considerations. The small time steps that are required to ensure numerical stability of these schemes also leads to reduced computational efficiency. The lack of development of fully implicit schemes has placed micro–macro methods at a disadvantage compared with their purely macroscopic counterparts. However, LASO, RAMÍREZ and PICASSO [2004] have made some progress in the development of efficient coupled (implicit) schemes for time-dependent micro/macro calculations. The feasibility of solving the very large nonlinear system of equations for both the macroscopic and microscopic degrees of freedom at each time step was achieved by means of a size reduction technique. The basis of this technique was to reduce the size of the system to one having the same size as the corresponding macroscopic description of the same problem using Schur's complement in which the microstructural degrees of freedom are eliminated from the system at the block level. The additional fill-in with respect to the macroscopic formulation is minor, and there is no deterioration of the conditioning of the Schur complement system with respect to the corresponding macroscopic system.

The nonlinear system at each time step is linearized using Newton's method, for example, giving rise to a system of the form

$$A\boldsymbol{x}^{n+1} = \boldsymbol{b}^n, \tag{2.42}$$

where the Jacobian matrix A has the partitioned form

$$A = \begin{pmatrix} A_{MM} & A_{Mm} \\ A_{mM} & A_{mm} \end{pmatrix}. \tag{2.43}$$

The first block of rows corresponds to the discretization of the conservation laws with the polymeric stress replaced by the approximation to the Kramers expression. Therefore, the diagonal block A_{MM} expresses the interconnectedness between the macroscopic degrees of freedom such as the nodal values of velocity and pressure, and the off-diagonal block A_{Mm} expresses the dependence of the macroscopic unknowns on the microscopic unknowns through the substitution of the polymeric stress in terms of the ensemble average (2.14). The last block of rows corresponds to the discretization of the stochastic differential equation. In this case, the diagonal block A_{mm} expresses the interconnectedness between the microscopic degrees of freedom such as the components of the dumbbell connector vectors, and the off-diagonal block A_{mM} expresses the dependence of the microscopic unknowns on the macroscopic unknowns through terms containing the velocity gradient in the Langevin equation, for example. The solution and right-hand side vectors can also be partitioned accordingly:

$$\boldsymbol{x} = \begin{pmatrix} \boldsymbol{x}_M \\ \boldsymbol{x}_m \end{pmatrix}, \quad \boldsymbol{b} = \begin{pmatrix} \boldsymbol{b}_M \\ \boldsymbol{b}_m \end{pmatrix}. \tag{2.44}$$

Generally, the size of $\boldsymbol{x}_m$ is much larger than $\boldsymbol{x}_M$ since there are many more microscopic than macroscopic degrees of freedom. Therefore, the size of the system (2.42) renders its solution by direct methods infeasible, in general.

The approach of LASO, RAMÍREZ and PICASSO [2004] exploits the structure of the diagonal block A_{mm} to ensure that the computational work required to solve the system (2.42) is comparable to that required to solve an equivalent macroscopic problem. The structure of A_{mm} can be arranged so that it has desirable properties.

If the microscopic degrees of freedom in CONNFFESSIT or LPM are ordered by element and then by model polymer molecules within an element, then A_{mm} is block diagonal. In this case, the number of blocks is equal to the total number of model polymer molecules, and the size of each block corresponds to the number of degrees of freedom associated with each model polymer molecule, i.e., the dimension of configuration space. For the method of Brownian configuration fields, a block diagonal structure for A_{mm} is also obtained if the degrees of freedom are ordered field-wise. In this case, the diagonal blocks are sparse, and the number of blocks is equal to the number of configuration fields.

Eliminating the microstructural degrees of freedom leads to the Schur complement system for the macroscopic degrees of freedom

$$\left(A_{MM} - A_{Mm}A_{mm}^{-1}A_{mM}\right)\boldsymbol{x}_M^{n+1} = \boldsymbol{b}_M^n - A_{Mm}A_{mm}^{-1}\boldsymbol{b}_m^n, \tag{2.45}$$

which is a modified discrete Navier–Stokes problem. Once $\boldsymbol{x}_M^{n+1}$ is known, the microstructural degrees of freedom are found from

$$\boldsymbol{x}_m^{n+1} = A_{mm}^{-1}\left(\boldsymbol{b}_m^n - A_{mM}\boldsymbol{x}_M^{n+1}\right). \tag{2.46}$$

LASO, RAMÍREZ and PICASSO [2004] argue that the systems (2.45) and (2.46) can be solved with a moderate amount of computational effort. They also show that the Schur complement $A_{MM} - A_{Mm}A_{mm}^{-1}A_{mM}$ remains sparse with bandwidth bounded by the bandwidth of A_{MM} and that there is no deterioration in the conditioning of the reduced system compared with the corresponding macroscopic system.

The very simple structure of the diagonal blocks in A_{mm} for fully implicit implementations of CONNFFESSIT and LPM means that, in general, the diagonal block A_{mm}^{-1} can be constructed analytically. A comparison of the CPU times for implicit and explicit CONNFFESSIT showed that the former was only two or three times slower per time step than the latter.

Analytical expressions for A_{mm}^{-1} are not available for the method of Brownian configuration fields resulting in a greater computational overhead to construct and solve the reduced system. However, in a later paper, RAMÍREZ and LASO [2005] propose an efficient iterative method for solving the modified Navier–Stokes problem (2.45) that avoids the explicit and costly construction of the Schur complement. The iterative method employed, which is a preconditioned GMRES method, is based on the use matrix-vector multiplications. Since the Jacobian is not available in explicit form, a preconditioner that is known to work well for the Navier–Stokes problem is used. This preconditioner is not optimal, however, so that there is scope for further improvements to be made to the algorithm.

LASO, RAMÍREZ and PICASSO [2004] have implemented the fully implicit micro–macro method to simulate the start-up of Couette flow and have shown that enhanced stability and accuracy are achieved compared with the decoupled scheme for both CONNFFESSIT and BCF. Since each model polymer molecule or configuration field is treated independently, these implicit micro–macro methods are well suited for efficient implementation on parallel computing systems.

2.5. Stochastic methods for reptation models

The Doi–Edwards model (see page 222) provides a description of the configuration pdf of the model polymer in terms of two variables: the unit orientation vector of a tube segment $\boldsymbol{u}$ and the normalized arc length of a polymer chain s (see Fig. 2.1).

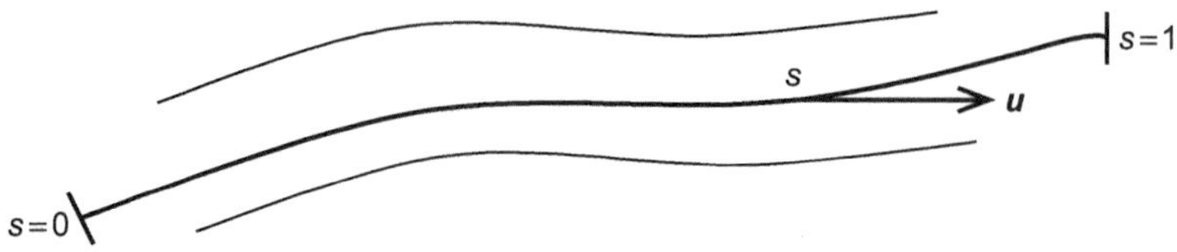

FIG. 2.1 The Doi–Edwards model polymer.

The stochastic process $\boldsymbol{U}_t$ satisfies a deterministic differential equation since there is only a first-order derivative with respect to $\boldsymbol{u}$ in (1.41):

$$\frac{D\boldsymbol{U}_t}{Dt} = (\boldsymbol{\delta} - \boldsymbol{U}_t\boldsymbol{U}_t) \cdot \nabla\boldsymbol{v}\boldsymbol{U}_t. \tag{2.47}$$

In this equation, the term $\nabla\boldsymbol{v}\boldsymbol{U}_t$ communicates the fact that $\boldsymbol{U}_t$ follows the flow field, while the operator $\boldsymbol{\delta} - \boldsymbol{U}_t\boldsymbol{U}_t$ ensures that the property $|\boldsymbol{U}_t| = 1$ is preserved.

The stochastic process S_t is a pure diffusion process that describes the reptational motion of the chain. It specifies the part of the polymer chain that is currently resident in the tube segment with the orientation $\boldsymbol{u}$.

Although the evolution equations (1.43) and (1.44) for the stochastic processes $\boldsymbol{U}_t$ and S_t are decoupled, the processes themselves are coupled through the boundary conditions (1.42) imposed when the chain escapes from its original tube. This provides the mechanism for the reptation process. The stochastic differential equation for S_t may be solved, for example, using either the Euler–Maruyama scheme

$$S_{i+1} = S_i + W_i \frac{1}{\pi}\sqrt{\frac{\Delta t_i}{\tau_d}}, \tag{2.48}$$

or the predictor-corrector scheme (2.17) described earlier in this chapter. If the value of S_{i+1} obtained by this scheme lies outside the interval [0, 1], it is replaced by the value obtained by reflection at the boundary it has just traversed, i.e., $S_{i+1} \to -S_{i+1}$ for $S_{i+1} < 0$, and $S_{i+1} \to 2 - S_{i+1}$ for $S_{i+1} > 1$, and $\boldsymbol{U}$ is reset to a randomly oriented unit vector. The stress tensor is calculated from the tube segment orientation tensor using (1.37).

Micro–macro simulations of polymer melts can be performed in much the same way as for polymer solutions combining a macroscopic treatment of the conservation equations with stochastic techniques for determining the polymer stress. Very few numerical simulations have been performed using the Doi–Edwards model in complex flows. An exception is the contribution of VAN HEEL, HULSEN and VAN DEN BRULE [1999] in which the Brownian configuration field and deformation gradient field methods were compared in simulations of the Doi–Edwards model with the independent alignment assumption in the start-up of two-dimensional flow past a cylinder confined between two parallel plates.

The Brownian configuration field method has already been described for dumbbell models for dilute polymer solutions (see Section 2.3). The application of the technique to the Doi–Edwards model follows a similar procedure except that it is based on the evolution of a number, N_f, of configuration fields $\boldsymbol{u}_k$, $k = 1, \ldots, N_f$, that represent the orientation distribution of the tube segments rather than the dumbbell connector vector. The configuration field approach avoids the need to follow the evolution of the orientation of tube segments associated with a number of discrete particles, a process that involves the determination of the trajectories of these particles. Each of the N_f configuration fields is a global and continuous representation of the tube segment orientation. These fields evolve according to

$$\frac{D\boldsymbol{u}_k}{Dt} = (\boldsymbol{\delta} - \boldsymbol{u}_k\boldsymbol{u}_k) \cdot \nabla \boldsymbol{v} \boldsymbol{u}_k, \tag{2.49}$$

for $k = 1, \ldots, N_f$. Associated with each field is a stochastic process or random walker s_k that satisfies

$$\mathrm{d}s_k = \frac{1}{\pi}\sqrt{\frac{1}{\tau_d}}\mathrm{d}W_k, \tag{2.50}$$

where s_k is a function of time but not of position. At the beginning of the simulation, the configuration fields are initialized using a spatially uniform configuration, i.e.,

$$\boldsymbol{u}_k(\mathbf{x}, 0) = \boldsymbol{u}_k^0,$$

where $\boldsymbol{u}_k^0$ is drawn from the isotropic distribution on the surface of the unit sphere. The initial orientation of the configuration fields is uncorrelated. The associated random walkers s_k are set to independent random numbers from a uniform distribution in the interval [0, 1]. During a simulation, whenever the random walker s_k is reflected, the associated configuration field is removed and replaced by a new and spatially uniform random configuration.

More recent reptation models such as the single segment reptation model of FANG, KRÖGER and ÖTTINGER [2000] have resolved, to a large extent, the failure of the Doi–Edwards model to predict the nonlinear rheology of entangled linear polymers. The model of FANG, KRÖGER and ÖTTINGER [2000] assumes uniform monomer density and isotropic tube cross section. In this model, the fields no longer evolve according to an equation that is purely deterministic as in the Doi–Edwards model (2.49). Furthermore, the associated random walkers evolve according to a stochastic differential equation that contains a deterministic drift term unlike (2.50), which is purely stochastic. The local velocity gradient influences the evolution of s_k through a stretch parameter that represents the ratio of the contour length of the chain to its equilibrium contour length. This means that reflections of s_k are dependent on both time and space. For this reason, it is not possible to replace the entire field upon reflection of s_k at a certain location, since it is highly unlikely that the tube completes its life span at the same time for all the locations in the flow domain. Since the tube survival probability is different at different locations in the flow domain, the method of Brownian configuration fields cannot be used for this model. However, GIGRAS and KHOMAMI [2002] have developed an adaptive configuration fields method (ACFM) that combines the essential aspects of the method of Brownian configuration fields and the deformation fields methods for reptation models with a stochastic strain measure and local variations of life span distribution.

CHAPTER 3

Fokker–Planck-Based Numerical Methods

3.1. Dilute solutions, locally homogeneous flows

As outlined in the Introduction, the polymer dynamics of a dilute polymer solution modeled by FENE dumbbells with the force law (1.26) can be described in a locally or globally homogeneous flow by the Fokker–Planck equation

$$\frac{D\psi}{Dt} = \mathcal{L}_{FP}(\nabla \boldsymbol{v})\psi = \frac{\partial}{\partial \boldsymbol{q}} \cdot \left(\frac{1}{2\lambda_H} \frac{\partial \psi}{\partial \boldsymbol{q}} + \frac{1}{2\lambda_H} \boldsymbol{F}(\boldsymbol{q})\psi - \nabla \boldsymbol{v} \boldsymbol{q} \psi \right). \tag{3.1}$$

Unlike in the previous chapter, we will here describe some numerical methods that solve (3.1) directly for the pdf ψ. Although this equation can be discretized in $\boldsymbol{q}$-space using any numerical method, we argue that spectral methods are especially appropriate in this case because in the whole simulation of a complex flow, (3.1) should be coupled with the conservation equations (1.1) and (1.2), and the dependence of ψ on position in physical space $\boldsymbol{x}$ should be taken into account, thus making the whole problem multidimensional. We therefore prefer to discretize the dependence of ψ on $\boldsymbol{q}$ in such a way as to have the least possible number of degrees of freedom, hence the choice of spectral methods. Another attractive choice can be a wavelet-based method as in ARMSTRONG, NAYAK, GHOSH and BROWN [1996], NAYAK [1998], SUEN, NAYAK, ARMSTRONG and BROWN [2003], for example.

We shall be interested in two-dimensional planar flows. However, even in this case, there is no physical reason to suppose that the dumbbells lie in the plane of the flow; hence, the configuration vector $\boldsymbol{q}$ should be three dimensional. For simplicity, one can also consider a simplified model where $\boldsymbol{q}$ is restricted to lie in the plane of the flow. In the latter case, we will denote the model as 2D FENE and in the former case as 3D FENE.

Collocation spectral method were applied by CHAUVIÈRE and LOZINSKI [2004b], LOZINSKI and CHAUVIÈRE [2003] to 2D FENE simulations and by CHAUVIÈRE and LOZINSKI [2004a] to 3D FENE simulations. Unlike these papers, we shall here construct spectral approximations based on the Galerkin method with numerical quadrature. Similar spectral methods were used in DU, LIU and YU [2005], KNEZEVIC [2008], KNEZEVIC and SÜLI [2009]. The last paper also contains a theoretical study of the convergence of the numerical method. Earlier spectral implementations are reported in FAN [1989a,b, 1985c]. Note also a finite difference scheme in YU, DU and LIU [2005].

3.1.1. A spectral discretization of the Fokker–Planck operator in the 2D FENE model

We consider first Eqn (1.28) alone for the probability density $\psi(t, \boldsymbol{q})$ with $(t, \boldsymbol{q}) \in [0, T] \times D$, $D = \{\boldsymbol{q} \in \mathbb{R}^2 : q < \sqrt{b}\}$, modeling a globally homogeneous flow with the imposed and fixed velocity gradient $\boldsymbol{\kappa} = \nabla \boldsymbol{v}$. Equation (1.28) is completed with the initial condition at time $t = 0$, which is usually chosen as the equilibrium stationary solution ($\boldsymbol{\kappa} = \boldsymbol{0}$) that takes the following form for 2D FENE dumbbells:

$$\psi_{\text{eq}} = \frac{b+2}{2\pi b}\left(1 - \frac{q^2}{b}\right)^{b/2}. \tag{3.2}$$

The behavior of ψ on the boundary ∂D is conditioned by the requirement that the random vector $\boldsymbol{q}(t)$, the solution to the corresponding SDE (1.25), of which ψ is the probability density, is supposed to stay inside D. This requirement is incorporated into the stochastic simulation scheme (cf. Section 2.2), and it has also been proved to be satisfied by the exact solution of the corresponding SDE, at least in some special cases provided $b > 2$ (see JOURDAIN and LELIÈVRE [2003]). Hence, dealing with the probability density function ψ of $\boldsymbol{q}(t)$, we need to make sure that the normal component $\boldsymbol{j} \cdot \boldsymbol{n}$ of the probability flux

$$\boldsymbol{j} = \frac{1}{2\lambda_H}\frac{\partial \psi}{\partial \boldsymbol{q}} + \frac{1}{2\lambda_H}\boldsymbol{F}(\boldsymbol{q})\psi - \boldsymbol{\kappa}\boldsymbol{q}\psi,$$

vanishes on the boundary $q = \sqrt{b}$. The weak formulation now reads: find $\psi(t, \boldsymbol{q})$ such that

$$\int_D \frac{\partial \psi}{\partial t}\phi \mathrm{d}\boldsymbol{q} + \int_D \left(-\boldsymbol{\kappa}\boldsymbol{q}\psi + \frac{1}{2\lambda_H}\boldsymbol{F}(\boldsymbol{q})\psi + \frac{1}{2\lambda_H}\nabla\psi\right) \cdot \nabla\phi \mathrm{d}\boldsymbol{q} = 0 \tag{3.3}$$

for any sufficiently smooth test function $\phi(\boldsymbol{q})$ for which all the integrals above make sense. Note that the boundary conditions specified above would imply both $\psi = 0$ and $\partial\psi/\partial\boldsymbol{n} = 0$ on ∂D for a solution smooth up to the boundary. We do not prescribe the behavior of ψ near the boundary for the moment, but rather perform the change of variable

$$\psi(t, \boldsymbol{q}) = \left(1 - \frac{q^2}{b}\right)^s \alpha(t, \boldsymbol{q}) \tag{3.4}$$

with some (as yet) unspecified parameter $s \geq 0$. Making a symmetric ansatz in the test function $\phi = \left(1 - \frac{q^2}{b}\right)^{-s} \beta$ in (3.3), we arrive at the weak problem in terms of α and any appropriate test function $\beta(\boldsymbol{q})$

$$\int_D \frac{\partial \alpha}{\partial t}\beta \mathrm{d}\boldsymbol{q} + a_\kappa(\alpha, \beta) = 0 \tag{3.5}$$

with

$$a_\kappa(\alpha, \beta) = \int_D \left(-\boldsymbol{\kappa}\boldsymbol{q}\alpha + \frac{b-2s}{2\lambda_H b}\boldsymbol{F}(\boldsymbol{q})\alpha + \frac{1}{2\lambda_H}\nabla\alpha\right) \cdot \left(\frac{2s}{b}\boldsymbol{F}(\boldsymbol{q})\beta + \nabla\beta\right)\mathrm{d}\boldsymbol{q}. \tag{3.6}$$

The bilinear form $a_\kappa(\alpha, \beta)$ is studied thoroughly by KNEZEVIC and SÜLI [2009] in the case $s = b/4$, i.e., for the change of variables $\psi \sim \sqrt{\psi_{\text{eq}}}\alpha$. The mathematical analysis is greatly simplified in this case because the form $a_0(\alpha, \beta)$ becomes symmetric. It is then natural to pose the problem in the weighted Sobolev space $\mathcal{H}$ with the norm $||\alpha||^2_{\mathcal{H}} = \int_D \left(\alpha^2 + \left|\frac{1}{2}\boldsymbol{F}(\boldsymbol{q})\alpha + \nabla\alpha\right|^2\right) \mathrm{d}\boldsymbol{q}$ since the bilinear form $a_\kappa(\alpha, \beta) + K \int_D \alpha\beta \mathrm{d}\boldsymbol{q}$ is bounded and coercive on $\mathcal{H}$ for a sufficiently large number K. Moreover, the space $\mathcal{H}$ is continuously and densely embedded into $L^2(D)$, thus proving that the parabolic problem (3.5) is well posed in $\mathcal{H}$. Some additional properties of the functional space $\mathcal{H}$ permit also an analysis of a spectral discretization of (3.5) with $s = b/4$. We prefer, however, to leave this parameter unspecified for the moment for the reasons explained below. Note that some results on the Fokker–Planck equation with the ansatz (3.4) with any $1/2 < s < b/2$ are also available in the above-cited paper, but they do not seem to be readily applicable to the analysis of the weak formulation (3.5) and (3.6).

Since the vector $\boldsymbol{q}$ in the 2D FENE model lies in the disc D, it is natural to represent it in polar coordinates, i.e,

$$q_1 = r\cos\theta, \ q_2 = r\sin\theta, \ \text{with } r \in [0, \sqrt{b}] \text{ and } \theta \in [0, 2\pi]. \tag{3.7}$$

The passage to polar coordinates in (3.5) necessitates boundary conditions at $r = 0$, which can be taken as $\frac{\partial\psi}{\partial r} = 0$. Indeed, changing the vector $\boldsymbol{q}$ to $-\boldsymbol{q}$ means physically just relabeling the beads of the dumbbell; hence, one should have $\psi(t, \boldsymbol{q}) = \psi(t, -\boldsymbol{q})$. In other words, the unknowns ψ and α can be considered functions of r^2 and θ. We map $r^2 \in (0, b)$ to $\eta \in (-1, 1)$ as is standard in spectral methods and summarize the change of variables from $\psi(t, r, \theta)$ to $\alpha(t, \eta, \theta)$ as

$$\psi(t, r, \theta) = \left(\frac{1-\eta}{2}\right)^s \alpha(t, \eta, \theta) \text{ with } r^2 = b\frac{1+\eta}{2}, \ \eta \in [-1, 1]. \tag{3.8}$$

We will search for an approximate solution $\alpha_N(t, \eta, \theta)$ to (3.5) in the finite dimensional space

$$V_N = \operatorname{span}\{h_k(\eta)\Phi_{il}(\theta), \ i = 0, 1, \ i \le l \le N_F, \ 1 \le k \le N_R\}. \tag{3.9}$$

Here, $\Phi_{il}(\theta) = (1 - i)\cos(2l\theta) + i\sin(2l\theta)$, $i = 0, 1$, $l = i, \ldots, N_F$ and $h_k(\eta)$, $1 \le k \le N_R$ are Lagrange interpolating polynomials based on the Gauss–Legendre (GL) points η_i, $i = 1, \ldots, N_R$ (see CANUTO, HUSSAINI, QUARTERONI and ZANG [2006] for details). Note that the set $\{\eta_i\}$ is chosen so that it does not include the endpoints $\eta = \pm 1$. Only the Fourier modes of even order are kept in the approximation space V_N because of the symmetry of α.

Let $(u, v)_{N_R, N_F}$ for any two functions u, v of η, θ denote the approximation of $\int_{-1}^{1}\int_0^{2\pi} uv \mathrm{d}\theta \mathrm{d}\eta$, in which the integrals with respect to η are evaluated using Gauss quadrature rule on GL points with weights ω_i on the nodes η_i, $i = 1, \ldots, N_R$, whereas the integrals with respect to θ are computed analytically:

$$(u, v)_{N_R, N_F} = \sum_{i=1}^{N_R} \omega_i \int_0^{2\pi} u(\eta_i, \theta)v(\eta_i, \theta)\mathrm{d}\theta. \tag{3.10}$$

Similarly, we denote by $a_{\kappa,N_R,N_F}(u,v)$ the approximation of the bilinear form $a_{\kappa}(u,v)$ defined in (3.6) in which the integrals are replaced by the same quadrature rule as in (3.10). A semidiscretization of (3.5) in $\boldsymbol{q}$-space can be then written as

$$\left(\frac{\partial \alpha_N}{\partial t}, \chi\right)_{N_R,N_F} = a_{\kappa,N_R,N_F}(\alpha_N, \chi), \quad \forall \chi \in V_N. \tag{3.11}$$

Note that $a_{\kappa,N_R,N_F}(\alpha,\beta)$ is well defined for any $\alpha, \beta \in V_N$ unlike $a_{\kappa}(\alpha,\beta)$ which becomes infinite if α or β does not vanish at $\eta = \pm 1$. The scheme (3.11) works well in practice although there is no mathematical justification for it for the moment. The alternative approach of KNEZEVIC and SÜLI [2009] circumvents this difficulty by modifying the approximation space (3.9) by taking only the functions vanishing at $\eta = 1$ and at $\eta = -1$, the latter only for the Fourier harmonics Φ_{il} with $l \geq 1$.

The spectral method (3.11) conserves the integral of the probability density $\psi_N = \left(\frac{1-\eta}{2}\right)^s \alpha_N$ reconstructed as in (3.8) provided the parameter s is an integer from 0 to N_R. Indeed, for such s, we can put $\chi = \left(\frac{1-\eta}{2}\right)^s$ in (3.11). Noting that GL quadrature on N_R points is exact for the polynomials of degree up to $2N_R - 1$ in η and our integration rule is exact in θ, this yields

$$\begin{aligned}\frac{\mathrm{d}}{\mathrm{d}t}\int_D \psi_N \mathrm{d}\boldsymbol{q} &= \frac{b}{4}\int_{-1}^{1}\int_{0}^{2\pi} \frac{\partial \alpha_N}{\partial t}\left(\frac{1-\eta}{2}\right)^s \mathrm{d}\theta \mathrm{d}\eta = \left(\frac{\partial \alpha_N}{\partial t}, \left(\frac{1-\eta}{2}\right)^s\right)_{N_R,N_F} \\ &= -a_{\kappa,N_R,N_F}\left(\alpha_N, \left(\frac{1-\eta}{2}\right)^s\right) = 0.\end{aligned}$$

The last equality here follows from the fact the χ can be written in the original variables as $(1 - q^2/b)^s$ so that $\nabla\chi + \frac{2s}{b}\boldsymbol{F}(\boldsymbol{q})\chi = 0$ and $a_{\kappa,N_R,N_F}(\alpha,\chi)$ vanishes for any α.

In view of an implementation of (3.11), we note that V_N is a linear space of dimension $N = N_R(2N_F + 1)$, and any $\alpha_N \in V_N$ can be represented by an N-tuple $\boldsymbol{\alpha} = \{\alpha^i_{kl}\}$ where α^i_{kl} are the coefficients in the expansion $\alpha_N = \sum_{k=1}^{N_R}\sum_{i=0}^{1}\sum_{l=i}^{N_F}\alpha^i_{kl}h_k(\eta)\Phi_{il}(\theta)$. With this representation, the problem (3.11) is rewritten as a system of linear ODEs

$$\frac{\mathrm{d}\boldsymbol{\alpha}}{\mathrm{d}t} = \mathcal{M}_{FP}(\boldsymbol{\kappa})\boldsymbol{\alpha}, \tag{3.12}$$

where the matrix $\mathcal{M}_{FP}(\boldsymbol{\kappa})$ is calculated by setting χ in (3.11) to all the basis functions in (3.9) and then multiplying by the inverse of the mass matrix obtained from the left-hand side of (3.11). By the above remark on the conservation of the integral of ψ_N, $\mathcal{M}_{FP}(\boldsymbol{\kappa})$ is singular for an appropriate choice of s so that the approximate stationary solution of (1.28) can be obtained by calculating the eigenvector of $\mathcal{M}_{FP}(\boldsymbol{\kappa})$ corresponding to zero eigenvalue and normalizing it, i.e., by solving the linear system with a matrix obtained from $\mathcal{M}_{FP}(\boldsymbol{\kappa})$ by modifying one row in it. If we are interested in the time-dependent solution, the ODE (3.12) can be discretized by, e.g., implicit Euler scheme

$$\frac{\boldsymbol{\alpha}^{j+1} - \boldsymbol{\alpha}^j}{\Delta t} = \mathcal{M}_{FP}(\boldsymbol{\kappa})\boldsymbol{\alpha}^{j+1}. \tag{3.13}$$

Either way, with an approximate solution $\alpha_N \in V_N$, one can compute an approximation to the elastic extra stress by Kramers expression (1.39), or in terms of α_N,

$$\boldsymbol{\tau} = \frac{\eta_p}{\lambda_H}\frac{b+4}{b}\left(-\boldsymbol{\delta} + \left(\alpha_N, \boldsymbol{qq}\left(1-\frac{q^2}{b}\right)^{s-1}\right)_{N_R,N_F}\right). \tag{3.14}$$

Let us now return to the choice of parameter s in (3.8). As already mentioned, an a priori attractive choice for s would be $s = b/4$, as in KNEZEVIC and SÜLI [2009]. One can also argue that an optimal choice should be $s = b/2$ since $\boldsymbol{F}(\boldsymbol{q})$ becomes unbounded near the boundary ∂D of the disc, and the term with $\boldsymbol{\kappa q}\psi$ in the governing equation (1.28) is negligible there. Therefore, ψ should behave in the vicinity of the boundary ∂D like the solution of (1.28) with $\boldsymbol{\kappa} = 0$, i.e., like (3.2). Moreover, it is proved in JOURDAIN, LE BRIS, LELIÈVRE and OTTO [2006] that if the initial condition satisfies $c_0(1-q^2/b)^{b/2} < \psi_0(\boldsymbol{q}) < C_0(1-q^2/b)^{b/2}$ with some positive constants c_0, C_0, then necessarily $c(t)(1-q^2/b)^{b/2} < \psi_0(t,\boldsymbol{q}) < C(t)(1-q^2/b)^{b/2}$ with positive $c(t)$, $C(t)$ varying exponentially in time. Some more estimates in the same spirit are available in LIU and LIU [2008]. A numerical scheme based on the ansatz (3.8) with $s = b/2$ is also implemented in DU, LIU and YU [2005].

However, our numerical experiments demonstrate that taking smaller values of s may be advantageous for the stability of the simulation. By stability, we mean here the exponential convergence of $\psi(t,\boldsymbol{q})$ as $t \to \infty$ to the stationary solution $\psi_\infty(q)$. Note that the existence and uniqueness of the stationary solution is an open problem, but supposing that such a solution exists and satisfies some natural hypotheses, one can prove the exponential convergence to it for the time-dependent problem with any initial condition (see JOURDAIN, LE BRIS, LELIÈVRE and OTTO [2006]). On the discrete level, this notion of stability translates simply to the requirement that all the nonzero eigenvalues of the matrix $-\mathcal{M}_{FP}(\boldsymbol{\kappa})$ in (3.12) have a positive real part. To verify this, we have calculated the spectrum of the matrix $\mathcal{M}_{FP}(\boldsymbol{\kappa}_e)$ with $\boldsymbol{\kappa}_e = \operatorname{diag}(\dot{\varepsilon}, -\dot{\varepsilon})$ corresponding to extensional flow. Equation (3.1) has in this case an analytical stationary solution of the form $C\psi_{\text{eq}}(\boldsymbol{q})\exp(\lambda_H\boldsymbol{\kappa}_e : \boldsymbol{qq})$ so that one can also check the convergence of the numerical method. In Table 3.1.1, we report the relative error in the polymeric stress for the stationary solution and the minimum nonzero eigenvalue of the discretization matrix $\Re\lambda_{\min} = \min_{\lambda\in sp(-\mathcal{M}_{FP}(\boldsymbol{\kappa})),\ \lambda\neq 0}\Re(\lambda)$ computed for the extensional flow of the FENE dumbbells with $b = 12$ on several meshes with $N_R = N_F = N$. We take three different values of $\dot{\varepsilon}$ and four values of s: 0, 1, $3 = b/4$, $6 = b/2$. The relative error is reported only if the corresponding discretization is stable, i.e., $\Re\lambda_{\min} > 0$ so that the stationary solution can be attained in a time-dependent simulation. We observe that the numerical method seems to be stable for any value of s and $\dot{\varepsilon}$ provided the mesh is sufficiently fine. However, the larger the value of s taken, the finer should be the mesh in order not to obtain spurious unstable modes. This phenomenon is particular pronounced in a strong flow (large $\dot{\varepsilon}$). One observes as well the influence of the parameter s on the convergence. While the method converges eventually for any choice of s, the convergence is generally the fastest under $s = 1$. We adopt therefore this value of s for all the simulations below.

The lack of stability with larger values of s can be explained by observing that the solution of the Fokker–Planck equation under a strong extensional flow (in fact, under any strong flow) tends to have high peaks near the boundary of the disc D, i.e., $\eta \approx 1$. Since $\alpha = \psi/(1-\eta)^s$, the peaks will be more pronounced for large values of s, and there is a

TABLE 3.1.1

The minimum real part of the eigenvalues of the matrix $\mathcal{M}(\kappa_e)$ discretizing the Fokker–Planck operator in extensional flows and the relative error ϵ_{rel} in the polymeric stress at steady state when the latter is attained

(a) $\dot{\varepsilon} = 1$:

	$N=4$		$N=8$		$N=12$		$N=16$	
s	$\Re\lambda_{min}$	ϵ_{rel}	$\Re\lambda_{min}$	ϵ_{rel}	$\Re\lambda_{min}$	ϵ_{rel}	$\Re\lambda_{min}$	ϵ_{rel}
0	0.62	$4.6\cdot10^{-2}$	0.69	$8.0\cdot10^{-3}$	0.69	$3.4\cdot10^{-4}$	0.69	$1.1\cdot10^{-5}$
1	0.55	0.11	0.69	$1.1\cdot10^{-4}$	0.69	$9.4\cdot10^{-9}$	0.69	$3.9\cdot10^{-13}$
3	$2.8\cdot10^{-2}$	0.80	0.68	$1.8\cdot10^{-3}$	0.69	$1.0\cdot10^{-6}$	0.69	$9.7\cdot10^{-12}$
6	−39	–	−0.27	–	0.69	$7.8\cdot10^{-5}$	0.69	$2.7\cdot10^{-10}$

(b) $\dot{\varepsilon} = 5$:

	$N=8$		$N=16$		$N=24$		$N=32$	
s	$\Re\lambda_{min}$	ϵ_{rel}	$\Re\lambda_{min}$	ϵ_{rel}	$\Re\lambda_{min}$	ϵ_{rel}	$\Re\lambda_{min}$	ϵ_{rel}
0	3.8	$4.2\cdot10^{-2}$	4.9	$1.1\cdot10^{-2}$	4.5	$1.5\cdot10^{-3}$	4.5	$4.2\cdot10^{-5}$
1	0.55	1.7	4.3	$3.1\cdot10^{-2}$	4.5	$1.2\cdot10^{-5}$	4.5	$3.0\cdot10^{-10}$
3	−12	–	−28	–	5.5	$4.8\cdot10^{-3}$	4.5	$5.6\cdot10^{-8}$
6	−313	–	−205	–	−2.5	–	4.4	$2.0\cdot10^{-4}$

(c) $\dot{\varepsilon} = 10$:

	$N=8$		$N=16$		$N=24$		$N=32$	
s	$\Re\lambda_{min}$	ϵ_{rel}	$\Re\lambda_{min}$	ϵ_{rel}	$\Re\lambda_{min}$	ϵ_{rel}	$\Re\lambda_{min}$	ϵ_{rel}
0	−3.7	–	5.1	$3.4\cdot10^{-2}$	10.3	$1.5\cdot10^{-2}$	9.5	$2.1\cdot10^{-3}$
1	−52	–	−3.1	–	6.8	$5.4\cdot10^{-2}$	9.4	$1.8\cdot10^{-4}$
3	−268	–	−48	–	−14	–	3.6	0.18
6	−4750	–	−2692	–	−589	–	−75	–

higher likelihood of instability if the peaks are not sufficiently well resolved by the mesh. We illustrate this by two pictures: in Fig. 3.1, we plot the stationary solution for ψ in the extensional flow with $\dot{\varepsilon} = 5$ computed on a sufficiently fine mesh with $N_R = N_F = 32$ and with $s = 1$. Note here the presence of two peaks near the boundary. In Fig. 3.2, we plot the results for α for the same flow on the same mesh but with three values of s. To see better the difference between the three simulations, we plot only the dependence on η with $\theta = 0$ fixed and rescale α so that its maximum is equal to 1.

3.1.2. *An alternative time discretization of the Fokker–Planck equation*

Let us now describe another time discretization of (3.11), which proves useful in complex flow simulations. The goal is to arrive at an implicit scheme where, unlike (3.13), the approximation matrix in the right-hand side will depend on the velocity gradient through only one scalar parameter. We decompose first the velocity gradient for a planar flow into

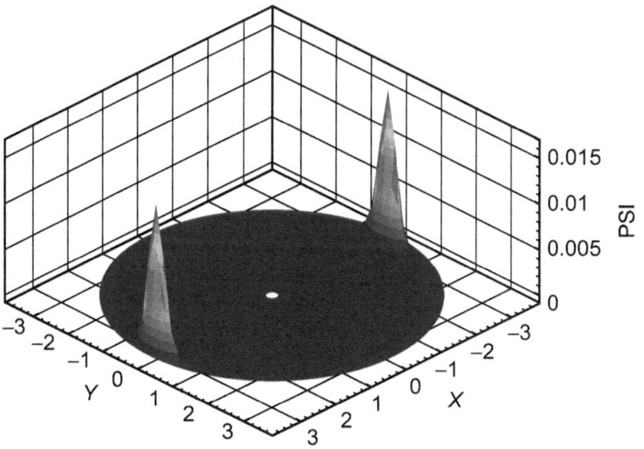

FIG. 3.1 The stationary solution of the Fokker–Planck equation for ψ in the extensional flow with $\dot{\varepsilon} = 5$ computed with $s = 1$ and $N_R = N_F = 32$.

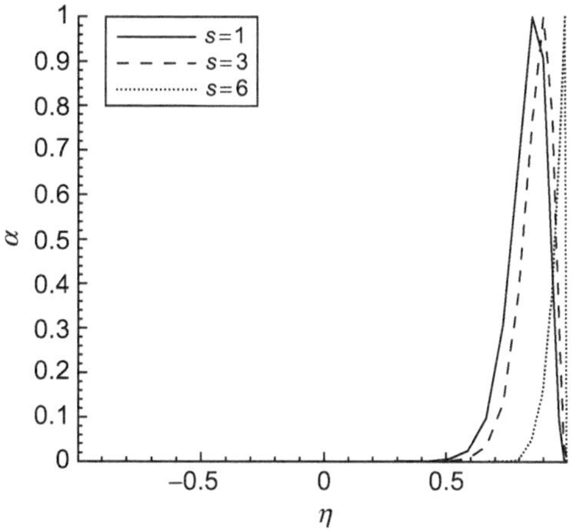

FIG. 3.2 The stationary solution in terms of α of (3.5) as a function of η for $\theta = 0$. Extensional flow with $\dot{\varepsilon} = 5$ computed on the mesh $N_R = N_F = 32$.

symmetric and antisymmetric parts and rotate the coordinates to the principal axes of the symmetric part. We thus have

$$\boldsymbol{\kappa} = k\boldsymbol{\pi}_{\phi}\boldsymbol{\delta}_1\boldsymbol{\pi}_{-\phi} + k_a\boldsymbol{\delta}_a, \tag{3.15}$$

where

$$\boldsymbol{\delta}_1 = \begin{pmatrix} 1 & 0 \\ 0 & -1 \end{pmatrix}, \qquad \boldsymbol{\delta}_a = \begin{pmatrix} 0 & 1 \\ -1 & 0 \end{pmatrix}, \tag{3.16}$$

and $\boldsymbol{\pi}_\phi$ is the matrix of the rotation by angle ϕ. The scalars k, ϕ, and k_a appearing in (3.15) are calculated from $\boldsymbol{\kappa}$ via

$$k = \sqrt{\kappa_{11}^2 + (\kappa_{12} + \kappa_{21})^2/4}, \quad k_a = \frac{\kappa_{12} - \kappa_{21}}{2} \tag{3.17}$$

$$\sin 2\phi = \frac{\kappa_{12} + \kappa_{21}}{2k}, \quad \cos 2\phi = \frac{\kappa_{11}}{k}, \tag{3.18}$$

Let Π_ϕ be the operator defined for an arbitrary 2π-periodic function $\Phi(\theta)$ by

$$\Pi_\phi \Phi(\theta) = \Phi(\theta + \phi), \tag{3.19}$$

i.e., the result of the rotation of coordinates $\boldsymbol{\pi}_{-\phi}$. By substituting (3.15) into (3.11), and using the relations $\boldsymbol{\delta}_a \boldsymbol{q} \cdot \frac{\partial}{\partial \boldsymbol{q}} = -\frac{\partial}{\partial \boldsymbol{\theta}}$ and $\Pi_\phi(\pi_{-\phi}\nabla\alpha) = \nabla(\Pi_\phi\alpha)$, we arrive at the following variational problem for α_N, which should be verified for all $\beta \in V_N$

$$\left(\frac{\partial\alpha}{\partial t} - k_a\frac{\partial\alpha}{\partial\theta}, \beta\right)_{N_R,N_F} + k a^1_{N_R,N_F}(\Pi_\phi\alpha, \Pi_\phi\beta) + a^0_{N_R,N_F}(\Pi_\phi\alpha, \Pi_\phi\beta) = 0, \tag{3.20}$$

where $a^0_{N_R,N_F}$ is the part of the bilinear form a_{κ,N_R,N_F} in (3.11) independent of $\boldsymbol{\kappa}$, i.e., $a^0_{N_R,N_F} = a_{\mathbf{0},N_R,N_F}$ and $a^1_{N_R,N_F} = a_{\delta_1,N_R,N_F} - a^0_{N_R,N_F}$. The form $a^0_{N_R,N_F}$ is invariant under the rotations Π_ϕ, which is why the rotations can be included into the last term in (3.20).

We discretize (3.20) in time by a Crank–Nicolson scheme on the interval $[t_n, t_{n+1}]$ treating the derivative in θ by a method of characteristics and approximating $\alpha((t_n + t_{n+1})/2)$ by

$$\widetilde{\alpha}^{n+\frac{1}{2}} = \frac{1}{2}\left(\Pi_{-\frac{1}{2}k_a\Delta t}\alpha^{n+1} + \Pi_{\frac{1}{2}k_a\Delta t}\alpha^n\right).$$

The time discretization reads

$$\left(\frac{\Pi_{-\frac{1}{2}k_a\Delta t}\alpha^{n+1} - \Pi_{\frac{1}{2}k_a\Delta t}\alpha^n}{\Delta t}, \beta\right)_{N_R,N_F} + k a^1_{N_R,N_F}\left(\Pi_\phi\widetilde{\alpha}^{n+\frac{1}{2}}, \Pi_\phi\beta\right)$$
$$+ a^0_{N_R,N_F}\left(\Pi_\phi\widetilde{\alpha}^{n+\frac{1}{2}}, \Pi_\phi\beta\right) = 0. \tag{3.21}$$

Noting that $\Pi_\phi V_N = V_N$ since

$$\Pi_\phi \Phi_{il}(\theta) = \cos(2l\phi)\Phi_{il}(\theta) - (-1)^i \sin(2l\phi)\Phi_{il}(\theta), \tag{3.22}$$

we can replace the test function β by $\Pi_{-\phi}\beta$ so that we obtain for any $\beta \in V_N$

$$\left(\frac{\Pi_{\varphi-\frac{1}{2}k_a\Delta t}\alpha^{n+1} - \Pi_{\varphi+\frac{1}{2}k_a\Delta t}\alpha^n}{\Delta t}, \beta\right)_{N_R,N_F} + k a^1_{N_R,N_F}\left(\Pi_\phi\widetilde{\alpha}^{n+1/2}, \beta\right)$$
$$+ a^0_{N_R,N_F}\left(\Pi_\phi\widetilde{\alpha}^{n+1/2}, \beta\right) = 0. \tag{3.23}$$

Going to matrix notation, we again represent $\alpha_N^j \in V_N$ by N-tuples $\boldsymbol{\alpha}^j$. The rotation operator Π_ϕ becomes, in these notations, the rotation matrix, which we still denote by the same symbol and which is very easy to implement in the basis (3.9) thanks to (3.22). Equation (3.23) now becomes

$$\boldsymbol{\alpha}^{n+1} = \Pi_{-\phi+\frac{1}{2}k_a\Delta t}\left(\boldsymbol{\delta} - \frac{\Delta t}{2}\mathcal{M}_0 - \frac{k\Delta t}{2}\mathcal{M}_1\right)^{-1}$$

$$\left(\boldsymbol{\delta} + \frac{\Delta t}{2}\mathcal{M}_0 + \frac{k\Delta t}{2}\mathcal{M}_1\right)\Pi_{\phi+\frac{1}{2}k_a\Delta t}\boldsymbol{\alpha}^n. \tag{3.24}$$

The matrices $\mathcal{M}_0$ and $\mathcal{M}_1$ here correspond to the discretizations of the forms $a^0_{N_R,N_F}$ and $a^1_{N_R,N_F}$, respectively.

Finally, supposing that the matrix $\boldsymbol{M} = \frac{\Delta t}{2}\left(\boldsymbol{\delta} - \frac{\Delta t}{2}\mathcal{M}_0\right)^{-1}\mathcal{M}_1$ is diagonalizable, we write it in the form $\boldsymbol{M} = \boldsymbol{P}\boldsymbol{D}\boldsymbol{P}^{-1}$ where $\boldsymbol{D}$ is the diagonal matrix formed with the eigenvalues of $\boldsymbol{M}$. Thus, we can express (3.24) in the form

$$\boldsymbol{\alpha}^{n+1} = \Pi_{\frac{1}{2}k_a\Delta t-\phi}\boldsymbol{P}\left(\boldsymbol{\delta} - k\boldsymbol{D}\right)^{-1}\boldsymbol{P}^{-1}(\boldsymbol{R} + k\boldsymbol{M})\Pi_{\phi+\frac{1}{2}k_a\Delta t}\boldsymbol{\alpha}^n, \tag{3.25}$$

with $\boldsymbol{M} = \left(\boldsymbol{\delta} - \frac{\Delta t}{2}\mathcal{M}_0\right)^{-1}\left(\boldsymbol{\delta} + \frac{\Delta t}{2}\mathcal{M}_0\right)$. Note that none of the matrices $\boldsymbol{D}$, $\boldsymbol{P}$, $\boldsymbol{P}^{-1}$, $\boldsymbol{M}$, and $\boldsymbol{R}$ depend on $\boldsymbol{\kappa}$.

3.1.3. Complex flow simulations

By a complex flow we mean here a locally (but not globally) homogeneous flow of a 2D FENE fluid governed by the Fokker–Planck equation (3.1) coupled with the momentum and mass conservation equations (1.1) and (1.2). As in the case of stochastic micro–macro simulations (cf. Section 2.2), it is natural to decouple numerically the solution of conservation equations (1.1) and (1.2) from that of the Fokker–Planck equation. This gives in particular the velocity gradient field that feeds (3.1). The last equation is then integrated from t to $t + \Delta t$ in order to compute the extra-stress $\boldsymbol{\tau}(t + \Delta t, \boldsymbol{x})$. The whole loop can be then repeated on the next time step.

Let us recapitulate the method. We suppose that some finite element spaces on a triangulation $\mathcal{T}$ of Ω are chosen to approximate the velocity and pressure in the conservation equations (1.1) and (1.2) and the probability density function ψ is represented by the N-tuple $\boldsymbol{\alpha}^n(\boldsymbol{x})$ for any grid point $\boldsymbol{x}$ in $\mathcal{T}$ via (3.8) and (3.9). On the $(n + 1)$-st time step, we perform the following:

- Update velocity and pressure $\boldsymbol{v}^n$ and p^n using a discretization of (1.1)–(1.2) supplying at the right-hand side of the momentum equation the known approximation of the elastic extra-stress $\boldsymbol{\tau}^n(\boldsymbol{x})$ at time $t_n = n\Delta t$.
- Update the approximation of $\alpha(t_n)$ solving the Fokker–Planck equation without the convective terms. We thus calculate $\boldsymbol{\alpha}^{n+1/2}(\boldsymbol{x})$ at each grid point $\boldsymbol{x}$ of the physical

domain Ω using the velocity gradient $\kappa_{kl} = \partial v_k^n / \partial x_l$ calculated from the latest available velocity field at the grid point $\boldsymbol{x}$.

$$\frac{\boldsymbol{\alpha}^{n+1/2}(\boldsymbol{x}) - \boldsymbol{\alpha}^n(\boldsymbol{x})}{\Delta t} = \mathcal{M}_{FP}(\boldsymbol{\kappa}(\boldsymbol{x}))\boldsymbol{\alpha}^{n+1/2}(\boldsymbol{x}). \tag{3.26}$$

- Update the approximation of $\alpha(t_n)$ accounting for the convective terms in the Fokker–Planck equation. We solve thus N copies of the transport equation in Ω:

$$\frac{\boldsymbol{\alpha}^{n+1} - \boldsymbol{\alpha}^{n+1/2}}{\Delta t} + \boldsymbol{v}^n \cdot \nabla \boldsymbol{\alpha}^{n+1} = 0. \tag{3.27}$$

- Use $\boldsymbol{\alpha}^{n+1}$ to calculate the elastic extra-stress $\boldsymbol{\tau}^{n+1}(\boldsymbol{x})$ at time t_{n+1} at each grid point $\boldsymbol{x}$.

In practice, the configuration step (3.26) can be a bottleneck in this algorithm since it requires the solution of a linear system of size N with the matrix varying from one grid point to another. This step can be implemented much more efficiently if we replace (3.26) by (3.25), which should be applied again at each grid point $\boldsymbol{x}$ with the parameters k_a, k, and ϕ calculated from the latest available velocity gradient at $\boldsymbol{x}$. Since all the matrices in (3.25) are independent from the velocity gradient, they can be constructed in a preprocessing stage. Thus, implementing (3.25) will amount only to some matrix-vector multiplications and rotations that can be cheaply calculated by (3.22).

3.2. Numerical methods for flows without the local homogeneity assumption

We turn now to situations when the flow domain is of a size comparable with that of the polymer molecules so that the effects of diffusion of polymer molecules outside from the streamlines cannot be neglected. Assuming again that the solution is dilute and that polymer molecules can be sufficiently well represented by FENE dumbbells, we recall that configuration of dumbbells is governed by the Fokker–Planck equation (1.34) for the probability density $\psi(t, \boldsymbol{r}_c, \boldsymbol{q})$ where $\boldsymbol{r}_c$ points to the center of mass of the dummbells and $\boldsymbol{q}$ is their end-to-end vector nondimensionalized as in (1.23). The main difficulty for the numerical treatment of this equation is of geometrical nature. The domain of definition for $\boldsymbol{r}_c, \boldsymbol{q}$ is given by the requirement that both beads of the dumbbell cannot leave the flow domain Ω and the length of the dumbbell cannot exceed $\sqrt{b}$, i.e., for a given $\boldsymbol{r}_c \in \Omega$ the vector $\boldsymbol{q}$ "lives" in the domain

$$D(\boldsymbol{r}_c) = \{\boldsymbol{q} : q < \sqrt{b}\} \bigcap \{\boldsymbol{q} : \boldsymbol{r}_c \pm \ell_0 \boldsymbol{q}/2 \in \Omega\}.$$

It is thus difficult to discretize in $\boldsymbol{q}$ using straightforward spectral methods like that of the previous section because $D(\boldsymbol{r}_c)$ is no longer of a simple form. We will therefore use a fictitious domain approach by embedding $D(\boldsymbol{r}_c)$ in a simple domain $\tilde{D}(\boldsymbol{r}_c)$ (typically a rectangle) and requiring that ψ vanishes in $\tilde{D}(\boldsymbol{r}_c) \setminus D(\boldsymbol{r}_c)$. Note that such an approach was used also in AMMAR, MOKDAD, CHINESTA and KEUNINGS [2006a], AMMAR, RYCKELYNCK, CHINESTA and KEUNINGS [2006b] in the context of finite element methods for the Fokker–Planck equation for a homogeneous flow of 2D FENE dumbbells, i.e., when the domain of $\boldsymbol{q}$ is a disc.

It is simpler to physically interpret the boundary conditions for the Fokker–Planck equation written in the variables $\boldsymbol{r}_1, \boldsymbol{r}_2$

$$\begin{aligned}\frac{\partial \psi}{\partial t} &= -\frac{\partial}{\partial \boldsymbol{r}_1} \cdot \left[\boldsymbol{v}(\boldsymbol{r}_1)\psi - \frac{1}{4\lambda_H} \boldsymbol{F}\left(\frac{\boldsymbol{r}_1 - \boldsymbol{r}_2}{\ell_0}\right)\psi \right] \\ &\quad - \frac{\partial}{\partial \boldsymbol{r}_2} \cdot \left[\boldsymbol{v}(\boldsymbol{r}_2)\psi - \frac{1}{4\lambda_H} \boldsymbol{F}\left(\frac{\boldsymbol{r}_2 - \boldsymbol{r}_1}{\ell_0}\right)\psi \right] \\ &\quad + \frac{\ell_0^2}{4\lambda_H} \frac{\partial^2 \psi}{\partial \boldsymbol{r}_1^2} + \frac{\ell_0^2}{4\lambda_H} \frac{\partial^2 \psi}{\partial \boldsymbol{r}_2^2},\end{aligned} \tag{3.28}$$

rather than for Eq (1.34) in $\boldsymbol{r}_c, \boldsymbol{q}$. Assume, for example, that the wall Γ is purely repulsive, and $\boldsymbol{v} \cdot \boldsymbol{n} = 0$ on Γ means that the normal component of the probability flux in terms of $\boldsymbol{r}_1, \boldsymbol{r}_2$ vanishes for configurations where one of the beads touches the wall:

$$\left(\ell_0^2 \frac{\partial \psi}{\partial \boldsymbol{r}_i} + \boldsymbol{F}\left(\frac{\boldsymbol{r}_i - \boldsymbol{r}_{3-i}}{\ell_0}\right)\psi \right) \cdot \boldsymbol{n} = 0 \quad \text{for } \boldsymbol{r}_i \in \Gamma,\ i = 1, 2. \tag{3.29}$$

Combining Eqn (3.28) with (3.29) yields the weak formulation of the problem:

$$\begin{aligned}\int\limits_{\boldsymbol{r}_1,\boldsymbol{r}_2\in\Omega}\int \frac{\partial \psi}{\partial t}\varphi \,\mathrm{d}\boldsymbol{r}_1\mathrm{d}\boldsymbol{r}_2 &= \int\limits_{\boldsymbol{r}_1,\boldsymbol{r}_2\in\Omega}\int \left(\boldsymbol{v}(\boldsymbol{r}_1) \cdot \frac{\partial \varphi}{\partial \boldsymbol{r}_1} + \boldsymbol{v}(\boldsymbol{r}_2) \cdot \frac{\partial \varphi}{\partial \boldsymbol{r}_2} \right) \psi \,\mathrm{d}\boldsymbol{r}_1\mathrm{d}\boldsymbol{r}_2 \\ &\quad - \frac{1}{4\lambda_H} \int\limits_{\boldsymbol{r}_1,\boldsymbol{r}_2\in\Omega}\int \left(\ell_0^2 \frac{\partial \psi}{\partial \boldsymbol{r}_1} + \boldsymbol{F}\left(\frac{\boldsymbol{r}_1 - \boldsymbol{r}_2}{\ell_0}\right)\psi \right) \cdot \frac{\partial \varphi}{\partial \boldsymbol{r}_1} \mathrm{d}\boldsymbol{r}_1\mathrm{d}\boldsymbol{r}_2 \\ &\quad - \frac{1}{4\lambda_H} \int\limits_{\boldsymbol{r}_1,\boldsymbol{r}_2\in\Omega}\int \left(\ell_0^2 \frac{\partial \psi}{\partial \boldsymbol{r}_2} + \boldsymbol{F}\left(\frac{\boldsymbol{r}_2 - \boldsymbol{r}_1}{\ell_0}\right)\psi \right) \cdot \frac{\partial \varphi}{\partial \boldsymbol{r}_2} \mathrm{d}\boldsymbol{r}_1\mathrm{d}\boldsymbol{r}_2,\end{aligned} \tag{3.30}$$

where $\varphi = \varphi(\boldsymbol{r}_1, \boldsymbol{r}_2)$ is a suitable test function. We note that this weak formulation is difficult to discretize efficiently since proper account should be taken of the localization of ψ in the subdomain of small $|\boldsymbol{r}_1 - \boldsymbol{r}_2|$. We prefer therefore to rewrite the problem again in terms of $\boldsymbol{r}_c$ and $\boldsymbol{q}$, cf. (1.34). After rescaling the force as in (1.26), this yields

$$\begin{aligned}&\int\limits_{\Omega}\int\limits_{D(\boldsymbol{r}_c)} \frac{\partial \psi}{\partial t}\varphi \,\mathrm{d}\boldsymbol{q}\mathrm{d}\boldsymbol{r}_c - \int\limits_{\Omega}\int\limits_{D(\boldsymbol{r}_c)} \frac{1}{2}(\boldsymbol{v}(\boldsymbol{r}_c + \ell_0\boldsymbol{q}/2) + \boldsymbol{v}(\boldsymbol{r}_c - \ell_0\boldsymbol{q}/2))\psi \cdot \frac{\partial \varphi}{\partial \boldsymbol{q}} \mathrm{d}\boldsymbol{q}\mathrm{d}\boldsymbol{r}_c \\ &\quad - \int\limits_{\Omega}\int\limits_{D(\boldsymbol{r}_c)} \frac{1}{\ell_0}(\boldsymbol{v}(\boldsymbol{r}_c + \ell_0\boldsymbol{q}/2) - \boldsymbol{v}(\boldsymbol{r}_c - \ell_0\boldsymbol{q}/2))\psi \cdot \frac{\partial \varphi}{\partial \boldsymbol{q}} \mathrm{d}\boldsymbol{q}\mathrm{d}\boldsymbol{r}_c \\ &\quad + \frac{1}{2\lambda_H} \int\limits_{\Omega}\int\limits_{D(\boldsymbol{r}_c)} \left(\frac{\partial \psi}{\partial \boldsymbol{q}} + \boldsymbol{F}(\boldsymbol{q})\psi \right) \cdot \frac{\partial \varphi}{\partial \boldsymbol{q}} \mathrm{d}\boldsymbol{q}\mathrm{d}\boldsymbol{r}_c \\ &\quad + \frac{\ell_0^2}{8\lambda_H} \int\limits_{\Omega}\int\limits_{D(\boldsymbol{r}_c)} \frac{\partial \psi}{\partial \boldsymbol{r}_c} \cdot \frac{\partial \varphi}{\partial \boldsymbol{r}_c} \mathrm{d}\boldsymbol{q}\mathrm{d}\boldsymbol{r}_c = 0.\end{aligned} \tag{3.31}$$

Let us now simplify the model by approximating the difference of velocities $\frac{1}{\ell_0}(\boldsymbol{v}(\boldsymbol{r}_c + \ell_0\boldsymbol{q}/2) - \boldsymbol{v}(\boldsymbol{r}_c - \ell_0\boldsymbol{q}/2))$ by the differential $\nabla\boldsymbol{v}(\boldsymbol{r}_c)\boldsymbol{q}$. This is reasonable because the most important features of the fully nonhomogeneous flows come from the diffusion of the centers of mass of the molecules, i.e., the last term in (3.31), which we keep unchanged. We can replace $\frac{1}{2}(\boldsymbol{v}(\boldsymbol{r}_c + \ell_0\boldsymbol{q}/2) + \boldsymbol{v}(\boldsymbol{r}_c - \ell_0\boldsymbol{q}/2))$ by $\boldsymbol{v}(\boldsymbol{r}_c)$ for the same reasons. A time discretization of (3.31) can be done by operator splitting, i.e., starting from ψ^n that approximates $\psi(t^n)$, we find an approximation for $\psi(t^{n+1})$ by solving two problems for $\psi^{n+1/2}$ and ψ^{n+1}, the first of which takes care of the differential operators with respect to $\boldsymbol{q}$ and the second one of those with respect to $\boldsymbol{r}_c$:

$$\int_\Omega \int_{D(\boldsymbol{r}_c)} \frac{\psi^{n+1/2} - \psi^n}{\Delta t} \varphi \, \mathrm{d}\boldsymbol{q} \mathrm{d}\boldsymbol{r}_c - \int_\Omega \int_{D(\boldsymbol{r}_c)} \nabla\boldsymbol{v}^n(r_c)\boldsymbol{q} \cdot \psi \frac{\partial\varphi}{\partial\boldsymbol{q}} \, \mathrm{d}\boldsymbol{q} \mathrm{d}\boldsymbol{r}_c + \frac{1}{2\lambda_H} \int_\Omega \int_{D(\boldsymbol{r}_c)} \left(\frac{\partial\psi}{\partial\boldsymbol{q}} + \boldsymbol{F}(\boldsymbol{q})\psi\right) \cdot \frac{\partial\varphi}{\partial\boldsymbol{q}} \, \mathrm{d}\boldsymbol{q}\mathrm{d}\boldsymbol{r}_c = 0, \tag{3.32}$$

$$\int_\Omega \int_{D(\boldsymbol{r}_c)} \frac{\psi^{n+1} - \psi^{n+1/2}}{\Delta t} \varphi \, \mathrm{d}\boldsymbol{q} \mathrm{d}\boldsymbol{r}_c - \int_\Omega \int_{D(\boldsymbol{r}_c)} \boldsymbol{v}^n(r_c)\psi \cdot \frac{\varphi}{\boldsymbol{r}_c} \, \mathrm{d}\boldsymbol{q} \mathrm{d}\boldsymbol{r}_c + \frac{\ell_0^2}{8\lambda_H} \int_\Omega \int_{D(\boldsymbol{r}_c)} \frac{\partial\psi}{\partial\boldsymbol{r}_c} \cdot \frac{\partial\varphi}{\partial\boldsymbol{r}_c} \mathrm{d}\boldsymbol{q}\mathrm{d}\boldsymbol{r}_c = 0. \tag{3.33}$$

We are now going to describe a discretization of problems (3.32) and (3.33) applying it to nonhomogeneous start-up plane Poiseuille flow of a 2D FENE fluid. The flow geometry is shown in Fig. 3.3(a) and consists of two plates $y = \pm d$ between which a dilute polymer solution flows under a constant pressure gradient. We assume that stress and velocity depend only on y so that the dependence on the position vector $\boldsymbol{r}_c$ will be denoted in the sequel of the section by dependence on y. Figure 3.3(b) illustrates the configuration spaces $D(y)$ for two different choices of y.

First step (3.32): discretization in configuration space Suppose that some grid $y_k \in [-\mathrm{d}, \mathrm{d}]$, $k = 1, \ldots, N_y$ in the physical space. Since problem (3.32) does not contain the derivatives in $\boldsymbol{r}_c$, it can be discretized separately at each grid point y_k. As in the case of homogeneous flows in Section 3.1, we introduce the change of variables (3.8) so that problem (3.32) can be rewritten as (cf. (3.5))

$$\int_{D(y_k)} \frac{\alpha_k^{n+1/2} - \alpha_k^n}{\Delta t} \beta \mathrm{d}\boldsymbol{q} + \int_{D(y_k)} \left(-\boldsymbol{\kappa}_k^n \boldsymbol{q} \alpha_k^{n+1/2} + \frac{b-2s}{2\lambda_H b} \boldsymbol{F}(\boldsymbol{q}) \alpha_k^{n+1/2} + \frac{1}{2\lambda_H} \nabla \alpha_k^{n+1/2}\right) \cdot \left(\frac{2s}{b} \boldsymbol{F}(\boldsymbol{q})\beta + \nabla\beta\right) \mathrm{d}\boldsymbol{q} = 0, \tag{3.34}$$

with $\boldsymbol{\kappa}_k^n = \nabla\boldsymbol{v}^n(y_k)$. Here, α_k^n is an approximation of $\alpha^n(y_k, \boldsymbol{q})$.

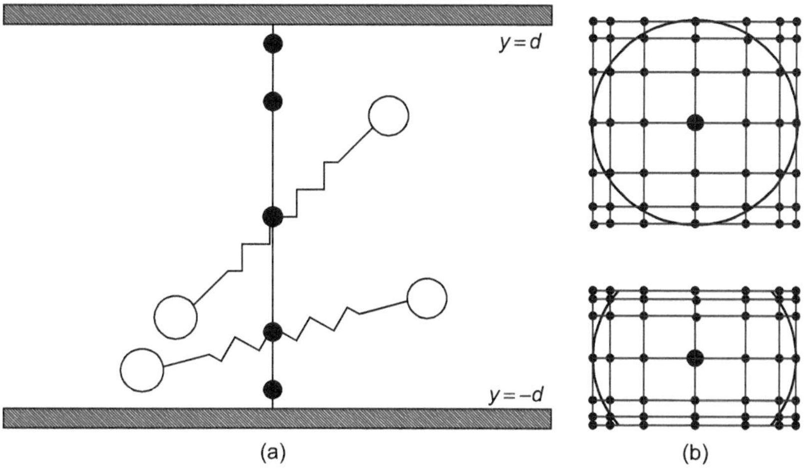

FIG. 3.3 (a) Flow between two parallel walls and GL collocation grid and (b) configuration spaces $D(y)$ for two different values y_k of y with superposed GLL grids.

For each y_k, the corresponding configuration space $D(y_k)$ is the intersection of the disc $q < \sqrt{b}$ and the rectangle $\tilde{D}(y_k) = (-\sqrt{b}, \sqrt{b}) \times (-d_k, d_k)$ where $d_k = \min(2(d - |y_k|), \sqrt{b})$. We introduce in this rectangle the Gauss–Lobatto–Legendre (GLL) points $(q_x^{k,i}, q_y^{k,j})$, $i = 0, \ldots, N_{q_x}^k, j = 0, \ldots, N_{q_y}^k$, and then expand α_k^n in terms of a tensorized basis consisting of Lagrange interpolating polynomials based on these points. The unknown α^n in (3.34) is then approximated by a linear combination of these polynomials requiring that it vanishes on $\tilde{D}(y_k) \setminus D(y_k)$. That is, we write

$$\alpha_k^n(\boldsymbol{q}) = \sum_{i=0}^{N_{q_x}^k} \sum_{j=0}^{N_{q_y}^k} \widehat{\alpha}_{ijk}^n H_i^k(q_x) H_j^k(q_y), \tag{3.35}$$

where the coefficients $\widehat{\alpha}_{ijk}^n$ are set to zero for polynomials corresponding to grid points outside the disc, i.e., such that $(q_x^{k,i})^2 + (q_y^{k,j})^2 \geq b$. In other words, the discrete space Σ_k to which both trial and test functions in configuration space belong is defined to consist of polynomials $H_i^k(q_x)H_j^k(q_y)$ with i, j such that $(q_x^{k,i})^2 + (q_y^{k,j})^2 < b$. The set of such nodes $(q_x^{k,i}, q_y^{k,j})$ will be denoted $\mathcal{T}_k$.

Let us verify numerically the validity of this approach and the influence of the parameter s in the simplest case when the configuration space $D(y_k)$ is the whole disc, and the results can be thus compared with those of Section 3.1. More specifically, we apply the discretization described above to the homogeneous Fokker–Planck equation with the constant velocity gradient $\boldsymbol{\kappa}_e = \operatorname{diag}(\dot{\varepsilon}, -\dot{\varepsilon})$. Some results are reported in Table 3.2.2. We take there the parameter $b = 12$, three increasing values of $\dot{\varepsilon}$, and the meshes with $N_{q_x} = N_{q_y} = N$. We present the relative error in the steady state extra stress (for which an analytical solution exists) only for stable simulations, i.e., for those that the steady state can be attained by a time-marching scheme. We observe that the fictitious domain approach seems to converge at all values of

TABLE 3.2.2
The relative error ϵ_{rel} in the polymeric stress at steady state in extensional flows discretized by the fictitious domain approach

(a) $\dot{\varepsilon} = 1$:

s	$N = 10$	$N = 20$	$N = 30$	$N = 40$	$N = 50$	$N = 60$
0	–	$8.4 \cdot 10^{-2}$	$5.8 \cdot 10^{-4}$	$9.4 \cdot 10^{-5}$	$2.5 \cdot 10^{-5}$	$9.5 \cdot 10^{-6}$
1	$7.8 \cdot 10^{-2}$	$5.5 \cdot 10^{-4}$	$5.5 \cdot 10^{-7}$	$7.5 \cdot 10^{-7}$	$2.3 \cdot 10^{-7}$	$8.9 \cdot 10^{-8}$
3	–	$3.8 \cdot 10^{-4}$	$8.8 \cdot 10^{-7}$	$2.3 \cdot 10^{-8}$	$1.8 \cdot 10^{-8}$	$1.1 \cdot 10^{-8}$
6	–	–	–	–	$2.4 \cdot 10^{-5}$	$5.6 \cdot 10^{-6}$

(b) $\dot{\varepsilon} = 5$:

s	$N = 20$	$N = 30$	$N = 40$	$N = 50$	$N = 60$	$N = 70$
0	–	–	–	–	–	$3.3 \cdot 10^{-4}$
1	–	–	$1.4 \cdot 10^{-3}$	$1.4 \cdot 10^{-3}$	$3.6 \cdot 10^{-5}$	$1.2 \cdot 10^{-5}$
3	–	–	–	$3.2 \cdot 10^{-3}$	$7.5 \cdot 10^{-5}$	$3.0 \cdot 10^{-6}$
6	–	–	–	–	–	–

(c) $\dot{\varepsilon} = 10$:

s	$N = 20$	$N = 30$	$N = 40$	$N = 50$	$N = 60$	$N = 70$
0	–	–	–	–	–	–
1	–	–	–	$1.4 \cdot 10^{-2}$	$1.8 \cdot 10^{-3}$	$1.0 \cdot 10^{-4}$
3	–	–	–	–	–	$5.1 \cdot 10^{-2}$
6	–	–	–	–	–	–

The dashes mean that the corresponding simulation is unstable so that the steady state is unattainable.

the parameter s, but the fastest convergence is achieved with $s = 1$. This is comparable with the results for the previous spectral method reported in Table 3.1.1, which is in general much more efficient.

Second step (3.33): discretization in physical space Since problem (3.33) does not contain the derivatives in $\boldsymbol{q}$, it can be discretized separately at any $\boldsymbol{q} : q < \sqrt{b}$. Taking into account the absence of convective derivatives in the case of a Couette flow problem (3.33) can be rewritten for any such $\boldsymbol{q}$ as

$$\int\limits_{\Omega(\boldsymbol{q})} \frac{\alpha^{n+1} - \alpha^{n+1/2}}{\Delta t} \beta(y)\, \mathrm{d}y + \frac{\ell_0^2}{8\lambda_H} \int\limits_{\Omega(\boldsymbol{q})} \frac{\partial \alpha^{n+1}}{\partial y} \cdot \frac{\mathrm{d}\beta}{\mathrm{d}y}(y)\, \mathrm{d}y = 0, \tag{3.36}$$

where $\Omega(\boldsymbol{q}) = (-d + \ell_0|q_y|/2, d - \ell_0|q_y|/2)$. We can now reinterpret (3.36) as the following differential boundary value problem

$$\frac{\alpha^{n+1}-\alpha^{n+1/2}}{\Delta t}=\frac{\ell_0^2}{8\lambda_H}\frac{\mathrm{d}^2\alpha^{n+1}}{\mathrm{d}y^2}\ \text{for } y\in\Omega(\boldsymbol{q}),\qquad (3.37)$$
$$\left.\frac{\mathrm{d}\alpha^{n+1}}{\mathrm{d}y}\right|_{y=\pm(d-|q_y|/2)}=0.$$

The discretization of the second derivative in y will depend on the point of configuration space $\boldsymbol{q}$ where it is applied. Indeed, if $1<k<N_y$ and $\boldsymbol{q}\in D(y_{k\pm 1})$, then we can use the standard central difference approximation:

$$\left.\frac{\partial^2\alpha^{n+1}}{\partial y^2}\right|_{y=y_k}=\frac{2}{h_{k+1}+h_k}\left(\frac{\alpha_{k+1}^{n+1}-\alpha_k^{n+1}}{h_{k+1}}-\frac{\alpha_k^{n+1}-\alpha_{k-1}^{n+1}}{h_k}\right)+O(h^2)\qquad (3.38)$$

with $h_k=y_k-y_{k-1}$, $h=\max_k h_k$. Otherwise, $\alpha_{k-1}^{n+1}(\boldsymbol{q})$ or $\alpha_{k+1}^{n+1}(\boldsymbol{q})$ is not defined, and we construct instead a first-order approximation for $\frac{\partial^2\alpha}{\partial y^2}(y_k,\boldsymbol{q},t_{n+1})$ using the boundary condition in (3.37). We detail this approximation only in the first case when $\alpha_{k-1}^{n+1}(\boldsymbol{q})$ is not defined, and the other case can be treated similarly. Let the point y_b be such that a dumbbell centered at y_b with the given end-to-end vector $\boldsymbol{q}$ touches the wall, i.e., $y_b=-d+\ell_0 q_y/2$. By assumption, $y_{k-1}<y_b\le y_k$ or simply $y_b\le y_k$ if $k=1$. Let $\delta=|y_b-y_k|$ so that $\delta<h$. Using the Taylor expansion, we can write

$$0=\frac{\mathrm{d}\alpha^{n+1}}{\mathrm{d}y}(y_b)=\frac{\mathrm{d}\alpha^{n+1}}{\mathrm{d}y}(y_k)-\delta\frac{\mathrm{d}^2\alpha^{n+1}}{\mathrm{d}y^2}(y_k)+O(h^2).\qquad (3.39)$$

Likewise we have

$$\alpha^{n+1}(y_{k+1})=\alpha^{n+1}(y_k)+h_k\frac{\mathrm{d}\alpha^{n+1}}{\mathrm{d}y}(y_k)+\frac{h_k^2}{2}\frac{\mathrm{d}^2\alpha^{n+1}}{\mathrm{d}y^2}(y_k)+O(h^3),$$

which in combination with (3.39) gives

$$\left.\frac{\partial^2\alpha^{n+1}}{\partial y^2}\right|_{y=y_k}=\frac{\alpha^{n+1}(y_{k+1})-\alpha^{n+1}(y_k)}{h_k(\delta+h_k/2)}+O(h).\qquad (3.40)$$

Discretization of (3.36) is now achieved by replacing the second derivative in y by its approximation (3.38) or (3.40). This discretization is so far described for any configuration vector, but in practice, it should be implemented for all $\boldsymbol{q}$ in the ensembles of nodes $\mathcal{T}(y_k)$. Since these ensembles differ from one grid point y_k to another and all the grid points are coupled by the approximations (3.38), (3.40) of the second derivative in y, their implementation requires an interpolation from one set of nodes $\mathcal{T}(y_k)$ to another. This is achieved by evaluating the Lagrange polynomials in (3.35) at the points from other nodal sets.

Computing the number density and the polymeric stress Unlike a locally homogeneous flow, the number density n of polymer molecules cannot be assumed to be constant in the fully nonhomogeneous case. Assuming that all the mass of the polymer is concentrated at the dumbbell beads, we should rather compute n as

$$n(\mathbf{r}, t) = \int\limits_{D(\mathbf{r}_c)} \psi(\mathbf{r} + \ell_0\mathbf{q}/2, \mathbf{q}, t)\, \mathrm{d}\mathbf{q}$$

where it is nondimensionalized by its average value n_{avg}. The Kramers expression for the stress should be also modified accordingly as developed by BILLER and PETRUCCIONE [1987]:

$$\tau(\boldsymbol{r}, t) = \frac{\eta_p}{\lambda}\frac{b + \mathrm{d} + 2}{b}\Big(- 2n(\boldsymbol{r}, t)kT\boldsymbol{\delta}$$

$$+ \int\limits_{D(\boldsymbol{r}_c)} \int\limits_{s=0}^{1} \boldsymbol{q}\boldsymbol{F}(\boldsymbol{q})\psi(\boldsymbol{r} + (s - 1/2)\ell_0\boldsymbol{q}, \boldsymbol{q}, t)\, \mathrm{d}s\, \mathrm{d}\boldsymbol{q}\Big).$$

3.3. Numerical methods for concentrated solutions

We describe here a spectral method for the simplest reptation model of Doi–Edwards written in the form of the Fokker–Planck equation (1.41) with the boundary conditions (1.42). We shall treat only the case of a homogeneous flow although our approach can be easily adapted to complex flow simulations as in FANG, LOZINSKI and OWENS [2004]. In fact, the last paper is devoted to a more elaborate model, namely Öttinger's simplified uniform model (see (ÖTTINGER [1999])), which is an example of a modern reptation theory that incorporates the stretching of polymer chains and the convective constraint release into the original picture of Doi–Edwards.

We approximate the configuration probability density ψ, expressing dependence on s and on a generic point $\boldsymbol{u}$ on the unit sphere, by

$$\psi(\boldsymbol{u}, s, t) \approx \sum_{i=0}^{1}\sum_{l=0}^{N_s}\sum_{n=0}^{N_u}\sum_{m=i}^{n} \psi_{i,\ell,n,m}(t)\Phi^i_{2n,2m}(\theta, \varphi)H_\ell(s). \tag{3.41}$$

In (3.41), $\Phi^i_{n,m} = P^m_n(\cos\theta)((1 - i)\cos m\varphi + i \sin m\varphi)$ $(i = 0, 1)$ are spherical harmonics defined in terms of the associated Legendre polynomials P^m_n and the spherical polar coordinates θ and φ. We note that only the spherical harmonics of even order appear in (3.41) because of the symmetry of ψ in $\boldsymbol{u}$. $H_\ell(s)$ in (3.41) are Lagrange interpolating polynomials of degree N_s based on the GLL points s_j, $j = 0, \ldots, N_s$ scaled to the interval $[-1, 1]$.

Inserting (3.41) into the Fokker–Planck equation (1.41) for ψ, we now seek to simplify the terms in the square parentheses appearing on the right-hand side of (1.41). It is shown by FAN [1989a] that

$$\frac{\partial}{\partial \boldsymbol{u}} \cdot \left[(\boldsymbol{\delta} - \boldsymbol{u}\boldsymbol{u}) \cdot \boldsymbol{\kappa} \cdot \boldsymbol{u}\Phi^i_{n,m}\right] = \sum_{k=m-2}^{m+2}\sum_{j=n-2}^{n+2} a^{m,k}_{n,j}\left(w^k_j\Phi^i_{j,k} + (-1)^{1-i}v^k_j\Phi^{1-i}_{j,k}\right), \tag{3.42}$$

where the coefficients $a^{m,k}_{n,j}$ and the linear combinations of velocity gradients w^k_j and v^k_j are supplied in tables 1-3 of the same paper.

Using (3.41) and (3.42), we form the product of (1.41) with a test function $\Phi^i_{2p,2q}(\theta, \varphi)L_k(s)$ ($i = 0, 1; p = 0, \dots, N_u; q = i, \dots, p; k = 0, \dots, N_s$) and integrate over configurational space $B(0, 1) \times (0, 1)$. The integral with respect to s is evaluated using a GLL quadrature rule, and orthogonality of the spherical harmonics over $B(0, 1)$ is exploited. Denoting the approximation of $\psi_{i,k,p,q}(j\Delta t)$ by $\psi^j_{i,k,p,q}$, a discretized Fokker–Planck equation can be written as

$$\begin{aligned}&\frac{\omega_k}{\Delta t}\left(\psi^{j+1}_{i,k,p,q} - \psi^j_{i,k,p,q}\right)\\&\quad + \omega_k \sum_{n=p-1}^{p+1}\sum_{m=q-1}^{q+1} a^{2q,2m}_{2p,2n}\left(w^{2m}_{2n}\psi^j_{i,k,n,m} + (-1)^i v^{2m}_{2n}\psi^j_{1-i,k,n,m}\right)\\&\quad + \frac{1}{\pi^2\tau_d}\sum_{\ell=0}^{N_s}\psi^j_{i,\ell,p,q}\left(L'_\ell(s), L'_k(s)\right)_{N_s} = 0, \qquad (3.43)\end{aligned}$$

where $(\cdot, \cdot)_{N_s}$ denotes the $(N_s + 1)$ point GLL quadrature evaluation of the L^2 inner product over $[0, 1]$

$$(u, v)_{N_s} = \sum_{k=0}^{N_s}\omega_k u(s_k)v(s_k)$$

with quadrature weights ω_k. Once we have the probability density in the form (3.41), the components of the orientation tensor $\boldsymbol{S} = \langle \boldsymbol{uu}\rangle$ may be easily computed using again the exact integration in $\boldsymbol{u}$ and the quadrature above for integration in s.

We note that we have used the explicit Euler time marching scheme in (3.43). Strangely enough, our numerical experiments indicate that passing to an implicit scheme does not enhance stability of simulations for this model. A much more important stabilization effect can be achieved by adding a diffusion term of the form $D\frac{\partial}{\partial \boldsymbol{u}} \cdot \frac{\partial \phi}{\partial \boldsymbol{u}}$ to (1.41). Fortunately, this term is present in more realistic reptation models like that of ÖTTINGER [1999]. Implementing this term is straightforward as the spherical harmonics are the eigenfunctions of the Laplace operator on the unit sphere; specifically $\frac{\partial}{\partial \boldsymbol{u}} \cdot \frac{\partial \Phi^i_{n,m}}{\partial \boldsymbol{u}} = -n(n+1)\Phi^i_{n,m}$, cf. FANG, LOZINSKI and OWENS [2004].

3.4. Models with high-dimensional configuration spaces

The number of configuration space dimensions in all the models we have considered up to now has been low (≤ 3). A more detailed description of the polymer molecules, however, often gives rise to models with significantly more degrees of freedom. A prototypical example of these is the bead-spring chain model. A linear polymer molecule is represented there by a chain consisting of $(d + 1)$ beads joined consecutively by d massless springs. Note that the dumbbell model (3.1) is just a special case of it with $d = 1$. The bead-spring chain is represented by d random vectors $\boldsymbol{q}_j$. Their dynamics in a globally homogeneous flow is given either by the SDEs (1.31) or by the Fokker–Planck equation (1.33) for the probability density function $\psi(t, \boldsymbol{q}_1, \dots, \boldsymbol{q}_d)$. The spring force is denoted here by $\boldsymbol{F}(\boldsymbol{q})$, and we consider again the case of FENE forces given by (1.26).

One can try to generalize the spectral method described in Section 3.1 for $d = 1$ to the chains of any length. We would then use the approximation space V_N defined by (3.9) to represent the dependence of ψ, after the change of variables $\psi \to \alpha$ as in (3.2), on each of the vectors $\boldsymbol{q}_j$ represented by (η_j, θ_j). More specifically, we should use the following ansatz:

$$\alpha \approx \sum_{\boldsymbol{i}:1\leq\|\boldsymbol{i}\|_\infty\leq N} \alpha_{\boldsymbol{i}}(t)\Phi_{i_1}(\boldsymbol{q}_1)\cdots\Phi_{i_d}(\boldsymbol{q}_d) \in V_N \otimes \cdots \otimes V_N. \tag{3.44}$$

Here, $\boldsymbol{i} = (i_1, \ldots, i_d)$ is a multi-index with the norm $||\boldsymbol{i}||_\infty = \max(i_1, \ldots, i_d)$, and $\{\Phi_i\}$, $i = 1, \ldots, N$, is some basis for V_N. We will refer to the approximation space $V_{\text{FTP}} = V_N \otimes \cdots \otimes V_N$ as the full tensor product space. Its dimension is N^d, which makes numerical methods based directly on (3.44) prohibitively expensive even for moderate values of d, say $d \geq 4$. This phenomenon is well known under the name "the curse of dimension": standard full tensor-product bases lead to computational effort that grows exponentially with total dimension. To keep the cost acceptable, one therefore needs to drastically reduce the number of terms in the sum approximating α in (3.44). One can conceive two general strategies to do this, which we can term, respectively, a priori and a posteriori. In an a priori strategy, one tries to represent α on only a small subspace $V_{\text{STP}} \subset V_{\text{FTP}}$, which we call a sparse tensor-product space and which is chosen so that it approximates any function of interest reasonably well. The idea for the construction of V_{STP} goes back to SMOLYAK [1963]. Very roughly speaking, one gets rid of the functions that oscillate too wildly in all the directions $\boldsymbol{q}_1, \ldots, \boldsymbol{q}_d$ (high harmonics), so that when $d = 2$, for example, one keeps in V_{STP} only the basis functions of the type "low harmonic$(\boldsymbol{q}_1)\times$ low harmonic$(\boldsymbol{q}_2)$" or "high harmonic$(\boldsymbol{q}_1)\times$ low harmonic$(\boldsymbol{q}_2)$" or "low harmonic$(\boldsymbol{q}_1)\times$ high harmonic$(\boldsymbol{q}_2)$" but not those of the type "high harmonic$(\boldsymbol{q}_1)\times$ high harmonic$(\boldsymbol{q}_2)$." In general, a high-dimensional basis is derived from a one-dimensional basis by forming tensor products of univariate formulae whose indices lie in an appropriate simplex. This approach was extensively studied in the last 20 years under the name Sparse Grids: see, for example, the review of BUNGARTZ and GRIEBEL [2004] and the references therein. In particular, this method was used for parabolic PDEs by VON, PETERSDORFF and SCHWAB [2004] and by GRIEBEL and OELTZ [2007]. A typical result on the convergence of these methods when they are based on a finite-element discretization of univariate functions on a mesh of size h is that one gets an optimal convergence rate with respect to h up to a logarithmic factor $\sim (\log_2 h)^d$ while keeping only $O(h^{-1}|\log h|^{d-1})$ basis functions in the sparse tensor product space, which should be contrasted to $O(h^{-d})$ basis functions in the full tensor product. In the present chapter, we report on an implementation of this approach for the Fokker–Planck equation of the bead-spring chain model, following essentially DELAUNAY, LOZINSKI and OWENS [2007].

An alternative a posteriori approach to overcome the "curse of dimension" is to try to approximate the unknown α by an m-term sum

$$\alpha \approx \sum_{i=1}^{m} \alpha_1^{(i)}(\boldsymbol{q}_1)\cdots\alpha_d^{(i)}(\boldsymbol{q}_d), \tag{3.45}$$

where the univariate functions $\alpha_i^{(1)}(\boldsymbol{q}) \in V_N$ are chosen so that the sum above gives the optimal or quasi-optimal approximation of α between all such m-term expressions for a fixed and hopefully not very big m. Following BEYLKIN and MOHLENKAMP [2002], we call the number m the separation rank of the approximation (3.45). Note that unlike the approach of Sparse

Grids above, we do not impose any a priori restrictions on the choice of $\alpha_1^{(i)}(\boldsymbol{q}_1)\cdots\alpha_d^{(i)}(\boldsymbol{q}_d)$. The ensemble of m-term sums in (3.45) does not form a linear space so that to find the best fitting sum one needs to solve a nonlinear optimization both in view of a theoretical justification and of a practical implementation. To perform the optimization in (3.45), BEYLKIN and MOHLENKAMP [2002, 2005] proposed the alternating least-squares algorithm, which assumes that the function α is represented already in the form of the sum of products as in (3.45) but with $\tilde{m} > m$ terms. Starting then from an initial approximation for the m-term sum, one iteratively refines it. One does so only in one direction k at a time by fixing the functions in the other directions $\alpha_i^{(j)}$, $j \neq k$, $i = 1, \ldots, m$ and minimizing the norm of the residual over the functions $\alpha_i^{(k)}$, $j \neq k$, $i = 1, \ldots, m$, sweeping over the directions $k = 1, \ldots, d$. Assuming that the minimization is done in the norm of some Hilbert space, the refinement in the direction k amounts to standard least square univariate problem. This procedure termed as the separation rank reduction was applied successfully to the ground-state multiparticle Schrödinger problem by BEYLKIN and MOHLENKAMP [2005].

An alternative and less expensive algorithm to the optimization in (3.45) is to find successively the representations of separation rank $1, 2, \ldots$ until the needed tolerance is reached. The mth step in this procedure assumes that we already have an $(m-1)$-term approximation and we do not touch it any longer. Rather we optimize only on the term number m, i.e., over the unknown functions $\{\alpha_m^{(k)},\ 1 \leq d\}$. This again leads to a nonlinear optimization problem, which can be solved by a fixed point or a Newton algorithm. This approach fits into the framework of greedy algorithms, i.e., algorithms that seek a global optimum by making the locally optimal choices at each iteration (see CORMEN, LEISERSON, RIVEST and STEIN [2001]). Of course, by doing this there is no hope of producing the best m-term approximation, even if the optimization problem on each iteration is solved exactly. However, DEVORE and TEMLYAKOV [1996], TEMLYAKOV [2000] prove the convergence of some greedy type algorithms for m-term approximations in Hilbert spaces. Note that similar algorithms were proposed and studied in MALLAT and ZHANG [1993] under the name of Matching pursuits.

This idea of using greedy algorithms to construct successive approximations of the type (3.45) was introduced by AMMAR, MOKDAD, CHINESTA and KEUNINGS [2006a], AMMAR, RYCKELYNCK, CHINESTA and KEUNINGS [2006b] in the context of the numerical solution of multidimensional PDEs. It is applied there to several problems including the stationary Fokker–Planck equation of the FENE bead-spring chain model. In AMMAR, MOKDAD, CHINESTA and KEUNINGS [2007], the approach is extended to time-dependent problems. One should note, however, that the method in the original papers of AMMAR et al. can be interpreted as a greedy algorithm for the error minimization only for elliptic symmetric problems since they can be rewritten in terms of minimization of an energy functional (see also a more complete theoretical study in LE BRIS, LELIÈVRE and MADAY [2009]). The extension of the method to more general problems like the Fokker–Planck equation is done merely by analogy. In particular, its convergence is not guaranteed. We prefer, therefore, to discuss in this chapter a more recent modification of the low-rank separation method, introduced in AMMAR, CHINESTA and FALCÓ [2010], where one minimizes on each iteration an L^2 norm of the residual rather than a norm of the error. Such a method can be linked with the greedy algorithms of DEVORE and TEMLYAKOV [1996] for essentially any PDE.

Yet another way to overcome the curse of dimension in **high-dimensional** systems may be based on quasi-Monte Carlo method (NIEDERREITER [1992]). This approach can be viewed as somewhere between stochastic and deterministic Fokker–Planck-based methods. It was employed for the bead-spring chain model undergoing simple shear flow by VENKITESWARAN and JUNK [2005a,b]. The error estimate in terms of the number of nodes M (say) behaves like $O(\sqrt{M})$, independently of the dimension. Although this is no better than for a standard Monte Carlo method, the method of these papers manifested less variance in the results, and for fixed accuracy, the authors achieved an improvement in the computational time over the simple Monte Carlo method.

3.4.1. *A high-order sparse tensor product Fokker–Planck-based method*

Let us first explain the idea of the method of sparse grids on the simplest example of the Poisson problem in the d-dimensional hypercube $\Omega = (0,1)^d$: given the function $f : \Omega \to \mathbb{R}$ find $u : \Omega \to \mathbb{R}$ such that

$$-\Delta u = f \text{ in } \Omega, \quad u|_{\partial\Omega} = 0. \tag{3.46}$$

We start by defining some nested finite element spaces $\mathcal{V}_l,\ l = 1, \ldots, L$ such that $\mathcal{V}_l \subset \mathcal{V}_{l+1}$. The simplest choice is to take $\mathcal{V}_l \subset H_0^1([0,1])$ as the set of piecewise linear functions on the uniform division $\mathcal{T}_l$ of the segment $[0,1]$ into 2^l segments so that $\dim \mathcal{V}_l = 2^l - 1$. We then introduce the increment spaces $\mathcal{W}_l,\ l = 1, \ldots, L$ such that $\mathcal{W}_1 = \mathcal{V}_1$ and $\mathcal{V}_l = \mathcal{V}_{l-1} \oplus \mathcal{W}_l$, $\mathcal{W}_l = (I - P_{l-1})\mathcal{V}_l$ for $l > 1$. Here, $P_l,\ l = 1, \ldots, L$, are some projectors $C^0([0,1]) \to \mathcal{V}_l$, which are usually taken either as a nodal interpolation on $\mathcal{T}_l$ or as the orthogonal projector in L^2. These univariate spaces are then used to construct a multi variate sparse tensor product space

$$V_{\mathrm{STP}} = \bigoplus_{k: d \le \|k\|_1 \le d+l-1} \mathcal{W}_{k_1} \bigotimes \cdots \bigotimes \mathcal{W}_{k_d}, \tag{3.47}$$

and the approximate solution $u_L \in V_{\mathrm{STP}}$ is sought by using a Galerkin method

$$\int_\Omega \nabla u \cdot \nabla v = \int_\Omega fv, \quad \forall v \in V_{\mathrm{STP}}.$$

It can be proved (see, for example, BUNGARTZ and GRIEBEL [2004]) that the error $||\nabla(u_L - u)||_{L^2(\Omega)}$ is $O(2^{-L}L^d)$, while the dimension of V_{STP} is only $O(2^L L^{d-1})$. Note that a standard Galerkin approximation on the full tensor product space $\mathcal{V}_L \bigotimes \cdots \bigotimes \mathcal{V}_L$ would lead to a slightly better error estimate of $O(2^{-L})$, but the dimension of the discrete problem would grow exponentially in d like $O(2^{Ld})$.

To clarify the construction of the sparse tensor product, we can construct a basis $\{\phi_1, \ldots, \phi_{2^L-1}\}$ such that $\mathcal{V}_1 = \mathcal{W}_1 = \mathrm{span}(\phi_1)$, $\mathcal{W}_2 = \mathrm{span}(\phi_2, \phi_3)$, and so on till $\mathcal{W}_L = \mathrm{span}(\phi_{2^{L-1}}, \phi_{2^L-1})$. Letting $\boldsymbol{i} = (i_1, \ldots, i_d)$ be a **d**-dimensional multi-index, we introduce the product functions

$$\Phi_{\boldsymbol{i}}(x_1, \ldots, x_d) = \prod_{j=1}^{d} \varphi_{i_j}(x_j). \tag{3.48}$$

A basis of V_{STP} is formed by functions $\Phi_{\boldsymbol{i}}$ with the multi-indices from the set

$$\mathcal{I}_{L,d} = \{\boldsymbol{i} \in \mathbb{N}^d : 2^{k_j-1} \le i_j \le 2^{k_j} - 1 \text{ for } \boldsymbol{k} \in \mathbb{N}^d : d \le \| \boldsymbol{k} \|_1 \le d + l - 1\}. \tag{3.49}$$

We would here like to adapt the ideas above to the solution of the Fokker–Planck equation (1.33) of the bead-spring chain model with 2D FENE connectors. We rewrite this equation for $\psi(t, \boldsymbol{q}_1, \ldots, \boldsymbol{q}_d)$, $\boldsymbol{q}_i \in D$ in terms of univariate operators acting on functions of a single vector $\boldsymbol{q} \in D$:

$$\frac{\partial \psi}{\partial t} = \mathcal{L}_{FP}^{(d)}(\psi) = -\sum_{k=1}^{d} L^k \psi + \frac{1}{4\lambda_H} \sum_{k=2}^{d} (N_x^{k-1} M_x^k + N_y^{k-1} M_y^k)\psi$$
$$+ \frac{1}{4\lambda_H} \sum_{k=1}^{d-1} (N_x^{k+1} M_x^k + N_y^{k+1} M_y^k)\psi, \tag{3.50}$$

where the operators L, $M_{x,y}$, and $N_{x,y}$ are defined by the following variational formulations for any appropriate functions $\psi(\boldsymbol{q})$, $\phi(\boldsymbol{q})$

$$\int_D (L\psi)\phi \mathrm{d}\boldsymbol{q} = \int_D \left(-\nabla \boldsymbol{v} \cdot \boldsymbol{q} + \frac{1}{2\lambda_H} \boldsymbol{F}(\boldsymbol{q}) + \frac{1}{2\lambda_H} \frac{\partial}{\partial \boldsymbol{q}} \right) \psi \cdot \frac{\partial \phi}{\partial \boldsymbol{q}} \mathrm{d}\boldsymbol{q}, \tag{3.51}$$

$$\int_D (M_x \psi)\phi \mathrm{d}\boldsymbol{q} = \int_D \psi \frac{\partial \phi}{\partial q_x} \mathrm{d}\boldsymbol{q}, \quad \int_D (M_y \psi)\phi \mathrm{d}\boldsymbol{q} = \int_D \psi \frac{\partial \phi}{\partial q_y} \mathrm{d}\boldsymbol{q},$$

$$\int_D (N_x \psi)\phi \mathrm{d}\boldsymbol{q} = \int_D \left(F_x(\boldsymbol{q}) + \frac{\partial}{\partial q_x} \right) \psi \phi \mathrm{d}\boldsymbol{q}, \quad \int_D (N_y \psi)\phi \mathrm{d}\boldsymbol{q} = \int_D \left(F_y(\boldsymbol{q}) + \frac{\partial}{\partial q_y} \right) \psi \phi \mathrm{d}\boldsymbol{q}.$$

The upper index k is used in (3.50) to indicate that the corresponding univariate operator acts in the direction of the dumbbell number k, for example, $L^k = I \otimes \cdots \otimes L \otimes \cdots \otimes I$ where the operator L is in the kth position. The operator L is precisely the operator from the Fokker–Planck equation of the dumbbell model (3.1). We would therefore like to use the spectral methods developed in Section 3.1 to discretize the univariate operators in (3.50) and to construct the univariate approximation spaces $\mathcal{V}_l$. However, it is not clear how to generalize the construction outlined above in the case of the Laplacian to the present problem since there is no straightforward choice for the nested triangulations $\mathcal{T}_l$ in the context of spectral methods on a two-dimensional disc. Alternatively, one can directly choose some basis functions $\{\psi_1(\boldsymbol{q}), \ldots, \psi_{2^L-1}(\boldsymbol{q})\}$ and define the spaces $\mathcal{V}_l$ and $\mathcal{W}_l$ as

$$\mathcal{V}_l = \mathrm{span}\{\psi_1, \ldots, \psi_{2^l-1}\}, \quad \mathcal{W}_l = \mathrm{span}\{\psi_{2^{l-1}}, \ldots, \psi_{2^l-1}\} \tag{3.52}$$

with $l = 1, \ldots, L$. Since the essential feature of the increment spaces $\mathcal{V}_l$ is that they contain more and more oscillating function when l increases, an appropriate choice for ψ_i can be to take them as the eigenfunctions of the univariate dumbbell operator $\mathcal{L}_{FP}$. More specifically, for the level of approximation L, we compute numerically the leading $2^L - 1$ eigenfunctions $\psi_1(\boldsymbol{q}), \ldots, \psi_{2^L-1}(\boldsymbol{q})$ corresponding to the eigenvalues $\lambda_1, \ldots, \lambda_{2^L-1}$ as

$$-\mathcal{L}_{FP}^{(1)} \psi_i = L\psi_i = \lambda_i \psi_i. \tag{3.53}$$

The eigenvalues are numbered so that $0 = \lambda_1 < \mathrm{Re}(\lambda_2) \leq \mathrm{Re}(\lambda_3) \leq \cdots$.

Construction of an eigenbasis The discretization of the univariate operator $L = \mathcal{L}_{FP}^{(1)}$ can be done along the same lines as in Section 3.1. We can no longer suppose, however, that $\psi_i(\boldsymbol{q})$ are even functions ($\psi_i(-\boldsymbol{q}) = \psi_i(\boldsymbol{q})$) since they will be used to represent the probability density of dumbbells in a chain, and the beads in the chain cannot be arbitrarily renumbered. Therefore, we should add to the approximation space (3.9) some functions to approximate the odd part of ψ, i.e., the products of odd Fourier harmonics in θ with the appropriate radial functions in r. The change of variables (3.8) does not fit to the odd harmonics because the corresponding radial function should vanish at $r = 0$. It is natural to approximate these functions by polynomials containing only the odd powers of r. We adapt (3.8) accordingly as follows. Any $\psi(\boldsymbol{q})$ representing a probability density of a dumbbell is first decomposed into the sum of its even $\psi_0(\boldsymbol{q}) = (\psi(\boldsymbol{q}) + \psi(-\boldsymbol{q}))/2$ and odd $\psi_1(\boldsymbol{q}) = (\psi(\boldsymbol{q}) - \psi(-\boldsymbol{q}))/2$ parts. We define then α_0 and α_1 as

$$\psi_0 = \left(1 - \frac{\rho^2}{b}\right)^s \alpha_0, \;\; \psi_1 = \rho\left(1 - \frac{\rho^2}{b}\right)^s \alpha_1, \tag{3.54}$$

with some real parameter s and finally recombine α_0 and α_1 into $\alpha = \alpha_0 + \alpha_1$. This change of variables $\psi \to \alpha$ will be denoted by the operator $\mathcal{R}$, i.e., $\psi = \mathcal{R}\alpha$. Analogously, we introduce the operator $\mathcal{R}^*$ to represent the test functions so that for any suitable function ϕ, we define $\beta = \mathcal{R}^*\phi = \beta_0 + \beta_1$ with

$$\phi_0 = \left(1 - \frac{\rho^2}{b}\right)^{-s} \beta_0, \;\; \phi_1 = \frac{1}{\rho}\left(1 - \frac{\rho^2}{b}\right)^{-s} \beta_1, \tag{3.55}$$

ϕ_0 and ϕ_1 being the even and the odd parts of ϕ. The finite dimensional space in which the new unknown α will be approximated is only slightly different from (3.9):

$$V_N = \text{span}\{h_k(\eta)\widetilde{\Phi}_{il}(\theta),\; i = 0, 1,\; i \le l \le 2N_F,\; 1 \le k \le N_R\}, \tag{3.56}$$

where $\{h_k(\eta)\}_{1\le k\le N_R}$ are Lagrange interpolating polynomials based on the GL points η_i, $i = 1, \ldots, N_R$, $\widetilde{\Phi}_{il}(\theta) = (1-i)\cos(l\theta) + i\sin(l\theta)$, $i = 0, 1$, $l = i, \ldots, N_F$, and $N = \text{dim}V_N = N_R(4N_F + 1)$. The eigenproblem (3.53) can now be discretized as follows: find $\psi_i = \mathcal{R}\alpha_i$ with $\alpha_i \in V_N$ such that

$$\begin{aligned}&\left(-\boldsymbol{\kappa}\boldsymbol{q}\mathcal{R}\alpha_i + \frac{1}{2\lambda_H}\boldsymbol{F}(\boldsymbol{q})\mathcal{R}\alpha_i + \frac{1}{2\lambda_H}\nabla\mathcal{R}\alpha_i, \nabla\mathcal{R}^*\beta\right)_{N_R,N_F} \\ &\quad = \lambda_i\,(\alpha_i, \beta)_{N_R,N_F}, \quad \forall\beta \in V_N.\end{aligned} \tag{3.57}$$

As in Section 3.1, the notation $(\cdot,\cdot)_{N_R,N_F}$ here stands for the numerical evaluation of $\int_D \cdot \mathrm{d}\boldsymbol{q}$ where the integrals with respect to η are evaluated using the standard Gauss quadrature rule based on the GL points, and integrals with respect to θ are evaluated analytically.

The bilinear forms (3.51) will also be used to discretize the operators L, $M_{x,y}$, and $N_{x,y}$ in the same manner as in (3.57). These discretization will be denoted as $\widehat{L}$, $\widehat{M}_{x,y}$, and $\widehat{N}_{x,y}$ so that, for instance, $\widehat{M}_x : \mathcal{R}V_N \to \mathcal{R}V_N$ is defined for any $\alpha \in V_N$ as

$$(\widehat{M}_x\mathcal{R}\alpha, \mathcal{R}^*\beta)_{N_R,N_F} = \left(\mathcal{R}\alpha_N, \frac{\partial}{\partial q_x}\mathcal{R}^*\beta_N\right)_{N_R,N_F}, \quad \forall\beta \in V_n.$$

Multidimensional problem Returning to the multidimensional problem (3.50), a straight-forward implementation of the sparse tensor product method would read: find $\psi(t) \in V_{\text{STP}}$, i.e., $\psi(t) = \sum_{i \in \mathcal{I}_{L,d}} \psi_i(t) \Phi_i$ with $\Phi_i(\boldsymbol{q}_1, \ldots, \boldsymbol{q}_d) = \psi_{i_1}(\boldsymbol{q}_1) \cdots \psi_{i_d}(\boldsymbol{q}_d) = \mathcal{R}\alpha_{i_1}(\boldsymbol{q}_1) \cdots \mathcal{R}\alpha_{i_d}(\boldsymbol{q}_d)$ such that

$$\sum_{i \in \mathcal{I}_{L,d}} \frac{\partial \psi_i}{\partial t} \left(\Phi_i, \widetilde{\Phi}_j\right)_{N_R,N_F,d} = \sum_{i \in \mathcal{I}_{L,d}} \psi_i \left(\mathcal{L}_{FP}^{(d)} \Phi_i, \widetilde{\Phi}_j\right)_{N_R,N_F,d} \quad \forall \boldsymbol{j} \in \mathcal{I}_{L,d}, \tag{3.58}$$

where $\widetilde{\Phi}_i(\boldsymbol{q}_1, \ldots, \boldsymbol{q}_d) = \mathcal{R}^* \alpha_{i_1}(\boldsymbol{q}_1) \cdots \mathcal{R}^* \alpha_{i_d}(\boldsymbol{q}_d)$ and $(\cdot, \cdot)_{N_R,N_F,d}$ stands for a multidimensional extension of the univariate quadrature $(\cdot, \cdot)_{N_R,N_F}$ so that, for instance,

$$\left(\Phi_i, \widetilde{\Phi}_j\right)_{N_R,N_F,d} = \prod_{k=1}^{d} \left(\alpha_{i_k}, \alpha_{j_k}\right)_{N_R,N_F}.$$

Numerically we have found however that the convergence of this approximation is very slow. Much better results are obtained if the trial functions in (3.58) are replaced by the functions $\Phi_j^* \in \mathcal{R}^* V_N$ that form the basis dual to the basis $\{\Phi_i\}$. The functions Φ_j^* are defined for any multi-index $\boldsymbol{j} \in \mathbb{N}^d$ as $\Phi_j^*(\boldsymbol{q}_1, \ldots, \boldsymbol{q}_d) = \prod_{k=1}^{d} \psi_{j_k}^*(\boldsymbol{q}_k)$ where $\psi_j^* = \mathcal{R}^* \beta_j$ and $\beta_j \in V_N$ are chosen so that

$$\langle \alpha_i, \beta_j \rangle_{N_R,N_F} = \delta_{ij}, \quad i, j = 1, \ldots, N. \tag{3.59}$$

In practice, the vectors in $\mathbb{R}^d$ representing β_j in the basis of the dual space V_N' can be cheaply constructed as the right eigenvectors of the matrix in the right-hand side of (3.57). Using Φ_j^* as the trial functions leads to the following Petrov–Galerkin approximation of (3.50): find $\psi(t) = \sum_{i \in \mathcal{I}_{L,d}} \psi_i(t) \Phi_i$ such that

$$\frac{\partial \psi_j}{\partial t} = \sum_{i \in \mathcal{I}_{L,d}} \psi_i \left(\mathcal{L}_{FP}^{(d)} \Phi_i, \Phi_j^*\right)_{N_R,N_F,d} \quad \forall \boldsymbol{j} \in \mathcal{I}_{L,d}. \tag{3.60}$$

A detailed form of the last equation reads

$$\begin{aligned}
\frac{\mathrm{d}\psi_j(t)}{\mathrm{d}t} &= -\sum_{k=1}^{d} \sum_{i_k=1}^{N_l} \psi_{j_1 \ldots j_{k-1} i_k j_{k+1} \ldots j_d} (\widehat{L}\psi_{i_k}, \psi_{j_k}^*)_{N_R,N_F} \\
&\quad + \frac{1}{4\lambda_H} \sum_{k=2}^{d} \sum_{i_{k-1}=1}^{N_l} \sum_{i_k=1}^{N_l} \psi_{j_1 \ldots j_{k-2} i_{k-1} i_k j_{k+1} \ldots j_d} \\
&\quad \times \Big[(\widehat{N}_x \psi_{i_{k-1}}, \psi_{j_{k-1}}^*)_{N_R,N_F} (\widehat{M}_x \psi_{i_k}, \psi_{j_k}^*)_{N_R,N_F} \\
&\quad + (\widehat{N}_y \psi_{i_{k-1}}, \psi_{j_{k-1}}^*)_{N_R,N_F} (\widehat{M}_y \psi_{i_k}, \psi_{j_k}^*)_{N_R,N_F} \Big]
\end{aligned}$$

$$+\frac{1}{4\lambda_H}\sum_{k=1}^{d-1}\sum_{i_k=1}^{N_l}\sum_{i_{k+1}=1}^{N_l}\psi_{j_1\cdots j_{k-1}i_k i_{k+1}j_{k+2}\cdots j_d}$$
$$\times\Big[(\widehat{M}_x\psi_{i_k},\psi^*_{j_k})_{N_R,N_F}(\widehat{N}_x\psi_{i_{k+1}},\psi^*_{j_{k+1}})_{N_R,N_F}$$
$$+(\widehat{M}_y\psi_{i_k},\psi^*_{j_k})_{N_R,N_F}(\widehat{N}_y\psi_{i_{k+1}},\psi^*_{j_{k+1}})_{N_R,N_F}\Big]. \tag{3.61}$$

Discretization in time is affected via a semi-implicit scheme where the operators L^k are treated implicitly and the others explicitly. We need thus to invert the matrix corresponding to the operator $\boldsymbol{\delta} - \Delta t\sum_{k=1}^{d} L^k$, which can be very expensive. We therefore replace this matrix by $\bigotimes_{k=1}^{d}\left(\boldsymbol{\delta} - \Delta t\widehat{L}^k\right)$ and thereby introduce an error of order $O(\Delta t^2)$. Only the matrices corresponding to one-dimensional operators need therefore to be inverted.

The elastic extra-stress tensor $\boldsymbol{\tau}$ may be written in the form $\boldsymbol{\tau} = \sum_{j=1}^{d}\boldsymbol{\tau}_j$, where $\boldsymbol{\tau}_j$ is the contribution of the jth segment in all the polymer chains of the solution to the total elastic extra stress. In the case of a homogeneous flow, the stress depends only on time. The Kramers expression for $\boldsymbol{\tau}_j$ at time $n\Delta t$ now reads

$$\boldsymbol{\tau}_j(n\Delta t) \approx \frac{\eta_p}{\lambda_H}\frac{b+4}{b}\left[\int_{D^d}\boldsymbol{q}_j\bigotimes\boldsymbol{F}(\boldsymbol{q}_j)\psi(n\Delta t,\boldsymbol{q}_1,\ldots,\boldsymbol{q}_d) - \boldsymbol{\delta}\right]. \tag{3.62}$$

3.4.2. *Low-rank separation algorithms*

We turn now to another class of methods suitable for high-dimensional problems, namely the methods introduced by AMMAR, MOKDAD, CHINESTA and KEUNINGS [2006a, 2007], which construct successive terms in the separated representation (3.45) in an iterative manner. A mathematical justification of this approach can be provided by the results of DEVORE and TEMLYAKOV [1996] on greedy algorithms. They treat a general problem of approximation in a Hilbert space H by elements from a prescribed set satisfying some quite general hypotheses. To be more precise, suppose that H is equipped with an inner product $\langle\cdot,\cdot\rangle$ so that the induced norm $||x|| := \langle x,x\rangle^{1/2}$. Let us call a system $\mathcal{D}$ of elements from H a dictionary if each $g \in \mathcal{D}$ has norm one ($||g|| = 1$) and its linear span is dense in H. We let Σ_m denote the collection of all functions in H, which can be expressed as a linear combination of at most m elements of $\mathcal{D}$

$$\Sigma_m = \left\{ s = \sum_{i=1}^{m} c_i w_i, \quad c_i \in \mathbb{R}, \quad w_i \in \mathcal{D}\right\}.$$

For a function $u \in H$, we seek for successive approximations $u_m \in \Sigma_m$ so that $u_m \to u$ in H as $m \to \infty$. The simplest way to do it is the Pure Greedy Algorithm: set $u_0 = 0$ and define inductively for each $m \geq 1$

$$u_m = u_{m-1} + G(u - u_{m-1}), \tag{3.63}$$

where $G(v) \in \Sigma_1$ for any $v \in H$ denotes an element from Σ_1, which minimizes $||v - g||$ over $g \in \Sigma_1$. The above algorithm is greedy in the sense that at each iteration, it approximates the residual $R_m = u - u_{m-1}$ as well as possible by a single function from $\mathcal{D}$. It is easy to see that each step of the algorithm provides the best m-term approximation to u in the case when $\mathcal{D}$ is an orthonormal basis of H. In general, it is not so, but it is proved in DEVORE and TEMLYAKOV [1996] that the error $||u_m - u||$ is bounded by $Mm^{-1/6}$ for any u from the closure of the set

$$\mathcal{A}_1^o(\mathcal{D}, M) = \left\{ u \in H : u = \sum_{i=1}^{s} c_i w_i,\ s < \infty,\ c_i \in \mathbb{R},\ w_i \in \mathcal{D},\ \sum_{i=1}^{s} |c_i| \le M \right\}, \quad (3.64)$$

for some $M \ge 0$. The theoretical rate of convergence can be improved if at each step of the greedy algorithm above we rearrange the coefficients before the terms composing $u_m = \sum_{i=1}^{m} \alpha_i u^{(i)}$, $\alpha_i \in \mathbb{R}$, $u^{(i)} \in \mathcal{D}$ so that it gives the best approximation to u. This idea gives rise to Orthogonal Greedy Algorithms in which one constructs a sequence $\widetilde{u}_m \in \Sigma_m$, setting $\widetilde{u}_0 = 0$ and then defining inductively

$$\begin{aligned} \widetilde{u}^{(m)} &= \frac{G(u - \widetilde{u}_{m-1})}{||G(u - \widetilde{u}_{m-1})||}, \\ \widetilde{u}_m &= \sum_{i=1}^{m} \widetilde{\alpha}_m^{(i)} \widetilde{u}^{(i)} \text{ such that } \left\langle \widetilde{u}_m, u^{(i)} \right\rangle = \left\langle u, u^{(i)} \right\rangle \text{ for } i = 1, \ldots, m. \end{aligned} \quad (3.65)$$

In other words, $\widetilde{u}_m$ is the orthogonal projection of u on $\text{span}\{\widetilde{u}^{(1)}, \ldots, \widetilde{u}^{(m)}\}$. The error $||\widetilde{u}_m - u||$ is bounded by $Mm^{-1/2}$ for any u from the closure of the set (3.64).

Example: Poisson problem Let us apply the algorithms described above to the approximation of the solution u to the Poisson problem (3.46) in the d-dimensional hypercube $\Omega = [0, 1]^d$. We set $H = H_0^1(\Omega)$ equipped with the inner product $< u, v >= a(u, v) = \int_\Omega \nabla u \cdot \nabla v d\mathbf{x}$ and the corresponding norm $||u||^2 = \int_\Omega |\nabla u|^2 d\mathbf{x}$. The weak formulation of the Poisson problem reads: find $u \in H$ such that

$$a(u, v) = \int_\Omega fv, \quad \forall v \in H. \quad (3.66)$$

We want to approximate the unknown solution u by m-term sums of the form

$$u_m = \sum_{i=1}^{m} \alpha_i u_1^{(i)}(x_1) \cdots u_d^{(i)}(x_d). \quad (3.67)$$

In the abstract language of the preceding paragraph, this means that we seek approximation in the linear spans of the functions from the dictionary

$$\mathcal{D} = \left\{ \prod_{k=1}^{d} u_k(x_k) \text{ for } u_k \in H_0^1(0, 1),\ \left\| \prod_{k=1}^{d} u_k(x_k) \right\| = 1 \right\}.$$

To implement the greedy algorithms, we should be able to find the best approximation in

$$\Sigma_1 = \mathbb{R}\mathcal{D} = \left\{ \prod_{k=1}^{d} u_k(x_k) \text{ for } u_k \in H_0^1(0, 1) \right\}$$

of any function $r \in H$, i.e., to minimize $||r - g||$ over $u \in \Sigma_1$. The necessary condition for the minimum is vanishing of the first variation

$$\delta||u - r||^2 = 0 \iff a(u - r, \delta u) = 0, \tag{3.68}$$

for all δu of the form

$$\begin{aligned}\delta u = {} & \delta u_1(x_1)u_2(x_2)\cdots u_d(x_d) + u_1(x_1)\delta u_2(x_2)\cdots u_d(x_d) \\ & + \cdots + u_1(x_1)u_2(x_2)\cdots \delta u_d(x_d)\end{aligned} \tag{3.69}$$

with any $\delta u_1, \delta u_2, \ldots, \delta u_d \in H$.

We can now write down the Pure Greedy Algorithm (3.63) for the Poisson problem (3.66). On the mth iteration, we have the approximation u_{m-1} of the form (3.67), and we want to find the best possible approximation $\hat{u} \in \Sigma_1$ of the residual

$$r = u - \sum_{i=1}^{m-1} \alpha_i \prod_{k=1}^{d} u_k^{(i)}(x_k).$$

The function $\hat{u} = \hat{u}_1(x_1)\cdots\hat{u}_d(x_d)$ should satisfy (3.68). Denoting δu_k as v_k and taking into account (3.66), we can rewrite (3.68) as

$$\begin{aligned} a\left(\prod_{s=1}^{d}\hat{u}_s(x_s), \prod_{\substack{t=1,\ldots,d\\ t\neq k}} \hat{u}_t(x_t)v_k(x_k)\right) = {} & \int_\Omega f \prod_{\substack{t=1,\ldots,d\\ t\neq k}} \hat{u}_t(x_t)v_k(x_k)\,\mathrm{d}\boldsymbol{x} \\ & - \sum_{i=1}^{m-1}\alpha_i a\left(\prod_{s=1}^{d} u_s^{(i)}(x_s), \prod_{\substack{t=1,\ldots,d\\ t\neq k}} \hat{u}_t(x_t)v_k(x_k)\right), \quad \forall v_k \in H(0,1). \end{aligned} \tag{3.70}$$

This can be rewritten more explicitly in a strong form as a set of d coupled boundary valued problems for $\hat{u}_k \in H_0^1(0,1)$, $k = 1, \ldots, d$:

$$\begin{aligned} & (-\hat{u}_k'') \prod_{\substack{t=1,\ldots,d\\ t\neq k}} (\hat{u}_t, \hat{u}_t) + \hat{u}_k \sum_{\substack{j=1,\ldots,d\\ j\neq k}}^{d} (\hat{u}_j', \hat{u}_j') \prod_{\substack{t=1,\ldots,d\\ t\neq k,j}} (\hat{u}_t, \hat{u}_t) \\ & = \int_\Omega f \prod_{\substack{t=1,\ldots,d\\ t\neq k}} \hat{u}_t(x_t)\,\mathrm{d}\boldsymbol{x}_k' \\ & \quad - \sum_{i=1}^{m-1}\alpha_i \left((-u_k^{(i)''}) \prod_{\substack{t=1,\ldots,d\\ t\neq k}} (u_t^{(i)}, \hat{u}_t) + u_k^{(i)} \sum_{\substack{j=1,\ldots,d\\ j\neq k}}^{d} (u_j^{(i)'}, \hat{u}_j') \prod_{\substack{t=1,\ldots,d\\ t\neq k,j}} (u_t^{(i)}, \hat{u}_t) \right) \end{aligned} \tag{3.71}$$

with the notation $d\mathbf{x}'_k = dx_1 \cdots dx_{k-1} dx_{k+1} \cdots dx_d$. Once a solution of this problem is known, one can update the iterate in the Pure Greedy Algorithm by setting

$$u_k^{(m)} = \frac{\hat{u}_k}{||\hat{u}_k||}, \quad k = 1, \ldots, d, \quad \alpha_m = ||\hat{u}_1|| \cdots ||\hat{u}_d||. \tag{3.72}$$

Turning to the Orthogonal Greedy Algorithms, we note that it also seeks successive approximation to u of the form (3.67) with coefficients α that are allowed to vary from one iteration to another:

$$u_m = \sum_{i=1}^{m} \alpha_m^{(i)} u_1^{(i)}(x_1) \cdots u_d^{(i)}(x_d). \tag{3.73}$$

The step number m of this algorithm is composed of two substeps. The first one is the same as before, i.e., the set of boundary value problems (3.71) for $\hat{u}_k$ with α_i replaced by $\alpha_{m-1}^{(i)}$. The second one adjusts the coefficients $\alpha_m^{(i)}$ by setting $u_k^{(m)} = \hat{u}_k/||\hat{u}_k||$ and solving for $\alpha_m^{(i)}$

$$a\left(\sum_{i=1}^{m} \alpha_m^{(i)} u_1^{(i)}(x_1) \cdots u_d^{(i)}(x_d), u_1^{(j)}(x_1) \cdots u_d^{(j)}(x_d)\right)$$
$$= \int f(\mathbf{x}) u_1^{(j)}(x_1) \cdots u_d^{(j)}(x_d)\, d\mathbf{x}, \qquad \text{for } j = 1, \ldots, m. \tag{3.74}$$

We recall that both Pure and Orthogonal algorithms are guaranteed to converge at least when the exact solution u is sufficiently smooth so that the constant M in (3.64) is finite. Note that this constant is evidently finite if the Hilbert space H is finite dimensional so that both greedy algorithms should converge on the discrete level, when all the problems in (3.71) are discretized by some numerical method. This results is also proved independently in AMMAR, CHINESTA and FALCÓ [2010].

Generalizations for a broad class of problems We first observe that the Laplace operator $A = -\Delta$ in the Poisson problem (3.66) above can be rewritten in a separated form

$$A = \sum_{n=1}^{N_{op}} \bigotimes_{j=1}^{d} A_j^n \tag{3.75}$$

with $N_{op} = d$ and the univariate operators $A_j^n : H_0^1(0,1) \to H_0^1(0,1)$ defined by $(A_j^j u, v) = (u', v')\ \forall v \in H_0^1(0,1)$, and $A_j^n u = u$ if $n \neq j$. Let us suppose for simplicity that the right-hand side is also of separated form

$$f = \sum_{n=1}^{N_f} \bigotimes_{j=1}^{d} f_j^n \tag{3.76}$$

with some $f_j^n \in L^2(0, 1)$. The problem (3.71) on the mth step of the Orthogonal Greedy Algorithm becomes in these notations after replacing α_i by $\alpha_{m-1}^{(i)}$

$$\left[\sum_{n=1}^{N_{op}} \prod_{\substack{t=1\ldots d\\ t\neq k}} (A_t^n \hat{u}_t, \hat{u}_t) A_k^n\right] \hat{u}_k = \sum_{n=1}^{N_f} \prod_{\substack{t=1\ldots d\\ t\neq k}} (f_t^n, \hat{u}_t) f_k^n$$
$$- \sum_{i=1}^{m-1} \alpha_{m-1}^{(i)} \left[\sum_{n=1}^{N_{op}} \prod_{\substack{t=1\ldots d\\ t\neq k}} (A_t^n u_t^{(i)}, \hat{u}_t) A_k^n\right] u_k^{(i)}. \tag{3.77}$$

We now note that the problem (3.77) makes sense for a general linear problem of the form $Au = f$ in which the unknown u is searched in a Hilbert space H that is a tensor product of univariate Hilbert spaces $H = H_1 \otimes \cdots \otimes H_d$, and the operator $A : H \to H$ has the form (3.75) with some $A_j^n : H_j \to H_j$. This idea is the basis of the method of a separated representations introduced in AMMAR, MOKDAD, CHINESTA and KEUNINGS [2006a, 2007]. More precisely, each iteration of this method aims at constructing the mth term in the approximation $u_m = \sum_{i=1}^m \alpha_m^{(i)} u_1^{(i)} \otimes \cdots \otimes u_d^{(i)}$ of the solution u to the problem $Au = f$ with a given f supposing that the first $m - 1$ terms are already known. It consists of two substeps. The first one, referred to as the basis enrichment step and corresponding to an iteration of the Pure Greedy Algorithm, consists in solving the nonlinear system of equations in (3.77) for $\hat{u}_k$, $k = 1, \ldots, d$ and setting $u_k^{(m)} = \hat{u}_k/||\hat{u}_k||$. The second substep is referred to as the projection step and is a natural generalization of (3.74) in the Orthogonal Greedy Algorithm. It is the standard Galerkin approximation over the basis $\{u_1^{(i)} \otimes \cdots \otimes u_d^{(i)},\ i = 1, \ldots, m\}$. One thus solves the linear system for $\alpha_m^{(i)}$, $i = 1, \ldots, m$ given by

$$\sum_{i=1}^{m} \left[\sum_{n=1}^{N_{op}} \prod_{t=1}^{d} (A_t^n u_t^{(i)}, u_t^{(j)}) A_k^n\right] \alpha_m^{(i)} = \sum_{n=1}^{N_f} \prod_{t=1}^{d} (f_t^n, u_t^{(j)}), \tag{3.78}$$

for $j = 1, \ldots, m$.

The method was applied in the original articles by AMMAR et al. to a variety of multidimensional problems including the Poisson problem, the time-dependent advection-diffusion problem, the 2D FENE dumbbell model, and 1D FENE bead-spring chain model (both in a homogeneous flow). The last model involves the time-dependent Fokker–Planck equation with the differential operator having a natural separated representation so that the application of the method (3.77) and (3.78) is mostly straightforward. The only difficulty is how to treat the dependence of the unknown on time. The paper of AMMAR, MOKDAD, CHINESTA and KEUNINGS [2007] proposes a nonincremental strategy for this, treating time as just one of the independent variables alongside $q_1, \ldots, q_d$ for a chain consisting of d springs. Since the initial conditions are not homogeneous (one usually takes the equilibrium solution as the initial condition), one should first make the change of variables $\psi = \psi_{\text{eq}} + \widetilde{\psi}$ so that

$\widetilde{\psi}|_{t=0} = 0$ and then approximate $\widetilde{\psi}$ by sums of the form

$$\widetilde{\psi}(q_1, \ldots, q_d, t) \approx \sum_{i=1}^{m} \alpha_m^{(i)} \psi_1^{(i)}(q_1) \cdots \psi_d^{(i)}(q_d) \psi_{d+1}^{(i)}(t), \tag{3.79}$$

where all the functions $\psi_0^{(d+1)}(t)$ vanish at $t = 0$. One applies thus the iterative algorithm (3.77) and (3.78) with u replaced by $\widetilde{\psi}$ and d replace by $d + 1$. An alternative, incremental strategy to treat the dependence on time is implemented in LEONENKO and PHILLIPS [2009] for the same problem of 1D FENE chains. One uses there a standard implicit Euler time marching scheme, which gives a multidimensional problem at each time step, and an approximate separated representation to it is searched by the algorithm (3.77) and (3.78) on each time step. Either way, the univariate problems in (3.77) can be discretized by any numerical methods. Finite elements in space and time were used in AMMAR, MOKDAD, CHINESTA and KEUNINGS [2006a, 2007], while the spectral methods were chosen in LEONENKO and PHILLIPS [2009]. These univariate problems are coupled in a nonlinear fashion so that one needs an iterative method to solve them. The simplest approach advocated in AMMAR, MOKDAD, CHINESTA and KEUNINGS [2006a, 2007] consists in solving the linear problems for $\hat{u}_k$ in (3.77) one after another for $k = 1, \ldots, d$ taking each time the latest available approximations for the other unknowns $\hat{u}_j$, $j \neq k$. One repeats these iterations until a fixed point is attained below some tolerance. It is proven in AMMAR, CHINESTA and FALCÓ [2010] that such a procedure, referred to as the Block Coordinated Descent Algorithm, converges to a critical point of the coupled nonlinear problem. Some results using Newton iterations were also obtained in AMMAR, MOKDAD, CHINESTA and KEUNINGS [2006a], but no comparison between the two approaches is reported.

Although accurate approximations of very low separation rank (usually $m \leq 5$) to the solutions of several multidimensional problems are reported in the above-cited papers, one should emphasize, however, that the reduced representation method as presented by (3.77) and (3.78) is equivalent to the theoretically substantiated greedy algorithms only if the operator A is symmetric positive definite as it is the case with the Laplace operator. This method is thus not guaranteed to converge (as $m \to \infty$) for nonsymetric and/or time-dependent problems as the Fokker–Planck equation. Fortunately, one can easily modify the method so that to recover the framework of a greedy approximation algorithm eventually for any operator A as proposed in AMMAR, CHINESTA and FALCÓ [2010]. The key idea there is to construct the approximations $u_m = \sum_{i=1}^{m} \alpha_i^{(m)} u_1^{(i)} \otimes \cdots \otimes u_d^{(i)}$ to the solution u of $Au = f$ that minimize a norm of the residual $f - Au_m = A(u - u_m)$ rather than that of the error $u_m - u$ as before. Indeed, given any invertible operator A whose image is in a Hilbert space with the inner product $(\cdot, \cdot)$, one can introduce the inner product $(A\cdot, A\cdot)$ on the domain of A and apply the Orthogonal Greedy Algorithm for the approximation with respect to the norm $||\cdot||_A$ induced by this product. Thus, to construct the best rank-one approximation $\hat{u} = \hat{u}_1 \otimes \cdots \otimes \hat{u}_d$ to a given function u that minimizes $||\hat{u} - u||_A = ||A\hat{u} - f||$ with $f = Au$, one should solve the following set of nonlinear problems

$$(A\hat{u} - f, A(\hat{u}^1 \otimes \cdots \otimes v^k \otimes \cdots \otimes \hat{u}^d)) = 0 \tag{3.80}$$

for $k = 1, \ldots, d$ and any v_k such that $\hat{u}^1 \otimes \cdots \otimes v^k \otimes \cdots \otimes \hat{u}^d$ is in the domain of A. Supposing again that operator A and the right-hand side f are written in the separated form (3.75) and (3.76), and noting that mth iteration of the Orthogonal Greedy Algorithm with respect to the norm $||\cdot||_A$ involves the minimization problem (3.80) with f replaced by the residual $r_{m-1} = f - A\left(\sum_{i=1}^{m} \alpha_i^{(m-1)} u_1^{(i)} \otimes \cdots \otimes u_d^{(i)}\right)$, we rewrite it in the strong form as

$$\left[\sum_{n=1}^{N_{op}} \sum_{n'=1}^{N_{op}} \prod_{\substack{t=1...d \\ t\neq k}} (A_t^n \hat{u}_t, A_t^{n'} \hat{u}_t)(A_k^n)^* A_k^{n'}\right] \hat{u}_k = \sum_{n=1}^{N_f} \sum_{n'=1}^{N_{op}} \prod_{\substack{t=1...d \\ t\neq k}} (f_t^n, A_t^{n'} \hat{u}_t) f_k^n$$

$$- \sum_{i=1}^{m-1} \alpha_i^{(m-1)} \left[\sum_{n=1}^{N_{op}} \sum_{n'=1}^{N_{op}} \prod_{\substack{t=1...d \\ t\neq k}} (A_t^n u_t^{(i)}, A_t^{n'} \hat{u}_t)(A_k^n)^* A_k^{n'}\right] u_k^{(i)}. \tag{3.81}$$

Each iteration of the modified algorithm proceeds thus by solving the last system of equations for $\hat{u}_k$, $k = 1, \ldots, d$ supposing that the preceding approximation u_{m-1} is already computed, setting $u_k^{(m)} = \hat{u}_k / ||\hat{u}_k||$ and finally adjusting the coefficients $\alpha_i^{(m)}$, $i = 1, \ldots, m$ by solving the linear system

$$\sum_{i=1}^{m} \left[\sum_{n=1}^{N_{op}} \sum_{n'=1}^{N_{op}} \prod_{t=1}^{d} (A_t^n u_t^{(i)}, A_t^{n'} u_t^{(j)}) A_k^n\right] \alpha_i^{(m)} = \sum_{n=1}^{N_f} \sum_{n'=1}^{N_{op}} \prod_{t=1}^{d} (f_t^n, A_t^{n'} u_t^{(j)}), \tag{3.82}$$

for $j = 1, \ldots, m$. We will refer to the algorithm (3.81) and (3.82) as the minimizing residual low-rank representation algorithm and will report some numerical results for it at the end of the next chapter.

CHAPTER 4

Numerical Results

We now use the numerical techniques outlined in the previous chapters in order to solve the governing equations for some popular benchmark problems. A dimensionless form of the equations of motion (1.1) may be written as

$$\mathrm{Re}\frac{D\boldsymbol{v}}{Dt} - \beta\nabla^2\boldsymbol{v} + \nabla p = \nabla\cdot\boldsymbol{\tau}, \tag{4.1}$$

$$\nabla\cdot\boldsymbol{v}, \tag{4.2}$$

where Re is a Reynolds number and β denotes the ratio between the Newtonian (solvent) viscosity and the sum of the Newtonian viscosity and a zero shear-rate polymeric viscosity. As explained in the Introduction, Eqns (4.1) and (4.2) are coupled with a constitutive equation for the stress field $\boldsymbol{\tau}$, where this is available, or with a kinetic theory model for the stress, if not. In Section 4.1, the Brownian configuration fields method described in Section 2.3 is used for the simulation of the flow of a dilute solution of both Hookean and FENE dumbbells around a confined cylinder, and comparison is made with computations of the same flow of, respectively, an Oldroyd-B and FENE-P fluid. For all computations involving solutions of dumbbells, the Kramers expression (see (1.39)) for the dimensionless elastic stress

$$\boldsymbol{\tau} = \frac{(1-\beta)\alpha_{b,d}}{\mathrm{We}}(-\boldsymbol{\delta} + \langle \boldsymbol{qF}\rangle), \tag{4.3}$$

is used for the stress calculator, where We denotes a dimensionless characteristic shear rate and is called the Weissenberg number. The parameter $\alpha_{b,d}$ has already been defined in (1.40). In Section 4.2, the Fokker–Planck-based spectral methods of Section 3.1 are compared with stochastic methods for the solution of start-up plane Couette flow and steady Poiseuille flow of a FENE fluid. Then, steady Poiseuille flow in a narrow channel is considered in Section 4.3 in order to showcase the Fokker–Planck-based spectral methods of Section 3.2 for nonhomogeneous flows of a dilute polymer solution. The flow of melts and concentrated polymeric solutions may be dealt with using the methods of Section 3.3, and these are here applied to the calculation of the evolution of the shear stress in a homogeneous shear flow. Finally, in Section 4.5, the approach of low-rank separation representations and of sparse tensor product Fokker–Planck-based spectral methods, both discussed in Section 3.4, is used for flows involving model polymers having higher dimensional configuration spaces. Comparisons are made with stochastic methods.

4.1. Second-generation micro–macro techniques

The complex problem of flow past a cylinder placed symmetrically in a channel is considered in this section. The aspect or blockage ratio is defined to be $\Lambda = \frac{R}{H}$, where R is the radius of the cylinder and H is the half-width of the channel. We consider the 50% blockage case, i.e., $\Lambda = 0.5$. This value has been chosen consistently as one of the benchmark problems in the field of computational rheology. The cylinder benchmark problem is acknowledged to be more difficult than the related sphere problem because, for the same aspect ratio Λ, the planar flow past a cylinder undergoes a stronger contraction and expansion than the axisymmetric flow past a sphere. A comprehensive discussion of these problems (flow past a cylinder and sphere) can be found in the monograph of OWENS and PHILLIPS [2002].

A transient scheme is used to solve the problem in which the solution of the conservation laws is decoupled from the solution of the evolution equation for the Brownian configuration fields within each time step. The coupling between the macroscopic and microscopic stages is achieved as follows: after the microscopic stage, the extra-stress tensor is evaluated by taking an arithmetic mean over the N_f configuration fields, and then, its divergence is computed and used to form the source term in the momentum equation. After the microscopic stage, the new velocity field is used to evolve the Brownian configuration fields forward in time over the next time step. In the macroscopic stage, all terms in the field equations are discretized implicitly except the divergence of the extra-stress tensor. In the microscopic stage, the stochastic differential equation for the configuration fields is discretized using second-order explicit schemes.

The basis of the numerical method employed for the stochastic simulations is the method of Brownian configuration fields. The spectral element method is used to discretize the governing equations in space. Full details of the method may be found in PHILLIPS and SMITH [2006], VARGAS, MANERO and PHILLIPS [2009].

Although the dimensionless drag coefficient is often used in the literature as a measure for testing the accuracy of numerical approximations to the solution of flow past a cylinder by comparing predictions with other results in the literature, it may not be sensitive to inaccuracies in the stress components away from the cylinder surface. Therefore, it is important to examine the behavior of the stress components globally since numerical oscillations or other mesh dependent features may be overseen by simply concentrating on the computation of the drag coefficient. The expression for a dimensionless drag on the cylinder is

$$F = 2\int_0^{\pi} \left\{ \left(-p + 2\beta \frac{\partial u}{\partial x} + \tau_{xx} \right) \cos\theta + \left(\beta \left(\frac{\partial v}{\partial x} + \frac{\partial u}{\partial y} \right) + \tau_{xy} \right) \sin\theta \right\} \mathrm{d}\theta. \tag{4.4}$$

4.1.1. Oldroyd-B and Hookean dumbbell models

Here, we compare macroscopic predictions using the Oldroyd-B model with micro–macro predictions based on the evolution of Brownian configuration fields using Hookean dumbbells. Since the Hookean dumbbell model is mathematically equivalent to the Oldroyd-B model, this comparison serves to demonstrate the reliability of the micro–macro approach in simulating a complex flow. A comparison of the evolution of the drag predicted by the two models is given in Fig. 4.1 for We = 1 and Re = 0.01. Very good agreement is obtained

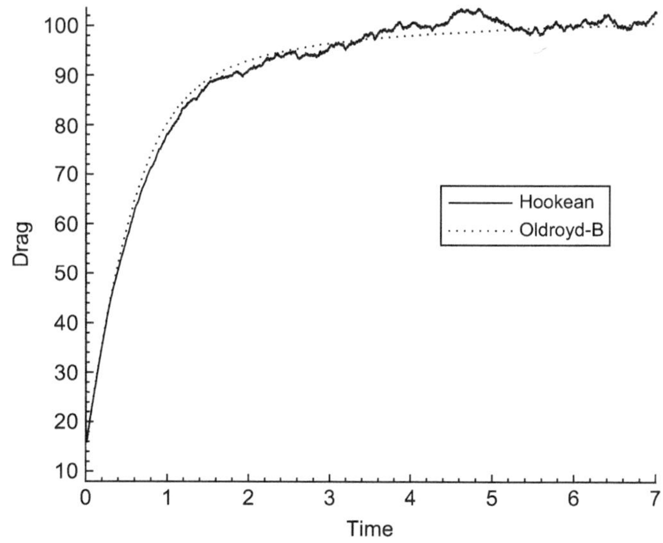

FIG. 4.1 Comparison of the dimensionless drag on the cylinder using the Oldroyd-B and Hookean dumbbell models for We $= 1$, Re $= 0.01$, and $\beta = 1/9$ with $N = 6$. For the Hookean dumbbell model $N_f = 2000$.

between the two approaches particularly as steady state is reached. In the micro–macro computation $N_f = 2000$ Brownian configuration fields are used, and for both sets of computations, the polynomial order N in each spectral element was taken to be 6.

However, good agreement of the two approaches as far as the calculation of the drag is concerned should not be interpreted as meaning that the corresponding global fields are necessarily in close agreement as well. We demonstrate that the corresponding global fields are also in agreement by comparing contours of the components of the extra-stress tensor generated by the macroscopic and micro-macro approaches. This comparison is shown in Fig. 4.2. Excellent quantitative agreement is obtained across the two approaches as evidenced by the fine scale features predicted and the location of contours of the same height.

4.1.2. FENE and FENE-P models

Our study of FENE models begins with a discussion of the influence of the discretization parameters on the evolution of the drag for We $= 3$, Re $= 0.01$, $\beta = 1/9$, and $b = 50$. In Fig. 4.3, we present the evolution of the drag on the cylinder as a function of polynomial order, N, for the FENE model. The number of spectral elements remains constant in these simulations. These results demonstrate that mesh convergence is obtained as the order of the approximation is increased. Very little difference is observed in the evolution of the drag for $N > 5$, and certainly as steady state is reached, the agreement is very good. The differences that exist can be explained by appealing to the noise in the stochastic simulations.

In Fig. 4.4, we show the influence of the number configuration fields, N_f, on the evolution of the drag. Steadily increasing the number of configuration fields, $N_f = 100, 500, 1000, 2000$, reduces the temporal fluctuations in the drag as expected. The large temporal fluctuations in the drag that are present when $N_f = 100$ are substantially reduced when $N_f = 500$.

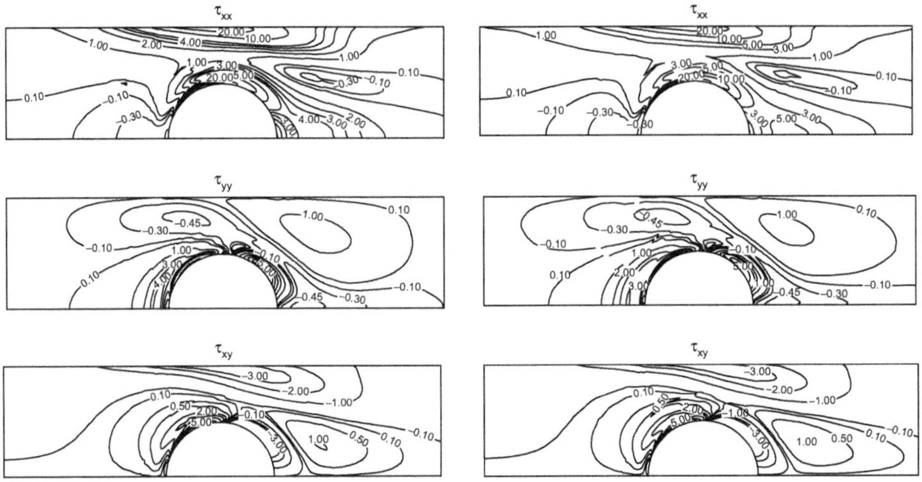

FIG. 4.2 Comparison of contour plots of the stress components generated using the Oldroyd-B model (left) and the Hookean dumbbell model (right) for We = 1, Re = 0.01, and $\beta = 1/9$ with $N = 6$. For the Hookean dumbbell model $N_f = 2000$.

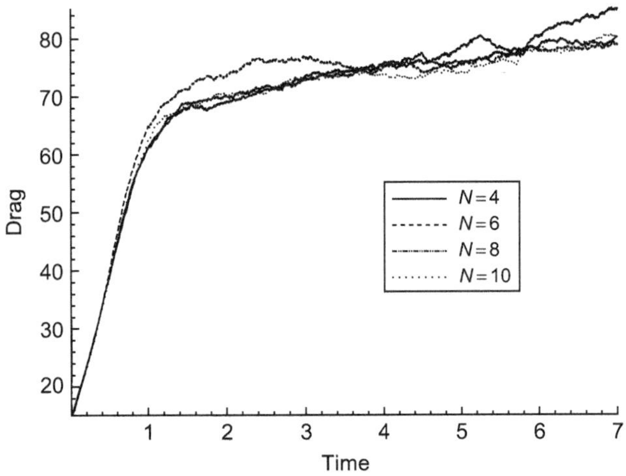

FIG. 4.3 Effect of mesh refinement: drag on the cylinder for the FENE model with We = 3, Re = 0.01, $\beta = 1/9$, and $b = 50$ for $4 \leq N \leq 10$ and $N_f = 2000$.

Again we see general improvement in the prediction of the drag with refinement in terms the microscopic part of the calculation, and there is very little difference when the number of configuration fields is doubled from 1000 to 2000.

A comparison of the predictions of the FENE and FENE-P models is shown in Fig. 4.5 where the polymeric stress components are plotted around the cylinder and along the downstream axis for Re = 0.01, $\beta = 1/9$, $N_f = 2000$, and $b = 50$. The influence of the

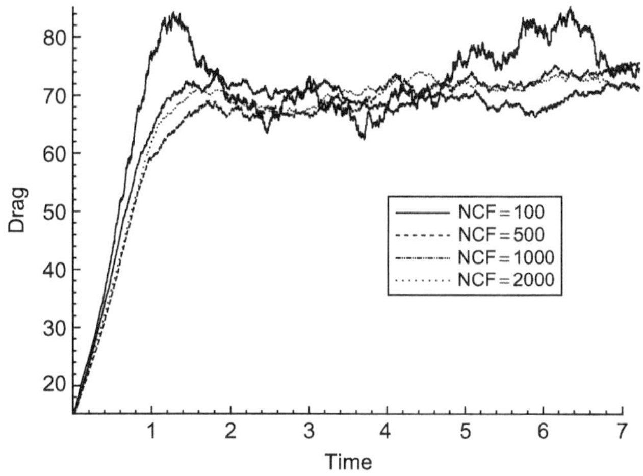

FIG. 4.4 Dependence of the evolution of the drag on the number of configuration fields for the FENE dumbbell model with We = 3, Re = 0.01, $\beta = 1/9$, and $b = 50$.

Weissenberg number on these stress profiles in this series of plots the dependence on Weissenberg number is also shown in this figure for We = 0.2, 0.4, 1, 2, and 4.

The axial stress component τ_{xx} dominates the other stress components, in terms of its magnitude, on the surface of the cylinder. The axial stress initially exhibits strong stress growth, and for the FENE model, its maximum value almost doubles as the Weissenberg number is increased from We = 0.2 to We = 1. This is followed by stress relaxation from around We = 1. This behavior is more accentuated for the FENE-P model than for the FENE model, and there is strong stress relaxation with the maximum value of the shear stress for We = 4 falling to about half its value for We = 1 and below that for We = 0.2. The maximum value of τ_{xx} in the rear wake remains small in comparison with peak on the cylinder and exhibits little variation with We.

Although the FENE-P model provides a good approximation to the FENE model in steady flows, large differences are expected for transient flows (see HERRCHEN and ÖTTINGER [1997], and KEUNINGS [1997], for example). The Peterlin approximation to the FENE model radically changes the statistical properties of the underlying kinetic theory in the sense that the configuration distribution for FENE-P dumbbells is always a Gaussian and thus is never localized, irrespective of the flow dynamics (KEUNINGS [1997]). A direct consequence of this is that nothing prevents individual FENE-P dumbbells from deforming beyond their maximum extensibility $\sqrt{b}$. It is only the average $\langle Q^2 \rangle$ that is bounded for FENE-P dumbbells. Therefore, drastic differences between the FENE and FENE-P models are to be expected when simulating complex flows. This clearly demonstrates that care should be exercised in deriving and using closure approximations. More physically realistic closure approximations for FENE kinetic theory do exist, however, such as the so-called FENE-L closure approximation that was developed by LIELENS, KEUNINGS and LEGAT [1999] using a two-parameter representation of the canonical radial distribution.

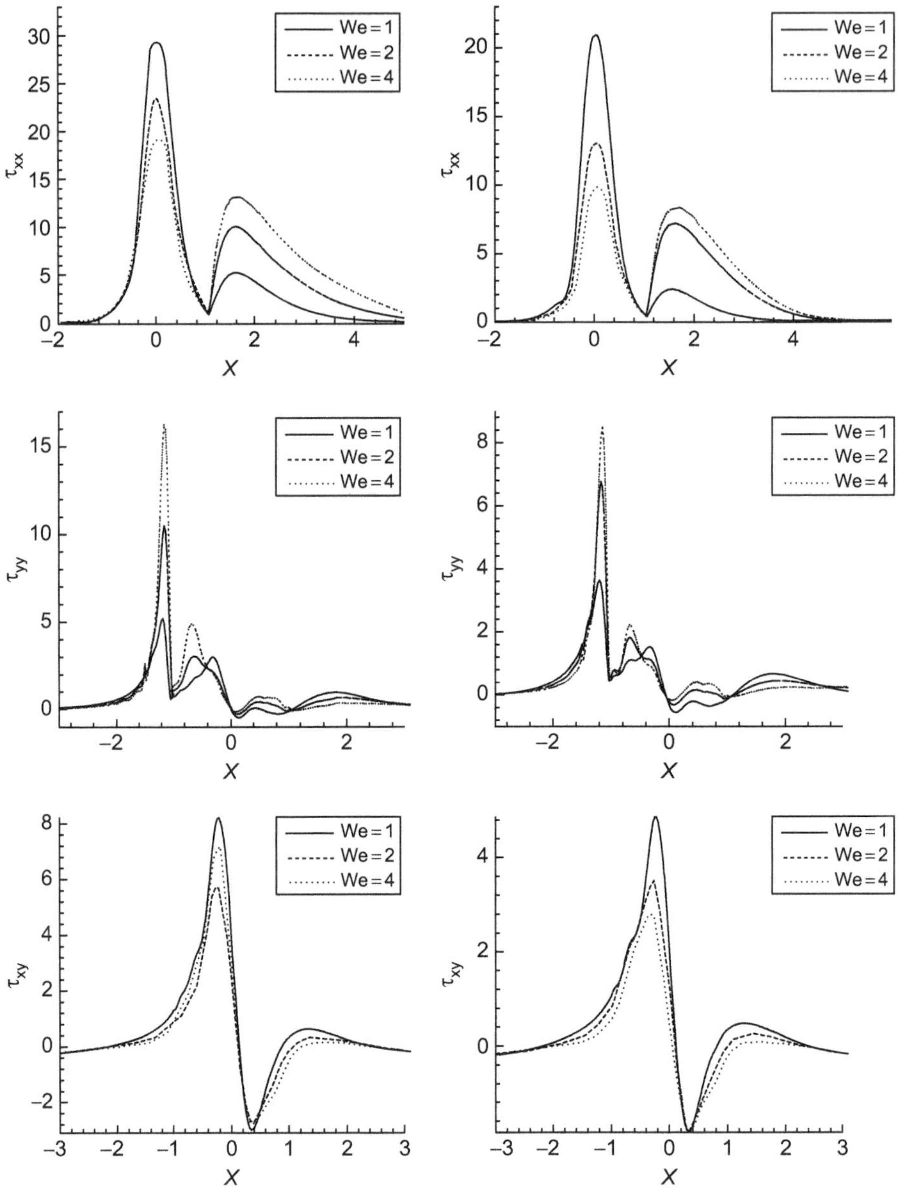

FIG. 4.5 Comparison of the dependence of the profiles of the polymeric stress components along the axis of symmetry line ($y = 0$) and around the cylinder on the Weissenberg number for the FENE model (left) and FENE-P model (right) with $\mathrm{Re} = 0.01$, $\beta = 1/9$, and $b = 50$ for $N = 6$ and $N_f = 2000$.

4.2. Fokker–Planck-based numerical methods for locally homogeneous flows of dilute polymeric solutions

4.2.1. *Start-up of plane Couette flow*

Consider the start-up of plane Couette flow in which a viscoelastic fluid is enclosed between two parallel plates of infinite length separated by a distance L (see Fig. 3.3(a)). For $t < 0$, the fluid and the two plates are at rest. At $t = 0$, the top plate begins to move in the positive x-direction with a speed U. The problem is to find the time development of u_x, the horizontal component of the velocity, for $t > 0$.

The velocity field is assumed to be of the form $u_x = u_x(y, t)$, $u_y = 0$. This velocity field automatically satisfies the incompressibility constraint. Therefore, the velocity at any moment in time can be determined from the horizontal component of the momentum equation:

$$\mathrm{Re}\frac{\partial u_x}{\partial t} = \beta\frac{\partial^2 u_x}{\partial y^2} - \frac{\partial \tau_{xy}}{\partial y}. \tag{4.5}$$

The polymeric contribution to the extra-stress tensor at each instant in time is computed using (1.39). In fact, for this one-dimensional problem, only the shear stress is required.

In our simulations of flow in a channel, we choose the discretization points y_k, $k = 1, \ldots, N_y$ to be the Gauss–Legendre–Lobatto points mapped onto the interval $(0, L)$.

We present the evolution of velocity in Fig. 4.6. First, we give the results for a Newtonian flow at Reynolds number $\mathrm{Re} = 1$ in the form of velocity evolution at three points $y = 0.25,\ 0.5,\ 0.75$ computed analytically. The results for a polymer fluid modeled by FENE dumbbells with $b = 50$, $\mathrm{We} = 1$, and $\beta = 1/9$ under the same Reynolds number are on the same figure to the right. We observe a qualitative difference in the time-dependent behavior between the two fluids as the velocity profiles cease to be monotone when passing

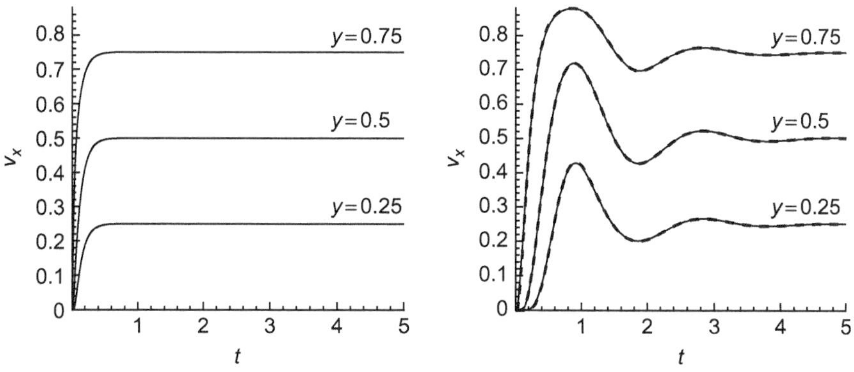

FIG. 4.6 Time evolution of velocity in the Couette flow at $\mathrm{Re} = 1$. Left – Newtonian fluid, right – FENE fluid with $\beta = 1/9$, $\mathrm{We} = 1$, $b = 50$. Dashed line represents the results on the mesh $N_y = 15$, $N_R = 10$, $N_F = 5$. Solid line represents the results on the mesh $N_y = 25$, $N_R = 24$, $N_F = 12$. The relative difference between the two curves is 0.06%.

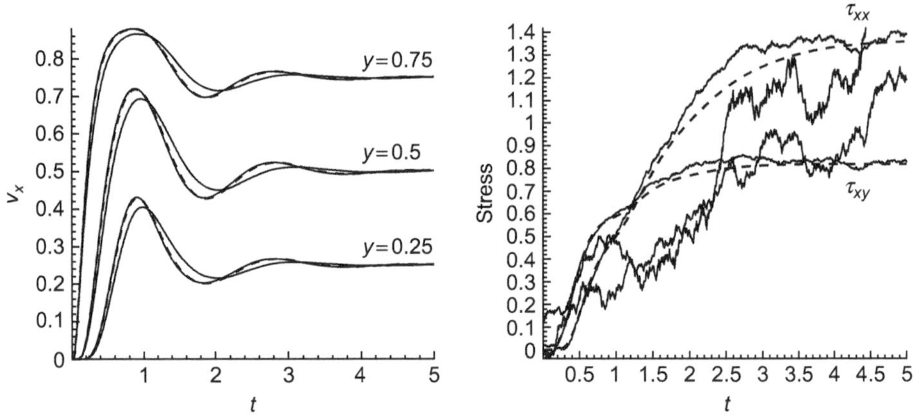

FIG. 4.7 Evolution of velocity (at $y = 0.25,\ 0.5,\ 0.75$) and stress (at $y = 0.5$) for the FENE fluid in the Couette flow with physical parameters as in Fig. 4.6. Dashed line computed by the Fokker–Planck-based method with $N_R = 10$, $N_F = 5$, light solid line by a stochastic simulation with $N_f = 200$, thick solid line by a stochastic simulation with $N_f = 4000$.

from a Newtonian fluid to a polymer one. The results for the FENE fluid in Fig. 4.6 are calculated by the Fokker–Planck-based method. We present them on two meshes to illustrate that excellent convergence is achieved already on the mesh $N_y = 15$, $N_R = 10$, $N_F = 5$. We now turn our attention to stochastic simulations in Fig. 4.7. On the left, the evolution of velocity is presented again as calculated by the converged Fokker–Planck method above and two stochastic simulations with $N_f = 200$ and $N_f = 4000$. The temporal fluctuations are hardly seen on the plots for the velocity, and a perfect match is achieved with $N_f = 4000$ (in fact, the agreement in the velocity is very good already for $N_f = 1000$). The situation is entirely different when we look at the plots of the stress. The results are extremely noisy, and we should increase the number of random realizations till 4000 in order to have reasonable agreement with the results of the Fokker–Planck simulations. To give an idea of the computing time on a laptop computer, we mention that the Fokker–Planck simulations take 2 s on the mesh $N_y = 15$, $N_R = 10$, $N_F = 5$ and 82 s on the mesh $N_y = 25$, $N_R = 24$, $N_F = 12$, while stochastic simulations with $N_f = 200$ take 3 s on the mesh $N_y = 15$ and 5 s on the mesh $N_y = 25$. The computing time scales approximately linearly with N_f. Taking into account that there is considerable error in the velocity and large temporal variations in the stress when stochastic simulations with a small N_f are used, we conclude that the Fokker–Planck simulations are much more efficient in this example. The advantage of Fokker–Planck simulations would be even more pronounced for the FENE dumbbells with smaller b and/or in a slower flow because then a less refined mesh would be sufficient to achieve an accurate approximation.

4.2.2. *Steady Poiseuille flow*

Consider now steady plane Poiseuille flow in which a viscoelastic fluid is enclosed between two parallel plates of infinite length separated by a distance $2L$ (the same geometry as in

Fig. 3.3(a) except for the width of the channel). Both walls are assumed to be at rest, and a pressure gradient P is applied in the horizontal direction so that the horizontal component of the momentum equation reads

$$\beta \frac{\mathrm{d}^2 u_x}{\mathrm{d}y^2} = \frac{\mathrm{d}\tau_{xy}}{\mathrm{d}y} + P. \tag{4.6}$$

Results for such a flow of a solution of FENE dumbbells with $b = 50$ under the nondimensionalized pressure gradient $P = 10$ are presented in Figs 4.8 and 4.9. In the first figure, we look at the velocity profiles under three values of the solvent viscosity ratio: $\beta = 1/9$, $1/2$, and $8/9$. The average velocity depends strongly on β. This indicates that the effective polymer viscosity in such a strong flow is much smaller than its value at zero shear rate

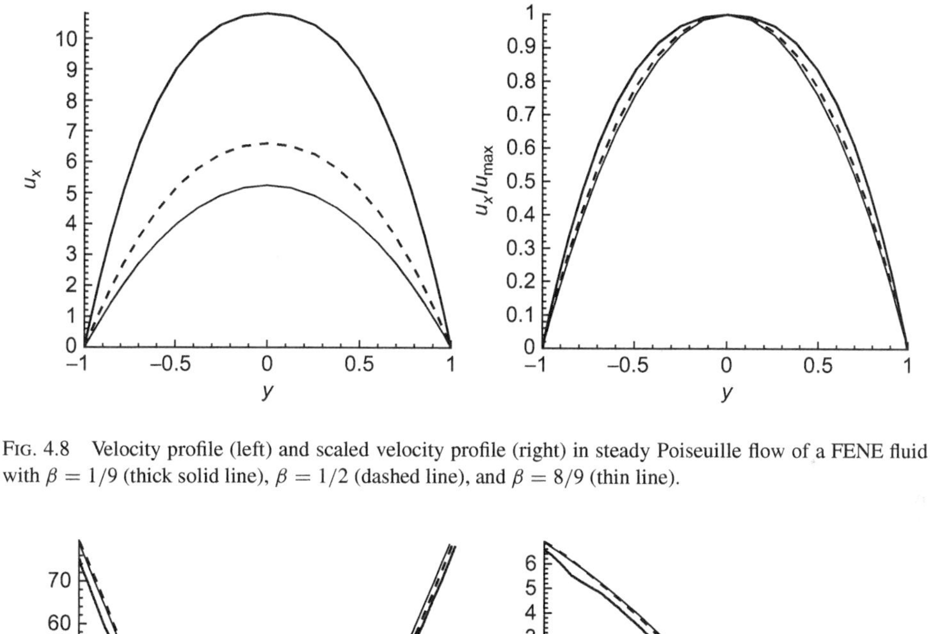

FIG. 4.8 Velocity profile (left) and scaled velocity profile (right) in steady Poiseuille flow of a FENE fluid with $\beta = 1/9$ (thick solid line), $\beta = 1/2$ (dashed line), and $\beta = 8/9$ (thin line).

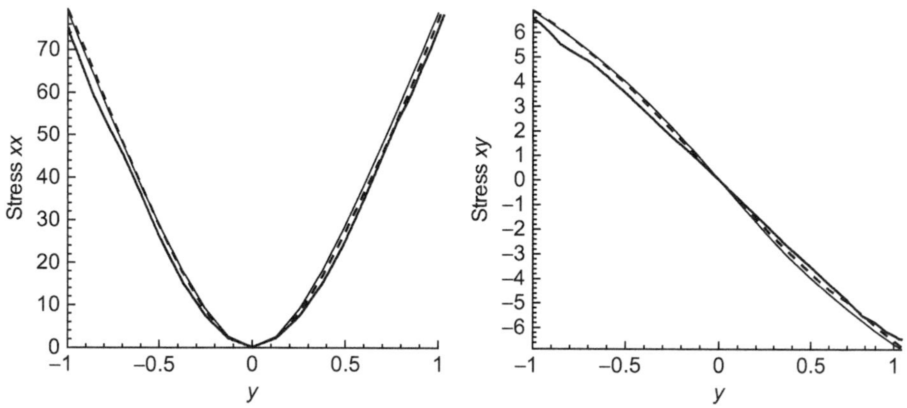

FIG. 4.9 Stress profile in the stationary Poiseuille flow. Dashed – Fokker–Planck (657 s CPU), thick solid – stochastic with $N_f = 200$ (16 s CPU), thin solid – stochastic with $N_f = 1000$ (80 s CPU).

(shear thinning). Another manifestation of this phenomenon is slight flattening of the velocity profile. To see it better, we rescale the velocity profiles by their maximum values on the right half of Fig. 4.8.

The predictions for the stress in the same flow are presented at Fig. 4.9. We compare there also the Fokker–Planck implementation with stochastic simulations. All the simulations were performed on the Gauss–Legendre–Lobatto grid with $N_y = 25$ points in the physical space. We had to take a rather fine grid for the probability density ($N_R = 30$, $N_F = 15$) in order to preserve the stability of the Fokker–Planck simulation so that it becomes very expensive. On the contrary, one obtains reasonably accurate results for such a strong flow using a stochastic simulation, even with a small number $N_f = 200$ of samples.

4.3. Fokker–Planck-based numerical methods for nonhomogeneous flows of dilute polymeric solutions: steady Poiseuille flow in a narrow channel

We now turn our attention to strongly nonhomogeneous flow modeled as in Section 3.2. In our simulations, we use two kinds of grids in physical space. The first one is to represent the stress and the probability density function, as explained in Section 3.2. We choose it to be the Gauss–Legendre (GL) grid of N_y points mapped onto the interval $(-L, L)$. Note that this set of collocation points does not include the end points $\pm$d since $\psi^c(y, \boldsymbol{q}, t)$ has no meaning for y lying on the boundary (configuration space has zero two-dimensional measure there). The second grid is the set of $N_y + 1$ Gauss–Legendre–Lobatto points mapped onto the interval $(-L, L)$ (and thus including the end points $-L, L$). It is used to represent the velocity u_x.

Let us present some results for Poiseuille flow with an applied dimensionless pressure gradient $P = 10$. In Fig. 4.10, the profiles of the steady dimensionless velocity $u_x(y)$ and of the polymer number density $n(y)$ are shown for different values of l_0/d. The case $l_0/\mathrm{d} = 0$ corresponds to the locally homogeneous FENE model (cf. Figs 4.8 and 4.9). The polymer number density is a constant in this case. However, we observe that as l_0/L increases from

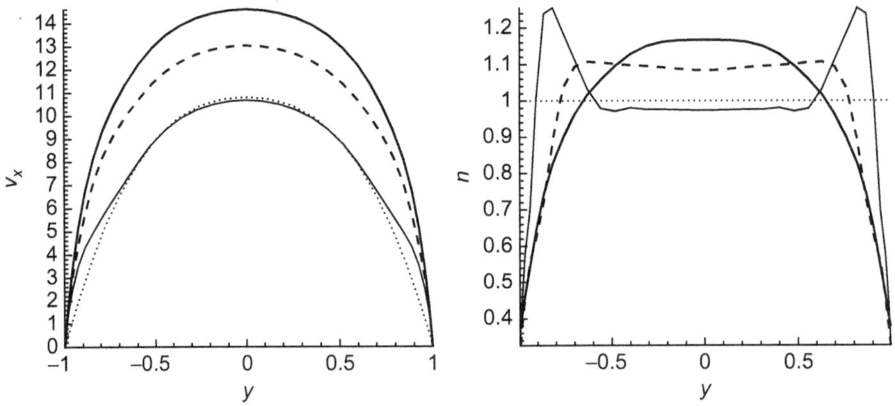

FIG. 4.10 Velocity v_x and polymer number density n for different ratios ℓ_0/L: dotted line $\ell_0/L = 0$, thin line $\ell_0/L = 0.05$, dashed line $\ell_0/\mathrm{d} = 0.L$, thick line $\ell_0/L = 0.2$. Mesh parameters: $N_y = 35$, $N^k_{q_x} = N^k_{q_y} = 30$.

0 to 0.2, the wall effects become stronger, and polymer migrates from the channel walls $y = \pm L$ toward the center of the channel. As a consequence, the velocity gradient steepens near the walls in order to maintain the total shear stress, and the profile flattens near the channel center since the total viscosity increases there.

4.4. Fokker–Planck-based numerical methods for melts and concentrated polymeric solutions: Couette flow of a Doi–Edwards fluid

Some simulation results for the Doi–Edwards model (discussed in Sections 1.1.2, 2.5, and 3.3) are presented in Fig. 4.11. We are looking there at the evolution of the stress in a homogeneous shear flow with velocity of the form $u_x = \dot{\gamma} y$, $u_y = u_z = 0$ starting from the equilibrium solution for the probability density $\psi_{eq}(\boldsymbol{u}, s) = \frac{1}{4\pi}\delta(|\boldsymbol{u}| - 1)$. All the quantities of interest are nondimensionalized, namely the time is rescaled by τ_d, the shear rate $\dot{\gamma}$ by $1/\tau_d$ and the stress by G_N^0. We present the results for two values of nondimensional shear rate: $\dot{\gamma} = 2$ on the left and $\dot{\gamma} = 10$ on the right of Fig. 4.11. In the first case, the reasonable convergence of the Fokker–Planck-based simulations is achieved on the mesh $N_u = N_s = 14$ and the time step $\Delta t = 2 \times 10^{-3}$, the method being unstable at bigger time steps. We compare these results with those of stochastic simulations as described in Section 2.5 with $N_f = 2500$ and $N_f = 10000$ random samples and the time step $\Delta t = 5 \times 10^{-3}$. A perfect matching is observed for $N_f = 10000$, while significant noise is noticeable with fewer number of samples. The Fokker–Planck-based calculation took 2.3 s of CPU time on a laptop computer, while the stochastic simulations with $N_f = 2500$ and $N_f = 10000$ took, respectively, 3.4 and 13.2 s. We conclude thus that the Fokker–Planck-based method allows one to obtain an accurate solution at a lower cost than the stochastic simulations. The situation is quite different for a higher shear rate of $\dot{\gamma} = 10$, however. In this case, we had to decrease the time step in a Fokker–Planck simulation down to $\Delta t = 2 \times 10^{-4}$ in order to preserve the stability

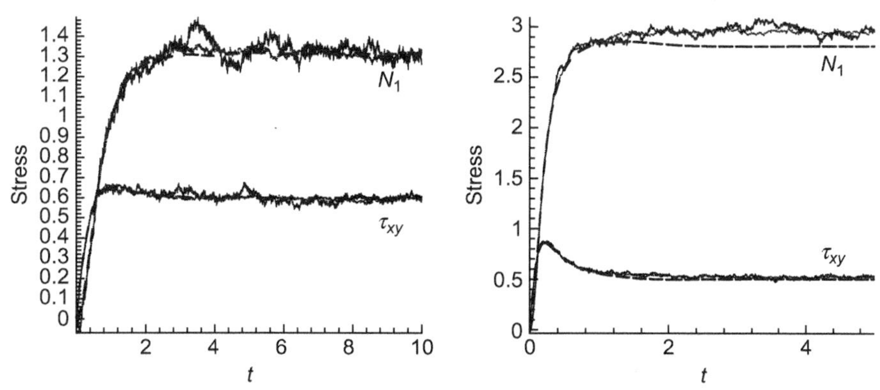

FIG. 4.11 Time evolution of the shear stress τ_{xy} and of the first normal stress difference $N_1 = \tau_{xx} - \tau_{yy}$ in the shear flow of a Doi–Edwards fluid at nondimensional shear rate $\dot{\gamma} = 2$ on the left and $\dot{\gamma} = 10$ on the right. The dashed line represents the results of a Fokker–Planck simulation ($N_u = N_s = 14$, $\Delta t = 2 \times 10^{-3}$ for $\dot{\gamma} = 1$ and $N_u = 20$, $N_s = 14$, $\Delta t = 2 \times 10^{-4}$ for $\dot{\gamma} = 1$), the solid line corresponds to a stochastic simulations with $N_f = 2500$, and the dotted line to that with $N_f = 10000$.

and to refine the mesh in $\mathbf{u}$ to $N_u = 20$, thus increasing the CPU time to 46 s. Meanwhile the stochastic simulations rest robust with the same parameters as before. They are thus much more efficient than the Fokker–Planck simulations at this relatively large value of $\dot{\gamma}$.

4.5. Fokker–Planck-based numerical methods for high-dimensional configuration spaces

4.5.1. *Sparse tensor product Fokker–Planck-based methods*

We present simulations of a homogeneous shear flow with velocity gradient $\kappa = \begin{pmatrix} 0 & 1 \\ 0 & 0 \end{pmatrix}$ for FENE chains with $b = 12$. We use both the sparse tensor product (STP) approach for the Fokker–Planck equation described in Section 3.4 and a stochastic method with $M = 15000$ and $M = 60000$ pseudo-random realizations. The STP simulations were done on a mesh with $N_R = 12, N_F = 6$ and the STP truncation parameter $L = 6$ and $L = 7$. A time step $\Delta t = 0.01$ was used in all the simulations. Some results for the evolution of the stress component τ_{xx} are presented in Fig. 4.12. We observe already excellent agreement between the two methods for chains of 2 springs but a considerable deterioration in the results for chains having d = 5, although the STP method is still convergent with increasing L. In Fig. 4.13(a), we give quantitative results on the discrepancy between the two approaches. We there plot the relative error in the value of τ_{xx} averaged on the time interval [10, 15] taking the stochastic results with $M = 60000$ as a reference solution, against which the Fokker–Planck-based calculations are compared. The CPU time in seconds on a Pentium IV machine is presented in Fig. 4.13(b). The x-axes on both plots in Fig. 4.13 represents the number of springs. In summary, our STP method may be seen to be competitive with the stochastic approach only for short chains consisting of up to 4 springs.

4.5.2. *Low-rank separation algorithms*

Finally, we present some results for the time-dependent simulations of the 2D FENE bead-spring chain model using the approach (3.81) and (3.82) of minimizing residual low-rank

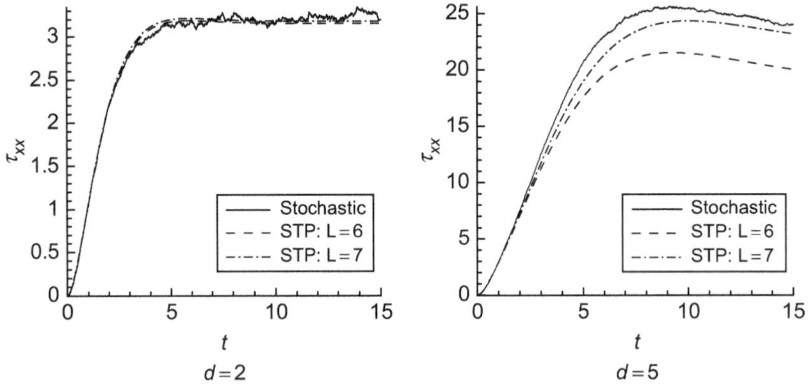

FIG. 4.12 Evolution of τ_{xx} in a shear flow calculated with STP and stochastic methods.

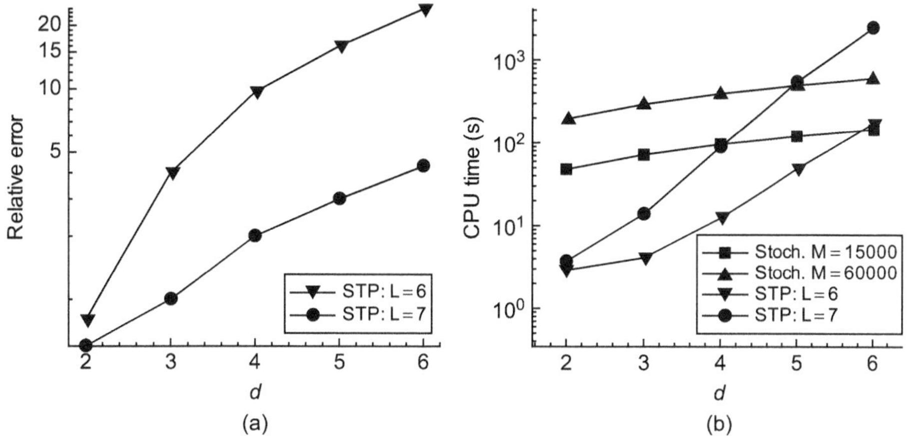

FIG. 4.13 (a) Relative error for τ_{xx} computed using STP method. (b) CPU time (s) of simulations with STP and stochastic methods.

separation representations (see Section 3.4.2). We consider the model with d chains so that the probability density function $\psi(\boldsymbol{q}_1, \ldots, \boldsymbol{q}_d, t)$ satisfies the Fokker–Planck equation (3.50)

$$\frac{\partial \psi}{\partial t} - \mathcal{L}_{FP}^{(d)}(\psi) = 0$$

with the differential operator $\mathcal{L}_{FP}^{(d)}$ having a separated representation involving the univariate operators L, $M_{x,y}$, and $N_{x,y}$ given by (3.51). We take the equilibrium solution ψ_{eq} as the initial condition and introduce the new unknown $\widetilde{\psi} = \psi - \psi_{\text{eq}}$ satisfying

$$\frac{\partial \widetilde{\psi}}{\partial t} - \mathcal{L}_{FP}^{(d)}(\widetilde{\psi}) = \mathcal{L}_{FP}^{(d)}(\psi_{\text{eq}}) \tag{4.7}$$

with the homogeneous initial conditions $\widetilde{\psi}|_{t=0} = 0$. We apply then the method (3.81) and (3.82) to construct successive approximations of the form (3.79) to the solution of the problem (4.7) viewed as a (d + 1)-dimensional problem with the operator $A = \frac{\partial}{\partial t} - \mathcal{L}_{FP}^{(\mathrm{d})}$.

We report on two series of numerical experiments for the 2D FENE chain model with the extensibility parameter $b = 12$ in a shear flow with the velocity gradient $\boldsymbol{\kappa} = \begin{pmatrix} 0 & \dot{\gamma} \\ 0 & 0 \end{pmatrix}$ with $\dot{\gamma} = 0.1$ and $\dot{\gamma} = 1$. We used the same spectral discretization of one-dumbbell functions $\psi_j^{(i)}(\boldsymbol{q}_j)$ as described in Section 3.4 on sparse tensor product methods. The time-dependent functions $\psi_{d+1}^{(i)}(t)$ and the corresponding operator $\frac{\partial}{\partial t}$ were discretized by the spectral collocation method on the GLL grid scaled to the interval $(0, T_{\text{fin}})$, T_{fin} being the final time of the computation. The nonlinear problems on iterations (3.81) were solved by a fixed-point method until the relative tolerance of 10^{-4} was reached. We performed 150 iterations of the minimizing residual low-rank separation representations method in all the cases using the grid with $N_R = 6$, $N_F = 3$ for the functions of $\boldsymbol{q}_j$ and the GLL grid of $N_T = 150$ points for

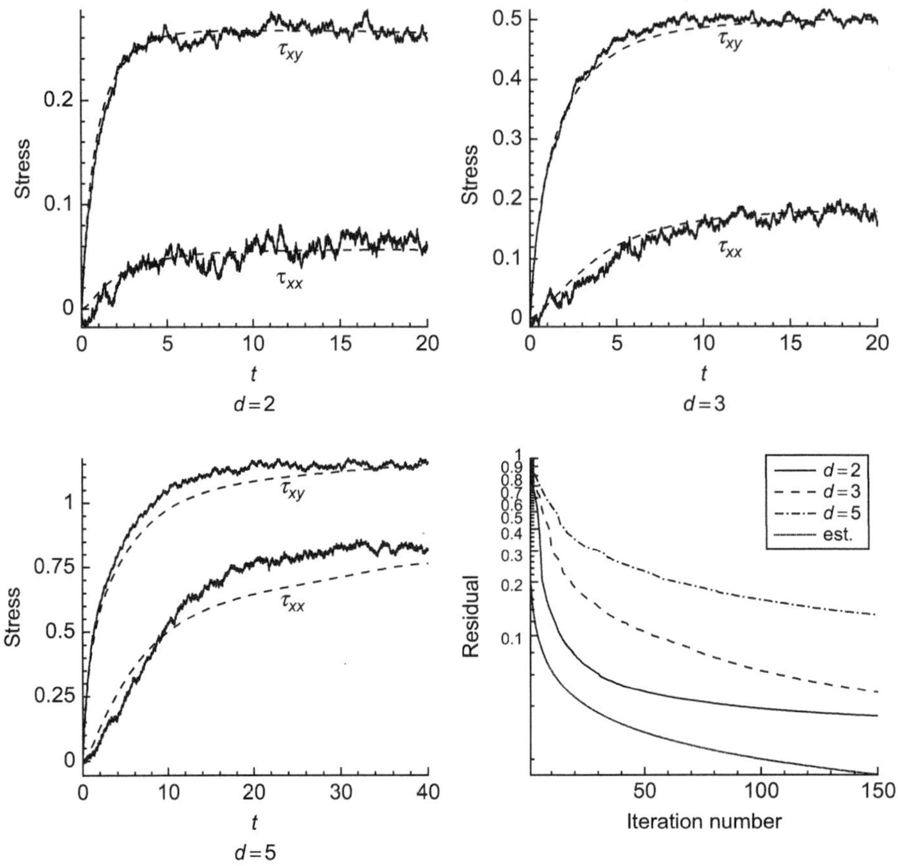

FIG. 4.14 Evolution of τ_{xx} and τ_{xy} in the 2D FENE chain model in a shear flow with the shear rate $\dot{\gamma} = 0.1$ calculated both by the minimizing residual separated representations algorithm with 150 iterations and by stochastic methods. The results are for the chains of d = 2, 3, and 5 springs. The graph on the bottom right represents the decrease of the residual on iterations in comparison with the theoretical estimate est $= C/\sqrt{m}$ taking $C = 0.2$.

the functions of time. The results for the evolution of the stress are presented in Figs 4.14 and 4.15 and compared against stochastic simulations with 100,000 samples for the chains consisting of d = 2, 3, and 5 springs. The first Fig. 4.14 contains the results at a small value of $\dot{\gamma} = 0.1$, and the second Fig. 4.15 is at a moderate value of $\dot{\gamma} = 1$. We observe a very good agreement in the case of low shear rate $\dot{\gamma} = 0.1$ and short chains d = 2, 3, and a qualitatively good agreement for a longer chain d = 5 as well as for a chain of d = 2 springs at a higher shear rate $\dot{\gamma} = 1$. On the other hand, even such a big number of iterations was not enough to produce acceptable results with $\dot{\gamma} = 1$ and the chains of 3 or 5 springs. The method in its present form seems thus limited to the simulations at small velocity gradients and with not very long chains. On a more positive side, we note that the theoretically predicted rate of decrease in the residual norm of order at worst $1/\sqrt{m}$ is observed in all the cases as demonstrated by the last graphs in both Figs 4.14 and 4.15.

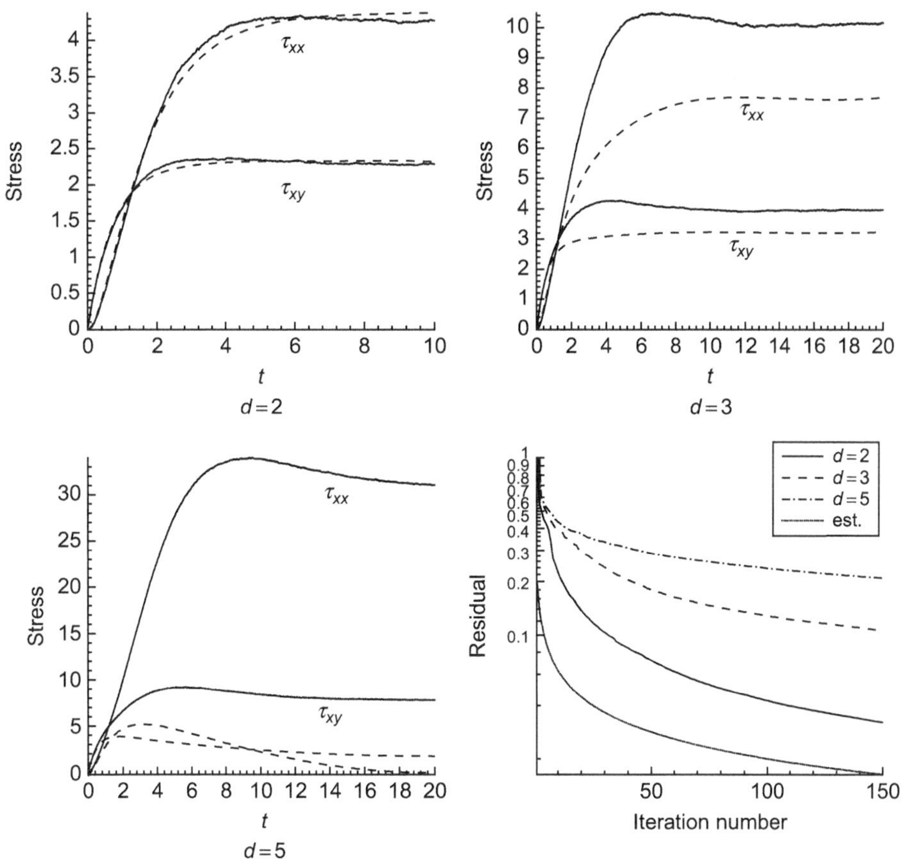

FIG. 4.15 Evolution of τ_{xx} and τ_{xy} in the 2D FENE chain model in a shear flow with the shear rate $\dot{\gamma} = 1$ calculated both by the minimizing residual separated representations algorithm with 150 iterations and by stochastic methods. The results are for the chains of d = 2, 3, and 5 springs. The graph on the bottom right represents the decrease of the residual on iterations in comparison with the theoretical estimate est $= C/\sqrt{m}$ taking $C = 0.2$.

Acknowledgments

The authors wish to acknowledge, with gratitude, useful and constructive discussions with Amine Ammar (Laboratoire de Rhéologie, Grenoble), Claude Le Bris (Ecole Nationale des Ponts et Chaussées), Francisco Chinesta (Ecole Centrale de Nantes), Antonio Falcó (Universidad CEU Cardenal Herrera), Jiannong Fang (Ecole Polytechnique Fédérale de Lausanne), David Knezevic (Oxford University), Tony Lelièvre (Ecole Nationale des Ponts et Chaussées), Ganna Leonenko (Swansea University), Jie Shen (Purdue University), Endre Süli (Oxford University), and Rene Vargas (Universidad Nacional Autónoma de México).

The work of the second author was supported by a Discovery Grant from the Natural Sciences and Engineering Research Council of Canada.

Bibliography

AMMAR, A., CHINESTA, F., FALCÓ, A. (2010). On the convergence of a greedy rank-one update algorithm for a class of linear systems. *Arch. Comput. Methods Eng.* In press.

AMMAR, A., MOKDAD, B., CHINESTA, F., KEUNINGS, R. (2006a). A new family of solvers for some classes of multidimensional partial differential equations encountered in kinetic theory modeling of complex fluids. *J. Non-Newton. Fluid Mech.* **139**, 153–176.

AMMAR, A., MOKDAD, B., CHINESTA, F., KEUNINGS, R. (2007). A new family of solvers for some classes of multidimensional partial differential equations encountered in kinetic theory modeling of complex fluids. Part II: Transient simulation using space-time separated representations. *J. Non-Newton. Fluid Mech.* **144**, 98–121.

AMMAR, A., RYCKELYNCK, D., CHINESTA, F., KEUNINGS, R. (2006b). On the reduction of kinetic theory models related to finitely extensible dumbbells. *J. Non-Newton. Fluid Mech.* **134**, 136–147.

ARMSTRONG, R.C., NAYAK, R., GHOSH, I., BROWN, R.A. (1996). The use of kinetic theory and microstructural models in the analysis of complex flows of viscoelastic liquids. In: Aït-Kadi, A., J.M.D., James, D.F., Williams, M.C. (eds.), *Proceedings of the XIIth International Congress on Rheology* (Laval University, Quebec City), pp. 307-310.

BAJAJ, M., BHAT, P.P., PRAKASH, J.R., PASQUALI, M. (2006). Multiscale simulation of viscoelastic free surface flows. *J. Non-Newton. Fluid Mech.* **140**, 87–107.

BARRETT, J.W., SCHWAB, C., SÜLI, E. (2005). Existence of global weak solutions for some polymeric flow models. *Math. Models Methods. Appl. Sci.* **15**, 939–983.

BARRETT, J.W., SÜLI, E. (2007). Existence of global weak solutions to some regularized kinetic models for dilute polymers. *Multiscale Model. Simul.* **6**, 506–546.

BEYLKIN, G., MOHLENKAMP, M.J. (2002). Numerical operator calculus in higher dimensions. *Proc. Natl. Acad. Sci. USA* **99**, 10246–10251.

BEYLKIN, G., MOHLENKAMP, M.J. (2005). Algorithms for numerical analysis in high dimensions. *SIAM J. Sci. Comput.* **26**, 2133–2159.

BILLER, P., PETRUCCIONE, F. (1987). The flow of dilute polymer solutions in confined geometries: a consistent numerical approach. *J. Non-Newton. Fluid Mech.* **25**, 347–364.

BIRD, R.B., ARMSTRONG, R.C., HASSAGER, O. (1987a). *Dynamics of Polymeric Liquids* Volume 1 (John Wiley and Sons, New York).

BIRD, R.B., CURTISS, C.F., ARMSTRONG, R.C., HASSAGER, O. (1987b). *Dynamics of Polymeric Liquids* Volume 2 (John Wiley and Sons, New York).

BIRD, R.B., DEAGUIAR, J.R. (1983). An encapsulated dumbbell model for concentrated polymer solutions and melts i. theoretical developments and constitutive equation. *J. Non-Newton. Fluid Mech.* **13**, 149–160.

BONVIN, J., PICASSO, M. (1999). Variance reduction methods for CONNFFESSIT-like simulations. *J. Non-Newton. Fluid Mech.* **84**, 191–215.

BONVIN, J.C. (2000). Numerical simulation of viscoelastic fluids with mesoscopic models, Ph.D. Thesis (Ecole Polytechnique Fédérale de Lausanne, Lausanne, Switzerland).

BUNGARTZ, H.-J., GRIEBEL, M. (2004). Sparse grids. *Acta Numer.* **13**, 147–269.

CANUTO, C., HUSSAINI, M.Y., QUARTERONI, A., ZANG, T.A. (2006). *Spectral Methods: Fundamentals in Single Domains. Scientific Computation* (Springer-Verlag, Berlin).

CHAUVIÈRE, C. (2002). A new method for micro-macro simulations of viscoelastic flows. *SIAM J. Sci. Comput.* **23** (6), 2123–2140.

CHAUVIÈRE, C., FANG, J., LOZINSKI, A., OWENS, R.G. (2003). On the numerical simulation of flows of polymer solutions using high-order methods based on the Fokker-Planck equation. *Int. J. Mod. Phys.* B **17**, 9–14.

CHAUVIÈRE, C., LOZINSKI, A. (2003). An efficient technique for simulations of viscoelastic flows, derived from the Brownian configuration field method. *SIAM J. Sci. Comput.* **24**, 1823–1837.

CHAUVIÈRE, C., LOZINSKI, A. (2004a). Simulation of complex viscoelastic flows using the Fokker-Planck equation: 3D FENE model. *J. Non-Newton. Fluid Mech.* **122**, 201–214.

CHAUVIÈRE, C., LOZINSKI, A. (2004b). Simulation of dilute polymer solutions using a Fokker-Planck equation. *Comput. Fluids* **33** (5-6), 687–696.

CORMEN, T.H., LEISERSON, C.E., RIVEST, R.L., STEIN, C. (2001). *Introduction to Algorithms, Second Edition* (MIT Press, Cambridge, MA).

CROCHET, M.J., DAVIES, A.R., WALTERS, K. (1984). *Numerical Simulation of Non-Newtonian Flow* (Elsevier, Amsterdam).

DE GENNES, P.G. (1971). Reptation of a polymer chain in the presence of fixed obstacles. *J. Chem. Phys.* **55**, 572–579.

DELAUNAY, P., LOZINSKI, A., OWENS, R.G. (2007). Sparse tensor-product Fokker-Planck-based methods for nonlinear bead-spring chain models of dilute polymer solutions. In: Bandrauk, A., Delfour, M.C., Le Bris C. (eds.), *High-dimensional Partial Differential Equations in Science and Engineering*. CRM Proceedings and Lecture Notes Volume 41 (American Mathematical Society, Providence, RI), pp. 73–89. http://www.ams.org/bookstore?fn=20&arg1=crmpseries&ikey=CRMP-41

DEVORE, R.A., TEMLYAKOV, V.N. (1996). Some remarks on greedy algorithms. *Adv. Comput. Math.* **5** (2-3), 173–187.

DOI, M., EDWARDS, S.F. (1978a). Dynamics of concentrated polymer systems: Brownian motion in the equilibrium state. *J. Chem. Soc. Faraday Trans.* **74**, 1789–1801.

DOI, M., EDWARDS, S.F. (1978b). Dynamics of concentrated polymer systems: molecular motion under flow. *J. Chem. Soc. Faraday Trans.* **74**, 1801–1817.

DOI, M., EDWARDS, S.F. (1978c). Dynamics of concentrated polymer systems: the constitutive equation. *J. Chem. Soc. Faraday Trans.* **74**, 1818–1832.

DOI, M., EDWARDS, S.F. (1988). *The Theory of Polymer Dynamics* (Oxford University Press, Oxford).

DU, Q., LIU, C., YU, P. (2005). FENE dumbbell model and its several linear and nonlinear closure approximations. Multiscale Model. *Simul.* **4**, 709–731.

WEINAN, E. W., LI, T., ZHANG, P. (2004). Well-posedness for the dumbbell model of polymeric fluids. *Commun. Math. Phys.* **248**, 409–427.

FAN, X. (1989a). Molecular model and flow calculation: I. The numerical solutions to multibead-rod models in homogeneous flows. *Acta Mech. Sin.* **5**, 49–59.

FAN, X. (1989b). Molecular models and flow calculations: II. Simulation of steady planar flow. *Acta Mech. Sin.* **5**, 216–226.

FAN, X.J. (1985a). A numerical investigation of the hydrodynamic interaction of Hookean dumbbell suspensions in steady state shear flow. *J. Chem. Phys.* **85**, 6237–6238.

FAN, X.J. (1985b). Viscosity, first and second normal-stress coefficients, and molecular stretching in concentrated solutions and melts. *J. Non-Newton. Fluid Mech.* **17**, 251–265.

FAN, X.J. (1985c). Viscosity, first normal-stress coefficient, and molecular stretching in dilute polymer solutions. *J. Non-Newton. Fluid Mech.* **17**, 125–144.

FANG, J., KRÖGER, M., ÖTTINGER, H.-C. (2000). A thermodynamically admissible reptation model for fast flows of entangled polymers. II. Model predictions for shear and extensional flows. *J. Rheol.* **44**, 1293–1317.

FANG, J., LOZINSKI, A., OWENS, R.G. (2004). Towards more realistic kinetic models for concentrated solutions and melts. *J. Non-Newton. Fluid Mech.* **122**, 79–90.

FEIGL, K., LASO, M., ÖTTINGER, H.-C. (1995). CONNFFESSIT approach for solving a two-dimensional viscoelastic flow problem. *Macromolecules* **28**, 3261–3274.

GALLEZ, X., HALIN, P., LIELENS, G., KEUNINGS, R., LEGAT, V. (1999). The adaptive Lagrangian particle method for macroscopic and micro-macro computations of time-dependent viscoelastic flows. *Comput. Methods Appl. Mech. Eng.* **180**, 345–364.

GARDINER, C.W. (2003). *Handbook of Stochastic Methods* (Springer-Verlag, Berlin, Heidelberg, New York).

GIGRAS, P.G., KHOMAMI, B. (2002). Adaptive configuration fields: a new multiscale simulation technique for reptation-based models with a stochastic strain measure and local variations of life span distribution. *J. Non-Newton. Fluid Mech.* **108**, 99–122.

GRIEBEL, M., OELTZ, D. (2007). A sparse grid space-time discretization scheme for parabolic problems. *Computing* **81**, 1–34.

GUILLOPÉ, C., SAUT, J.C. (1990). Existence results for the flow of viscoelastic fluids with a differential constitutive law. *Non. Anal. Theory Methods Appl.* **15**, 849–869.

HALIN, P., LIELENS, G., KEUNINGS, R., LEGAT, V. (1998). The Lagrangian particle method for macroscopic and micro-macro viscoelastic flow computations. *J. Non-Newton. Fluid Mech.* **79**, 387–403.

HELZEL, C., OTTO, F. (2006). Multiscale simulations for suspensions of rod-like molecules. *J. Comput. Phys.* **216**, 52–75.

HERRCHEN, M., ÖTTINGER, H.-C. (1997). A detailed comparison of various FENE dumbbell models. *J. Non-Newton. Fluid Mech.* **68**, 17–42.

HULSEN, M.A., PETERS, E.A.J.F., VAN DEN BRULE, B.H.A.A. (2001). A new approach to the deformation fields method for solving complex flows using integral constitutive equations. *J. Non-Newton. Fluid Mech.* **98**, 201–221.

HULSEN, M.A., VAN HEEL, A.P.G., VAN DEN BRULE, B.H.A.A. (1997). Simulation of viscoelastic flows using Brownian configuration fields. *J. Non-Newton. Fluid Mech.* **70**, 79–101.

JENDREJACK, R.M., DE PABLO, J.J., GRAHAM, M.D. (2002). A method for multiscale simulation of flowing complex fluids. *J. Non-Newton. Fluid Mech.* **108**, 123–142.

JOURDAIN, B., LE BRIS, C., LELIÈVRE, T. (2004a). On a variance reduction technique for micro-macro simulations of polymeric fluids. *J. Non-Newton. Fluid Mech.* **122**, 91–106.

JOURDAIN, B., LE BRIS, C., LELIÈVRE, T., OTTO, F. (2006). Long-time asymptotics of a multiscale model for polymeric fluid flows. *Arch. Ration. Mech. Anal.* **181**, 97–148.

JOURDAIN, B., LELIÈVRE, T. (2003). Mathematical analysis of a stochastic differential equation arising in the micro-macro modelling of polymeric fluids. In: Davies, I.M., Jacob, N., Truman, A., Hassan, O., Morgan, K., Weatherill, N.P., (eds.), *Probabilistic Methods in Fluids* (World Scientific Publishing, River Edge, NJ), pp. 205–223. http://www.worldscibooks.com/mathematics/5156.html

JOURDAIN, B., LELIÈVRE, T., LE BRIS, C. (2002). Numerical analysis of micro-macro simulations of polymeric fluid flows: a simple case. *Math. Models Methods Appl. Sci.* **12**, 1205–1243.

JOURDAIN, B., LELIÈVRE, T., LE BRIS, C. (2004b). Existence of solution for a micro-macro model of polymeric fluid: the FENE model. *J. Funct. Anal.* **209**, 162–193.

KEUNINGS, R. (1997). On the Peterlin approximation for finitely extensible dumbbells. *J. Non-Newton. Fluid Mech.* **68**, 85–100.

KEUNINGS, R. (2004).Micro-macro methods for the multiscale simulation of viscoelastic flow using molecular models of kinetic theory. In: Binding, D.M., Walters, K. (eds.), *Rheology Reviews* (British Society of Rheology, Aberystwyth, UK), pp. 67–83.

KLOEDEN, P.E., PLATEN, E. (1992). Numerical solution of stochastic differential equations. *Applications of Mathematics* Volume 23 (Springer, Berlin).

KNEZEVIC, D.J. (2008). Analysis and implementation of numerical methods for simulating dilute polymeric fluids, Ph.D. Thesis (University of Oxford, Oxford, UK).

KNEZEVIC, D.J., SÜLI, E. (2009). Spectral Galerkin approximation of Fokker-Planck equations with unbounded drift. M2AN Math. *Model. Numer. Anal.* **43** (3), 445–485.

KOPPOL, A.P., SURESHKUMAR, R., KHOMAMI, B. (2007). An efficient algorithm for multiscale flow simulations of dilute polymeric solutions using bead-spring chains. *J. Non-Newton. Fluid Mech.* **141**, 180–192.

KRAMERS, H.A. (1944). Het gedrag van macromoleculen in een stroomende vloeistof. *Physica* **11**, 1–19.

LASO, M., ÖTTINGER, H.-C. (1993). Calculation of viscoelastic flow using molecular models: the CONNFFESSIT approach. *J. Non-Newton. Fluid Mech.* **47**, 1–20.

LASO, M., RAMÍREZ, J., PICASSO, M. (2004). Implicit micro-macro methods. *J. Non-Newton. Fluid Mech.* **122**, 215–226.

LE BRIS, C., LELIÈVRE, T., MADAY, Y. (2009). Results and questions on a nonlinear approximation approach for solving high-dimensional partial differential equations. *Constr. Approx.* **30** (3), 621–651.

LELIÈVRE, T. (2004). Optimal error estimate for the CONNFFESSIT approach in a simple case. *Comput. Fluids* **33**, 815–820.

LEONENKO, G.M., PHILLIPS, T.N. (2009). On the solution of the Fokker-Planck equation using a high-order reduced basis approximation. Comput. *Methods Appl. Mech. Eng.* **199**, 158–168.

LI, T., ZHANG, H., ZHANG, P.-W. (2004). Local existence for the dumbbell model of polymeric fluids. *Comm. Partial Differ. Equ.* **29**, 903–923.

LIELENS, G., KEUNINGS, R., LEGAT, V. (1999). The FENE-L and FENE-LS closure approximations to the kinetic theory of finitely extensible dumbbells. *J. Non-Newton. Fluid Mech.* **87**, 179–196.

LIN, F.-H., LIU, C., ZHANG, P. (2005). On hydrodynamics of viscoelastic fluids. Commun. *Pure Appl. Math.* **58**, 1437–1471.

LIN, F.-H., ZHANG, P., ZHANG, Z. (2008). On the global existence of smooth solution to the 2-D FENE dumbbell model. *Commun. Math. Phys.* **277**, 531–553.

LIONS, P.-L., MASMOUDI, N. (2000). Global solutions for some oldroyd models of Non-Newtonian. flows. *Chinese Ann. Math. Ser. B* **21**, 131–146.

LIONS, P.-L., MASMOUDI, N. (2007). Global existence of weak solutions to some micro-macro models. *C.R. Acad. Sci. Paris Ser. I* **345**, 15–20.

LIU, C., LIU, H. (2008). Boundary conditions for the microscopic fene models. *SIAM J. Appl. Math.* **68**, 1304–1315.

LOZINSKI, A., CHAUVIÈRE, C. (2003). A fast solver for Fokker-Planck equation applied to viscoelastic flows calculations: 2D FENE model. *J. Comput. Phys.* **189** (2), 607–625.

LOZINSKI, A., CHAUVIÈRE, C., FANG, J., OWENS, R.G. (2003). Fokker–Planck simulations of fast flows of melts and concentrated polymer solutions in complex geometries. *J. Rheol.* **47**, 535–561.

LOZINSKI, A., OWENS, R.G., FANG, J. (2004). A Fokker-Planck-based numerical method for modelling non-homogeneous flows of dilute polymeric solutions. *J. Non-Newton. Fluid Mech.* **122**, 273–286.

MALLAT, S., ZHANG, Z. (1993). Matching pursuits with time-frequency dictionaries. *IEEE Trans. Signal Process.* **41** (12), 3397–3415.

MASMOUDI, N. (2008). Well-posedness for the FENE dumbbell model of polymeric flows. *Commun. Pure Appl. Math.* **61** (12), 1685–1714.

MELCHIOR, M., ÖTTINGER, H.-C. (1995). Variance reduced simulations of stochastic differential equations. *J. Chem. Phys.* **103**, 9506–9509.

MELCHIOR, M., ÖTTINGER, H.-C. (1996). Variance reduced simulations of polymer dynamics. *J. Chem. Phys.* **105**, 3316–3331.

MIL'SHTEIN, G.N. (1974). Approximate integration of stochastic differential equations. *Theory Probab. Appl.* **19**, 557–562.

NAYAK, R. (1998). Molecular simulation of liquid crystal polymer flow: a wavelet-finite element analysis, Ph.D. Thesis (MIT, Cambridge, MA).

NIEDERREITER, H. (1992). Random number generation and quasi-Monte Carlo methods. *CBMS-NSF Regional Conference Series in Applied Mathematics* Volume 63 (Society for Industrial and Applied Mathematics (SIAM), Philadelphia, PA).

ÖTTINGER, H.C. (1996). *Stochastic Processes in Polymeric Fluids* (Springer-Verlag, Berlin).

ÖTTINGER, H.-C. (1999). A thermodynamically admissible reptation model for fast flows of entangled polymers. *J. Rheol.* **43**, 1461–1493.

ÖTTINGER, H.-C., VAN DEN BRULE, B.H. A.A., HULSEN, M.A. (1997). Brownian configuration fields and variance reduced CONNFFESSIT. *J. Non-Newton. Fluid Mech.* **70**, 255–261.

OTTO, F., TZAVARAS, A.E. (2008). Continuity of velocity gradients in suspensions of rod-like molecules. *Commun. Math. Phys.* **277**, 729–758.

OWENS, R.G., PHILLIPS, T.N. (2002). *Computational Rheology* (Imperial College Press, London).

PETERLIN, A. (1966). Hydrodynamics of macromolecules in a velocity field with longitudinal gradient. *J. Polym. Sci. B: Polym. Lett.* **4**, 287–291.

PETERS, E.A.J.F., HULSEN, M.A., VAN DEN BRULE, B.H.A.A. (2000a). Instationary Eulerian viscoelastic flow simulations using time separable Rivlin-Sawyers constitutive equations. *J. Non-Newton. Fluid Mech.* **89**, 209–228.

PETERS, G.W.M., VAN HEEL, A.P.G., HULSEN, M.A., VAN DE BRULE, B.H.A.A. (2000b). Generalization of the deformation field method to simulate advanced reptation models in complex flow. *J. Rheol.* **44**, 811–829.

PHILLIPS, T.N., SMITH, K.D. (2006). A spectral element approach to the simulation of viscoelastic flows using Brownian configuration fields. *J. Non-Newton. Fluid Mech.* **138**, 98–110.

RAMÍREZ, J., LASO, M. (2005). Size reduction methods for the implicit time-dependent simulation of micro-macro viscoelastic flow problems. *J. Non-Newton. Fluid Mech.* **127**, 41–49.

RENARDY, M. (1991). An existence theorem for model equations resulting from kinetic theories of polymer solutions. *SIAM J. Math. Anal.* **22**, 313–327.

SCHIEBER, J.D. (1992). Do internal viscosity models satisfy the fluctuation-dissipation theorem? *J. Non-Newton. Fluid Mech.* **45**, 47–61.

SCHIEBER, J.D. (2006). Generalized Brownian configuration fields for Fokker-Planck equations including center-of-mass diffusion. *J. Non-Newton. Fluid Mech.* **135**, 179–181.

SCHIEBER, J.D., ÖTTINGER, H.C. (1988). The effects of bead inertia on the Rouse model. *J. Chem. Phys.* **89**, 6972–6981.

SHAQFEH, E.S.G., JAGADEESHAN, R.P. (2008). International Workshop on Mesoscale and Multiscale Description of Complex Fluids (IWMMCOF'06), Prato, Italy, July 5–8, 2006. *J. Non-Newton. Fluid Mech.* **149**, 1–2.

SMOLYAK, S.A. (1963). Quadrature and interpolation formulas for tensor products of certain classes of functions. *Dokl. Akad. Nauk SSSR* **4**, 240–243.

SOMASI, M., KHOMAMI, B. (2000). Linear stability and dynamics of viscoelastic flows using time-dependent stochastic simulation techniques. *J. Non-Newton. Fluid Mech.* **93**, 339–362.

SOMASI, M., KHOMAMI, B., WOO, N.J., HUR, J.S., SHAQFEH, E.S.G. (2002). Brownian dynamics simulations of bead-rod and bead-spring chains: numerical algorithms and coarse-graining issues. *J. Non-Newton. Fluid Mech.* **108**, 227–255.

SUEN, J.K., NAYAK, R., ARMSTRONG, R.C., BROWN, R.A. (2003). A wavelet-galerkin method for simulating the doi model with orientation-dependent rotational diffusivity. *J. Non-Newton. Fluid Mech.* **114** (2-3), 197–228.

SUEN, J.K.C., JOO, Y.L., ARMSTRONG, R.C. (2002). Molecular orientation effects in viscoelasticity. *Annu. Rev. Fluid Mech.* **34**, 417–444.

TEMLYAKOV, V.N. (2000). Weak greedy algorithms. *Adv. Comput. Math.* **12**, 213–227.

VAN HEEL, A.P.G., HULSEN, M.A., VAN DEN BRULE, B.H.A.A. (1999). Simulation of the Doi-Edwards model in complex flow. *J. Rheol.* **43**, 1239–1260.

VARGAS, R.O., MANERO, O., PHILLIPS, T.N. (2009). Viscoelastic flow past confined objects using a micro-macro approach. *Rheol. Acta* **48**, 373–395.

VENKITESWARAN, G., JUNK, M. (2005a). A QMC approach for high dimensional Fokker-Planck equations modelling polymeric liquids. *Math. Comput. Simul.* **68**, 43–56.

VENKITESWARAN, G., JUNK, M. (2005b). Quasi-Monte Carlo algorithms for diffusion equations in high dimensions. *Math. Comput. Simul.* **68**, 23–41.

VON PETERSDORFF, T., SCHWAB, C. (2004). Numerical solution of parabolic equations in high dimensions. *M2AN Math. Model. Numer. Anal.* **38**, 93–127.

WAPPEROM, P., KEUNINGS, R. (2000). Simulation of linear polymer melts in transient complex flow. *J. Non-Newton. Fluid Mech.* **95**, 67–83.

WAPPEROM, P., KEUNINGS, R., LEGAT, V. (2000). The backward-tracking Lagrangian particle method for transient viscoelastic flows. *J. Non-Newton. Fluid Mech.* **91**, 273–295.

WARNER, H.R. (1972). Kinetic theory and rheology of dilute suspensions of finitely extendible dumbbells. *Ind. Eng. Chem. Fundam.* **11**, 379–387.

YU, P., DU, Q., LIU, C. (2005). From micro to macro dynamics via a new closure approximation to the FENE model of polymeric fluids. *Multiscale Model. Simul.* **3**, 895–917.

ZHANG, H., ZHANG, P. (2006). Local existence for the FENE-dumbbell model of polymeric fluids. *Arch. Rat. Mech. Anal.* **181**, 373–400.

Viscoelastic Flows with Complex Free Surfaces: Numerical Analysis and Simulation

Andrea Bonito

Department of Mathematics, Texas A&M University, College Station, TX 77843, USA
E-mail: bonito@math.tamu.edu

Philippe Clément

Analysis Group, TU Delft, Mekelweg 4, 2628 CD Delft, The Netherlands
E-mail: Ph.P.J.E.Clement@ewi.tudelft.nl

Marco Picasso

Institut d'Analyse et Calcul Scientifique, Ecole Polytechnique Fédérale de Lausanne, 1015 Lausanne, Switzerland
E-mail: marco.picasso@epfl.ch

Numerical Methods for Non-Newtonian Fluids
Special Volume (R. Glowinski and Jinchao Xu, Guest Editors) of
HANDBOOK OF NUMERICAL ANALYSIS, VOL. XVI
P.G. Ciarlet (Editor)

ISSN 1570-8659
DOI 10.1016/B978-0-444-53047-9.00003-4

Contents

CHAPTER 1

Modeling of Viscoelastic Flows with Complex Free Surfaces

Viscoelastic flows with complex free surfaces are considered. Such flows are involved not only in several industrial processes involving paints, plastics, food, or adhesives but also in geophysical applications such as mud flows or avalanches.

Viscoelastic fluids are viscous fluids having elastic properties. They cannot be described with the classical theories of fluid or continuum mechanics. Additional laws have to be added in order to relate the stress to the velocity, this being the scope of rheology.

The rheology of viscoelastic flows depends on the microscopic details of the fluid. As a consequence, an accurate mathematical modeling should consider all the physical scales involved. Consider for instance the case of a polymeric liquid, say polyethylene $(C_2H_4)_n$, where $n = 10^4$ is the number of monomers. Since the size of the C–C bond is 10^{-10} m, then the size of the fully extended molecule is $10^{-10} \times 10^4 = 10^{-6}$ m, whereas the size of the macroscopic workpiece – a car bumper for instance – is about 1 m. Clearly, all these microscopic details cannot be included in a macroscopic numerical simulation, but intermediate – mesoscopic – models can be considered. Only the simplest macroscopic and mesoscopic models are considered here. More realistic and complex models can be found in classical textbooks of non-Newtonian flows (see BIRD, CURTISS, ARMSTRONG and HASSAGER [1987], LARSON [1999], ÖTTINGER [1996], and RENARDY [2000] for instance). We also refer to SINGH, JOSEPH, HESLAB, GLOWINSKI and PAN [2000] for the description of suspended particles in a viscoelastic flow.

Consider a cavity Λ of $\mathbb{R}^d$, $d = 2$ or 3, partially filled with a viscoelastic fluid. We are interested in computing the fluid shape between time 0 and time T. The notations are reported in Fig. 1.1 and are the following. Let $D(t) \subset \Lambda$ be the liquid region at time t, and let $\varphi :$ $\Lambda \times (0, T)$ be the characteristic function of the liquid, that is

$$\varphi(x,t) = 1 \quad \text{if } x \in D(t),$$
$$= 0 \quad \text{if not.}$$

Then, the space-time domain D_T containing the fluid is defined by

$$D_T = \{(x,t) \in \Lambda \times (0,T); \varphi(x,t) = 1\}. \tag{1.1}$$

In the liquid region D_T, the momentum equation is

$$\rho\frac{\partial u}{\partial t} + \rho(u\cdot\nabla)u - \operatorname{div}\sigma^{\text{tot}} = f. \tag{1.2}$$

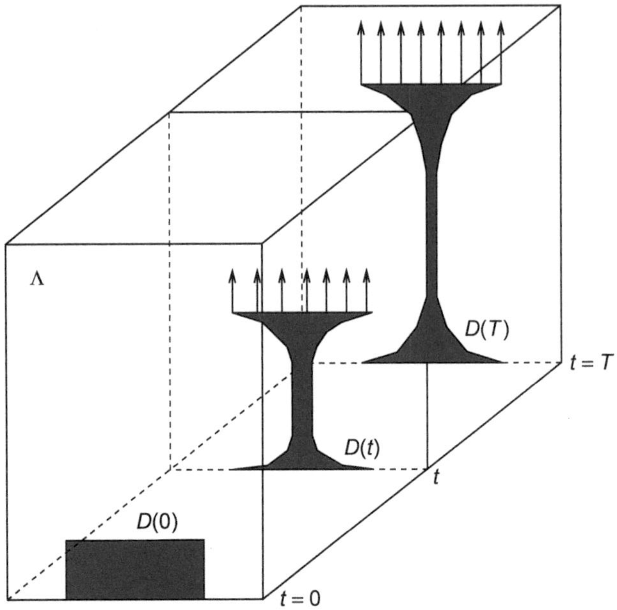

FIG. 1.1 Notations: the stretching of a filament in two space dimensions is considered. At initial time, the viscoelastic fluid is at rest and occupies the domain $D(0)$, which is part of the cavity Λ. At $t > 0$, the upper part of the liquid domain moves at given velocity, and the fluid domain is $D(t)$.

Here, ρ is the fluid density, $u : D_T \to \mathbb{R}^d$ is the fluid velocity, $\sigma^{\text{tot}} : D_T \to \mathbb{R}^{d\times d}$ is the total stress tensor of the fluid, and $f : D_T \to \mathbb{R}^d$ are volume forces, for instance gravity forces $f = \rho g$. Consider the case of a polymeric fluid, that is, a Newtonian solvent plus polymer chains. Then, the total stress is the sum of a Newtonian contribution and a non-Newtonian one

$$\sigma^{\text{tot}} = 2\eta_s \epsilon(u) - pI + \sigma, \tag{1.3}$$

where $\eta_s \geq 0$ is the solvent viscosity, $\epsilon(u) = \frac{1}{2}(\nabla u + \nabla u^T)$ is the symmetric part of the velocity gradient with $(\nabla u)_{ij} = \partial u_i/\partial x_j$, $p : D_T \to \mathbb{R}$ denotes the pressure, I is the unit tensor in $\mathbb{R}^{d\times d}$, and σ is the extra stress (the non-Newtonian part of the stress) due to the polymer chains for instance. Inserting (1.3) into (1.2) and assuming incompressibility yields the following mass and momentum equations

$$\rho\frac{\partial u}{\partial t} + \rho(u\cdot\nabla)u - 2\eta_s \text{div}\, \epsilon(u) + \nabla p - \text{div}\, \sigma = f, \tag{1.4}$$

$$\text{div}\, u = 0, \tag{1.5}$$

in the liquid domain D_T.

In order to obtain the space-time liquid domain D_T defined by (1.1), Lagrangian or Eulerian methods can be advocated. Since our aim is to solve flows in complex domains such as jet buckling or fingering instabilities, we shall consider Eulerian methods so that an

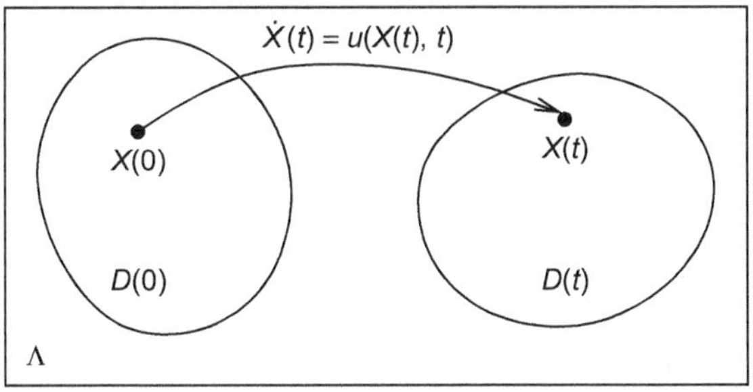

FIG. 1.2 Trajectories of a fluid particle from time 0 to time t. The liquid domain at time t is $D(t)$, and the cavity containing the liquid is Λ.

equation is needed for the characteristic function φ of the liquid region. Again, we have the choice between level set OSHER and FEDKIW [2003], SETHIAN and SMEREKA [2003] methods or volume of fluid (VOF) RIDER and KOTHE [1998], SCARDOVELLI and ZALESKI [1999] methods. We select here the VOF formulation and obtain this equation by assuming that all the fluid particles move with the fluid velocity u. Therefore, given the liquid domain $D(0)$ at time 0, the liquid domain at time t is given by

$$D(t) = \{X(t) \in \Lambda \text{ such that } \dot{X}(t) = u(X(t), t) \text{ with } X(0) \in D(0)\}; \tag{1.6}$$

see Fig. 1.2. Hereabove, it is understood that the velocity u is smooth enough so that the differential equation $\dot{X}(t) = u(X(t), t)$ has a unique solution, for instance u continuous and Lipschitz with respect to the space variable. We now claim that if the function φ satisfies

$$\frac{\partial \varphi}{\partial t} + u \cdot \nabla \varphi = 0 \qquad \text{in } \Lambda \times (0, T), \tag{1.7}$$

in a weak sense and if $\varphi(\cdot, 0)$ is the characteristic function of $D(0)$, then $\varphi(\cdot, t)$ is the characteristic function of $D(t)$. Indeed, the solution of (1.7) is given by

$$\varphi(X(t), t) = \varphi(X(0), 0), \quad \text{where } \dot{X}(t) = u(X(t), t), \quad 0 \le t \le T.$$

Since $\varphi(\cdot, 0)$ is the characteristic function of $D(0)$, we therefore have

$$\varphi(X(t), t) = \varphi(X(0), 0) = 1 \qquad \text{for all } X(0) \in D(0).$$

Using (1.6), we finally obtain

$$\varphi(X(t), t) = 1 \qquad \text{for all } X(t) \in D(t);$$

thus, $\varphi(\cdot, t)$ is the characteristic function of $D(t)$.

Let us summarize the situation. Our goal is to find the characteristic function of the liquid φ in the whole cavity Λ, the velocity u, the pressure p, and the extra stress σ in the liquid region $D(t)$ and satisfying (1.4), (1.5), and (1.7). We still need to provide a relation between u and σ. This can be done by considering either macroscopic or mesoscopic models.

1.1. Macroscopic models

When considering viscoelastic flows at macroscopic scale, one has to choose between differential and integral models. In Chapter 3, numerical results will be presented when considering the simplest of the differential models presented here, namely the Oldroyd-B model. However, a brief presentation of differential and integral models is proposed hereafter. We refer again to BIRD, CURTISS, ARMSTRONG and HASSAGER [1987], LARSON [1999], ÖTTINGER [1996], RENARDY [2000] for classical textbooks and to the contribution of Lozinski and Phillips in this book.

1.1.1. Differential models

The simplest differential model is the so-called Oldroyd-B constitutive equation

$$\sigma + \lambda \left(\frac{\partial \sigma}{\partial t} + u \cdot \nabla \sigma - \nabla u \sigma - \sigma \nabla u^T \right) = 2\eta_p \epsilon(u), \tag{1.8}$$

where $\lambda \geq 0$ is the fluid relaxation time and $\eta_p \geq 0$ is the polymer viscosity. The term $\nabla u \sigma$ denotes the matrix–matrix product between ∇u and σ, and the expression within the parenthesis is the upper convected derivative of σ. When the solvent viscosity η_s vanishes, (1.4), (1.5), and (1.8) are the upper convected Maxwell model. Many other models are available in the literature BIRD, CURTISS, ARMSTRONG and HASSAGER [1987], LARSON [1999]; for instance, the extra stress of the eight modes Oldroyd-B model is defined by $\sigma = \sigma_1 + \cdots + \sigma_8$, where each σ_i, $i = 1, 8$, satisfies (1.8) with λ replaced by the ith relaxation time λ_i. Also, the corotational Oldroyd-B model is obtained by replacing the terms $-\nabla u \sigma - \sigma \nabla u^T$ in (1.8) by

$$\frac{1}{2} \left(\sigma \left(\nabla u - \nabla u^T \right) - \left(\nabla u - \nabla u^T \right) \sigma \right).$$

The Oldroyd-B model can be generalized to models involving more derivatives, for instance the third-order retarded motion model (see BIRD, CURTISS, ARMSTRONG and HASSAGER [1987]). Finally, nonlinear extensions of the Oldroyd-B model have been proposed, for instance the Giesekus, Leonov, and Phan-Thien Tanner models are contained in the general formulation

$$f(\sigma)\sigma + \lambda \left(\frac{\partial \sigma}{\partial t} + u \cdot \nabla \sigma - \nabla u \sigma - \sigma \nabla u^T \right) = 2\eta_p \epsilon(u),$$

where $f(\sigma)$ is a scalar function depending on σ and $tr(\sigma)$. Constitutive equations can be specialized to particular viscoelastic fluids. For instance, the Rolie–Poly model is designed for entangled polymer melts LIKHTMAN and GRAHAM [2003], and the Extended Pom-Pom model VERBEETEN, PETERS and BAAIJENS [2004] has been developed in order to take into account the morphology of branched polymer melts.

Although the Oldroyd-B model (1.8) is too simple to reproduce some of the viscoelastic effects reported in experiments – shear thinning in shear flows for instance – it already contains mathematical difficulties absent in Newtonian flows. Moreover, as we will see in

Section 1.2, this model is linked to the simplest mesoscopic model, namely the Hookean dumbbell model.

1.1.2. Integral models

Following LIN, LIU and ZHANG [2005], the integral (Lagrangian) formulation of the Oldroyd-B model (1.8) is the following. Let $x \in D(0)$ be the initial position of a particle moving with the fluid velocity u. The position of this particle at time t is denoted by $X(t)$ and is the solution at time t of

$$\dot{X}(s) = u(X(s), s) \qquad 0 \le s \le t,$$
$$X(0) = x,$$

or equivalently

$$X(t) = x + \int_0^t u(X(s), s)\mathrm{d}s. \tag{1.9}$$

Let $F : D(0) \times (0, T) \to \mathbb{R}^{d\times d}$ be the deformation tensor defined by

$$F(x, t) = \frac{\partial X}{\partial x}(t) \quad x \in D(0), \quad t \ge 0.$$

Then, the integral formulation of the extra stress for an Oldroyd-B fluid is, in Lagrangian coordinates:

$$\tilde{\sigma}(x, t) = \frac{\eta_p}{\lambda^2}\left(\int_0^t \mathrm{e}^{-(t-s)/\lambda} F(x, t)F^{-1}(x, s)F^{-T}(x, s)F^{T}(x, t)\mathrm{d}s + \lambda(F(x, t)F^{T}(x, t) - I)\right), \tag{1.10}$$

for all $x \in D(0)$ and $t \ge 0$. We now check that (1.10) indeed coincides with the Oldroyd-B model (1.8). Differentiating (1.9) with respect to x and t yields

$$\frac{\partial F}{\partial t}(x, t) = \nabla u(X(t), t)F(x, t) \qquad x \in D(0), \quad t \ge 0,$$

so that $\tilde{\sigma}$ defined by (1.10) satisfies

$$\frac{\partial \tilde{\sigma}}{\partial t}(x, t) + \frac{1}{\lambda}\tilde{\sigma}(x, t) = \nabla u(X(t), t)\tilde{\sigma}(x, t) + \tilde{\sigma}(x, t)\nabla u(X(t), t)^{T} + \frac{\eta_p}{\lambda}\left(\nabla u(X(t), t) + \nabla u(X(t), t)^{T}\right).$$

Finally, we introduce the extra stress in Eulerian coordinates $\sigma : D_T \to \mathbb{R}^{d\times d}$ defined by

$$\sigma(X(t), t) = \tilde{\sigma}(x, t) \qquad x \in D(0), \quad t \geq 0,$$

and check that σ satisfies (1.8).

Several extensions of the integral Oldroyd-B model have been proposed, for instance the famous K-BKZ model (see BIRD, CURTISS, ARMSTRONG and HASSAGER [1987] Section 8.3). Implementing integral models in complex flows requires some additional effort; see for instance KEUNINGS [2003] for a review. Indeed, the particles path has to be stored – at least during some time proportional to the relaxation time λ – which is expensive to implement. Therefore, integral formulations will not be considered in this contribution.

1.2. Mesoscopic models

Consider a polymeric liquid that is a newtonian solvent and polymer chains. Polymers chains are long molecules made out of many identical blocks called monomers. The modeling of liquid polymers ranges from the atomic to the mesoscopic scale.

At the atomic scale, molecular dynamics can be considered in order to study specific problems such as single chains in a flow, rupture of a filament or nanodrops. For instance in KOPLIK and BANAVAR [2003], a polymer melt is considered. A finitely extensible non-linear elastic (FENE) potential applies between two monomers of a given chain, whereas a Lennard–Jones potential acts between two monomers not belonging to the same chain. Such simulations are of interest in localized regions but cannot be performed at the macroscopic level. Another possibility is to consider a kinetic theory of liquid polymers.

At the mesoscopic level, polymer chains can be modeled by a collection of beads connected with springs, the Rouse chain (see Fig. 1.3). When considering a dilute solution of polymers (a spaghetti soup), the chains do not interact each other, but the interaction is only through the Newtonian solvent. When considering a polymer melt (a plate of spaghetti), the chains are entangled, and the movement of the beads is possible only along the chain: the chains reptates. From the industrial viewpoint, most processes involve polymer melts rather than dilute solutions. However, dilute solutions are better understood from the mathematical viewpoint, and we will focus in this contribution on dilute solutions. Moreover, we will even simplify the Rouse chain model and consider the dumbbell model, that is, two beads connected with an elastic spring (see again Fig. 1.3). Due to the increase of computer power, realistic numerical simulations can be nowadays performed on chains; see KEUNINGS [2004] and the references therein.

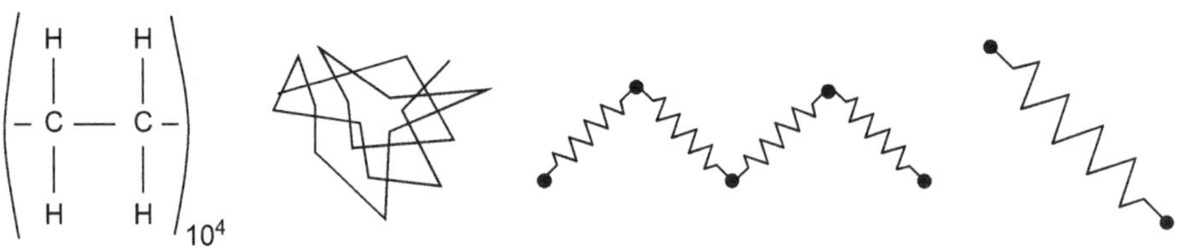

FIG. 1.3 Modeling of polymer chains from nanoscale to mesoscale. From left to right: polyethylene, polymer chain, the Rouse chain, a dumbbell.

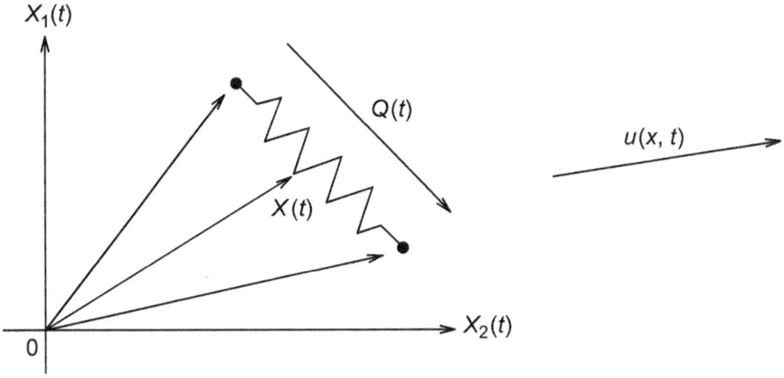

FIG. 1.4 A dumbbell placed in a flow field $u(x,t)$. The beads positions are $X_1(t)$ and $X_2(t)$, the spring elongation is $Q(t) = X_2(t) - X_1(t)$, and the center of mass is $X(t) = \frac{1}{2}(X_1(t) + X_2(t))$.

1.2.1. The Dumbbell model

Consider as in Fig. 1.4 a dumbbell. The two beads positions are denoted by $X_1(t)$, $X_2(t)$, the spring elongation is $Q(t) = X_2(t) - X_1(t)$, and the center of mass is $X(t) = \frac{1}{2}(X_1(t) + X_2(t))$. We now derive a stochastic differential equation for the elongation $Q(t)$. The forces acting on each bead are (i) the drag force that is proportional to the relative velocity $\dot{X}_i(t) - u(X_i(t), t)$ between the velocity of bead i and the fluid velocity, $i = 1, 2$ (ii) the elastic force due to the spring elongation $X_2(t) - X_1(t)$ (iii) the random forces due to thermal agitation and collisions with the solvent $R_i(t)$, $i = 1, 2$. Writing Newton's equations on the beads yields

$$
\begin{aligned}
m\ddot{X}_1(t) &= \xi\left(u(X_1(t), t) - \dot{X}_1(t)\right) + F\left(X_2(t) - X_1(t)\right) + R_1(t),\\
m\ddot{X}_2(t) &= \xi\left(u(X_2(t), t) - \dot{X}_2(t)\right) - F\left(X_2(t) - X_1(t)\right) + R_2(t),
\end{aligned}
$$

where m is the mass of each bead, ξ is the drag coefficient, and F is the force due to spring elongation. Adding and subtracting the two above equations and neglecting inertia leads to

$$
\begin{aligned}
\dot{X}(t) &= \frac{1}{2}\left(u(X_1(t), t) + u(X_2(t), t)\right) + \frac{1}{2\xi}\left(R_1(t) + R_2(t)\right),\\
\dot{Q}(t) &= u(X_2(t), t) - u(X_1(t), t) - \frac{2}{\xi}F\left(Q(t)\right) + \frac{1}{\xi}\left(R_2(t) - R_1(t)\right).
\end{aligned}
$$

The stochastic term $R_1(t) + R_2(t)$ is then neglected, while $R_2(t) - R_1(t)$ is assumed to be proportional to white noise (the formal derivative of Brownian motion). An order one Taylor expansion

$$
u(X_i(t), t) \simeq u(X(t), t) + \nabla u(X(t), t)\,(X_i(t) - X(t)) \qquad i = 1, 2,
$$

yields

$$
\begin{aligned}
\dot{X}(t) &= u(X(t), t),\\
\dot{Q}(t) &= \nabla u(X(t), t) Q(t) - \frac{2}{\xi}F\left(Q(t)\right) + \frac{2k\theta}{\xi}\dot{B}(t),
\end{aligned}
$$

where k is Boltzmann's constant, θ is the absolute temperature, and B is a Wiener process (see for instance REVUZ and YOR [1994] for a definition). We now specialize the spring force F into:

$$\text{Hookean springs: } F(q) = Hq \qquad \forall q \in \mathbb{R}^d,$$

$$\text{FENE springs: } F(q) = H\frac{q}{1 - \dfrac{|q|^2}{Q_0}} \qquad \forall q \in \mathbb{R}^d,\ |q| < \sqrt{Q_0},$$

where H is the spring stiffness. The case of finitely extensible nonlinear elastic (FENE) dumbbells prevents the springs to have an elongation greater than $\sqrt{Q_0}$, which corresponds to the length of the fully elongated chain. A scaling of the spring elongation $Q(t)$ by $\sqrt{k\theta/H}$ yields the following stochastic differential equations

$$\mathrm{d}X(t) = u(X(t), t)\mathrm{d}t, \tag{1.11}$$

$$\mathrm{d}Q(t) = \left(\nabla u(X(t), t)Q(t) - \frac{1}{2\lambda}F(Q(t))\right)\mathrm{d}t + \frac{1}{\sqrt{\lambda}}\mathrm{d}B(t), \tag{1.12}$$

where $\lambda = \xi/4H$ is the relaxation time and the spring force F is now defined by

$$\text{Hookean springs: } F(q) = q \qquad \forall q \in \mathbb{R}^d,$$

$$\text{FENE springs: } F(q) = \frac{q}{1 - \dfrac{|q|^2}{b}} \qquad \forall q \in \mathbb{R}^d,\ |q| < \sqrt{b}.$$

According to ÖTTINGER [1996], the parameter b is linked to the number of monomer units in the polymer chains. Typical values range from $b = 10$ to $b = 1000$. When b is large, FENE dumbbells behave as Hookean dumbbells. The Eulerian equation corresponding to (1.11), (1.12) is

$$\begin{aligned}\mathrm{d}q(x, t, \omega) = \Big(&-u(x, t)\cdot\nabla q(x, t, \omega) + \nabla u(x, t)q(x, t, \omega)\\ &-\frac{1}{2\lambda}F(q(x, t, \omega))\Big)\mathrm{d}t + \frac{1}{\sqrt{\lambda}}\mathrm{d}B(t, \omega),\end{aligned} \tag{1.13}$$

for each (x, t) belonging to the liquid domain D_T and for each event ω in Ω, the space of events. It now remains to provide an expression for the extra-stress tensor σ; see BIRD, CURTISS, ARMSTRONG and HASSAGER [1987] Section 13.3 for details. This expression is

$$\sigma = \frac{\eta_p}{\lambda}\left(\mathbb{E}(F(q)q^T) - I\right), \tag{1.14}$$

where η_p is the polymer viscosity, $\mathbb{E}(\cdot)$ is the mathematical expectation, and $F(q)q^T$ is the symmetric tensor with coefficients $F_i(q)q_j$, $i, j = 1, d$, that is:

$$\text{Hookean springs: } F_i(q)q_j = q_iq_j \qquad \forall q \in \mathbb{R}^d,$$

$$\text{FENE springs: } F_i(q)q_j = \frac{q_iq_j}{1 - \dfrac{|q|^2}{b}} \qquad \forall q \in \mathbb{R}^d,\ |q| < \sqrt{b}.$$

The deterministic formulation corresponding to (1.13), (1.14) is obtained by considering the probability density f for the dumbbell elongations. Here, $f(x, t, q)\mathrm{d}x\mathrm{d}q$ denotes the probability of finding a dumbbell at time t, located between x and $x + \mathrm{d}x$ having elongation between q and $x + \mathrm{d}q$. Following KLOEDEN and PLATEN [1992], ÖTTINGER [1996], REVUZ and YOR [1994] for instance, the probability density f must satisfy the Fokker–Planck equation

$$\frac{\partial f}{\partial t} + \mathrm{div}_x(uf) + \mathrm{div}_q\left((\nabla_x u)qf - \frac{1}{2\lambda}F(q)f\right) = \frac{1}{2\lambda}\mathrm{div}_q(\nabla_q f). \tag{1.15}$$

The deterministic counterpart of (1.14) is then

$$\text{Hookean springs:}\quad \sigma(x, t) = \frac{\eta_p}{\lambda}\left(\int\limits_{q\in\mathbb{R}^d} qq^T f(x, t, q)\mathrm{d}q - I\right),$$

$$\text{FENE springs:}\quad \sigma(x, t) = \frac{\eta_p}{\lambda}\left(\int\limits_{q\in\mathbb{R}^d, |q|<\sqrt{b}} \frac{qq^T}{1 - \frac{|q|^2}{b}} f(x, t, q)\mathrm{d}q - I\right).$$

Finally, we mention the reflected dumbbell BONITO, LOZINSKI and MOUNTFORD [To appear] model, which, roughly speaking, stands in between the Hookean and FENE models. In this new model, the dumbbells are subject to a linear spring force as long as the spring elongation does not exceed $\sqrt{b}$. When this value is reached, the force is modified to prevent further elongation. The latter nonlinear force is mathematically expressed as the subdifferential of the convex potential

$$\Pi(q) = \begin{cases} \frac{1}{2}|q|^2, & \text{if } |q| < \sqrt{b} \\ +\infty, & \text{otherwise} \end{cases}$$

so that the corresponding stochastic PDE for the elongation dumbbells becomes, in fact, a reflected stochastic PDE. A comparison of different numerical algorithms in this context is proposed in BONITO, LOZINSKI and MOUNTFORD [To appear].

1.2.2. Link between Hookean dumbbells and the Oldroyd-B model

One of the striking properties of the Hookean dumbbell model is that it leads to the Oldroyd-B model. Indeed, let q be a solution of (1.13) with $F(q) = q$. Ito's formula (see for instance REVUZ and YOR [1994]) applied to $V = \mathbb{E}(qq^T)$ yields

$$V + \lambda\left(\frac{\partial V}{\partial t} + u \cdot \nabla V - \nabla u V - V \nabla u^T\right) = I. \tag{1.16}$$

Inserting into (1.14), we obtain that the extra-stress σ satisfies exactly (1.8). The same formal calculation can be performed using the Fokker–Planck equation. Indeed, let

$F(q) = q$, multiply (1.15) by qq^T and integrate with respect to the q variable, and then we obtain that

$$V = \int_{q \in \mathbb{R}^d} qq^T f(x, t, q) \mathrm{d}q$$

satisfies (1.16). It should be noted that this formal computation can be justified rigorously and can be extended to the Rouse chain. For instance, the Rouse chain with nine beads and eight Hookean springs is equivalent to the eight modes Oldroyd-B model with appropriate relaxation times λ_i, $i = 1, 8$; see Section 15.3 in BIRD, CURTISS, ARMSTRONG and HASSAGER [1987].

The formal equivalence between Hookean dumbbells and the Oldroyd-B model has been historically used in order to derive macroscopic models arising from this kinetic theory. For instance, the FENE-P model is obtained by setting the springs forces F and the extra-stress σ to

$$F(q) = \frac{q}{1 - \dfrac{\mathbb{E}(|q|^2)}{b}} \qquad \text{and} \qquad \sigma = \frac{\eta_p}{\lambda}\left(\frac{\mathbb{E}(qq^T)}{1 - \dfrac{\mathbb{E}(|q|^2)}{b}} - I\right).$$

Let $V = \mathbb{E}(qq^T)$, $tr(V) = \mathbb{E}(|q|^2)$. Using again formal stochastic calculus, we obtain

$$\frac{V}{1 - \dfrac{tr(V)}{b}} + \lambda\left(\frac{\partial V}{\partial t} + u \cdot \nabla V - \nabla u V - V \nabla u^T\right) = I,$$

the extra stress being now defined by

$$\sigma = \frac{\eta_p}{\lambda}\left(\frac{V}{1 - \dfrac{tr(V)}{b}} - I\right).$$

The FENE model has no macroscopic counterpart. However, using expansions of the probability density f in powers of λ yields the retarded motion model; see for instance Section 13.5 in BIRD, CURTISS, ARMSTRONG and HASSAGER [1987] or DEGOND, LEMOU and PICASSO [2002]. We also refer to DU, LIU and YU [2005] for recent, high-order approximations of FENE dumbbells.

Due to the increase of computers power, mesoscopic models have been solved numerically in order to obtain more realistic results; see KEUNINGS [2004] for a review. Both the deterministic and stochastic formulations of this kinetic theory have been considered. We now discuss which of the two formulations should be used when performing numerical simulations of viscoelastic flows with dumbbells or chains.

For dumbbells, the kinetic variable $q(.,.,.) \in \mathbb{R}^d$, $d = 2, 3$, and it is not clear which of the deterministic or stochastic formulations is in principle more efficient from the computational point of view. For chains, that is to say when considering several beads connected by springs, the stochastic formulation should be more efficient than the deterministic one, for the reasons detailed hereafter.

Consider a chain with N springs and $N+1$ beads, then the kinetic variable $q(.,.,.) \in \mathbb{R}^{d\times N}$, $d = 2, 3$. The stochastic formulation of a chain leads to a stochastic differential equation similar to (1.13) and to an expression of the extra-stress similar to (1.14). The Monte Carlo method is used to approach the expectation

$$\mathbb{E}\left(F(q)q^T\right) \simeq \frac{1}{M}\sum_{m=1}^{M} F(q_m)q_m^T,$$

where the q_m, $m = 1, \ldots, M$, are independent copies of the stochastic process q. Assuming the velocity $u(x, t)$ to be a known quantity at a given point (x, t) in the space-time domain, both the number of degrees of freedom and the computational cost required to solve (1.13) and (1.14) are $O(dNM)$, the convergence rate being $O(M^{-1/2})$ from the central limit theorem. On the other side, the deterministic formulation (1.15) requires a grid of $\mathbb{R}^{dN}$. If the grid is uniform with mesh size h, then the number of degrees of freedom is $O(h^{-dN})$, the computational cost is at least the same, and the rate of convergence of the error is $O(h^r)$ depending on the method used (for instance, $r = 2$ when using standard order two centered finite differences). Therefore, the rate of convergence is the same in both methods provided $h^r = O(M^{-1/2})$, and the number of degrees of freedom of the deterministic method is then $O(M^{dN/2r})$). We thus conclude that the Monte Carlo method is favorable if $dN \geq 2r$, that is to say with long chains.

Recently, the sparse tensor product method has been proposed in order to solve parabolic equations in high dimensions GRIEBEL [2006], VON PETERSDORFF and SCHWAB [2004]. When applied to (1.15), the number of degrees of freedom could be reduced from $O(h^{-dN})$ to $O(h^{-1}|\log h|^{dN-1})$ without error increase so that the method could be competitive with long chains (see for instance DELAUNAY, LOZINSKI and OWENS [2007]).

1.3. Initial and boundary conditions

Let us go back to the free-surface problem described at the beginning of this chapter, namely (1.4), (1.5), (1.7), supplemented by the macroscopic model (1.8) or by the mesoscopic model (1.13) (1.14). We now discuss initial and boundary conditions for these two problems.

At initial time, the characteristic function of the liquid region $\varphi(0) : \Lambda \to \mathbb{R}$ is prescribed, which defines the initial liquid region $D(0)$. The velocity field in the liquid region $u(0) : D(0) \to \mathbb{R}^d$ is then prescribed. When considering macroscopic models (the Oldroyd-B model (1.8) for instance), the initial extra-stress $\sigma(0) : D(0) \to \mathbb{R}^d$ is also prescribed. When dumbbells are considered, Eqns (1.13) and (1.14), the initial elongation $q(0)$ – a stochastic variable – is also prescribed. In general, initial conditions correspond to zero extra-stress that is a $\mathcal{N}(0, I)$ random variable (a normal distribution with zero mean and unit variance) for $q(0)$.

Let us now consider the boundary conditions. For the sake of clarity, two test cases are considered, namely the filling of a cavity with a viscoelastic jet and the stretching of a viscoelastic filament (see Fig. 1.5).

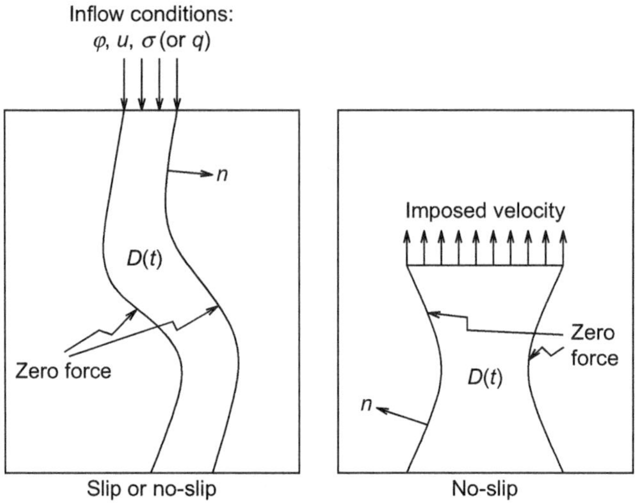

FIG. 1.5 Boundary conditions. Left: filling of a cavity with a viscoelastic jet. Right: stretching of a viscoelastic filament. The liquid domain at time t is $D(t)$.

The boundary conditions for the velocity field are the following. It is assumed that no external forces act on the liquid–air free surface, and effects of surface tension are neglected so that

$$-pn + (2\eta_s\epsilon(u) + \sigma)\,n = 0, \tag{1.17}$$

where n is the unit outer normal of the liquid–air free surface. On the boundary of the liquid domain being in contact with the walls, either slip, imposed, or no-slip (that is $u = 0$) boundary conditions apply. Slip boundary conditions correspond to zero normal velocity $u \cdot n = 0$ and zero tangent stress

$$(-pn + (2\eta_s\epsilon(u) + \sigma)\,n) \cdot t = 0,$$

where t is the unit outer vector tangent to the boundary of the cavity (two tangent vectors are to be considered in three space dimensions).

Let us now consider a macroscopic model, for instance the Oldroyd-B model (1.8). We define the inflow boundary

$$\{x \in \partial\Lambda \cap \partial D(t);\, u(x,t) \cdot n(x,t) < 0\}.$$

Since Eqns (1.7) and (1.8) are transport equations, both φ and σ are to be imposed at the inflow boundary. Similarly, when considering the mesoscopic dumbbell model (1.13) and (1.14), both φ and q are to be imposed at the inflow boundary.

1.4. Summary

The mathematical models considered in this contribution are the (macroscopic) Oldroyd-B model and the (mesoscopic) FENE dumbbell model.

Given a cavity containing an Oldroyd-B fluid, the free-surface model consists in finding the characteristic function of the liquid φ, the velocity u, the pressure p, and the extra-stress σ in the liquid satisfying (1.4), (1.5), (1.7), and (1.8), with appropriate initial and boundary conditions.

Alternatively, the free-surface FENE dumbbell model consists in finding the characteristic function of the liquid φ, the velocity u, the pressure p, the extra-stress σ, and the dumbbell elongations q in the liquid satisfying (1.4), (1.5), (1.7), (1.13), and (1.14), with appropriate initial and boundary conditions.

CHAPTER 2

Numerical Analysis of Simplified Problems

Our goal is now to design a numerical method for solving viscoelastic flows with complex free surfaces. Following GLOWINSKI [2003], an operator splitting method is used for the time discretization. The prediction step consists in solving convection problems only. Then, the new liquid domain is obtained, and the correction step consists in solving a viscoelastic flow problem (either macroscopic or mesoscopic) without convection and in a prescribed domain.

In the next section, we propose a review of numerical methods for viscoelastic flows, with emphasis on finite elements. Then, the splitting algorithm is proposed in Section 2.2, which allows convection to be decoupled from the other physical phenomena. In Section 2.3 we present some results pertaining to the so-called three fields Stokes problem. Finally, we propose an existence and convergence result for the problem involved in the correction step of the splitting algorithm. The corresponding Oldroyd-B problem is considered in Section 2.4, and the Hookean dumbbell problem in Section 2.5.

2.1. Numerical models for viscoelastic flows: a chronological review

2.1.1. Numerical computations

Macroscopic models Following CROCHET and WALTERS [1983], the first papers reporting numerical computations of viscoelastic flows in two space dimensions were published in the mid 1970s. The finite element method became rapidly a method of choice in order to compute flows past submerged obstacles or contractions, and flows with simple free surfaces, die swell for instance. In the early 1980s, the increase of computer power allowed mesh refinement, but numerical oscillations were soon reported BERIS, ARMSTRONG and BROWN [1984], CROCHET and KEUNINGS [1982], MENDELSON, YEH, BROWN and ARMSTRONG [1982]. A typical choice at the time was to use Galerkin finite elements on quadrilateral (no upwinding), with continuous functions, piecewise quadratic/linear for the velocity/pressure (the famous $Q_2 - P_1$ stable element, see GIRAULT and RAVIART [1986]) and continuous, piecewise quadratic stresses.

Rapidly, researchers came across the so-called high Deborah/Weissenberg number problem. Here, the Deborah/Weissenberg number We is a dimensionless number measuring the elastic behavior of the flow $\mathrm{We} = \lambda V/L$, where λ is the relaxation time present in the

Oldroyd-B constitutive equation (1.8), V is a characteristic velocity, and L is a characteristic length. Quoting CROCHET and WALTERS [1983] in 1983, "There is no doubt that presently the outstanding problem in the numerical simulation of viscoelastic flows concerns the upper limit on the nondimensional parameter W (found in all published works) above which the numerical algorithms fail to converge (...) The limit is relatively low, so low in fact that many of the important and dramatic experimental results fall outside the present scope of numerical simulation (...) The limit on W is common to all published works. It applies to finite-difference or finite-element techniques, to differential and integral constitutive models, and to flows with and without abrupt changes in geometry. Some suggestions for possible causes of the W barrier are the bifurcation phenomena, the unsuitability of the constitutive models, and the failure of the iterative numerical schemes."

In KEUNINGS [1986], the 4:1 axisymmetric contraction flow of upper convected Maxwell and Leonov fluids is computed, and the influence of mesh refinement is discussed, five meshes being used. The limit on We was clearly depending on the mesh size. In BROWN, ARMSTRONG, BERIS and YEH [1986], on the same test case, the limit on We was decreasing with mesh size. This phenomena disappeared when smoothing the reentrant corner of the contraction. Still in BROWN, ARMSTRONG, BERIS and YEH [1986], on the eccentric rotating cylinder test case, a bifurcation point was clearly identified for large We numbers. The use of upwinding techniques was advocated in order to discretize the transport term $u \cdot \nabla\sigma$ in the constitutive equation (1.8).

The SUPG method was introduced in MARCHAL and CROCHET [1987] in the framework of viscoelastic flows. In the same paper, the $Q_2 - P_1$ finite element was used for the velocity-pressure approximation, and each quadrangle was cut into 4×4 smaller quadrangles, and the extra stress was continuous, piecewise linear in each of these small quadrangles. Later, this finite element was proved to be stable and convergent (in the sense of the inf-sup condition) for the three fields Stokes problem FORTIN and PIERRE [1989]; see also Section 2.3 hereafter.

Discontinuous stresses were introduced in FORTIN and FORTIN [1989] for quadrangular elements. The velocity was continuous, piecewise quadratic, and the pressure was discontinuous, piecewise linear so that the element satisfies the inf-sup condition for the Stokes problem. The extra stress was discontinuous, piecewise quadratic, and the Lesaint-Raviart upwinding technique was used LESAINT and RAVIART [1974].

The so-called elastic viscous split stress (EVSS) method enabled the use of low-order finite elements for the extra stress, and we refer for instance to BAAIJENS [1998] for a review paper. The idea of EVSS was to add to the set of equations a new unknown d such that $d = \epsilon(u)$, for stability purposes. The analysis of FORTIN, GUÉNETTE and PIERRE [2000] proved that low-order finite elements could be used for the extra stress, while keeping inf-sup stable elements for the velocity-pressure only. The link with stabilized (Galerkin least square) formulations was proposed in BONVIN, PICASSO and STENBERG [2001].

High-order elements were also considered to compute viscoelastic flows. For instance, high-order methods were considered in CHAUVIÈRE and OWENS [2000], FAN [2003] for the flow of a falling sphere in a cylindrical tube, and the same limit on We was found in both papers. Spectral elements for time-dependent viscoelastic flows were studied in FIÉTIER and DEVILLE [2002a]. Finite volume methods have also been successfully employed for solving viscoelastic flows, with the same limitations of the We number (see for instance MOMPEAN and DEVILLE [2000], and ABOUBACAR and WEBSTER [2003]).

Nowadays, the high Deborah/Weissenberg number problem is still under debate. For instance, the high-resolution parallel computations performed in KIM et al. [2005] for the planar 4:1 contraction flow still report decreasing We with decreasing mesh size. Apparently, this phenomena seems to disappear with the corotational Maxwell model SANDRI [2005]. This numerical observation is consistent with the fact that from the mathematical viewpoint, the corotational Maxwell model is particular (see Section 2.1.3. Also, when the re-entrant corner is rounded-off with a small radius, accurate spectral computations FIÉTIER and DEVILLE [2002b] have shown that instabilities disappear. A numerical scheme satisfying an L^1 estimate and positive definiteness of the extra stress was proposed in LEE and XU [2006], LOZINSKI and OWENS [2003]. However, numerical results CHAUVIÈRE and LOZINSKI [2003] confirmed a limiting Deborah/Weissenberg number when using the scheme proposed in LOZINSKI and OWENS [2003]. Computations with high Deborah/Weissenberg numbers were recently performed in HULSEN, FATTAL and KUPFERMAN [2005] using a log-based evolution equation, although mesh convergence was not certified. In WAPPEROM and RENARDY [2005], the flow past a cylinder is considered at high Weissenberg numbers, using a prescribed velocity field. Boundary layers of size $O(\mathrm{We}^{-5})$ are obtained, showing that extremely small mesh size is required in order to compute the stress with sufficient accuracy. Also, a stabilization by jump of the gradients coupled with a nonlinear artificial viscosity shock-capturing type has been introduced in BONITO and BURMAN [2008] increasing the Weissenberg number limit. Finally, a defect correction method was advocated in ERVIN and HOWELL [2008], TREBOTICH, COLELLA and MILLER [2005] in order to reach high Weissenberg numbers for the planar 4:1 contraction flow.

Mesoscopic models Until 1993, the kinetic theory of liquid polymers was evaluated on simple flows (shear, extension), the velocity gradient being a known, prescribed quantity. In 1993, the stochastic formulation of FENE dumbbell model was solved numerically for planar shear flow LASO and ÖTTINGER [1993], the velocity field and dumbbell elongations being coupled for the first time. The goal of solving the kinetic theory was to obtain more realistic results with mesoscopic models and, eventually, to circumvent the high Deborah/Weissenberg number problem. In FEIGL, LASO and ÖTTINGER [1995], LASO, PICASSO and ÖTTINGER [1997], two-dimensional Lagrangian computations were presented, and Eulerian computations were proposed in HALIN, LIELENS, KEUNINGS and LEGAT [1998], HULSEN, VAN HEEL and VAN DEN BRULE [1997] and showed to be more efficient. The use of variance reduction techniques was clearly demonstrated in BONVIN and PICASSO [1999, 2001, 2002], JOURDAIN, LELIÈVRE and LE BRIS [2004b], ÖTTINGER, VAN DEN BRULE and HULSEN [1997]. The heterogeneous multiscale method was applied in LI, VANDEN-EIJNDEN and ZHANG [2004] to the framework of FENE dumbbells in one and two space dimensions.

The deterministic formulation of FENE dumbbells was also considered. An efficient spectral element method was used in LOZINSKI and CHAUVIÈRE [2003] for coupling the mass and momentum equations to the Fokker–Planck equation (1.15).

Mesoscopic computations have been extended to more complex models such as chains KOPPOL, SURESHKUMAR and KHOMAMI [2007], reptation models (see the review paper KEUNINGS [2004] and the references therein). These methods are first attempts towards more realistic models. We refer to KRÖGER [2004] for a general physical picture of "micro–macro" models for polymers.

2.1.2. *Mathematical analysis*

Notations For simplicity, the notations will be abridged as follow whenever there is no possible confusion. Let D be a domain of $\mathbb{R}^d$, $d = 2, 3$. For a real number $1 \leq p < +\infty$ (resp. $p = \infty$), $L^p(D)$ denotes the space of $p-$power integrable functions (resp. essentially bounded functions) defined on D with values in $\mathbb{R}$, $\mathbb{R}^d$ or $\mathbb{R}^{d\times d}$. Also, for a positive integer m, a real number $1 \leq p \leq +\infty$, $W^{m,p}(D)$ denotes the usual corresponding Sobolev space, that is, the space of functions defined on D with derivative up to mth order in $L^p(D)$. When $p = 2$, these spaces are the Hilbert spaces denoted as $H^m(D)$. As usual, the space $H_0^1(D)$ denotes the space of $H^1(D)$ velocities vanishing on the boundary $\partial\Omega$, whereas $L_0^2(D)$ denotes the space of $L^2(D)$ pressures with zero mean. The dual of $H_0^1(D)$ is denoted by $H^{-1}(D)$. Then, the notation $(\cdot,\cdot)_D$ stands for the $L^2(D)$ scalar product for scalar, vectors, or tensors, with induced norm $\|\cdot\|_{L^2(D)}$.

Given $T > 0$, a Banach space B and a positive integer m, the space of functions defined on $[0, T]$ with values in B, continuous, with (time) derivatives up to mth order also continuous is denoted by $C^m([0, T]; B)$. Also, for a positive integer m and a real number $0 < \mu < 1$, $C^{m+\mu}([0, T]; B)$ stands for the corresponding Hölder space, whereas $h^{m+\mu}([0, T]; B)$ stands for the little Hölder space; see for instance LUNARDI [1995] for a definition. For a real number $1 \leq p \leq +\infty$, $L^p(0, T; B)$ denotes the standard Bochner space. Finally, for a positive integer m, a real number $1 \leq p \leq +\infty$, $W^{m,p}(0, T; B)$ is the space of functions defined on $[0, T]$ with values in B having (time) derivatives up to mth order in $L^p(0, T; B)$.

Macroscopic models From the mathematical point of view, the high Deborah/ Weissenberg number problem translates into the fact that no a priori estimates are available in adequate norms. Indeed, consider the Oldroyd-B problem (1.4), (1.5), and (1.8), in a prescribed domain D with $u = 0$ on the boundary, take the weak formulation, and choose u, p, and σ as test functions. The following formal estimate is then obtained

$$\begin{aligned}\frac{\rho}{2}\frac{\mathrm{d}}{\mathrm{d}t}\|u\|^2_{L^2(D)} + 2\eta_s\|\epsilon(u)\|^2_{L^2(D)} + \frac{\lambda}{2\eta_p}\frac{\mathrm{d}}{\mathrm{d}t}\|\sigma\|^2_{L^2(D)} + \frac{1}{2\eta_p}\|\sigma\|^2_{L^2(D)}\\ = \frac{\lambda}{2\eta_p}\int_D tr\left(\left(\nabla u\sigma + \sigma\nabla u^T\right)\sigma\right),\end{aligned} \tag{2.1}$$

which is not sufficient to obtain global existence for any data using energy methods.

The mathematical analysis of macroscopic viscoelastic flows started in 1985 with the study of the change of type (elliptic/hyperbolic) in the upper convected Maxwell model JOSEPH, RENARDY and SAUT [1985]. The existence of a strong solution to the steady flow of an upper convected Maxwell fluid in differential form was proved in RENARDY [1985a]. More precisely, given a smooth bounded domain D of $\mathbb{R}^3$, given an integer $m \geq 1$, and given $f \in H^m(D)$ sufficiently small, there exists a stationary solution

$$u \in H^{m+2}(D), \quad p \in H^{m+1}(D), \quad \sigma \in H^{m+1}(D),$$

of (1.4), (1.5), and (1.8) with $\eta_s = 0$. Extensions to Oldroyd-B (with several relaxation modes), Giesekus, and Phan-Thien Tanner fluids were also proposed. The same technique was used in RENARDY [1985b] to prove existence of the K-BKZ integral model.

Existence for the time-dependent flow of an Oldroyd-B fluid has been first addressed in GUILLOPÉ and SAUT [1990]. Local existence of strong solutions was proved, so as global existence for small data. More precisely, given a smooth bounded domain D of $\mathbb{R}^3$, given the final time $T > 0$, and given initial velocity $u_0 \in H^2(D) \cap H_0^1(D)$, initial extra-stress $\sigma_0 \in H^2(D)$, and source term $f \in L^\infty(0, T; H_0^1(D))$, $\partial f/\partial f \in L^\infty(0, T; H^{-1}(D))$, sufficiently small in their respective spaces, there exists a solution

$$u \in L^2(0, T; H^3(D)), \quad \frac{\partial u}{\partial t} \in L^2(0, T; H^1(D)), \tag{2.2}$$

$$p \in L^2(0, T; H^2(D)), \quad \sigma \in \mathcal{C}^1(0, T; H^2(D)) \tag{2.3}$$

of (1.4), (1.5), and (1.8) when $\eta_s > 0$. Extensions to Jeffreys fluids can be found in HAKIM [1994]. The case of exterior problems is considered in NOVOTNÝ, SEQUEIRA and VIDEMAN [1999]. Generalizations to Banach spaces and a review can be found in FERNÁNDEZ-CARA, GUILLÉN and ORTEGA [2002]. A necessary condition for blow up is provided in CHEMIN and MASMOUDI [2001].

Existence of a weak solution for any data has been proved in LIONS and MASMOUDI [2000], but for the corotational Oldroyd-B model only. The case of the corotational Oldroyd-B model is particular since the right-hand side in (2.1) must be replaced by

$$\frac{\lambda}{4\eta_p} \int_D tr\left(\left(\sigma\left(\nabla u - \nabla u^T\right) - \left(\nabla u - \nabla u^T\right)\sigma\right)\sigma\right),$$

which cancels. We also refer to LIN, LIU and ZHANG [2005], LIU and WALKINGTON [2001] for related work.

In LEE and XU [2006], LOZINSKI and OWENS [2003], it is noted that taking the trace of (1.8) yields an $L^1(D)$ estimate for the extra stress. However, this estimate does not seem to be sufficient to prove the well-posedness of the Oldroyd-B problem for any data. Estimates involving log-Sobolev inequalities can be found in HU and LELIEVRE [2007]. An example of nonintegrable extra stress for high Deborah/Weissenberg numbers can be found in SANDRI [1999].

Mesoscopic models The mathematical analysis of mesoscopic models started in 1991. Existence of a solution for a deterministic nonlinear dumbbell problem was obtained in RENARDY [1991]. The solvent viscosity was zero, and the FENE formulation was not included in the theory. The complete analysis and numerical analysis of a one-dimensional (stochastic) Hookean dumbbell problem were proposed in JOURDAIN, LELIÈVRE and LE BRIS [2002]; see also E, LI and ZHANG [2002] for a similar study.

Existence of a solution for FENE dumbbells (still in one space dimension) was proposed in JOURDAIN, LELIÈVRE and LE BRIS [2004a]. Existence of nonlinear (stochastic) dumbbells problem in $[0, 1]^d$ with periodic boundary conditions was proposed in E, LI and ZHANG [2004]; however, the analysis does not apply to FENE dumbbells. A similar result was obtained in LI, ZHANG and ZHANG [2004] for the corresponding deterministic formulation. The analysis of the transport term in the dumbbell equations was proposed in LE BRIS and LIONS [2004]. The well posedness of the FENE deterministic equation (without coupling with the mass and momentum equations) was considered in DU, LIU and YU [2005]. The well posedness of a modified deterministic dumbbell problem (including the FENE formulation) was considered in BARRETT, SCHWAB and SÜLI [2005], BARRETT and SÜLI [2007]. Existence of a weak solution could be proved provided the gradient of the velocity field in the Fokker–Planck equation (1.15) was mollified. See also ZHANG, ZHANG and ZHANG [2008] where same techniques are applied to the Hookean dumbbell model. Local existence of the deterministic FENE dumbbell model was obtained in ZHANG and ZHANG [2006]. In JOURDAIN, LE BRIS, LELIEVRE and OTTO [2006], it was proved that convergence to a stationary solution could be obtained for FENE dumbbells, whereas Hookean dumbbells are unstable. In HU and LELIEVRE [2007], new entropy estimates involving log-Sobolev inequalities are proved for FENE dumbbells. Existence of a weak solution of the deterministic corotational FENE dumbbell model has been proved for any data in LIN, ZHANG and ZHANG [2008], LIONS and MASMOUDI [2007].

2.1.3. Numerical analysis

Macroscopic models To the author's knowledge, the first finite element analysis pertaining to viscoelastic flows in two space dimensions was published in 1989 FORTIN and PIERRE [1989]. The so-called three fields Stokes problem was considered, setting $\partial\sigma/\partial t = 0$, $\eta_s = 0$, and $\lambda = 0$ in (1.4), (1.5), and (1.8) to obtain

$$-\text{div}\,\sigma + \nabla p = f, \quad \text{div}\,u = 0, \quad \sigma - 2\eta_p\epsilon(u) = 0.$$

It was proved that the continuous $Q_2 - P_1 - 16Q_1$ finite element proposed in MARCHAL and CROCHET [1987] was stable (in the sense of Brezzi's inf-sup condition) and convergent with optimal order. More precisely, let h be the typical mesh size, and let u_h, p_h, σ_h be the finite element approximations of u, p, σ, respectively. Then, the following a priori error estimate holds

$$\begin{aligned}&\|u - u_h\|_{H^1(D)} + \|p - p_h\|_{L^2(D)} + \|\sigma - \sigma_h\|_{L^2(D)}\\&\quad \le Ch^2\left(\|u\|_{H^3(D)} + \|p\|_{H^2(D)} + \|\sigma\|_{H^2(D)}\right),\end{aligned}$$

where C is independent of the mesh size h and of the exact solution u, p, σ.

Convergence of a finite element discretization for the Oldroyd-B (nonlinear) stationary problem corresponding to (1.4), (1.5), and (1.8) was first proved in BARANGER and SANDRI [1992a], the transport term being disregarded in the momentum equation. Triangular elements were considered, the velocity/pressure being continuous piecewise quadratic/linear, and the extra stress was discontinuous piecewise linear so that the element was stable for the three fields Stokes problem. Moreover, the transport term in the extra-stress constitutive equation was discretized using the method of LESAINT and RAVIART [1974]. Assuming that $\eta_s > 0$ and that the solution (u, p, σ) of the continuous problem was small in the $H^3(D) \times H^2(D) \times H^2(D)$ norm, the authors proved existence and uniqueness of a finite element solution in a $O(h^{3/2})$ neighbourhood of (u, p, σ), so as an optimal $O(h^{3/2})$ convergence rate in the $H^1(D) \times L^2(D) \times L^2(D)$ norm.

Other finite element spaces were considered in RUAS, CARNEIRO DE ARAÚJO and SILVA RAMOS [1993], SANDRI [1993] for the three fields Stokes problem and in SANDRI [1994] for the Oldroyd-B stationary problem. A numerical algorithm decoupling velocity/pressure and extra-stress computations was analyzed in NAJIB and SANDRI [1995].

Convergence of a space-time discretization for the time-dependent Oldroyd-B model was first considered in BARANGER and WARDI [1995]. An implicit Euler scheme was considered for the time discretization, together with triangular finite elements (continuous, piecewise quadratic/linear velocity/pressure; discontinuous, piecewise linear extra stress). Assuming that $\eta_s > 0$ and that the solution (u, p, σ) of the continuous problem was small in the norm

$$\left(\mathcal{C}^1(0,T;H^3(D)) \cap \mathcal{C}^2(0,T;L^2(D))\right) \times \mathcal{C}^0(0,T;H^2(D))$$
$$\times \left(\mathcal{C}^1(0,T;H^2(D)) \cap \mathcal{C}^2(0,T;L^2(D))\right),$$

assuming the stability condition $\Delta t \leq C_1 h^{3/2}$ between the time step and the mesh size, the authors proved existence and an optimal convergence rate

$$\left(\Delta t \sum_{n=0}^{N} \|u(t^n) - u_h^n\|_{H^1(D)}^2\right)^{1/2} + \left(\Delta t \sum_{n=0}^{N} \|p(t^n) - p_h^n\|_{L^2(D)}^2\right)^{1/2}$$
$$+ \left(\Delta t \sum_{n=0}^{N} \|\sigma(t^n) - \sigma_h^n\|_{L^2(D)}^2\right)^{1/2} \leq C(h^{3/2} + \Delta t),$$

with C independent of the mesh size h and time step Δt. Other results pertaining to the numerical analysis of time-dependent problems were obtained in BENSAADA and ESSELAOUI [2005], ERVIN and HEUER [2004], ERVIN and MILES [2003], MACHMOUM and ESSELAOUI [2001], SARAMITO [1994].

The convergence of high-order methods, more precisely *hp* methods, for the three fields Stokes problem was performed in SCHWAB and SURI [1999].

A posteriori error estimates have been proposed for instance in ERVIN and NTASIN [2005], JIN and TANNER [1994], NAJIB, SANDRI and ZINE [2004], OWENS [1998], PICASSO and RAPPAZ [2001].

Mesoscopic models The numerical analysis of mesoscopic models is recent. Since the Hookean dumbbell model is formally equivalent to the Oldroyd-B problem (see Section 1.2.2), it is expected that a space discretization, which is convergent for macroscopic models, should also be convergent for mesoscopic models. This conjecture has been observed in numerical computations, but, up to the author's knowledge, there is no convergence proof for FENE dumbbells in two or three space dimensions.

In JOURDAIN, LELIÈVRE and LE BRIS [2002], LELIÈVRE [2004], the complete analysis and numerical analysis of the Hookean dumbbell problem are performed in the framework of a one-dimensional shear flow. The error due to time and space discretization is considered, and the analysis of the Monte Carlo method is also included. A similar study can be found in E, LI and ZHANG [2002]. For the sake of clarity, we briefly report hereafter some of the results obtained in JOURDAIN, LELIÈVRE and LE BRIS [2002], LELIÈVRE [2004].

Consider the shear flow of an Hookean dumbbell fluid between two parallel infinite planes, the lower plane being at rest, the upper plane moving at imposed velocity (see Fig. 2.1). Let $u(x, t)$ be the horizontal velocity and $P(t, \omega)$ and $Q(x, t, \omega)$ be the horizontal

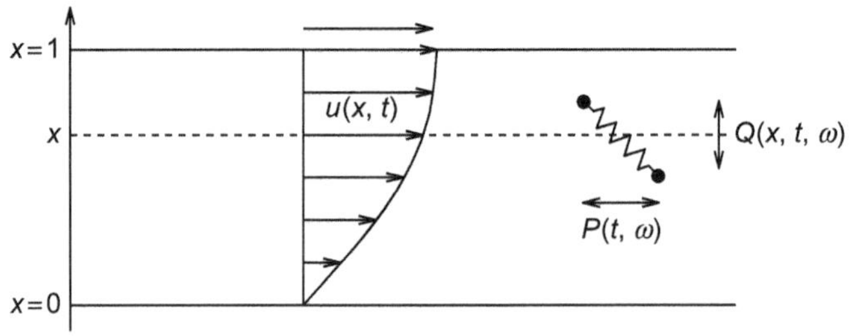

FIG. 2.1 Hookean dumbbells in a Couette flow.

and vertical dumbbell elongation ($\omega \in \Omega$ the space of events). Then, after a lifting of the boundary conditions, Eqns (1.4), (1.5), (1.13), and (1.14) with $F(q) = q$ reduce to

$$\rho \frac{\partial u}{\partial t} - \eta_s \frac{\partial^2 u}{\partial x^2} - \frac{\eta_p}{\lambda} \frac{\partial}{\partial x} \mathbb{E}(PQ) = f, \tag{2.4}$$

$$\mathrm{d}P(t, \omega) = -\frac{1}{2\lambda} P(t, \omega)\mathrm{d}t + \frac{1}{\sqrt{\lambda}} \mathrm{d}V(t, \omega), \tag{2.5}$$

$$\mathrm{d}Q(x, t, \omega) = \left(\frac{\partial u}{\partial x}(x, t) P(t, \omega) - \frac{1}{2\lambda} Q(x, t, \omega) \right) \mathrm{d}t + \frac{1}{\sqrt{\lambda}} \mathrm{d}W(t, \omega), \tag{2.6}$$

where V and W are two independent Wiener processes and f is due to the lifting of the boundary conditions. From the mathematical viewpoint, the shear flow simplifies considerably the model since the quadratic terms analogous to $\nabla u \sigma + \sigma \nabla u^T$ in (1.8) disappeared. Indeed, since P is given by

$$P(t, \omega) = \mathrm{e}^{-t/2\lambda} P(0, \omega) + \int_0^t \mathrm{e}^{-(t-s)/2\lambda} \mathrm{d}V(s, \omega),$$

the terms $\partial/\partial x\, \mathbb{E}(PQ)$ in (2.4) and $\partial u/\partial x\, P$ in (2.6) are linear rather than quadratic. Formal a priori estimates can be obtained. Indeed, taking the weak formulation corresponding to (2.4), choosing u as a test function, we obtain

$$\frac{\rho}{2} \frac{\mathrm{d}}{\mathrm{d}t} \|u\|^2_{L^2(0,1)} + \eta_s \|\frac{\partial u}{\partial x}\|^2_{L^2(0,1)} + \frac{\eta_p}{\lambda} \int_0^1 \mathbb{E}(PQ) \frac{\partial u}{\partial x} \mathrm{d}x = \int_0^1 fu\mathrm{d}x.$$

We then invoke Ito's formula and obtain after computing the expectation

$$\frac{1}{2} \frac{\mathrm{d}}{\mathrm{d}t} \|\mathbb{E}(Q^2)\|^2_{L^2(0,1)} + \frac{1}{2\lambda} \|\mathbb{E}(Q^2)\|^2_{L^2(0,1)} = \int_0^1 \mathbb{E}(PQ) \frac{\partial u}{\partial x} \mathrm{d}x + \frac{1}{2\lambda}.$$

Multiplying the above equation by η_p/λ and summing with the previous one yield an a priori estimate for

$$\|u\|_{L^\infty(0,T;L^2(0,1))} + \left\|\frac{\partial u}{\partial x}\right\|_{L^2(0,T;L^2(0,1))} + \|Q\|_{L^\infty(0,T;L^2(0,1;L^2(\Omega)))}.$$

Existence of a weak solution can be proved in using the Faedo–Galerkin method. The authors of JOURDAIN, LELIÈVRE and LE BRIS [2002], LELIÈVRE [2004] then consider a space, time, and Monte Carlo discretization of (2.4)–(2.6). Assuming sufficient regularity of the data and that the time step Δt is small enough, they prove the following convergence rate:

$$\begin{aligned}
&\|u(t^N) - u_h^N 1_{\mathcal{A}_N}\|_{L^2(0,1;L^2(\Omega))} \\
&\quad + \left\| \mathbb{E}\left(P\left(t^N\right) Q\left(t^N\right)\right) - \frac{1}{M}\sum_{j=1}^{M} P^{N,j} Q_h^{N,j} 1_{\mathcal{A}_N} \|_{L^1(0,1;L^1(\Omega))} \right\| \\
&= O\left(h + \Delta t + \frac{1}{\sqrt{M}}\right).
\end{aligned}$$

Here, $t^N = N\Delta t = T$ and $\mathcal{A}_N$ is the set defined by

$$\mathcal{A}_N = \left\{ \forall k \le N, \frac{1}{M}\sum_{j=1}^{M} (P^{N,j})^2 < \frac{13}{20}\frac{1}{\Delta t} \right\}.$$

Similar results have been obtained in E, LI and ZHANG [2002].

To the author's knowledge, the numerical analysis of mesoscopic models in more than one space dimensions has been addressed only in BONITO, CLÉMENT and PICASSO [2006a], LI and ZHANG [2006]. In LI and ZHANG [2006], a priori error estimates are obtained for Hookean dumbbells and a finite difference method in $[0, 1]^d$ with periodic boundary conditions. The space, time, and Monte Carlo discretizations are considered. Assuming $u \in \mathcal{C}^5([0, T] \times D)$, $\Delta t = h^2$ and the Monte Carlo parameter $M = h^{-\alpha}$, $\alpha > d$, it was proved that the velocity error in the $L^\infty(0, T; L^2([0, 1]^d)$ norm was of order $O(h^2 + \Delta t + 1/M^{(1-\epsilon)/2})$, after excluding an event with probability depending on

$$\frac{1}{h^d \Delta t} e^{-M} \quad \text{and} \quad \frac{1}{h^d \Delta t} e^{-M^\epsilon},$$

where $0 < \epsilon < 1$ is an arbitrary small number.

In BONITO, CLÉMENT and PICASSO [2006a, 2006b], Hookean dumbbells are considered in a bounded smooth domain D, and a finite element discretization is considered in space. Pathwise results are obtained. It should be noted that the convective terms in (1.4), (1.13) are removed in order to perform the analysis. The reason for disregarding convective terms is motivated by the use of an operator splitting scheme, which will be presented in the next section. This analysis will be detailed in Section 2.5.

2.2. Time discretization: an operator splitting scheme

Let us consider the free-surface Oldroyd-B model (1.4), (1.5), (1.7), (1.8) or alternatively the free-surface FENE dumbbell model (1.4), (1.5), (1.7), (1.13), (1.14). Following BONITO, PICASSO and LASO [2006], CABOUSSAT [2005, 2006], CABOUSSAT, PICASSO and RAPPAZ [2005], MARONNIER, PICASSO and RAPPAZ [1999, 2003], an order one operator splitting scheme is used for the time discretization, which allows advection and diffusion phenomena to be decoupled. We refer for instance GLOWINSKI [2003] chapters 2 and 6.30 for a general description of operator splitting methods.

Let $0 = t^0 < t^1 < t^2 < \cdots < t^N = T$ be a subdivision of the time interval $[0, T]$, define $\Delta t^n = t^n - t^{n-1}$ the nth time step, $n = 1, 2, \ldots, N$, Δt the largest time step. At time t^{n-1}, assume that an approximation $\varphi^{n-1} : \Lambda \to \mathbb{R}$ of the volume fraction of liquid is known, which defines the approximation D^{n-1} of the liquid region at time t^{n-1}:

$$D^{n-1} = \{x \in \Lambda;\, \varphi^{n-1}(x) = 1\}.$$

2.2.1. *The free-surface Oldroyd-B model*

Consider the free-surface Oldroyd-B model (1.4), (1.5), (1.7), (1.8), and assume that approximations of the velocity $u^{n-1} : D^{n-1} \to \mathbb{R}^d$ and the extra-stress $\sigma^{n-1} : D^{n-1} \to \mathbb{R}^{d\times d}$ are available. Then, φ^n, D^n, u^n, σ^n are computed by means of a splitting algorithm as illustrated in Fig. 2.2. The prediction step consists in solving three advection problems, which yields the new volume fraction of liquid φ^n, the new liquid region D^n, the predicted velocity $u^{n-\frac{1}{2}} : D^n \to \mathbb{R}^d$, and the predicted extra-stress $\sigma^{n-\frac{1}{2}} : D^n \to \mathbb{R}^{d\times d}$. Then, the correction step is performed, and an Oldroyd-B problem without convection is solved in the liquid region D^n, which yields the new velocity $u^n : D^n \to \mathbb{R}^d$, pressure $p^n : D^n \to \mathbb{R}$, and extra-stress $\sigma^n : D^n \to \mathbb{R}^{d\times d}$.

Prediction step The prediction step consists in solving between time t^{n-1} and t^n the three advection problems:

$$\frac{\partial \tilde{u}}{\partial t} + (\tilde{u} \cdot \nabla)\tilde{u} = 0, \tag{2.7}$$

$$\frac{\partial \tilde{\sigma}}{\partial t} + (\tilde{u} \cdot \nabla)\tilde{\sigma} = 0, \tag{2.8}$$

$$\frac{\partial \tilde{\varphi}}{\partial t} + \tilde{u} \cdot \nabla\tilde{\varphi} = 0, \tag{2.9}$$

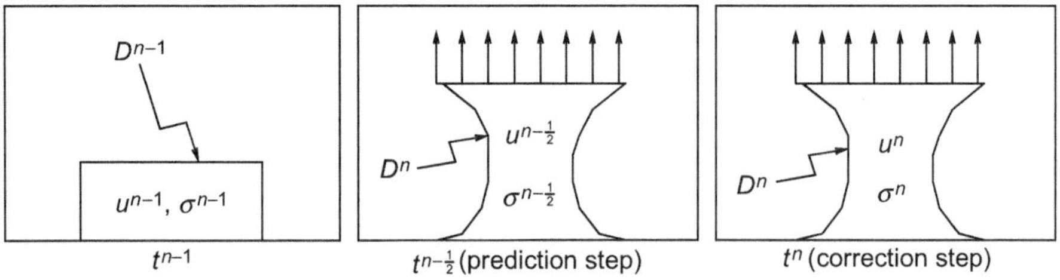

FIG. 2.2 The splitting algorithm.

with initial conditions

$$\tilde{u}\left(t^{n-1}\right) = u^{n-1},$$
$$\tilde{\sigma}\left(t^{n-1}\right) = \sigma^{n-1},$$
$$\tilde{\varphi}\left(t^{n-1}\right) = \varphi^{n-1}.$$

These three problems can be solved exactly using the method of characteristics (see for instance PIRONNEAU [1989], PIRONNEAU, LIOU and TEZDUYAR [1992], and QUARTERONI and VALLI [1991]), the trajectories of the velocity field being straight lines. Indeed, the trajectories are given by $X'(t) = \tilde{u}(X(t), t)$, but since $\tilde{u}$ is constant along the trajectories, we have $X'(t) = \tilde{u}(X(t^{n-1}), t^{n-1}) = u^{n-1}(X(t^{n-1}))$. Let $u^{n-\frac{1}{2}}$, $\sigma^{n-\frac{1}{2}}$, φ^n denote the solution at time t^n of (2.7), (2.8), (2.9), respectively. We thus have

$$u^{n-\frac{1}{2}}(x + \Delta t^n u^{n-1}(x)) = u^{n-1}(x), \tag{2.10}$$
$$\sigma^{n-\frac{1}{2}}(x + \Delta t^n u^{n-1}(x)) = \sigma^{n-1}(x), \tag{2.11}$$
$$\varphi^{n}(x + \Delta t^n u^{n-1}(x)) = \varphi^{n-1}(x), \tag{2.12}$$

for all x belonging to D^{n-1}. Once φ^n is known in the cavity Λ, then the liquid region at time t^n is defined by

$$D^n = \left\{y \in \Lambda;\ \varphi^n(y) = 1\right\}. \tag{2.13}$$

Correction step The new liquid region D^n being known, the predicted velocity $u^{n-\frac{1}{2}} : D^n \to \mathbb{R}^d$ and the extra-stress $\sigma^{n-\frac{1}{2}} : D^n \to \mathbb{R}^{d\times d}$ being also known, an Oldroyd-B problem without convection is solved:

$$\rho\frac{\partial \hat{u}}{\partial t} - 2\eta_s \operatorname{div} \epsilon(\hat{u}) + \nabla\hat{p} - \operatorname{div}\hat{\sigma} = f, \tag{2.14}$$
$$\operatorname{div}\hat{u} = 0, \tag{2.15}$$
$$\hat{\sigma} + \lambda\left(\frac{\partial\hat{\sigma}}{\partial t} - \nabla\hat{u}\sigma - \sigma\nabla\hat{u}^T\right) = 2\eta_p\epsilon(\hat{u}), \tag{2.16}$$

in the slab $D^n \times (t^{n-1}, t^n)$, with initial conditions

$$\hat{u}(t^{n-1}) = u^{n-\frac{1}{2}},$$
$$\hat{\sigma}(t^{n-1}) = \sigma^{n-\frac{1}{2}}.$$

Then, the corrected velocity $u^n : D^n \to \mathbb{R}^d$ and the extra-stress $\sigma^n : D^n \to \mathbb{R}^{d\times d}$ are defined by

$$u^n = \hat{u}(t^n), \quad \sigma^n = \hat{\sigma}(t^n).$$

In Section 2.4 we discuss the well posedness of the correction step (2.14), (2.15), and (2.16). Also, we prove convergence of a finite element discretization in space.

2.2.2. *The free-surface FENE dumbbell model*

In the case of the free-surface FENE dumbbell model (1.4), (1.5), (1.7), (1.13), (1.14), given approximations of the velocity $u^{n-1} : D^{n-1} \to \mathbb{R}^d$ and the dumbbell elongations $q^{n-1} : D^{n-1} \times \Omega \to \mathbb{R}^d$, the prediction step consists in solving three advection problems, which yields the new volume fraction of liquid φ^n, the new liquid region D^n, the predicted velocity $u^{n-\frac{1}{2}} : D^n \to \mathbb{R}^d$, and the predicted dumbbell elongations $q^{n-\frac{1}{2}} : D^n \times \Omega \to \mathbb{R}^d$. We thus keep (2.10), (2.12) and replace (2.11) by

$$q^{n-\frac{1}{2}}(x + \Delta t^n u^{n-1}(x), \omega) = q^{n-1}(x, \omega), \tag{2.17}$$

for all x belonging to D^{n-1}, for all event ω in Ω. The new liquid region D^n is then obtained using (2.13), and the predicted extra-stress $\sigma^{n-\frac{1}{2}} : D^n \to \mathbb{R}^{d\times d}$ is defined as

$$\sigma^{n-\frac{1}{2}} = \frac{\eta_p}{\lambda}\left(\mathbb{E}\left(F\left(q^{n-\frac{1}{2}}\right)\left(q^{n-\frac{1}{2}}\right)^T\right) - I\right). \tag{2.18}$$

The correction step consists in solving (2.14), (2.15) and replacing (2.16) with

$$d\hat{q} = \left(\nabla u\hat{q} - \frac{1}{2\lambda}F(\hat{q})\right)dt + \frac{1}{\sqrt{\lambda}}dB, \tag{2.19}$$

$$\hat{\sigma} = \frac{\eta_p}{\lambda}\left(\mathbb{E}\left(F(\hat{q})\hat{q}^T\right) - I\right), \tag{2.20}$$

with initial conditions

$$\hat{u}\left(t^{n-1}\right) = u^{n-\frac{1}{2}},$$

$$\hat{q}\left(t^{n-1}\right) = q^{n-\frac{1}{2}}.$$

Then, the corrected velocity $u^n : D^n \to \mathbb{R}^d$, dumbbell elongations $q^n : D^n \times \Omega \to \mathbb{R}^d$, and extra-stress $\sigma^n : D^n \times \Omega \to \mathbb{R}^{d\times d}$ are defined by

$$u^n = \hat{u}(t^n), \quad q^n = \hat{q}(t^n), \quad \sigma^n = \frac{\eta_p}{\lambda}\left(\mathbb{E}\left(F(q^n)(q^n)^T\right) - I\right).$$

In Section 2.5 we discuss the well posedness of the correction step (2.14), (2.15), (2.19), and (2.20) when $F(q) = q$ (Hookean dumbbells). Also, we prove convergence of a finite element discretization in space.

2.3. The three fields stokes problem

2.3.1. *The continuous problem*

The simplest problem when solving viscoelastic flow problems with finite elements is the three fields Stokes problem obtained by considering the correction step (2.14)–(2.16) setting formally $\rho = 0$ and $\lambda = 0$.

Then, given a domain $D \subset \mathbb{R}^d$, $d = 2$ or 3, given $f : D \to \mathbb{R}^d$, we are looking for $u : D \to \mathbb{R}^d$, $p : D \to \mathbb{R}$, and $\sigma : D \to \mathbb{R}^{d\times d}$ such that

$$- 2\eta_s \text{div } \epsilon(u) + \nabla p - \text{div } \sigma = f \quad \text{in } D, \tag{2.21}$$

$$\text{div } u = 0 \quad \text{in } D, \tag{2.22}$$

$$\sigma - 2\eta_p \epsilon(u) = 0 \quad \text{in } D, \tag{2.23}$$

$$u = 0 \quad \text{on } \partial D. \tag{2.24}$$

Formally, a smooth solution to this problem satisfies, after elimination of σ:

$$- 2(\eta_s + \eta_p)\text{div } \epsilon(u) + \nabla p = f,$$

$$\text{div } u = 0.$$

Therefore, the problem should be well posed even the solvent viscosity $\eta_s = 0$, provided $\eta_p > 0$. Indeed, existence of a unique weak solution and continuous dependence on the data f can be proved for $\eta_s \geq 0$ and $\eta_p > 0$ using the inf-sup framework BABUŠKA and AZIZ [1972]. The details can be found in BARANGER and SANDRI [1992b], BONVIN, PICASSO and STENBERG [2001], but, for the sake of clarity, we briefly report the arguments hereafter.

As usual, the space $H_0^1(D)$ denotes the space of $H^1(D)$ velocities vanishing on the boundary $\partial\Omega$, whereas $L_0^2(D)$ denotes the space of $L^2(D)$ pressures with zero mean. Recall that $(\cdot, \cdot)_D$ stands for the $L^2(D)$ scalar product for scalar, vectors, or tensors, with induced norm $\| \cdot \|_{L^2(D)}$. The weak formulation corresponding to (2.21)–(2.24) writes : find $u \in H_0^1(D)$, $p \in L_0^2(D)$, $\sigma \in L^2(D)$ such that

$$2\eta_s(\epsilon(u), \epsilon(v))_D - (p, \text{div } v)_D + (\sigma, \epsilon(v))_D = (f, v)_D \quad \forall v \in H_0^1(D), \tag{2.25}$$

$$(\text{div } u, q)_D = 0 \quad \forall q \in L_0^2(D), \tag{2.26}$$

$$(\sigma - 2\eta_p \epsilon(u), \tau)_D = 0 \quad \forall \tau \in L^2(D). \tag{2.27}$$

Setting $W = H_0^1(D)^d \times L_0^2(D) \times L^2(D)^{d\times d}$, we can rewrite this problem as finding $(u, p, \sigma) \in W$ such that

$$B(u, p, \sigma; v, q, \tau) = F(v, q, \tau) \qquad \forall (v, q, \tau) \in W, \tag{2.28}$$

where $B : W \times W \to \mathbb{R}$ is the symmetric bilinear form defined by

$$\begin{aligned} B(u, p, \sigma; v, q, \tau) = {} & 2\eta_s(\epsilon(u), \epsilon(v))_D - (p, \text{div } v)_D + (\sigma, \epsilon(v))_D \\ & - (\text{div} u, q)_D - \frac{1}{2\eta_p}(\sigma, \tau)_D + (\epsilon(u), \tau)_D, \end{aligned}$$

and $F : W \to \mathbb{R}$ is the linear form defined by

$$F(v, q, \tau) = (f, v)_D.$$

The space W is equipped with the norm $\|\cdot\|_W$ defined for all $(v, q, \tau) \in W$ by

$$\|v, q, \tau\|_W^2 = 2(\eta_s + \eta_p)\|\epsilon(v)\|_{L^2(D)}^2 + \frac{1}{2(\eta_s + \eta_p)}\|q\|_{L^2(D)}^2 + \frac{1}{2\eta_p}\|\tau\|_{L^2(D)}^2.$$

Then, the well posedness of (2.28) is a consequence of the following Lemma.

LEMMA 2.1. *The symmetric bilinear form B satisfies the inf-sup condition, uniformly with respect to* $\eta_s \geq 0$ *and* $\eta_p > 0$: $\exists C > 0, \forall \eta_s \geq 0, \forall \eta_p > 0, \forall (u, p, \sigma) \in W$

$$\sup_{(v,q,\tau)\in W\setminus\{0\}} \frac{B(u, p, \sigma; v, q, \tau)}{\|v, q, \tau\|_W} \geq C\|u, p, \sigma\|_W.$$

PROOF. In order to prove that B satisfies the Babuška inf-sup conditions BABUŠKA and AZIZ [1972], it suffices to prove that $\exists C_1, C_2 > 0, \forall \eta_s \geq 0, \forall \eta_p > 0, \forall (u, p, \sigma) \in W, \exists (v, q, \tau) \in W$ such that

$$B(u, p, \sigma; v, q, \tau) \geq C_1\|u, p, \sigma\|_W^2 \quad \text{and} \quad \|v, q, \tau\|_W \leq C_2\|u, p, \sigma\|_W. \tag{2.29}$$

Let $(u, p, \sigma) \in W$. Clearly, we have

$$B(u, p, \sigma; u, -p, -\sigma) = 2\eta_s\|\epsilon(u)\|_{L^2(D)}^2 + \frac{1}{2\eta_p}\|\sigma\|_{L^2(D)}^2,$$
$$B(u, p, \sigma; 0, 0, 2\eta_p\epsilon(u)) = 2\eta_p\|\epsilon(u)\|_{L^2(D)}^2 - (\sigma, \epsilon(u))_D.$$

On the other side, the classical inf-sup condition between pressure and velocity implies $\exists C_3 > 0, \forall p \in L_0^2(D), \exists \tilde{v} \in H_0^1(D)$ such that

$$\|p\|_{L^2(D)}^2 = (p, \operatorname{div} \tilde{v})_D \quad \text{and} \quad \|\epsilon(\tilde{v})\|_{L^2(D)} \leq C_3\|p\|_{L^2(D)};$$

thus, we have

$$B(u, p, \sigma; -\tilde{v}, 0, 0) = -2\eta_s(\epsilon(u), \epsilon(\tilde{v}))_D + \|p\|_{L^2(D)}^2 + (\sigma, \epsilon(\tilde{v}))_D.$$

Therefore, for $\delta > 0$, we have

$$\begin{aligned}
&B(u, p, \sigma; u - \delta\tilde{v}, -p, -\sigma + 2\delta\eta_p\epsilon(u)) \\
&\quad = 2(\eta_s + \delta\eta_p)\|\epsilon(u)\|_{L^2(D)}^2 + \frac{1}{2\eta_p}\|\sigma\|_{L^2(D)}^2 + \delta\|p\|_{L^2(D)}^2 \\
&\qquad - (\sigma, \epsilon(u))_D - 2\delta\eta_s(\epsilon(u), \epsilon(\tilde{v}))_D + \delta(\sigma, \epsilon(\tilde{v}))_D,
\end{aligned}$$

and we can prove that (2.29) holds with

$$(v, q, \tau) = (u - \delta\tilde{v}, -p, -\sigma + 2\delta\eta_p)$$

provided $\delta > 0$ is chosen sufficiently small. Please note that δ does not depend on η_s or η_p; we refer to BARANGER and SANDRI [1992b], BONVIN, PICASSO and STENBERG [2001] for details. □

From the above lemma, we deduce that if $f \in H^{-1}(D)$, then the problem is well posed: there exists a unique solution to (2.28) and $C > 0$ independent of f such that

$$\|u, p, \sigma\|_W \leq C\|f\|_{H^{-1}(D)}.$$

REMARK 2.1. The interested reader should note that a different weak formulation was used in FORTIN and PIERRE [1989], SANDRI [1993]. Indeed, (2.25)–(2.27) was considered with $\eta_s = 0$ and rewritten in the framework of Brezzi inf-sup theorem BREZZI [1974] as finding $(u, p, \sigma) \in W$ such that

$$\begin{aligned} &a(\sigma, p; \tau, q) + b(\tau, q; u) = 0 \qquad && \forall (\tau, q) \in L^2(D) \times L_0^2(D), \\ &b(\sigma, p; v) = -(f, v)_D && \forall v \in H_0^1(D), \end{aligned}$$

where the bilinear forms a and b are defined by

$$a(\sigma, p; \tau, q) = \frac{1}{2\eta_p}(\sigma, \tau)_D \quad \text{and} \quad b(\tau, q; u) = -(\epsilon(u), \tau)_D + (\operatorname{div} u, q)_D.$$

Then, a and b satisfy Brezzi's inf-sup conditions, i.e., a is coercive onto

$$K = \{(\sigma, p) \in L^2(D) \times L_0^2(D) \text{ such that } b(\sigma, p; v) = 0, \forall v \in H_0^1(D)\},$$

and b satisfies the inf-sup condition

$$\exists C_2 > 0, \forall v \in H_0^1(D) \qquad \sup_{(\sigma,p) \in L^2(D)\times L^2(D)\backslash\{0\}} \frac{b(\sigma, p; v)}{\|\sigma, p\|_{L^2(D)\times L^2(D)}} \geq C_2\|v\|_{H^1(D)};$$

thus, the problem is well posed.

2.3.2. *Finite element discretizations*

We are now interested in computing finite elements approximations of (2.25)–(2.27) or equivalently (2.28). For any $h > 0$, let $\mathcal{T}_h$ be a finite element mesh of D into triangles ($d = 2$) or tetrahedrons ($d = 3$), regular in the sense of CIARLET and LIONS [1991].

Galerkin methods Let $V_h \subset H_0^1(D)^d$, $Q_h \subset L_0^2(D)$, and $M_h \subset L^2(D)^{d\times d}$ be finite element subspaces for the velocity, pressure, and extra stress, respectively, and let $W_h = V_h \times Q_h \times M_h$. A Galerkin method corresponding to (2.28) writes : find $(u_h, p_h, \sigma_h) \in W_h$ such that

$$B(u_h, p_h, \sigma_h; v_h, q_h, \tau_h) = F(v_h, q_h, \tau_h) \qquad \forall (v_h, q_h, \tau_h) \in W_h. \tag{2.30}$$

Clearly, if the finite element spaces satisfy the conditions div $V_h \subset M_h$ and

$$\exists C > 0, \forall h > 0, \forall q_h \in Q_h \qquad \sup_{v_h \in V_h\backslash\{0\}} \frac{(q_h, \operatorname{div} v_h)}{\|\nabla v_h\|_{L^2(D)}} \geq C\|q_h\|_{L^2(D)}, \tag{2.31}$$

then the discrete analog of Lemma 2.1 holds since the proof can be reproduced in the discrete space W_h instead of W. Thus, the problem is well posed and optimal a priori error estimates hold.

An example of spaces V_h, Q_h, and M_h satisfying the two above conditions when $d = 2$ is the following. The velocity is continuous, piecewise quadratic; thus the velocity divergence is discontinuous piecewise linear, so as the extra stress. Moreover, the pressure is continuous, piecewise linear so that (2.31) is satisfied. This finite element discretizations has been used in BARANGER and SANDRI [1992a].

REMARK 2.2. Let us consider the framework introduced in Remark 2.1.Following FORTIN and PIERRE [1989], SANDRI [1993], optimal a priori error estimates can be recovered provided the finite element spaces V_h, Q_h, and M_h are such that the bilinear forms a and b satisfy Brezzi's discrete inf-sup conditions, uniformly with respect to h.

An example of spaces V_h, Q_h, and M_h satisfying these two conditions when $d = 2$ is the following FORTIN and PIERRE [1989]. The velocity, pressure, and extra stress are continuous on quadrangles. The velocity is piecewise quadratic, and the pressure is piecewise linear. Each quadrangle is cut into 4×4 quadrangles, and the extra stress is piecewise linear on these smaller quadrangles. We refer to SANDRI [1993] for a similar example on triangles.

Stabilized Galerkin least-square formulations Stabilized Galerkin least square formulations can be considered in order to avoid compatibility conditions between the finite element spaces V_h, Q_h, and M_h. The simplest stabilized scheme consists in considering continuous, piecewise linear spaces for the velocity, pressure, and extra stress together with the following Galerkin least-square formulation: find $(u_h, p_h, \sigma_h) \in W_h = V_h \times Q_h \times M_h$ such that

$$B_h(u_h, p_h, \sigma_h; v_h, q_h, \tau_h) = F_h(v_h, q_h, \tau_h) \qquad \forall (v_h, q_h, \tau_h) \in W_h. \tag{2.32}$$

Here, B_h is the bilinear form defined by

$$\begin{aligned} B_h(u_h, p_h, \sigma_h; v_h, q_h, \tau_h) &= B(u_h, p_h, \sigma_h; v_h, q_h, \tau_h) \\ &- \sum_{K \in \mathcal{T}_h} \frac{\alpha h_K^2}{2\eta_p} (-2\eta_s \operatorname{div} \epsilon(u_h) + \nabla p_h - \operatorname{div} \sigma_h, \nabla q_h)_K \\ &+ 2\eta_p \left(\frac{1}{2\eta_p} \sigma_h - \epsilon(u_h), -\epsilon(v_h) \right)_D, \end{aligned} \tag{2.33}$$

and F_h is the linear form defined by

$$F_h(v_h, q_h, \tau_h) = F(v_h, q_h, \tau_h) - \sum_{K \in \mathcal{T}_h} \frac{\alpha h_K^2}{2\eta_p} (f, \nabla q)_K,$$

where $\alpha > 0$ is a dimensionless parameter and $(\cdot, \cdot)_K$ denotes the $L^2(K)$ scalar product. The stabilized scheme (2.32) is designed so that it is stable and consistent. Consistency means that if the solution (u, p, σ) of (2.28) is smooth enough, then

$$B_h(u, p, \sigma; v_h, q_h, \tau_h) = F_h(v_h, q_h, \tau_h) \qquad \forall (v_h, q_h, \tau_h) \in W_h.$$

The key point for proving stability is stated in the following Lemma.

LEMMA 2.2. *Let C_I be the largest constant involved in the following inverse inequality*

$$C_I \sum_{K\in\mathcal{T}_h} h_K^2 \|\text{div } \sigma_h\|^2_{L^2(K)} \leq \|\sigma_h\|^2_{L^2(D)} \qquad \forall \sigma_h \in M_h, \tag{2.34}$$

and let $0 < \alpha < C_I$. Then, $\exists C > 0$, $\forall \eta_s \geq 0$, $\forall \eta_p > 0$, $\forall h > 0$, $\forall (u_h, p_h, \sigma_h) \in W_h$

$$B_h(u_h, p_h, \sigma_h; u_h, -p_h, -\sigma_h) \geq C\|u_h, p_h, \sigma_h\|_h^2,$$

where $\|\cdot\|_h$ is the discrete norm defined by

$$\|u_h, p_h, \sigma_h\|_h^2 = 2(\eta_s + \eta_p)\|\epsilon(u_h)\|^2_{L^2(D)} + \frac{1}{2(\eta_s+\eta_p)} \sum_{K\in\mathcal{T}_h} h_K^2 \|\nabla p_h\|^2_{L^2(K)} + \frac{1}{2\eta_p}\|\sigma_h\|^2_{L^2(D)}.$$

PROOF. Let $(u_h, p_h, \sigma_h) \in W_h$; we have

$$B_h(u_h, p_h, \sigma_h; u_h, -p_h, -\sigma_h) = 2(\eta_s + \eta_p)\|\epsilon(u_h)\|^2_{L^2(D)} + \frac{1}{2\eta_p}\|\sigma_h\|^2_{L^2(D)} + \sum_{K\in\mathcal{T}_h} \frac{\alpha h_K^2}{2\eta_p}\left(\|\nabla p_h\|^2_{L^2(D)} - (\text{div } \sigma_h, \nabla p_h)_D\right) - (\sigma_h, \epsilon(u_h))_D.$$

It then suffices to use (2.34) and Young's inequality to obtain the result. We again refer to BONVIN, PICASSO and STENBERG [2001] for details. □

From this Lemma, the discrete analog of Lemma 2.1 can be proved for the bilinear form B_h : $\exists C > 0$, $\forall \eta_s \geq 0$, $\forall \eta_p > 0$, $\forall h > 0$, $\forall (u_h, p_h, \sigma_h) \in W_h$

$$\sup_{(v_h,q_h,\tau_h)\in W_h\setminus\{0\}} \frac{B_h(u_h, p_h, \sigma_h; v_h, q_h, \tau_h)}{\|v_h, q_h, \tau_h\|_W} \geq C\|u_h, p_h, \sigma_h\|_W$$

so that optimal a priori error estimates hold.

EVSS stabilization In FORTIN, GUÉNETTE and PIERRE [2000], an elastic viscous split stress (EVSS) scheme was analyzed. The EVSS scheme consists in adding to the three fields Stokes problem (2.21)–(2.23) a new field d, for stability purposes, as following:

$$-2(\eta_s + \eta_p)\text{div } \epsilon(u) + \nabla p - \text{div } (\sigma - 2\eta_p d) = f \quad \text{in } D, \tag{2.35}$$

$$\text{div } u = 0 \quad \text{in } D, \tag{2.36}$$

$$\sigma - 2\eta_p \epsilon(u) = 0 \quad \text{in } D, \tag{2.37}$$

$$d - \epsilon(u) = 0 \quad \text{in } D. \tag{2.38}$$

Equal order finite elements were used to approach d and σ; thus, the Galerkin finite element formulation corresponding to (2.35)–(2.38) consists in finding $(u_h, p_h, \sigma_h, d_h) \in W_h = V_h \times$

$Q_h \times M_h \times M_h$ such that

$$B(u_h, p_h, \sigma_h, d_h; v_h, q_h, \tau_h, e_h) = F(v_h, q_h, \tau_h, e_h) \qquad \forall (v_h, q_h, \tau_h, e_h) \in W_h. \quad (2.39)$$

Here, B is the bilinear form defined by

$$\begin{aligned} B(u_h, p_h, \sigma_h, d_h; v_h, q_h, \tau_h, e_h) &= 2(\eta_s + \eta_p)(\epsilon(u_h), \epsilon(v_h))_D - (p_h, \operatorname{div} v_h)_D \\ &+ (\sigma_h - 2\eta_p d_h, \epsilon(v_h))_D - (\operatorname{div} u_h, q_h)_D - \frac{1}{2\eta_p}(\sigma_h, \tau_h)_D + (\epsilon(u_h), \tau_h)_D \\ &+ 2\eta_p(d_h - \epsilon(u_h), e_h)_D. \end{aligned}$$

Since σ_h and d_h belong to same finite element space M_h, it is clear that $\sigma_h = 2\eta_p d_h$ so that solving (2.39) is equivalent to finding $(u_h, p_h) \in V_h \times Q_h$ such that

$$\begin{aligned} &2(\eta_s + \eta_p)(\epsilon(u_h), \epsilon(v_h))_D - (p_h, \operatorname{div} v_h)_D \\ &\quad - (\operatorname{div} u_h, q_h)_D = (f, v_h)_D \qquad \forall (v_h, q_h) \in V_h \times Q_h, \end{aligned} \quad (2.40)$$

and then finding $\sigma_h \in M_h$ such that

$$(\sigma_h, \tau_h)_D = 2\eta_p(\epsilon(u_h), \tau_h)_D \qquad \forall \tau_h \in M_h.$$

Therefore, (2.39) is well posed whenever the finite element spaces V_h and Q_h satisfy the classical discrete inf-sup condition (2.31).

The connection between stabilized Galerkin least square formulations and EVSS stabilization has been studied for the three fields Stokes problem in BONVIN, PICASSO and STENBERG [2001]. An extension to a simplified stationary Oldroyd-B problem has been considered in PICASSO and RAPPAZ [2001]. It should be noted that when considering the Oldroyd-B model or FENE dumbbells, both stabilized Galerkin least square and EVSS schemes differ. However, the EVSS formulation is much simpler to implement; therefore, it is usually preferred. Numerical simulations of FENE dumbbells using the EVSS scheme have been proposed in BONVIN and PICASSO [2001, 2002].

2.4. A simplified Oldroyd-B problem

This section is devoted to the study of the correction step of the free-surface algorithm for the Oldroyd-B problem presented in Section 2.2. This is (2.14)–(2.16) that are recalled hereafter for the convenience of the reader. Two different considerations are discussed here. First, existence and uniqueness results with small data are presented. Second, the well posedness of a stabilized finite element discretization in space is obtained, so as optimal convergence results. The results presented here can be found in more detail in BONITO, CLÉMENT and PICASSO [2007].

Let D be a bounded, connected open set of $\mathbb{R}^d$, $d \geq 2$ with boundary ∂D of class C^2, and let $T > 0$ be the final time. We consider the following problem. Given initial conditions $u_0 : D \to \mathbb{R}^d$, $\sigma_0 : D \to \mathbb{R}^{d\times d}$, a force term f, a constant density $\rho > 0$, constant solvent and polymer viscosities $\eta_s > 0$, $\eta_p > 0$, and a constant relaxation time $\lambda > 0$, find the velocity $u : D \times (0, T) \to \mathbb{R}^d$, pressure $p : D \times (0, T) \to \mathbb{R}$, and extra stress $\sigma : D \times (0, T) \to \mathbb{R}^{d\times d}$ such that

$$\rho\frac{\partial u}{\partial t} - 2\eta_s \operatorname{div} \epsilon(u) + \nabla p - \operatorname{div} \sigma = f \qquad \text{in } D \times (0, T), \tag{2.41}$$

$$\operatorname{div} u = 0 \qquad \text{in } D \times (0, T), \tag{2.42}$$

$$\frac{1}{2\eta_p}\sigma + \frac{\lambda}{2\eta_p}\left(\frac{\partial \sigma}{\partial t} - (\nabla u)\sigma - \sigma(\nabla u)^T\right) - \epsilon(u) = 0 \quad \text{in } D \times (0, T), \tag{2.43}$$

$$u = 0 \qquad \text{on } \partial D \times (0, T), \tag{2.44}$$

$$u(\cdot, 0) = u_0, \qquad \sigma(\cdot, 0) = \sigma_0 \qquad \text{in } D. \tag{2.45}$$

Note that when comparing to (2.14)–(2.16), the hat symbols have been omitted for clarity purpose.

When D is of class $\mathcal{C}^2$, the implicit function theorem has been used in BONITO, CLÉMENT and PICASSO [2007] to prove that the above problem admits a unique solution

$$\begin{aligned} &u \in W^{1,q}(0, T; L^r(D)) \cap L^q(0, T; W^{2,r}(D)),\\ &p \in L^q(0, T; W^{1,r}(D)),\\ &\sigma \in W^{1,q}(0, T; W^{1,r}(D)), \end{aligned} \tag{2.46}$$

with $1 < q < \infty$, $d < r < \infty$, for any data f, u_0, σ_0 small enough in appropriate spaces. We assume that such a result also holds when D is a convex polygon; see PICASSO and RAPPAZ [2001] for a proof in the framework of the corresponding stationary problem. Then, the above regularity is sufficient to ensure the existence and convergence of a stabilized, continuous, piecewise linear finite element discretization in space.

The finite element approximation in space is now introduced. For any $h > 0$, let $\mathcal{T}_h$ be a decomposition of the computational domain D into triangles K with diameter h_K less than h, regular in the sense of CIARLET and LIONS [1991]. We consider as in Section 2.3.2 the finite element spaces V_h, Q_h, and M_h corresponding to continuous, piecewise linear velocity, pressure, and extra stress. We denote i_h the $L^2(D)$ projection onto V_h, Q_h, or M_h and introduce the following stabilized finite element discretization in space of (2.41)–(2.45). Given f, u_0, σ_0 find

$$(u_h, p_h, \sigma_h) : t \to (u_h(t), p_h(t), \sigma_h(t)) \in V_h \times Q_h \times M_h$$

such that $u_h(0) = i_h u_0$, $\sigma_h(0) = i_h \sigma_0$ and such that the following weak formulation holds in $]0, T[$:

$$\begin{aligned} &\rho\left(\frac{\partial u_h}{\partial t}, v_h\right)_D + 2\eta_s\,(\epsilon(u_h), \epsilon(v_h))_D - (p_h, \operatorname{div} v_h)_D + (\sigma_h, \epsilon(v_h))_D\\ &\quad - (f, v_h)_D + (\operatorname{div} u_h, q_h)_D + \sum_{K\in\mathcal{T}_h} \frac{\alpha h_K^2}{2\eta_p}(\nabla p_h, \nabla q_h)_K\\ &\quad + \frac{1}{2\eta_p}(\sigma_h, \tau_h)_D + \frac{\lambda}{2\eta_p}\left(\frac{\partial \sigma_h}{\partial t} - (\nabla u_h)\sigma_h - \sigma_h(\nabla u_h)^T, \tau_h\right)_D\\ &\quad - (\epsilon(u_h), \tau_h)_D = 0, \end{aligned} \tag{2.47}$$

for all $(v_h, q_h, \tau_h) \in V_h \times Q_h \times M_h$. Here, $\alpha > 0$ is a dimensionless stabilization parameter.

In order to prove that the solution of the nonlinear finite element discretization (2.47) exists and converges to that of (2.41)–(2.45), we shall use the abstract Theorem 2.1 of CALOZ and RAPPAZ [1997]. For this purpose, we introduce X_h defined by

$$X_h = L^2(0, T; V_h) \times L^\infty(0, T; M_h)$$

equipped with the norm $||\cdot||_{X_h}$ defined for all $x_h = (u_h, \sigma_h) \in X_h$ by

$$\|x_h\|^2_{X_h} = 2\eta_s \int_0^T \|\epsilon(u_h(t))\|^2_{L^2(\Omega)} \mathrm{d}t + \frac{\lambda}{4\eta_p} \sup_{t \in [0,T]} \|\sigma_h(t)\|^2_{L^2(\Omega)}.$$

Then, we rewrite the solution of (2.47) as the following fixed point problem. Given $y = (f, u_0, \sigma_0) \in Y$, find $x_h = (u_h, \sigma_h) \in X_h$ such that

$$x_h = \mathsf{T}_h(y, S(x_h)). \tag{2.48}$$

Here, Y is the functional space corresponding to the data (f, u_0, σ_0) (see BONITO, CLÉMENT and PICASSO [2007] for details), and S is defined by

$$S(x_h) = \frac{\lambda}{2\eta_p}\left((\nabla u_h)\sigma_h + \sigma_h(\nabla u_h)^T\right)_D.$$

Given $y = (f, u_0, \sigma_0) \in Y$ and $g \in L^2(0, T; L^2(D))$, computing $\mathsf{T}_h(y, g)$ consists in solving a time-dependent three fields Stokes problem discretized in space, namely

$$\begin{aligned}\mathsf{T}_h : Y \times L^2(0, T; L^2(D)) &\to X_h \\ (f, u_0, \sigma_0, g) &\to \mathsf{T}_h(f, u_0, \sigma_0, g) := (\tilde{u}_h, \tilde{\sigma}_h),\end{aligned}$$

where for $t \in (0, T)$

$$(\tilde{u}_h, \tilde{p}_h, \tilde{\sigma}_h) : t \to (\tilde{u}_h(t), \tilde{p}_h(t), \tilde{\sigma}_h(t)) \in V_h \times Q_h \times M_h$$

satisfies $\tilde{u}_h(0) = i_h u_0$, $\tilde{\sigma}_h(0) = i_h \sigma_0$, and

$$\begin{aligned}&\rho\left(\frac{\partial \tilde{u}_h}{\partial t}, v_h\right)_D + 2\eta_s(\epsilon(\tilde{u}_h), \epsilon(v_h))_D - (\tilde{p}_h, \operatorname{div} v_h)_D + (\tilde{\sigma}_h, \epsilon(v_h))_D - (f, v_h)_D \\ &\quad + (\operatorname{div} \tilde{u}_h, q_h)_D + \sum_{K \in \mathcal{T}_h} \frac{\alpha h_K^2}{2\eta_p}(\nabla \tilde{p}_h, \nabla q_h)_K \\ &\quad + \frac{1}{2\eta_p}(\tilde{\sigma}_h, \tau_h)_D + \frac{\lambda}{2\eta_p}\left(\frac{\partial \tilde{\sigma}_h}{\partial t}, \tau_h\right)_D - (\epsilon(\tilde{u}_h), \tau_h)_D - \frac{\lambda}{2\eta_p}(g, \tau_h)_D = 0\end{aligned} \tag{2.49}$$

for all $(v_h, q_h, \tau_h) \in V_h \times Q_h \times M_h$, a.e in $(0, T)$.

In BONITO, CLÉMENT and PICASSO [2007], it is proved that (2.48) has a unique solution converging to that of (2.41)–(2.45). Indeed, following PICASSO and RAPPAZ [2001], (2.48) is written as the following nonlinear problem: given $y = (f, u_0, \sigma_0) \in Y$, find $x_h = (u_h, \sigma_h) \in X_h$ such that

$$F_h(y, x_h) = 0, \tag{2.50}$$

where $F_h : Y \times X_h \to X_h$ is defined by

$$F_h(y, x_h) = x_h - \mathsf{T}_h(y, S(x_h)). \tag{2.51}$$

The abstract Theorem 2.1 of CALOZ and RAPPAZ [1997] can be then used in order to prove existence and convergence of a solution to (2.50). The mapping $F_h : Y \times X_h \to X_h$ is $\mathcal{C}^1$. Moreover, the scheme is consistent, $D_x F_h$ has bounded inverse at $i_h x$, and $D_x F_h$ is locally Lipschitz at $i_h x$. Here, i_h is the $L^2(D)$ projection onto the finite element space X_h, and $x = (u, \sigma)$ is the solution of the continuous problem (2.41)–(2.45), with regularity (2.46). Therefore, applying Theorem 2.1 of CALOZ and RAPPAZ [1997], existence of a semidiscrete solution x_h can be proved in the neighborhood of $i_h x$ provided the data y is small enough in Y, the space of data. Note that this regularity implies that the trajectories $(x, t) \to q_h(x, t, \omega)$ are continuous, for almost each event $\omega \in \Omega$. Moreover, optimal error estimates hold for $\|x - x_h\|_{X_h}$, that is:

$$\|u - u_h\|_{L^2(0,T;H^1(D))} + \|\sigma - \sigma_h\|_{L^\infty(0,T;L^2(D))} = O(h).$$

We refer to BONITO, CLÉMENT and PICASSO [2007] for details.

2.5. A simplified Hookean dumbbells problem

As a first step toward the analysis of stochastic models for viscoelastic fluids, this section is devoted to the study of the correction step of the free-surface algorithm for Hookean dumbbell model (see Section 2.2). This is (2.14), (2.15) supplemented by (2.19) and (2.20) with $F(q) = q$, which are recalled hereafter for the convenience of the reader. A pathwise existence will be provided with enough regularity to ensure the convergence of the finite element scheme proposed. The results presented here can be found in more detail in BONITO, CLÉMENT and PICASSO [2006a, 2006b].

We refer to BONVIN and PICASSO [1999], JOURDAIN, LELIÈVRE and LE BRIS [2004b] for presentations related to the Monte Carlo discretization and the use of variance reduction techniques.

Let $D \subset \mathbb{R}^d$ be the "physical" space, $T > 0$ be the final time, and $(\Omega, \mathcal{F}, \mathcal{P})$ be a complete filtered probability space. Given $f : D \times [0, T] \to \mathbb{R}^d$, $u_0 : D \to \mathbb{R}^d$, and $q_0 : \Omega \to \mathbb{R}^d$, we are seeking for the velocity $u : D \times [0, T] \to \mathbb{R}^d$, the pressure $p : D \times [0, T] \to \mathbb{R}$, and the dumbbell elongation $q : D \times [0, T] \times \Omega \to \mathbb{R}^d$ such that

$$\begin{aligned} &\rho \frac{\partial u}{\partial t} - 2\eta_s \operatorname{div} \epsilon(u) + \nabla p \\ &\qquad - \frac{\eta_p}{\lambda} \operatorname{div} \left(\mathbb{E}(qq^T) - I\right) = f \qquad \text{in } D \times (0, T), \end{aligned} \tag{2.52}$$

$$\operatorname{div} u = 0 \qquad \text{in } D \times (0, T), \tag{2.53}$$

$$\mathrm{d}q = \left(\nabla u q - \frac{1}{2\lambda} q\right) \mathrm{d}t + \frac{1}{\sqrt{\lambda}} \mathrm{d}B \qquad \text{in } D \times (0, T) \times \Omega, \tag{2.54}$$

$$u = 0 \qquad \text{on } \partial D \times (0, T), \tag{2.55}$$

$$u(\cdot, 0) = u_0 \qquad \text{in } D, \tag{2.56}$$

$$q(\cdot, 0, \cdot) = q_0 \qquad \text{in } D \times \Omega, \tag{2.57}$$

where q_0 satisfies

$$\mathbb{E}(q_0) = 0 \qquad \text{and} \qquad \mathbb{E}(q_0 q_0^T) = I. \tag{2.58}$$

Note that comparing to (2.14), (2.15), (2.19), and (2.20), the hat symbols have been omitted for clarity purpose. Also note that Eqns (2.54) and (2.57) are notations for

$$q(x, t, \omega) - q_0(\omega) = \int_0^t \left(\nabla u(x, s) q(x, s, \omega) - \frac{1}{2\lambda} q(x, s, \omega)\right) \mathrm{d}s + \frac{1}{\sqrt{\lambda}} B(t, \omega),$$

where $(x, t, \omega) \in D \times [0, T] \times \Omega$.

When D is of class $\mathcal{C}^2$, the implicit function theorem has been used in BONITO, CLÉMENT and PICASSO [2006b] to prove that the above problem admits a unique solution (u, p, q) satisfying

$$\begin{aligned} &u \in h^{1+\mu}([0, T]; L^r(D)) \cap h^{\mu}([0, T]; W^{2,r}(D)) \\ &p \in h^{\mu}([0, T]; W^{1,r}(D)) \\ &q \in L^{\gamma}(\Omega; h^{\mu}([0, T]; W^{1,r}(D))) \end{aligned} \tag{2.59}$$

with $r > d$, $0 < \mu < 1/2$, and $\gamma \geq 2$, for any data f, u_0, small enough in appropriate spaces. Note that this regularity implies that the trajectories $(x, t) \to q(x, t, \omega)$ are continuous, for almost each event ω.

We now consider the finite element discretization in space. First, we assume that the existence result presented hereabove still holds when D is a convex polygon in $\mathbb{R}^2$. For any $h > 0$, let $\mathcal{T}_h$ be a decomposition of D into triangles K with diameter h_K less than h, regular in the sense of CIARLET and LIONS [1991]. We consider the finite element spaces V_h, Q_h, and R_h corresponding to continuous, piecewise linear velocity, pressure, and dumbbell elongations. We denote i_h the $L^2(D)$ projection onto V_h, Q_h, or R_h and introduce the following stabilized finite element discretization in space of (2.52)–(2.57). Given f, u_0, q_0 find

$$\begin{aligned} &(u_h, p_h, q_h) : (0, T) \times \Omega \to V_h \times Q_h \times R_h, \\ &\qquad (t, \omega) \to (u_h(t), p_h(t), q_h(t, \omega)), \end{aligned}$$

such that $u_h(0) = i_h u_0$, $q_h(0, \omega) = q_0(\omega)$ and such that the following weak formulation holds in $(0, T) \times \Omega$:

$$\begin{aligned}
&\rho \left(\frac{\partial u_h}{\partial t}, v_h \right)_D + 2\eta_s \left(\epsilon(u_h), \epsilon(v_h)\right)_D - (p_h, \operatorname{div} v_h)_D \\
&+ \frac{\eta_p}{\lambda} \left(\mathbb{E}\left(q_h (q_h)^T\right) - I, \epsilon(v_h)\right)_D - (f, v_h)_D \\
&+ (\operatorname{div} u_h, s_h)_D + \sum_{K \in \mathcal{T}_h} \frac{\alpha h_K^2}{2\eta_p} (\nabla p_h, \nabla s_h)_K \\
&+ (q_h(t), r_h)_D - (q_0, r_h)_D + \int_0^t \left(\frac{1}{2\lambda} q_h(k) - \nabla u_h(k) q_h(k), r_h \right)_D \mathrm{d}k \\
&\qquad - \frac{1}{\sqrt{\lambda}} (B(t), r_h)_D = 0, \qquad (2.60)
\end{aligned}$$

for all $(v_h, s_h, r_h) \in V_h \times Q_h \times R_h$. Here, $\alpha > 0$ is a dimensionless stabilization parameter.

In order to avoid complications when considering stochastic processes with value in Banach spaces, the following decomposition is introduced

$$q = q^{\mathrm{eq}} + q^d.$$

Here, $q^{\mathrm{eq}} : [0, T] \times \Omega \to \mathbb{R}^d$ corresponds to physical equilibrium and is the so-called *Ornstein–Uhlenbeck* stochastic process satisfying

$$\mathrm{d}q^{\mathrm{eq}} = -\frac{1}{2\lambda} q^{\mathrm{eq}} \mathrm{d}t + \frac{1}{\sqrt{\lambda}} \mathrm{d}B, \quad q^{\mathrm{eq}}(0) = q_0, \qquad (2.61)$$

while $q^d : D \times [0, T] \times \Omega \to \mathbb{R}^d$ satisfies a deterministic differential equation with a stochastic forcing term

$$\frac{\partial q^d}{\partial t} + \frac{1}{2\lambda} q^d - (\nabla u) q^d = (\nabla u) q^{\mathrm{eq}}, \quad q^d(0) = 0. \qquad (2.62)$$

Then, using the fact that

$$\mathbb{E}(q^{\mathrm{eq}}(s) q^{\mathrm{eq}}(t)^T) = \mathrm{e}^{-\frac{|t-s|}{2\lambda}} I, \quad s, t \in [0, T], \qquad (2.63)$$

the momentum equation (2.52) is

$$\begin{aligned}
&\rho \frac{\partial u}{\partial t} - 2\eta_s \operatorname{div} \epsilon(u) + \nabla p \\
&- \frac{\eta_p}{\lambda} \operatorname{div} \left(\mathbb{E}(q^d (q^d)^T + q^d (q^{\mathrm{eq}})^T + q^{\mathrm{eq}} (q^d)^T) \right) = f. \qquad (2.64)
\end{aligned}$$

As for the continuous problem, we use the decomposition

$$q_h = q^{\mathrm{eq}} + q_h^d;$$

thus, we are finally looking for (u_h, p_h, q_h^d) such that

$$\begin{aligned}
&\rho\left(\frac{\partial u_h}{\partial t}, v_h\right)_D + 2\eta_s\,(\epsilon(u_h), \epsilon(v_h))_D - (p_h, \operatorname{div} v_h)_D \\
&+ \frac{\eta_p}{\lambda}\mathbb{E}\left(q_h^d\left(q_h^d\right)^T + q_h^d\left(q^{\mathrm{eq}}\right)^T + q^{\mathrm{eq}}\left(q_h^d\right)^T, \epsilon(v_h)\right)_D - (f, v_h)_D \\
&+ (\operatorname{div} u_h, s_h)_D + \sum_{K\in\mathcal{T}_h} \frac{\alpha h_K^2}{2\eta_p}(\nabla p_h, \nabla s_h)_K \\
&+ (q_h^d(t), r_h)_D + \int_0^t \left(\frac{1}{2\lambda} q_h^d(k) - \nabla u_h(k)(q^{\mathrm{eq}}(k) + q_h^d(k)), r_h\right)_D \mathrm{d}k \\
&- \frac{1}{\sqrt{\lambda}}(B(t), r_h)_D \mathrm{d}k = 0
\end{aligned} \tag{2.65}$$

for all $(v_h, s_h, r_h) \in V_h \times Q_h \times R_h$.

As in the previous subsection, we write the above nonlinear problem as an abstract fixed point problem. Given $y = (f, u_0) \in Y$, find $x_h = (u_h, q_h^d) \in X_h$ such that

$$x_h = \mathsf{T}_h\left(y, S_1(x_h), S_2(x_h)\right). \tag{2.66}$$

Here, Y is the functional space for the data (f, u_0) (see BONITO, CLÉMENT and PICASSO [2006a] for details), and

$$X_h = L^2(0, T; V_h) \times L^2(\Omega; L^\infty(0, T; R_h)),$$

provided with the norm $||.||_{X_h}$ defined for all $x_h = (u_h, q_h) \in X_h$ by

$$||x_h||_{X_h}^2 = 2\eta_s \int_0^T ||\epsilon(u_h(t))||_{L^2(D)}^2 \mathrm{d}t + \int_\Omega \sup_{t\in[0,T]} ||q_h(\omega, t)||_{L^2(D)}^2 \,\mathrm{d}\mathcal{P}(\omega).$$

Also, the operators S_1 and S_2 are defined by

$$S_1(x_h) = \mathbb{E}\left(q_h^d (q_h^d)^T\right) \quad \text{and} \quad S_2(x_h) = \nabla u_h\, q_h^d,$$

while the linear operator T_h is defined by

$$\mathsf{T}_h : Y \times L^2(0, T; L^2(D)) \times L^2(\Omega; L^2(0, T; L^2(D))) \to X_h$$

$$(f_1, u_0, f_2, w) \to \mathsf{T}_h(f_1, u_0, f_2, w) = (\tilde{u}_h, \tilde{q}_h^d) \in X_h,$$

where for $(t, \omega) \in (0, T) \times \Omega$

$$(\tilde{u}_h, \tilde{p}_h, \tilde{q}_h^d) : (t, \omega) \to (\tilde{u}_h(t), \tilde{p}_h(t), \tilde{q}_h^d(t, \omega)) \in V_h \times Q_h \times R_h$$

satisfies $\tilde{u}_h(0) = i_h u_0$ and

$$\rho \left(\frac{\partial \tilde{u}_h}{\partial t}, v_h \right)_D + 2\eta_s \left(\epsilon(\tilde{u}_h), \epsilon(v_h)\right)_D - (\tilde{p}_h, \operatorname{div} v_h)_D$$
$$+ \frac{\eta_p}{\lambda} \left(\mathbb{E}(\tilde{q}_h^d (q^{\mathrm{eq}})^T + q^{\mathrm{eq}} (\tilde{q}_h^d)^T) + f_2, \epsilon(v_h) \right)_D - (f_1, v_h)_D$$
$$+ (\operatorname{div} \tilde{u}_h, s_h)_D + \sum_{K \in \mathcal{T}_h} \frac{\alpha h_K^2}{2\eta_p} (\nabla \tilde{p}_h, \nabla s_h)_K$$
$$\left(\tilde{q}_h^d(t), r_h\right)_D + \int_0^t \left(\frac{1}{2\lambda} \tilde{q}_h^d(k) - \nabla \tilde{u}_h(k) q^{\mathrm{eq}}(k) - w, r_h \right)_D \mathrm{d}k = 0, \tag{2.67}$$

for all $(v_h, s_h, r_h) \in V_h \times Q_h \times R_h$, a.e. in $(0, T)$ and a.e. in Ω.

In BONITO, CLÉMENT and PICASSO [2006a], it is proved that (2.67) has a unique solution converging to that of (2.64), (2.62). As in the previous subsection, (2.66) is rewritten as the following nonlinear problem: given $y = (f, u_0) \in Y$, find $x_h = (u_h, q_h^d) \in X_h$ such that

$$F_h(y, x_h) = 0, \tag{2.68}$$

where $F_h : Y \times X_h \to X_h$ is defined by

$$F_h(y, x_h) = x_h - \mathsf{T}_h\left(y, S_1(x_h), S_2(x_h)\right). \tag{2.69}$$

The abstract Theorem 2.1 of CALOZ and RAPPAZ [1997] can be then used in order to prove existence and convergence of a solution to (2.50). The mapping $F_h : Y \times X_h \to X_h$ is $\mathcal{C}^1$. Moreover, the scheme is consistent, $D_x F_h$ has bounded inverse at $i_h x$, and $D_x F_h$ is locally Lipschitz at $i_h x$. Here, i_h is the $L^2(D)$ projection onto the finite element space X_h, and $x = (u, q^d)$ is the solution of the continuous problem (2.62), (2.64), with regularity (2.59). Therefore, applying Theorem 2.1 of CALOZ and RAPPAZ [1997], existence of a semidiscrete solution x_h can be proved in the neighborhood of $i_h x$ provided the data y is small enough in Y. Moreover, optimal error estimates hold for $\|x - x_h\|_{X_h}$, that is:

$$\|u - u_h\|_{L^2(0,T;H^1(D))} + \|q^d - q_h^d\|_{L^2(\Omega;L^\infty(0,T;L^2(D)))} = O(h).$$

Note that the convergence result obtained here ensures the convergence of almost all trajectories. Also, a posteriori error estimates can be derived, and we refer to BONITO, CLÉMENT and PICASSO [2006a] for details.

CHAPTER 3

Numerical Simulation of Viscoelastic Flows with Complex Free Surfaces

3.1. Space discretization: structured cells and finite elements

We now come back to the splitting scheme described in Section 2.2.1 to solve the free-surface Oldroyd-B model. Recall that during the prediction step, three advection problems have to be solved, which leads to formula (2.10)–(2.12), whereas during the correction step, the Oldroyd-B problem without convection (2.14)–(2.16) has to be solved.

Two distinct grids are used to solve the prediction and correction steps (see Fig. 3.1). Since the shape of the cavity Λ can be complex (this is for instance the case in mold filling or extrusion processes), finite element techniques are well suited for solving (2.14)–(2.16) using an unstructured mesh. On the other hand, a structured grid of cubic cells is used to implement (2.10)–(2.12). The reasons for using a structured grid is the following. First, the method of characteristics can be easily implemented on structured grids. Second, the size of the cells can be tuned in order to control numerical diffusion when projecting (2.10)–(2.12) on the structured grid. Numerical experiments reported in CABOUSSAT, PICASSO and RAPPAZ [2005], MARONNIER, PICASSO and RAPPAZ [1999, 2003] have shown that choosing the cells spacing three to five times smaller than the mesh spacing is a good trade-off between numerical diffusion and computational cost or memory storage.

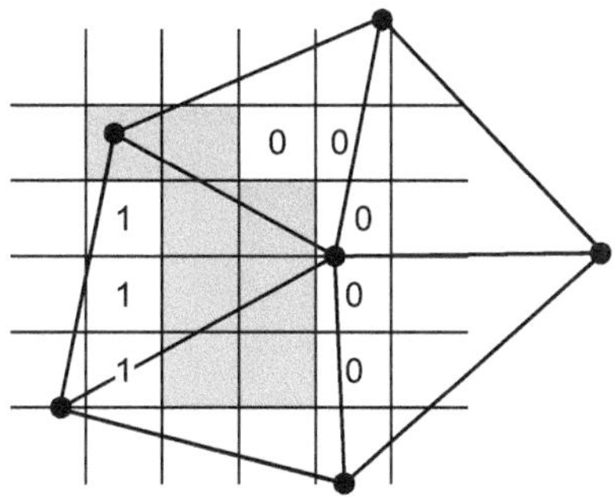

FIG. 3.1 Two grids are used for the computations. In order to reduce numerical diffusion and to simplify the implementation, the volume fraction of liquid is computed on a structured grid of small cells. The velocity, pressure, and extra stress are computed on an unstructured finite element mesh with larger size. The symbol 1 (resp. 0) denotes a cell completely filled (resp. empty). The cells that are partially filled are shaded. The goal is to reduce the width of the partially filled region to a value smaller than the finite element spacing.

We also refer to TOMÉ, CASTELO and CUMINATO [2008] for similar numerical simulations using finite difference methods.

3.1.1. Advection step: structured grid of cubic cells

The implementation of (2.10)–(2.12) is now discussed. Assume that the grid is made out of cubic cells C_{ijk} of size h. Let φ_{ijk}^{n-1}, u_{ijk}^{n-1}, and σ_{ijk}^{n-1} be the approximate value of φ, u, and σ at center of cell number (ijk) and time t^{n-1}. According to (2.10)–(2.12), the advection step on cell number (ijk) consists in advecting φ_{ijk}^{n-1}, u_{ijk}^{n-1} and σ_{ijk}^{n-1} by $\Delta t^n u_{ijk}^{n-1}$ and then projecting the values onto the structured grid. An example of cell advection and projection is presented in Fig. 3.2 in two space dimensions.

This advection algorithm is unconditionally stable with respect to the CFL condition – velocity times the time step divided by the cells spacing h – and $O(\Delta t + h^2/\Delta t)$ convergent, according to the theoretical results available for the characteristics-Galerkin method PIRONNEAU [1989], PIRONNEAU, LIOU and TEZDUYAR [1992], QUARTERONI and VALLI [1991]. However, this algorithm has two drawbacks. Indeed, numerical diffusion is introduced when projecting the values of the advected cells on the grid (recall that the volume fraction of liquid is discontinuous across the interface). Moreover, if the time step is too large, two cells may arrive at the same place, producing numerical (artificial) compression.

In order to enhance the quality of the volume fraction of liquid, two postprocessing procedures have been implemented. We refer to CABOUSSAT [2005], MARONNIER, PICASSO and RAPPAZ [1999, 2003] for a description in two and three space dimensions. The first procedure reduces numerical diffusion and is a simplified implementation of the simple

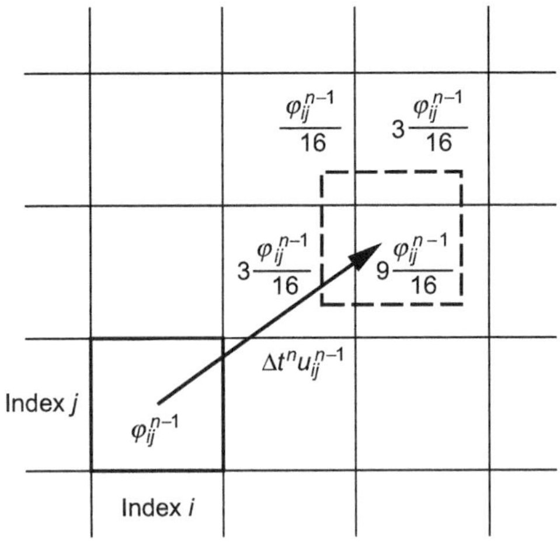

FIG. 3.2 An example of two-dimensional advection of φ_{ij}^{n-1} by $\Delta t^n u_{ij}^{n-1}$ and projection on the grid. The advected cell is represented by the dashed lines. The four cells containing the advected cell receive a fraction of φ_{ij}^{n-1}, according to the position of the advected cell. In this example, the new values of the volume fraction of liquid φ^n are updated as follows: $\varphi_{i+1,j+1}^n = \varphi_{i+1,j+1}^n + 3/16\varphi_{ij}^{n-1}$; $\varphi_{i+2,j+1}^n = \varphi_{i+2,j+1}^n + 9/16\varphi_{ij}^{n-1}$; $\varphi_{i+1,j+2}^n = \varphi_{i+1,j+2}^n + 1/16\varphi_{ij}^{n-1}$; $\varphi_{i+2,j+2}^n = \varphi_{i+2,j+2}^n + 3/16\varphi_{ij}^{n-1}$.

linear interface calculation (SLIC) algorithm CHORIN [1980], NOH and WOODWARD [1976], SCARDOVELLI and ZALESKI [1999]; see Figs 3.3 and 3.4 for a simple example. In the SLIC procedure, if a cell is partially filled with liquid, then the volume fraction of liquid is condensed along the cells faces, edges, or corners (see Fig. 3.5), according to the volume fraction of liquid of the neighboring cells (see Fig. 3.6).

The second procedure removes artificial compression (that is, values of the volume fraction of liquid greater than one), which may happen when the volume fraction of liquid advected in two cells arrives at the same place (see Fig. 3.7). The aim of this procedure is to produce new values φ_{ijk}^{n} that are between zero and one and is as follows. At each time step, all the cells having values φ_{ijk}^{n} greater than one (strictly) or between zero and one (strictly)

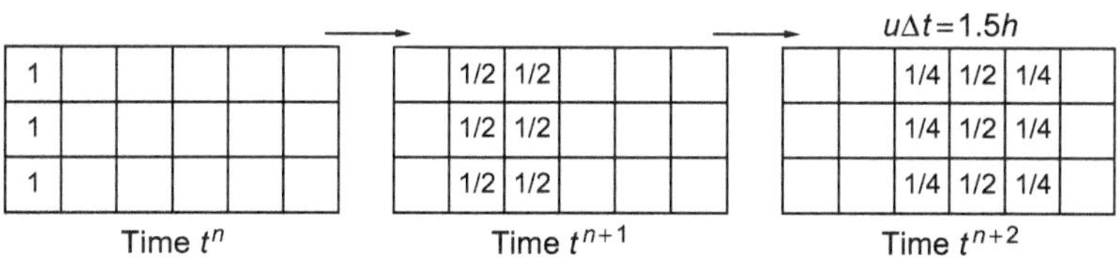

FIG. 3.3 Numerical diffusion during the advection step. At time t^n, the cells have volume fraction of liquid one or zero. The velocity u is horizontal, and the time step Δt is chosen so that $u\Delta t = 1.5h$ where h is the cells spacing.

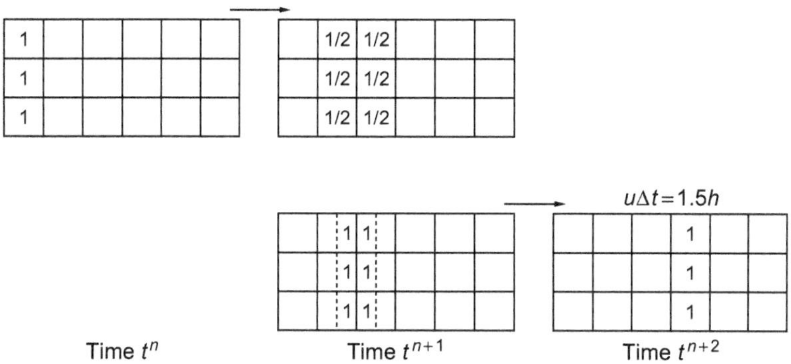

FIG. 3.4 Reducing numerical diffusion using the SLIC algorithm. Before advecting a cell partially filled with liquid, the volume fraction of liquid is condensed along the cells boundaries, according to the neighboring cells.

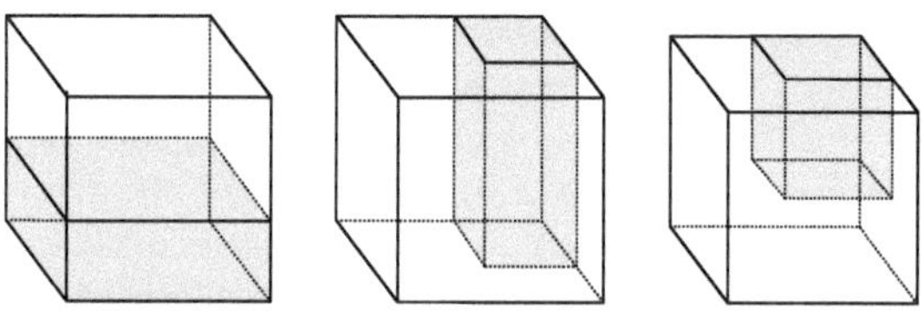

FIG. 3.5 SLIC algorithm. If the cell is partially filled with liquid, the liquid is pushed along a face, an edge, or a vertex of the cell, according to the neighbors volume fraction of liquid.

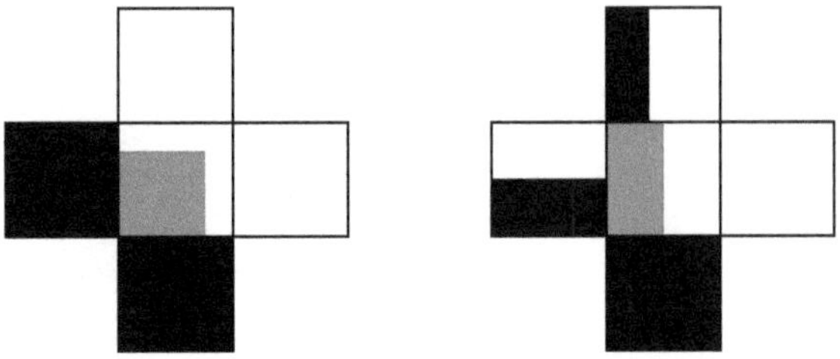

FIG. 3.6 SLIC algorithm. The volume fraction of liquid in a cell partially filled with liquid is pushed according to the volume fraction of liquid of the neighboring cells. Two examples are proposed. Left: the left and bottom neighboring cells are full of liquid, the right and top neighboring cells are empty, and the liquid is pushed at the bottom-left corner of the cell. Right: the bottom neighboring cell is full of liquid, the right neighboring cell is empty, the other two neighboring cells are partially filled with liquid, and the volume fraction of liquid is pushed along the left side of the cell.

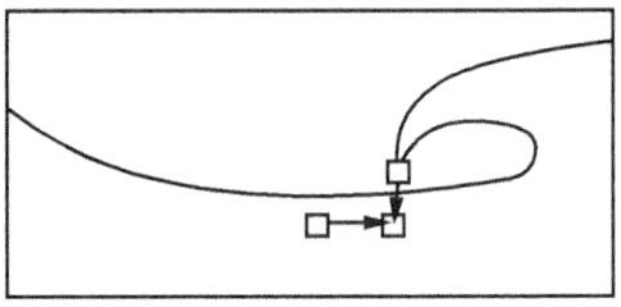

FIG. 3.7 An example of numerical (artificial) compression.

are sorted according to their values φ_{ijk}^n. This can be done in an efficient way using quick sort algorithms. The cells having values φ_{ijk}^n greater than one are called the dealer cells, whereas the cells having values φ_{ijk}^n between zero and one are called the receiver cells. The second procedure then consists in moving the fraction of liquid in excess in the dealer cells to the receiver cells; see MARONNIER, PICASSO and RAPPAZ [1999, 2003] for details.

Validation of these procedures using standard two-dimensional test cases taken from AULISA, MANSERVISI and SCARDOVELLI [2003], RIDER and KOTHE [1998] have been performed in CABOUSSAT [2005]. Translation, rotation, and stretching of a circular region of fluid are shown in Fig. 3.8. For more details, we refer to Section 5.1 of CABOUSSAT [2005].

In a number of industrial applications, the shape of the cavity containing the liquid is complex. Therefore, a special data structure has been implemented in order to reduce the memory requirements used to store the cell data. An example is proposed in Fig. 3.9. The cavity containing the liquid is meshed into tetrahedrons. Without any particular cells data structure, a great number of cells would be stored in the memory without ever being used. The data structure makes use of three hierarchical levels to define the cells. At the coarsest level, the cavity is meshed into windows, which can be glued together. Each window is then subdivided into blocks. Finally, a block is cut into smaller cubes, namely the cells (ijk). When a block is free of liquid ($\varphi = 0$), it is switched off, that is to say the memory corresponding to the cells is not allocated. When liquid enters a block, the block is switched on, that is to say the memory corresponding to the cells is allocated.

Once values φ_{ijk}^n, $u_{ijk}^{n-\frac{1}{2}}$ and $\sigma_{ijk}^{n-\frac{1}{2}}$ have been computed on the cells (ijk), values are interpolated at the vertices P of the finite element mesh. More precisely, the volume fraction of

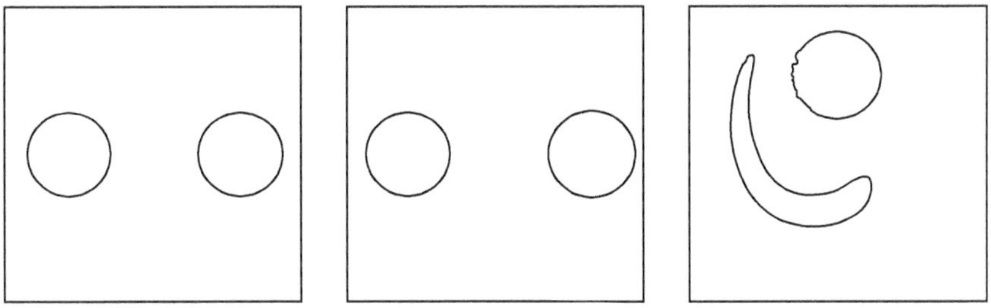

FIG. 3.8 Validation of the advection step. Left: translation of a circular region of liquid, the interface is shown at initial and final time. Middle: rotation of a circular region of liquid, the interface is shown at initial and final time. Right: single vortex test case, the interface is shown at time $t = 1$ (maximal deformation) and $t = 2$ s (return to initial circular shape).

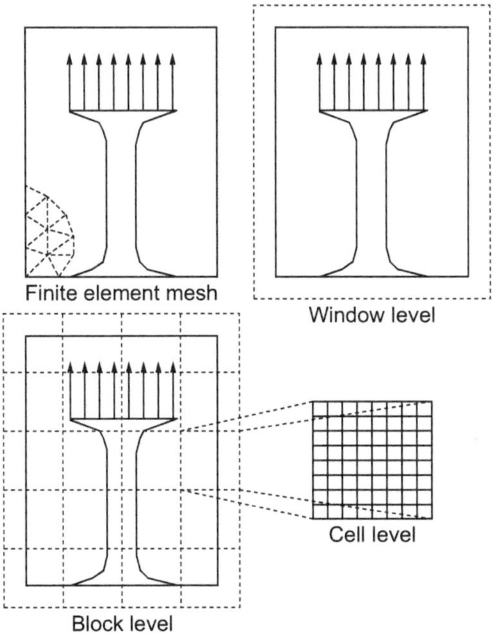

FIG. 3.9 The hierarchical window-block-cell data structure used to implement cells advection in the framework of the 2D filament stretching.

liquid at vertex P is computed by considering all the cells (ijk) contained in the triangles K containing vertex P (see Fig. 3.10), using the following formula:

$$\varphi^n(P) = \frac{\displaystyle\sum_{\substack{K \\ P\in K}} \sum_{(ijk)\subset K} \phi_P(x_{ijk})\varphi^n_{ijk}}{\displaystyle\sum_{\substack{K \\ P\in K}} \sum_{(ijk)\subset K} \phi_P(x_{ijk})}. \tag{3.1}$$

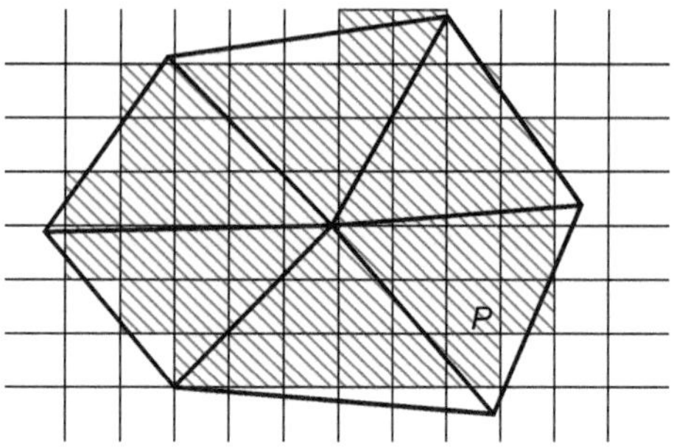

FIG. 3.10 Interpolation of the volume fraction of liquid from the structured cells to the unstructured finite element mesh. The volume fraction of liquid at vertex P depends on the volume fraction of liquid in the shaded cells.

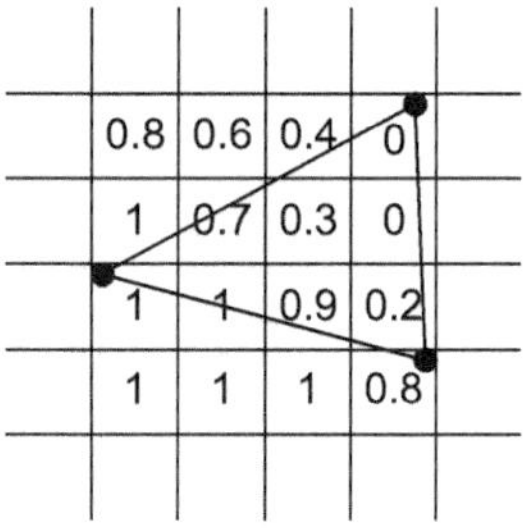

FIG. 3.11 A two-dimensional example of liquid element. The values of the volume fraction of liquid φ at the center of the cells are known. A value φ is then interpolated at the vertices of the finite element mesh. The displayed triangle has at least one vertex with value φ greater than 0.5. Therefore, the triangle is liquid, and the velocity, the pressure, and the extra stress will be computed at the three vertices of the triangle.

Here, x_{ijk} denotes the center of cell (ijk), and ϕ_P is the finite element basis function attached to vertex P. Similar formula hold for the velocity and extra stress. Then, the liquid region is defined as follows. An element (tetrahedron) of the mesh is said to be liquid if (at least) one of its vertices has a volume fraction of liquid $\varphi^n > 0.5$ (see Fig. 3.11). The computational domain D^n used for solving (1.2)–(1.8) is then defined to be the union of all liquid elements. At this point, we would like to stress that the values of the volume fraction of liquid on the unstructured finite element mesh are only used in order to define the liquid region. Again, advection of the volume fraction of liquid only occurs on the structured cells and not on the unstructured finite element mesh. Also, the volume constraint is not directly enforced in the numerical model. However, if numerical diffusion of the volume fraction of liquid is small, then the volume constraint will be satisfied. This is precisely the goal of the two post processing procedures that have been added. In all the computations, we have observed that the (numerical) diffusion layer of the volume fraction of liquid $(0 < \varphi < 1)$ is of the order of one or two cells and that the volume constraint is satisfied up to 1%. In order to achieve this goal, the two post processing procedures must be switched on and the cells spacing must be three to five times smaller than the mesh spacing.

3.1.2. Correction step: stokes and Oldroyd-B with finite elements

Let us now turn to the finite element techniques used for solving (2.14)–(2.16). Given the new liquid domain D^n (remember that D^n is the union of liquid elements belonging to the mesh), let V_h, Q_h, and M_h be the finite element subspaces of continuous, piecewise linear velocity, pressure, and extra stress defined on D^n. We follow Section 2.3.2 and use an elastic viscous split stress (EVSS) formulation with continuous, piecewise linear stabilized finite elements. More precisely, given the predicted velocity $u_h^{n-1/2} \in V_h$, the extra-variable $d_h^{n-1/2} \in M_h$ defined by

$$\left(d_h^{n-1/2}, e_h\right)_{D^n} = \left(\epsilon\left(u_h^{n-1/2}\right), e_h\right)_{D^n} \qquad \forall e_h \in M_h,$$

is introduced for stability purposes. Solving this equation results in solving a diagonal linear system provided a mass lumping quadrature formula is used. Since the mass lumping quadrature formula is order two accurate in space, the global accuracy of the method is not affected. Once $d_h^{n-1/2}$ is computed, the predicted extra-stress $\sigma_h^{n-1/2}$ being known, the new velocity $u_h^n \in V_h$ and pressure $p_h^n \in Q_h$ are obtained by solving the following Stokes problem

$$\begin{aligned}
&\frac{\rho}{\Delta t^n}\left(u_h^n, v_h\right)_{D^n} + 2(\eta_s+\eta_p)\left(\epsilon(u_h^n), \epsilon(v_h)\right)_{D^n} - \left(p^n, \operatorname{div} v_h\right)_{D^n}\\
&\quad = \frac{\rho}{\Delta t^n}\left(u_h^{n-1/2}, v_h\right)_{D^n}\left(2\eta_p d_h^{n-1/2} - \sigma^{n-1/2}, \epsilon(v_h)\right)_{D^n} + (\rho g, v_h)_{D^n},\\
&\left(\operatorname{div} u_h^n, q_h\right)_{D^n} + \sum_{K\subset D^n}\alpha_K\left(\frac{\rho}{\Delta t^n}u_h^n + \nabla p_h^n, \nabla q_h\right)_{D^n}\\
&\quad = \sum_{K\subset D^n}\alpha_K\left(\frac{\rho}{\Delta t^n}u_h^{n-1/2} + \operatorname{div}\sigma_h^{n-1/2} + \rho g, \nabla q_h\right)_{D^n},
\end{aligned} \tag{3.2}$$

for all test functions $v_h \in V_h$ and $q_h \in Q_h$. Here, α_K is the local stabilization coefficient defined by

$$\alpha_K = \begin{cases} \dfrac{|K|^{2/3}}{12(\eta_s+\eta_p)} & \text{if } \mathrm{Re}_K \le 3,\\[2ex] \dfrac{|K|^{2/3}}{4\mathrm{Re}_K(\eta_s+\eta_p)} & \text{else,}\end{cases}$$

where, following FRANCA and FREY [1992], the local Reynolds number Re_K is defined by

$$\mathrm{Re}_K = \frac{\rho|K|^{\frac{1}{3}}\|u_h^{n-1/2}\|_{L^\infty(K)}}{2(\eta_s+\eta_p)}.$$

Note that in (3.2), the corrected velocity u_h^n can be prescribed on the boundary of the cavity Λ whenever needed; see Fig. 1.5 for a discussion related to boundary conditions. Also note that the boundary condition (1.17) is implicitly contained in the above variational formulation. All the degrees of freedom corresponding to velocity and pressure are stored in a single matrix, and the linear system is solved using the GMRES algorithm with a classical incomplete LU preconditioner and no restart.

It then remains to update the extra-stress $\sigma_h^n \in M_h$ from Oldroyd-B constitutive equation:

$$\left(1+\frac{\lambda}{\Delta t^n}\right)(\sigma^n, \tau)_{D^n} = \left(\frac{\lambda}{\Delta t^n}\sigma^{n-1/2} + \lambda \nabla u_h^n \sigma^{n-1/2} + \lambda \sigma^{n-1/2}(\nabla u_h^n)^T \right.$$
$$\left. + 2\eta_p(\epsilon(u_h^n), \tau\right)_{D^n} \qquad \forall \tau \in M_h.$$

Here, σ_h^n must be prescribed at the inflow boundary, if there is one (see Fig. 1.5). Again, this equation results in solving a diagonal linear system whenever a mass lumping quadrature formula is used.

Finally, once the new velocity u_h^n and extra-stress σ_h^n are computed at the vertices of the finite element mesh, values are interpolated at the center of the cells (ijk):

$$u_{ijk}^n = \sum_P \phi_P(x_{ijk}) u_P^n, \tag{3.3}$$

where P denotes a mesh vertex, x_{ijk} denotes the center of cell (ijk), ϕ_P denotes the finite element basis function corresponding to vertex P, and u_P^n is the velocity at vertex P. A similar formula is used for the extra-stress σ_{ijk}^n. Please note that the volume fraction of liquid is not interpolated from the finite element mesh to the cells. Indeed, the volume fraction of liquid is only computed on the structured cells. It is interpolated on the unstructured finite element mesh only in order to define the liquid region after the prediction step; see Fig. 3.11.

3.1.3. Implementations details

The memory storage is the following. For each cubic cell, the volume fraction of liquid, the velocity, and the extra stress must be stored in order to implement (2.10)–(2.12), therefore $1 + 3 + 6 = 10$ values. For each vertex of the finite element mesh, the velocity, the pressure, the extra stress, and the EVSS field $d_h^{n-1/2}$ must be stored, therefore $3 + 1 + 6 + 6 = 16$ values. The code is written in the C++ programming language, and the finite element data structure is classical. The data structure of the cells is as follows. Each cell is labeled by indices (ijk) within a block. Also, each block is labeled by indices (ijk) within a window (see Fig. 3.9).

In order to perform efficient interpolation between the two grids (structured cells/unstructured finite elements), the following data structure is needed. In order to implement interpolation from the finite element mesh to the cells, Eqn (3.3), the index of the finite element (tetrahedron) containing each cell is needed. Alternatively, in order to implement interpolation from the cells to the finite element mesh, Eqn (3.1), the list of the cells contained in each finite element (tetrahedron) is required. This additional data structure is built at the beginning of each computation. It can be stored in case several computations are performed with the same grids. The additional CPU time required to build this data structure is small (less than 1%) compared to the total CPU time.

3.2. Extension to mesoscopic models

Computations with the FENE free-surface algorithm presented in Section 2.2.2 have been performed in GRANDE, LASO and PICASSO [2003] in two space dimensions and are not reported here. The use of variance reduction techniques is advocated (see for instance BONVIN and PICASSO [1999], and JOURDAIN, LELIÈVRE and LE BRIS [2004b]).

3.3. Numerical results

In this section, numerical results pertaining to Oldroyd-B three-dimensional flows with complex free surfaces are presented. First, the method is validated on test cases for which an exact solution is available. Then, numerical simulations are proposed on two test cases involving flows with complex free surfaces, namely jet buckling and the stretching of a filament.

3.3.1. Numerical validation

Elongational flow At initial time, liquid at rest occupies a cylinder with radius $R_0 = 0.0034$ m and height $L_0 = 0.0019$ m. Then, the velocity field on the top and bottom sides of the cylinder is imposed to be

$$u(x, y, z, t) = \begin{pmatrix} -\dfrac{1}{2}\dot{\epsilon}_0 x \\ -\dfrac{1}{2}\dot{\epsilon}_0 y \\ \dot{\epsilon}_0 z \end{pmatrix},$$

with $\dot{\epsilon}_0 = 4.68\ \text{s}^{-1}$, whereas (1.17) applies on the lateral sides. Since there is no inflow velocity, no boundary conditions have to be enforced for the extra stress. A simple calculation shows that for all time t, the above velocity field satisfies the momentum equations, that the extra-stress tensor is homogeneous, for instance

$$\sigma_{zz}(x, y, z, t) = \frac{2\eta_p \dot{\epsilon}_0}{1 - 2\dot{\epsilon}_0 \lambda}\left(1 - \mathrm{e}^{-\left(\frac{1}{\lambda} - 2\dot{\epsilon}_0\right)t}\right),$$

and that the liquid region remains a cylinder with radius $R(t) = R_0 \mathrm{e}^{-\frac{1}{2}\dot{\epsilon}_0 t}$. Indeed, the trajectories of the fluid particles are defined by $X'(t) = u(X(t), t)$, which yields

$$\begin{pmatrix} X(t) = X(0)\mathrm{e}^{-\frac{1}{2}\dot{\epsilon}_0 t} \\ Y(t) = Y(0)\mathrm{e}^{-\frac{1}{2}\dot{\epsilon}_0 t} \\ Z(t) = Z(0)\mathrm{e}^{\dot{\epsilon}_0 t} \end{pmatrix}.$$

Two meshes are used for the computations. The computational domain is the block $[-0.004\,\text{m}, 0.004\,\text{m}] \times [-0.004\,\text{m}, 0.004\,\text{m}] \times [0\,\text{m}, 0.03\,\text{m}]$ in the xyz directions. The 3D meshes are obtained by extruding the 2D meshes shown in Fig. 3.12, from $z = 0$ to $z = 0.03$ m, and then cutting the prisms into tetrahedrons. The coarse (resp. fine) mesh has

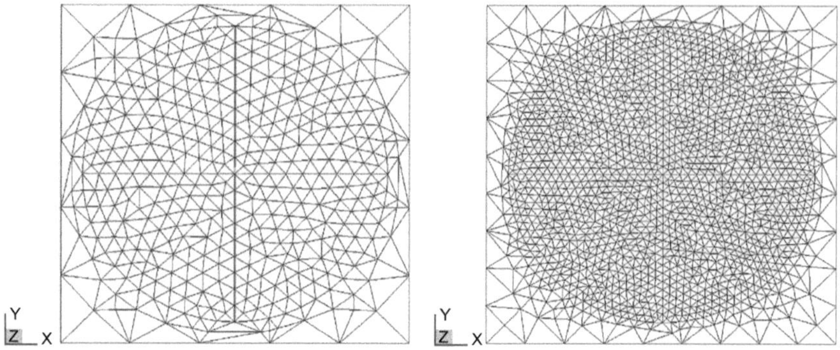

FIG. 3.12 Elongational flow: 2D cut of the mesh at $z = 0$; left: coarse mesh; right: fine mesh.

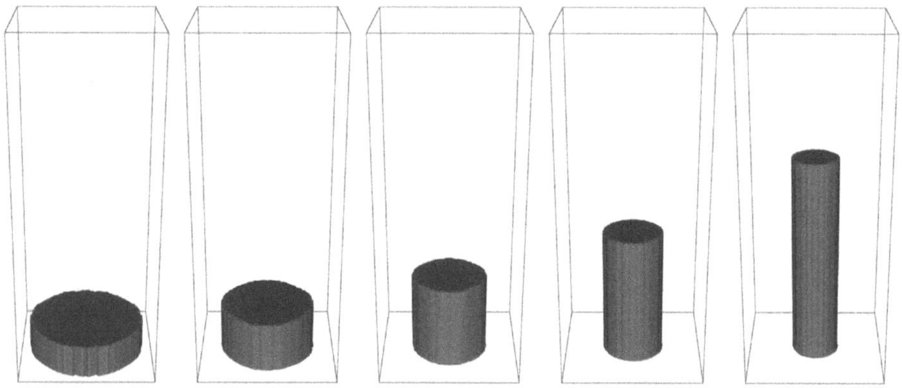

FIG. 3.13 Elongational flow: shape of the liquid region (the volume corresponding to volume fraction of liquid $\varphi > 0.5$ is shown); simulation at different times (from left to right): $t = 0, 0.1, 0.2, 0.3, 0.4$ s.

62000 (resp. 462000) vertices and mesh size 0.00035 m (resp. 0.000175 m). When using the coarse (resp. fine) mesh, the cell size is 0.0001 m (resp. 0.00005 m). The time step was $\Delta t = 0.01$ s for the coarse mesh (resp. $\Delta t = 0.005$ s for the fine mesh) so that the CFL number of the cells – velocity times the time step divided by the cells spacing – equals 0.9 at time $t = 0$ and 3.7 at time $t = 0.3$.

Numerical results corresponding to 0.05 % by weight Polystyrene (the parameter values are taken from CORMENZANA, LEDD, LASO and DEBBAUT [2001], $\rho = 1030\,\text{kg/m}^3$, $\eta_s = 9.15\,\text{Pa}\cdot\text{s}$, $\eta_p = 25.8\,\text{Pa}\cdot\text{s}$, $\lambda = 0.421$ s, thus $\text{De} = \lambda\dot{\epsilon}_0 = 1.97$) are reported in Figs 3.13 and 3.14. Clearly, the computed velocity agrees perfectly with the exact velocity, whereas the error for the extra stress is within 10% on the fine grid. The fact that the velocity is more precise than the extra stress is not surprising since the finite element method is expected to be of order two (in the L^2 norm and in a fixed domain) for the velocity but only of order one for the extra stress.

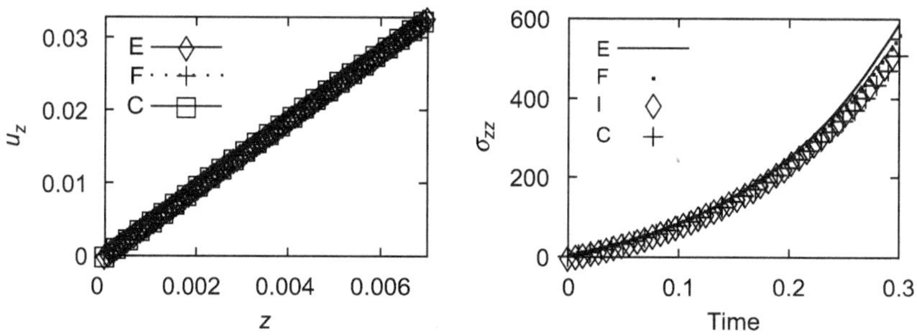

FIG. 3.14 Elongational flow (E = exact solution, F = fine mesh, I = intermediate mesh, C = coarse mesh); left: vertical velocity u_z along the vertical axis Oz at final time $t = 0.3$ s; right: extra-stress σ_{zz} at $z = 0.0006$ m as a function of time.

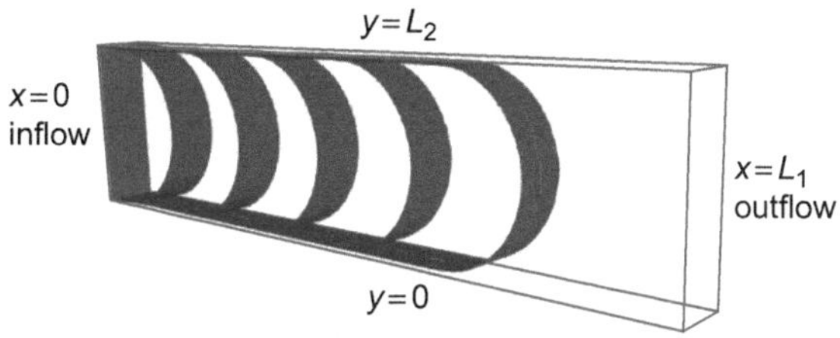

FIG. 3.15 Filling of a pipe; notations and isovalue $\varphi = 0.5$ for a Newtonian fluid at times $t = 0,\ 0.6,\ 1.2,\ 1.8,\ 2.4,\ 3.0$ s.

Filling of a straight pipe Consider a rectangular pipe of dimensions $[0, L_1] \times [0, L_2] \times [0, L_3]$ in the xyz directions, where $L_1 = 4$ m, $L_2 = 1$ m, $L_3 = 0.3$ m. At the initial time, the pipe is empty. Then, fluid enters from the left side ($x = 0$) with velocity and extra stress given by

$$u(x, y, z, t) = \begin{pmatrix} u_x \\ 0 \\ 0 \end{pmatrix}, \quad \sigma(x, y, z, t) = \begin{pmatrix} \sigma_{xx} & \sigma_{xy} & 0 \\ \sigma_{xy} & 0 & 0 \\ 0 & 0 & 0 \end{pmatrix}, \tag{3.4}$$

with $u_x(y) = 6y(L_2 - y)$, $\sigma_{xx}(y) = 72\eta_s\lambda(2y - L_2)^2$, and $\sigma_{xy}(y) = -6\eta_p(2y - L_2)$. The boundary conditions are detailed in Fig. 3.15 and are the following. On the top and bottom sides ($y = 0$ and $y = L_2$), no-slip boundary conditions apply. On the front and rear sides ($z = 0$ and $z = L_3$), slip boundary conditions apply. On the right side ($x = L_1$), the fluid is free to exit the pipe with zero vertical velocity. The parameter values are taken from TOMÉ, MANGIAVACCHI, CUMINATO, CASTELO and MCKEE [2002] Subsection 6.1 and are the following: $\rho = 1\,\text{kg/m}^3$, $\eta_s = \eta_p = 0.5\,\text{Pa}\cdot\text{s}$. Three finite element meshes are used in this subsection (see Table 3.1 for details). The cells spacing is five times smaller than the finite element mesh spacing.

We first consider the filling of the pipe, starting from an empty pipe. This experiment has been considered in PICHELI and COUPEZ [1998], TOMÉ, MANGIAVACCHI, CUMINATO,

TABLE 3.1
Filling of a pipe; the three mesh used to check convergence

Mesh	Subdivisions (radius × height)	Vertices	Tetrahedrons
Coarse	40 × 10 × 3	1804	7200
Intermediate	80 × 20 × 6	11900	57600
Fine	160 × 40 × 12	85813	460800

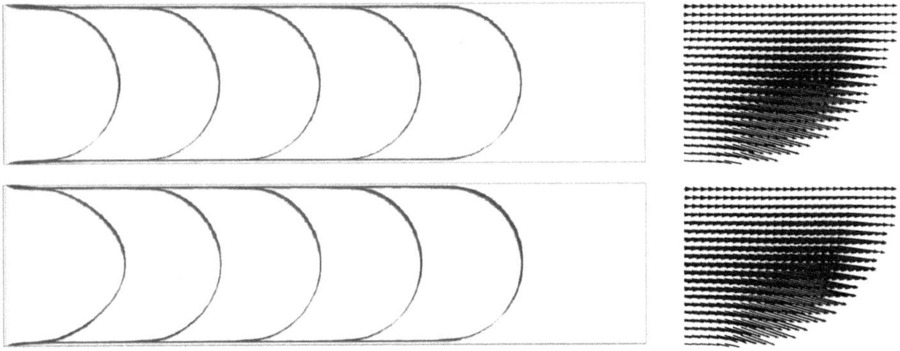

FIG. 3.16 Filling of a pipe. Left: position of the free surface at time $t = 0,\ 0.6,\ 1.2,\ 1.8,\ 2.4,\ 3.0$ s. Right: velocity field close to the free surface at time $t = 1.8$ s. Top : Newtonian flow; bottom: viscoelastic flow ($\lambda = 5$ s thus De $= 5$).

CASTELO and MCKEE [2002] and is sometimes called fountain flow. The imposed velocity and extra-stress profile at the inlet are those corresponding to Poiseuille flow (see (3.4)). Following GRAMBERG, VAN VROONHOVEN and VAN DE VEN [2004], after some time, the shape of free surface should be close to a half circle. In Fig. 3.15, the velocity and the shape of the free surface is shown at several times. The mesh is the finest one, and the time step is $\Delta t = 0.03$ s so that the CFL number of the cells – velocity times the time step divided by the cells spacing – equals 4.5. Away from the inlet, the position of the free surface is the same for both Newtonian and viscoelastic flows (see Fig. 3.16). As predicted theoretically GRAMBERG, VAN VROONHOVEN and VAN DE VEN [2004], the shape is almost circular. Details of the fountain flow at the free surface is provided in Fig. 3.16.

Once totally filled with liquid, the velocity and extra stress must satisfy (3.4) in the whole pipe. Convergence of the stationary solution is checked with $\lambda = 1$ s, thus De $= \lambda U/L_2 = 1$, where $U = 1$ m/s is the average velocity. In Fig. 3.17, σ_{xx}, σ_{xy}, and u_x are plotted along the vertical line $x = L_1/2,\ 0 \leq y \leq L_2,\ z = L_3/2$. Convergence can be observed even though boundary layer effects are present, this being classical with low-order finite elements. In Fig. 3.18, the error in the L^2 norm of σ_{xx}, σ_{xy}, and u_x is plotted versus the mesh size. Clearly, order one convergence rate is observed for the extra stress, order two for the velocity, this being consistent with theoretical predictions on simplified problems.

3.3.2. *Jet buckling*

The transient flow of a 3D jet injected into a parallelepiped cavity is now reproduced. The cavity is a parallelepiped of width 0.05 m, depth 0.05 m, and height 0.1 m, the diameter of

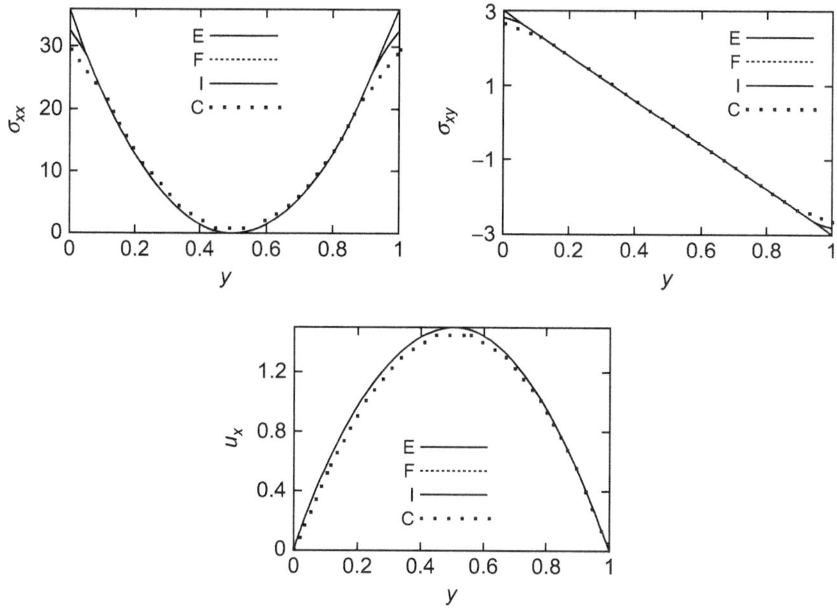

FIG. 3.17 Filling of a pipe; in all the plots E = exact solution, F = fine mesh, I = intermediate mesh, and C = coarse mesh; top left: $\sigma_{xx} = 72\eta_s\lambda(2y - L_2)^2$ along the vertical line $x = L_1/2$, $0 \leq y \leq L_2$, $z = L_3/2$; middle: $\sigma_{xy} = -6\eta_p(2y - L_2)$, bottom: horizontal velocity $u_x = 6y(L_2 - y)$.

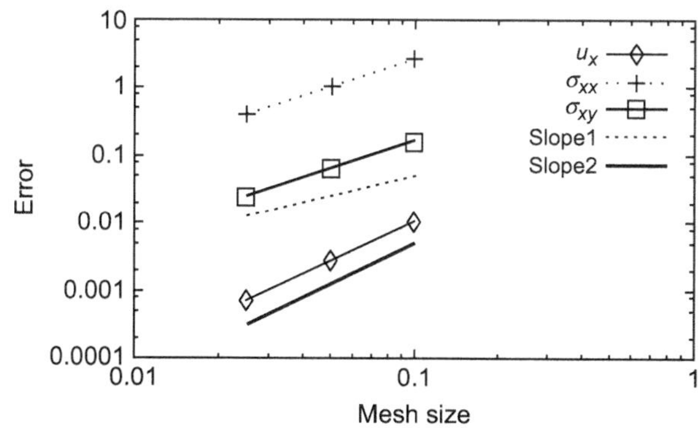

FIG. 3.18 Filling of a pipe; error in the L^2 norm with respect to the mesh size.

the jet being $D = 0.005\,\text{m}$. Liquid enters from the top of the cavity with vertical velocity $U = 0.5\,\text{m/s}$. The fluid parameters are $\rho = 1030\,\text{kg/m}^3$; in the Newtonian case, the viscosity is $\eta_s + \eta_p = 10.3\,\text{Pa} \cdot \text{s}$, and $\lambda = 0\,\text{s}$; in the viscoelastic case, the viscosities are $\eta_s = 1.03$, $\eta_p = 9.27\,\text{Pa} \cdot \text{s}$, and the relaxation time $\lambda = 1\,\text{s}$ so that $\text{De} := \lambda U/D = 100$. The finite element mesh has 503171 vertices and 2918760 tetrahedrons. The cells size is 0.0002 m, and the time step is 0.001 s; thus, the CFL number of the cells – velocity times the time step divided by the cells spacing – is 2.5. The shape of the jet is shown in

FIG. 3.19 Jet buckling in a thick cavity. Shape of the jet at time $t = 0.125$ s (col. 1), $t = 0.45$ s (col. 2), $t = 0.6$ s (col. 3), $t = 0.9$ s (col. 4), $t = 1.15$ s (col. 5), $t = 1.6$ s (col. 6), Newtonian fluid (row 1), viscoelastic fluid De $= 100$ (row 2).

Figs 3.19–3.20 for Newtonian and viscoelastic flows. This computation took 64 h on a AMD opteron CPU with 8-Gb memory.

In TOMÉ and MCKEE [1999], Tomé and McKee provided an empirical threshold on the Reynolds number for a Newtonian jet to buckle. Our experiments indicates that this relation does not hold for viscoelastic flow and that the Weissenberg number should be taken into account (see BONITO, PICASSO and LASO [2006]).

3.3.3. *Filament stretching*

The flow of an Oldroyd-B fluid contained between two parallel coaxial circular disks with radius $R_0 = 0.003$ m is considered. At the initial time, the distance between the two end-plates is $L_0 = 0.0019$ m, and the liquid is at rest. Then, the top end-plate is moved vertically with velocity $L_0\dot{\epsilon_0}e^{\dot{\epsilon_0}t}$. The model data ($\rho$, η_s, η_p, λ, $\dot{\epsilon_0}$) and the fine mesh described in the *Elongational flow* test case above are used here. The initial time step is $\Delta t^0 = 0.005$ s yielding an initial CFL number of the cells – velocity times the time step divided by the cells spacing – close to one. Moreover, the time step at time t^n is chosen so that the distance of the moving end-plate between two time steps is constant, that is,

$$\Delta t^n = \Delta t^{n-1} e^{-\dot{\epsilon_0}\Delta t^{n-1}}.$$

Therefore, the CFL number remains constant throughout the simulation. The shape of the liquid region at time $t = 0.5$ s is represented in Fig. 3.21, for both Newtonian and

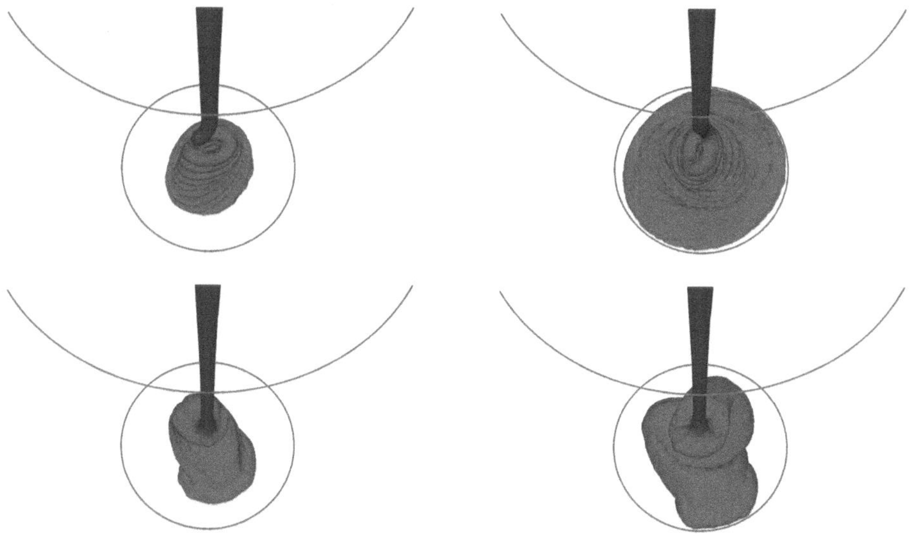

FIG. 3.20 Jet buckling in a thick cavity. View from the top. Shape of the jet at time $t = 0.6$ s (col. 1), $t = 1.15$ s (col. 2), Newtonian fluid (row 1), viscoelastic fluid De = 100 (row 2).

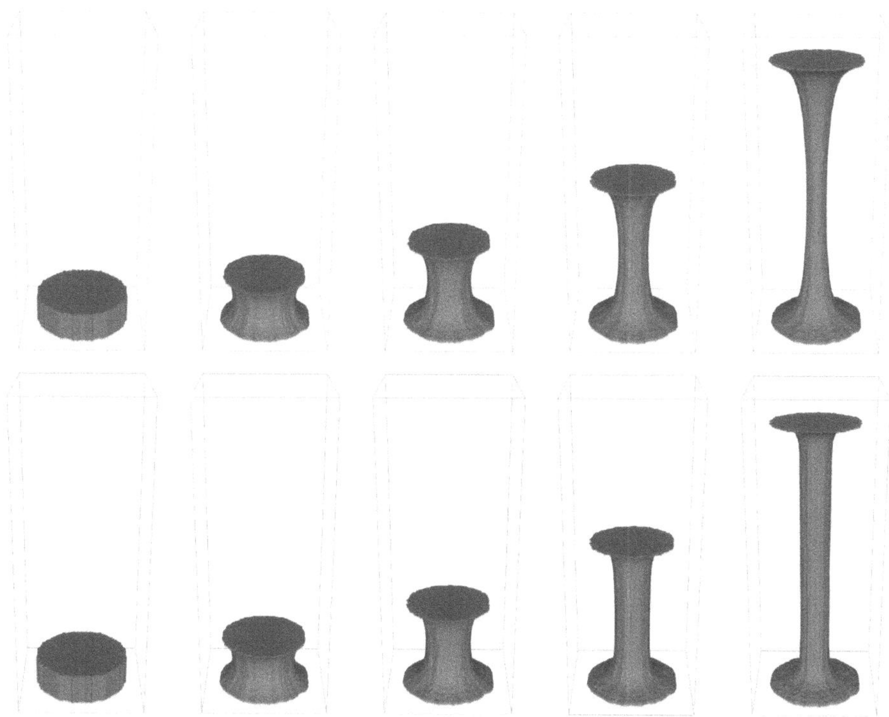

FIG. 3.21 Filament stretching. Aspect ratio $L_0/R_0 = 19/30$. The Hencky strains $\epsilon := \dot{\epsilon}_0 t$ are (column 1) 0; (column 2) 0.57; (column 3) 1.12; (column 4) 2.25; (column 5) 4.49; (top row) Newtonian fluid; (bottom row) Viscoelastic fluid with $\lambda = 0.421$ s (We = 1.97).

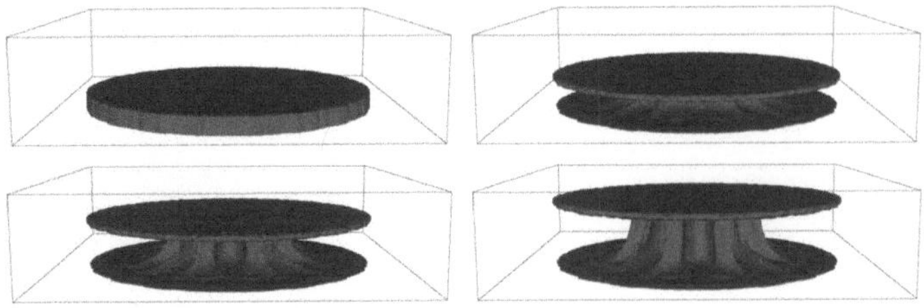

FIG. 3.22 Filament stretching, $\lambda = 0.421$ s (We $= 1.97$), aspect ratio $L_0/R_0 = 1/20$. Shape of the liquid region at time: (top left) $t = 0$ s, (top right) $t = 0.33$ s, (bottom left) $t = 0.66$ s, (bottom right) $t = 1$ s.

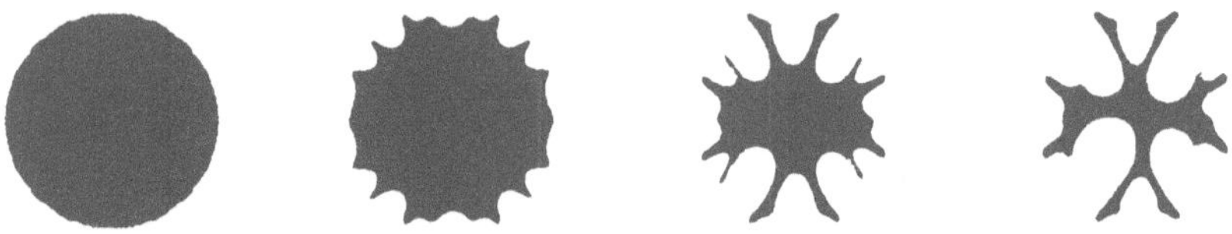

FIG. 3.23 Filament stretching, $\lambda = 0.421$ s (We $= 1.97$), aspect ratio $L_0/R_0 = 1/20$. Horizontal cut through the middle of the liquid region at time: (from left to right) $t = 0$ s, $t = 0.33$ s, $t = 0.66$ s, $t = 1$ s.

non-Newtonian computations. As reported in YAO and MCKINLEY [1998], the "necking" phenomena occurring in the central part of the liquid for Newtonian fluids is not observed for viscoelastic fluids, due to strain hardening. This calculation requires 2 h (resp. 24 h) on the coarse mesh (resp. fine mesh) using a single user Pentium 4 CPU 2.8 Ghz, with 2-Gb memory, under the Linux operating system. Most of the time is spent in solving the associated Stokes problem. The memory usage is 200 Mb for the coarse mesh, resp. 1.6 Gb for the fine mesh.

We now show that our numerical model is capable to reproduce fingering instabilities reported in BACH, RASMUSSEN, LONGIN and HASSAGER [2002], DERKS, LINDNER, CRETON and BONN [2003], MCKINLEY and SRIDHAR [2002], RASMUSSEN and HASSAGER [1999] for non-Newtonian flows. Following Section 4.4 in MCKINLEY and SRIDHAR [2002], we take an aspect ratio $L_0/R_0 = 1/20$ ($R_0 = 0.003$ m, $L_0 = 0.00015$ m) so that the Weissenberg number We $= DeR_0^2/L_0^2$ is large. The finite element mesh has 50 vertices along the radius and 25 vertices along the height; thus, the mesh size is 0.00006 m. The cells size is 0.00001 m, and the initial time step is $\Delta t^0 = 0.01$ s; thus, the CFL number of the cells – velocity times the time step divided by the cells spacing – is close to one. The shape of the filament is reported in Figs 3.22 and 3.23. Fingering instabilities can be observed from the very beginning of the stretching, leading to branched structures, as described in BACH, RASMUSSEN, LONGIN and HASSAGER [2002], DERKS, LINDNER, CRETON and BONN [2003], MCKINLEY and SRIDHAR [2002]. Clearly, such complex shapes cannot be obtained using Lagrangian models, and the mesh distortion would be too large.

Acknowledgment

We wish to acknowledge funding support. Bonito was partially supported by NSF grant DMS-0914977 and by the Swiss National Science Foundation.

Bibliography

ABOUBACAR, M., WEBSTER, M.F. (2003). Development of an optimal hybrid finite volume/element method for viscoelastic flows. *Int. J. Numer. Methods Fluids* **41** (11), 1147–1172.

AULISA, E., MANSERVISI, S., SCARDOVELLI, R. (2003). A mixed markers and volume-of-fluid method for the reconstruction and advection of interfaces in two-phase and free-boundary flows. *J. Comput. Phys.* **188** (2), 611–639.

BAAIJENS, F.P.T. (1998). Mixed finite element methods for viscoelastic flow analysis: a review. *J. Non-Newton. Fluid Mech.* **79**, 361–385.

BABUŠKA, I., AZIZ, A.K. (1972). Survey lectures on the mathematical foundations of the Finite Element Method. In: *The Mathematical Foundations of the Finite Element Method with Applications to Partial Differential Equations (Proceedings of a Symposium, University of Maryland, Baltimore, MD, 1972)*(Academic Press, New York), pp. 1–359. (With the collaboration of Fix, G. and Kellogg, R.B.).

BACH, A., RASMUSSEN, H.K., LONGIN, P.-Y., HASSAGER, O. (2002). Growth of non-axisymmetric disturbances of the free surface in the filament stretching rheometer: experiments and simulation. *J. Non-Newton. Fluid Mech.* **108**, 163–186.

BARANGER, J., SANDRI, D. (1992a). Finite element approximation of viscoelastic fluid flow: existence of approximate solutions and error bounds. I. Discontinuous constraints. *Numer. Math.* **63** (1), 13–27.

BARANGER, J., SANDRI, D. (1992b). A formulation of Stokes's problem and the linear elasticity equations suggested by the Oldroyd model for viscoelastic flow. *RAIRO Modél. Math. Anal. Numér.* **26** (2), 331–345.

BARANGER, J., WARDI, S. (1995). Numerical analysis of a FEM for a transient viscoelastic flow. *Comput. Methods Appl. Mech. Eng.* **125** (1–4), 171-185.

BARRETT, J.W., SCHWAB, C., SÜLI, E. (2005). Existence of global weak solutions for some polymeric flow models. *Math. Models Methods Appl. Sci.* **15** (6), 939–983.

BARRETT, J.W., SÜLI, E. (2007). Existence of global weak solutions to some regularized kinetic models for dilute polymers. *Multiscale Model. Simul.* **6** (2), 506–546. (Electronic).

BENSAADA, M., ESSELAOUI, D. (2005). Numerical analysis of stabilized method for transient viscoelastic flows. *Int. J. Pure Appl. Math.* **21** (4), 441–473.

BERIS, A.N., ARMSTRONG, R.C., BROWN, R.A. (1984). Finite element calculation of viscoelastic flow in a journal bearing: I. small eccentricities. *J. Non-Newton. Fluid Mech.* **16** (1–2), 141–172.

BIRD, R., CURTISS, C., ARMSTRONG, R., HASSAGER, O. (1987). *Dynamics of Polymeric Liquids* Volume 1 and 2 (John Wiley & Sons, NewYork).

BONITO, A., BURMAN, E. (2008). A continuous interior penalty method for viscoelastic flows. *SIAM J. Sci. Comput.* **30** (3), 1156–1177.

BONITO, A., CLÉMENT, Ph., PICASSO, M. (2006a). Finite element analysis of a simplified stochastic Hookean dumbbells model arising from viscoelastic flows. *M2AN Math. Model. Numer. Anal.* **40** (4), 785–814.

BONITO, A., CLÉMENT, Ph., PICASSO, M. (2006b). Mathematical analysis of a simplified Hookean dumbbells model arising from viscoelastic flows. *J. Evol. Equ.* **6** (3), 381–398.

BONITO, A., CLÉMENT, Ph., PICASSO, M. (2007). Mathematical and numerical analysis of a simplified time-dependent viscoelastic flow. *Numer. Math.* **107** (2), 213–255.

Bonito, A., Lozinski, A., Mountford, Th. (2010). Modeling viscoelastic flows using reflected stochastic differential equations. *Commun. Math. Sci.* **8** (3), 655–670.

Bonito, A., Picasso, M., Laso. M. (2006). Numerical simulation of 3D viscoelastic flows with free surfaces. *J. Comput. Phys.* **215** (2), 691–716.

Bonvin, J., Picasso, M. (1999). Variance reduction methods for connffessit-like simulations. *J. Non-Newton. Fluid Mech.* **84**, 191–215.

Bonvin, J., Picasso, M. (2001). A finite element/Monte-Carlo method for polymer dilute solutions. *Comput. Vis. Sci.* **4** (2), 93–98. Second AMIF International Conference (Il Ciocco, 2000).

Bonvin, J., Picasso, M. (2002). Mesoscopic models for viscoelastic flows: coupling finite element and Monte Carlo methods. *Monte Carlo Methods Appl.* **8** (1), 73–81.

Bonvin, J., Picasso, M., Stenberg, R. (2001). GLS and EVSS methods for a three-field Stokes problem arising from viscoelastic flows. *Comput. Methods Appl. Mech. Eng.* **190** (29–30), 3893–3914.

Brezzi, F. (1974). On the existence, uniqueness and approximation of saddle-point problems arising from Lagrangian multipliers. *Rev. Française Automat. Informat. Recherche Opérationnelle Sér. Rouge* **8** (R-2), 129–151.

Brown, R.A., Armstrong, R.C., Beris, A.N., Yeh, P.W. (1986). Galerkin finite element analysis of viscoelastic flows. *Comput. Methods Appl. Mech. Eng.* **58**, 201–226.

Caboussat, A. (2005). Numerical simulation of two-phase free surface flows. *Arch. Comput. Methods Eng., State of the Art Reviews* **12** (2), 165–224.

Caboussat, A. (2006). A numerical method for the simulation of free surface flows with surface tension. *Comput. Fluids* **35** (10), 1205–1216.

Caboussat, A., Picasso, M., Rappaz, J. (2005). Numerical simulation of free surface incompressible liquid flows surrounded by compressible gas. *J. Comput. Phys.* **203** (2), 626–649.

Caloz, G., Rappaz, J. (1997). Numerical analysis for nonlinear and bifurcation problems. In: *Handbook of Numerical Analysis*, Volume V (North-Holland, Amsterdam), pp. 487–693.

Chauvière, C., Lozinski, A. (2003). An efficient technique for simulations of viscoelastic flows, derived from the Brownian configuration field method. *SIAM J. Sci. Comput.* **24** (5), 1823–1837. (Electronic).

Chauvière, C., Owens, R.G. (2000). How accurate is your solution?: Error indicators for viscoelastic flow calculations. *J. Non-Newton. Fluid Mech.* **95** (1), 1–33.

Chemin, J.-Y., Masmoudi, N. (2001). About lifespan of regular solutions of equations related to viscoelastic fluids. *SIAM J. Math. Anal.* **33** (1), 84–112.

Chorin, A.J. (1980). Flame advection and propagation algorithms. *J. Comput. Phys.* **35**, 1–11.

Ciarlet, P.G., Lions, J.-L. (eds.) (1991). *Handbook of Numerical Analysis*, Volume II (North-Holland, Amsterdam). Finite Element Methods. Part 1.

Cormenzana, J., Ledd, A., Laso, M., Debbaut, B. (2001). Calculation of free surface flows using CONNFFESSIT. *J. Rheol.* **45** (1), 237–258.

Crochet, M.J., Keunings, R. (1982). On numerical die swell calculation. *J. Non-Newton. Fluid Mech.* **10** (1–2), 85–94.

Crochet, M.J., Walters, K. (1983). Numerical methods in non-Newtonian fluid mechanics. *Annu. Rev. Fluid. Mech.* **15**, 241–260.

Degond, P., Lemou, M., Picasso, M. (2002). Viscoelastic fluid models derived from kinetic equations for polymers. *SIAM J. Appl. Math.* **62** (5), 1501–1519.

Delaunay, P., Lozinski, A., Owens R.G. (2007). Sparse tensor product Fokker-Planck based methods for nonlinear bead spring chain models of dilute polymer solutions. *CRM Proc. Lect. Notes* **41**, 73–90.

Derks, D., Lindner, A., Creton, C., Bonn, D. (2003). Cohesive failure of thin layers of soft model adhesives under tension. *J. Appl. Phys.* **93** (3), 1557–1566.

Du, Q., Liu, C., Yu, P. (2005). FENE dumbbell model and its several linear and nonlinear closure approximations. *Multiscale Model, Simul.* **4** (3), 709–731.

E, W., Li, T., Zhang, P. (2002). Convergence of a stochastic method for the modeling of polymeric fluids. *Acta Math. Appl. Sin. Engl. Ser.* **18** (4), 529–536.

E, W., Li, T.J., Zhang, P.W. (2004). Well-posedness for the dumbbell model of polymeric fluids. *Comm. Math. Phys.* **248** (2), 409–427.

ERVIN, V.J., HEUER, N. (2004). Approximation of time-dependent, viscoelastic fluid flow: Crank-Nicolson, finite element approximation. *Numer. Methods Partial Differ. Equ.* **20** (2), 248–283.

ERVIN, V.J., HOWELL, J.S. (2008). A two-parameter defect-correction method for computation of steady-state viscoelastic fluid flow. *Appl. Math. Comp.* (To appear).

ERVIN, V.J., MILES, W.W. (2003). Approximation of time-dependent viscoelastic fluid flow: SUPG approximation. *SIAM J. Numer. Anal.* **41** (2), 457–486. (Electronic).

ERVIN, V.J., NTASIN, L.N. (2005). A posteriori error estimation and adaptive computation of viscoelastic fluid flow. *Numer. Methods Partial Differ. Equ.* **21** (2), 297–322.

FAN, Y. (2003). Limiting behavior of the solutions of a falling sphere in a tube filled with viscoelastic fluids. *J. Non-Newton. Fluid Mech.* **110** (2-3), 77–102.

FEIGL, K., LASO, M., ÖTTINGER, H.C. (1995). The CONNFFESSIT approach for solving a two-dimensional viscoelastic fluid problem. *Macromolecules* **28**, 3261–3274.

FERNÁNDEZ-CARA, E., GUILLÉN, F., ORTEGA, R.R. (2002). Mathematical modeling and analysis of viscoelastic fluids of the Oldroyd kind. In: *Handbook of Numerical Analysis*, Volume VIII (North-Holland, Amsterdam), pp. 543–661.

FIÉTIER, N., DEVILLE, M. (2002a). Simulations of time-dependent flows of viscoelastic fluids with spectral element methods. *J. Sci. Comput.* **17** (1–4), 649–657.

FIÉTIER, N., DEVILLE, M.O. (2002b). Simulations of time-dependent flows of viscoelastic fluids with spectral element methods. *J. Sci. Comput.* **17** (1–4), 649–657.

FORTIN, A., GUÉNETTE, R., PIERRE, R. (2000). On the discrete EVSS method. *Comput. Methods Appl. Mech. Eng.* **189** (1), 121–139.

FORTIN, M., FORTIN, A. (1989). A new approach for the fem simulation of viscoelastic flows. *J. Non-Newton. Fluid Mech.* **32** (3), 295–310.

FORTIN, M., PIERRE, R. (1989). On the convergence of the mixed method of Crochet and Marchal for viscoelastic flows. *Comput. Methods Appl. Mech. Eng.* **73** (3), 341–350.

FRANCA, L., FREY, S. (1992). Stabilized finite element methods. II. The incompressible Navier-Stokes equations. *Comput. Methods Appl. Mech. Eng.* **99** (2–3), 209–233.

GIRAULT, V., RAVIART, P.-A. (1986). Finite element methods for Navier-Stokes equations. *Springer Series in Computational Mathematics* Volume 5 (Springer-Verlag, Berlin). Theory and Algorithms.

GLOWINSKI, R. (2003). Finite element methods for incompressible viscous flow. In: *Handbook of Numerical Analysis*, Volume IX (North-Holland, Amsterdam), pp. 3–1176.

GRAMBERG, H.J.J., VAN VROONHOVEN, J.C.W., VAN DE VEN, A.A.F. (2004). Flow patterns behind the free flow front for a Newtonian fluid injected between two infinite parallel plates. *Eur. J. Mech. B Fluids* **23** (4), 571–585.

GRANDE, E., LASO, M., PICASSO, M. (2003). Calculation of variable-topology free surface flows using connffessit. *J. Non-Newton. Fluid Mech.* **113** (2), 127–145.

GRIEBEL, M. (2006). Sparse grids and related approximation schemes for higher dimensional problems. In: *Foundations of Computational Mathematics, Santander 2005, London Mathematical Society Lecture Note Ser.*, Volume 331 (Cambridge University Press, Cambridge), pp. 106–161.

GUILLOPÉ, C., SAUT, J.-C. (1990). Existence results for the flow of viscoelastic fluids with a differential constitutive law. *Nonlinear Anal.* **15** (9), 849–869.

HAKIM, A. (1994). Mathematical analysis of viscoelastic fluids of White-Metzner type. *J. Math. Anal. Appl.* **185** (3), 675–705.

HALIN, P., LIELENS, G., KEUNINGS, LEGAT, V. (1998). The Lagrangian particle method for macroscopic and micromacro viscoelastic flow computations. *J. Non-Newton. Fluid Mech.* **79** (2–3), 387–403.

HU, D., LELIEVRE, T. (2007). New entropy estimates for the Oldroyd-B model, and related models. *Comm. Math. Sci.* **5** (4), 909–916.

Hulsen, M.A., Fattal, R., Kupferman, R. (2005). Flow of viscoelastic fluids past a cylinder at high Weissenberg number: stabilized simulations using matrix logarithms. *J. Non-Newton. Fluid Mech.* **127** (1), 27–39.

HULSEN, M.A., VAN HEEL, A.P.G., VAN DEN BRULE, B.H.A.A. (1997). Simulation of viscoelastic flows using Brownian configuration fields. *J. Non-Newton. Fluid Mech.* **70**, 79–101.

Jin, H., Tanner, R.I. (1994). An a posteriori error estimate and adaptive refinement procedure for viscoelastic fluid flows: the modified upper-convected Maxwell model. *Comput. Mech.* **13** (6), 400–413.

Joseph, D.D., Renardy, M., Saut, J.-C. (1985). Hyperbolicity and change of type in the flow of viscoelastic fluids. *Arch. Ration. Mech. Anal.* **87** (3), 213–251.

Jourdain, B., Le Bris, C., Lelievre, T., Otto, F. (2006). Long-time asymptotics of a multiscale model for polymeric fluid flows. *Arch. Ration. Mech. Anal.* **181** (1), 94–148.

Jourdain, B., Lelièvre, T., Le Bris, C. (2002). Numerical analysis of micro-macro simulations of polymeric fluid flows: a simple case. *Math. Models Methods Appl. Sci.* **12** (9), 1205–1243.

Jourdain, B., Lelièvre, T., Le Bris, C. (2004a). Existence of solution for a micro-macro model of polymeric fluid: the FENE model. *J. Funct. Anal.* **209** (1), 162–193.

Jourdain, B., Lelièvre, T., Le Bris, C. (2004b). On a variance reduction technique for micromacro simulations of polymeric fluids. *J. Non-Newton. Fluid Mech.* **122** (1–3), 91–106.

Keunings, R. (1986). On the high Weissenberg number problem. *J. Non-Newton. Fluid Mech.* **20**, 209–226.

Keunings, R. (2003). Finite element methods for integral viscoelastic fluids. *Rheol. Rev.* 167–195.

Keunings, R. (2004). Micro-marco methods for the multi-scale simulation of viscoelastic flow using molecular models of kinetic theory. *Rheol. Rev.* 67–98.

Kim, J.M., Kim, C., Kim, J.H., Chung, C., Ahn, K.H., Lee, S.J. (2005). High-resolution finite element simulation of 4:1 planar contraction flow of viscoelastic fluid. *J. Non-Newton. Fluid Mech.* **129** (1), 23–37.

Kloeden, P.E., Platen, E. (1992). Numerical solution of stochastic differential equations. *Applications of Mathematics (New York)* Volume 23 (Springer-Verlag, Berlin).

Koplik, J., Banavar, J.R. (2003). Extensional rupture of model non-Newtonian fluid filaments. *Phys. Rev. E* 67(011502).

Koppol, A.P., Sureshkumar, R., Khomami, B. (2007). An efficient algorithm for multiscale flow simulation of dilute polymeric solutions using bead-spring chains. *J. Non-Newton. Fluid Mech.* **141** (2–3), 180–192.

Kröger, M. (2004). Simple models for complex nonequilibrium fluids. *Phys. Rep.* **390** (6), 453–551.

Larson, R.G. (1999). *The Structure and Rheology of Complex Fluids*(Oxford University Press, Oxford).

Laso, M., Öttinger, H.-C. (1993). Calculation of viscoelastic flow using molecular models. *J. Non-Newton. Fluid Mech.* **47**, 1–20.

Laso, M., Picasso, M., Öttinger H.C. (1997). 2d time-dependent viscoelastic flow calculations using CONNFFESSIT. *AIChE. J.* **43** (4), 877–892.

Le Bris, C., Lions, P.-L. (2004). Renormalized solutions of some transport equations with partially $W^{1,1}$ velocities and applications. *Ann. Mat. Pura Appl. (4)* **183** (1), 97–130.

Lee, Y.-J., Xu, J. (2006). New formulations, positivity preserving discretizations and stability analysis for non-Newtonian flow models. *Comput. Methods Appl. Mech. Eng.* **195** (9–12), 1180–1206.

Lelièvre, T. (2004). Optimal error estimate for the CONNFFESSIT approach in a simple case. *Comput. Fluids* **33** (5–6), 815–820.

Lesaint, P., Raviart, P.-A. (1974). On a finite element method for solving the neutron transport equation. In: *Mathematical Aspects of Finite Elements in Partial Differential Equations (Proceedings of a Symposium Mathematical Research Center, University Wisconsin, Madison, WI, 1974)*, Publication No. 33 (Mathematical Research Center, University of Wisconsin-Madison, Academic Press, New York), pp. 89–123.

Li, T., Vanden-Eijnden, E., Zhang, P., E.W. (2004). Stochastic models of polymeric fluids at small Deborah number. *J. Non-Newton. Fluid Mech.* **121** (2–3), 117–125.

Li, T., Zhang, H., Zhang, P. (2004). Local existence for the dumbbell model of polymeric fluids. *Comm. Partial Differ. Equ.* **29** (5–6), 903–923.

Li, T. Zhang, P. (2006). Convergence analysis of BCF method for Hookean dumbbell model with finite difference scheme. *Multiscale Model. Simul.* **5** (1), 205–234. (Electronic).

Likhtman, A.E., Graham, R.S. (2003). Simple constitutive equation for linear polymer melts derived from molecular theory: Roliepoly equation. *J. Non-Newton. Fluid Mech.* **114** (1), 1–12.

Lin, F., Zhang, P., Zhang, Z. (2008). On the global existence of smooth solution to the 2-D FENE dumbbell model. *Comm. Math. Phys.* **277** (2), 531–553.

LIN, F.H., LIU, C., ZHANG, P. (2005). On hydrodynamics of viscoelastic fluids. *Comm. Pure Appl. Math.* **58** (11), 1437–1471.

LIONS, P.L., MASMOUDI, N. (2000). Global solutions for some Oldroyd models of non-Newtonian flows. *Chinese Ann. Math. Ser. B* **21** (2), 131–146.

LIONS, P.L., MASMOUDI, N. (2007). Global existence of weak solutions to some micro-macro models. *C. R. Math. Acad. Sci. Paris* **345** (1), 15–20.

LIU, C., WALKINGTON, N.J. (2001). An Eulerian description of fluids containing visco-elastic particles. *Arch. Ration. Mech. Anal.* **159** (3), 229–252.

LOZINSKI, A., CHAUVIÈRE, C. (2003). A fast solver for Fokker–Planck equation applied to viscoelastic flows calculations: 2D FENE model. *J. Comput. Phys.* **189** (2), 607–625.

LOZINSKI, A., OWENS, R.G. (2003). An energy estimate for the Oldroyd B model: theory and applications. *J. Non-Newton. Fluid Mech.* **112**, 161–176.

LUNARDI, A. (1995). Analytic semigroups and optimal regularity in parabolic problems. *Progress in Nonlinear Differential Equations and Their Applications* Volume 16. (Birkhuser Verlag, Basel).

MACHMOUM, A., ESSELAOUI, D. (2001). Finite element approximation of viscoelastic fluid flow using characteristics method. *Comput. Methods Appl. Mech. Eng.* **190** (42), 5603–5618.

MARCHAL, J.M., CROCHET, M.J. (1987). A new mixed finite element for calculating viscoelastic flow. *J. Non-Newton. Fluid Mech.* **28** (1), 77–114.

MARONNIER, V., PICASSO, M., RAPPAZ, J. (1999). Numerical simulation of free surface flows. *J. Comput. Phys.* **155** (2), 439–455.

MARONNIER, V., PICASSO, M., RAPPAZ, J. (2003). Numerical simulation of 3d free surface flows. *Int. J. Numer. Methods Fluids* **42** (7), 697–716.

MCKINLEY, G.H., SRIDHAR, T. (2002). Filament-stretching rheometry of complex fluids. *Annu. Rev. Fluid Mech.* **34**, 375–415.

MENDELSON, M.A., YEH, P.W., BROWN, R.A., ARMSTRONG, R.C. (1982). Approximation error in finite element calculation of viscoelastic fluid flows. *J. Non-Newton. Fluid Mech.* **10** (1–2), 31–54.

MOMPEAN, G., DEVILLE, M. (2000). Unsteady finite volume simulation of Oldroyd-B fluid through a three-dimensional planar contraction. *J. Non-Newton. Fluid Mech.* **72** (2–3), 253–279.

NAJIB, K., SANDRI, D. (1995). On a decoupled algorithm for solving a finite element problem for the approximation of viscoelastic fluid flow. *Numer. Math.* **72** (2), 223–238.

NAJIB, K., SANDRI, D., ZINE, A.-M. (2004). On a posteriori estimates for a linearized Oldroyd's problem. *J. Comput. Appl. Math.* **167** (2), 345–361.

NOH, W.F., WOODWARD, P. (1976). *SLIC (Simple Line Interface Calculation)*, Lectures Notes in Physics Volume 59 (Springer-Verlag) pp. 330–340.

NOVOTNÝ, A., SEQUEIRA, A., VIDEMAN, J.H. (1999). Steady motions of viscoelastic fluids in three-dimensional exterior domains. Existence, uniqueness and asymptotic behaviour. *Arch. Ration. Mech. Anal.* **149** (1), 49–67.

OSHER, S., FEDKIW, R. (2003). *Level Set Methods and Dynamic Implicit Surfaces*. Applied Mathematical Sciences Volume 153 (Springer-Verlag, New York).

ÖTTINGER, H.-C. (1996). *Stochastic Processes in Polymeric Fluids* (Springer-Verlag, Berlin).

ÖTTINGER, H.-C., VAN DEN BRULE, B.H.A.A., HULSEN, M.A. (1997). Brownian configuration fields and variance reduced CONNFFESSIT. *J. Non-Newton. Fluid Mech.* **70**, 255–261.

OWENS, R.G. (1998). A posteriori error estimates for spectral element solutions to viscoelastic flow problems. *Comput. Methods Appl. Mech. Eng.* **164** (3–4), 375–395.

PICASSO, M., RAPPAZ, J. (2001). Existence, a priori and a posteriori error estimates for a nonlinear three-field problem arising from Oldroyd-B viscoelastic flows. *M2AN Math. Model. Numer. Anal.* **35** (5), 879–897.

PICHELI, E., COUPEZ, T. (1998). Finite element solution of the 3D mold filling problem for viscous incompressible fluid. *Comput. Methods Appl. Mech. Eng.* **163** (1–4), 359–371.

PIRONNEAU, O. (1989). *Finite Element Methods for Fluids* (Wiley, Chichester).

PIRONNEAU, O., LIOU, J., TEZDUYAR, T. (1992). Characteristic-Galerkin and Galerkin/least-squares space-time formulations for the advection-diffusion equation with time-dependent domains. *Comput. Methods Appl. Mech. Eng.* **100**, 117–141.

Quarteroni, A., Valli, A. (1991). *Numerical Approximation of Partial Differential Equations*. Number 23 in Springer Series in Computational Mathematics (Springer-Verlag, Berlin).

Rasmussen, H.K., Hassager, O. (1999). Three-dimensional simulations of viscoelastic instability in polymeric filaments. *J. Non-Newton. Fluid Mech.* **82**, 189–202.

Renardy, M. (1985a). Existence of slow steady flows of viscoelastic fluids with differential constitutive equations. *Z. Angew. Math. Mech.* **65** (9), 449–451.

Renardy, M. (1985b). A local existence and uniqueness theorem for a K-BKZ-fluid. *Arch. Ration. Mech. Anal.* **88** (1), 83–94.

Renardy, M. (1991). An existence theorem for model equations resulting from kinetic theories of polymer solutions. *SIAM J. Math. Anal.* **22** (2), 313–327.

Renardy, M. (2000). Mathematical analysis of viscoelastic flows. *CBMS-NSF Regional Conference Series in Applied Mathematics* Volume 73 (Society for Industrial and Applied Mathematics (SIAM), Philadelphia, PA).

Revuz, D., Yor, M. (1994). Continuous martingales and Brownian motion. *Grundlehren der Mathematischen Wissenschaften [Fundamental Principles of Mathematical Sciences]* Volume 293 (Springer-Verlag, Berlin).

Rider, W.J., Kothe, D.B. (1998). Reconstructing volume tracking. *J. Comput. Phys.* **141** (2), 112–152.

Ruas, V., Carneiro de Araújo, J.H., Silva Ramos, M.A.M. (1993). Approximation of the three-field Stokes system via optimized quadrilateral finite elements. *RAIRO Modél. Math. Anal. Numér.* **27** (1), 107–127.

Sandri, D. (1993). Analyse d'une formulation à trois champs du problème de Stokes. *RAIRO Modél. Math. Anal. Numér.* **27** (7), 817–841.

Sandri, D. (1994). Finite element approximation of viscoelastic fluid flow: existence of approximate solutions and error bounds. Continuous approximation of the stress. *SIAM J. Numer. Anal.* **31** (2), 362–377.

Sandri, D. (1999). Non-integrable extra stress tensor solution for a flow in a bounded domain of an Oldroyd fluid. *Acta Mech.* **135** (1–2), 95–99.

Sandri, D. (2005). Numerical study around the corotational Maxwell model for the viscoelastic fluid flows. *Eur. J. Mech. B Fluids* **24** (6), 733–750.

Saramito, P. (1994). A new θ-scheme algorithm and incompressible FEM for viscoelastic fluid flows. *RAIRO Modél. Math. Anal. Numér.* **28** (1), 1–35.

Scardovelli, R., Zaleski, S. (1999). Direct numerical simulation of free-surface and interfacial flow. In: *Annual Review of Fluid Mechanics*, Volume 31. Volume 31 of Annu. Rev. Fluid Mech. (Annual Reviews, Palo Alto, CA), pp. 567–603.

Schwab, C., Suri, M. (1999). Mixed *hp* finite element methods for Stokes and non-Newtonian flow. *Comput. Methods Appl. Mech. Eng.* **175** (3–4), 217–241.

Sethian, J.A., Smereka, P. (2003). Level set methods for fluid interfaces. In: *Annual Review of Fluid Mechanics*, Volume 35. Volume 35 of Annu. Rev. Fluid Mech. (Annual Reviews, Palo Alto, CA), pp. 341–372.

Singh, P., Joseph, D.D., Heslab, T.I., Glowinski, R., Pan, T.-W. (2000). A distributed Lagrange multiplier/fictitious domain method for viscoelastic particulate flows. *J. Non-Newton. Fluid Mech.* **91** (2–3), 165–188.

Tomé, M.F., Castelo, A., Cuminato, J.A. (2008). A numerical method for solving the Oldroyd-B model for 3d free surface flows. *J. Non-Newton. Fluid Mech.*

Tomé, M.F., Mangiavacchi, N., Cuminato, J.A., Castelo, A., McKee, S. (2002). A finite difference technique for simulation unsteady viscoelastic free surface flows. *J. Non-Newton. Fluid Mech.* **106**, 61–106.

Tomé, M.F., McKee, S. (1999). Numerical simulation of viscous flow: buckling of planar jets. *Int. J. Numer. Methods Fluids.* **29**, 705–718.

Trebotich, A., Colella, P., Miller, G.H. (2005). A stable and convergent scheme for viscoelastic flow in contraction channels. *J. Comput. Phys.* **205**, 315–342.

Verbeeten, W.M.H., Peters, G.W.M., Baaijens, F.P.T. (2004). Numerical simulations of the planar contraction flow for a polyethylene melt using the xpp model. *J. Non-Newton. Fluid Mech.* **117** (2–3), 73–84.

von Petersdorff, T., Schwab, C. (2004). Numerical solution of parabolic equations in high dimensions. *M2AN Math. Model. Numer. Anal.* **38** (1), 93–127.

Wapperom, P., Renardy, M. (2005). Numerical prediction of the boundary layers in the flow around a cylinder using a fixed velocity field. *J. Non-Newton. Fluid Mech.* **125** (1), 35–48.

Yao, M., McKinley, G.H. (1998). Numerical simulation of extensional deformations of viscoelastic liquid bridges in filament stretching devices. *J. Non-Newton. Fluid Mech.* **74** (1–3), 47–88.

Zhang, H., Zhang, P. (2006). Local existence for the FENE-Dumbbells model of polymeric liquids. *Arch. Ration. Mech. Anal.* **181**, 373–400.

Zhang, L.Y., Zhang, H., Zhang, P.W. (2008). Global existence of weak solutions to the regularized Hookean dumbbell model. *Commun. Math. Sci.* **6** (1), 85–124.

Stable Finite Element Discretizations for Viscoelastic Flow Models

Young-Ju Lee

Department of Mathematics, Rutgers, The State University of New Jersey, NJ 08901, USA
E-mail: leeyoung@math.rutgers.edu

Jinchao Xu

Department of Mathematics and Center for Computational Mathematics and Application, the Pennsylvania State University, PA 16802, USA
E-mail: xu@math.psu.edu

Chen-Song Zhang

Department of Mathematics and Center for Computational Mathematics and Application, the Pennsylvania State University, PA 16802, USA
E-mail: zhangchensong@gmail.com

Numerical Methods for Non-Newtonian Fluids
Special Volume (R. Glowinski and Jinchao Xu, Guest Editors) of
HANDBOOK OF NUMERICAL ANALYSIS, VOL. XVI
P.G. Ciarlet (Editor)

ISSN 1570-8659
DOI 10.1016/B978-0-444-53047-9.00004-6

1. Introduction

A complex fluid, also called a non-Newtonian fluid, is "a fluid made up of a lot of different kinds of stuff" as described by GELBART and BEN-SHAUL [1996]. This high number of complexities and their interactions can produce a variety of new nontrivial physical phenomena (BIRD, CURTISS, ARMSTRONG and HASSAGER [1987]), including, for example, the rod-climbing Weissenberg effect (DEALY and VU [1977]) and the die swell (CLERMONT and PIERRARD [1976]). Among the phenomena that have been of particular research interest in recent years are the flow behaviors in highly elastic complex fluid with a vanishingly small Reynolds number (GROISMAN and STEINBERG [1998], THOMASES and SHELLEY [2009]). It is agreed that the peculiar behavior of the highly elastic fluids flow, known as "elastic turbulence," originates in the strong nonlinear mechanical properties of the polymer solutions (GROISMAN and STEINBERG [1998]), and it is similar to the phenomena observed from the strong inertial effects in Newtonian fluids. During the past decade, these phenomena have been the subject of many theoretical and experimental studies.

The controlling parameter of the strength of the nonlinearity of complex fluid models is the Weissenberg number or the Deborah number. Roughly speaking, the larger the Weissenberg number, the stronger the elasticity of the polymer solutions (see GROISMAN and STEINBERG [1998]). One crucial outstanding problem in computational rheology is that computations for complex fluid models with a high Weissenberg number have encountered great difficulty due to a breakdown in the convergence of the algorithms at critical values of the Weissenberg number. Although some significant progress has been made in recent years (e.g., FATTAL and KUPFERMAN [2004]), the level of fundamental correctness in the relevant regimes of large Weissenberg number has yet to be obtained. The main aim of this article is to focus on the issues that arise in simulating highly elastic and high Weissenberg number flows.

After briefly reviewing recent progress regarding the theoretical and numerical study of generic polymeric fluids in high Weissenberg number regimes, we will give a detailed presentation of a family of algorithms originally proposed by LEE and XU's [2006] and some new results developed in the last few years. In particular, we will give a more refined presentation of the positivity-preserving discretization schemes proposed in LEE and XU's [2006] and present some preliminary numerical experiments. In our discussion, we will

- demonstrate how a general macroscopic viscoelastic fluid model can be reformulated, in terms of the conformation tensor, as a Riccati differential equation;
- use this reformulation to establish the positive definiteness of the conformation tensor;
- use key numerical methods based on the Eulerian–Lagrangian method, which discretizes the momentum equation and constitutive equations by solving the nonlinear ordinary differential equations that define the characteristics related to the transport part of the equation;
- discuss how the resulting discrete system can be effectively solved iteratively by combining multigrid and parallel computing techniques; and
- show that the nonlinear iterations uniformly converge and the computational costs of the methods are uniformly optimal with respect to relevant physical parameters (such as the Reynolds number and the Weissenberg number) as well as time step and mesh sizes (see LEE, XU and ZHANG [To appear]).

The rest of the article is organized as follows. In Section 2, we review the basic properties of the flow maps, the generalized Lie derivatives, and the algebraic Riccati differential equations. In Section 3, we introduce the connection between the algebraic Riccati differential equations and the macroscopic constitutive relations for viscoelastic fluids. In Section 4, we discuss the properties of various macroscopic models for viscoelastic fluids. In Section 5, we briefly review the existing numerical schemes designed for simulating viscoelastic fluids at high Weissenberg number regimes. In Section 6, we discuss the numerical discretization schemes that preserve the positive definiteness of the conformation tensor. In Section 7, we briefly consider the solution techniques for the resulting discrete systems. In Section 8, we show the energy estimate, the long-term stability, and the well posedness, of the discrete solution, and then in Section 9, we present the implementation details and corresponding numerical results. Finally, in Section 10, we offer concluding remarks.

2. Flow maps, generalized Lie derivatives, and Riccati equations

2.1. *Notation*

Throughout this article, we use the standard notation for Sobolev spaces: $H^k(\Omega)$ denotes the classical Sobolev space of scalar functions on a bounded domain $\Omega \subset \mathbb{R}^d$ whose derivatives up to order k are square integrable, with the full norm $\|\cdot\|_k$ and the corresponding seminorm $|\cdot|_k$. The symbol $H_0^1(\Omega)$ denotes the subspace of $H^1(\Omega)$ whose trace vanishes on the boundary $\partial\Omega$. We will also discuss the corresponding spaces restricted to the subdomain of Ω. For any $\omega \subset \Omega$, we denote $\|\cdot\|_{k,\omega}$ and $|\cdot|_{k,\omega}$ as the norm and the seminorm, respectively, on the domain ω. The usual L^∞-norm and L^2-norm will be denoted by $\|\cdot\|_\infty$ and $\|\cdot\|_0$, respectively. The symbol $L_0^2(\Omega)$ denotes a subspace of $L^2(\Omega)$ consisting of functions that have a zero average. $(\cdot,\cdot)$ and $\langle\cdot,\cdot\rangle$ denote the classical L^2-inner product and the dual pairing, respectively. The space $L^p(0,T;H^1(\Omega))$ for $1 \le p < \infty$ is the Hilbert space consisting of functions $f(x,t) : \Omega \times [0,T] \mapsto \mathbb{R}$ such that

$$\left(\int_0^T \|f(\cdot,\nu)\|_1^p \,d\nu\right)^{1/p} < \infty.$$

The symbol $\mathbb{M}$ denotes the space of matrix-valued functions whose ranges are in $\mathbb{R}^{d\times d}$, and $\mathbb{S}$ denotes the subspace of $\mathbb{M}$ consisting of the symmetric matrices. In addition, $\mathbb{S}^+$ denotes the subset of $\mathbb{S}$ consisting of positive-definite matrices. Finally, following (XU [1992]), the symbol $A \lesssim B$ means $A \le CB$ with a constant C independent of space mesh size h and time step k, and $A \lesssim B$ is an abbreviation of $A \lesssim B \lesssim A$.

2.2. *Flow maps and the deformation tensor*

Consider a bounded domain $\Omega \subset \mathbb{R}^d$ ($d = 2$ or 3) and a velocity field of flow $\mathbf{u} = (u_i) \in \mathbb{R}^d$. The motion of a particle can be described by the flow map $\phi_{t,s} : \Omega \mapsto \Omega$ such that

$$\frac{\partial}{\partial s}\phi_{t,s}(x) = \mathbf{u}(\phi_{t,s}(x), s), \quad \phi_{t,t}(x) = x. \tag{2.1}$$

We note that by $\phi_{t,t}(x) = x$, we mean that the Eulerian coordinate is coincident with the Lagrangian (or material) coordinate at time t. As in classical mechanics, the flow map $\phi_{t,s}$ is assumed to be a diffeomorphism. Moreover, the flow map satisfies the composition rule, i.e., $\phi_{t_1,t_2}\phi_{t,t_1} = \phi_{t,t_2}$, for any $t_1, t_2 \geq 0$.

By means of this flow map, with an abuse of notation, we define

$$v(t, s) = v(x, t; s) := v(\phi_{t,s}(x), s) \quad \text{and} \quad v(t, t) = v(t) = v(x, t; t) = v(x, t),$$

where v can be any (scalar, vector, or tensor) function. Furthermore, for any $v(x, t)$, we have the following definition of the material derivative:

$$\frac{Dv}{Dt}(x, t) := \frac{\partial}{\partial s} v(\phi_{t,s}(x), s)\Big|_{s=t} = \big(v_t + (\mathbf{u} \cdot \nabla)v\big)(x, t). \tag{2.2}$$

Of the different conventions to define the gradient of velocity $\mathbf{u}$, denoted by $\nabla\mathbf{u}$ (or $\nabla_x\mathbf{u}$), we use the convention that $(\nabla\mathbf{u})_{i,j} = (\partial_j u_i)_{i,j}$, i.e.,

$$\nabla\mathbf{u} := \begin{pmatrix} \nabla u_1^T \\ \nabla u_2^T \\ \vdots \\ \nabla u_d^T \end{pmatrix} = \begin{pmatrix} \partial_1 u_1 & \partial_2 u_1 & \cdots & \partial_d u_1 \\ \partial_1 u_2 & \partial_2 u_2 & \cdots & \partial_d u_2 \\ \vdots & \vdots & \ddots & \vdots \\ \partial_1 u_d & \partial_2 u_d & \cdots & \partial_d u_d \end{pmatrix}.$$

For any two time variables, t_1 and t_2, we define the relative deformation gradient $\boldsymbol{F}(x, t; t_1, t_2)$ ($\boldsymbol{F}(t; t_1, t_2)$ in short) as follows:

$$\boldsymbol{F}(t; t_1, t_2) := \nabla_z \phi_{t_1,t_2}(z), \quad \text{with } z = \phi_{t,t_1}(x). \tag{2.3}$$

In case $t_1 = t$, we have $\boldsymbol{F}(t; t, t_2) = \nabla_x \phi_{t,t_2}(x)$. Geometrically, the deformation gradient $\boldsymbol{F}(t; t_1, t_2)$ measures the relative deformation between two configurations at t_1 and t_2.

From the definition of $\boldsymbol{F}(t; t_1, t_2)$ and the chain rule, we can derive the following ordinary differential equation:

$$\begin{aligned} \frac{\partial \boldsymbol{F}(t; t_1, t_2)}{\partial t_2} &= \frac{\partial}{\partial t_2} \nabla_z \phi_{t_1,t_2}(z) = \nabla_z \mathbf{u}(\phi_{t,t_2}(x), t_2) = \nabla_z \mathbf{u}(\phi_{t_1,t_2}(z), t_2) \\ &= \nabla\mathbf{u}(\phi_{t_1,t_2}(z), t_2)\boldsymbol{F}(t; t_1, t_2) = \nabla\mathbf{u}(t, t_2)\boldsymbol{F}(t; t_1, t_2) \end{aligned} \tag{2.4}$$

and the initial condition $\boldsymbol{F}(t; t_1, t_1) = \boldsymbol{\delta}$, where $\boldsymbol{\delta}$ is the identity tensor.

Throughout this article, we will only consider incompressible fluids, namely, $\nabla \cdot \mathbf{u} = 0$, which implies the determinant of $\boldsymbol{F}(t; t_1, t_2)$ is one, i.e., $\det \boldsymbol{F}(t; t_1, t_2) = 1$. Therefore, it is invertible. Furthermore, the inverse of $\boldsymbol{F}(t; t_1, t_2)$ is given by $\boldsymbol{F}(t; t_2, t_1)$ unambiguously because we have the following relation:

$$\boldsymbol{F}(t; t_2, t_1) = \nabla_{z'} \phi_{t_2,t_1}(z'), \quad \text{with } z' = \phi_{t,t_2}(x). \tag{2.5}$$

Using (2.5), we can derive the following relation that

$$\begin{aligned} 0 = \frac{\partial\big(\boldsymbol{F}(t; t_1, t_2)\boldsymbol{F}(t; t_2, t_1)\big)}{\partial t_2} &= \frac{\partial \boldsymbol{F}(t; t_1, t_2)}{\partial t_2}\boldsymbol{F}(t; t_2, t_1) + \boldsymbol{F}(t; t_1, t_2)\frac{\partial \boldsymbol{F}(t; t_2, t_1)}{\partial t_2} \\ &= \nabla\mathbf{u}(t, t_2) + \boldsymbol{F}(t; t_1, t_2)\frac{\partial \boldsymbol{F}(t; t_2, t_1)}{\partial t_2}. \end{aligned}$$

Therefore, we obtain

$$\frac{\partial \boldsymbol{F}(t;t_2,t_1)}{\partial t_2} = -\boldsymbol{F}(t;t_2,t_1)\nabla\mathbf{u}(t,t_2) \quad \text{and} \quad \boldsymbol{F}(t;t_1,t_1)=\boldsymbol{\delta}. \tag{2.6}$$

2.3. *Generalized Lie derivatives*

We will now introduce the generalized Lie derivative. For any given continuous function $\boldsymbol{\Phi}(t)=\boldsymbol{\Phi}(x,t)\colon \Omega\times(0,+\infty)\mapsto \mathbb{M}$, we define $\boldsymbol{L}(t;t_1,t_2)$ to be the transition matrix (or evolution matrix) that satisfies the following ordinary differential equation:

$$\frac{\partial \boldsymbol{L}(t;t_1,t_2)}{\partial t_2} = \boldsymbol{\Phi}(t,t_2)\boldsymbol{L}(t;t_1,t_2) \quad \text{and} \quad \boldsymbol{L}(t;t_1,t_1)=\boldsymbol{\delta}. \tag{2.7}$$

We can view this transition matrix $\boldsymbol{L}(t;t_1,t_2)$ as a generalization of the deformation gradient $\boldsymbol{F}(t;t_1,t_2)$; when $\boldsymbol{\Phi}(t,t_2)=\nabla\mathbf{u}(t,t_2)$, $\boldsymbol{L}(t;t_1,t_2)$ reduces to $\boldsymbol{F}(t;t_1,t_2)$.

The following lemma gives a fundamental property of the transition matrices (see, for example, BROCKETT [1970, theorem 2, section 1.4].)

LEMMA 2.1 (Composition Rule). *For any time levels, $t, t_0, t_1, t_2 \geq 0$, we have the following property*

$$\boldsymbol{L}(t;t_1,t_2)\boldsymbol{L}(t;t_0,t_1)=\boldsymbol{L}(t;t_0,t_2). \tag{2.8}$$

In particular, we also have

$$\boldsymbol{L}(t;t_1,t_0)\boldsymbol{L}(t;t_0,t_1)=\boldsymbol{\delta}.$$

Furthermore, we have that $\mathbf{L}(t;t_2,t_1)$ is the inverse of $\mathbf{L}(t;t_1,t_2)$ and it satisfies:

$$\frac{\partial \boldsymbol{L}(t;t_2,t_1)}{\partial t_2} = -\boldsymbol{L}(t;t_2,t_1)\boldsymbol{\Phi}(t,t_2) \quad \textit{and} \quad \boldsymbol{L}(t;t_1,t_1)=\boldsymbol{\delta}. \tag{2.9}$$

PROOF. Given any point $y\in\Omega$, we consider the ordinary differential equation:

$$\frac{\partial y(s)}{\partial s} = \boldsymbol{\Phi}(t,s)y(s) \quad \text{and} \quad y(t_1)=y. \tag{2.10}$$

Then, by definition (2.7), we obtain $y(s)=\boldsymbol{L}(t;t_1,s)y$. Similarly, let $z(s)$ satisfy the following ODE:

$$\frac{\partial z(s)}{\partial s} = \boldsymbol{\Phi}(t,s)z(s) \quad \text{and} \quad z(t_0)=y(t_0). \tag{2.11}$$

We have the relation:

$$z(s)=\boldsymbol{L}(t;t_0,s)y(t_0)=\boldsymbol{L}(t;t_0,s)\boldsymbol{L}(t;t_1,t_0)y.$$

It follows that $z(t_1)=\boldsymbol{L}(t;t_0,t_1)\boldsymbol{L}(t;t_1,t_0)y$, which implies that

$$z(s)=\boldsymbol{L}(t;t_1,s)z(t_1)=\boldsymbol{L}(t;t_1,s)\boldsymbol{L}(t;t_0,t_1)\boldsymbol{L}(t;t_1,t_0)y.$$

Consequently, by the definition and the uniqueness of the transition matrix $\boldsymbol{L}$, we have

$$\boldsymbol{L}(t;t_0,s)\boldsymbol{L}(t;t_1,t_0)y = \boldsymbol{L}(t;t_1,s)\boldsymbol{L}(t;t_0,t_1)\boldsymbol{L}(t;t_1,t_0)y, \quad \forall\, y \in \Omega,\ s \geq 0.$$

Hence, we can get the composition rule. Furthermore, by simply taking $t_2 = t_0$, we obtain the second equation in this lemma.

Hence, we immediately see that $\boldsymbol{L}(t;t_1,t_2)\boldsymbol{L}(t;t_2,t_1) = \boldsymbol{L}(t;t_2,t_2) = \boldsymbol{\delta}$. By taking derivatives with respect to t_2 on both sides, we can obtain

$$\frac{\partial \boldsymbol{L}(t;t_1,t_2)}{\partial t_2}\boldsymbol{L}(t;t_2,t_1) + \boldsymbol{L}(t;t_1,t_2)\frac{\partial \boldsymbol{L}(t;t_2,t_1)}{\partial t_2} = 0.$$

By plugging (2.7) into the equation above, we can see that $\boldsymbol{L}(t;t_2,t_1)$ is the inverse of $\boldsymbol{L}(t;t_1,t_2)$ and it satisfies the following ODE:

$$\frac{\partial \boldsymbol{L}(t;t_2,t_1)}{\partial t_2} = -\boldsymbol{L}(t;t_2,t_1)\big(\boldsymbol{\Phi}(t,t_2)\boldsymbol{L}(t;t_1,t_2)\big)\boldsymbol{L}(t;t_2,t_1) = -\boldsymbol{L}(t;t_2,t_1)\boldsymbol{\Phi}(t,t_2)$$

and the initial condition $\boldsymbol{L}(t;t_1,t_1) = \boldsymbol{\delta}$. □

Now, we are ready to introduce to the definition and the properties of the generalized Lie derivative.

DEFINITION 2.1 (Generalized Lie Derivative). We define the generalized Lie derivative of a symmetric tensor with respect to $\boldsymbol{\Phi}$ in the Lagrangian frame as follows: for $t, s \geq 0$,

$$\mathcal{L}_{\mathbf{u},\boldsymbol{\Phi}}\boldsymbol{\zeta}(t,s) := \boldsymbol{L}(t;t,s)\frac{\partial\left(\boldsymbol{L}(t;s,t)\boldsymbol{\zeta}(t,s)\boldsymbol{L}(t;s,t)^T\right)}{\partial s}\boldsymbol{L}(t;t,s)^T. \tag{2.12}$$

In the Eulerian coordinates, we let $s = t$ and

$$\mathcal{L}_{\mathbf{u},\boldsymbol{\Phi}}\boldsymbol{\zeta}(t) := \boldsymbol{L}(t;t,s)\frac{\partial\left(\boldsymbol{L}(t;s,t)\boldsymbol{\zeta}(t,s)\boldsymbol{L}(t;s,t)^T\right)}{\partial s}\boldsymbol{L}(t;t,s)^T\bigg|_{s=t}. \tag{2.13}$$

The following lemma then gives the generalized Lie derivative defined above in the Eulerian frame:

LEMMA 2.2 (The Generalized Lie Derivative in the Eulerian Frame). *For any* $\boldsymbol{\zeta} = \boldsymbol{\zeta}(x,t)\colon \Omega \times (0,+\infty) \mapsto \mathbb{S}$, *we have*

$$\mathcal{L}_{\mathbf{u},\boldsymbol{\Phi}}\boldsymbol{\zeta}(t) = \frac{D\boldsymbol{\zeta}(t)}{Dt} - \boldsymbol{\Phi}(t)\boldsymbol{\zeta}(t) - \boldsymbol{\zeta}(t)\boldsymbol{\Phi}(t)^T. \tag{2.14}$$

PROOF. Using Eqn (2.9) and the product rule, we have

$$\frac{\partial\big(\boldsymbol{L}(t;s,t)\boldsymbol{\zeta}(t,s)\boldsymbol{L}(t;s,t)^T\big)}{\partial s} = \frac{\partial \boldsymbol{L}(t;s,t)}{\partial s}\boldsymbol{\zeta}(t,s)\boldsymbol{L}(t;s,t)^T + \boldsymbol{L}(t;s,t)\frac{\partial \boldsymbol{\zeta}(t,s)}{\partial s}\boldsymbol{L}(t;s,t)^T + \boldsymbol{L}(t;s,t)\boldsymbol{\zeta}(t,s)\frac{\partial \boldsymbol{L}(t;s,t)^T}{\partial s}.$$

Hence, we can immediately obtain

$$\begin{aligned}\mathcal{L}_{\mathbf{u},\boldsymbol{\Phi}}\boldsymbol{\zeta}(t,s) &= \boldsymbol{L}(t;t,s)\frac{\partial(\boldsymbol{L}(t;s,t)\boldsymbol{\zeta}(t,s)\boldsymbol{L}(t;s,t)^T)}{\partial s}\boldsymbol{L}(t;t,s)^T\\ &= \boldsymbol{L}(t;t,s)\frac{\partial \boldsymbol{L}(t;s,t)}{\partial s}\boldsymbol{\zeta}(t,s) + \frac{\partial\boldsymbol{\zeta}(t,s)}{\partial s} + \boldsymbol{\zeta}(t,s)\frac{\partial \boldsymbol{L}(t;s,t)^T}{\partial s}\boldsymbol{L}(t;t,s)^T.\end{aligned}$$

Using the relation (2.9), we observe that

$$\mathcal{L}_{\mathbf{u},\boldsymbol{\Phi}}\boldsymbol{\zeta}(t,s) = \frac{\partial\boldsymbol{\zeta}(t,s)}{\partial s} - \boldsymbol{\Phi}(t,s)\boldsymbol{\zeta}(t,s) - \boldsymbol{\zeta}(t,s)\boldsymbol{\Phi}(t,s)^T. \tag{2.15}$$

Now, by letting $s = t$, we get Eqn (2.14) in this lemma. □

The derivative $\mathcal{L}_{\mathbf{u},\boldsymbol{\Phi}}\boldsymbol{\zeta}$ is known as the Truesdell stress rate (SIMO and HUGHES [1998]). The notion of generalized Lie derivatives makes it possible to treat many complicated time derivatives in a unified way. This observation has been used in developing numerical schemes effectively in the pioneering work by HUGHES and WINGET [1980]. The main advantage that had been obtained was that the temporal discretization induced from this type of Lie derivative-based algorithms can have the numerical frame indifference, which is called the *incrementally objective discretization.* A key observation in LEE and XU's [2006] is that many macroscopic constitutive equations can be reformulated into a well-known symmetric matrix Riccati differential equation in terms of the aforementioned generalized Lie derivatives. Many new numerical methods can be obtained based on this observation. This will be further explored in Sections 3 and 6.

2.4. *A few examples of generalized Lie derivatives*

With some appropriate choices of transition matrices, one can consider many types of generalized Lie derivatives; see especially HUGHES [1984] and SIMO and HUGHES [1998]. For example, if $\boldsymbol{\Phi}(t)$ is the zero matrix, then the transition matrix $\boldsymbol{L}(t;s,t) \equiv \boldsymbol{\delta}$ and the corresponding generalized Lie derivative are reduced to the usual material derivative (2.2).

We now give a few more examples that will be used in the next section.

EXAMPLE 2.1 (Upper Convective Maxwell Derivative). If $\boldsymbol{\Phi}(t) = \nabla\mathbf{u}(t)$, then the transition matrix $\boldsymbol{L}(t;s,t) = \boldsymbol{F}(t;s,t)$ is the deformation gradient. From Lemma 2.2, the generalized Lie derivative with respect to $\boldsymbol{\Phi}(t)$ is just the upper convective Maxwell derivative (OLDROYD [1950]):

$$\mathcal{L}_{\mathbf{u},\boldsymbol{\Phi}}\boldsymbol{\zeta}(t) = \frac{D\boldsymbol{\zeta}(t)}{Dt} - \nabla\mathbf{u}(t)\,\boldsymbol{\zeta}(t) - \boldsymbol{\zeta}(t)\nabla\mathbf{u}(t)^T, \qquad \forall\boldsymbol{\zeta}(t)\in\mathbb{M}. \tag{2.16}$$

EXAMPLE 2.2 (Lower Convective Maxwell Derivative). If $\boldsymbol{\Phi}(t) = -\nabla\mathbf{u}(t)^T$, then the transition matrix $\boldsymbol{L}(t;s,t) = \boldsymbol{F}(t;t,s)$, the inverse of $\boldsymbol{F}(t;s,t)$ (cf. Eqn (2.6)). In this case, we have that

$$\mathcal{L}_{\mathbf{u},\boldsymbol{\Phi}}\boldsymbol{\zeta}(t) = \frac{D\boldsymbol{\zeta}(t)}{Dt} + \nabla\mathbf{u}(t)^T\,\boldsymbol{\zeta}(t) + \boldsymbol{\zeta}(t)\,\nabla\mathbf{u}(t), \qquad \forall\boldsymbol{\zeta}(t)\in\mathbb{M}. \tag{2.17}$$

This is the well-known lower convective Maxwell derivative.

These two examples above have been studied in terms of the Lie derivative by THIFFEAULT [2001].

EXAMPLE 2.3 (Gordon–Schowalter Derivative). We can also take $\boldsymbol{\Phi}(t) = \boldsymbol{R}(t)$ with

$$\boldsymbol{R}(t) := \frac{\mathsf{a}+1}{2}\nabla\mathbf{u}(t) + \frac{\mathsf{a}-1}{2}\nabla\mathbf{u}^T(t),$$

where $\mathsf{a} \in [-1, 1]$ is some parameter. In this case, the generalized Lie derivative with respect to $\boldsymbol{R}(t)$ can be given as follows:

$$\mathcal{L}_{\mathbf{u},\boldsymbol{R}}\boldsymbol{\zeta}(t) = \frac{D\boldsymbol{\zeta}(t)}{Dt} - \boldsymbol{R}(t)\,\boldsymbol{\zeta}(t) - \boldsymbol{\zeta}(t)\,\boldsymbol{R}(t)^T. \tag{2.18}$$

This is known as the Gordon–Schowalter derivative (GORDON and SCHOWALTER [1972]), in which the transition matrix is often denoted by $\boldsymbol{E}$ (JOHNSON and SEGALMAN [1977]), and we also use this convention in the rest of this article.

2.5. *Riccati differential equations*

The classical symmetric matrix Riccati differential equation (ABOU-KANDIL, FREILING, IONESCU and JANK [2003]) for a symmetric tensor $\boldsymbol{\zeta} : \Omega \times (0, +\infty) \mapsto \mathbb{S}$ has this form:

$$\frac{D\boldsymbol{\zeta}(t)}{Dt} = \boldsymbol{A}(t)\boldsymbol{\zeta}(t) + \boldsymbol{\zeta}(t)\boldsymbol{A}(t)^T - \boldsymbol{\zeta}(t)\boldsymbol{B}(t)\boldsymbol{\zeta}(t)^T + \boldsymbol{G}(t), \tag{2.19}$$

with a symmetric positive semidefinite initial condition $\boldsymbol{\zeta}(t, 0) = \boldsymbol{\zeta}_0$. Typically, it is assumed that the coefficient matrices $\boldsymbol{A}$, $\boldsymbol{B}$, and $\boldsymbol{G}$ are bounded and that the matrices $\boldsymbol{B}$ and $\boldsymbol{G}$ are both symmetric and positive semidefinite.

In particular, in this study, we are interested in two important properties of the Riccati differential equation (2.19):

- Equation (2.19) has a certain closed-form solution, from which the solution $\boldsymbol{\zeta}$ can be proved to be symmetric positive definite under certain conditions.
- The positivity-preserving schemes for such equations can easily be devised, especially in time, as investigated in the literature, as well as in terms of the solution to the Riccati form of the ODE (DIECI and EIROLA [1996]).

The following theorem shows further how this view can be exploited to establish the property of the solution to a symmetric matrix Riccati differential equation.

THEOREM 2.1 (Solution of Riccati Equations). *Equation (2.19) is equivalent to*

$$\mathcal{L}_{\mathbf{u},\boldsymbol{\Phi}}\boldsymbol{\zeta}(t) = \boldsymbol{G}(t), \quad \textit{with} \quad \boldsymbol{\Phi}(t) = \boldsymbol{A}(t) - \frac{1}{2}\boldsymbol{B}(t)\boldsymbol{\zeta}(t). \tag{2.20}$$

Furthermore, we can write $\boldsymbol{\zeta}$ in a closed form as follows: for any $t, s \geq 0$,

$$\boldsymbol{\zeta}(t) = \boldsymbol{L}(t; s, t)\boldsymbol{\zeta}(t, s)\boldsymbol{L}(t; s, t)^T + \int_s^t \boldsymbol{L}(t; \nu, t)\boldsymbol{G}(t, \nu)\boldsymbol{L}(t; \nu, t)^T \mathrm{d}\nu, \tag{2.21}$$

where the transition matrix $\boldsymbol{L}$ *satisfies the following ODE:*

$$\frac{\partial \boldsymbol{L}(t; t_1, t_2)}{\partial t_2} = \boldsymbol{\Phi}(t, t_2)\boldsymbol{L}(t; t_1, t_2) \quad \textit{and} \quad \boldsymbol{L}(t; t_1, t_1) = \boldsymbol{\delta}.$$

PROOF. We first rewrite Eqn (2.19) in the Lagrangian frame:

$$\frac{\partial \boldsymbol{\zeta}(t, s)}{\partial s} = \boldsymbol{A}(t, s)\boldsymbol{\zeta}(t, s) + \boldsymbol{\zeta}(t, s)\boldsymbol{A}(t, s)^T - \boldsymbol{\zeta}(t, s)\boldsymbol{B}(t, s)\boldsymbol{\zeta}(t, s)^T + \boldsymbol{G}(t, s). \tag{2.22}$$

Hence, the equivalence between two Eqns (2.19) and (2.20) is straightforward from the definition of generalized Lie derivatives and (2.15). We can write Eqn (2.20) as follows:

$$\boldsymbol{L}(t; t, s)\frac{\partial(\boldsymbol{L}(t; s, t)\boldsymbol{\zeta}(t, s)\boldsymbol{L}(t; s, t)^T)}{\partial s}\boldsymbol{L}(t; t, s)^T = \boldsymbol{G}(t, s). \tag{2.23}$$

This relation can also be cast into the following form:

$$\frac{\partial(\boldsymbol{L}(t; \nu, t)\boldsymbol{\zeta}(t, \nu)\boldsymbol{L}(t; \nu, t)^T)}{\partial \nu} = \boldsymbol{L}(t; \nu, t)\boldsymbol{G}(t, \nu)\boldsymbol{L}(t; \nu, t)^T. \tag{2.24}$$

By taking integration (from s to t) with respect to ν on both sides of the equality above, we obtain the desired result. This completes the proof. □

REMARK 2.1 (Positive Definiteness of the Solution). Notice that the expression of $\boldsymbol{\zeta}$ given in Eqn (2.21) suggests that $\boldsymbol{\zeta}$ is always positive definite if $\boldsymbol{G}$ and $\boldsymbol{\zeta}_0$ are positive definite.

In the rest of this article, we drop the first variable of the transition matrix $\boldsymbol{L}(t; t_1, t_2)$ when t_1 or t_2 is equal to t. For example, $\boldsymbol{L}(t; s, t)$ is denoted simply by $\boldsymbol{L}(s, t)$. Same notation applies to the deformation gradient $\boldsymbol{F}(s, t) = \boldsymbol{F}(t; s, t)$ as well.

3. General macroscopic viscoelastic models

Most macroscopic complex fluid models are given by three fundamental equations: the momentum balance equation, the continuity equation, and a constitutive law. In this section, as stated earlier, we will reformulate various constitutive equations from viscoelastic fluid models into symmetric matrix Riccati differential equations (ABOU-KANDIL, FREILING, IONESCU and JANK [2003]). This new formulation will be a key ingredient in understanding viscoelastic fluid models and in developing new numerical algorithms. The link between viscoelastic fluid models and symmetric matrix Riccati differential equations was first established by LEE and XU'S [2006]. It is then successfully used by LEE [2004] to compute the falling sphere simulation through the Johnson–Segalman fluids. The close relation between the general macroscopic viscoelastic fluid models and the symmetric matrix Riccati differential equations in this section leads to a number of important numerical schemes for solving non-Newtonian equations in a unified framework, and it opens new doors to further development.

3.1. *Basic fluid models*

Consider fluids that occupy a bounded domain $\Omega \subset \mathbb{R}^d$. Define the Reynolds number $\mathsf{Re} := \bar{U}\bar{L}/\eta_0$ where η_0 is the zero shear viscosity and $\bar{U}$ and $\bar{L}$ are the characteristic velocity scale and the length scale, respectively. The dimensionless form of the momentum balance and continuity equations in these models can be written as follows:

$$\mathsf{Re}\left(\frac{\partial \mathbf{u}}{\partial t} + \mathbf{u}\cdot\nabla\mathbf{u}\right) = -\nabla p + \nabla\cdot \boldsymbol{T}, \tag{3.1}$$

and

$$\nabla\cdot\mathbf{u} = 0, \tag{3.2}$$

where $\mathbf{u}$ is the velocity of the fluids, p is the pressure, and $\boldsymbol{T}$ is the extra-stress tensor that can be decomposed into two parts (GROISMAN and STEINBERG [1998]) in the dilute polymeric fluids as

$$\boldsymbol{T} = 2\eta_s\mathcal{D}(\mathbf{u}) + \boldsymbol{\tau}, \tag{3.3}$$

where η_s is the Newtonian viscosity and $\mathcal{D}(\mathbf{u})$ is the symmetric part of the gradient of velocity,

$$\mathcal{D}(\mathbf{u}) = \frac{\nabla\mathbf{u} + (\nabla\mathbf{u})^T}{2}. \tag{3.4}$$

We remark that $2\eta_s\mathcal{D}(\mathbf{u})$ is the solvent contribution of the stress. We note also that the tensor $\boldsymbol{\tau}$ is the polymeric contribution of the stress, which arises from the high-molecular-weight viscoelastic macromolecules and enters the equation of motion linearly.

3.2. *The Oldroyd-B model*

Most complex fluid models share the same mathematical form for the momentum and continuum equations as (3.1) and (3.2); different constitutive equations for the polymeric stress $\boldsymbol{\tau}$ lead to different complex fluid models. One basic model for complex fluids that introduces the outstanding challenge for high Weissenberg number regimes is called the Oldroyd-B model (OLDROYD [1950]).

The Oldroyd-B model (OLDROYD [1958]) obeys the following constitutive relation for $\boldsymbol{\tau}$:

$$\boldsymbol{\tau} + \mathsf{Wi}\left(\frac{\partial \boldsymbol{\tau}}{\partial t} + \mathbf{u}\cdot\nabla\boldsymbol{\tau} - \nabla\mathbf{u}\,\boldsymbol{\tau} - \boldsymbol{\tau}(\nabla\mathbf{u})^T\right) = 2\eta_p\mathcal{D}(\mathbf{u}), \tag{3.5}$$

where η_p is the polymeric viscosity and the Weissenberg number $\mathsf{Wi} = \lambda\bar{U}/\bar{L}$, where λ, $\bar{U}$, and $\bar{L}$ are the relaxation time, the characteristic velocity scale, and the length scale, respectively. The Weissenberg number is proportional to the material relaxation time.

The Oldroyd-B constitutive model (3.5) can be viewed as the simplest nonlinear extension of Maxwell's idea of formulating a system of ordinary differential equations to determine the stress in terms of the velocity gradient and the time derivative. It is easy to see that the upper convective Maxwell time derivative $\frac{\partial \boldsymbol{\tau}}{\partial t} + \mathbf{u} \cdot \nabla \boldsymbol{\tau} - \nabla \mathbf{u} \boldsymbol{\tau} - \boldsymbol{\tau} \nabla \mathbf{u}^T$ that appears in the model can be identified to be $\mathcal{L}_{\mathbf{u},\nabla\mathbf{u}}$ (see Example 2.1); therefore, the Oldroyd-B constitutive law can simply be written as

$$\boldsymbol{\tau} + \mathsf{Wi}\, \mathcal{L}_{\mathbf{u},\nabla\mathbf{u}} \boldsymbol{\tau} = 2\eta_p \mathcal{D}(\mathbf{u}). \tag{3.6}$$

It is well known that the Oldroyd-B model reduces to the upper convected Maxwell (UCM) model for the special case in which $\eta_s = 0$. It has been proved that Eqns (3.1), (3.2), and (3.6) are stable in the sense of Hadamard (OWENS and PHILLIPS [2002]).

Writing the constitutive equation as in Eqn. (3.6) is an attempt to relate the stress $\boldsymbol{\tau}$ and the rate of strain $\mathcal{D}(\mathbf{u})$. For instance, when Wi becomes zero, the stress is linearly proportional to the rate of strain, which is the Newtonian constitutive relation; in this case, the Eqns (3.1), (3.2), and (3.3) become the Navier–Stokes equations. The Weissenberg number Wi is thus the characteristic constant that distinguishes the polymeric fluids from the Newtonian fluids.

3.3. *A reformulation of the Oldroyd-B model in terms of the conformation tensor*

We now take the Oldroyd-B model (OLDROYD [1950]) as an illustrative example to show how the Oldroyd-B constitutive law can be reformulated in terms of the conformation tensor and viewed as a symmetric matrix Riccati differential equation.

The constitutive law (3.5) is frequently written in terms of the conformation tensor

$$\boldsymbol{c} := \boldsymbol{\tau} + \frac{\eta_p}{\mathsf{Wi}} \boldsymbol{\delta}. \tag{3.7}$$

From a physical point of view, the conformation tensor can be regarded as a molecular deformation tensor on a continuum level. More precisely, the conformation tensor is the ensemble average of the dyadic product of the end-to-end vector of the dumbbell. It is, therefore, symmetric and positive definite, and it is often used as a primary variable in viscoelastic flow calculations (CARREAU and GRMELA [1987]).

We recall that the rate of the strain tensor $\mathcal{D}(\mathbf{u})$ can be expressed in terms of the upper convected derivative of the identity tensor $\boldsymbol{\delta}$:

LEMMA 3.1 (Lie Derivative of the Identity). *Let $\boldsymbol{\delta}$ be the identity tensor. Then, we have*

$$\mathcal{L}_{\mathbf{u},\nabla\mathbf{u}} \boldsymbol{\delta} = -2\mathcal{D}(\mathbf{u}). \tag{3.8}$$

This is a direct consequence of the definition of $\mathcal{L}_{\mathbf{u},\nabla\mathbf{u}}$, and this simple identity plays a significant role in understanding various constitutive equations. We can reformulate the model (3.6) as follows:

$$\boldsymbol{\tau} + \mathsf{Wi} \left(\mathcal{L}_{\mathbf{u},\nabla\mathbf{u}} \boldsymbol{\tau} + \frac{\eta_p}{\mathsf{Wi}} \mathcal{L}_{\mathbf{u},\nabla\mathbf{u}} \boldsymbol{\delta} \right) = 0. \tag{3.9}$$

By adding $\frac{\eta_p}{\mathsf{Wi}} \boldsymbol{\delta}$ to both sides of the equation above and using the fact that the operator $\mathcal{L}_{\mathbf{u},\nabla\mathbf{u}}$ is linear, we obtain

$$\left(\boldsymbol{\tau} + \frac{\eta_p}{\mathsf{Wi}} \boldsymbol{\delta} \right) + \mathsf{Wi}\, \mathcal{L}_{\mathbf{u},\nabla\mathbf{u}} \left(\boldsymbol{\tau} + \frac{\eta_p}{\mathsf{Wi}} \boldsymbol{\delta} \right) = \frac{\eta_p}{\mathsf{Wi}} \boldsymbol{\delta}. \tag{3.10}$$

In terms of the conformation tensor, the constitutive equation (3.6) becomes

$$\boldsymbol{c} + \mathsf{Wi}\, \mathcal{L}_{\mathbf{u},\nabla\mathbf{u}}\boldsymbol{c} = \frac{\eta_p}{\mathsf{Wi}}\boldsymbol{\delta}, \tag{3.11}$$

and we observe that Eqn (3.11) can be written such that for $\mathsf{Wi} \neq 0$,

$$\frac{D\boldsymbol{c}}{Dt} - \nabla\mathbf{u}\,\boldsymbol{c} - \boldsymbol{c}\,\nabla\mathbf{u}^T + \frac{1}{\mathsf{Wi}}\boldsymbol{c} = \frac{\eta_p}{\mathsf{Wi}^2}\boldsymbol{\delta}. \tag{3.12}$$

REMARK 3.1 (Algebraic Riccati Form of Oldroyd-B). Equation (3.12) can be further reformulated into the following form:

$$\frac{D\boldsymbol{c}}{Dt} + \left(\frac{1}{2\mathsf{Wi}} - \nabla\mathbf{u}\right)\boldsymbol{c} + \boldsymbol{c}\left(\frac{1}{2\mathsf{Wi}} - \nabla\mathbf{u}\right)^T = \frac{\eta_p}{\mathsf{Wi}^2}\boldsymbol{\delta}. \tag{3.13}$$

This form can be immediately identified with the symmetric matrix Riccati differential equation for $\boldsymbol{c}$ as introduced in the general form (2.19) with the choice of the coefficient matrices that

$$\boldsymbol{A}(t) = \frac{1}{2\mathsf{Wi}}\boldsymbol{\delta} - \nabla\mathbf{u}, \quad \boldsymbol{B}(t) \text{ is a zero matrix, and } \boldsymbol{G}(t) = \frac{\eta_p}{\mathsf{Wi}^2}\boldsymbol{\delta}. \tag{3.14}$$

REMARK 3.2 (Positivity of the Conformation Tensor for the Oldroyd-B Model). The positive definiteness of $\boldsymbol{c}$ is thought to have been first established by HULSEN [1990] directly from the differential model (3.11). From the Riccati form of the Oldroyd-B constitutive law (3.13) and Lemma 2.1, it is easy to establish that if $\boldsymbol{c}(0)$ is given to be a positive definite tensor, then the conformation tensor $\boldsymbol{c}$ is always positive definite since $\boldsymbol{G}$ is non-negative. In fact, this technique will allow us to provide an integral equivalent equation of the differential equation given by Eqn (3.12) and establish the positive definiteness of the conformation tensor $\boldsymbol{c}$ in a transparent manner as well. See Eqn (3.30).

3.4. Conformation tensor formulation of the Johnson–Segalman model

The Oldroyd-B model (3.6) is a basic constitutive model for complex fluids. Many improvements for constitutive equations have been developed from the Oldroyd-B model, e.g., by modifying the upper convective derivative or by adding additional terms to better fit the rheological property of the fluids. A few such examples will be discussed in this section.

Let us first consider the Johnson–Segalman model (JOHNSON and SEGALMAN [1977]):

$$\boldsymbol{\tau} + \mathsf{Wi}\, \mathcal{L}_{\mathbf{u},\boldsymbol{R}}\boldsymbol{\tau} = 2\eta_p \mathcal{D}(\mathbf{u}), \tag{3.15}$$

where $\mathcal{L}_{\mathbf{u},\boldsymbol{R}}$ is the Gordon–Schowalter derivative as in Example 2.3. The Johnson–Segalman model is often the model of choice for studying material instability, such as shark-skin and spurt, which have been the subject of considerable research interest in recent years.

Similar to our approach with the Oldroyd-B model, we first reformulate Eqn (3.15) in terms of the conformation tensor $\boldsymbol{c}$. We obtain

$$\mathcal{L}_{\mathbf{u},\boldsymbol{R}}\boldsymbol{\delta} = -\left(\frac{\mathsf{a}+1}{2}\nabla\mathbf{u} + \frac{\mathsf{a}-1}{2}\nabla\mathbf{u}\right) - \left(\frac{\mathsf{a}+1}{2}\nabla\mathbf{u} + \frac{\mathsf{a}-1}{2}\nabla\mathbf{u}\right)^T = -2\mathsf{a}\mathcal{D}(\mathbf{u}).$$

Therefore, for nonzero Wi and a, the model (3.15) can be written as follows:

$$\boldsymbol{\tau} + \mathsf{Wi}\,\mathcal{L}_{\mathbf{u},\boldsymbol{R}}\boldsymbol{\tau} = -\frac{\eta_p}{\mathsf{a}}\mathcal{L}_{\mathbf{u},\boldsymbol{R}}\boldsymbol{\delta} \Longrightarrow \boldsymbol{\tau} + \frac{\eta_p}{\mathsf{aWi}} + \mathsf{Wi}\,\mathcal{L}_{\mathbf{u},\boldsymbol{R}}\left(\boldsymbol{\tau} + \frac{\eta_p}{\mathsf{aWi}}\boldsymbol{\delta}\right) = \frac{\eta_p}{\mathsf{aWi}}\boldsymbol{\delta}.$$

Defining $\boldsymbol{c} = \boldsymbol{\tau} + \frac{\eta_p}{\mathsf{aWi}}\boldsymbol{\delta}$, we arrive at the following reformulation of the Johnson–Segalman model (3.15):

$$\boldsymbol{c} + \mathsf{Wi}\,\mathcal{L}_{\mathbf{u},\boldsymbol{R}}\boldsymbol{c} = \frac{\eta_p}{\mathsf{aWi}}\boldsymbol{\delta}. \tag{3.16}$$

Recall that the generalized Lie derivative $\mathcal{L}_{\mathbf{u},\boldsymbol{R}}$ is determined by the transition matrix $\boldsymbol{E}(s,t)$ that satisfies the following ODE:

$$\frac{D\boldsymbol{E}(s,t)}{Dt} = \left(\frac{\mathsf{a}+1}{2}\nabla\mathbf{u} + \frac{\mathsf{a}-1}{2}\nabla\mathbf{u}^T\right)\boldsymbol{E}(s,t) \quad \text{and} \quad \boldsymbol{E}(s,s) = \boldsymbol{\delta}. \tag{3.17}$$

The tensor $\boldsymbol{E}(s,t)$ obeying (3.17) was first introduced by JOHNSON and SEGALMAN [1977] as a deformation tensor for viscoelastic fluids that have certain degree of nonaffinity. The parameter a is related to the so-called slippage parameter $\xi = 1 - \mathsf{a}$, which measures the nonaffinity in the reaction of macromolecules under the exerted force from the surrounding fluids.

3.5. *Conformation reformulation for a general viscoelastic model*

To summarize, we now discuss macroscopic models that can be written in the following general form:

$$\mathcal{L}_{\mathbf{u},\boldsymbol{R}}\boldsymbol{c} + \alpha\boldsymbol{c} = \beta\boldsymbol{\delta}, \tag{3.18}$$

where $\alpha \geq 0$ and $\beta > 0$ may depend on t and/or $\boldsymbol{c}$. For instance, the Johnson–Segalman model can be recovered from Eqn (3.18) by choosing $\alpha = \frac{1}{\mathsf{Wi}}$ and $\beta = \frac{\eta_p}{\mathsf{aWi}^2}$. It would be of interest to consider Eqn (3.18) as it is, which is because the generalized Lie derivative $\mathcal{L}_{\mathbf{u},\boldsymbol{R}}$ is ubiquitous in general macroscopic equations. We can, in fact, derive the solution expression $\boldsymbol{c}$ in terms of the transition matrix $\boldsymbol{E}$ as follows.

THEOREM 3.1 (Explicit Solution of the Constitutive Equation). *The solution to the constitutive equation in Riccati form (3.18) satisfies*

$$\boldsymbol{c}(t) = \exp\left(-\int_s^t \alpha(\varsigma)\mathrm{d}\varsigma\right)\boldsymbol{E}(s,t)\boldsymbol{c}(t,s)\boldsymbol{E}(s,t)^T + \int_s^t \exp\left(-\int_\nu^t \alpha(\varsigma)\mathrm{d}\varsigma\right)\beta(\nu)\boldsymbol{E}(\nu,t)\boldsymbol{E}(\nu,t)^T\mathrm{d}\nu. \tag{3.19}$$

PROOF. Note that Eqn (3.18) can be reformulated as follows:

$$\mathcal{L}_{\mathbf{u},\boldsymbol{\Phi}}\boldsymbol{c} = \beta\boldsymbol{\delta},$$

where the generalized Lie derivative is with respect to

$$\boldsymbol{\Phi}(t) := \frac{\mathsf{a}+1}{2}\nabla\mathbf{u} + \frac{\mathsf{a}-1}{2}\nabla\mathbf{u}^T - \frac{\alpha(t)}{2}.$$

From the Lemma 2.1, we arrive at the following expression for $\boldsymbol{c}$:

$$\boldsymbol{c}(t) = \boldsymbol{L}(s,t)\boldsymbol{c}(t,s)\boldsymbol{L}(s,t)^T + \int_s^t \beta(\nu)\boldsymbol{L}(\nu,t)\boldsymbol{L}(\nu,t)^T \mathrm{d}\nu. \tag{3.20}$$

On the other hand, the matrix $\boldsymbol{L}(s,t)$ can be expressed as follows:

$$\boldsymbol{L}(s,t) = \exp\left(-\int_s^t \frac{\alpha(\nu)}{2}\mathrm{d}\nu\right)\boldsymbol{E}(s,t). \tag{3.21}$$

To see this, we note that $\boldsymbol{L}_1(s,t) = \boldsymbol{E}(s,t)$ is the solution to the following ODE:

$$\frac{\partial \boldsymbol{L}_1(s,t)}{\partial t} = \left(\frac{\mathsf{a}+1}{2}\nabla\mathbf{u}(t) + \frac{\mathsf{a}-1}{2}\nabla\mathbf{u}(t)^T\right)\boldsymbol{L}_1(s,t)$$

and the solution to the equation

$$\frac{\partial \boldsymbol{L}_2(s,t)}{\partial t} = -\frac{\alpha(t)}{2}\boldsymbol{L}_2(s,t)$$

is given by

$$\boldsymbol{L}_2(s,t) = \exp\left(-\int_s^t \frac{\alpha(\nu)}{2}\mathrm{d}\nu\right)\boldsymbol{\delta}. \tag{3.22}$$

The simple observation that $\boldsymbol{L}(s,t) = \boldsymbol{L}_1(s,t)\boldsymbol{L}_2(s,t)$ completes the proof. □

The simple formulation (3.18) can, in fact, represent many existing models. For example, it can represent the well-known Phan-Thien and Tanner (PTT) model THIEN and TANNER [1977] and other models that belong to the finitely extensible nonlinear elastic (FENE) models (CHILCOTT and RALLISON [1988], ILG, KARLIN and ÖTTINGER [2002], LIELENS, HALIN, JAUMAIN, KEUNINGS and LEGAT [1998], REMMELGAS, SINGH and LEAL [1999], SZERI [2000]).

EXAMPLE 3.1 (The Phan-Thien and Tanner Model). The Phan-Thien and Tanner (PTT) model can be given in the following expression:

$$\mathcal{F}(\boldsymbol{\tau})\boldsymbol{\tau} + \mathsf{Wi}\,\mathcal{L}_{\mathbf{u},\boldsymbol{R}}\boldsymbol{\tau} = 2\eta_p\mathcal{D}(\mathbf{u}), \tag{3.23}$$

where $\mathcal{F}$ is a scalar function defined by

$$\mathcal{F}(\boldsymbol{\tau}) = \exp\left(\frac{\epsilon \mathsf{Wi}}{\eta_p}\,\mathrm{tr}(\boldsymbol{\tau})\right), \tag{3.24}$$

where ϵ is a parameter. The model (3.23) can easily be transformed in terms of the conformation tensor $\boldsymbol{c}$ as follows:

$$\mathcal{L}_{\mathbf{u},\boldsymbol{R}}\boldsymbol{c} + \frac{\mathcal{G}(\boldsymbol{c})}{\mathsf{Wi}}\boldsymbol{c} = \eta_p \frac{\mathcal{G}(\boldsymbol{c})}{\mathsf{aWi}^2}\boldsymbol{\delta}, \quad \text{with } \mathcal{G}(\boldsymbol{c}) = \mathcal{F}\Big(\boldsymbol{c} - \frac{\eta_p}{\mathsf{aWi}}\boldsymbol{\delta}\Big). \tag{3.25}$$

Therefore, the PTT model belongs to the class of models that can be represented by Eqn (3.18).

EXAMPLE 3.2 (General Single-Variable Models). We note that the general single-variable models as introduced by Hulsen (e.g., BERIS and EDWARDS [1994], HULSEN [1990]) can be given in terms of the conformation tensor $\boldsymbol{c}$ as follows:

$$\frac{D\boldsymbol{c}}{Dt} = \boldsymbol{a}(t)\,\boldsymbol{c} + \boldsymbol{c}\,\boldsymbol{a}(t)^T + g_0\boldsymbol{\delta} + g_1\boldsymbol{c} + g_2\boldsymbol{c}^2, \tag{3.26}$$

where g_i's $(i = 0, 1, 2)$ are given functions that may depend on time and/or $\boldsymbol{c}$. HULSEN [1990] provided a sufficient condition that $g_0 > 0$ for which the conformation tensor $\boldsymbol{c}$ for models of the form (3.26) remains positive definite for all time if it is positive initially. His arguments were based on the rate of change in the determinant of $\boldsymbol{c}$ along the trajectory. Our framework cast (3.26) into the general Riccati equation

$$\frac{D\boldsymbol{c}}{Dt} - \boldsymbol{A}(t)\boldsymbol{c} - \boldsymbol{c}\boldsymbol{A}(t)^T = \boldsymbol{G}(t), \tag{3.27}$$

with the coefficient matrices

$$\boldsymbol{A}(t) := \boldsymbol{a}(t) + \frac{g_1}{2}\boldsymbol{\delta} + \frac{g_2\boldsymbol{c}}{2} \quad \text{and} \quad \boldsymbol{G}(t) := g_0\boldsymbol{\delta}.$$

An alternative reformulation of (3.27) can be given by

$$\mathcal{L}_{\mathbf{u},\boldsymbol{A}}\boldsymbol{c} = \boldsymbol{G}(t). \tag{3.28}$$

This reformulation in terms of the generalized Lie derivative with respect to $\boldsymbol{\Phi}$ immediately proves the positivity of $\boldsymbol{c}$ under the assumption that $g_0 > 0$.

It should be note here that the analytic expression (3.19) of the conformation tensor $\boldsymbol{c}$ can be used to derive the corresponding integral models. Indeed, under some appropriate assumption (such as $\alpha \geq \alpha_0$ for some positive constant α_0 and $\boldsymbol{E}(s,t)$ is bounded for $s \leq t$), we formally obtain the following integral models by taking $s \to -\infty$,

$$\boldsymbol{c}(t) = \int_{-\infty}^{t} \exp\left(-\int_{\nu}^{t} \frac{\alpha(\nu)}{2}\mathrm{d}\nu\right)\beta(\nu)\boldsymbol{E}(\nu,t)\boldsymbol{E}(\nu,t)^T\mathrm{d}\nu. \tag{3.29}$$

In particular, this includes the Johnson–Segalman integral model, which does not seem to be known in the literature; see JOSEPH [1990]. Furthermore, as an immediate consequence of $\mathsf{a} = 1$, we obtain the well-known integral expression for the Oldroyd-B model (3.12) as follows:

$$c(t) = \frac{\eta_p}{\mathsf{Wi}^2} \int_{-\infty}^{t} \exp\left(-\frac{t-\nu}{\mathsf{Wi}}\right) \boldsymbol{F}(\nu,t)\boldsymbol{F}(\nu,t)^T \mathrm{d}\nu. \tag{3.30}$$

Although it has been widely believed that the integral expression (3.30) of the conformation tensor is equivalent to (3.11) (see, e.g., RENARDY [2000b]), a rigorous justification for this equivalence is missing in the literature (see relevant remarks made both by JOSEPH [1990, p.15], RENARDY [2000b, p.18]). Note that it is easy to establish that the integral model can result in the differential model (3.11) by taking the (material) time derivative. With the help of the Riccati formulation, the justification that the differential model results in the integral model is completed with ease, which would have been difficult otherwise.

4. Basic mathematical and physical properties of the models

In this section, we give a brief overview of the existing mathematical analysis of basic theoretical issues such as the existence and stability of the solution to complex fluid models. While these theoretical works are obviously of interest themselves, they are also instrumental to designing of appropriate numerical methods for these models.

4.1. Existence theory

On the mathematical theory for complex fluid models, many fundamental questions, such as whether (weak) solutions exist, are still open (CHEMIN and MASMOUDI [2001], LIN, LIU and ZHANG [2007], LIONS and MASMOUDI [2000]). On the Oldroyd-B model, the existence of global weak solutions even at regimes of low Weissenberg number has not been fully understood yet. The global existence of weak solutions was established by BARRETT and SÜLI [2008] for the corotational models, under the assumption that the velocity field is regularized. For the noncorotational models, like the Oldroyd-B model, both the velocity and the extra-stress fields are assumed to be mollified in the weak formulations in order to obtain the global existence of weak solutions (BARRETT and SÜLI [2008]).

Some studies have been published on short-time existence (GUILLOPE and SAUT [1990a,b], JOURDAIN, LELIVRE and BRIS [2004], LI and ZHANG [2004], RENARDY [1991]) and global existence with small initial data (GUILLOPE and SAUT [1990a,b], LIN, LIU and ZHANG [2005]) of the solutions. In particular, LIN, LIU and ZHANG [2005] established the existence of classical solutions for the Oldroyd-B model at infinite Weissenberg number with small initial data. In another notable work, LIONS and MASMOUDI [2000] proved the existence of global weak solutions for the corotational Jeffreys model based on the L^2-norm energy estimate for both velocity and stress fields. This type of energy estimate does not, however, hold for the Oldroyd-B model; therefore, the global existence of the Oldroyd-B model is still an open problem for general initial data. A stability result has also been obtained by HE and ZHANG [2009]: if the initial data is sufficiently close to the equilibrium, the solution approaches to the equilibrium with a certain decaying rate measured in the

L^2-norm. The main idea behind the existence and stability proof in HE and ZHANG [2009] and LIN, LIU and ZHANG [2005] is to take full advantage of the *Divergence-free condition* imposed at the velocity field, which is shown to generate a dissipation mechanism and hence stabilize the equation. This work strongly indicates that incompressibility plays an important role in the stability of the system. It hints at the importance of preserving incompressibility on the discrete level in order to obtain stable numerical schemes.

While the global existence of the Oldroyd-B model for general "large" data is missing, there are several global-in-time existence results of some complex fluid models for simple shear flows. ENGLER [1987] and GUILLOPE and SAUT [1990a,b] obtained global existence results for shear flows obeying a class of nonlinear integro-differential models and the Johnson–Segalman model, respectively. This problem has recently been revisited by RENARDY [2009], in which the global existence for shear flow under the PTT (GIESEKUS [1982], THIEN and TANNER [1977]), and Johnson–Segalman models (JOHNSON and SEGALMAN [1977]) has been established for some parameters, although not for the Oldroyd-B model. As Renardy stated in RENARDY [2009] that the positive definiteness of the conformation tensor is crucial to his proof.

In addition, several studies indicate that the Oldroyd-B model might produce nonsmooth stress fields; for example, see RENARDY [2006] and BAJAJ, PASQUALI and PRAKASH [2008]. Renardy's (RENARDY [2006]) results have been correlated with the numerical results of THOMASES and SHELLEY [2007], where certain numerical evidence of singularity formation is provided. The latter study tried to explain why the flow-past-cylinder benchmark problem presents numerical challenges. We note that these singular solutions are obtained for the steady-state Oldroyd-B model, and it is unclear whether or not singularity will form for the time-dependent equations.

While global-in-time existence remains illusive for continuous problems, we will establish the global existence of the discrete solutions for macroscopic viscoelastic models in this article; see Section 8. Similar to the theories on the continuous level, the strong divergence-free condition and the positivity of the conformation tensors play critical roles in our analysis. As suggested by these successful theoretical efforts, our guiding principle is that the positivity of the conformation tensors and the strong divergence-free condition for the velocity fields should be both preserved in the fully discrete level. Both ingredients are crucial in deriving the discrete energy estimates and global existence for the numerical solutions.

4.2. *Energy estimates*

Energy estimates are basic ingredients in the analysis of well posedness of the equations, and they are also crucially important in designing well-posed numerical discretization schemes as well. We will present some basic energy estimates (LEE and XU's [2006], LOZINSKI and OWENS [2003]) for the continuous equations in this section, and we will later extend these estimates to the discrete level.

To state the energy estimate, let us first introduce an energy norm for the conformation tensor:

$$\|\boldsymbol{\sigma}\|_{L^1} := \int_{\Omega} \mathrm{tr}(\boldsymbol{\sigma})\,\mathrm{d}x, \qquad \forall \boldsymbol{\sigma} \in \boldsymbol{S}_h. \tag{4.1}$$

This obviously defines a norm on the space of positive-definite tensors. We note that for the conformation tensor $\boldsymbol{c}$, the norm $\|\boldsymbol{c}\|_{L^1}$ has its own physical meaning as well. The trace of the conformation tensor $\boldsymbol{c}$ can be viewed as the length from the tail to the head of a macromolecule: the longer the length, the more elastic energy it stores. We can view $\|\boldsymbol{c}\|_{L^1}$ as the total elastic energy due to the interaction between the macromolecules and surrounding fluids.

Based on this, we can define the total energy (kinetic and elastic) of the fluid at time level t to be

$$\mathcal{E}(t) := \mathsf{Re}\|\mathbf{u}(\cdot,t)\|_0^2 + \frac{1}{2}\|\boldsymbol{c}(\cdot,t)\|_{L^1}. \tag{4.2}$$

For all the estimates presented in this section, the following *bridging identity* is crucial:

$$(\boldsymbol{c}(\cdot,t) : \mathcal{D}(\mathbf{u})(\cdot,t)) = \int_\Omega \operatorname{tr}(\nabla\mathbf{u}(\cdot,t)\boldsymbol{c}(\cdot,t))\,\mathrm{d}x. \tag{4.3}$$

This identity plays a role in bridging the energy term in the momentum equation and its counterpart in the constitutive equation.

Now, we take the Oldroyd-B model (in terms of the conformation tensor) in the Riccati form as an example:

$$\mathsf{Re}\left(\frac{\partial\mathbf{u}}{\partial t} + \mathbf{u}\cdot\nabla\mathbf{u}\right) = -\nabla p + 2\mu_s\nabla\cdot\mathcal{D}(\mathbf{u}) + \nabla\cdot\boldsymbol{c}, \tag{4.4}$$

$$\nabla\cdot\mathbf{u} = 0, \tag{4.5}$$

$$\frac{\partial\boldsymbol{c}}{\partial t} + \mathbf{u}\cdot\nabla\boldsymbol{c} - \nabla\mathbf{u}\,\boldsymbol{c} - \boldsymbol{c}\,\nabla\mathbf{u}^T + \frac{1}{\mathsf{Wi}}\boldsymbol{c} = \frac{\eta_p}{\mathsf{Wi}^2}\boldsymbol{\delta}. \tag{4.6}$$

From (4.3), we can easily establish the following energy law for Eqns (4.4)–(4.6):

$$\frac{\mathrm{d}}{\mathrm{d}t}\mathcal{E}(t) = -\eta_s\|\mathcal{D}(\mathbf{u}(\cdot,t))\|_0^2 - \frac{1}{2\mathsf{Wi}}\|\boldsymbol{c}(\cdot,t)\|_{L^1} + \frac{d}{2}\frac{1-\eta_s}{\mathsf{Wi}^2}|\Omega|, \tag{4.7}$$

where $|\Omega|$ and d are the measure and spatial dimension of the domain Ω, respectively. From the energy law (4.7), we can derive the following energy estimate for the Oldroyd-B model.

THEOREM 4.1 (Continuous Energy Estimate). *For* $\mathsf{Wi} \neq 0$, *the Oldroyd-B model* (4.4)–(4.6) *admits the following energy estimate:*

$$\mathcal{E}(t) \le \mathrm{e}^{-\mathsf{C}_1 t}\mathcal{E}(0) + \frac{\mathsf{C}_2}{\mathsf{C}_1}\left(1 - \mathrm{e}^{-\mathsf{C}_1 t}\right) \tag{4.8}$$

$$\eta_s\int_0^t \|\mathcal{D}(\mathbf{u}(\cdot,\nu))\|_0^2\,\mathrm{d}\nu \le \mathcal{E}(0) + \mathsf{C}_2\,t, \tag{4.9}$$

with the constants

$$\mathsf{C}_1 = \min\left(\frac{\mathsf{C}_\Omega\eta_s}{\mathsf{Re}}, \frac{1}{\mathsf{Wi}}\right) \quad \textit{and} \quad \mathsf{C}_2 = \frac{d}{2}\frac{1-\eta_s}{\mathsf{Wi}^2}|\Omega|, \tag{4.10}$$

where C_Ω *is a positive constant depending on* Ω *only. For the special case when* $\mathsf{Wi} = \infty$, *we have*

$$\mathcal{E}(t) \le \mathcal{E}(0) \qquad \textit{and} \qquad \eta_s \int_0^t \|\mathcal{D}(\mathbf{u}(\cdot, \nu))\|_0^2 \le \mathcal{E}(0). \tag{4.11}$$

PROOF. From Korn's inequality, we have the following inequality $C_\Omega \|\mathbf{u}\|_0 \le \|\mathcal{D}(\mathbf{u})\|_0$, where C_Ω depends only on Ω. The energy law (4.7) then leads to

$$\frac{d}{dt}\Big(\mathsf{Re}\|\mathbf{u}(\cdot, t)\|_0^2 + \frac{1}{2}\|\boldsymbol{c}(\cdot, t)\|_{L^1}\Big) \le -\eta_s C_\Omega \|\mathbf{u}(\cdot, t)\|_0^2 - \frac{1}{2\mathsf{Wi}}\|\boldsymbol{c}(\cdot, t)\|_{L^1} + \frac{d}{2}\frac{1-\eta_s}{\mathsf{Wi}^2}|\Omega|.$$

We define $C_1 = \min(C_\Omega \eta_s/\mathsf{Re}, 1/\mathsf{Wi})$ and $C_2 = \frac{d}{2}\frac{1-\eta_s}{\mathsf{Wi}^2}|\Omega|$ to obtain the following inequality:

$$\frac{d}{dt}\Big(\mathsf{Re}\|\mathbf{u}(\cdot, t)\|_0^2 + \frac{1}{2}\|\boldsymbol{c}(\cdot, t)\|_{L^1}\Big) \le -C_1\Big(\mathsf{Re}\|\mathbf{u}(\cdot, t)\|_0^2 + \frac{1}{2}\|\boldsymbol{c}(\cdot, t)\|_{L^1}\Big) + C_2. \tag{4.12}$$

This estimate gives the desired estimate (4.8) immediately by Gronwall's inequality. We take integration with respect to time on both sides of (4.7) to get

$$\begin{aligned} &\mathsf{Re}\|\mathbf{u}(\cdot, t)\|_0^2 + \frac{1}{2}\|\boldsymbol{c}(\cdot, t)\|_{L^1} + \eta_s \int_0^t \|\mathcal{D}(\mathbf{u}(\cdot, \nu))\|_0^2 \,d\nu \\ &\quad \le \mathsf{Re}\|\mathbf{u}(\cdot, 0)\|_0^2 + \frac{1}{2}\|\boldsymbol{c}(\cdot, 0)\|_{L^1} - \frac{1}{2\mathsf{Wi}} \int_0^t \|\boldsymbol{c}(\cdot, \nu)\|_{L^1}\,d\nu + C_2 t \\ &\quad \le \mathsf{Re}\|\mathbf{u}(\cdot, 0)\|_0^2 + \frac{1}{2}\|\boldsymbol{c}(\cdot, 0)\|_{L^1} + C_2 t. \end{aligned}$$

This completes the proof. □

REMARK 4.1 (The Effect of the Weissenberg Number). The parameter Wi is the ratio between the relaxation time of the macromolecules and the characteristic time. The longer the relaxation time, the longer it takes for the macromolecules to return to their original states; this can be interpreted as showing that the fluid is less dependent on its initial state. This has been correlated in the energy estimate (4.8); that is, the coefficient function multiplied to the initial data decays slowly when Wi becomes larger.

5. Existing numerical methods for viscoelastic fluid models

In this section, we offer a brief overview of numerical methods for solving viscoelastic fluid models, especially in regard to studies focused on the well-known high Weissenberg number problem (HWNP). The problem is associated with the breakdown of the numerical solutions to the complex fluid models when the Weissenberg numbers are *moderately* large. This outstanding problem has been one of the driving forces for developing new numerical techniques for complex fluids (see OWENS and PHILLIPS [2002]).

5.1. Mixed formulations

To numerically achieve mesh convergence and long-time stability beyond certain critical Weissenberg numbers for various viscoelastic models has proven difficult. Numerous attempts have been made to overcome the high Weissenberg number problem with mixed finite element methods. Most of the early work on viscoelastic flow analysis is based on the mixed finite element formulations for $(\mathbf{u}, p, \boldsymbol{\tau})$; see Baaijens's BAAIJENS [1998] review for more details. Two basic problems have been encountered with the above formulations. First, as the value of the Weissenberg number increases, the importance of the convective term grows, which makes Galerkin discretizations not suitable. Second, the discretization spaces for the three variables must be carefully selected with respect to each other in order to satisfy appropriate stability conditions for the three fields.

Numerical success in the early stage of computational rheology can be found in MARCHAL and CROCHET [1987], which introduced a new mixed finite element method for the numerical simulation of viscoelastic flows. In MARCHAL and CROCHET [1987], the authors showed that the streamline-upwinding (SU) method by HUGHES and BROOKS [1982] could be used for viscoelastic fluid simulation in order to stabilize the hyperbolic constitutive equation. Further, FORTIN and PIERRE [1989] analyzed the finite element spaces employed in MARCHAL and CROCHET [1987]. Another approach, introduced by FORTIN and FORTIN [1989], used the discontinuous Galerkin (DG) method by LESAINT RAVIART [1979] for the constitutive equation combined with the element-wise streamline-upwinding method for the momentum equation.

Much work has been done in this line of research. To stabilize the numerical simulation, these algorithms focus on adding more diffusion to the momentum equation in order to make the ellipticity of the equation explicit. SUN, SMITH, ARMSTRONG and BROWN [1999] summarized the main ideas as follows:

(1) reformulating the momentum and the constitutive equation to make the elliptic character of this equation explicit with respect to the velocity field;
(2) splitting the formulation into the solution of the velocity-pressure saddle point problem equations for an incompressible fluid and the calculation of the extra-stress field from the hyperbolic constitutive equation;
(3) applying numerically stable and accurate methods, like SUPG or DG methods, for solution of the hyperbolic constitutive equations; and
(4) introducing accurate and smooth interpolation of velocity gradients for additional numerical stability in solution of the constitutive equation.

KING, APELIAN, ARMSTRONG and BROWN [1988] made the first effort in this direction; they introduced the explicitly elliptic momentum equation (EEME) and gave a reformulation of the momentum equation that makes its ellipticity explicit for the upper convected Maxwell (UCM) models. This method was later generalized by BERIS, ARMSTRONG and BROWN [1984, 1986]: their elastic-viscous split-stress (EVSS) formulation splits the extra-stress $\boldsymbol{T}$ into a viscous part and an elastic part; i.e., $\boldsymbol{T} = \boldsymbol{\tau}_v + \boldsymbol{\tau}_e$, where $\boldsymbol{\tau}_v = 2\eta_a \mathcal{D}(\mathbf{u})$ and η_a is a parameter for viscosity; in this way, the formulation introduces another variable, the rate-of-strain $\mathcal{D}(\mathbf{u})$. The application of this method has been limited to a few models; furthermore, it introduced a new term containing convected derivatives of $\mathcal{D}(\mathbf{u})$. RAJAGOPALAN,

ARMSTRONG and BROWN [1990] modified this method by using a least square approximation of $\mathcal{D}(\mathbf{u})$ and generalized it to the Oldroyd-B model, as well as the Giesekus models.

The EEME and EVSS formulations are significant improvements over previous methods based on the standard viscous model in terms of numerical stability. They allow numerically stable and accurate calculations at moderately high Wi values. However, almost every flow problem has levels of elasticity that cannot be calculated with these methods for any given finite element mesh. SUN, PHAN-THIEN and TANNER [1996] argued that the failure is due to a steep stress gradient and introduced an adaptive way for choosing the viscosity parameter η_a to tackle the difficulty; this is known as the adaptive viscoelastic stress splitting (AVSS) method. An alternative EVSS method was proposed in BROWN, SZADY, NORTHEY and ARMSTRONG [1993] and SZADY et al. [1995], who applied least square approximation for the gradient $\nabla\mathbf{u}$ instead of $\mathcal{D}(\mathbf{u})$; this type of methods is called EVSS-G. GUENETTE and FORTIN [1995] and LIU, BORNSIDE, ARMSTRONG and BROWN [1998] applied the stress splitting at the discrete level, which gives the discrete elastic-viscous split-stress (DEVSS) and DEVSS-G, respectively. SUN, SMITH, ARMSTRONG and BROWN [1999] combined all these techniques to create the discrete adaptive viscoelastic stress splitting–discontinuous Galerkin (DAVSS-G/DG) method, which does exactly what the name suggests.

Besides the finite element formulation, other related discretization schemes, such as finite volume method and the spectral methods, have been applied to viscoelastic fluids. Just to mention a few, HU and JOSEPH [1990] designed a finite volume (FV) discretization based on the semi-implicit method for pressure-linked equations revised (SIMPLER) for the UCM model on orthogonal staggered grids. OLIVEIRA, PINHO and PINTO [1998] generalized this method to nonorthogonal collocated grids. ALVES, OLIVEIRA and PINHO [2003] and ALVES, PINHO and OLIVEIRA [2001] discussed FVM on nonorthogonal grids combined with a high-resolution scheme (HRS) instead of usual upwind difference schemes for the Oldroyd-B model and the PTT model for the planar contraction benchmark problem. CHAUVIÉRE and OWENS [2001] applied the spectral method for viscoelastic flows and introduced the streamline-upwind Petrov/Galerkin (SUPG) for constitutive equations.

5.2. Steep stress layers and log-conformation formulation

The breakdown of the numerical algorithms for high Weissenberg numbers has often been associated with the steep stress gradients in the narrow regions of the flow domain. Even for the well-known flow-past-cylinder problem, which has no geometric singularities and generates smooth flows, the numerical simulation is still difficult. Observed in many numerical simulations is that this difficulty inheres in the thin stress layers that develop around the cylinder and in the wake along the centerline, where the flow is purely elongational; see the review by BAAIJENS [1998] for example. BERIS, ARMSTRONG and BROWN [1983, 1987] have demonstrated the formation of elastic boundary layers in both asymptotic analysis and spectral/finite-element calculations for the flow between two eccentric rotating cylinders. RENARDY [2000a] analyzed the width of the boundary layer and the wake for the UCM model with fixed Newtonian kinematics. Based on this work, WAPPEROM and RENARDY [2005] applied a Lagrangian technique to simulate viscoelastic flow past a cylinder benchmark problem; they showed that for an ultradilute solution, the governing equations for the Oldroyd-B model can be solved for arbitrarily large values of Wi under the assumption that the underlying velocity field is fixed to be Newtonian.

By far the most successful method is (FATTAL and KUPFERMAN's [2004]) matrix-logarithm formulation of the conformation tensor for the constitutive laws; this method introduces a new variable $\boldsymbol{\Psi} = \ln \boldsymbol{c}$ and rewrite the constitutive equations in terms of $\boldsymbol{\Psi}$ for numerical calculations. The main motivation relies on the fact that the stress tensor is exponential in regions of high deformation rates or stagnation points; numerical instabilities are caused by failure to balance exponential growth with convection. In FATTAL and KUPFERMAN [2004], the authors pointed out the inappropriateness of polynomial-based approximations to represent the stress.

Let the conformation tensor be $\boldsymbol{c} = \boldsymbol{\delta} + \frac{\mu_p}{\mathsf{Wi}}\boldsymbol{\tau}$. Notice that the conformation tensor is different from that defined in (3.7) by a constant multiplier. We can then write the Oldroyd-B constitutive equation (3.5) in terms of $\boldsymbol{c}$:

$$\frac{\partial \boldsymbol{c}}{\partial t} + \mathbf{u} \cdot \nabla \boldsymbol{c} - \nabla \mathbf{u}\, \boldsymbol{c} - \boldsymbol{c}(\nabla \mathbf{u})^T = \frac{1}{\mathsf{Wi}}(\boldsymbol{\delta} - \boldsymbol{c}). \tag{5.1}$$

The core feature of the transformation is the decomposition of the velocity gradient into a traceless extensional component $\boldsymbol{B}$ and a pure rotational component $\boldsymbol{R}$:

$$\nabla \mathbf{u} = \boldsymbol{R} + \boldsymbol{B} + \boldsymbol{N}\boldsymbol{c}^{-1}, \tag{5.2}$$

where $\boldsymbol{N}$ is antisymmetric. By plugging (5.2) in the Oldroyd-B constitutive relation, we obtain

$$\frac{\partial \boldsymbol{c}}{\partial t} + (\mathbf{u} \cdot \nabla)\boldsymbol{c} - (\boldsymbol{R}\boldsymbol{c} - \boldsymbol{c}\boldsymbol{R}) - 2\boldsymbol{B}\boldsymbol{c} = \frac{1}{\mathsf{Wi}}(\boldsymbol{\delta} - \boldsymbol{c}). \tag{5.3}$$

Because of the symmetric positive-definite (SPD) nature of the conformation tensor, we can have the factorization $\boldsymbol{c} = \boldsymbol{U}\boldsymbol{\Lambda}\boldsymbol{U}^T$, where $\boldsymbol{U}$ is an orthogonal matrix that consists of the eigenvectors of $\boldsymbol{c}$ and $\boldsymbol{\Lambda}$ is a diagonal matrix made with the corresponding eigenvalues of $\boldsymbol{c}$. Therefore, we obtain $\boldsymbol{\Psi} = \boldsymbol{U}(\ln \boldsymbol{\Lambda})\boldsymbol{U}^T$. Then, we can write the Oldroyd-B constitutive relation in terms of $\boldsymbol{\Psi}$ and solve it numerically:

$$\frac{\partial \boldsymbol{\Psi}}{\partial t} + (\mathbf{u} \cdot \nabla)\boldsymbol{\Psi} - (\boldsymbol{R}\boldsymbol{\Psi} - \boldsymbol{\Psi}\boldsymbol{R}) - 2\boldsymbol{B}\boldsymbol{\Psi} = \frac{1}{\mathsf{Wi}} \exp(-\boldsymbol{\Psi})(\boldsymbol{\delta} - \exp(\boldsymbol{\Psi})). \tag{5.4}$$

Note that the positivity of $\boldsymbol{c}$ is guaranteed automatically in this way.

FATTAL and KUPFERMAN [2004] reported numerical results on the lid-driven cavity benchmark for a finitely extensible Chilcott–Rallison (FENE-CR) fluid (CHILCOTT and RALLISON [1988]) with a Weissenberg number of up to 5.0. FATTAL and KUPFERMAN [2005] made a break-through in HWNP with this idea on the Oldroyd-B model for the lid-driven cavity benchmark using finite difference methods. Recently, this method has been further investigated by PAN, HAO and GLOWINSKI [2009] based on the finite element method and an operator-splitting Lie's scheme. HULSEN, FATTAL and KUPFERMAN [2005] applied the log-conformation formulation combined with the DEVSS/DG to the Oldroyd-B as well as the Giesekus model for the flow-past-cylinder benchmark in the finite element context. CORONADO, ARORA, BEHR and PASQUALI [2007], on the other hand, gave an alternative log-conformation formulation and applied the DEVSS-TG/SUPG method (PASQUALI and SCRIVEN [2002]) for the flow-past-cylinder benchmark. Their reported numerical results are for Weissenberg numbers of up to 1.8 (HULSEN, FATTAL and KUPFERMAN [2005]) and 1.05 (CORONADO, ARORA, BEHR and PASQUALI [2007]). Recently, AFONSO, OLIVEIRA, PINHO and ALVES [2009] applied a finite volume method on the log-conformation and reported computational results for the flow-past-cylinder benchmark up to a Weissenberg number of 2.5; but mesh convergence was not confirmed for Weissenberg numbers larger than around 0.9.

The stability of the algorithm in FATTAL and KUPFERMAN [2004] has been analyzed by BOYAVAL, LELIEVRE and MANGOUBI [2009]. The key ingredients used in the analysis were the divergence-free condition and the positivity of the conformation tensor. We note that the log-formulation of the conformation tensor preserves positivity in the discrete sense naturally. In fact, preserving the positivity of the conformation tensor is regarded as one of the main issues in developing stable numerical schemes for viscoelastic flows. We next discuss several attempts to address this issue.

5.3. *Positivity-preserving schemes*

The HWNP has been closely investigated in correlation with the loss of the positivity-preserving property of the conformation tensor $\boldsymbol{c}$ on the discrete level (BERIS and EDWARDS [1994], DUPRET, MARCHAL and CROCHET [1985], HULSEN [1988], JOSEPH and SAUT [1986], OWENS and PHILLIPS [2002]). Although there have been many attempts to construct positivity-preserving schemes, only a handful of schemes are available that preserve the positive-definite character of the conformation tensor on the discrete level. These include the log-conformation schemes discussed above. Another notable example is given by LOZINSKI and OWENS [2003]. Using the fact that the conformation tensor $\boldsymbol{c}$ is positive, they wrote the conformation tensor $\boldsymbol{c}$ as $\boldsymbol{c} = \boldsymbol{C}\boldsymbol{C}^T$ and defined the matrix $\boldsymbol{C}$ by the solution to some semidiscrete equation. More precisely, let $\mathbf{u}^n$ and $\boldsymbol{c}^n = \boldsymbol{C}^n(\boldsymbol{C}^n)^T$ be the nth time step approximate solution to the velocity field $\mathbf{u}$ and the conformation tensor $\boldsymbol{c}$, respectively. Then, $\boldsymbol{C}^n$ is defined as the solution to the equation given as follows:

$$\boldsymbol{C}^n + k\Big(\frac{1}{2\mathrm{Wi}}\boldsymbol{C}^n + (\mathbf{u}^n \cdot \nabla)\boldsymbol{C}^n - \nabla\mathbf{u}^n\boldsymbol{C}^n\Big) = \sqrt{\boldsymbol{c}^{n-1} + \frac{\eta_p k}{\mathrm{Wi}^2}\boldsymbol{\delta}}, \tag{5.5}$$

where k is the time step size. It is further shown that this semidiscretization is consistent in LOZINSKI and OWENS [2003]. This approach has been further explored by HAO, PAN, GLOWINSKI and JOSEPH [2009] as well.

VAITHIANATHAN, ROBERT, BRASSEUR and COLLINS [2006, 2007] have developed a positivity-preserving algorithm that takes into account that the conformation tensor $\boldsymbol{c}$ can be decomposed as $\boldsymbol{c} = \boldsymbol{U}\boldsymbol{\Lambda}\boldsymbol{U}^T$, where $\boldsymbol{U}$ is the orthogonal matrix that consists of eigenvectors of $\boldsymbol{c}$ and $\boldsymbol{\Lambda}$ is the diagonal matrix consisting of the eigenvalues of $\boldsymbol{c}$. They wrote equations for both $\boldsymbol{\Lambda}$ and $\boldsymbol{U}$. These are evolved by solving equations that define these unknowns, which have been successfully applied for the turbulent flow of a viscoelastic polymer solution (VAITHIANATHAN, ROBERT, BRASSEUR and COLLINS [2006, 2007]).

LEE and XU's's [2006] made an attempt to tackle the high Weissenberg number problems by preserving the positivity of the conformation tensors on the fully discrete level. The idea relies on the link between the constitutive equations and the symmetric matrix Riccati differential equations (ABOU-KANDIL, FREILING, IONESCU and JANK [2003], REID [1972]). We will demonstrate why this is crucial in the stability of the solutions and prove that the discrete solution exists in time without break-down in Section 8. More importantly, we will show how this approach can be generalized so that the proposed method will be able to handle most existing macroscopic constitutive equations in a unified and efficient way.

LEE and XU's [2006] method is closely related to the method proposed by PETERA [2002], which is based on a conformation tensor formulation of the constitutive laws and the method of characteristics for the upper convected time derivative directly. PETERA [2002] developed an Eulerian–Lagrangian discretization based on the direct discretization of the generalized Lie derivative introduced by Hughes and Winget in their pioneering work (FORTIN and ESSELAOUI [1987], HUGHES and WINGET [1980]). However, this method does not preserve the strong divergence-free condition. While the method by PETERA [2002] can be shown to preserve the positivity of the conformation tensor, due to the lack of the strong divergence-free condition in his scheme, stability could not be proven; the relevant energy estimates were missing as well. Moreover, the fact that the conformation tensor formulations can be identified with the Riccati equations, as discussed in Section 3, was not noticed there. In particular, the techniques introduced by FORTIN and ESSELAOUI [1987] have been further investigated by KABANEMI, BERTRAND, TANGUY and AIT-KADI [1994].

5.4. *Constitutive equations with diffusion*

Other approaches attempt to stabilize the viscoelastic models by adding a small diffusion term to the constitutive equations. Considering the fact that the difficulty of simulating and proving the global-in-time solutions to the general complex fluids lies in the hyperbolic nature of the constitutive equations, this is a natural choice. Not only does this addition stabilize the equation, it also eases the proof of the global-in-time existence of solutions (see LIN, LIU and ZHANG [2005], for example.) Another existence proof can be found in EL-KAREH and LEAL [1989]. Furthermore, SURESHKUMAR and BERIS [1995] investigated this approach for the Poiseuille flow of the Oldroyd-B model and concluded that a small stress diffusivity can be introduced so that enhanced stability can be achieved without altering the flow rheology.

The main issue, which is still open here, is how to impose the boundary conditions upon adding the diffusivity of the stress fields to the constitutive equations (BERIS and EDWARDS [1994]). The pure Neumann boundary or the Robin boundary conditions are often given (see ADAMS, FIELDING and OLMSTED [2008], BLACK and GRAHAM [2001]). It is, however, impossible to know how macromolecules react near the boundary in general and the construction of the right boundary conditions still remains elusive. In fact, the molecular derivation of the Oldroyd-B model has the diffusion terms although the diffusivity constant is small, and it has been shown that, generally, the multiscale approach is more stable (BAJAJ, BHAT, PRAKASH and PASQUALI [2006]). The addition of dissipation, therefore, may help achieve the stability of the numerical calculations. This technique has been widely used for turbulent drag reduction (SURESHKUMAR and BERIS [1997]). Experiments in SURESHKUMAR and BERIS [1995] showed that adding the diffusion term in the constitutive equations has definite positive effects without altering the flow pattern significantly.

6. A family of Eulerian–Lagrangian finite element methods

In this section, we present our numerical methods to solve the viscoelastic flow models introduced in Section 3. Typically, the viscoelastic fluids are described by the time-dependent models, and even steady-state computations are generally performed using time marching (ALVES, OLIVEIRA and PINHO [2003]) for the corresponding time-dependent problem. Therefore, our interest lies in developing time-dependent non-Newtonian models and their time and space discretization. Our aim here is to introduce, in a systematic way, a class of positivity-preserving discretizations of the Riccati formulation of the constitutive equations in terms of the conformation tensor. In particular, we will demonstrate that our schemes possess the important stability property, and we refer to a recent work by BOYAVAL, LELIEVRE and MANGOUBI [2009], where similar stability results have been presented as well.

6.1. *Temporal discretization*

In the Lagrangian frame, it has been established that the general macroscopic constitutive equation can be cast into (3.18). Therefore, it is natural to employ the Lagrangian approach, and there are two main ways to use this approach.

The first idea, the pure Lagrangian approach or the method of characteristics, is to follow the particle trajectories in time and to use the initial positions of the particles as nodes at which the solutions are evaluated. This approach has an inherent disadvantage in that grid points can be severely distorted, and therefore, the accuracy of long-time calculations can easily be degraded. Further, the relocation of particle positions is unavoidable, in which case it is necessary to interpolate the solutions at the new positions. And, this in turn introduces additional numerical error (BAAIJENS [1993]).

In order to avoid mesh distortion, we can view the fixed discretization at any time level as the particle positions and consider the characteristic foot (or departure foot) as the previous position of this given particle. This method, known as the Eulerian–Lagrangian method (ELM) or the semi-Lagrangian method (SLM), was introduced to the finite element community in the early 1980s; see DOUGLAS and RUSSELL [1982] and PIRONNEAU [1982].

6.1.1. *Eulerian–Lagrangian method for the momentum equation*

The ELM begins by establishing the characteristic foot of any given particle at the current time step. Note that similar approaches have been applied to the computation of viscoelastic flows in BONITO, PICASSO and LASO [2006] and PHILLIPS and WILLIAMS [1999] as well.

Let x be the position of any material particle at the current time t; let x also be used to refer to the particle itself. Suppose that the particle x moves with the velocity $\mathbf{u}(\phi_{t,s}(x), s)$ at time s. The characteristic foot (or the departure foot) $y = \phi_{t,s}(x)$ of the particle x at any previous time $s \leq t$ can be found by solving the following flow map equation:

$$\frac{\partial}{\partial s}\phi_{t,s}(x) = \mathbf{u}(\phi_{t,s}(x), s), \quad \phi_{t,t}(x) = x. \tag{6.1}$$

A straightforward approximation scheme for (6.1) is the first-order forward Euler method:

$$\frac{x - y}{k} = \mathbf{u}(y, s) + O(k), \tag{6.2}$$

where k is the time step size $k = t - s$. We denote the approximate solution to Eqn (6.2) by $\widetilde{y}$, which satisfies the equation

$$\widetilde{y} = x - k\mathbf{u}(\widetilde{y}, s). \tag{6.3}$$

We note that the characteristic foot $\widetilde{y}$ is defined implicitly. Hence, we must apply certain nonlinear iterations to obtain the solution to Eqn (6.3).

Unfortunately, the explicit Euler scheme does not preserve volume, which is crucial for the stability of numerical simulations and the convergence of nonlinear iterative methods; see Section 8 for more detail. FENG and SHANG [1995] discussed volume-preserving numerical schemes for the ordinary differential equation (6.1) and noted that the simplest one replaces $\mathbf{u}(\widetilde{y}, s)$ in (6.3) with $\mathbf{u}\big((\widetilde{y}+x)/2, s\big)$, i.e.,

$$\widetilde{y} = x - k\mathbf{u}\Big(\frac{\widetilde{y}+x}{2}, s\Big). \tag{6.4}$$

Next, we offer a simple discussion to demonstrate why this scheme is volume-preserving.

LEMMA 6.1 (First-Order Volume-Preserving Scheme). *Let* $\Omega \subset \mathbb{R}^2$ *Suppose* $\mathbf{u}(s) \in (H^1(\Omega))^d$ *and* $\nabla \cdot \mathbf{u}(s) = 0$. *If the time step size k is small enough, then the scheme* (6.4) *is well defined and volume preserving, i.e.,* $\det(\nabla\widetilde{y}) = 1$.

PROOF. First, if k is small enough, Eqn (6.4) is solvable, and the scheme is well defined. Now let $\boldsymbol{J} := \nabla\widetilde{y}$ be the Jocobian matrix. Taking derivative with respect to x on both sides of (6.4), we can obtain

$$\boldsymbol{J} = \boldsymbol{\delta} - \frac{k}{2}\nabla\mathbf{u}\Big(\frac{\widetilde{y}+x}{2}, s\Big)(\boldsymbol{J}+\boldsymbol{\delta}).$$

This immediately implies that

$$\Big[\boldsymbol{\delta} + \frac{k}{2}\nabla\mathbf{u}\Big(\frac{\widetilde{y}+x}{2}, s\Big)\Big]\boldsymbol{J} = \boldsymbol{\delta} - \frac{k}{2}\nabla\mathbf{u}\Big(\frac{\widetilde{y}+x}{2}, s\Big).$$

Hence, if k is small enough, $\boldsymbol{\delta} + \frac{k}{2}\nabla\mathbf{u}\big(\frac{\widetilde{y}+x}{2}, s\big)$ is invertible, and we can solve for $\boldsymbol{J}$. So the determinate of the Jacobian matrix is

$$\det\boldsymbol{J} = \det\Big[\boldsymbol{\delta} + \frac{k}{2}\nabla\mathbf{u}\Big(\frac{\widetilde{y}+x}{2}, s\Big)\Big]^{-1}\det\Big[\boldsymbol{\delta} - \frac{k}{2}\nabla\mathbf{u}\Big(\frac{\widetilde{y}+x}{2}, s\Big)\Big].$$

To show $\det\boldsymbol{J} = 1$, we assume that

$$\frac{k}{2}\nabla\mathbf{u}\Big(\frac{\widetilde{y}+x}{2}, s\Big) = \begin{pmatrix} a_{11} & a_{12} \\ a_{21} & a_{22} \end{pmatrix} \quad \text{and} \quad a_{11} + a_{22} = 0.$$

Therefore, by direct calculation, we obtain

$$\det\boldsymbol{J} = \frac{1 - a_{11} - a_{22} + a_{11}a_{22} - a_{12}a_{21}}{1 + a_{11} + a_{22} + a_{11}a_{22} - a_{12}a_{21}} = 1,$$

which completes the proof. □

REMARK 6.1 (Alternative Volume-Preserving Schemes). The scheme (6.4) is only of first order. An alternative is this second-order scheme:

$$\widetilde{y} = x - \frac{k}{2}\Big(\mathbf{u}\Big(\frac{\widetilde{y}+x}{2}, s\Big) + \mathbf{u}\Big(\frac{\widetilde{y}+x}{2}, t\Big)\Big). \tag{6.5}$$

The aforementioned two schemes (6.4) and (6.5) preserve volume in $\mathbb{R}^2$. For three-dimensional problems, constructing volume-preserving schemes is possible, but more complicated; see FENG and SHANG [1995] for details.

We will now apply this idea to discretizing the momentum equation of the Oldroyd-B model. More specifically, we will assume that the solution at time level $s = t^{\text{old}}$ is known; that is, $(\mathbf{u}^{\text{old}}, p^{\text{old}}, \boldsymbol{T}^{\text{old}})$ is given, and for any given mesh points at time level $t = t^{\text{new}}$, we let $\widetilde{y}$ be solutions to the discrete flow map equation (6.3). Namely, $\widetilde{y}$ is an approximation of the departure foot $y = \phi_{t^{\text{new}},\ t^{\text{old}}}(x)$. In the Lagrangian view, the momentum equation (3.1) can be viewed as an ODE; therefore, it can be discretized by using the flow map solutions as follows:

$$\mathsf{Re}\left(\frac{\mathbf{u}^{\text{new}} - \mathbf{u}^{\text{old}} \circ \widetilde{y}}{k}\right) = \eta_s \Delta \mathbf{u}^{\text{new}} - \nabla p^{\text{new}} + \nabla \cdot \boldsymbol{T}^{\text{new}}.$$

Therefore, we arrive at the following semidiscrete equation (continuous in space variable):

$$\frac{\mathsf{Re}}{k}\mathbf{u}^{\text{new}} - \eta_s \Delta \mathbf{u}^{\text{new}} + \nabla p^{\text{new}} - \nabla \cdot \boldsymbol{T}^{\text{new}} = \frac{\mathsf{Re}}{k}\mathbf{u}^{\text{old}} \circ \widetilde{y}. \tag{6.6}$$

6.1.2. Eulerian–Lagrangian method for constitutive equations

The particle-following approach (6.6) can be naturally applied to approximate generalized Lie derivatives as well. We now explain it using the model equation (3.18) with positive constant parameters α and β as an example.

Approximations based on the generalized Lie derivative First, we consider the numerical discretization of the generalized Lie derivative $\mathcal{L}_{u,\boldsymbol{R}}\boldsymbol{\zeta}$ at time t^{new}. By Definition 2.1, we can employ the first-order difference approximation for the time derivative to obtain

$$\begin{aligned} &\left.\frac{\boldsymbol{E}(s,t)\boldsymbol{\zeta}(s,t)\boldsymbol{E}(s,t)^T - \boldsymbol{E}(s-k,t)\boldsymbol{\zeta}(s-k,t)\boldsymbol{E}(s-k,t)^T}{k}\right|_{s=t} \\ &\quad = \frac{\boldsymbol{\zeta}(t,t) - \boldsymbol{E}(t-k,t)\boldsymbol{\zeta}(t-k,t)\boldsymbol{E}(t-k,t)^T}{k}. \end{aligned} \tag{6.7}$$

Let $\widetilde{\boldsymbol{E}}$ be an approximate solution to the ODE for the transition matrix, namely,

$$\frac{\partial \boldsymbol{E}(s,t)}{\partial t} = \boldsymbol{R}(s,t)\boldsymbol{E}(s,t) \quad \text{and} \quad \boldsymbol{E}(s,s) = \boldsymbol{\delta}. \tag{6.8}$$

For example, we can apply the explicit Euler method:

$$\frac{\widetilde{\boldsymbol{E}} - \boldsymbol{\delta}}{k} = \boldsymbol{R}(t^{\text{old}})\boldsymbol{\delta} \Longrightarrow \widetilde{\boldsymbol{E}} = \boldsymbol{\delta} + k\boldsymbol{R}(t^{\text{old}}). \tag{6.9}$$

We can also apply the implicit Euler method:

$$\frac{\widetilde{\boldsymbol{E}} - \boldsymbol{\delta}}{k} = \boldsymbol{R}(t^{\text{new}})\widetilde{\boldsymbol{E}} \Longrightarrow \widetilde{\boldsymbol{E}} = \left(\boldsymbol{\delta} - k\boldsymbol{R}(t^{\text{new}})\right)^{-1}. \tag{6.10}$$

Using either (6.9) or (6.10) for approximating the transition matrix $\widetilde{\boldsymbol{E}}$ and the approximate solution $\widetilde{y}$ to the flow map equation, we derive a numerical discretization of the generalized Lie derivative as follows:

$$\mathcal{L}_{u,\boldsymbol{R}}\zeta(t^{\text{new}}) \approx \frac{\boldsymbol{\zeta}^{\text{new}} - \widetilde{\boldsymbol{E}}(\boldsymbol{\zeta}^{\text{old}} \circ \widetilde{y})\widetilde{\boldsymbol{E}}^T}{k}. \tag{6.11}$$

This approximation can be easily shown to satisfy the desired property that the conformation tensor is positive definite in the semidiscrete level when applied to approximate the general Riccati form of the constitutive law (3.18).

Lemma 6.2 (Positivity-Preserving Semidiscretization). *Consider the semidiscrete scheme (6.11) for the model equation (3.18) with positive parameters α and β, namely,*

$$\frac{\boldsymbol{c}^{\text{new}} - \widetilde{\boldsymbol{E}}(\boldsymbol{c}^{\text{old}} \circ \widetilde{y})\widetilde{\boldsymbol{E}}^T}{k} + \alpha \boldsymbol{c}^{\text{new}} = \beta\boldsymbol{\delta}. \tag{6.12}$$

If $\boldsymbol{c}^{\text{old}}$ is positive definite, then the numerical scheme preserves positivity, namely, $\boldsymbol{c}^{\text{new}}$ is still positive definite.

Proof. We can solve the Eqn (6.12) in terms of $\boldsymbol{c}^{\text{new}}$ to obtain

$$(1 + k\alpha)\boldsymbol{c}^{\text{new}} = \widetilde{\boldsymbol{E}}(\boldsymbol{c}^{\text{old}} \circ \widetilde{y})\widetilde{\boldsymbol{E}}^T + k\beta\boldsymbol{\delta}. \tag{6.13}$$

As an immediate consequence, if $\boldsymbol{c}^{\text{old}}$ is positive definite, then so is $\boldsymbol{c}^{\text{new}}$. □

Approximations for the Riccati form of constitutive laws However, the above discretization of the generalized Lie derivative is not the only way to obtain positive-definite discrete conformation tensors. We can simply apply the discretization of the material derivative to obtain the following semidiscrete systems:

$$\frac{\boldsymbol{c}^{\text{new}} - \boldsymbol{c}^{\text{old}} \circ \widetilde{y}}{k} - \boldsymbol{R}^{\text{new}}\boldsymbol{c}^{\text{new}} - \boldsymbol{c}^{\text{new}}(\boldsymbol{R}^{\text{new}})^T + \alpha\boldsymbol{c}^{\text{new}} = \beta\boldsymbol{\delta}. \tag{6.14}$$

Known as the Lyapunov equation, this equation can be reformulated as a symmetric algebraic Riccati equation by a simple change of variable:

$$\left(\frac{\alpha k + 1}{2k}\boldsymbol{\delta} - \boldsymbol{R}^{\text{new}}\right)\boldsymbol{c}^{\text{new}} + \boldsymbol{c}^{\text{new}}\left(\frac{\alpha k + 1}{2k}\boldsymbol{\delta} - \boldsymbol{R}^{\text{new}}\right)^T = \frac{\boldsymbol{c}^{\text{old}} \circ \widetilde{y}}{k} + \beta\boldsymbol{\delta}. \tag{6.15}$$

In fact, the solution $\boldsymbol{c}^{\text{new}}$ is positive definite if $\boldsymbol{c}^{\text{old}} \circ \widetilde{y}$ is positive definite.

Approximations based on the explicit solution We can also design numerical schemes based on the explicit solution of the Riccati form of the constitutive equations. In general, the numerical schemes based on the analytic solution (3.19) require approximations of $\boldsymbol{E}(s, t)$, (3.22), and the time integral in (3.19); see Lemma 3.1. Approximations of $\boldsymbol{E}(s, t)$ do not affect the approximated conformation tensor's positivity property. The integral expression in (3.19) should be computed by using numerical quadratures with positive weights in order to maintain the positivity property, which is the approach taken by DIECI [1994] and DIECI and EIROLA [1994, 1996].

There are many possible methods based on explicit integral expression (3.19) of the solution; an extensive list of schemes can be found in LEE and XU's [2006]. Here, we present only one such example: if α and β are both constants, the explicit form of the solution can be approximated by the left-point Euler method:

$$\boldsymbol{c}^{\mathrm{new}} = \exp(-k\alpha)\widetilde{\boldsymbol{E}}(\boldsymbol{c}^{\mathrm{old}} \circ \widetilde{y})\,\widetilde{\boldsymbol{E}}^{T} + k\beta\boldsymbol{\delta}. \tag{6.16}$$

If we use the first-order Taylor expansion for the exponential function and drop higher-order terms with respect to k on the right-hand side, then we get the exact same scheme as (6.12).

6.2. *Spatial discretization*

It is now clear that the Eulerian–Lagrangian framework provides semidiscrete equations, which preserves the positivity of the conformation tensor. The main goal of this section is to introduce spatial discretizations so that positivity can be realized in the fully discrete sense as well. It is worth noting that the Eulerian–Lagrangian approach follows the particle trajectory and the interpolated solution may not be positive even if the solution is positive at mesh points. This will restrict the choice of the approximation spaces, in particular the approximate stress field. In this section, we introduce various approximation spaces for which the positivity of the conformation tensors can be preserved.

6.2.1. *Stokes-like saddle point problems*

We begin by introducing the equations that will be discretized in space. After applying the ELM to the model problem (3.1), (3.2), and (3.18), we obtain the following semidiscrete problem: find $(\mathbf{u}^{\mathrm{new}}, p^{\mathrm{new}}, \boldsymbol{c}^{\mathrm{new}})$ such that

$$\frac{\mathrm{Re}}{k}\mathbf{u}^{\mathrm{new}} - \eta_s\Delta\mathbf{u}^{\mathrm{new}} + \nabla p^{\mathrm{new}} = \frac{\mathrm{Re}}{k}\mathbf{u}^{\mathrm{old}} \circ \widetilde{y} + \nabla\cdot\boldsymbol{c}^{\mathrm{new}} \tag{6.17}$$

$$\nabla\cdot\mathbf{u}^{\mathrm{new}} = 0 \tag{6.18}$$

$$(1 + k\alpha)\boldsymbol{c}^{\mathrm{new}} = \widetilde{\boldsymbol{E}}(\boldsymbol{c}^{\mathrm{old}} \circ \widetilde{y})\widetilde{\boldsymbol{E}}^{T} + \beta\boldsymbol{\delta}. \tag{6.19}$$

We note that Eqn (6.17) is nonlinear and coupled together through the conformation tensor with (6.19); the nonlinearity also lies in the computation of $\widetilde{y}$ and $\widetilde{\boldsymbol{E}}$. The finite element spaces for the unknowns $\mathbf{u}$, p, and $\boldsymbol{c}$ should be constructed carefully based on the stability conditions to keep the solution from blowing up as the mesh size reduces.

In this section, we identify the ingredients necessary to achieving this goal by considering the momentum equation and continuity equation with the conformation tensor being given explicitly. We consider the most straightforward linearization method here: the conformation tensor $\boldsymbol{c}$ explicitly given at each iteration. In each nonlinear iteration, Eqns (6.17) and (6.18) can be written as the following system of equations up to a simple rescaling:

$$\begin{cases} \rho^2\mathbf{u} - \kappa^2\Delta\mathbf{u} + \nabla p = \mathbf{g} \\ \nabla\cdot\mathbf{u} = 0, \end{cases} \tag{6.20}$$

where $\mathbf{g}$ depends on $\mathbf{u}$ and $\rho^2, \kappa^2 \lesssim 1$.

The main goal now is to identify stable finite element pairs for the velocity and pressure so that accuracy is independent of all relevant parameters ρ^2 and κ^2. We begin by casting Eqn (6.20) into a weak formulation as follows: find $(\mathbf{u}, p) \in (H_0^1(\Omega))^d \times L_0^2(\Omega)$ such that

$$\begin{cases} a_p(\mathbf{u}, \mathbf{v}) + b(\mathbf{v}, p) = \langle \mathbf{g}, \mathbf{v} \rangle & \forall \mathbf{v} \in (H_0^1(\Omega))^d \\ b(\mathbf{u}, q) \qquad\qquad\; = 0 & \forall q \in L_0^2(\Omega), \end{cases} \tag{6.21}$$

where the bilinear forms $a_p(\cdot,\cdot) : (H_0^1)^d \times (H_0^1)^d \mapsto \mathbb{R}$ and $b(\cdot,\cdot) : (H_0^1)^d \times L_0^2 \mapsto \mathbb{R}$ are defined as

$$a_p(\mathbf{u}, \mathbf{v}) := \rho^2(\mathbf{u}, \mathbf{v}) + \kappa^2(\nabla\mathbf{u} : \nabla\mathbf{v}) \quad \text{and} \quad b(\mathbf{v}, p) := -\int_\Omega (\nabla \cdot \mathbf{v})\, p \,\mathrm{d}x.$$

Here, we use the standard notation that $(\cdot : \cdot)$ acting on two matrix-valued functions $\boldsymbol{A} = (a_{i,j}) \in \mathbb{M}$ and $\boldsymbol{B} = (b_{i,j}) \in \mathbb{M}$ denotes

$$(\boldsymbol{A} : \boldsymbol{B}) := \int_\Omega \mathrm{tr}(\boldsymbol{AB}) \,\mathrm{d}\boldsymbol{x} = \int_\Omega \sum_{i,j=1}^{d} a_{i,j} b_{i,j} \,\mathrm{d}\boldsymbol{x}, \tag{6.22}$$

where $\mathrm{tr} : \mathbb{M} \mapsto \mathbb{R}$ is the standard trace operator of a matrix.

Apparently, the bilinear form $a_p(\cdot,\cdot)$ induces a norm

$$\|\mathbf{u}\|_{a_p}^2 := a_p(\mathbf{u}, \mathbf{u}) = \rho^2 \|\mathbf{u}\|_0^2 + \kappa^2 |\mathbf{u}|_1^2.$$

We now introduce the energy norm $|||\cdot|||$ on $(H_0^1(\Omega))^d$ as follows:

$$|||\mathbf{u}|||^2 := \|\mathbf{u}\|_{a_p}^2 + \|\nabla \cdot \mathbf{u}\|_0^2 \qquad \forall \mathbf{u} \in (H_0^1(\Omega))^d.$$

It is then clear that the bilinear forms $a_p(\cdot,\cdot)$ and $b(\cdot,\cdot)$ are continuous, i.e.,

$$\begin{aligned} a_p(\mathbf{u}, \mathbf{v}) &\lesssim |||\mathbf{u}|||\; |||\mathbf{v}||| && \forall \mathbf{u}, \mathbf{v} \in (H_0^1)^d \\ b(\mathbf{v}, q) &\lesssim |||\mathbf{v}|||\; \|q\|_0 && \forall \mathbf{v} \in (H_0^1)^d,\, q \in L_0^2. \end{aligned}$$

We also note that $|||\mathbf{u}||| = \|\mathbf{u}\|_{a_p}$ for any $\mathbf{u} \in \mathcal{N} := \{\mathbf{v} \in (H_0^1)^d : \nabla \cdot \mathbf{v} = 0\}$. Therefore, $a_p(\cdot,\cdot)$ is also elliptic on $\mathcal{N}$.

Furthermore, using the following inf-sup condition (or the Brezzi condition) that

$$\sup_{\mathbf{v} \in (H_0^1)^d} \frac{b(\mathbf{v}, q)}{\|\mathbf{v}\|_1} \gtrsim \|q\|_0 \qquad \forall q \in L_0^2, \tag{6.23}$$

and the fact that $\|\nabla \cdot \mathbf{u}\|_0 \le \|\mathbf{u}\|_1$, we can easily obtain that for $\rho^2, \kappa^2 \lesssim 1$,

$$\sup_{\mathbf{v} \in (H_0^1)^2} \frac{b(\mathbf{v}, q)}{|||\mathbf{v}|||} \gtrsim \sup_{\mathbf{v} \in (H_0^1)^2} \frac{b(\mathbf{v}, q)}{\|\mathbf{v}\|_1} \gtrsim \|q\|_0 \qquad \forall q \in L_0^2. \tag{6.24}$$

This means that the Eqn (6.21) is uniformly stable with respect to the norm $|||\cdot|||$ for the velocity and $\|\cdot\|_0$ for the pressure.

6.2.2. Stable discretizations of the generalized Stokes equation

Similar considerations can be directly transferred to the discrete case as well. We assume that the domain $\Omega \subset \mathbb{R}^d$ has been partitioned into triangular/tetrahedral elements $\mathcal{T}_h = \{E\}$ and that the conforming and quasi-uniform partition $\mathcal{T}_h$ satisfies

$$\overline{\Omega} = \bigcup_{E \in \mathcal{T}_h} \overline{E}. \tag{6.25}$$

Based on the partitions $\mathcal{T}_h$, we will choose appropriate approximation spaces $\mathbf{V}_h$ and W_h for the primitive variables $\mathbf{u}$ and p, respectively.

Consider a discrete weak formulation that is formulated by making the appropriate choice of space $\mathbf{V}_h \subset (H_0^1(\Omega))^d$ for the velocity and $W_h \subset L_0^2(\Omega)$ for the pressure: find $(\mathbf{u}_h, p_h) \in \mathbf{V}_h \times W_h$ such that

$$\begin{cases} a_p(\mathbf{u}_h, \mathbf{v}_h) + b(\mathbf{v}_h, p_h) = \langle \mathbf{g}, \mathbf{v}_h \rangle & \forall \mathbf{v}_h \in \mathbf{V}_h \\ \qquad\qquad\qquad b(\mathbf{u}_h, q_h) = 0 & \forall q_h \in W_h. \end{cases} \tag{6.26}$$

As demonstrated by XIE, XU and XUE [2008], the uniform well-posedness and error analysis for the finite element pairs $\mathbf{V}_h \times W_h$ can be achieved if we can show that the following two conditions are satisfied from the well-known (Brezzi theory BREZZI [1974], BREZZI and FORTIN [1991]), namely,

$$\sup_{\mathbf{v}_h \in \mathbf{V}_h} \frac{b(\mathbf{v}_h, q_h)}{\|\mathbf{v}_h\|_1} \gtrsim \|q_h\|_0 \qquad \forall q_h \in W_h, \tag{6.27}$$

and

$$\nabla \cdot \mathbf{V}_h \subseteq W_h. \tag{6.28}$$

We define $a(\mathbf{u}, \mathbf{v}) := a_p(\mathbf{u}, \mathbf{v}) + a_s(\mathbf{u}, \mathbf{v})$ with $a_s(\mathbf{u}, \mathbf{v}) := (\nabla \cdot \mathbf{u}, \nabla \cdot \mathbf{v})$. Under the two afore stated conditions (6.27) and (6.28), we can immediately see that

$$\sup_{\mathbf{v}_h \in \mathbf{V}_h} \frac{b(\mathbf{v}_h, q_h)}{|||\mathbf{v}_h|||} \gtrsim \|q_h\|_0 \qquad \forall q_h \in W_h \tag{6.29}$$

and

$$a(\mathbf{u}_h, \mathbf{u}_h) \gtrsim |||\mathbf{u}_h|||^2 \qquad \forall \mathbf{u}_h \in \mathcal{N}_h, \tag{6.30}$$

where $\mathcal{N}_h := \{\mathbf{v}_h \in \mathbf{V}_h : \nabla \cdot \mathbf{v}_h = 0\}$.

Now, we give some examples of conforming finite element methods that satisfy both conditions, (6.27) and (6.28).

EXAMPLE 6.1 (Scott–Vogelius Finite Elements). The $P_0^4 - P_{-1}^3$ Scott–Vogelius element (SCOTT and VOGELIUS [1985a,b]) is important in fluid mechanics computation. It uses the piecewise continuous polynomials on triangles of a degree up to 4 to approximate the velocity field and uses the piecewise discontinuous polynomials of a degree up to 3 for the pressure. In $\mathbb{R}^2$, on each triangle, the P_0^4 space has 15 degrees of freedom (DOFs) determined by values at three vertices, three quartering points on each edge, and three interior points inside each triangle. The P^3_{-1} space has 10 DOFs on each triangle, all of which are given inside the triangle independently. This element is stable in the sense that it satisfies the inf-sup condition, if the triangulation is singular-point free (a vertex is called singular if all edges meeting at the vertex fall into two crossing straight lines). This kind of element is of key importance because it can preserve the incompressible condition, i.e., the discrete velocity is divergence-free pointwise.

EXAMPLE 6.2 (Austin–Manteuffel–McCormick Finite Elements). The tensor-product finite element in two-dimensional domains given by AUSTIN, MANTEUFFEL and MCCORMICK [2004] can easily be extended into three-dimensional domains. In each reference triangle, for the horizontal component of the velocity fields, the product of the cubic Hermite polynomial in the x_1 variable and the quadratic polynomial in the x_2 variable are used. For the vertical component of the velocity fields, the product of the quadratic polynomial in the x_1 variable and the cubic Hermite polynomial in the x_2 variable are used. For the pressure, the product of the quadratic polynomial in the x_1 and x_2 variable is used. This element has been shown to be uniformly stable by LEE, WU and CHEN [2009].

6.2.3. *Approximation Space for Stress*

Our guiding principle in choosing the approximation space for the stress fields is to preserve the positivity of the conformation tensor. The main bottleneck in this process is to evaluate the conformation tensor at any points, which may not necessarily be the mesh points. This can be reinterpreted as constructing an interpolation operator $\Pi_h^S : \mathbb{M} \mapsto \mathbb{M}$, which preserves positivity in the following sense:

$$\boldsymbol{\sigma} > 0 \quad \Longrightarrow \quad \Pi_h^S(\boldsymbol{\sigma}) > 0, \quad \forall \boldsymbol{\sigma} \in \mathbb{M}, \tag{6.31}$$

where $\boldsymbol{\sigma} > 0$ means $\boldsymbol{\sigma}$ is positive definite.

For this purpose, we first consider the scalar positivity-preserving interpolations. We start with two simple examples:

EXAMPLE 6.3 (Piecewise Constant Interpolation). The simplest finite element space that preserves the positivity of the scalar functions is, of course, the space of the piecewise constant functions. In this case, the existence of the positivity-preserving interpolation operator Π_h is obvious. For example, we can take, on each element $E \in \mathcal{T}_h$,

$$\Pi_h(g)(x) := \frac{1}{|E|} \int_E g \, \mathrm{d}x \qquad \forall x \in E, \tag{6.32}$$

where $|E|$ is the area of E. It is easy to see that

$$\|\Pi_h(g)\|_\infty = \max_E \frac{1}{|E|} \int_E g \leq \max_E |E|^{-1/2} \|g\|_0 \tag{6.33}$$

and

$$\|\Pi_h(g)\|_{L^1} = \sum_E \int_E |\Pi_h(g)| \leq \sum_E \int_E |g| = \|g\|_{L^1}. \tag{6.34}$$

Notice that these two inequalities are sharp. To see this, we can take a function g that is 1 on an element E and 0 elsewhere.

EXAMPLE 6.4 (Piecewise Linear Interpolation). The other choices can be given by continuous or discontinuous piecewise linear finite element spaces. For cases in which we choose a globally continuous piecewise linear finite element space, the standard pointwise nodal value interpolant would be positivity preserving. In case the solution is not smooth or the point values of g are not well defined, we can define the nodal value of $\Pi_h(g)(x_i)$ as the local mean value as follows:

$$\Pi_h(g)(x_i) := \frac{1}{|B_i|} \int_{B_i} g \, \mathrm{d}x, \tag{6.35}$$

where $B_i = B(x_i, r_i(x_i))$ and where the ball centered at x_i and with radius $r_i(x_i)$ with $r_i(x_i)$ small enough so that B_i is contained in the union of closed elements containing x_i. This interpolation can be shown to be of second-order accuracy (NOCHETTO and WAHLBIN [2002]). For a case in which we choose a discontinuous piecewise linear finite element space, the construction of positivity-preserving operator Π_h for the above continuous piecewise linear element can be applied similarly.

REMARK 6.2 (High-Order Interpolations). We note that it is well known that the positivity-preserving interpolant cannot be made for a polynomial of degree 2 or higher. To summarize, we can choose the approximation space S_h for the conformation tensor as either piecewise constant or piecewise linear polynomial spaces in case the positivity preserving is the main restriction and, therefore, the accuracy of the approximations for such choices is either first-order or second-order.

Now, we introduce a lemma that though simple, is useful, as it allows us to construct positivity-preserving interpolation operators for tensors based on simple scalar interpolations.

LEMMA 6.3 (Positivity-Preserving Interpolations). *Let Π_h be a positivity-preserving interpolation operator for scalar functions, that is, if $g > 0$ on Ω, then $\Pi_h(g) > 0$ on Ω. Then, the interpolation operator Π_h induces Π_h^S such that*

$$\Pi_h^S(\boldsymbol{\sigma}) = (\Pi_h(\boldsymbol{\sigma}_{i,j}))_{i,j=1,\ldots,d}. \tag{6.36}$$

And Π_h^S is a positivity-preserving interpolation in $\mathbb{M}$.

PROOF. We note that the operator Π_h defined on scalar functions preserves the positivity in the sense that $g > 0$ implies $\Pi_h(g) > 0$. We choose any positive-definite tensor $\boldsymbol{\sigma} = (\sigma_{i,j})_{i,j=1,\ldots,d} \in \mathbb{M}$ and any nonzero vector $\xi = (\xi_i)_{i=1,\ldots,d} \in \mathbb{R}^d$ and observe that

$$0 < \xi^T \boldsymbol{\sigma} \xi = \sum_{i,j=1}^{d} \xi_i \sigma_{i,j} \xi_j \quad \Longrightarrow \quad 0 < \Pi_h\left(\xi^T \boldsymbol{\sigma} \xi\right).$$

We exploit the fact that the operator Π_h is linear to see that

$$\Pi_h\left(\xi^T \boldsymbol{\sigma} \xi\right) = \sum_{i,j=1}^{d} \xi_i \Pi_h(\sigma_{i,j}) \xi_j. \tag{6.37}$$

Therefore, the operator Π_h^S is positivity preserving. □

The following simple lemma is useful for deriving the discrete analog of the bridging identity (4.3) that has been crucially used to obtain the energy estimate in the continuous level.

Lemma 6.4 (Bridging Lemma). *For matrix-valued functions $\boldsymbol{A}$ and $\boldsymbol{B} : \mathbb{R}^d \to \mathbb{M}$, the following identities hold true:*

$$\left(\Pi_h^S(\boldsymbol{A}) : \boldsymbol{B}\right) = \left(\Pi_h^S(\boldsymbol{A}) : \Pi_h^S(\boldsymbol{B})\right) = \left(\boldsymbol{A} : \Pi_h^S(\boldsymbol{B})\right). \tag{6.38}$$

Proof. These equalities can be obtained by noticing that Π_h^S is an L^2 projection to the space of constant matrices. We show a more direct proof here. It is enough to show that for any scalar functions f and g, we have

$$\int_\Omega f\, \Pi_h(g)\, \mathrm{d}x = \int_\Omega \Pi_h(f)\, \Pi_h(g)\, \mathrm{d}x = \int_\Omega \Pi_h(f)\, g\, \mathrm{d}x. \tag{6.39}$$

And it can be shown from the following relation:

$$\begin{aligned}\int_\Omega f\, \Pi_h(g)\, \mathrm{d}x &= \int_\Omega f \sum_E \left(\frac{1}{|E|}\int_E g\, \mathrm{d}x\right) \varphi_E\, \mathrm{d}y = \sum_E \left(\int_\Omega f\, \varphi_E\, \mathrm{d}y\right)\left(\frac{1}{|E|}\int_E g\, \mathrm{d}x\right)\\ &= \sum_E \int_\Omega \left(\frac{1}{|E|}\int_E f\, \mathrm{d}y\right) \varphi_E\, \mathrm{d}z \left(\frac{1}{|E|}\int_E g\, \mathrm{d}x\right) = \int_\Omega \Pi_h(f)\, \Pi_h(g)\, \mathrm{d}x.\end{aligned}$$

The second equality follows using the same argument. □

Remark 6.3 (Connecting the Momentum Balance with the Constitutive Laws). Lemma 6.4 can be used to establish the discrete analog of the important bridging identity (4.3) in the continuous level, namely,

$$\left(\Pi_h^S(\nabla \mathbf{u}) : \boldsymbol{c}\right) = \left(\Pi_h^S(\nabla \mathbf{u}) : \Pi_h^S(\boldsymbol{c})\right) = \left(\nabla \mathbf{u} : \Pi_h^S(\boldsymbol{c})\right). \tag{6.40}$$

We will apply this to obtain the discrete energy estimate in Section 8.

6.3. *Full discretizations*

In this section, we will conclude our discussion on discretization by combining time and space discretizations. We choose the approximation spaces $\mathbf{V}_h \times W_h \in (H_0^1(\Omega))^d \times L_0^2(\Omega)$

so that they satisfy both the inf-sup condition and the strong divergence-free condition. In addition, we choose $\boldsymbol{S}_h$ to be a symmetric tensor space whose entries belong to the piecewise polynomial spaces with a degree less than or equal to one.

The weak formulation of the semidiscrete system of Eqns (6.17)–(6.19) can be written as follows: given $\mathbf{u}_h^{\mathrm{old}}, p_h^{\mathrm{old}}$, and $\boldsymbol{c}_h^{\mathrm{old}}$, find $(\mathbf{u}_h^{\mathrm{new}}, p_h^{\mathrm{new}}, \boldsymbol{c}_h^{\mathrm{new}}) \in \mathbf{V}_h \times W_h \times \boldsymbol{S}_h$ such that for any $(\mathbf{v}_h, q_h, \boldsymbol{\sigma}_h) \in \mathbf{V}_h \times W_h \times \boldsymbol{S}_h$,

$$\mathsf{Re}\left(\frac{\mathbf{u}_h^{\mathrm{new}}}{k}, \mathbf{v}_h\right) + \eta_s\left(\mathcal{D}(\mathbf{u}_h^{\mathrm{new}}) : \mathcal{D}(\mathbf{v}_h)\right) - \left(p_h^{\mathrm{new}}, \nabla \cdot \mathbf{v}_h\right) = \mathsf{Re}\left(\frac{\Pi_h^V(\mathbf{u}_h^{\mathrm{old}} \circ \widetilde{y})}{k}, \mathbf{v}_h\right) - \left(\boldsymbol{c}_h^{\mathrm{new}} : \mathcal{D}(\mathbf{v}_h)\right), \tag{6.41}$$

$$\left(\nabla \cdot \mathbf{u}_h^{\mathrm{new}}, q_h\right) = 0, \tag{6.42}$$

$$(1 + k\alpha)\left(\boldsymbol{c}_h^{\mathrm{new}} : \boldsymbol{\sigma}_h\right) = \left(\widetilde{\boldsymbol{E}}_h \Pi_h^S(\boldsymbol{c}_h^{\mathrm{old}} \circ \widetilde{y})\widetilde{\boldsymbol{E}}_h^T : \boldsymbol{\sigma}_h\right) + \beta\left(\boldsymbol{\delta} : \boldsymbol{\sigma}_h\right). \tag{6.43}$$

Based on the various approximations for the constitutive equation in Section 6.1.2, we can devise many approximations for the constitutive equation (3.18). There are a number of approaches to handling the constitutive laws. The weak formulation (6.43) leads us to the following discrete equation:

$$\boldsymbol{A} C_h = F_h, \tag{6.44}$$

where $\boldsymbol{A} = (a_{i,j}) \in \mathbb{M}$, and $a_{i,j} = \int_\Omega (1 + k\alpha)\boldsymbol{\varphi}_j \boldsymbol{\varphi}_i \,\mathrm{d}x$ and $\{\boldsymbol{\varphi}_i\}_i$ are the basis functions for each entry of the stress approximation fields, the entries of C_h are the components of the expression of the tensor $\boldsymbol{c}_h^{\mathrm{new}}$ in terms of the finite element basis, and F_h is the force terms due to the right-hand side in Eqn (6.43).

REMARK 6.4 (Discretization Based on the Algebraic Riccati Form). Note that the discretization of the material derivative in the constitutive equation (6.14) leads to the following discrete constitutive equation:

$$\frac{\boldsymbol{c}_h^{\mathrm{new}} - \Pi_h^S(\boldsymbol{c}_h^{\mathrm{old}} \circ \widetilde{y})}{k} - \boldsymbol{R}_h \boldsymbol{c}_h^{\mathrm{new}} - \boldsymbol{c}_h^{\mathrm{new}} \boldsymbol{R}_h^T + \alpha \boldsymbol{c}_h^{\mathrm{new}} = \beta \boldsymbol{\delta}. \tag{6.45}$$

The equation can be recast into the well-known algebraic Riccati equation called the Lyapunov equation given as follows:

$$\left(\frac{\alpha k + 1}{2k} - \boldsymbol{R}_h\right)\boldsymbol{c}_h^{\mathrm{new}} + \boldsymbol{c}_h^{\mathrm{new}}\left(\frac{\alpha k + 1}{2k} - \boldsymbol{R}_h\right)^T = \frac{\Pi_h^S(\boldsymbol{c}_h^{\mathrm{old}} \circ \widetilde{y})}{k} + \beta \boldsymbol{\delta}. \tag{6.46}$$

EXAMPLE 6.5 (A Fully Discrete Scheme for the Oldroyd-B Model). We would like to give a fully discrete scheme for the Oldroyd-B model, (3.1), (3.2), and (3.11), which we will discuss in later sections; for various other schemes, we refer interested readers to LEE and XU's [2006].

Algorithm 1 Full Discretization–One Time Step

Step 0: Given $\mathbf{u}_h^n$ and $\boldsymbol{c}_h^n$.

Step 1: For any particle x, compute the departure feet

$$\widetilde{y}^n = x - k\,\mathbf{u}_h^n\left(\frac{\widetilde{y}^n + x}{2}\right).$$

Step 2: Solve the following nonlinear system:

$$\begin{cases} \mathrm{Re}\,\mathbf{u}_h^{n+1} - k\Delta_h \mathbf{u}_h^{n+1} + k\nabla_h p_h^{n+1} = k\nabla_h \cdot \boldsymbol{c}_h^{n+1} + \mathrm{Re}\,\Pi_h^V(\mathbf{u}_h^n \circ \widetilde{y}^n), \\ \nabla \cdot \mathbf{u}_h^{n+1} = 0, \\ (1 + k\alpha)\boldsymbol{c}_h^{n+1} = \boldsymbol{F}_h^{n+1}\Pi_h^S(\boldsymbol{c}_h^n \circ \widetilde{y}^n)(\boldsymbol{F}_h^{n+1})^T + k\beta\boldsymbol{\delta}, \end{cases}$$

where $\boldsymbol{F}_h^{n+1} := \left(\boldsymbol{\delta} - k\Pi_h^S(\nabla \mathbf{u}_h^{n+1})\right)^{-1}$.

7. Fast and robust solvers for Stokes-type systems

As discussed in Section 6, by applying the Eulerian–Lagrangian method (ELM) to the non-Newtonian models, we reduce the task of solving nonlinear systems of equations to solving symmetric linear systems of Stokes type at each iteration. Therefore, the optimal solution methods and multilevel preconditioners for non-Newtonian fluids can be devised, if we can solve the following Stokes-type equation defined in Ω:

$$\rho^2\,\mathbf{u} - \kappa^2\Delta\mathbf{u} + \nabla p = \mathbf{g} \quad \text{and} \quad \nabla \cdot \mathbf{u} = 0, \tag{7.1}$$

where $\mathbf{g}$ is a function that depends on the conformation tensor from the constitutive equation of the underlying model. We note that ρ^2 and κ^2 in Eqn (7.1) are material-dependent parameters. (The uniformly stable finite elements with respect to the parameters ρ and κ were discussed in Section 6.)

7.1. *Discrete Stokes-type system*

The purpose of this section is to consider the fast solution techniques for such parameter-dependent problems as the Stokes-type equation given in (7.1). We begin by writing the discrete weak formulation of the Stokes-type equation (7.1) given as follows: find $(\mathbf{u}_h, p_h) \in \mathbf{V}_h \times W_h$ such that

$$\begin{cases} a_p(\mathbf{u}_h, \mathbf{v}_h) + b(\mathbf{v}_h, p_h) = \langle \mathbf{g}, \mathbf{v}_h \rangle & \forall \mathbf{v}_h \in \mathbf{V}_h \\ b(\mathbf{u}_h, q_h) \qquad\qquad\quad = 0 & \forall q_h \in W_h, \end{cases} \tag{7.2}$$

where the bilinear forms are defined as

$$a_p(\mathbf{u}_h, \mathbf{v}_h) := \rho^2(\mathbf{u}_h, \mathbf{v}_h) + \kappa^2(\nabla\mathbf{u}_h : \nabla\mathbf{v}_h) \quad \text{and} \quad b(\mathbf{v}_h, p_h) := -\int_\Omega \nabla \cdot \mathbf{v}_h\, p_h\, \mathrm{d}x.$$

Throughout this section, for convenience of the presentation, we will consider the operator form of Eqn (7.2) given as follows:

$$\begin{pmatrix} \mathcal{A}_p & \mathcal{B}^* \\ \mathcal{B} & 0 \end{pmatrix} \begin{pmatrix} \mathbf{u}_h \\ p_h \end{pmatrix} = \begin{pmatrix} \mathbf{g} \\ 0 \end{pmatrix}, \tag{7.3}$$

where $\mathcal{A}_p = \rho^2 I - \kappa^2 \Delta_h$, $\mathcal{B} = -\nabla\cdot$, and $\mathcal{B}^* = \nabla_h$.

Our goal here is to discuss two types of iterative methods for solving the discrete version of Eqn (7.1): one is the augmented Lagrangian method, and the other is the preconditioned minimal residual (MinRes) method. For a comparison of the computational costs of solving techniques for Stokes-type systems, we refer to the recent work of LARIN and REUSKEN [2008] and references therein; see XU [2009, 2010] as well.

7.2. Augmented Lagrangian method

Augmented Lagrangian methods for Stokes problems have been introduced by Fortin and Glowinski in FORTIN and GLOWINSKI [1982] and FORTIN and GLOWINSKI [1983]. They have been further discussed in GLOWINSKI and LE TALLEC [1989] and GLOWINSKI [2003]. In this section, we discuss the augmented Lagrangian Uzawa method that can be shown to be fast and robust with respect to parameters ρ, κ as well as the mesh size h.

We assume that the mixed finite elements employed here satisfy the uniform accuracy for the aforementioned Stokes-type equation (7.1). Namely, the pair of finite element spaces $\mathbf{V}_h$ and W_h for velocity fields and pressure satisfy the classical inf-sup conditions and the strong divergence-free condition, namely, $\nabla \cdot \mathbf{V}_h \subseteq W_h$ as discussed in XIE, XU and XUE [2008]. For conforming finite elements, it is well known that the Scott–Vogelius elements SCOTT and VOGELIUS [1985a,b] enjoy the optimal approximation property for the problem (7.1). And it has recently been established that finite elements introduced by AUSTIN, MANTEUFFEL and MCCORMICK [2004] have such a property as well LEE [2009]; see Example 6.1 and Example 6.2.

The Augmented Lagrangian iterative method for the operator form of Stokes-type equation (7.3) can be viewed as the Uzawa method for the following penalized problem:

$$\begin{pmatrix} \mathcal{A}_p + \mu^2 \mathcal{B}^* \mathcal{B} & \mathcal{B}^* \\ \mathcal{B} & 0 \end{pmatrix} \begin{pmatrix} \mathbf{u}_h \\ p_h \end{pmatrix} = \begin{pmatrix} \mathbf{g} \\ 0 \end{pmatrix}, \tag{7.4}$$

where $\mu^2 \geq 0$ is an arbitrary parameter. Note that due to the fact that the strong divergence-free condition holds for the finite element pair, the formulations (7.3) and (7.4) are equivalent. The optimal choice of damping parameter for the Uzawa method has been discussed by NOCHETTO and PYO [2004].

An application of the Uzawa method with damping parameter μ^2 reads as follows: given $(\mathbf{u}_h^i, p_h^i)$, the new iterate $(\mathbf{u}_h^{i+1}, p_h^{i+1})$ is obtained by solving the following equations in an alternating way:

$$(\mathcal{A}_p + \mu^2 \mathcal{B}^* \mathcal{B}) \mathbf{u}_h^{i+1} = \mathbf{g} - B^* p_h^i \tag{7.5}$$

$$p_h^{i+1} = p_h^i + \mu^2 B \mathbf{u}_h^{i+1}. \tag{7.6}$$

The contraction factor of the Uzawa iterations (7.5) can be shown to be $O(\mu^{-2})$ when $\mu^2 \gg 1$ (LEE, WU, XU and ZIKATANOV [2007]). Therefore, if μ^2 is big enough, the Uzawa iteration converges very fast. However, as discussed in LEE, WU and CHEN [2009], the trade-off for achieving such a fast convergence is the inversion of a nearly singular operator $\mathcal{A}_p + \mu^2\mathcal{B}^*\mathcal{B}$.

It should be noted here that while the construction of robust multilevel methods for the operator $\mathcal{A}_p + \mu^2\mathcal{B}^*\mathcal{B}$ is well-known, theoretical analysis on this point is missing from the literature. In fact, AUSTIN, MANTEUFFEL and MCCORMICK [2004] posed the theoretical justification of their numerical experiments on the multilevel method for the operator $\mathcal{A}_p + \mu^2\mathcal{B}^*\mathcal{B}$ as an open problem. Recent papers by LEE [2009] and LEE, WU and CHEN [2009], however, have addressed this question. In this article, we will not attempt to reproduce this work. Instead, we focus on algorithmic details for the robust multigrid methods for the operator $\mathcal{A}_p + \mu^2\mathcal{B}^*\mathcal{B}$ in terms of μ^2 as well as the mesh size, thereby introducing fast and robust solvers for Stokes-type equations.

We now present the robust multigrid algorithm for $\mathcal{A}_p + \mu^2\mathcal{B}^*\mathcal{B}$ in an abstract framework. Let $\mathbf{V}$ be a real Hilbert space with the inner product $a(\cdot,\cdot)$ and the induced norm $\|\cdot\|_a = a(\cdot,\cdot)^{1/2}$. We begin by constructing *multilevel* finite element spaces on which our multigrid method is based. We assume that Ω has been triangulated by nested triangulations $\mathcal{T}_1 \subset \mathcal{T}_2 \subset \cdots \subset \mathcal{T}_L$, where $\mathcal{T}_L$ forms the finest triangulation of Ω. For each $1 \le l \le L$, we let $\{x_l^i\}_i$ be vertices of the triangulation $\mathcal{T}_l$ and denote $\mathcal{T}_l^i$ by the set of triangles in $\mathcal{T}_l$ meeting at the vertex x_l^i. We define the local patch as follows:

$$\Omega_l^i = \bigcup_{E \in \mathcal{T}_l^i} E. \tag{7.7}$$

These patches form an overlapping covering of Ω for each k. We then build the finite element spaces on Ω_l^i as follows:

$$\mathbf{V}_l^i = \{\mathbf{v}_l \in \mathbf{V}_l : \operatorname{supp}(\mathbf{v}_l) \subset \overline{\Omega_l^i}\}. \tag{7.8}$$

Correspondingly, we also define W_l^i, the subspace of W_l, which is supported on Ω_l^i. It is then clear that

$$\mathbf{V} = \sum_{l=1}^{L} \mathbf{V}_l = \sum_{l=1}^{L} \sum_{i=1}^{N_l} \mathbf{V}_l^i,$$

where N_l is the number of vertices for the triangulation $\mathcal{T}_l$.

We further introduce additional notation that the space $\mathcal{N}_l^i$ for $1 \le l \le J$ and $1 \le i \le N_l$,

$$\begin{aligned} \mathcal{N}_l^i &= \{\mathbf{u} \in \mathbf{V}_l^i : (\nabla \cdot \mathbf{u}, q) = 0, \quad \forall q \in \nabla \cdot \mathbf{V}_l^i\} \\ &= \{\mathbf{u} \in \mathbf{V}_l : (\nabla \cdot \mathbf{u}, \nabla \cdot \mathbf{v}) = 0, \quad \forall \mathbf{v} \in \mathbf{V}_l\}. \end{aligned}$$

The robust multigrid method will be constructed by the successive subspace correction (SSC) method with local exact solvers in each subspace $\mathbf{V}_l^i$. In this setting, it is easy to demonstrate that the finite element spaces $\mathbf{V}_l$ and W_l generated based on the triangulations $\mathcal{T}_l$ are nested, especially for Scott–Vogelius finite elements and Austin–Mantueffel and McCormick finite elements, namely, we have

$$\mathbf{V}_1 \subset \cdots \mathbf{V}_l \subset \cdots \mathbf{V}_J, \quad \text{and} \quad W_1 \subset \cdots W_l \subset \cdots W_J. \tag{7.9}$$

Furthermore, in this setting, the following assumptions hold true:

$$\text{A1:} \quad \mathbf{V} = \sum_{l=1}^{L} \sum_{i=1}^{N_l} \mathbf{V}_l^i \quad \text{and} \quad \text{A2:} \quad \mathcal{N} = \sum_{l=1}^{L} \sum_{i=1}^{N_l} (\mathbf{V}_l^i \cap \mathcal{N}).$$

Under these assumptions, we can establish that the following subspace correction algorithm possesses the optimal convergence property; see LEE [2009] and LEE, WU and CHEN [2009].

ALGORITHM 2 Successive Subspace Correction Method

Give the initial guess $\mathbf{u}^0 \in \mathbf{V}$ and let $m = 0$.
while The residual is bigger than the given tolerance **do**
  $\mathbf{u}_0^m = \mathbf{u}^m$;
  for $l = 1, \ldots, L$ **do**
    for $i = 1, \ldots, N_l$ **do**
      Find $\mathbf{e}_i \in \mathbf{V}_l^i$, s.t. $a(\mathbf{e}_i, \mathbf{v}_i) = \mathbf{g}(\mathbf{v}_i) - a(\mathbf{u}_{i-1}^m, \mathbf{v}_i), \quad \forall \mathbf{v}_i \in \mathbf{V}_l^i$;
      $\mathbf{u}_i^m = \mathbf{u}_{i-1}^m + \mathbf{e}_i$;
    end for
  end for
  $\mathbf{u}^{m+1} = \mathbf{u}_L^m$;
  $m = m + 1$;
end while

7.3. *Preconditioned MinRes method*

In this section, we will introduce another algorithm; for this one, it is not necessary to assume the strong divergence-free condition. For instance, we can employ the well-known Taylor–Hood finite elements (TAYLOR and HOOD [1973]) to approximate velocity/pressure fields. To solve the discrete saddle point problem, we can use the preconditioned minimal residual (MinRes) method by PAIGE and SAUNDERS [1975]. Like the conjugate gradient method, the efficiency of MinRes depends heavily on the construction of preconditioners, which should be spectrally equivalent to the inverse of the original operator.

The time-dependent Stokes system has the coefficient matrix in the following form

$$\mathcal{A} = \begin{pmatrix} \mathcal{A}_p & \mathcal{B}^* \\ \mathcal{B} & 0 \end{pmatrix}. \tag{7.10}$$

For this system, we apply the MinRes method with the block diagonal preconditioners by RUSTEN and WINTHER [1992] and BRAMBLE and PASCIAK [1997], namely,

$$\mathcal{P} = \begin{pmatrix} \mathcal{P}_A & 0 \\ 0 & \mathcal{P}_S \end{pmatrix},$$

where $\mathcal{P}_L$ is a multigrid preconditioner for the Laplace-like matrix $\mathcal{A}_p$ and $\mathcal{P}_S$ is a preconditioner corresponding to the Schur complement. The matrix $\mathcal{A}_p$ has a block-diagonal form with each diagonal block corresponding to a scalar reaction-diffusion problem. And the Schur complement preconditioner can be chosen to be

$$\mathcal{P}_S = \max(\kappa^2, \rho^2 h^2) M^{-1} + \rho^2(-\Delta_N)^{-1},$$

where M is the mass matrix for the pressure space and $-\Delta_N$ is the auxiliary Laplace operator with the Neumann boundary condition. This preconditioner is shown to be uniform with respect to ρ, κ, and h; see BRAMBLE and PASCIAK [1997] and OLSHANSKII, PETERS and REUSKEN [2006]. Since fast solvers, like multigrid method (BRAMBLE [1993], BRANDT [1977], HACKBUSCH [1985]), for scalar reaction-diffusion problems are available (see XU [2010]), we can solve the Stokes-type system efficiently.

8. Stability analysis and existence of discrete solutions

In this section, we show that our discretization schemes as discussed in Section 6 are stable. The stability will then be used to establish the existence of the discrete solutions in time evolution. For simplicity and clarity in presenting the main ideas of the proof, we only discuss the Oldroyd-B model:

$$\begin{cases} \mathsf{Re}\left(\frac{\partial \mathbf{u}}{\partial t} + \mathbf{u} \cdot \nabla \mathbf{u}\right) = \eta_s \Delta \mathbf{u} - \nabla p + \nabla \cdot \boldsymbol{c}, & \text{in } \Omega \times \mathbb{R}^+, \\ \nabla \cdot \mathbf{u} = 0, & \text{in } \Omega \times \mathbb{R}^+, \\ \alpha \boldsymbol{c} + \mathcal{L}_{\mathbf{u}, \nabla \mathbf{u}} \boldsymbol{c} = \beta \boldsymbol{\delta}, & \text{in } \Omega \times \mathbb{R}^+, \\ \mathbf{u}(x, t) = 0, & \text{in } \partial\Omega \times \mathbb{R}^+, \\ \mathbf{u}(x, 0) = \mathbf{u}_0(x), & \text{in } \Omega, \\ \boldsymbol{c}(x, 0) = \boldsymbol{c}_0(x), & \text{in } \Omega, \end{cases} \tag{8.1}$$

with $\alpha = 1/\mathsf{Wi}$ and $\beta = \eta_p/\mathsf{Wi}^2$, in a polygonal domain $\Omega \subset \mathbb{R}^2$ here. The extension of these stability and convergence results to more general cases is straightforward.

8.1. Stability analysis for discrete solutions

We first consider the discrete analog of the continuous energy estimate in Theorem 4.1. We are not aiming to presenting the stability analysis for all the schemes introduced in Section 6; instead, we focus on a particular scheme and demonstrate unambiguously the critical role played in the analysis by the positivity of the conformation tensor and the volume preservation of flow maps.

Let us now start to investigate the discrete scheme in Algorithm 1: given the solution $(\mathbf{u}_h^n, p_h^n, \boldsymbol{c}_h^n) \in \mathbf{V}_h \times W_h \times \boldsymbol{S}_h$ from the previous time level, find $(\mathbf{u}_h^{n+1}, p_h^{n+1}, \boldsymbol{c}_h^{n+1}) \in \mathbf{V}_h \times W_h \times \boldsymbol{S}_h$ by the following relation equations:

$$\mathsf{Re}\Big(\frac{\mathbf{u}_h^{n+1} - \Pi_h^V(\mathbf{u}_h^n \circ \widetilde{y}^n)}{k}, \mathbf{v}_h\Big) - (p_h^{n+1}, \nabla \cdot \mathbf{u}_h) + \eta_s \left(\mathcal{D}(\mathbf{u}_h^{n+1}) : \mathcal{D}(\mathbf{v}_h)\right) = -\left(\boldsymbol{c}_h^{n+1} : \mathcal{D}(\mathbf{v}_h)\right) \tag{8.2}$$

$$(\nabla \cdot \mathbf{u}_h^{n+1}, q_h) = 0 \tag{8.3}$$

$$(1 + k\alpha)\left(\boldsymbol{c}_h^{n+1} : \boldsymbol{\sigma}_h\right) = \left(\widetilde{\boldsymbol{F}}_h^{n+1} \Pi_h^S(\boldsymbol{c}_h^n \circ \widetilde{y}^n)(\widetilde{\boldsymbol{F}}_h^{n+1})^T : \boldsymbol{\sigma}_h\right) + k\beta\,(\boldsymbol{\delta} : \boldsymbol{\sigma}_h), \tag{8.4}$$

for all $(\mathbf{v}_h, q_h, \boldsymbol{\sigma}_h) \in \mathbf{V}_h \times W_h \times \boldsymbol{S}_h$.

Here, $\widetilde{\boldsymbol{F}}_h^{n+1} = \left(\boldsymbol{\delta} - k\Pi_h^S(\nabla \mathbf{u}_h^{n+1})\right)^{-1}$ is an approximation to the deformation tensor $\boldsymbol{F}$; see Section 6.3 for more details. And we use the interpolation operator $\Pi_h^S : L^2(\Omega) \mapsto L^2(\Omega)$ introduced in Section 6.1, i.e.,

$$\Pi_h^S(\boldsymbol{\sigma}) = \left(\Pi_h(\sigma_{i,j})\right)_{i,j=1,\ldots,d} \quad \text{with } \Pi_h(g) := \sum_{E \in \mathcal{T}_h} \left(\frac{1}{|E|} \int_E g\, \mathrm{d}x\right) \phi_E(\cdot), \tag{8.5}$$

where $\mathcal{T}_h = \{E\}$ is a quasi-uniform triangular partition of the physical domain Ω with characteristic mesh size h, and ϕ_E is a characteristic function that is one on $\overline{E}$ and zero elsewhere.

As discussed in Section 6.2.2, we assume that the pair of spaces $\mathbf{V}_h$ and W_h satisfy the inf-sup condition as well as

$$\nabla \cdot \mathbf{u}_h \in W_h \qquad \forall\, \mathbf{u}_h \in \mathbf{V}_h. \tag{8.6}$$

The property (8.6) is crucial to constructing the volume-preserving flow map in both two- and three-dimensional domains as discussed in FENG and SHANG [1995].

The flow map $\phi_{t,s} : \Omega \mapsto \Omega$ can be obtained so that the approximate flow map $\widetilde{y}$ satisfies the following identity:

$$\int_\Omega g \circ \widetilde{y}\, \mathrm{d}x = \int_\Omega g\, \mathrm{d}x \qquad \forall\, g \in L^1(\Omega). \tag{8.7}$$

And (8.7) is the key to deriving uniform energy estimates for the solution to the discrete model equations (8.2)–(8.4). Recall that the discrete scheme (6.4) preserves volume in $\mathbb{R}^2$. We can easily show that this scheme satisfies (8.7) by simple change of variables.

Now, we are ready to present the discrete analog of the energy estimate.

THEOREM 8.1 (Discrete Energy Estimate). *The discrete solution to (8.2)–(8.4) admits the following estimate: if* $\mathsf{Wi} < \infty$ *and* $n \geq 1$, *then*

$$\mathsf{Re}\|\mathbf{u}_h^n\|_0^2 + \|\boldsymbol{c}_h^n\|_{L^1} \leq c_1 \mathrm{e}^{-\mathsf{C}_1 t^n}\left(\mathsf{Re}\|\mathbf{u}_h^0\|_0^2 + \|\boldsymbol{c}_h^0\|_{L^1}\right) + c_2, \tag{8.8}$$

$$2\eta_s \sum_{\ell=1}^{n} k\|\mathcal{D}(\mathbf{u}_h^\ell)\|_0^2 \leq \mathsf{Re}\,\|\mathbf{u}_h^0\|_0^2 + \|\boldsymbol{c}_h^0\|_{L^1} + c_2 t^n. \tag{8.9}$$

Here, c_1 *and* c_2 *are generic constants.*

PROOF. From (8.4), we have the following relation:

$$
\begin{aligned}
&(1+k\alpha)\, \boldsymbol{c}_h^{n+1} \\
&= \left(\boldsymbol{\delta} - k\Pi_h^S(\nabla \mathbf{u}_h^{n+1})\right)^{-1} \Pi_h^S(\boldsymbol{c}_h^n \circ \widetilde{\mathcal{Y}}^n) \left(\boldsymbol{\delta} - k\Pi_h^S(\nabla \mathbf{u}_h^{n+1})\right)^{-T} + k\beta\boldsymbol{\delta}.
\end{aligned} \tag{8.10}
$$

We first multiply $\boldsymbol{\delta} - k\Pi_h^S(\nabla \mathbf{u}_h^{n+1})$ to the left and $\boldsymbol{\delta} - k\Pi_h^S(\nabla \mathbf{u}_h^{n+1})^T$ to the right of Eqn (8.10) to obtain that

$$
\begin{aligned}
&(1+k\alpha) \left(\boldsymbol{\delta} - k\Pi_h^S(\nabla \mathbf{u}_h^{n+1})\right) \boldsymbol{c}_h^{n+1} \left(\boldsymbol{\delta} - k\Pi_h^S(\nabla \mathbf{u}_h^{n+1})\right)^T \\
&= \Pi_h^S(\boldsymbol{c}_h^n \circ \widetilde{\mathcal{Y}}^n) + k\beta \left(\boldsymbol{\delta} - k\Pi_h^S(\nabla \mathbf{u}_h^{n+1})\right) \left(\boldsymbol{\delta} - k\Pi_h^S(\nabla \mathbf{u}_h^{n+1})\right)^T.
\end{aligned} \tag{8.11}
$$

Hence, by taking trace and then integration on both sides of the equation above, we get that

$$
\begin{aligned}
&k(1+k\alpha) \int_\Omega \operatorname{tr}\left(\Pi_h^S(\nabla \mathbf{u}_h^{n+1}) \boldsymbol{c}_h^{n+1} + \boldsymbol{c}_h^{n+1} \Pi_h^S(\nabla \mathbf{u}_h^{n+1})^T\right) \mathrm{d}x \\
&= (1+k\alpha)\, \|\boldsymbol{c}_h^{n+1}\|_{L^1} - \|\boldsymbol{c}_h^n \circ \widetilde{\mathcal{Y}}^n\|_{L^1} - \mathrm{d}\beta|\Omega|k \\
&\quad + k^2\beta \int_\Omega \operatorname{tr}\left(\Pi_h^S(\nabla \mathbf{u}_h^{n+1}) + \Pi_h^S(\nabla \mathbf{u}_h^{n+1})^T\right) \mathrm{d}x \\
&\quad + k^2(1+k\alpha) \int_\Omega \operatorname{tr}\left(\Pi_h^S(\nabla \mathbf{u}_h^{n+1}) \Big(\boldsymbol{c}_h^{n+1} - \frac{k\beta}{1+k\alpha}\boldsymbol{\delta}\Big) \Pi_h^S(\nabla \mathbf{u}_h^{n+1})^T\right) \mathrm{d}x.
\end{aligned} \tag{8.12}
$$

We note that from (8.10), the approximate conformation tensor $\boldsymbol{c}_h^{n+1}$ satisfies

$$
(1+k\alpha)\boldsymbol{c}_h^{n+1} - k\beta\boldsymbol{\delta} \geq 0 \qquad \forall n \geq 1, \tag{8.13}
$$

if the initial condition $\boldsymbol{c}_h^0 \geq 0$. Since $\boldsymbol{c}_h^{n+1}$ is symmetric, by Lemma 6.4 and the discrete divergence-free condition, we can easily see that

$$
\begin{aligned}
&\int_\Omega \operatorname{tr}\left(\Pi_h^S(\nabla \mathbf{u}_h^{n+1}) + \Pi_h^S(\nabla \mathbf{u}_h^{n+1})^T\right) \mathrm{d}x \\
&= \int_\Omega \operatorname{tr}\left(\nabla \mathbf{u}_h^{n+1} + (\nabla \mathbf{u}_h^{n+1})^T\right) \mathrm{d}x = 2\int_\Omega \nabla \cdot \mathbf{u}_h^{n+1} \mathrm{d}x = 0
\end{aligned} \tag{8.14}
$$

and

$$
\begin{aligned}
&\int_\Omega \operatorname{tr}\left(\Pi_h^S(\nabla \mathbf{u}_h^{n+1}) \boldsymbol{c}_h^{n+1}\right) \mathrm{d}x = \int_\Omega \operatorname{tr}\left(\boldsymbol{c}_h^{n+1} \Pi_h^S(\nabla \mathbf{u}_h^{n+1})^T\right) \mathrm{d}x \\
&\qquad = \left(\boldsymbol{c}_h^{n+1} : \Pi_h^S(\mathcal{D}(\mathbf{u}_h^{n+1}))\right) = \left(\boldsymbol{c}_h^{n+1} : \mathcal{D}(\mathbf{u}_h^{n+1})\right).
\end{aligned} \tag{8.15}
$$

Finally, based on the volume-preserving property of $\widetilde{y}^n$, we have $\|c_h^n \circ \widetilde{y}^n\|_{L^1} = \|c_h^n\|_{L^1}$. Taking the facts (8.13), (8.14), and (8.15) into account, we can derive the following inequality from (8.12):

$$\left(c_h^{n+1} : \mathcal{D}(\mathbf{u}_h^{n+1})\right) \geq \frac{1}{2k}\|c_h^{n+1}\|_{L^1} - \frac{1}{2k(1+k\alpha)}\|c_h^n\|_{L^1} - \frac{\mathrm{d}\beta|\Omega|}{2(1+k\alpha)}. \tag{8.16}$$

We now consider the momentum equation (8.2). Using the energy method, together with the discrete divergence-free condition and (8.16), we can obtain

$$\begin{aligned}
&\frac{\mathsf{Re}}{k}\|\mathbf{u}_h^{n+1}\|_0^2 + \eta_s\|\mathcal{D}(\mathbf{u}_h^{n+1})\|_0^2 \\
&\quad = \frac{\mathsf{Re}}{k}\left(\Pi_h^V(\mathbf{u}_h^n \circ \widetilde{y}^n), \mathbf{u}_h^{n+1}\right) - \left(c_h^{n+1} : \mathcal{D}(\mathbf{u}_h^{n+1})\right) \\
&\quad \leq \frac{\mathsf{Re}}{k}\left(\mathbf{u}_h^n \circ \widetilde{y}^n, \mathbf{u}_h^{n+1}\right) - \frac{1}{2k}\|c_h^{n+1}\|_{L^1} + \frac{1}{2k(1+k\alpha)}\|c_h^n\|_{L^1} + \frac{\mathrm{d}\beta|\Omega|}{2(1+k\alpha)}.
\end{aligned} \tag{8.17}$$

Applying the Cauchy–Schwarz inequality and the standard kick-back argument, we obtain the following relation:

$$\begin{aligned}
&\frac{\mathsf{Re}}{2k}\|\mathbf{u}_h^{n+1}\|_0^2 + \eta_s\|\mathcal{D}(\mathbf{u}_h^{n+1})\|_0^2 + \frac{1}{2k}\|c_h^{n+1}\|_{L^1} \\
&\quad \leq \frac{\mathsf{Re}}{2k}\|\mathbf{u}_h^n\|_0^2 + \frac{1}{2k(1+k\alpha)}\|c_h^n\|_{L^1} + \frac{\mathrm{d}\beta|\Omega|}{2(1+k\alpha)}.
\end{aligned} \tag{8.18}$$

We are now in the position to show the first estimate (8.8). Multiplying $2k$ to both sides of (8.18) and using Korn's inequality, we obtain that

$$\begin{aligned}
\kappa\|\mathbf{u}_h^{n+1}\|_0^2 + \|c_h^{n+1}\|_{L^1} &\leq \mathsf{Re}\|\mathbf{u}_h^n\|_0^2 + \frac{1}{1+k\alpha}\|c_h^n\|_{L^1} + \frac{k\mathrm{d}\beta|\Omega|}{1+k\alpha} \\
&\leq \exp(-\mathsf{C}_1 k)\left(\kappa\|\mathbf{u}_h^n\|_0^2 + \|c_h^n\|_{L^1}\right) + \frac{k\mathrm{d}\beta|\Omega|}{1+k\alpha},
\end{aligned} \tag{8.19}$$

where $\kappa = \mathsf{Re} + 2k\eta_s\mathsf{C}_\Omega$ and C_Ω is a positive constant depending only on Ω. Here, $\mathsf{C}_1 > 0$ is chosen to be a constant such that

$$\max\left(\frac{\mathsf{Re}}{\mathsf{Re} + 2k\eta_s\mathsf{C}_\Omega}, \frac{1}{1+k\alpha}\right) \leq \exp(-\mathsf{C}_1 k), \quad 0 \leq k \leq 1. \tag{8.20}$$

Now, we use the induction argument to obtain:

$$\begin{aligned}
\kappa\|\mathbf{u}_h^n\|_0^2 + \|c_h^n\|_{L^1} &\leq \exp(-\mathsf{C}_1 t^n)\left(\kappa\|\mathbf{u}_h^0\|_0^2 + \|c_h^0\|_{L^1}\right) + \frac{k\mathrm{d}\beta|\Omega|}{1+k\alpha}\sum_{l=0}^{n}\exp(-\mathsf{C}_1 t^l) \\
&\leq \exp(-\mathsf{C}_1 t^n)\left(\kappa\|\mathbf{u}_h^0\|_0^2 + \|c_h^0\|_{L^1}\right) + \mathsf{C}_2^n,
\end{aligned} \tag{8.21}$$

where

$$\mathsf{C}_2^n = k\mathrm{d}\beta|\Omega|\frac{1 - \exp(-\mathsf{C}_1 t^n)}{1 - \exp(-\mathsf{C}_1 k)}. \tag{8.22}$$

It is clear that we can choose generic constants c_1 and c_2 such that

$$\kappa \|\mathbf{u}_h^0\|_0^2 + \|\boldsymbol{c}_h^0\|_{L^1} \leq c_1 \left(\mathsf{Re}\|\mathbf{u}_h^0\|^2 + \|\boldsymbol{c}_h^0\|_{L^1}\right) \quad \text{and} \quad \mathsf{C}_2^n \leq c_2. \tag{8.23}$$

We then obtain the desired result (8.8).

We now drive the other estimate (8.9). First, we multiply $2k$ to both sides of (8.18) and take summation for $l = 1, 2, \ldots, n$ for both sides to obtain:

$$\begin{aligned} 2\eta_s \sum_{l=1}^{n} k\|\mathcal{D}(\mathbf{u}_h^l)\|_0^2 &\leq c_1 \left(\mathsf{Re}\|\mathbf{u}_h^0\|_0^2 + \|\boldsymbol{c}_h^0\|_{L^1}\right) + \sum_{l=1}^{n} k d\beta|\Omega|, \\ &\leq c_1 \left(\mathsf{Re}\|\mathbf{u}_h^0\|_0^2 + \|\boldsymbol{c}_h^0\|_{L^1}\right) + c_2 t^n. \end{aligned}$$

This completes the proof. □

We will now consider the limiting case in which $\mathsf{Wi} = \infty$; in this case, $\alpha = \beta = 0$.

COROLLARY 8.1. *Assume that* $\mathsf{Wi} = \infty$ *and* $\alpha = \beta = 0$. *Then, the following estimates hold true for any* $n \geq 1$*:*

$$\mathsf{Re}\|\mathbf{u}_h^n\|_0^2 + \|\boldsymbol{c}_h^n\|_{L^1} \leq \mathsf{Re}\|\mathbf{u}_h^0\|_0^2 + \|\boldsymbol{c}_h^0\|_{L^1} \tag{8.24}$$

and

$$\eta_s \sum_{l=0}^{n} k\|\mathcal{D}(\mathbf{u}_h^l)\|_0^2 \leq \mathsf{Re}\|\mathbf{u}_h^0\|_0^2 + \|\boldsymbol{c}_h^0\|_{L^1}. \tag{8.25}$$

PROOF. Note that in the limiting case, $\alpha = \beta = 0$ and $\boldsymbol{c}_h^n$ is itself a conformation tensor for $n \geq 0$. The result then immediately follows from two estimates (8.8) and (8.9) since $\mathsf{C}_1 = 0$ and $\mathsf{C}_2^n = 0$ for all $n \geq 0$. This completes the proof. □

REMARK 8.1 (Effects of Load). When there is a nonzero external force term $\boldsymbol{f}$ on the right-hand side of the momentum equation (8.2), it can be shown that the energy estimates in Theorem 8.1 are still valid as long as $\eta_s > 0$. In this case, the L^2 norm of $\boldsymbol{f}$ will enter into the constant c_2 in the inequality (8.8).

8.2. Existence of the discrete solutions

The discrete model equations (8.2)–(8.4) are fully nonlinear, and the well posedness of this model is not trivial. The main purpose of this section is to prove the existence of the discrete solution. We will show that the solution to the discrete problem exists for sufficiently small time step size k; furthermore, the discrete solution is unique. These will, in turn, confirm that the discrete problem, (8.2)–(8.4), is well defined. Theoretically, the restriction of k is only given by the mesh size h.

Let tol be the tolerance for the nonlinear iteration. We assume that $\mathbf{u}_h^n$, p_h^n, and $\boldsymbol{c}_h^n$ at the time level t^n are available. Then, we have the following algorithm for time marching:

The Algorithm 3 is a single-step time-marching algorithm. Once the initial condition $(\mathbf{u}_h^0, p_h^0, \boldsymbol{c}_h^0)$ is given, we can proceed to the evolution process. Note that the presence of $\boldsymbol{f}$ on the right-hand side of the Stokes-type equation is due to the Dirichlet boundary condition.

ALGORITHM 3 Nonlinear Iteration

Step 0: Given $\mathbf{u}_h^n$, p_h^n, and $\boldsymbol{c}_h^n$. Set

$$\mathbf{u}_h^{n,0} := \mathbf{u}_h^n, \quad p_h^{n,0} := p_h^n, \quad \text{and} \quad \boldsymbol{c}_h^{n,0} := \boldsymbol{c}_h^n.$$

Step 1: For any particle x, compute the departure feet

$$\widetilde{y}^n = x - k\,\mathbf{u}_h^{n,0}\Big(\frac{\widetilde{y}^n + x}{2}\Big).$$

Step 2: For $\ell = 0, 1, 2, \ldots,$ do

(1) Solve the Stokes-type system

$$\begin{cases} \mathsf{Re}\,\mathbf{u}_h^{n,\ell+1} - k\Delta_h \mathbf{u}_h^{n,\ell+1} + k\nabla_h p_h^{n,\ell+1} \\ \qquad = \mathsf{Re}\,\Pi_h^V(\mathbf{u}_h^{n,0} \circ \widetilde{y}^n) + k\nabla_h \cdot \boldsymbol{c}_h^{n,\ell} + k\boldsymbol{f}, \\ \nabla \cdot \mathbf{u}_h^{n,\ell+1} = 0. \end{cases}$$

(2) Update the conformation tensor

$$(1 + k\alpha)\boldsymbol{c}_h^{n,\ell+1} = \boldsymbol{F}_h^{n,\ell+1}\Pi_h^S(\boldsymbol{c}_h^{n,0} \circ \widetilde{y}^n)(\boldsymbol{F}_h^{n,\ell+1})^T + k\beta\boldsymbol{\delta},$$

where $\boldsymbol{F}_h^{n,\ell+1} := \big(\boldsymbol{\delta} - k\Pi_h^S(\nabla \mathbf{u}_h^{n,\ell+1})\big)^{-1}$.

(3) If $\|\mathbf{u}_h^{n,\ell+1} - \mathbf{u}_h^{n,\ell}\|_1 \le \mathsf{tol}$ and $\|p_h^{n,\ell+1} - p_h^{n,\ell}\|_0 \le \mathsf{tol}$, then break.

Step 3: Update solution: $\mathbf{u}_h^{n+1} := \mathbf{u}_h^{n,\ell+1}$, $p_h^{n+1} := p_h^{n,\ell+1}$, and $\boldsymbol{c}_h^{n+1} := \boldsymbol{c}_h^{n,\ell+1}$.

REMARK 8.2 (Solving the Flow Map Equations). From the energy estimate (8.9), we know that $\|\mathbf{u}_h^n\|_1$ is bounded. Hence, if the time step size k is small enough, the nonlinear equation for the flow map in Step 1 of Algorithm 3 is solvable by the inverse function theorem. We will discuss iterative methods for solving the flow map equation in Section 9.

REMARK 8.3 (Feet Searching). We remark that one of the key ingredients for the Eulerian–Lagrangian method here used to solve the Riccati form of the constitutive equation is to find the function values at the departure feet, $\mathbf{u}_h^{n,0} \circ \widetilde{y}^n$ and $\boldsymbol{c}_h^{n,0} \circ \widetilde{y}^n$, as quickly and accurately as possible. In Algorithm 3, we did not provide referents for the points x. It can have different meanings in different discretization methods. But, of course, it is not practical to trace back through all points. We will explain the implementation details in Section 9.2.1. Usually, for the finite difference method, x is any grid point; for the finite element method, x is any quadrature point.

REMARK 8.4 (Parallel Computing for Solving Ricatti Equations). We note that the constitutive equations can be solved in a fully parallel way. This is because all the coefficients are defined locally by a $d \times d$ matrix equation in each of the nodes, all of which are completely independent.

The main goal in this section is to show that the Algorithm 3 is convergent in each time step under certain conditions on the time step size k. This has been discussed by LEE, XU and ZHANG [To appear]. Before introducing the main existence result, we would like to make a few comments:

(1) The discrete model, (8.2)–(8.4), is a highly coupled nonlinear system of equations. Therefore, we need to apply certain iterative methods that lead to the solution implicitly given in order to satisfy the discrete models. There are many such nonlinear iterative schemes, and we focus on one of them in Algorithm 3. Moreover, extending the existence proof to other methods is possible.
(2) This result can be achieved from the uniform stability estimate established in the previous section. In addition, we note that the notion we used here as the well posedness for the solution to the discrete models (8.2)–(8.4) should be distinguished from the one introduced by KREISS [2001]. In particular, our result does not necessarily imply stability with respect to the perturbation of the data.
(3) Our analysis fully exploits the finite dimensionality of the solution space; therefore, technically, it will be difficult to extend this analysis to the existence analysis for the continuous level.

Our proof is based on the induction argument. Specially, we will assume that at time level t^n, the discrete solutions $\mathbf{u}_h^n$ and $\boldsymbol{c}_h^n$ are well defined and generate a sequence of iterates according to Algorithm 3 and show that the nonlinear iteration converges and defines $\mathbf{u}_h^{n+1}$ and $\boldsymbol{c}_h^{n+1}$. More precisely, we will show that the solutions at the time level t^{n+1} can be obtained uniquely by the Algorithm 3. We note that if $\mathbf{u}_h^n$ and $\boldsymbol{c}_h^n$ at the time level t^n satisfy the uniform bounds

$$\|\mathbf{u}_h^n\|_0 \lesssim 1 \quad \text{and} \quad \|\boldsymbol{c}_h^n\|_{L^1} \lesssim 1, \tag{8.26}$$

the fixed-point iteration (Algorithm 3) converges. We, therefore, conclude our proof by a simple recursive argument.

REMARK 8.5 (Inverse Inequalities). We recall the well-known inverse inequalities (cf. BRENNER and SCOTT [2002, chapter 4]) that

$$\|\mathbf{v}\|_\infty \lesssim h^{-1}\|\mathbf{v}\|_0 \quad \text{and} \quad \|\nabla \mathbf{v}\|_0 \lesssim h^{-1}\|\mathbf{v}\|_0, \qquad \forall\, \mathbf{v} \in \mathbf{V}_h. \tag{8.27}$$

Let us first establish that the sequence generated from Algorithm 3 is bounded uniformly.

LEMMA 8.1 (Uniform Boundedness). *Suppose that* $\boldsymbol{f} \in \left(L^2(\Omega)\right)^2$ *For sufficiently small* k, *the sequence generated by Algorithm 3 is uniformly bounded in* L^2 *norm for the velocity and* L^1 *norm for the stress field, respectively.*

PROOF. Using the strong divergence-free finite elements as in Section 6, we have $\nabla \cdot \mathbf{u}_h^{n,\ell} = 0$, for $\ell = 0, 1, 2, \ldots,$ By Lemma 6.1, it holds that $\det(\nabla \widetilde{\mathbf{y}}^n) = 1$. Employing the energy method and the inverse inequality (8.27), we derive that

$$\begin{aligned}\mathsf{Re}\,\|\mathbf{u}_h^{n,\ell+1}\|_0^2 + k\|\nabla \mathbf{u}_h^{n,\ell+1}\|_0^2 \le \frac{\mathsf{Re}}{2}\|\mathbf{u}_h^{n,0}\|_0^2 + \frac{\mathsf{Re}}{2}\|\mathbf{u}_h^{n,\ell+1}\|_0^2 + k\|\boldsymbol{f}\|_0\|\mathbf{u}_h^{n,\ell+1}\|_0 \\ + k\|\boldsymbol{c}_h^{n,\ell}\|_{L^1}\|\nabla \mathbf{u}_h^{n,\ell+1}\|_\infty.\end{aligned}$$

Therefore, using the Cauchy–Schwarz inequality, we obtain

$$\begin{aligned}\mathsf{Re}\,\|\mathbf{u}_h^{n,\ell+1}\|_0^2 + k\|\nabla \mathbf{u}_h^{n,\ell+1}\|_0^2 &\le \frac{\mathsf{Re}}{2}\|\mathbf{u}_h^{n,0}\|_0^2 + \frac{\mathsf{Re}}{2}\|\mathbf{u}_h^{n,\ell+1}\|_0^2 \\ &+ k\Big(\frac{\nu^{-2}}{2}\|\boldsymbol{f}\|_0^2 + \frac{\nu^2}{2}\|\mathbf{u}_h^{n,\ell+1}\|_0^2\Big) + k\Big(\frac{C_1 h^{-2}}{2}\|\boldsymbol{c}_h^{n,\ell}\|_{L^1}^2 + \frac{1}{2}\|\nabla \mathbf{u}_h^{n,\ell+1}\|_0^2\Big),\end{aligned}$$

where ν is chosen such that $|\mathbf{u}_h^{n,\ell+1}|_1^2 \ge \nu^2\|\mathbf{u}_h^{n,\ell+1}\|_0^2$. We can then get, for all $\ell = 0, 1, 2, \ldots$, that

$$\mathsf{Re}\,\|\mathbf{u}_h^{n,\ell+1}\|_0^2 \le \mathsf{Re}\,\|\mathbf{u}_h^{n,0}\|_0^2 + C_1 k h^{-2}\|\boldsymbol{c}_h^{n,\ell}\|_{L^1}^2 + C_0 k\|\boldsymbol{f}\|_0^2, \tag{8.28}$$

where C_0 and C_1 are generic constants independent of k and h. Without loss of generality, we can assume that both C_0 and C_1 are greater than 1.

The equation (3) in Step 1 for updating the conformation tensor reveals the following inequality

$$\|\boldsymbol{c}_h^{n,\ell+1}\|_{L^1} \le \|\boldsymbol{F}_h^{n,\ell+1}\|_\infty^2\|\boldsymbol{c}_h^{n,0}\|_{L^1} + d|\Omega|\beta k. \tag{8.29}$$

Combining the last two inequalities, (8.28) and (8.29), we obtain

$$\begin{aligned}\mathsf{Re}\,\|\mathbf{u}_h^{n,\ell+1}\|_0^2 \le \mathsf{Re}\,\|\mathbf{u}_h^{n,0}\|_0^2 + 2C_1 k h^{-2}\|\boldsymbol{F}_h^{n,\ell}\|_\infty^4\|\boldsymbol{c}_h^{n,0}\|_{L^1}^2 \\ + C_0 k\|\boldsymbol{f}\|_0^2 + 2C_1 d^2|\Omega|^2\beta^2 k^2.\end{aligned} \tag{8.30}$$

Now, we define

$$\overline{C} := \left(\mathsf{Re}\,\|\mathbf{u}_h^{n,0}\|_0^2 + 4C_1\|\boldsymbol{c}_h^{n,0}\|_{L^1}^2 + C_0\|\boldsymbol{f}\|_0^2 + 2C_1 d^2|\Omega|^2\beta^2\right)^{\frac{1}{2}}.$$

And we will show that, if k is small enough, $\overline{C}$ is a uniform upper bound for $\|\mathbf{u}_h^{n,\ell}\|_0$ and $\|\boldsymbol{c}_h^{n,\ell}\|_{L^1}$. This is apparently true for $\ell = 0$. Now, suppose that this is also true for ℓ, and we can now prove that it is true for $\ell + 1$ with a fixed time step size k.

Using the inequality (6.33) and the inverse inequality, we have

$$\|\Pi_h^S(\nabla \mathbf{u}_h^{n,\ell})\|_\infty \le h^{-1}\|\nabla \mathbf{u}_h^{n,\ell}\|_0 \le C_2 h^{-2}\|\mathbf{u}_h^{n,\ell}\|_0 \le \overline{C}C_2 h^{-2}.$$

Hence, we can choose $2\overline{C}C_2 k \le h^2$, which implies that $\boldsymbol{\delta} - k\Pi_h^S(\nabla \mathbf{u}_h^{n,\ell})$ is invertible and that $\boldsymbol{F}_h^{n,\ell}$ is well defined. Furthermore, we also have that

$$\|\boldsymbol{F}_h^{n,\ell}\|_\infty \le \frac{1}{1 - kh^{-1}\|\nabla \mathbf{u}_h^{n,\ell}\|_0} \le \frac{1}{1 - \overline{C}C_2 k h^{-2}} \le 2.$$

Plugging the inequality above to (8.30), we obtain

$$\mathsf{Re}\,\|\mathbf{u}_h^{n,\ell+1}\|_0^2 \le \mathsf{Re}\,\|\mathbf{u}_h^{n,0}\|_0^2 + 32C_1kh^{-2}\|\boldsymbol{c}_h^{n,0}\|_{L^1}^2 + C_0k\|\boldsymbol{f}\|_0^2 + 2C_1d^2|\Omega|^2\beta^2k^2 \le \overline{C}^2,$$

if $k \le \min(1, h^2/8)$. And then (8.29) immediately gives that $\|\boldsymbol{c}_h^{n,\ell+1}\|_{L^1}$ is bounded uniformly, which completes the proof. □

REMARK 8.6 (Condition on Time Step Size). From the proof, we can see that there is an upper bound for the time step size:

$$k \le \min\left(1, \frac{h^2}{8}, \frac{h^2}{2\overline{C}C_2}\right),$$

which depends on neither nonlinear iteration step ℓ nor time level n. Furthermore, this still holds for the infinite Weissenberg number case where $\alpha = \beta = 0$.

We would like to remark that the Lemma 8.1 also shows that the discrete conformation tensor given by Algorithm 3 is always symmetric and positive definite.

THEOREM 8.2 (Positivity of the Discrete Conformation Tensor). *If the initial condition $\boldsymbol{c}_h^0$ is symmetric positive definite and the time step size k is small enough, then the discrete conformation tensor $\boldsymbol{c}_h^n$ is always symmetric positive definite for all $n = 1, 2, 3, \ldots,$*

PROOF. It is trivial to the symmetry is kept for all n. We only need to check the positivity here. From the proof of Theorem 8.1, we have seen that if k is small enough, $\boldsymbol{F}_h^{n,\ell}$ is well defined. Since Π_h^S is a positivity preserving interpolation and $\beta > 0$, we can obtain that $\boldsymbol{c}_h^{n,\ell}$ is positive definite by induction. As this is true for all ℓ and n, it completes the proof. □

We are now ready to show the existence of the solution from the compactness argument. We will show that the sequence converges to a unique limit and conclude our main result in this section. To begin with, it is helpful to notice that for any invertible matrices $\boldsymbol{A}$ and $\boldsymbol{B}$, we have that

$$\boldsymbol{A}^{-1} - \boldsymbol{B}^{-1} = \boldsymbol{A}^{-1}(\boldsymbol{B} - \boldsymbol{A})\boldsymbol{B}^{-1}. \tag{8.31}$$

We arrive at the main result for this section as below.

THEOREM 8.3 (Convergence of Algorithm 3). *The nonlinear iteration in Algorithm 3 converges if k small enough.*

PROOF. For ease of our presentation, we define

$$\boldsymbol{e}_{\mathbf{u}}^{\ell+1} := \mathbf{u}_h^{n,\ell+1} - \mathbf{u}_h^{n,\ell} \quad \text{and} \quad \boldsymbol{e}_{\boldsymbol{c}}^{\ell+1} := \boldsymbol{c}_h^{n,\ell+1} - \boldsymbol{c}_h^{n,\ell}.$$

By subtracting the momentum equation for $\mathbf{u}_h^{n,\ell+1}$ from the equation for $\mathbf{u}_h^{n,\ell}$ in Algorithm 3 and taking integration by parts, we obtain that

$$\mathsf{Re}\,\|\boldsymbol{e}_{\mathbf{u}}^{\ell+1}\|_0^2 + k\|\nabla\boldsymbol{e}_{\mathbf{u}}^{\ell+1}\|_0^2 \le k\|\nabla\boldsymbol{e}_{\mathbf{u}}^{\ell+1}\|_\infty\|\boldsymbol{e}_{\boldsymbol{c}}^{\ell}\|_{L^1}.$$

Therefore, by the inverse inequality (8.27) and the Cauchy–Schwarz inequality, we conclude that

$$\mathrm{Re}\,\|\boldsymbol{e}_{\mathbf{u}}^{\ell+1}\|_0^2 \le C_3 k h^{-2} \|\boldsymbol{e}_{\boldsymbol{c}}^{\ell}\|_{L^1}^2, \quad \forall \ell = 0, 1, 2, \ldots. \tag{8.32}$$

By subtracting the constitutive equations for $\boldsymbol{c}_h^{n,\ell+1}$ and $\boldsymbol{c}_h^{n,\ell}$, we obtain the following inequality:

$$\|\boldsymbol{e}_{\boldsymbol{c}}^{\ell+1}\|_{L^1} \le \|\boldsymbol{F}_h^{n,\ell+1} \Pi_h^S(\boldsymbol{c}_h^{n,0} \circ \widetilde{y}^n)(\boldsymbol{F}_h^{n,\ell+1})^T - \boldsymbol{F}_h^{n,\ell} \Pi_h^S(\boldsymbol{c}_h^{n,0} \circ \widetilde{y}^n)(\boldsymbol{F}_h^{n,\ell})^T\|_{L^1}.$$

In the proof of Lemma 8.1, we have seen that $\boldsymbol{F}_h^{n,\ell}$ is well defined and also that the $\|\boldsymbol{F}_h^{n,\ell}\|_0 \le 2$ for $\ell = 0, 1, 2, \ldots$.

Therefore, we have

$$\begin{aligned}\|\boldsymbol{e}_{\boldsymbol{c}}^{\ell+1}\|_{L^1} \le\ & \|\boldsymbol{F}_h^{n,\ell+1} \Pi_h^S(\boldsymbol{c}_h^{n,0} \circ \widetilde{y}^n)(\boldsymbol{F}_h^{n,\ell+1})^T - \boldsymbol{F}_h^{n,\ell} \Pi_h^S(\boldsymbol{c}_h^{n,0} \circ \widetilde{y}^n)(\boldsymbol{F}_h^{n,\ell+1})^T\|_{L^1} \\ & + \|\boldsymbol{F}_h^{n,\ell} \Pi_h^S(\boldsymbol{c}_h^{n,0} \circ \widetilde{y}^n)(\boldsymbol{F}_h^{n,\ell+1})^T - \boldsymbol{F}_h^{n,\ell} \Pi_h^S(\boldsymbol{c}_h^{n,0} \circ \widetilde{y}^n)(\boldsymbol{F}_h^{n,\ell})^T\|_{L^1}.\end{aligned}$$

Using an argument similar to that in the proof of Lemma 8.1, we obtain

$$\begin{aligned}\|\boldsymbol{e}_{\boldsymbol{c}}^{\ell+1}\|_{L^1} &\le 2(\|\boldsymbol{F}_h^{n,\ell+1}\|_\infty + \|\boldsymbol{F}_h^{n,\ell}\|_\infty)\|\boldsymbol{c}_h^{n,0}\|_{L^1}\|\boldsymbol{F}_h^{n,\ell+1} - \boldsymbol{F}_h^{n,\ell}\|_\infty \\ &\le 8\|\boldsymbol{c}_h^{n,0}\|_{L^1}\|\boldsymbol{F}_h^{n,\ell+1} - \boldsymbol{F}_h^{n,\ell}\|_\infty \le 32 k h^{-1}\|\boldsymbol{c}_h^{n,0}\|_{L^1}\|\nabla \mathbf{u}_h^{n,\ell+1} - \nabla \mathbf{u}_h^{n,\ell}\|_0,\end{aligned}$$

where the last inequality is from the inverse inequality and the following fact:

$$\begin{aligned}\boldsymbol{F}_h^{n,\ell+1} - \boldsymbol{F}_h^{n,\ell} &= \boldsymbol{F}_h^{n,\ell+1}\left((\boldsymbol{F}_h^{n,\ell})^{-1} - (\boldsymbol{F}_h^{n,\ell+1})^{-1}\right)\boldsymbol{F}_h^{n,\ell} \\ &= \boldsymbol{F}_h^{n,\ell+1}\left(\boldsymbol{\delta} - k\Pi_h(\nabla \mathbf{u}_h^{n,\ell}) - \boldsymbol{\delta} + k\Pi_h(\nabla \mathbf{u}_h^{n,\ell+1})\right)\boldsymbol{F}_h^{n,\ell} \\ &= k\boldsymbol{F}_h^{n,\ell+1}\Pi_h(\nabla \mathbf{u}_h^{n,\ell+1} - \nabla \mathbf{u}_h^{n,\ell})\boldsymbol{F}_h^{n,\ell}.\end{aligned}$$

By invoking the inverse inequality again, we conclude that

$$\mathrm{Re}\,\|\boldsymbol{e}_{\mathbf{u}}^{\ell+1}\|_0^2 \le C_3 k h^{-2}\|\boldsymbol{e}_{\boldsymbol{c}}^{\ell}\|_{L^1}^2 \le C_4 k^3 h^{-4}\|\boldsymbol{c}_h^{n,0}\|_{L^1}^2\|\boldsymbol{e}_{\mathbf{u}}^{\ell}\|_0^2. \tag{8.33}$$

For sufficiently small k, more specifically $C_4\overline{C}^2 k^3 h^{-4} \le 1/2$, Eqn (8.33) implies that the sequences $\{\|\boldsymbol{e}_{\mathbf{u}}^{\ell}\|_0\}$ and $\{\|\boldsymbol{e}_{\boldsymbol{c}}^{\ell}\|_{L^1}\}$ are contractions. Hence, $\mathbf{u}_h^{n,\ell}$ converges in the L^2 sense and $\boldsymbol{c}_h^{n,\ell}$ converges in the L^1 sense. □

THEOREM 8.4 (Global Existence of the Discrete Solution). *For any initial guess $\mathbf{u}_h^0$ and $\boldsymbol{c}_h^0$, there is a positive constant κ_0, such that the discrete systems* (8.2)–(8.4) *have a unique solution for all $n \ge 0$ as long as $k \le \kappa_0 h^2$.*

PROOF. This theorem follows directly from Theorem 8.2 by noting that, in its proof, the time step size k and all other constants appearing in the proof of Theorem 8.2 are independent of the time level t^n. □

8.3. Computational complexity

So far, we have discussed all the details of the fully discrete scheme, Algorithm 3. We can now further investigate the computational complexity of Algorithm 3.

First, if k is small enough, then the nonlinear equation for the flow map in Step 1, Algorithm 3 is solvable. Specifically, Step 1 can be solved in a fixed number of iterations. Second, Theorem 8.3 guarantees that the fixed-point iteration for solving the coupled system in Step 2 can also be terminated in a finite number of iterations to any given tolerance tol. Finally, we have seen that Stokes-type systems can be solved by optimal multilevel methods independent of h, k, Re, and η_s. Based on these observations, we can easily see the following result:

COROLLARY 8.2 (Computation Complexity). *If the time step size k is small enough, Algorithm 1 converges uniformly with respect to* Re *and* Wi *and the computational complexity is* $O(N \log N)$, *where N is the total spatial degrees of freedom.*

9. Implementation details and numerical experiments

In this section, we will give details of our implementation of Algorithm 3. We will also offer some preliminary numerical experiments.

9.1. A benchmark problem

We consider the Poiseuille flow between two parallel plates around a cylinder with circular cross section for the numerical tests. The problem is well suited as a benchmark problem for understanding the viscoelastic models in a smooth flow without geometric singularity; see, for example, AFONSO, OLIVEIRA, PINHO and ALVES [2009], CORONADO, ARORA, BEHR and PASQUALI [2007], and SUN, SMITH, ARMSTRONG and BROWN [1999]. We start by describing the geometry and boundary conditions.

9.1.1. A two-dimensional model problem

Consider the computational domain $\Omega \subset \mathbb{R}^2$ as described in Fig. 9.1.

We use a symmetric domain with $R = 1$, $H_1 = H_2 = 2$, and $L_1 = L_2 = 15$. The ratio of the distance between the two plates and the diameter of the circular hole is 2.

As discussed in Section 3, the nondimensional Oldroyd-B model can be written as follows: find $(\mathbf{u}, p, \boldsymbol{c})$ for $x \in \Omega$ and $t \in (0, +\infty)$ such that

$$\left\{ \begin{aligned} \mathsf{Re}\left(\frac{\partial \mathbf{u}}{\partial t} + \mathbf{u}\cdot\nabla\mathbf{u}\right) + \nabla p - \eta_s \Delta \mathbf{u} &= \nabla \cdot \boldsymbol{c} \\ \nabla \cdot \mathbf{u} &= 0 \\ \tfrac{1}{\mathsf{Wi}}\boldsymbol{c} + \mathcal{L}_{\mathbf{u},\nabla\mathbf{u}}\boldsymbol{c} &= \tfrac{(1-\eta_s)}{\mathsf{Wi}^2}\boldsymbol{\delta}. \end{aligned} \right. \tag{9.1}$$

On the top and bottom walls, we impose the no-slip boundary condition for the flow velocity $\mathbf{u}$; at the outflow boundary, we give the Neumann boundary condition for $\mathbf{u}$. And the inflow

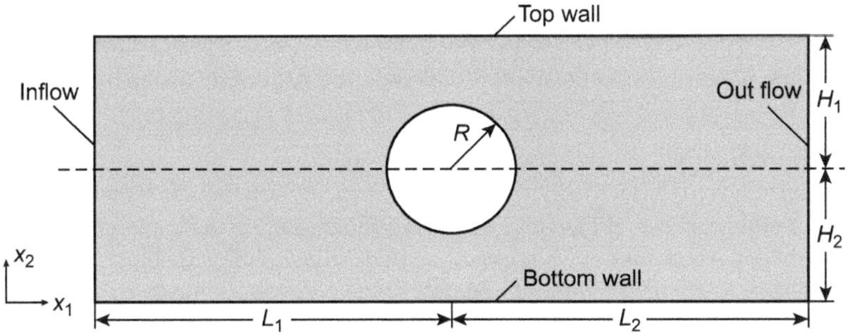

FIG. 9.1 Flow-past-cylinder domain

boundary condition for velocity is given by

$$\mathbf{u} = \begin{pmatrix} 1.5\left(1 - \left(\frac{x_2}{H}\right)^2\right) \\ 0 \end{pmatrix}.$$

Therefore, the average speed of the background inflow fluids is 1.0 in the horizontal direction and 0.0 in the vertical direction.

9.1.2. *Drag coefficient*

In order to compare with the established benchmark results in the literature, we focus on the dimensionless drag coefficient. The definition of the drag coefficient can be given as follows:

$$F_D = \frac{1}{U}\int_{\partial B} \left(-p\delta + \eta_s(\nabla\mathbf{u} + \nabla\mathbf{u}^T) + \boldsymbol{c}\right)\mathbf{n}\cdot\mathbf{e}_1 \,\mathrm{d}\Gamma, \tag{9.2}$$

where $\mathbf{n}$ is the outer unit normal vector for the boundary of circle ∂B (vector pointing outward from the circle) and $\mathbf{e}_1 = (1, 0)^T$ and U is the mean background flow velocity.

There are two standard ways of computing the drag coefficient: one way is to compute the line integral directly on the curved boundary as in (9.2); an alternative is to do integration by parts and transform the integral into a volume integral (see JOHN [2004], for example.) Let $\boldsymbol{\varphi} = (\varphi_1, 0)^T$ be a smooth function in Ω in which φ_1 equals one on ∂B and vanishes on $\partial\Omega \setminus \partial B$. Multiplying $\boldsymbol{\varphi}$ on both ends of (9.1), we obtain that

$$\int_{\Omega} \mathrm{Re}\left(\frac{\partial\mathbf{u}}{\partial t} + \mathbf{u}\cdot\nabla\mathbf{u}\right)\cdot\boldsymbol{\varphi}\,\mathrm{d}x = \int_{\Omega} (-\nabla p + \eta_s\Delta\mathbf{u} + \nabla\cdot\boldsymbol{c})\cdot\boldsymbol{\varphi}\,\mathrm{d}x. \tag{9.3}$$

Applying integration by parts, we get

$$F_D = \frac{1}{U}\int_{\Omega} \left\{\mathrm{Re}\left(\frac{\partial\mathbf{u}}{\partial t} + \mathbf{u}\cdot\nabla\mathbf{u}\right) - \nabla p + \eta_s\Delta\mathbf{u} + \nabla\cdot\boldsymbol{c}\right\}\cdot\boldsymbol{\varphi}\,\mathrm{d}x. \tag{9.4}$$

9.2. *Implementation details*

In this section, we discuss the details for implementing Algorithm 3 step by step.

9.2.1. *Flow map*

We first discuss how to numerically approximate the departure foot of any point x in Step 1 of Algorithm 3 and how to find the value of certain functions at $\widetilde{y}^n$ using interpolation. Notice that in Step 1 of Algorithm 3, $\widetilde{y}^n$ appears on both sides of the equation for the midpoint rule. We can use a simple fixed-point iteration or the Newton–Raphson method to solve the nonlinear equation

$$G(y) = 0 \quad \text{with } G(y) := y + k\mathbf{u}_h^n\Big(\frac{x+y}{2}\Big) - x.$$

In our experiments, we employ the Newton–Raphson method, and we stop the Newton–Raphson's iteration once the residual is less than 10^{-10}. In our experiments, the Newton–Raphson method usually converges in 2 to 4 iterations.

9.2.2. *Feet searching*

Once the coordinates of the departure foot $\widetilde{y}^n$ are computed as discussed in the previous section, we need to find the element in which $\widetilde{y}^n$ is located in order to perform interpolations in Step 2. Note that the time step size is usually small and the departure foot should not be too far away from the corresponding arrival point x. So it is natural to start searching for the host element by beginning with the element that contains x and then following the characteristics to locate the host element of each characteristic foot (ALLIEVI and BERMEJO [1997]). In order to describe the algorithm, we introduce two data structures first:

(i) PATCH(:,x) gives all elements that share a given point x, and PATCH(i,x) is the local index for the ith element in the patch.
(ii) NEIGH(:,E) gives the neighboring elements of a given element E, and NEIGH(s,E) is the neighbor of E opposite the side s.

Now, we are ready to describe the feet-searching algorithm.

ALGORITHM 4 Finding the Host Elements of Departure Feet

Step 0. Set the current element $E = \text{PATCH}(1, x)$ and $i = 1$.

Step 1. Find the reference coordinate r of y in E. If $r_1 + r_2 > 1$, then $s = 1$; else if $r_1 < 0$, then $s = 2$; else if $r_2 < 0$, then $s = 3$; otherwise, return E as the host element of y and stop.

Step 2. If NEIGH(s, E) has not yet been visited and it is not out-of-boundary, set $E =$ NEIGH(s, E); else if PATCH($i+1$, x) is not empty, set $E = \text{PATCH}(i+1, x)$ and $i = i+1$. Go back to Step 1.

Step 3. Let E be the next nonvisited element and go to Step 1.

9.2.3. Stokes solvers

As stated in Section 7, the main computational cost at each iteration in Algorithm 3 is to solve the Stokes-like systems in Step 2. To date, we have only tested the Taylor–Hood $P_0^2 - P_0^1$ element; our implementation of the Scott–Vogelius $P_0^4 - P_{-1}^3$ is on going. We discussed two types of Stokes solvers in Section 7. Here, we test the preconditioned MinRes method using the flow-past-cylinder benchmark problem; we, therefore, (i) test the Stokes solvers for steady-state problems and (ii) test the time-marching scheme as a smoother to solve steady-state problems.

We report the dimensionless drag coefficient for various meshes in Tables 9.1 and 9.2 and for iteration numbers in Table 9.3. From Table 9.2, we note that the steady-state solution does not depend on time step size k. The differences between drag coefficients using different time step sizes k are less than 10^{-4}.

9.2.4. Using subdivisions to improve accuracy

In order to further improve accuracy, we divide each element $E \in \mathcal{T}_h$ into several subelements and then define the degree of freedom on each subelement for the piecewise constant

TABLE 9.1

Drag coefficient for steady-state Stokes flow between parallel plates

	DOF	$h_{\min}$	Drag coefficient	Difference
Mesh 1	33001	1.8e-3	132.24599	
Mesh 2	58979	7.1e-4	132.30130	5.5315e-02
Mesh 3	109729	3.2e-4	132.33079	2.9488e-02
Mesh 4	213575	2.0e-4	132.34571	1.4917e-02
Mesh 5	416409	7.1e-5	132.35255	6.8414e-03

The convergence rate for the drag coefficient is DOF^{-1} where DOF is the degrees of freedom. Reference value $= 132.34 \sim 132.36$.

TABLE 9.2

Steady limit of the drag coefficient for the time-dependent Stokes flow between parallel plates

	DOF	$h_{\min}$	$k = 0.5$	$k = 0.1$	$k = 0.01$
Mesh 1	33001	1.8e-3	132.24577	132.24577	132.24577
Mesh 2	58979	7.1e-4	132.30126	132.30125	132.30125
Mesh 3	109729	3.2e-4	132.33078	132.33078	132.33078
Mesh 4	213575	2.0e-4	132.34571	132.34571	132.34571
Mesh 5	416409	7.1e-5	132.35256	132.35256	132.35255

TABLE 9.3

Number of iterations for the MinRes solver with zero initial guess for the steady-state Stokes system

	Mesh 1	Mesh 2	Mesh 3	Mesh 4	Mesh 5
Number of iteration	106	106	109	112	113

The stopping criteria is that the relative residual is smaller than 10^{-8}.

tensor $\boldsymbol{c}_h^n$. We notice that MARCHAL and CROCHET [1987] have employed a similar technique, which has been used to enhance stability.

Here, we have some different considerations due to the difficulties that inheres in the Eulerian–Lagrangian method:

- The integrand on the right-hand side of the momentum equation is usually nonsmooth (piecewise polynomial), and using subelements can improve the accuracy of numerical quadrature.
- When the velocity field is nonconstant, the deformed element $E(y)$ changes its shape, and using subelements can describe the shape of deformed triangles much better.

The extra cost of this approach is that, after locating the host element of y_q, we need to find in the subelement in which it is located and then evaluate interpolation on each subelement.

9.3. *Numerical experiments*

For the benchmark problem (the flow past a cylinder in a two-dimensional setting), the Newtonian viscosity η_s chosen is 0.59, and the Reynolds number is assumed to be 0. So the polymeric viscosity is $\eta_p = 1 - \eta_s = 0.41$. For the computational domain in Fig. 9.1, we take $R = 1$, $H = 2R$, and $L_1 = L_2 = 15$. Under this setting, many research groups have obtained results for Wi up to about 1.2, and they have agreed on the problems for $\mathsf{Wi} \leq 0.7$; see AFONSO, OLIVEIRA, PINHO and ALVES [2009], for example. In this section, we test our algorithms using the two-dimensional benchmark problem above on different meshes, with various Weissenberg numbers. The main purpose is to validate the convergence of the proposed algorithms and the long-term stability of the computation.

First, we fix $\mathsf{Wi} = 0.1$ and 0.5; we use different meshes to test the convergence of the algorithm. We report drag coefficients in Table 9.4. Notice that the drag coefficients converge when we refine the mesh.

In order to reduce the error introduced by the interpolation, we employ the subelement technique introduced in Section 9.2.4. We divide each element into 4 and 16 congruent elements by applying regular refinement (by dividing each triangle into four smaller congruent triangles) once and twice, respectively. The numerical results for $\mathsf{Wi} = 0.1$ are reported in Table 9.5. We find that accuracy is improved by using more accurate interpolation.

We test the proposed algorithm using 16 subelements on a mildly refined mesh (Mesh 2) for various Weissenberg numbers. As discussed in Section 8, our algorithms remain stable as the time steps increase. The drag coefficients are reported in Fig. 9.2, and the results are consistent with the literature at least for Wi less than or equal to 0.75.

We find that the positivity of the conformation tensor can be preserved in the discrete sense. But we have yet to implement a discretization scheme that will maintain a strong

TABLE 9.4
Mesh dependence of the drag coefficient for low Weissenberg numbers

	Spatial DOF	k	$\mathsf{Wi} = 0.1$	$\mathsf{Wi} = 0.5$
Mesh 1	6269	2.5×10^{-3}	127.60	115.41
Mesh 2	25471	1.25×10^{-3}	129.33	117.23
Mesh 3	102674	6.25×10^{-4}	129.94	117.99
Mesh 4	412277	3.125×10^{-4}	130.11	118.35

TABLE 9.5
Comparison of drag coefficient for Wi = 0.1 numbers using different number of subelements

	Spatial DOF	1 subelement	4 subelements	16 subelements
Mesh 1	3477	121.96931	124.68434	125.62053
Mesh 2	15509	127.13945	128.40418	128.94619
Mesh 3	25763	128.08089	129.00947	129.28447
Mesh 4	50068	128.41062	129.19577	129.53881

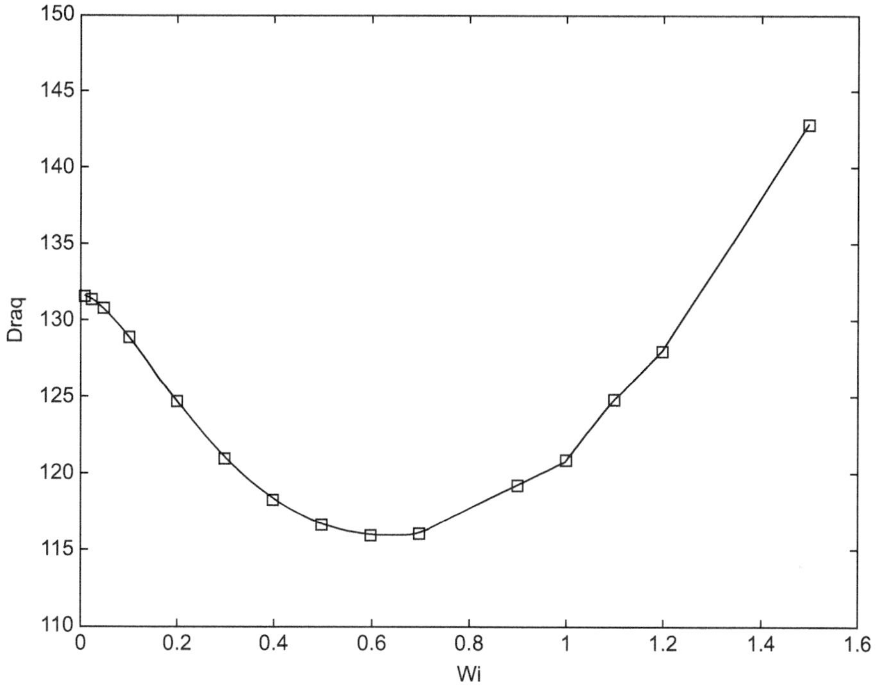

FIG. 9.2 Drag coefficients for various Weissenberg numbers.

divergence-free condition. Though the algorithm has not been fully realized, we obtained a stable numerical solution in time. We checked the grid convergence for different Weissenberg numbers, and we observed that the numerical solutions for Weissenberg numbers larger than 1.0 exhibit difficulties in grid convergence. In future research work, we will work to fully implement the proposed algorithms and study mesh convergence in high Weissenberg number regimes.

10. Concluding remarks

In this article, we reviewed the link between various constitutive equations from viscoelastic fluid models and symmetric matrix Riccati differential equations. We presented several building blocks for the unified and stable numerical treatment of viscoelastic fluid models. We provided the proof that the resulting discrete problem admits a globally unique solution. We discussed how Stokes-type linear systems can be solved effectively using multigrid methods. We also presented some of our recent efforts to implement the designed algorithms in order to demonstrate some of our theoretical results.

Acknowledgments

The first author was supported in part by NSF DMS-0753111, DMS-0915028, and the Startup fund from the Rutgers University. The second and third authors were supported in part by NSF DMS-0749202, DMS-0915153, and the Center for Computational Mathematics and Application, the Pennsylvania State University.

References

ABOU-KANDIL, H., FREILING, G., IONESCU, V., JANK, G. (2003). *Matrix Riccati Equations: In Control and Systems Theory*, Systems and Control (Birkhäuser Verlag, Boston).

ADAMS, J., FIELDING, S., OLMSTED, P. (2008). The interplay between boundary conditions and flow geometries in shear banding: Hysteresis, band configurations, and surface transitions. *J. Non-Newton. Fluid Mech.* **151**, 101–118. 4th Annual European Rheology Conference.

AFONSO, A., OLIVEIRA, P., PINHO, F., ALVES, M. (2009). The log-conformation tensor approach in the finite-volume method framework. *J. Non-Newton. Fluid Mech.* **157**, 55–65.

ALLIEVI, A., BERMEJO, R. (1997). A generalized particle search-locate algorithm for arbitrary grids. *J. Comput. Phys.* **132**, 157–166.

ALVES, M.A., OLIVEIRA, P.J., PINHO, F.T. (2003). Benchmark solutions for the flow of Oldroyd-B and PTT fluids in planar contractions. *J. Non-Newton. Fluid Mech.* **110**, 45–75.

ALVES, M.A., PINHO, F.T., OLIVEIRA, P.J. (2001). The flow of viscoelastic fluids past a cylinder: Finite-volume high-resolution methods. *J. Non-Newton. Fluid Mech.* **97**, 207–232.

AUSTIN, T.M., MANTEUFFEL, T.A., MCCORMICK, S. (2004). A robust multilevel approach for minimizing $\mathbf{H}(\mathrm{div})$-dominated functionals in an $\mathbf{H}^1$-conforming finite element space. *Numer. Linear Algebra Appl.* **11**, 115–140.

BAAIJENS, F. (1993). An U-ALE formulation of 3-D unsteady viscoelastic flow. *Int. J. Numer. Methods Eng.* **36**, 1115–1143.

BAAIJENS, F.P. (1998). Mixed finite element methods for viscoelastic flow analysis: A review. *J. Non-Newton. Fluid Mech.* **79**, 361–385.

BAJAJ, M., BHAT, P., PRAKASH, J.R., PASQUALI, M. (2006). Multiscale simulation of viscoelastic free surface flows. *J. Non-Newton. Fluid Mech.* **140**, 87–107.

BAJAJ, M., PASQUALI, M., PRAKASH, J. (2008). Coil-stretch transition and the breakdown of computations for viscoelastic fluid flow around a confined cylinder. *J. Rheol.* **52**, 197–223.

BARRETT, J., SÜLI, E. (2008). Existence of global weak solutions to dumbbell models for dilute polymers with microscopic cut-off. *M3AS.* **18**, 935–971.

BERIS, A., ARMSTRONG, R., BROWN, R. (1984). Finite element calculation of viscoelastic flow in a journal bearing: I. small eccentricities. *J. Non-Newton. Fluid Mech.* **16**, 141–172.

BERIS, A., ARMSTRONG, R., BROWN, R. (1986). Finite element calculation of viscoelastic flow in a journal bearing: II. moderate eccentricity. *J. Non-Newton. Fluid Mech.* **19**, 323–347.

BERIS, A., ARMSTRONG, R.C., BROWN, R.A. (1983). Perturbation theory for viscoelastic fluids between eccentric rotating cylinders. *J. Non-Newton. Fluid Mech.* **13**, 109–148.

BERIS A., EDWARDS, B. (1994). *Thermodynamics of Flow Systems, with Internal Mocrostructure* (Oxford Science Publication, New York).

BERIS, A.N., ARMSTRONG, R.C., BROWN, R.A. (1987). Spectral/finite-element calculations of the flow of a Maxwell fluid between eccentric rotating cylinders. *J. Non-Newton. Fluid Mech.* **22**, 129–167.

BIRD, R., CURTISS, C., ARMSTRONG, R., HASSAGER, O. (1987). *Dynamics of Polymeric Liquids, Kinetic Theory* Volume 2 (Weiley Interscience, New York).

BLACK, W., GRAHAM, M. (2001). Slip, concentration fluctuations, and flow instability in sheared polymer solutions. Macromolecules. **34**, 5731–5733.

Bonito, A., Picasso, M., Laso, M. (2006). Numerical simulation of 3D viscoelastic flows with free surfaces. *J. Comput. Phys.* **215**, 691–716.

Boyaval, S., Lelievre, T., Mangoubi, C. (2009). Free-energy-dissipative schemes for the Oldroyd-B model. *ESAIM:M2AN.* **43**, 523–561.

Bramble, J., Pasciak, J. (1997). Iterative techniques for time dependent Stokes problems. *Comput. Methods Appl. Mech. Eng.* **1–2**, 13–30.

Bramble, J.H. (1993). *Multigrid Methods.* Pitman Research Notes in Mathematical Sciences Volume 294. (Longman Scientific & Technical, Essex, England).

Brandt, A. (1977). Multi–level adaptive techniques (MLAT) for partial differential equations: ideas and software. In: Rice, J.R. (ed.), (Academic Press, New York), 277–318.

Brenner, S.C., Scott, L.R. (2002). *The mathematical theory of finite element methods.* Texts in Applied Mathematics Volume 15, (Springer-Verlag, New York), second ed.

Brezzi, F. (1974). On the existence, uniqueness and approximation of saddle point problems arising from Lagrange multipliers. *RAIRO Numer. Anal.* **8**, 129–151.

Brezzi, F., Fortin, M. (1991). *Mixed and hybrid finite element methods* (Springer-Verlag, New York).

Brockett, R.W. (1970). *Finite dimensional linear systems* (Wiley, New York).

Brown, R., Szady, M., Northey, P., Armstrong, R. (1993). On the numerical stability of mixed finite element methods for viscoelastic flows governed by differential constitutive equations. *Theor. Comput. Fluid Dyn.* **5**, 77–106.

Carreau, P., Grmela, M. (1987). Conformation tensor rheological models. *J. Non-Newton. Fluid Mech.* **23**, 271–294.

Chauvière, C., Owens, R. G. (2001). A new spectral element method for the reliable computation of viscoelastic flow. *Comput. Methods Appl. Mech. Eng.* **190**, 3999–4018.

Chemin, J.Y., Masmoudi, N. (2001). About lifespan of regular solutions of equations related to viscoelastic fluids. *SIAM J. Math. Anal.* **33**, 84–112.

Chilcott, M., Rallison, J. (1988). Creeping flow of dilute polymer solutions past cylinders and spheres. *J. Non-Newton. Fluids Mech.* **29**, 381–432.

Clermont, J., Pierrard, J. (1976). Experimental study of a non-viscometric flow: Kinematics of a viscoelastic fluid at the exit of a cylindrical tube. *J. Non-Newton. Fluid Mech.* **1**, 175–182.

Coronado, O.M., Arora, D., Behr, M., Pasquali, M. (2007). A simple method for simulating general viscoelastic fluid flows with an alternate log-conformation formulation. *J. Non-Newton. Fluid Mech.* **147**, 189–199.

Dealy, J., Vu, T. (1977). The Weissenberg effect in molten polymers. *J. Nonnewt. Fluid Mech.* **3**, 127–140.

Dieci, L. (1994). Structure preserving piecewise polynomial interpolation for definite matrices. *Lin. Alg. Appl.* **202**, 25–32.

Dieci, L., Eirola, T. (1994). Positive deffiniteness in the numerical solution of Riccati differential equations. *Numer. Math.* **67**, 303–313.

Dieci, L., Eirola, T. (1996). Preserving monotonicity in the numerical solution of Riccati differential equations. *Numer. Math.* **74**, 35–47.

Douglas, J., Russell, T.F. (1982). Numerical methods for convection-dominated diffusion problems based on combining the method of characteristics with finite element or finite difference procedures. *SIAM J. Numer. Anal.* **19**, 871–885.

Dupret, F., Marchal, J., Crochet, M. (1985). On the consequence of discretization errors in the numerical calculation of viscoelastic flow. *J. Non-Newton. Fluid Mech.* **18**, 173–186.

El-Kareh, A.W., Leal, L. (1989). Existence of solutions for all Deborah numbers for a non-Newtonian model modified to include diffusion. *J. Non-Newton. Fluid Mech.* **33**, 257–287.

Engler, H. (1987). On the dynamic shear flow problem for viscoelastic liquids. *SIAM J. Math. Anal.* **18**, 972–990.

Fattal, R., Kupferman, R. (2004). Constitutive laws for the matrix-logarithm of the conformation tensor. *J. Non-Newton. Fluid Mech.* **124**, 281–285.

Fattal, R., Kupferman, R. (2005). Time-dependent simulation of viscoelastic flows at high Weissenberg number using the log-conformation representations. *J. Non-Newton. Fluid Mech.* **126**, 23–37.

Feng, K., Shang, Z.-J. (1995). Volume-preserving algorithms for source-free dynamical systems. *Numer. Math.* **71**, 451–463.

Fortin, M., Esselaoui, D. (1987). A finite element procedure for viscoelastic flows. *Inter. J. Numer. Methods in Fluids.* **7**, 1035–1052.

Fortin, M., Fortin, A. (1989). A new approach for the FEM simulation of viscoelastic flows. *J. Non-Newton. Fluid Mech.* **32**, 295–310.

Fortin, M., Glowinski, R. (1982). *Méthodes de lagrangien augmenté*. Méthodes Mathématiques de l'Informatique Volume 9 (Gauthier-Villars, Paris). Applications à la résolution numérique de problèmes aux limites.

Fortin, M., Glowinski, R. (1983). *Augmented Lagrangian methods*. Studies in Mathematics and its Applications Volume 15. (North-Holland Publishing Co., Amsterdam). Applications to the numerical solution of boundary value problems, Translated from the French by B. Hunt and D. C. Spicer.

Fortin, M., Pierre, R. (1989). On the convergence of the mixed method of Crochet and Marchal for viscoelastic flows. *Comput. Methods Appl. Mech. Engrg.* **73**, 341–350.

Gelbart, W., Ben-Shaul, (1996). The "new" science of "complex fluids". *J. Phys. Chem.* **100**, 13169–13189.

Giesekus, H. (1982). A simple constitutive equation for polymer fluids based on the concept of deformation-dependent tensorial mobility. *J. Non-Newton. Fluid Mech.* **11**, 69–109.

Glowinski, R. (2003). Finite element methods for incompressible viscous flow. In: *Handbook of Numerical Analysis*, Volume IX (North-Holland, Amsterdam), 3–1176.

Glowinski, R., Le Tallec, P. (1989). *Augmented Lagrangian and operator-splitting methods in nonlinear mechanics*. SIAM Studies in Applied Mathematics Volume 9 (Society for Industrial and Applied Mathematics (SIAM), Philadelphia, PA).

Gordon, R., Schowalter, W. (1972). Anisotropic fluid theory: A different approach to the dumbbell theory of dilute polymer solutions. *Trans. Soc. Rheo.* **16**, 79–97.

Groisman, A., Steinberg, V. (1998). Mechanism of elastic instability in Couette flow of polymer solutions. *Experiment Phys. Fluids*, **10**, 2451–2463.

Guénette, R., Fortin, M. (1995). A new mixed finite element method for computing viscoelastic flows. *J. Non-Newton. Fluid Mech.* **60**, 27–52.

Guillope, C., Saut, J.-C. (1990a). Existence results for the flow of viscoelastic fluids with a differential constitutive law. *Nonlinear Anal.* **15**, 849–869.

Guillope, C., Saut, J.-C. (1990b). Global existence and one-dimensional nonlinear stability of shearing motions of viscoelastic fluids of Oldroyd type. *RAIRO Anal. Numér.* **24**, 369–401.

Hackbusch, W. (1985). *Multigrid Methods and Applications*. Computational Mathematics Volume 4. (Springer-Verlag, Berlin).

Hao, J., Pan, T.-W., Glowinski, R., Joseph, D.D. (2009). A fictitious domain/distributed Lagrange multiplier method for the particulate flow of Oldroyd-B fluids: A positive definiteness preserving approach. *J. Non-Newton. Fluid Mech.* **156**, 95–111.

He, L., Zhang, P. (2009). L2 decay of solutions to a micro-macro model for polymeric fluids near equilibrium. *SIAM J. Math. Anal.* **40**, 1905–1922.

Hu, H.H., Joseph, D.D. (1990). Numerical simulation of viscoelastic flow past a cylinder. *J. Non-Newton. Fluid Mech.* **37**, 347–377.

Hughes, T. (1984). Numerical Implementation of Constitutive Models: Rate-Independent Deviatoric Plasticity: Theoretical Foundation for Large-Scale Computations for Nonlinear Material Behavior. (Martinus Nijhoff Publisher, Dordrecht, The Netherlands).

Hughes, I., Brooks, A. (1982). A theoretical framework for Petrov-Galerkin methods with discontinuous weighting functions: Applications to the streamline upwind procedure. In: Gallagher, R., Norrie, D., Oden, D., Zienkiewicz, J. (eds.), *Finite Element in Fluids* Volume 4 (Wiley-Interscience Publisher, John Wiley and Sons, Inc., New York), pp. 47–65.

HUGHES, T., WINGET, J. (1980). Finite rotation effects in numerical integration of rate constitutive equations arising in large-deformation analysis. *Inter. J. Numer. Meth. Eng.* **15**, 1862–1867.

HULSEN, M. (1988). Some properties and analytical expressions for plane flow of Leonov and Giesekus models. *J. Non-Newton Fluid Mech.* **30**, 85–92.

HULSEN, M. (1990). A sufficient condition for a positive definite configuration tensor in differential models. *J. Non-Newton Fluid Mech.* **38**, 93–100.

HULSEN, M.A., FATTAL, R., KUPFERMAN, R. (2005). Flow of viscoelastic fluids past a cylinder at high Weissenberg number: Stabilized simulations using matrix logarithms. *J. Non-Newton. Fluid Mech.* **127**, 27–39.

ILG, P., KARLIN, I., ÖTTINGER, H. (2002). Canonical distribution functions in polymer dynamics.(I). Dilute solutions of flexible polymers. *Physica A: Stat. Mech. Appl.* **315**, 367–385.

JOHN, V. (2004). Reference values for drag and lift of a two-dimensional time-dependent flow around a cylinder. *Int. J. Numer. Methods Fluids.* **44**, 777–788.

JOHNSON, M., SEGALMAN, D. (1977). A model for viscoelastic fluid behaviour which allows non-Newtonian deformation. *J. Non-Newton. Fluid Mech.* **2**, 255–270.

JOSEPH, D. (1990a). Fluid dynamics of viscoelastic liquids. *Applied Mathematical Sciences* Volume 84. (Springer, New York), p. 755.

JOSEPH, D., SAUT, J. (1986). Change of type and loss of evolution in the flow of viscoelastic fluids. *J. Non-Newton. Fluid Mech.* **20**, 117–141.

JOSEPH, D.D. (1990b). Fluid dynamics of two miscible liquids with diffusion and gradient stresses. *Eur. J. Mech. B Fluids* **9**, 565–596.

JOURDAIN, B., LELIVRE, T., BRIS, C.L. (2004). Existence of solution for a micro-macro model of polymeric fluid: the FENE model. *J. Funct Anal.* **209**, 162–193.

KABANEMI, K., BERTRAND, F., TANGUY, P., AIT-KADI, A. (1994). A pseudo-transient finite element method for the resolution of viscoelastic fluid flow problems by the method of characteristics. *J. Non-Newt. Fluid Mech.* **55**, 283–305.

KING, R.C., APELIAN, M.N., ARMSTRONG, R.C., BROWN, R.A. (1988). Numerically stable finite element techniques for viscoelastic calculations in smooth and singular geometries. *J. Non-Newton. Fluid Mech.* **29**, 147–216.

KREISS, H.-O. (2001). *Time-dependent partial differential equations and their numerical solution.* (Birkhuser Verlag, Basel).

LARIN, M., REUSKEN, A. (2008). A comparative study of efficient iterative solvers for generalized Stokes equations. *Numer. Linear Algebra Appl.* **15**, 13–34.

LEE, Y.-J. (2004). Modeling and Simulations of Non-Newtonian Fluid Flows, PhD thesis (State College, Pennsylvania).

LEE, Y.-J. (2009). Uniform stability analysis of Austin, Manteuffel and McCormick finite elements and fast and robust iterative methods for the Stokes-like equations. *Numer. Linear Algebra Appl.* **17** (1), 109–138, January 2010.

LEE, Y.-J., WU, J., CHEN, J. (2009). Robust multigrid method for the planar linear elasticity problems. *Numer. Math.* **113**, 473–496.

LEE, Y.-J., WU, J., XU, J., ZIKATANOV, L. (2007). Robust subspace correction methods for nearly singular systems. *Math. Models Methods Appl. Sci.* **17**, 1937–1963.

LEE, Y.-J., XU, J. (2006). New formulations, positivity preserving discretizations and stability analysis for non-Newtonian flow models. *Comput. Methods Appl. Mech. Engrg.* **195**, 1180–1206.

LEE, Y.-J. XU, J., ZHANG, C-S. On the global existence and uniqueness of solutions to discretized viscoelastic flow models. *Math. Models Methods Appl. Sci.* In press.

LESAINT, P., RAVIART, P.A. (1979). Finite element collocation methods for first order systems. *Math. Comput.* **33**, 891–918.

LI, W.E.T., ZHANG, P. (2004). Well-posedness for the dumbbell model of polymeric fluids. *Commun. Math. Phys.* **248**, 409–427.

LIELENS, G., HALIN, P., JAUMAIN, I., KEUNINGS, R., LEGAT, V. (1998). New closure approximations for the kinetic theory of finitely extensible dumbbells. *J. Non-Newton. Fluid Mech.* **76** 249–279.

LIN, F.-H., LIU, C., ZHANG, P. (2005). On hydrodynamics of viscoelastic fluids. *Comm. Pure Appl. Math.* **58**, 1437–1471.

LIN, F.-H., LIU, C., ZHANG, P. (2007). On a micro-macro model for polymeric fluids near equilibrium. *Commun. Pure Appl. Math.* LX, 838–866.

LIONS, P.L., MASMOUDI, N. (2000). Global solutions for some Oldroyd models of non-Newtonian flows. *Chinese Ann. Math. Ser. B.* **21**, 131–146.

LIU, A.W., BORNSIDE, D.E., ARMSTRONG, R.C., BROWN, R.A. (1998). Viscoelastic flow of polymer solutions around a periodic, linear array of cylinders: comparisons of predictions for microstructure and flow fields. *J. Non-Newton. Fluid Mech.* **77**, 153–190.

LOZINSKI, A., OWENS, R.G. (2003). An energy estimate for the Oldroyd B model: Theory and applications. *J. Non-Newton. Fluid Mech.* **112**, 161–176.

MARCHAL, J., CROCHET, M. (1987). A new mixed finite element for calculating viscoelastic flow. *J. Non-Newton. Fluid Mech.* **26**, 77–114.

NOCHETTO, R.H., PYO, J.-H. (2004). Optimal relaxation parameter for the Uzawa method. *Numer. Math.* **98**, 695–702.

NOCHETTO, R.H., WAHLBIN, L.B. (2002). Positivity preserving finite element approximation. *Math. Comp.* **71**, 1405–1419.

OLDROYD, J. (1950). On the formulation of rheological equations of state. *Proc. Roy. Soc.* **A200**, 523–541.

OLDROYD, J. (1958). Non-Newtonian effects in steady motion of some idealized elasto-viscous liquids. *Proc. R. Soc. A.* **245**, 278–297.

OLIVEIRA, P.J., PINHO, F.T., PINTO, G.A. (1998). Numerical simulation of non-linear elastic flows with a general collocated finite-volume method. *J. Non-Newton. Fluid Mech.* **79**, 1–43.

OLSHANSKII, M.A., PETERS, J., REUSKEN, A. (2006). Uniform preconditioners for a parameter dependent saddle point problem with application to generalized Stokes interface equations. *Numer. Math.* **105**, 159–191.

OWENS, R., PHILLIPS, T. (2002). *Computational Rheology* (Imperial College Press, London).

PAIGE, C.C., SAUNDERS, M.A. (1975). Solutions of sparse indefinite systems of linear equations. *SIAM J. Numer. Anal.* **12**, 617–629.

PAN, T.-W., HAO, J., GLOWINSKI, R. (2009). On the simulation of a time-dependent cavity flow of an Oldroyd-B fluid. *Int. J. Numer. Methods Fluids.* **60**, 791–808.

PASQUALI, M., SCRIVEN, L.E. (2002). Free surface flows of polymer solutions with models based on the conformation tensor. *J. Non-Newton. Fluid Mech.* **108**, 363–409.

PETERA, J. (2002). A new finite element scheme using the Lagrangian framework for simulation of viscoelastic fluid flows. *J. Non-Newton. Fluid Mech.* **103**, 1–43.

PHILLIPS, T.N., WILLIAMS, A.J. (1999). Viscoelastic flow through a planar contraction using a semi-lagrangian nite volume method. *J. Non-Newton. Fluid Mech.* **87**, 215–246.

PIRONNEAU, O. (1981/82). On the transport-diffusion algorithm and its applications to the Navier-Stokes equations. *Numer. Math.* **38**, 309–332.

RAJAGOPALAN, D., ARMSTRONG, R., BROWN, R. (1990). Finite-element methods for calculation of steady, viscoelastic flow using constitutive-equations with a Newtonian viscosity. *J. Non-Newton. Fluid Mech.* **36**, 159–192.

REID, W. (1972). Riccati differential equations. Mathematics in Science and Engineering Voloume 86 (Academic Press, New York).

REMMELGAS, J., SINGH, P., LEAL, L. (1999). Computational studies of nonlinear elastic dumbbell models of Boger fluids in a cross-slot flow. *J. Non-Newton. Fluid Mech.* **88**, 31–61.

RENARDY, M. (1991). An existence theorem for model equations resulting from kinetic theories of polymer solutions. *SIAM J. Math. Anal.* **22**, 313–327.

RENARDY, M. (2000a). Asymptotic structure of the stress field in flow past a cylinder at high Weissenberg number. *J. Non-Newton. Fluid. Mech.* **90**, 13–23.

RENARDY, M. (2000b). *Mathematical Analysis of Viscoelastic Flows.* CBMS-NSF regional conference series in applied mathematics Volume 73 (SIAM, Philadelphia).

RENARDY, M. (2006). A comment on smoothness of viscoelastic stresses. *J. Non-Newton. Fluid Mech.* **138**, 204–205.

RENARDY, M. (2009). Global existence of solutions for shear flow of certain viscoelastic fluids. *J. Math. Fluid Mech.* **11**, 91–99.

RUSTEN, T., WINTHER, R. (1992). A preconditioned iterative method for saddlepoint problems. *SIAM J. Matrix Anal. Appl.* **13** (3), 887–904.

SCOTT, L., VOGELIUS, M. (1985a). *Conforming finite element methods for incompressible and nearly incompressible continua.* Lectures in Appl. Math., Large-scale computations in fluid mechanics, Part 2 (La Jolla, Calif., 1983) Volume 22-2 (Amer. Math. Soc., Providence, RI).

SCOTT, L., VOGELIUS, M. (1985b). Norm estimates for a maximal right inverse of the divergence operator in spaces of piecewise polynomials. *RAIRO Modél. Math. Anal. Numér.* **19**, 111–143.

SIMO, J., HUGHES, T. (1998). *Computational inelasticity.* Interdisciplinary Applied Mathematics Volume 7 (Springer, New York).

SUN, J., PHAN-THIEN, N., TANNER, R.I. (1996). An adaptive viscoelastic stress splitting scheme and its applications: AVSS/SI and AVSS/SUPG. *J. Non-Newton. Fluid Mech.* **65**, 75–91.

SUN, J., SMITH, M.D., ARMSTRONG, R.C., BROWN, R.A. (1999). Finite element method for viscoelastic flows based on the discrete adaptive viscoelastic stress splitting and the discontinuous Galerkin method: DAVSS-G/DG. *J. Non-Newton. Fluid Mech.* **86**, 281–307.

SURESHKUMAR, R., BERIS, A. (1995). Effect of artificial stress diffusivity on the stability of numerical calculations and the flow dynamics of time-dependent viscoelastic flows. *J. Non-Newton. Fluid. Mech.* **60**, 53–80.

SURESHKUMAR, R., BERIS, A. (1997). Direct numerical simulation of the turbulent channel flow of a polymer solution. *Phys. Fluids* **9**, 743–755.

SZADY, M., SALAMON, T., LIU, A., BORNSIDE, D., ARMSTRONG, R., BROWN, R. (1995). A new mixed finite-element method for viscoelastic flows governed by differential constitutive-equations. *J. Non-Newton. Fluid. Mech.* **59**, 215–243.

SZERI, A. (2000). A deformation tensor model for nonlinear rheology of FENE polymer solutions. *J. Non-Newton. Fluid Mech.* **92**, 1–25.

TAYLOR, C., HOOD, P. (1973). A numerical solution of the Navier-Stokes equations using the finite element technique. *Int. J. Comput. Fluids.* **1**, 73–100.

THIEN N., TANNER, R. (1977). A new constitutive equation derived from network theory. *J. Non-Newton. Fluid Mech.* **2**, 353–365.

THIFFEAULT, J.-L. (2001). Covariant time derivatives for dynamical systems. *J. Phys. A* **34**, 5875–5885.

THOMASES, B., SHELLEY, M. (2007). Emergence of singular structures in Oldroyd-B fluids. *Phys. Fluids* **19**, 103103.

THOMASES, B., SHELLEY, M. (2009). A transition to mixing and oscillations in a Stokesian viscoelastic flow. *Phys Rev. Lett.* **103**, 094501.

VAITHIANATHAN, T., ROBERT, A., BRASSEUR, J.G., COLLINS, L.R. (2006). An improved algorithm for simulating three-dimensional viscoelastic turbulence. *J. Non-Newton. Fluid Mech.* **140**, 3–22.

VAITHIANATHAN, T., ROBERT, A., BRASSEUR, J.G., COLLINS, L.R. (2007). Polymer mixing in shear-driven turbulence. *J. Fluid Mech.* **585**, 487–497.

WAPPEROM, P., RENARDY, M. (2005). Numerical prediction of the boundary layers in the flow around a cylinder using a fixed velocity field. *J. Non-Newton. Fluid Mech.* **125**, 35–48.

XIE, X., XU, J., XUE, G. (2008). Uniformly-stable finite element methods for Darcy-Stokes-Brinkman models. *J. Comput. Math.* **26**, 437–455.

XU, J. (1992). Iterative methods by space decomposition and subspace correction. *SIAM Rev.* **34**, 581–613.

XU, J. (2009). Optimal algorithms for discretized partial differential equations. In: ICIAM 07—6th International Congress on Industrial and Applied Mathematics. *Eur. Math. Soc*, (Zürich), pp. 409–444.

XU, J. (2010). Fast Poisson-based solvers for linear and nonlinear PDEs. In: *Proceedings of International Congress of Mathematicians* (World Scientific Publishing Company, Hyderabad, India).

Positive Definiteness Preserving Approaches for Viscoelastic Flow of Oldroyd-B Fluids: Applications to a Lid-Driven Cavity Flow and a Particulate Flow

Tsorng-Whay Pan, Jian Hao, and Roland Glowinski

Department of Mathematics, University of Houston, Houston, TX 77204, USA

E-mail: pan@math.uh.edu, jnhao@nscu.edu, roland@math.uh.edu

1. Introduction

One of the difficulties (e.g., see BAAIJENS [1998], and KEUNINGS [2000]) for simulating viscoelastic flows is the breakdown of the numerical methods. It has been widely believed that the lack of positive definiteness preserving property of the conformation tensor at the discrete level during the *entire time integration* is one of the reasons for this breakdown. To preserve the positive definiteness property of the conformation tensor, a sophisticated third-order upwind positive only scheme was developed in SINGH and LEAL [1993] and was used in SINGH, JOSEPH, HELSA, GLOWINSKI and PAN [2000] when simulating the sedimentation of disks in an Oldroyd-B fluid. The following methods published recently also preserve the positive definiteness of the conformation tensor: In LOZINSKI and OWENS [2003], the authors factorized the conformation tensor to get $\boldsymbol{\sigma} = \mathbf{A}\mathbf{A}^T$ and then try to write down the equations for $\mathbf{A}$ approximately at the discrete level. Hence, the positive definiteness of the conformation tensor is forced with such an approach. In LEE and XU [2006], the authors have developed a unified numerical discretization framework that can be used for simulating most

Numerical Methods for Non-Newtonian Fluids
Special Volume (R. Glowinski and Jinchao Xu, Guest Editors) of
HANDBOOK OF NUMERICAL ANALYSIS, VOL. XVI
P.G. Ciarlet (Editor)

ISSN 1570-8659
DOI 10.1016/B978-0-444-53047-9.00005-8

of existing constitutive equations so that the positiveness of the conformation tensor at the continuous level can be extended to its discrete analog. In FATTAL and KUPFERMAN [2004], the constitutive equations were reformulated as equations for the matrix logarithm of the conformation tensor to preserve its positive definiteness. The main advantage of using the log-conformation tensor is that one can better resolve the exponential behavior of the conformation tensor in the region where there are singularities and boundary layers.

In this article, we discuss two numerical methods for simulating the time-dependent flow of Oldroyd-B fluids. Both methods preserve the positive definiteness of the conformation tensor at the discrete time level. In the first method, we have combined the factorization approach developed in LOZINSKI and OWENS [2003] with a fictitious domain/distributed Lagrange multiplier method (see, e.g., GLOWINSKI, PAN, HESLA, JOSEPH and PERIAUX [2001]) to simulate particulate flow in Oldroyd-B fluids. In the second method, we have combined the technique of log-conformation tensor in FATTAL and KUPFERMAN [2004] with an operator splitting scheme of the Lie type to simulate a time-dependent lid-driven cavity Stokes flow. Even though the lid-driven cavity flow has closed planar streamlines in a simple confined geometry, its conformation tensor does have sharp boundary layers attached to the lid at high Weissenberg numbers. In the following two sections, we present first the formulation of the problem and then the computational methodology and numerical results. The numerical results presented in this article show that both methods are stable and robust.

2. Particulate flow

2.1. Generalities

The motion of particles in non-Newtonian fluids is not only of fundamental theoretical interest but also of importance in many applications to industrial processes involving particle-laden materials (see, e.g., CHHABRA [1993], and MCKINLEY [2002]). Although numerical methods for simulating particulate flows in Newtonian fluids have been very successful, numerically simulating the motion of particulate flows in a viscoelastic fluid is quite complicated and challenging. There have been recent works on the simulation of the sedimentation of particles in viscoelastic fluids, such as Oldroyd-B fluids in, e.g., FENG, HUANG and JOSEPH [1996], HU, PATANKAR and ZHU [2001], HUANG, HU and JOSEPH [1998], SINGH, JOSEPH, HELSA, GLOWINSKI and PAN [2000], YU, PHAN-THIEN, FAN and TANNER [2002]; Oldroyd-B fluids with shear thinning in, e.g., HUANG, HU and JOSEPH [1998], YU, WACHS and PEYSSON [2006]: and viscoelastic fluids of the FENE-Dumbbells type in BINOUS and PHILLIPS [1999]. In FENG, HUANG and JOSEPH [1996], one used the finite element solver POLYFLOW to study the two-dimensional sedimentation of circular particles in an Oldroyd-B fluid; one obtained chains of two particles aligned with the direction of sedimentation, which is precisely the microstructure observed in actual experiments in JOSEPH, LIU, POLETTO and FENG [1994]. In HUANG, HU and JOSEPH [1998], an arbitrary Lagrangian–Eulerian (ALE) moving mesh technique (see also HU, PATANKAR and ZHU [2001]) was used to investigate the cross-stream migration and orientations of elliptic particles in Oldroyd-B fluids (with and without shear thinning); in this article, Huang, Hu and Joseph found that the orientation of elliptic particles depends on two critical numbers, namely the elasticity and Mach numbers. In SINGH, JOSEPH, HELSA, GLOWINSKI and PAN [2000], a fictitious domain/distributed Lagrange multiplier (FD/DLM) method for particulate flow of Oldroyd-B fluids was developed for fixed structured mesh by generalizing the FD/DLM methodologies developed for simulating particulate flows of Newtonian fluids (e.g., see GLOWINSKI [2003], GLOWINSKI, PAN, HESLA and JOSEPH [1999], and GLOWINSKI, PAN, HESLA, JOSEPH and PERIAUX [2001]). A sophisticated third-order upwind positive only scheme (see SINGH and LEAL [1993]) was used to keep the positivity of the conformation tensor in SINGH, JOSEPH, HELSA, GLOWINSKI and PAN [2000]; chains of two particles aligned with the direction of sedimentation were obtained, and for the case of multiple circular particles, many chains of two particles were found next to the channel walls. YU, PHAN-THIEN, FAN and TANNER [2002] modified the above DLM methods by using rectangular finite elements (with globally continuous piecewise bilinear (resp. piecewise constant) approximations for the velocity (resp. the pressure)); they used a third-order upwind-biased finite difference scheme to discretize the constitutive equation for the conformation tensor. In YU, WACHS and PEYSSON [2006], a different FD/DLM method with finite difference methods was developed and used to investigate the sedimentation of particles in an Oldroyd-B fluid with shear thinning.

One of the difficulties (e.g., see BAAIJENS [1998], and KEUNINGS [2000]) for simulating viscoelastic flows is the breakdown of the numerical methods. It has been widely believed that the lack of positive definiteness preserving property of the conformation tensor at the discrete level during the *entire time integration* is one of the reasons for the breakdown. In FENG, HUANG and JOSEPH [1996], HU, PATANKAR and ZHU [2001], HUANG, HU and JOSEPH [1998], YU, PHAN-THIEN, FAN and TANNER [2002], YU, WACHS and PEYSSON [2006], all dedicated to the simulation of particulate flow, no specific treatments have been used to preserve the positive definiteness property of the conformation tensor (or at least nothing was mentioned in these articles). To preserve the positive definiteness property of the conformation tensor, several methods have been published recently, and some can be combined easily with the FD/DLM method through operator splitting techniques. In LOZINSKI and OWENS [2003], one factorized the conformation tensor to get $\boldsymbol{\sigma} = \mathbf{A}\mathbf{A}^T$ and then try to write down the equations for $\mathbf{A}$ approximately at the discrete level. Hence, the positive definiteness of the conformation tensor is forced with such an approach. In LEE and XU [2006], one has developed a unified numerical discretization framework that can be used for simulating most of existing constitutive equations in a way that the positiveness of the conformation tensor at the continuous level can be extended to its discrete analog. In FATTAL and KUPFERMAN [2005], one reformulated the constitutive equations as equations for the matrix logarithm of the conformation tensor to preserve the property of the positive definiteness of the conformation tensor. The main advantage of using the log-conformation tensor is that one can better resolve the exponential behavior of the conformation tensor in the region where there are singularities and boundary layers.

In this article, we consider the numerical simulation of the sedimentation of circular particles in a two-dimensional channel filled with an Oldroyd-B fluid. A fictitious domain/distributed Lagrange multiplier method preserving positive definiteness of the conformation tensor has been developed. The fluid-particle system is treated implicitly using a combined weak formulation. The governing equations for the Oldroyd-B fluid are solved everywhere, including inside the particles, via a fictitious domain method. We use distributed Lagrange multipliers to force the flow inside the particles to be a rigid-body motion. An operator-splitting technique called the Lie's scheme in CHORIN, HUGHES, MARSDEN and

McCracken [1978] has been used to decouple the difficulties associated with the incompressibility, advection, rigid-body motion enforcement, and the terms in the constitutive equation. The resulting method is easy to implement and quite modular implying that different space and time approximations can be used to treat the various steps. By factoring the conformation tensor, which is the technique developed in Lozinski and Owens [2003], we solve the equivalent equations for the conformation tensor. The new scheme preserves the positive definiteness of the conformation tensor at the discrete time level. The advection terms have been decoupled from the rest and solved by a wave-like equation method, which does not introduce numerical dissipation. In the numerical simulations, we have considered the cases of one particle, two particles, and then several particles sedimenting in an Oldroyd-B fluid. Our results agree with the ones in the literature and with experimental observations for the cases of one or two particles as in Feng, Huang and Joseph [1996], Yu, Phan-Thien, Fan and Tanner [2002]. For the case of multiple particles, our results agree well with the observations that the particles form *stable long* chains parallel to the flow direction when the *Mach number* $M = \sqrt{\text{ReDe}}$ is less than 1 and the *elasticity number* $E = \text{De}/\text{Re}$ is greater than a critical value, which depends on the blockage ratio (see Huang, Hu and Joseph [1998]), while the chains of multiple particles were not obtained in Yu, Phan-Thien, Fan and Tanner [2002] and not stable enough in Singh, Joseph, Helsa, Glowinski and Pan [2000]. Here, $\text{Re} = \rho_f UD/\eta$ is the *Reynolds number*, $\text{De} = U\lambda_1/D$ is the *Deborah number*, ρ_f being the fluid density, U is the particle velocity, D is the particle diameter, η is the viscosity of the fluid, and λ_1 is the relaxation time of the fluid.

The remainder of Section 2 is organized as follows: in Section 2.2, we present a FD/DLM formulation for particulate flows in an Oldroyd-B fluid. Then, in Section 2.3, we discuss the operator splitting technique, the space and time discretization of the FD/DLM formulation, and how we apply the Lozinski and Owens' method to get the equivalent equations for the conformation tensor. In Section 2.4, the algorithms for solving the subproblems obtained from the operator splitting are discussed. In Section 2.5, numerical results for the cases of the sedimentation of one, two, three, and six particles are shown and commented.

2.2. *Mathematical formulation*

2.2.1. *The governing equations*

Let Ω be a bounded domain in $\mathbb{R}^2$, and let Γ be the boundary $\partial\Omega$ of Ω. We suppose that Ω is filled with a viscoelastic fluid of the Oldroyd-B type and density ρ_f and contains also N moving circular particles of density ρ_s (see Fig. 1). Let $B(t) = \cup_{i=1}^{N} B_i(t)$ where $B_i(t)$ is the ith solid particle in the fluid for $i = 1, \ldots, N$. We denote by $\gamma_i(t)$ the boundary $\partial B_i(t)$ of $B_i(t)$ for $i = 1, \ldots, N$. For $T > 0$, the governing equations for the fluid-particle system are

$$\rho_f\left(\frac{\partial \mathbf{u}}{\partial t} + (\mathbf{u}\cdot\nabla)\mathbf{u}\right) = \rho_f\mathbf{g} - \nabla p + 2\mu\nabla\cdot\mathbf{D}(\mathbf{u}) + \nabla\cdot\mathbf{T} \text{ in } \Omega\backslash\overline{B(t)},\ t\in(0,T), \quad (2.1)$$

$$\nabla\cdot\mathbf{u} = 0 \text{ in } \Omega\backslash\overline{B(t)},\ t\in(0,T), \quad (2.2)$$

$$\mathbf{u}(\mathbf{x},0) = \mathbf{u}_0(\mathbf{x}),\quad \forall\mathbf{x}\in\Omega\backslash\overline{B(0)},\quad \text{with } \nabla\cdot\mathbf{u}_0 = 0, \quad (2.3)$$

$$\mathbf{u} = \mathbf{g}_0 \quad \text{on } \Gamma\times(0,T),\ \text{with} \int_\Gamma \mathbf{g}_0\cdot\mathbf{n}\,d\Gamma = 0, \quad (2.4)$$

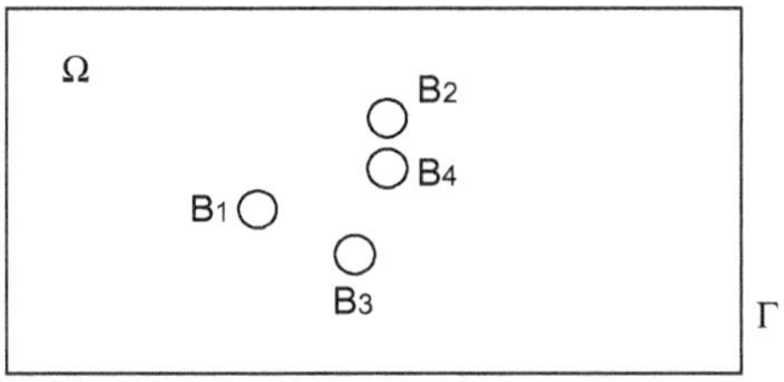

FIG. 1 Example of a two-dimensional flow region with four particles.

$$\mathbf{u} = \mathbf{V}_{p,i} + \omega_i \times \overrightarrow{\mathbf{G}_i\mathbf{x}}, \ \forall \mathbf{x} \in \partial B_i, \ i = 1, \ldots, N, \tag{2.5}$$

$$\frac{\partial \mathbf{T}}{\partial t} + (\mathbf{u} \cdot \nabla)\mathbf{T} - (\nabla \mathbf{u})\mathbf{T} - \mathbf{T}(\nabla \mathbf{u})^t + \frac{1}{\lambda_1}\mathbf{T} = 2\frac{\eta}{\lambda_1}\mathbf{D}(\mathbf{u}) \text{ in } \Omega\backslash\overline{B(t)}, \ t \in (0, T), \tag{2.6}$$

$$\mathbf{T}(\mathbf{x}, 0) = \mathbf{T}_0(\mathbf{x}), \ \mathbf{x} \in \Omega\backslash\overline{B(0)}, \tag{2.7}$$

$$\mathbf{T} = \mathbf{T}_L, \ \text{on } \Gamma^-, \tag{2.8}$$

where $\mathbf{u} = \{u_i\}_{i=1}^2$ is the flow velocity, p is the pressure, $\mathbf{g}$ is the gravity, $\mathbf{T}$ is the extra-stress tensor, $\mu = \eta_1\lambda_2/\lambda_1$ is the Newtonian viscosity of the fluid, $\eta = (\eta_1 - \mu)$ is the elastic viscosity of the fluid, η_1 is the fluid viscosity, λ_1 is the relaxation time of the fluid, λ_2 is the retardation time of the fluid, $\mathbf{n}$ is the outer normal unit vector at Γ, Γ^- is the upstream portion of Γ, $2\mathbf{D}(\mathbf{u}) = \nabla \mathbf{u} + (\nabla \mathbf{u})^t$, $(\mathbf{v} \cdot \nabla)\mathbf{w} = \left\{\sum_{j=1}^2 v_j \frac{\partial w_i}{\partial x_j}\right\}_{i=1}^2$. In (2.5) (which represents the no-slip condition on the boundary of the ith particle), $\mathbf{V}_{p,i}$ is the translation velocity, $\omega_i \times \overrightarrow{\mathbf{G}_i\mathbf{x}} = (-\omega_i(x_2 - G_{i,2}), \omega_i(x_1 - G_{i,1}))$ where ω_i is the angular velocity, $\mathbf{G}_i = (G_{i,1}, G_{i,2})$ is the center of mass, and $\mathbf{x} = \{x_i\}_{i=1}^2$ is the generic point on the boundary of the particle.

The motion of the particles is modeled by the Newton's laws:

$$M_i \frac{d\mathbf{V}_{p,i}}{dt} = M_i\mathbf{g} + \mathbf{F}_i + \mathbf{F}_i^r, \tag{2.9}$$

$$I_i \frac{d\omega_i}{dt} = F_i^t, \tag{2.10}$$

$$\frac{d\mathbf{G}_i}{dt} = \mathbf{V}_{p,i}, \tag{2.11}$$

$$\mathbf{G}_i(0) = \mathbf{G}_i^0, \ V_{p,i}(0) = V_{p,i}^0, \ \omega_i(0) = \omega_i^0, \tag{2.12}$$

for $i = 1, \ldots, N$, where in (2.9)–(2.12), M_i and I_i are the mass and inertia of the ith particle, respectively; $\mathbf{F}_i$ and F_i^t denote, respectively, the force and the torque imposed on the ith particle by the fluid, and $\mathbf{F}_i^r$ is a short-range repulsion force imposed on the ith particle by other particles and the wall to prevent particle/particle and particle/wall penetration.

REMARK 2.1. Let us define the conformation tensor $\mathbf{T}'$ as $\mathbf{T}' = \mathbf{T} + (\eta/\lambda_1)\mathbf{I}$, where $\eta, \lambda_1 > 0$, and $\mathbf{I}$ is the identity matrix. The conformation tensor $\mathbf{T}'$ is symmetric and positive definite (see JOSEPH [1990]). Then, the constitutive equation can be written in terms of $\mathbf{T}'$ as

$$\frac{\partial \mathbf{T}'}{\partial t} + (\mathbf{u} \cdot \nabla)\mathbf{T}' - (\nabla \mathbf{u})\mathbf{T}' - \mathbf{T}'(\nabla \mathbf{u})^t + \frac{1}{\lambda_1}\mathbf{T}' = \frac{\eta}{\lambda_1^2}\mathbf{I}. \tag{2.13}$$

There is advantage at using (2.13) when applying the factorization approach.

REMARK 2.2. The force $\mathbf{F}_i$ and the torque F_i^t imposed on the ith particle by the fluid are given by

$$\mathbf{F}_i = -\int_{\partial B_i} \boldsymbol{\sigma}\mathbf{n}\, ds,$$

$$F_i^t = -\int_{\partial B_i} \mathbf{Gx} \times \boldsymbol{\sigma}\mathbf{n}\, ds,$$

where $\boldsymbol{\sigma}$ is the stress tensor, $\mathbf{x}$ is the generic point on the boundary of the ith particle, $\mathbf{n}$ is the unit normal vector on the boundary of the particle, pointing to the center of the particle, and $\mathbf{a} \times \mathbf{b} = (a_1b_2 - a_2b_1)\ \vec{\mathbf{e}}_3$ (where $\vec{\mathbf{e}}_3 = (0, 0, 1)$).

REMARK 2.3. It is almost impossible to simulate the motion of multiple particles or particles close to a wall without repulsive forces to prevent the particle/particle and particle/wall penetration. A simple way is to define a safe zone around a particle such that when the particle/particle or particle/wall gap is smaller than some threshold a repulsive force is activated. To prevent particles from penetrating each other or the four walls $\Gamma_1, \Gamma_2, \Gamma_3, \Gamma_4$, we adopt the following collision strategy (see, e.g., GLOWINSKI, PAN, HESLA and JOSEPH [1999], GLOWINSKI, PAN, HESLA, JOSEPH and PERIAUX [2001], and GLOWINSKI [2003, chapter 8]). Assume that the particles are circular. For the particle–particle repulsive force, we take

$$\mathbf{F}_{i,j}^p = \begin{cases} 0, & \text{if } d_{i,j} > R_i + R_j + \rho_0, \\ \dfrac{1}{\epsilon_p}(\mathbf{G}_i - \mathbf{G}_j)(R_i + R_j + \rho_0 - d_{i,j})^2, & \text{if } d_{i,j} \le R_i + R_j + \rho_0, \end{cases}$$

where $d_{i,j} = |\mathbf{G}_i - \mathbf{G}_j|$ is the distance between the center of the ith particle and that of the jth particle, R_i is the radius of the ith particle, ρ_0 is the force range, and ϵ_p is a given small "stiffness" parameter.

For the particle-wall repulsive force, we take

$$\mathbf{F}_{i,j}^w = \begin{cases} 0, & \text{if } d_{i,j} > 2R_i + \rho_0, \\ \dfrac{1}{\epsilon_w}(G_i - G_i')(2R_i + \rho_0 - d_i)^2, & \text{if } d_{i,j} \le 2R_i + \rho_0, \end{cases}$$

where $d_{i,j} = |G_i - G_i'|$ is the distance between the center of the ith particle and that of the virtual particle which is on the other side of the wall Γ_j and tangent to the wall so that the line segment joining two centers is perpendicular to the wall, and ϵ_w is an another small stiffness parameter. We sum up all above forces to get $\mathbf{F}_i^r$ in (2.9).

For those readers wondering how to adjust the stiffness parameters and the force range ρ_0, see GLOWINSKI, PAN, HESLA, JOSEPH and PERIAUX [2001], GLOWINSKI [2003, chapter 8], for details.

2.2.2. A fictitious domain formulation

To derive a *fictitious domain–based variational formulation* for the governing equations of the particulate flow described in Section 2.2.1, we consider, for simplicity, only one solid particle in the fluid. The principle of this derivation is relatively simple; it relies on the following steps (see, e.g., GLOWINSKI, PAN, HESLA and JOSEPH [1999], and GLOWINSKI [2003, chapter 8]):

(a) Start from the following combined weak formulation (of the *virtual power* type):

$$\begin{cases} \rho_f \displaystyle\int_{\Omega\backslash\overline{B(t)}} \left[\frac{\partial \mathbf{u}}{\partial t} + (\mathbf{u}\cdot\nabla)\mathbf{u}\right]\cdot \mathbf{v}\,d\mathbf{x} + 2\mu \int_{\Omega\backslash\overline{B(t)}} \mathbf{D}(\mathbf{u}) : \mathbf{D}(\mathbf{v})\,d\mathbf{x} - \\ \displaystyle\int_{\Omega\backslash\overline{B(t)}} p\nabla\cdot\mathbf{v}\,d\mathbf{x} - \int_{\Omega\backslash\overline{B(t)}} \mathbf{v}\cdot(\nabla\cdot\mathbf{T}')\,d\mathbf{x} + \mathbf{M}\frac{d\mathbf{V}}{dt}\cdot\mathbf{Y} + I\frac{d\omega}{dt}\theta - \mathbf{F}^r\cdot\mathbf{Y} \\ = \rho_f \displaystyle\int_{\Omega\backslash\overline{B(t)}} \mathbf{g}\cdot\mathbf{v}\,d\mathbf{x} + \mathbf{M}\mathbf{g}\cdot\mathbf{Y}, \\ \forall\{\mathbf{v},\mathbf{Y},\theta\} \in (H^1(\Omega\setminus\overline{B(t)}))^2 \times \mathbb{R}^2 \times \mathbb{R}, \text{ and verifying} \\ \mathbf{v} = 0 \text{ on } \Gamma,\ \mathbf{v}(\mathbf{x}) = \mathbf{Y} + \theta \times \overrightarrow{\mathbf{G}(t)\mathbf{x}},\ \forall \mathbf{x}\in\partial B(t),\ t\in(0,T), \end{cases} \tag{2.14}$$

$$\int_{\Omega\backslash\overline{B(t)}} q\nabla\cdot\mathbf{u}(t)\,d\mathbf{x} = 0,\ \forall q \in L^2(\Omega\setminus\overline{B(t)}),\ \ t\in(0,T), \tag{2.15}$$

$$\mathbf{u} = \mathbf{g}_0 \text{ on } \Gamma, \tag{2.16}$$

$$\mathbf{u}(\mathbf{x},t) = \mathbf{V}(t) + \omega \times \overrightarrow{\mathbf{G}(t)\mathbf{x}},\ \forall\mathbf{x}\in\partial B(t),\ t\in(0,T), \tag{2.17}$$

$$\int_{\Omega\backslash\overline{B(t)}} \left(\frac{\partial \mathbf{T}'}{\partial t} + (\mathbf{u}\cdot\nabla)\mathbf{T}' - (\nabla\mathbf{u})\mathbf{T}' - \mathbf{T}'(\nabla\mathbf{u})^t + \frac{1}{\lambda_1}\mathbf{T}'\right) : \mathbf{s}\,d\mathbf{x} \tag{2.18}$$

$$= \frac{\eta}{\lambda_1^2}\int_{\Omega\backslash\overline{B(t)}} \mathbf{I} : \mathbf{s}\,d\mathbf{x},\ \forall\mathbf{s}\in(H^1(\Omega\setminus\overline{B(t)}))^{2\times 2},\quad \mathbf{s} = \mathbf{0} \text{ on } \Gamma^-,$$

$$\mathbf{T}' = \mathbf{T}'_L \text{ on } \Gamma^-, \tag{2.19}$$

$$\frac{d\mathbf{G}}{dt} = \mathbf{V}, \tag{2.20}$$

$$\mathbf{T}'(\mathbf{x},0) = \mathbf{T}'_0(x),\ \forall\mathbf{x}\in\Omega, \tag{2.21}$$

$$\mathbf{u}(\mathbf{x},0) = \mathbf{u}_0(\mathbf{x}),\ \forall\mathbf{x}\in\Omega\setminus\overline{B_0}, \tag{2.22}$$

$$\mathbf{G}(0) = \mathbf{G}_0,\ \mathbf{V}(0) = \mathbf{V}_0,\ \omega(0) = \omega_0,\ B(0) = B_0. \tag{2.23}$$

(b) Fill the particle B with the surrounding fluid.
(c) Impose a rigid-body motion to the fluid inside B.
(d) Modify the global weak formulation (2.14)–(2.23) accordingly, taking advantage of the fact that if $\mathbf{v}$ is a rigid-body motion velocity field, then $\nabla \cdot \mathbf{v} = 0$ and $\mathbf{D}(\mathbf{v}) = \mathbf{0}$, and the conformation tensor $\mathbf{T}'$ is constant inside the particle.
(e) Use a *Lagrange multiplier* defined over B to force the rigid-body motion inside B.

Assuming that B is made of a homogeneous material of density ρ_s, the "program" above leads to a fictitious domain formulation. To obtain such a fictitious domain formulation, we define first the following functional spaces

$$\mathbf{V}_{\mathbf{g}_0(t)} = \{\mathbf{v} \mid \mathbf{v} \in (H^1(\Omega))^2, \mathbf{v} = \mathbf{g}_0(t) \text{ on } \Gamma\},$$

$$L_0^2(\Omega) = \left\{q \mid q \in L^2(\Omega), \int_\Omega q \, d\mathbf{x} = 0\right\},$$

$$\Lambda(t) = H^1(B(t))^2,$$

$$\mathbf{V}_{\mathbf{T}'_L(t)} = \{\mathbf{T}' \mid \mathbf{T}' \in (H^1(\Omega))^{2\times 2}, \mathbf{T}' = \mathbf{T}'_L(t) \text{ on } \Gamma^-\},$$

$$\mathbf{V}_{\mathbf{T}'_0} = \{\mathbf{T}' \mid \mathbf{T}' \in (H^1(\Omega))^{2\times 2}, \mathbf{T}' = 0 \text{ on } \Gamma^-\}.$$

The fictitious domain formulation of the governing equations reads as follows:

For a.e. $t > 0$, find $\mathbf{u}(t) \in \mathbf{V}_{\mathbf{g}_0(t)}$, $p(t) \in L_0^2(\Omega)$, $\mathbf{T}'(t) \in \mathbf{V}_{\mathbf{T}'_L(t)}$, $\mathbf{V}(t) \in \mathbb{R}^2$, $\mathbf{G}(t) \in \mathbb{R}^2$, $\omega(t) \in \mathbb{R}$, $\boldsymbol{\lambda}(t) \in \Lambda(t)$ such that

$$\begin{cases} \rho_f \int_\Omega \left[\frac{\partial \mathbf{u}}{\partial t} + (\mathbf{u} \cdot \nabla)\mathbf{u}\right] \cdot \mathbf{v} \, d\mathbf{x} + 2\mu \int_\Omega \mathbf{D}(\mathbf{u}) : \mathbf{D}(\mathbf{v}) \, d\mathbf{x} - \int_\Omega p \nabla \cdot \mathbf{v} \, d\mathbf{x} \\ - \int_\Omega \mathbf{v} \cdot (\nabla \cdot \mathbf{T}') \, d\mathbf{x} + (1 - \rho_f/\rho_s) \left[\mathbf{M}\frac{d\mathbf{V}}{dt} \cdot \mathbf{Y} + I\frac{d\omega}{dt}\theta\right] \\ - < \boldsymbol{\lambda}, \mathbf{v} - \mathbf{Y} - \theta \times \overrightarrow{\mathbf{G}(t)\mathbf{x}} >_{B(t)} - \mathbf{F}^r \cdot \mathbf{Y} \\ = \rho_f \int_\Omega \mathbf{g} \cdot \mathbf{v} d\mathbf{x} + (1 - \rho_f/\rho_s)\mathbf{M}\mathbf{g} \cdot \mathbf{Y}, \\ \forall \{\mathbf{v}, \mathbf{Y}, \theta\} \in (H_0^1(\Omega))^2 \times \mathbb{R}^2 \times \mathbb{R}, \quad \text{a.e. } t \in (0, T), \end{cases} \tag{2.24}$$

(2.25)

$$\int_\Omega q \nabla \cdot \mathbf{u}(t) \, d\mathbf{x} = 0, \forall q \in L^2(\Omega), \quad \text{a.e. } t \in (0, T), \tag{2.26}$$

$$< \boldsymbol{\mu}, \mathbf{u}(\mathbf{x}, t) - \mathbf{V}(t) - \omega \times \overrightarrow{\mathbf{G}(t)\mathbf{x}} >_{B(t)} = 0, \quad \forall \boldsymbol{\mu} \in \boldsymbol{\Lambda}(t), \text{ a.e. } t \in (0, T), \tag{2.27}$$

$$\int_\Omega \left(\frac{\partial \mathbf{T}'}{\partial t} + (\mathbf{u} \cdot \nabla)\mathbf{T}' - (\nabla \mathbf{u})\mathbf{T}' - \mathbf{T}'(\nabla \mathbf{u})^t + \frac{1}{\lambda_1}\mathbf{T}'\right) : \mathbf{s} \, d\mathbf{x} \tag{2.28}$$

$$= \frac{\eta}{\lambda_1^2} \int_\Omega \mathbf{I} : \mathbf{s} \, d\mathbf{x}, \forall \mathbf{s} \in \mathbf{V}_{\mathbf{T}'_0}, \quad \text{and} \quad \mathbf{T}' = (\eta/\lambda_1)\mathbf{I} \quad \text{in} \quad B(t),$$

$$\frac{d\mathbf{G}}{dt} = \mathbf{V}, \tag{2.29}$$

$$\mathbf{T}'(\mathbf{x}, 0) = \mathbf{T}_0'(x), \forall \mathbf{x} \in \Omega, \ \mathbf{G}(0) = \mathbf{G}_0, \ \mathbf{V}(0) = \mathbf{V}_0, \ \omega(0) = \omega_0, \ B(0) = B_0, \tag{2.30}$$

$$\mathbf{u}(\mathbf{x}, 0) = \begin{cases} \mathbf{u}_0(\mathbf{x}), \forall \mathbf{x} \in \Omega \setminus \overline{B_0}, \\ \mathbf{V}_0 + \omega_0 \times \overrightarrow{\mathbf{G}_0\mathbf{x}}, \ \forall \mathbf{x} \in \overline{B_0}. \end{cases} \tag{2.31}$$

From a theoretical point of view, a natural choice for $< \cdot, \cdot >_{B(t)}$ is provided by, e.g.,

$$< \boldsymbol{\mu}, \mathbf{v} >_{B(t)} = \int_{B(t)} [\boldsymbol{\mu} \cdot \mathbf{v} + l^2 \mathbf{D}(\boldsymbol{\mu}) : \mathbf{D}(\mathbf{v})] \, d\mathbf{x}, \tag{2.32}$$

where l is a characteristic length (the diameter of B, for example). From a practical point of view, when it come, to space discretization, a simple and efficient strategy is discussed in the following section (also see, e.g., GLOWINSKI, PAN, HESLA, JOSEPH and PERIAUX [2001], and GLOWINSKI [2003, chapter 8]).

REMARK 2.4. Since, in the Eqn (2.24), $\mathbf{u}$ is divergence free and satisfies Dirichlet boundary conditions on Γ, we have

$$2 \int_{\Omega} \mathbf{D}(\mathbf{u}) : \mathbf{D}(\mathbf{v}) d\mathbf{x} = \int_{\Omega} \nabla \mathbf{u} : \nabla \mathbf{v} d\mathbf{x}, \ \forall \mathbf{v} \in (H_0^1(\Omega))^2. \tag{2.33}$$

This is a substantial simplification from a computational point of view, which is another advantage of the fictitious domain approach. With this simplification, we can use fast solvers for the elliptic problems in order to speed up computations, as shown in the following section. Also the gravity $\mathbf{g}$ in (2.24) can be absorbed in the pressure.

2.3. *Numerical methods and operator splitting scheme*

2.3.1. *Finite element approximation*

In order to solve problem (2.24)–(2.31) numerically, we shall discretize Ω using a regular finite element triangulation $\mathcal{T}_h$ for the velocity and conformation tensor, where h is the mesh size, and a twice coarser triangulation $\mathcal{T}_{2h}$ for the pressure. Practically, we should construct first the coarse triangulation, $\mathcal{T}_{2h}$, and then construct the finer triangulation, $\mathcal{T}_h$, by joining the midpoints of each triangle in $\mathcal{T}_{2h}$ as shown in Fig. 2. Thus, each element of $\mathcal{T}_{2h}$ contains four elements of $\mathcal{T}_h$.

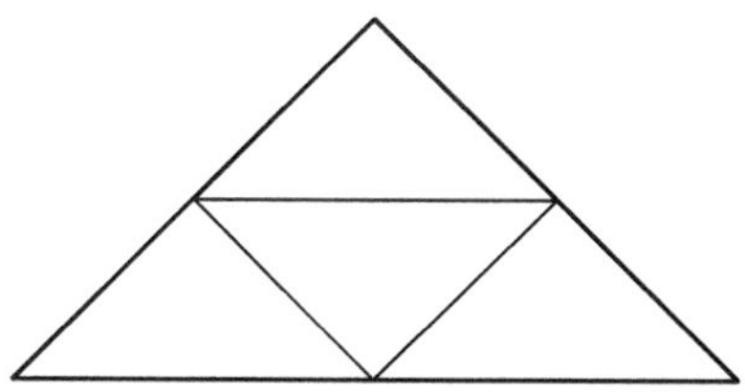

FIG. 2 Subdivision of a triangle of $\mathcal{T}_{2h}$.

The following finite dimensional spaces are defined for approximating $\mathbf{V}_{\mathbf{g}_0(t)}$, $(H_0^1(\Omega))^2$, $L^2(\Omega)$, $L_0^2(\Omega)$, $\mathbf{V}_{\mathbf{T}_L'(t)}$, $\mathbf{V}_{\mathbf{T}_0'}$, respectively,

$$\mathbf{V}_{\mathbf{g}_{0h}(t)} = \{\mathbf{v}_h \mid \mathbf{v}_h \in (C^0(\overline{\Omega}))^2, \mathbf{v}_h|_E \in (P_1)^2, \forall E \in \mathcal{T}_h, \mathbf{v}_h|_\Gamma = \mathbf{g}_{0h}(t)\},$$

$$\mathbf{V}_{0h} = \{\mathbf{v}_h \mid \mathbf{v}_h \in (C^0(\overline{\Omega}))^2, \mathbf{v}_h|_E \in (P_1)^2, \forall E \in \mathcal{T}_h, \mathbf{v}_h|_\Gamma = 0\},$$

$$L_h^2 = \{q_h \mid q_h \in C^0(\overline{\Omega}), q_h|_E \in P_1, \forall E \in \mathcal{T}_{2h}\},$$

$$L_{0h}^2 = \{q_h \mid q_h \in L_h^2, \int_\Omega q_h \, \mathbf{dx} = 0\},$$

$$\mathbf{V}_{\mathbf{T}_{Lh}'(t)} = \{s_h \mid s_h \in (C^0(\overline{\Omega}))^{2\times 2}, s_h|_E \in (P_1)^{2\times 2}, \forall E \in \mathcal{T}_h, s_h|_{\Gamma_h^-} = \mathbf{T}_{Lh}'(t)\},$$

$$\mathbf{V}_{\mathbf{T}_{0h}'} = \{s_h \mid s_h \in (C^0(\overline{\Omega}))^{2\times 2}, s_h|_E \in (P_1)^{2\times 2}, \forall E \in \mathcal{T}_h, s_h|_{\Gamma_h^-} = 0\},$$

where P_1 is the space of the polynomials in two variables of degree ≤ 1, $\mathbf{g}_{0h}(t)$ is an approximation of $\mathbf{g}_0$ satisfying $\int_\Gamma \mathbf{g}_{0h}(t) \cdot \mathbf{n} d\Gamma = 0$, and $\Gamma_h^- = \{\mathbf{x} \mid \mathbf{x} \in \Gamma,\ \mathbf{g}_{0h}(\mathbf{x}, t) \cdot \mathbf{n}(\mathbf{x}) < 0\}$.

We "approximate" $\mathbf{\Lambda}(t)$ by $\mathbf{\Lambda}_h(t)$ defined as follows: let $\{\mathbf{x}_i\}_{i=1}^K$ be a set of points from $\overline{B(t)}$ that covers $\overline{B(t)}$ "evenly" (see Fig. 3), and then, we define

$$\mathbf{\Lambda}_h(t) = \{\boldsymbol{\mu} \mid \boldsymbol{\mu} = \sum_{j=1}^K \boldsymbol{\mu}_j \delta(\mathbf{x} - \mathbf{x}_j),\ \boldsymbol{\mu}_j \in \mathbb{R}^2,\ \forall j = 1, \ldots, K\}, \tag{2.34}$$

where $\mathbf{x} \to \delta(\mathbf{x} - \mathbf{x}_j)$ is the Dirac measure at $\mathbf{x}_j$. Then, instead of the scalar product of $(H^1(B_h(t)))^2$, we shall use $< \boldsymbol{\mu}, \mathbf{v} >_{B_h(t)}$ defined by

$$< \boldsymbol{\mu}, \mathbf{v} >_{B_h(t)} = \sum_{j=1}^K \boldsymbol{\mu}_j \cdot \mathbf{v}(\mathbf{x}_j), \forall \boldsymbol{\mu} \in \mathbf{\Lambda}_h(t),\ \mathbf{v} \in \mathbf{V}_{\mathbf{g}_{0h}(t)} \text{ or } \mathbf{V}_{0h}. \tag{2.35}$$

Using the "scalar product" defined by (2.35) implies that the rigid-body motion of $B(t)$ is forced via a *collocation method*, which is also easier to implement than using finite element subspace to approximate $\mathbf{\Lambda}(t)$. In (2.34), $\mathbf{x} \to \delta(\mathbf{x} - \mathbf{x}_j)$ is the Dirac measure at $\mathbf{x}_j$, and the set $\{\mathbf{x}_j\}_{j=1}^K$ is the union of two subsets, namely: (i) The set of the points of the velocity grid contained in $B(t)$ and whose distance at $\partial B(t)$ is no less than ch, h being a space discretization step and c a constant ≈ 1. (ii) A set of control points located on $\partial B(t)$ and forming a mesh whose step size is of the order of h.

Using the finite dimensional spaces above leads to the following approximation of problem (2.24)–(2.31):

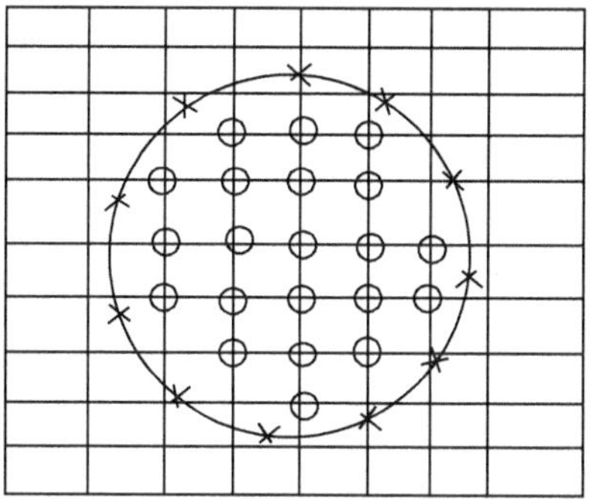

FIG. 3 Example of a set of collocation points: an union of a set of interior points (∘) and of a set of points on the boundary of the particle (×).

For a.e. $t > 0$, find $\mathbf{u}_h(t) \in \mathbf{V}_{\mathbf{g}_{0h}(t)}$, $p(t) \in L^2_{0h}$, $\mathbf{T}'_h(t) \in \mathbf{V}_{\mathbf{T}'_{Lh}(t)}$, $\mathbf{V}(t) \in \mathbb{R}^2$, $\mathbf{G}(t) \in \mathbb{R}^2$, $\omega(t) \in \mathbb{R}$, $\boldsymbol{\lambda}_h(t) \in \boldsymbol{\Lambda}_h(t)$ such that

$$\begin{cases} \rho_f \int_\Omega \left[\frac{\partial \mathbf{u}_h}{\partial t} + (\mathbf{u}_h \cdot \nabla)\mathbf{u}_h\right] \cdot \mathbf{v}\mathrm{d}\mathbf{x} + \mu \int_\Omega \nabla \mathbf{u}_h : \nabla \mathbf{v}\mathrm{d}\mathbf{x} - \int_\Omega p\nabla \cdot \mathbf{v}\mathrm{d}\mathbf{x} \\ -\int_\Omega \mathbf{v} \cdot (\nabla \cdot \mathbf{T}'_h)\mathrm{d}\mathbf{x} + (1 - \rho_f/\rho_s)\left[\mathbf{M}\frac{\mathrm{d}\mathbf{V}}{\mathrm{d}t} \cdot \mathbf{Y} + I\frac{\mathrm{d}\omega}{\mathrm{d}t}\theta\right] \\ -\mathbf{F}^r \cdot \mathbf{Y} = (1 - \rho_f/\rho_s)\mathbf{M}\mathbf{g} \cdot \mathbf{Y} + < \boldsymbol{\lambda}_h, \mathbf{v} - \mathbf{Y} - \theta \times \overrightarrow{\mathbf{G}(t)\mathbf{x}} >_{B_h(t)}, \\ \forall \{\mathbf{v}, \mathbf{Y}, \theta\} \in \mathbf{V}_{0h} \times \mathbb{R}^2 \times \mathbb{R}, \end{cases} \tag{2.36}$$

$$\int_\Omega q\nabla \cdot \mathbf{u}_h(t)\,\mathrm{d}\mathbf{x} = 0, \forall q \in L^2_h, \tag{2.37}$$

$$< \boldsymbol{\mu}, \mathbf{u}_h(t) - \mathbf{V}(t) - \omega(t) \times \overrightarrow{\mathbf{G}(t)\mathbf{x}} >_{B_h(t)} = 0, \forall \boldsymbol{\mu} \in \boldsymbol{\Lambda}_h(t), \tag{2.38}$$

$$\int_\Omega \left(\frac{\partial \mathbf{T}'_h}{\partial t} + (\mathbf{u}_h \cdot \nabla)\mathbf{T}'_h - (\nabla \mathbf{u}_h)\mathbf{T}'_h - \mathbf{T}'_h(\nabla \mathbf{u}_h)^t + \frac{1}{\lambda_1}\mathbf{T}'_h\right) : \mathbf{s}_h\mathrm{d}\mathbf{x} \tag{2.39}$$

$$= \frac{\eta}{\lambda_1^2}\int_\Omega \mathbf{I} : \mathbf{s}_h\mathrm{d}\mathbf{x}, \forall \mathbf{s}_h \in \mathbf{V}_{\mathbf{T}'_{0h}}, \ \textit{and}\ \mathbf{T}'_h = (\eta/\lambda_1)\mathbf{I}\ \textit{in}\ B_h(t),$$

$$\frac{\mathrm{d}\mathbf{G}}{\mathrm{d}t} = \mathbf{V}, \tag{2.40}$$

$$\mathbf{T}'_h(\mathbf{x}, 0) = \mathbf{T}'_{0h}(x), \forall \mathbf{x} \in \Omega, \tag{2.41}$$

$$\mathbf{G}(0) = \mathbf{G}_0,\ \mathbf{V}(0) = \mathbf{V}_0,\ \omega(0) = \omega_0, \tag{2.42}$$

$$\mathbf{u}_h(\mathbf{x}, 0) = \mathbf{u}_{0h}(\mathbf{x}), \forall \mathbf{x} \in \Omega, \tag{2.43}$$

where $\mathbf{u}_{0h}$ is an approximation of $\mathbf{u}_0$ so that $\int_\Omega q\nabla \cdot \mathbf{u}_{0h}\mathrm{d}\mathbf{x} = 0,\ \forall q \in L^2_h$.

2.3.2. An operator splitting scheme

Consider the following initial value problem:

$$\frac{d\phi}{dt} + A(\phi) = 0 \text{ on } (0, T), \quad \phi(0) = \phi_0 \tag{2.44}$$

with $0 < T < +\infty$. We suppose that operator A has a decomposition such as $A = \sum_{j=1}^{J} A_j$ with $J \geq 2$.

Let $\tau (> 0)$ be a time-discretization step; we denote $n\tau$ by t^n. With ϕ^n denoting an approximation of $\phi(t^n)$, the *Lie's scheme* reads as follows:

$$\phi^0 = \phi_0; \tag{2.45}$$

then, for $n \geq 0$, assuming that ϕ^n is known, compute ϕ^{n+1} via

$$\begin{cases} \dfrac{d\phi}{dt} + A_j(\phi) = 0 \text{ on } (t^n, t^{n+1}), \\ \phi(t^n) = \phi^{n+(j-1)/J}; \ \phi^{n+j/J} = \phi(t^{n+1}), \end{cases} \tag{2.46}$$

for $j = 1, \ldots, J$.

The Lie's operator splitting scheme allows us to decouple the following difficulties:

(1) The incompressibility condition and the related unknown pressure
(2) The advection terms
(3) The rigid-body motion in $B_h(t)$ and the related DLM $\boldsymbol{\lambda}_h$.

Since the conformation tensor $\mathbf{T}'$ is symmetric and positive definite, by Cholesky factorization there exists a 2×2 lower triangular matrix $\mathbf{A}$ such that $\mathbf{T}' = \mathbf{A}\mathbf{A}^t$, with $\mathbf{A}^t$ the transpose of $\mathbf{A}$. Similarly, we can define finite dimensional spaces $\mathbf{V}_{\mathbf{A}_{Lh}(t)}$ and $\mathbf{V}_{\mathbf{A}_{0h}}$ for $\mathbf{A}$. We take advantage of the following lemma, when applying operator splitting.

LEMMA 2.1. *For the above $\mathbf{T}'$ and $\mathbf{A}$, given $\mathbf{u} \in \mathbb{R}^2$ and $\lambda_1 (> 0)$*

(a) If $\mathbf{A}$ satisfies the equation $\dfrac{d\mathbf{A}}{dt} + (\mathbf{u} \cdot \nabla)\mathbf{A} = \mathbf{0}$, then $\mathbf{T}'$ satisfies the equation $\dfrac{d\mathbf{T}'}{dt} + (\mathbf{u} \cdot \nabla)\mathbf{T}' = \mathbf{0}$;

(b) if $\mathbf{A}$ satisfies the equation $\dfrac{d\mathbf{A}}{dt} + \dfrac{1}{2\lambda_1}\mathbf{A} - (\nabla\mathbf{u})\mathbf{A} = \mathbf{0}$, then $\mathbf{T}'$ satisfies the equation $\dfrac{d\mathbf{T}'}{dt} + \dfrac{1}{\lambda_1}\mathbf{T}' - (\nabla\mathbf{u})\mathbf{T}' - \mathbf{T}'(\nabla\mathbf{u})^T = \mathbf{0}$.

PROOF. (a) Multiplying the equation by $\mathbf{A}^t$ to the right, and the transpose of the equation by $\mathbf{A}$ to the left, we have

$$\frac{d\mathbf{A}}{dt}\mathbf{A}^t + (\mathbf{u}\cdot\nabla)\mathbf{A}\mathbf{A}^t = \mathbf{0}, \quad (L1)$$

$$\mathbf{A}\frac{d\mathbf{A}^t}{dt} + \mathbf{A}(\mathbf{u}\cdot\nabla)\mathbf{A}^t = \mathbf{0}, \quad (L2)$$

Adding (L1) and (L2) gives

$$\frac{d(\mathbf{A}\mathbf{A}^t)}{dt} + (\mathbf{u}\cdot\nabla)(\mathbf{A}\mathbf{A}^t) = \mathbf{0}; \text{ that is, } \frac{d\mathbf{T}'}{dt} + (\mathbf{u}\cdot\nabla)(\mathbf{T}') = \mathbf{0}.$$

(b) Multiplying the equation by $\mathbf{A}^t$ to the right, and the transpose of the equation by $\mathbf{A}$ to the left, we have

$$\frac{d\mathbf{A}}{dt}\mathbf{A}^t + \frac{1}{2\lambda_1}\mathbf{A}\mathbf{A}^t - (\nabla\mathbf{u})\mathbf{A}\mathbf{A}^t = \mathbf{0}, \quad (L3)$$

$$\mathbf{A}\frac{d\mathbf{A}^t}{dt} + \frac{1}{2\lambda_1}\mathbf{A}\mathbf{A}^t - \mathbf{A}\mathbf{A}^t(\nabla\mathbf{u})^t = \mathbf{0}. \quad (L4)$$

Adding (L3) and (L4) gives

$$\frac{d(\mathbf{A}\mathbf{A}^t)}{dt} + \frac{1}{\lambda_1}\mathbf{A}\mathbf{A}^t - (\nabla\mathbf{u})\mathbf{A}\mathbf{A}^t - \mathbf{A}\mathbf{A}^t(\nabla\mathbf{u})^t = \mathbf{0},$$

or,

$$\frac{d(\mathbf{T}')}{dt} + \frac{1}{\lambda_1}\mathbf{T}' - (\nabla\mathbf{u})\mathbf{T}' - \mathbf{T}'(\nabla\mathbf{u})^t = \mathbf{0}.$$

□

Applying the Lie's scheme to the problem (2.36)–(2.43) with the above factorization and equations for $\mathbf{A}$, we obtain

$$\mathbf{u}^0 = \mathbf{u}_{0h}, \mathbf{T}'^0 = \mathbf{T}'_{0h}, \mathbf{G}^0 = \mathbf{G}_0, \mathbf{V}^0 = \mathbf{V}_0, \omega^0 = \omega_0 \text{ given,} \tag{2.47}$$

for $n \geq 0$, $\mathbf{u}^n$, $\mathbf{T}'^n$, $\mathbf{G}^n$, $\mathbf{V}^n$, ω^n being known, we compute $\mathbf{u}^{n+\frac{1}{5}}$, and $p^{n+\frac{1}{5}}$ via the solution of

$$\begin{cases} \rho_f \int\limits_\Omega \frac{\mathbf{u}^{n+\frac{1}{5}} - \mathbf{u}^n}{\Delta t} \cdot \mathbf{v}d\mathbf{x} - \int\limits_\Omega p^{n+\frac{1}{5}}\nabla\cdot\mathbf{v}d\mathbf{x} = 0, \forall\mathbf{v}\in\mathbf{V}_{0h}, \\ \int\limits_\Omega q\nabla\cdot\mathbf{u}^{n+\frac{1}{5}}\,d\mathbf{x} = 0, \forall q \in L_h^2; \mathbf{u}^{n+\frac{1}{5}} \in \mathbf{V}_{\mathbf{g}_{0h}}^{n+1}, p^{n+\frac{1}{5}} \in L_{0h}^2. \end{cases} \tag{2.48}$$

Next, we compute $\mathbf{u}^{n+\frac{2}{5}}$ and $\mathbf{A}^{n+\frac{2}{5}}$ via the solution of

$$\begin{cases} \rho_f \int_\Omega \frac{d\mathbf{u}(t)}{dt} \cdot \mathbf{v} d\mathbf{x} + \int_\Omega (\mathbf{u}^{n+\frac{1}{5}} \cdot \nabla)\mathbf{u}(t) \cdot \mathbf{v} d\mathbf{x} = 0, \forall \mathbf{v} \in \mathbf{V}_{0h}^{n+1,-}, \\ \mathbf{u}(t^n) = \mathbf{u}^{n+\frac{1}{5}}, \\ \mathbf{u}(t) \in \mathbf{V}_h, \mathbf{u}(t) = \mathbf{g}_{0h}(t^{n+1}) \text{ on } \Gamma^{n+1,-} \times [t^n, t^{n+1}], \end{cases} \tag{2.49}$$

$$\begin{cases} \int_\Omega \frac{d\mathbf{A}(t)}{dt} : \mathbf{s} d\mathbf{x} + \int_\Omega (\mathbf{u}^{n+\frac{1}{5}} \cdot \nabla)\mathbf{A}(t) : \mathbf{s} d\mathbf{x} = 0, \forall \mathbf{s} \in \mathbf{V}_{\mathbf{A}_{0h}}, \\ \mathbf{A}(t^n) = \mathbf{A}^n, \quad \text{where } \mathbf{A}^n(\mathbf{A}^n)^t = \mathbf{T}'^n, \\ \mathbf{A}(t) \in \mathbf{V}_{\mathbf{A}_{Lh}}^{n+1}, \ t \in [t^n, t^{n+1}], \end{cases} \tag{2.50}$$

and set $\mathbf{u}^{n+\frac{2}{5}} = \mathbf{u}(t^{n+1})$ and $\mathbf{A}^{n+\frac{2}{5}} = \mathbf{A}(t^{n+1})$, where $\Gamma^{n+1,-} = \{\mathbf{x} \in \Gamma, \mathbf{g}_{0h}(t^{n+1})(\mathbf{x}) \cdot \mathbf{n}(\mathbf{x}) < 0\}$, $\mathbf{V}_h = \{\mathbf{v}_h | \mathbf{v}_h \in (C^0(\overline{\Omega}))^2, \mathbf{v}_h|_E \in (P_1)^2, \forall E \in \mathcal{T}_h\}$, and $\mathbf{V}_{0h}^{n+1,-} = \{\mathbf{v} \in \mathbf{V}_h, \mathbf{v} = 0, \text{ on } \Gamma^{n+1,-}\}$.

Then, compute $\mathbf{u}^{n+\frac{3}{5}}$ and $\mathbf{A}^{n+\frac{3}{5}}$ via the solution of

$$\begin{cases} \rho_f \int_\Omega \frac{\mathbf{u}^{n+\frac{3}{5}} - \mathbf{u}^{n+\frac{2}{5}}}{\Delta t} \cdot \mathbf{v} d\mathbf{x} + \alpha\mu \int_\Omega \nabla \mathbf{u}^{n+\frac{3}{5}} : \nabla \mathbf{v} \, d\mathbf{x} = 0, \\ \forall \mathbf{v} \in \mathbf{V}_{0h}; \mathbf{u}^{n+\frac{3}{5}} \in \mathbf{V}_{\mathbf{g}_{0h}}^{n+1}, \end{cases} \tag{2.51}$$

$$\begin{cases} \int_\Omega \left(\frac{\mathbf{A}^{n+\frac{3}{5}} - \mathbf{A}^{n+\frac{2}{5}}}{\Delta t} - (\nabla \mathbf{u}^{n+\frac{2}{5}})\mathbf{A}^{n+\frac{3}{5}} + \frac{1}{2\lambda_1} \mathbf{A}^{n+\frac{3}{5}} \right) : \mathbf{s} \, d\mathbf{x} = 0, \\ \forall \mathbf{s} \in \mathbf{V}_{\mathbf{A}_{0h}}; \mathbf{A}^{n+\frac{3}{5}} \in \mathbf{V}_{\mathbf{A}_{Lh}}^{n+1}, \end{cases} \tag{2.52}$$

and set

$$\mathbf{T}'^{n+\frac{3}{5}} = \mathbf{A}^{n+\frac{3}{5}}(\mathbf{A}^{n+\frac{3}{5}})^t + \frac{\eta \Delta t}{\lambda_1^2}\mathbf{I}.$$

We predict the position and the translation velocity of the center of mass as follows:

Take $\mathbf{V}^{n+\frac{3}{5},0} = \mathbf{V}^n$ and $\mathbf{G}^{n+\frac{3}{5},0} = \mathbf{G}^n$; then predict the new position and translation velocity via the following subcycling and predicting-correcting technique:

For $k = 1, 2, \ldots, N$, compute

$$\hat{\mathbf{V}}^{n+\frac{3}{5},k} = \mathbf{V}^{n+\frac{3}{5},k-1} + (1 - \rho_f/\rho_s)^{-1} M^{-1} \mathbf{F}^r(\mathbf{G}^{n+\frac{3}{5},k-1})\Delta t/2N, \tag{2.53}$$

$$\hat{\mathbf{G}}^{n+\frac{3}{5},k} = \mathbf{G}^{n+\frac{3}{5},k-1} + (\Delta t/4N)(\hat{\mathbf{V}}^{n+\frac{3}{5},k} + \mathbf{V}^{n+\frac{3}{5},k-1}), \tag{2.54}$$

$$\mathbf{V}^{n+\frac{3}{5},k} = \mathbf{V}^{n+\frac{3}{5},k-1} \tag{2.55}$$
$$+ (1-\rho_f/\rho_s)^{-1}M^{-1}(\mathbf{F}^r(\hat{\mathbf{G}}^{n+\frac{3}{5},k}) + \mathbf{F}^r(\mathbf{G}^{n+\frac{3}{5},k-1}))\Delta t/4N,$$
$$\mathbf{G}^{n+\frac{3}{5},k} = \mathbf{G}^{n+\frac{3}{5},k-1} + (\Delta t/4N)(V^{n+\frac{3}{5},k} + V^{n+\frac{3}{5},k-1}), \tag{2.56}$$

end do;
let $\mathbf{V}^{n+\frac{3}{5}} = \mathbf{V}^{n+\frac{3}{5},N}$, $\mathbf{G}^{n+\frac{3}{5}} = \mathbf{G}^{n+\frac{3}{5},N}$.
Next compute $\{\mathbf{u}^{n+\frac{4}{5}}, \boldsymbol{\lambda}^{n+\frac{4}{5}}, \mathbf{V}^{n+\frac{4}{5}}, \omega^{n+\frac{4}{5}}\}$ via the solution of

$$\begin{cases} \rho_f \displaystyle\int_\Omega \frac{\mathbf{u}^{n+\frac{4}{5}} - \mathbf{u}^{n+\frac{3}{5}}}{\Delta t} \cdot \mathbf{v}d\mathbf{x} + \beta\mu \int_\Omega \nabla\mathbf{u}^{n+\frac{4}{5}} : \nabla\mathbf{v}d\mathbf{x} \\ +(1-\dfrac{\rho_f}{\rho_s})\left[M\dfrac{\mathbf{V}^{n+\frac{4}{5}} - \mathbf{V}^{n+\frac{3}{5}}}{\Delta t}\cdot\mathbf{Y} + I\dfrac{\omega^{n+\frac{4}{5}} - \omega^n}{\Delta t}\theta\right] \\ =< \boldsymbol{\lambda}^{n+\frac{4}{5}}, \mathbf{v} - \mathbf{Y} - \theta \times \overrightarrow{\mathbf{G}^{n+\frac{3}{5}}\mathbf{x}} >_{B_h^{n+\frac{3}{5}}} +(1-\rho_f/\rho_s)\mathbf{Mg}\cdot\mathbf{Y}, \\ \forall \mathbf{v} \in \mathbf{V}_{0h}, \mathbf{Y} \in \mathbb{R}^2, \theta \in \mathbb{R}, \\ < \boldsymbol{\mu}, \mathbf{u}^{n+\frac{4}{5}} - \mathbf{V}^{n+\frac{4}{5}} - \omega^{n+\frac{4}{5}} \times \overrightarrow{\mathbf{G}^{n+\frac{3}{5}}\mathbf{x}} >_{B_h^{n+\frac{3}{5}}} = 0, \forall \boldsymbol{\mu} \in \boldsymbol{\Lambda}_h^{n+\frac{3}{5}}, \\ \mathbf{u}^{n+\frac{4}{5}} \in \mathbf{V}_{\mathbf{g}_{0h}}^{n+1}, \boldsymbol{\lambda}^{n+\frac{4}{5}} \in \boldsymbol{\Lambda}_h^{n+\frac{3}{5}}, \end{cases} \tag{2.57}$$

and set $\mathbf{T}'^{n+\frac{4}{5}} = \mathbf{T}'^{n+\frac{3}{5}}$, and then let $\mathbf{T}'^{n+\frac{4}{5}} = (\eta/\lambda_1)\mathbf{I}$ in $B_h^{n+\frac{3}{5}}$.

Finally, take $\mathbf{V}^{n+1,0} = \mathbf{V}^{n+\frac{4}{5}}$ and $\mathbf{G}^{n+1,0} = \mathbf{G}^{n+\frac{3}{5}}$; then predict the final position and translation velocity as follows:

For $k = 1, 2, \ldots, N$, compute

$$\hat{\mathbf{V}}^{n+1,k} = \mathbf{V}^{n+1,k-1} + (1-\rho_f/\rho_s)^{-1}M^{-1}\mathbf{F}^r(\mathbf{G}^{n+1,k-1})\Delta t/2N, \tag{2.58}$$
$$\hat{\mathbf{G}}^{n+1,k} = \mathbf{G}^{n+1,k-1} + (\Delta t/4N)(\hat{\mathbf{V}}^{n+1,k} + \mathbf{V}^{n+1,k-1}), \tag{2.59}$$
$$\mathbf{V}^{n+1,k} = \mathbf{V}^{n+1,k-1} \tag{2.60}$$
$$+ (1-\rho_f/\rho_s)^{-1}M^{-1}(\mathbf{F}^r(\hat{\mathbf{G}}^{n+1,k}) + \mathbf{F}^r(\mathbf{G}^{n+1,k-1}))\Delta t/4N,$$
$$\mathbf{G}^{n+1,k} = \mathbf{G}^{n+1,k-1} + (\Delta t/4N)(\mathbf{V}^{n+1,k} + \mathbf{V}^{n+1,k-1}), \tag{2.61}$$

end do;
let $\mathbf{V}^{n+1} = \mathbf{V}^{n+1,N}$, $\mathbf{G}^{n+1} = \mathbf{G}^{n+1,N}$.

Then, compute $\mathbf{u}^{n+1}$ via the solution of

$$\begin{cases} \rho_f \displaystyle\int_\Omega \frac{\mathbf{u}^{n+1} - \mathbf{u}^{n+\frac{4}{5}}}{\Delta t}\cdot\mathbf{v}d\mathbf{x} + \gamma\mu\int_\Omega \nabla\mathbf{u}^{n+1} : \nabla\mathbf{v}d\mathbf{x} \\ = \displaystyle\int_\Omega \mathbf{v}\cdot(\nabla\cdot\mathbf{T}'^{n+\frac{4}{5}})d\mathbf{x}, \forall\mathbf{v} \in \mathbf{V}_{0h}; \mathbf{u}^{n+1} \in \mathbf{V}_{\mathbf{g}_{0h}}^{n+1}. \end{cases} \tag{2.62}$$

We complete the final step by setting $\mathbf{T}'^{n+1} = \mathbf{T}'^{n+\frac{4}{5}}$, and $\omega^{n+1} = \omega^{n+\frac{4}{5}}$.

In the above, $\mathbf{V}_{\mathbf{g}_{0h}}^{n+1} = \mathbf{V}_{\mathbf{g}_{0h}(t^{n+1})}$, $\mathbf{\Lambda}_h^{n+s} = \mathbf{\Lambda}_h(t^{n+s})$, $\mathbf{V}_{\mathbf{A}_{Lh}}^{n+1} = \mathbf{V}_{\mathbf{A}_{Lh}(t^{n+1})}$, $B_h^{n+s} = B_h(t^{n+s})$, $\alpha + \beta + \gamma = 1$, where $\alpha, \beta, \gamma \geq 0$.

REMARK 2.5. In the scheme above, at each time step n, the discretized conformation tensor $\mathbf{T}'^n$ is clearly symmetric, and positive definite, by construction.

REMARK 2.6. In the scheme (2.47)–(2.62), we have applied the backward Euler's method for the time discretization of (2.48), (2.51), (2.52), and (2.57). In (2.53)–(2.56) and (2.58)–(2.61), we have used a predicting-correcting scheme to obtain the position of the mass center and the translation velocity of the particle. In order to let the short-range repulsive discussed in Remark 2.3 be activated effectively, we have used N sub-time steps to move the particle during one time step.

2.4. *On the solutions of the subproblems from operator splitting*

Problem (2.48) is a "degenerated" quasi-Stokes problem; problems (2.49) and (2.50) are advection problems; problem (2.57) concerns the rigid-body motion enforcement. Problems (2.51) and (2.62) are classical elliptic problems, which can be solved by a matrix-free fast solver from FISHPAK by ADAMS, SWARZTRAUBER and SWEET [1980]. Problem (2.52) gives a simple equation at each grid point, which can be solved easily if we use trapezoidal quadrature rule to compute the integrals.

2.4.1. *Solution of the degenerated quasi-stokes subproblems*

Subproblem (2.48) can be viewed as a degenerated quasi-Stokes problem of the following form (some h and n have been dropped):

$$\alpha \int_\Omega \mathbf{u} \cdot \mathbf{v} d\mathbf{x} - \int_\Omega p \nabla \cdot \mathbf{v} d\mathbf{x} = \int_\Omega \mathbf{f} \cdot \mathbf{v} d\mathbf{x}, \forall \mathbf{v} \in \mathbf{V}_{0h}, \tag{2.63}$$

$$\int_\Omega q \nabla \cdot \mathbf{u} \, d\mathbf{x} = 0, \forall q \in L_h^2, \tag{2.64}$$

with $\{\mathbf{u}, p\} \in \mathbf{V}_{\mathbf{g}_{0h}} \times L_{0h}^2$, where $\alpha > 0$.

In (2.63)–(2.64) $\mathbf{u}$ can be interpreted as the L^2projection of $\mathbf{f}/\alpha$ on the subspace of $\mathbf{V}_{\mathbf{g}_{0h}}$ consisting of those functions satisfying

$$\int_\Omega q \nabla \cdot \mathbf{v} \, d\mathbf{x} = 0, \forall q \in L_h^2. \tag{2.65}$$

The pressure p is the Lagrange multiplier associated to the linear constraint in (2.64); p is not unique unless we specify an additional relation, for example, $p \in L_{0h}^2$.

The saddle point problem (2.63) and (2.64) can be solved by a Uzawa/preconditioned conjugate gradient algorithm operating in the space L_{0h}^2 (see, e.g., GLOWINSKI, PAN and PERIAUX [1998], and GLOWINSKI [2003]); this algorithm reads as follows:

Step 0: Initialization

$$p^0 \in L^2_{0h} \text{ is given;} \tag{2.66}$$

solve the projection problem:

$$\begin{cases} \alpha \int_\Omega \mathbf{u}^0 \cdot \mathbf{v} d\mathbf{x} = \int_\Omega \mathbf{f} \cdot \mathbf{v} d\mathbf{x} + \int_\Omega p^0 \nabla \cdot \mathbf{v} d\mathbf{x}, \forall \mathbf{v} \in \mathbf{V}_{0h}, \\ \mathbf{u}^0 \in \mathbf{V}_{\mathbf{g}_{0h}}, \end{cases} \tag{2.67}$$

then

$$\begin{cases} \int_\Omega r^0 q d\mathbf{x} = \int_\Omega q \nabla \cdot \mathbf{u}^0 d\mathbf{x}, \forall q \in L^2_h, \\ r^0 \in L^2_h, \end{cases} \tag{2.68}$$

and finally

$$\begin{cases} \int_\Omega \nabla g^0 \cdot \nabla q d\mathbf{x} = \int_\Omega r^0 q d\mathbf{x}, \forall q \in L^2_h, \\ g^0 \in L^2_{0h}. \end{cases} \tag{2.69}$$

Take

$$w^0 = g^0. \tag{2.70}$$

Step 1: Descent

Then, for $k \geq 0$, assuming that $\mathbf{u}^k$, p^k, r^k, g^k, w^k are known, compute $\mathbf{u}^{k+1}$, p^{k+1}, r^{k+1}, g^{k+1}, w^{k+1} as follows:
solve

$$\begin{cases} \alpha \int_\Omega \bar{\mathbf{u}}^k \cdot \mathbf{v} d\mathbf{x} = \int_\Omega w^k \nabla \cdot \mathbf{v} d\mathbf{x}, \forall \mathbf{v} \in \mathbf{V}_{0h}, \\ \bar{\mathbf{u}}^k \in \mathbf{V}_{0h}, \end{cases} \tag{2.71}$$

then

$$\begin{cases} \int_\Omega \bar{r}^k q d\mathbf{x} = \int_\Omega q \nabla \cdot \bar{\mathbf{u}}^k d\mathbf{x}, \forall q \in L^2_h, \\ \bar{r}^k \in L^2_h, \end{cases} \tag{2.72}$$

and finally

$$\begin{cases} \int_\Omega \nabla \bar{g}^k \cdot \nabla q d\mathbf{x} = \int_\Omega \bar{r}^k q d\mathbf{x}, \forall q \in L^2_h, \\ \bar{g}^k \in L^2_{0h}. \end{cases} \tag{2.73}$$

Compute

$$\rho_k = \frac{\int_\Omega r^k g^k \mathbf{dx}}{\int_\Omega \bar{r}^k w^k \mathbf{dx}}, \tag{2.74}$$

and then

$$\mathbf{u}^{k+1} = \mathbf{u}^k - \rho_k \bar{\mathbf{u}}^k, \tag{2.75}$$
$$p^{k+1} = p^k - \rho_k w^k, \tag{2.76}$$
$$g^{k+1} = g^k - \rho_k \bar{g}^k, \tag{2.77}$$
$$r^{k+1} = r^k - \rho_k \bar{r}^k. \tag{2.78}$$

Step 2: Convergence test and new descent direction
If

$$\frac{\int_\Omega r^{k+1} g^{k+1} \mathbf{dx}}{\int_\Omega r^0 g^0 \mathbf{dx}} \leq \epsilon,$$

take $p = p^{k+1}$ and $\mathbf{u} = \mathbf{u}^{k+1}$; otherwise, compute

$$\gamma_k = \frac{\int_\Omega r^{k+1} g^{k+1} \mathbf{dx}}{\int_\Omega r^k g^k \mathbf{dx}}, \tag{2.79}$$

and set

$$w^{k+1} = g^{k+1} + \gamma_k w^k. \tag{2.80}$$

Do $k = k + 1$ and go back to (2.71).

REMARK 2.7. In the above algorithm, problems (2.69) and (2.73) (preconditioned steps) are classical elliptic problems and have been solved by a matrix-free fast solver from FISHPAK.

2.4.2. *Solution of the advection subproblems*

Solving the pure advection problem (2.49) is a more delicate issue. We solve this advection problem by a wave-like equation method (as in DEAN and GLOWINSKI [1997], and DEAN, GLOWINSKI and PAN [1998]). After translation and dilation on the time axis, each component of the velocity vector **u** and of the tensor **A** is solution of a transport equation of the following type:

$$\begin{cases} \dfrac{\partial \varphi}{\partial t} + (\mathbf{U} \cdot \nabla)\varphi = 0 \text{ in } \Omega \times (0, 1), \\ \varphi(0) = \varphi_0, \ \varphi = g \text{ on } \Gamma^- \times (0, 1), \end{cases} \tag{2.81}$$

with $\nabla \cdot \mathbf{U} = 0$ and $\partial \mathbf{U}/\partial t = 0$ on $\Omega \times (0, 1)$. Thus, (2.81) is equivalent to the (formally) well-posed problem:

$$\begin{cases} \dfrac{\partial^2 \varphi}{\partial t^2} - \nabla \cdot ((\mathbf{U} \cdot \nabla \varphi)\mathbf{U}) = 0 \text{ in } \Omega \times (0, 1), \\ \varphi(0) = \varphi_0, \ \dfrac{\partial \varphi}{\partial t}(0) = -\mathbf{U} \cdot \nabla \varphi_0, \\ \varphi = g \text{ on } \Gamma^- \times (0, 1), (\mathbf{U} \cdot \mathbf{n})\left(\dfrac{\partial \varphi}{\partial t} + (\mathbf{U} \cdot \nabla)\varphi\right) = 0 \text{ on } (\Gamma \backslash \Gamma^-) \times (0, 1). \end{cases} \tag{2.82}$$

Solving the wave-like equation (2.82) by a classical finite element/time stepping method is quite easy since a variational formulation of (2.82) is given by

$$\begin{cases} \displaystyle\int_\Omega \frac{\partial^2 \varphi}{\partial t^2} v \, d\mathbf{x} + \int_\Omega (\mathbf{U} \cdot \nabla \varphi)(\mathbf{U} \cdot \nabla v) \, d\mathbf{x} \\ \displaystyle\quad + \int_{\Gamma \backslash \Gamma^-} \mathbf{U} \cdot \mathbf{n} \frac{\partial \varphi}{\partial t} v d\Gamma = 0, \ \forall v \in W_0, \\ \varphi(0) = \varphi_0, \ \dfrac{\partial \varphi}{\partial t}(0) = -\mathbf{U} \cdot \nabla \varphi_0, \\ \varphi = g \text{ on } \Gamma^- \times (0, 1), \end{cases} \tag{2.83}$$

with the test function space W_0 defined by

$$W_0 = \{v | \ v \in H^1(\Omega), \ v = 0 \text{ on } \Gamma^-\}.$$

Let H_h^1 be a C^0- conforming finite element subspace of $H^1(\Omega)$ as discussed in, e.g., CIARLET [1978], CIARLET [1991]. We define $W_{0h} = H_h^1 \cap W_0$; we suppose that $\lim_{h \to 0} W_{0h} = W_0$ in the usual finite element sense. Next, we define $\tau_1 > 0$ by $\tau_1 = \Delta t/Q$, where Q is a positive integer, and we discretize problem (2.83) by

$$\varphi^0 = \varphi_{0h} (\approx \varphi_0), \tag{2.84}$$

$$\begin{cases} \displaystyle\int_\Omega (\varphi^{-1} - \varphi^1) v \, d\mathbf{x} = 2\tau_1 \int_\Omega (\mathbf{U}_h \cdot \nabla \varphi^0) v \, d\mathbf{x}, \ \forall v \in W_{0h}, \\ \varphi^{-1} - \varphi^1 \in W_{0h}, \end{cases} \tag{2.85}$$

and for $q = 0, 1, \ldots, Q - 1$,

$$\begin{cases} \varphi^{q+1} \in H_h^1, \ \varphi^{q+1} = g_h \text{ on } \Gamma^-, \\ \displaystyle\int_\Omega \frac{\varphi^{q+1} + \varphi^{q-1} - 2\varphi^q}{\tau_1^2} v \, d\mathbf{x} + \int_\Omega (\mathbf{U}_h \cdot \nabla \varphi^q)(\mathbf{U}_h \cdot \nabla v) \, d\mathbf{x} \\ \displaystyle\quad + \int_{\Gamma \backslash \Gamma^-} \mathbf{U}_h \cdot \mathbf{n} \left(\frac{\varphi^{q+1} - \varphi^{q-1}}{2\tau_1} \right) v d\Gamma = 0, \ \forall v \in W_{0h}, \end{cases} \tag{2.86}$$

where $\mathbf{U}_h$ and g_h are the approximates of $\mathbf{U}$ and g, respectively.

REMARK 2.8. Scheme (2.84)–(2.86) is a centered scheme, which is formally second-order accurate with respect to space and time discretizations. To be stable, scheme (2.84)–(2.86) has to verify a condition such as

$$\tau_1 \leq ch,$$

with c of order of $1/||\mathbf{U}||$. Since the advection problem is decoupled from the other ones, we can choose a proper time step here so that the above condition is satisfied. If one uses the trapezoidal rule to compute the first and the third integrals in (2.86), the above scheme becomes explicit, i.e., φ^{q+1} is obtained via the solution of a linear system with diagonal matrix.

REMARK 2.9. Scheme (2.84)–(2.86) does not introduce numerical dissipation, unlike the *upwinding* schemes commonly used to solve transport problems like (2.81).

REMARK 2.10. Let us consider the homogeneous boundary condition, $\mathbf{U}_h|_\Gamma = \mathbf{0}$, and set $Q = 1$ in (2.84)–(2.86). Then, we have the following:

$$\begin{cases} \int\limits_\Omega \dfrac{\varphi^1 - \varphi^0}{\Delta t}\, v\,d\mathbf{x} + \int\limits_\Omega (\mathbf{U}_h \cdot \nabla\varphi^0)\, v\,d\mathbf{x} \\ = -\dfrac{\Delta t}{2}\int\limits_\Omega (\mathbf{U}_h \cdot \nabla\varphi^0)(\mathbf{U}_h \cdot \nabla v)\,d\mathbf{x},\ \forall v \in H_h^1;\ \varphi^1 \in H_h^1. \end{cases} \tag{2.87}$$

The right-hand-side term in (2.87) is a naturally built-in diffusion term only acting in the direction of streamlines. This extra term is also close to the one introduced in streamline-diffusion methods (e.g., see JOHNSON [1986]).

2.4.3. *Solution of the rigid-body motion enforcement problems*

Problem (2.57) has the following form:

Find $\mathbf{u} \in \mathbf{V}_{\mathbf{g}_{0h}}$, $\mathbf{V} \in \mathbb{R}^2$, $\omega \in \mathbb{R}$, $\boldsymbol{\lambda} \in \boldsymbol{\Lambda}_h$ satisfying

$$\begin{cases} \alpha \int\limits_\Omega \mathbf{u}\cdot\mathbf{v}\,d\mathbf{x} + \mu\int\limits_\Omega \nabla\mathbf{u} : \nabla\mathbf{v}\,d\mathbf{x} \\ +\left(1 - \dfrac{\rho_f}{\rho_s}\right)\left[M\dfrac{\mathbf{V} - \mathbf{V}_0}{\Delta t}\cdot\mathbf{Y} + I\dfrac{\omega - \omega_0}{\Delta t}\theta\right] \\ = \int\limits_\Omega \mathbf{f}\cdot\mathbf{v}\,d\mathbf{x} + (1 - \rho_f/\rho_s)M\mathbf{g}\cdot\mathbf{Y} + <\boldsymbol{\lambda}, \mathbf{v} - \mathbf{Y} - \theta\times\mathbf{r}>_{B_h}, \\ \forall\mathbf{v} \in \mathbf{V}_{0h}, \mathbf{Y} \in \mathbb{R}^2, \theta \in \mathbb{R}, \\ <\boldsymbol{\mu}, \mathbf{u} - \mathbf{V} - \omega\times\mathbf{r}>_{B_h} = 0, \forall\boldsymbol{\mu} \in \boldsymbol{\Lambda}_h, \end{cases} \tag{2.88}$$

where the center $\mathbf{G}$ of the mass of the particle B_h is assumed known and $\mathbf{r} = \overrightarrow{\mathbf{Gx}}$. A conjugate gradient method for solving (2.88) has been discussed in GLOWINSKI, PAN, HESLA and JOSEPH [1999]; this algorithm reads as follows:

Step 0: Initialization

Assume $\boldsymbol{\lambda}^0 \in \boldsymbol{\Lambda}_h$ is given.

Find $\mathbf{u}^0 \in \mathbf{V}_{\mathbf{g}_{0h}}$, $\mathbf{V}^0 \in \mathbb{R}^2$, and $\omega^0 \in \mathbb{R}$ satisfying

$$\alpha \int_\Omega \mathbf{u}^0 \cdot \mathbf{v}\,d\mathbf{x} + \mu \int_\Omega \nabla \mathbf{u}^0 : \nabla \mathbf{v}\,d\mathbf{x} = \int_\Omega \mathbf{f} \cdot \mathbf{v}\,d\mathbf{x} + < \boldsymbol{\lambda}^0, \mathbf{v} >_{B_h}, \forall \mathbf{v} \in \mathbf{V}_{0h}, \quad (2.89)$$

$$\left(1 - \frac{\rho_f}{\rho_s}\right) M \frac{\mathbf{V}^0 - \mathbf{V}_0}{\Delta t} \cdot \mathbf{Y} + < \boldsymbol{\lambda}^0, \mathbf{Y} >_{B_h} = (1 - \rho_f/\rho_s)\mathbf{M}\mathbf{g} \cdot \mathbf{Y}, \forall \mathbf{Y} \in \mathbb{R}^2, \quad (2.90)$$

$$\left(1 - \frac{\rho_f}{\rho_s}\right) I \frac{\omega^0 - \omega_0}{\Delta t} \theta + < \boldsymbol{\lambda}^0, \theta \times \mathbf{r} >_{B_h} = 0, \forall \theta \in \mathbb{R}. \quad (2.91)$$

Find $\mathbf{g}^0 \in \boldsymbol{\Lambda}_h$ satisfying

$$< \boldsymbol{\mu}, \mathbf{g}^0 >_{B_h} = < \boldsymbol{\mu}, \mathbf{u}^0 - \mathbf{V}^0 - \omega^0 \times \mathbf{r} >_{B_h}, \forall \boldsymbol{\mu} \in \boldsymbol{\Lambda}_h. \quad (2.92)$$

Set $\mathbf{w}^0 = \mathbf{g}^0$.
For $m = 0, 1, \ldots,$ assuming $\mathbf{u}^m$, $\mathbf{V}^m$, ω^m, $\boldsymbol{\lambda}^m$, $\mathbf{g}^m$, and $\mathbf{w}^m$ are known, compute $\mathbf{u}^{m+1}$, $\mathbf{V}^{m+1}$, ω^{m+1}, $\boldsymbol{\lambda}^{m+1}$, $\mathbf{g}^{m+1}$, and $\mathbf{w}^{m+1}$ as follows:

Step 1: Descent

Find $\bar{\mathbf{u}}^m \in \mathbf{V}_{0h}$, $\bar{\mathbf{V}}^m \in \mathbb{R}^2$, and $\bar{\omega}^m \in \mathbb{R}$ satisfying

$$\alpha \int_\Omega \bar{\mathbf{u}}^m \cdot \mathbf{v}\,d\mathbf{x} + \mu \int_\Omega \nabla \bar{\mathbf{u}}^m : \nabla \mathbf{v}\,d\mathbf{x} = < \mathbf{w}^m, \mathbf{v} >_{B_h}, \forall \mathbf{v} \in \mathbf{V}_{0h}, \quad (2.93)$$

$$\left(1 - \frac{\rho_f}{\rho_s}\right) M \frac{\bar{\mathbf{V}}^m}{\Delta t} \cdot \mathbf{Y} + < \mathbf{w}^m, \mathbf{Y} >_{B_h} = 0, \forall \mathbf{Y} \in \mathbb{R}^2, \quad (2.94)$$

$$\left(1 - \frac{\rho_f}{\rho_s}\right) I \frac{\bar{\omega}^m}{\Delta t} \theta + < \mathbf{w}^m, \theta \times \mathbf{r} >_{B_h} = 0, \forall \theta \in \mathbb{R}. \quad (2.95)$$

Find $\bar{\mathbf{g}}^m \in \boldsymbol{\Lambda}_h$ satisfying

$$< \boldsymbol{\mu}, \bar{\mathbf{g}}^m >_{B_h} = < \boldsymbol{\mu}, \bar{\mathbf{u}}^m - \bar{\mathbf{V}}^m - \bar{\omega}^m \times \mathbf{r} >_{B_h}, \forall \boldsymbol{\mu} \in \boldsymbol{\Lambda}_h. \quad (2.96)$$

Set

$$\rho_m = \frac{< \mathbf{g}^m, \mathbf{g}^m >_{B_h}}{< \mathbf{w}^m, \bar{\mathbf{g}}^m >_{B_h}}, \quad (2.97)$$

$$\boldsymbol{\lambda}^{m+1} = \boldsymbol{\lambda}^m - \rho_m \mathbf{w}^m, \quad (2.98)$$

$$\mathbf{u}^{m+1} = \mathbf{u}^m - \rho_m \bar{\mathbf{u}}^m, \quad (2.99)$$

$$\mathbf{V}^{m+1} = \mathbf{V}^m - \rho_m \bar{\mathbf{V}}^m, \quad (2.100)$$

$$\omega^{m+1} = \omega^m - \rho_m \bar{\omega}^m, \quad (2.101)$$

$$\mathbf{g}^{m+1} = \mathbf{g}^m - \rho_m \bar{\mathbf{g}}^m, \quad (2.102)$$

Step 2: Convergence test and new descent direction

If

$$\frac{< \mathbf{g}^{m+1}, \mathbf{g}^{m+1} >_{B_h}}{< \mathbf{g}^0, \mathbf{g}^0 >_{B_h}} \leq \epsilon,$$

take $\mathbf{u} = \mathbf{u}^{m+1}$, $\mathbf{V} = \mathbf{V}^{m+1}$, $\omega = \omega^{m+1}$, and $\boldsymbol{\lambda} = \boldsymbol{\lambda}^{m+1}$.
Otherwise, set

$$\gamma_m = \frac{< \mathbf{g}^{m+1}, \mathbf{g}^{m+1} >_{B_h}}{< \mathbf{g}^m, \mathbf{g}^m >_{B_h}}, \tag{2.103}$$

$$\mathbf{w}^{m+1} = \mathbf{g}^{m+1} + \gamma_m \mathbf{w}^m, \tag{2.104}$$

$$m = m + 1, \tag{2.105}$$

and go to (2.93).

The above algorithm, as it stands, cannot be used for simulating the cases of neutrally buoyant particles. A modified algorithm for neutrally buoyant particles can be found in PAN and GLOWINSKI [2002], PAN and GLOWINSKI [2005].

2.5. Numerical results

2.5.1. Sedimentation of a single particle

The test case of a single circular particle sedimenting in a channel filled with an Oldroyd-B fluid is considered. The channel is infinitely long and has a width of 1. The computational domain is $\Omega = (0, 1) \times (0, 6)$ initially and then moves down with the mass center of the particle (see HU [1996] for more details). The density of the fluid ρ_f is 1, the density of the disk ρ_s is 1.01, the viscosity of the fluid η_1 is 0.2, the relaxation time λ_1 is 1, the retardation time λ_2 is $0.25\lambda_1$, and the diameter of the disk D is 0.25 (the radius r is 0.125). The initial position of the disk center is (0.35, 2.5).

We first conducted convergence tests for the numerical solutions obtained from our algorithms. Some results are shown in Fig. 4. We can see that the numerical solutions obtained with different values of h and Δt basically converge to the same limit; a finer mesh and a smaller time step would result in more accurate solutions. A large overshoot of the settling velocities and the angular velocity, followed by a damped oscillation, can be observed in our simulations, which is consistent with previously reported results (e.g., see FENG, HUANG and JOSEPH [1996], and YU, PHAN-THIEN, FAN and TANNER [2002]). Also notice that the particle rotates counter-clockwised in an anomalous manner, as if rolling up the nearby wall (see SINGH and JOSEPH [2000]).

We then investigate the wall effects by releasing the particle at different distances from the wall. We have chosen the parameters used in FENG, HUANG and JOSEPH [1996]. The density of the fluid ρ_f is 1, the density of the disk ρ_s is 1.0007, the viscosity of the fluid η_1 is 0.034, the relaxation time λ_1 is 2.025, the retardation time λ_2 is $0.125\lambda_1$, and the diameter of the disk D is 0.25. Figure 5 shows the trajectories of a particle released from three different lateral positions $x_1 = 0.25$, 0.3, and 0.375. The particles from all three releases reached precisely the same lateral equilibrium position at $x_1 = 0.343$ (2.75 radii). The particles in

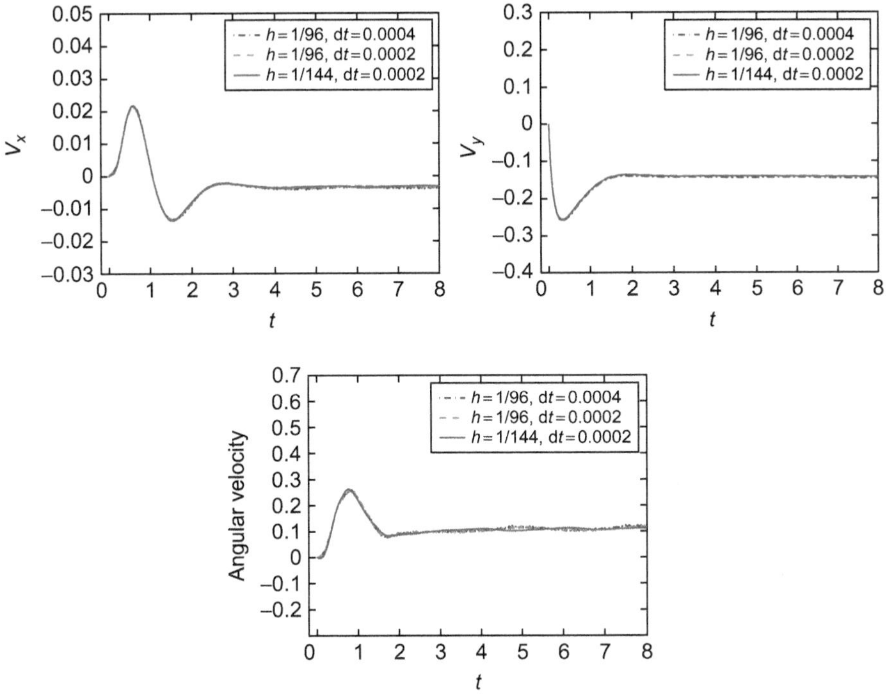

FIG. 4 Histories of the x_1-component (upper left) and x_2-component (upper right) of the translational velocity and of the angular velocity (bottom) of a circular particle in a channel for different mesh sizes and time steps; blockage ratio = 4; Re $= 0.18$, De $= 0.57$, $E = 3.2$, $M = 0.32$.

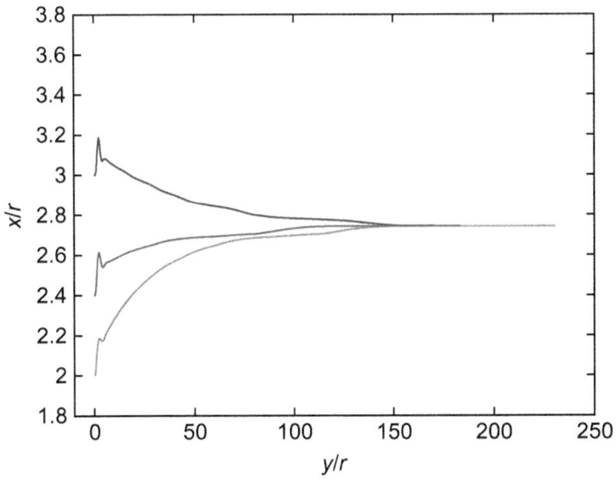

FIG. 5 Trajectory of a circular particle released from three different lateral positions computed with $h = 1/128$ and $\Delta t = 0.0004$; blockage ratio $= 4$; Re $= 0.42$, De $= 0.46$, $E = 1.1$, $M = 0.44$.

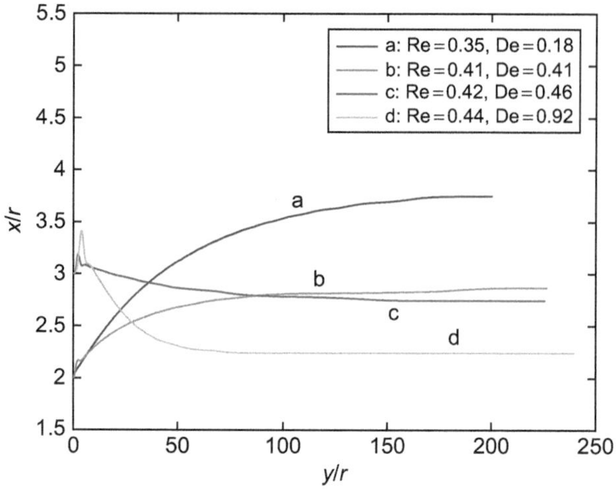

FIG. 6 The effect of the elasticity number E on the trajectory of a particle settling in a channel. $E = 0.5$, 1.0, 1.1, and 2.2 from (a) to (d), respectively, by changing the relaxation time; blockage ratio = 4.

all the cases were observed to be pushed away from the wall initially. At $t \simeq 1.6$, the farther particle was attracted to drift toward the wall and gradually approached the eccentric equilibrium position. These observations agree with the ones observed by FENG, HUANG and JOSEPH [1996]. The direction of the initial drift is opposite to the one observed by YU, PHAN-THIEN, FAN and TANNER [2002]. We then varied the relaxation time to investigate the effects of elasticity on the lateral equilibrium position. As shown in Fig. 6, the smaller the elasticity number E is, the closer to the centerline the equilibrium position is. On the other hand, the larger the elasticity number E is, the closer to the wall the equilibrium position is.

2.5.2. *Sedimentation of two particles*

The second test case concerns two circular particles sedimenting side by side in a channel filled with an Oldroyd-B fluid. The computational domain is $\Omega = (0, 1) \times (0, 6)$ initially and then moves down with the lower mass center of two particles. The density of the fluid ρ_f is 1, the density of the disk ρ_s is 1.01, the viscosity of the fluid η_1 is 0.2, the relaxation time λ_1 is 1, the retardation time λ_2 is $0.25\lambda_1$, and the diameter of the disk D is 0.25. The repulsion parameter ε is 2.5×10^{-5}. The safe zone parameter ρ_0 is h. The initial positions of the disks are (0.35, 2.5) and (0.65, 2.5). The mesh size for velocity and stress tensor is $h = 1/96$, and the time step is $\Delta t = 0.0004$. It is well known that the particles in this case will attract and approach each other, and the doublet rotates until the line of the centers is aligned with the falling direction (see, JOSEPH, LIU, POLETTO and FENG [1994], SINGH, JOSEPH, HELSA, GLOWINSKI and PAN [2000], and YU, PHAN-THIEN, FAN and TANNER [2002]), which is significantly unlike the phenomenon of *drafting, kissing, and tumbling* in Newtonian fluids.

Figure 7 gives the snapshots of the doublet at various moments of time, showing the phenomenon of *drafting, kissing, and chaining* for two particles in a viscoelastic fluid (here since the computational domain Ω moves with the lower mass center of the two particles by adding nodes to the bottom and removing them from the top of Ω during the simulation, the

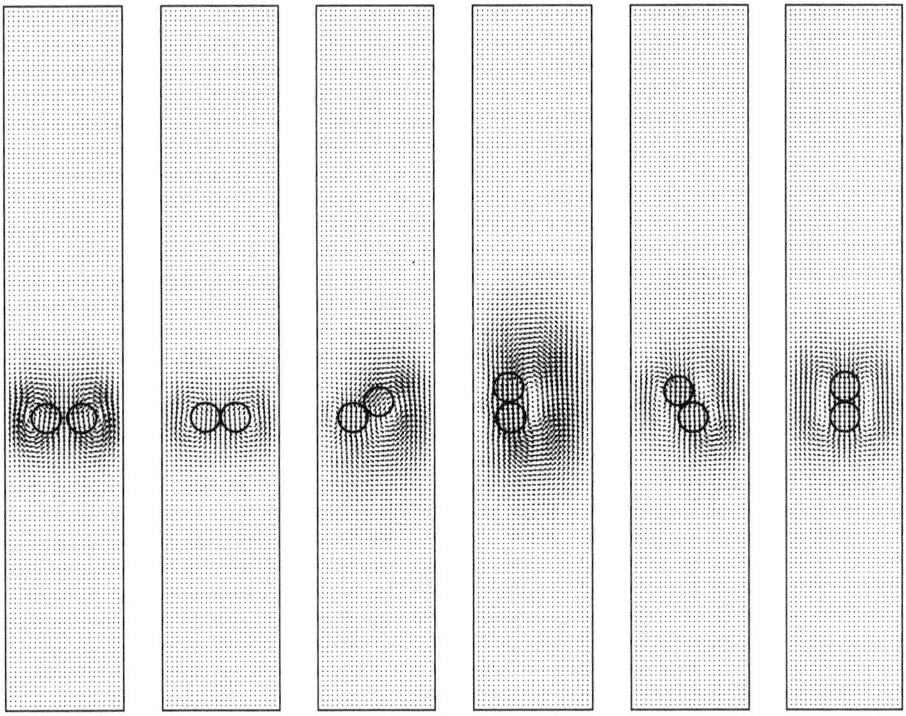

FIG. 7 Snapshots of the positions of the two particles at $t = 0.2$, 10, 19, 23, 32, and 90 (from left to right); $h = 1/96$ and $\triangle t = 0.0004$, showing the phenomenon of *drafting, kissing, and chaining* in a viscoelastic fluid; blockage ratio = 4; Re = 0.21, De = 0.66, $E = 3.2$, $M = 0.37$.

two particles seem stationary in those snapshots). The average terminal velocity is 0.166, the Reynolds number Re is 0.21, the Deborah number De is 0.66, the elasticity number E is 3.2, and the Mach number M is 0.37. Histories of the x_1 and x_2 coordinates of centers of the two particles are given in Fig. 8; histories of the x_1 and x_2 components of translational velocities and angular velocities of the two particles are given in Fig. 9. From the figures, we can see that after $t = 80$, the translational velocity $V_1 \approx 0$, the angular velocity $\omega \approx 0$, and the translational velocity V_2 is approximately a constant. The chain of the two particles was approximately on the centerline.

For the case of two particles sedimenting side by side in a viscoelastic fluid with initial separation short enough, the particles attract, kiss, and chain only when the elasticity number E is larger than the critical value, which depends on the blockage ratio, while the two particles behave like in a Newtonian fluid when the elasticity number E is less than the critical value, as pointed out in HUANG, HU and JOSEPH [1998]. The results of our simulations shown in Figs 10 and 11 have confirmed this claim. In the simulations, we used the above parameters except for the diameter of particles and the relaxation time. The diameter of the two particles is 0.125 so that the blockage ratio is 8. For the case in Fig. 10 with the relaxation time $\lambda_1 = 0.1$, its elasticity number $E = 0.16$ is less than the critical value; we can see that the two particles just break away and never touch each other during the simulation. They

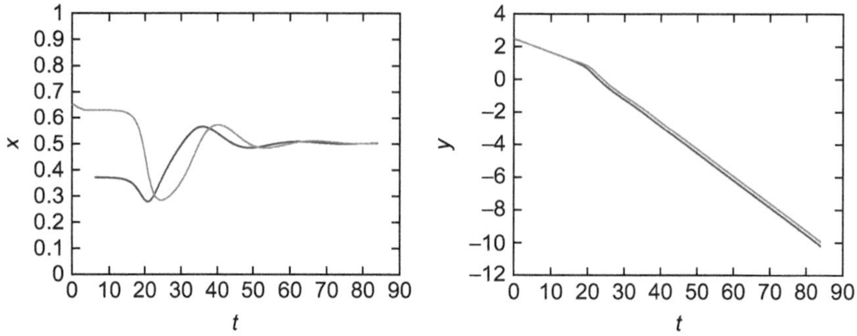

FIG. 8 Histories of the x_1 (left) and x_2 (right) components of centers of the two particles; $h = 1/96$ and $\triangle t = 0.0004$; blockage ratio = 4; Re = 0.21, De = 0.66, $E = 3.2$, $M = 0.37$.

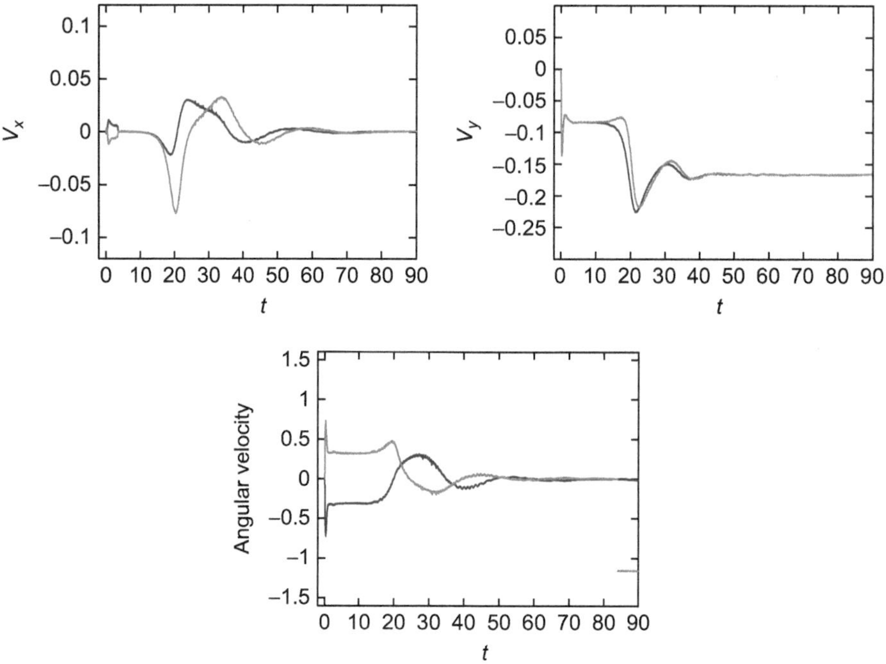

FIG. 9 Histories of the x_1 (upper left) and x_2 (upper right) components of velocities and angular velocity (bottom) of the two particles; $h = 1/96$ and $\triangle t = 0.0004$; blockage ratio = 4; Re = 0.21, De = 0.66, $E = 3.2$, $M = 0.37$.

behave as in a Newtonian fluid. For the case in Fig. 11 with the relaxation time $\lambda_1 = 1$, its elasticity number $E = 1.6$ is larger than the critical value; the two particles attract, kiss, and chain to form a doublet, and then, the doublet rotates until the line of the centers is aligned with the direction of sedimentation.

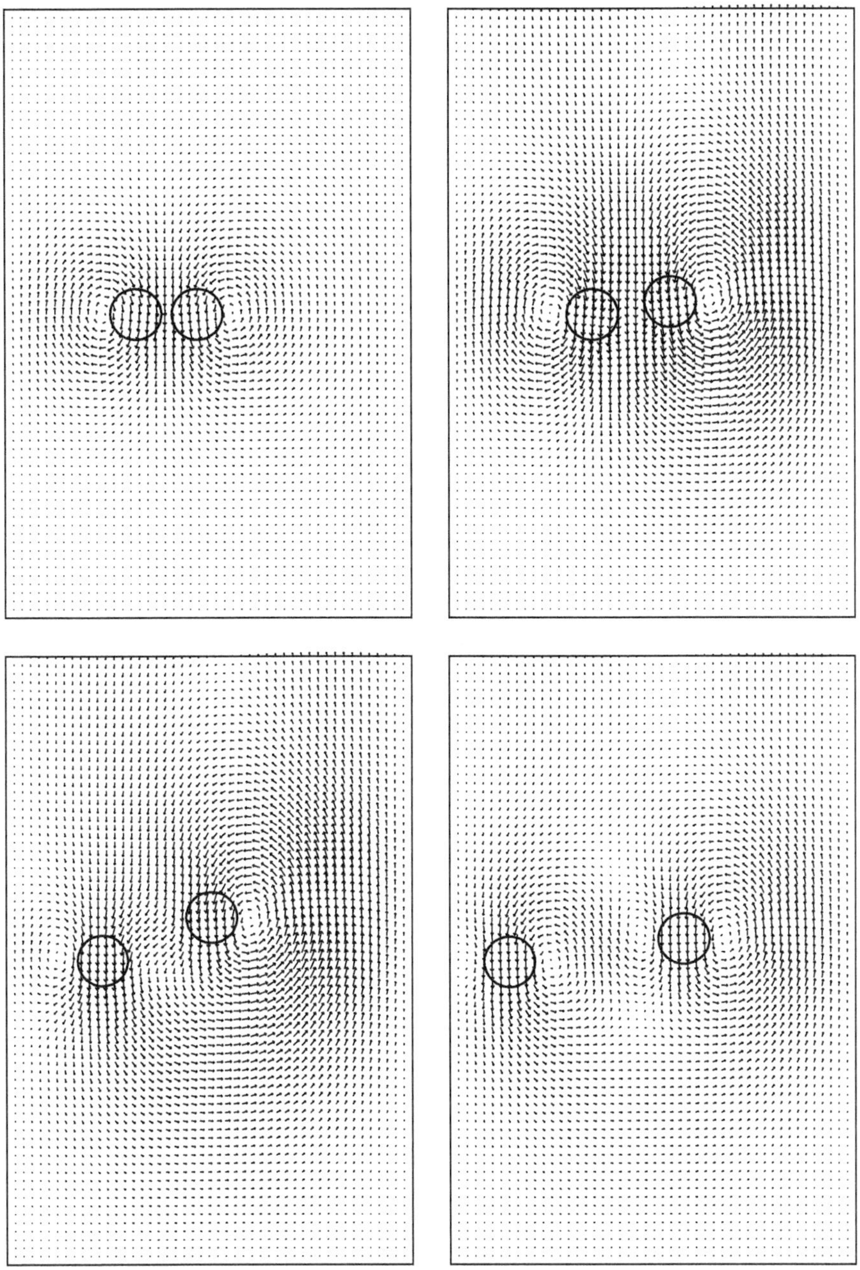

FIG. 10 Positions of the two particles at $t = 0.2$, 8, 24, 100 (from left to right and from top to bottom), with the elasticity number $E = 0.16$, which is less than the critical value, blockage ratio $= 8$; Re $= 0.37$, De $= 0.059$, and $M = 0.15$. The two particles behave like two circular particles sedimenting in a Newtonian fluid.

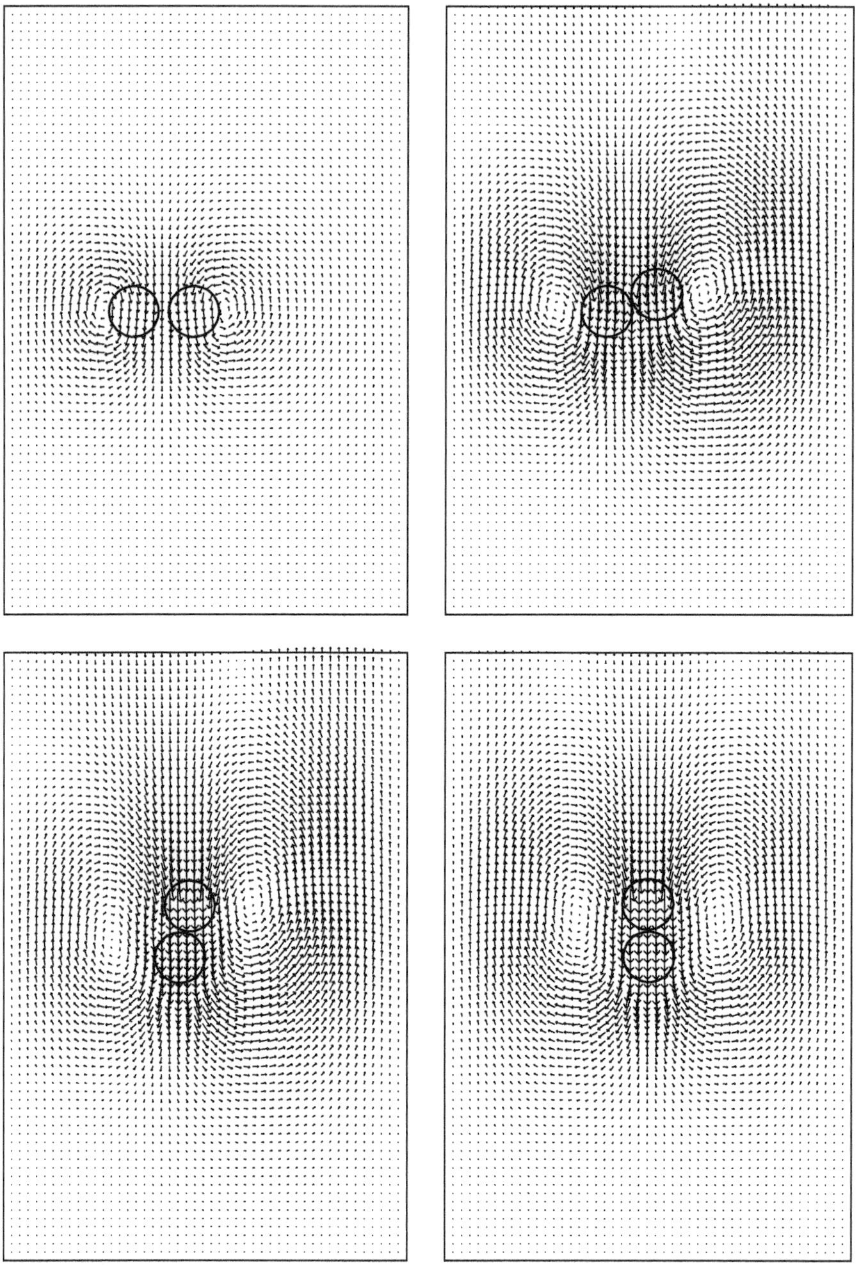

FIG. 11 Positions of the two particles at $t = 0.2$, 8, 22, 100 (from left to right and from top to bottom), with the elasticity number $E = 1.6$, which is larger than the critical value, blockage ratio = 8; Re = 0.85, De = 1.36, and $M = 1.07$; showing the phenomenon of *drafting, kissing, and chaining* in a viscoelastic fluid.

2.5.3. Sedimentation of three particles

The third test case concerns three circular particles sedimenting in a channel filled with an Oldroyd-B fluid. The channel is infinitely long and has a width of 2. The computational domain is $\Omega = (0, 2) \times (0, 5)$ initially and then moves down with the lowest mass center of three particles. The diameter of the particles is 0.25. The blockage ratio is 8. The density of the fluid is $\rho_f = 1$, the density of the disks is $\rho_s = 1.005$, the viscosity of the fluid is $\eta_1 = 0.26$, the relaxation time is $\lambda_1 = 1.5$, and the retardation time is $\lambda_2 = 0.125\lambda_1$. The initial positions of the disks are (0.6, 2.0), (0.88, 2.0), and (1.16, 2.0). The repulsion parameter is $\varepsilon = 1.0 \times 10^{-6}$, the safe zone parameter is $\rho_0 = h$, the mesh size for velocity and stress tensor is $h = 1/96$, and the time step is $\Delta t = 0.0004$. In our simulation, the three particles formed a chain along the flow direction, which verifies the known observations and experiments. Figure 12 gives the snapshots of the particles forming a chain at various moments of time. We can see that the particles approximately form a chain at $t = 26$. At $t = 102$, the chain is almost on the center line. The average terminal velocity is 0.17, the Reynolds number is $\mathrm{Re} = 0.16$, the Deborah number is $\mathrm{De} = 1.03$, the elasticity number is $E = 6.24$, and the Mach number is $M = 0.41$.

2.5.4. Sedimentation of six particles

The fourth test case concerns six circular particles sedimenting in a channel filled with an Oldroyd-B fluid. The channel is infinitely long and has a width of 1. The computational domain is $\Omega = (0, 1) \times (0, 7)$ initially and moves down with the lowest mass center of the six particles. The diameter of the particles is 0.25. The density of the fluid is $\rho_f = 1$, the density of the disk is $\rho_s = 1.01$, the viscosity of the fluid is $\eta_1 = 0.26$, the relaxation time is $\lambda_1 = 1.3$, the retardation time is $\lambda_2 = 0.125\lambda_1$, and the diameter of the disk is $D = 0.25$. The initial positions of the disks are (0.23, 2.0),(0.5, 2.0), (0.78, 2.0), (0.22, 2.30),(0.5, 2.3), and (0.77, 2.3). The other parameters are the same as in the second test case. The mesh size for the velocity and stress tensor is $h = 1/96$, and the time step is $\Delta t = 0.0004$. We know that when the elasticity number E is larger than the critical value ($O(1)$) and the Mach number M is less than the critical value ($O(1)$), the particles in this case will form chains that are parallel to the flow (see, SINGH, JOSEPH, HELSA, GLOWINSKI and PAN [2000], and YU, PHAN-THIEN, FAN and TANNER [2002]). In our simulations, all the six particles are lined up along the flow direction, which verifies the known observations and experiments. Figure 13 gives the snapshots of the lining up of particles at various moments of time. We can see that the six particles form approximately a straight line at $t = 20$; at $t = 30$, the trailing particle has been separated from the leading five particles. This observation agrees with experiments showing that, sometimes, the last particle in the chain gets detached in PATANKAR and HU [2000]. It is known that a long chain falls faster than a single particle in the fluid. This long body effect tends to detach the last particle from the chain. The average terminal velocity is 0.147, the Reynolds number is $\mathrm{Re} = 0.14$, the Deborah number is $\mathrm{De} = 0.76$, the elasticity number is $E = 5.4$, and the Mach number is $M = 0.33$.

3. Cavity flow

Generally, viscoelastic computation in complex flows at high Weissenberg number has proven to be a tremendous challenge, in particular for systems where singularities are

FIG. 12 Positions of the three particles at $t = 0.2$, 15, 26, 50, 67, 102 (from left to right and from top to bottom), forming a chain in a viscoelastic fluid; blockage ratio = 8; Re = 0.16, De = 1.03, $E = 6.24$, $M = 0.41$.

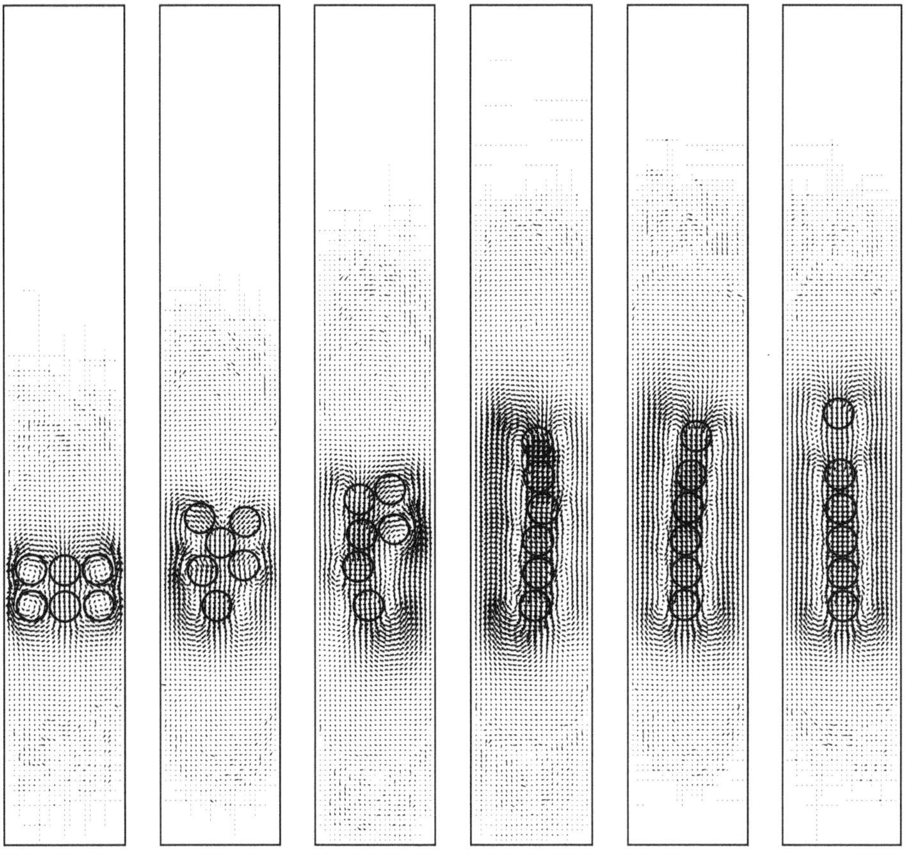

FIG. 13 Snapshots of positions of six particles lining up at $t = 0.2, 7, 11, 20, 24, 30$; $h = 1/96$ and $\triangle t = 0.0004$; blockage ratio $= 4$; Re $= 0.14$, De $= 0.76$, $E = 5.4$, $M = 0.33$.

present. Examples include cavity flows with a steadily moving lid, and only a limited number of computational methods provide satisfactory results (see, e.g., BAAIJENS [1998]). There have been few numerical studies of cavity flows of viscoelastic fluids. PHELAN, MALONE and WINTER [1989] implemented a hyperbolic numerical solution method and tested their method by considering the cavity flow of a shearing-thinning fluid. GRILLET and SHAQFEH [1996] used a perturbation technique to investigate the first effects of elasticity on the flow geometry of the semicavity flow problem. GRILLET, YANG, KHOMAMI and SHAQFEH [1999] studied the numerical modeling of two-dimensional steady lid-driven cavity flow. They introduced leakage to relieve the corner singularity in the simulations by including small rounded channels at the corners where the fluid can leak through and used a mixed finite element with streamline upwind/Petrov-Galerkin (SUPG) stabilization in the discretization of the constitutive equation for the conformation tensor. In FATTAL and KUPFERMAN [2005], the authors used a second-order finite difference scheme to simulate the Stokes flow of an Oldroyd-B fluid in a lid-driven cavity. They reformulated the constitutive equation as an equation for the matrix logarithm of the conformation tensor to preserve the property of the positive definiteness of the conformation tensor, which was developed in the earlier work of FATTAL and KUPFERMAN [2004]. To discretize the advection term in the constitutive equation, they have applied the Kurganov–Tadmor scheme in KURGANOV and TADMOR [2000] with min-mod limiter (see, e.g., LEVEQUE [1992] for an extensive reference on limiters).

From previous work, we have found that there are two important considerations when trying to simulate time-dependent viscoelastic flows at high Weissenberg number. First, the positive definiteness of the conformation tensor has to be preserved at the *discrete level* during the *entire time integration*. Besides the technique developed in FATTAL and KUPFERMAN [2005], another attempt for obtaining the positive definiteness preserving scheme when discretizing the constitutive equation is a recent work by LOZINSKI and OWENS [2003], where the authors factorize the conformation tensor to get $\boldsymbol{c} = \mathbf{A}\mathbf{A}^T$ and then try to write down the equations for $\mathbf{A}$ approximately at the discrete level. Hence, the positive definiteness of the conformation tensor is forced with such an approach. In a most recent work of LEE and XU [2006], one has developed a unified numerical discretization framework that can be used for simulating most of existing constitutive equations so that the positiveness of the conformation tensor of the continuous level can be extended to its discrete analog. However, the main advantage of using the log-conformation tensor is that we can better resolve the exponential behavior of the conformation tensor in the region where there are boundary layers. In this article, we have incorporated the log-conformation tensor technique developed in FATTAL and KUPFERMAN [2005] with an operator splitting technique to preserve the positive definiteness of the conformation tensor. Second, the constitutive equation is a hyperbolic equation and lacks diffusion term. In SURESHKUMAR and BERIS [1995], an additional diffusion term added to the constitutive equation for the Oldroyd-B fluid did stabilize the computations. SUPG methods have been used widely with finite element methods (see BAAIJENS [1998] and the references therein for details) to stabilize the numerical schemes used for solving the constitutive equation. The min-mod limiter used in FATTAL and KUPFERMAN [2005] is known to be very stable but introduces additional diffusion; indeed the additional diffusion obtained directly or indirectly from the above numerical techniques does stabilize to some extent the numerical schemes used for solving the constitutive equation. It is the opinion of the authors that additional (but not too much) diffusion smooths out some of the high-frequency modes from the discrete conformation tensor so that the numerical scheme is stabilized. To reduce the number of high-frequency modes in the first place, we have chosen a finite element approach for discretizing the conformation tensor defined on a coarser mesh (compared to the mesh for the velocity field); actually in, e.g., GRILLET, YANG, KHOMAMI and SHAQFEH [1999], FATTAL and KUPFERMAN [2005], SARAMITO [1995], the discrete conformation tensor was also defined on coarser meshes.

In HULSEN, FATTAL and KUPFERMAN [2005], CORONADO, ARORA, BEHR, PASQUALI [2007], the technique of log-conformation tensor has been used with finite element methods to simulate viscoelastic fluid flows past a cylinder. This article is a follow up of the work by PAN and HAO [2007] in which the points mentioned above have been taken into account to develop a stable scheme for the solution of a two-dimensional lid-driven cavity Stokes flow for an Oldroyd-B fluid at high Weissenberg numbers. In PAN and HAO [2007], the advection was treated with the first-order upwind scheme which, just like the min-mod limiter used in FATTAL and KUPFERMAN [2005], produces too much artificial diffusion. Even though the lid-driven cavity flow has closed planar streamlines in a simple confined geometry, its conformation tensor does have sharp boundary layers attached to the lid at high Weissenberg numbers. The numerical results in FATTAL and KUPFERMAN [2005], PAN and HAO [2007] were obtained with uniform meshes, and hence, the boundary layer has not been well resolved. In both FATTAL and KUPFERMAN [2005], PAN and HAO [2007], some convergent results have been shown, but the ones like the convergence of the conformation tensor on the lid have not been shown. In this article, we have improved the methodology developed in PAN and HAO [2007] by applying a second-order upwinding scheme to treat the advection and using nonuniform meshes with very fine mesh close to the lid and the left and right sides of the cavity. With these modifications, we have obtained convergent numerical results for Weissenberg number up to 1.25. For higher Weissenberg number cases, it is very difficult to resolve the boundary layer of the conformation tensor since its maximum value has shown an exponential relation to the Weissenberg number as discussed in Section 3.3.3 unless we use extremely fine meshes close to the lid. In the following section, we first introduce the formulation of the problem. Then, we discuss how to apply the Lie's scheme to split the constitutive equation into subproblems and how to reformulate those subproblems via the technique developed in FATTAL and KUPFERMAN [2005]. In Section 3.2, we discuss the space and time discretizations together with numerical methods for solving the subproblems. Numerical results are presented in Section 3.3.

3.1. Formulation of the problem

We consider a two-dimensional lid-driven cavity Stokes flow for an Oldroyd-B fluid. Let $\Omega = (0, 1) \times (0, 1)$ be the region occupied by the fluid, Γ the boundary of Ω and $T > 0$ (see Fig. 14). The flow model problem is governed by

$$-\nabla p + \mu \Delta \mathbf{u} + \frac{\eta}{\lambda_1} \nabla \cdot \boldsymbol{c} = \mathbf{0} \text{ in } \Omega \times (0, T), \tag{3.1}$$

$$\nabla \cdot \mathbf{u} = 0 \text{ in } \Omega \times (0, T), \tag{3.2}$$

$$\frac{\partial \boldsymbol{c}}{\partial t} + (\mathbf{u} \cdot \nabla) \boldsymbol{c} - (\nabla \mathbf{u}) \boldsymbol{c} - \boldsymbol{c} (\nabla \mathbf{u})^T = \frac{1}{\lambda_1} (\mathbf{I} - \boldsymbol{c}) \text{ in } \Omega \times (0, T), \tag{3.3}$$

$$\boldsymbol{c}(0) = \boldsymbol{c}_0 \text{ in } \Omega, \tag{3.4}$$

$$\mathbf{u} = \mathbf{g}(t) \text{ on } \Gamma \times (0, T) \text{ with } \int_\Gamma \mathbf{g}(t) \cdot \mathbf{n} \, d\Gamma = 0 \text{ on } (0, T). \tag{3.5}$$

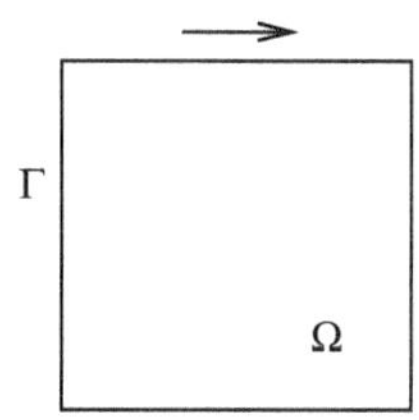

FIG. 14 Lid-driven flow in a square cavity.

Here, $\mathbf{u}$ and p are the flow velocity and pressure, $\boldsymbol{c}$ is the conformation tensor (with $\mathbf{T}' = \frac{\eta}{\lambda_1}\boldsymbol{c}$), μ and η are the solvent and polymer viscosities, λ_1 is a characteristic relaxation time for the fluid, while $\mathbf{n}$ is the unit outward normal vector at the boundary Γ. We use the notation $v(t)$ to denote the function $\mathbf{x} \to v(\mathbf{x}, t)$ in (3.4), (3.5) and below.

For $\mathbf{g}(t)$ in (3.5), we have chosen the same regularized boundary condition given in FATTAL and KUPFERMAN [2005]:

$$\mathbf{g}(\mathbf{x}, t) = \begin{cases} (g(x,t), 0)^T & \text{on } \{\mathbf{x} | \mathbf{x} = (x, 1)^T,\ 0 < x < 1\}, \\ (0, 0)^T & \text{otherwise on } \Gamma, \end{cases} \tag{3.6}$$

with $g(x, t) = 8(1 + \tanh 8(t - 0.5))x^2(1 - x)^2$. The discontinuity of the velocity field at the two upper corners has been removed in (3.6). The inflow boundary conditions for the conformation tensor are not needed since there is no inflow boundary for this case. The Weissenberg number is $\mathrm{Wi} = \lambda_1 U/L$ where U and L are the characteristic velocity and length scale. With $U = 1$ as the speed from the lid and $L = 1$ as the width of the cavity, $\mathrm{Wi} = \lambda_1$.

We have applied an operator-splitting technique, namely the Lie's scheme similar to the one in CHORIN, HUGHES, MARSDEN and MCCRACKEN [1978], to solve (3.1)–(3.5). The Lie's scheme is *first-order* accurate, but its low order accuracy is compensated by easy implementation, less cost in computational time, good stability, and robustness properties. For example, it has been successfully applied to develop numerical methods for simulating the interaction of solid particles and fluid (see, e.g., GLOWINSKI [2003], GLOWINSKI, PAN, HESLA, JOSEPH and PERIAUX [2001], and PAN and GLOWINSKI [2005]). Let Δt be a time discretization step and $t^n = n\Delta t$. Applying the operator-splitting technique to (3.1)–(3.5) yields the following.

For $n \geq 0$, $\boldsymbol{c}^n$ being known, we compute first $\mathbf{u}^{n+1} (\approx \mathbf{u}(t^{n+1}))$ and $p^{n+1} (\approx p(t^{n+1}))$ via the solution of the following problem

$$-\boldsymbol{\nabla} p^{n+1} + \mu \Delta\, \mathbf{u}^{n+1} = -\frac{\eta}{\lambda_1} \boldsymbol{\nabla} \cdot \boldsymbol{c}^n \ in\ \Omega, \tag{3.7}$$

$$\boldsymbol{\nabla} \cdot \mathbf{u}^{n+1} = 0\ in\ \Omega, \tag{3.8}$$

$$\mathbf{u}^{n+1} = \mathbf{g}(t^{n+1})\ on\ \Gamma. \tag{3.9}$$

Next, we compute $\boldsymbol{c}^{n+1}$ via the following steps: first solve

$$\frac{\partial \boldsymbol{c}}{\partial t} + (\mathbf{u}^{n+1} \cdot \boldsymbol{\nabla})\, \boldsymbol{c} = \mathbf{0}\ in\ \Omega \times (t^n, t^{n+1}), \tag{3.10}$$

$$\boldsymbol{c}(t^n) = \boldsymbol{c}^n\ in\ \Omega, \tag{3.11}$$

and set $\boldsymbol{c}^{n+1/2} = \boldsymbol{c}(t^{n+1})$. Then solve

$$\frac{\partial \boldsymbol{c}}{\partial t} - (\boldsymbol{\nabla} \mathbf{u}^{n+1})\, \boldsymbol{c} - \boldsymbol{c}\, (\boldsymbol{\nabla} \mathbf{u}^{n+1})^T + \frac{1}{\lambda_1} \boldsymbol{c} = \frac{1}{\lambda_1} \mathbf{I}\ in\ \Omega \times (t^n, t^{n+1}), \tag{3.12}$$

$$\boldsymbol{c}(t^n) = \boldsymbol{c}^{n+1/2}\ in\ \Omega, \tag{3.13}$$

and set $\boldsymbol{c}^{n+1} = \boldsymbol{c}(t^{n+1})$.

To keep $\boldsymbol{c}$ positive definite, we have combined in the following the matrix logarithm formulation of the conformation tensor developed in FATTAL and KUPFERMAN [2005], FATTAL and KUPFERMAN [2004] with the above operator splitting scheme. But first for a symmetric positive definite matrix $\boldsymbol{c}$, we have that $\boldsymbol{\psi} = \log \boldsymbol{c}$. (Recall that a symmetric positive definite matrix $\mathbf{A}$ can always be diagonalized as $\mathbf{A} = \mathbf{R}\,\boldsymbol{\Lambda}\,\mathbf{R}^T$ and that $\log \mathbf{A} = \mathbf{R}\,\log \boldsymbol{\Lambda}\,\mathbf{R}^T$.) In FATTAL and KUPFERMAN [2004], it was shown that with $\mathbf{u}^n$ being a divergence-free velocity field and $\boldsymbol{c}$ a symmetric positive definite tensor field, the velocity gradient $\nabla \mathbf{u}^n$ can be decomposed as

$$\nabla \mathbf{u}^n = \boldsymbol{\omega}_n + \mathbf{B}_n + \mathbf{N}_n \boldsymbol{c}^{-1}, \tag{3.14}$$

where $\boldsymbol{\omega}_n$ and $\mathbf{N}_n$ are skew-symmetric and $\mathbf{B}_n$ is symmetric, trace-free, and commutes with $\boldsymbol{c}$. Using these matrices, we obtain the following variant of scheme (3.7)–(3.13):

For $n \geq 0$, $\boldsymbol{c}^n$ (and $\boldsymbol{\psi}^n = \log \boldsymbol{c}^n$) being known, we compute first $\mathbf{u}^{n+1}$ and p^{n+1} via the solution of the following problem

$$-\nabla p^{n+1} + \mu \Delta\, \mathbf{u}^{n+1} = -\frac{\eta}{\lambda_1}\, \nabla \cdot \boldsymbol{c}^n \ \textit{in}\ \Omega, \tag{3.15}$$

$$\nabla \cdot \mathbf{u}^{n+1} = 0 \ \textit{in}\ \Omega, \tag{3.16}$$

$$\mathbf{u}^{n+1} = \mathbf{g}(t^{n+1}) \ \textit{on}\ \Gamma. \tag{3.17}$$

Next, we compute $\boldsymbol{\psi}^{n+1}$ via the following steps: first solve

$$\frac{\partial \boldsymbol{\psi}}{\partial t} + (\mathbf{u}^{n+1} \cdot \nabla)\, \boldsymbol{\psi} = \mathbf{0} \ \textit{in}\ \Omega \times (t^n, t^{n+1}), \tag{3.18}$$

$$\boldsymbol{\psi}(t^n) = \boldsymbol{\psi}^n \ \textit{in}\ \Omega, \tag{3.19}$$

and set $\boldsymbol{\psi}^{n+1/2} = \boldsymbol{\psi}(t^{n+1})$. Then, solve

$$\frac{\partial \boldsymbol{\psi}}{\partial t} - [\boldsymbol{\omega}_{n+1}\, \boldsymbol{\psi} - \boldsymbol{\psi}\, \boldsymbol{\omega}_{n+1}] - 2\mathbf{B}_{n+1} = \frac{1}{\lambda_1}\,(\mathrm{e}^{-\boldsymbol{\psi}} - \mathbf{I}) \ \textit{in}\ \Omega \times (t^n, t^{n+1}), \tag{3.20}$$

$$\boldsymbol{\psi}(t^n) = \boldsymbol{\psi}^{n+1/2} \ \textit{in}\ \Omega, \tag{3.21}$$

and set $\boldsymbol{\psi}^{n+1} = \boldsymbol{\psi}(t^{n+1})$ and $\boldsymbol{c}^{n+1} = \mathrm{e}^{\boldsymbol{\psi}^{n+1}}$.

REMARK 3.1. To compute $\boldsymbol{\omega}$, $\mathbf{B}$, and $\mathbf{N}$ from a divergence-free velocity field $\mathbf{u}$ for a two-dimensional case, we can use the following formulas given in FATTAL and KUPFERMAN [2004]: (i) If $\boldsymbol{c}$ is proportional to the unit tensor then set $\mathbf{B} = (\nabla \mathbf{u} + (\nabla \mathbf{u})^T)/2$ and $\boldsymbol{\omega} = \mathbf{0}$. (ii) Otherwise, diagonalize $\boldsymbol{c}$ via

$$\boldsymbol{c} = \mathbf{R} \begin{pmatrix} \lambda_1 & 0 \\ 0 & \lambda_2 \end{pmatrix} \mathbf{R}^T, \tag{3.22}$$

and set

$$\begin{pmatrix} m_{11} & m_{12} \\ m_{21} & m_{22} \end{pmatrix} = \mathbf{R}^T\, (\nabla \mathbf{u})\, \mathbf{R}. \tag{3.23}$$

Then,

$$\mathbf{N} = \mathbf{R}\begin{pmatrix} 0 & n \\ -n & 0 \end{pmatrix}\mathbf{R}^T, \quad \mathbf{B} = \mathbf{R}\begin{pmatrix} m_{11} & 0 \\ 0 & m_{22} \end{pmatrix}\mathbf{R}^T, \quad \omega = \mathbf{R}\begin{pmatrix} 0 & s \\ -s & 0 \end{pmatrix}\mathbf{R}^T, \tag{3.24}$$

with $n = (m_{12} + m_{21})/(\lambda_1^{-1} - \lambda_2^{-1})$, and $s = (\lambda_2 m_{12} + \lambda_1 m_{21})/(\lambda_2 - \lambda_1)$. □

REMARK 3.2. The subproblem (3.18), (3.19) is an advection problem. To solve it, we advocate the wave-like equation method discussed in Section 2.3.2. Each entry of the matrix $\boldsymbol{\psi}$ satisfies a *transport equation* of the following type:

$$\begin{cases} \dfrac{\partial \varphi}{\partial t} + \mathbf{V} \cdot \nabla \varphi = 0 \text{ in } \Omega \times (t^n, t^{n+1}), \\ \varphi(t^n) = \varphi_0 \text{ in } \Omega, \end{cases} \tag{3.25}$$

with $\nabla \cdot \mathbf{V} = 0$ and $\partial \mathbf{V}/\partial t = \mathbf{0}$ on (t^n, t^{n+1}) with empty Γ^-.

Using the properties $\nabla \cdot \mathbf{V} = 0$ and $\partial \mathbf{V}/\partial t = \mathbf{0}$ on $\Omega \times (t^n, t^{n+1})$, we have that problem (3.25) is "equivalent" to the (formally) well-posed problem:

$$\begin{cases} \dfrac{\partial^2 \varphi}{\partial t^2} - \nabla \cdot ((\mathbf{V} \cdot \nabla \varphi)\mathbf{V}) = 0 \text{ in } \Omega \times (t^n, t^{n+1}), \\ \varphi(t^n) = \varphi_0, \ \dfrac{\partial \varphi}{\partial t}(t^n) = -\mathbf{V} \cdot \nabla \varphi_0, \\ \mathbf{V} \cdot \mathbf{n}\left(\dfrac{\partial \varphi}{\partial t} + \mathbf{V} \cdot \nabla \varphi\right) = 0 \text{ on } \Gamma \times (t^n, t^{n+1}). \end{cases} \tag{3.26}$$

Solving the wave-like equation (3.26) by a classical finite element/time stepping method is quite easy since a variational formulation of (3.26) is given by

$$\begin{cases} \displaystyle\int_\Omega \frac{\partial^2 \varphi}{\partial t^2} v \, d\mathbf{x} + \int_\Omega (\mathbf{V} \cdot \nabla \varphi)(\mathbf{V} \cdot \nabla v) \, d\mathbf{x} = 0, \ \forall v \in H^1(\Omega), \ \text{a.e. on } (t^n, t^{n+1}), \\ \varphi(t^n) = \varphi_0, \ \dfrac{\partial \varphi}{\partial t}(t^n) = -\mathbf{V} \cdot \nabla \varphi_0. \end{cases} \tag{3.27}$$

Since for the driven cavity problem, we have $\mathbf{V} \cdot \mathbf{n} = 0$, the boundary condition in (3.26) is satisfied automatically. A solution method for problem (3.27) will be described in the following section. □

REMARK 3.3. Actually subproblem (3.20) and (3.21) can be solved directly using a further splitting, namely

$$\frac{\partial \boldsymbol{\psi}}{\partial t} - [\omega_{n+1} \boldsymbol{\psi} - \boldsymbol{\psi} \omega_{n+1}] - 2\mathbf{B}_{n+1} = \mathbf{0} \text{ in } \Omega \times (t^n, t^{n+1}), \tag{3.28}$$

$$\boldsymbol{\psi}(t^n) = \boldsymbol{\psi}^{n+1/2} \text{ in } \Omega, \tag{3.29}$$

and set $\tilde{\boldsymbol{\psi}}^{n+1} = \boldsymbol{\psi}(t^{n+1})$ and $\tilde{\boldsymbol{c}}^{n+1} = e^{\tilde{\boldsymbol{\psi}}^{n+1}}$. Then solve

$$\frac{\partial \boldsymbol{c}}{\partial t} = \frac{1}{\lambda_1}(\mathbf{I} - \boldsymbol{c}) \text{ in } \Omega \times (t^n, t^{n+1}), \tag{3.30}$$

$$\boldsymbol{c}(t^n) = \tilde{\boldsymbol{c}}^{n+1} \text{ in } \Omega, \tag{3.31}$$

and set $\boldsymbol{c}^{n+1} = \boldsymbol{c}(t^{n+1})$ and $\boldsymbol{\psi}^{n+1} = \log(\boldsymbol{c}^{n+1})$.
The closed form solutions of the above two subproblems can be obtained easily (at least for two-dimensional flows). □

3.2. Space and time discretizations

Concerning the *space approximation*, we use P_1-*iso*-P_2 and P_1 finite elements for the velocity field and pressure, respectively (as, e.g., in BERCOVIER and PIRONNEAU [1979], BRISTEAU, GLOWINSKI and PERIAUX [1987], and GLOWINSKI [2003, chapter 5]). More precisely with h as *space discretization step*, we introduce a finite element triangulation $\mathcal{T}_h$ of $\overline{\Omega}$ and then $\mathcal{T}_{2h}$ a triangulation twice coarser (in practice, we should construct $\mathcal{T}_{2h}$ first and then $\mathcal{T}_h$ by joining the midpoints of the edges of $\mathcal{T}_{2h}$, dividing thus each triangle of $\mathcal{T}_{2h}$ into four similar subtriangles, as shown already in Fig. 2).

Next, we define the following finite dimensional spaces:

$$V_{\mathbf{g}_h(t)} = \{\mathbf{v}_h \mid \mathbf{v}_h \in (C^0(\overline{\Omega}))^2,\ \mathbf{v}_h|_T \in P_1 \times P_1,\ \forall T \in \mathcal{T}_h,\ \mathbf{v}_h|_\Gamma = \mathbf{g}_h(t)\}, \tag{3.32}$$

$$V_{0h} = \{\mathbf{v}_h \mid \mathbf{v}_h \in (C^0(\overline{\Omega}))^2,\ \mathbf{v}_h|_T \in P_1 \times P_1,\ \forall T \in \mathcal{T}_h,\ \mathbf{v}_h|_\Gamma = \mathbf{0}\}, \tag{3.33}$$

$$L_h^2 = \{q_h \mid q_h \in C^0(\overline{\Omega}),\ q_h|_T \in P_1,\ \forall T \in \mathcal{T}_{2h}\}, \tag{3.34}$$

$$L_{0h}^2 = \{q_h \mid q_h \in L_h^2,\ \int_\Omega q_h \,\mathrm{d}\mathbf{x} = 0\}; \tag{3.35}$$

in (3.32)–(3.35), $\mathbf{g}_h(t)$ is an approximation of $\mathbf{g}(t)$ verifying $\int_\Gamma \mathbf{g}_h(t) \cdot \mathbf{n} \,\mathrm{d}\Gamma = 0$, and P_1 is the space of the polynomials in two variables of degree ≤ 1. The discrete conformation tensor belongs to

$$\mathbf{W}_2 = \left\{\mathbf{A}_h \mid \mathbf{A}_h = \begin{pmatrix} A_{1,h} & A_{2,h} \\ A_{2,h} & A_{3,h} \end{pmatrix},\ A_{i,h} \in L_h^2,\ i = 1, 2, 3\right\}. \tag{3.36}$$

Using these finite element spaces, we obtain the following realization of scheme (3.15)–(3.21) (after dropping some of the subscripts h):

For $n \geq 0$, $\boldsymbol{c}^n$ (and $\boldsymbol{\psi}^n$) being known, we compute first $\mathbf{u}^{n+1}$ and p^{n+1} via the solution of the following problem

$$\begin{cases} \int_\Omega p^{n+1}\nabla\cdot\mathbf{v}\,d\mathbf{x} - \mu\int_\Omega \nabla\mathbf{u}^{n+1}:\nabla\mathbf{v}\,d\mathbf{x} = -\dfrac{\eta}{\lambda_1}\int_\Omega(\nabla\cdot\boldsymbol{c}^n)\cdot\mathbf{v}\,d\mathbf{x},\ \forall\mathbf{v}\in V_{0h}, \\ \int_\Omega q\nabla\cdot\mathbf{u}^{n+1}\,d\mathbf{x} = 0,\ \forall q\in L_h^2, \\ \mathbf{u}^{n+1}\in V_{\mathbf{g}_h}^{n+1},\ p^{n+1}\in L_{0h}^2. \end{cases} \tag{3.37}$$

Next, we compute $\boldsymbol{\psi}^{n+1} = \begin{pmatrix}\psi_1^{n+1} & \psi_2^{n+1}\\ \psi_2^{n+1} & \psi_3^{n+1}\end{pmatrix}$ via the following steps: first solve

$$\begin{cases} \int_\Omega \dfrac{\partial^2\psi_i}{\partial t^2}v\,d\mathbf{x} + \int_\Omega(\mathbf{u}^{n+1}\cdot\nabla\psi_i)(\mathbf{u}^{n+1}\cdot\nabla v)\,d\mathbf{x} = 0,\ \forall v\in L_h^2,\ \text{on }(t^n,t^{n+1}), \\ \psi_i(t^n) = \psi_i^n,\ \dfrac{\partial\psi_i}{\partial t}(t^n) = -\mathbf{u}^{n+1}\cdot\nabla\psi_i^n;\ \psi_i(t)\in L_h^2, \end{cases} \tag{3.38}$$

for i =1, 2, 3, and set $\boldsymbol{\psi}^{n+1/2} = \begin{pmatrix}\psi_1(t^{n+1}) & \psi_2(t^{n+1})\\ \psi_2(t^{n+1}) & \psi_3(t^{n+1})\end{pmatrix}$. Then solve

$$\int_\Omega\left[\frac{\partial\boldsymbol{\psi}}{\partial t} - (\boldsymbol{\omega}_{n+1}\boldsymbol{\psi} - \boldsymbol{\psi}\boldsymbol{\omega}_{n+1}) - 2B_{n+1}\right]:\mathbf{T}\,d\mathbf{x}$$

$$= \int_\Omega\left[\frac{1}{\lambda_1}(e^{-\boldsymbol{\psi}} - \mathbf{I})\right]:\mathbf{T}\,d\mathbf{x},\ \forall\mathbf{T}\in\mathbf{W}_2,\ \text{on}\ (t^n,t^{n+1}), \tag{3.39}$$

$$\boldsymbol{\psi}(t^n) = \boldsymbol{\psi}^{n+1/2};\ \boldsymbol{\psi}(t)\in\mathbf{W}_2, \tag{3.40}$$

and set $\boldsymbol{\psi}^{n+1} = \boldsymbol{\psi}(t^{n+1})$ and $\boldsymbol{c}^{n+1} = e^{\boldsymbol{\psi}^{n+1}}$.
In (3.37), $V_{\mathbf{g}_h}^{n+1} = V_{\mathbf{g}_h(t^{n+1})}$, and

$$\mathbf{A}:\mathbf{B} = a_{11}b_{11} + a_{12}b_{12} + a_{21}b_{21} + a_{22}b_{22},\ \text{for}\ \mathbf{A} = \begin{pmatrix}a_{11} & a_{12}\\ a_{21} & a_{22}\end{pmatrix},\ \mathbf{B} = \begin{pmatrix}b_{11} & b_{12}\\ b_{21} & b_{22}\end{pmatrix}.$$

At each step in the above scheme, we encounter simpler subproblems, which can be solved by *simple and standard* numerical methods. First, the Stokes problem (3.37) is a classical problem and has been solved by an Uzawa/conjugate gradient algorithm in GLOWINSKI [2003], in which a sequence of elliptic problems has been solved by a red-black SOR iterative method. The wave-like equation (3.38) is solved by the following time-stepping method, which is a special case of the one discussed in Section 2.3.2 due to the different boundary condition and discretization considered here.

We first define a sub-time-step, $\tau_1 > 0$, by $\tau_1 = \Delta t/Q_1$, where Q_1 is a positive integer and we discretize problem (3.38) in time by

$$\varphi^0 = \varphi_0, \tag{3.41}$$

$$\begin{cases} \int_\Omega (\varphi^{-1} - \varphi^1) v \, d\mathbf{x} = 2\tau_1 \int_\Omega (\mathbf{V} \cdot \nabla \varphi^0) v \, d\mathbf{x}, \ \forall v \in L_h^2, \\ \varphi^{-1} - \varphi^1 \in L_h^2, \end{cases} \tag{3.42}$$

and for $q = 0, \ldots, Q_1 - 1$,

$$\begin{cases} \varphi^{q+1} \in L_h^2, \\ \int_\Omega \dfrac{\varphi^{q+1} + \varphi^{q-1} - 2\varphi^q}{\tau_1^2} v \, d\mathbf{x} + \int_\Omega (\mathbf{V} \cdot \nabla \varphi^q)(\mathbf{V} \cdot \nabla v) \, d\mathbf{x} = 0, \ \forall v \in L_h^2, \end{cases} \tag{3.43}$$

where, in (3.41) and (3.43), φ_0 is the initial value and $\mathbf{V} = \mathbf{u}^{n+1}$.
Scheme (3.41)–(3.43) is a centered scheme, which is formally second-order accurate with respect to space and time discretizations. To be stable, scheme (3.41)–(3.43) has to verify a condition such as

$$\tau_1 \leq ch, \tag{3.44}$$

with c of the order of $1/||\mathbf{V}||$. If one chooses an appropriate numerical integration method to compute the first integral in (3.43), the above scheme becomes *explicit*, i.e., φ^{q+1} is obtained via the solution of a linear system with a *diagonal* matrix (e.g., the trapezoidal rule has been used for the results reported in the article). When computing $\mathbf{V} \cdot \nabla f$ in (3.42) and (3.43), we have applied a *second-order upwind scheme* to compute $\dfrac{\partial f}{\partial x_1}$ and $\dfrac{\partial f}{\partial x_2}$ when solving with scheme (3.37)–(3.40). The first-order upwind scheme used in PAN and HAO [2007] produces too much artificial diffusion. Another detail is that at each time step in scheme (3.41)–(3.43), we do not update the values of φ^q at the boundary grid points at which we have $\mathbf{V} = \mathbf{0}$. When solving subproblem (3.39) and (3.40), we have applied the trapezoidal rule to get pointwise differential equation at each grid point and then solve it via the further splitting discussed in Remark 3.3. In (3.39), $\nabla \mathbf{u}^{n+1}$ is computed via second-order difference scheme on the mesh points used for the conformation tensor, and then, $\boldsymbol{\omega}_{n+1}$ and $\mathbf{B}_{n+1}$ are computed according to (3.22)–(3.24).

3.3. Numerical results

In this section, we consider the numerical results for the lid-driven cavity Stokes flow by the numerical schemes described in the above sections. The boundary condition for the velocity field in (3.6) is given by

$$\mathbf{g}(\mathbf{x}, t) = \begin{cases} (g(x_1, t), 0)^T & \text{on } \{\mathbf{x} | \mathbf{x} = (x_1, 1)^T, \ 0 < x_1 < 1\}, \\ (0, 0)^T & \text{otherwise on } \Gamma, \end{cases}$$

FIG. 15 An example of mesh $\mathcal{T}_h$ for $N = 32$.

with $g(x_1, t) = 8(1 + \tanh 8(t - 0.5))x_1^2(1 - x_1)^2$; this choice gives a smooth start, and for $t \gg \frac{1}{2}$, the lid velocity attains its maximum, $\mathbf{u} = (1, 0)^T$, at the center, $x_1 = 1/2$. The initial condition for $\boldsymbol{c}$ is $\boldsymbol{c}_0 = \mathbf{I}$. The viscosities, μ and η, in all calculations are both equal to 1. The relaxation times λ_1 considered here are 0.75 and 1.25 (so the Weissenberg numbers are 0.75 and 1.25, respectively).

The mesh $\mathcal{T}_{2h}$ for the pressure is a triangular mesh obtained by the following way. We have chosen points $x_{2,j} = 1 - (1 - \frac{2j}{N})^2$, for $j = 0, 1, \ldots, N/2$, in the x_2 direction. In the x_1 direction, we first choose $x_{1,i} = 2(2i/N)^2$, for $i = 0, 1, \ldots, N/4$, and then set $x_{1,i} = 1 - x_{1,\frac{N}{2}-i}$, for $i = N/4 + 1, \ldots, N/2$. Using the lines $x_1 = x_{1,i}$ and $x_2 = x_{2,j}$, for $i, j = 1, \ldots, N/2 - 1$, we divide the unit square into smaller rectangles, and each rectangle is divided into two triangles. After obtaining the triangular mesh for the pressure, we join the midpoints of the edges of each triangle to divide it into four smaller triangles as shown in Fig. 2 to obtain the mesh for the velocity field. In Fig. 15, an example mesh for the velocity field for $N = 32$ is shown. With nonuniform triangular meshes, the discrete elliptic problems arising from the Uzawa/conjugate gradient algorithm at each iteration have been solved by a red-black SOR iterative method. We have parallelized the code via OpenMP and run it on quad-core CPUs to speed up the computation.

3.3.1. The case Wi $= 0.75$

This is a "nice" test case since the Weissenberg number is not too high. The results obtained with $N = 288$, 320, and 352 are computed with the time steps $\Delta t = 0.0015$, 0.0012, and 0.001, respectively. The kinetic energy grows as the lid accelerates, reaches a maximum at the end of the acceleration, and decreases toward a steady value as the elastic energy builds up. The history of the kinetic energy, $\frac{1}{2}\|\mathbf{u}_h\|_2^2$, and the history of the elastic energy, $\int_\Omega (c_{11} + c_{22})\,d\mathbf{x}$, are shown in Fig. 16. We have obtained a steady state solution for Wi $=$ 0.75 as

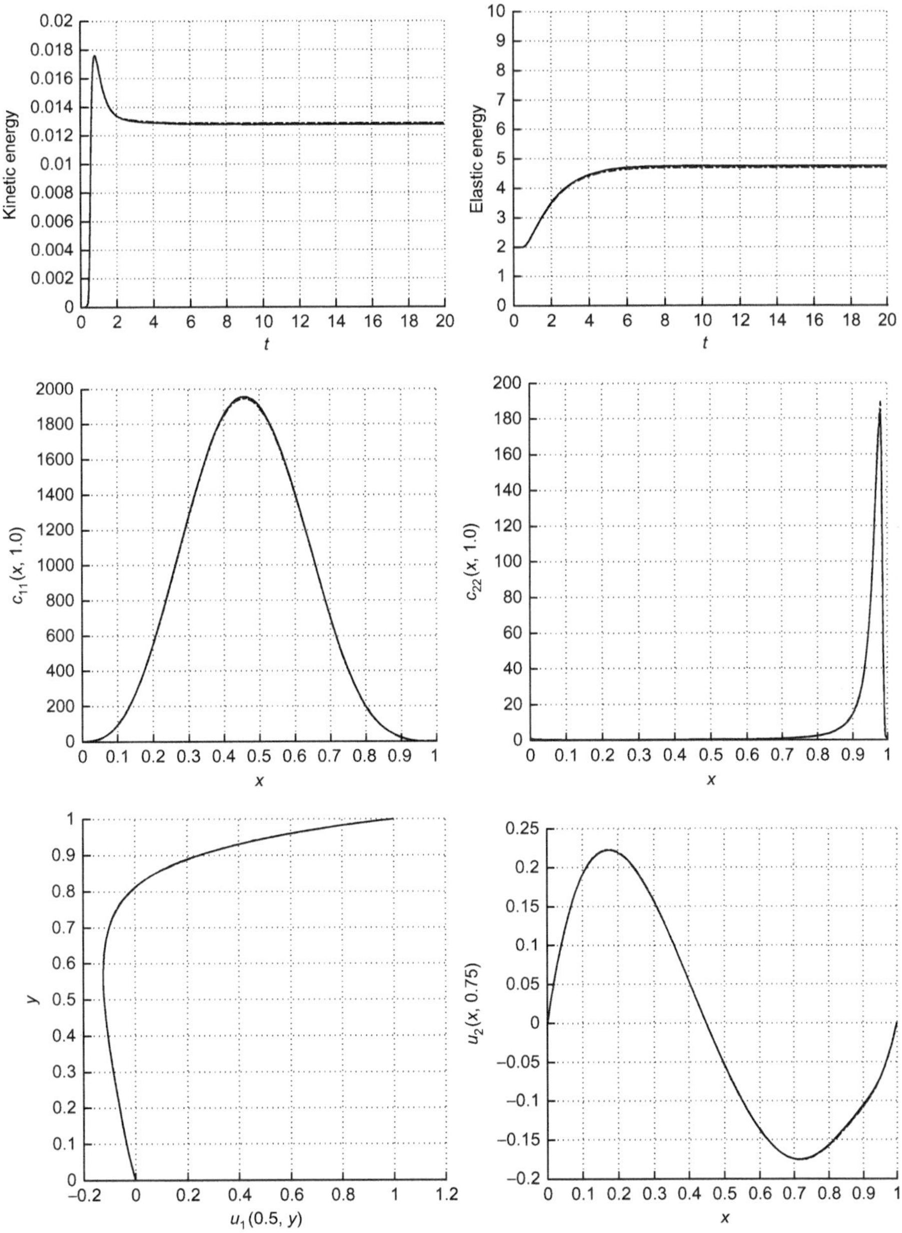

FIG. 16 Histories of the kinetic energy (upper left) and the elastic energy (upper right), and the cross section of $c_{11}(x_1, 1)$ (middle left), $c_{22}(x_1, 1)$ (middle right), $u_1(x_2, 0.5)$ (lower left), and $u_2(x_1, 0.75)$ (lower right) at $t = 20$ obtained with $N = 288$ (dashed line), $N = 320$ (dash-dotted line), and $N = 352$ (solid line) for $\mathrm{Wi} = 0.75$.

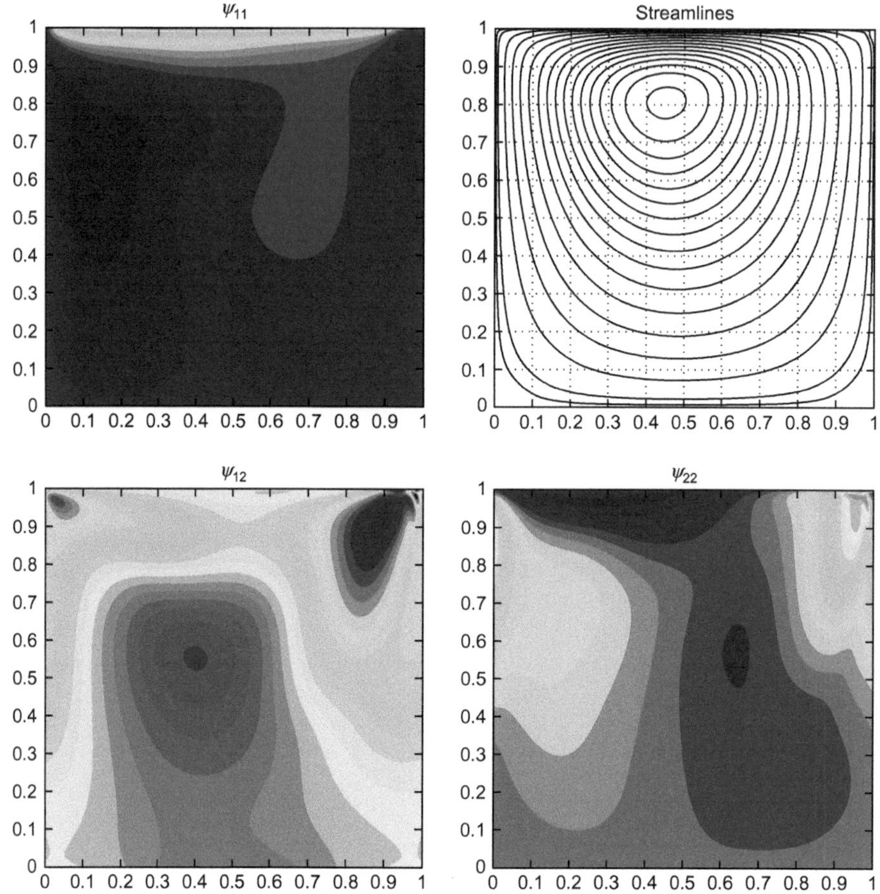

FIG. 17 The contour plots of ψ_{11} (upper left), ψ_{12} (lower left), ψ_{22} (lower right), and the streamlines (upper right) obtained with $N = 256$ at $t = 20$ for Wi $= 0.75$.

shown by the kinetic and elastic energy in Fig. 16 at $t = 20$. The streamlines and the contour plots of ψ_{ij} obtained with $N = 288$ at $t = 20$ are shown in Fig. 17. The minimal value of the stream function obtained with $N = 256$ and $\Delta t = 0.0015$ is –0.06646064 at (0.4525945, 0.8085937). The cross sections of ψ_{ij} at $x_1 = 0.5$ and at $x_2 = 1$, c_{ij} at $x_2 = 1$, u_1 at $x_1 = 0.5$, and u_2 at $x_2 = 0.75$ are shown in Figs 16 and 18. These results show the convergence when reducing the mesh size and time step. As shown in Figs 16 and 18, c_{11} and c_{22} do have sharp boundary layer attached to the lid. The center of the core vortex region shifts in the upstream direction as observed in the experiments of PAKDEL, SPIEGELBERG and MCKINLEY [1997].

3.3.2. *The case* Wi $= 1.25$

In this section, the results obtained with $N = 352$, 386, and 412 are computed with the time steps $\Delta t = 0.0014$, 0.00117, and 0.001, respectively. The kinetic energy and elastic energy behave like those of the case Wi $= 0.75$. Their histories are shown in Fig. 19. We have obtained a steady state solution for Wi $= 1.25$ at $t = 40$. The smallest value of the stream

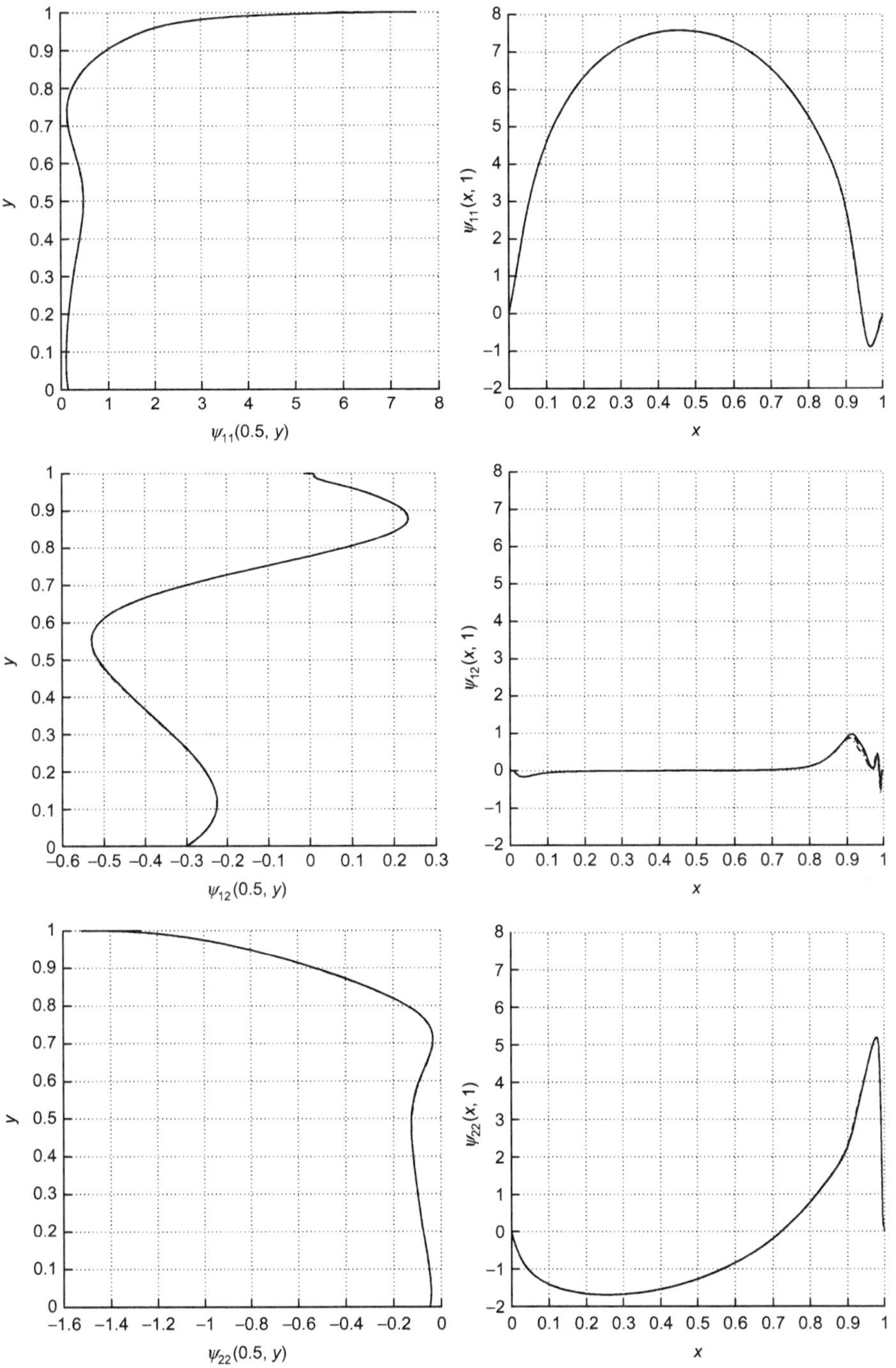

FIG. 18 The cross section of ψ_{11}, ψ_{12}, and ψ_{22} (from top to bottom) at $x_1 = 0.5$ (left) and $x_2 = 1$ (right) obtained with $N = 288$ (dashed line), $N = 320$ (dash-dotted line), and $N = 352$ (solid line) at $t = 20$ for $\mathrm{Wi} = 0.75$.

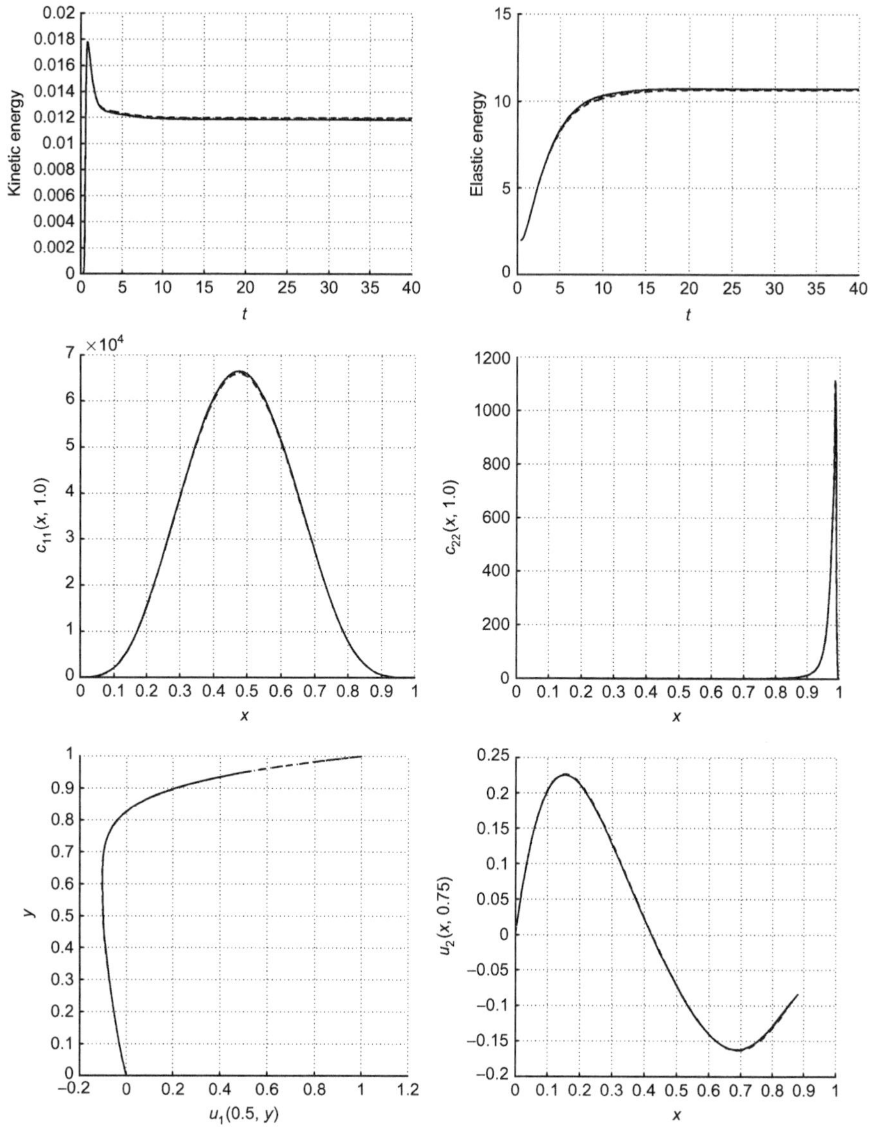

FIG. 19 Histories of the kinetic energy (upper left) and the elastic energy (upper right) and the cross section of $c_{11}(x_1, 1)$ (middle left), $c_{22}(x_1, 1)$ (middle right), $u_1(x_2, 0.5)$ (lower left), and $u_2(x_1, 0.75)$ (lower right) at $t = 40$ obtained with $N = 352$ (dashed line), $N = 386$ (dash-dotted line), and $N = 412$ (solid line) for $\mathrm{Wi} = 1.25$.

function obtained with $N = 352$ and $\Delta t = 0.001$ is –0.06228973 at (0.4288804, 0.8208129). The cross sections of ψ_{ij} at $x_1 = 0.5$ and at $x_2 = 1$, c_{11} and c_{22} at $x_2 = 1$, u_1 at $x_1 = 0.5$, and u_2 at $x_2 = 0.75$ are shown in Figs 19 and 20. These results show the convergence when reducing the mesh size and time step. As shown in Figs 19 and 20, the boundary layer of c_{11}

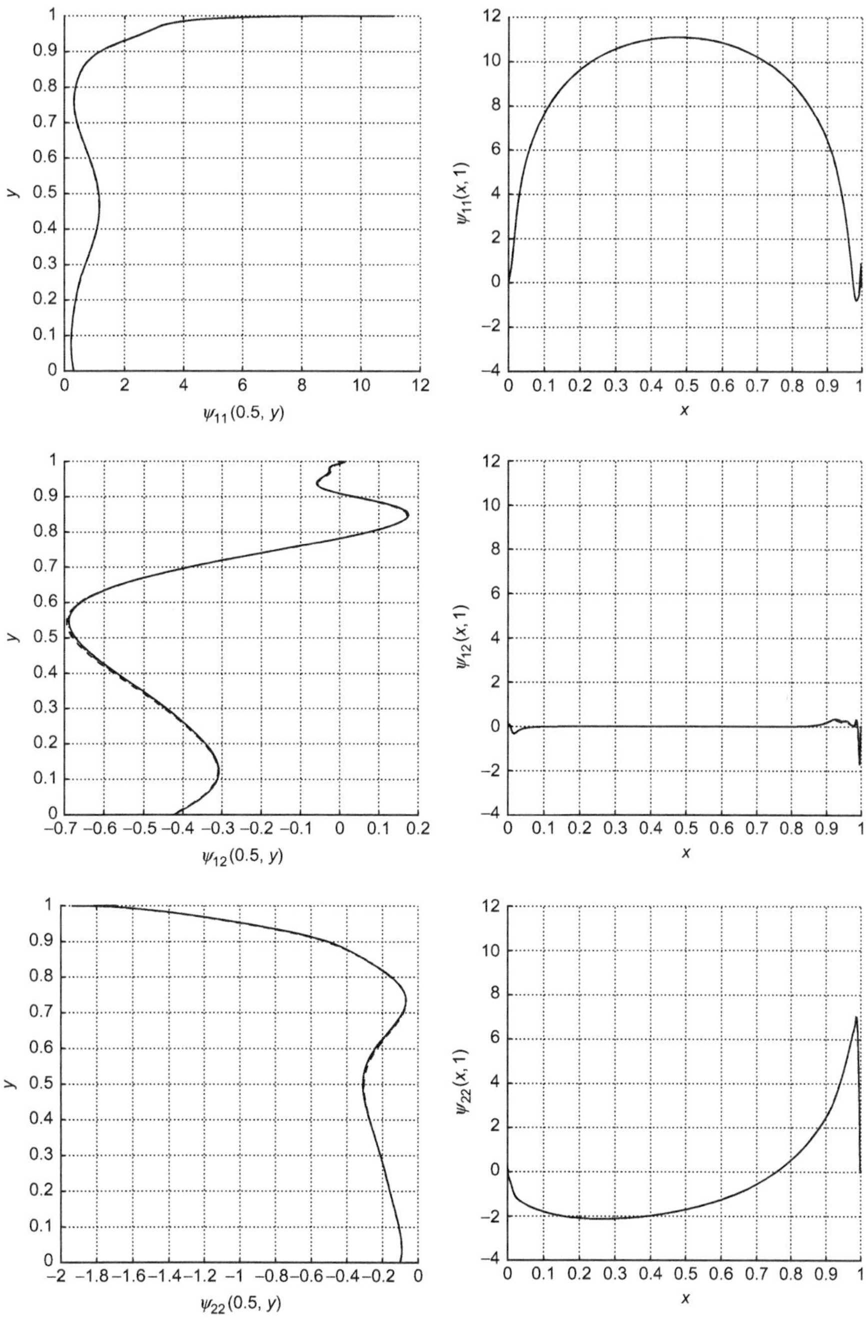

FIG. 20 The cross section of ψ_{11}, ψ_{12}, and ψ_{22} (from top to bottom) at $x_1 = 0.5$ (left) and $x_2 = 1$ (right) obtained with $N = 356$ (dashed line), $N = 386$ (dash-dotted line), and $N = 412$ (solid line) at $t = 40$ for $\mathrm{Wi} = 1.25$.

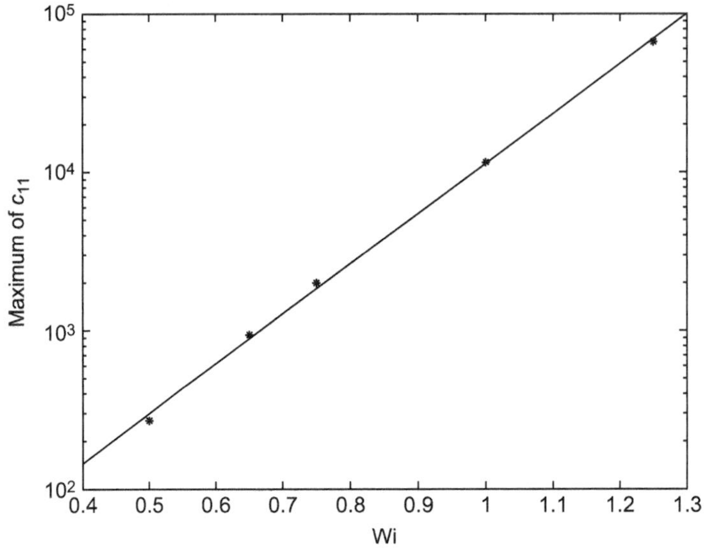

FIG. 21 The exponential curve fitting for the maximum of c_{11} and Wi.

attached to the lid becomes much higher. The maximum of c_{11} is 68882.98 obtained with $N = 352$ at $t = 40$.

The largest values of ψ_{11} at $x_1 = 0.5$ shown in FATTAL and KUPFERMAN [2005] for Wi = 1 (resp. Wi = 2) are less than 7 (resp. 8). Our values shown in Fig. 20 are about 11.1. We believe that those values obtained in FATTAL and KUPFERMAN [2005] for Wi = 1 and 2 were not well resolved due to the use of uniform meshes, which are not fine enough mesh close to the lid. Also they might be smoothed out by the numerical diffusion produced by the Kurganov–Tadmor scheme with min-mod limiter used in FATTAL and KUPFERMAN [2005].

3.3.3. The growth of c_{11}

Using the curve fitting for the maximum of c_{11} obtained at Wi = 0.5, 0.65, 0.75, 1, and 1.25, we have obtained the relation $c_{11}(\mathrm{Wi}) = \mathrm{e}^{2.0625+7.2734\,\mathrm{Wi}}$, and its plot is shown in Fig. 21. The growth of c_{11} indicates that extremely fine meshes are needed to resolve the boundary layer of c_{11} and that the cavity flow is a difficult problem at high Weissenberg number.

Acknowledgments

We acknowledge the helpful comments and suggestions of R. Bai, S. Canic, E.J. Dean, J. He, H.H. Hu, P.Y. Huang, G.P. Galdi, D.D. Joseph, A. Lozinski, M. Pasquali, and P. Singh. We acknowledge also the support of NSF (grants ECS-9527123, CTS-9873236, DMS-9973318, CCR-9902035, DMS-0209066, DMS-0443826) and DOE/LASCI (grant R71700K-292-000-99).

Bibliography

ADAMS, J., SWARZTRAUBER, P., SWEET, R. (1980). FISHPAK: A package of FORTRAN subprograms for the solution of separable elliptic partial differential equations (National Center for Atmospheric Research, Boulder, CO).

BAAIJENS, F.P.T. (1998). Mixed finite element methods for viscoelastic flow analysis: a review. *J. Non-Newton. Fluid Mech.* **79**, 361–385.

BERCOVIER, M., PIRONNEAU, O. (1979). Error estimates for finite element method solution of the Stokes problem in the primitive variables. *Numer. Math.* **33**, 211–224.

BINOUS, H., PHILLIPS, R.J. (1999). Dynamic simulation of one and two particles sedimenting in viscoelastic suspensions of FENE dumbbells. *J. Non-Newton. Fluid Mech.* **83**, 93–130.

BRISTEAU, M.O., GLOWINSKI, R., PERIAUX, J. (1987). Numerical methods for the Navier-Stokes equations. Applications to the simulation of compressible and incompressible viscous flow. *Comput. Phys. Rep.* **6**, 73–187.

CHHABRA, R.P. (1993). *Bubbles, Drops, and Particles in Non-Newtonian Fluids* (CRC Press, Boca Raton, FL).

CHORIN, A.J., HUGHES, T.J.R., MARSDEN, J.E., MCCRACKEN, M. (1978). Product formulas and numerical algorithms. *Comm. Pure Appl. Math.* **31**, 205–256.

CIARLET, P.G. (1978). *The Finite Element Method for Elliptic Problems* (North-Holland, Amsterdam).

CIARLET, P.G. (1991). Basic error estimates for elliptic problems. In: Ciarlet, P.G., Lions, J.L. (eds.), *Handbook of Numerical Analysis II* (North-Holland, Amsterdam), pp. 17–351.

CORONADO, O.M., ARORA, D., BEHR, M., PASQUALI, M. (2007). A simple method for simulating general viscoelastic fluid flows with an alternate log-conformation formulation. *J. Non-Newton. Fluid Mech.* **147**, 189–199.

DEAN, E.J., GLOWINSKI, R. (1997). A wave equation approach to the numerical solution of the Navier-Stokes equations for incompressible viscous flow. *C.R. Acad. Sci. Paris, Série 1* **325**, 783–791.

DEAN, E.J., GLOWINSKI, R., PAN, T.-W. (1998). A wave equation approach to the numerical solution of incompressible viscous fluid flow modeled by the Navier-Stokes equations. In: De Santo, J.A. (ed.), *Mathematical and Numerical Aspects of Wave Propagation* (SIAM, Philadelphia, PA), pp. 65–74.

FATTAL, R., KUPFERMAN, R. (2004). Constitutive laws for the matrix-logarithm of the conformation tensor. *J. Non-Newton. Fluid Mech.* **123**, 281–285.

FATTAL, R., KUPFERMAN, R. (2005). Time-dependent simulation of viscoelastic flows at high Weissenberg number using the log-conformation representation. *J. Non-Newton. Fluid Mech.* **126**, 23–37.

FENG, J., HUANG, P.Y., JOSEPH, D.D. (1996). Dynamic simulation of sedimentation of solid particles in an Oldroyd-B fluid. *J. Non-Newton. Fluid Mech.* **63**, 63–88.

GLOWINSKI, R. (2003). Finite element methods for incompressible viscous flows. In: Ciarlet, P.G., Lions, J.L. (eds.), *Handbook of Numerical Analysis IX* (North-Holland, Amsterdam), pp. 3–1176.

GLOWINSKI, R., PAN, T.-W., PERIAUX, J. (1998). Distributed Lagrange multiplier methods for incompressible flow around moving rigid bodies. *Comput. Methods Appl. Mech. Eng.* **151**, 181–194.

GLOWINSKI, R., PAN, T.-W., HESLA, T., JOSEPH, D.D. (1999). A distributed Lagrange multiplier/fictitious domain method for particulate flows. *Int. J. Multiphase Flow* **25**, 755–794.

GLOWINSKI, R., PAN, T.-W., HESLA, T., JOSEPH, D.D., PERIAUX, J. (2001). A fictitious domain approach to the direct numerical simulation of incompressible viscous flow past moving rigid bodies: application to particulate flow. *J. Comput. Phys.* **169**, 363–426.

GRILLET, A.M., SHAQFEH, E.S.G. (1996). Observations of viscoelastic instabilities in recirculation flows of Boger fluids. *J. Non-Newton. Fluid Mech.* **64**, 141–155.

GRILLET, A.M., YANG, B., KHOMAMI, B., SHAQFEH, E.S.G. (1999). Modeling of viscoelastic lid driven cavity flow using finite element simulations. *J. Non-Newton. Fluid Mech.* **88**, 99–131.

HU, H.H. (1996). Direct simulation of flows of solid-liquid mixtures. *Int. J. Multiphase Flow* **22**, 335.

HU, H.H., PATANKAR, N.A., ZHU, M.Y. (2001). Direct numerical simulations of fluid-solid systems using the arbitrary Lagrangian-Eulerian technique. *J. Comput. Phys.* **169**, 427–462.

HUANG, P.Y., HU, H.H., JOSEPH, D.D. (1998). Direct simulation of the sedimentation of elliptic particles in Oldroy-B fluids. *J. Fluid Mech.* **362**, 297–325.

HULSEN, M.A., FATTAL, R., KUPFERMAN, R. (2005). Flow of viscoelastic fluids past a cylinder at high Weissenberg number: stabilized simulations using matrix logarithms. *J. Non-Newton. Fluid Mech.* **127**, 27–39.

JOHNSON, C. (1986). Streamline diffusion methods for problems in fluid mechanics. In: Gallagher, R. (ed.), *Finite Element in Fluids IV* (Wiley-Interscience, Chichester, England and New York), pp. 251–261.

JOSEPH, D.D. (1990). *Fluid Dynamics of Viscoelastic Liquids* (Springer, New York).

JOSEPH, D.D., LIU, Y.J., POLETTO, M., FENG, J. (1994). Aggregation and dispersion of spheres falling in viscoelastic liquids. *J. Non-Newton. Fluid Mech.* **54**, 45–86.

KEUNINGS, R. (2000). A survey of computational rheology. In: Binding, D.M., Hudson, N.E., Mewis, J., Piau, J-M, Petrie, C.J.S., Townsend, P., Wagner, M.H., and Walters, K. (eds.), *13th International Congress on Rheology,* Volume 1 (British Society of Rheology, Glasgow), p. 7.

KURGANOV, A., TADMOR, E. (2000). New high-resolution central schemes for nonlinear conservation laws and convection-diffusion equations. *J. Comput. Phys.* **160**, 241–282.

LEE, Y.J., XU, J. (2006). New formulations, positivity preserving discretizations and stability analysis for non-Newtonian flow models. *Comput. Methods Appl. Mech. Eng.* **195**, 1180–1206.

LEVEQUE, R. (1992). *Numerical Methods for Conservation Laws* (Birkhauser Verlag, Basel).

LOZINSKI, A., OWENS, R.G. (2003). An energy estimate for the Oldroyd B model: theory and applications. *J. Non-Newton. Fluid Mech.* **112**, 161–176.

MCKINLEY, G.H. (2002). Steady and transient motion of spherical particles in viscoelastic liquids. In: De Kee, D., Chhabra, R.P. (eds.), *Transport Processes in Bubbles, Drops & Particles* (Taylor & Francis, New York), pp. 338–375.

PAKDEL, P., SPIEGELBERG, S.H., MCKINLEY, G.H. (1997). Cavity flows of elastic liquids: two-dimensional flows. *Phys. Fluids* **9**, 3123–3140.

PAN, T.-W., GLOWINSKI, R. (2002). Direct simulation of the motion of neutrally buoyant circular cylinders in plane Poiseuille flow. *J. Comput. Phys.* **181**, 260–279.

PAN, T.-W., GLOWINSKI, R. (2005). Direct simulation of the motion of neutrally buoyant balls in a three-dimensional Poiseuille flow. *C.R.M écanique Acad. Sci. Paris* **333**, 884–895.

PAN, T.-W., HAO, J. (2007). Numerical simulation of a lid-driven cavity viscoelastic flow at high Weissenberg numbers. *C. R. Acad. Sci. Paris, Série I* **344**, 283–286.

PATANKAR, N.A., HU, H.H. (2000). A numerical investigation of the Detachment of the trailing particle from a chain sedimenting in Newtonian and Viscoelastic Fluids. *J. Fluids Eng.* **122**, 517–521.

PHELAN, F.R., MALONE, M.F., WINTER, H.H. (1989). A purely hyperbolic model for unsteady viscoelastic flow. *J. Non-Newton. Fluid Mech.* **32**, 197–224.

SARAMITO, P. (1995). Efficient simulation of nonlinear viscoelastic fluid flows. *J. Non-Newton. Fluid Mech.* **60**, 199–223.

SINGH, P., LEAL, L.G. (1993). Finite-Element simulation of the start-up problem for a viscoelastic fluid in an eccentric rotating cylinder geometry using a third-order upwind scheme. *Theor. Comput. Fluid Dyn.* **5**, 107–137.

SINGH, P., JOSEPH, D.D., HELSA, T.I., GLOWINSKI, R., PAN, T.-W. (2000). A distributed Lagrange multiplier/fictitious domain method for viscoelastic particulate flows. *J. Non-Newton. Fluid Mech.* **91**, 165–188.

SINGH, P., JOSEPH, D.D. (2000). Sedimentation of a sphere near a vertical wall in an Oldroyd-B fluid. *J. Non-Newton. Fluid Mech.* **94**, 179–203.

SURESHKUMAR, R., BERIS, A.N. (1995). Effect of artificial stress diffusivity on the stability of numerical calculations and the flow dynamics of time-dependent viscoelastic flows. *J. Non-Newton. Fluid Mech.* **60**, 53–80.

YU, Z., PHAN-THIEN, N., FAN, Y., TANNER, R.I. (2002). Viscoelastic mobility problem of a system of particles. *J. Non-Newton. Fluid Mech.* **104**, 87–124.

YU, Z., WACHS, A., PEYSSON, Y. (2006). Numerical simulation of particle sedimentation in shear-shinning fluids with a fictitious domain method. *J. Non-Newton. Fluid Mech.* **136**, 126–139.

On the Numerical Simulation of Viscoplastic Fluid Flow

Roland Glowinski

Department of Mathematics, University of Houston, Houston, TX, USA
E-mail:roland@math.uh.edu

Anthony Wachs

Fluid Mechanics Department, IFP Energies Nouvelles, 1 & 4 avenue de Bois Préau, 92852 Rueil Malmaison, France
E-mail: anthony.wachs@ifpenergiesnouvelles.fr

Numerical Methods for Non-Newtonian Fluids
Special Volume (R. Glowinski and Jinchao Xu, Guest Editors) of
HANDBOOK OF NUMERICAL ANALYSIS, VOL. XVI
P.G. Ciarlet (Editor)

ISSN 1570-8659
DOI 10.1016/B978-0-444-53047-9.00006-X

Contents

CHAPTER 1

Viscoplastic Fluid Flow: A Review

1. Introduction

Among the various classes of *non-Newtonian* materials, those exhibiting *viscoplastic* properties are particularly interesting in accordance with their ability to strain only if the stress intensity exceeds a minimum value (called the *yield stress*, usually). Many industrial processes involve viscoplastic fluids. We will mention only a few of them, namely: mud, cement slurries, food stuff, waxy crude oils, suspensions, emulsions, foams, ... As mentioned in HUILGOL and YOU [2005], even if the presence of a true yield stress is widely debated, the yield stress concept is clearly useful from an engineering standpoint. As a result, in a viscoplastic fluid flow, the flow pattern highlights two kinds of regions: (1) the regions where the stress intensity exceeds the yield stress, and (2) the regions where it does not. The former and latter regions are usually called the *yielded* and *unyielded regions*, respectively. The most commonly encountered viscoplastic model is the *Bingham fluid* (BINGHAM [1922]). This model may be ranked at the top of the list of models in the mind of practitioners, when they think about a "yield stress fluid"; this is primarily due to its simplicity, from both the experimental and numerical standpoints. Although this model contains the primary feature needed to be called viscoplastic, which is the presence of a nonzero yield stress, it is deemed as a crude simplification of the true rheological behavior because its predictions may depart, quite significantly, from experimental data.

Due to the significant number of industrial applications, and the still unresolved fundamental issues, associated with the specific behavior of this class of materials, there has been a spurt in research activity relating to the mechanics of viscoplastic fluids. The lack of understanding, or the inability to predict simple phenomena like the critical density ratio of spheres suspended in a yield stress fluid, has motivated the investigations of a fair number of scientists and engineers around the world. Thanks to this significant effort and, in particular, to the contributions of OLDROYD [1947], PRAGER [1961], MOSSOLOV and MIASNIKOV [1965], DUVAUT and LIONS [1972a, 1976], BARNES and WALTERS [1985], BIRD, DAI and YARUSSO [1983], PAPANASTASIOU [1987] (see also GLOWINSKI [1974, 1984], BRISTEAU and GLOWINSKI [1974], and GLOWINSKI, LIONS and TRÉMOLIÈRES [1976, 1981]), some "dull" areas of viscoplastic fluid mechanics are getting clearer. Over the past years, advances in the use of asymptotic techniques or variational methods are a strong testimony that our understanding is gradually improving in this area. Recent efforts in terms of computational methods extend the research activity to more complex geometries and/or to the combination with other complex features of the flow: multifluid (FRIGAARD and SCHERZER [1998, 2000]), free surfaces

(VOLA, BOSCARDIN and LATCHÉ [2003], and DIMAKOPOULOS and TSAMOPOULOS [2003]), compressibility (DAVIDSON, NGUYEN, CHANG and RONNINGSEN [2004], and VINAY, WACHS and AGASSANT [2006]), and stability issues (FRIGAARD, HOWISON and SOBEY [1994]). The main mathematical difficulty when computing the solution of viscoplastic flow problems is the *nondifferentiability* of the constitutive law at the yield point, which is also pointed by some practitioners, in the specialized literature, as the ability to track yield surfaces. The most straightforward and convenient way to circumvent the difficulty associated with the nondifferentiability is to approach the true equation by an approximated one in which the material never truly yields, but instead behaves like a very viscous material. This way of dealing with a yield stress fluid is called a *regularization method* (BERCOVIER and ENGELMAN [1980], PAPANASTASIOU [1987], and ALLOUCHE, FRIGAARD and SONA [2000]) because the equation modeling the yield stress behavior is regularized, that is made differentiable. Many investigators have used this procedure to obtain trustworthy and useful results. However, this approach has also come under criticism (FRIGAARD and NOUAR [2005]). A third group of investigators (cf., e.g., GLOWINSKI, LIONS and TRÉMOLIÈRES [1976, 1981]) have rejuvenated the *augmented Lagrangian* approach and applied it to a wide range of problems in mechanics and physics, including problems from viscoplasticity. This class of methods has the advantage of considering the true constitutive equation, in contrast to regularization procedures. We think that the main reason behind the fact that the augmented Lagrangian approach has been overlooked for the last 20 years, at least by a part of the Rheology community, stems from the underlying *variational inequality* approach, which may be difficult to grasp for some nonapplied mathematicians. Recently, computational specialists realized that although the theoretical issues associated with the method may be "sharp," its practical implementation is fairly easy. This has led to a significant amount of numerical investigations, and associated results, relying on augmented Lagrangian-based algorithms. Anyway, irrespective of the numerical methods used to simulate yield stress fluid flows and to track yield surfaces, most often, they end up as solvers for fixed-point or saddle-point problems, which may be costly to solve. In addition, although the convergence of these iterative methods has been proved theoretically, it may be very slow for very "tough" problems. This observation has motivated several scientists, researchers, and engineers to investigate the following issue: how to speedup the convergence of fixed point iterations for viscoplastic flows? We will revisit this challenge in the following parts of this article, starting with the present chapter.

The ability to accurately compute the flow of a viscoplastic material, thanks to an efficient algorithm, combined with an appropriate space-time discretization, has drawn the attention of applied mathematicians and other practitioners for many years. However, although it is generally assumed that the so-called Bingham and Herschel–Bulkley models are appropriate ones, it should be also acknowledged that these widely used models are largely empirical. In fact, these models, and their generalizations are merely the results of attempts at fitting rheometrical flow curves, without connection to any microscopic theory, unlike the approach taken nowadays in viscoelasticity. Apparently, the lack of any structural theory to support the existence of a yield stress is not a subject of worry (although the whole viscoplasticity community would welcome such a theory), and simple physical arguments, highlighted by observations and rheological measurements, seem to suit a majority of rheologists. However, the existence of a reliable theory would help to close the long-standing debate on the existence of a true yield stress (BARNES and WALTERS [1985], and BARNES [1999]), a concept that relies mainly on the accuracy of our measuring capabilities at low-deformation rates.

The determination of the yield stress is usually based on the extrapolation of the stress–strain rate curve to the zero strain rate axis. This procedure has been at the heart of the debate on the existence of a true yield stress for many years. In fact, the nonbelievers of the "true yield stress" theory claim that this extrapolation technique is flawed and is used primarily because of the limited capabilities of the rheometers (including the most modern ones) to measure properly strain rates less than $10^{-5}/s$.The second part of their argument assumes that, considering the continuous improvement of rheometers, we will be able to show, eventually, that the stress–strain rate curve passes through the origin. Let us emphasize again that it seems that, up to now, the yield stress concept has been very useful to the rheology community in order to describe materials that barely strain, or do not strain at all, below a critical value of the stress intensity. This also raises another interesting issue, that any viscoplasticity specialist should wonder when investigating her/his problem, namely: *what happens below the yield stress*? This seems to be an issue that the viscoplastic community is more willing to address (than, say, devising a proper theory of fluid viscoplasticity), and some members of this community have already started investigating this topic. Observations and characterization of yield stress materials have shown that below the yield stress, they behave as either rigid solids or elastic solids or very viscous fluids. Actually, it seems that very few materials behave as rigid solids, which supports the idea that Bingham or Herschel–Bulkley models are idealizations. But, honestly, at very-low strain rates, it is highly problematic to distinguish between reversible elastic strains and nonreversible creep. The elastic solid behavior looks promising and is often associated with the formation of gel structures in the bulk of the material. Therefore, at very-low shear rates, the material deforms and, as the stress that has been applied is released, the material recovers its initial state without damage to its internal structure. If the stress intensity is gradually increased, there is a critical value beyond which the internal structure of the material begins to break down, and the material starts flowing. This type of behavior suggests the merging of viscoelastic and viscoplastic models into viscoelastoplastic ones that would encompass a wide range of rheologically complex material behaviors; we will return to this, with more details, later in this chapter. Similarly, the existence of an abrupt transition between reversible elastic strain and irreversible viscous strain appears also to be a subject of controversy, or at least of debate. In particular, recent results, obtained by many experimentalists, strongly suggest that *thixotropic* effects (also known as *time* or *memory* effects), related to the microstructure of viscoplastic materials, affect their behavior dramatically; these evidences support the concept of a gradual transition, which may be somehow incompatible with the true yield stress approach. In any case, little is known about the stress field below the yield stress (assuming the existence of such a yield). From an experimental standpoint, measurement techniques for the stress field, or for the tracking of yield surfaces, are quite complicated, keeping in mind that many viscoplastic fluids are not transparent. Finally, since thixotropy, that is aging, matters, it is obvious that at the time of measurement, what we measure is highly dependent of the material history: What stress and strain has the material undergone, to which temperatures has it been exposed, . . . ? In many cases, the measurements are protocol-dependent and, without extra care, may be nonreproducible. This knowledge is shared and widely accepted by those rheologists dealing with pastes, emulsions, gels and so on. Eventually, it is fair to wonder if a viscoplastic model without thixotropy, or with a yield stress value measured experimentally without following a precise protocol, makes sense or is representative of the material behavior. Clearly, some very simple viscoplastic materials behave as predicted by the Bingham model, but most of them possess also some degree of thixotropy.

Our objective in this review article is to address *the numerical simulation of viscoplastic fluid flow*. Although the issues raised above are highly interesting and important for the overall growth of the field, we will assume here that the yield stress notion is a worthwhile idealization, and that the *Bingham model* and its derivatives are a good starting point. At any rate, from the mathematical and computational standpoints, the main difficulties are rather associated with the discontinuity at the yield point than with any sophistication that we may add to the model to deal with thixotropy, temperature dependence, compressibility, and so on.

2. Applications

Among the various industrial applications involving viscoplastic fluids, we choose to focus on the energy supply industry (in relation to the background of the second author of this article), although we are aware that in other industries and scientific areas, there is a keen interest on yield stress fluids as well (food processing, geophysical flows, and so on.)

2.1. Waxy crude oils

In the *Oil & Gas industry*, one has extensively used pipelines to transport large amounts of crude oil over short or long distances. The transportation of conventional (that is Newtonian, lowly viscous, steady physical properties, single-phase, ...) crude oils is a relatively easy to handle task, however, pipelining crude oils containing large proportions of high molecular weight compounds, like *paraffin*, may cause many specific difficulties (UHDE and KOPP [1971], SMITH and RAMSDEN [1978], and MODI, KISWANTO and MERRILL [1994]). Most of the complexity comes from paraffin crystals forming an interlocking gel-like structure that modifies some of the crude oils rheological properties (CAZAUX [1998]). The above crystallization mechanism is mainly controlled by temperature. These oils, known as *waxy crude oils*, usually exhibit high "wax appearance temperature" (WAT) and high "pour point." Using a standardized test (*ASTM D* 97), the pour point correspond to an experimentally measured temperature, below which the oil has a tendency to freeze, or not to pour while being cooled. The word "high" applies to those situations where the pour point temperature is higher than the temperature of the external conditions surrounding the pipeline. Below the pour point, the oil rheological behavior is characterized by thixotropic, temperature-dependent and shear-dependent yield stress and viscosity (ECONOMIDES and CHANEY [1983], WARDAUGH and BOGER [1987], RONNINGSEN [1992], and HÉNAUT and BRUCY [2001]).

From a viscoplastic flow point of view, the main concern with waxy crude oil transportation is the issue of restarting (PERKINS and TURNER [1971], and SMITH and RAMSDEN [1978]). From an operational standpoint, transporting waxy crude oil under steady flowing conditions is not a too complex operation, but it is still of primary interest for both practical and fundamental reasons. However, the situation becomes more "tricky" if a shutdown occurs. Flow shutdowns may occur for different reasons, such as maintenance, emergency situations, pumps failures, and so on. Under nonflowing conditions, when the pipeline is subjected to severe external temperature conditions (especially in Artic regions, sub sea installations, and so on), the temperature of the crude oil in the pipeline starts to drop. This temperature decrease leads to the crystallization of the paraffin compounds and, eventually, as the temperature drops below the pour point, to the buildup of a gel-like structure in the crude oil bulk. If the temperature decrease lasts long enough, the waxy crude oil undergoes a *thermal shrinkage* related to the formation of gaseous cavities that impart a kind of *compressibility* to the material. Eventually, the waxy crude oil restarting issue consists in resuming the flow of a compressible gel-like material, usually by injecting some fresh warm oil (expected to be Newtonian and incompressible) at the pipe inlet (CAWKWELL and CHARLES [1987], and CHANG, NGUYEN and RONNINGSEN [1999]). It clearly appears that the *temperature* is the key factor of the whole shutdown and restart process. Actually, waxy crude oils are usually transported in pipelines under steady flow conditions, far below the pour point; this implies that for this kind of flow, waxy crude oils already exhibit some of their specific rheological properties, such as shear thinning viscosity, yield stress, and temperature-dependence.

Predicting, accurately, the restart of a waxy crude oil flow requires a fairly precise description of the initial state (that is at the time of restart) of the material in the pipeline. Because the rheological properties are temperature- and temperature-history-dependent, knowing the evolution of the temperature field in the pipeline during the shutdown is required (COOPER, SMITH, CHARLES, RYAN and ALEXANDER [1978]). Similarly, in order to predict the evolution of the temperature during the shutdown (that is the temperature history), the temperature field at the time of the shutdown has to be known; this field corresponds to the temperature under steady flowing conditions. In other words, the survey of a waxy crude oil flow restart implies to consider the whole process: the steady flowing conditions (namely the production conditions), the flow stopping (maintenance, emergency, ...), and the resuming of the production (the flow restart itself). In the three phases we mentioned just earlier, the yield stress properties of the material play a key role, in the following way:

1. *In production conditions*, the flow corresponds to the steady flow of a temperature-dependent viscoplastic fluid. Essentially, the fluid enters the pipe at a warm temperature, and is gradually cooled down while flowing toward the pipe outlet, due to the outside temperature conditions, usually lower than the inlet temperature. The temperature dependence of the rheological properties of the crude oil (yield stress and viscosity) leads to a nontrivial flow pattern in terms of yielded/unyielded regions. This is equivalent to the flow of a fluid whose viscosity and yield stress are continuously varying along the length of the pipeline.
2. *Oil at rest: shutdown time*. Once the flow stops, the temperature starts dropping until it matches the outside temperature. Under these conditions, heat transfer takes place by conduction and natural convection. The yield stress property of the waxy crude oil usually limits the effects of natural convection because a minimal temperature gradient is necessary to create a flow, that is, the shear stress associated with the temperature gradient has to overcome the yield stress, otherwise no flow is possible.
3. *Resuming the production*. The restart of a waxy crude oil flow consists in injecting a fresh warm oil at the pipe inlet, in order to flush, out of the pipeline, the gelled oil which is a compressible, viscoplastic, thixotropic, and temperature-dependent material. The combined effects of compressibility, viscoplasticity, thixotropy, and temperature led to a complex flow dynamics. Undoubtedly, the prediction of the restart pressure is a difficult task, and the combination of all the particular properties of the material leads, usually, to a prediction fairly below the conservative relation

$P = 4\tau_y \frac{L}{D}$, where τ_y denotes the yield stress, L the length of the pipeline, and D its diameter.

As the flow restarts, a structural breakdown mechanism comes into play. Indeed, the thixotropic properties of the material are essentially related to the gel structure, consisting of crystallized paraffin compounds. This solid gel-like structure can be destroyed by shearing effects in the material. As a consequence, the values of the yield stress and of the viscosity decrease. In a pipeline, because the pressure is imposed at the inlet section, these decreases of the yield stress and viscosity lead to an increase of the flow rate, which in turn increases the shear rate at the pipe-wall. This speeds-up the destruction of the gel structure until steady flow conditions are approached. In the restart process, it is appropriate to say that compressibility acts like a triggering factor. Indeed, the flow created, at the first stages of the restart process, by the compressibility properties of the waxy crude oil, induces a structural breakdown mechanism, leading to a fast drop of the yield stress and viscosity. Therefore, thanks to the compressibility, one can restart the flow using a lower pressure than one required for resuming an incompressible viscoplastic yield stress flow. Finally, the fresh warm oil entering the pipeline is cooled down while flowing downstream to the outlet section. Consequently, a structure buildup occurs; it contributes to the increase of the values of the yield stress and viscosity. This phenomenon is opposite to the shear-driven structure breakdown mechanism. The waxy crude oil restart issue is indeed a relevant illustration of the coupled phenomena, which stem from the yield stress property of the material.

2.2. *Complex fluids in drilling operations*

Yield stress fluids are also common in the Oil & Gas industry for drilling operations (GUILLOT, HENDRIKS, CALLET and VIDICK [1990], SMITH and RAVI [1991], and PEYSSON [2004]). A first example is the drilling muds used today; they are complex fluids made of various components. There exist two main mud families, characterized by the base fluid: *water-based* muds and *oil-based* muds. Water-based muds are essentially clay suspensions with polymer additive, whereas oil-based muds are emulsions of brine (around 30% in volume) in an oil phase with surfactants, polymers, and various chemical additives. These fluids are designed in such a way that they exhibit properties beneficial to the efficient achievement of drilling operations. First, *rock cuttings* need to be removed from the well to the surface (see Figure 2.1, below). Thus, on the one hand, the mud apparent viscosity has to be high enough to assure a sufficiently large drag on the solids.

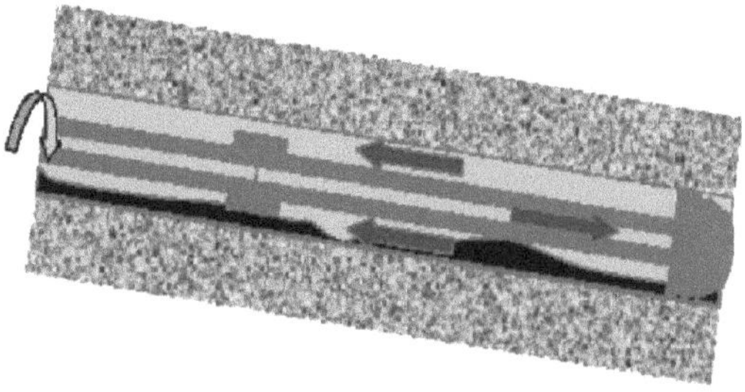

FIG. 2.1 Rock cutting removal.

On the other hand, the power of the surface-located pumping facilities provides an upper bound that should not be exceeded for the viscosity of the drilling mud, otherwise the pumps cannot maintain the flow. Furthermore, ideally those drilling muds must remain homogeneous without any settling of the solid phase when the circulation (visualized in Fig. 2.1) is stopped. Thus, a *yield stress* is needed at rest, to prevent the solid phase sedimentation. Again, the design of this yield stress is a delicate balance between the efficient prevention of the sedimentation and pumping system capabilities (because the higher the yield stress, the higher the apparent viscosity). The amount and nature of clay and polymer additives control these properties.

The drilling of naturally fractured reservoirs may lead to significant mud losses, a real inconvenience for drilling operators. Mud losses in natural fractures are characterized by a typical mud-loss-history curve, obtained by observing the mud pit level evolution. This curve begins with a high mud-loss rate (quick decrease of the mud pit level) as the drilling bit hits the fracture and the mud starts invading the fracture; then the mud-loss rate decreases and, finally, stops because of the yield stress of the mud. This "self-plugging" property can be very convenient to limit mud losses in difficult areas.

Like drilling mud, cement slurry is essential in drilling operations. It is the material used to seal the annular space between the casing ring and the borehole wall. The major goal of the cementing operation is to prevent any circulation of gas, oil or water between different rock layers and to mechanically fix the tubing in the well. The cement slurry is a paste-like material that evolves in time due to chemical reactions; its mechanical characteristics, "between" solid and liquid, make it close to a Bingham material. The yield stress of the slurry is an important parameter because it controls the flow of the cement in the well and, therefore, the accuracy and efficiency of the sealing operation. Indeed, the final position of the cement slurry in the annular space, and the way the slurry displaces the initial mud are crucial for a good mechanical sealing of the well.

2.3. Nuclear energy applications

Nowadays, the energy produced by nuclear power plants worldwide amounts to about 7% of the totally produced energy. The power plant production of this type of energy relies to the controlled use of *nuclear fission* (of uranium, mainly). Nuclear energy is produced by a controlled nuclear chain reaction that produces heat, which is used to produce steam and drive a steam turbine. The turbine is then usually used to produce electricity, but it can also be used to produce mechanical work. In a nuclear reactor, accidents involving the reactor usually mean that the nuclear fission reaction has gone out of control. In such severe situations, the reactor may blaze and the nuclear fuel melts with its steel environment to form a compound called *corium* in nuclear industry. In practice, corium is a blend of nuclear fuel with whatever has melted due to the very-high temperature of the material flowing out of the reactor. Intense heat transfer accompanied by phase transition and chemical reactions all occur in corium. The free-surface flow of corium in a nuclear accident is thus of primary importance, and it is crucial to be able to predict how far the corium layer may flow (PIAR, MICHEL, BABIK, LATCHÉ, GUILLARD and RUGGIERI [1999]).

Focusing on the fluid mechanics aspects of the problem only (that is, heat transfer and chemical reactions are not considered), the corium can be modeled as a yield stress fluid. As a result of this particular rheological feature, the gravity-induced corium flow will actually stop at some point (in contrast to a purely viscous fluid), enabling the engineers to predict the width and thickness of the final layer as well as how far it has reached (VOLA, BABIK and LATCHÉ [2004]). The cooling of the corium layer, due to heat transfer with the surrounding environment, promotes the solidification of the material, which in turn slows down the expansion of the layer (mainly because its yield stress and viscosity increase). An additional issue is the heat transfer with the surface on which the corium is flowing (made of concrete, usually). This surface is quite often severely altered as a result of melting and erosion. Therefore, the prediction of the area covered by the corium layer, as it stops, allows us also to improve the design and safety of nuclear plants, like applying a coating on the concrete surface, or building this surface with another material than simple concrete (like, for example, ceramic materials that withstand more efficiently the rough contact with the corium material).

3. Constitutive laws

3.1. Generalities

A variety of constitutive equations of various form and complexity have been proposed in the literature for modeling the rheological behavior of viscoplastic materials. Our objective is not to discuss all of them, but (1) just to show how to start with the most basic law that mimics the behavior of a yield stress fluid, and then (2) to derive a complex constitutive equation that involves other properties beyond a nonzero yield stress.

The simplest, and at the same time the most relevant yield stress model is the *Bingham* one (BINGHAM [1922]). This model has the following features:

- A nonzero yield stress, which is the threshold value beyond which the material yields.
- A constant plastic viscosity at stress levels beyond the yield stress.

The Bingham model reads as follows:

$$\begin{cases} \boldsymbol{\tau} = 2\mu \mathbf{D} + \tau_y \dfrac{\mathbf{D}}{\|\mathbf{D}\|} & \text{if } \|\boldsymbol{\tau}\| > \tau_y, \\ \mathbf{D} = \mathbf{0} & \text{if } \|\boldsymbol{\tau}\| \leq \tau_y, \end{cases} \tag{3.1}$$

where $\boldsymbol{\tau}$ denotes the extra-stress tensor, $\mathbf{D} = \frac{1}{2}[\nabla \mathbf{u} + (\nabla \mathbf{u})^t]$ is the rate of strain tensor, $\mathbf{u}$ is the medium velocity, τ_y is the yield stress, and μ is the plastic viscosity (that is, the fluid phase viscosity). The norm $\|.\|$ is the *Euclidian* one defined (with obvious notation) by

$$\|\boldsymbol{\chi}\| = \sqrt{\frac{1}{2} \sum_{1 \leq i,j \leq d} |\chi_{ij}|^2}, \quad \forall\, \boldsymbol{\chi} \in \mathbf{R}^{d \times d}, \quad d = 1,\ 2,\ \text{or } 3. \tag{3.2}$$

Starting from the Bingham model (3.1), the first additional feature is the *shear-thinning* property of the plastic viscosity. This leads to the following *Herschel–Bulkley* model:

$$\begin{cases} \boldsymbol{\tau} = 2\mu_0 \dot{\gamma}^{n-1} \mathbf{D} + \tau_y \dfrac{\mathbf{D}}{\|\mathbf{D}\|} & \text{if } \|\boldsymbol{\tau}\| > \tau_y, \\ \mathbf{D} = \mathbf{0} & \text{if } \|\boldsymbol{\tau}\| \leq \tau_y, \end{cases} \tag{3.3}$$

where $n \in [0, 2]$ stands for the shear-thinning coefficient, $\dot{\gamma} = 2\|\mathbf{D}\|$ is the generalized shear rate and μ_0 is the plastic viscosity at zero shear rate. Actually, when $n \in [0, 1)$ (resp., $n \in (1, 2]$) the viscosity is shear-thinning (resp., shear-thickening).

Another simple two-parameter model, which implicitly shows a shear-thinning property is due to *Casson* (BIRD, DAI and YARUSSO [1983]); this model (initially developed to model the viscous behavior of blood) reads as follows:

$$\sqrt{\boldsymbol{\tau}} = \sqrt{\tau_y \frac{\mathbf{D}}{\|\mathbf{D}\|}} + \sqrt{2\mu\mathbf{D}} \quad \text{if } \|\boldsymbol{\tau}\| > \tau_y, \; \mathbf{D} = \mathbf{0} \quad \text{if } \|\boldsymbol{\tau}\| \leq \tau_y, \tag{3.4}$$

which can be written as

$$\boldsymbol{\tau} = \tau_y \frac{\mathbf{D}}{\|\mathbf{D}\|} + 2(\mu + 2\sqrt{\tau_y \mu}\dot{\gamma}^{-1/2})\mathbf{D} \quad \text{if } \|\boldsymbol{\tau}\| > \tau_y, \; \mathbf{D} = \mathbf{0} \quad \text{if } \|\boldsymbol{\tau}\| \leq \tau_y. \tag{3.5}$$

Compared with the Herschel–Bulkley model, the viscosity of the Casson model converges to a nonzero asymptotic value (namely μ) as $\dot{\gamma} \to +\infty$, and the total plastic viscosity depends of the yield stress.

In order to emphasize the other possible properties of yield stress materials, we will go back to the *waxy crude oils*. Indeed, these oils are good representatives of this class of fluids, despite their well-known highly complex rheological behavior. Above the *wax appearance temperature* (WAT), they behave like a simple *Newtonian fluid*. As the temperature drops below the WAT, the viscosity starts to increase sharply and becomes sensitive to mechanical constraints, in relation to the presence of paraffin crystals and the gel-like structure of the materials. The mechanical properties of two Canadian artic crude oils (Cape Allison and Bent Horn) are discussed in CAWKWELL and CHARLES [1989]; this study shows high thixotropic properties and a strong temperature and temperature history dependence. Similarly, the mechanical properties of Australian (Jabiru, Johnson and McKee) and Chinese (Da Qing) crude oils are discussed in WARDAUGH and BOGER [1987]; this study shows that the rheological behavior of these oils is also strongly affected by the shear rate and temperature history. Waxy crude oil can usually be modeled by a nonisothermal thixotropic and viscoplastic constitutive equation. The models encountered in the literature consist of generalized standard viscoplastic models (Bingham's or Herschel–Bulkley's). In RONNINGSEN [1992], one introduces the time dependence by merely allowing the yield stress and plastic velocity to be functions of time. In HOUSKA [1981] and SESTAK, CHARLES, CAWKWELL and HOUSKA [1987], one introduces, in the standard viscoplastic model, a scalar variable that describes the structure of the material. This structure parameter obeys a first-order transient differential equation, akin to a first-order chemical advection-reaction equation; moreover, in the Houska's model, one assumes that the yield stress and plastic viscosity are *affine* functions of the structure parameter. In the early 1970s, Perkins and Turner (see PERKINS and TURNER [1971]) developed a thixotropic model, in which the time effect is considered through the cumulative strain undergone by the material. In this model, the yield stress and the viscosity vary with the reciprocal of the cumulative strain, which, in turn, is a function of time. From a temperature view point, in CAWKWELL and CHARLES [1989] one extends the Houska's model to nonisothermal situations, by simply replacing the constant rheological parameters of the model by temperature-dependent ones. In HÉNAUT [2002], the presence of gas voids in the oil bulk is evidenced; this imparts a kind of compressibility to the material. Recently, in VINAY, WACHS and AGASSANT [2006], one has extended the standard Herschel–Bulkley model, in order to account for compressible effects.

Because of the lack of general thermodynamic theory of viscoplastic fluids, the thermal effects in the constitutive equations rely on the temperature dependence of the rheological parameters. In practice, the temperature dependence of the yield stress, and of the viscosity, is provided by experimental data. The *nonisothermal* Bingham model can be written as follows:

$$\begin{cases} \boldsymbol{\tau} = 2\mu(\theta)\mathbf{D} + \tau_y(\theta)\dfrac{\mathbf{D}}{\|\mathbf{D}\|} & \text{if } \|\boldsymbol{\tau}\| > \tau_y(\theta), \\ \mathbf{D} = \mathbf{0} & \text{if } \|\boldsymbol{\tau}\| \leq \tau_y(\theta), \end{cases} \tag{3.6}$$

where θ denotes the temperature.

The inclusion of *compressible effects* requires, in principle, the introduction of a *second viscosity* coefficient, denoted by ξ, here. Then, the compressible Bingham model reads as follows:

$$\begin{cases} \boldsymbol{\tau} = 2\mu\mathbf{D} + \left[\left(\xi - \dfrac{2}{3}\mu\right)\nabla\cdot\mathbf{u}\right]\mathbf{I} + \tau_y\dfrac{\mathbf{D}}{\|\mathbf{D}\|} & \text{if } \|\boldsymbol{\tau}\| > \tau_y, \\ \mathbf{D} = \mathbf{0} & \text{if } \|\boldsymbol{\tau}\| \leq \tau_y, \end{cases} \tag{3.7}$$

where $\mathbf{I}$ denotes the identity tensor. If the material is assumed to be a *Stokes fluid*, that is the viscosity forces are due to the shear only, but not from volume variation, then the second viscosity coefficient vanishes, that is $\xi = 0$. In such a situation, the constitutive law of a compressible Bingham fluid reads as follows:

$$\begin{cases} \boldsymbol{\tau} = 2\mu\left[\mathbf{D} - \dfrac{1}{3}(\nabla\cdot\mathbf{u})\mathbf{I}\right] + \tau_y\dfrac{\mathbf{D}}{\|\mathbf{D}\|} & \text{if } \|\boldsymbol{\tau}\| > \tau_y, \\ \mathbf{D} = \mathbf{0} & \text{if } \|\boldsymbol{\tau}\| \leq \tau_y. \end{cases} \tag{3.8}$$

The introduction of time effects, that is of the thixotropy, in the constitutive equation has been, and still is, a subject of extensive research. The reason of such investigations, derives from the sheer nature of viscoplastic materials, which are often suspensions, soft materials, polymers, and so on. The fine description and clear understanding of their microscopic structure should enable us to improve the derivation of constitutive equations at the mathematical level. The yield stress and viscosity decays, observed experimentally, are essentially related to a structure breakdown mechanism. Usually, this breakdown mechanism relies on a strain and/or strain rate; various models have been proposed in the literature to describe this effect, several of them being briefly discussed below.

3.2. The Houska's model

In the Houska's model (introduced in HOUSKA [1981]), both the yield stress and viscosity are divided into two parts:

1. A permanent (time-independent) part.
2. A thixotropic (time-dependent) part.

Both the yield stress and viscosity are *affine functions* of a *structure parameter* λ, as shown by

$$\tau_y = \tau_{y0} + \lambda \tau_{y1}, \tag{3.9}$$

$$\mu = \mu_0 + \lambda \Delta \mu_0, \tag{3.10}$$

λ being a function of x and t taking its values in the closed interval $[0, 1]$. We have, accordingly:

At $\lambda = 0$, the structure is completely broken down and the yield stress and viscosity reach their minimal values.
At $\lambda = 1$, the material is fully structured and the yield stress and viscosity have their maximal values.

The structure breakdown mechanism, which describes the yield stress and viscosity decays, is driven by shear rate. The space-time distribution of the structure parameter relies on a first-order equation (of the advection-reaction type), while the shear-thinning viscosity is assumed to obey a power law. Finally, the *complete Houska's model* reads as follows:

$$\begin{cases} \boldsymbol{\tau} = 2(\mu_0 + \lambda \Delta \mu_0)\dot{\gamma}^{n-1}\mathbf{D} + (\tau_{y0} + \lambda \tau_{y1})\dfrac{\mathbf{D}}{\|\mathbf{D}\|} & \text{if } \|\boldsymbol{\tau}\| > \tau_{y0} + \lambda \tau_{y1}, \\ \mathbf{D} = \mathbf{0} \quad \text{if } \|\boldsymbol{\tau}\| \leq \tau_{y0} + \lambda \tau_{y1}, \end{cases} \tag{3.11}$$

$$\frac{\partial \lambda}{\partial t} + \mathbf{u} \cdot \nabla \lambda = a(1 - \lambda) - b\dot{\gamma}^m, \tag{3.12}$$

where μ_0 denotes the permanent viscosity, $\Delta \mu_0$ the thixotropic viscosity, τ_{y0} the permanent yield stress, τ_{y1} the thixotropic yield stress, n the shear-thinning coefficient, a the buildup parameter, b the breakdown parameter, and m an adjustable parameter.

Actually, a close inspection reveals that the Houska's model is a generalized Herschel–Bulkley model (see (3.3)), in which the rheological parameters, namely yield stress and viscosity, are affine functions of the structure parameter λ.

3.3. The Perkins and Turner model

The *Perkins and Turner model* was introduced in the early 1970s (PERKINS and TURNER [1971]) to mimic the thixotropic behavior of waxy crude oils. The thixotropy is handled by allowing the rheological parameters to be functions of the cumulative strain. The underlying idea implies that the more strained is the material, the more broken down should be the microscopic structure. Consequently, yield stress and viscosity vary with the reciprocal of the cumulative strain. The complete Perkins and Turner model reads as follows:

$$\begin{cases} \boldsymbol{\tau} = 2\mu(\varepsilon)\mathbf{D} + \tau_y(\varepsilon)\dfrac{\mathbf{D}}{\|\mathbf{D}\|} & \text{if } \|\boldsymbol{\tau}\| > \tau_y(\varepsilon), \\ \mathbf{D} = \mathbf{0} & \text{if } \|\boldsymbol{\tau}\| \leq \tau_y(\varepsilon), \end{cases} \tag{3.13}$$

$$\frac{\partial \varepsilon}{\partial t} + \mathbf{u} \cdot \nabla \varepsilon = \dot{\gamma}, \tag{3.14}$$

$$\tau_y(\varepsilon) = \frac{\tau_y}{(1 + \varepsilon)^{b_1}}, \quad \mu(\varepsilon) = \mu_0 + \frac{b_2 \dot{\gamma}^{b_3 - 1}}{(1 + \varepsilon)^{b_4}}, \tag{3.15}$$

where b_1, b_2, b_3, and b_4 are obtained by fitting experimental data.

A simple analysis easily reveals that the models of *Houska* [(3.11), (3.12)] and *Perkins and Turner* [(3.13)–(3.15)] are qualitatively very similar. The role of the structure parameter λ in the Houska's model is equivalent to the one of the cumulative strain ε in the Perkins and Turner model. The main difference concerns the range in which these parameters vary: $\lambda(x, t) \in [0, 1]$, whereas $\varepsilon(x, t) \in [0, +\infty]$.

3.4. The phenomenological model of Chang, Nguyen, and Ronningsen

In CHANG, NGUYEN and RONNINGSEN [1999], one provides a descriptive model involving thixotropy, creeping behavior, and elasticity. In fact, many investigators carried out experiments that revealed the possibility of a creep motion of the material below the yield stress. If the experiments are carried long enough, the material finally yields. This behavior is known as the creeping below the yield stress and it considerably complicates the meaning given to the yield stress. Actually, it follows from these observations that a timescale has to be introduced in the model; moreover, after a prolonged mechanical constraint, the final yielding may also be considered as a form of thixotropy. From these various considerations, Chang et al proposed a *two-yield stress concept*, where

1. τ_{ye} is the *elastic limit yield stress*, corresponding to the limit of elastic reversible deformation in the material.
2. τ_{ys} is the *static yield stress*, corresponding to the classical definition of the yield stress, that is, a level of stress that needs to be exceeded to entail a flow.

Accordingly, if one applies a stress τ_w, Chang et al proposed three types of behavior, namely:

1. If $\tau_w < \tau_{ye}$, the material is subjected to reversible elastic deformation and there is no flow.
2. If $\tau_{ye} < \tau_w < \tau_{ys}$, the material undergoes a very small plastic deformation (creep behavior) up to the yielding of the material, either because the applied stress τ_w slowly increases with time, or because the yield stress τ_{ys} slowly decreases with time as a result of thixotropy (a slow structure breakdown mechanism).
3. If $\tau_w > \tau_{ys}$, the applied stress overcomes the yield stress and the material starts flowing.

The creeping behavior proposed by Chang et al also suggests that the yielding of the material may be controlled by a critical strain or strain rate, instead of a critical stress as the yield stress. But this issue goes far beyond the scope of this article.

4. Numerical methods

4.1. Generalities

No matter how sophisticated the yield stress rheological model may be at the mathematical level, we still need to use a suitable method to deal with the two main difficulties associated with yield stress models, namely:

1. The *nondifferentiability* of the stress–strain rate relation at the yield point.
2. The indeterminate nature of the stress field below the yield point.

Because the simplest yield stress model, namely (3.1), exhibits these two features, we will focus on the *Bingham model* in this section. Irrespective of the rheological model, the flow of the material is governed by the following conservation equations:

- *The mass conservation equation.* We assume here that the material is *incompressible*. It follows from this property that the mass conservation equation reduces to the *divergence-free* condition for the velocity field, that is

$$\nabla \cdot \mathbf{u} = 0. \tag{4.1}$$

- *The momentum equation.* This a well-known equation in Fluid Mechanics. Usually, for the visco-plastic flows that we consider the advection terms are small compared with the other terms and may be neglected. The momentum equation takes then the following form:

$$\rho \frac{\partial \mathbf{u}}{\partial t} - \nabla \cdot \boldsymbol{\tau} + \nabla p = \mathbf{f}, \tag{4.2}$$

where ρ (> 0) stands for the *material density*, p for the *isotropic pressure*, and where $\mathbf{f}$ models the external forces (situations where $\mathbf{f} = \mathbf{0}$ are fairly common).

Suppose now that the Bingham fluid is filling a bounded region (domain) Ω of $\mathbf{R}^d$, $d = 1, 2$, or 3 and that $(0, T)$ is a time interval. Let's assume that Dirichlet boundary conditions hold for the velocity field $\mathbf{u}$ on the boundary Γ of Ω. Assuming that the data $\mathbf{f}$, $\mathbf{g}$, and $\mathbf{u}_0$ are smooth enough, the *transient creeping flow* of a Bingham material satisfies the following system:

$$\{\mathbf{u}(t), p(t)\} \in (H^1(\Omega))^d \times L^2(\Omega), \quad \text{a.e. on } (0, T),$$

$$\rho \frac{\partial \mathbf{u}}{\partial t} - \nabla \cdot \boldsymbol{\tau} + \nabla p = \mathbf{f}, \quad \text{in } \Omega \times (0, T) \tag{4.3}$$

$$\nabla \cdot \mathbf{u} = 0 \text{ in } \Omega \times (0, T) \tag{4.4}$$

$$\begin{cases} \boldsymbol{\tau} = 2\mu \mathbf{D} + \tau_y \dfrac{\mathbf{D}}{\|\mathbf{D}\|} & \text{if } \|\boldsymbol{\tau}\| > \tau_y, \\ \mathbf{D} = \mathbf{0} & \text{if } \|\boldsymbol{\tau}\| \leq \tau_y, \end{cases} \tag{4.5}$$

$$\mathbf{u} = \mathbf{g} \text{ on } \Gamma \times (0, T), \quad \text{with } \int_\Gamma \mathbf{g}(t).\mathbf{n}\, d\Gamma = 0, \text{ a.e. on } (0, T), \tag{4.6}$$

$$\mathbf{u}|_{t=0} = \mathbf{u}_0 \text{ with } \nabla \cdot \mathbf{u}_0 = 0. \tag{4.7}$$

In the Bingham model (4.3)–(4.7):

1. The *Sobolev space* $H^1(\Omega)$ is defined by

$$H^1(\Omega) = \left\{ v | v \in L^2(\Omega), \frac{\partial v}{\partial x_i} \in L^2(\Omega), \forall i = 1, \ldots, d \right\}, \tag{4.8}$$

the derivatives in (4.8) being in the sense of *distributions* (see, e.g., TARTAR [2007]).

2. We have used the notation $\varphi(t)$ for the function $x \to \varphi(x, t)$.
3. $\mathbf{n}$ denotes the unit outward normal vector at the boundary Γ of Ω.

With the exception of Szabo and Hassager (see SZABO and HASSAGER [1992]) who computed the flow in the yielded regions only, and then tracked the yield surfaces with a remeshing approach, there are mainly two families of computational methods. The methods of the *first family* rely on *variational inequality* formulations, an approach pioneered in DUVAUT and LIONS [1972a, 1976] for the mathematical analysis of Bingham flow problems, and followed by others from a computational point of view (as in, e.g., BRISTEAU and GLOWINSKI [1974], GLOWINSKI, LIONS and TRÉMOLIÈRES [1976, 1981], FORTIN and GLOWINSKI [1982, 1983], and GLOWINSKI and LE TALLEC [1989]). The associated computational methodology relies on a variety of *multiplier* functions (Lagrange's and others), the corresponding solution algorithms being of the *Uzawa* type. As already mentioned, the underlying mathematical developments may be difficult to follow for some, but the practical implementation is fairly easy because of the modularity of the methodology. The methods of the *second family* rely on *regularization*, this approach stemming from the recognition that the flow can not be computed directly, and requires thus a mean to circumvent the nondifferentiability of the constitutive law. Regularization methods have been applied, in GLOWINSKI, LIONS and TRÉMOLIÈRES [1976, 1981] to the numerical simulation of Bingham flow in cylinders; generalizations to Bingham flow in cavities can be found in BERCOVIER and ENGELMAN [1980], and PAPANASTASIOU [1987]. Various regularization methods are available; their common feature is the fact that they approximate the true, discontinuous constitutive law by a not only continuous but differentiable one. The resulting approximate problem involves an operator that is continuously differentiable, thus, allowing the use of the Newton's method for the computation of the solution. Regularization is achieved through the introduction of a parameter. This regularization parameter may be tuned so that the approximate model is as close as desired to the exact one. In practice, the smaller the regularization parameter, the closer the approximate model to the exact one.

The two approaches, briefly described above, have generated a large number of publications and have been widely used by practitioners, providing valuable results. The related methods will be described below.

4.2. Multipliers techniques

For the methods of this family, the starting point is the availability of a *variational formulation* of the mechanical problem under consideration. To the best of our knowledge, the *variational inequality formulation* of the Bingham flow problem (4.3)–(4.7) is due to Duvaut and Lions (see DUVAUT and LIONS [1972a, 1976]); this formulation can be found in Chapter 2 of this article. Basically, there are two families of *multiplier-based algorithms*, allowing the numerical solution of the Bingham flow problem, through its variational inequality formulation. One has, more precisely (and chronologically):

1. *Projection-like algorithms* based on the introduction of a (kind of) multiplier field (not exactly of the Lagrange's type).
2. *Augmented Lagrangian algorithms* based on the introduction of a Lagrange multiplier field and of an additional strain rate tensor field.

In order to show how to apply the variational methods discussed in DUVAUT and LIONS [1976] (see also GLOWINSKI, LIONS and TRÉMOLIÈRES [1976, 1981]) we consider the Bingham model constitutive law defined by (3.1), assuming that both $\boldsymbol{\tau}$ and $\mathbf{D}$ belong to $(L^2(\Omega))^{d\times d}$. Using relatively simple results of *Convex Analysis* (available, for example, in EKELAND and TEMAM [1999], and IONESCU and SOFONEA [1986]), one can prove that (3.1) has the following equivalent variational formulation (with $\mathrm{d}x = \mathrm{d}x_1 \ldots \mathrm{d}x_d$):

$$2\int_\Omega \mu \mathbf{D} : (\mathbf{q} - \mathbf{D})\mathrm{d}x + \sqrt{2}\int_\Omega \tau_y(|\mathbf{q}| - |\mathbf{D}|)\mathrm{d}x \geq \int_\Omega \boldsymbol{\tau} : (\mathbf{q} - \mathbf{D})\mathrm{d}x, \ \forall\, \mathbf{q} \in (L^2(\Omega))^{d\times d}, \tag{4.9}$$

where $\mathbf{S}:\mathbf{T}$ (resp., $|\mathbf{S}|$) denotes the *Fröbenius* scalar product (resp., norm) of the two $d \times d$ tensors $\mathbf{S}$ and $\mathbf{T}$ (resp., of the $d \times d$ tensor $\mathbf{S}$), that is,

$$\mathbf{S} : \mathbf{T} = \sum_{1\leq i,j\leq d} s_{ij}\, t_{ij} \ \text{and} \ |\mathbf{S}| = \sqrt{\mathbf{S}:\mathbf{S}} = \sqrt{\sum_{1\leq i,j\leq d} s_{ij}^2}, \ \forall \quad \mathbf{S} = (s_{ij})_{1\leq i,j\leq d},\ \mathbf{T} = (t_{ij})_{1\leq i,j\leq d} \tag{4.10}$$

(we have thus $|\,.\,| = \sqrt{2}\|\cdot\|$, with the tensorial norm $\|\cdot\|$ defined by (3.2)).

From the symmetry of $\boldsymbol{\tau}$, a variational formulation corresponding to the conservation equations (4.1) and (4.2) reads as follows:

$$\int_\Omega \nabla\cdot\mathbf{u}\, q\mathrm{d}x = 0, \ \forall\, q \in L^2(\Omega), \tag{4.11}$$

$$\rho\int_\Omega \frac{\partial \mathbf{u}}{\partial t}\cdot\mathbf{v}\mathrm{d}x + \int_\Omega \boldsymbol{\tau} : \mathbf{D}(\mathbf{v})\mathrm{d}x - \int_\Omega p\nabla\cdot\mathbf{v}\mathrm{d}x = \int_\Omega \mathbf{f}\cdot\mathbf{v}\mathrm{d}x, \ \forall \mathbf{v} \in (H_0^1(\Omega))^d, \tag{4.12}$$

with $H_0^1(\Omega) = \{v | v \in H^1(\Omega), v = 0 \text{ on } \Gamma\}$. Combining (4.9), (4.11), and (4.12), we obtain the following *variational inequality formulation* of the unsteady flow of an incompressible Bingham fluid:

Find $\{\mathbf{u}, p\} \in (H^1(\Omega))^d \times L^2(\Omega)$ such that

$$\int_\Omega \nabla\cdot\mathbf{u}\, q\mathrm{d}x = 0, \ \forall\, q \in L^2(\Omega), \tag{4.13}$$

$$\rho\int_\Omega \frac{\partial \mathbf{u}}{\partial t}\cdot(\mathbf{v}-\mathbf{u})\mathrm{d}x - \int_\Omega p\nabla\cdot(\mathbf{v}-\mathbf{u})\mathrm{d}x + 2\mu\int_\Omega \mathbf{D}(\mathbf{u}) : \mathbf{D}(\mathbf{v}-\mathbf{u})\mathrm{d}x + \sqrt{2}\int_\Omega \tau_y(|\mathbf{D}(\mathbf{v})| - |\mathbf{D}(\mathbf{u})|)\mathrm{d}x \geq \int_\Omega \mathbf{f}\cdot(\mathbf{v}-\mathbf{u})\mathrm{d}x, \ \forall\, \mathbf{v} \in (H_0^1(\Omega))^d, \tag{4.14}$$

to be completed by (4.6) and (4.7); above, we have used the dot product notation for the canonical Euclidian scalar product of two vectors of $\mathbf{R}^d$. Applying, for example, the *backward Euler scheme* to the time-discretization of (4.6), (4.7), (4.13), (4.14), we obtain (with $\Delta t(>0)$ a time discretization step that we suppose constant for simplicity):

$$\mathbf{u}^0 = \mathbf{u}_0, \tag{4.15}$$

and for $n \geq 1$, $\mathbf{u}^{n-1}$ being known, we obtain $\mathbf{u}^n$ and p^n from the solution of the following system:

$$\int_\Omega q\nabla \cdot \mathbf{u}^n \, \mathrm{d}x = 0, \ \forall\, q \in L^2(\Omega), \tag{4.16}$$

$$\begin{aligned}
&\rho \int_\Omega \frac{\mathbf{u}^n - \mathbf{u}^{n-1}}{\Delta t} \cdot (\mathbf{v}-\mathbf{u}^n)\mathrm{d}x + 2\mu \int_\Omega \mathbf{D}(\mathbf{u}^n) : \mathbf{D}(\mathbf{v} - \mathbf{u}^n)\mathrm{d}x \\
&\quad - \int_\Omega p^n \nabla \cdot (\mathbf{v} - \mathbf{u}^n)\mathrm{d}x + \sqrt{2} \int_\Omega \tau_y(|\mathbf{D}(\mathbf{v})| - |\mathbf{D}(\mathbf{u}^n)|)\mathrm{d}x \\
&\quad \geq \int_\Omega \mathbf{f}(n\Delta t) \cdot (\mathbf{v} - \mathbf{u}^n)\mathrm{d}x, \quad \forall \mathbf{v} \in (H_0^1(\Omega))^d,
\end{aligned} \tag{4.17}$$

$$\mathbf{u}^n = \mathbf{g}(n\Delta t) \text{ on } \Gamma. \tag{4.18}$$

Assuming that the functions $\mathbf{u_0}$, $\mathbf{f}$, and $\mathbf{g}$ are sufficiently smooth, it can be proved (see, e.g., DUVAUT and LIONS [1972a, 1976]) that the system (4.16)–(4.18) has a unique solution $\{\mathbf{u}^n, p^n\}$ in $\mathbf{V}_{\mathbf{g}^n} \times L_0^2(\Omega)$, where $\mathbf{V}_{\mathbf{g}^n} = \{\mathbf{v} | \mathbf{v} \in (H^1(\Omega))^d,\ \mathbf{v} = \mathbf{g}(n\Delta t) \text{ on } \mathbf{\Gamma}\}$ and $L_0^2(\Omega) = \{q \in L^2(\Omega) | \int_\Omega q \ \mathrm{d}x = 0\}$. Actually, the above pair $\{\mathbf{u}^n,\ p^n\}$ is the unique solution of the following *saddle-point problem*:

$$\begin{aligned}
&\{\mathbf{u}^n,\ p^n\} \in \mathbf{V}_{\mathbf{g}^n} \times L_0^2(\Omega), \\
&\mathcal{G}_n(\mathbf{u}^n,\ q) \leq \mathcal{G}_n(\mathbf{u}^n,\ p^n) \leq \mathcal{G}_n(\mathbf{v},\ p^n), \ \ \forall\ \{\mathbf{v}, q\} \in \mathbf{V}_{\mathbf{g}^n} \times L_0^2(\Omega)
\end{aligned} \tag{4.19}$$

with

$$\begin{aligned}
\mathcal{G}_n(\mathbf{v},\ q) = {} & \frac{1}{2}\rho \int_\Omega |\mathbf{v}|^2 \mathrm{d}x + \Delta t\mu \int_\Omega |\mathbf{D}(\mathbf{v})|^2 \mathrm{d}x + \Delta t\sqrt{2} \int_\Omega \tau_y |\mathbf{D}(\mathbf{v})| \mathrm{d}x \\
& - \Delta t \int_\Omega q\nabla \cdot \mathbf{v}\mathrm{d}x - \int_\Omega (\Delta t \mathbf{f}^n + \rho\, \mathbf{u}^{n-1}) \cdot \mathbf{v}\mathrm{d}x.
\end{aligned} \tag{4.20}$$

Both formulations (either as *variational inequality* or as a *saddle-point problem*) are clearly an improvement when compared with the original formulation (4.3)–(4.7). The next step is to derive a tractable computational method with good convergence properties. At this stage, we see two main approaches:

1. *The decomposition-coordination approach*

Here, one introduces an auxiliary additional variable **d** representing the strain rate tensor **D(u)** (*decomposition step*). The saddle-point problem (4.19) is clearly equivalent to

$$\begin{aligned}&\{\{\mathbf{u}^n, \mathbf{d}^n\}, p^n\} \in \mathbf{W}_{\mathbf{g}^n} \times L_0^2(\Omega),\\ &\mathcal{G}_n^*(\mathbf{u}^n, \mathbf{d}^n, q) \leq \mathcal{G}_n^*(\mathbf{u}^n, \mathbf{d}^n, p^n) \leq \mathcal{G}_n^*(\mathbf{v}, \mathbf{q}, p^n), \forall\ \{\{\mathbf{v}, \mathbf{q}\}, q\} \in \mathbf{W}_{\mathbf{g}^n}\\ &\quad \times L_0^2(\Omega),\end{aligned} \tag{4.21}$$

another saddle-point problem, where

$$\begin{aligned}\mathcal{G}_n^*(\mathbf{v}, \mathbf{q}, q) = \frac{1}{2}\rho \int_\Omega |\mathbf{v}|^2 \mathrm{d}x + \Delta t \mu \int_\Omega |\mathbf{D}(\mathbf{v})|^2 \mathrm{d}x + \Delta t \sqrt{2} \int_\Omega \tau_y |\mathbf{q}| \mathrm{d}x\\ - \Delta t \int_\Omega q \nabla \cdot \mathbf{v} \mathrm{d}x - \int_\Omega (\Delta t \mathbf{f}^n + \rho\, \mathbf{u}^{n-1}) \cdot \mathbf{v} \mathrm{d}x.\end{aligned} \tag{4.22}$$

and

$$\mathbf{W}_{\mathbf{g}^n} = \{\{\mathbf{v}, \mathbf{q}\} | \mathbf{v} \in \mathbf{V}_{\mathbf{g}^n},\ \mathbf{q} \in (L^2(\Omega))^{d\times d},\ \mathbf{D}(\mathbf{v}) - \mathbf{q} = \mathbf{0}\}. \tag{4.23}$$

The *coordination step* consists in introducing a *Lagrange multiplier field* $\boldsymbol{\lambda}^n \in (L^2(\Omega))^{d\times d}$ (homogeneous to a plastic stress tensor), to relax the constraint $\mathbf{D}(\mathbf{u}^n) - \mathbf{d}^n = \mathbf{0}$. This leads to the following saddle-point problem:

$$\begin{aligned}&\mathbf{u}^n \in \mathbf{V}_{\mathbf{g}^n}, \mathbf{d}^n \in (L^2(\Omega))^{d\times d}, \boldsymbol{\lambda}^n \in (L^2(\Omega))^{d\times d}, p^n \in L_0^2(\Omega),\\ &\tilde{\mathcal{L}}_n(\mathbf{u}^n, \mathbf{d}^n, \boldsymbol{\mu}, q) \leq \tilde{\mathcal{L}}_n(\mathbf{u}^n, \mathbf{d}^n, \boldsymbol{\lambda}^n, p^n) \leq \tilde{\mathcal{L}}_n(\mathbf{v}, \mathbf{q}, \boldsymbol{\lambda}^n, p^n),\\ &\forall\, \mathbf{v} \in \mathbf{V}_{\mathbf{g}^n}, \mathbf{q} \in (L^2(\Omega))^{d\times d}, \boldsymbol{\mu} \in (L^2(\Omega))^{d\times d}, q \in L_0^2(\Omega),\end{aligned} \tag{4.24}$$

where the Lagrangian $\tilde{\mathcal{L}}_n$ is defined by

$$\tilde{\mathcal{L}}_n(\mathbf{v}, \mathbf{q}, \boldsymbol{\mu}, q) = \mathcal{G}_n^*(\mathbf{v}, \mathbf{q}, q) + \int_\Omega \boldsymbol{\mu} : (\mathbf{D}(\mathbf{v}) - \mathbf{q}) \mathrm{d}x. \tag{4.25}$$

The Lagrangian functional $\tilde{\mathcal{L}}_n$ can be *augmented* as follows:

$$\tilde{\mathcal{L}}_{nr}(\mathbf{v}, \mathbf{q}, \boldsymbol{\mu}, q) = \tilde{\mathcal{L}}_n(\mathbf{v}, \mathbf{q}, \boldsymbol{\mu}, q) + \frac{1}{2} r \int_\Omega |\mathbf{D}(\mathbf{v}) - \mathbf{q}|^2 \mathrm{d}x, \tag{4.26}$$

where r (>0) is the *augmentation* (in fact a *penalty*) parameter.

To solve the saddle-point problem (4.24), we advocate the *Uzawa's algorithms* discussed in, e.g., FORTIN and GLOWINSKI [1982, 1983], GLOWINSKI and LE TALLEC [1989] (actually, we used the algorithm called ALG2 in the above references). The advantage of this approach is that it decouples, in some sense, the computation of $\mathbf{u}^n$ from that of $\mathbf{d}^n$. In particular, for $\{\mathbf{v}, \boldsymbol{\mu}, q\}$ given, the problem consisting in the minimization, with respect to $\mathbf{q}$ in $(L^2(\Omega))^{d\times d}$, of the augmented Lagrangian $\tilde{\mathcal{L}}_{nr}$ has a unique solution whose *closed form* is easy to compute, thus, circumventing the difficulties associated with the nondifferentiability of the functional $\mathbf{v} \to \Delta t\sqrt{2}\int_\Omega \tau_y|\mathbf{D}(\mathbf{v})|\mathrm{d}x$. In Chapters 2 and 3, we will return on the application of augmented Lagrangian algorithms (such as ALG2) to the solution of viscoplastic flow problems.

2. *The orthogonal projection approach*

This approach relies on the equivalence (proved for the first time in DUVAUT and LIONS [1972a]; see also DUVAUT and LIONS [1976]) between the system (4.3)–(4.7) and the following one

$$\{\mathbf{u}(t), \boldsymbol{\lambda}(t), p(t)\} \in (H^1(\Omega))^d \times (L^2(\Omega))^{d\times d} \times L^2(\Omega), \text{ a.e. on } (0, T)$$

$$\rho\frac{\partial \mathbf{u}}{\partial t} - 2\mu\nabla\cdot\mathbf{D}(\mathbf{u}) - \sqrt{2}\tau_y\nabla\cdot\boldsymbol{\lambda} + \nabla p = \mathbf{f} \text{ in } \Omega\times(0, T), \tag{4.27}$$

$$\nabla\cdot\mathbf{u} = 0 \text{ in } \Omega\times(0, T), \tag{4.28}$$

$$\boldsymbol{\lambda} = \boldsymbol{\lambda}^t, \tag{4.29}$$

$$|\boldsymbol{\lambda}(x, t)| \le 1, \text{ a.e. in } \Omega\times(0, T), \tag{4.30}$$

$$\boldsymbol{\lambda}.\mathbf{D}(\mathbf{u}) = |\mathbf{D}(\mathbf{u})|, \tag{4.31}$$

$$\mathbf{u} = \mathbf{g} \text{ on } \Gamma\times(0, T), \tag{4.32}$$

$$\mathbf{u}(0) = \mathbf{u}_0. \tag{4.33}$$

Relations (4.29)–(4.31) imply that, a.e. on $(0, T)$,

$$\boldsymbol{\lambda}(t) = P_{\boldsymbol{\Lambda}}(\boldsymbol{\lambda}(t) + r\mathbf{D}(\mathbf{u}(t))), \quad \forall r > 0 \tag{4.34}$$

and conversely; in (4.34): (i) $\boldsymbol{\Lambda}$ is the closed convex subset of $(L^2(\Omega))^{d\times d}$ (of $(L^\infty(\Omega))^{d\times d}$, actually) defined by

$$\boldsymbol{\Lambda} = \{\boldsymbol{\mu} | \boldsymbol{\mu} \in (L^2(\Omega))^{d\times d},\ \boldsymbol{\mu} = \boldsymbol{\mu}^t,\ |\boldsymbol{\mu}(x)| \le 1, \text{ a.e. on } \Omega\}. \tag{4.35}$$

(ii) $P_{\boldsymbol{\Lambda}}$ is the *orthogonal projection operator* from $(L^2(\Omega))^{d\times d}$ onto $\boldsymbol{\Lambda}$. One has

$$P_{\boldsymbol{\Lambda}}(\boldsymbol{\mu}(x)) = \frac{\boldsymbol{\mu}(x) + \boldsymbol{\mu}^t(x)}{\max(2, |\boldsymbol{\mu}(t) + \boldsymbol{\mu}^t(x)|)} \text{ a.e. on } \Omega,\ \forall\, \boldsymbol{\mu} \in (L^2(\Omega))^{d\times d}. \tag{4.36}$$

Numerical simulation methods for Bingham flow, based on the equivalence between (4.3)–(4.7) and (4.27)–(4.33), and taking advantage of (4.34), will be discussed in Chapter 2 (see also DEAN, GLOWINSKI and GUIDOBONI [2007], and the references therein).

The computational methods derived from either the *augmented Lagrangian* or *orthogonal projection* approaches are fairly *modular*, making them relatively easy to implement.

4.3. *Regularization methods*

The multiplier-based solution methods discussed in Section 4.2 rely on relatively sophisticated tools from *Convex Duality theory*; moreover, the convergence of the associated algorithms (of the Uzawa's type, essentially) may be slow (although several techniques to speed up their convergence can be found in Chapter 2, and also in DEAN, GLOWINSKI and GUIDOBONI [2007]). These facts explain why some practitioners have looked for conceptually simpler methods in order to simulate yield stress flows. Among these methods, one finds the *regularization methods*, briefly mentioned already in Section 4.1. The idea behind the regularization methods is pretty simple: it consists in approximating the nonsmooth constitutive law modeling yield stress flows by a smoother one. A convenient approach to regularization is to write the original constitutive law in terms of apparent (or equivalent) viscosity, and then to approximate this viscosity by a smoother one making computations possible without recourse to multipliers (through the Newton's method, for example). For simplicity, we will choose the Bingham model to describe some of these regularization procedures (the generalization to more complicated viscoplastic models is straightforward). Thus, considering the constitutive law (3.1) of the Bingham model, we observe that (3.1) is equivalent to the following system:

$$\boldsymbol{\tau} = 2\mu_e \mathbf{D}(\mathbf{u}) \tag{4.37}$$

$$\mu_e = \mu + \frac{\tau_y}{2\|\mathbf{D}(\mathbf{u})\|}. \tag{4.38}$$

The above system is obviously well-suited to yielded regions, that is, regions where the strain-rate tensor $\mathbf{D}(\mathbf{u})$ is nonzero. However, if $\|\mathbf{D}(\mathbf{u})\| \to 0$, that is $\|\boldsymbol{\tau}\| \to \tau_y$, then $\mu_e \to +\infty$, preventing the use of standard computational methods. The basic idea behind regularization is to approximate the above nonsmooth equivalent viscosity μ_e by $\mu_{e,\varepsilon}$ which is finite everywhere, although necessarily "very" large in those regions where $\mu_e = +\infty$. Let us give our readers some justification for the use of regularization procedures:

1. From a *practical* point of view, *regularization* is a way to make computations possible (via the Newton's method, for example), and to obtain, at the same time solutions close to the actual one if the regularization parameter ε is "small enough" (actually, this requires also the usual space-time discretization parameters to be also small enough, but there is nothing new with this requirement).
2. From a *rheological* point of view, some scientists claim that regularized models are in fact closer to the physical reality since *true yield stress fluids do not exist*, and that in any case the material under consideration will strain, though very slightly. This argument takes us back to the *yield stress* "myth" and the wild debate that many rheologists have "enjoyed" for the last thirty years (see Section 1); in this article we will not further enter in this debate (the interested readers should look at BARNES and WALTERS [1985], and BARNES [1999]).
3. From a *mathematical* point of view, regularization provides procedures to approximate nondifferentiable functionals and operators by smoother ones. Some 'classical' regularization procedures (formulated in term of *equivalent viscosity*) are described below:

   a. The *exponential* model of PAPANASTASIOU [1987]:

$$\mu_{e,\varepsilon} = \mu + \frac{\tau_y}{2\|\mathbf{D}(\mathbf{u})\|}\left(1 - e^{\frac{-2\|\mathbf{D}(\mathbf{u})\|}{\varepsilon}}\right). \tag{4.39}$$

b. The model of BERCOVIER and ENGELMAN [1980]:

$$\mu_{e,\varepsilon} = \mu + \frac{\tau_y}{\sqrt{\varepsilon^2 + 4\|\mathbf{D}(\mathbf{u})\|^2}}. \tag{4.40}$$

Actually, the regularization procedure associated with (4.40) is discussed in, e.g., CEA and GLOWINSKI [1972], GLOWINSKI, LIONS and TRÉMOLIÈRES [1976, 1981].

c. The simple *algebraic* model employed in ALLOUCHE, FRIGAARD and SONA [2000] (and likely elsewhere, previously):

$$\mu_{e,\varepsilon} = \mu + \frac{\tau_y}{\varepsilon + 2\|\mathbf{D}(\mathbf{u})\|}. \tag{4.41}$$

Of these three models the easiest one to employ is clearly the second one since it provides a C^∞ approximation of μ_e (making it thus very suitable for solution methods *à la* Newton's). In Chapter 2, we will discuss some of the approximation properties of (4.40).

REMARK 4.1. A related, but slightly different approach has been proposed in BEVERLY and TANNER [1992]. It is known as the *biviscosity model*. Essentially, it consists in introducing a critical shear rate $\dot{\gamma}_c$, as small as possible, and a second viscosity $\bar{\mu}$, as large as possible, so that

$$\boldsymbol{\tau} = \begin{cases} 2\bar{\mu}\mathbf{D}(\mathbf{u}), & \text{if } \|\mathbf{D}(\mathbf{u})\| \le \dot{\gamma}_c, \\ 2\mu\mathbf{D}(\mathbf{u}) + \dfrac{\mathbf{D}(\mathbf{u})}{\|\mathbf{D}(\mathbf{u})\|}\tau_y, & \text{otherwise.} \end{cases} \tag{4.42}$$

In other words, unyielded regions are replaced by very viscous ones delineated by the critical shear rate $\dot{\gamma}_c$. This model does not regularize the constitutive equation (which is still nondifferentiable) it improves, however, the computational capabilities of the original Bingham model because the apparent viscosity can be computed whatever is the strain rate.

REMARK 4.2. In Chapter 2 we will discuss, and investigate computationally, a novel regularization method. This method is more sophisticated than the ones discussed just above because it relies on a regularization of the dual problem, namely of the problem which has as solution the tensor-valued function $\boldsymbol{\lambda}$ encountered in (4.27)–(4.33), after (formal) elimination of $\mathbf{u}$. To the best of our knowledge, this new regularization method was introduced in DEAN, GLOWINSKI and GUIDOBONI [2007].

The two approaches we just discussed (namely, the one based on *multipliers* and the one based on *regularization*) have been extensively used for actual computations. In the next section, we will report on the results, we are aware of, obtained by viscoplasticity practitioners using the methods briefly described earlier.

5. A brief history of computational viscoplasticity

Our goal in this section is to present the computational results obtained by the viscoplasticity community during the last three decades. The list of references below is obviously incomplete and we apologize to those colleagues whose contributions are not quoted below, although they deserve to be.

The simplest problem that is conceivable is the simple *shear flow*. In particular, the *Poiseuille flow* in a plane channel or in an axi-symmetric duct has received a great attention. The main reason for the particular attention given to these test problems is the possibility to obtain exact analytical solutions, which is highly valuable in order to "measure" the accuracy of computational methods and to validate their computer implementations. Besides, pipeline flows are fairly common in industry, *Oil & Gas* and *Mining*, in particular. The analytical solution of the Poiseuille flow can be found in many publications (e.g., GLOWINSKI, LIONS and TRÉMOLIÈRES [1976, 1981], and GLOWINSKI [1984, 2008]). In BRISTEAU [1975], BRISTEAU and GLOWINSKI [1974] the authors provide an error analysis of the finite-element approximation of the Bingham flow problem. The Poiseuille flow of a Bingham fluid in a cylinder has been recently revisited by Saramito and Roquet in the particular case of a square cross section (see ROQUET [2000], and SARAMITO and ROQUET [2001]). In the two above references, the flow simulator relies on an augmented Lagrangian method combined with an adaptive mesh generator allowing a highly accurate tracking of the yield surface. The methodology developed by the two above practitioners has proved to be quite efficient at providing accurate yielded/unyielded region patterns. Even more recently, the augmented Lagrangian methodology has been extended, by Huilgol and You to Herschel–Bulkley and Casson fluid flow in pipes of both circular and square cross-section (cf. HUILGOL and YOU [2005]).

Another test problem that has been extensively investigated concerns the flow of a Bingham fluid in a *two-dimensional lid-driven cavity*. Publications dealing with this specific test problem are quite numerous; let us mention among many others FORTIN and GLOWINSKI [1982, 1983], MITSOULIS and ZISIS [2001], DEAN and GLOWINSKI [2002], VOLA, BOSCARDIN and LATCHÉ [2003], GLOWINSKI [2003], DEAN, GLOWINSKI and GUIDOBONI [2007] (see also Chapter 2, Section 17 of this article). Both, multipliers techniques and regularization methods have been applied successfully to the numerical simulation of such a flow. It is the opinion of the authors of the present article that the main reasons, explaining why this problem has motivated so many investigators, are as follows: (1) It is a very common problem in Fluid Mechanics. (2) Unlike Poiseuille flows in channels and cylinders whose velocity is a scalar quantity, it leads to a genuine two-dimensional flow problem (albeit, the simplest one) where the unknown velocity is a two component vector-valued function. (3) Last but not least, this problem is a natural generalization of the most documented test problem in Computational Fluid Mechanics, namely, the simulation of a lid-driven incompressible Newtonian viscous flow in a square cavity; looking at the changes brought by the extra-viscous term associated with the non-Newtonian behavior is an interesting issue in itself. Let us describe what various numerical simulations have shown concerning this test problem: At low Reynolds numbers, the flow is fully governed by the *Bingham number*, that is the magnitude of the yield stress. For a Bingham material, the flow pattern is well-known: assuming that the upper side of the cavity moves tangentially with uniform velocity while the other parts of the boundary do not move, unyielded regions are located in the two lower corners and in the center of the recirculation region (as shown in, e.g., Fig. 17.5 of Chapter 2); the former regions are dead zones, that is regions where the flow velocity is zero, while the latter enjoys a rigid body motion. The size of the unyielded regions grows with the Bingham number. The results available in the literature are fairly close to each others, with however some slight variations in the intensity and position of the central vortex, as well as in the precise shape of the unyielded regions. Nowadays, the hydrodynamics of the two-dimensional lid-driven Bingham flow in a square cavity seems to be well understood; this explains why this particular flow problem is commonly used to validate new computational methods (indeed, in Chapter 3, we will use this flow problem to validate the *finite-volume/augmented Lagrangian* simulation method discussed there).

Another popular benchmark problem is provided by the *flow past an obstacle*, either a *disk* in two dimensions or a *sphere* in three dimensions. Similarly to the two-dimensional lid-driven flow in a square cavity, this type of flow has received a considerable attention from the Computational Fluid Dynamics community, for all types of fluids and flow regimes. The case of low Reynolds numbers viscoplastic flow is particularly interesting for the following reasons:

1. The yield surfaces can not be intuitively guessed, and this has led to some erroneous interpretation of the flow kinematics.
2. The prediction of the obstacle drag coefficient provides a good benchmark problem to compare and validate numerical methods and their computer implementations.
3. Solving this problem may help solving a reciprocal one, namely the settling of a rigid body in a quiescent viscoplastic fluid, an important issue being to know if the rigid body will move under the effect of its own weight (buoyancy forces). Introducing (as in CHHABRA [2006]) a *yield gravity parameter* Y_G defined by $Y_G = \frac{\tau_y}{gd\Delta\rho}$, where g is the gravity acceleration, d is the sphere diameter, and $\Delta\rho$ is the density difference; there exists a critical value of Y_G below which the sphere will start moving. Various laboratory experiments have produced critical values of Y_G in the interval [0.04, 0.08] (for more details, see CHHABRA [2006], an excellent book indeed). From a computational point of view, this flow problem has been addressed by BERIS, TSAMOPOULOS, ARMSTRONG and BROWN [1985], BLACKERY and MITSOULIS [1997], BEAULNE and MITSOULIS [1997], LIU, MULLER and DENN [2002], all using a *regularization* method based on the *Papanastasiou* approximation (4.39). They all predict a similar yielded/unyielded region pattern, namely: rigid zones attached to the sphere at the leading and trailing edges and a yielded region surrounding the sphere that, in a cross plane, looks like two disks partially superposed. Concerning the critical value of Y_G, all these authors found $Y_G \approx 0.048$. Surprisingly, the flow of a viscoplastic fluid past a disk in either a bounded or unbounded region (clearly, the two-dimensional analog of the flow past a sphere) has received less attention (likely because less physical). This problem has been investigated in ZISIS and MITSOULIS [2002], MITSOULIS [2004] using *regularization* methods, and in ROQUET [2000], ROQUET and SARAMITO [2003] using an *augmented Lagrangian* method combined with an adaptive mesh generator, which refines the mesh in the neighborhood of the yield surfaces (as in SARAMITO and ROQUET [2001]), allowing thus an accurate tracking of the yield/unyielded interface. The settling of rigid disks in a bounded cavity containing

a Bingham fluid is discussed in DEAN, GLOWINSKI and PAN [2003]; the numerical experiments presented there suggest that if a two-dimensional analog of Y_G is large enough, the disks stop settling in finite time before reaching the bottom of the cavity.

A last type of test problems, favored by computational rheologists, is provided by *contraction* or *expansion* flows, either plane or axisymmetric, also referred as *entry* or *exit* flows, respectively. These problems have been thoroughly investigated when the fluid is *viscoelastic* because in that case, the viscoelastic extra-stress tensor develops a singularity at re-entrant corners; less attention has been given to their viscoplastic analogs. Let us mention, however, the following contributions: ABDALI, MITSOULIS and MARKATOS [1992], MITSOULIS, ABDALI and MARKATOS [1993], MITSOULIS and HUILGOL [2004], all using *regularization*, and COUPEZ, ZINE and AGASSANT [1994] using an *augmented Lagrangian* method. The flow pattern is pretty simple: upstream and downstream of the contraction the flow is of the Poiseuille type with a central plug region, however the recirculation region of the Newtonian case becomes a stagnant zone with zero velocity. When the flow is driven by a pressure difference ΔP between the inlet and outlet, a challenge is to identify the critical value of ΔP below which there is no flow and the whole domain is unyielded. It is worth noticing that this type of test problems has motivated one of the few successful three-dimensional viscoplastic flow simulations, namely, the one discussed in BURGOS and ALEXANDROU [1999] for a Herschel–Bulkley flow in a square duct with a 1:2 expansion.

Many industrial applications involve *pipe flows* through circular or annular cross-sections. In particular, in the Oil & Gas industry, pipelining is extensively used either for *drilling operations* or for *crude transportation*. In drilling operations, mud or cement slurry flow through a circular pipe downward, and through an annular one upward (as shown in Fig. 2.1). This type of flow has been investigated in WALTON and BITTLESTON [1991], SZABO and HASSAGER [1992], NOURI, UMUR and WHITELAW [1993], NOURI and WHITELAW [1994], HUSSAIN and SHARIF [2000], for either Bingham or Herschel–Bulkley viscoplastic fluids. The two main objectives of these investigations were as follows: (1) The description of the unyielded regions. (2) To find the influence of the possible eccentricity between the circular and annular pipes on the relation *pressure drop* $\rightarrow$ *flow rate*. There exists, in particular, a critical value of the eccentricity-depending of the Bingham number-above for which the narrow side of the annular pipe is a stagnant zone. The flow of two Bingham fluid mixtures has been investigated in FRIGAARD and SCHERZER [1998, 2000], MOYERS-GONZALEZ and FRIGAARD [2004], in order to mimic *cementing* processes in drilling operations. These authors explored the zero flow limit, both with *regularization* and *augmented Lagrangian* methods and showed the advantages of the second approach for computing this limit. Also, in the particular case of two Bingham fluids in an inclined pipe, with the heavier fluid on top of the lighter one, the above authors were able to provide some criterion on the *stability* of the configuration, that is whether the yield stress will be able to sustain the density difference (and no transverse flow will occur) or not; this is an important issue concerning the efficiency of cementing processes.

Beyond the relatively simple situations which have been considered so far (incompressible, isothermal, mostly single-phase) these past few years have witnessed some original and more advanced work in an attempt to simulate situations involving a more complex physics. Indeed, *computational viscoplasticity* seems to have reached a turning point where, the basic computational methods being well understood (despite the fact that there is still room for progress for incompressible single-phase flows, such as speeding up the convergence of the solution algorithms), many practitioners are eager to consider more complex and realistic situations. In that direction, let us mention MITSOULIS, ABDALI and MARKATOS [1993], NOUAR, DESAUBRY and ZENAIDI [1998], NOUAR, BENAOUDA-ZOUAOUI and DESAUBRY [2000], VINAY, WACHS and AGASSANT [2005], ZHANG, VOLA and FRIGAARD [2006] for flow with heat transfer; DIMAKOPOULOS and TSAMOPOULOS [2003], VOLA, BABIK and LATCHÉ [2004] for free surface flows; DAVIDSON, NGUYEN, CHANG and RONNINGSEN [2004], VINAY, WACHS and AGASSANT [2006] for compressible flows; and DEAN, GLOWINSKI and PAN [2003], YU and WACHS [2007] for particulate flows.

In all fairness, we have to also mention that significant progresses have been made in viscoplasticity, at the theoretical, analytical, and experimental levels; for conciseness, they will not be addressed in this article. From a computational standpoint, the more complex problems discussed in the references given just above, do not seem to cause any particular difficulties, as testified by the numerical experiments reported in these references; this suggests that the regularization-based and multiplier-based methods used there apply on problems more complicated than those which motivated their introduction. Although the regularization and multiplier methods seem to differ substantially, they are all of the *fixed-point* type and have shown slow convergence when yield stress effects dominate the flow. In Chapter 2, we will report on various techniques (some introduced in HE and GLOWINSKI [2000] and further improved in DEAN, GLOWINSKI and GUIDOBONI [2007]), which speed up the convergence of the multiplier/projection methods. In Chapter 2, we will also report on a method combining regularization and multipliers; this method (introduced in the reference just above) has the interesting property that its speed of convergence improves as τ_y increases, everything else being the same. Another important computational issue is the accurate tracking of the yield surfaces; in FRIGAARD and NOUAR [2005], the authors point to some drawbacks of the regularization methods concerning this issue, particularly when compared with multiplier methods in some specific situations (such as the stability of the yield surfaces). Having said that, regularization methods are still widely used and provide valuable results. Actually, a part of this debate between regularization and multiplier methods concerns the dependence of the computed solutions, and therefore of the yield surfaces, on the magnitude of the regularization parameter, which, ideally, should be taken as small as possible, which is not realistic for reasons that have to do with the condition number of the approximate problem. However, the convergence criterion of augmented Lagrangian method relies on the satisfaction of the constraint $\mathbf{d} - \mathbf{D}(\mathbf{u}) = \mathbf{0}$, which at a computational level translates as $\|\mathbf{d} - \mathbf{D}(\mathbf{u})\| \le \varepsilon$, implying that the computed solution is ε-dependent. The compared sensitivities of regularization and multiplier methods to their respective small parameters is a complicated and largely open issue. However, what is clear is the superior ability of multiplier-based methods to simulate the flow cessation or no-flow limit; this is a significant advantage over regularization methods (let us mention again that the multiplier/regularization method discussed in Chapter 2 seems to combine the best of both worlds).

To conclude this long section, we have reported in Table 5.1 below some useful references from Computational Viscoplasticity:

The list of references in Table 5.1 is far being exhaustive; more references will be given in the following chapters (see also the article by R. Hoppe and W. Litvinov, on the simulation of electro-rheological flow, in this volume of the *Handbook of Numerical Analysis*).

TABLE 5.1
Some useful references from Computational Viscoplasticity

Type of problems	References
Poiseuille flow in a cylinder	GLOWINSKI [1974], GLOWINSKI, LIONS and TRÉMOLIÈRES [1976, 1981], BRISTEAU and GLOWINSKI [1974], ROQUET [2000], SARAMITO and ROQUET [2001], HUILGOL and YOU [2005], WALTON and BITTLESTON [1991], HUSSAIN and SHARIF [2000], SZABO and HASSAGER [1992], NOURI, UMUR and WHITELAW [1993], NOURI and WHITELAW [1994]
Two-dimensional lid-driven cavity flow	FORTIN and GLOWINSKI [1982, 1983], GLOWINSKI and LE TALLEC [1989], DEAN and GLOWINSKI [2002], GLOWINSKI [2003], MITSOULIS and ZISIS [2001], VOLA, BOSCARDIN and LATCHÉ [2003]
Flow past an obstacle	BERIS, TSAMOPOULOS, ARMSTRONG and BROWN [1985], BLACKERY and MITSOULIS [1997], ZISIS and MITSOULIS [2002], MITSOULIS [2004], LIU, MULLER and DENN [2002], ROQUET [2000], ROQUET and SARAMITO [2003]
Contraction or expansion flow	ABDALI, MITSOULIS and MARKATOS [1992], MITSOULIS, ABDALI and MARKATOS [1993], COUPEZ, ZINE and AGASSANT [1994], BURGOS and ALEXANDROU [1999], MITSOULIS and HUILGOL [2004]
Two-fluid flow	FRIGAARD and SCHERZER [1998, 2000], MOYERS-GONZALEZ and FRIGAARD [2004]
Flow with heat transfer	MITSOULIS, ABDALI and MARKATOS [1993], NOUAR, DESAUBRY and ZENAIDI [1998], NOUAR, BENAOUDA-ZOUAOUI and DESAUBRY [2000], VINAY, WACHS and AGASSANT [2005], ZHANG, VOLA and FRIGAARD [2006]
Free surface flow	DIMAKOPOULOS and TSAMOPOULOS [2003], VOLA, BABIK and LATCHÉ [2004]
Compressible flow	VINAY, WACHS and AGASSANT [2006], DAVIDSON, NGUYEN, CHANG and RONNINGSEN [2004]
Particulate flow	DEAN, GLOWINSKI and PAN [2003], YU and WACHS [2007]
Nonisothermal electrorheological flow	HOPPE and LITVINOV [2004]

6. Conclusion

The main objective of this chapter was to give the reader an overview of the main mathematical and computational aspects of viscoplasticity. Viscoplastic materials are involved in a wide range of industrial processes and geological flows, from tooth-paste to avalanches of granular matter. Many breakthroughs have been achieved since (1) the pioneering work of Bingham on the description of the material rheological behavior, (2) the investigations of, e.g., Duvaut, J.L. Lions, Glowinski, Trémolières, and Papanastasiou on the construction of efficient computational methods for the simulation of viscoplastic flows. On-going theoretical and experimental studies constantly improve the understanding of the specific behavior of this class of materials, including modifications of the basic models (such as Bingham's) in order to include effects such as thixotropy, temperature dependence, compressibility, electro-magnetism, and so on. Two families of numerical methods are available to the practitioners, both fairly easy to implement. The application of these methods has led to an abundant literature, describing these methods, the test problems, and the related numerical results (some of the corresponding references are given in Table 5.1).

In the three following chapters, we will focus on some of the issues mentioned in the present chapter. The discussion will include convergence studies, the description of new algorithms, applications going much beyond the basic Bingham's model, etc. A particular attention will be given to multiplier methods.

CHAPTER 2

Bingham Flow In Cylinders and Cavities

7. Introduction and Synopsis

As already mentioned in Chapter 1, the *numerical simulation of Bingham fluid flow* has been, for many years, the subject of intensive scrutiny. Among the reasons explaining this situation, let us mention:

1. The fact that materials as diverse as fresh concrete, tortilla dough, fruit-syrup mixtures, blood in the capillaries, mud used in drilling technologies, tooth paste, etc. . . , have a Bingham medium behavior, namely: below a certain stress yield, the material enjoys rigidity; above this yield, the above material behaves like an incompressible viscous fluid.
2. For applied mathematicians and numerical analysts, Bingham flow modeling has been a permanent source of challenging problems for many decades already, the main breakthrough in this direction being the *variational inequality* formulation due to G. Duvaut and J.L. Lions (see, e.g., DUVAUT and LIONS [1972a, 1976]).
3. Bingham media are, in some sense, the simplest *viscoplastic* materials encountered in Continuum Mechanics, and the various aspects of their analysis has proved helpful when considering viscoplastic flow with more complicated constitutive law (like, for example, the *nonisothermal compressible and thixotropic viscoplastic flow* discussed in Chapter 3).
4. The following quotation from BALMFORTH and FRIGAARD [2007a] (listing the most important developments in viscoplasticity in recent years):

   "*For good or bad, the Bingham model is the theoretical paradigm in the field.*"

Our goal in this chapter (which follows closely DEAN, GLOWINSKI and GUIDOBONI [2007]) is to review several of the approaches we are aware of, concerning the *numerical simulation* of Bingham flow. Roughly speaking, there exist two main approaches: one based on *regularization* procedures and, the other based on the use of *multipliers*. There is no way that we can describe all the related methods in this chapter; we will discuss, nevertheless, quite a few of them, considering first the case of *Bingham flow in cylindrical pipes* and then the more general case of *Bingham flow in multidimensional cavities*.

It has become practically impossible to give all the references related to the modeling and simulation of Bingham fluid flow (more than 18,000 entries in *Google Scholar* as of

October 12, 2007); in addition to DUVAUT and LIONS [1972a, 1976], DEAN, GLOWINSKI and GUIDOBONI [2007], let us mention, among many others, PRAGER [1961], GERMAIN [1973], GLOWINSKI and LE TALLEC [1989], GUYON, HULIN and PETIT [2001], BALMFORTH and FRIGAARD [2007b] (see also the references therein and those given in Chapter 1).

8. On the modeling of Bingham viscoplastic flow

The material in this section is pretty classical. It has been introduced here to fix the notation and to remind of some basic facts concerning the mathematical modeling of Bingham flow. Let thus Ω be a domain (i.e., an open and connected region) of $\mathbf{R}^d$ ($d = 2$ or 3 in applications); we denote by Γ the boundary of Ω. The *isothermal flow* of an *incompressible Bingham viscoplastic medium*, during the time interval $(0,\ T)$, is modeled by the following system of equations (clearly of the Navier–Stokes type):

$$\rho\left[\frac{\partial \mathbf{u}}{\partial t} + (\mathbf{u}\cdot\nabla)\mathbf{u}\right] = \nabla\cdot\boldsymbol{\sigma} + \mathbf{f} \text{ in } \Omega\times(0,T), \tag{8.1}$$

$$\nabla\cdot\mathbf{u} = 0 \text{ in } \Omega\times(0,T), \tag{8.2}$$

$$\boldsymbol{\sigma} = -p\mathbf{I} + \sqrt{2}\tau_y\frac{\mathbf{D}(\mathbf{u})}{|\mathbf{D}(\mathbf{u})|} + 2\mu\mathbf{D}(\mathbf{u}), \tag{8.3}$$

$$\mathbf{u}(0) = \mathbf{u}_0 \text{ (with } \nabla\cdot\mathbf{u}_0 = 0). \tag{8.4}$$

For simplicity, we shall consider only *Dirichlet boundary conditions*, namely:

$$\mathbf{u} = \mathbf{u}_\Gamma \text{ on } \Gamma\times(0,\ T) \text{ with } \int_\Gamma \mathbf{u}_\Gamma(t)\cdot\mathbf{n}\,d\Gamma = 0, \text{ a.e. in } (0,\ T). \tag{8.5}$$

In system (8.1)–(8.5):

- ρ (resp., μ and τ_y) is the *density* (resp., are the *viscosity* and *plasticity yield*) of the Bingham medium; we have $\rho > 0$, $\mu > 0$, and $\tau_y > 0$.
- $\mathbf{f}$ is a density of external forces.
- $\mathbf{D}(\mathbf{v}) = \frac{1}{2}[\nabla\mathbf{v} + (\nabla\mathbf{v})^t](= (D_{ij}(\mathbf{v}))_{1\le i,j\le d})$, $\forall\, \mathbf{v} \in (H^1(\Omega))^d$, and $|\mathbf{D}(\mathbf{v})|$ is the Fröbenius norm of tensor $\mathbf{D}(\mathbf{v})$, that is

$$|\mathbf{D}(\mathbf{v})| = \left(\sum_{1\le i,j\le d} |D_{ij}(\mathbf{v})|^2\right)^{\frac{1}{2}}.$$

- $\mathbf{n}$ is the outward unit normal vector at Γ.
- We have denoted (and will denote later on) by $\phi(t)$ the function $x \to \phi(x,\ t)$.

We observe that if $\tau_y = 0$, system (8.1)–(8.5) reduces to the Navier–Stokes equations modeling isothermal incompressible Newtonian viscous fluid flow. Having said that, if $\tau_y > 0$ the above model makes no sense on the (rigid) set

$$Q_0 = \{\{x,t\}|\{x,t\} \in \Omega\times(0,T),\ \mathbf{D}(\mathbf{u})(x,t) = \mathbf{0}\}.$$

Following DUVAUT and LIONS [1972a, chapter 6], [1976, chapter 6], we eliminate the above difficulty by considering instead of the (doubly) nonlinear system (8.1)–(8.5) the following *variational inequality model* (where $dx = dx_1 \ldots dx_d$):

Find $\{\mathbf{u}(t), p(t)\} \in (H^1(\Omega))^d \times L^2(\Omega)$ such that a.e. in $(0,\ T)$ we have

$$\begin{aligned}
&\rho \int_\Omega \frac{\partial \mathbf{u}}{\partial t}(t) \cdot (\mathbf{v} - \mathbf{u}(t))dx + \rho \int_\Omega (\mathbf{u}(t) \cdot \nabla)\mathbf{u}(t) \cdot (\mathbf{v} - \mathbf{u}(t))dx \\
&\quad + \mu \int_\Omega \nabla \mathbf{u}(t) : \nabla(\mathbf{v} - \mathbf{u}(t))dx + \sqrt{2}\tau_y[j(\mathbf{v}) - j(u(t))] \\
&\quad - \int_\Omega p(t)\nabla \cdot (\mathbf{v} - \mathbf{u}(t))dx \geq \int_\Omega \mathbf{f}(t) \cdot (\mathbf{v} - \mathbf{u}(t))dx, \ \forall\, \mathbf{v} \in \mathbf{V}_\Gamma(t),
\end{aligned} \tag{8.6}$$

$$\nabla \cdot \mathbf{u}(t) = 0 \ \text{ in } \Omega, \tag{8.7}$$

$$\mathbf{u}(0) = \mathbf{u}_0, \tag{8.8}$$

$$\mathbf{u}(t) = \mathbf{u}_\Gamma(t) \text{ on } \Gamma, \tag{8.9}$$

with, in (8.6),

$$j(\mathbf{v}) = \int_\Omega |\mathbf{D}(\mathbf{v})|dx, \quad \forall\, \mathbf{v} \in (H^1(\Omega))^d, \tag{8.10}$$

$$\mathbf{V}_\Gamma(t) = \{\mathbf{v} | \mathbf{v} \in (H^1(\Omega))^d, \ \mathbf{v} = \mathbf{u}_\Gamma(t) \text{ on } \Gamma\} \tag{8.11}$$

and $\mathbf{S} : \mathbf{T} = \sum_{i=1}^d \sum_{j=1}^d s_{ij}t_{ij}, \forall\, \mathbf{S} = (s_{ij})_{1 \leq i,j \leq d}, \mathbf{T} = (t_{ij})_{1 \leq i,j \leq d} \in \mathbf{R}^{d \times d}$.

Various comments concerning formulation (8.6)–(8.9) can be found in, e.g., [2003, chapter 10]. The *variational inequality* formulation of *temperature-dependent* Bingham flow can be found in, e.g., DUVAUT and LIONS [1972b] (see also VINAY, WACHS and AGASSANT [2005] (and Chapter 3) for a discussion of another type of temperature-dependent viscoplastic flow).

In the following sections, we are going to review a variety of computational techniques, which have been developed during the last four decades for the solution of problems (8.1)–(8.5) and (8.6)–(8.9). For simplicity, we will start our discussion with *Bingham flow in cylinders*, and then consider *flow in bounded multidimensional cavities*.

REMARK 8.1. It follows from DUVAUT and LIONS [1972a, 1976] that there exists a *tensor-valued function* $\boldsymbol{\lambda}$ such that the formulation (8.6)–(8.9) is equivalent to

$$\begin{aligned}
&\rho\left[\frac{\partial \mathbf{u}}{\partial t} + (\mathbf{u} \cdot \nabla)\mathbf{u}\right] = \nabla \cdot \boldsymbol{\sigma} + \mathbf{f} \ \text{ in } \Omega \times (0, T), \quad \nabla \cdot \mathbf{u} = 0 \ \text{ in } \Omega \times (0, T), \\
&\qquad\qquad \mathbf{u} = \mathbf{u}_\Gamma \ \text{ on } \Gamma \times (0,\ T), \ \ \mathbf{u}(0) = \mathbf{u}_0
\end{aligned} \tag{8.12}$$

with

$$\boldsymbol{\sigma} = -p\mathbf{I} + 2\mu\mathbf{D}(\mathbf{u}) + \sqrt{2}\tau_y\boldsymbol{\lambda}, \tag{8.13}$$

$$\boldsymbol{\lambda} : \mathbf{D}(\mathbf{u}) = |\mathbf{D}(\mathbf{u})|, \ \ \boldsymbol{\lambda} = \boldsymbol{\lambda}^t, \ \ |\boldsymbol{\lambda}| \leq 1. \tag{8.14}$$

We can take advantage of the above formulation to solve (8.6)–(8.9) numerically, as shown in Section 17. Incidentally, assuming that $\mathbf{u}$, p, and $\boldsymbol{\lambda}$ arc known, relation (8.13) provides the stress tensor $\boldsymbol{\sigma}$ everywhere in $\Omega \times (0, T)$ (this includes the rigid set Q_0). The tensor-valued function $\sqrt{2}\,\tau_y\,\boldsymbol{\lambda}$ can be viewed as the *extra-stress tensor* associated to the viscoplastic behavior of the medium.

9. Bingham flow in cylinders: (I) Formulation

The *isothermal* and *unsteady axial flow* of an *incompressible viscoplastic Bingham fluid* in an infinitely long cylinder of (bounded) cross-section Ω is (formally) modeled by the following *nonlinear parabolic equation* (where Γ is the boundary of Ω):

$$\rho\frac{\partial u}{\partial t} - \mu\nabla^2 u - \tau_y\nabla\cdot\left(\frac{\nabla u}{|\nabla u|}\right) = C \text{ in } \Omega\times(0,\ T), \quad u = 0 \text{ on } \Gamma\times(0,\ T),$$
$$u(0) = u_0. \tag{9.1}$$

In system (9.1), (1) u is the *axial velocity* of the flow, i.e., $\mathbf{u} = \{0, 0, u\}$, assuming that the fluid flows in the Ox_3-direction, Ω being parallel to the (Ox_1, Ox_2)-plane. (2) C is the pressure drop per unit length (it is a function of t only, and possibly a constant). System (9.1) is a particular case of (8.1)–(8.5).

Before going further, let us observe that (as in Section 8, for (8.1)–(8.5)), model (9.1) makes no sense in the (space-time) rigid region

$$Q_0 = \{\{x, t\} | \{x, t\} \in \Omega\times(0, T),\ \nabla u(x, t) = \mathbf{0}\}.$$

There are classically two approaches to overcome the above difficulty, namely the *regularization* and the *multiplier* approaches; both will be discussed hereafter.

10. Bingham flow in cylinders: (II) the regularization approach

Let ε be a *small positive* parameter. The idea here is to replace system (9.1) by the following *well-posed nonlinear parabolic* problem:

$$\rho\frac{\partial u_\varepsilon}{\partial t} - \mu\nabla^2 u_\varepsilon - \tau_y\nabla\cdot\left(\frac{\nabla u_\varepsilon}{\sqrt{\varepsilon^2 + |\nabla u_\varepsilon|^2}}\right) = C \text{ in } \Omega\times(0,\ T),$$
$$u_\varepsilon = 0 \text{ on } \Gamma\times(0,\ T),$$
$$u_\varepsilon(0) = u_0. \tag{10.1}$$

We will return on the approximation properties of u_ε in Section 11. Let us mention that the above regularization procedure has been widely used, not only in *Viscoplasticity*, but also in *Image Processing* (see, e.g., CHAN, GOLUB and MULET [1999] and the references therein). It has however some drawbacks, a major one being that if $C = 0$, the well-known property that $u(t) \to 0$ in *finite time*, as t increases, is lost (as is the interface between the rigid and fluid regions). An alternative to regularization is provided by the *multiplier* approach, to be discussed in the following section.

REMARK 10.1. Other regularization procedures can be found in PAPANASTASIOU [1987] (see also Chapter 1, Section 4.3).

11. Bingham flow in cylinders: (III) variational inequality formulation. The multiplier approach

It follows from DUVAUT and LIONS [1972a, 1976] that a mechanically and mathematically correct formulation of (9.1) is provided by the following *variational inequality* type model:

$$\begin{aligned} & u(t) \in H_0^1(\Omega), \quad \text{a.e. in } (0,\ T), \\ & \rho \int_\Omega \frac{\partial u}{\partial t}(v - u(t))\mathrm{d}x + \mu \int_\Omega \nabla u(t) \cdot \nabla (v - u(t))\mathrm{d}x \\ & \qquad + \tau_y[j(v) - j(u(t))] \geq C \int_\Omega (v - u(t))\mathrm{d}x, \quad \forall v \in H_0^1(\Omega),\ u(0) = u_0 \end{aligned} \tag{11.1}$$

with

$$j(v) = \int_\Omega |\nabla v| \mathrm{d}x. \tag{11.2}$$

It follows from, e.g., DUVAUT and LIONS [1972a, 1976] that problem (11.1) has a *unique solution.*

REMARK 11.1. In order to be rigorous, we should replace in (11.1) the term $\int_\Omega \frac{\partial u}{\partial t}(t)(v - u(t))\mathrm{d}x$ by $\left\langle \frac{\partial u}{\partial t}(t),\ v - u(t) \right\rangle$, where $\langle .\ ,\ . \rangle$ denotes the *duality pairing* between $H^{-1}(\Omega)$ (the dual space of $H_0^1(\Omega)$) and $H_0^1(\Omega)$, which coincides with the canonical $L^2(\Omega)$-scalar product when the first argument is sufficiently smooth so that it belongs to $L^2(\Omega)$; this observation applies also to the first integral in (8.6). However, following in that a well-established tradition (initialized very likely by J.L. Lions), we will keep the integral notation used in (11.1) (and (8.6)).

The following mathematical results hold, all important from a computational point of view. The *first one* concerns the approximation properties of the *regularization procedure* defined by (10.1); it reads as follows:

THEOREM 11.1. *Let u and u_ε be the respective solutions of problems (11.1) and (10.1); we have then, if $u_0 \in L^2(\Omega)$,*

$$\|u_\varepsilon(t) - u(t)\|_{L^2(\Omega)} \leq \sqrt{\frac{\tau_y|\Omega|}{\mu\lambda_0}} \sqrt{1 - \exp\left(-\frac{2\mu\lambda_0}{\rho}t\right)}\ \sqrt{\varepsilon}, \quad \forall\, t \in [0,\ T], \tag{11.3}$$

with $|\Omega|$ = meas.(Ω) and $\lambda_0(>0)$ the smallest eigenvalue of $-\nabla^2$ operating in $H_0^1(\Omega)$.

PROOF. Let us define $j_\varepsilon : H^1(\Omega) \to \mathbf{R}$ by

$$j_\varepsilon(v) = \int_\Omega \sqrt{\varepsilon^2 + |\nabla v|^2}\,\mathrm{d}x. \tag{11.4}$$

The functional j_ε is clearly convex and C^∞ over $H^1(\Omega)$; moreover,

$$0 < j_\varepsilon(v) - j(v) = \varepsilon^2 \int_\Omega \frac{\mathrm{d}x}{\sqrt{\varepsilon^2 + |\nabla v|^2} + |\nabla v|} \leq \varepsilon|\Omega|, \ \forall\, v \in H^1(\Omega). \tag{11.5}$$

It follows from DUVAUT and LIONS [1972a, 1976] (and from the *convexity* and *differentiability* properties of the functional j_ε) that there is *equivalence* between the *nonlinear parabolic problem* (10.1) and the *parabolic variational inequality problem*

$$\begin{aligned}
&u_\varepsilon(t) \in H_0^1(\Omega) \text{ a.e. in } (0,\ T),\\
&\rho \int_\Omega \frac{\partial u_\varepsilon}{\partial t}(v - u_\varepsilon(t))\mathrm{d}x + \mu \int_\Omega \nabla u_\varepsilon(t) \cdot \nabla(v - u_\varepsilon(t))\mathrm{d}x + \tau_y[j_\varepsilon(v) - j_\varepsilon(u_\varepsilon(t))]\\
&\quad \geq C \int_\Omega (v - u_\varepsilon(t))\mathrm{d}x, \forall v \in H_0^1(\Omega),\ u_\varepsilon(0) = u_0.
\end{aligned} \tag{11.6}$$

Take $v = u_\varepsilon(t)$ in (11.1) (resp., $v = u(t)$ in (11.6)). We obtain then by addition that

$$\begin{aligned}
&\rho \int_\Omega \frac{\partial}{\partial t}(u_\varepsilon - u)(t)(u_\varepsilon - u)(t)\mathrm{d}x + \mu \int_\Omega |\nabla(u_\varepsilon - u)(t)|^2 \mathrm{d}x\\
&\quad + \tau_y[j_\varepsilon(u_\varepsilon(t)) - j(u_\varepsilon(t))] \leq \tau_y[j_\varepsilon(u(t)) - j(u(t))], \ \text{ a.e. in } (0,\ T).
\end{aligned} \tag{11.7}$$

Combining (11.7) with (11.5) and $\lambda_0 \|v\|^2_{L^2(\Omega)} \leq \int_\Omega |\nabla v|^2 \mathrm{d}x, \forall v \in H_0^1(\Omega)$ (*Poincaré inequality*), we obtain

$$\frac{\rho}{2}\frac{\mathrm{d}}{\mathrm{d}t}\|(u_\varepsilon - u)(t)\|^2_{L^2(\Omega)} + \lambda_0\mu\|(u_\varepsilon - u)(t)\|^2_{L^2(\Omega)} \leq \tau_y|\Omega|\varepsilon, \ \text{ a.e. in } (0, T) \tag{11.8}$$

Denote $\|(u_\varepsilon - u)(t)\|^2_{L^2(\Omega)}$ by $z(t)$; we have then $z(0) = 0$ and (from (11.8))

$$\frac{\mathrm{d}}{\mathrm{d}t}z(t) + \frac{2\lambda_0\mu}{\rho}z(t) \leq \frac{2\tau_y}{\rho}|\Omega|\varepsilon, \quad \text{a.e. in } (0, T).$$

The above inequality implies in turn that

$$\mathrm{e}^{-\frac{2\lambda_0\mu}{\rho}t}\frac{\mathrm{d}}{\mathrm{d}t}\mathrm{e}^{\frac{2\lambda_0\mu}{\rho}t}z(t) \leq \frac{2\tau_y}{\rho}|\Omega|\varepsilon, \ \text{ a.e. in } (0,\ T),$$

that is

$$\frac{\mathrm{d}}{\mathrm{d}t}\mathrm{e}^{\frac{2\lambda_0\mu}{\rho}t}z(t) \leq \mathrm{e}^{\frac{2\lambda_0\mu}{\rho}t}\frac{2\tau_y}{\rho}|\Omega|\varepsilon, \ \text{ a.e. in } (0,\ T). \tag{11.9}$$

Integrating from 0 to t the (ordinary) *differential inequality* (11.9) yields

$$z(t) \leq \frac{\tau_y}{\lambda_0\mu}|\Omega|\left[1 - e^{-\frac{2\lambda_0\mu}{\rho}t}\right]\varepsilon, \text{ in } [0,\ T]. \tag{11.10}$$

Relation (11.3) follows from (11.10) and from the definition of $z(t)$. □

REMARK 11.2. It follows from relation (11.3) that

$$\lim_{\varepsilon\to 0} u_\varepsilon = u \text{ in } C^0([0,\ T];\, L^2(\Omega)) \tag{11.11}$$

(in (11.11), $C^0([0,\ T];\, L^2(\Omega))$ denotes the space of the functions continuous over $[0,\ T]$ with values in $L^2(\Omega)$). The convergence properties (11.3) and (11.11) provide quite clearly a justification of the regularization procedure described in Section 10. Actually, relation (11.3) implies

$$\|u_\varepsilon - u\|_{C^0([0,\ T];L^2(\Omega))} \leq \sqrt{\frac{\tau_y|\Omega|}{\mu\lambda_0}}\,\varepsilon^{\frac{1}{2}}. \tag{11.12}$$

We do not claim that the order of convergence $\frac{1}{2}$ encountered in (11.12) is optimal.

The *second* result concerns the behavior of $u(t)$ when $t \to +\infty$. We have then the following:

THEOREM 11.2. *Suppose that C does not depend of t and that $T = +\infty$. We have then, if $u_0 \in L^2(\Omega)$, the following asymptotic behavior:*

$$u(t) = 0, \quad \forall\, t \geq T_c, \quad \text{if } C < \tau_y\gamma|\Omega|^{-\frac{1}{2}}, \tag{11.13}$$

where

$$\gamma = \inf_v \left[\frac{\int_\Omega |\nabla v|\mathrm{d}x}{\|v\|_{L^2(\Omega)}}\right], \qquad v \in H_0^1(\Omega)\backslash\{0\},$$

and

$$T_c = \frac{\rho}{\mu\lambda_0}\ln\left[1 + \frac{\mu\lambda_0}{\gamma\tau_y - C|\Omega|^{1/2}}\|u_0\|_{L^2(\Omega)}\right]. \tag{11.14}$$

If $C \geq \gamma\tau_y|\Omega|^{-\frac{1}{2}}$, then the following estimate holds:

$$\|u(t) - u_\infty\|_{L^2(\Omega)} \leq \|u_0 - u_\infty\|_{L^2(\Omega)} e^{-\frac{\mu\lambda_0}{\rho}t}, \quad \forall t \geq 0, \tag{11.15}$$

with u_∞ the corresponding steady-state solution.

PROOF. (i) Suppose that $u_0(=u(0))=0$ and that $C \leq \tau_y \gamma |\Omega|^{-\frac{1}{2}}$; it follows then from the *Schwarz inequality* in $L^2(\Omega)$ and from the *Nirenberg–Strauss inequality*

$$\gamma \|v\|_{L^2(\Omega)} \leq \int_\Omega |\nabla v| \mathrm{d}x, \quad \forall\, v \in H_0^1(\Omega) \tag{11.16}$$

(with $\gamma > 0$; cf. STRAUSS [1973]) that $u=0$ is the *unique solution* of (11.1); indeed if we take $u=0$ in (11.1), the only thing left to show is that

$$\tau_y \int_\Omega |\nabla v| \mathrm{d}x \geq C \int_\Omega v \mathrm{d}x, \quad \forall\, v \in H_0^1(\Omega);$$

the above inequality follows directly from $C \leq \tau_y \gamma |\Omega|^{-\frac{1}{2}}$, from (11.16) and from $|\int_\Omega v \,\mathrm{d}x| \leq |\Omega|^{\frac{1}{2}} \|v\|_{L^2(\Omega)}, \quad \forall\, v \in L^2(\Omega)$.

(ii) Taking $v = 2u(t)$ and $v = 0$ in (11.1) shows that

$$\begin{aligned}
&\frac{\rho}{2}\frac{\mathrm{d}}{\mathrm{d}t}\|u(t)\|^2_{L^2(\Omega)} + \mu \int_\Omega |\nabla u(t)|^2 \mathrm{d}x + \tau_y \int_\Omega |\nabla u(t)| \mathrm{d}x \\
&\quad = C \int_\Omega u(t) \mathrm{d}x, \quad \text{a.e. in } (0, +\infty).
\end{aligned} \tag{11.17}$$

From the *Schwarz inequality* in $L^2(\Omega)$, and from the inequalities $\lambda_0 \|v\|^2_{L^2(\Omega)} \leq \int_\Omega |\nabla v|^2 \mathrm{d}x, \quad \forall\, v \in H_0^1(\Omega)$ (Poincaré's) and (11.16) (Nirenberg–Strauss'), it follows from (11.17) that

$$\begin{aligned}
&\frac{\rho}{2}\frac{\mathrm{d}}{\mathrm{d}t}\|u(t)\|^2_{L^2(\Omega)} + \mu\lambda_0 \|u(t)\|^2_{L^2(\Omega)} + (\tau_y \gamma - C|\Omega|^{\frac{1}{2}})\|u(t)\|_{L^2(\Omega)} \leq 0 \\
&\text{a.e. in } (0, +\infty).
\end{aligned} \tag{11.18}$$

Suppose that $u_0 \neq 0$. If $u(t) \neq 0$, $\forall\, t \in [0, \infty)$, the function $t \to \|u(t)\|_{L^2(\Omega)}$ is absolutely continuous; it is, therefore, differentiable almost everywhere and

$$\frac{\mathrm{d}}{\mathrm{d}t}\|u(t)\|^2_{L^2(\Omega)} = 2\|u(t)\|_{L^2(\Omega)} \frac{\mathrm{d}}{\mathrm{d}t}\|u(t)\|_{L^2(\Omega)} \quad \text{a.e. in } (0, +\infty). \tag{11.19}$$

Because $\|u(t)\|_{L^2(\Omega)} > 0$, $\forall\, t \in [0, +\infty)$, it follows from (11.18) and (11.19) that

$$\rho \frac{\mathrm{d}}{\mathrm{d}t}\|u(t)\|_{L^2(\Omega)} + \mu\lambda_0 \|u(t)\|_{L^2(\Omega)} + (\tau_y \gamma - C|\Omega|^{\frac{1}{2}}) \leq 0 \text{ a.e. in } (0, +\infty). \tag{11.20}$$

Denote $\|u(t)\|_{L^2(\Omega)} + \frac{\tau_y \gamma - C|\Omega|^{\frac{1}{2}}}{\mu\lambda_0}$ by $z(t)$; we have then $z(t) \geq 0, \ \forall\, t \geq 0$, and (from (11.20))

$$\frac{\mathrm{d}}{\mathrm{d}t} z(t) + \frac{\mu\lambda_0}{\rho} z(t) \leq 0, \quad \text{a.e. in } (0, +\infty).$$

The above inequality implies in turn that

$$\frac{\mathrm{d}}{\mathrm{d}t}\left[z(t)\mathrm{e}^{\frac{\mu\lambda_0}{\rho}t}\right] \leq 0, \quad \text{a.e. in } (0, \infty). \tag{11.21}$$

Integrating the differential inequality (11.21) from 0 to t, we obtain

$$z(t) \leq z(0)\mathrm{e}^{-\frac{\mu\lambda_0}{\rho}t}, \ \forall\, t \geq 0. \tag{11.22}$$

Because $z(t) \geq 0, \ \forall\, t \geq 0$, relation (11.22) implies

$$\lim_{t\to+\infty} z(t) = 0. \tag{11.23}$$

However, we have

$$z(t) = \|u(t)\|_{L^2(\Omega)} + \frac{\tau_y\gamma - C|\Omega|^{\frac{1}{2}}}{\mu\lambda_0} \geq \frac{\tau_y\gamma - C|\Omega|^{\frac{1}{2}}}{\mu\lambda_0} > 0, \ \forall\, t \geq 0. \tag{11.24}$$

Because (11.24) contradicts (11.23), there exists t^*, with $0 < t^* < +\infty$, such that

$$u(t^*) = 0. \tag{11.25}$$

It follows from Part (i) that $u(t) = 0, \ \forall\, t \geq t^*$, and from (11.22), (11.24) that $t^* \leq T_c$, with T_c the solution of

$$z(0)\mathrm{e}^{-\frac{\mu\lambda_0}{\rho}T_c} = \frac{\tau_y\gamma - C|\Omega|^{\frac{1}{2}}}{\mu\lambda_0},$$

namely (because $z(0) = \|u(0)\|_{L^2(\Omega)} + \frac{\tau_y\gamma - C|\Omega|^{\frac{1}{2}}}{\mu\lambda_0}$), $T_c = \frac{\rho}{\mu\lambda_0}\ln\left[1 + \frac{\mu\lambda_0}{\gamma\tau_y - C|\Omega|^{\frac{1}{2}}}\,\|u_0\|_{L^2(\Omega)}\right]$. The above relation validates (11.14).

(iii) Denote by u_∞ the *steady state solution* associated with the parabolic problem (11.1); u_∞ is, thus, the *unique solution* of the following *elliptic variational inequality*

$$\begin{cases} u_\infty \in H_0^1(\Omega); \ \forall\, v \in H_0^1(\Omega) \\ \mu\displaystyle\int_\Omega \nabla u_\infty \cdot \nabla(v - u_\infty)\mathrm{d}x + \tau_y \int_\Omega [|\nabla v| - |\nabla u_\infty|]\mathrm{d}x \geq C\int_\Omega (v - u_\infty)\mathrm{d}x. \end{cases} \tag{11.26}$$

Taking $v = u_\infty$ in (11.1) (resp., $v = u(t)$ in (11.26)), we obtain by addition

$$\frac{\rho}{2}\frac{\mathrm{d}}{\mathrm{d}t}\|u(t) - u_\infty\|^2_{L^2(\Omega)} + \mu\int_\Omega |\nabla(u(t) - u_\infty)|^2\mathrm{d}x \leq 0, \quad \text{a.e. in } (0, +\infty). \tag{11.27}$$

The estimate (11.15) follows easily from (11.27) and from the *Poincaré inequality*

$$\lambda_0\|v\|^2_{L^2(\Omega)} \leq \int_\Omega |\nabla v|^2\mathrm{d}x, \ \ \forall\, v \in H_0^1(\Omega).$$

□

REMARK 11.3. The estimate (11.15) *is not optimal* since, in the case where one has $C < \tau_y\gamma|\Omega|^{-\frac{1}{2}}$, it does not predict the convergence of $u(t)$ to u_∞ (= 0, here) in finite time. Actually, we doubt of the optimality of the above estimate when $C \geq \tau_y\gamma|\Omega|^{-\frac{1}{2}}$ because the inequality in (11.15) does not show any explicit dependence in τ_y (the *plasticity yield*), other than the dependence in u_∞, an implicit function of τ_y.

We will conclude this section with the following:

THEOREM 11.3. *The solution of problem (11.1) is characterized by the existence of a vector-valued function* $\boldsymbol{\lambda}(= \{\lambda_1,\ \lambda_2\})$, *such that*

$$\rho\frac{\partial u}{\partial t} - \mu\nabla^2 u - \tau_y\nabla\cdot\boldsymbol{\lambda} = C \ \textit{in}\ \Omega\times(0,\ T),\quad u = 0\ \textit{on}\ \Gamma\times(0,\ T),$$
$$u(0) = u_0,\ |\boldsymbol{\lambda}(x,t)| \leq 1\quad \textit{a.e. in}\ \Omega\times(0,\ T)\ \textit{and}\ \boldsymbol{\lambda}\cdot\nabla u = |\nabla u|, \tag{11.28}$$

with $|\mathbf{q}| = \sqrt{q_1^2 + q_2^2}$.

Theorem 11.3 is proved in, e.g., DUVAUT and LIONS [1972a, 1976]. Actually, there are several ways to prove the above theorem: nonconstructive ones relying on the *Hahn–Banach Theorem*, and more constructive ones based on the *regularization procedure* briefly discussed in Section 10. Indeed, using the notation of Section 10, it is relatively easy to show that when $\varepsilon \to 0$, the vector $\lambda_\varepsilon \left(= \frac{\nabla u_\varepsilon}{\sqrt{\varepsilon^2+|\nabla u_\varepsilon|^2}}\right)$ encountered in (10.1) converges weakly $-*$ in $(L^\infty(\Omega\times(0,T)))^2$ to a limit verifying the properties of $\boldsymbol{\lambda}$ specified in the statement of Theorem 11.3. Actually, the regularization approach has been used in GLOWINSKI [1984] to prove the *steady state analog* of Theorem 11.3. Further observations are in order; among them:

REMARK 11.4. Without being (strictly speaking) a *Lagrange* (or *Kuhn–Tucker*) *multiplier*, the vector $\boldsymbol{\lambda}$ shares many properties with such vectors, explaining why we call it a *multiplier* in the sequel. Among its properties, let us emphasize that *the multiplier* $\boldsymbol{\lambda}$ is *nonunique* (as shown in, e.g., GLOWINSKI, LIONS and TRÉMOLIÈRES [1981], and HE and GLOWINSKI [2000]), however $\nabla\cdot\boldsymbol{\lambda}$ is unique.

REMARK 11.5. The last two relations in (11.28) are equivalent to

$$\boldsymbol{\lambda}(t) = \mathrm{P}_{\boldsymbol{\Lambda}}[\boldsymbol{\lambda}(t) + r\,\tau_y\,\nabla u(t)],\quad \forall\ r \geq 0,\quad \text{a.e. in } (0\ T), \tag{11.29}$$

where, in (11.29), the *closed convex* set $\boldsymbol{\Lambda}$ and the *projection operator* $\mathbf{P}_{\boldsymbol{\Lambda}}(:\ (L^2(\Omega))^2 \to \boldsymbol{\Lambda})$ are defined by

$$\boldsymbol{\Lambda} = \{\mathbf{q}|\mathbf{q} \in (L^2(\Omega))^2,\ |\mathbf{q}(x)| \leq 1,\quad \text{a.e. in } \Omega\} \tag{11.30}$$

and

$$\mathrm{P}_{\boldsymbol{\Lambda}}(\mathbf{q})(x) = \frac{\mathbf{q}(x)}{\max(1,\ |\mathbf{q}(x)|)},\quad \text{a.e. in } \Omega,\quad \forall\ \mathbf{q} \in (L^2(\Omega))^2, \tag{11.31}$$

respectively. The present remark has important *computational implications*, as shown in Section 13.

REMARK 11.6. One can prove relatively easily that the constant γ occurring in the *Nirenberg–Strauss inequality* (11.16) *is independent of the size and shape of* Ω, a most remarkable property indeed. Actually, it can be shown (this is a little more complicated (see TALENTI [1976])) that $\gamma = 2\sqrt{\pi}(= 3.5449077\ldots)$.

12. Bingham flow in cylinders: (IV) time-discretization of problem (11.1)

To the best of our knowledge, the *backward Euler scheme*, described below, is the only scheme preserving the *asymptotic behavior* of the solution of the continuous problem (namely, problem (11.1)), including the *return to rest in finite time* (if the plasticity yield τ_y is large enough). The scheme reads as follows (with Δt (> 0) a *time-discretization step* that we suppose constant, for simplicity):

$$u^0 = u_0; \tag{12.1}$$

then, for $n \geq 1$, compute u^n from u^{n-1} through the solution of

$$\begin{aligned} &u^n \in H_0^1(\Omega),\\ &\rho \int_\Omega (u^n - u^{n-1})(v - u^n)\mathrm{d}x + \mu\Delta t \int_\Omega \nabla u^n \cdot \nabla(v - u^n)\mathrm{d}x\\ &\quad + \tau_y \Delta t[j(v) - j(u^n)] \geq \Delta t C^n \int_\Omega (v - u^n)\mathrm{d}x, \quad \forall\, v \in H_0^1(\Omega) \end{aligned} \tag{12.2}$$

with $C^n = C(n\Delta t)$. It follows from, e.g., GLOWINSKI [1984, chapter 1] that (12.2) is an *elliptic variational inequality* (of the *second kind*) problem, which has a *unique* solution. Concerning scheme (12.1), (12.2) we have the following *stability*:

THEOREM 12.1. *Suppose that $C(t)$ is bounded in $(0, T)$ and that $u_0 \in L^2(\Omega)$; then the scheme (12.1), (12.2) is unconditionally stable. Moreover, if $T = +\infty$ and if the upper bound of $|C(t)|$ is small enough, there exists an integer n_c such that*

$$u^n = 0, \quad \forall\, n \geq n_c. \tag{12.3}$$

PROOF. Taking $v = 2u^n$, and then $v = 0$ in the inequality in (12.2), we obtain, by comparison, the following relation:

$$\begin{aligned} &\rho \int_\Omega (u^n - u^{n-1})u^n\mathrm{d}x + \mu\Delta t \int_\Omega |\nabla u^n|^2\mathrm{d}x + \tau_y \Delta t j(u^n)\\ &\quad = \Delta t\, C^n \int_\Omega u^n\mathrm{d}x, \quad \forall\, n \geq 1. \end{aligned} \tag{12.4}$$

Let us denote $\sup_{t\in(0,\,T)} |C(t)|$ by $\|C\|_\infty$. Using the *Schwarz inequality* in $L^2(\Omega)$ and the *Nirenberg–Strauss inequality* (11.16), it follows from (12.4) that

$$\|u^n\|^2_{L^2(\Omega)} - \|u^{n-1}\|^2_{L^2(\Omega)} + \frac{2\mu\Delta t}{\rho}\|\nabla u^n\|^2_{(L^2(\Omega))^2} \leq \frac{2\Delta t}{\rho}(\|C\|_\infty|\Omega|^{\frac{1}{2}} - \tau_y\gamma)^+\|u^n\|_{L^2(\Omega)}, \quad \forall\, n \geq 1, \tag{12.5}$$

where $z^+ = \max(0, z)$, $\forall\, z \in \mathbf{R}$. If $\tau_y \geq \|C\|_\infty|\Omega|^{\frac{1}{2}}\gamma^{-1}$, relation (12.5) implies $\|u^n\|_{L^2(\Omega)} \leq \|u^0\|_{L^2(\Omega)}$ $\forall\, n \geq 1$, i.e., the *unconditional stability* of the scheme. Suppose now that $\tau_y < \|C\|_\infty|\Omega|^{\frac{1}{2}}\gamma^{-1}$ and denote by K the (positive) quantity $\|C\|_\infty|\Omega|^{\frac{1}{2}} - \tau_y\gamma$.

Combining (12.5) and the relations

$$2ab \leq \alpha a^2 + \frac{b^2}{\alpha}, \quad \forall\, a,\ b \in \mathbf{R}, \quad \text{and } \alpha > 0$$

and

$$\lambda_0\|v\|^2_{L^2(\Omega)} \leq \int_\Omega |\nabla v|^2 \mathrm{d}x, \quad \forall\, v \in H^1_0(\Omega) \ \text{(Poincaré inequality)},$$

we obtain

$$\left(1 + \frac{\mu\Delta t}{\rho}\lambda_0\right)\|u^n\|^2_{L^2(\Omega)} \leq \frac{\Delta t K^2}{\rho\lambda_0\mu} + \|u^{n-1}\|^2_{L^2(\Omega)}, \quad \forall\, n \geq 1. \tag{12.6}$$

Let us denote by θ the quantity $\left(1 + \frac{\mu\Delta t}{\rho}\lambda_0\right)$; it follows from (12.6) that

$$\|u^n\|^2_{L^2(\Omega)} \leq \frac{\Delta t K^2}{\rho\lambda_0\mu}\sum_{j=1}^{n}\theta^{-j} + \|u^0\|^2_{L^2(\Omega)}\theta^{-n}, \quad \forall\, n \geq 1. \tag{12.7}$$

Because $\theta > 1$, (12.7) implies that

$$\|u^n\|^2_{L^2(\Omega)} \leq \frac{\Delta t K^2}{\rho\lambda_0\mu}\frac{\theta^{-1}}{1-\theta^{-1}} + \|u^0\|^2_{L^2(\Omega)}, \quad \forall\, n \geq 1.$$

The above relation implies in turn $\left(\text{because } \frac{\theta^{-1}}{1-\theta^{-1}} = \frac{1}{\theta-1} = \frac{\rho}{\mu\Delta t\lambda_0}\right)$ that

$$\|u^n\|^2_{L^2(\Omega)} \leq \left(\frac{K}{\lambda_0\mu}\right)^2 + \|u^0\|^2_{L^2(\Omega)}, \quad \forall\, n \geq 1,$$

that is the *unconditional stability* of scheme (12.1), (12.2) when $\tau_y < \|C\|_\infty\,|\Omega|^{\frac{1}{2}}\gamma^{-1}$.

To complete the proof of the theorem, we still have to show that property (12.3) holds if $|C(t)|$ is sufficiently small and $T = +\infty$. Suppose, indeed, that

$$\|C\|_\infty < \gamma\tau_y|\Omega|^{-\frac{1}{2}}; \tag{12.8}$$

it follows then from (12.4), and from the various inequalities already used earlier, that

$$\rho(\|u^n\|_{L^2(\Omega)} - \|u^{n-1}\|_{L^2(\Omega)})\|u^n\|_{L^2(\Omega)} + \Delta t\lambda_0\mu\|u^n\|^2_{L^2(\Omega)}$$
$$+ \Delta t(\gamma\tau_y - \|C\|_\infty|\Omega|^{\frac{1}{2}})\|u^n\|_{L^2(\Omega)}, \quad \forall\, n \geq 1. \tag{12.9}$$

Because (from (12.8)) $\tau_y\gamma - \|C\|_\infty|\Omega|^{\frac{1}{2}} > 0$, it follows from (12.9) that if there exists n_0 such that $u^{n_0} = 0$, then $u^n = 0, \ \forall\, n \geq n_0$. Suppose now that $u^n \neq 0, \ \forall\, n \geq 0$. We have then, from (12.9),

$$(\|u^n\|_{L^2(\Omega)} - \|u^{n-1}\|_{L^2(\Omega)}) + \frac{\Delta t\lambda_0\mu}{\rho}\|u^n\|_{L^2(\Omega)} + \frac{\Delta t}{\rho}(\tau_y\gamma - \|C\|_\infty||\Omega|^{\frac{1}{2}}) \leq 0,$$
$$\forall\, n \geq 1. \tag{12.10}$$

Relation (12.10) can be rewritten as

$$(\|u^n\|_{L^2(\Omega)} - \|u^{n-1}\|_{L^2(\Omega)}) + \frac{\Delta t\lambda_0\mu}{\rho}\left[\|u^n\|_{L^2(\Omega)} + \frac{\tau_y\gamma - \|C\|_\infty||\Omega|^{\frac{1}{2}}}{\lambda_0\mu}\right] \leq 0,$$
$$\forall\, n \geq 1. \tag{12.11}$$

Introduce now $y^n = \|u^n\|_{L^2(\Omega)} + \frac{\tau_y\gamma - \|C\|_\infty||\Omega|^{\frac{1}{2}}}{\lambda_0\mu}$; it follows then from (12.11) that

$$\left(1 + \Delta t\frac{\lambda_0\mu}{\rho}\right)y^n \leq y^{n-1}, \quad \forall\, n \geq 1,$$

which implies that

$$y^n \leq \left(1 + \Delta t\frac{\lambda_0\mu}{\rho}\right)^{-n} y_0, \quad \forall\, n \geq 1. \tag{12.12}$$

Relation (12.12) implies in turn that $\lim_{n\to+\infty} y^n = 0$, i.e.,

$$\lim_{n\to+\infty}\left[\|u^n\|_{L^2(\Omega)} + \frac{\tau_y\gamma - \|C\|_\infty|\Omega|^{\frac{1}{2}}}{\lambda_0\mu}\right] = 0,$$

which is impossible because $\tau_y\gamma - \|C\|_\infty|\Omega|^{\frac{1}{2}} > 0$. Thus, there exists an index n_c such that (12.3) holds; this completes the proof of the theorem. □

REMARK 12.1. From relation (12.12), we can derive an *upper bound* for the above index n_c. Indeed, it follows from the definition of y^n that (12.12) can not hold for those n verifying

$$n \geq \frac{\ln\left[1 + \frac{\lambda_0\mu\|u_0\|_{L^2(\Omega)}}{\tau_y\gamma - \|C\|_\infty||\Omega|^{\frac{1}{2}}}\right]}{\ln\left(1 + \Delta t\frac{\lambda_0\mu}{\rho}\right)}, \tag{12.13}$$

implying that the corresponding $u^n = 0$. Assuming that $\{u^n\}_{n\geq 0}$ converges in some sense to the solution u of the continuous problem (11.1), when $\Delta t \to 0$ (a result not too difficult to prove), we observe (after multiplying both sides of the inequality (12.13) by Δt) that

$$u(t) = 0, \qquad \forall\, t \geq \frac{\rho}{\lambda_0 \mu} \ln\left[1 + \frac{\lambda_0 \mu \|u_0\|_{L^2(\Omega)}}{\tau_y \gamma - \|C\|_\infty |\Omega|^{\frac{1}{2}}}\right]. \tag{12.14}$$

Relation (12.14) is consistent with the "cut-off" relations (11.13), (11.14). To the best of our knowledge, the *backward Euler scheme* is the *only* time-discretization scheme to enjoy those properties, mimicking those of the continuous model.

REMARK 12.2. Concerning the solutions of problem (11.1), there are many situations of practical interest where $u(t)$ lacks the $C^2(\bar{\Omega})$-regularity property and where, moreover, $u \notin C^1([0,\ T];\ L^2(\Omega))$. This lack of regularity, with respect to both the space and time variables, suggests that there are less advantages at using approximations of higher order than *backward Euler* for the *time-discretization* and *piecewise linear finite-elements* for the *space one*. The numerical experiments and comparisons reported in BRISTEAU and GLOWINSKI [1974] validate this prediction; what was compared in the above references were computational methods and results based, on the one hand, on

$$u_h^0 = u_{0h} (\in V_h); \tag{12.15}$$

then, for $n \geq 1$, compute u_h^n from u_h^{n-1} through the solution of

$$\begin{aligned} &u_h^n \in V_{0h}, \\ &\rho \int_{\Omega_h} \left(u_h^n - u_h^{n-1}\right)(v - u_h^n)\mathrm{d}x + \Delta t \mu \int_{\Omega_h} \nabla u_h^n \cdot \nabla (v - u_h^n)\mathrm{d}x \\ &\quad + \Delta t \tau_y \left[j_h(v) - j_h\left(u_h^n\right)\right] \geq \Delta t C^n \int_{\Omega_h} (v - u_h^n)\mathrm{d}x, \quad \forall\, v \in V_{0h}, \end{aligned} \tag{12.16}$$

and, on the other hand, on

$$u_h^0 = u_{0h} (\in V_h); \tag{12.17}$$

then compute u_h^1 from

$$u_h^1 = 2u_h^{1/2} - u_h^0, \tag{12.18}$$

where, in (12.18), $u_h^{1/2}$ is the solution of

$$\begin{aligned} &u_h^{1/2} \in V_{0h}, \\ &\rho \int_{\Omega_h} \left(u_h^{1/2} - u_h^0\right)\left(v - u_h^{1/2}\right)\mathrm{d}x + \frac{1}{2}\Delta t \mu \int_{\Omega_h} \nabla u_h^{1/2} \cdot \nabla \left(v - u_h^{1/2}\right)\mathrm{d}x \\ &\quad + \Delta t \tau_y \left[j_h(v) - j_h\left(u_h^{1/2}\right)\right] \geq \frac{1}{2}\Delta t C^{1/2} \int_{\Omega_h} \left(v - u_h^{1/2}\right)\mathrm{d}x, \quad \forall\, v \in V_{0h} \end{aligned} \tag{12.19}$$

and next, for $n \geq 2$, u_h^n is obtained from u_h^{n-1} and u_h^{n-2} via the solution of

$$
\begin{aligned}
& u_h^n \in V_{0h}, \\
& \rho \int_{\Omega_h} \left(\frac{3}{2} u_h^n - 2u_h^{n-1} + \frac{1}{2} u_h^{n-2} \right) (v - u_h^n) \mathrm{d}x + \Delta t \mu \int_{\Omega_h} \nabla u_h^n \cdot \nabla (v - u_h^n) \mathrm{d}x \\
& \quad + \Delta t \tau_y \left[j_h(v) - j_h\left(u_h^n\right) \right] \geq \Delta t C^n \int_{\Omega_h} (v - u_h^n) \mathrm{d}x, \quad \forall\, v \in V_{0h}.
\end{aligned}
\tag{12.20}
$$

In BRISTEAU and GLOWINSKI [1974], we had:

- The *finite-element spaces* V_h and V_{0h}, in (12.15) and (12.16), defined by

$$V_h = \{v | v \in C^0(\overline{\Omega}_h),\ v|_K \in P_1, \quad \forall\, K \in \mathcal{T}_h\} \tag{12.21}$$

  and

$$V_{0h} = \{v | v \in V_h,\ v = 0 \text{ on } \Gamma\}, \tag{12.22}$$

  respectively, with $\mathcal{T}_h$ a *triangulation* of Ω, $\overline{\Omega}_h = \bigcup_{K \in \mathcal{T}_h} K$ (assuming that the triangles K are closed), Ω_h is the interior of $\overline{\Omega}_h$, and P_1 is the space of the polynomials in two variables of degree ≤ 1.
- In (12.17)–(12.20), the finite-element space V_{0h} defined by

$$V_{0h} = \{v | v \in C^0(\overline{\Omega}_h),\ v|_K \in P_2(K), \quad \forall\, K \in \mathcal{T}_h,\ v = 0 \text{ on } \Gamma\} \tag{12.23}$$

  with $\mathcal{T}_h$ a triangulation containing possibly curved triangles (to better follow the curved parts of the boundary, if such parts exist), $\overline{\Omega}_h = \bigcup_{K \in \mathcal{T}_h} K$ (assuming that the triangles K are closed), $\Omega_h =$ the interior of $\overline{\Omega}_h$, $P_2(K) = P_2$ (the space of the polynomials in two variables of degree ≤ 2) if K is a rectilinear triangle and, if K is a curved triangle, $P_2(K)$ is obtained from P_2 via the quadratic-mapping-based isoparametric methodology discussed in, e.g., CIARLET [1978, 1991], GLOWINSKI [2003].
- $\lim_{h \to 0} u_{0h} = u_0$ in $L^2(\Omega)$.
- In (12.16),

$$j_h(v) = \sum_{K \in \mathcal{T}_h} \int_K |\nabla v| \mathrm{d}x,$$

  while, in (12.19) and (12.20), $j_h(v)$ is obtained from $\sum_{K \in \mathcal{T}_h} \int_K |\nabla v| \mathrm{d}x$ by using a *Simpson rule* related *numerical integration method* to compute the integrals over the triangles K of the triangulation $\mathcal{T}_h$.

Observe that step (12.19) is of the *Crank–Nicolson* type, while the scheme used in (12.20) is *fully implicit* and *two-step backward*. Scheme (12.17)–(12.20) is *second-order accurate* (when applied to the solution of smooth problems (which is not the case here)) and *stiff A-stable* (like scheme (12.15), (12.16)). Other schemes are possible.

It is worth mentioning that an *adaptive finite-element method* for the solution of the steady state variant of problem (11.1) is discussed in SARAMITO and ROQUET [2001]; it relies on *piecewise quadratic approximations* similar to those in (12.23). *Adaptivity* provides here a way to overcome the low accuracy of the approximate solutions resulting from the lack of regularity of the solution (H^2-regularity at most).

13. Bingham flow in cylinders: (V) steady flow

13.1. Formulation of the problem: synopsis

Suppose that the *pressure drop* C is *independent* of t. It follows then from, e.g., GLOWINSKI [1984] that the *steady-state* problem associated with (11.1), namely

$$u_\infty \in H_0^1(\Omega), \quad \mu \int_\Omega \nabla u_\infty \cdot \nabla (v - u_\infty) \mathrm{d}x + \tau_y [j(v) - j(u_\infty)] \geq$$
$$C \int_\Omega (v - u_\infty) \mathrm{d}x, \quad \forall\, v \in H_0^1(\Omega), \tag{13.1}$$

has a *unique solution*. In order to solve (13.1) (an *elliptic variational inequality* problem), several approaches are possible, several of them discussed in, e.g., BRISTEAU and GLOWINSKI [1974], GLOWINSKI, LIONS and TRÉMOLIÈRES [1981], GLOWINSKI [1984], HE and GLOWINSKI [2000]; among them

1. Solve the corresponding problem (11.1) on the time interval $(0, +\infty)$ until one reaches a steady-state solution.
2. Apply directly to (13.1) the "old-fashioned" *Uzawa method* introduced a very long time ago in CEA and GLOWINSKI [1972].
3. Use some of the time dependent methods advocated in HE and GLOWINSKI [2000] which provide short-cuts to u_∞.
4. Use *augmented Lagrangian* methods associated with the *linear constraint* $\mathbf{p} = \nabla u$ (this approach seems to be increasingly popular and has been used in, e.g., COUPEZ, ZINE and AGASSANT [1994], SARAMITO and ROQUET [2001], ROQUET and SARAMITO [2003], VOLA, BOSCARDIN and LATCHÉ [2003], MOYERS-GONZALEZ and FRIGAARD [2004], HUILGOL and YOU [2005], VINAY, WACHS and AGASSANT [2005], LITVINOV and HOPPE [2005], the last reference concerning the numerical simulation of *Electro-Rheological fluid flow*).

The four above approaches will be discussed below. We will discuss also a novel approach, recently introduced in DEAN, GLOWINSKI and GUIDOBONI [2007]; it combines *penalty techniques*, the *Newton's method* and *conjugate gradient algorithms*.

13.2. Computing u_∞ via the solution of the time-dependent problem (13.1)

Relation (11.15) in the statement of Theorem 11.2 (see Section 11), shows that integrating (11.1) from 0 to $+\infty$ provides u_∞ with *exponential speed* of convergence in $L^2(\Omega)$; actually, this property still holds if one applies the *backward Euler scheme* to the solution of problem (11.1). Let us prove this property: assuming that the pressure drop C is time independent, the backward Euler scheme takes here the following form:

$$u^0 = u_0; \tag{13.2}$$

then, for $n \geq 1$, compute u^n from u^{n-1} through the solution of

$$u^n \in H_0^1(\Omega), \quad \rho \int_\Omega (u^n - u^{n-1})(v - u^n)\mathrm{d}x + \Delta t \mu \int_\Omega \nabla u^n \cdot \nabla (v - u^n)\mathrm{d}x$$

$$+ \Delta t \tau_y [j_h(v) - j_h(u^n)] \geq \Delta t C^n \int_\Omega (v - u^n)\mathrm{d}x, \quad \forall\, v \in H_0^1(\Omega). \tag{13.3}$$

Denote $u^n - u_\infty$ by $\bar{u}^n$, taking $v = u^n$ (resp., $v = u_\infty$) in (13.1) (resp., (13.3)) and adding (after multiplying by Δt both sides of the inequality in (13.1)) we obtain

$$\rho \int_\Omega (\bar{u}^n - \bar{u}^{n-1})\bar{u}^n \mathrm{d}x + \Delta t \mu \|\nabla \bar{u}^n\|^2_{(L^2(\Omega))^2} \leq 0, \quad \forall\, n \geq 1. \tag{13.4}$$

Combining (13.4) with the *Schwarz inequality* in $L^2(\Omega)$ and the *Poincaré inequality* in $H_0^1(\Omega)$, we obtain

$$\frac{\rho}{2}\left(\|\bar{u}^n\|^2_{L^2(\Omega)} - \|\bar{u}^{n-1}\|^2_{L^2(\Omega)}\right) + \Delta t \lambda_0 \mu \|\bar{u}^n\|^2_{L^2(\Omega)} \leq 0, \quad \forall\, n \geq 1 \tag{13.5}$$

with λ_0 the smallest eigenvalue of $-\nabla^2$ operating in $H_0^1(\Omega)$. It follows from (13.5) that

$$\left(1 + 2\frac{\lambda_0 \mu}{\rho}\Delta t\right)\|\bar{u}^n\|^2_{L^2(\Omega)} \leq \|\bar{u}^{n-1}\|^2_{L^2(\Omega)}, \quad \forall\, n \geq 1,$$

which implies in turn that

$$\|\bar{u}^n\|^2_{L^2(\Omega)} \leq \left(1 + 2\frac{\lambda_0 \mu}{\rho}\Delta t\right)^{-\frac{n}{2}} \|\bar{u}^0\|_{L^2(\Omega)}, \quad \forall\, n \geq 0. \tag{13.6}$$

The *exponential convergence* (in $L^2(\Omega)$) of u^n to u_∞ follows from (13.6) (which is nothing but a *time-discrete analog* of relation (11.15)). Actually, relation (13.6) still holds for the *finite-element* analogs of (13.1) and (13.2), (13.3).

Of course, when applying the fully implicit scheme (13.2), (13.3) to the computation of u_∞, we still have to address the solution of the *elliptic variational inequality* problems (13.3); this important issue will be discussed in the following section.

13.3. An iterative method à la Uzawa for the solution of problems (13.1) and (13.3)

Both problems (13.1) and (13.3) are particular cases of

$$u \in H_0^1(\Omega); \quad \alpha \int_\Omega u(v - u)\mathrm{d}x + \mu \int_\Omega \nabla u \cdot \nabla (v - u)\mathrm{d}x + \tau_y [j(v) - j(u)]$$

$$\geq \int_\Omega f(v - u)\mathrm{d}x, \quad \forall\, v \in H_0^1(\Omega) \tag{13.7}$$

with $\alpha \geq 0$ and $f \in L^2(\Omega)$.

A classical method to solve problem (13.7) is the one introduced many years ago in CEA and GLOWINSKI [1972]; it reduces the solution of the above problem to a sequence of *linear Dirichlet problems* for the operator $\alpha\mathbf{I} - \mu\nabla^2$ and simple projections operations. This method relies on the (now classical) equivalence between problem (13.7) and the following system:

$$\alpha u - \mu\nabla^2 u - \tau_y \nabla\cdot\boldsymbol{\lambda} = f \text{ in } \Omega, \quad u = 0 \text{ on } \Gamma, \quad \boldsymbol{\lambda}\cdot\nabla u = |\nabla u|, \ \boldsymbol{\lambda}\in\boldsymbol{\Lambda} \tag{13.8}$$

with (cf. (11.30)) $\boldsymbol{\Lambda} = \{\mathbf{q}|\mathbf{q}\in(L^2(\Omega))^2, |\mathbf{q}(x)|\le 1, \text{ a.e. in } \Omega\}$ in (13.8). The last two relations in (13.8) are equivalent to

$$\boldsymbol{\lambda} = P_{\boldsymbol{\Lambda}}[\boldsymbol{\lambda} + r\tau_y\nabla u], \quad \forall\, r\ge 0 \tag{13.9}$$

with the operator $P_{\boldsymbol{\Lambda}}$ defined by (11.31). In order to solve (13.7), through (13.8) and (13.9), we advocate, following CEA and GLOWINSKI [1972], the *fixed point algorithm* below:

$$\boldsymbol{\lambda}^0 \text{ is given in } \boldsymbol{\Lambda}(\,\boldsymbol{\lambda}^0 = \mathbf{0}, \text{ for example}); \tag{13.10}$$

then, for $n\ge 0$, $\boldsymbol{\lambda}^n$ being known, compute u^n and $\boldsymbol{\lambda}^{n+1}$ as follows:
Solve (in $H_0^1(\Omega)$)

$$\alpha u^n - \mu\nabla^2 u^n = \tau_y\nabla\cdot\boldsymbol{\lambda}^n + f \text{ in } \Omega, \quad u^n = 0 \text{ on } \Gamma, \tag{13.11}$$

and update $\boldsymbol{\lambda}^n$ through

$$\boldsymbol{\lambda}^{n+1} = P_{\boldsymbol{\Lambda}}[\boldsymbol{\lambda}^n + r\tau_y\nabla u^n]. \tag{13.12}$$

REMARK 13.1. Suppose that the system (13.8) has a solution $\{u,\boldsymbol{\lambda}\}\in H_0^1(\Omega)\times\boldsymbol{\Lambda}$ (which is indeed the case from, e.g., GLOWINSKI, LIONS and TRÉMOLIÈRES [1976, 1981], and GLOWINSKI [1984]). It can be shown (see, again the above two references) that the above pair is necessarily a *saddle-point* over $H_0^1(\Omega)\times\boldsymbol{\Lambda}$ of the *Lagrangian functional* $\mathcal{L}: H^1(\Omega)\times(L^2(\Omega))^2\to\mathbf{R}$ defined by

$$\mathcal{L}(v,\boldsymbol{\mu}) = \frac{1}{2}\left[\alpha\int_\Omega |v|^2 dx + \mu\int_\Omega |\nabla v|^2 dx\right] + \tau_y\int_\Omega \boldsymbol{\mu}\cdot\nabla v\, dx - \int_\Omega f v\, dx, \tag{13.13}$$

that is the pair $\{u,\boldsymbol{\lambda}\}$ verifies (from the definition of a saddle-point; see, e.g., GLOWINSKI [2003, chapter 4])

$$\begin{aligned}&\{u,\boldsymbol{\lambda}\}\in H_0^1(\Omega)\times\boldsymbol{\Lambda}, \ \mathcal{L}(u,\boldsymbol{\mu})\le\mathcal{L}(u,\boldsymbol{\lambda})\le\mathcal{L}(v,\boldsymbol{\lambda}),\\ &\forall\,\{v,\boldsymbol{\mu}\}\in H_0^1(\Omega)\times\boldsymbol{\Lambda}.\end{aligned} \tag{13.14}$$

Conversely, any solution of (13.14) is solution of system (13.8). It follows from, e.g., the above reference that algorithm (13.10)–(13.12) is nothing but an *Uzawa algorithm* applied to the solution of the saddle-point problem (13.14) with $\mathcal{L}$ defined by (13.13); for a systematic study of Uzawa algorithms, see, e.g., GLOWINSKI [2003, chapter 4] and the references therein.

Proving the convergence of algorithm (13.10)–(13.12) (for $r > 0$ sufficiently small) is a relatively simple exercise; owing to the importance of these topics (in order to investigate, in the following sections, the convergence of variants of algorithm (13.10)–(13.12)), we thought that it was worth to give a proof of the convergence of the above algorithm. We have, thus, the following convergence:

THEOREM 13.1. *Suppose that*

$$0 < r < \frac{2\mu}{\tau_y^2} \tag{13.15}$$

in (13.12). Then, $\forall\, \boldsymbol{\lambda}^0 \in \boldsymbol{\Lambda}$, *the sequence* $\{u^n, \boldsymbol{\lambda}^n\}$ *generated by algorithm (13.10)–(13.12) verifies*

$$\lim_{n\to+\infty} \{u^n, \boldsymbol{\lambda}^n\} = \{u^n, \boldsymbol{\lambda}^*\} \text{ in } H_0^1(\Omega) \times ((L^\infty(\Omega))^2 \text{ weak } -*), \tag{13.16}$$

where $\{u, \boldsymbol{\lambda}^*\}$ *is a solution of (13.8) in* $H_0^1(\Omega) \times \boldsymbol{\Lambda}$.

PROOF. Let $\{u, \boldsymbol{\lambda}\}$ be a solution of (13.8) in $H_0^1(\Omega) \times \boldsymbol{\Lambda}$ and let us denote $u^n - u$ and $\boldsymbol{\lambda}^n - \boldsymbol{\lambda}$ by $\bar{u}^n$ and $\bar{\lambda}^n$, respectively. Taking into account the fact that the operator $P_{\boldsymbol{\Lambda}}$ is a *contraction* of $(L^2(\Omega))^2$, we obtain by subtraction between (13.8), (13.9) and (13.11), (13.12) that, $\forall\, n \geq 0$,

$$\begin{aligned} &\alpha\bar{u}^n - \mu\nabla^2\bar{u}^n = \tau_y\nabla\cdot\bar{\lambda}^n + f \text{ in } \Omega, \quad \bar{u}^n = 0 \text{ on } \Gamma, \\ &\|\bar{\lambda}^{n+1}\|^2_{(L^2(\Omega))^2} \leq \|\bar{\lambda}^n + r\tau_y\nabla\bar{u}^n\|^2_{(L^2(\Omega))^2}, \end{aligned} \tag{13.17}$$

which implies in turn

$$\|\bar{\lambda}^n\|^2_{(L^2(\Omega))^2} - \|\bar{\lambda}^{n+1}\|^2_{(L^2(\Omega))^2} \geq -2r\tau_y \int_\Omega \bar{\lambda}^n\cdot\nabla\bar{u}^n \mathrm{d}x - r^2\tau_y^2\|\nabla\bar{u}^n\|^2_{(L^2(\Omega))^2}. \tag{13.18}$$

We observe, next, that, after integration by parts, the first two relations in (13.17) imply that

$$\alpha\|\bar{u}^n\|^2_{(L^2(\Omega))^2} + \mu\|\nabla\bar{u}^n\|^2_{(L^2(\Omega))^2} = -\tau_y \int_\Omega \bar{\lambda}^n\cdot\nabla\bar{u}^n \mathrm{d}x. \tag{13.19}$$

Combining relations (13.18) and (13.19), we obtain

$$\begin{aligned} \|\bar{\lambda}^n\|^2_{(L^2(\Omega))^2} - \|\bar{\lambda}^{n+1}\|^2_{(L^2(\Omega))^2} &\geq 2r\left[\alpha\|\bar{u}^n\|^2_{(L^2(\Omega))^2} + \mu\|\nabla^2\bar{u}^n\|^2_{(L^2(\Omega))^2}\right] \\ &\quad - r^2\tau_y^2\|\nabla\bar{u}^n\|^2_{(L^2(\Omega))^2} \geq r\left(2 - \frac{r\tau_y^2}{\mu}\right) \\ &\quad \left[\alpha\|\bar{u}^n\|^2_{(L^2(\Omega))^2} + \mu\|\nabla^2\bar{u}^n\|^2_{(L^2(\Omega))^2}\right]. \end{aligned} \tag{13.20}$$

Suppose that the condition (13.15) holds; it implies that $r\left(2-\frac{r\tau_y^2}{\mu}\right) > 0$. It follows then from (13.20) that the sequence $\{\|\bar{\lambda}^n\|^2_{(L^2(\Omega))^2}\}_{n\geq 0}$ is decreasing; this sequence being bounded from below by 0 converges to some (non-negative) limit, which implies that

$$\lim_{n\to+\infty}\left(\|\bar{\lambda}^n\|^2_{(L^2(\Omega))^2} - \|\bar{\lambda}^{n+1}\|^2_{(L^2(\Omega))^2}\right) = 0. \tag{13.21}$$

Combining (13.20) and (13.21), we obtain $\lim_{n\to+\infty}\bar{u}^n = 0$ in $H_0^1(\Omega)$ namely the convergence of $\{u^n\}_{n\geq 0}$ to u in $H_0^1(\Omega)$. Proving the convergence of $\{\boldsymbol{\lambda}^n\}_{n\geq 0}$ is a more complicated issue that we will not address here (it is discussed in, e.g., GLOWINSKI [2003, chapter 4], and GLOWINSKI, LIONS and TRÉMOLIÈRES [1981]). □

REMARK 13.2. All the solutions of system (13.8) share the same u. Suppose now that $\{u, \boldsymbol{\lambda}\}$ and $\{u, \boldsymbol{\lambda}'\}$ are solutions of (13.8); we have then

$$\nabla\cdot(\boldsymbol{\lambda}' - \boldsymbol{\lambda}) = 0. \tag{13.22}$$

Keeping in mind that

$$(L^2(\Omega))^2 = \nabla H_0^1(\Omega) \oplus \mathbf{S}_0 \tag{13.23}$$

with $\mathbf{S}_0 = \{\mathbf{q}|\mathbf{q}\in (L^2(\Omega))^2, \nabla\cdot\mathbf{q} = 0\}$, it follows from (13.22) that all the pairs $\{u, \boldsymbol{\lambda}\}$, solutions of (13.8), not only share the same argument u, but all the $\boldsymbol{\lambda}$s have the same component in $\nabla H_0^1(\Omega)$ when decomposed according to (13.23). Another consequence is the following:

Consider $\mathbf{q} \in (L^2(\Omega))^2$; it follows from (13.23) that $\mathbf{q} = \mathbf{q}_1 + \mathbf{q}_2$, with $\mathbf{q}_1 \in \nabla H_0^1(\Omega)$ and $\mathbf{q}_2 \in \mathbf{S}_0$, respectively, the above decomposition being unique. If $\{u, \boldsymbol{\lambda}\}$ is solution of (13.8), all the $\boldsymbol{\lambda}$s have $\boldsymbol{\lambda}_1$ in common in the decomposition (13.23) of $(L^2(\Omega))^2$. Suppose now that r verifies the condition (13.15); it follows then from Theorem 13.1 and from (13.17), (13.23) that

$$\lim_{n\to+\infty}\boldsymbol{\lambda}_1^n = \boldsymbol{\lambda}_1 \text{ in } (L^2(\Omega))^2. \tag{13.24}$$

The frequently observed slow convergence of algorithm (13.10)–(13.12), particularly when τ_y is large, seems to be related to the relative importance of $\boldsymbol{\lambda}_2$ compared to $\boldsymbol{\lambda}_1$. The more important is $\boldsymbol{\lambda}_2$, the slower is the convergence, everything else being the same.

REMARK 13.3. By formal elimination of u in (13.8), we can show that $\boldsymbol{\lambda}$ is in fact solution of a kind of *elliptic variational inequality* of the *obstacle type*. To show this property, we observe that any pair $\{u, \boldsymbol{\lambda}\}$ solution of (13.8) in $H_0^1(\Omega)\times\boldsymbol{\Lambda}$ verifies

$$\begin{aligned}&\boldsymbol{\lambda}\in\boldsymbol{\Lambda}, \int_\Omega(-\nabla u)\cdot(\boldsymbol{\mu}-\boldsymbol{\lambda})\,dx \geq 0, \quad \forall\,\boldsymbol{\mu}\in\boldsymbol{\Lambda},\\ &\alpha u - \mu\nabla^2 u = f + \tau_y\nabla\cdot\boldsymbol{\lambda} \text{ in } \Omega, \ u = 0 \text{ on } \Gamma.\end{aligned} \tag{13.25}$$

Next, we introduce the continuous and linear operator $\mathbf{A}$ from $(L^2(\Omega))^2$ into $(L^2(\Omega))^2$, defined as follows:

$$\mathbf{A}\mathbf{q} = -\nabla u_{\mathbf{q}}, \quad \forall\, \mathbf{q} \in (L^2(\Omega))^2, \tag{13.26}$$

where $u_{\mathbf{q}}$ is the unique solution in $H_0^1(\Omega)$ of the Dirichlet problem

$$\alpha u_{\mathbf{q}} - \mu\nabla^2 u_{\mathbf{q}} = \tau_y \nabla\cdot\mathbf{q} \text{ in } \Omega, \; u_{\mathbf{q}} = 0 \text{ on } \Gamma. \tag{13.27}$$

We have then, $\forall \mathbf{q},\ \mathbf{q}' \in (L^2(\Omega))^2$,

$$\begin{aligned}\int_\Omega (\mathbf{A}\mathbf{q})\cdot\mathbf{q}'\mathrm{d}x &= -\int_\Omega \nabla u_{\mathbf{q}}\cdot\mathbf{q}'\mathrm{d}x = \langle\nabla\cdot\mathbf{q}', u_q\rangle \\ &= \frac{1}{\tau_y}\langle\alpha u_{\mathbf{q}'} - \mu\nabla^2 u_{\mathbf{q}'}, u_{\mathbf{q}}\rangle = \frac{1}{\tau_y}\int_\Omega [\alpha u_{\mathbf{q}} u_{\mathbf{q}'} + \mu\nabla u_{\mathbf{q}}\cdot\nabla u_{\mathbf{q}'}]\mathrm{d}x,\end{aligned} \tag{13.28}$$

where in (13.28), $\langle .,.\rangle$ denotes the pairing between $H^{-1}(\Omega)$ (the dual space of $H_0^1(\Omega)$) and $H_0^1(\Omega)$. It follows from (13.28) that operator $\mathbf{A}$ is *symmetric* and *positive semidefinite*; $\mathbf{A}$ is not positive definite because we clearly have $Ker(\mathbf{A}) = \mathbf{S}_0$, with $\mathbf{S}_0$ as in (13.23). Finally, define u_f as the unique solution in $H_0^1(\Omega)$ of the Dirichlet problem

$$\alpha u_f - \mu\nabla^2 u_f = f \text{ in } \Omega, \; u_f = 0 \text{ on } \Gamma \tag{13.29}$$

(if Γ is *smooth* and/or Ω is *convex*, then $u_f \in H^2(\Omega)\cap H_0^1(\Omega)$, implying that $\nabla u_f \in (H^1(\Omega))^2 \subset (L^s(\Omega))^2$, $\forall\, s \in [1, +\infty)$. It follows then from (13.25)–(13.27) and (13.29) that

$$\nabla u = \nabla u_f + \nabla u_{\boldsymbol{\lambda}} = \nabla u_f - \mathrm{A}\boldsymbol{\lambda},$$

which combined with (13.25) implies that the vector-valued function $\boldsymbol{\lambda}$ is a solution of the following "elliptic" *variational inequality* (in the sense of LIONS and STAMPACCHIA [1967]):

$$\begin{aligned}&\boldsymbol{\lambda} \in \boldsymbol{\Lambda}, \\ &\int_\Omega \mathbf{A}\boldsymbol{\lambda}\cdot(\boldsymbol{\mu}-\boldsymbol{\lambda})\mathrm{d}x \geq \int_\Omega \nabla u_f\cdot(\boldsymbol{\mu}-\boldsymbol{\lambda})\mathrm{d}x, \quad \forall\, \boldsymbol{\mu} \in \boldsymbol{\Lambda}\end{aligned} \tag{13.30}$$

which, from the very nature of the *convex* set $\boldsymbol{\Lambda}$, is definitely an *obstacle problem*. Incidentally, an *equivalent* formulation of algorithm (13.10)–(13.12) is given by

$$\boldsymbol{\lambda}^0 \text{ is given in } \boldsymbol{\Lambda}; \tag{13.31}$$

then, for $n \geq 0$, assuming that $\boldsymbol{\lambda}^n$ is known, compute $\boldsymbol{\lambda}^{n+1}$ as follows

$$\boldsymbol{\lambda}^{n+1} = P_{\boldsymbol{\Lambda}}[\boldsymbol{\lambda}^n - r\,\tau_y(\mathbf{A}\boldsymbol{\lambda}^n - \nabla u_f)]. \tag{13.32}$$

From the *symmetry* of operator $\mathbf{A}$ (see (13.28)), algorithm (13.31), (13.32) is clearly a *gradient method with projection.*

Two other *iterative methods* for the solution of problem (13.7), (13.8) will be discussed in the following sections; more methods are discussed in HE and GLOWINSKI [2000].

13.4. A (pseudo-) time relaxation approach for the solution of problem (13.7), (13.8)

To the best of our knowledge, the method to be discussed now has been introduced in DEAN, GLOWINSKI and GUIDOBONI [2007]; in fact, it improves on related methods discussed in HE and GLOWINSKI [2000] and has some similarities with methods recently introduced in *Image Processing*. The idea is pretty simple and quite general: it consists in associating with (13.7) and (13.8) a well-chosen *initial value problem* (*flow* in the *Dynamical System* terminology) that we integrate from 0 to $+\infty$ in order to capture the related *steady-state solutions*, if such solutions exist, which is the case here. The initial value problem that we consider is a dynamical variant of problem (13.30); it is defined as follows (with τ a pseudo-time):

$$\boldsymbol{\lambda}(0) = \boldsymbol{\lambda}_0 \ (\in \boldsymbol{\Lambda}); \tag{13.33}$$

$$\boldsymbol{\lambda}(\tau) \in \boldsymbol{\Lambda}, \ \tau \in [0, +\infty),$$
$$\int_\Omega \left(\frac{\partial \boldsymbol{\lambda}}{\partial \tau} + \mathbf{A}\boldsymbol{\lambda}\right)(\tau)\cdot(\boldsymbol{\mu} - \boldsymbol{\lambda}(\tau))\mathrm{d}x \geq \tau_y \int_\Omega \nabla u_f \cdot (\boldsymbol{\mu} - \boldsymbol{\lambda}(\tau))\mathrm{d}x, \quad \forall\, \boldsymbol{\mu} \in \boldsymbol{\Lambda}. \tag{13.34}$$

An equivalent, but more explicit formulation of the initial value problem (13.33), (13.34) is obtained by replacing (13.34) by

$$\begin{aligned}
&\alpha u(\tau) - \mu\nabla^2 u(\tau) - \tau_y \nabla\cdot\boldsymbol{\lambda}(\tau) = f \text{ in } \Omega, \ u(\tau) = 0 \text{ on } \Gamma,\\
&\boldsymbol{\lambda}(\tau) \in \boldsymbol{\Lambda}, \ \ \tau \in [0, +\infty), \ \int_\Omega \left(\frac{\partial \boldsymbol{\lambda}}{\partial \tau} - \tau_y \nabla u\right)(\tau)\cdot(\boldsymbol{\mu} - \boldsymbol{\lambda}(\tau))\mathrm{d}x \geq 0, \ \ \forall\, \boldsymbol{\mu} \in \boldsymbol{\Lambda}.
\end{aligned} \tag{13.35}$$

To *time-discretize* (13.33), (13.35), we advocate the following *backward Euler scheme*:

$$\boldsymbol{\lambda}^0 = \boldsymbol{\lambda}_0; \tag{13.36}$$

then, for $n \geq 1$, compute $\boldsymbol{\lambda}^n$ from $\boldsymbol{\lambda}^{n-1}$ via the solution of

$$\alpha u^n - \mu\nabla^2 u^n - \tau_y \nabla\cdot\boldsymbol{\lambda}^n = f \text{ in } \Omega, \ u^n = 0 \text{ on } \Gamma, \tag{13.37}$$

$$\boldsymbol{\lambda}^n \in \boldsymbol{\Lambda}, \ \int_\Omega \left(\frac{\boldsymbol{\lambda}^n - \boldsymbol{\lambda}^{n-1}}{\Delta\tau} - \tau_y \nabla u^n\right)\cdot(\boldsymbol{\mu} - \boldsymbol{\lambda}^n)\mathrm{d}x \geq 0, \ \ \forall\, \boldsymbol{\mu} \in \boldsymbol{\Lambda}, \tag{13.38}$$

with $\Delta\tau (> 0)$ a (pseudo) time-discretization step. From the properties of the linear operator $\mathbf{A}$, problem (13.37), (13.38) has a *unique* solution. In order to solve system (13.37), (13.38) we observe that (13.38) is *equivalent* to

$$\boldsymbol{\lambda}^n \in \boldsymbol{\Lambda}, \quad \int_\Omega \left[\boldsymbol{\lambda}^n + r\frac{\boldsymbol{\lambda}^{n-1} - \boldsymbol{\lambda}^n}{\Delta\tau} + r\tau_y \nabla u^n - \boldsymbol{\lambda}^n\right] \cdot (\boldsymbol{\mu} - \boldsymbol{\lambda}^n)\mathrm{d}x \leq 0,$$
$$\forall\, \boldsymbol{\mu} \in \boldsymbol{\Lambda}, \quad \forall\, r \geq 0. \tag{13.39}$$

Because $\mathbf{P}_\Lambda$ is a *projection operator*, there is equivalence between (13.39) and

$$\boldsymbol{\lambda}^n = P_{\boldsymbol{\Lambda}}\left[\boldsymbol{\lambda}^n + r\frac{\boldsymbol{\lambda}^{n-1} - \boldsymbol{\lambda}^n}{\Delta\tau} + r\tau_y \nabla u^n\right], \quad \forall\, r \geq 0. \tag{13.40}$$

Inspired by (13.37), (13.40), we suggest the following *fixed point algorithm* to compute $\{u^n, \boldsymbol{\lambda}^n\}$ from $\boldsymbol{\lambda}^{n-1}$:

$$\boldsymbol{\lambda}_0^n \text{ is given in } \boldsymbol{\Lambda} (\text{ a most natural choice being } \boldsymbol{\lambda}_0^n = \boldsymbol{\lambda}^{n-1}); \tag{13.41}$$

for $k \geq 0$, $\boldsymbol{\lambda}_k^n$ being known, compute u_k^n and $\boldsymbol{\lambda}_{k+1}^n$ as follows:
Solve first

$$\alpha u_k^n - \mu \nabla^2 u_k^n = f + \tau_y \nabla \cdot \boldsymbol{\lambda}_k^n \text{ in } \Omega, \ u_k^n = 0 \text{ on } \Gamma, \tag{13.42}$$

and update $\boldsymbol{\lambda}_k^n$ via

$$\boldsymbol{\lambda}_{k+1}^n = P_{\boldsymbol{\Lambda}}\left[\boldsymbol{\lambda}_k^n + r\frac{\boldsymbol{\lambda}^{n-1} - \boldsymbol{\lambda}_k^n}{\Delta\tau} + r\tau_y \nabla u_k^n\right]. \tag{13.43}$$

Concerning the convergence of algorithm (13.41)–(13.43) to the unique solution of system (13.37), (13.38) in $H_0^1(\Omega) \times (L^2(\Omega))^2$, we have the following:

THEOREM 13.2. *Suppose that*

$$0 < r \leq \frac{2\mu}{2\mu + \tau_y^2 \Delta\tau}\Delta\tau; \tag{13.44}$$

we have then, $\forall \boldsymbol{\lambda}_0^n$ *in (13.41),*

$$\lim_{k\to+\infty} \{u_k^n, \ \boldsymbol{\lambda}_k^n\} = \{u^n, \ \boldsymbol{\lambda}^n\} \text{ in } H_0^1(\Omega) \times (L^2(\Omega))^2, \tag{13.45}$$

the convergence being geometric.

PROOF. We proceed as in the proof of Theorem 13.1. Denoting thus $u_k^n - u^n$ and $\boldsymbol{\lambda}_k^n - \boldsymbol{\lambda}^n$ by $\bar{u}_k^n$ and $\bar{\boldsymbol{\lambda}}_k^n$, respectively, we obtain by subtraction between (13.37), (13.40) and (13.42), (13.43) that, $\forall k \geq 0$,

$$\alpha \bar{u}_k^n - \mu \nabla^2 \bar{u}_k^n = \tau_y \nabla \cdot \bar{\boldsymbol{\lambda}}_k^n \text{ in } \Omega, \ \bar{u}_k^n = 0 \text{ on } \Gamma, \tag{13.46}$$

and (from the contraction properties of operator $P_{\boldsymbol{\Lambda}}$)

$$\|\bar{\boldsymbol{\lambda}}_{k+1}^{n}\|_{(L^2(\Omega))^2} \leq \left\| \left(1 - \frac{r}{\Delta\tau}\right) \bar{\boldsymbol{\lambda}}_{k}^{n} + r\tau_y \nabla \bar{u}_k^n \right\|_{(L^2(\Omega))^2}. \tag{13.47}$$

Combining (13.46) with (13.47) yields

$$\|\bar{\boldsymbol{\lambda}}_{k+1}^{n}\|^2_{(L^2(\Omega))^2} + r\left[2\mu - r\left(\frac{2\mu}{\Delta\tau} + \tau_y^2\right)\right] \|\nabla \bar{u}_{k+1}^n\|^2_{(L^2(\Omega))^2}$$
$$\leq \left\| \left(1 - \frac{r}{\Delta\tau}\right) \bar{\boldsymbol{\lambda}}_{k}^{n} \right\|^2_{(L^2(\Omega))^2}. \tag{13.48}$$

It follows from (13.48) that the convergence property (13.45) will take place if r verifies

$$\left|1 - \frac{r}{\Delta\tau}\right| < 1 \quad \text{and} \quad r\left[2\mu - r\left(\frac{2\mu}{\Delta\tau} + \tau_y^2\right)\right] \geq 0, \quad \text{i.e.,}$$

$$0 < r < 2\Delta\tau \quad \text{and} \quad r \leq \frac{2\mu\Delta\tau}{2\mu + \tau_y^2\Delta\tau},$$

respectively. Because $\frac{2\mu\Delta\tau}{2\mu+\tau_y^2\Delta\tau} \leq \Delta\tau < 2\Delta\tau$ the condition (13.44) implies, clearly, the convergence property (13.45); indeed, we have more because the above discussion shows the existence of a constant K such that

$$\|u_k^n - u^n\|_{(L^2(\Omega))^2} \leq K\|\boldsymbol{\lambda}_0^n - \boldsymbol{\lambda}^n\|_{(L^2(\Omega))^2} \left|1 - \frac{r}{\Delta\tau}\right|^k, \quad \forall\, k \geq 0, \tag{13.49}$$

and

$$\|\boldsymbol{\lambda}_k^n - \boldsymbol{\lambda}^n\|_{(L^2(\Omega))^2} \leq K\|\boldsymbol{\lambda}_0^n - \boldsymbol{\lambda}^n\|_{(L^2(\Omega))^2} \left|1 - \frac{r}{\Delta\tau}\right|^k, \quad \forall\, k \geq 0. \tag{13.50}$$

Relations (13.49) and (13.50) complete the proof of the theorem. □

REMARK 13.4. If one takes $\Delta\tau = \frac{2\mu}{\tau_y^2}$ and $r = \frac{\Delta\tau}{2} (= \frac{\mu}{\tau_y^2})$, the convergence condition (13.44) is verified, the *contraction factor* being $\frac{1}{2}$ in relations (13.49) and (13.50).

13.5. A penalty-Newton-Uzawa-conjugate gradient method for the solution of problem (13.7), (13.8)

The *dual problem* (13.30) is essentially (see Section 13.3) an *obstacle problem* associated with the *point-wise constraint*

$$|\boldsymbol{\lambda}(x)| \leq 1, \quad \text{a.e. in } \Omega. \tag{13.51}$$

An alternative to the *projection methods* discussed so far is provided by a variant of the *penalty-Newton-conjugate gradient method* applied in GLOWINSKI, KUZNETSOV and PAN

[2003], DACOROGNA, GLOWINSKI, KUZNETSOV and PAN [2004], GLOWINSKI, SHIAU, KUO and NASSER [2006] to the solution of *time-dependent variational inequalities* of the *obstacle type*. Let ε be a *small positive* parameter; we approximate the dual problem (13.30) by

$$\mathbf{A}\boldsymbol{\lambda}_\varepsilon + \frac{1}{\varepsilon}(|\boldsymbol{\lambda}_\varepsilon|^2 - 1)^{+2}\boldsymbol{\lambda}_\varepsilon = \nabla u_f, \tag{13.52}$$

with $\xi^+ = \max(0, \xi)$, $\forall \xi \in \mathbf{R}$. The nonlinearity in (13.52) is reminiscent of the *Ginzburg–Landau's* one (see, e.g., BETHUEL, BREZIS and HELEIN [1994]). Using *convexity* arguments and the fact that (13.52) is the *Euler–Lagrange equation* of the following problem from *Calculus of Variations*:

$$\boldsymbol{\lambda}_\varepsilon \in (L^6(\Omega))^2, \quad j_\varepsilon(\boldsymbol{\lambda}_\varepsilon) \le j_\varepsilon(\boldsymbol{\mu}), \qquad \forall\, \boldsymbol{\mu} \in (L^6(\Omega))^2, \tag{13.53}$$

with

$$j_\varepsilon(\boldsymbol{\mu}) = \frac{1}{2}\int_\Omega \mathbf{A}\boldsymbol{\mu}\cdot\boldsymbol{\mu}\,\mathrm{d}x + \frac{1}{6\varepsilon}\int_\Omega (|\boldsymbol{\mu}|^2 - 1)^{+3}\mathrm{d}x - \int_\Omega \nabla u_f\cdot\boldsymbol{\mu}\,\mathrm{d}x,$$

we can easily prove that problem (13.52), (13.53) has a solution in $(L^6(\Omega))^2$. From the definition of operator $\mathbf{A}$ (see Section 13.3), problem (13.52), (13.53) is *equivalent* to the following *nonlinear system*:

$$\alpha u_\varepsilon - \mu\nabla^2 u_\varepsilon - \tau_y \nabla\cdot\boldsymbol{\lambda}_\varepsilon = f \text{ in } \Omega, \;\; u_\varepsilon = 0 \text{ on } \Gamma, \tag{13.54}$$

$$-\nabla u_\varepsilon + \frac{1}{\varepsilon}(|\boldsymbol{\lambda}_\varepsilon|^2 - 1)^{+2}\boldsymbol{\lambda}_\varepsilon = \mathbf{0} \tag{13.55}$$

(easier to handle than (13.52), (13.53), in practice). Using relatively simple variants of the methods discussed in GLOWINSKI [1984] (concerning the solution of *elliptic variational inequalities*), we can prove the following convergence properties:

$$\lim_{\varepsilon\to 0} u_\varepsilon = u \text{ in } H_0^1(\Omega), \quad \lim_{\varepsilon\to 0} \boldsymbol{\lambda}_\varepsilon = \boldsymbol{\lambda} \;\text{ weakly in }\; (L^6(\Omega))^2, \tag{13.56}$$

where, in (13.56), the pair $\{u, \boldsymbol{\lambda}\}$ is a solution of problem (13.7), (13.8) (implying, in turn, that $\boldsymbol{\lambda}$ is a solution of the *obstacle problem* (13.30)). Having thus justified the introduction of the approximate (by *penalization*) problem (13.52), we still have to address its solution. The *Newton's method* is an obvious candidate to achieve such a goal; this leads to the following algorithm (after multiplying by ε both sides of (13.52) and dropping the subscript ε):

$$\boldsymbol{\lambda}^0 \text{ is given in } (L^6(\Omega))^2 \; (\boldsymbol{\lambda}^0 = \mathbf{0}, \text{ for example}); \tag{13.57}$$

for $n \ge 0$, compute $\boldsymbol{\lambda}^{n+1}$ from $\boldsymbol{\lambda}^n$ via

$$\boldsymbol{\lambda}^{n+1} = \boldsymbol{\lambda}^n + \delta\boldsymbol{\lambda}^n, \tag{13.58}$$

$\delta\boldsymbol{\lambda}^n$ being the solution of the following *linear problem*:

$$\varepsilon \mathbf{A}\delta\boldsymbol{\lambda}^n + (|\boldsymbol{\lambda}^n|^2 - 1)^{+2}\delta\boldsymbol{\lambda}^n + 4(|\boldsymbol{\lambda}^n|^2 - 1)^{+}\boldsymbol{\lambda}^n(\boldsymbol{\lambda}^n \cdot \delta\boldsymbol{\lambda}^n) = \\ -[\varepsilon(\mathbf{A}\boldsymbol{\lambda}^n - \nabla u_f) + (|\boldsymbol{\lambda}^n|^2 - 1)^{+2}\boldsymbol{\lambda}^n]; \tag{13.59}$$

we stop iterating when, typically, $\|\delta\boldsymbol{\lambda}^n\|_{(L^2(\Omega))^2} \leq tol._1$.

The linear operator in the left-hand side of (13.59) is clearly *symmetric* and *positive semidefinite*; these properties suggest solving (13.59) by a *conjugate gradient algorithm*. From a *practical point of view*, it is preferable to consider directly the *equivalent system* (13.54), (13.55); we are going to solve it by a *Newton's algorithm* operating in $H_0^1(\Omega) \times (L^6(\Omega))^2$; this algorithm (equivalent to (13.57)–(13.59)) reads as follows:

$$\boldsymbol{\lambda}^0 \text{ is given in } (L^6(\Omega))^2 \ (\boldsymbol{\lambda}^0 = \mathbf{0}, \text{ for example}); \tag{13.60}$$

solve

$$\alpha u^0 - \mu\nabla^2 u^0 = \tau_y \nabla \cdot \boldsymbol{\lambda}^0 + f \text{ in } \Omega, \ u^0 = 0 \text{ on } \Gamma, \tag{13.61}$$

(the above elliptic problem has a unique solution in $H_0^1(\Omega)$ (in fact in $W_0^{1,6}(\Omega)$)). Then, for $n \geq 0$, compute $\{u^{n+1}, \boldsymbol{\lambda}^{n+1}\}$ from $\{u^n, \boldsymbol{\lambda}^n\}$ via

$$\{u^{n+1}, \boldsymbol{\lambda}^{n+1}\} = \{u^n + \delta u^n, \boldsymbol{\lambda}^n + \delta\boldsymbol{\lambda}^n\} \tag{13.62}$$

where, in (13.62), $\{\delta u^n, \delta\boldsymbol{\lambda}^n\}$ is solution to

$$\alpha\delta u^n - \mu\nabla^2\delta u^n - \tau_y\nabla \cdot \delta\boldsymbol{\lambda}^n = 0 \text{ in } \Omega, \quad \delta u^n = 0 \text{ on } \Gamma, \tag{13.63}$$

$$-\varepsilon\nabla\delta u^n + (|\boldsymbol{\lambda}^n|^2 - 1)^{+2}\delta\boldsymbol{\lambda}^n + 4(|\boldsymbol{\lambda}^n|^2 - 1)^{+}\boldsymbol{\lambda}^n(\boldsymbol{\lambda}^n \cdot \delta\boldsymbol{\lambda}^n) \\ = \varepsilon\nabla u^n - (|\boldsymbol{\lambda}^n|^2 - 1)^{+2}\boldsymbol{\lambda}^n. \tag{13.64}$$

We are going to discuss now the solution of system (13.63), (13.64) by an *Uzawa-conjugate gradient algorithm* operating in the *Hilbert space* $H_0^1(\Omega) \times (L^2(\Omega))^2$. To further simplify the notation we denote δu^n by ψ, $\delta\boldsymbol{\lambda}^n$ by $\mathbf{p}$, and by $\mathbf{Q}$ the space $(L^2(\Omega))^2$; the system (13.63), (13.64) takes then the following form:

$$\alpha\psi - \mu\nabla^2\psi - \tau_y\nabla \cdot \mathbf{p} = 0 \text{ in } \Omega, \ \psi = 0 \text{ on } \Gamma, \tag{13.65}$$

$$-\varepsilon\nabla\psi + (|\boldsymbol{\lambda}^n|^2 - 1)^{+2}\mathbf{p} + 4(|\boldsymbol{\lambda}^n|^2 - 1)^{+}\boldsymbol{\lambda}^n(\boldsymbol{\lambda}^n \cdot \mathbf{p}) \\ = \varepsilon\nabla u^n - (|\boldsymbol{\lambda}^n|^2 - 1)^{+2}\boldsymbol{\lambda}^n, \tag{13.66}$$

leading to the following algorithm:

Step 0. *Initialization*

$$\mathbf{p}^0 \text{ is given in } \mathbf{Q} \ (\mathbf{p}^0 = \mathbf{0} \text{ is a natural choice here}); \tag{13.67}$$

solve the following *Dirichlet problem*:

$$\psi^0 \in H_0^1(\Omega), \tag{13.68}$$

$$\alpha \int_\Omega \psi^0 \phi \mathrm{d}x + \mu \int_\Omega \nabla\psi^0 \cdot \nabla\phi \mathrm{d}x = -\tau_y \int_\Omega \mathbf{p}^0 \cdot \nabla\phi \mathrm{d}x, \quad \forall\, \phi \in H_0^1(\Omega), \tag{13.69}$$

and then

$$\begin{aligned} &\mathbf{g}^0 \in \mathbf{Q}, \\ &\int_\Omega \mathbf{g}^0 \cdot \mathbf{q} \mathrm{d}x = -\varepsilon \int_\Omega \nabla(\psi^0 + u^n) \cdot \mathbf{q} \mathrm{d}x + \int_\Omega (|\boldsymbol{\lambda}^n|^2 - 1)^{+2} (\mathbf{p}^0 + \boldsymbol{\lambda}^n) \cdot \mathbf{q} \mathrm{d}x \\ &\qquad + 4 \int_\Omega (|\boldsymbol{\lambda}^n|^2 - 1)^+ (\boldsymbol{\lambda}^n \cdot \mathbf{p}^0)(\boldsymbol{\lambda}^n \cdot \mathbf{q}) \mathrm{d}x, \quad \forall\, \mathbf{q} \in \mathbf{Q}. \end{aligned} \tag{13.70}$$

Set

$$\mathbf{w}^0 = \mathbf{g}^0. \tag{13.71}$$

For $m \geq 0$, assuming that $\mathbf{p}^m$, $\mathbf{g}^m$ and $\mathbf{w}^m$ are known, the last two different from $\mathbf{0}$, we proceed as follows to compute $\{\psi, \mathbf{p}\}$:

Step 1. *Descent*

Solve

$$\begin{aligned} &\bar{\psi}^m \in H_0^1(\Omega), \\ &\alpha \int_\Omega \bar{\psi}^m \phi \mathrm{d}x + \mu \int_\Omega \nabla\bar{\psi}^m \cdot \nabla\phi \mathrm{d}x = -\tau_y \int_\Omega \mathbf{w}^m \cdot \nabla\phi \mathrm{d}x, \quad \forall\, \phi \in H_0^1(\Omega), \end{aligned} \tag{13.72}$$

and then

$$\begin{aligned} &\bar{\mathbf{g}}^m \in \mathbf{Q}, \\ &\int_\Omega \bar{\mathbf{g}}^m \cdot \mathbf{q} \mathrm{d}x = -\varepsilon \int_\Omega \nabla\bar{\psi}^m \cdot \mathbf{q} \mathrm{d}x + \int_\Omega (|\boldsymbol{\lambda}^n|^2 - 1)^{+2} \mathbf{w}^m \cdot \mathbf{q} \mathrm{d}x \\ &\qquad + 4 \int_\Omega (|\boldsymbol{\lambda}^n|^2 - 1)^+ (\boldsymbol{\lambda}^n \cdot \mathbf{w}^m)(\boldsymbol{\lambda}^n \cdot \mathbf{q}) \mathrm{d}x, \quad \forall\, \mathbf{q} \in \mathbf{Q}. \end{aligned} \tag{13.73}$$

Compute

$$\rho_m = \frac{\int_\Omega |\mathbf{g}^m|^2 \mathrm{d}x}{\int_\Omega \bar{\mathbf{g}}^m \cdot \mathbf{w}^m \mathrm{d}x} \tag{13.74}$$

and

$$\mathbf{p}^{m+1} = \mathbf{p}^m - \rho_m \mathbf{w}^m, \quad \mathbf{g}^{m+1} = \mathbf{g}^m - \rho_m \bar{\mathbf{g}}^m. \tag{13.75}$$

Step 2. *Testing the convergence and construction of* $\mathbf{w}^{m+1}$

If $\frac{\int_\Omega |\mathbf{g}^{m+1}|^2 \mathrm{d}x}{\int_\Omega |\mathbf{g}^0|^2 \mathrm{d}x} \le tol._2$ take $\mathbf{p} = \mathbf{p}^{m+1}$ and compute ψ from the solution of

$$\alpha\psi - \mu\nabla^2\psi = \tau_y \nabla \cdot \mathbf{p} \text{ in } \Omega, \quad \psi = 0 \text{ on } \Gamma;$$

else compute

$$\gamma_m = \frac{\int_\Omega |\mathbf{g}^{m+1}|^2 \mathrm{d}x}{\int_\Omega |\mathbf{g}^m|^2 \mathrm{d}x} \tag{13.76}$$

and

$$\mathbf{w}^{m+1} = \mathbf{g}^{m+1} + \gamma_m \mathbf{w}^{m+1}. \tag{13.77}$$

Do $m = m + 1$ and return to (13.72).

Algorithm (13.67)–(13.77) is less complicated than it looks like; it requires essentially the solution of *one Dirichlet problem per iteration.*

REMARK 13.5. Algorithm (13.67)–(13.77) has been written in *variational form* in order to facilitate its *finite-element implementation*, an issue to be addressed in Section 15.

REMARK 13.6. The *conjugate gradient* solution of *linear* and *nonlinear* problems in *Hilbert spaces* is discussed in, e.g., GLOWINSKI [2003, chapter 4] (see also the references therein and KŘÍZEK, NEITTAANMÄKI, GLOWINSKI and KOROTOV [2004]).

14. Bingham flow in cylinders: (VI) an augmented Lagrangian approach to the solution of problem (13.7)

Problem (13.7) is *equivalent* to the following *minimization* one

$$\begin{aligned} &u \in H_0^1(\Omega), \\ &J(u) \le J(v), \quad \forall\, v \in H_0^1(\Omega), \end{aligned} \tag{14.1}$$

with

$$J(v) = \frac{1}{2}\int_\Omega [\alpha|v|^2 + \mu|\nabla v|^2]\mathrm{d}x + \tau_y \int_\Omega |\nabla v|\mathrm{d}x - \int_\Omega fv\,\mathrm{d}x.$$

The idea behind the *augmented Lagrangian* method which follows is to decouple nonlinearity and derivatives; this will be done by treating ∇v as an independent variable $\mathbf{q}$ and then by forcing the relation $\nabla v - \mathbf{q} = \mathbf{0}$ by *penalization* and the use of a *Lagrange multiplier*. In order to implement the above idea, we will proceed as follows:

1. We denote $(L^2(\Omega))^2$ by $\mathbf{Q}$ and define $\mathbf{W}$ and $j(.,.)$ by

$$\mathbf{W} = \{\{v, \mathbf{q}\} \mid v \in H_0^1(\Omega), \quad \mathbf{q} \in \mathbf{Q}, \quad \nabla v - \mathbf{q} = \mathbf{0}\} \tag{14.2}$$

and

$$j(v, \mathbf{q}) = \frac{1}{2}\int_\Omega [\alpha|v|^2 + \mu|\nabla v|^2]\mathrm{d}x + \tau_y \int_\Omega |\mathbf{q}|\mathrm{d}x - \int_\Omega fv\mathrm{d}x. \tag{14.3}$$

2. We observe that problem (14.1) is equivalent to

$$\begin{aligned}&\{u, \mathbf{p}\} \in \mathbf{W},\\ &j(u, \mathbf{p}) \le j(v, \mathbf{q}), \quad \forall\, \{v, \mathbf{q}\} \in \mathbf{W}.\end{aligned} \tag{14.4}$$

3. With $r > 0$, we define an *augmented Lagrangian* functional $\mathcal{L}_r : (H_0^1(\Omega) \times \mathbf{Q}) \times \mathbf{Q} \to \mathbf{R}$ by

$$\mathcal{L}_r(\{v, \mathbf{q}\}, \boldsymbol{\mu}) = j(v, \mathbf{q}) + \frac{1}{2}r\int_\Omega |\nabla v - \mathbf{q}|^2\mathrm{d}x + \int_\Omega \boldsymbol{\mu} \cdot (\nabla v - \mathbf{q})\mathrm{d}x \tag{14.5}$$

and observe that if $\{\{u, \mathbf{p}\}, \boldsymbol{\lambda}\}$ is a *saddle-point* of $\mathcal{L}_r$ over $(H_0^1(\Omega) \times \mathbf{Q}) \times \mathbf{Q}$, that is verifies

$$\begin{aligned}&\{\{u, \mathbf{p}\}, \boldsymbol{\lambda}\} \in (H_0^1(\Omega) \times \mathbf{Q}) \times \mathbf{Q},\\ &\mathcal{L}_r(\{u, \mathbf{p}\}, \boldsymbol{\mu}) \le \mathcal{L}_r(\{u, \mathbf{p}\}, \boldsymbol{\lambda}) \le \mathcal{L}_r(\{v, \mathbf{q}\}, \boldsymbol{\lambda}),\\ &\forall\, \{\{v, \mathbf{q}\}, \boldsymbol{\mu}\} \in (H_0^1(\Omega) \times \mathbf{Q}) \times \mathbf{Q},\end{aligned} \tag{14.6}$$

then, the pair $\{u, \mathbf{p}\}$ is solution of problem (14.4), which implies, in turn, that u is the solution of problem (14.1) and that $\mathbf{p} = \nabla u$ (augmented Lagrangians, other than the one defined by (14.5), can be used; we can, in particular, replace $\mu|\nabla v|^2$ in (14.3) by $\mu|\mathbf{q}|^2$, but it does not seem to make a significant difference from an algorithmic point of view (concerning the speed of convergence, in particular)).

4. In order to solve the saddle-point problem (14.6), we advocate (following, e.g., GLOWINSKI, LIONS and TRÉMOLIÈRES [1981], FORTIN and GLOWINSKI [1982, 1983], GLOWINSKI [1984], and GLOWINSKI and LE TALLEC [1989]) the following *Uzawa* type algorithm (called *ALG2* in the above references):

$$\{u^{-1}, \boldsymbol{\lambda}^0\} \text{ is given in } H_0^1(\Omega) \times \mathbf{Q}; \tag{14.7}$$

for $n \ge 0$, u^{n-1} and $\boldsymbol{\lambda}^n$ being known, solve

$$\begin{aligned}&\mathbf{p}^n \in \mathbf{Q},\\ &\mathcal{L}_r(\{u^{n-1}, \mathbf{p}^n\}, \boldsymbol{\lambda}^n) \le \mathcal{L}_r(\{u^{n-1}, \mathbf{q}\}, \boldsymbol{\lambda}^n), \quad \forall\, \mathbf{q} \in \mathbf{Q},\end{aligned} \tag{14.8}$$

then

$$\begin{aligned}&u^n \in H_0^1(\Omega),\\ &\mathcal{L}_r(\{u^n, \mathbf{p}^n\}, \boldsymbol{\lambda}^n) \le \mathcal{L}_r(\{v, \mathbf{p}^n\}, \boldsymbol{\lambda}^n), \quad \forall\, v \in H_0^1(\Omega),\end{aligned} \tag{14.9}$$

and update $\boldsymbol{\lambda}^n$ by

$$\boldsymbol{\lambda}^{n+1} = \boldsymbol{\lambda}^n + r(\nabla u^n - \mathbf{p}^n). \tag{14.10}$$

It follows from, e.g., GLOWINSKI, LIONS and TRÉMOLIÈRES [1981], FORTIN and GLOWINSKI [1982, 1983], GLOWINSKI [1984], GLOWINSKI and LE TALLEC [1989] that the following *convergence result* holds

$$\forall \{u^{-1}, \boldsymbol{\lambda}^0\} \in H_0^1(\Omega) \times \mathbf{Q}, \text{ one has } \lim_{n\to+\infty} \{u^n, \mathbf{p}^n\} = \{u, \nabla u\} \in H_0^1(\Omega) \times \mathbf{Q} \tag{14.11}$$

where, in (14.11), u is the solution of problem (13.7), (14.1). Concerning the *implementation* of algorithm (14.7)–(14.10), a close inspection shows that problem (14.8) reduces to

$$\mathbf{p}^n = \arg\min_{\mathbf{q}\in\mathbf{Q}} \left[\frac{1}{2}\int_\Omega |\mathbf{q}|^2 \mathrm{d}x + \tau_y \int_\Omega |\mathbf{q}| \mathrm{d}x - \int_\Omega (r\nabla u^{n-1} + \boldsymbol{\lambda}^n) \cdot \mathbf{q} \mathrm{d}x \right]. \tag{14.12}$$

The minimization problem in (14.12) can be solved *point-wise*, leading to the following *closed form* solution for $\mathbf{p}^n$:

$$\text{a.e. in } \Omega, \quad \mathbf{p}^n(x) = \frac{1}{r}\left(1 - \frac{\tau_y}{|\mathbf{X}^n(x)|}\right)^+ \mathbf{X}^n(x), \tag{14.13}$$

with $\mathbf{X}^n = r\nabla u^{n-1} + \boldsymbol{\lambda}^n$ (and $\xi^+ = \max(0, \xi)$). Moreover, (14.9) reduces to the following *linear Dirichlet problem* (written here in *variational* form):

$$\begin{aligned} &u^n \in H_0^1(\Omega), \\ &\alpha \int_\Omega u^n v \mathrm{d}x + (\mu + r) \int_\Omega \nabla u^n . \nabla v \mathrm{d}x = \int_\Omega f v \mathrm{d}x + \int_\Omega (r\mathbf{p}^n - \boldsymbol{\lambda}^n) \cdot \nabla v \mathrm{d}x, \\ &\forall v \in H_0^1(\Omega); \end{aligned} \tag{14.14}$$

the numerical solution of elliptic problems such as (14.14) is routine nowadays.

REMARK 14.1. By updating $\boldsymbol{\lambda}^n$ after step (14.8), we obtain the following variant of algorithm (14.7)–(14.10):

$$\{u^{-1}, \boldsymbol{\lambda}^0\} \text{ is given in } H_0^1(\Omega) \times \mathbf{Q}; \tag{14.15}$$

for $n \geq 0$, u^{n-1} and $\boldsymbol{\lambda}^n$ being known, solve

$$\begin{aligned} &\mathbf{p}^n \in \mathbf{Q}, \\ &\mathcal{L}_r(\{u^{n-1}, \mathbf{p}^n\}, \boldsymbol{\lambda}^n) \leq \mathcal{L}_r(\{u^{n-1}, \mathbf{q}\}, \boldsymbol{\lambda}^n), \quad \forall\, \mathbf{q} \in \mathbf{Q}, \end{aligned} \tag{14.16}$$

update $\boldsymbol{\lambda}^n$ by

$$\boldsymbol{\lambda}^{n+1/2} = \boldsymbol{\lambda}^n + r(\nabla u^{n-1} - \mathbf{p}^n), \tag{14.17}$$

solve

$$u^n \in H_0^1(\Omega),$$
$$\mathcal{L}_r(\{u^n, \mathbf{p}^n\}, \boldsymbol{\lambda}^{n+1/2}) \leq \mathcal{L}_r(\{v, \mathbf{p}^n\}, \boldsymbol{\lambda}^{n+1/2}), \quad \forall\, v \in H_0^1(\Omega), \tag{14.18}$$

finally, update $\boldsymbol{\lambda}^{n+1/2}$ by

$$\boldsymbol{\lambda}^{n+1} = \boldsymbol{\lambda}^{n+1/2} + r(\nabla u^n - \mathbf{p}^n). \tag{14.19}$$

The above algorithm (called ALG3 in GLOWINSKI, LIONS and TRÉMOLIÈRES [1981], FORTIN and GLOWINSKI [1982, 1983], GLOWINSKI [1984, 2008], and GLOWINSKI and LE TALLEC [1989]) verifies also the convergence properties given in (14.11). The choice of r is for both algorithms a critical issue for which we refer to, e.g., FORTIN and GLOWINSKI [1982, 1983], GLOWINSKI and LE TALLEC [1989]. Actually, both algorithms have very close relation with *operator-splitting schemes* such as *Peaceman–Rachford's* and *Douglas–Rachford's* (see the above references for details). Concerning the relative merits of *ALG2* and *ALG3*, let us say that it seems (see, e.g., FORTIN and GLOWINSKI [1982, 1983], and GLOWINSKI and LE TALLEC [1989]) that *ALG3* is faster for *smooth problems*, whereas *ALG2* is more robust; because the problem under consideration involves the non differentiable functional $v \to \int_\Omega |\nabla v| \mathrm{d}x$, we favor *ALG2*.

REMARK 14.2. To the best of our knowledge, the particular *augmented Lagrangian* methodology discussed in this section has been introduced in GLOWINSKI and MARROCCO [1974] and [1975], for the solution of the *Dirichlet problem* for the s – *Laplacian* operator; namely,

$$\begin{aligned} -\nabla \cdot (|\nabla u|^{s-2} \nabla u) &= f \text{ in } \Omega \\ u &= 0 \text{ on } \Gamma, \end{aligned} \tag{14.20}$$

with Ω a bounded domain of $\mathbf{R}^d$, Γ its boundary and $1 < s < +\infty$. If $\Omega \subset \mathbf{R}^2$ and f is a constant, (14.20) models the flow of a non-Newtonian incompressible viscous fluid, of the *power law* type, in a cylinder of cross-section Ω, the flow being induced by a constant drop of pressure (proportional to f) per unit length. As shown in the above two references, *ALG2* has been quite effective at solving problem (14.20), even for s close to 1 ($s = 1.1$, for example). Other applications to *Fluid Mechanics* and *Nonlinear Elasticity* can be found in, e.g., FORTIN and GLOWINSKI [1982, 1983], GLOWINSKI and LE TALLEC [1989], GLOWINSKI and HOLMSTRÖM [1995] (see also the references therein). Other applications of the *augmented Lagrangian* methodology include the solution of the *Monge–Kantorovich optimal transportation problem* (see BENAMOU and BRENIER [2000]), of the *two-dimensional elliptic Monge–Ampère equation* (see DEAN and GLOWINSKI [2003, 2006a,b], and DEAN, GLOWINSKI and PAN [2005]), and of *inverse problems in seismic reflection tomography* (see DELBOS, GILBERT, GLOWINSKI and SINOQUET [2006]).

15. Bingham flow in cylinders: (VII) finite-element approximation

In Section 12, Remark 12.2, we have been advocating the use of *low-order* space-time approximations for Bingham flow in cylinders, the main reason being the relatively low regularity of the solution. From these considerations, the *backward Euler scheme* (discussed in

Section 12) will be our method of choice for the *time-discretization*. Similarly, we will rely on *globally continuous, piecewise affine approximations* for the *space-discretization*. This combination leads us to the scheme (12.15), (12.16) described in Section 12, Remark 12.2. We have, thus, to solve at each time step a *finite-dimensional* problem of the following type:

$$\begin{aligned} & u_h \in V_{0h}, \\ & \alpha \int_{\Omega_h} u_h(v - u_h)\mathrm{d}x + \mu \int_{\Omega_h} \nabla u_h \cdot \nabla(v - u_h)\mathrm{d}x + \tau_y[j_h(v) - j_h(u_h)] \\ & \quad \geq \int_{\Omega_h} f_h(v - u_h)\mathrm{d}x, \quad \forall\, v \in V_{0h}, \end{aligned} \tag{15.1}$$

with

$$j_h(v) = \int_{\Omega_h} |\nabla v|\mathrm{d}x = \sum_{K \in \mathcal{T}_h} \int_K |\nabla v|\mathrm{d}x. \tag{15.2}$$

The finite-dimensional problem (15.1) has a *unique* solution characterized by the existence of $\boldsymbol{\lambda}_h$ such that

$$\begin{aligned} & \{u_h, \boldsymbol{\lambda}_h\} \in V_{0h} \times \boldsymbol{\Lambda}_h, \\ & \alpha \int_{\Omega_h} u_h v \mathrm{d}x + \mu \int_{\Omega_h} \nabla u_h \cdot \nabla v \mathrm{d}x + \tau_y \int_{\Omega_h} \boldsymbol{\lambda}_h . \nabla v \mathrm{d}x = \int_{\Omega_h} f_h v \mathrm{d}x, \quad \forall\, v \in V_{0h}, \\ & \boldsymbol{\lambda}_h \cdot \nabla u_h = |\nabla u_h|, \end{aligned} \tag{15.3}$$

with

$$\boldsymbol{\Lambda}_h = \{\boldsymbol{\mu} \mid \boldsymbol{\mu} \in (L^2(\Omega))^2, \quad \forall\, K \in \mathcal{T}_h,\ \boldsymbol{\mu}|_K = \boldsymbol{\mu}_K \in \mathbf{R}^2,\ |\boldsymbol{\mu}_K| \leq 1\}. \tag{15.4}$$

System (15.3) takes various *equivalent* forms, among them

$$\begin{aligned} & u_h \in V_{0h}, \\ & \alpha \int_{\Omega_h} u_h v \mathrm{d}x + \mu \int_{\Omega_h} \nabla u_h \cdot \nabla v \mathrm{d}x + \tau_y \int_{\Omega_h} \boldsymbol{\lambda}_h . \nabla v \mathrm{d}x = \int_{\Omega_h} f_h v \mathrm{d}x, \quad \forall\, v \in V_{0h}, \\ & \boldsymbol{\lambda}_h = \mathbf{P}_{\boldsymbol{\Lambda}_h}(\boldsymbol{\lambda}_h + r\tau_y \nabla u_h), \quad \forall\, r \geq 0, \end{aligned} \tag{15.5}$$

and

$$\begin{aligned} & \{u_h, \boldsymbol{\lambda}_h\} \in V_{0h} \times \boldsymbol{\Lambda}_h, \\ & \alpha \int_{\Omega_h} u_h v \mathrm{d}x + \mu \int_{\Omega_h} \nabla u_h \cdot \nabla v \mathrm{d}x + \tau_y \int_{\Omega_h} \boldsymbol{\lambda}_h \cdot \nabla v \mathrm{d}x = \int_{\Omega_h} f_h v \mathrm{d}x, \quad \forall\, v \in V_{0h}, \\ & - \int_{\Omega_h} \nabla u_h \cdot (\boldsymbol{\mu} - \boldsymbol{\lambda}_h)\mathrm{d}x \geq 0, \quad \forall\, \boldsymbol{\mu} \in \boldsymbol{\Lambda}_h; \end{aligned} \tag{15.6}$$

above, $\mathbf{P}_{\boldsymbol{\Lambda}_h}$ is the *orthogonal projection operator* from $\mathbf{L}_h$ onto $\boldsymbol{\Lambda}_h$ (with

$$\mathbf{L}_h = \{\boldsymbol{\mu} \mid \boldsymbol{\mu} \in (L^2(\Omega_h))^2, \quad \forall\, K \in \mathcal{T}_h,\, \boldsymbol{\mu}|_K = \boldsymbol{\mu}_K \in \mathbf{R}^2\});$$

operator $\mathbf{P}_{\boldsymbol{\Lambda}_h}$ verifies

$$\mathbf{P}_{\boldsymbol{\Lambda}_h}(\boldsymbol{\mu})|_K = \frac{\boldsymbol{\mu}_K}{\max(1, |\boldsymbol{\mu}_K|)}, \quad \forall\, K \in \mathcal{T}_h, \quad \forall\, \boldsymbol{\mu} \in \mathbf{L}_h. \tag{15.7}$$

From these formulations, deriving the *fully discrete* analogs of the various iterative methods discussed in the preceding sections is straightforward.

16. Bingham flow in cylinders: (VIII) numerical experiments

In this section, we will focus on the solution of the *steady flow problem* (13.1), in the particular case, in which Ω is the *disk* of radius R centered at $\{0, 0\}$, that is

$$\Omega = \{x \mid x = \{x_1, x_2\} \in \mathbf{R}^2,\ x_1^2 + x_2^2 < R^2\}. \tag{16.1}$$

Assume that $C > 0$, then for the above cross-section Ω, the solution of (13.1) is given by

$$u_\infty = \begin{cases} \left(\dfrac{R - r}{2\mu}\right)\left[\dfrac{C}{2}(R + r) - 2\tau_y\right] & \text{if } R' \le r \le R, \\[2ex] \left(\dfrac{R - R'}{2\mu}\right)\left[\dfrac{C}{2}(R + R') - 2\tau_y\right] & \text{if } 0 \le r \le R', \end{cases} \tag{16.2}$$

with $r = \sqrt{x_1^2 + x_2^2}$ and $R' = \frac{2\tau_y}{C}$.

For the numerical experiments described below, we took $R = \frac{1}{4}$, $C = 16$, and $\mu = \frac{1}{4}$, implying that $u_\infty = 0$ if $\tau_y \ge 2$. To approximate the related problem (13.1), we have used the *finite-element spaces* described in Section 12, Remark 12.2, and in Section 15; these spaces being defined from *triangulations* of Ω like the one shown in Fig. 16.1, below (with h the length of the largest edge(s) of the triangulation).

In Table 16.1, we have reported, for various values of τ_y and h, some of the numerical results obtained by applying, to the solution of problem (13.1), the discrete variant of algorithm (13.10)–(13.12) (with $\alpha = 0$ and $f = C$), associated with the triangulation $\mathcal{T}_h$. The above algorithm has been initialized with $\boldsymbol{\lambda}_h^0 = \mathbf{0}$, and we stopped iterating as soon as $\|\boldsymbol{\lambda}_h^{n+1} - \boldsymbol{\lambda}_h^n\|_{L^2} \le 10^{-4}$, *nit* being the corresponding number of iterations; for r, we took $\frac{\mu}{\tau_y^2}$ (which is consistent with the convergence condition (13.15)). For $h = \frac{1}{64}$ (resp., $\frac{1}{128}$ and $\frac{1}{256}$), $\mathcal{T}_h$ consists of 1,976 (resp., 7,945 and 31,690) triangles and has 1,039 (resp., 4,074 and 16,047) vertices. The results shown in Table 16.1 (and Fig. 16.2) suggest that $\|u_h - u\|_{L^2(\Omega)} \approx O(h^2)$, while $\|\nabla(u_h - u)\|_{(L^2(\Omega))^2} \approx O(h)$; these results are consistent with error estimates proved in, e.g., GLOWINSKI [1984].

The results shown in Table 16.2 concern the same test problem and have been obtained using the same triangulations than above. However, for these computations, we have used

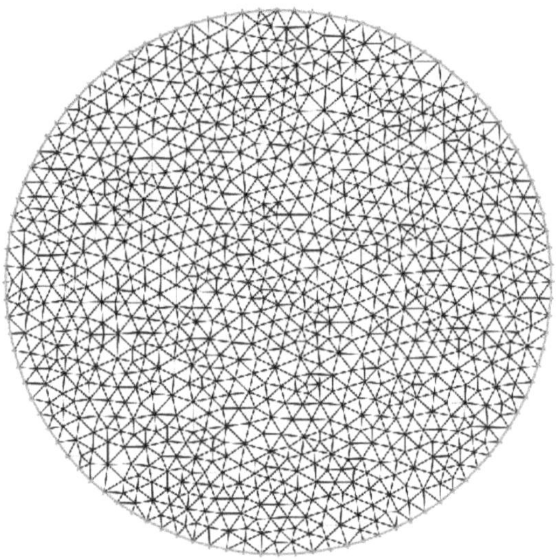

FIG. 16.1 A triangulation $\mathcal{T}_h$ of the disk Ω (courtesy of G. Guidoboni).

TABLE 16.1
Numerical results obtained by the discrete variant of algorithm (13.10)–(13.12) (courtesy of G. Guidoboni)

τ_y	h	*nit*	$\|u_h - u\|_{L^2(\Omega)}$	$\|\nabla(u_h - u)\|_{(L^2(\Omega))^2}$
0.2	1/64	4	1.2206×10^{-4}	1.0964×10^{-2}
	1/128	4	3.0895×10^{-5}	4.9999×10^{-3}
	1/256	3	7.6938×10^{-6}	2.7501×10^{-3}
1.0	1/64	15	1.0071×10^{-4}	2.1055×10^{-2}
	1/128	8	2.4395×10^{-5}	1.0090×10^{-2}
	1/256	5	6.1040×10^{-6}	5.1368×10^{-3}
1.7	1/64	20	1.3162×10^{-4}	1.8784×10^{-2}
	1/128	7	2.8520×10^{-5}	8.0858×10^{-3}
	1/256	2	5.3213×10^{-6}	4.0120×10^{-3}
1.9	1/64	2	6.3682×10^{-5}	1.1223×10^{-2}
	1/128	2	1.5250×10^{-5}	4.8720×10^{-3}
	1/256	2	4.5102×10^{-6}	2.4759×10^{-3}
2.1	1/64	2	5.1227×10^{-15}	5.2260×10^{-14}
	1/128	2	3.2679×10^{-14}	3.1831×10^{-13}
	1/256	2	2.0857×10^{-13}	2.0147×10^{-12}

the *nested* iterative method obtained by combining the (pseudo-) time discretization scheme (13.36)–(13.38) (initialized with $\boldsymbol{\lambda}_h^0 = \mathbf{0}$) with algorithm (13.41)–(13.43) (initialized with $\boldsymbol{\lambda}_{h,0}^n = \boldsymbol{\lambda}_h^{n-1}$). We took $\Delta\tau = \frac{2\mu}{\tau_y^2}$, $r = \frac{\Delta\tau}{2}$, and stop iterating as soon as $\|\boldsymbol{\lambda}_h^{n+1} - \boldsymbol{\lambda}_h^n\|_{L^2}$

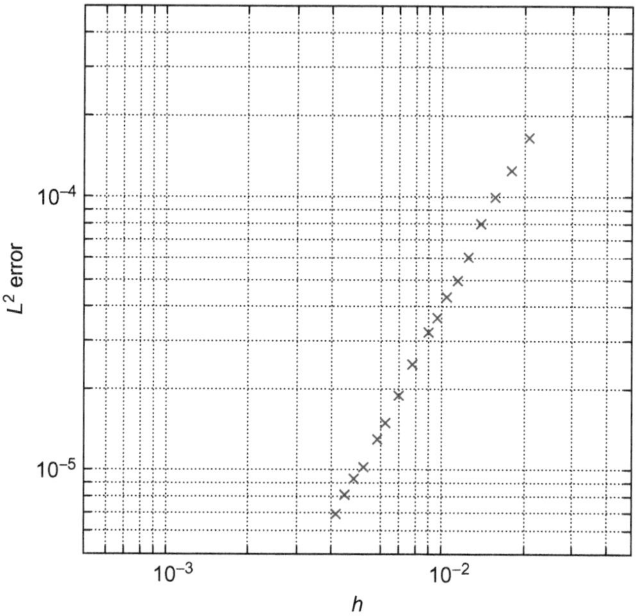

FIG. 16.2 Variation of $\|u_h - u\|_{L^2(\Omega)}$ versus h for $\tau_y = 1$ (log scales) (courtesy of G. Guidoboni).

TABLE 16.2
Numerical results obtained by the discrete variant of algorithm (13.36)–(13.38), (13.41)–(13.43) (courtesy of G. Guidoboni)

τ_y	h	*nit*	$\|u_h - u\|_{L^2(\Omega)}$	$\|\nabla(u_h - u)\|_{(L^2(\Omega))^2}$
0.2	1/64	3	1.2213×10^{-4}	1.0959×10^{-2}
	1/128	3	3.0822×10^{-5}	4.9971×10^{-3}
	1/256	3	7.6864×10^{-6}	2.7502×10^{-3}
1.0	1/64	14	1.0055×10^{-4}	2.1043×10^{-2}
	1/128	7	2.3587×10^{-5}	1.0061×10^{-2}
	1/256	5	5.8432×10^{-6}	5.1390×10^{-3}
1.7	1/64	20	1.2672×10^{-4}	1.8807×10^{-2}
	1/128	7	1.9826×10^{-5}	8.1015×10^{-3}
	1/256	6	4.8495×10^{-6}	4.1762×10^{-3}
1.9	1/64	7	1.0990×10^{-4}	1.1936×10^{-2}
	1/128	7	1.7497×10^{-5}	5.3066×10^{-3}
	1/256	7	4.6899×10^{-6}	2.6584×10^{-3}
2.1	1/64	7	3.3682×10^{-17}	3.8609×10^{-16}
	1/128	7	2.1175×10^{-16}	2.0748×10^{-15}
	1/256	7	2.3071×10^{-15}	2.2257×10^{-14}

$\leq 10^{-4}$ (*outer iterations*) and $\|\boldsymbol{\lambda}^n_{h,k+1} - \boldsymbol{\lambda}^n_{h,k}\|_{L^2} \leq 10^{-5}$(*inner iterations*) (other stopping strategies are possible). The numbers in the *nit* column correspond to outer iterations. We observe that the approximation errors given in Table 16.2 are of the same order than those in Table 16.1 (except for $\tau_y = 2.1$, where the solution u_h being 0, what we have obtained, with both algorithms, is (a kind of) numerical noise). Finally, we have shown in Table 16.3 (and Figs. 16.3 and 16.4) the results obtained using a discrete variant of the *penalty-Newton–Uzawa-conjugate gradient* method discussed in Section 13.5. The computations have been done taking $h = \frac{1}{128}$ and varying τ_y and the *penalty* parameter ε. The *Newton's* (resp., *Uzawa-conjugate gradient*) iterations have been initialized with $\boldsymbol{\lambda}^0_h = \mathbf{0}$ (resp., $\mathbf{p}^0_h = \mathbf{0}$) and (using Section 13.5 notation) we took $tol._1 = 10^{-6}$ and $tol._2 = 10^{-4}$ in the stopping criteria. In Table 16.3, *nit* is the number of iterations necessary to achieve convergence. We observe that the number of Newton's iterations decreases as τ_y increases; this property was expected because the size of the fluid region (where $|\boldsymbol{\lambda}(x)| = 1$) is a decreasing function of τ_y, everything else being the same; on the other hand, the number of Uzawa-conjugate gradient iterations stays around 5, for the values of τ_y and ε considered here. Figure 16.3, which corresponds to $\tau_y = 1$, suggests that for ε moderately small $\|u_{h,\varepsilon} - u^*_h\|_{(L(\Omega))^2}$ varies like $\sqrt{\varepsilon}$ (which is what we were expecting; here, u^*_h is the approximate solution computed through the discrete variant of algorithm (13.10)–(13.12)), while it stays constant for smaller

TABLE 16.3

Numerical results obtained by a discrete variant of the penalty-Newton–Uzawa-conjugate gradient method of Section 13.5 for $h = 1/128$ (u^*_h is the corresponding solution obtained by the related discrete variant of algorithm (13.10)–(13.12)) (courtesy of G. Guidoboni)

τ_y	ε	*nit*	$\|u_{h,\varepsilon} - u^*_h\|_{L^2(\Omega)}$	$\|u_{h,\varepsilon} - u\|_{L^2(\Omega)}$	$\|\nabla(u_{h,\varepsilon} - u)\|_{(L^2(\Omega))^2}$
0.2	10^{-3}	19	5.4330×10^{-4}	5.4523×10^{-4}	7.2570×10^{-3}
	10^{-5}	19	6.7749×10^{-5}	7.5271×10^{-5}	5.0353×10^{-3}
	10^{-7}	19	4.5020×10^{-5}	5.5144×10^{-5}	5.0137×10^{-3}
1.0	10^{-3}	11	3.2531×10^{-3}	3.2621×10^{-3}	3.6440×10^{-2}
	10^{-5}	11	3.9400×10^{-4}	4.0331×10^{-4}	1.0732×10^{-2}
	10^{-7}	11	2.4255×10^{-4}	2.5211×10^{-4}	1.0257×10^{-2}
1.7	10^{-3}	9	1.3728×10^{-3}	1.3978×10^{-3}	2.6178×10^{-2}
	10^{-5}	6	3.8398×10^{-4}	4.0890×10^{-4}	1.0300×10^{-2}
	10^{-7}	6	3.6026×10^{-4}	3.8519×10^{-4}	1.0092×10^{-2}
1.9	10^{-3}	11	3.8816×10^{-4}	4.0174×10^{-4}	1.4630×10^{-2}
	10^{-5}	4	1.4912×10^{-4}	1.6300×10^{-4}	6.4455×10^{-3}
	10^{-7}	4	1.4636×10^{-4}	1.6024×10^{-4}	6.4008×10^{-3}
2.1	10^{-3}	2	3.2679×10^{-14}	2.6906×10^{-28}	2.7906×10^{-27}
	10^{-5}	2	3.2679×10^{-14}	2.6728×10^{-28}	2.7729×10^{-27}
	10^{-7}	2	3.2679×10^{-14}	2.6958×10^{-28}	2.7919×10^{-27}

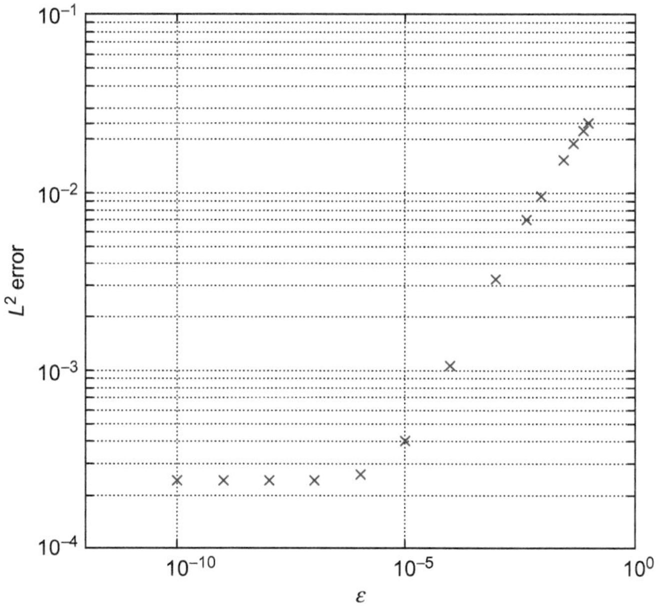

FIG. 16.3 Variation of $\|u_{h,\varepsilon} - u_h^*\|_{L^2(\Omega)}$ versus ε for $h = 1/128$ and $\tau_y = 1$ (courtesy of G. Guidoboni).

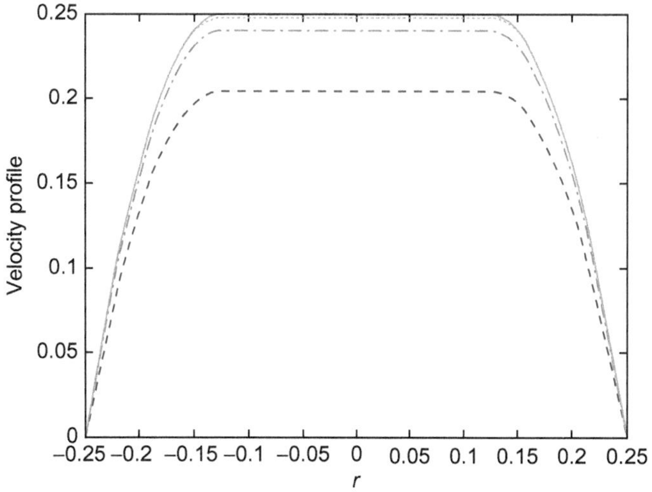

FIG. 16.4 Graphs of the exact solution (—) and of the approximated solutions restricted to a diameter of Ω for $\varepsilon = 3 \times 10^{-2}$ (– – ––), 10^{-3} (–· –· –·), and 10^{-5}(. . .) ($h = 1/128$ and $\tau_y = 1$) (courtesy of G. Guidoboni).

values of ε. We suspect that to recover the $\sqrt{\varepsilon}$ behavior for the very small values of ε, we should use smaller tolerances in the stopping criteria of the various iterative methods used to compute $u_{h,\varepsilon}$ and u_h^*. In Fig. 16.4, we compare for $\tau_y = 1$, the exact solution with the approximated ones, obtained with $h = \frac{1}{128}$ and various values of ε; we observe that the condition $\nabla u = \mathbf{0}$ is well approximated in the rigidity region.

REMARK 16.1. The last three rows of Table 16.3 suggest that for $\tau_y = 2.1$, we have (with obvious notation) $u_\varepsilon = 0$ ($= u$, since $u = 0$ if $\tau_y \geq 2$) for the three values of ε considered here.

REMARK 16.2. We will conclude this discussion, concerning the numerical simulation of Bingham flow in cylinders by mentioning that, if one is interested by the *steady-state solution* only, it may be advantageous to consider the following (non physical) initial value problem:

$$u(0) = u_0 (\in H_0^1(\Omega)),$$

$$\int_\Omega \nabla\left(\frac{\partial u}{\partial t}\right) \cdot \nabla(v-u)\,\mathrm{d}x + \mu \int_\Omega \nabla u \cdot \nabla(v-u)\,\mathrm{d}x + \tau_y \left[\int_\Omega |\nabla v|\,\mathrm{d}x - \int_\Omega |\nabla u|\,\mathrm{d}x\right]$$

$$\geq C \int_\Omega (v-u)\,\mathrm{d}x, \quad \forall\, v \in H_0^1(\Omega), \tag{16.3}$$

where $\frac{\mathrm{d}C}{\mathrm{d}t} = 0$. Indeed, suppose that u_∞ is the solution of the corresponding steady-state problem (13.1), it is then fairly easy to prove that

$$\|u(t) - u_\infty\|_{H_0^1(\Omega)} \leq \|u_0 - u_\infty\|_{H_0^1(\Omega)} \mathrm{e}^{-\mu t}, \quad \forall\, t \geq 0, \tag{16.4}$$

a stronger convergence result than the one given by (11.15) in Section 11. We refer to HE and GLOWINSKI [2000] concerning the practical implementation of the above approach (which is no more complicated to implement than the one based on (11.1)).

17. Bingham flow in cavities

17.1. Generalities

The modeling of *multidimensional Bingham flow* can be found in Section 8. From now on, we will assume that Ω is a *bounded* region of $\mathbf{R}^d$ (with $d = 2$ or 3). The *numerical simulation* of such flow has been addressed in, e.g., SANCHEZ [1998], DEAN and GLOWINSKI [2002], GLOWINSKI [2003, chapter 10, section 50] (see also the references therein). In this article, we will review, briefly, the *operator-splitting*-based methodology discussed in the three above references and present some numerical results obtained using it. Before describing the above methodology, let us mention that a popular approach to overcome the difficulties associated with the *non differentiability* of the functional

$$\mathbf{v} \to \int_\Omega |\mathbf{D}(\mathbf{v})|\,\mathrm{d}x$$

is to approximate the problems (8.1)–(8.5) and (8.6)–(8.9) by *regularization*; among the various regularization procedures, the one below is classical:

$$\rho\left[\frac{\partial \mathbf{u}_\varepsilon}{\partial t} + (\mathbf{u}_\varepsilon \cdot \nabla)\mathbf{u}_\varepsilon\right] = \nabla \cdot \boldsymbol{\sigma}_\varepsilon + \mathbf{f} \text{ in } \Omega \times (0,\ T), \tag{17.1}$$

$$\nabla \cdot \mathbf{u}_\varepsilon = 0 \text{ in } \Omega \times (0,\ T), \tag{17.2}$$

$$\boldsymbol{\sigma}_\varepsilon = -p_\varepsilon \mathbf{I} + \tau_y \sqrt{2}\frac{\mathbf{D}(\mathbf{u}_\varepsilon)}{\sqrt{\varepsilon^2 + |\mathbf{D}(\mathbf{u}_\varepsilon)|^2}} + 2\mu\mathbf{D}(\mathbf{u}_\varepsilon), \tag{17.3}$$

$$\mathbf{u}_\varepsilon(0) = \mathbf{u}_0, \tag{17.4}$$

$$\mathbf{u}_\varepsilon = \mathbf{u}_\Gamma \text{ on } \Gamma \times (0,\ T). \tag{17.5}$$

A drawback of the above regularization procedure is that it does not have the property that $\mathbf{u}(t)$ reaches the value $\mathbf{0}$ in *finite time* if $\mathbf{u}_\Gamma = \mathbf{0}$ and $\mathbf{f} = \mathbf{0}$, unlike the solutions of (8.6)–(8.9). We will not investigate further the regularization approach associated with (17.1)–(17.5). Instead, in the following sections, we will discuss the solution of problem (8.6)–(8.9) through a *time discretization by operator-splitting*; with this approach (already investigated in, e.g., SANCHEZ [1998], DEAN and GLOWINSKI [2002], GLOWINSKI [2003, chapter 10, section 50]), we will be able to solve problem (8.6)–(8.9) by a methodology closely related to various methods used for the solution of the *Navier–Stokes equations* modeling *incompressible Newtonian viscous flow* (that is, those equations obtained by taking $\tau_y = 0$ in (8.1)–(8.5) and (8.6)–(8.9)).

REMARK 17.1. Above, we have mentioned that if $\mathbf{u}_\Gamma = \mathbf{0}$ and $\mathbf{f} = \mathbf{0}$, then $\mathbf{u}(t) = \mathbf{0}$, for t large enough. Owing to the importance of this property, we are going to prove it hereafter; we have thus the following:

THEOREM 17.1. *Suppose that $T = +\infty$ in (8.6)–(8.9); if $\mathbf{u}_\Gamma = \mathbf{0}$ and $\mathbf{f} = \mathbf{0}$, we have $\mathbf{u}(t) = \mathbf{0}$ for t large enough.*

PROOF. The proof which follows is a variant of the one used to prove relation (11.13) in Section 11. Take $\mathbf{v} = \mathbf{0}$ and $\mathbf{v} = 2\mathbf{u}(t)$ in (8.6). It follows then from (8.6) and (8.7) that

$$\frac{1}{2}\rho\frac{\mathrm{d}}{\mathrm{d}t}\int_\Omega |\mathbf{u}|^2 \mathrm{d}x + \rho\int_\Omega (\mathbf{u}(t)\cdot\nabla)\mathbf{u}(t)\cdot\mathbf{u}(t)\mathrm{d}x + \mu\int_\Omega |\nabla\mathbf{u}(t)|^2\mathrm{d}x$$
$$+ \sqrt{2}\tau_y j(u(t)) = 0, \text{ a.e. } t \in (0, +\infty). \tag{17.6}$$

Taking advantage of the following (classical) relations

$$\int_\Omega (\mathbf{w}\cdot\nabla)\mathbf{v}\cdot\mathbf{v}\mathrm{d}x = 0, \quad \forall\, \mathbf{v},\ \mathbf{w} \in (H_0^1(\Omega))^d, \quad \nabla\cdot\mathbf{w} = 0, \tag{17.7}$$

$$\int_\Omega |\nabla\mathbf{v}|^2\mathrm{d}x \geq \lambda_0 \int_\Omega |\mathbf{v}|^2\mathrm{d}x, \quad \forall\, \mathbf{v} \in (H_0^1(\Omega))^d, \tag{17.8}$$

$$j(\mathbf{v}) \left(= \int_\Omega |\mathbf{D}(\mathbf{v})|\mathrm{d}x\right) \geq \gamma\|\mathbf{v}\|_{(L^2(\Omega))^d}, \quad \forall\, \mathbf{v} \in (H_0^1(\Omega))^d \tag{17.9}$$

(where in (17.8), (17.9), $\lambda_0 (> 0)$ is the *smallest eigenvalue* of $-\nabla^2$ operating over $H_0^1(\Omega)$, and γ is a *positive* constant (cf. STRAUSS [1973])), (17.6) implies

$$\frac{1}{2}\rho\frac{\mathrm{d}}{\mathrm{d}t}\|\mathbf{u}\|^2_{(L^2(\Omega))^d} + \mu\lambda_0\|\mathbf{u}\|^2_{(L^2(\Omega))^d} + \sqrt{2}\tau_y\gamma\|\mathbf{u}\|_{(L^2(\Omega))^d} \leq 0,$$
$$\text{a.e. } t \in (0, +\infty). \tag{17.10}$$

Suppose that $\mathbf{u}(t) \neq \mathbf{0}, \quad \forall\, t \in [0, +\infty)$; we have then

$$\frac{\mathrm{d}}{\mathrm{d}t}\|\mathbf{u}\|^2_{(L^2(\Omega))^d} = 2\|\mathbf{u}\|_{(L^2(\Omega))^d}\frac{\mathrm{d}}{\mathrm{d}t}\|\mathbf{u}\|_{(L^2(\Omega))^d},$$

which combined with (17.10) implies

$$\rho\frac{\mathrm{d}}{\mathrm{d}t}\|\mathbf{u}\|_{(L^2(\Omega))^d} + \mu\lambda_0\|\mathbf{u}\|_{(L^2(\Omega))^d} + \sqrt{2}\tau_y\gamma \leq 0,$$
$$\text{a.e. } t \in (0, +\infty). \tag{17.11}$$

Relation (17.11) can be rewritten as

$$\frac{\mathrm{d}}{\mathrm{d}t}\left[\|\mathbf{u}\|_{(L^2(\Omega))^d} + \sqrt{2}\frac{\tau_y\gamma}{\mu\lambda_0}\right] + \frac{\mu\lambda_0}{\rho}\left[\|\mathbf{u}\|_{(L^2(\Omega))^d} + \sqrt{2}\frac{\tau_y\gamma}{\mu\lambda_0}\right] \leq 0,$$
$$\text{a.e. } t \in (0, +\infty). \tag{17.12}$$

By time-integration of the differential inequality in (17.12), we obtain

$$\|\mathbf{u}(t)\|_{(L^2(\Omega))^d} + \sqrt{2}\frac{\tau_y\gamma}{\mu\lambda_0} \leq \left[\|\mathbf{u}_0\|_{(L^2(\Omega))^d} + \sqrt{2}\frac{\tau_y\gamma}{\mu\lambda_0}\right]\mathrm{e}^{-\frac{\mu\lambda_0}{\rho}t}, \quad \forall\, t \geq 0. \tag{17.13}$$

Because $\lim_{t\to+\infty} \mathrm{e}^{-\frac{\mu\lambda_0}{\rho}t} = 0$, relation (17.13) makes no sense "as soon as" $t > T_c$ with

$$T_c = \frac{\rho}{\lambda_0\mu}\ln\left(1 + \frac{\mu\lambda_0}{\sqrt{2}g\tau_y}\|\mathbf{u}_0\|_{(L^2(\Omega))^d}\right). \tag{17.14}$$

We have thus

$$\mathbf{u}(t) = \mathbf{0}, \quad \forall\, t \geq T_c. \tag{17.15}$$

Relation (17.15) concludes the proof of the theorem and provides, in adition, an estimate of the "cutoff" time. □

REMARK 17.2. The estimate of the cutoff time given by (17.14) is not optimal. Actually, a more accurate upper bound T_c can be obtained by taking for λ_0 and γ in (17.10) and (17.14), the quantities defined by

$$\lambda_0 = \inf_{\mathbf{v}\in\mathbf{V}_0\setminus\{\mathbf{0}\}} \frac{\int_\Omega |\nabla\mathbf{v}|^2 \mathrm{d}x}{\int_\Omega |\mathbf{v}|^2 \mathrm{d}x} \tag{17.16}$$

and

$$\gamma = \inf_{\mathbf{v}\in\mathbf{V}_0\setminus\{\mathbf{0}\}} \frac{\int_\Omega |\mathbf{D}(\mathbf{v})|\mathrm{d}x}{\sqrt{\int_\Omega |\mathbf{v}|^2 \mathrm{d}x}}, \tag{17.17}$$

respectively, with, in (17.16) and (17.17), $\mathbf{V}_0 = \{\mathbf{v}|\mathbf{v} \in (H_0^1(\Omega))^d,\ \nabla\cdot\mathbf{v} = 0\}$. The quantity λ_0 defined by (17.16) is, clearly, the *smallest eigenvalue* of the *Stokes operator* acting on $(H_0^1(\Omega))^d$, i.e., the smallest λ such that

$$\begin{aligned} &\{\mathbf{w},\ p\} \in (H_0^1(\Omega))^d \times L^2(\Omega),\\ &-\nabla^2\mathbf{w} + \nabla p = \lambda\mathbf{w} \text{ in } \Omega,\quad \nabla\cdot\mathbf{w} = 0 \text{ in } \Omega,\quad \int_\Omega |\mathbf{w}|^2 \mathrm{d}x = 1. \end{aligned} \tag{17.18}$$

If $\Omega = (0, 1)^2$, we have $\lambda_0 = 2\pi^2 (= 19.739\ldots)$ in (17.8), while we have, on the other hand, $\lambda_0 = 52.3\ldots$ in (17.16) (as shown in, e.g., GLOWINSKI [2003, chapter 7]); this implies that taking $\nabla\cdot\mathbf{v} = 0$ into account, in the definition of λ_0, leads to an estimate of the cutoff time which is more than 2.5 smaller than the one given by (17.8), (17.9), and (17.14).

REMARK 17.3. Suppose that $d = 2$, then as in Remark 11.6, the value of γ in (17.9) (and (17.17) *is independent of the size and of the shape of* Ω (an upper bound of the constant γ in (17.9) is given by $2^{3/4}\sqrt{\pi} = 2.9809001\ldots$; an upper bound of the one in (17.17) is given by $2\sqrt{\pi}$).

REMARK 17.4. Bingham fluid type models have been used to describe *soil mechanics* phenomena such as *landslides*. A basic reference in that direction is HILD, IONESCU, LACHAND-ROBERT and ROSCA [2002].

17.2. Time-discretization of problem (8.6)–(8.9) by operator-splitting

There are many ways to discretize problem (8.6)–(8.9) by operator-splitting. Among the many possible schemes, we will discuss only one, of the *Marchuk–Yanenko* type (see GLOWINSKI [2003, chapter 6] and the references therein); this scheme reads as follows (with, as usual, $t^{n+\alpha} = (n+\alpha)\Delta t$):

$$\mathbf{u}^0 = \mathbf{u}_0; \tag{17.19}$$

then, for $n \geq 0$, $\mathbf{u}^n$ being known, we compute $\{\mathbf{u}^{n+1/3}, p^{n+1}\}$, $\mathbf{u}^{n+2/3}$, and $\mathbf{u}^{n+1}$ as follows:

1. Solve the *generalized Stokes* problem

$$\begin{aligned} &\frac{\rho}{\Delta t}(\mathbf{u}^{n+1/3} - \mathbf{u}^n) - \frac{\mu}{2}\nabla^2\mathbf{u}^{n+1/3} + \nabla p^{n+1} = \mathbf{f}^{n+1}\ (= \mathbf{f}(t^{n+1})) \text{ in } \Omega,\\ &\nabla\cdot\mathbf{u}^{n+1/3} = 0 \text{ in } \Omega,\\ &\mathbf{u}^{n+1/3} = \mathbf{u}_\Gamma^{n+1}\ (= \mathbf{u}_\Gamma(t^{n+1})) \text{ on } \Gamma. \end{aligned} \tag{17.20}$$

2. Solve the transport problem

$$\frac{\partial \mathbf{u}}{\partial t} + (\mathbf{u}^{n+1/3} \cdot \nabla)\mathbf{u} = \mathbf{0} \text{ in } \Omega \times (t^n, t^{n+1}),$$
$$\mathbf{u}(t^n) = \mathbf{u}^{n+1/3},$$
$$\mathbf{u} = \mathbf{u}_\Gamma^{n+1} \text{ on } \Gamma_-^{n+1} \times (t^n, t^{n+1}) \tag{17.21}$$

(with $\Gamma_-^{n+1} = \{x \mid x \in \Gamma, \ (\mathbf{u}_\Gamma^{n+1} \cdot \mathbf{n})(x) < 0\}$, $\mathbf{n}$ being the outward unit normal vector at Γ) and set

$$\mathbf{u}^{n+2/3} = \mathbf{u}(t^{n+1}). \tag{17.22}$$

3. Solve, finally, the following elliptic variational inequality:

$$\mathbf{u}^{n+1} \in (H^1(\Omega))^d, \quad \mathbf{u}^{n+1} = \mathbf{u}_\Gamma^{n+1} \text{ on } \Gamma,$$
$$\rho \int_\Omega \left(\mathbf{u}^{n+1} - \mathbf{u}^{n+2/3}\right) \cdot \left(\mathbf{v} - \mathbf{u}^{n+1}\right) \mathrm{d}x + \frac{\mu \Delta t}{2} \int_\Omega \nabla \mathbf{u}^{n+1} : \nabla(\mathbf{v} - \mathbf{u}^{n+1}) \mathrm{d}x$$
$$+ \tau_y \sqrt{2} \Delta t \left[\int_\Omega |\mathbf{D}(\mathbf{v})| \mathrm{d}x - \int_\Omega |\mathbf{D}(\mathbf{u}^{n+1})| \mathrm{d}x \right] \geq 0,$$
$$\forall\, \mathbf{v} \in (H^1(\Omega))^d, \quad \mathbf{v} = \mathbf{u}_\Gamma^{n+1} \text{ on } \Gamma. \tag{17.23}$$

Closely related operator-splitting techniques have been used in SANCHEZ [1998] for the simulation of Bingham flow in two-dimensional square cavities.

REMARK 17.5. It follows from, e.g., GLOWINSKI [1984, chapters 1 & 2], that the variational inequality problem (17.23) has a *unique solution* characterized by the existence of a $d \times d$ *tensor-valued function* $\boldsymbol{\lambda}^{n+1}$ such that

$$\mathbf{u}^{n+1} \in (H^1(\Omega))^d, \quad \mathbf{u}^{n+1} = \mathbf{u}_\Gamma^{n+1} \text{ on } \Gamma, \quad \boldsymbol{\lambda}^{n+1} \in (L^\infty(\Omega))^{d \times d}, \boldsymbol{\lambda}^{n+1} = (\boldsymbol{\lambda}^{n+1})^t,$$
$$\frac{\rho}{\Delta t}(\mathbf{u}^{n+1} - \mathbf{u}^{n+2/3}) - \frac{\mu}{2} \nabla^2 \mathbf{u}^{n+1} - \tau_y \sqrt{2} \nabla \cdot \boldsymbol{\lambda}^{n+1} = 0 \text{ in } \Omega,$$
$$|\boldsymbol{\lambda}^{n+1}(x)| \leq 1 \quad \text{a.e. in } \Omega, \ \boldsymbol{\lambda}^{n+1}, (x) : \mathbf{D}(\mathbf{u}^{n+1})(x) = |\mathbf{D}(\mathbf{u}^{n+1})(x)| \quad \text{a.e. in } \Omega. \tag{17.24}$$

The multiplier $\boldsymbol{\lambda}^{n+1}$ is not necessarily unique.

The computer *implementation* of the operator-splitting scheme (17.19)–(17.23) will be discussed in the following sections.

17.3. On the finite-element approximation of problem (8.6)–(8.9)

In this section (assuming that Ω is a bounded polygonal domain of $\mathbf{R}^2$), we are going to *space-approximate* problem (8.6)–(8.9) by a variant of the well-known *Bercovier–Pironneau finite-element approximation* of the Stokes and Navier–Stokes equations (see, e.g., BERCOVIER and PIRONNEAU [1979], PIRONNEAU [1989], and GLOWINSKI [1984, 2003, 2008] for the theory and practice of the Bercovier–Pironneau approximation). The notation being essentially the same as in Section 15, the fundamental discrete spaces are thus:

$$\mathbf{V}_h = \{\mathbf{v} \mid \mathbf{v} \in (C^0(\overline{\Omega}))^2,\ \mathbf{v}|_K \in (P_1)^2,\ \ \forall\, K \in \mathcal{T}_{h/2}\}, \tag{17.25}$$

$$\mathbf{V}_{0h} = \{\mathbf{v} \mid \mathbf{v} \in \mathbf{V}_h, \mathbf{v} = \mathbf{0} \text{ on } \Gamma\}\ \ (= \mathbf{V}_h \cap (H_0^1(\Omega))^2), \tag{17.26}$$

and

$$P_h = \{q \mid q \in C^0(\overline{\Omega}),\ q|_K \in P_1,\ \ \forall\, K \in \mathcal{T}_h\}, \tag{17.27}$$

where, as usual, P_1 denotes the space of the two variables polynomials of degree ≤ 1, and where $\mathcal{T}_{h/2}$ is the triangulation of Ω, obtained from the pressure triangulation $\mathcal{T}_h$ by joining the midpoints of the edges of its elements (as shown in Fig. 17.1, below).

The *continuous in time* approximation of problem (8.6)–(8.9), associated with the above finite-element spaces, is defined as follows:

For $t \in (0, T)$ find $\{\mathbf{u}_h(t), p_h(t)\} \in \mathbf{V}_h \times P_h$ such that

$$\begin{aligned}
&\rho \int_\Omega \left[\frac{\partial \mathbf{u}_h}{\partial t}(t) + (\mathbf{u}_h(t) \cdot \boldsymbol{\nabla})\mathbf{u}_h(t)\right] \cdot (\mathbf{v} - \mathbf{u}_h(t))\mathrm{d}x \\
&\quad + \mu \int_\Omega \boldsymbol{\nabla}\mathbf{u}_h(t) : \boldsymbol{\nabla}(\mathbf{v} - \mathbf{u}_h(t))\mathrm{d}x \\
&\quad - \int_\Omega p_h(t)\boldsymbol{\nabla} \cdot (\mathbf{v} - \mathbf{u}_h(t))\mathrm{d}x + \tau_y\sqrt{2}\left[\int_\Omega |\mathbf{D}(\mathbf{v})|\mathrm{d}x - \int_\Omega |\mathbf{D}(\mathbf{u}_h(t))|\mathrm{d}x\right] \\
&\quad \geq \int_\Omega \mathbf{f}_h(t) \cdot (\mathbf{v} - \mathbf{u}_h(t))\mathrm{d}x, \quad \forall\, \mathbf{v} \in \mathbf{V}_h,\ \mathbf{v} = \mathbf{u}_{\Gamma h}(t) \text{ on } \Gamma,
\end{aligned} \tag{17.28}$$

$$\int_\Omega \boldsymbol{\nabla} \cdot \mathbf{u}_h(t)\, q\mathrm{d}x = 0, \quad \forall\, q \in P_h, \tag{17.29}$$

$$\mathbf{u}_h(0) = \mathbf{u}_{0h}, \tag{17.30}$$

$$\mathbf{u}_h(t) = \mathbf{u}_{\Gamma h}(t) \text{ on } \Gamma; \tag{17.31}$$

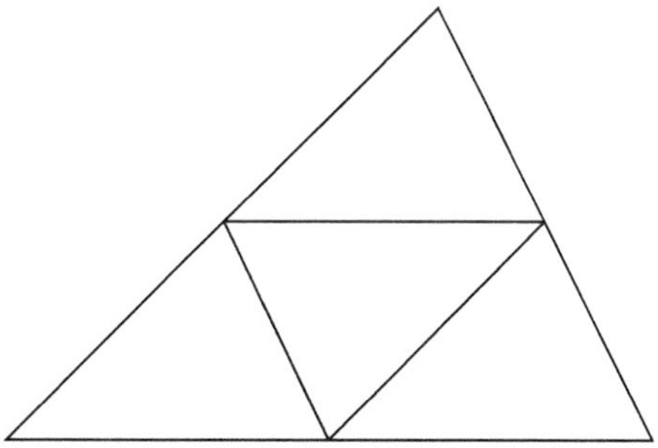

FIG. 17.1 Dividing $K \in \mathcal{T}_h$ to construct $\mathcal{T}_{h/2}$.

in (17.28)–(17.31):

- $\mathbf{f}_h$ is an approximation of $\mathbf{f}$.
- $\mathbf{u}_{\Gamma h}$ is an approximation of $\mathbf{u}_\Gamma$ such that

$$\int_\Gamma \mathbf{u}_{\Gamma h}(t)\cdot \mathbf{n}\,\mathrm{d}\Gamma = 0, \quad \forall\, t \in (0,\ T),$$

$$\mathbf{u}_{\Gamma h}(t) \in \gamma\, \mathbf{V}_h = \{\boldsymbol{\mu} | \boldsymbol{\mu} = \mathbf{v}|_\Gamma,\ \mathbf{v} \in \mathbf{V}_h\}.$$

- $\mathbf{u}_{0h}$ is an approximation of $\mathbf{u}_0$ so that $\mathbf{u}_{0h} \in \mathbf{V}_h$ and $(\mathbf{u}_{0h} - \mathbf{u}_{\Gamma h}(0))|_\Gamma = 0$.
- It is easy to compute $\int_\Omega |\mathbf{D}(\mathbf{v})|\mathrm{d}x$, $\forall \mathbf{v} \in \mathbf{V}_h$, because (17.25) implies that, $\forall\ K \in \mathcal{T}_{h/2}$, we have $\mathbf{D}(\mathbf{v}|_K) \in \mathbf{R}^{2\times 2}$ and therefore $|\mathbf{D}(\mathbf{v}|_K)| \in \mathbf{R}$, which implies in turn that

$$\int_\Omega |\mathbf{D}(\mathbf{v})|\mathrm{d}x\ (= j(\mathbf{v})) = \sum_{K\in\mathcal{T}_{h/2}} \text{meas.}(K)|\mathbf{D}(\mathbf{v}|_K)|, \quad \forall\ \mathbf{v} \in \mathbf{V}_h.$$

There is thus no need for numerical integration to compute $j(\mathbf{v})$, if $\mathbf{v} \in \mathbf{V}_h$. The convergence, as $h \to 0$, of the approximate solution $\{\mathbf{u}_h, p_h\}$ to its continuous counterpart $\{\mathbf{u}, p\}$ is discussed in, e,g., FORTIN [1972], GLOWINSKI, LIONS and TRÉMOLIÈRES [1976, chapter 6], [1981, chapter 6].

REMARK 17.6. Suppose that A_1, A_2 and A_3 are the vertices of a triangle K; we suppose that ∂K is oriented counterclockwise. We have then

$$\text{meas.}(K) = \frac{1}{2}|\overrightarrow{A_1A_2} \times \overrightarrow{A_1A_3}|. \tag{17.32}$$

Moreover, if $v \in P_1$, with $v(A_i) = v_i$, $\forall i = 1, 2$, and 3, we have (see, e.g., GLOWINSKI [2003, chapter 5] for details):

$$\frac{\partial v}{\partial x_1} = -\frac{1}{2\text{meas.}(K)}(v_1\overrightarrow{A_2A_3} + v_2\overrightarrow{A_3A_1} + v_3\overrightarrow{A_1A_2})\cdot \mathbf{e}_2, \tag{17.33}$$

$$\frac{\partial v}{\partial x_1} = \frac{1}{2\text{meas.}(K)}(v_1\overrightarrow{A_2A_3} + v_2\overrightarrow{A_3A_1} + v_3\overrightarrow{A_1A_2})\cdot \mathbf{e}_1, \tag{17.34}$$

with $\mathbf{e}_1 = \{1, 0, 0\}$ and $\mathbf{e}_2 = \{0, 1, 0\}$ (assuming that K is contained in the plane $(0x_1, 0x_2)$).

Using formulae (17.32)–(17.34) (and, if necessary the two-dimensional *trapezoidal* and *Simpson* rules; see, e.g., GLOWINSKI [2003]), we can derive a formulation of (17.28)–(17.31) more suitable for computations.

17.4. *Solution of the generalized Stokes subproblem (17.20)*

Combining scheme (17.19)–(17.23) with the finite-element spaces described in Section 17.3 leads to the following approximation of the *generalized Stokes problem* (17.20):

Find $\{\mathbf{u}_h^{n+1/3}, p_h^{n+1}\} \in \mathbf{V}_h \times P_h$ such that

$$\rho \int_{\Omega} \frac{\mathbf{u}_h^{n+1/3} - \mathbf{u}_h^n}{\Delta t} \cdot \mathbf{v} dx + \frac{\mu}{2} \int_{\Omega} \nabla \mathbf{u}_h^{n+1/3} : \nabla \mathbf{v} dx - \int_{\Omega} p_h^{n+1} \nabla \cdot \mathbf{v} dx$$
$$= \int_{\Omega} \mathbf{f}_h^{n+1} \cdot \mathbf{v} dx, \qquad \forall\, \mathbf{v} \in \mathbf{V}_{0h}, \tag{17.35}$$

$$\int_{\Omega} \nabla \cdot \mathbf{u}_h^{n+1/3} q dx = 0, \quad \forall\, q \in P_h, \tag{17.36}$$

$$\mathbf{u}_h^{n+1/3} = \mathbf{u}_{\Gamma h}^{n+1} \text{ on } \Gamma. \tag{17.37}$$

The *fully discrete generalized Stokes problem* (17.35)–(17.37) is of the *Bercovier–Pironneau* type; it can be solved using a discrete analog of the *preconditioned conjugate gradient algorithms* discussed in GLOWINSKI [2003, chapter IV] (algorithm (21.46)–(21.60), in particular).

17.5. Solution of the transport subproblems (17.21)

To solve the *transport problem* (17.21), we shall combine the finite-element spaces described in Section 17.3 with the *wave-like equation* method discussed in GLOWINSKI [2003, chapter VI]; we obtain then the following semidiscrete wave-like equation problem:

Find $\mathbf{u}_h(t) \in \mathbf{V}_h$ such that, $\forall\, t \in (t^n, t^{n+1})$,

$$\int_{\Omega} \frac{\partial^2 \mathbf{u}_h(t)}{\partial t^2} \cdot \mathbf{v} dx + \int_{\Omega} (\mathbf{u}_h^{n+1/3} \cdot \nabla) \mathbf{u}_h(t) \cdot (\mathbf{u}_h^{n+1/3} \cdot \nabla) \mathbf{v} dx$$
$$+ \int_{\Gamma \setminus \Gamma_-^{n+1}} \mathbf{u}_h^{n+1/3} \cdot \mathbf{n} \frac{\partial \mathbf{u}_h}{\partial t}(t) \cdot \mathbf{v} d\Gamma = 0, \quad \forall\, \mathbf{v} \in \mathbf{V}_{0h,-}^{n+1}, \tag{17.38}$$

$$\mathbf{u}_h(t^n) = \mathbf{u}_h^{n+1/3}, \tag{17.39}$$

$$\begin{cases} \dfrac{\partial \mathbf{u}_h}{\partial t}(t^n) \in \mathbf{V}_{0h,-}^{n+1}, \\ \displaystyle\int_{\Omega} \frac{\partial \mathbf{u}_h}{\partial t}(t^n) \cdot \mathbf{v} dx = -\int_{\Omega} (\mathbf{u}_h^{n+1/3} \cdot \nabla) \mathbf{u}_h^{n+1/3} \cdot \mathbf{v} dx, \quad \forall\, \mathbf{v} \in \mathbf{V}_{0h,-}^{n+1}, \end{cases} \tag{17.40}$$

$$\mathbf{u}_h(t) = \mathbf{u}_{\Gamma h}^{n+1} \text{ on } \Gamma_-^{n+1}, \tag{17.41}$$

with, in (17.38)–(17.41),

$$\Gamma_-^{n+1} = \{x \mid x \in \Gamma,\ (\mathbf{u}_h^{n+1/3} \cdot \mathbf{n})(x) < 0\},$$

and

$$\mathbf{V}_{0h,-}^{n+1} = \{\mathbf{v} \mid \mathbf{v} \in \mathbf{V}_h,\ \mathbf{v} = \mathbf{0} \text{ on } \Gamma_-^{n+1}\}.$$

The solution of problem (17.38)–(17.41) has been discussed at length in GLOWINSKI [2003, chapter VI], where it has been validated by the results of numerical experiments for a large variety of two- and three-dimensional test problems (see also GLOWINSKI, GUIDOBONI and PAN [2006]).

17.6. Solution of the elliptic variational inequality (17.23)

We approximate problem (17.23) by the following fully discrete elliptic variational inequality:

$$\text{Find } \mathbf{u}_h^{n+1} \in \mathbf{V}_h, \mathbf{u}_h^{n+1} = \mathbf{u}_{\Gamma h}^{n+1} \text{ on } \Gamma \text{ such that}$$
$$\rho \int_\Omega \frac{\mathbf{u}_h^{n+1} - \mathbf{u}_h^{n+2/3}}{\Delta t} \cdot (\mathbf{v} - \mathbf{u}_h^{n+1}) \mathrm{d}x + \frac{\mu}{2} \int_\Omega \nabla \mathbf{u}_h^{n+1} : \nabla(\mathbf{v} - \mathbf{u}_h^{n+1}) \mathrm{d}x$$
$$+ \tau_y \sqrt{2} \left[\int_\Omega |\mathbf{D}(\mathbf{v})| \mathrm{d}x - \int_\Omega |\mathbf{D}(\mathbf{u}_h^{n+1})| \mathrm{d}x \right] \geq 0, \quad \forall\, \mathbf{v} \in \mathbf{V}_h,\ \mathbf{v} = \mathbf{u}_{\Gamma h}^{n+1} \text{ on } \Gamma. \tag{17.42}$$

Problem (17.42) has a *unique* solution. To solve the above problem, we are going to take advantage of its equivalence with:

$$\text{Find } \{\mathbf{u}_h^{n+1}, \boldsymbol{\lambda}_h^{n+1}\} \in \mathbf{V}_h \times \mathbf{L}_h, \quad \mathbf{u}_h^{n+1} = \mathbf{u}_{\Gamma h}^{n+1} \text{ on } \Gamma, \quad \boldsymbol{\lambda}_h^{n+1} = (\boldsymbol{\lambda}_h^{n+1})^t$$

such that

$$\rho \int_\Omega \frac{\mathbf{u}_h^{n+1} - \mathbf{u}_h^{n+2/3}}{\Delta t} \cdot \mathbf{v} \mathrm{d}x + \frac{\mu}{2} \int_\Omega \nabla \mathbf{u}_h^{n+1} : \nabla \mathbf{v} \mathrm{d}x$$
$$+ \tau_y \sqrt{2} \int_\Omega \boldsymbol{\lambda}_h^{n+1} : \mathbf{D}(\mathbf{v}) \mathrm{d}x = 0, \quad \forall\, \mathbf{v} \in \mathbf{V}_{0h}, \tag{17.43}$$

$$|\boldsymbol{\lambda}_h^{n+1}| \leq 1 \quad \text{a.e. in } \Omega, \quad \boldsymbol{\lambda}_h^{n+1} : \mathbf{D}(\mathbf{u}_h^{n+1}) = |\mathbf{D}(\mathbf{u}_h^{n+1})| \quad \text{a.e. in } \Omega \tag{17.44}$$

with the space $\mathbf{L}_h$ defined by

$$\mathbf{L}_h = \{\mathbf{q} \mid \mathbf{q} \in (L^\infty(\Omega))^{2\times 2}, \ \ \mathbf{q}|_K \in \mathbf{R}^{2\times 2}, \forall K \in \mathcal{T}_{h/2}\}; \tag{17.45}$$

we have, thus, $\nabla \mathbf{v}$ and $\mathbf{D}(\mathbf{v})$ belonging to $\mathbf{L}_h$, $\forall\, \mathbf{v} \in \mathbf{V}_h$. It follows from the symmetry of $\boldsymbol{\lambda}_h^{n+1}$ that

$$\int_\Omega \boldsymbol{\lambda}_h^{n+1} : \mathbf{D}(\mathbf{v}) \mathrm{d}x = \int_\Omega \boldsymbol{\lambda}_h^{n+1} : \nabla \mathbf{v}\, \mathrm{d}x, \quad \forall\, \mathbf{v} \in \mathbf{V}_h \tag{17.46}$$

and from relation (17.44) that

$$\boldsymbol{\lambda}_h^{n+1} = \mathbf{P}_{\boldsymbol{\Lambda}_h}(\boldsymbol{\lambda}_h^{n+1} + r\tau_y \sqrt{2} \mathbf{D}(\mathbf{u}_h^{n+1})), \quad \forall\, r \geq 0, \tag{17.47}$$

with

$$\boldsymbol{\Lambda}_h = \boldsymbol{\Lambda} \cap \mathbf{L}_h (\text{that is, } \ \boldsymbol{\Lambda}_h = \{\mathbf{q} \mid \mathbf{q} \in \mathbf{L}_h, \ |\mathbf{q}|_K| \leq 1, \ \ \forall\, K \in \mathcal{T}_{h/2}\}), \tag{17.48}$$

and

$$\mathbf{P}_{\mathbf{\Lambda}_h}(\mathbf{q})|_K = \begin{cases} \mathbf{q}|_K & \text{if } |\mathbf{q}|_K| \leq 1, \\ \mathbf{q}|_K/|\mathbf{q}|_K| & \text{if } |\mathbf{q}|_K| > 1. \end{cases} \tag{17.49}$$

Denote by $\mathbf{\Lambda}_h^\sigma$ the (closed convex) subset of $\mathbf{\Lambda}_h$ defined by

$$\mathbf{\Lambda}_h^\sigma = \{\mathbf{q} \mid \mathbf{q} \in \mathbf{\Lambda}_h, \ \mathbf{q} = \mathbf{q}^t\}; \tag{17.50}$$

it is a simple exercise to show that (with obvious notation)

$$\mathbf{P}_{\mathbf{\Lambda}_h^\sigma}(\mathbf{q}) = \mathbf{P}_{\mathbf{\Lambda}_h}\left(\frac{\mathbf{q}+\mathbf{q}^t}{2}\right), \quad \forall\, \mathbf{q} \in \mathbf{L}_h. \tag{17.51}$$

Combining (17.47) with (17.51) yields

$$\boldsymbol{\lambda}_h^{n+1} = \mathbf{P}_{\mathbf{\Lambda}_h^\sigma}(\boldsymbol{\lambda}_h^{n+1} + r\tau_y\sqrt{2}\nabla\mathbf{u}_h^{n+1}), \quad \forall\, r \geq 0. \tag{17.52}$$

We have, thus, shown that problem (17.43), (17.44) is equivalent to

Find $\{\mathbf{u}_h^{n+1}, \boldsymbol{\lambda}_h^{n+1}\} \in \mathbf{V}_h \times \mathbf{L}_h, \ \mathbf{u}_h^{n+1} = \mathbf{u}_{\Gamma h}^{n+1}$ on Γ,

$$\rho\int_\Omega \frac{\mathbf{u}_h^{n+1} - \mathbf{u}_h^{n+2/3}}{\Delta t}\cdot\mathbf{v}dx + \frac{\mu}{2}\int_\Omega \nabla\mathbf{u}_h^{n+1} : \nabla\mathbf{v}dx + \tau_y\sqrt{2}\int_\Omega \boldsymbol{\lambda}_h^{n+1} : \mathbf{v}dx = 0,$$

$$\forall\, \mathbf{v} \in \mathbf{V}_{0h}, \tag{17.53}$$

$$\boldsymbol{\lambda}_h^{n+1} = \mathbf{P}_{\mathbf{\Lambda}_h^\sigma}(\boldsymbol{\lambda}_h^{n+1} + r\tau_y\sqrt{2}\nabla\mathbf{u}_h^{n+1}), \quad \forall\, r \geq 0. \tag{17.54}$$

Following Section 17.3, we shall use the following iterative method *à la Uzawa* to solve problem (17.42):

$$\boldsymbol{\lambda}_h^{n+1,0} \text{ is given in } \mathbf{\Lambda}_h^\sigma; \tag{17.55}$$

then, for $k \geq 0$, assuming that $\boldsymbol{\lambda}_h^{n+1,k} \in \mathbf{\Lambda}_h^\sigma$ is known, solve

$$\mathbf{u}_h^{n+1,k} \in \mathbf{V}_h, \ \mathbf{u}_h^{n+1,k} = \mathbf{u}_{\Gamma h}^{n+1} \text{ on } \Gamma,$$

$$\rho\int_\Omega \mathbf{u}_h^{n+1,k}\cdot\mathbf{v}dx + \frac{\mu\Delta t}{2}\int_\Omega \nabla\mathbf{u}_h^{n+1,k} : \nabla\mathbf{v}dx = \rho\int_\Omega \mathbf{u}_h^{n+1/3}\cdot\mathbf{v}dx$$

$$-\tau_y\sqrt{2}\Delta t\int_\Omega \boldsymbol{\lambda}_h^{n+1,k} : \nabla\mathbf{v}dx \quad \forall\, \mathbf{v} \in \mathbf{V}_{0h}, \tag{17.56}$$

and compute

$$\boldsymbol{\lambda}_h^{n+1,k} = \mathbf{P}_{\mathbf{\Lambda}_h^\sigma}\left(\boldsymbol{\lambda}_h^{n+1,k} + r\tau_y\sqrt{2}\nabla\mathbf{u}_h^{n+1,k}\right). \tag{17.57}$$

Concerning the convergence of algorithm (17.55)–(17.57), we have the following:

THEOREM 17.2. *Suppose that*

$$0 < r < \frac{\mu}{2\tau_y^2}; \tag{17.58}$$

we have then, $\forall\ \boldsymbol{\lambda}_h^{n+1,0} \in \boldsymbol{\Lambda}_h^{\sigma}$,

$$\lim_{k\to+\infty} \{\mathbf{u}_h^{n+1,k}, \boldsymbol{\lambda}_h^{n+1,k}\} = \{\mathbf{u}_h^{n+1}, \boldsymbol{\lambda}_h^{n+1,*}\}, \tag{17.59}$$

where, in (17.59), the pair $\{\mathbf{u}_h^{n+1}, \boldsymbol{\lambda}_h^{n+1,*}\}$ *is a solution of problem (17.43), (17.44),* $\mathbf{u}_h^{n+1}$ *being then the unique solution of problem (17.42).*

PROOF. Proving the convergence of $\{\mathbf{u}_h^{n+1,k}\}_{k\geq 0}$ is fairly easy; we just proceed as in the proof of Theorem 13.1 in Section 13.3. Suppose that $\mathbf{q} \in \mathbf{L}_h$; we shall denote by $\|\mathbf{q}\|_0$ the $L^2(\Omega)$-norm of $\mathbf{q}$ defined by $\|\mathbf{q}\|_0 = (\int_\Omega |\mathbf{q}|^2 \mathrm{d}x)^{\frac{1}{2}}$; operator $\mathbf{P}_{\boldsymbol{\Lambda}_h^{\sigma}}$ is a contraction for the above norm. Next, we denote by $\bar{\mathbf{u}}_h^{n+1,k}$ and $\bar{\boldsymbol{\lambda}}_h^{n+1,k}$ the differences $\mathbf{u}_h^{n+1,k} - \mathbf{u}_h^{n+1}$ and $\boldsymbol{\lambda}_h^{n+1,k} - \boldsymbol{\lambda}_h^{n+1}$, where $\{\mathbf{u}_h^{n+1}, \boldsymbol{\lambda}_h^{n+1}\} \in \mathbf{V}_h \times \boldsymbol{\Lambda}_h^{\sigma}$ is a solution of problem (17.43), (17.44). By subtraction, we clearly obtain

$$\bar{\mathbf{u}}_h^{n+1,k} \in \mathbf{V}_{0h},$$

$$\rho \int_\Omega \bar{\mathbf{u}}_h^{n+1,k} \cdot \mathbf{v} \mathrm{d}x + \frac{\mu \Delta t}{2} \int_\Omega \boldsymbol{\nabla} \bar{\mathbf{u}}_h^{n+1,k} \cdot \boldsymbol{\nabla} \mathbf{v} \mathrm{d}x = -\tau_y \sqrt{2} \Delta t \int_\Omega \bar{\boldsymbol{\lambda}}_h^{n+1,k} : \boldsymbol{\nabla} \mathbf{v} \mathrm{d}x,$$

$$\forall\, \mathbf{v} \in \mathbf{V}_{0h}, \tag{17.60}$$

$$\|\bar{\boldsymbol{\lambda}}_h^{n+1,k+1}\|_0 \leq \|\bar{\boldsymbol{\lambda}}_h^{n+1,k} + \tau_y r \sqrt{2} \boldsymbol{\nabla} \bar{\mathbf{u}}_h^{n+1,k}\|_0. \tag{17.61}$$

Taking $\mathbf{v} = \bar{\mathbf{u}}_h^{n+1,k}$ in (17.60) and combining with (17.61), we obtain

$$\begin{aligned}
&\|\bar{\boldsymbol{\lambda}}_h^{n+1,k}\|_0^2 - \|\bar{\boldsymbol{\lambda}}_h^{n+1,k+1}\|_0^2 \\
&\quad \geq -2r\tau_y\sqrt{2} \int_\Omega \bar{\boldsymbol{\lambda}}_h^{n+1,k} : \boldsymbol{\nabla} \bar{\mathbf{u}}_h^{n+1,k} \mathrm{d}x - 2r^2\tau_y^2 \|\boldsymbol{\nabla} \bar{\mathbf{u}}_h^{n+1,k}\|_0^2 \\
&\quad \geq r\,\mu\left(\frac{2\rho}{\mu\Delta t}\|\bar{\mathbf{u}}_h^{n+1,k}\|_{(L^2(\Omega))^2}^2 + \|\boldsymbol{\nabla} \bar{\mathbf{u}}_h^{n+1,k}\|_0^2\right) - 2r^2\tau_y^2 \|\boldsymbol{\nabla} \bar{\mathbf{u}}_h^{n+1,k}\|_0^2 \\
&\quad \geq r(\mu - 2r\tau_y^2)\left(\frac{2\rho}{\mu\Delta t}\|\bar{\mathbf{u}}_h^{n+1,k}\|_{(L^2(\Omega))^2}^2 + \|\boldsymbol{\nabla} \bar{\mathbf{u}}_h^{n+1,k}\|_0^2\right).
\end{aligned} \tag{17.62}$$

Suppose that the double inequality (17.58) holds; it follows then from (17.62) that the sequence $\{\|\bar{\boldsymbol{\lambda}}_h^{n+1,k}\|_0^2\}_{k\geq 0}$ is *decreasing*. Because it is bounded from below by 0, it converges to some limit, implying that

$$\lim_{k\to\infty} (\|\bar{\boldsymbol{\lambda}}_h^{n+1,k}\|_0^2 - \|\bar{\boldsymbol{\lambda}}_h^{n+1,k+1}\|_0^2) = 0; \tag{17.63}$$

as (17.58) implies $r(\mu - 2r\tau_y^2) > 0$, combining (17.62) with (17.63) shows that $\lim_{k\to+\infty} \bar{\mathbf{u}}_h^{n+1,k} = \mathbf{0}$, i.e., $\lim_{k\to+\infty} \mathbf{u}_h^{n+1,k} = \mathbf{u}_h^{n+1}$. To prove the convergence of $\{\boldsymbol{\lambda}_h^{n+1,k}\}_{k\geq 0}$, we should proceed as in GLOWINSKI [2003, chapter 4] or GLOWINSKI, LIONS and TRÉMOLIÈRES [2001]. □

REMARK 17.7. Actually, the upper bound in (17.58) is pessimistic. Indeed, from relation (17.60), we can easily show that the convergence result (17.59) still holds if r verifies

$$0 < r < \left(1 + \frac{2\rho}{\mu \Delta t \beta_h^M}\right) \frac{\mu}{2\tau_y^2}, \tag{17.64}$$

where, in (17.64), β_h^M is the *largest eigenvalue* of the following discrete *eigenvalue problem*:

$$\{\mathbf{w}_h, \beta\} \in \mathbf{V}_{0h} \times \mathbf{R},$$
$$\int_\Omega \nabla \mathbf{w}_h : \nabla \mathbf{v} \mathrm{d}x = \beta \int_\Omega \mathbf{w}_h \cdot \mathbf{v} \mathrm{d}x, \quad \forall\, \mathbf{v} \in \mathbf{V}_{0h},$$
$$\int_\Omega |\mathbf{w}_h|^2 \mathrm{d}x = 1.$$

We recall that $\beta_h^M = \mathbf{O}(h^{-2})$.

17.7. Numerical experiments

The computational methodology discussed in Sections 17.2–17.6 has been applied to the solution of problem (8.6)–(8.9), assuming that:

1. $\Omega = (0,\ 1) \times (0,\ 1),\ \ \Gamma = \partial\Omega$.
2. $\rho = 1,\ \mu = 1,\ \tau_y = 1$.
3. $\Gamma_N = \{x | x = \{x_1,\ x_2\}, x_2 = 1, 0 < x_1 < 1\}$ and (sliding upper boundary)

$$\mathbf{u}_\Gamma(x) = \begin{cases} \mathbf{0} & \text{if } x \in \Gamma \backslash \Gamma_N, \\ 16\{x_1^2(1 - x_1)^2, 0\} & \text{if } x \in \Gamma_N. \end{cases} \tag{17.65}$$

4. $\mathbf{u}_0 = \mathbf{0}$.

REMARK 17.8. The methodology discussed in Sections 17.2–17.6 is robust enough to handle without additional difficulties the case where $\mathbf{u}_\Gamma$ is defined by

$$\mathbf{u}_\Gamma(x) = \begin{cases} \mathbf{0} & \text{if } x \in \Gamma \backslash \Gamma_N, \\ \{1, 0\} & \text{if } x \in \Gamma_N. \end{cases} \tag{17.66}$$

The regularization associated with (17.65) is a classical one in the context of wall driven cavity flow, which is the main reason why we used it here.

The results shown below have been obtained using for $\mathcal{T}_h$ a uniform triangulation like the one shown in Fig. 17.2, below, but for a smaller space discretization step, namely $\Delta x_1 = \Delta x_2 = \frac{1}{64}$; for the time discretization, we took $\Delta t = 10^{-3}$.

On Fig. 17.3, we have visualized the time variation of the *computed kinetic energy*; it is clear from this figure that we have fast convergence to a steady flow. On Fig. 17.4, we have shown the streamlines of the computed (quasi) steady-state solution. The *rigidity* (black) and *plastic* (white) regions have been visualized on Fig. 17.5. The rigidity region is the one where $\mathbf{D}(\mathbf{u}) = \mathbf{0}$; it reconnects *tangentially* with the boundary Γ of Ω in the two lower corners, as shown in the above figure, and in Fig. 17.6 in which the graph of the function $x \to |\boldsymbol{\lambda}_h(x)|$ has been visualized (we recall that $|\boldsymbol{\lambda}(x)| = 1$ in the plastic region). The above results are in good agreement with those reported in SANCHEZ [1998].

To conclude the presentation of the results associated with the test problem under consideration, we will report on the following numerical experiments: with $\mathbf{u}_0$, μ, g, and $\mathbf{u}_\Gamma$ as above, we solved-approximately-problem (8.6)–(8.9) up to $t = 0.05$; let us denote by $\mathbf{u}_h(0.05)$ the approximate velocity at $t = 0.05$. At $t = 0.05$, we froze the motion of the upper wall implying that for $t > 0.05$, the Bingham flow is modeled by (8.6)–(8.8) completed by the following boundary conditions:

$$\mathbf{u}(t) = \mathbf{0} \text{ on } \Gamma, \text{ if } t > 0.05,$$

with $\mathbf{u}_h(0.05)$ as initial condition at $t = 0.05$. Figure 17.7 suggests that the flow behaves as expected, namely, the fluid returns to rest in *finite time*. The kinetic energy behavior observed in the above figure is consistent with the one reported in GLOWINSKI and LE TALLEC [1989], GLOWINSKI, LIONS and TRÉMOLIÈRES [1981], BÉGIS and GLOWINSKI [1983], concerning the

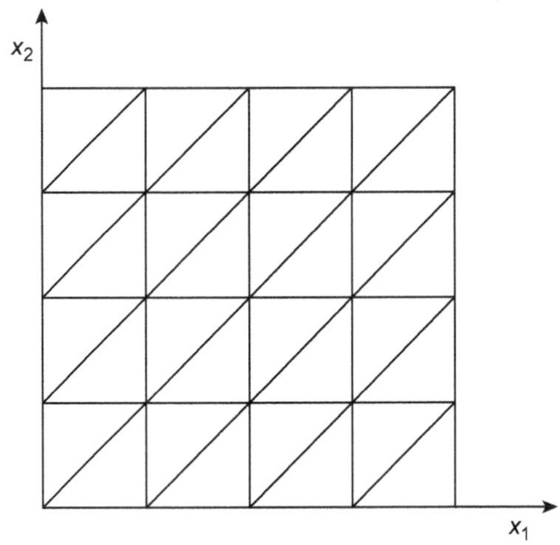

FIG. 17.2 A uniform triangulation of Ω.

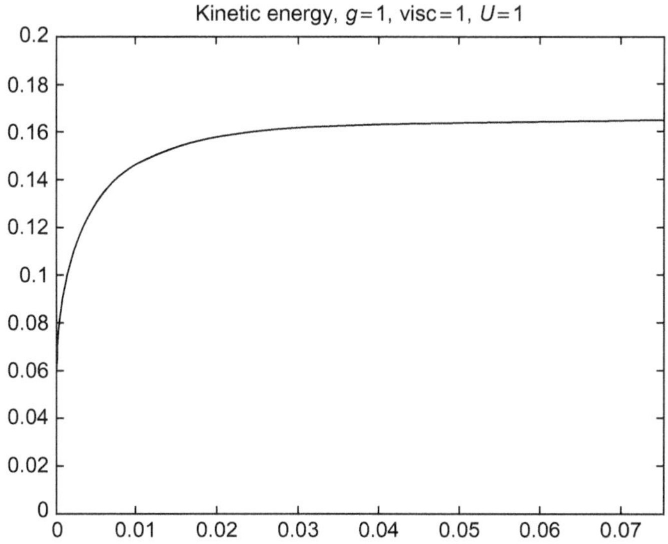

FIG. 17.3 Variation of the kinetic energy (courtesy of E.J. Dean).

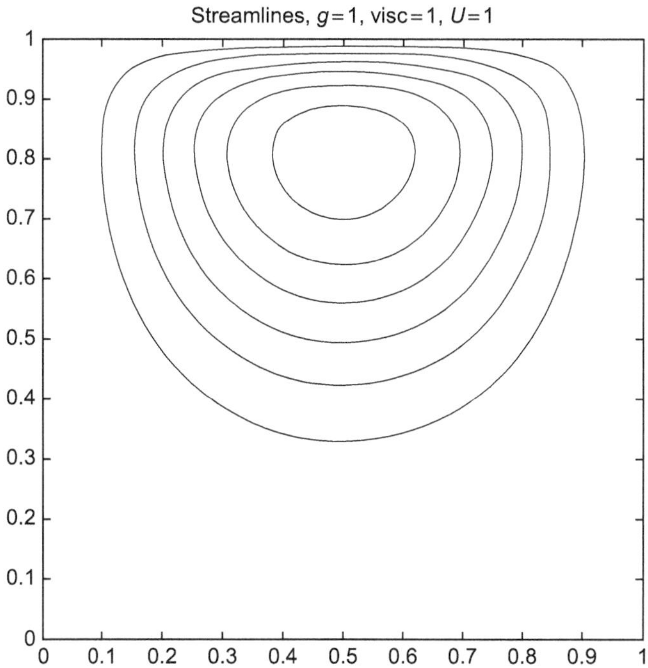

FIG. 17.4 Streamlines of the computed solution at steady state (courtesy of E.J. Dean).

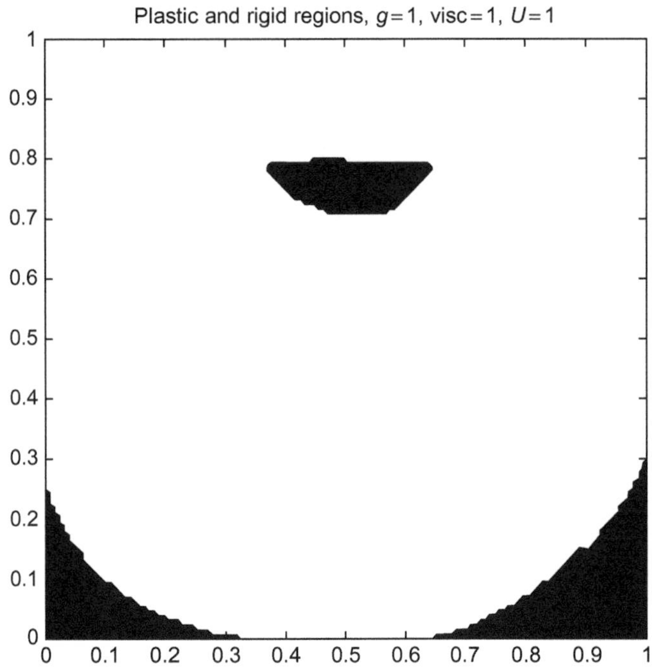

FIG. 17.5 Visualization of the computed rigid (black) and plastic (white) regions (courtesy of E.J. Dean).

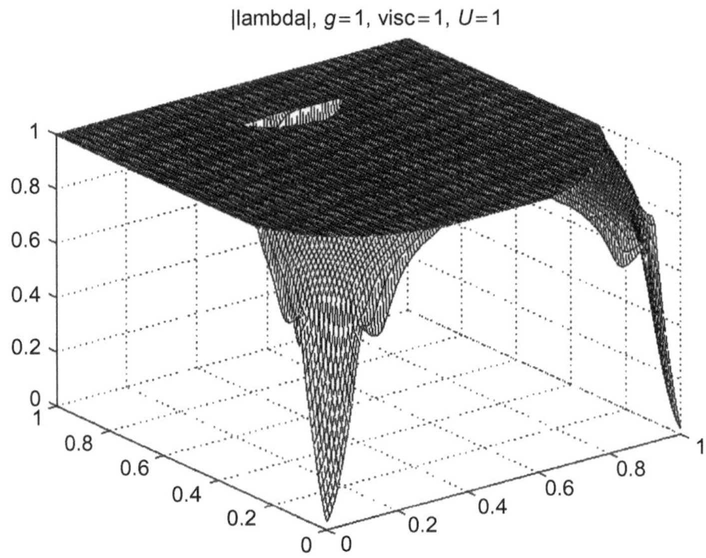

FIG. 17.6 Graph of $|\boldsymbol{\lambda}_h|$ (courtesy of E.J. Dean).

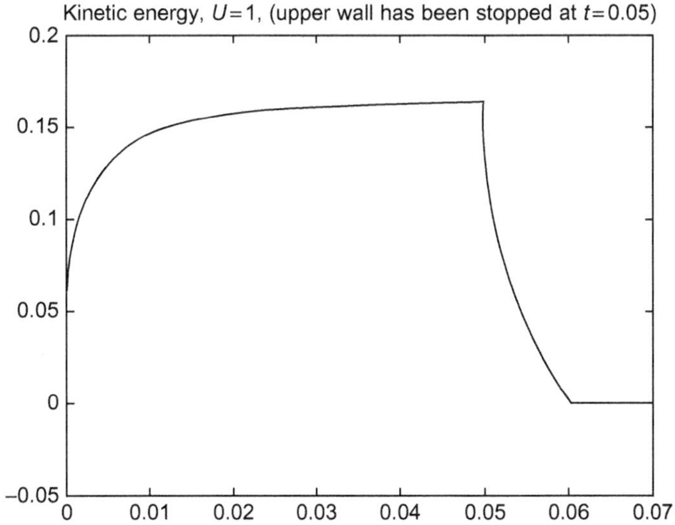

FIG. 17.7 Variation of the kinetic energy (courtesy of E.J. Dean).

solution of a closely related test problem (in the above three references the solution was computed using an *equivalent stream-function formulation* of problem (8.6)–(8.9), coupled with a variant of the augmented Lagrangian algorithm (14.7)–(14.10)).

17.8. Further comments on the asymptotic behavior of the time-discretization schemes

The numerical results shown in Section 17.7 (Fig. 17.7, particularly) strongly suggest that the operator-splitting based time-discretization scheme that we used is able to reproduce the return to rest in finite time, a property enjoyed by the solution of the continuous problem if $\mathbf{f} = \mathbf{0}$ and $\mathbf{u}_\Gamma = \mathbf{0}$ (see Theorem 17.1 of Section 17.1 for details). Actually, the numerical results reported in DEAN and GLOWINSKI [2002], GLOWINSKI [2003, chapter 10] (obtained by applying the methodology discussed in this chapter on the same test problem, but with $\tau_y = 0.1$, instead of 1) clearly show an example where the return to rest property does not hold for the discrete problem. We are going to show that this property is not lost if one uses the following *backward Euler scheme* for the time-discretization of problem (8.6)–(8.9):

$$\mathbf{u}^0 = \mathbf{u}_0; \tag{17.67}$$

then, for $n \geq 0$, $\mathbf{u}^{n-1}$ being known, find $\{\mathbf{u}^n, p^n\} \in (H_0^1(\Omega))^d \times L^2(\Omega)$ such that

$$\rho \int_\Omega \frac{\mathbf{u}^n - \mathbf{u}^{n-1}}{\Delta t} \cdot (\mathbf{v} - \mathbf{u}^n) \mathrm{d}x + \rho \int_\Omega (\mathbf{u}^n \cdot \nabla)\mathbf{u}^n \cdot (\mathbf{v} - \mathbf{u}^n) \mathrm{d}x$$
$$+ \mu \int_\Omega \nabla \mathbf{u}^n : \nabla(\mathbf{v} - \mathbf{u}^n) \mathrm{d}x$$

$$\tau_y\sqrt{2}(j(\mathbf{v}) - j(\mathbf{u}^n)) - \int_\Omega p^n \nabla \cdot (\mathbf{v} - \mathbf{u}^n) dx \geq 0, \quad \forall\, \mathbf{v} \in (H_0^1(\Omega))^2, \tag{17.68}$$

$$\nabla \cdot \mathbf{u}^n = 0 \text{ in } \Omega. \tag{17.69}$$

Assuming that problem (17.68), (17.69) has a solution, $\forall n \geq 1$ (it is not difficult to prove that it is indeed the case), take $\mathbf{v} = 0$ and $\mathbf{v} = 2\mathbf{u}^n$ in (17.68); because $\int_\Omega \mathbf{u}^n \cdot \mathbf{u}^{n-1} dx \leq \|\mathbf{u}^n\|_{0,\Omega}\, \|\mathbf{u}^{n-1}\|_{0,\Omega}$ (with $\|\mathbf{v}\|_{0,\Omega} = \|\mathbf{v}\|_{(L^2(\Omega))^d}$), it follows from (17.68), (17.69) that

$$\frac{\rho}{\Delta t}\|\mathbf{u}^n\|_{0,\Omega}(\|\mathbf{u}^n\|_{0,\Omega} - \|\mathbf{u}^{n-1}\|_{0,\Omega}) + \rho\int_\Omega (\mathbf{u}^n \cdot \nabla)\mathbf{u}^n \cdot \mathbf{u}^n dx + \mu \int_\Omega |\nabla \mathbf{u}^n|^2 dx$$
$$+ \tau_y\sqrt{2}j(\mathbf{u}^n) \leq 0, \quad \forall\, n \geq 1. \tag{17.70}$$

Relation (17.70) enjoys further simplifications because (a) relation (17.69) implies that $\rho \int_\Omega (\mathbf{u}^n \cdot \nabla)\mathbf{u}^n \cdot \mathbf{u}^n dx = 0$, and (b) Remark 17.2 implies (see Section 17.1) $\int_\Omega |\nabla \mathbf{u}^n|^2 dx \geq \lambda_0 \int_\Omega |\mathbf{u}^n|^2 dx$ and $\int_\Omega |\nabla \mathbf{u}^n| dx \geq \gamma \|\mathbf{u}^n\|_{0,\Omega}$ (with both λ_0 and $\gamma > 0$). Combining the inequalities in (a) and (b) with (17.70) we obtain, $\forall\, n \geq 1$,

$$\frac{\rho}{\Delta t}\|\mathbf{u}^n\|_{0,\Omega}(\|\mathbf{u}^n\|_{0,\Omega} - \|\mathbf{u}^{n-1}\|_{0,\Omega}) + \mu\lambda_0\|\mathbf{u}^n\|_{0,\Omega}^2 + \gamma\tau_y\sqrt{2}\|\mathbf{u}^n\|_{0,\Omega} \leq 0. \tag{17.71}$$

The above inequality shows that if $\mathbf{u}^{n-1} = \mathbf{0}$, then $\mathbf{u}^{n+k} = \mathbf{0}$, $\forall\, k \geq 0$. Suppose now that $\mathbf{u}^n \neq \mathbf{0}$ $\forall\, n \geq 0$. We have then, from (17.71),

$$\frac{\rho}{\Delta t}(\|\mathbf{u}^n\|_{0,\Omega} - \|\mathbf{u}^{n-1}\|_{0,\Omega}) + \mu\lambda_0\|\mathbf{u}^n\|_{0,\Omega} + \gamma\tau_y\sqrt{2} \leq 0, \quad \forall\, n \geq 1. \tag{17.72}$$

It follows from (17.72) that

$$\|\mathbf{u}^n\|_{0,\Omega} + \frac{\gamma\tau_y\sqrt{2}}{\lambda_0\mu} \leq \left(1 + \frac{\mu\lambda_0}{\rho}\Delta t\right)^{-1}\left[\|\mathbf{u}^{n-1}\|_{0,\Omega} + \frac{\gamma\tau_y\sqrt{2}}{\lambda_0\mu}\right], \quad \forall\, n \geq 1,$$

which implies in turn that

$$\|\mathbf{u}^n\|_{0,\Omega} + \frac{\gamma\tau_y\sqrt{2}}{\lambda_0\mu} \leq \left(1 + \frac{\mu\lambda_0}{\rho}\Delta t\right)^{-n}\left[\|\mathbf{u}_0\|_{0,\Omega} + \frac{\gamma\tau_y\sqrt{2}}{\lambda_0\mu}\right], \quad \forall\, n \geq 1. \tag{17.73}$$

Because $\lim_{n\to+\infty}\left(1 + \frac{\mu\lambda_0}{\rho}\Delta t\right)^{-n} = 0$, relation (17.73) makes no sense if $n > n_c$, with

$$n_c = \frac{\ln\left(1 + \frac{\mu\lambda_0}{\gamma\tau_y\sqrt{2}}\|\mathbf{u}_0\|_{0,\Omega}\right)}{\ln\left(1 + \frac{\mu\lambda_0}{\rho}\Delta t\right)}; \tag{17.74}$$

we have thus $\mathbf{u}^n = \mathbf{0}$, $\forall\, n > n_c$. Relation (17.74) is a discrete analog of $\mathbf{u}(t) = \mathbf{0}, \forall t \geq T_c$. It is worth noticing that, as expected,

$$\lim_{\Delta t \to 0} n_c \Delta t = \frac{\rho}{\mu\lambda_0} \ln\left(1 + \frac{\mu\lambda_0}{\gamma\tau_y\sqrt{2}}\|\mathbf{u}_0\|_{0,\Omega}\right) = T_c.$$

We have shown thus that the solution $\{\mathbf{u}^n\}_{n\geq 1}$ of problem (17.67)–(17.69) behaves "discretely" like the solution of problem (8.6)–(8.9). To prove (and have) the same result after *space-discretization*, it will definitely help to have

$$\rho \int_\Omega (\mathbf{u}_h^n \cdot \nabla)\mathbf{u}_h^n \cdot \mathbf{u}_h^n \mathrm{d}x = 0, \quad \forall\, n \geq 1. \tag{17.75}$$

This will not be the case in general, if one uses the *Hood–Taylor* or *Bercovier–Pironneau* finite-element methods to approximate problem (8.6)–(8.9). An easy way to overcome this difficulty, and recover the convergence to zero in finite discrete time would be to replace $\rho \int_\Omega (\mathbf{u}_h^n \cdot \nabla)\mathbf{u}_h^n \cdot (\mathbf{v} - \mathbf{u}_h^n)\mathrm{d}x$ by $\int_\Omega [(\mathbf{u}_h^n \cdot \nabla)\mathbf{u}_h^n + \frac{1}{2}(\nabla \cdot \mathbf{u}_h^n)\mathbf{u}_h^n] \cdot (\mathbf{v} - \mathbf{u}_h^n)\mathrm{d}x$, an idea due to R.Temam (for the Newtonian incompressible Navier–Stokes equations; see MARION and TEMAM [1998] and the references therein).

CHAPTER 3

Numerical Simulation of Nonisothermal, Compressible and Thixotropic Viscoplastic Flow: An Augmented Lagrangian Finite-Volume Approach

18. Generalities: synopsis

In Chapter 1, we mentioned several industrial applications in which viscoplastic materials play a major role. We also emphasized the fact that in real-life situations, the viscoplastic properties of a given material are usually associated with additional complex properties such as temperature dependence, thixotropy, and compressibility. The combined effect of these rheological properties leads to intricate phenomena. Because both authors are associated with the Oil & Gas industry, we found natural to focus on waxy crude oil flow because waxy crude oils are very good candidates to exhibit unusual properties beyond the existence of a nonzero yield stress.

In order to give our readers more insight on waxy crude oil flow problems, let us recall briefly the features of this type of problems, and also the industrial context where they take place. In the Oil & Gas industry, the most convenient way to transport large quantities of oil over short or long distances is clearly by pipelines. Pipelines have been massively used, for many decades already, implying that mastering the way one uses them is of primary importance for oil companies. From industrial and business standpoints, the motivations are clear, but what field engineers are really looking for is to obtain a controlled steady flow, which does not damage or destroy installations, assuming that the pressure drop along the pipe stays above a critical value (that one would like as low as possible).

Transient operations like shutdowns and restarts should be avoided as much as possible. However, unpredictable accidents may (and will) occur and the lack of knowledge or methods to handle them is highly detrimental because the resulting consequences may be catastrophic. This is where research engineers and other scientists may (and should) help. At this stage, the motivations are twofold: (1) improve the understanding of this type of flow at a fundamental level, and (2) derive methods to improve the practical handling of field pipelines.

The transportation of conventional (that is Newtonian, slightly viscous, single-phase, with steady physical properties, etc.) crude oils is a task relatively easy to handle; however, pipelining crude oils that contain large amounts of high molecular weight compounds (such as paraffin) may cause many specific difficulties (see UHDE and KOPP [1971], SMITH and RAMSDEN [1978], and MODI, KISWANTO and MERRILL [1994]). As mentioned already in Chapter 1, most of the complexity is related to formation by the paraffin crystals of an interlocking gel-like structure, which modifies the rheological properties of the crude oil (CAZAUX [1998]). This crystallization mechanism is mostly controlled by temperature. These oils (commonly known as waxy crude oils) exhibit, usually, high wax appearance temperature (WAT) and high pour point. Using a standardized test (ASMT D97), the pour point corresponds to an experimentally measured temperature below which the oil under consideration has a tendency to gel or to stop pouring. In this context, the word *high* applies to situations where this temperature is higher than the external temperature in the vicinity of the pipeline. This observation is mandatory because some readers may believe that waxy crude oil gelling occurs only in very cold climates like those encountered in Alaska or Northern Siberia, while in fact this phenomenon may take place also in Central Africa and Australia, usually considered as warm regions. Below the pour point, the rheological behavior of these oils is characterized by yield stress and viscosity coefficients which are thixotropy, temperature, and shear dependent (see ECONOMIDES and CHANEY [1983], WARDAUGH and BOGER [1987, 1991], WARDAUGH, BOGER and TONNER [1988], RONNINGSEN [1992], and HÉNAUT and BRUCY [2001]). If the temperature drop lasts long enough, the waxy crude oil undergoes a thermal shrinkage related to the occurrence of gas-filled cavities (bubbles), which confer to the material some kind of compressibility.

There are two major issues related to waxy crude oil flows in pipelines. These two issues, namely the transient regime and the steady flow, are not specifically related to the particular behavior of waxy crude oils; indeed, they are commonly encountered when considering pipeline flows. Actually, in practical situations, transporting waxy crude oil in steady flowing conditions is not a too complex operation. However, it is a good starting point for those researchers willing to understand and highlight the particular characteristics of this class of flows. It is generally agreed that the primary concern with waxy crude oil transportation is the restarting issue (see, e.g., PERKINS and TURNER [1971], and SMITH and RAMSDEN [1978]). Flow shutdowns may occur for various reasons, such as maintenance, emergency situations, and so on. Under nonflowing conditions, when the temperature in the core of the pipeline is higher than the external temperature, the pipeline temperature starts dropping; this temperature drop causes the crystallization of the paraffin compound and eventually, as the temperature drops below the pour point, there is a build up of a gel-like structure in the crude oil bulk. If the temperature drop lasts long enough, the waxy crude oil undergoes, as mentioned earlier, a thermal shrinkage following the occurrence of gas-filled bubbles, which confer to the material some kind of compressibility. Taking all these facts into account, the waxy crude oil restarting issue consists in resuming the flow of a compressible gel-like material, usually by injecting some fresh warm oil (expected to be Newtonian and incompressible) at the pipe inlet (see CAWKWELL and CHARLES [1987], and CHANG, NGUYEN and RONNINGSEN [1999]). Nevertheless, it appears that the temperature is the key parameter of the whole shutdown and restart process. Moreover, waxy crude oils are usually transported in pipelines under steady flowing conditions far below the pour point. Actually, even for this kind of flow, waxy crude oils already exhibit some of their specific rheological features, such as shear-thinning viscosity, yield stress, temperature-dependent properties, and so on.

From what has been told above, waxy crude oils have clearly a very complex rheological behavior; more specifically:

1. Above the WAT, they behave as simple incompressible Newtonian viscous fluids.
2. As the temperature drops below the WAT, the viscosity starts to increase sharply and then becomes stress dependent, in relation with the presence of paraffin crystals and of the related gel-like structure of the material.

In CAWKWELL and CHARLES [1989], the properties of two North Canadian crude oils, namely Cape Allison and Bent Horn, are discussed. These crude oils exhibit high thixotropic properties and strong temperature and temperature history dependences. Two related publications are WARDAUGH and BOGER [1987], WARDAUGH, BOGER and TONNER [1988] which deal, respectively, with Australian crude oils (Jabiru, Jackson and McKee) and a Chinese one (Da Quing); the rheological behavior of these crude oils is strongly affected by shear rate and temperature history. Waxy crude oils can be modeled, usually, by a nonisothermal thixotropic and viscoplastic constitutive equation. The models proposed in the literature consist of generalized standard viscoplastic models (e.g., Bingham or Herschel–Bulkley). In RONNINGSEN [1992], the time dependence is introduced by allowing the yield stress and the viscosity to be functions of time. In HOUSKA [1981], SESTAK, CHARLES, CAWKWELL and HOUSKA [1987], one proposed to introduce in the standard viscoplastic model a scalar variable which describes the structure of the material. This structure parameter obeys a first-order, time-dependent partial differential equation, the yield stress and the viscosity being prescribed as affine functions of the structure parameter (see Chapter 1, Section 3). In CAWKWELL and CHARLES [1989], the Houska's model is extended to nonisothermal situations by simply replacing the constant rheological parameters of the model by temperature-dependent ones. The thermal shrinkage undergone by the oil is related to the occurrence of gas-filled bubbles. The gel formation, governed by the paraffin crystallization, is controlled by thermal and flow mechanisms. Thus, the location and volume of those bubbles embedded in the gelled crude oil depend of the cooling and flow rate in the pipe (see HÉNAUT [2002]). These bubbles occupying 4–8% of the total volume of the pipe, depending of the shutdown conditions, provide thus a global compressibility to the fluid.

Before starting investigating the numerical simulation of waxy crude oil flows, let us summarize the (very interesting, in our opinion) rheological features of waxy crude oils, beyond the mere existence of a nonzero yield stress:

- *Shear-thinning viscosity.*
- *Temperature-dependent viscosity and yield stress.*
- *Thixotropic viscosity and yield stress.*
- *Slight compressibility.*

These properties justify our claim that waxy crude oils are perfect candidates to investigate the numerical simulation of yield stress fluid flows, clearly more complicated than the Bingham ones discussed in Chapter 2.

It is likely that, initially, our readers will be convinced that the main interest in dealing with waxy crude oils lies with their particular rheological features, additional to the yield stress. However, practitioners know that the main difficulty for the numerical simulation of this class of fluid flows is related to the nondifferentiability of the constitutive law and the inability to evaluate the stress in those regions where the material has not yielded. As already pointed out in Chapters 1 and 2, two main approaches have been advocated in order to overcome the computational difficulties associated with the nonsmoothness of the constitutive law.

The first approach relies on *regularization*; it has been widely used for many years by many practitioners (see, among many others, BERIS, TSAMOPOULOS, ARMSTRONG and BROWN [1985], PAPANASTASIOU [1987], ABDALI, MITSOULIS and MARKATOS [1992], MITSOULIS, ABDALI and MARKATOS [1993], BURGOS and ALEXANDROU [1999], MITSOULIS and ZISIS [2001], LIU, MULLER and DENN [2002], and MITSOULIS and HUILGOL [2004]). The key idea behind regularization methods is to approximate the (nonsmooth) constitutive law by a differentiable one. To the best of our knowledge, the most popular regularization procedure is the exponential one proposed in PAPANASTASIOU [1987] (see also Chapter 1, Section 4). The method is easy to implement because the regularized problem involves a differentiable nonlinear viscosity operator. Unfortunately, the criterion to decide whether a flow region is yielded or unyielded has become less clear-cut as pointed out in ABDALI, MITSOULIS and MARKATOS [1992], MITSOULIS, ABDALI and MARKATOS [1993]; indeed, if the regularization related strain-rate tensor (namely $\mathbf{D}(\mathbf{u}_\varepsilon)$, $\mathbf{u}_\varepsilon$ being the velocity solution of the regularized problem) vanishes, it is on a set of measure zero. Thus, with regularization methods, the determination of the yielded and unyielded regions relies on a Von Mises stress criterion (namely, x belongs to the unyielded region if $|\mathbf{D}(\mathbf{u}_\varepsilon(x))|$ is "small enough").

The second approach is mathematically more complicated; it relies on the use of *multipliers* and on the theory of variational inequalities. Concerning viscoplastic flow, these notions were introduced in DUVAUT and LIONS [1972a, 1976] for purely mathematical reasons (to prove the existence of solutions and to characterize these solutions), but they proved to be very useful computationally as shown in, e.g., FORTIN and GLOWINSKI [1982, 1983]. GLOWINSKI and LE TALLEC [1989], GLOWINSKI [2003], DEAN, GLOWINSKI and GUIDOBONI [2007] (see also Chapters 1 and 2 of the present article). Among those multiplier-based methods, the ones using *augmented Lagrangian* techniques (as in Chapter 1, Section 4, and Chapter 2, Section 14) seem to have, at the moment, the favor of the computational viscoplasticity community. Among the reasons explaining this trend, we see (1) The robustness and modularity of this methodology. (2) The fact that with the augmented Lagrangian approach one deals with the genuine constitutive laws, implying that the yielded and unyielded regions are the true ones (modulo their finite-element or finite-volume approximation). These facts justify our choice of an augmented Lagrangian-based methodology for the numerical simulation of those waxy crude oil flows discussed in the following sections of this chapter.

In addition to the above references, let us mention the following publications, in which augmented Lagrangian methods have been used for the simulation of viscoplastic flow: VOLA, BOSCARDIN and LATCHÉ [2003] for lid-driven cavity flow; ROQUET [2000], ROQUET and SARAMITO [2003] for flow around cylinders; COUPEZ, ZINE and AGASSANT [1994] for flow in convergent geometries; and YU and WACHS [2007] for the sedimentation of particles in Bingham fluids.

In this chapter, we would like to address two different types of flow: (1) nonisothermal steady flow, and (2) transient compressible and thixotropic flow (a part of the material presented in this chapter can be found in VINAY, WACHS and AGASSANT [2005, 2006]). The geometry corresponds to a three-dimensional axisymmetric pipe. The solution is obtained through a time-dependent approach which allows one to decouple velocity-pressure and temperature in the case of nonisothermal flow problems, while it allows the decoupling of the above variables from the structure parameter in the case of a thixotropic flow. At each time step, the pressure-velocity and temperature problems are solved sequentially. Most of the computational time is dedicated to the solution of the velocity-pressure problems; for solving these problems, we advocate various Lagrange multiplier-based iterative methods associated with well-chosen augmented Lagrangian functionals; these algorithms will be detailed in Section 20.

At first glance, the variational formulations and methods used to model and solve the viscoplastic flow problems, at the continuous level, call for *finite-element*-based space approximations. Indeed, the results reported in, e.g., FORTIN and GLOWINSKI [1982, 1983], GLOWINSKI and LE TALLEC [1989] (see also Chapter 2, Sections 16 and 17) show the feasibility of this approach. However, the use of finite-element approximations is not mandatory and other space discretization methods can be considered as well. In this chapter (see Section 21), the space discretization of the governing equations will be achieved by a *finite-volume* method, "operating" on a staggered grid. This finite-volume approach makes (relatively) easy the discretization of the convection terms by explicit TVD (Total Variation Diminishing) schemes (for a detailed description of TVD schemes see, e.g., YEE, WARMING and HARTEN [1985], VINCENT [1999], WANG and HUTTER [2001], and LEVEQUE [2002]). Our first (resp., second) test problem will be the simulation of the lid-driven flow in a square cavity (resp., of the axisymmetric Poiseuille flow). Our computational methods will be validated by comparison with data available in the literature (numerical results for the first test problem, exact analytical solutions for the second one). These two test cases will show that the finite-volume-based methodology that we advocate in this chapter is easy to implement and provides good quality numerical results.

The accurate prediction of the restart of a waxy crude oil flow requires a fairly precise knowledge of the initial state (that is, of the state at the time of restart) of the material in the pipeline. Because the rheological properties are temperature- and temperature-history dependent, one needs to know the evolution of the temperature field in the pipeline during the time interval when the shutdown occurred (see, e.g., COOPER, SMITH, CHARLES, RYAN and ALEXANDER [1978]). Similarly, in order to predict the temperature drop that occurred during the shutdown (that is, the temperature history), one needs to know the temperature field at the beginning of the shutdown; this field corresponds to the temperature under the steady flowing conditions. It follows from these observations that the first waxy crude oil flow problem to be considered will be the nonisothermal steady flow of a viscoplastic fluid. We retain the Bingham's model to describe the viscoplastic feature of the fluid, the temperature dependence being introduced in the model by allowing the rheological parameters (namely, the viscosity and the yield stress) to be function of the temperature. The steady state solution is obtained as the stationary solution of a system of time-dependent equations. The objective of this first study is to obtain the description of the flow pattern, when a yield stress fluid flows through a pipeline and cools down due to temperatures on the pipeline external boundary, which are lower than the inlet temperature; the yield stress temperature-dependent case will be of particular interest. Concerning the temperature dependence of the rheological parameters, let us mention that in NOUAR, DESAUBRY and ZENAIDI [1998], NOUAR, BENAOUDA-ZOUAOUI and DESAUBRY [2000], one has investigated, computationally and experimentally, the thermal convection phenomena for non-Newtonians fluids in a horizontal annular duct. However, in the above publications, only the temperature dependence of the viscosity coefficient has been taken into account.

Next, we will focus on the transient corresponding to the restart of a waxy crude oil flow in a pipeline. Here, the heat transfer is not significant; on the other hand, the oil compressibility, related to the presence of bubbles, plays a prevailing role. In CAWKWELL and CHARLES [1987, 1989] one has simulated a 1-D compressible thixotropic viscoplastic flow, using a model from SESTAK, CHARLES, CAWKWELL and HOUSKA [1987]. The compressibility effect is taken into account through the pressure dependence of the density, thanks to an isothermal compressibility coefficient. Because the initial state is a no-flow condition, the numerical simulations correspond to the restart of the flow. The time required to clear the pipe will be computed, assuming first the incompressibility of the fluid, and then its compressibility. Comparisons show that the corresponding clearing times are different. Indeed, the computed clearing time associated with the compressible case is 42% shorter compared with the incompressible one. Among the various publications related to the flow of slightly compressible non-Newtonian fluids, let us mention: CAWKWELL and CHARLES [1989] for 1-D compressible thixotropic viscoplastic flows, GOLAY and HELLUY [1998] for viscous compressible flows, SILVA and COUPEZ [2002], KESHTIBAN, BELBLIDIA and WEBSTER [2004, 2005] for compressible viscoelastic flow. Recently, in DAVIDSON, NGUYEN, CHANG and RONNINGSEN [2004], one has proposed a semianalytical 1-D approach for the restarting of a pipeline filled with a compressible gelled waxy crude oil. The crucial point of our implementation is the adaptation of Lagrangian functional based methods, developed for incompressible viscoplastic flows, to situations where compressibility occurs. The compressibility is introduced in the continuity equation using the isothermal compressibility coefficient, so that the continuity equation will be expressed in term of pressure instead of density; the augmented Lagrangian functionals encountered in the incompressible case will be modified accordingly. Finally, the compressible Stokes type subproblems associated with the operator-splitting time-discretization scheme will be solved by a modified Uzawa algorithm as well.

Our goal in this chapter is to apply the augmented Lagrangian/finite-volume methodology, briefly sketched earlier, to a variety of problems, starting with the following two well-known flow problems: (1) the two-dimensional lid-driven square cavity flow, and (2) the axisymmetric flow in a pipeline with circular cross-section.

This chapter, dedicated to the numerical simulation of temperature-dependent steady flows and unsteady compressible (possibly thixotropic) flows, should be seen as an attempt to analyze the waxy crude oil restart issue and provide a better understanding of this class of flows.

19. Governing equations

19.1. Conservation equations

We suppose that the flow region is a bounded domain of $\mathbf{R}^d$ (with $1 \leq d \leq 3$) and that $(0, T)$ is a time interval. The unsteady nonisothermal flow of a compressible thixotropic viscoplastic fluid is governed by the following conservation equations:

- *Continuity equation:*

$$\frac{\mathrm{d}\rho}{\mathrm{d}t} + \rho \nabla \cdot \mathbf{u} = 0 \text{ in } \Omega \times (0,\ T), \tag{19.1}$$

where: ρ is the *fluid density*, $\frac{\mathrm{d}}{\mathrm{d}t}$ is the *convective time derivative* (that is, $\frac{\mathrm{d}}{\mathrm{d}t} = \frac{\partial}{\partial t} + \mathbf{u} \cdot \boldsymbol{\nabla}$) and $\mathbf{u}$ is the flow velocity vector with $\mathbf{u} = \{u_x, u_y, u_z\}$. In (19.1), $\boldsymbol{\nabla}$ denotes the gradient operator.

The compressibility dependence is taken into account through the pressure dependence of the density (that is, $\rho = \rho(p)$). Actually, the isothermal compressibility measuring the pressure variation–induced compressibility is defined by

$$\chi_\Theta = \frac{1}{\rho}\left(\frac{\partial \rho}{\partial p}\right)_\Theta, \tag{19.2}$$

p being the pressure. It follows from (19.2) that the continuity equation (19.1) can be reformulated as

$$\chi_\Theta\left(\frac{\partial p}{\partial t} + \mathbf{u} \cdot \nabla p\right) + \nabla \cdot \mathbf{u} = 0 \text{ in } \Omega \times (0,\ T). \tag{19.3}$$

Taking $\rho_0 \bar{U}^2$ as characteristic pressure, the dimensionless number χ' related to the compressibility is defined by

$$\chi' = \chi_\Theta \rho_0 \bar{U}^2; \tag{19.4}$$

in (19.4), $\bar{U}$ denotes a characteristic velocity and ρ_0 the fluid density at atmospheric pressure.

- *Momentum equation:*

$$\rho\left[\frac{\partial \mathbf{u}}{\partial t} + (\mathbf{u} \cdot \nabla)\mathbf{u}\right] + \boldsymbol{\nabla} p = \boldsymbol{\nabla} \cdot \boldsymbol{\tau} \text{ in } \Omega \times (0,\ T), \tag{19.5}$$

with $\boldsymbol{\tau}$ standing for the extra-stress tensor. We define the (dimensionless) Reynolds number $\mathcal{R}e$ by

$$\mathcal{R}e = \frac{\rho_0 \bar{U} L_c}{\mu}; \tag{19.6}$$

in relation (19.6), μ denotes the dynamic viscosity, and L_c a characteristic length (we can also define the Mach number $\mathcal{M}_a$ by $\mathcal{M}_a = \frac{\bar{U}}{c}$, where c is the speed of sound).

- *Energy equation:*

$$\rho C_p\left(\frac{\partial \Theta}{\partial t} + \mathbf{u} \cdot \nabla \Theta\right) = \lambda_f \nabla^2 \Theta + \boldsymbol{\tau} : \mathbf{D}(\mathbf{u}) \text{ in } \Omega \times (0,\ T), \tag{19.7}$$

where C_p is the heat capacity, Θ is the temperature, and λ_f is the thermal conductivity. A dimensional analysis of the energy equation (19.7) provides two additional dimensionless numbers, namely:

1. The Brinkman number $\mathcal{B}r$, used to measure the influence of the viscous dissipation; $\mathcal{B}r$ is defined by

$$\mathcal{B}r = \frac{\mu \bar{U}^2}{\lambda_f(\Theta_{\text{ext}} - \Theta_{\text{fluid}})}, \tag{19.8}$$

   where Θ_{fluid}(resp., Θ_{ext}) stands for a characteristic temperature in the flow region (resp., on its boundary).
2. The Peclet number $\mathcal{P}\text{e}$, to quantify the importance of the convection comparatively to the diffusion; $\mathcal{P}\text{e}$ is defined by

$$\mathcal{P}\text{e} = \frac{\rho C_p \bar{U} L_c}{\lambda_f}. \tag{19.9}$$

19.2. Constitutive equations

Our objective is to use a rheological model that can handle nonisothermal, compressible, and thixotropic effects. The approach used to derive such a model is to start from the basic Bingham model and then add to it the features required in order to include all the desired effects. If one assumes that the fluid is incompressible and that its flow is isothermal, the corresponding Bingham model has been defined in Chapter 1, Section 3, by (3.1) and (3.2), and was further discussed in Chapter 2. We recall that the Bingham constitutive equation is given by

$$\begin{cases} \boldsymbol{\tau} = 2\mu \mathbf{D}(\mathbf{u}) + \tau_y \dfrac{\mathbf{D}(\mathbf{u})}{\|\mathbf{D}(\mathbf{u})\|} & \text{if } \|\boldsymbol{\tau}\| > \tau_y, \\ \mathbf{D}(\mathbf{u}) = \mathbf{0} & \text{if } \|\boldsymbol{\tau}\| \le \tau_y, \end{cases}$$

where, as in Chapters 1 and 2, τ_y is the yield stress, μ is the plastic viscosity coefficient, and $\mathbf{D}(\mathbf{u}) = \frac{1}{2}[\boldsymbol{\nabla}\mathbf{u} + (\boldsymbol{\nabla}\mathbf{u})^t]$ is the strain rate tensor ($\|.\|$ corresponds to the tensor norm defined by (3.2) in Chapter 1, Section 3). We define the dimensionless Bingham number $\mathcal{B}\text{n}$, representative of the viscoplastic effects, by

$$\mathcal{B}\text{n} = \frac{\tau_y L_c}{\mu \bar{\bar{U}}}. \tag{19.10}$$

In order to take into account the thermal effects in the constitutive equation, we simply allow the rheological parameters (namely viscosity and yield stress) to be functions of the temperature, this temperature dependence being provided by experimental data. We also assume the shear-thinning of the dynamic viscosity. The compressibility contributes to an additional term in the extra-stress tensor. Finally, the viscosity and yield-stress breakdown and build up mechanisms associated with the thixotropic properties are taken into account via a parameter λ_s, which describes the internal structure of the fluid (see Chapter 1, Section 3). This structure parameter λ_s measures the degree of gelling of the waxy crude oil: λ_s is defined so that it belongs to the closed interval [0, 1], $\lambda_s = 0$ (resp., $\lambda_s = 1$) corresponding to a fully broken down gel (resp., to a fully gelled material). The magnitude of the rheological parameters (viscosity and yield stress) is a function of λ_s. In the Houska model, viscosity and yield stress are affine functions of λ_s (see relations (3.9) and (3.10) in Chapter 1, Section 3).

Taking into account all the above assumptions, the complete constitutive equation (system, in fact) reads as follows:

$$\begin{cases} \boldsymbol{\tau} = 2\mu(\Theta, \lambda_s)\mathbf{D}(\mathbf{u}) + \left[\left[\xi - \frac{2}{3}\mu(\Theta, \lambda_s)\right] \nabla \cdot \mathbf{u}\right]\mathbf{I} & \\ \quad + \tau_y(\Theta, \lambda_s)\dfrac{\mathbf{D}(\mathbf{u})}{\|\mathbf{D}(\mathbf{u})\|} & \text{if } \|\boldsymbol{\tau}\| > \tau_y(\Theta, \lambda_s), \\ \mathbf{D}(\mathbf{u}) = \mathbf{0} & \text{if } \|\boldsymbol{\tau}\| \leq \tau_y(\Theta, \lambda_s), \end{cases} \tag{19.11}$$

$$\frac{\partial \lambda_s}{\partial t} + \mathbf{u} \cdot \nabla \lambda_s = a(1 - \lambda_s) - b\lambda_s \dot{\gamma}^m, \tag{19.12}$$

$$\mu(\Theta, \lambda_s) = [\mu_0(\Theta) + \lambda_s \mu_1(\Theta)]\dot{\gamma}^{n-1}, \tag{19.13}$$

$$\tau_y(\Theta, \lambda_s) = \tau_{y0}(\Theta) + \lambda_s \tau_{y1}(\Theta). \tag{19.14}$$

System (19.11)–(19.14) includes temperature dependence, compressibility, and thixotropy; it is the compressible counterpart of the incompressible Houska model described by relations (3.9)–(3.12) in Chapter 1, Section 3. In (19.11)–(19.14): (1) ξ (the only additional parameter with respect to (3.9)–(3.12)) denotes the second viscosity. (2) a denotes the build up coefficient. (3) b denotes the break down coefficient. (4) m is an adjusting parameter. (5) μ_0 and μ_1 denote the constant viscosity and thixotropic viscosity, respectively. (6) τ_{y0} and τ_{y1} denote the constant yield stress and thixotropic yield stress, respectively.

20. Augmented Lagrangian-based solution algorithms

20.1. Synopsis

The continuity equation (19.3), the momentum equation (19.5) and the energy equation (19.7), together with the constitutive system (19.11)–(19.14), form the problem whose solution we need to address. The actual mathematical challenge lies in the formulation and solution of the velocity-pressure system (19.3), (19.5) because solving the energy and structure parameter equations (19.7) and (19.12) is straightforward. In the next paragraphs, we will discuss an augmented Lagrangian-based solution method for the velocity-pressure problem. In Section 20.2, we will consider the incompressible case and present an augmented Lagrangian solution method well-suited to this type of situations. Then, in Section 20.3, we will show how to modify the algorithm discussed in Section 20.2, in order to handle the compressibility. On the basis of the results of the numerical experiments presented in Sections 24–26, we claim that our methodology has the capability to efficiently simulate the unsteady and temperature-dependent flow of a compressible Bingham fluid.

20.2. The incompressible case

We assume in this paragraph that the flow is incompressible and that the exponent n in relations (19.13) and (19.14) is equal to 1. It follows from these assumptions that the continuity equation and the constitutive law reduce to

$$\nabla \cdot \mathbf{u} = 0, \tag{20.1}$$

$$\begin{cases} \boldsymbol{\tau} = 2\mu \mathbf{D}(\mathbf{u}) + \tau_y \dfrac{\mathbf{D}(\mathbf{u})}{\|\mathbf{D}(\mathbf{u})\|} & \text{if } \|\boldsymbol{\tau}\| > \tau_y, \\ \mathbf{D}(\mathbf{u}) = \mathbf{0} & \text{if } \|\boldsymbol{\tau}\| \leq \tau_y, \end{cases} \tag{20.2}$$

respectively. The time integration of the above system requires the knowledge of initial and boundary conditions; concerning the velocity field, we will assume that $\mathbf{u}$ verifies a Dirichlet boundary condition. If one uses one of the time-discretization schemes discussed in Chapter 1, Section 4 and Chapter 2, Section 17 (or a variant of these schemes), we have to solve at each time step a variational problem of the following type (with μ and τ_y possibly varying with x):

$$\begin{aligned} &\mathbf{u}^{\text{new}} \in \mathbf{S}_\Gamma(\Omega), \\ &\rho \int_\Omega \left(\frac{\mathbf{u}^{\text{new}} - \mathbf{u}^{\text{old}}}{\Delta t} \right) \cdot (\mathbf{v} - \mathbf{u}^{\text{new}}) \mathrm{d}x + 2 \int_\Omega \mu \mathbf{D}(\mathbf{u}^{\text{new}}) : \mathbf{D}(\mathbf{v} - \mathbf{u}^{\text{new}}) \mathrm{d}x \\ &\quad + \sqrt{2} \int_\Omega \tau_y [|\mathbf{D}(\mathbf{v})| - |\mathbf{D}(\mathbf{u}^{\text{new}})|] \mathrm{d}x \geq \int_\Omega \mathbf{f} \cdot (\mathbf{v} - \mathbf{u}^{\text{new}}) \mathrm{d}x, \forall \mathbf{v} \in \mathbf{S}_\Gamma(\Omega); \end{aligned} \tag{20.3}$$

in (20.3), the space $\mathbf{S}_\Gamma(\Omega)$ is defined by

$$\mathbf{S}_\Gamma(\Omega) = \{\mathbf{v} | \mathbf{v} \in (H^1(\Omega))^d, \nabla \cdot \mathbf{v} = 0, \mathbf{v} = \mathbf{u}_\Gamma \text{ on } \Gamma\}. \tag{20.4}$$

If $\mathbf{u}_\Gamma$ verifies

$$\int_\Gamma \mathbf{u}_\Gamma \cdot \mathbf{n} \, \mathrm{d}\Gamma = 0, \tag{20.5}$$

and is the trace on Γ of a vector-valued function belonging to $(H^1(\Omega))^d$, then the affine space (linear if $\mathbf{u}_\Gamma = \mathbf{0}$) $\mathbf{S}_\Gamma(\Omega)$ is not empty (in (20.5), $\mathbf{n}$ denotes the outward unit normal vector at Γ). It follows from, e.g., GLOWINSKI, LIONS and TRÉMOLIÈRES [1981], GLOWINSKI and LE TALLEC [1989] that the variational problem (20.3) has a unique solution and is equivalent to the following minimization problem:

$$\begin{aligned} &\mathbf{u}^{\text{new}} \in \mathbf{S}_\Gamma(\Omega), \\ &J_{\Delta t}(\mathbf{u}^{\text{new}}) \leq J_{\Delta t}(\mathbf{v}), \forall \mathbf{v} \in \mathbf{S}_\Gamma(\Omega), \end{aligned} \tag{20.6}$$

with $J_{\Delta t}(.)$ defined by

$$\begin{aligned} J_{\Delta t}(\mathbf{v}) = {} & \frac{\rho}{2\Delta t} \int_\Omega |\mathbf{v}|^2 \mathrm{d}x + \int_\Omega \mu |\mathbf{D}(\mathbf{v})|^2 \mathrm{d}x + \sqrt{2} \int_\Omega \tau_y |\mathbf{D}(\mathbf{v})| \mathrm{d}x \\ & - \int_\Omega \left(\frac{\rho}{\Delta t} \mathbf{u}^{\text{old}} + \mathbf{f} \right) \cdot \mathbf{v} \mathrm{d}x. \end{aligned} \tag{20.7}$$

As mentioned quite a few times in the preceding chapters and sections, the main difficulty when attempting to solve the minimization problem (20.6) is the nondifferentiability of the third term in the functional $J_{\Delta t}(\cdot)$ defined by (20.7). In order to overcome the above difficulty, we advocate a decomposition-coordination approach taking advantage of the equivalence between problem (20.6) and the following problem (where the superscript *new* has been dropped)

$$\begin{aligned}&\{\mathbf{u}, \mathbf{p}\} \in \mathbf{W}_\Gamma(\Omega),\\&j_{\Delta t}(\mathbf{u}, \mathbf{p}) \leq j_{\Delta t}(\mathbf{v}, \mathbf{q}), \forall \{\mathbf{v}, \mathbf{q}\} \in \mathbf{W}_\Gamma(\Omega),\end{aligned} \tag{20.8}$$

where

$$\mathbf{W}_\Gamma(\Omega) = \{\{\mathbf{v}, \mathbf{q}\} | \mathbf{v} \in \mathbf{S}_\Gamma(\Omega), \mathbf{q} \in \mathbf{Q}, \mathbf{D}(\mathbf{v}) - \mathbf{q} = \mathbf{0}\}, \tag{20.9}$$

$$\mathbf{Q} = \{\mathbf{q} | \mathbf{q} \in (L^2(\Omega))^{d\times d}, \mathbf{q} = \mathbf{q}^t\}, \tag{20.10}$$

$$\begin{aligned}j_{\Delta t}(\mathbf{v}, \mathbf{q}) = {}& \frac{\rho}{2\Delta t}\int_\Omega |\mathbf{v}|^2 \mathrm{d}x + \int_\Omega \mu |\mathbf{D}(\mathbf{v})|^2 \mathrm{d}x + \sqrt{2}\int_\Omega \tau_y |\mathbf{q}| \mathrm{d}x\\&- \int_\Omega \left(\frac{\rho}{\Delta t}\mathbf{u}^{\mathrm{old}} + \mathbf{f}\right).\mathbf{v}\mathrm{d}x.\end{aligned} \tag{20.11}$$

To relax the linear constraints $\nabla \cdot \mathbf{v} = 0$ and $\mathbf{D}(\mathbf{v}) - \mathbf{q} = \mathbf{0}$, we introduce two Lagrange multipliers functions, namely p and $\boldsymbol{\lambda}$, associated with $\nabla \cdot \mathbf{v} = 0$ and $\mathbf{D}(\mathbf{v}) - \mathbf{q} = \mathbf{0}$, respectively. The multipliers p and $\boldsymbol{\lambda}$ can be interpreted as the flow pressure and a plastic stress-tensor, respectively. This leads to associate with (20.8)–(20.11) the following augmented Lagrangian functional (with r a positive parameter):

$$\begin{aligned}\mathcal{L}_r(\mathbf{v}, \mathbf{q}; q, \boldsymbol{\mu}) = {}& j_{\Delta t}(\mathbf{v}, \mathbf{q}) + \frac{r}{2}\int_\Omega |\mathbf{D}(\mathbf{v}) - \mathbf{q}|^2 \mathrm{d}x - \int_\Omega q\nabla.\mathbf{v}\mathrm{d}x\\&+ \int_\Omega \boldsymbol{\mu} : (\mathbf{D}(\mathbf{v}) - \mathbf{q})\mathrm{d}x,\end{aligned} \tag{20.12}$$

and then the following saddle-point problem

$$\begin{aligned}&\{\{\mathbf{u}, \mathbf{p}\}, \{p, \boldsymbol{\lambda}\}\} \in (\mathbf{V}_\Gamma(\Omega) \times \mathbf{Q}) \times (L^2(\Omega) \times \mathbf{Q}),\\&\mathcal{L}_r(\mathbf{u}, \mathbf{p}; q, \boldsymbol{\mu}) \leq \mathcal{L}_r(\mathbf{u}, \mathbf{p}; p, \boldsymbol{\lambda}) \leq \mathcal{L}_r(\mathbf{v}, \mathbf{q}; p, \boldsymbol{\lambda}),\\&\forall \{\{\mathbf{v}, \mathbf{q}\}, \{q, \boldsymbol{\mu}\}\} \in (\mathbf{V}_\Gamma(\Omega) \times \mathbf{Q}) \times (L^2(\Omega) \times \mathbf{Q}),\end{aligned} \tag{20.13}$$

where

$$\mathbf{V}_\Gamma(\Omega) = \{\mathbf{v} \mid \mathbf{v} \in (H^1(\Omega))^d, \mathbf{v} = \mathbf{u}_\Gamma \text{ on } \Gamma\}. \tag{20.14}$$

By partial differentiation of the augmented Lagrangian $\mathcal{L}_r$ at $\{\{\mathbf{u}, \mathbf{p}\}, \{p, \boldsymbol{\lambda}\}\}$, we obtain the following necessary and sufficient optimality conditions (for more information on

Lagrangian functionals and saddle-point problems see, e.g., GLOWINSKI [2003, chapter 4] and the references therein):

1. Differentiating with respect to the pair $\{\mathbf{v}, q\}$ provides

$$\{\mathbf{u}, p\} \in \mathbf{V}_\Gamma(\Omega) \times L^2(\Omega), \tag{20.15}$$

$$\frac{\rho}{\Delta t}\int_\Omega \mathbf{u}\cdot\mathbf{v}\,dx + \int_\Omega (r+2\mu)\mathbf{D}(\mathbf{u}) : \mathbf{D}(\mathbf{v})\,dx - \int_\Omega p\nabla\cdot\mathbf{v}\,dx + \int_\Omega (\boldsymbol{\lambda} - r\mathbf{p}) : \mathbf{D}(\mathbf{v})\,dx = \int_\Omega \left(\frac{\rho}{\Delta t}\mathbf{u}^{\text{old}} + \mathbf{f}\right)\cdot\mathbf{v}\,dx,\ \forall\mathbf{v} \in (H_0^1(\Omega))^d, \tag{20.16}$$

$$\int_\Omega \nabla\cdot\mathbf{u}\,q\,dx = 0,\ \forall q \in L^2(\Omega). \tag{20.17}$$

It follows from (20.15)–(20.17) that the quadruple $\{\mathbf{u}, \mathbf{p}, p, \boldsymbol{\lambda}\}$ verifies

$$\begin{aligned} &\frac{\rho}{\Delta t}\mathbf{u} - \nabla\cdot[(r+2\mu)\mathbf{D}(\mathbf{u})] + \nabla p - \nabla\cdot(\boldsymbol{\lambda} - r\mathbf{p}) = \frac{\rho}{\Delta t}\mathbf{u}^{\text{old}} + \mathbf{f} \text{ in } \Omega,\\ &\nabla\cdot\mathbf{u} = 0 \text{ in } \Omega,\\ &\mathbf{u} = \mathbf{u}_\Gamma \text{ on } \Gamma. \end{aligned} \tag{20.18}$$

For $\mathbf{p}$ and $\boldsymbol{\lambda}$ given, the pair $\{\mathbf{u}, p\}$ is solution of a Stokes type system; the solution of such systems has been discussed at length in GLOWINSKI [2003, chapter 4].

2. Differentiating with respect to the pair $\{\mathbf{q}, \boldsymbol{\mu}\}$ provides

$$\{\mathbf{p}, \boldsymbol{\lambda}\} \in \mathbf{Q} \times \mathbf{Q}, \tag{20.19}$$

$$r\int_\Omega \mathbf{p} : (\mathbf{q} - \mathbf{p})\,dx + \sqrt{2}\left[\int_\Omega \tau_y|\mathbf{q}|\,dx - \int_\Omega \tau_y|\mathbf{p}|\,dx\right] - \int_\Omega [r\mathbf{D}(\mathbf{u}) + \boldsymbol{\lambda}] : (\mathbf{q} - \mathbf{p})\,dx \geq 0, \quad \forall\mathbf{q} \in \mathbf{Q}, \tag{20.20}$$

$$\mathbf{D}(\mathbf{u}) - \mathbf{p} = \mathbf{0}. \tag{20.21}$$

The vector-valued function $\mathbf{u}$ in (20.13) is the solution of problem (20.6). In order to solve the saddle-point problem (20.13) (or equivalently the optimality system (20.15)–(20.21)), we advocate the algorithm ALG2 discussed in, e.g., GLOWINSKI and LE TALLEC [1989] (see also Chapter 2, Section 14). When applying ALG2 to the solution of problem (20.13), we obtain:

$$\{\mathbf{u}^{-1}, \boldsymbol{\lambda}^0\} \text{ is given in } \mathbf{V}_\Gamma(\Omega) \times \mathbf{Q}; \tag{20.22}$$

then, for $m \geq 0$, assuming that $\{\mathbf{u}^{m-1}, \boldsymbol{\lambda}^m\}$ is known, compute $\mathbf{p}^m$, $\{\mathbf{u}^m, p^m\}$ and $\boldsymbol{\lambda}^{m+1}$ as follows: Solve first

$$\begin{aligned}
&\mathbf{p}^m \in \mathbf{Q}, \\
&r\int_\Omega \mathbf{p}^m : (\mathbf{q} - \mathbf{p}^m)\mathrm{d}x + \sqrt{2}\left[\int_\Omega \tau_y|\mathbf{q}|\mathrm{d}x - \int_\Omega \tau_y|\mathbf{p}^m|\mathrm{d}x\right] \\
&\quad - \int_\Omega [r\mathbf{D}(\mathbf{u}^{m-1}) + \boldsymbol{\lambda}^m] : (\mathbf{q} - \mathbf{p}^m)\mathrm{d}x \geq 0, \forall \mathbf{q} \in \mathbf{Q},
\end{aligned} \tag{20.23}$$

and then

$$\begin{aligned}
&\frac{\rho}{\Delta t}\mathbf{u}^m - \nabla \cdot [(r + 2\mu)\mathbf{D}(\mathbf{u}^m)] + \nabla p^m = \nabla \cdot (\boldsymbol{\lambda}^m - r\mathbf{p}^m) + \frac{\rho}{\Delta t}\mathbf{u}^{\text{old}} + \mathbf{f} \text{ in } \Omega, \\
&\nabla \cdot \mathbf{u}^m = 0 \text{ in } \Omega, \\
&\mathbf{u}^m = \mathbf{u}_\Gamma \text{ on } \Gamma.
\end{aligned} \tag{20.24}$$

Finally, update $\boldsymbol{\lambda}^m$ by

$$\boldsymbol{\lambda}^{m+1} = \boldsymbol{\lambda}^m + r[\mathbf{D}(\mathbf{u}^m) - \mathbf{p}^m]. \tag{20.25}$$

The Stokes problem (20.24) can be solved by those algorithms discussed in, e.g., GLOWINSKI [2003, chapter 4]. Concerning the solution of problem (20.23), let us observe that this problem is equivalent to

$$\mathbf{p}^m = \arg\min_{\mathbf{q} \in \mathbf{Q}}\left[\frac{r}{2}\int_\Omega |\mathbf{q}|^2\,\mathrm{d}x + \sqrt{2}\int_\Omega \tau_y|\mathbf{q}|\mathrm{d}x - \int_\Omega [r\mathbf{D}(\mathbf{u}^{m-1}) + \boldsymbol{\lambda}^m] : \mathbf{q}\mathrm{d}x\right]. \tag{20.26}$$

The minimization problem in (20.26) has a closed form solution given by

$$\mathbf{p}^m = \frac{1}{r}\left(1 - \frac{\sqrt{2}\tau_y}{|\mathbf{X}^m|}\right)^+ \mathbf{X}^m, \tag{20.27}$$

with $\mathbf{X}^m = r\mathbf{D}(\mathbf{u}^{m-1}) + \boldsymbol{\lambda}^m$.

REMARK 20.1. The augmented Lagrangian functional defined by (20.11) and (20.12) has the properties required for the convergence of algorithm (20.22)–(20.25). This follows from the results given in, e.g., FORTIN and GLOWINSKI [1982, 1983], GLOWINSKI [1984], GLOWINSKI and LE TALLEC [1989] concerning the convergence of augmented Lagrangian algorithms in Hilbert spaces.

REMARK 20.2. Remark 14.1 of Chapter 2, Section 14, still applies to algorithm (20.22)–(20.25).

20.3. The compressible case

Taking into account the constitutive law (19.11), the momentum equation (19.5) reads as follows, after an appropriate time-discretization (here, we use directly a variational formulation, the notation being as in Section 20.2):

$$\mathbf{u}^{\text{new}} \in \mathbf{V}_\Gamma(\Omega), \tag{20.28}$$

$$\begin{aligned}
&\frac{1}{\Delta t}\int_\Omega \mathbf{u}^{\text{new}}\cdot(\mathbf{v}-\mathbf{u}^{\text{new}})\,dx + 2\int_\Omega \mu\mathbf{D}(\mathbf{u}^{\text{new}}):\mathbf{D}(\mathbf{v}-\mathbf{u}^{\text{new}})\,dx\\
&\quad+\int_\Omega\left(\xi-\frac{2}{3}\mu\right)\nabla\cdot\mathbf{u}^{\text{new}}\nabla\cdot(\mathbf{v}-\mathbf{u}^{\text{new}})\,dx\\
&\quad+\sqrt{2}\left[\int_\Omega \tau_y|\mathbf{D}(\mathbf{v})|\,dx-\int_\Omega \tau_y|\mathbf{D}(\mathbf{u}^{\text{new}})|\,dx\right]-\int_\Omega p\nabla\cdot(\mathbf{v}-\mathbf{u}^{\text{new}})\,dx\\
&\geq\int_\Omega\left(\frac{\rho}{\Delta t}\mathbf{u}^{\text{old}}+\mathbf{f}\right)\cdot(\mathbf{v}-\mathbf{u}^{\text{new}})\,dx,\quad\forall\mathbf{v}\in\mathbf{V}_\Gamma(\Omega).
\end{aligned}\tag{20.29}$$

Relations (20.28) and (20.29) have to be completed by the time-discrete analog of the continuity equation (19.3), namely

$$\frac{\chi_\Theta}{\Delta t}p^{\text{new}}+\nabla\cdot\mathbf{u}^{\text{new}}=g. \tag{20.30}$$

As in Section 2.2 for the incompressible case, the main difficulty is, from a computational standpoint, the presence of a nondifferentiable functional of $\mathbf{v}$ in the variational inequality (20.28), (20.29). As in Section 20.2, we introduce $\mathbf{q}=\mathbf{D}(\mathbf{v})$ and $\mathbf{p}^{\text{new}}=\mathbf{D}(\mathbf{u}^{\text{new}})$. After dropping the superscript *new*, there is equivalence between (20.28), (20.29) and

$$\{\mathbf{u},\mathbf{p},\boldsymbol{\lambda}\}\in\mathbf{V}_\Gamma(\Omega)\times\mathbf{Q}\times\mathbf{Q}, \tag{20.31}$$

$$\begin{aligned}
&\frac{1}{\Delta t}\int_\Omega \rho\mathbf{u}\cdot(\mathbf{v}-\mathbf{u})\,dx+2\int_\Omega\mu\mathbf{D}(\mathbf{u}):\mathbf{D}(\mathbf{v}-\mathbf{u})\,dx\\
&\quad+\int_\Omega\left(\xi-\frac{2}{3}\mu\right)\nabla\cdot\mathbf{u}\nabla\cdot(\mathbf{v}-\mathbf{u})\,dx+\sqrt{2}\left[\int_\Omega\tau_y|\mathbf{q}|\,dx-\int_\Omega\tau_y|\mathbf{p}|\,dx\right]\\
&\quad-\int_\Omega p\nabla\cdot(\mathbf{v}-\mathbf{u})\,dx+r\int_\Omega[\mathbf{D}(\mathbf{u})-\mathbf{p}]:[\mathbf{D}(\mathbf{v}-\mathbf{u})-(\mathbf{q}-\mathbf{p})]\,dx\\
&\quad+\int_\Omega\boldsymbol{\lambda}:[\mathbf{D}(\mathbf{v}-\mathbf{u})-(\mathbf{q}-\mathbf{p})]\,dx\geq\int_\Omega\left(\frac{\rho}{\Delta t}\mathbf{u}^{\text{old}}+\mathbf{f}\right)\cdot(\mathbf{v}-\mathbf{u})\,dx,\\
&\forall\{\mathbf{v},\mathbf{q}\}\in\mathbf{V}_\Gamma(\Omega)\times\mathbf{Q},
\end{aligned}\tag{20.32}$$

$$\mathbf{D}(\mathbf{u})-\mathbf{p}=\mathbf{0}. \tag{20.33}$$

Taking advantage of the fact that $\mathbf{D}(\mathbf{v}) - \mathbf{D}(\mathbf{u}) + \mathbf{p} - \mathbf{q} = \mathbf{D}(\mathbf{v}) - \mathbf{q}$, we can easily show that the system (20.32), (20.33) is equivalent to

$$
\begin{aligned}
&\frac{1}{\Delta t}\int_\Omega \rho\mathbf{u}\cdot\mathbf{v}\mathrm{d}x + \int_\Omega \mu\mathbf{D}(\mathbf{u}) : \mathbf{D}(\mathbf{v})\mathrm{d}x + \int_\Omega \left(\xi - \frac{2}{3}\mu\right)\nabla\cdot\mathbf{u}\nabla\cdot\mathbf{v}\mathrm{d}x \\
&\quad - \int_\Omega p\nabla\cdot\mathbf{v}\mathrm{d}x + r\int_\Omega [\mathbf{D}(\mathbf{u}) - \mathbf{p}] : \mathbf{D}(\mathbf{v})\mathrm{d}x + \int_\Omega \boldsymbol{\lambda} : \mathbf{D}(\mathbf{v})\mathrm{d}x \\
&\quad = \int_\Omega \left(\frac{\rho}{\Delta t}\mathbf{u}^{\mathrm{old}} + \mathbf{f}\right)\cdot\mathbf{v}\mathrm{d}x, \quad \forall\mathbf{v}\in(H_0^1(\Omega))^d, \\
&\sqrt{2}\left[\int_\Omega \tau_y|\mathbf{q}|\mathrm{d}x - \int_\Omega \tau_y|\mathbf{p}|\mathrm{d}x\right] + r\int_\Omega [\mathbf{p} - \mathbf{D}(\mathbf{u})] : (\mathbf{q} - \mathbf{p})\mathrm{d}x \\
&\quad - \int_\Omega \boldsymbol{\lambda} : (\mathbf{q} - \mathbf{p})\mathrm{d}x \geq 0, \quad \forall\,\mathbf{q}\in\mathbf{Q}, \\
&\mathbf{D}(\mathbf{u}) - \mathbf{p} = \mathbf{0}.
\end{aligned}
\tag{20.34}
$$

Finally, the system that one has to solve at each time step reads as follows:

$$\{\mathbf{u}, \mathbf{p}, p, \boldsymbol{\lambda}\} \in \mathbf{V}_\Gamma(\Omega) \times \mathbf{Q} \times L^2(\Omega) \times \mathbf{Q}, \tag{20.35}$$

$$
\begin{aligned}
&\frac{1}{\Delta t}\int_\Omega \rho\mathbf{u}\cdot\mathbf{v}\mathrm{d}x + 2\int_\Omega \mu\mathbf{D}(\mathbf{u}) : \mathbf{D}(\mathbf{v})\mathrm{d}x + \int_\Omega \left(\xi - \frac{2}{3}\mu\right)\nabla\cdot\mathbf{u}\nabla\cdot\mathbf{v}\mathrm{d}x \\
&\quad - \int_\Omega p\nabla\cdot\mathbf{v}\mathrm{d}x + r\int_\Omega [\mathbf{D}(\mathbf{u}) - \mathbf{p}] : \mathbf{D}(\mathbf{v})\mathrm{d}x + \int_\Omega \boldsymbol{\lambda} : \mathbf{D}(\mathbf{v})\mathrm{d}x \\
&\quad = \int_\Omega \left(\frac{\rho}{\Delta t}\mathbf{u}^{\mathrm{old}} + \mathbf{f}\right)\cdot\mathbf{v}\mathrm{d}x, \forall\mathbf{v}\in(H_0^1(\Omega))^d,
\end{aligned}
\tag{20.36}
$$

$$\frac{\chi_\Theta}{\Delta t}\mathbf{p} + \nabla\cdot\mathbf{u} = g, \tag{20.37}$$

$$
\begin{aligned}
&\sqrt{2}\left[\int_\Omega \tau_y|\mathbf{q}|\mathrm{d}x - \int_\Omega \tau_y|\mathbf{p}|\mathrm{d}x\right] + r\int_\Omega [\mathbf{p} - \mathbf{D}(\mathbf{u})] : (\mathbf{q} - \mathbf{p})\mathrm{d}x \\
&\quad - \int_\Omega \boldsymbol{\lambda} : (\mathbf{q} - \mathbf{p})\mathrm{d}x \geq 0, \quad \forall\mathbf{q}\in\mathbf{Q},
\end{aligned}
\tag{20.38}
$$

$$\mathbf{D}(\mathbf{u}) - \mathbf{p} = \mathbf{0}. \tag{20.39}$$

In order to solve the nonlinear system (20.35)–(20.39), we advocate the following variant of algorithm (20.22)–(20.25):

$$\{\mathbf{u}^{-1}, \boldsymbol{\lambda}^0\} \text{ is given in } \mathbf{V}_\Gamma(\Omega) \times \mathbf{Q}; \tag{20.40}$$

then, for $m \geq 0$, assuming that $\{\mathbf{u}^{m-1}, \boldsymbol{\lambda}^m\}$ is known, compute $\mathbf{p}^m$, $\{\mathbf{u}^m, p^m\}$ and $\boldsymbol{\lambda}^{m+1}$ as follows:
Solve the two following variational problems:

$$\begin{aligned}
&\mathbf{p}^m \in \mathbf{Q},\\
&\sqrt{2}\left[\int_\Omega \tau_y|\mathbf{q}|\mathrm{d}x - \int_\Omega \tau_y|\mathbf{p}^m|\mathrm{d}x\right] + r\int_\Omega \mathbf{p}^m : (\mathbf{q} - \mathbf{p}^m)\mathrm{d}x\\
&\quad \geq \int_\Omega [r\mathbf{D}(\mathbf{u}^{m-1}) + \boldsymbol{\lambda}^m] : (\mathbf{q} - \mathbf{p}^m)\mathrm{d}x, \forall \mathbf{q} \in \mathbf{Q},
\end{aligned} \tag{20.41}$$

$$\begin{aligned}
&\{\mathbf{u}^m, p^m\} \in \mathbf{V}_\Gamma(\Omega) \times L^2(\Omega),\\
&\frac{1}{\Delta t}\int_\Omega \rho\mathbf{u}^m \cdot \mathbf{v}\mathrm{d}x + \int_\Omega (2\mu + r)\mathbf{D}(\mathbf{u}^m) : \mathbf{D}(\mathbf{v})\mathrm{d}x + \int_\Omega \left(\xi - \frac{2}{3}\mu\right)\nabla \cdot \mathbf{u}^m \nabla \cdot \mathbf{v}\mathrm{d}x\\
&\quad - \int_\Omega p^m \nabla \cdot \mathbf{v}\mathrm{d}x = \int_\Omega [r\mathbf{p}^m - \boldsymbol{\lambda}^m] : \mathbf{D}(\mathbf{v})\mathrm{d}x + \int_\Omega \left(\frac{\rho}{\Delta t}\mathbf{u}^{\text{old}} + \mathbf{f}\right) \cdot \mathbf{v}\mathrm{d}x,\\
&\forall \mathbf{v} \in (H_0^1(\Omega))^d,
\end{aligned} \tag{20.42}$$

$$\frac{\chi_\Theta}{\Delta t}p^m + \nabla \cdot \mathbf{u}^m = g \tag{20.43}$$

and update $\boldsymbol{\lambda}^m$ by

$$\boldsymbol{\lambda}^{m+1} = \boldsymbol{\lambda}^m + r[\mathbf{D}(\mathbf{u}^m) - \mathbf{p}^m]. \tag{20.44}$$

We will denote algorithm (20.40)–(20.44) by MUA (for Modified Uzawa Algorithm).

REMARK 20.3. Remark 14.1 of Chapter 2, Section 14, and Remark 20.2 of Section 20.2 still apply to MUA.

The solution of the generalized Stokes problem (20.42), (20.43) will be discussed in Section 22. On the other hand, problem (20.41) has a closed form solution given by

$$\mathbf{p}^m = \frac{1}{r}\left(1 - \frac{\sqrt{2}\tau_y}{|\mathbf{X}^m|}\right)^+ \mathbf{X}^m, \tag{20.45}$$

where $\mathbf{X}^m = r\mathbf{D}(\mathbf{u}^{m-1}) + \boldsymbol{\lambda}^m$. Actually, MUA is part of a relatively complex solution process, to be described in Section 20.4.

20.4. Solution method for the unsteady flow of a temperature-dependent compressible thixotropic viscoplastic flow

The numerical simulation of a time-dependent, nonisothermal, thixotropic, compressible viscoplastic flow requires the solution of a relatively complicated system coupling the hydrodynamic equations to the energy equation and to the equation modeling the evolution of the structure parameter λ_s. In other words, after a well-chosen time-discretization, the temperature, structure parameter, and shear dependent rheological properties (essentially, viscosity, and yield stress) are updated, based on fields computed at the previous time step. At each time step, a compressible viscoplastic problem (whose general formulation is given by (20.28)–(20.30)) will be solved by algorithm (20.40)–(20.44) (denoted by MUA in Section 20.3). The solution of the generalized Stokes problem (20.42), (20.43), encountered at each iteration of MUA, will be discussed in Section 22. The velocity field obtained from MUA will be introduced in the time-discrete equations approximating the continuous energy and structure parameter equations. This type of coupling (often referred to as a weak coupling) is very easy to implement and has proved efficient for all the situations considered in this chapter.

The solution method resulting of the above strategy reads as follows (with $\Delta t (> 0)$ a time-discretization step):

$$\mathbf{u}^0, p^0, \mathbf{p}^0, \boldsymbol{\lambda}^0, \Theta^0, \text{ and } \lambda_s^0 \text{ are given.} \tag{20.46}$$

For $k \geq 1$, we denote $k\Delta t$ by t^k; assuming that $\mathbf{u}^{k-1}, p^{k-1}, \mathbf{p}^{k-1}, \boldsymbol{\lambda}^{k-1}, \Theta^{k-1}$ and λ_s^{k-1} are known, we compute as follows the approximate solution $\{\mathbf{u}^k, p^k, \mathbf{p}^k, \boldsymbol{\lambda}^k, \Theta^k, \lambda_s^k\}$ at t^k:

Update the rheological parameters μ and τ_y by

$$\tau_y^k = \tau_y(\dot{\gamma}^{k-1}, \Theta^{k-1}, \lambda_s^{k-1}), \text{ and } \mu^k = \mu(\dot{\gamma}^{k-1}, \Theta^{k-1}, \lambda_s^{k-1}). \tag{20.47}$$

Use MUA (actually, a close variant of it) to solve the time-discrete, transient, compressible viscoplastic problem, that is (from (20.40)–(20.44)):

$$\mathbf{u}_{-1}^k = \mathbf{u}^{k-1}, \ p_{-1}^k = p^{k-1}, \boldsymbol{\lambda}_0^k = \boldsymbol{\lambda}^{k-1}; \tag{20.48}$$

for $m \geq 0$, $\mathbf{u}_{m-1}^k$, p_{m-1}^k, and $\boldsymbol{\lambda}_m^k$ being known, compute $\mathbf{p}_m^k$, $\mathbf{u}_m^k$, p_m^k, and $\boldsymbol{\lambda}_{m+1}^k$ as follows:

1. Compute the strain rate tensor $\mathbf{p}_m^k$ via

$$\mathbf{p}_m^k = \frac{1}{r}\left(1 - \frac{\sqrt{2}\tau_y^k}{|\mathbf{X}_m^k|}\right)^+ \mathbf{X}_m^k, \tag{20.49}$$

with

$$\mathbf{X}_m^k = r\mathbf{D}(\mathbf{u}_{m-1}^k) + \boldsymbol{\lambda}_m^k. \tag{20.50}$$

2. Solve the following generalized Stokes problem:

$$\frac{\rho_{m-1}^k}{\Delta t}\mathbf{u}_m^k - \nabla \cdot \left[(2\mu^k + r)\mathbf{D}(\mathbf{u}_m^k) - \frac{2}{3}\mu^k(\nabla \cdot \mathbf{u}_m^k)\mathbf{I}\right] =$$

$$\nabla \cdot (\boldsymbol{\lambda}_m^k - r\mathbf{p}_m^k) + \mathbf{f}(\mathbf{u}^{k-1}, \rho_{m-1}^k), \tag{20.51}$$

$$\frac{\chi_\Theta}{\Delta t} p_m^k + \nabla \cdot \mathbf{u}_m^k = g(\mathbf{u}^{k-1}, p^{k-1}), \tag{20.52}$$

$$\mathbf{u}_m^k = \mathbf{u}_\Gamma \text{ on } \boldsymbol{\Gamma}, \tag{20.53}$$

with

$$\mathbf{f}(\mathbf{u}^{k-1}, \rho_{m-1}^k) = \rho_{m-1}^k \left[\frac{\mathbf{u}^{k-1}}{\Delta t} - (\mathbf{u}^{k-1} \cdot \nabla)\mathbf{u}^{k-1}\right], \tag{20.54}$$

$$g(\mathbf{u}^{k-1}, p^{k-1}) = \chi_\Theta \left[\frac{p^{k-1}}{\Delta t} - \mathbf{u}^{k-1} \cdot \nabla p^{k-1}\right]. \tag{20.55}$$

3. Update $\boldsymbol{\lambda}_m^k$ by

$$\boldsymbol{\lambda}_{m+1}^k = \boldsymbol{\lambda}_m^k + r[\mathbf{D}(\mathbf{u}_m^k) - \mathbf{p}_m^k]. \tag{20.56}$$

4. Update the density by

$$\rho_m^k = \rho_0 e^{\chi_\Theta p_m^k}. \tag{20.57}$$

5. Define $\Delta\{\mathbf{u}, p\}_m$ and $\Delta\mathbf{D}_m$ by

$$\Delta\{\mathbf{u}, p\}_m = \|\mathbf{u}_m^k - \mathbf{u}_{m-1}^k\|_\infty + \|p_m^k - p_{m-1}^k\|_\infty, \tag{20.58}$$

$$\Delta\mathbf{D}_m = \|\mathbf{D}(\mathbf{u}_m^k) - \mathbf{p}_k^m\|_\infty. \tag{20.59}$$

If $\Delta\{\mathbf{u}, p\}_m \le tol_1$ and $\Delta\mathbf{D}_m \le tol_2$, take $\mathbf{u}^k = \mathbf{u}_m^k$ and $p^k = p_m^k$; else,

Do $m = m + 1$ and return to (20.49).

- Solve the time-discrete energy equation

$$\rho C_p \left(\frac{\Theta^k - \Theta^{k-1}}{\Delta t} + \mathbf{u}^k \cdot \nabla\Theta^{k-1}\right) - \lambda_f \nabla^2\Theta^k = 0 \text{ in } \Omega. \tag{20.60}$$

- Solve the time-discrete structure parameter equation

$$\frac{\lambda_s^k - \lambda_s^{k-1}}{\Delta t} + \mathbf{u}^k \cdot \nabla\lambda_s^{k-1} = a(1 - \lambda_s^k) - b\lambda_s^k \dot{\gamma}^m \text{ in } \Omega. \tag{20.61}$$

If one is looking for a steady-state solution, a reasonable (actually, quite demanding) stopping criterion is provided by

$$\frac{\|\mathbf{u}^k - \mathbf{u}^{k-1}\|_\infty}{\Delta t} \le tol_3. \tag{20.62}$$

21. A finite-volume scheme

21.1. Synopsis

Our primary objective in this chapter is to apply our numerical methodology to the simulation of pipeline flows. Therefore, for our computations, we consider a three-dimensional axisymmetric pipe, whose geometry is described using a cylindrical coordinate system $\{r, \theta, z\}$. If we denote by $\{u_r, u_\theta, u_z\}$, the corresponding components of the velocity, we will assume that $u_\theta = 0$. The finite-volume method (FVM) that we are going to discuss is well suited for this type of situation; it can be easily modified to handle the Cartesian representations of two and three dimensional geometries. In order to implement our finite-volume method, we proceed as follows:

1. We divide the computational domain Ω into a finite number of control volumes, as shown in Fig. 21.1.
2. The values of the unknown functions p, Θ, and λ_s are prescribed at the centers of these control volumes.
3. The values of u_r and u_z are prescribed at the centers of the cell faces, as shown on Fig. 21.1. These face centers are thus the nodes of a staggered grid.
4. The components p_{rr}, p_{zz}, and $\lambda_{rr}, \lambda_{zz}$ of the strain rate $\mathbf{p}$ and Lagrange multiplier $\boldsymbol{\lambda}$ are prescribed at the center of these control volumes, whereas the components $p_{\theta\theta}$ and $\lambda_{\theta\theta}$ are prescribed at the cell faces.
5. The nondiagonal components p_{rz} and λ_{rz} of $\mathbf{p}$ and $\boldsymbol{\lambda}$ are prescribed at the cell faces.

This space discretization, of the Marker & Cell (MAC) type, will allow the approximation of the first-order derivatives by second-order accurate centered schemes.

REMARK 21.1. For an introduction and a thorough discussion of finite-volume methods for the approximation of partial differential equations, see, e.g., EYMARD, GALLOUET and HERBIN [2000].

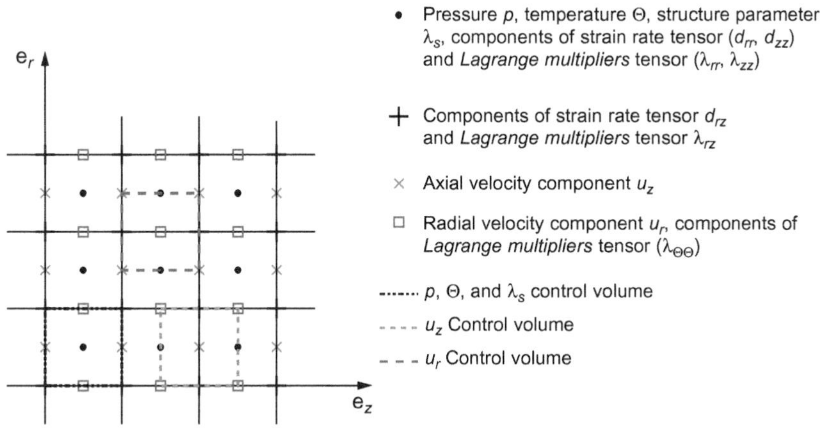

FIG. 21.1 Staggered grid.

21.2. Treatment of the convective terms

It follows from Section 20 that all the conservation equations, encountered in the flow model under consideration in this chapter, contain a convective term. It is highly recommended, if not required, to discretize this term with a carefully chosen scheme, in order to obtain a fair accuracy and avoid the spurious oscillations that are customary with standard centered schemes. It is well-known, from the literature, that upwind schemes are quite stable but not very accurate, due to the associated high level of numerical diffusion.

The search for schemes exhibiting simultaneously the robustness properties of monotone schemes close to the discontinuities, and second-order accuracy away from the discontinuities, lead to the concept of total variation (TV), defined as follows:

$$TV(\varphi^k) = \Sigma_i \left|\varphi_i^k - \varphi_i^{k-1}\right|, \tag{21.1}$$

and to the introduction of the so-called TVD schemes because they decrease the total variation, that is

$$TV(\varphi^{k+1}) \leq TV(\varphi^k), k \geq 0. \tag{21.2}$$

Among the various TVD schemes available in the literature, we selected one advocated in YEE, WARMING and HARTEN [1985]. This scheme is based on an explicit second-order *Lax-Wendroff* scheme and contains a *Superbee* type slope limiter. Let us discuss briefly the construction of such a scheme, when applied to the discretization of the following pure advection equation

$$\frac{\partial \varphi}{\partial t} + \mathbf{V} \cdot \nabla \varphi = 0 \tag{21.3}$$

The integral form of the advection equation in the C-centered control volume (see Fig. 21.2) reads as follows:

$$\int_{\Omega_C} \frac{\partial \varphi}{\partial t} r dr dz + \int_{\Omega_C} \mathbf{V} \cdot \nabla \varphi r dr dz = 0. \tag{21.4}$$

Let us introduce the flux $\mathbf{F}$ defined by $\mathbf{F} = \varphi \mathbf{V}$. Because $\mathbf{V} \cdot \nabla \varphi = \nabla \cdot \mathbf{F} - \varphi \nabla \cdot \mathbf{V}$, relation (21.4) becomes

$$\int_{\Omega_C} \frac{\partial \varphi}{\partial t} r dr dz + \int_{\Omega_C} \nabla \cdot \mathbf{F} r dr dz = \int_{\Omega_C} \varphi \nabla \cdot \mathbf{V} r dr dz. \tag{21.5}$$

We recall that in the $\{r, z\}$ system of coordinates we have (with obvious notation)

$$\nabla \cdot \mathbf{F} = \frac{1}{r} \frac{\partial}{\partial r}(r F_r) + \frac{\partial F_z}{\partial z}. \tag{21.6}$$

Assuming that one uses a forward Euler scheme for the time discretization of equation (21.3), it follows from (21.5) and (21.6) that a finite-volume discretization of (21.3) at

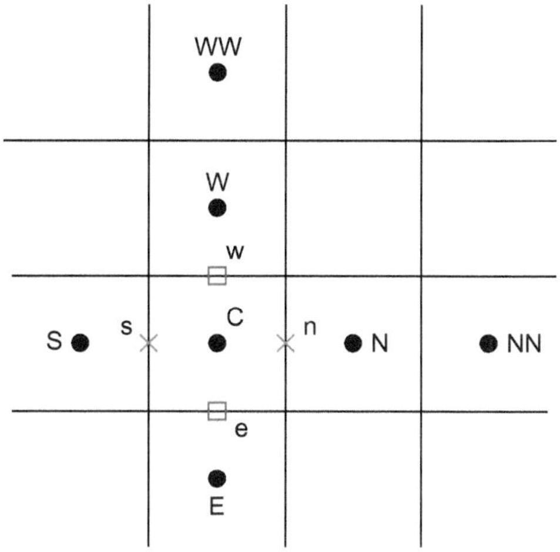

FIG. 21.2 A C-centered control volume Ω_C for the advection equation (21.3).

$t^k = k\Delta t$ reads as:

$$\frac{\varphi_C^k - \varphi_C^{k-1}}{\Delta t} + \frac{1}{r_C}\frac{r_w F_{rw}^{k-1} - r_e F_{re}^{k-1}}{\Delta r_C} + \frac{F_{zn}^{k-1} - F_{zs}^{k-1}}{\Delta z_C} = \varphi_C^{k-1}[\nabla \cdot \mathbf{V}^k]_C, \tag{21.7}$$

where in (21.7):

1. Δr_C and Δz_C denote $r_w - r_e$ and $z_n - z_s$, respectively.
2. F_{rw}^{k-1}, F_{re}^{k-1}, F_{zn}^{k-1}, and F_{zs}^{k-1} denote the approximate convective fluxes, obtained at t^{k-1} by a TVD Lax–Wendroff Superbee scheme, to be detailed shortly.
3. The subscripts w, e, n, and s are associated with the midpoints of the faces of the control volume Ω_C, whereas C, S, E, N, W, NN, WW, and so on, are associated with control volume centers.
4. The right-hand side in (21.7) is computed (approximately) at C using

$$\begin{aligned}[\nabla \cdot \mathbf{V}^k]_C &= \frac{\partial V_r^k}{\partial r}(C) + \frac{V_r^k(C)}{r_C} + \frac{\partial V_z^k}{\partial z}(C) \\ &\approx \frac{V_{rw}^k - V_{re}^k}{\Delta r_C} + \frac{V_{rw}^k + V_{re}^k}{2r_C} + \frac{V_{zn}^k - V_{zs}^k}{\Delta z_C}.\end{aligned} \tag{21.8}$$

An alternative to (21.8) is provided by

$$[\nabla \cdot \mathbf{V}^k]_C = \frac{1}{r_C}\frac{\partial (rV_r^k)}{\partial r}(C) + \frac{\partial V_z^k}{\partial z}(C) \approx \frac{r_w V_{rw}^k - r_e V_{re}^k}{r_C \Delta r_C} + \frac{V_{zn}^k - V_{zs}^k}{\Delta z_C}. \tag{21.9}$$

Collecting the above results we obtain, after discretizing at C, the advection equation (21.3):

$$\varphi_C^k = \left[1 + \Delta t\left(\frac{V_{rw}^k - V_{re}^k}{\Delta r_C} + \frac{V_{rw}^k + V_{re}^k}{2r_C} + \frac{V_{zn}^k - V_{zs}^k}{\Delta z_C}\right)\right]\varphi_C^{k-1} - \Delta t\left[\frac{r_w F_w^{k-1} - r_e F_e^{k-1}}{r_C \Delta r_C} + \frac{F_n^{k-1} - F_s^{k-1}}{\Delta z_C}\right], \tag{21.10}$$

with

$$F_w^{k-1} = \max(0, V_{rw}^k)\left[\varphi_C^{k-1} + \frac{1}{2}\phi_w^{k-1}\left(1 - \frac{\Delta t}{\Delta r_C}V_{rw}^k\right)(\varphi_W^{k-1} - \varphi_C^{k-1})\right] + \min(0, V_{rw}^k)\left[\varphi_W^{k-1} - \frac{1}{2}\phi_w^{k-1}\left(1 + \frac{\Delta t}{\Delta r_C}V_{rw}^k\right)(\varphi_W^{k-1} - \varphi_C^{k-1})\right], \tag{21.11}$$

$$F_e^{k-1} = \max(0, V_{re}^k)\left[\varphi_E^{k-1} + \frac{1}{2}\phi_e^{k-1}\left(1 - \frac{\Delta t}{\Delta r_C}V_{re}^k\right)(\varphi_C^{k-1} - \varphi_E^{k-1})\right] + \min(0, V_{re}^k)\left[\varphi_C^{k-1} - \frac{1}{2}\varphi_e^{k-1}\left(1 + \frac{\Delta t}{\Delta r_C}V_{re}^k\right)(\varphi_C^{k-1} - \varphi_E^{k-1})\right], \tag{21.12}$$

$$F_n^{k-1} = \max(0, V_{zn}^k)\left[\varphi_C^{k-1} + \frac{1}{2}\phi_n^{k-1}\left(1 - \frac{\Delta t}{\Delta z_C}V_{zn}^k\right)(\varphi_N^{k-1} - \varphi_C^{k-1})\right] + \min(0, V_{zn}^k)\left[\varphi_N^{k-1} - \frac{1}{2}\phi_n^{k-1}\left(1 + \frac{\Delta t}{\Delta z_C}V_{zn}^k\right)(\varphi_N^{k-1} - \varphi_C^{k-1})\right], \tag{21.13}$$

$$F_s^{k-1} = \max(0, V_{zs}^k)\left[\varphi_S^{k-1} + \frac{1}{2}\phi_s^{k-1}\left(1 - \frac{\Delta t}{\Delta z_C}V_{zs}^k\right)(\varphi_C^{k-1} - \varphi_S^{k-1})\right] + \min(0, V_{zs}^k)\left[\varphi_C^{k-1} - \frac{1}{2}\phi_s^{k-1}\left(1 + \frac{\Delta t}{\Delta z_C}V_{zs}^k\right)(\varphi_C^{k-1} - \varphi_S^{k-1})\right]; \tag{21.14}$$

in (21.11)–(21.14), $\phi_w^{k-1}, \phi_e^{k-1}, \phi_n^{k-1}$, and ϕ_s^{k-1} denote the *Superbee slope limiters* defined by

$$\phi_w^{k-1} = \phi(\xi_w^{k-1}),\ \phi_e^{k-1} = \phi(\xi_e^{k-1}),\ \phi_n^{k-1} = \phi(\xi_n^{k-1}),\ \text{and } \phi_s^{k-1} = \phi(\xi_s^{k-1}) \tag{21.15}$$

with

$$\xi_w^{k-1} = \frac{\varphi_C^{k-1} - \varphi_E^{k-1}}{\varphi_W^{k-1} - \varphi_C^{k-1}},\ \xi_e^{k-1} = \frac{\varphi_E^{k-1} - \varphi_{EE}^{k-1}}{\varphi_C^{k-1} - \varphi_E^{k-1}},\ \xi_n^{k-1} = \frac{\varphi_C^{k-1} - \varphi_S^{k-1}}{\varphi_N^{k-1} - \varphi_C^{k-1}},$$

$$\xi_s^{k-1} = \frac{\varphi_S^{k-1} - \varphi_{SS}^{k-1}}{\varphi_C^{k-1} - \varphi_S^{k-1}},\ \text{and}$$

$$\phi(\xi) = \max[0,\ \min(2\xi, 1),\ \min(\xi, 2)]. \tag{21.16}$$

21.3. Finite-volume discretization of the continuity equation

The integral form of the *continuity equation* (20.30), in the C-centered control volume Ω_C (see Fig. 21.3), reads as follows:

$$\frac{1}{\Delta t}\int_{\Omega_C} \chi_\Theta p^k \mathrm{d}\Omega_C + \int_{\Omega_C} \nabla\cdot\mathbf{u}^k \mathrm{d}\Omega_C = \int_{\Omega_C} g(\mathbf{u}^{k-1}, p^{k-1})\mathrm{d}\Omega_C. \tag{21.17}$$

Let us introduce $\mathbf{F}^{k-1} = p^{k-1}\mathbf{u}^{k-1}$; because (from (19.3))

$$g(\mathbf{u}^{k-1}, p^{k-1}) = \chi_\Theta\left(\frac{1}{\Delta t}p^{k-1} - \mathbf{u}^{k-1}\cdot\nabla p^{k-1}\right),$$

it follows from (21.17) and from the *divergence theorem* that

$$\begin{aligned}\frac{1}{\Delta t}\int_{\Omega_C} \chi_\Theta p^k \mathrm{d}\Omega_C + \int_{\partial\Omega_C} \mathbf{u}^k\cdot\mathbf{n}\mathrm{d}(\partial\Omega_C) &= \frac{1}{\Delta t}\int_{\Omega_C} \chi_\Theta p^{k-1}\mathrm{d}\Omega_C\\ - \int_{\partial\Omega_C} \chi_\Theta \mathbf{F}^{k-1}\cdot\mathbf{n}\mathrm{d}(\partial\Omega_C) &+ \int_{\Omega_C} \chi_\Theta p^{k-1}\nabla\cdot\mathbf{u}^{k-1}\mathrm{d}\Omega_C\end{aligned} \tag{21.18}$$

(in relation (21.18), it was assumed that χ_Θ is a constant). Calculation of the flux on each face of the control volume Ω_C yields

$$\begin{aligned}&\frac{1}{\Delta t}\chi_\Theta p_C^k + \frac{r_w u_{rw}^k - r_e u_{re}^k}{r_C \Delta r_C} + \frac{u_{zn}^k - u_{zs}^k}{\Delta z_C} =\\ &\chi_\Theta\left[\frac{1}{\Delta t}p_C^{k-1} - \frac{r_w F_{rw}^{k-1} - r_e F_{re}^{k-1}}{r_C\Delta r_C} - \frac{F_{zn}^{k-1} - F_{zs}^{k-1}}{\Delta z_C} + p_C^{k-1}(\nabla\cdot\mathbf{u}^{k-1})_C\right],\end{aligned} \tag{21.19}$$

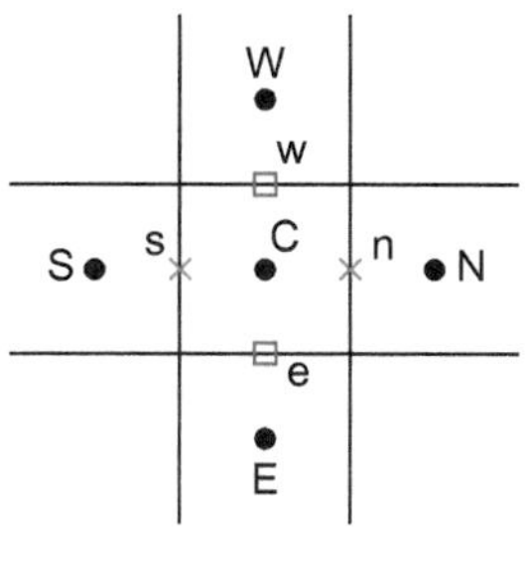

FIG. 21.3 A C-centered control volume Ω_C for the continuity equation.

the notation in (21.19) being as in Section 21.2. Assuming that $(\nabla \cdot \mathbf{u}^{k-1})_C$ is still defined by (21.8), the fully discrete form of the continuity equation reads as follows:

$$\frac{1}{\Delta t}\chi_\Theta p_C^k + \frac{r_w u_{rw}^k - r_e u_{re}^k}{r_C \Delta r_C} + \frac{u_{zn}^k - u_{zs}^k}{\Delta z_C} =$$
$$\chi_\Theta \left[p_C^{k-1}\left(\frac{1}{\Delta t} + \frac{u_{rw}^{k-1} - u_{re}^{k-1}}{\Delta r_C} + \frac{u_{rw}^{k-1} + u_{re}^{k-1}}{2r_C} + \frac{u_{zn}^{k-1} - u_{zs}^{k-1}}{\Delta z_C}\right)\right.$$
$$\left. - \left(\frac{r_w F_{rw}^{k-1} - r_e F_{re}^{k-1}}{r_C \Delta r_C} + \frac{F_{zn}^{k-1} - F_{zs}^{k-1}}{\Delta z_C}\right)\right], \tag{21.20}$$

where the convective flux $\mathbf{F}$ is discretized using the *Lax–Wendroff TVD Superbee scheme* defined by (21.11)–(21.16), that is

$$F_w^{k-1} = \max\left(0, u_{rw}^{k-1}\right)\left[p_C^{k-1} + \frac{1}{2}\phi_w^{k-1}\left(1 - \frac{\Delta t}{\Delta r_C}u_{rw}^{k-1}\right)\left(p_W^{k-1} - p_C^{k-1}\right)\right]$$
$$+ \min\left(0, u_{rw}^{k-1}\right)\left[p_W^{k-1} - \frac{1}{2}\phi_w^{k-1}\left(1 + \frac{\Delta t}{\Delta r_C}u_{rw}^{k-1}\right)\left(p_W^{k-1} - p_C^{k-1}\right)\right], \tag{21.21}$$

$$F_e^{k-1} = \max\left(0, u_{re}^{k-1}\right)\left[p_E^{k-1} + \frac{1}{2}\phi_e^{k-1}\left(1 - \frac{\Delta t}{\Delta r_C}u_{re}^{k-1}\right)\left(p_C^{k-1} - p_E^{k-1}\right)\right]$$
$$+ \min\left(0, u_{re}^{k-1}\right)\left[p_C^{k-1} - \frac{1}{2}\phi_e^{k-1}\left(1 + \frac{\Delta t}{\Delta r_C}u_{re}^{k-1}\right)\left(p_C^{k-1} - p_E^{k-1}\right)\right], \tag{21.22}$$

$$F_n^{k-1} = \max\left(0, u_{zn}^{k-1}\right)\left[p_C^{k-1} + \frac{1}{2}\phi_n^{k-1}\left(1 - \frac{\Delta t}{\Delta z_C}u_{zn}^{k-1}\right)\left(p_N^{k-1} - p_C^{k-1}\right)\right]$$
$$+ \min\left(0, u_{zn}^{k-1}\right)\left[p_N^{k-1} - \frac{1}{2}\phi_n^{k-1}\left(1 + \frac{\Delta t}{\Delta z_C}u_{zn}^{k-1}\right)\left(p_N^{k-1} - p_C^{k-1}\right)\right], \tag{21.23}$$

$$F_s^{k-1} = \max\left(0, u_{zs}^{k-1}\right)\left[p_S^{k-1} + \frac{1}{2}\phi_s^{k-1}\left(1 - \frac{\Delta t}{\Delta z_C}u_{zs}^{k-1}\right)\left(p_C^{k-1} - p_S^{k-1}\right)\right]$$
$$+ \min\left(0, u_{zs}^{k-1}\right)\left[p_C^{k-1} - \frac{1}{2}\phi_s^{k-1}\left(1 + \frac{\Delta t}{\Delta z_C}u_{zs}^{k-1}\right)\left(p_C^{k-1} - p_S^{k-1}\right)\right], \tag{21.24}$$

where in (21.21)–(21.24), ϕ_w^{k-1}, ϕ_e^{k-1}, ϕ_n^{k-1}, and ϕ_s^{k-1} denote the *Superbee slope limiters* defined by

$$\phi_w^{k-1} = \phi\left(\xi_w^{k-1}\right),\ \phi_e^{k-1} = \phi\left(\xi_e^{k-1}\right),\ \phi_n^{k-1} = \phi\left(\xi_n^{k-1}\right),\ \text{and } \phi_s^{k-1} = \phi\left(\xi_s^{k-1}\right) \tag{21.25}$$

with $\phi(\xi) = \max[0,\ \min(2\xi, 1),\ \min(\xi, 2)]$, and

$$\xi_w^{k-1} = \frac{p_C^{k-1} - p_E^{k-1}}{p_W^{k-1} - p_C^{k-1}},\ \xi_e^{k-1} = \frac{p_E^{k-1} - p_{EE}^{k-1}}{p_C^{k-1} - p_E^{k-1}},\ \xi_n^{k-1}$$
$$= \frac{p_C^{k-1} - p_S^{k-1}}{p_N^{k-1} - p_C^{k-1}},\ \xi_s^{k-1} = \frac{p_S^{k-1} - p_{SS}^{k-1}}{p_C^{k-1} - p_S^{k-1}}.$$

21.4. Finite-volume discretization of the momentum equation

At time $k\Delta t$, the integral vector form of the momentum equation (20.42) reads as follows (after omitting the superscript m and replacing the *augmented Lagrangian* parameter r by r_{AL} to avoid confusion with the cylindrical coordinate r):

$$\frac{1}{\Delta t}\int_D \rho^k \mathbf{u}^k \mathrm{d}D - \int_D \nabla\cdot\left[\left(2\mu^{k-1}+r_{AL}\right)\mathbf{D}(\mathbf{u}^k) - \frac{2}{3}\mu^{k-1}(\nabla\cdot\mathbf{u}^k)\mathbf{I}\right]\mathrm{d}D$$
$$+\int_D \nabla p^k \mathrm{d}D = \int_D \nabla\cdot\left(\boldsymbol{\lambda}^k - r_{AL}\mathbf{p}^k\right)\mathrm{d}D + \int_D \mathbf{f}\left(\mathbf{u}^{k-1},\rho^k\right)\mathrm{d}D, \tag{21.26}$$

where D is an arbitrary subdomain of Ω. For clarity, we introduce the tensors $\boldsymbol{\phi}$ and $\mathcal{T}$ defined by

$$\boldsymbol{\phi} = \left(2\mu^{k-1}+r_{AL}\right)\mathbf{D}(\mathbf{u}) - \frac{2}{3}\mu^{k-1}(\nabla\cdot\mathbf{u})\mathbf{I} = \begin{pmatrix}\phi_{rr} & 0 & \phi_{rz}\\ 0 & \phi_{\theta\theta} & 0\\ \phi_{rz} & 0 & \phi_{zz}\end{pmatrix} \tag{21.27}$$

and

$$\mathcal{T} = \boldsymbol{\lambda} - r_{AL}\mathbf{p} = \begin{pmatrix}\mathcal{T}_{rr} & 0 & \mathcal{T}_{rz}\\ 0 & \mathcal{T}_{\theta\theta} & 0\\ \mathcal{T}_{rz} & 0 & \mathcal{T}_{zz}\end{pmatrix}, \tag{21.28}$$

respectively (in (21.27), (21.28) (and below) we have dropped, for clarity, the superscripts k); we can easily show that the nonzero coefficients of the 3×3 tensor $\boldsymbol{\phi}$ are given by

$$\phi_{rr} = \left(\frac{4}{3}\mu^{k-1}+r_{AL}\right)\frac{\partial u_r}{\partial r} - \frac{2}{3}\mu^{k-1}\left(\frac{u_r}{r}+\frac{\partial u_z}{\partial z}\right), \tag{21.29}$$

$$\phi_{\theta\theta} = \left(\frac{4}{3}\mu^{k-1}+r_{AL}\right)\frac{u_r}{r} - \frac{2}{3}\mu^{k-1}\left(\frac{\partial u_r}{\partial r}+\frac{\partial u_z}{\partial z}\right), \tag{21.30}$$

$$\phi_{zz} = \left(\frac{4}{3}\mu^{k-1}+r_{AL}\right)\frac{\partial u_z}{\partial z} - \frac{2}{3}\mu^{k-1}\left(\frac{u_r}{r}+\frac{\partial u_r}{\partial r}\right), \tag{21.31}$$

$$\phi_{rr} = \left(\mu^{k-1}+\frac{1}{2}r_{AL}\right)\left(\frac{\partial u_r}{\partial z}+\frac{\partial u_z}{\partial r}\right). \tag{21.32}$$

Finally, we introduce the following column-vectors:

$$\boldsymbol{\phi}_r = \begin{pmatrix}\phi_{rr}\\ 0\\ \phi_{rz}\end{pmatrix}, \boldsymbol{\phi}_z = \begin{pmatrix}\phi_{rz}\\ 0\\ \phi_{zz}\end{pmatrix} \tag{21.33}$$

and

$$\mathcal{T}_r = \begin{pmatrix}\mathcal{T}_{rr}\\ 0\\ \mathcal{T}_{rz}\end{pmatrix}, \mathcal{T}_z = \begin{pmatrix}\mathcal{T}_{rz}\\ 0\\ \mathcal{T}_{zz}\end{pmatrix} \tag{21.34}$$

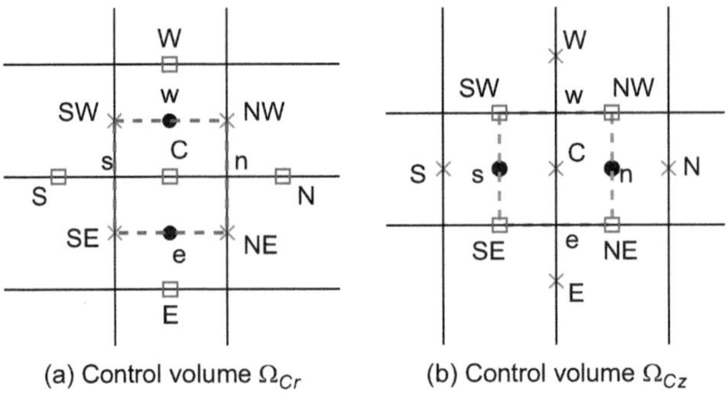

FIG. 21.4 Control volumes Ω_{Cr} (a) and Ω_{Cz} (b) for the momentum equation.

If one projects the momentum equation on the r-axis and z-axis, and integrates the resulting equations on the control volumes Ω_{Cr} and Ω_{Cz} shown in Fig. 21.4, one obtains, from (21.27)–(21.34), the following variants of (21.26):

$$\int_{\Omega_{Cr}} \frac{\rho}{\Delta t} u_r r dr d\theta + \int_{\Omega_{Cr}} \frac{\partial p}{\partial r} r dr d\theta - \int_{\Omega_{Cr}} \left(\nabla\cdot\boldsymbol{\phi}_r - \frac{\phi_{\theta\theta}}{r}\right) r dr d\theta = \int_{\Omega_{Cr}} \left(\nabla\cdot\boldsymbol{\mathcal{T}}_r - \frac{\mathcal{T}_{\theta\theta}}{r}\right) r dr d\theta + \int_{\Omega_{Cr}} \frac{\rho}{\Delta t} u_r^{k-1} r dr d\theta - \int_{\Omega_{Cr}} \rho\mathbf{u}^{k-1}\cdot\nabla u_r^{k-1} r dr d\theta \tag{21.35}$$

and

$$\int_{\Omega_{Cz}} \frac{\rho}{\Delta t} u_z r dr d\theta + \int_{\Omega_{Cz}} \frac{\partial p}{\partial z} r dr d\theta - \int_{\Omega_{Cz}} \nabla\cdot\boldsymbol{\phi}_z r dr d\theta = \int_{\Omega_{Cz}} \nabla\cdot\boldsymbol{\mathcal{T}}_z r dr d\theta + \int_{\Omega_{Cz}} \frac{\rho}{\Delta t} u_z^{k-1} r dr d\theta - \int_{\Omega_{Cz}} \rho\mathbf{u}^{k-1}\cdot\nabla u_z^{k-1} r dr d\theta. \tag{21.36}$$

First, let us introduce the fluxes $\mathbf{F}_r^{k-1} = u_r^{k-1}\mathbf{u}^{k-1}$ and $\mathbf{F}_z^{k-1} = u_z^{k-1}\mathbf{u}^{k-1}$; next, we approximate $\int_{\Omega_{Cr}} \rho\mathbf{u}^{k-1}\cdot\nabla u_r^{k-1} r dr d\theta$ in (21.35) (resp., $\int_{\Omega_{Cz}} \rho\mathbf{u}^{k-1}\cdot\nabla u_z^{k-1} r dr d\theta$, in (21.36)) by $\rho_C \int_{\Omega_{Cr}} \mathbf{u}^{k-1}\cdot\nabla u_r^{k-1} r dr d\theta$ (resp., $\rho_C \int_{\Omega_{Cz}} \mathbf{u}^{k-1}\cdot\nabla u_z^{k-1} r dr d\theta$,) with $\rho_C = \frac{1}{2}(\rho_e + \rho_w)$ (resp. $\rho_C = \frac{1}{2}(\rho_n + \rho_s)$); it follows then from (21.35), (21.36), and from the divergence theorem that

$$\int_{\Omega_{Cr}} \frac{\rho}{\Delta t} u_r r dr d\theta + \int_{\Omega_{Cr}} \frac{\partial p}{\partial r} r dr d\theta - \int_{\Omega_{Cr}} \boldsymbol{\phi}_r\cdot\mathbf{n} d(\partial\Omega_{Cr}) + \int_{\Omega_{Cr}} \frac{\phi_{\theta\theta}}{r} r dr d\theta$$

$$= \int\limits_{\Omega_{Cr}} \mathcal{T}_r \cdot \mathbf{n} d(\partial\Omega_{Cr}) - \int\limits_{\Omega_{Cr}} \frac{\mathcal{T}_{\theta\theta}}{r} r dr d\theta + \int\limits_{\Omega_{Cr}} \frac{\rho}{\Delta t} u_r^{k-1} r dr d\theta$$
$$- \rho_C \int\limits_{\Omega_{Cr}} \mathbf{F}_r^{k-1} \cdot \mathbf{n} d(\partial\Omega_{Cr}) + \rho_C \int\limits_{\Omega_{Cr}} u_r^{k-1} \nabla \cdot \mathbf{u}^{k-1} r dr d\theta, \tag{21.37}$$

$$\int\limits_{\Omega_{Cz}} \frac{\rho}{\Delta t} u_z r dr d\theta + \int\limits_{\Omega_{Cz}} \frac{\partial p}{\partial z} r dr d\theta - \int\limits_{\Omega_{Cz}} \boldsymbol{\phi}_z \cdot \mathbf{n} d(\partial\Omega_{Cz})$$
$$= \int\limits_{\Omega_{Cz}} \mathcal{T}_z \cdot \mathbf{n} d(\partial\Omega_{Cz}) + \int\limits_{\Omega_{Cz}} \frac{\rho}{\Delta t} u_z^{k-1} r dr d\theta$$
$$- \rho_C \int\limits_{\Omega_{Cz}} \mathbf{F}_z^{k-1} \cdot \mathbf{n} d(\partial\Omega_{Cz}) + \rho_C \int\limits_{\Omega_{Cz}} u_z^{k-1} \nabla \cdot \mathbf{u}^{k-1} r dr d\theta. \tag{21.38}$$

In (21.37) and (21.38), the fluxes are computed at the centers of the faces of the control volumes Ω_{Cr} and Ω_{Cz}, the notation in Fig. 21.4 being as in Fig. 21.1. After space discretization, Eqn (21.37) becomes

$$\frac{1}{2\Delta t}(\rho_e + \rho_w) u_{rC} + \frac{p_w - p_e}{\Delta r_C} - \frac{r_w \phi_{rrw} - r_e \phi_{rre}}{r_C \Delta r_C} - \frac{\phi_{rzn} - \phi_{rzs}}{\Delta z_C} + \frac{\phi_{\theta\theta C}}{r_C}$$
$$= \frac{r_w \mathcal{T}_{rrw} - r_e \mathcal{T}_{rre}}{r_C \Delta r_C} + \frac{\mathcal{T}_{rzn} - \mathcal{T}_{rzs}}{\Delta z_C} - \frac{\mathcal{T}_{\theta\theta C}}{r_C} + \frac{1}{2\Delta t}(\rho_e + \rho_w) u_{rC}^{k-1}$$
$$- \frac{1}{2}(\rho_e + \rho_w) \frac{r_w F_{rw}^{k-1} - r_e F_{re}^{k-1}}{r_C \Delta r_C} - \frac{1}{2}(\rho_e + \rho_w) \frac{F_{rn}^{k-1} - F_{rs}^{k-1}}{\Delta z_C}$$
$$+ \frac{1}{2}(\rho_e + \rho_w) u_{rC}^{k-1} [\nabla \cdot \mathbf{u}^{k-1}]_C, \tag{21.39}$$

with $\Delta r_C = r_w - r_e$, $\Delta z_C = z_n - z_s$, and $[\nabla \cdot \mathbf{u}^{k-1}]_C = \left(\frac{\partial u_r^{k-1}}{\partial r_C}\right)_C + \frac{u_r^{k-1}}{r_C} + \left(\frac{\partial u_z^{k-1}}{\partial z}\right)_C$.

Replacing, in (21.39), the component of the tensor $\boldsymbol{\Phi}$ by their actual values (see (21.29)–(21.32)), we obtain

$$\frac{1}{2\Delta t}(\rho_e + \rho_w) u_{rC} + \frac{p_w - p_e}{\Delta r_C} + \left(\frac{4}{3}\mu_C^{k-1} + r_{AL}\right) \frac{u_{rC}}{r_C^2}$$
$$- \frac{2}{3 r_C} \mu_C^{k-1} \left(\frac{\partial u_r}{\partial r} + \frac{\partial u_z}{\partial z}\right)_C - \frac{1}{r_C \Delta r_C} \left(r_w \left[\left(\frac{4}{3}\mu_w^{k-1} + r_{AL}\right) \left(\frac{\partial u_r}{\partial r}\right)_w \right.\right.$$
$$\left. - \frac{2}{3}\mu_w^{k-1} \left(\frac{u_{rw}}{r_w} + \left(\frac{\partial u_z}{\partial z}\right)_w \right) \right] - r_e \left[\left(\frac{4}{3}\mu_w^{k-1} + r_{AL}\right) \left(\frac{\partial u_r}{\partial r}\right)_e \right.$$
$$\left.\left. - \frac{2}{3}\mu_e^{k-1} \left(\frac{u_{re}}{r_e} + \left(\frac{\partial u_z}{\partial z}\right)_e \right) \right] \right) - \frac{1}{\Delta z_C} \left(\left(\mu_n^{k-1} + \frac{1}{2} r_{AL}\right) \left(\frac{\partial u_r}{\partial z} + \frac{\partial u_z}{\partial r}\right)_n \right.$$
$$\left. - \left(\mu_s^{k-1} + \frac{1}{2} r_{AL}\right) \left(\frac{\partial u_r}{\partial z} + \frac{\partial u_z}{\partial r}\right)_s \right)$$

$$
\begin{aligned}
&= \frac{r_w \mathcal{T}_{rrw} - r_e \mathcal{T}_{rre}}{r_C \Delta r_C} + \frac{\mathcal{T}_{rzn} - \mathcal{T}_{rzs}}{\Delta z_C} - \frac{\mathcal{T}_{\theta\theta C}}{r_C} + \frac{1}{2\Delta t}(\rho_e + \rho_w) u_{rC}^{k-1} \\
&\quad - \frac{1}{2}(\rho_e + \rho_w) \frac{r_w F_{rw}^{k-1} - r_e F_{re}^{k-1}}{r_C \Delta r_C} - \frac{1}{2}(\rho_e + \rho_w) \frac{F_{rn}^{k-1} - F_{rs}^{k-1}}{\Delta z_C} \\
&\quad + \frac{1}{2}(\rho_e + \rho_w) u_{rC}^{k-1} [\nabla \cdot \mathbf{u}^{k-1}]_C. \qquad (21.40)
\end{aligned}
$$

In (21.40), the velocities and their derivatives are evaluated as follows (the notation is as in Fig. 21.4 (a)):

$$
\begin{aligned}
&\left(\frac{\partial u_r}{\partial r}\right)_w = \frac{u_{rW} - u_{rC}}{r_W - r_C}, \ u_{rw} = \frac{1}{2}(u_{rW} + u_{rC}), \ \left(\frac{\partial u_z}{\partial z}\right)_w = \frac{u_{zNW} - u_{zSW}}{z_n - z_s}, \\
&\left(\frac{\partial u_r}{\partial r}\right)_e = \frac{u_{rC} - u_{rE}}{r_C - r_E}, \ u_{re} = \frac{1}{2}(u_{rC} + u_{rE}), \ \left(\frac{\partial u_z}{\partial z}\right)_e = \frac{u_{zNE} - u_{zSE}}{z_n - z_s}, \\
&\left(\frac{\partial u_r}{\partial z}\right)_n = \frac{u_{rN} - u_{rC}}{z_N - z_C}, \ \left(\frac{\partial u_r}{\partial z}\right)_s = \frac{u_{rC} - u_{rS}}{z_C - z_S}, \\
&\left(\frac{\partial u_z}{\partial r}\right)_n = \frac{u_{zNW} - u_{zNE}}{r_w - r_e}, \ \left(\frac{\partial u_z}{\partial r}\right)_s = \frac{u_{zSW} - u_{rSE}}{r_w - r_e}, \\
&\left(\frac{\partial u_r}{\partial r}\right)_C = \frac{u_{rW} - u_{rE}}{r_W - r_E}, \ \left(\frac{\partial u_z}{\partial z}\right)_C = \frac{(u_{zNW} + u_{zNE}) - (u_{zSW} + u_{zSE})}{2(z_n - z_s)}.
\end{aligned}
$$

Assuming that, in (21.40), the fluxes have been computed using a *TVD Lax–Wendroff scheme* with a *Superbee slope limiter*, we have the following:

$$
\begin{aligned}
F_{rw}^{k-1} &= \max\left(0, u_{rw}^{k-1}\right)\left[u_{rC}^{k-1} + \frac{1}{2}\phi_{rw}^{k-1}\left(1 - \frac{\Delta t}{\Delta r_C} u_{rw}^{k-1}\right)\left(u_{rW}^{k-1} - u_{rC}^{k-1}\right)\right] \\
&\quad + \min\left(0, u_{rw}^{k-1}\right)\left[u_{rW}^{k-1} - \frac{1}{2}\phi_{rw}^{k-1}\left(1 + \frac{\Delta t}{\Delta r_C} u_{rw}^{k-1}\right)\left(u_{rW}^{k-1} - u_{rC}^{k-1}\right)\right], \qquad (21.41) \\
F_{re}^{k-1} &= \max\left(0, u_{re}^{k-1}\right)\left[u_{rE}^{k-1} + \frac{1}{2}\phi_{re}^{k-1}\left(1 - \frac{\Delta t}{\Delta r_C} u_{re}^{k-1}\right)\left(u_{rC}^{k-1} - u_{rE}^{k-1}\right)\right] \\
&\quad + \min\left(0, u_{re}^{k-1}\right)\left[u_{rC}^{k-1} - \frac{1}{2}\phi_{re}^{k-1}\left(1 + \frac{\Delta t}{\Delta r_C} u_{re}^{k-1}\right)\left(u_{rC}^{k-1} - u_{rE}^{k-1}\right)\right], \qquad (21.42) \\
F_{zn}^{k-1} &= \max\left(0, u_{zn}^{k-1}\right)\left[u_{zC}^{k-1} + \frac{1}{2}\phi_{zn}^{k-1}\left(1 - \frac{\Delta t}{\Delta z_C} u_{zn}^{k-1}\right)\left(u_{zN}^{k-1} - u_{zC}^{k-1}\right)\right] \\
&\quad + \min\left(0, u_{zn}^{k-1}\right)\left[u_{zN}^{k-1} - \frac{1}{2}\phi_{zn}^{k-1}\left(1 + \frac{\Delta t}{\Delta z_C} u_{zn}^{k-1}\right)\left(u_{zN}^{k-1} - u_{zC}^{k-1}\right)\right], \qquad (21.43) \\
F_{zs}^{k-1} &= \max\left(0, u_{zs}^{k-1}\right)\left[u_{zS}^{k-1} + \frac{1}{2}\phi_{zs}^{k-1}\left(1 - \frac{\Delta t}{\Delta z_C} u_{zs}^{k-1}\right)\left(u_{zC}^{k-1} - u_{zS}^{k-1}\right)\right] \\
&\quad + \min\left(0, u_{zs}^{k-1}\right)\left[u_{zC}^{k-1} - \frac{1}{2}\phi_{zs}^{k-1}\left(1 + \frac{\Delta t}{\Delta z_C} u_{zs}^{k-1}\right)\left(u_{zC}^{k-1} - u_{zS}^{k-1}\right)\right], \qquad (21.44)
\end{aligned}
$$

with

$$u_{zn} = \frac{1}{2}(u_{zNW} + u_{zNE}), \quad u_{zs} = \frac{1}{2}(u_{zSW} + u_{sSE}),$$

$$\phi_{rw}^{k-1} = \phi\left(\frac{u_{rC}^{k-1} - u_{rE}^{k-1}}{u_{rW}^{k-1} - u_{rC}^{k-1}}\right),$$

$$\phi_{re}^{k-1} = \phi\left(\frac{u_{rE}^{k-1} - u_{rEE}^{k-1}}{u_{rC}^{k-1} - u_{rE}^{k-1}}\right), \quad \phi_{zn}^{k-1} = \phi\left(\frac{u_{zC}^{k-1} - u_{zS}^{k-1}}{u_{zN}^{k-1} - u_{sC}^{k-1}}\right),$$

$$\phi_{zs}^{k-1} = \phi\left(\frac{u_{zS}^{k-1} - u_{zSS}^{k-1}}{u_{rC}^{k-1} - u_{rS}^{k-1}}\right),$$

and $\phi(\xi) = \max[0, \min(2\xi, 1), \min(\xi, 2)]$.

In a similar fashion, we can derive the fully discrete analog of the z-momentum equation (21.38); this is left as an exercise to the reader (see VINAY [2005, chapter 2] for details).

21.5. Finite-volume discretization of the energy equation

Because a dimensional analysis performed in VINAY [2005, chapter 2] shows that the viscous dissipation is quite small, the term $\boldsymbol{\tau} : \mathbf{D}(\mathbf{u})$ has been dropped out from the energy equation (19.7); the above equation reduces then to

$$\rho C_p \left(\frac{\partial \Theta}{\partial t} + \mathbf{u} \cdot \nabla \Theta\right) = \lambda_f \nabla^2 \Theta. \tag{21.45}$$

Integrating the energy equation (21.45) over the C-centered control volume Ω_C (see Fig. 21.5), we obtain

$$\int_{\Omega_C} \rho C_p \left(\frac{\partial \Theta}{\partial t} + \mathbf{u} \cdot \nabla \Theta\right) r dr d\theta = \int_{\Omega_C} \lambda_f \nabla^2 \Theta r dr d\theta. \tag{21.46}$$

Let us introduce the flux $\mathbf{F} = \Theta \mathbf{u}$. Approximating ρ by ρ_C on Ω_C, it follows from the relations $\mathbf{u} \cdot \nabla \Theta = -\Theta \nabla \cdot \mathbf{u} + \nabla \cdot \mathbf{F}$ and $\nabla^2 \Theta = \nabla \cdot \nabla \Theta$, and from the divergence theorem,

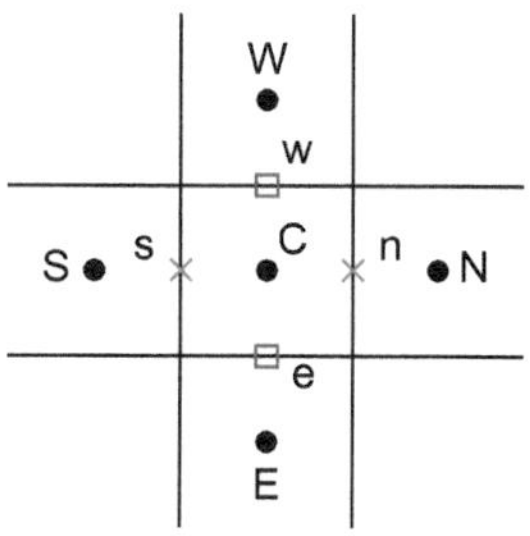

FIG. 21.5 A C-centered control volume Ω_C for the energy equation.

that (21.46) implies

$$\int_{\Omega_C} \frac{\partial \Theta}{\partial t} r\mathrm{d}r\mathrm{d}\theta + \int_{\partial\Omega_C} \mathbf{F}\cdot\mathbf{n}\mathrm{d}(\partial\Omega_C) = \frac{\lambda_f}{\rho_C C_p} \int_{\partial\Omega_C} \frac{\partial\Theta}{\partial n}\mathrm{d}(\partial\Omega_C) + \int_{\Omega_C} \Theta\nabla\cdot\mathbf{u}r\mathrm{d}r\mathrm{d}\theta. \tag{21.47}$$

After computing on each face of the control volume Ω_C the fluxes in (21.47), we obtain

$$\frac{\Theta_C^k - \Theta_C^{k-1}}{\Delta t} + \frac{r_w F_w^{k-1} - r_e F_e^{k-1}}{r_C \Delta r_C} + \frac{F_n^{k-1} - F_s^{k-1}}{\Delta z_C} = \frac{\lambda_f}{\rho_C^k C_p}\left[\frac{1}{r_C\Delta r_C}\left(r_w\left(\frac{\partial\Theta}{\partial r}\right)_w^k - r_e\left(\frac{\partial\Theta}{\partial r}\right)_e^k\right) + \frac{1}{\Delta z_C}\left(\left(\frac{\partial\Theta}{\partial z}\right)_n^k - \left(\frac{\partial\Theta}{\partial z}\right)_s^k\right)\right] + \Theta_C^{k-1}[\nabla\cdot\mathbf{u}^{k-1}]_C, \tag{21.48}$$

with $\Theta_C^{k-1}[\nabla\cdot\mathbf{u}^{k-1}]_C$ still given by (21.8), while the temperature derivatives are given by

$$\left(\frac{\partial\Theta}{\partial r}\right)_w^k = \frac{\Theta_W^k - \Theta_C^k}{\Delta r_w}, \quad \left(\frac{\partial\Theta}{\partial r}\right)_e^k = \frac{\Theta_C^k - \Theta_E^k}{\Delta r_e},$$
$$\left(\frac{\partial\Theta}{\partial z}\right)_n^k = \frac{\Theta_N^k - \Theta_C^k}{\Delta z_n}, \quad \left(\frac{\partial\Theta}{\partial z}\right)_s^k = \frac{\Theta_C^k - \Theta_S^k}{\Delta z_s}, \tag{21.49}$$

with, in (21.48) and (21.49), $\Delta r_C = r_w - r_e$, $\Delta z_C = z_n - z_s$, $\Delta r_w = r_W - r_C$, $\Delta r_e = r_C - r_E$, $\Delta z_n = z_N - z_C$, and $\Delta z_s = z_C - z_S$. The four components of the flux $\mathbf{F}^{k-1}$ have been obtained using, as in Sections 21.2–21.4, a Lax–Wendroff TVD scheme with a Superbee slope limiter, that is

$$F_w^{k-1} = \max\left(0, u_w^k\right)\left[\Theta_C^{k-1} + \frac{1}{2}\phi_w^{k-1}\left(1 - \frac{\Delta t}{\Delta r_C}u_w^k\right)\left(\Theta_W^{k-1} - \Theta_C^{k-1}\right)\right] + \min\left(0, u_w^k\right)\left[\Theta_W^{k-1} - \frac{1}{2}\phi_w^{k-1}\left(1 + \frac{\Delta t}{\Delta r_C}u_w^k\right)\left(\Theta_W^{k-1} - \Theta_C^{k-1}\right)\right], \tag{21.50}$$

$$F_e^{k-1} = \max\left(0, u_e^k\right)\left[\Theta_E^{k-1} + \frac{1}{2}\phi_e^{k-1}\left(1 - \frac{\Delta t}{\Delta r_C}u_e^k\right)\left(\Theta_C^{k-1} - \Theta_E^{k-1}\right)\right] + \min\left(0, u_e^k\right)\left[\Theta_C^{k-1} - \frac{1}{2}\phi_e^{k-1}\left(1 + \frac{\Delta t}{\Delta r_C}u_e^k\right)\left(\Theta_C^{k-1} - \Theta_E^{k-1}\right)\right], \tag{21.51}$$

$$F_n^{k-1} = \max\left(0, u_n^k\right)\left[\Theta_C^{k-1} + \frac{1}{2}\phi_n^{k-1}\left(1 - \frac{\Delta t}{\Delta z_C}u_n^k\right)\left(\Theta_N^{k-1} - \Theta_C^{k-1}\right)\right] + \min\left(0, u_n^k\right)\left[\Theta_N^{k-1} - \frac{1}{2}\phi_n^{k-1}\left(1 + \frac{\Delta t}{\Delta z_C}u_n^k\right)\left(\Theta_N^{k-1} - \Theta_C^{k-1}\right)\right], \tag{21.52}$$

$$F_s^{k-1} = \max\left(0, u_s^k\right)\left[\Theta_S^{k-1} + \frac{1}{2}\phi_s^{k-1}\left(1 - \frac{\Delta t}{\Delta z_C}u_s^k\right)\left(\Theta_C^{k-1} - \Theta_S^{k-1}\right)\right]$$
$$+ \min\left(0, u_s^k\right)\left[\Theta_C^{k-1} - \frac{1}{2}\phi_s^{k-1}\left(1 + \frac{\Delta t}{\Delta z_C}u_s^k\right)\left(\Theta_C^{k-1} - \Theta_S^{k-1}\right)\right], \quad (21.53)$$

with

$$\phi_w^{k-1} = \phi\left(\frac{\Theta_C^{k-1} - \Theta_E^{k-1}}{\Theta_W^{k-1} - \Theta_C^{k-1}}\right), \quad \phi_e^{k-1} = \phi\left(\frac{\Theta_E^{k-1} - \Theta_{EE}^{k-1}}{\Theta_C^{k-1} - \Theta_E^{k-1}}\right),$$
$$\phi_n^{k-1} = \phi\left(\frac{\Theta_C^{k-1} - \Theta_S^{k-1}}{\Theta_N^{k-1} - \Theta_C^{k-1}}\right), \quad \phi_s^{k-1} = \phi\left(\frac{\Theta_S^{k-1} - \Theta_{SS}^{k-1}}{\Theta_C^{k-1} - \Theta_S^{k-1}}\right),$$

and $\phi(\xi) = \max[0,\ \min(1,\ 2\xi),\ \min(2,\ \xi)]$.

21.6. Finite-volume discretization of the structure parameter equation

The integral form of the *structure parameter equation* (19.12), for the control volume Ω_C in Fig. 21.2, reads as follows:

$$\int_{\Omega_C}\left(\frac{\partial\lambda_s}{\partial t} + \mathbf{u}\cdot\nabla\lambda_s\right) r dr d\theta = \int_{\Omega_C}\left[a(1-\lambda_s) - b\lambda_s\dot{\gamma}^m\right] r dr d\theta. \quad (21.54)$$

Let us introduce $\mathbf{F} = \lambda_s\mathbf{u}$; it follows then from the divergence theorem that

$$\int_{\Omega_C}\frac{\partial\lambda_s}{\partial t} r dr d\theta + \int_{\Omega_C}\mathbf{F}\cdot\mathbf{n}d(\partial\Omega_C) =$$
$$\int_{\Omega_C}\left[a(1-\lambda_s) - b\lambda_s\dot{\gamma}^m\right] r dr d\theta + \int_{\Omega_C}\lambda_S\nabla\cdot\mathbf{u} r dr d\theta \quad (21.55)$$

Computing the fluxes in (21.55) on the faces of the control volume Ω_C, we obtain

$$\frac{\lambda_{sC}^k - \lambda_{sC}^{k-1}}{\Delta t} + \frac{r_w F_w^{k-1} - r_e F_e^{k-1}}{r_C\Delta r_C} + \frac{F_n^{k-1} - F_s^{k-1}}{\Delta z_C} =$$
$$a - \left[a + b(\dot{\gamma}_c^{k-1})^m\right]\lambda_{sC}^{k-1} + \lambda_{sC}^{k-1}[\nabla\cdot\mathbf{u}^k]_C \quad (21.56)$$

where $[\nabla\cdot\mathbf{u}^k]_C$ is still given by (21.8) and the four components of the flux $\mathbf{F}^{k-1}$ by

$$F_w^{k-1} = \max\left(0, u_w^k\right)\left[\lambda_{sC}^{k-1} + \frac{1}{2}\phi_w^{k-1}\left(1 - \frac{\Delta t}{\Delta r_C}u_w^k\right)\left(\lambda_{sW}^{k-1} - \lambda_{sC}^{k-1}\right)\right]$$
$$+ \min\left(0, u_w^k\right)\left[\lambda_{sW}^{k-1} - \frac{1}{2}\phi_w^{k-1}\left(1 + \frac{\Delta t}{\Delta r_C}u_w^k\right)\left(\lambda_{sW}^{k-1} - \lambda_{sC}^{k-1}\right)\right], \quad (21.57)$$

$$F_e^{k-1} = \max\left(0, u_e^k\right)\left[\lambda_{sE}^{k-1} + \frac{1}{2}\phi_e^{k-1}\left(1 - \frac{\Delta t}{\Delta r_C}u_e^k\right)\left(\lambda_{sC}^{k-1} - \lambda_{sE}^{k-1}\right)\right]$$
$$+ \min\left(0, u_e^k\right)\left[\lambda_{sC}^{k-1} - \frac{1}{2}\phi_e^{k-1}\left(1 + \frac{\Delta t}{\Delta r_C}u_e^k\right)\left(\lambda_{sC}^{k-1} - \lambda_{sE}^{k-1}\right)\right], \tag{21.58}$$

$$F_n^{k-1} = \max\left(0, u_n^k\right)\left[\lambda_{sC}^{k-1} + \frac{1}{2}\phi_n^{k-1}\left(1 - \frac{\Delta t}{\Delta z_C}u_n^k\right)\left(\lambda_{sN}^{k-1} - \lambda_{sC}^{k-1}\right)\right]$$
$$+ \min\left(0, u_n^k\right)\left[\lambda_{sN}^{k-1} - \frac{1}{2}\phi_n^{k-1}\left(1 + \frac{\Delta t}{\Delta z_C}u_n^k\right)\left(\lambda_{sN}^{k-1} - \lambda_{sC}^{k-1}\right)\right], \tag{21.59}$$

$$F_s^{k-1} = \max\left(0, u_s^k\right)\left[\lambda_{sS}^{k-1} + \frac{1}{2}\phi_s^{k-1}\left(1 - \frac{\Delta t}{\Delta z_C}u_s^k\right)\left(\lambda_{sC}^{k-1} - \lambda_{sS}^{k-1}\right)\right]$$
$$+ \min\left(0, u_s^k\right)\left[\lambda_{sC}^{k-1} - \frac{1}{2}\phi_s^{k-1}\left(1 + \frac{\Delta t}{\Delta z_C}u_s^k\right)\left(\lambda_{sC}^{k-1} - \lambda_{sS}^{k-1}\right)\right], \tag{21.60}$$

with

$$\phi_w^{k-1} = \phi\left(\frac{\lambda_{sC}^{k-1} - \lambda_{sE}^{k-1}}{\lambda_{sW}^{k-1} - \lambda_{sC}^{k-1}}\right), \quad \phi_e^{k-1} = \phi\left(\frac{\lambda_{sE}^{k-1} - \lambda_{sEE}^{k-1}}{\lambda_{sC}^{k-1} - \lambda_{sE}^{k-1}}\right),$$
$$\phi_n^{k-1} = \phi\left(\frac{\lambda_{sC}^{k-1} - \lambda_{sS}^{k-1}}{\lambda_{sN}^{k-1} - \lambda_{sC}^{k-1}}\right), \quad \phi_s^{k-1} = \phi\left(\frac{\lambda_{sS}^{k-1} - \lambda_{sSS}^{k-1}}{\lambda_{sC}^{k-1} - \lambda_{sS}^{k-1}}\right),$$

and $\phi(\xi) = \max[0, \min(1, 2\xi), \min(2, \xi)]$.

21.7. Evaluation of the strain-rate-related tensors

In this paragraph, we describe how the strain-rate tensor $\mathbf{p}$ is evaluated at the step (20.41) of the solution algorithm (20.40)–(20.44) (see Section 20.3). As mentioned in Section 21.1, and visualized in Fig. 21.1: (1) the p_{rr} and p_{zz} components are evaluated at the cell centers, (2) the $p_{\theta\theta}$ components are evaluated at the cell faces, and (3) the p_{rz} components are evaluated at the grid nodes. Actually, the computation of $\mathbf{p}$ does not require the integration of the equation in (20.41) over the control volumes; instead, each component p_{ij} is evaluated at the corresponding mesh location. For each component p_{ij}, the corresponding components λ_{ij} and $D_{ij}(\mathbf{u})$ of the tensors components $\boldsymbol{\lambda}$ and $\mathbf{D}(\mathbf{u})$ are computed at the same mesh location (see Fig. 21.1), implying that the term $\lambda_{ij} + r_{AL}D_{ij}(\mathbf{u})$ is easy to evaluate. However, the computation of $\|\boldsymbol{\lambda} + r_{AL}\mathbf{D}(\mathbf{u})\|$ is more delicate because some components have to be obtained by interpolation; this will be illustrated just below by the evaluation of p_{rr} at the center C of the control volume Ω_C in Figs. 21.1 and 21.6. Because p_{rr} is attached to C, all the components of $\boldsymbol{\lambda}$ and $\mathbf{D}(\mathbf{u})$ must also be evaluated at C. We have thus

$$p_{rrC} = \frac{1}{r_{AL}}\left[1 - \frac{\tau_y(\Theta_C)}{\|\boldsymbol{\lambda} + r_{AL}\mathbf{D}(\mathbf{u})\|_{rrC}}\right]^+ [\lambda_{rrC} + r_{AL}D_{rrC}(\mathbf{u})], \tag{21.61}$$

with

$$\|\boldsymbol{\lambda} + r_{AL}\mathbf{D}(\mathbf{u})\|_{rrC}$$
$$= \sqrt{\frac{[\lambda_{rrC}+r_{AL}D_{rrC}(\mathbf{u})]^2+[\lambda_{\theta\theta C}+r_{AL}D_{\theta\theta C}(\mathbf{u})]^2+\left[\lambda_{zzC}+r_{AL}D_{zzC}(\mathbf{u})\right]^2+2\left[\lambda_{rzC}+r_{AL}D_{rzC}(\mathbf{u})\right]^2}{2}}$$

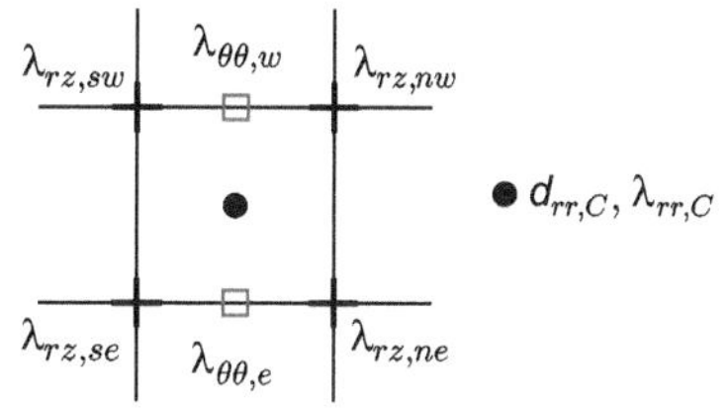

FIG. 21.6 Location of the Lagrange multipliers $\lambda_{\theta\theta}$ and λ_{rz} for the computation of p_{rr}.

where, as visualized in Fig. 21.6, we have

$$\lambda_{\theta\theta C} = \frac{1}{2}(\lambda_{\theta\theta w} + \lambda_{\theta\theta e}), \quad \lambda_{rzC} = \frac{1}{4}(\lambda_{rzsw} + \lambda_{rzse} + \lambda_{rznw} + \lambda_{rzne}).$$

Using a similar interpolation-based approach, one can compute the other three components of the tensor $\mathbf{p}$, namely p_{zz}, p_{rz}, and $p_{\theta\theta}$.

22. Solution of the linear systems

22.1. Solution of the generalized Stokes problems

The matrix form of the compressible Stokes problem (20.42), (20.43) reads as follows:

$$\begin{pmatrix} \mathbf{A} & \mathbf{B} \\ -\mathbf{B}^t & \mathbf{C} \end{pmatrix} \begin{pmatrix} \mathbf{u} \\ \mathbf{p} \end{pmatrix} = \begin{pmatrix} \mathbf{f} \\ \mathbf{g} \end{pmatrix}, \tag{22.1}$$

where $\mathbf{A}$ is an $N \times N$ symmetric and positive definite matrix, $\mathbf{B}$ a $M \times N$ matrix, $\mathbf{C}$ a $M \times M$ positive diagonal matrix, $\mathbf{u}$ and $\mathbf{f}$ belong to $\mathbb{R}^N$, and $\mathbf{p}$ and $\mathbf{g}$ belong to $\mathbb{R}^M$. The dimensionless form of the linear system (22.1) reads as

$$\begin{pmatrix} \dfrac{\mathbf{A}}{\bar{\mu} L_c} & \dfrac{\mathbf{B}}{L_c^2} \\ -\dfrac{\mathbf{B}^t}{L_c^2} & \dfrac{\mathbf{C}}{L_c^3/\bar{\mu}} \end{pmatrix} \begin{pmatrix} \dfrac{\mathbf{u}}{\bar{U}} \\ \dfrac{\mathbf{p}}{\bar{\mu}\bar{U}/L_c} \end{pmatrix} = \begin{pmatrix} \dfrac{\mathbf{f}}{\bar{\mu}\bar{U}L_c} \\ \dfrac{\mathbf{g}}{\bar{U}L_c^2} \end{pmatrix}, \tag{22.2}$$

where $\bar{\mu} = \mu + r + \rho_0 \bar{U} L_c$ is an augmented viscosity coefficient (fluid viscosity coefficient + augmented Lagrangian coefficient + unsteady term coefficient). Finally, with $\bar{\mathbf{X}}$ denoting a dimensionless matrix or vector, we obtain from (22.2)

$$\begin{pmatrix} \bar{\mathbf{A}} & \bar{\mathbf{B}} \\ -\bar{\mathbf{B}}^t & \bar{\mathbf{C}} \end{pmatrix} \begin{pmatrix} \bar{\mathbf{u}} \\ \bar{\mathbf{p}} \end{pmatrix} = \begin{pmatrix} \bar{\mathbf{f}} \\ \bar{\mathbf{g}} \end{pmatrix}. \tag{22.3}$$

In order to solve the linear system (22.3), we advocate a simple variant of the Uzawa/conjugate gradient algorithm discussed in FORTIN and GLOWINSKI [1982, 1983],

GLOWINSKI and LE TALLEC [1989], for the solution of linear systems such as (22.1), when matrix $\mathbf{C} = \mathbf{0}$. This modified algorithm (denoted by MSA) reads as follows:

- **Initialization:** $i = 0$
  - *Compute* $\mathbf{r}^0 = \mathbf{\Lambda}\bar{\mathbf{p}}^0 - \mathbf{b}$ *where* $\mathbf{\Lambda} = \bar{\mathbf{B}}^t(\bar{\mathbf{A}})^{-1}\bar{\mathbf{B}} + \bar{\mathbf{C}}$ *and* $\mathbf{b} = \bar{\mathbf{B}}^t(\bar{\mathbf{A}})^{-1}\bar{\mathbf{f}} + \bar{\mathbf{g}}$
    - *Solve* $\bar{\mathbf{A}}\bar{\mathbf{u}}^0 = \bar{\mathbf{f}} - \bar{\mathbf{B}}\bar{\mathbf{p}}^0$,
    - *Compute* $\mathbf{r}^0 = -\bar{\mathbf{B}}^t\bar{\mathbf{u}}^0 + \bar{\mathbf{C}}\bar{\mathbf{p}}^0 - \bar{\mathbf{g}}$.
  - *Compute the descent direction* $\mathbf{w}^0 = \mathbf{r}^0$.
- **Iterative process:** $i \geq 1$
  - *Computation of* $\bar{\mathbf{r}}^{i-1} = \mathbf{\Lambda}\mathbf{w}^{i-1}$
    - *Solve* $\bar{\mathbf{A}}\bar{\bar{\mathbf{u}}}^{i-1} = -\bar{\mathbf{B}}\mathbf{w}^{i-1}$,
    - *Compute* $\bar{\mathbf{r}}^{i-1} = -\bar{\mathbf{B}}^t\bar{\bar{\mathbf{u}}}^{i-1} + \bar{\mathbf{C}}\mathbf{w}^{i-1}$.
  - *Compute* $\alpha_{i-1} = \dfrac{|\mathbf{r}^{i-1}|^2}{\bar{\mathbf{r}}^{i-1} \cdot \mathbf{w}^{i-1}}$.
  - *Compute*
    - $\bar{\mathbf{p}}^i = \bar{\mathbf{p}}^{i-1} - \alpha_{i-1}\mathbf{w}^{i-1}$,
    - $\mathbf{r}^i = \mathbf{r}^{i-1} - \alpha_{i-1}\bar{\mathbf{r}}^{i-1}$.
  - *Testing the convergence and updating the descent direction*
    - *If* $|\mathbf{r}^i| \leq tol.|\mathbf{r}^0|$ *take* $\bar{\mathbf{p}} = \bar{\mathbf{p}}^i$ *and compute* $\bar{\mathbf{u}}$ *from (22.3); else*
    - *Compute* $\beta_i = \dfrac{|\mathbf{r}^i|^2}{|\mathbf{r}^{i-1}|^2}$,
    - *Set* $\mathbf{w}^i = \mathbf{r}^i + \beta_i\mathbf{w}^{i-1}$.

REMARK 22.1. The main difference between a classical *discrete incompressible Stokes problem* and the *discrete compressible Stokes problem* (22.1) is associated with matrix $\mathbf{C}$; indeed, we have $\mathbf{C} = \mathbf{0}$ in the incompressible case, while $\mathbf{C}$ is diagonal positive definite in the compressible case. Actually, the MSA algorithm, we described just above, still converges if $\mathbf{C} = \mathbf{0}$ (as shown in, e.g., FORTIN and GLOWINSKI [1982, 1983], GLOWINSKI and LE TALLEC [1989]). In the incompressible case, it has been suggested in the three above-mentioned references (see also GLOWINSKI [2003, chapter 4]) to replace the resulting system

$$\begin{pmatrix} \mathbf{A} & \mathbf{B} \\ -\mathbf{B}^t & \mathbf{0} \end{pmatrix} \begin{pmatrix} \mathbf{u} \\ \mathbf{p} \end{pmatrix} = \begin{pmatrix} \mathbf{f} \\ \mathbf{g} \end{pmatrix}, \tag{22.4}$$

by the following equivalent one

$$\begin{pmatrix} \mathbf{A} + \tilde{r}\mathbf{B}\mathbf{B}^t & \mathbf{B} \\ -\mathbf{B}^t & \mathbf{0} \end{pmatrix} \begin{pmatrix} \mathbf{u} \\ \mathbf{p} \end{pmatrix} = \begin{pmatrix} \mathbf{f} - \tilde{r}\mathbf{B}\mathbf{g} \\ \mathbf{g} \end{pmatrix}, \tag{22.5}$$

where $\tilde{r}$ is a positive parameter associated with an augmented Lagrangian functional (see GLOWINSKI [2003, chapter 4] for details).

REMARK 22.2. Theoretically, the greater $\tilde{r}$ the faster is the convergence of the variant of MSA associated with (22.5). However (see, e.g., WACHS [2000], and GLOWINSKI [2003]), the speed of convergence depends also of the condition number of the matrix $\mathbf{A} + \tilde{r}\mathbf{B}\mathbf{B}^t$. At each iteration of the above algorithm, we have to solve a linear system such as

$$(\mathbf{A} + \tilde{r}\mathbf{B}\mathbf{B}^t)\mathbf{U} = \mathbf{RHS}. \tag{22.6}$$

The matrix $\mathbf{A}$ in (22.6) being symmetric and positive definite, we can solve (22.6) using either a direct method *à la Cholesky* (with $\mathbf{L}$ such that $\mathbf{L}\mathbf{L}^t = \mathbf{A} + \tilde{r}\mathbf{B}\mathbf{B}^t$ computed once for all) or a preconditioned conjugate gradient algorithm. The numerical results presented hereafter have been obtained, using for the solution of (22.6) a conjugate gradient algorithm with SSOR preconditioning; with this approach, because the condition number of matrix $\mathbf{A} + \tilde{r}\mathbf{B}\mathbf{B}^t$ increases with $\tilde{r}$, the larger is $\tilde{r}$ the more expensive is the solution of (22.6). Finally, the speed of convergence of the resulting nested algorithm is a trade-off between the number of outer iterations (which is a decreasing function of $\tilde{r}$) and the number of inner iterations (which is an increasing function of $\tilde{r}$). For the class of problems considered in this chapter, the optimum value of $\tilde{r}$ lies (after scaling) in the range $[10^2, 10^4]$, as shown in WACHS [2000].

22.2. *Solution of the discrete energy and structure parameter equations*

The full discretization of the energy and structure equations (that is (19.7) and (19.12)) leads to the solution at each time step of two linear systems of the following form:

$$\mathbf{M}\mathbf{X}^k = \mathbf{RHS}^{k-1}, \tag{22.7}$$

where $\mathbf{M}$ is a $J \times J$ matrix and where $\mathbf{X}^k$ and $\mathbf{RHS}^{k-1}$ belong both to $\mathbb{R}^J$. In both cases, we have (with obvious notation)

$$\mathbf{RHS}^{k-1} = \mathbf{M}\mathbf{X}^{k-1} - \sum_{i,j} \mathbf{F}_{\text{convection}}(X_{ij}^{k-1}). \tag{22.8}$$

The matrix $\mathbf{M}$ associated with the energy equation (19.7) is symmetric and positive definite; in this chapter, we have used an SSOR preconditioned conjugate gradient algorithm to solve the related linear systems (22.7). The matrix $\mathbf{M}$ associated with the structure equation (19.12) being diagonal positive, the solution of the related systems (22.7) is a trivial operation. Actually, concerning the solution of the linear systems (22.7), a substantial amount of computational time is spent at computing $\mathbf{RHS}^{k-1}$, particularly the flux related part of it (that is $-\sum_{i,j} \mathbf{F}_{\text{convection}}(X_{ij}^{k-1})$).

Because the convection part of the continuity, momentum, energy, and structure equations is treated explicitly, a stability condition is required. The numerical results presented in this chapter have been obtained using as stability condition

$$\text{CFL} = \max_{\Omega_j} \left[\frac{|u_{rj}| + |u_{zj}|}{\min(\Delta r_j, \Delta z_j)} \right] \Delta t < \frac{1}{2}, \tag{22.9}$$

where Δr_j and Δz_j denote the sizes of the control volume Ω_j in the Or and Oz directions, respectively.

23. Numerical experiments: wall-driven cavity creeping flow

23.1. Synopsis

To validate our augmented Lagrangian/finite-volume methodology, the first test problem that we consider is a rather classical one, namely the two-dimensional wall-driven square cavity problem already considered in Chapter 2, Section 17. Our goal in this chapter is to validate our methodology by comparing the numerical results obtained with it to results available in the literature. Indeed, the two-dimensional wall-driven cavity problem has received a broad attention in the literature, first for Newtonian incompressible viscous fluids, and then for non-Newtonian fluids such as Bingham's. In the particular case of viscoplastic material (the case which interests us, here), let us mention, among several others, the contributions of BERCOVIER and ENGELMAN [1980], FORTIN and GLOWINSKI [1982, 1983], GLOWINSKI and LE TALLEC [1989], MITSOULIS and ZISIS [2001], DEAN and GLOWINSKI [2002], VOLA, BOSCARDIN and LATCHÉ [2003], GLOWINSKI [2003] (see also the references therein) and Chapter 2, Section 17, where we tackled this cavity flow problem with a methodology quite different from the one we use in this chapter. Actually, the contribution of MITSOULIS and ZISIS [2001] is a very good candidate for comparison purpose, for the following two main reasons: (1) The above authors are using a regularization/finite-element-based solution method, quite different from the augmented Lagrangian/finite-volume one that we use in this chapter. (2) The above reference contains a thorough discussion of the Bingham number dependence of the steady-state solutions.

The geometry of the flow region has been visualized in Fig. 23.1, in a Cartesian system of coordinates.

23.2. Governing equations

In order to compare our numerical results with those in MITSOULIS and ZISIS [2001], we need to address exactly the same test problem than these two authors, namely, the isothermal, steady, incompressible, inertia-less flow of a Bingham material. If we neglect the inertia in

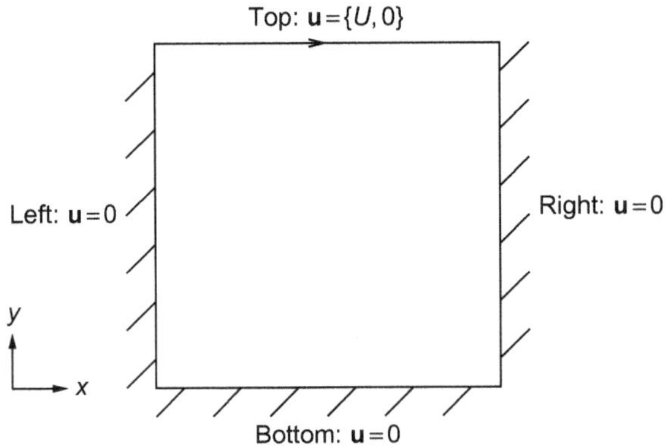

FIG. 23.1 Flow region geometry and boundary conditions.

the momentum equation, the creeping steady flow is modeled by the following system of equations and inequalities:

$$\nabla \cdot \mathbf{u} = 0, \tag{23.1}$$

$$\nabla p - \nabla \cdot \boldsymbol{\tau} = \mathbf{0}, \tag{23.2}$$

$$\begin{cases} \boldsymbol{\tau} = 2\mu\mathbf{D}(\mathbf{u}) + \tau_y \dfrac{\mathbf{D}(\mathbf{u})}{\|\mathbf{D}(\mathbf{u})\|} & \text{if } \|\boldsymbol{\tau}\| > \tau_y, \\ \mathbf{D}(\mathbf{u}) = \mathbf{0} & \text{if } \|\boldsymbol{\tau}\| \le \tau_y. \end{cases} \tag{23.3}$$

Assume that $\Omega = (0, a) \times (0, a)$; then we take a (resp., U) as characteristic length (resp., as characteristic velocity). The flow can be described by a single dimensionless number: the *Bingham number* $\mathcal{B}\mathrm{n} = \dfrac{\tau_y a}{\mu U}$.

23.3. Boundary conditions

The boundary conditions are those indicated in Fig. 23.1, that is:

- We assume that $\mathbf{u} = \mathbf{0}$ on the left, bottom, and right walls.
- We assume that $\mathbf{u} = \{U, 0\}$ on the top wall.

23.4. Results and discussion

We are going to investigate the steady, isothermal, incompressible creeping flow of a Bingham fluid in a lid-driven cavity for various Bingham numbers. In practice, we vary $\mathcal{B}n$ by changing the magnitude of the yield stress τ_y.

The computational results are presented and discussed in terms of the dimensionless number $\mathcal{B}\mathrm{n}$, of the dimensionless coordinates x/a and y/a, of the size and location of the yielded and unyielded regions, of the streamlines, of the dimensionless intensity $\psi^*_{\max}$ of the main vortex, and of the vertical position y^*_{eye} of the center of the main vortex.

23.4.1. Dimensionless parameters and meshes

The single dimensionless parameter, which governs the flow is the Bingham number $\mathcal{B}\mathrm{n}$. For this type of flow, $\mathcal{B}\mathrm{n}$ may vary from 0 (Newtonian case where the whole flow region is yielded) to $+\infty$ (which corresponds to a total blockage). In our computations, we have considered Bingham numbers varying from 0 to 1000; actually, we took for $\mathcal{B}\mathrm{n}$ the values 0, 0.1, 1, 2, 5, 20, 50, 200, 500, 1000. In practice, with the exception of τ_y, the various parameters in the problem, namely a, μ, and U were set to 1, implying that $\mathcal{B}\mathrm{n} = \tau_y$.

A uniform Cartesian grid was generated for the finite-volume discretization, with $N_x = N_y$ standing for the number of finite-volume cells in each direction. Several meshes have been considered, their characteristics being shown in Table 23.1.

23.4.2. Convergence properties of the Uzawa algorithm

The convergence of the Uzawa algorithm for any positive value of the augmentation parameter r is a basic result that (assuming several reasonable assumptions) has been proved mathematically in, e.g., GLOWINSKI, LIONS and TRÉMOLIÈRES [1976, 1981], FORTIN and

TABLE 23.1
Mesh characteristics

Meshes	$N_x = N_y$	*Number of cells*
Mesh 1	10	100
Mesh 2	20	400
Mesh 3	50	2,500
Mesh 4	100	10,000
Mesh 5	200	40,000

GLOWINSKI [1982, 1983], GLOWINSKI [1984], GLOWINSKI and LE TALLEC [1989], GLOWINSKI [2003]; in the context of viscoplasticity computations, this convergence property has been verified in the above references and in COUPEZ, ZINE and AGASSANT [1994], ROQUET and SARAMITO [2003]. One of our goals in this chapter is to confirm that the above convergence property still holds for the approximate problems derived from our finite-volume space-discretization scheme, for all the cases investigated here, whatever is the mesh size, the Bingham number, or the Lagrangian augmentation parameter r. Concerning the stopping criteria for the incompressible isothermal variant of algorithm (20.46)–(20.61), we have taken $tol_1 = tol_2 = 10^{-5}$ (since numerical experiments show that taking smaller values for tol_1 and tol_2 does not modify, in practice, the computed results, but may increase substantially the number of iterations necessary for convergence). It is worth noticing that the interpolation method we used in Section 21.7 to compute the components of the strain-rate-related tensor $\mathbf{p}$ does not affect the convergence properties of the Uzawa algorithm. Actually, the main challenge with this algorithm is the "good" choice of r, which is a nontrivial issue. The main difficulty stems from the fact that theory does not provide any good estimate or suitable guideline for the good choice of r (see however DELBOS, GILBERT, GLOWINSKI and SINOQUET [2006] for a strategy to vary r in order to speed up the convergence of a particular augmented Lagrangian algorithm). In practice, some preliminary tests are performed in order to assess which range of r provides the best speed of convergence. We illustrate the convergence property of our augmented Lagrangian algorithm by computing for $\mathcal{B}\mathrm{n} = 2$ the approximate solution associated with Mesh 4, for various values of r.

Let us highlight the influence of the augmented Lagrangian parameter r on the global convergence. Because to achieve convergence we need to verify $\Delta\{\mathbf{u}, p\}_i \leq tol_1$ and $\Delta D_i \leq tol_2$, we have plotted in Fig. 23.2 the behavior of these two convergence indicators, and in Fig. 23.3 the behavior of the maximum of these two indicators. It is quite obvious that an optimal value of r exists and that this optimum lies in the interval [2, 20] if $\mathcal{B}\mathrm{n} = 2$. In other words, choosing r in the interval $[\mathcal{B}\mathrm{n}, 10\mathcal{B}\mathrm{n}]$ seems to be a reasonable choice. In practice, we took r in the smaller interval $[2\mathcal{B}\mathrm{n}, 5\mathcal{B}\mathrm{n}]$ because computations performed for various values of $\mathcal{B}\mathrm{n}$ confirm the soundness of this choice. Finally, we wish to point out that for a given value of $\mathcal{B}\mathrm{n}$, the computed velocity and pressure fields, and the streamlines, are essentially independent of r. Actually, the computed yielded and unyielded regions can be slightly affected by the choice of r, because of their high sensitivity to the values of the numerical parameter and the demanding test we use to identify those regions where the discrete analog of $\mathbf{D}(\mathbf{u})$ vanishes. In Fig. 23.4, we have visualized the yielded and unyielded regions obtained with Mesh 4, for various values of r, when $\mathcal{B}\mathrm{n} = 2$. It is clear that the small discrepancies affecting the location of the yielded/unyielded regions are nearly unnoticeable

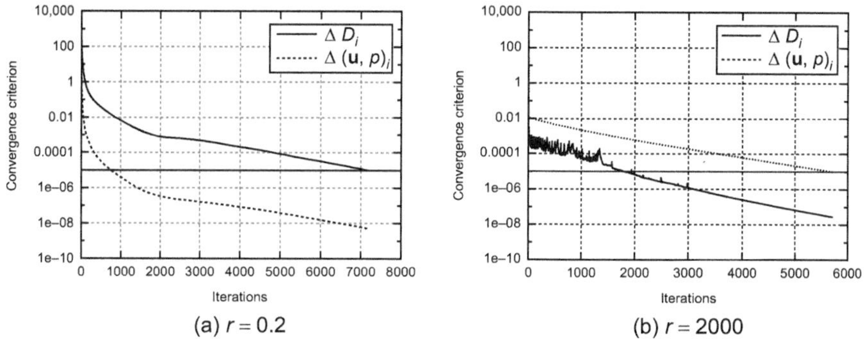

FIG. 23.2 Convergence of the Uzawa algorithm for $\mathcal{B}n = 2$ and Mesh 4: (a) $r = 0.2$, and (b) $r = 2000$.

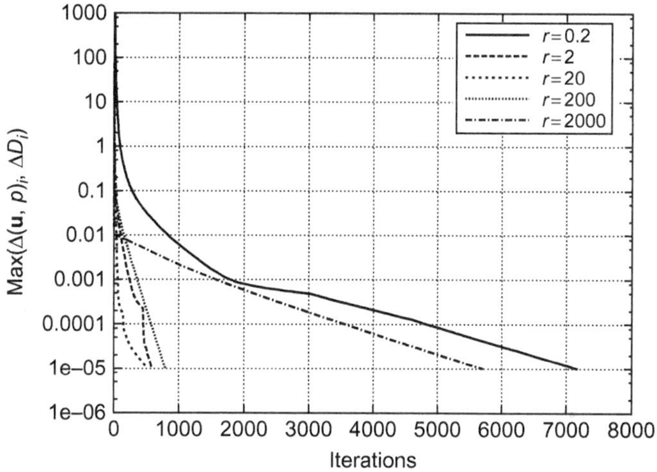

FIG. 23.3 Convergence of the Uzawa algorithm for $\mathcal{B}n = 2$ and Mesh 4, and for various values of r in the interval [0.2, 2000].

(they are of the order of the mesh size). These various results prove without much ambiguity that the computed solutions are essentially independent of r.

23.4.3. Convergence properties of the finite-volume approximation

In this paragraph, we are going to investigate the convergence of the approximate solutions as $\{\Delta x, \Delta y\} \to 0$, in the particular case where $\mathcal{B}n = 20$. The choice $\mathcal{B}n = 20$ is motivated by the fact that for this value of the Bingham number, the unyielded region occupies a large part of Ω. The convergence will be verified on the intensity $\psi^*_{\max}$ of the main vortex and on the vertical position y^*_{eye} of the center of the main vortex. Once again, we have used for our computations the five meshes described in Table 23.1; the numerical results have been reported in Table 23.2.

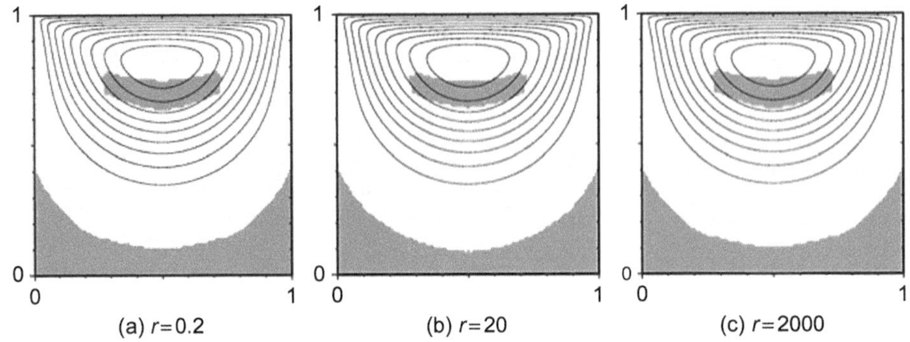

FIG. 23.4 Yielded (white) and unyielded (gray) regions, and streamlines for various values of the augmentation parameter r (Mesh 4, $\mathcal{B}n = 2$).

TABLE 23.2
Variations, as functions of the mesh size, of the main vortex intensity (ψ^*_{max}) and of the height of its center (y^*_{eye}) ($\mathcal{B}n = 20$)

Meshes	*Grid size* h^*	y^*_{eye}	ψ^*_{max}
Mesh 1	1/10	0.8999	1.57×10^{-3}
Mesh 2	1/20	0.9	3.16×10^{-2}
Mesh 3	1/50	0.9	3.63×10^{-2}
Mesh 4	1/100	0.9	4.00×10^{-2}
Mesh 5	1/200	0.9	4.00×10^{-2}

Table 23.2 shows that the height of the main vortex center is almost independent of the mesh size; however, $h^* < 1/50$ is required to obtain the stabilization of the intensity of the main vortex. Concerning the topology and size of the computed yielded and unyielded regions, let us say that, from Fig. 23.5, they look close to each others for $h^* \leq 1/50$ and quasi-identical for $h^* = 1/100$ and $1/200$. It is worth mentioning that, unlike for the intensity of the main vortex, the information, given by the coarsest mesh, on the topology and size of the yielded and unyielded regions, is pretty close to the one obtained with much finer meshes.

The above results provide a clear evidence of the convergence of the computed solutions as $\{\Delta x, \Delta y\} \to \mathbf{0}$. They show, also, that the geometrical parameters (including the pattern of the streamlines) have a fast convergence, compared with more quantitative ones, like the main vortex intensity. Because the numerical results associated with $h^* = 1/100$ and $h^* = 1/200$ are practically identical, we will use Mesh 4 to investigate further properties of the solutions, such as the influence of $\mathcal{B}n$ on the pattern of the yielded and unyielded regions.

23.4.4. Flow pattern as a function of the Bingham number

In this paragraph, we are going to discuss the influence of the Bingham number $\mathcal{B}n$ on the flow pattern. From its relative simplicity, the lid-driven cavity flow of a Bingham fluid has motivated a fairly abundant literature. Our main goal here is not to confirm the numerical

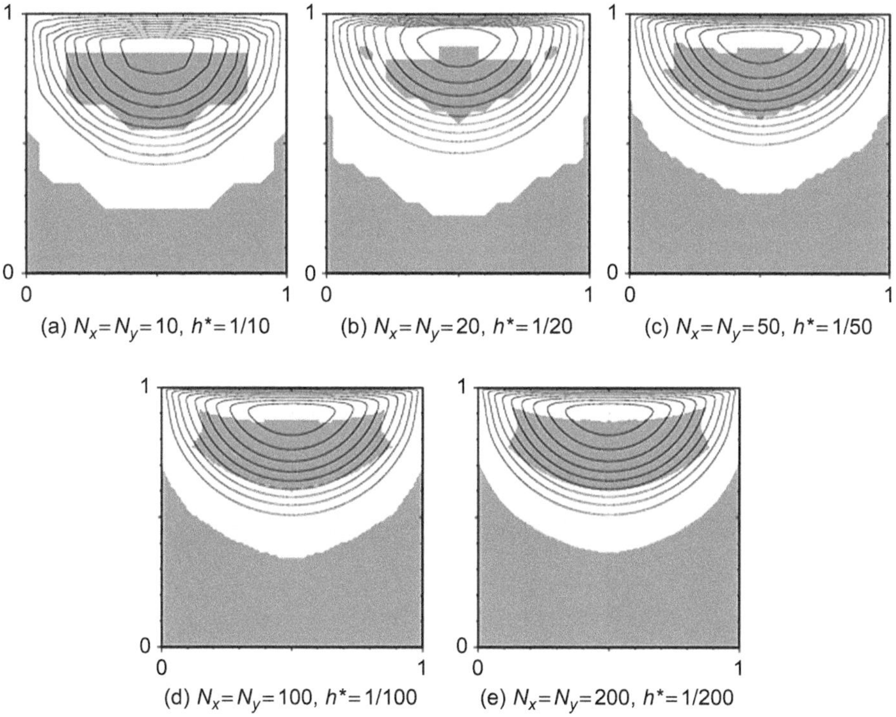

FIG. 23.5 Visualization of the yielded and unyielded regions for various values of the mesh size ($\mathcal{B}n = 20$)

results found by other scientists, but, actually, to validate our finite volume/augmented Lagrangian methodology by comparing our results with those, obtained by other methods, available in the literature.

The flow streamlines and the yielded and unyielded regions (computed with Mesh 4) have been visualized in Fig. 23.6. Because inertia has been neglected, the flow is, as expected, symmetric with respect to the line $x^* = 1/2$. Another expected property, obvious from these figures, is the growth of the unyielded region as $\mathcal{B}n$ increases. As soon as $\mathcal{B}n$ is nonzero, the flow recirculation regions in the bottom corners associated with a Newtonian fluid ($\mathcal{B}n = 0$) become unyielded flow stagnant regions and a moving unyielded region appear around the center of the main vortex (see Fig. 23.6(a)). As $\mathcal{B}n$ increases, the two flow stagnant unyielded regions in the bottom corners grow up, and finally merge to form a single flow stagnant unyielded region at the bottom of the cavity. Similarly, the moving unyielded region contained in the main vortex grows with $\mathcal{B}n$ as shown in Fig. 23.6. The shapes of the unyielded regions are in qualitative agreement with the ones in MITSOULIS and ZISIS [2001]; the small discrepancies one can observe between the results in the above publication and ours can be explained by the following facts: the methodology used in MITSOULIS and ZISIS [2001] combines finite-element approximation with regularization, the computations being done on a 40×40 mesh, while our method combines a finite-volume approximation with an augmented Lagrangian treatment of the nonsmoothness of the constitutive law, the

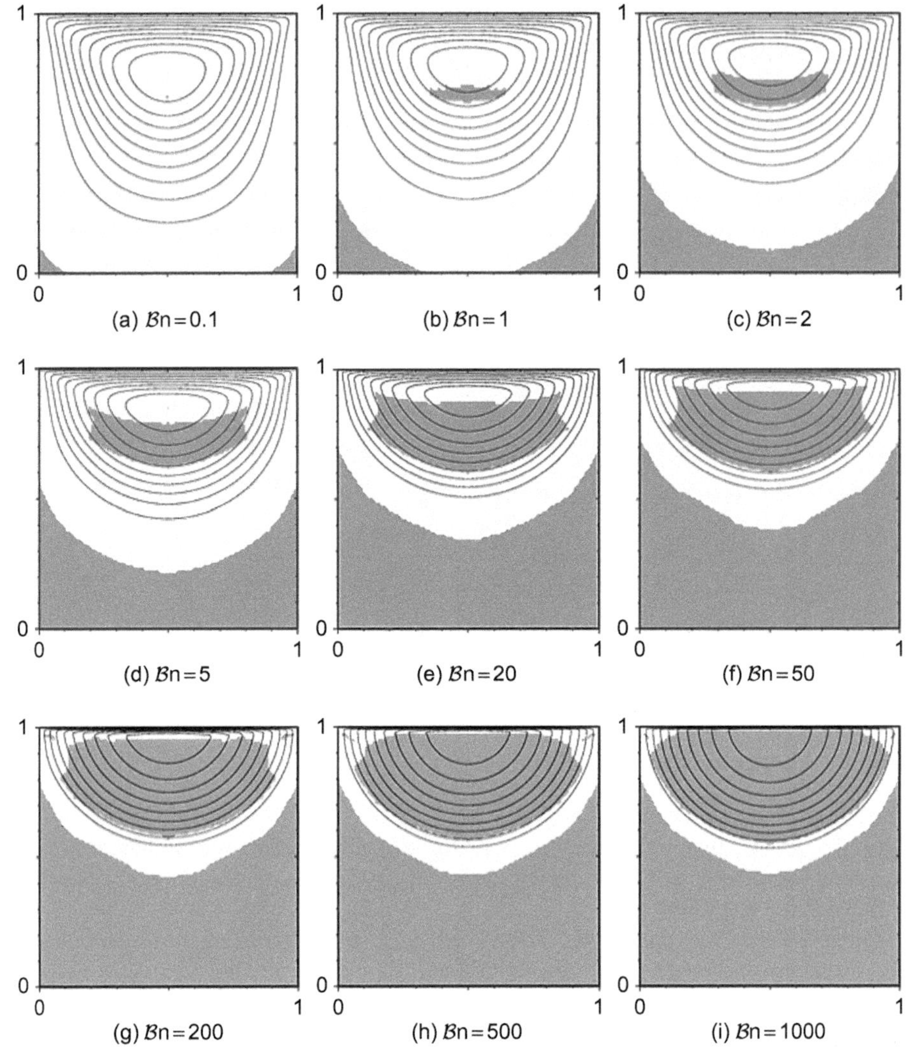

FIG. 23.6 Visualization of the flow streamlines and of the yielded (white) and unyielded (gray) regions for $\mathcal{B}$n varying from 0.1 to 1000 (Mesh 4).

computations being done on a 100×100 mesh. From these facts, we strongly believe that our methodology is more accurate than the one used in MITSOULIS and ZISIS [2001]. Actually, we have also compared (see Fig. 23.7) the values we obtained for y^*_{eye} and $\psi^*_{\max}$ with those in MITSOULIS and ZISIS [2001]. As shown by the above figure, there is a very good agreement between our results and those in the above reference up to $\mathcal{B}\text{n} = 200$; as for the yielded and unyielded regions, we think that the discrepancies observed for $\mathcal{B}\text{n} > 200$ follow from the fact that Mitsoulis and Zisis used a regularization-based approximation of

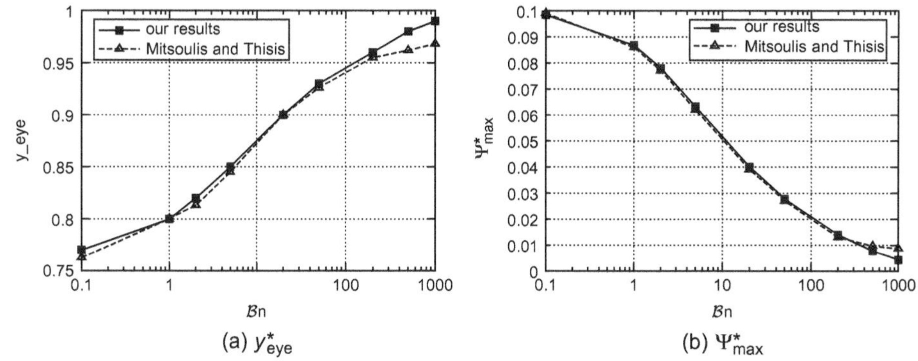

FIG. 23.7 (a) Variation as a function of $\mathcal{B}$n of the height of the center of the main vortex. (b) Variation as a function of $\mathcal{B}$n of the main vortex intensity. Comparison with MITSOULIS and ZISIS [2001] (Mesh 4).

the Bingham model, while our augmented Lagrangian-based methodology operates on the exact Bingham model (modulo a space discretization associated with a grid finer than the one used in MITSOULIS and ZISIS [2001]).

23.5. Further comments and conclusion on the two-dimensional lid-driven creeping flow problem

In the preceding parts of Section 23, we addressed the numerical simulation of the lid-driven isothermal, incompressible, steady creeping flow of a Bingham fluid in a square cavity. Our primary goal was the validation of our numerical methodology through a simple geometry-related test problem for which numerical results were already available in the literature.

Our numerical experiments have shown the convergence of the augmented Lagrangian/Uzawa algorithm we used whatever were the Bingham number $\mathcal{B}$n and the augmentation parameter r. We observed also that, as predicted by the theory, the computed solutions were independent of r. However, identifying the optimal value of r (that is, the one providing the fastest convergence of the augmented Lagrangian algorithm) is not an easy task. For this particular flow problem, we advocate choosing r in the interval [2$\mathcal{B}$n, 5$\mathcal{B}$n], but it is very likely that for another class of flow problems, this rule of thumb may be not valid anymore. We see the difficulty at identifying the optimal value of r as the main drawback of the augmented Lagrangian approach. From a space discretization point of view, our numerical experiments, and comparisons with the results in MITSOULIS and ZISIS [2001], show good convergence properties of our finite-volume approximation as $h \to 0$. It is worth noticing that the results reported in the above reference have been obtained by a method combining the regularization of the constitutive law with a finite-element approximation, that is a computational approach quite different from ours, thus providing a significant basis for comparisons.

In the next section, we will investigate a class of more complex flow problems, all related to the transportation of crude oils in pipelines. For this class of flows, the viscoplastic properties of the fluid are combined with less common features, such as temperature dependence, thixotropy, and compressibility. This will give us the opportunity to further validate our numerical methodology and to provide solutions to rather intricate (and indeed fascinating) problems of practical interest.

24. Study of nonisothermal incompressible flow in pipelines

24.1. Synopsis

In this section, we are going to investigate the numerical simulation of waxy crude oil flow in pipelines. Our first priority is to assess the capabilities of our methodology to take into account heat transfer and temperature-dependent rheology in a steady flow because such a flow is a representative of what is taking place in pipelines in steady production conditions. We assume in this section that the fluid is incompressible; for production situations, this is a reasonable assumption because (as our simulations will show in Section 25) compressibility will play a significant role only in the early transients of the restart. The waxy crude oil obeys a simple temperature-dependent Bingham constitutive law in which yield stress and/or viscosity are affine functions of the temperature. Our primary goal here is to investigate the influence of the temperature on the yielded and unyielded regions of the flow.

For our computations, we consider an axi-symmetric pipeline whose axial direction is parallel to Oz in an $\{r, \theta, z\}$ system of cylindrical coordinates (see Fig. 24.1). We assume from now on that we have $u_\theta = 0$, u_θ being the ortho-radial component of the velocity. In Fig. 24.1, R, L, and L_e denote, respectively, the radius of the pipe, the length of the pipe, and the distance downstream at which the temperature of the pipe changes abruptly.

24.2. Governing equations

The nonisothermal flow of an incompressible Bingham fluid is modeled by the following system of equations:

$$\nabla \cdot \mathbf{u} = 0, \tag{24.1}$$

$$\rho \left[\frac{\partial \mathbf{u}}{\partial t} + (\mathbf{u} \cdot \nabla)\mathbf{u} \right] + \nabla p = \nabla \cdot \boldsymbol{\tau}, \tag{24.2}$$

$$\begin{cases} \boldsymbol{\tau} = 2\mu(\Theta)\mathbf{D}(\mathbf{u}) + \tau_y \dfrac{\mathbf{D}(\mathbf{u})}{\|\mathbf{D}(\mathbf{u})\|} & \text{if } \|\boldsymbol{\tau}\| > \tau_y, \\ \mathbf{D}(\mathbf{u}) = \mathbf{0} & \text{if } \|\boldsymbol{\tau}\| \leq \tau_y, \end{cases} \tag{24.3}$$

$$\rho C_p \left(\frac{\partial \Theta}{\partial t} + \mathbf{u} \cdot \nabla \Theta \right) = \lambda_f \nabla^2 \Theta + \boldsymbol{\tau} : \mathbf{D}(\mathbf{u}). \tag{24.4}$$

An obvious candidate for the characteristic length L_c is the radius R of the pipe. In the particular case of those high viscosity and slow flowing oils that we consider, the Reynolds number $\mathcal{R}e$ is very small, as is the convection term in the momentum equation (24.2). Such flows are usually called creeping flows by practitioners. Moreover, the Brinkman number $\mathcal{B}r$ being also very small, the viscous dissipation term $\boldsymbol{\tau} : \mathbf{D}(\mathbf{u})$ will be discarded from the energy equation (24.4). From these simplifications, the remaining relevant dimensionless

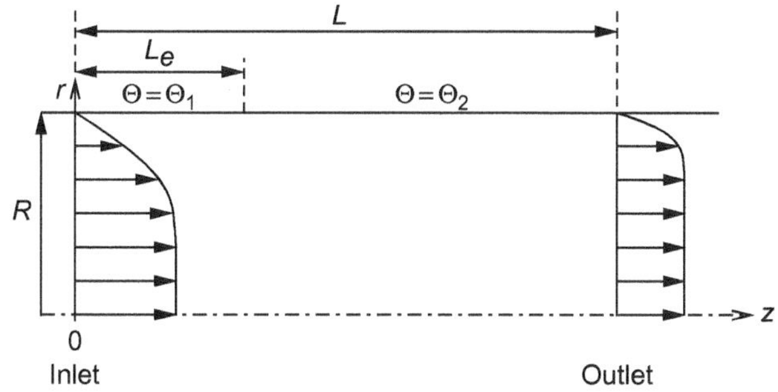

FIG. 24.1 Flow region geometry and boundary conditions for the steady nonisothermal flow of a Bingham fluid.

numbers are $\mathcal{R}\mathrm{e}$, $\mathcal{B}\mathrm{n}$, $\mathcal{P}\mathrm{e}$, and $\mathcal{C}\mathrm{a}$. The dimensionless number $\mathcal{C}\mathrm{a}$ is known as the *Cameron number*. It is defined by

$$\mathcal{C}\mathrm{a} = \frac{\lambda_f L}{\rho C_p \overline{U} R^2} = \frac{L}{R\mathcal{P}\mathrm{e}}; \tag{24.5}$$

$\mathcal{C}\mathrm{a}$ is used to estimate geometrical characteristics of the temperature field.

24.3. *Boundary conditions*

The equations modeling the flow, that is (24.1)–(24.4), have to be completed by boundary conditions. For the problem under consideration, these boundary conditions read as follows:

- *At the inlet*
  Fully developed Dirichlet boundary conditions are prescribed for $\mathbf{u}$ and Θ. We suppose that the length L_e of the pipe entry is large enough to prevent the velocity and temperature at the inlet to be affected by the steep change of temperature taking place downstream on the external boundary (wall) of the pipe. These boundary conditions are

$$u_r = 0, \;\; u_z = u_{z\text{-fully developed}} \tag{24.6}$$

$$\Theta = \Theta_{\text{fully developed}} \tag{24.7}$$

  Prescribing the profile of u_z at the inlet is like imposing, implicitly, the pressure drop.
- *At the wall*
  We assume a no-slip boundary condition for the velocity at the wall. The temperature verifies Dirichlet boundary conditions, corresponding to a steep cooling of the flow. More explicitly, we have

$$u_r = u_z = 0, \tag{24.8}$$

$$\Theta = \begin{cases} \Theta_1 & \text{if } z < L_e, \\ \Theta_2 & \text{if } z \geq L_e, \end{cases} \quad \text{with } \Theta_2 < \Theta_1. \tag{24.9}$$

- *Along the symmetry axis*

$$u_r = 0, \ \tau_{rz} = 0, \tag{24.10}$$

$$\frac{\partial \Theta}{\partial r} = 0. \tag{24.11}$$

- *At the outlet*

$$u_r = 0, \ \tau_{zz} = 0, \tag{24.12}$$

$$\frac{\partial \Theta}{\partial z} = 0. \tag{24.13}$$

The pressure being defined modulo an arbitrary additive constant, we will set it, arbitrarily, to 0 at the outlet.

24.4. Problem description

The problem that we consider is investigating the cooling of a viscoplastic fluid flow, assuming that the fluid rheological properties are temperature dependent. The cooling follows from the fact that the temperature on the wall of the pipe is lower than the entry temperature (as shown in (24.9)). The above scenario models the transportation of waxy crude oil in a pipeline, when the crude oil exits a pumping station at a warm temperature, and then cools down due to extreme temperature conditions along the pipe.

The strong temperature dependence of waxy crude oil rheological properties has been shown by many experimental surveys (see, e.g., CAWKWELL and CHARLES [1989], and HÉNAUT and BRUCY [2001]). However, to the best of our knowledge, one is still lacking a general framework for the description of the thermo-rheological properties of viscoplastic materials. As a first step in this direction, we are going to assume that the rheological parameters in our model are functions of the temperature; the "shape" of these functions can be obtained by data curve fitting or by classical WLF (Williams–Landel–Ferry) or Arrhenius equations. The good news are that from a qualitative point of view, the precise shape of the above functions is not crucial, implying that their approximation by simple affine functions of Θ will be sufficient to bring useful information for the temperature range that we consider.

In the first part of the pipeline (of length L_e), the wall temperature is maintained equal to the entry (inlet) temperature, implying that, in practice, the temperature is constant (and equal to the wall temperature) in this particular region of the flow. The flow pattern in the entry zone does not vary much with z and corresponds to a Poiseuille flow with a plug (solid) region of constant radius. Sufficiently far downstream from the location of the temperature discontinuity, the flow recovers a fully developed feature à la Poiseuille, but with a plug of larger radius because lower temperatures imply larger values for the yield stress τ_y. The fundamental questions that arise concern the transition zone: What is the flow pattern in this zone, particularly in terms of yielded and unyielded regions? Does the size and shape of the plug region vary continuously, as suggested in Fig. 24.2?

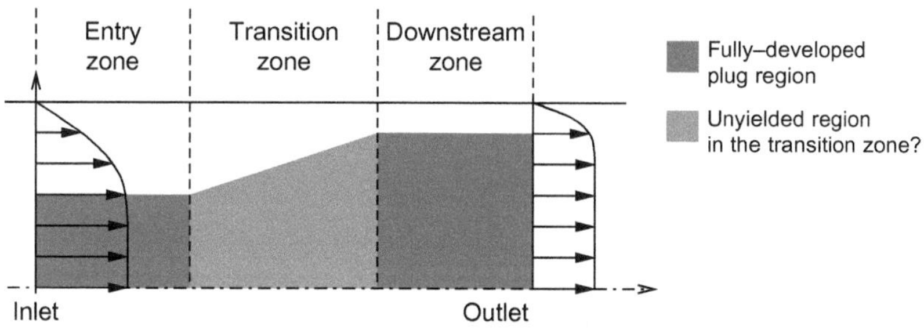

FIG. 24.2 Flow pattern in the transition region: a naïve first sketch.

24.5. Results and discussion

24.5.1. Synopsis

We are going to investigate the flow of a viscoplastic Bingham fluid in an axi-symmetric pipe for three different situations:

1. An isothermal case.
2. A nonisothermal case with a temperature-dependent viscosity and a constant yield stress.
3. A nonisothermal case with a temperature-dependent yield stress and a constant viscosity.

Actually, our numerical methodology can handle easily the simulation of the nonisothermal flow of a viscoplastic fluid whose viscosity and yield stress are both temperature dependent. However, we thought that investigating separately how the temperature dependence of the viscosity and yield stress affect the flow, may give a clearer view of the flow temperature dependence.

The computations performed via the transient decoupled algorithm (20.46)–(20.61), correspond to the real (in the physical sense) temporal evolution of the solution. However, within the scope of this section, we are interested in the steady flow only. Therefore, all the numerical results to be presented correspond to steady-state solutions. To perform transient calculations, we need to provide $\mathbf{u}|_{t=0}(= \mathbf{u}_0)$ and $\Theta|_{t=0}(= \Theta_0)$; for our computations, we used $\mathbf{u}_0 = \mathbf{0}$ and $\Theta_0 = \Theta_1$.

The results of our numerical experiments are presented and discussed in terms of dimensionless physical quantities: namely, the dimensionless Bingham number $\mathcal{B}n$, the dimensionless coordinates r/R and z/L, the dimensionless axial velocity $u_z/u_{z\text{-max}}$, the dimensionless temperature $\dfrac{\Theta - \Theta_2}{\Delta\Theta}$ (where $\Delta\Theta = \Theta_1 - \Theta_2$) and the dimensionless pressure $p/p_{\max}$ (where $p_{\max}$ is the maximum pressure obtained for $\Delta\Theta = 20^\circ K$).

24.5.2. Pipe dimensions, dimensionless parameters, and meshes

As already mentioned in Section 24.4, the flow region is divided into three parts:

1. An entry zone where the temperature field is fully developed, constant, and equal to Θ_1.

2. A transition zone where the temperature varies from Θ_1 to Θ_2.
3. A downstream zone where the temperature field is fully developed, constant, and equal to Θ_2.

The length of the transition zone depends on the Cameron and Peclet numbers. We can evaluate the length L_{fully} required for the downstream temperature field to recover a fully developed profile as

$$L_{\text{fully}} = \mathcal{C}a\mathcal{P}eR. \tag{24.14}$$

In order to limit the size of the computational domain, to lower the computational time, and to obtain a mesh ratio R_c (see its definition below) as close to 1 as possible, we chose a short pipe ($L/R = 40$). Because the temperature field is fully developed if $\mathcal{C}a \geq 1$, we set $\mathcal{C}a = 1$, thus (from (24.14)) the greater $\mathcal{P}e$, the greater L_{fully}. We take $L_e = L/4$, and, in order to match the pipe length and the requirements to obtain a fully developed temperature field at the pipe exit, we set the physical parameters so that $\mathcal{P}e = 10$ (which yields $L_{\text{fully}} = L/4$).

A uniform Cartesian grid is generated in the $\{r, z\}$ meridian rectangle associated with the flow region, with N_r (resp., N_z) cells in the Or (resp., Oz) direction. The mesh ratio is defined as

$$R_c = \frac{\Delta z}{\Delta r} = \frac{L/N_z}{R/N_r}. \tag{24.15}$$

Several meshes have been considered, with their characteristics being reported in Table 24.1.

24.5.3. Convergence properties of the iterative methods

The steady-state solution of the isothermal test problem was computed using all the 11 meshes described in Table 24.1. However, for the solution of the nonisothermal test problems, we used the six meshes from Mesh 4 to Mesh 9. For all the cases that we investigated, the convergence of the Uzawa algorithm was achieved at every time step and the

TABLE 24.1
Mesh characteristics and mesh size effects for a steady Bingham flow

Meshes	$N_r \times N_z$	*Mesh ratio* R_c	ε_r	ε_Θ
Mesh 1	10×50	8	2.56×10^{-2}	
Mesh 2	10×100	4	2.59×10^{-2}	
Mesh 3	10×200	2	2.57×10^{-2}	
Mesh 4	20×25	32	0.99×10^{-2}	6.31×10^{-4}
Mesh 5	20×50	16	1.03×10^{-2}	3.34×10^{-4}
Mesh 6	20×100	8	1.03×10^{-2}	1.72×10^{-4}
Mesh 7	20×200	4	1.02×10^{-2}	0.58×10^{-4}
Mesh 8	20×300	2.66	1.02×10^{-2}	0.29×10^{-4}
Mesh 9	20×400	2	1.02×10^{-2}	
Mesh 10	30×100	12	0.51×10^{-2}	
Mesh 11	30×200	6	0.51×10^{-2}	

whole transient algorithm (namely, the finite-volume analog of algorithm (20.46)–(20.61)) converged to a steady-state solution, the number of iterations and time steps necessary to achieve convergence depending, as expected, of the values chosen for the stopping criteria tol_1, tol_2, and tol_3. The unconditional (with respect to the value of the augmentation parameter r_{AL}) convergence property that was observed testifies of the robustness of the whole transient algorithm. As mentioned in Section 23, an appropriate choice for the value of the augmentation parameter r_{AL} reduces substantially the number of iterations necessary to achieve convergence, without modifying the computed solution. Numerical tests performed for various values of r_{AL} show that the optimal value of this parameter lies in the interval $[\mu, \rho/\Delta t]$ (in practice, we used $r_{AL} = 50$).

The generalization to nonisothermal situations (with viscosity and/or yield stress coefficients varying in space with Θ) is fairly straightforward, the algorithm still keeping its robust convergence properties. However, as expected, the speed of convergence decreases slightly if the magnitude of the spatial variations of the viscosity and yield stress increases. For example, the values of the other parameters being the same, for nonisothermal situations the Uzawa algorithm requires more iterations per time step for $\Delta\Theta = 20°K$ than for $\Delta\Theta = 1°K$ (but the difference, although noticeable, is not that large, as shown in Fig. 24.3 for the first time step).

Numerical tests were performed in order to determine suitable values for the stopping criteria tol_1, tol_2, and tol_3. Concerning the capture of the steady-state solution, we noticed that 10^{-5} was an appropriate value for tol_3; indeed, computations performed with $tol_3 \in [10^{-7}, 10^{-5})$ showed that the computed steady-state solutions are essentially identical for all $tol_3 \leq 10^{-5}$ (actually, $tol_3 = 10^{-4}$ provides steady-state solutions very close to those obtained with $tol_3 \leq 10^{-5}$). Concerning the convergence of the Uzawa

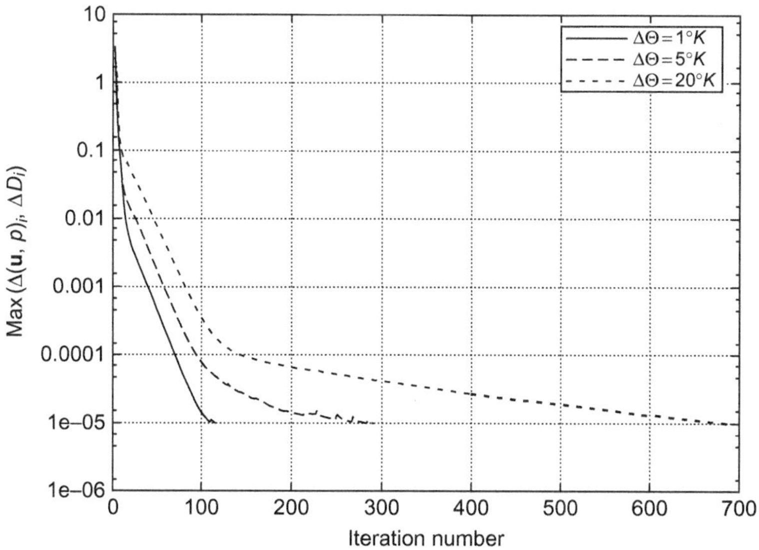

FIG. 24.3 Influence of the magnitude of the viscosity and yield stress spatial variations on the speed of convergence of the Uzawa algorithm at the first time step for several temperature drops ($\Delta\Theta = 1°K$, $\Delta\Theta = 5°K$, and $\Delta\Theta = 20°K$).

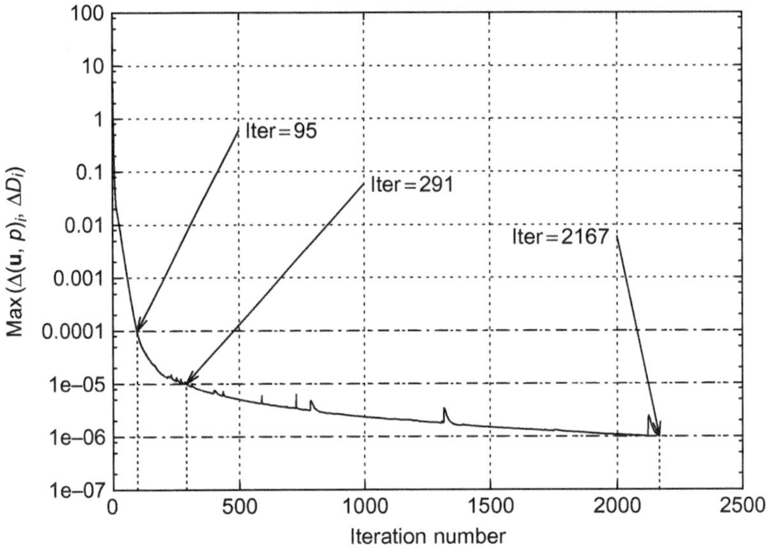

FIG. 24.4 Influence of the stopping criteria tol_1 and tol_2 on the number of iterations required for the convergence of the Uzawa algorithm ($\Delta\Theta = 5°K$).

algorithm, we performed calculations with the following three realizations of $\{tol_1, tol_2\}$: $\{10^{-4}, 10^{-4}\}$, $\{10^{-5}, 10^{-5}\}$ and $\{10^{-6}, 10^{-6}\}$; the computed results being quasi-identical, we retained $\{tol_1, tol_2\} = \{10^{-5}, 10^{-5}\}$ to be on the safe side. The ability of the Uzawa algorithm to converge for $tol_1 = tol_2 = 10^{-6}$ underlines its robustness, but the computational cost in that case is quite high because there is a steep deterioration of the speed of convergence between $tol_1 = tol_2 = 10^{-5}$ and $tol_1 = tol_2 = 10^{-6}$, as shown in Fig. 24.4 above.

24.5.4. Influence of the mesh size

Influence of Δr: In order to assess the influence of Δr on the approximate solutions, we compared the computed axial component of the velocity to the exact one for an isothermal situation with $\mathcal{B}\text{n} = 5$. For this particular case, the steady-state solution verifies $u_r = 0$ and u_z does not depend of z. The closed form solution of this (kind of) Poiseuille flow is known and can be found in Chapter 2, Section 16. Actually, for this test problem, $\mathbf{u}(r, z) = \mathbf{u}(r, 0) = \{0, u_{z\text{-inlet}}(r)\}$, $\forall z \in [0, L]$. The error ε_r between the computed and exact axial velocity solutions is defined as

$$\varepsilon_r = \frac{\|u_{z\text{-computed}} - u_{z\text{-exact}}\|_\infty}{\|u_{z\text{-exact}}\|_\infty} \tag{24.16}$$

with $\|\varphi\|_\infty = \max_{\{r,z\}\in(0,R)\times(0,L)} |\varphi(r, z)|$. The approximation error has been computed for the 11 meshes described in Table 24.1 and reported in the fourth column of the above table. As expected for the steady-state solution of the problem under consideration, ε_r is essentially independent of N_z, that is of Δz. A close inspection of Table 24.1 shows that $\varepsilon_r \approx O(|\Delta r|^{3/2})$, which is quite satisfactory for the L^∞-norm of the approximation error of

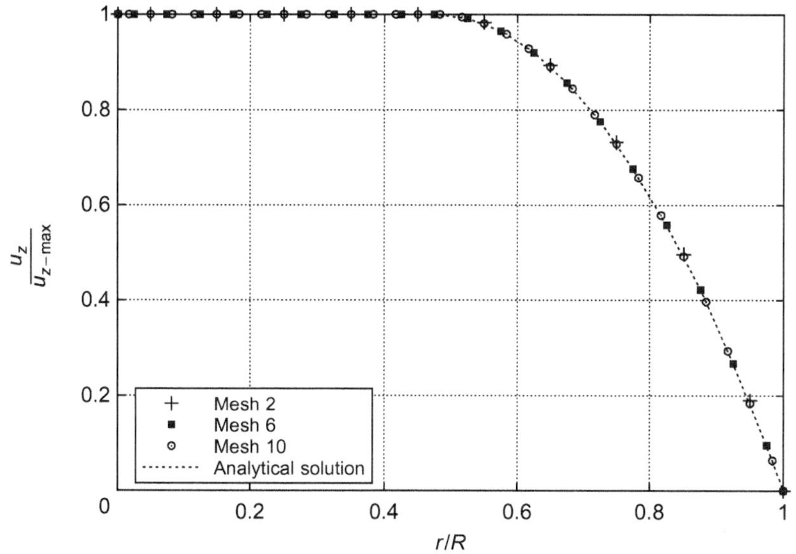

FIG. 24.5 Axial velocity profiles and comparisons with the exact steady-state solution of an isothermal test problem ($\mathcal{B}$n = 5; Meshes 2, 6, and 10).

the solution of a nonsmooth problem (for a smooth problem one would expect $O(|\Delta r|^2)$). An additional comparison between exact and computed solutions can be found in Fig. 24.5. Incidentally, these comparisons provide a validation of the interpolation method, described in Section 21.7, used for the finite volume approximation of the components of the strain tensors $\mathbf{D}(\mathbf{u})$ and $\mathbf{p}$. On the basis of the above results, we retained $N_r = 20$ as a reasonable compromise between the contradictory requirements of a good accuracy and a low computational time.

Influence of Δz*:* Assessing the influence of Δz on the space approximation error is more difficult from the lack of closed form solutions in the nonisothermal case. To estimate the contribution of Δz to the approximation error, we used the following estimator

$$\varepsilon_\Theta = \frac{|\Theta_{\text{average}}(N_z) - \Theta_{\text{average}}(\tilde{N}_z)|}{\Theta_{\text{average}}(\tilde{N}_z)} \tag{24.17}$$

where N_z takes the values 25, 50, 100, 200, and 300, the associated values of $\tilde{N}_z$ being 50, 100, 200, 300, and 400; in (24.17), Θ_{average} denotes the mean value of the steady-state temperature over the flow region, that is $\Theta_{\text{average}} = \frac{2}{LR^2}\int_{\Omega_{rz}} \Theta r dr dz$, where $\Omega_{rz} = (0, R) \times (0, L)$. The computations we have performed, with N_r fixed at 20 and $\Delta\theta = 20°K$, lead to the results reported in Table 24.1 and visualized in Fig. 24.6. The fifth column of Table 24.1 suggests that, roughly speaking, $\varepsilon_\Theta = O(\Delta z)$.

On the basis of the results of the above numerical experiments, dedicated to the assessment of the influence of Δr and Δz on the approximation error of our finite volume method,

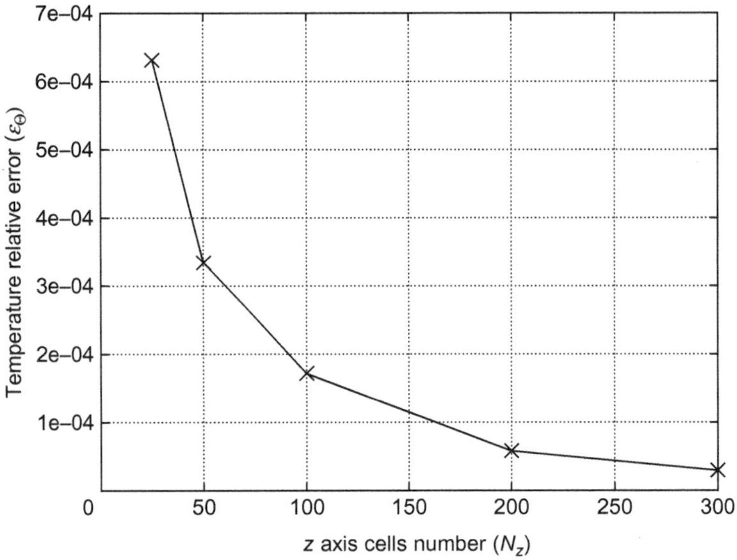

FIG. 24.6 Influence of $N_z(= L/\Delta z)$ on $\varepsilon_\Theta(N_r = 20, \Delta\Theta = 20°K)$.

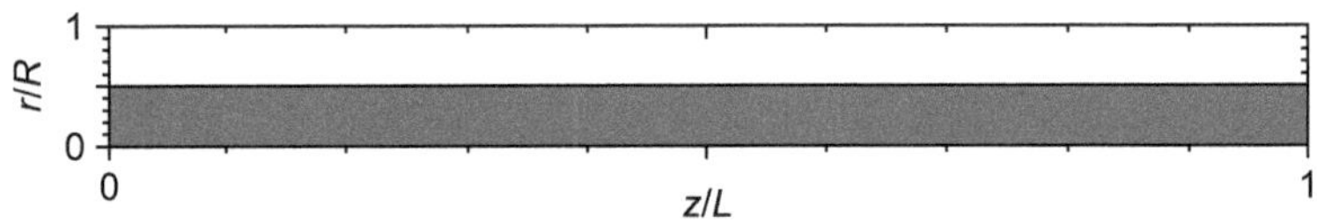

FIG. 24.7 Yielded (white) and unyielded (black) regions for an isothermal incompressible Bingham flow ($\mathcal{B}n = 5$).

we retained Mesh 7 ($N_r = 20, N_z = 200$) for further computations; this mesh offers a reasonable trade-off between good accuracy and low computational time.

24.5.5. Further comments on the approximation of isothermal Bingham flow

In this (short) section, we complete the above results, concerning the computation of the approximate steady-state solution when the Bingham fluid is incompressible and the flow isothermal. The results we focus on are the computed yielded and unyielded regions of the flow at $\mathcal{B}n = 5$. These regions have been visualized in Fig. 24.7 with the yielded (resp., unyielded) region in white (resp., black). The unyielded region is characterized by $\mathbf{D(u)} = \mathbf{0}$. The radius of the computed unyielded region is very close to $\frac{R}{2}$ (which is the theoretical value corresponding to $\mathcal{B}n = 5$).

24.5.6. Incompressible Bingham flow with constant yield stress and temperature-dependent viscosity

For the test problem discussed in this paragraph, we assume that the yield stress τ_y is independent of the temperature. However, we suppose that the fluid viscosity is an affine function of the temperature, that is $\mu = \mu(\Theta) = \mu_0 + \mu_1\Theta$, where μ_0 and μ_1 are two constants. Assuming that the temperature drop at the wall is $20°K$, μ_0 and μ_1 are chosen so that the outlet/inlet viscosity ratio (that is $\frac{\mu(\Theta_2)}{\mu(\Theta_1)}$) is 20; the corresponding inlet and outlet Bingham numbers are thus $\mathcal{B}n_{\text{Inlet}} = 10$ and $\mathcal{B}n_{\text{Outlet}} = 0.5$.

We have visualized in Fig. 24.8 the computed temperature distribution associated with the above yield stress and viscosity; in this figure, we clearly see the entry, transition, and downstream regions of the flow.

In Fig. 24.9, we have visualized the yielded and unyielded parts of the flow region. Here too, three subregions appear very clearly, consistent with those in Fig. 24.8: the entry zone on the left, the transition zone in the middle, and the downstream zone on the right. In the entry zone (where the temperature is quasi-constant) the flow has the features of a Poiseuille flow at temperature $\Theta = \Theta_1$ with a constant radius plug region in the center of the pipe. In the region downstream, the flow is again of the Poiseuille type, but at the temperature $\Theta = \Theta_2 (< \Theta_1)$. The radius of the outlet plug region is smaller than the radius of the inlet one, which makes sense because $\mathcal{B}n_{\text{Outlet}} < \mathcal{B}n_{\text{Inlet}}$. Actually, the main result is the one concerning the distribution yielded/unyielded in the transition zone: based on our computations, we observe that for this particular nonisothermal incompressible flow problem (where the viscosity is temperature dependent, whereas the yield stress τ_y is independent of Θ), the Bingham material is yielded in the whole transition region.

24.5.7. Incompressible Bingham flow with temperature-dependent yield stress and constant viscosity

We consider now a Bingham flow with temperature-dependent yield stress and constant viscosity. The constitutive law (24.3) still holds with $\mu =$ constant and the yield stress τ_y verifying

$$\tau_y = \tau_y(\Theta) = \tau_0 + \tau_1\Theta \tag{24.18}$$

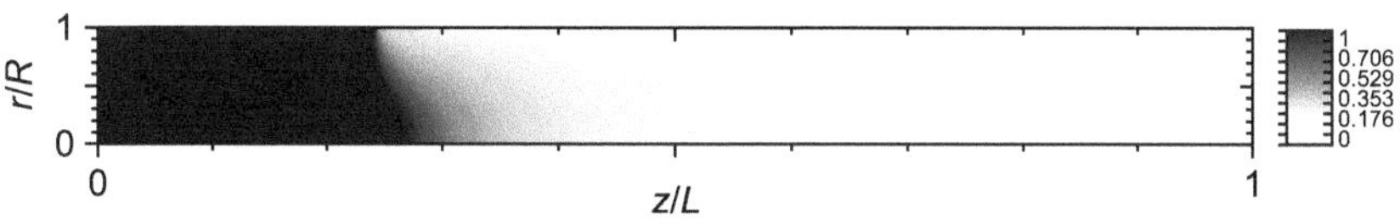

FIG. 24.8 Temperature distribution for a nonisothermal Bingham flow with temperature-independent yield stress and temperature-dependent viscosity ($\mathcal{B}n_{\text{Inlet}} = 10$, $\mathcal{B}n_{\text{Outlet}} = 0.5$, $\Delta\Theta = 20°K$).

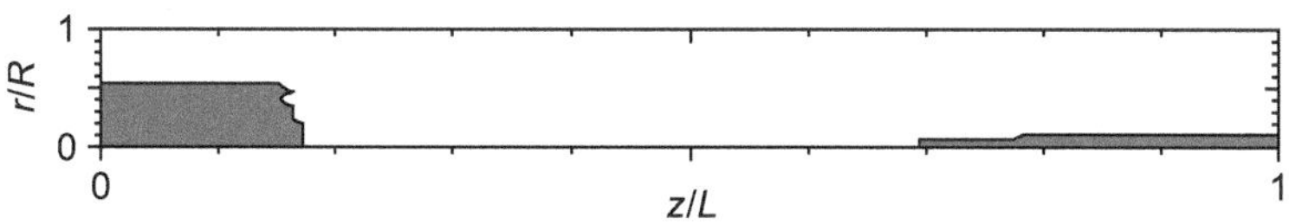

FIG. 24.9 Yielded (white) and unyielded (black) regions for a nonisothermal Bingham flow with temperature-independent yield stress and temperature-dependent viscosity ($\mathcal{B}n_{\text{Inlet}} = 10$, $\mathcal{B}n_{\text{Outlet}} = 0.5$, $\Delta\Theta = 20°K$).

where in (24.18), τ_0 and τ_1 are two constants. For the numerical experiments which follow, we assume that τ_0 and τ_1 have been chosen so that $\mathcal{B}n_{\text{Inlet}} = 1.5$ and $\mathcal{B}n_{\text{Outlet}} = 50$ if $\Delta\Theta = 20°K$. The results presented here correspond to three different temperature drops, namely $\Delta\Theta = 1°K$, $\Delta\Theta = 5°K$, and $\Delta\Theta = 20°K$; the corresponding Bingham numbers and outlet/inlet yield stress ratios have been reported in Table 24.2. The temperature distributions associated with the three above values of $\Delta\Theta$ have been visualized in Fig. 24.10. The entry, transition, and downstream regions look essentially the same, however, the time t_{steady} required to reach the steady-state solution (according to $tol_3 = 10^{-5}$ in the fully discrete analog of algorithm (20.45)–(20.61)) increases with $\Delta\Theta$, as reported in Table 24.2 and visualized in Fig. 24.11.

We have visualized in Fig. 24.12 the yielded and unyielded regions associated with the above three values of $\Delta\Theta$. Once again, we observe fully developed Poiseuille–Bingham flows in the entry and downstream regions, with the downstream region plug radius increasing with $\Delta\Theta$, while the entry region plug radius stays the same. Such a behavior makes sense, since in the three cases considered here the inlet Bingham numbers $\mathcal{B}n_{\text{Inlet}}$ are

TABLE 24.2
Flow parameters for an incompressible Bingham fluid with constant viscosity and temperature-dependent yield stress

$\Delta\Theta = \Theta_1 - \Theta_2 (°K)$	$\dfrac{\tau_y(\Theta_2)}{\tau_y(\Theta_1)}$	$\mathcal{B}n_{\text{Inlet}}$	$\mathcal{B}n_{\text{Outlet}}$	$t_{\text{steady}}\,(\text{s})$
1	2.617	1.5	3.92	54.60
5	9.083	1.5	13.62	67.84
20	33.333	1.5	50	75.34

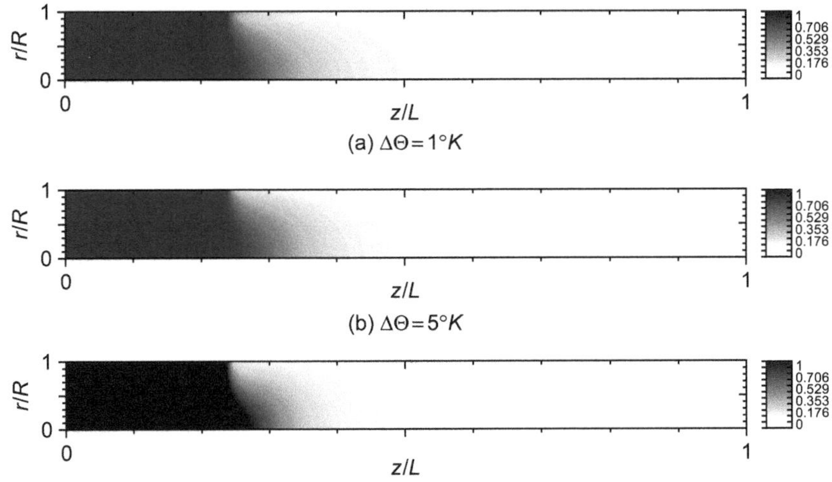

(a) $\Delta\Theta = 1°K$

(b) $\Delta\Theta = 5°K$

(c) $\Delta\Theta = 20°K$

FIG. 24.10 Temperature distribution for a nonisothermal incompressible Bingham flow with temperature-independent viscosity and temperature-dependent yield stress: (a) $\Delta\Theta = 1°K$, (b) $\Delta\Theta = 5°K$, and (c) $\Delta\Theta = 20°K$.

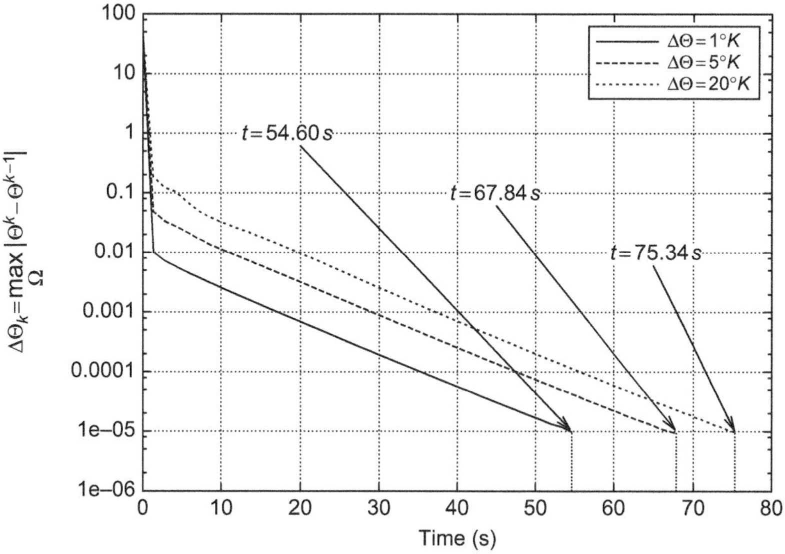

FIG. 24.11 Influence of the temperature drop $\Delta\Theta$ on the time evolution of the computed temperature for a nonisothermal incompressible Bingham flow with temperature-independent viscosity and temperature-dependent yield stress.

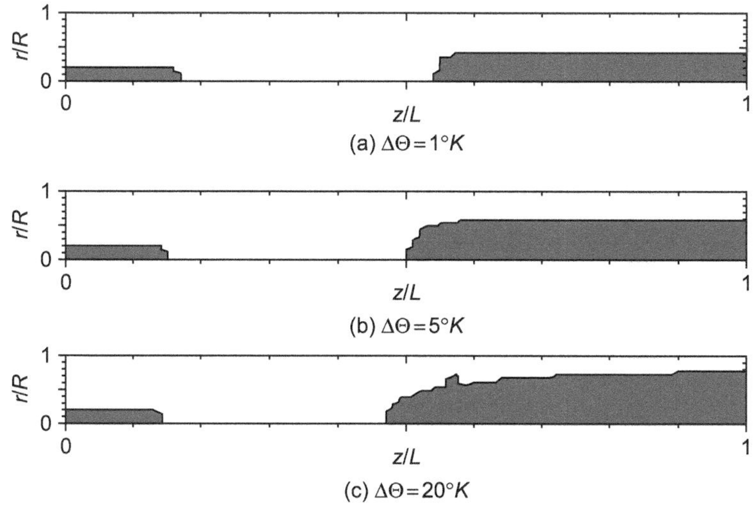

FIG. 24.12 Influence of the temperature drop on the yielded (white) and unyielded (black) regions for a nonisothermal incompressible Bingham flow with temperature-independent viscosity and temperature-dependent yield stress: (a) $\Delta\Theta = 1°K$, (b) $\Delta\Theta = 5°K$, and (c) $\Delta\Theta = 20°K$.

identical (all equal to 1.5), while the outlet Bingham number $\mathcal{B}n_{\text{Outlet}}$ increases with $\Delta\Theta$, according to the fourth column of Table 24.2. Figure 24.12 shows also that (1) the larger $\Delta\Theta$, the sooner (in the flow direction) begins the downstream plug region, (2) the length required for the downstream plug region to reach its fully developed radius increases with $\Delta\Theta$, and finally (3) the flow is yielded everywhere in the transition region. All these observations attest: (1) of the sensitivity of the flow to temperature changes when the yield stress is temperature-dependent, and (2) of the yielded character of the transition region. In Fig. 24.13, we have plotted for $\Delta\Theta = 1°K,\ 5°K$, and $20°K$, the profiles of the computed axial velocity u_z in the cross-sections of the pipe located at $z/L = 0$, 0.23, 0.25, 0.30, 0.35, 0.60, and 1. A striking result (consistent, however, with the fact that the transition region is essentially yielded) is the bell shape of these velocity profiles in the transition region (the region in which the temperature varies significantly); this bell shape is more pronounced for large $\Delta\Theta$. We observe also that the velocity along the pipe axis (that is at $r = 0$) is higher in the transition region than in the entry or downstream plug regions.

In Fig. 24.14, we have visualized the variations of the computed averaged pressure along the pipe (as a function of z/L); we recall that we prescribed $p = 0$ at the outlet and took as pressure of reference the maximal value of p associated with $\Delta\Theta = 20°K$. The averaged pressure is (within a reasonable accuracy) an affine function of z in the entry and downstream regions; however, the three graphs show some curvature in the transition region. We observe also that the larger is $\Delta\Theta$, the larger is the pressure drop between inlet and outlet. Having said all that, we have to acknowledge the fact that the pressure transition region is much smaller than the ones associated with the changes in temperature, velocity, and yielding/unyielding regions; indeed, the pressure is practically a piecewise affine function of z from 0 to L. To quantify this property, we have done the following comparison: (1) Using relation (16.2) from Chapter 2, Section 16, we can easily compute the pressure drops per unit length associated with $\mathcal{B}n_{\text{Inlet}}$ and $\mathcal{B}n_{\text{Outlet}}$. (2) From these values, it is a very simple exercise to find the pressure distribution $p_{\text{isothermal}}$ associated with an incompressible flow of the Poiseuille–Bingham type with $\mathcal{B}n = \mathcal{B}n_{\text{Inlet}}$ in the entry region (that is for $0 \leq z \leq L_e$), followed by another Poiseuille–Bingham flow associated this time with $\mathcal{B}n = \mathcal{B}n_{\text{Outlet}}$, immediately downstream of the entry region (that is for $L_e \leq z \leq L$); $p_{\text{isothermal}}$ is a piecewise affine function of z, continuous over $[0, L]$ and vanishing at $z = L$. (3) We define ε_p as $\max_{z\in[0,L]} [p_{\text{non-isothermal}}(z) - p_{\text{isothermal}}(z)]$. The values of ε_p reported in Table 24.3 show that $p_{\text{isothermal}}$ is a good approximation (from below) of $p_{\text{nonisothermal}}$. Finally, focusing on Fig. 24.13(a), it is reasonable to assume (from the flatness of all the graphs close to the z/L axis) that, for $\Delta\Theta = 1°K$, the transition region contains a plug whose radius varies continuously. Actually, Fig. 24.12(a) shows the irrelevance of the above assumption because, even for $\Delta\Theta = 1°K$, the transition region is fully yielded.

REMARK 24.1. The *unyielded regions* are characterized by $\mathbf{D}(\mathbf{u}) = \mathbf{0}$, with

$$\mathbf{D}(\mathbf{u}) = \begin{pmatrix} \dfrac{\partial u_x}{\partial x} & \dfrac{1}{2}\left(\dfrac{\partial u_x}{\partial y} + \dfrac{\partial u_y}{\partial x}\right) & \dfrac{1}{2}\left(\dfrac{\partial u_x}{\partial z} + \dfrac{\partial u_z}{\partial x}\right) \\ \dfrac{1}{2}\left(\dfrac{\partial u_x}{\partial y} + \dfrac{\partial u_y}{\partial x}\right) & \dfrac{\partial u_y}{\partial y} & \dfrac{1}{2}\left(\dfrac{\partial u_y}{\partial z} + \dfrac{\partial u_z}{\partial y}\right) \\ \dfrac{1}{2}\left(\dfrac{\partial u_x}{\partial z} + \dfrac{\partial u_z}{\partial x}\right) & \dfrac{1}{2}\left(\dfrac{\partial u_y}{\partial z} + \dfrac{\partial u_z}{\partial y}\right) & \dfrac{\partial u_z}{\partial z} \end{pmatrix}. \tag{24.19}$$

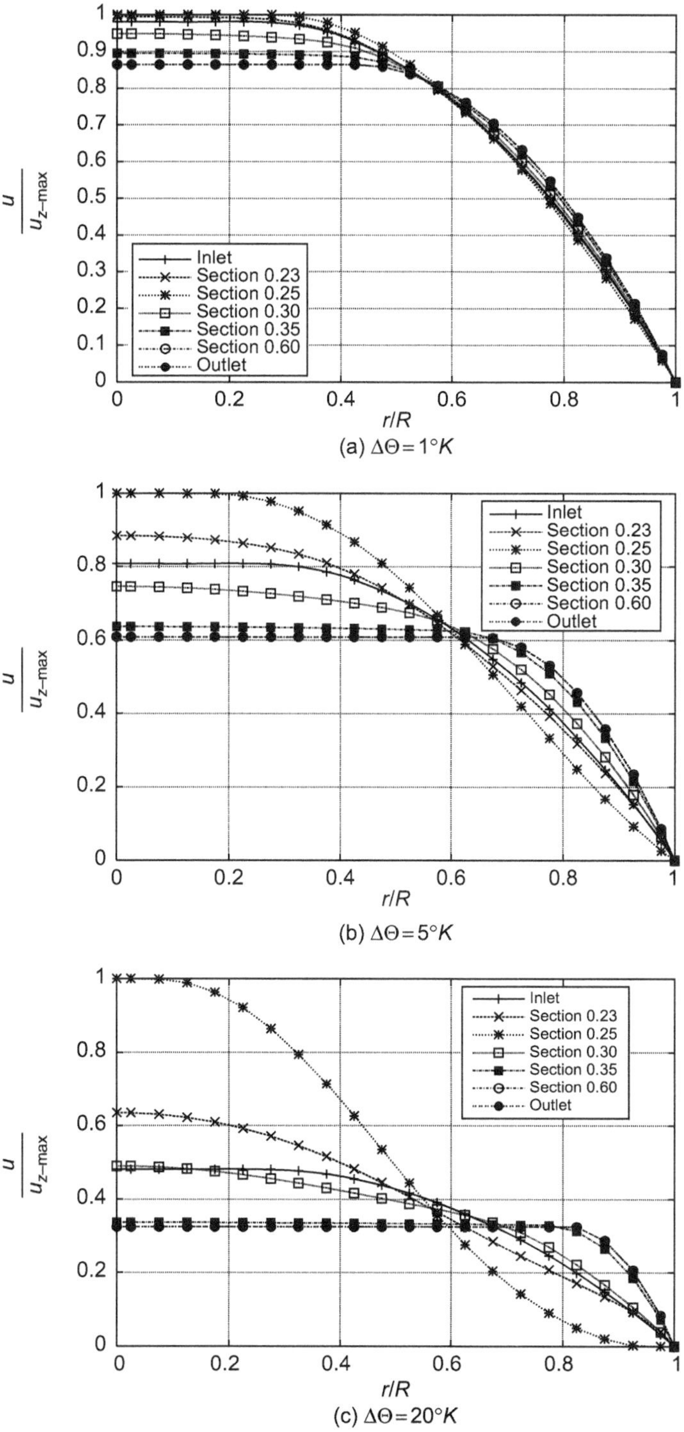

FIG. 24.13 Axial velocity profiles at various cross-sections for a nonisothermal incompressible Bingham flow with temperature-independent viscosity and temperature-dependent yield stress: (a) $\Delta\Theta = 1°K$, (b) $\Delta\Theta = 5°K$, and (c) $\Delta\Theta = 20°K$.

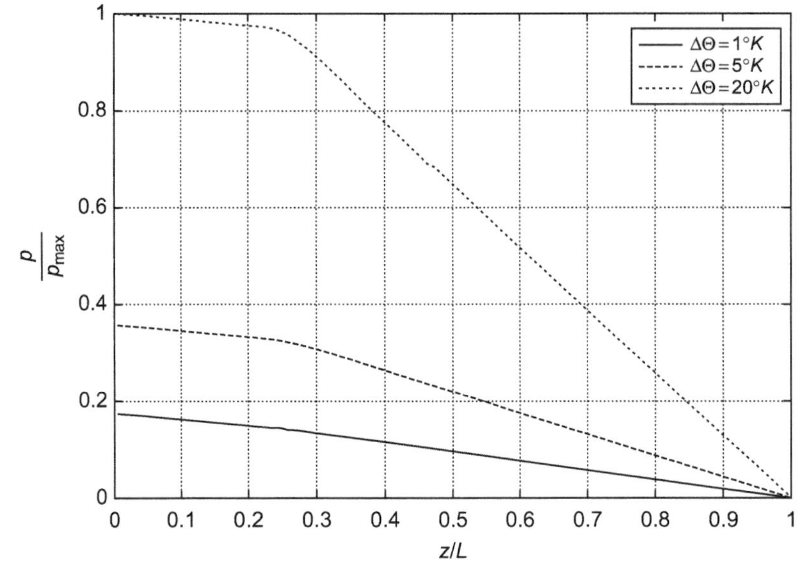

FIG. 24.14 Influence of the temperature drop magnitude $\Delta\Theta$ on the pressure drop for a nonisothermal Bingham flow with a temperature-dependent yield stress.

TABLE 24.3
Values of ε_p for $\Delta\Theta = 1°K$, $5°K$, and $20°K$

$\Delta\Theta$	ε_p
1	1.05×10^{-2}
5	0.98×10^{-2}
20	1.45×10^{-2}

It follows from (24.19) that to have a plug region, in the transition zone, which contains (parts of) the pipe axis, we need to have $\frac{\partial u_z}{\partial z} \equiv 0$ over an interval of the pipe axis contained in the transition zone. Our computations do not show this phenomenon, even for $\Delta\Theta = 1°K$. They show that in the transition zone, $u_z(0, \theta, z)$ increases with z, then reaches a maximal value and then decreases as z increases; we have first $\frac{\partial u_z}{\partial z}(0, \theta, z) > 0$, then $\frac{\partial u_z}{\partial z}(0, \theta, z_{\max}) = 0$, $\{0, \theta, z_{\max}\}$ being the point of the axis where the function $z \to u_z(0, \theta, z)$ reaches its maximal value and, finally, $\frac{\partial u_z}{\partial z}(0, \theta, z) < 0$ until the transition region encounters the plug region downstream. This behavior of $\frac{\partial u_z}{\partial z}$ along the pipe axis in the transition zone prevents the formation of a plug containing the axis and is consistent with the yielded feature of this transition region.

24.6. Further comments on the simulation of a steady nonisothermal incompressible Bingham flow in a pipeline

In the preceding paragraphs of Section 24, we have addressed, through a transient approach, the numerical simulation of a steady incompressible nonisothermal viscoplastic flow in a pipeline. The rheological model that we considered was of the Bingham type with viscosity or yield stress dependent of the temperature. In addition to the numerical experiments related to the above nonisothermal situations, we also carried out (see Section 24.5) the numerical simulation of incompressible isothermal Bingham flow in pipelines, in order to validate our finite volume/augmented Lagrangian methodology from an accuracy standpoint. The nonisothermal results discussed in Sections 24.5.6 and 24.5.7 highlight the strong sensitivity of the flow pattern to temperature changes. It is our opinion that the main result of these investigations can be stated as follows:

In pipe flow situations where the rheological parameters (viscosity and yield stress) of the Bingham viscoplastic model depend of the temperature, the fluid is yielded in those flow regions where the temperature varies.

Indeed, we have shown that small variations of the temperature (e.g., $\Delta\Theta = 1°K$) are sufficient to trigger the full yielding of the transition region. From a practical and experimental standpoint, it is likely that the distinction between unyielded and slightly yielded is not as clear-cut as we would like, but it is an important notion from a conceptual point of view, as it is mathematically and computationally. It seems, in particular, that our augmented Lagrangian-based methodology is better suited than regularization methods when it comes to the identification of the yielded and unyielded regions of the flow.

Those readers, looking for more realistic situations, may be tempted by the simulation of viscoplastic flow where viscosity and yield stress are both temperature dependent, the temperature dependence being described by relations more realistic than those in Sections 24.5.6 and 24.5.7 (such as WLF or Arrhenius equations). Although an experimental survey would be difficult to carry out, comparisons with experimental results for nonisothermal viscoplastic flow would be a most interesting achievement in order to verify if indeed the flow is yielded in those regions where $\boldsymbol{\nabla}\Theta \neq \mathbf{0}$.

Returning to the waxy crude oil flow problem, the next step will be the restarting issue, that is, investigating the early transients of the flow, starting from a quiescent state. To restart problems, the key feature is no longer the thermal dependence, but rather the compressibility of the fluid.

25. Transient isothermal compressible viscoplastic flow in a pipeline

We focus now on the restart problem, a problem in which the oil compressibility plays a significant role. We will assume that the flow is isothermal and that the waxy crude oil rheological behavior can be properly described by a compressible Bingham model. The geometry of the flow region is like the one encountered in Section 24.

25.1. Governing equations

We assume that the fluid under consideration is Stokesian, that is the viscosity forces are due to shear only, and not to volume variations, implying that $\xi = 0$, ξ being the second viscosity. From these hypotheses, the isothermal flow of a compressible Bingham fluid is governed by the following system of equations and inequalities:

$$\chi_\Theta \left(\frac{\partial p}{\partial t} + \mathbf{u} \cdot \nabla p \right) + \nabla \cdot \mathbf{u} = 0, \tag{25.1}$$

$$\rho \left[\frac{\partial \mathbf{u}}{\partial t} + (\mathbf{u} \cdot \nabla)\mathbf{u} \right] + \nabla p = \nabla \cdot \boldsymbol{\tau}, \tag{25.2}$$

$$\begin{cases} \boldsymbol{\tau} = 2\mu \left[\mathbf{D}(\mathbf{u}) - \dfrac{1}{3}(\nabla \cdot \mathbf{u})\mathbf{I} \right] + \tau_y \dfrac{\mathbf{D}(\mathbf{u})}{\|\mathbf{D}(\mathbf{u})\|} & \text{if } \|\boldsymbol{\tau}\| > \tau_y, \\ \mathbf{D}(\mathbf{u}) = \mathbf{0} & \text{if } \|\boldsymbol{\tau}\| \le \tau_y. \end{cases} \tag{25.3}$$

From a physical point of view, the flow is characterized by the following three dimensionless numbers: the Bingham number $\mathcal{B}\mathrm{n}$, the Reynolds number $\mathcal{R}\mathrm{e}$, and the compressibility coefficient χ'. For characteristic velocity, we take $\bar{U} = u_{z\text{-max}}$, $u_{z\text{-max}}$ being the maximal value of u_z. In the case of pipeline flows, χ' varies usually in the range $10^{-9} - 10^{-3}$.

25.2. *Flow geometry and boundary conditions*

The flow region is as in Section 24, that is, is axisymmetric; it has been visualized in Fig. 25.1. As before, we use a system $\{r, \theta, z\}$ of cylindrical coordinates to describe the flow region and we assume that $u_\theta \equiv 0$, u_θ being the ortho-radial component of the velocity. The boundary conditions verified by the flow read as follows:

- *At the inlet*
  Fully developed Dirichlet conditions are prescribed for the pressure and for the radial component of the velocity; we assume also that the axial component of the extra-stress tensor vanishes at $z = 0$. To summarize:

$$u_r|_{z=0} = 0, \;\; \tau_{zz}|_{z=0} = 0, \tag{25.4}$$

  and

$$p|_{z=0} = P_{\text{Inlet}}. \tag{25.5}$$

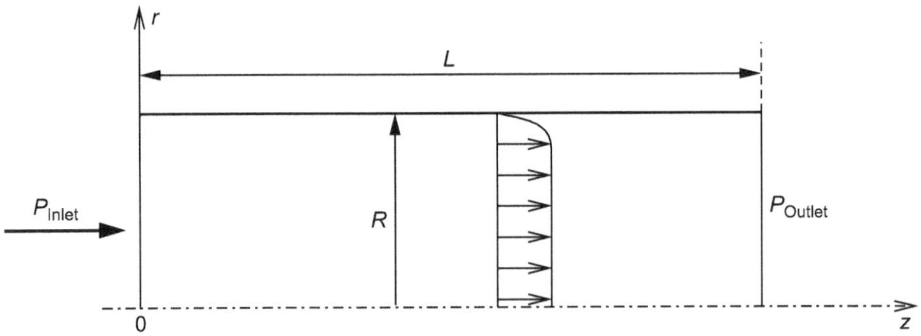

FIG. 25.1 Flow geometry and some boundary conditions.

- *At the wall*
  A no-slip boundary condition is prescribed for the velocity at the wall, that is

$$u_r|_{r=R} = u_z|_{r=R} = 0. \tag{25.6}$$

- *Along the axis of the pipe*

$$u_r|_{r=0} = 0, \ \ \tau_{rz}|_{r=0} = 0. \tag{25.7}$$

- *At the outlet*
  Dirichlet conditions are prescribed for the pressure and for the radial component of the velocity; we assume also that the axial component of the extra-stress tensor vanishes at $z = L$. To summarize:

$$u_r|_{z=L} = 0, \ \ \tau_{zz}|_{z=L} = 0, \tag{25.8}$$

  and

$$p|_{z=L} = P_{\text{Outlet}}. \tag{25.9}$$

The pressure being defined modulo an additive constant, we assume from now on that $P_{\text{Outlet}} = 0$.

25.3. *Numerical results and discussion*

25.3.1. *Generalities*

The flow simulations have been performed using the methodology discussed in Section 20.3; they correspond to the real-time evolution (in the physical sense) of the flow. For all the computations we have assumed that at $t = 0$, the flow is at rest ($\boldsymbol{u} = \mathbf{0}$ and $p = 0$ in the flow region). Our numerical results are presented and discussed in terms of dimensionless physical variables, namely, the dimensionless coordinates $r^* = r/R$ and $z^* = z/L$, the dimensionless time $t^* = t/t_{\text{ref}}$ (with $t_{\text{ref}} = R/u_{z\text{-max}}$), the dimensionless axial velocity $u_z/u_{z\text{-max}}$, the dimensionless pressure $\dfrac{p}{\rho_0 u_{z\text{-max}}^2}$ and, finally, the dimensionless compressibility coefficient χ'. Concerning the compressibility, our computations have been done with χ' varying from 1.38×10^{-9} to 1.45×10^{-4} because these values agree with those mentioned in Section 25.1 for pipeline flows. We observe that the definition of several of the above dimensionless variables requires dividing by the maximal value $u_{z\text{-max}}$ of the axial velocity; if the flow stops, then $u_{z\text{-max}} = 0$, implying that χ' and the characteristic pressure vanish, making impossible the definition of t_{ref}, $\mathcal{B}\text{n}$, and $\mathcal{R}\text{e}$. To avoid those difficulties associated with $u_{z\text{-max}} = 0$, we will take from now on $p^* = p/P_{\text{Inlet}}$ as dimensionless pressure. Similarly, we introduce new Bingham and compressibility numbers; these are the dimensionless quantities defined, respectively, by

$$\mathcal{B}\text{n}^* = \frac{2\tau_y}{\varpi R} \tag{25.10}$$

and

$$\chi^* = \chi_\Theta(p_{\text{Inlet}} - P_{\text{Outlet}}), \tag{25.11}$$

where $\varpi = \dfrac{P_{\text{Inlet}} - P_{\text{Outlet}}}{L}$, that is the pressure drop per unit length. Under steady flowing conditions, the Bingham fluid flows in the pipe if $0 < \mathcal{B}n^* < 1$, whereas there is no flow if $\mathcal{B}n^* \geq 1$. If there is no flow due to $\mathcal{B}n^* \geq 1$, we arbitrarily set $t_{\text{ref}} = 1$ and use χ^* as compressibility coefficient.

25.3.2. Pipe dimensions and influence of the mesh size

The pipe geometry is such that $L/R = 200$, the notation being as in Fig. 25.1. For the implementation of our finite volume method, we use a Cartesian grid in the $\{r, z\}$ space, with constant grid size in each direction. We have then $\Delta r = R/N_r$ and $\Delta z = L/N_z$, N_r, and N_z being two positive integers; we denote by R_c the ratio $\dfrac{\Delta r}{\Delta z}$. Several meshes have been tested, with their characteristics being reported in Table 25.1.

Influence of Δr: The influence of Δr has been reported in the fourth column of Table 25.1, with ε_r defined as in Section 24.5.4, that is, by comparing the computed axial velocity of a fully developed incompressible Bingham flow to its exact value given by relation (16.2) in Chapter 2, Section 16. These results suggest that $\varepsilon_r \approx O(\Delta r)$ and that $N_r = 20$ provides a reasonable trade-off between accuracy and computational time (see also Section 24.5.4).

Influence of Δz: To assess the influence of Δz, we have considered a transient compressible Bingham flow and proceeded essentially like in Section 24.5.4, the difference being that the pressure plays here the role the temperature played there. All the simulations have been done with $N_r = 20$, and N_z varying from 30 to 300 (that is with Meshes 2–6). At a given time (whose exact value is irrelevant), we have computed the following error indicator

$$\varepsilon_z = \frac{|P_{\text{average}}(N_z) - P_{\text{average}}(N_z')|}{P_{\text{average}}(N_z')}, \tag{25.12}$$

TABLE 25.1
Mesh characteristics and influence of the mesh size for transient compressible Bingham flow

Meshes	$N_r \times N_z$	*Mesh ratio* R_c	ε_r	ε_Θ
Mesh 1	10 × 50	40	2.56×10^{-2}	
Mesh 2	20 × 30	133	0.99×10^{-2}	1.60×10^{-2}
Mesh 3	20 ×50	80	1.03×10^{-2}	1.01×10^{-2}
Mesh 4	20 × 100	40	1.03×10^{-2}	0.55×10^{-2}
Mesh 5	20× 200	20	1.02×10^{-2}	0.31×10^{-2}
Mesh 6	20 × 300	13	1.02×10^{-2}	
Mesh 7	30 × 50	120	0.51×10^{-2}	
Mesh 8	40 ×50	160	0.51×10^{-2}	

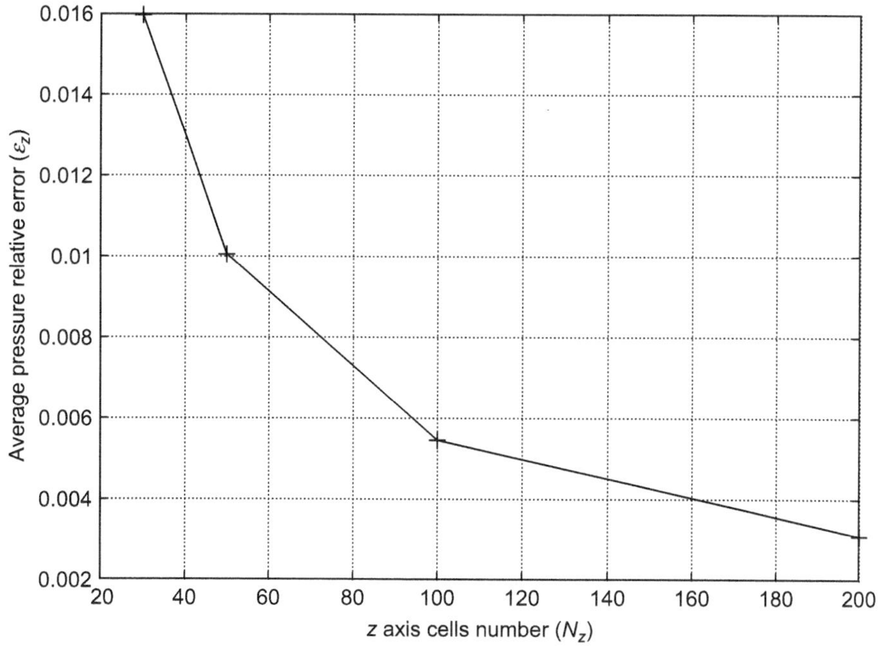

FIG. 25.2 Influence of N_z on ε_z.

where $P_{\text{average}} = \dfrac{2}{LR^2}\int_{(0,R)\times(0,L)} pr\,dr\,dz$ denotes the average pressure in the pipeline and N_z and N_z' the number of control volumes in the Oz-direction for two consecutive meshes (for example, for Mesh 2 we have $N_z = 30$ and $N_z' = 50$). The fifth column of Table 25.1 and Fig. 25.2 show that the smaller is Δz, the smaller is ε_z. However, the improvement is not significant for $N_z \geq 100$, as shown by Fig. 25.2, explaining why we have taken Mesh 4 ($N_r = 20$ and $N_z = 100$) for the numerical experiments to be discussed hereafter.

25.3.3. Convergence properties of the iterative methods

In order to compute the approximate solutions, we rely on two nested algorithms, namely MUA (for **M**odified **U**zawa **A**lgorithm) described in Section 20, and MSA (for **M**odified **S**tokes **A**lgorithm) described in Section 22. For our test problems, MUA converges for every time step, and the fully discrete analog of scheme (20.46)–(20.59) provides a steady-state solution; our computations have been performed with $tol_1 = tol_2 = tol_3 = 10^{-5}$, as recommended in Section 24.5.3. In order to assess the convergence properties, all the results presented in this section have been obtained at the first time step of scheme (20.46)–(20.59), this step being by far the most demanding from a computational point of view.

Generalizing the usual Uzawa transient algorithm (that is the one associated with incompressible isothermal Bingham flows), in order to handle compressible flows, does not bring any particular complication. In particular, MUA still has good convergence properties and it seems, actually, that compressibility enhances convergence as shown in Fig. 25.3.

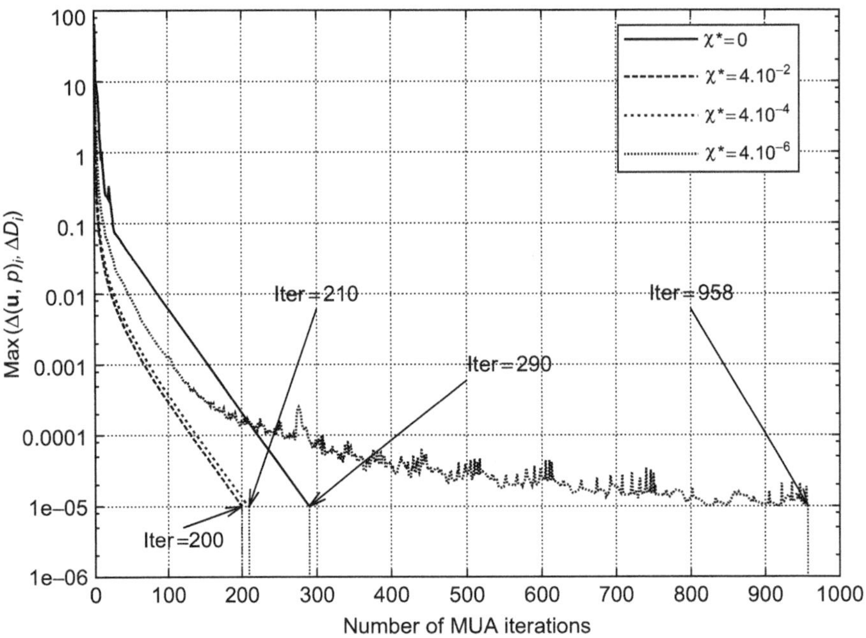

FIG. 25.3 Influence of the compressibility on the speed of convergence of MUA.

For all our test problems, the MSA algorithm discussed in Section 22 always converged at the first step of MUA, according to the value of *tol* in the stopping criterion; actually, we can chose *tol* as small as 10^{-10} and still have good convergence properties, explaining why we took precisely $tol = 10^{-10}$. In Fig. 25.4, we have visualized the influence of the compressibility on the speed of convergence of MSA "inside" the first MUA iteration. The compressibility coefficient plays here the role of a relaxation factor, because the larger is χ^*, the faster is the convergence: if the fluid is incompressible ($\chi^* = 0$), the Stokes system is not relaxed and the number of iterations required for the convergence is large. From these observations, we can claim that compressibility improves the convergence properties of MSA (an Uzawa/conjugate gradient-like algorithm).

As mentioned earlier, we observe that for the first iteration of MUA, the smaller is χ^*, the slower is the convergence of MSA (495 MSA iterations at $\chi^* = 0$, versus 35 iterations at $\chi^* = 4 \times 10^{-2}$ ($\chi' = 1.38 \times 10^{-5}$)). What about the second iteration of MUA? If $\chi^* = 4 \times 10^{-2}$, the number of iterations of MSA reduces to 20; however, if $\chi^* = 0$ only one iteration of MSA is needed to achieve convergence. This phenomenon can be explained as follows: the pressure is imposed at the pipe inlet and outlet; MSA used to solve the generalized Stokes problem (20.51)–(20.53) is a pressure-driven algorithm, which, in the incompressible case, identifies the pressure solution at the first iteration of MUA. From this property, in the incompressible case, only one iteration of MSA is needed for the kth iteration of MUA if $k \geq 2$, a remarkable property, indeed.

Having verified in the present section and in the preceding one the properties of our finite-volume/augmented Lagrangian methodology, we are going to use it, in the following sections, to explore various scenarios concerning compressible viscoplastic flows in pipelines.

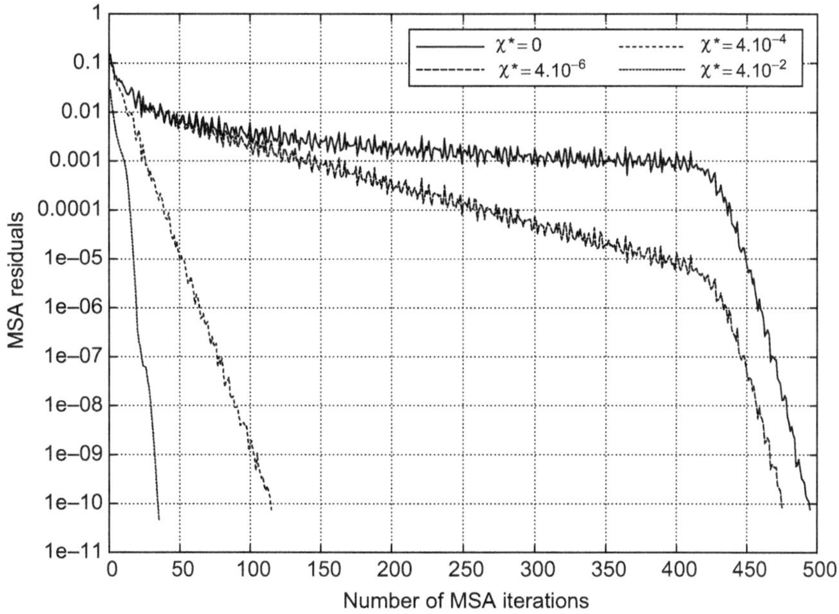

FIG. 25.4 Influence of the compressibility on the speed of convergence of MSA at the first iteration of MUA.

25.3.4. *Simulation of a Newtonian compressible flow with* $\chi^* = 8 \times 10^{-2} (\chi' = 8.84 \times 10^{-6})$ *and* $\mathcal{R}\mathrm{e} = 8.84 \times 10^{-2}$

In this paragraph, the results of the numerical simulation of a compressible Newtonian flow are presented, assuming that $\chi^* = 8 \times 10^{-2}$ $(\chi' = 8.84 \times 10^{-6})$ and $\mathcal{R}\mathrm{e} = 8.84 \times 10^{-2}$. We suppose that at $t^* = 0$, we have $\mathbf{u} = \mathbf{0}$ and $p = 0$.

The time variations of the computed dimensionless inlet and outlet mass flow rates have been represented on Fig. 25.5, using Q_{Steady} as reference mass flow rate (Q_{Steady} being the steady mass flow rate at the pipe inlet). In Fig. 25.5, we observe that, because of the fluid compressibility, the inlet mass flow rate enjoys, just after the start, a strong peak, exceeding by far the steady mass flow rate.

In Fig. 25.6, we have represented, for various values of t^*, the variations of the dimensionless pressure $p^* (= p/P_{\mathrm{Inlet}})$ as a function of z^*. The exponential and purely convex form of the pressure profiles is typical of a classical Newtonian compressible flow. A steady state has been reached when the inlet and outlet mass flow rates are equal and when p^* becomes an affine function of z^* (at $t^* = 46.16$).

25.3.5. *Simulation of a compressible Bingham flow for* $\chi^* = 4 \times 10^{-2}$ $(\chi' = 1.38 \times 10^{-5})$, $\mathcal{B}\mathrm{n}^* = 0.5$ $(\mathcal{B}\mathrm{n} = 4)$, *and* $\mathcal{R}\mathrm{e} = 1.11$

In this paragraph, we consider the numerical simulation of the flow of a compressible Bingham fluid for $\chi^* = 4 \times 10^{-2} (\chi' = 1.38 \times 10^{-5})$, $\mathcal{B}\mathrm{n}^* = 0.5 (\mathcal{B}\mathrm{n} = 4)$, and $\mathcal{R}\mathrm{e} = 1.11$. We assume that $\mathbf{u} = \mathbf{0}$ and $p = 0$ at $t^* = 0$.

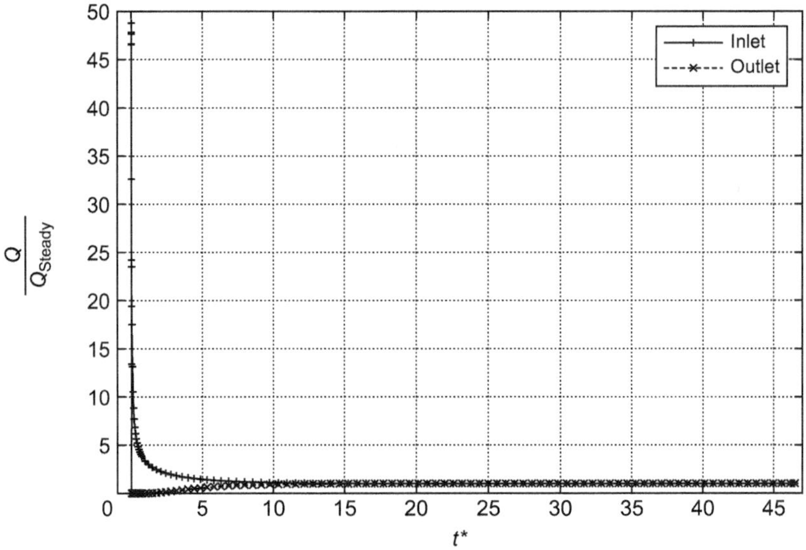

FIG. 25.5 Time evolution of the inlet (top) and outlet (bottom) mass flow rates for a Newtonian compressible flow [$\chi^* = 8 \times 10^{-2}(\chi^* = 8.84 \times 10^{-6})$ and $\mathcal{R}e = 8.84 \times 10^{-2}$].

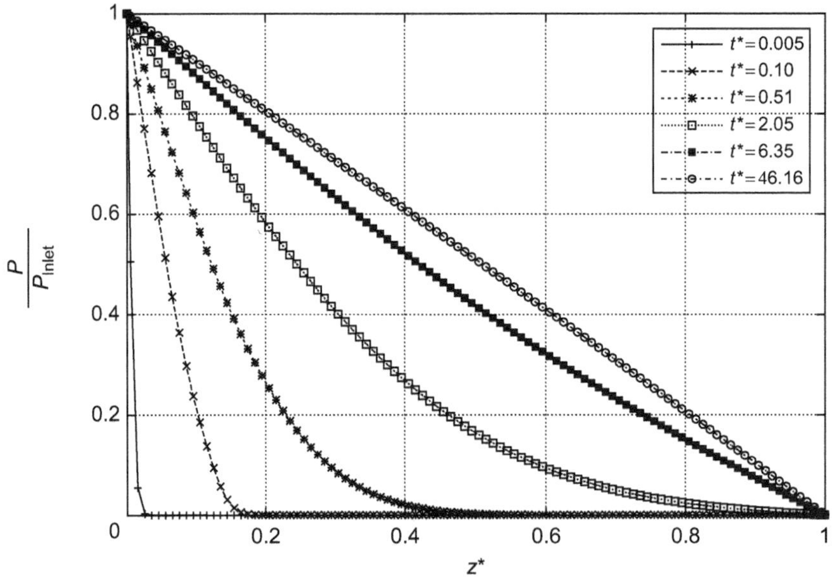

FIG. 25.6 Space-time evolution of the pressure for a Newtonian compressible flow [$\chi^* = 8 \times 10^{-2}(\chi^* = 8.84 \times 10^{-6})$ and $\mathcal{R}e = 8.84 \times 10^{-2}$].

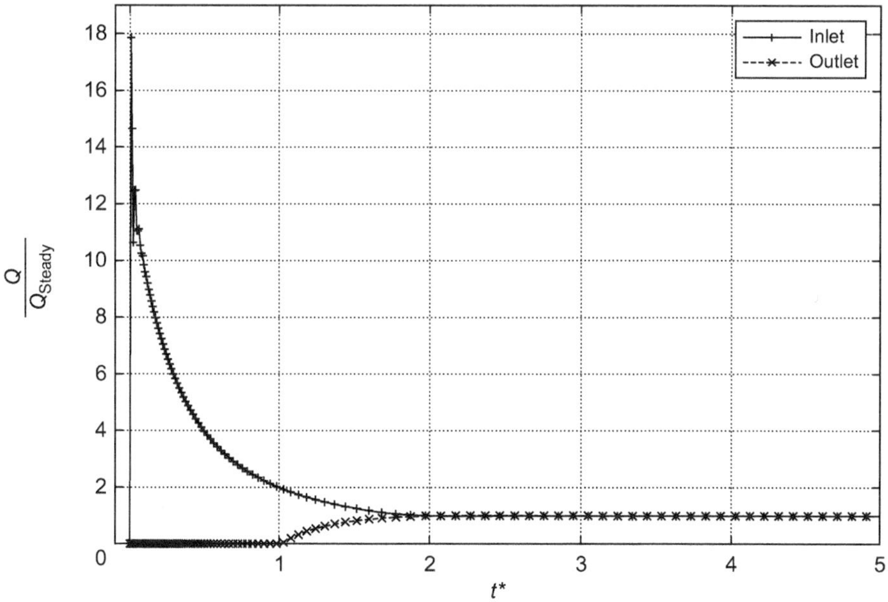

FIG. 25.7 Time evolution of the inlet (top) and outlet (bottom) mass flow rates for a compressible Bingham flow for $\chi^* = 4 \times 10^{-2}(\chi' = 1.38 \times 10^{-5})$, $\mathcal{B}\text{n}^* = 0.5$ ($\mathcal{B}\text{n} = 4$), and $\mathcal{R}\text{e} = 1.11$.

In Fig. 25.7, we have visualized the dimensionless inlet (top) and outlet (bottom) mass flow rates as functions of t^*, using Q_{Steady} as mass flow rate of reference. For this test problem, we observe that, as in the preceding Newtonian compressible case, the inlet mass flow rates shows a strong peak just after the start; both mass flow rates coincide at steady state.

In Fig. 25.8, we have represented, for various values of t^*, the dimensionless pressure p^* as a function of z^*. We observe a pressure peak close to the inlet at $t^* = 0.023$; actually, such a peak has been reported in CAWKWELL and CHARLES [1987] and put it down to compressibility and inertia. Moreover, contrary to the Newtonian compressible case, one observes that for t^* small enough, the pressure profile may exhibit an inflexion point, as shown in Fig. 25.8 for $t^* = 0.189$; for t^* large enough, the pressure profiles is convex and ends up affine at steady state.

Let us call compression front the point of the pipe axis at the interface between the positive pressure region and the zero pressure region. As shown on Fig. 25.8, the pressure is positive upstream the compression front, and is zero downstream. The pressure profiles exhibit at the compression front a slope discontinuity which disappears, only when this front reaches the outlet. This particular form of the pressure profile results, very likely, from the combined effects of the compressibility and viscoplasticity of the fluid. A close inspection of Fig. 25.8 shows that at $t^* = 1.975$, the pressure has practically reached its steady-state values because it coincides quite accurately with the pressure at $t^* = 5.013$, the value of t^* at which the velocity reaches its steady state, according to the value of tol_3 in (20.62) ($tol_3 = 10^{-5}$ here).

The time evolution of the axial velocity profile at the pipe outlet has been visualized on Fig. 25.9. Comparing the profiles at $t^* = 1.975$ and $t^* = 5.013$ strongly suggests that the

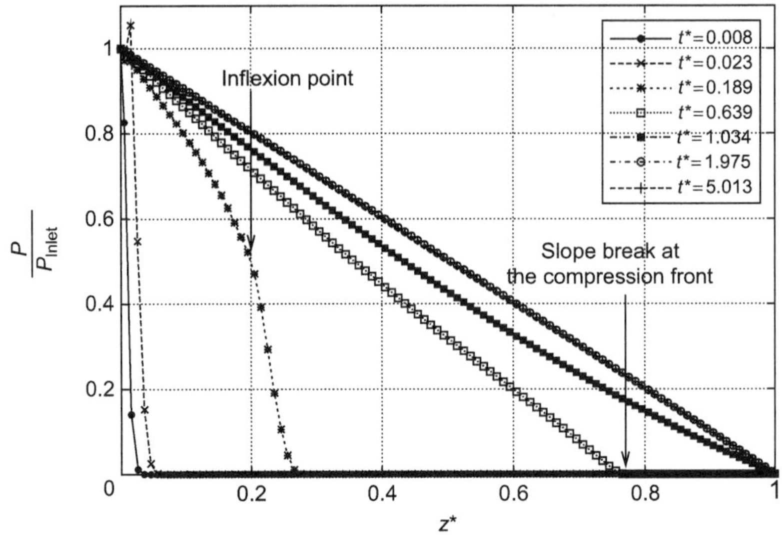

FIG. 25.8 Space-time evolution of the pressure for a compressible Bingham flow for $\chi^* = 4 \times 10^{-2} (\chi' = 1.38 \times 10^{-5})$, $\mathcal{B}n^* = 0.5$ ($\mathcal{B}n = 4$), and $\mathcal{R}e = 1.11$.

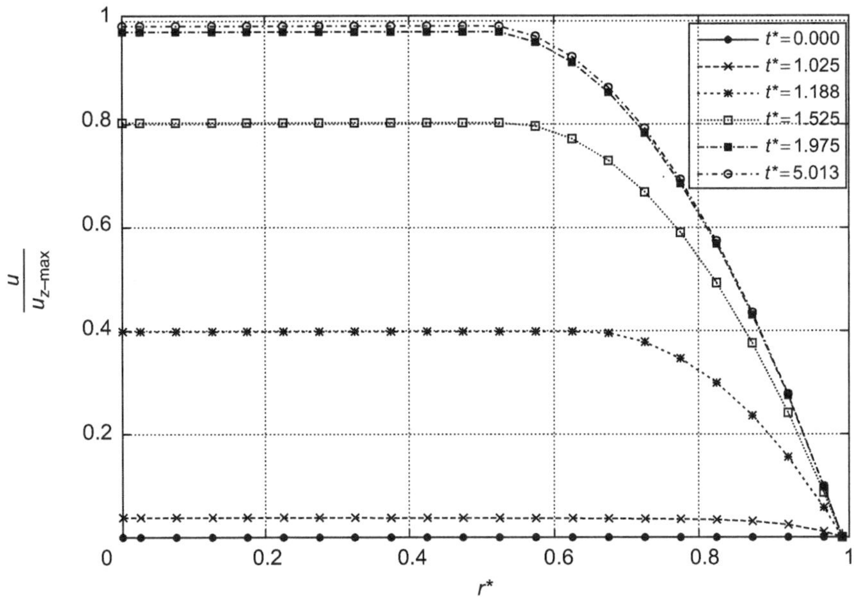

FIG. 25.9 Time evolution of the axial velocity profile at the pipe outlet for a compressible Bingham flow for $\chi^* = 4 \times 10^{-2} (\chi' = 1.38 \times 10^{-5})$, $\mathcal{B}n^* = 0.5$ ($\mathcal{B}n = 4$), and $\mathcal{R}e = 1.11$.

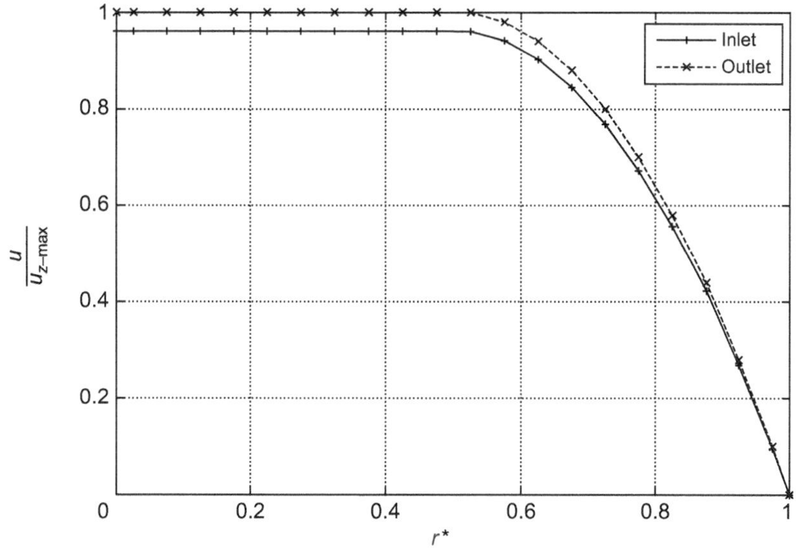

FIG. 25.10 Inlet and outlet axial velocity profiles for a compressible Bingham flow for $\chi^* = 4 \times 10^{-2}(\chi' = 1.38 \times 10^{-5})$, $\mathcal{B}n^* = 0.5$ ($\mathcal{B}n = 4$), and $\mathcal{R}e = 1.11$.

flow has practically reached its steady state at $t^* \approx 5$. The axial velocity profile at $t^* = 5.013$ has the features of a Bingham–Poiseuille one, with a plug region in the center whose radius is half the pipe radius. However, according to Fig. 25.10, at steady state, the inlet and outlet axial velocity profiles do not match (this follows from (19.3) which implies that at steady state $\nabla \cdot \mathbf{u} \neq 0$); actually, $\nabla \cdot \mathbf{u} \neq 0$ implies that $\mathbf{D}(\mathbf{u}) \neq 0$, that is, as a result of the compressibility, the flow is yielded at steady state. This analysis is confirmed by Fig. 25.11, which shows the time evolution of the yielded (white) and unyielded (black) regions. Actually, the above figures show that the compression front is also at the interface separating the yielded and unyielded regions.

25.3.6. Influence of the compressibility on the restart of a Bingham flow

In this paragraph, we are going to compare the results of the numerical simulation of three compressible Bingham flows which share $\mathcal{B}n* = 0.5$ ($\mathcal{B}n = 4$) and $\mathcal{R}e = 1.11$, but differ by the compressibility that takes here the following values: $\chi^* = 4 \times 10^{-6}, 4 \times 10^{-4}$, and 4×10^{-2} (corresponding to $\chi' = 1.38 \times 10^{-9}$, 1.38×10^{-7}, and 1.38×10^{-5}). In Fig. 25.12, we have visualized the time evolution of the inlet and outlet mass flow rates for the above three values of χ^*. The influence of the compressibility can be evaluated at the pipe inlet by observing the magnitude of the mass flow rate peak as a function of χ^*. In fact, the more compressible is the flow, the higher is the peak of the mass flow rate at the inlet. Moreover, the smaller is χ^*, the sooner the flow restarts at the outlet (for example, the outlet restart time at $\chi^* = 4 \times 10^{-2}$ is four times larger than it is at $\chi^* = 4 \times 10^{-4}$). More generally, the smaller is χ^*, the sooner the steady flow is reached. As mentioned earlier, one can find in DAVIDSON, NGUYEN, CHANG and RONNINGSEN [2004] the discussion of a global one-dimensional approach to deal with the restart of waxy crude oil flows in pipelines; our

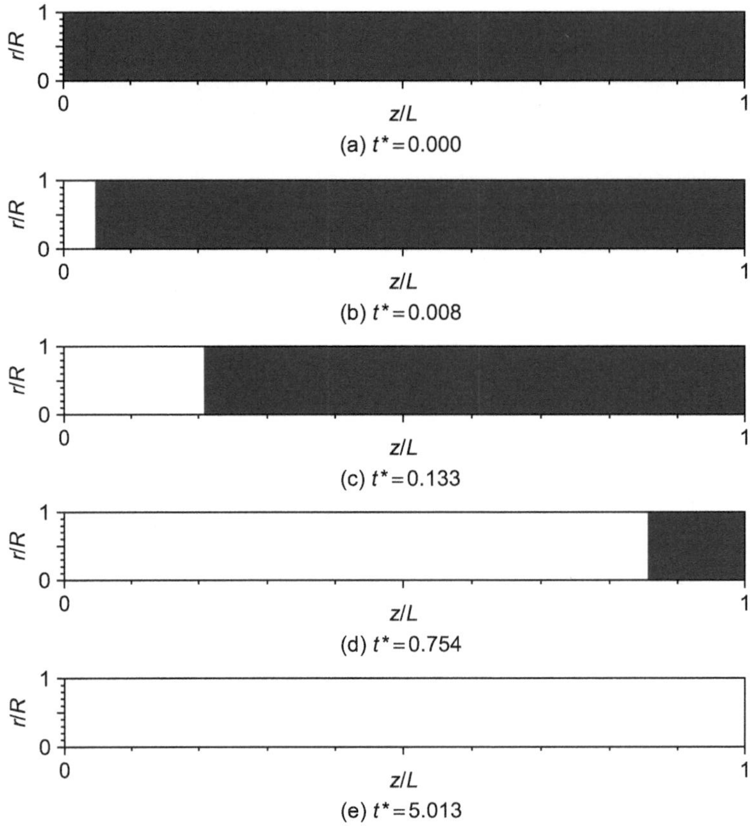

FIG. 25.11 Time evolution of the yielded (white) and unyielded (black) regions for a compressible Bingham flow for $\chi^* = 4 \times 10^{-2}(\chi' = 1.38 \times 10^{-5})$, $\mathcal{B}n^* = 0.5$ ($\mathcal{B}n = 4$), and $\mathcal{R}e = 1.11$.

approach is quite different because, in the above publication, the authors used a sophisticated one-dimensional model to take into account all the known properties of waxy crude oils (compressibility, viscoplasticity, thixotropy, and so on). Concerning, however, the influence of the compressibility on the restart time of a viscoplastic flow, the results we obtained, showing that the restart time increases with the compressibility, are in good agreement with those in the above reference.

In Fig. 25.13, we have visualized the time evolution of the pressure p^* for $\chi^* = 4 \times 10^{-6}$ and 4×10^{-4}. For these values of χ^*, the pressure peak observed for $\chi^* = 4 \times 10^{-2}$ does not exist. Concerning the inflexion point, observed in Fig. 25.8 for $t^* = 0.189$, this feature is still present for $\chi^* = 4 \times 10^{-4}$, but is gone for 4×10^{-6}. However, as it was the case for $\chi^* = 4 \times 10^{-2}$, we still observe a breaking of the slope at the compression front for the two above values of χ^* (as shown at $t^* = 0.023$ and $z^* \approx 0.5$ in Fig. 25.13(a), and at $t^* = 6 \times 10^{-4}$ and $z^* \approx 0.3$ in Fig. 25.13(b)). Actually, Fig. 25.13(a,b) shows that before stabilizing, the pressure oscillates around its final steady state (an affine function of z^*) until the oscillations are finally damped out.

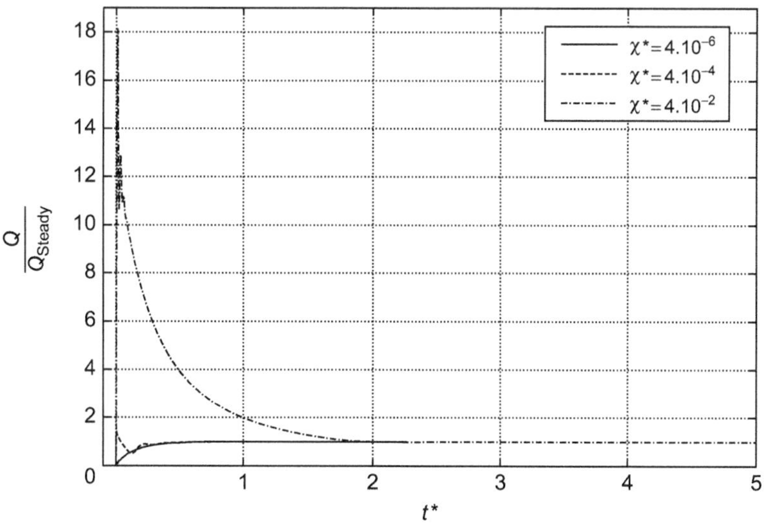

(a) Inlet mass flow rate

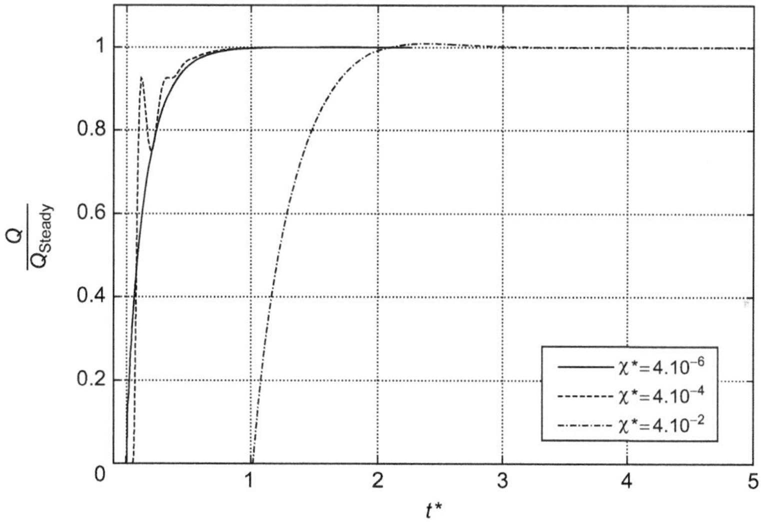

(b) Outlet mass flow rate

FIG. 25.12 Time evolution of the inlet (a) and outlet (b) mass flow rates for $\chi^* = 4 \times 10^{-6}$, 4×10^{-4}, 4×10^{-2} (compressible viscoplastic flow with $\mathcal{B}n^* = 0.5$ and $\mathcal{R}e = 1.11$).

Schematically, we have identified three regimes concerning the family of compressible viscoplastic flows we just investigated:

1. *The low-compressibility case* ($\chi^* = 4 \times 10^{-6}$)

   It follows from Fig. 25.12 that it takes a very short time before the flow restart taking place at the inlet reaches the outlet. Moreover, the pressure along the pipe

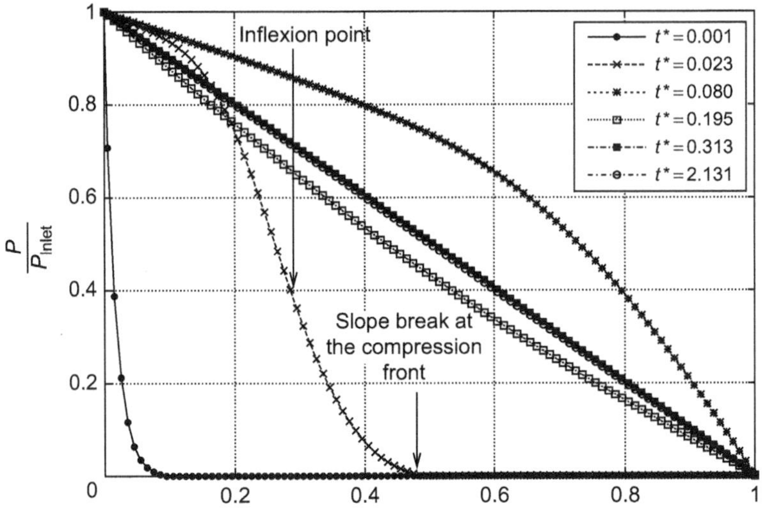

(a) $\chi^*=4.10^{-4}$

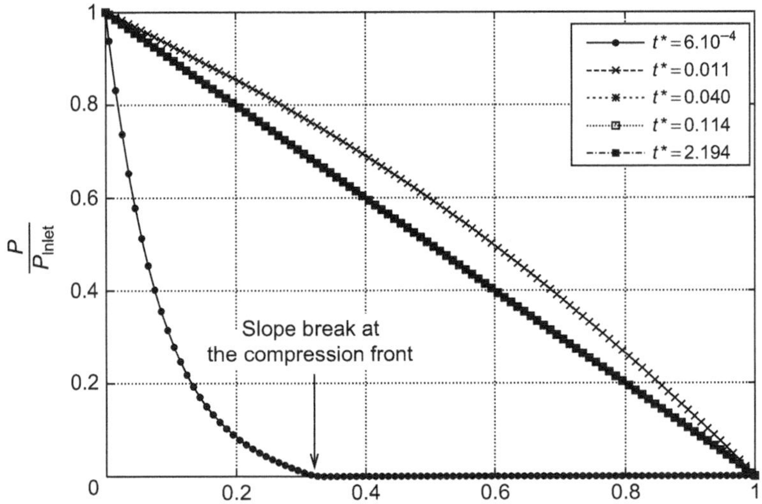

(b) $\chi^*=4.10^{-6}$

FIG. 25.13 Time evolution of the *pressure* for $\chi^* = 4 \times 10^{-6},\ 4 \times 10^{-4}$, (compressible viscoplastic flow with $\mathcal{B}n^* = 0.5$ and $\mathcal{R}e = 1.11$).

reaches its final distribution with one oscillation at most. Indeed, this case is very close to an incompressible one.

2. *The moderate compressibility case* ($\chi^* = 4 \times 10^{-4}$)

   Figure 25.12(b) shows that the time necessary for the flow restart at the inlet, to reach the outlet, is of the order of 5% of the time necessary to reach the steady state.

The above figure shows also that significant oscillations of the mass flow rate take place at the outlet before steady state is reached. Accordingly, Fig. 25.13(a) shows significant oscillations of the pressure profile before it reaches its steady state, the amplitude of these oscillations being larger than the one in Fig. 25.13(b) for $\chi^* = 4 \times 10^{-6}$.

3. *The high-compressibility case* ($\chi^* = 4 \times 10^{-2}$)
   It follows from Fig. 25.12(b) that the time it takes for the flow restart at the inlet, to reach the outlet, is of the order of 25% of the time necessary to reach steady state. The above figure shows also that the mass flow rate reaches its steady-state value without oscillating. Accordingly, Fig. 25.8 shows that the pressure reaches its steady state without oscillating.

25.3.7. Influence of the Bingham number on a compressible Bingham flow

In this paragraph, we are going to investigate the influence of the Bingham number on the flow of a compressible fluid; concerning the choice of Bingham, Reynolds and compressibility numbers we have retained the following triples for $\{\mathcal{B}\text{n}^*, \mathcal{R}\text{e}, \chi^*\}$: $\{0.1, 3.58, 4 \times 10^{-2}\}$ and $\{0.5, 1.11, 4 \times 10^{-2}\}$; actually, the second case has been already investigated in Section 25.3.5.

In Fig. 25.14, we have visualized the time evolution of the dimensionless inlet (a) and outlet (b) mass flow rates. For the two cases considered here, the dimensionless mass flow rate is defined using Q_{Steady} as reference mass flow rate, Q_{Steady} being the mass flow rate at steady state; however, to define t^* we have used as reference time the time t_{ref} associated with $\mathcal{B}\text{n}^* = 0.5$ (see Section 25.3.1 for the definition of t_{ref}). The inlet mass flow rates peak just after the start; because χ^* is the same for both flows, the peak values are also the same. However, at steady state, the mass flow rate depends on the Bingham number. The yield stress being smaller at $\mathcal{B}\text{n}^* = 0.1$ than at $\mathcal{B}\text{n}^* = 0.5$, the apparent viscosity is also smaller implying that the steady-state mass flow rate is larger.

Moreover, Fig. 25.14 shows also that the flow restarts at the outlet earlier for the smaller value of $\mathcal{B}\text{n}^*$. Actually, in VERSCHUUR, VERHEUL and DEN HARTOG [1971], one can find experimental results concerning the prediction of the pressure required to restart a pipeline containing a gelled waxy crude oil. In the above publication, it is shown that the time necessary to restart a pilot pipeline depends of the value of the yield stress: the smaller is the yield stress, the easier and quicker is the flow restart. From the comparison of these experimental results with our computational ones, we can conclude that our numerical methodology has provided a simulator able to predict properly real trends.

25.3.8. Numerical simulation of two compressible Bingham flows for $\mathcal{B}\text{n}^* = 1.1$ *(*$\chi^* = 4 \times 10^{-6}$ *and* $\chi^* = 4 \times 10^{-2}$*)*

In this paragraph, we are going to investigate the restart of two compressible Bingham flows sharing the same Bingham number ($\mathcal{B}\text{n}^* = 1.1$, here) but differing by their compressibility coefficient ($\chi^* = 4 \times 10^{-6}$ and $\chi^* = 4 \times 10^{-2}$, here). If the fluid was incompressible (that is if $\chi^* = 0$), we should be in a no-flow situation because $\mathcal{B}\text{n}^* > 1$.

Let us consider first the higher compressibility case ($\chi^* = 4 \times 10^{-2}$): on Fig. 25.15, we have visualized the time evolution of the inlet and outlet dimensionless mass flow rates on the time interval [0, 50], assuming that at $t^* = 0$, the flow is at rest. Due to the compressibility, the inlet mass flow rate reaches a peak immediately after the restart, and then decreases

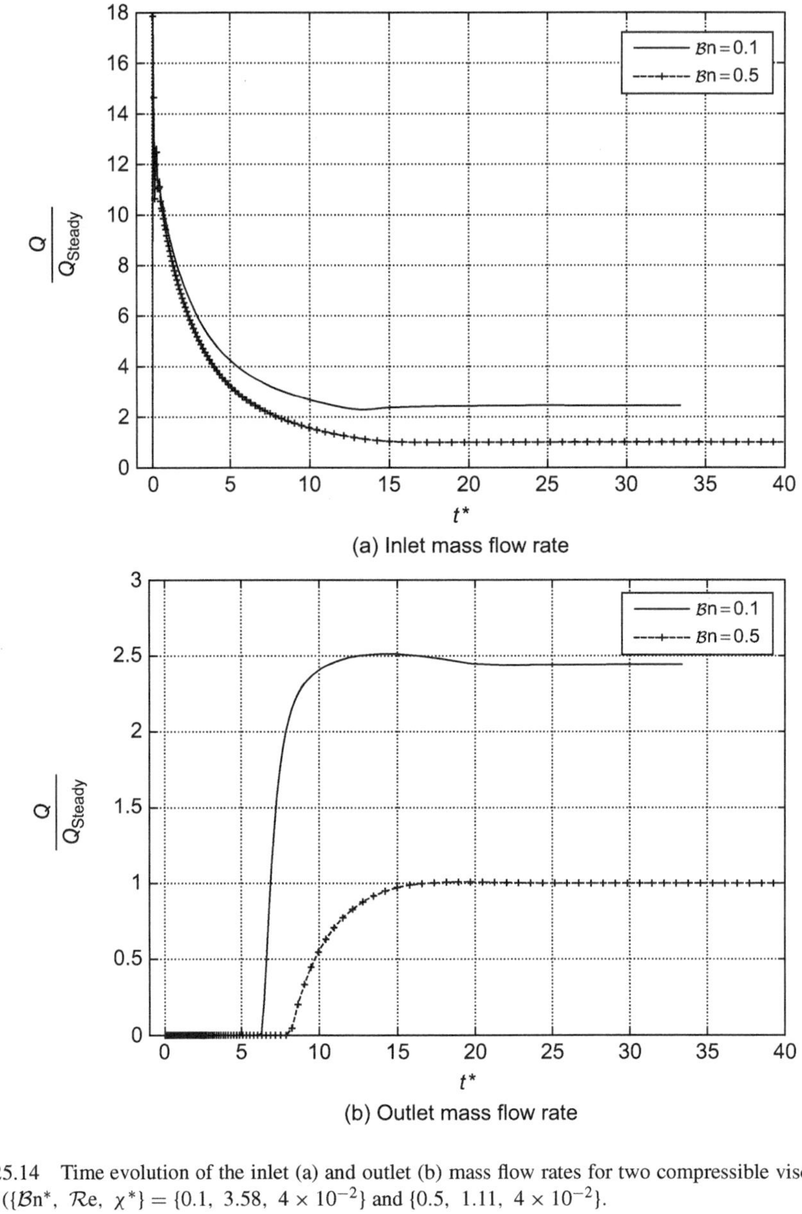

FIG. 25.14 Time evolution of the inlet (a) and outlet (b) mass flow rates for two compressible viscoplastic flows ($\{\mathcal{B}n^*, \mathcal{R}e, \chi^*\} = \{0.1, 3.58, 4 \times 10^{-2}\}$ and $\{0.5, 1.11, 4 \times 10^{-2}\}$.

very quickly to zero. However, the outlet mass flow rate never moves away from zero. A closer inspection shows that the no-flow situation prevails in the whole pipe at $t^* \approx 760$ because the inlet pressure is not high enough to maintain the flow. A further illustration of the above phenomenon is provided by Fig. 25.16, which describes the time evolution of the pressure; on this figure, we observe the progression of the compression front with the usual associated slope breaking of the pressure profile. For t^* small enough, the compressibility

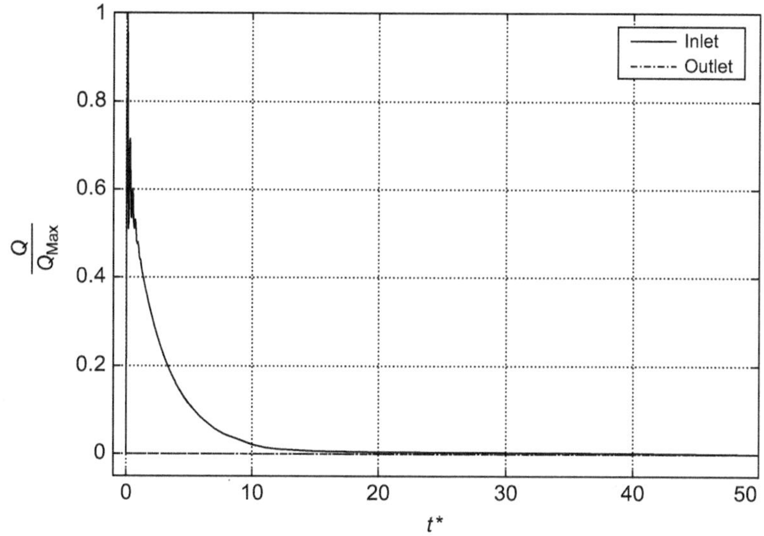

FIG. 25.15 Time evolution of the inlet and outlet dimensionless mass flow rates for a compressible viscoplastic flow ($\mathcal{B}n^* = 1.1,\ \chi^* = 4 \times 10^{-2}$).

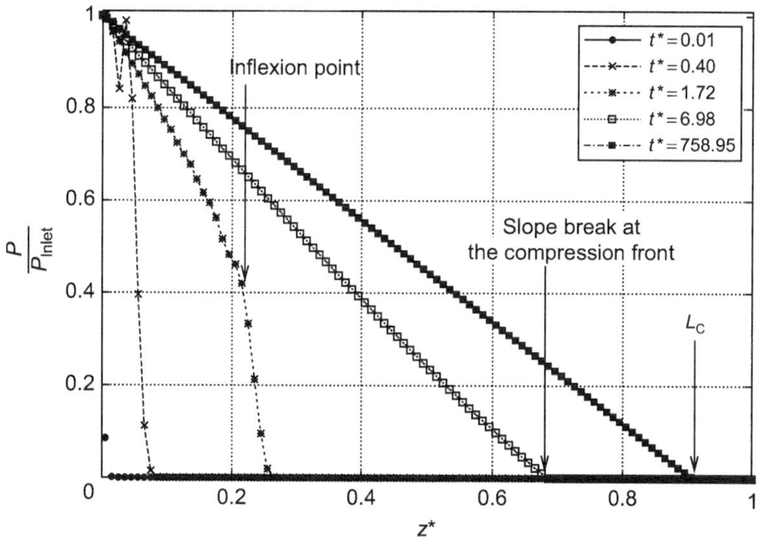

FIG. 25.16 Time evolution of the pressure for a compressible viscoplastic flow ($\mathcal{B}n^* = 1.1, \chi^* = 4 \times 10^{-2}$).

is the main factor of the flow restart and leads to strong pressure drops upstream the compression front; indeed, the pressure decreases very quickly from P_{Inlet} to 0 in the upstream region and then stays at 0 downstream of the compression front. The strong pressure drop taking place upstream of the compression front implies a local Bingham number lower than

1, allowing thus the fluid to flow in this region of the pipeline. When the compression front reaches the critical value L_C, the fluid stops flowing. Actually, this critical length corresponds to the critical pressure drop $\varpi_C = \dfrac{P_{\text{Inlet}} - P_{\text{Outlet}}}{L_C}$ for which $\mathcal{B}n^* = 1$ $\left(\text{where } \mathcal{B}n^* = \dfrac{2\tau_y}{\varpi_C}\right)$. The critical length L_C reached by the compression front when the fluid stops flowing is given by

$$L_C = \frac{R(P_{\text{Inlet}} - P_{\text{Outlet}})}{2\tau_y} \tag{25.13}$$

For the case considered here, we have (analytically) $L_C/L = 0.9$, which is in good agreement with the computed value of L_C observed on Fig. 25.16 (see the pressure profile associated with $t^* = 758.95$ (which from a practical point of view can be considered as the pressure profile at steady state)).

Let us consider now the lower compressibility case ($\chi^* = 4 \times 10^{-6}$): in Fig. 25.17, we have visualized the time evolution of the inlet and outlet dimensionless mass flow rates on the time interval [0, 0.25]. Contrary to the higher compressibility case, the flow at the outlet starts up and both inlet and outlet flow rates oscillate before dropping to zero. Further information is provided by Fig. 25.18, where the time evolution of the pressure has been visualized; the pressure oscillations appear clearly in this figure, as is the inflexion point associated with the unusual shape of the velocity profile, more pronounced than in the higher compressibility case. Moreover, in the present case, the compression front reaches the outlet; this was not the case for $\chi^* = 4 \times 10^{-2}$. To summarize, for the two compressible cases, which have been investigated in this paragraph, the fluid stops flowing ultimately because the inlet pressure is not large enough. However, the time required to reach the no-flow steady state, and the way this steady state is reached, are highly dependent of χ^*. Indeed, for $\chi^* = 4 \times 10^{-2}$,

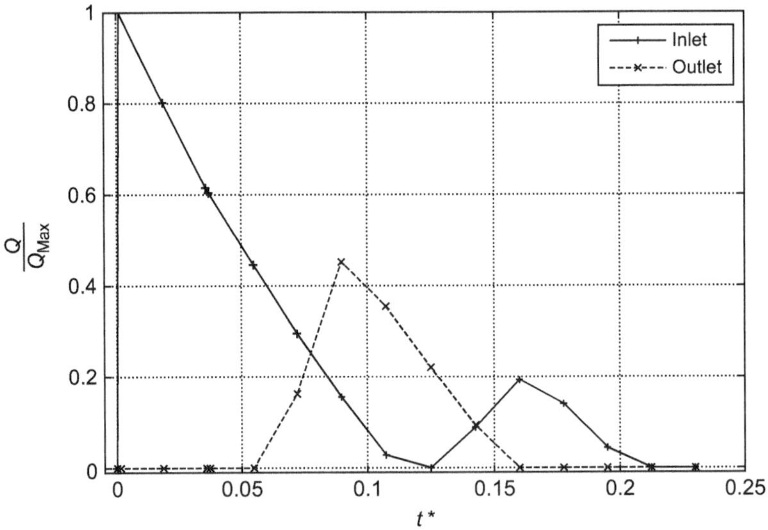

FIG. 25.17 Time evolution of the inlet and outlet dimensionless mass flow rates for a compressible viscoplastic flow for $\mathcal{B}n^* = 1.1$, $\chi^* = 4 \times 10^{-6}$)

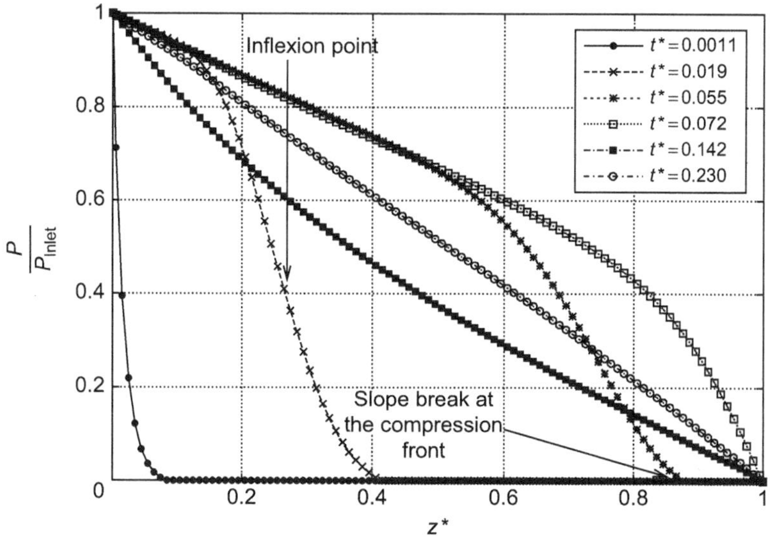

FIG. 25.18 Time evolution of the pressure for a compressible viscoplastic flow ($\mathcal{B}n^* = 1.1, \chi^* = 4 \times 10^{-6}$).

the steady state is reached around $t^* = 760$, while for $\chi^* = 4 \times 10^{-6}$ the steady state is reached at $t^* \approx 0.23$; moreover, for the smaller value of χ^*, we observe significant oscillations of the flow rates and of the pressure.

Our final statement will be:

If $\mathcal{B}n^* > 1$, the larger is χ^*, the larger is the time required for the fluid to stop flowing.

25.4. *Final comments on the transient, isothermal, and compressible flow of a viscoplastic fluid in a pipeline*

In Sections 25.1–25.3, we addressed the numerical simulation of the transient flow of a weakly compressible viscoplastic fluid in a pipeline; the rheological model that we considered was a compressible Bingham model. From a computational point of view, the main challenge was generalizing to compressible situations a Lagrange multiplier-based methodology dedicated to the simulation of incompressible viscoplastic flows. In order to achieve this objective, we had to modify the augmented Lagrangian functional and the various Uzawa algorithms that we use for the solution of the incompressible viscoplastic flow problems. The modified transient algorithm has proved to be very efficient and robust since convergence has been obtained (1) for all the problems considered in Section 25, and (2) for the augmentation parameter r (denoted also by r_{AL} in some paragraphs) varying in very large intervals of positive values. However, the speed of convergence varies with the compressibility coefficient: everything else being the same, the larger this coefficient, the faster the convergence. Concerning the physical properties of the flow, the following ones have been observed:

1. Compressibility induces oscillations of the inlet and outlet flow rates, and of the pressure as well.

2. For steady flows with $\mathbf{u} \neq \mathbf{0}$, unyielded regions cannot exist. Indeed, because $\frac{\partial u_z}{\partial z} \neq 0$ along the axis of the pipeline, plug regions cannot exist for such flows.
3. The values of the Bingham number and of the compressibility coefficient influence the restarting time of the flow at the pipe outlet. Indeed, high compressibility and/or large Bingham number increase the flow restarting time at the outlet, assuming that this flow restart does occur.
4. Just after the time of restart ($t = 0$, here), a high compressibility implies, at the inlet, a high peak for the mass flow rate, and a strong drop of the pressure.
5. If the flow restart is unsuccessful, the time it takes to have $\mathbf{u} = \mathbf{0}$ in the pipeline increases with the compressibility.

The main problem concerning the restart of waxy crude oil flows is to estimate the critical pressure at the inlet above which the oil will flow at the outlet; the numerical methods discussed in this chapter can contribute to the identification of this critical pressure. Actually, real-life waxy crude oils enjoy thixotropic properties; the numerical simulation of transient isothermal compressible thixotropic viscoplastic flow will be discussed in the following section.

26. Transient isothermal compressible and thixotropic flow in a pipeline: the isothermal restart of waxy crude oil flow

26.1. *Generalities*

The last problem that we wish to investigate in this chapter is very similar to the one discussed in Section 25, except that it will be assumed that the fluid to be considered is thixotropic, in addition to being viscoplastic. The thixotropic properties of waxy crude oils are related to a gel breakdown mechanism due to shear; basically, this breakdown entails the decrease of both viscosity and yield stress. Because compressibility provides high shear rates, the yield stress (and the corresponding Bingham number) decreases sharply, immediately after the restart of the flow, implying that at the end of the compression period, we may have $\mathcal{B}n^* < 1$ and $\frac{L_C}{L} > 1$ that is the oil is flowing at the outlet. This will be further discussed in the following parts of this section.

From a physical standpoint, thixotropy implies time-dependent rheological properties due to the ability of the gel-like structure of the material to break down or build up. From a modeling point of view, one more field, the structure parameter $\lambda_s (\in [0, 1])$, and an equation describing its evolution are included in the constitutive model, in order to describe the time evolution of both yield stress and viscosity. Therefore, the problem to be addressed is a slightly modified version of the one discussed in Section 25; however, the numerical results to be presented later in this section are of great interest because they illustrate the combined effects of compressibility and thixotropy on the restart of oil flow in pipelines. We will see, in particular, that the thixotropy property may allow the restarting of a compressible oil flow under conditions that will prevent the flow restart if the oil was not thixotropic.

The geometry of the flow region and the boundary conditions are similar to those encountered in Section 25, namely, we consider an axisymmetric pipeline with imposed pressure at the inlet and outlet cross sections.

26.2. Governing equations

In order to take *thixotropy* into account, we replace the system (25.1)–(25.3) by the following modified one:

$$\chi_\Theta\left(\frac{\partial p}{\partial t}+\mathbf{u}\cdot\nabla p\right)+\nabla\cdot\mathbf{u}=0, \tag{26.1}$$

$$\rho\left[\frac{\partial \mathbf{u}}{\partial t}+(\mathbf{u}\cdot\nabla)\mathbf{u}\right]+\nabla p=\nabla\cdot\boldsymbol{\tau}, \tag{26.2}$$

$$\begin{cases}\boldsymbol{\tau}=2\mu(\lambda_s)\left[\mathbf{D}(\mathbf{u})-\frac{1}{3}(\nabla\cdot\mathbf{u})\mathbf{I}\right]+\tau_y(\lambda_s)\dfrac{\mathbf{D}(\mathbf{u})}{\|\mathbf{D}(\mathbf{u})\|} & \text{if } \|\boldsymbol{\tau}\|>\tau_y(\lambda_s),\\ \mathbf{D}(\mathbf{u})=\mathbf{0} & \text{if } \|\boldsymbol{\tau}\|\le\tau_y(\lambda_s),\end{cases} \tag{26.3}$$

$$\frac{\partial\lambda_s}{\partial t}+\mathbf{u}\cdot\nabla\lambda_s=a(1-\lambda_s)-b\lambda_s\dot{\gamma}^m, \tag{26.4}$$

$$\mu(\lambda_s)=\mu_0+\lambda_s\mu_1, \tag{26.5}$$

$$\tau(\lambda_s)=\tau_{y0}+\lambda_s\tau_{y1}. \tag{26.6}$$

As mentioned already in Section 3, this thixotropic-viscoplastic model has been introduced by *Houska* in the early 1980s (see HOUSKA [1981]); a particular feature of the Houska's model is the affine relations existing between the structure parameter and the viscosity and yield stress of the fluid (see relations (26.5) and (26.6)). The build up coefficient a is usually quite small, compared with the break down coefficient b; thus, we will assume from now on that $a=0$. Actually, in restart problems, the time scale of the compressible period is usually small compared with the whole restart time, implying that the phenomenon is mostly governed by the structure breakdown rate $b\lambda_s\dot{\gamma}^m$. In addition to the compressibility, Reynolds and Bingham numbers, we introduce a new number $\mathcal{B}$d, related to thixotropic breakdown effects and defined as the product of the characteristic time scale of breakdown by a characteristic shear rate, namely

$$\mathcal{B}\mathrm{d}=b\left(\frac{\bar{U}}{L_C}\right)^{m-1}. \tag{26.7}$$

Moreover, since the yield stress $\tau(\lambda_s)$ is the sum of the constant term τ_{y0} with the thixotropy-dependent term $\lambda_s\tau_{y1}$, it is convenient to define additional Bingham numbers, such as:

- A space-time dependent Bingham number, associated with the total yield stress τ_y; it is defined by

$$\mathcal{B}n^*(\tau_{y0}+\lambda_s\tau_{y1}) = 2\frac{\tau_{y0}+\lambda_s\tau_{y1}}{\varpi R} \tag{26.8}$$

$\left(\text{as in Section 25, } \varpi = \dfrac{P_{\text{Inlet}} - P_{\text{Outlet}}}{L}\right)$.

- A Bingham number, associated with τ_{y0}; it is defined by

$$\mathcal{B}n_0^* = \frac{2\tau_{y0}}{\varpi R}. \tag{26.9}$$

- A Bingham number, associated with τ_{y1}; it is defined by

$$\mathcal{B}n_1^* = \frac{2\tau_{y1}}{\varpi R}. \tag{26.10}$$

- A maximal Bingham number, associated with $\tau_{y0}+\tau_{y1}$ it is defined by

$$\mathcal{B}n_{\max}^* = \mathcal{B}n^*(\tau_{y0}+\tau_{y1}) = 2\frac{\tau_{y0}+\tau_{y1}}{\varpi R} = \mathcal{B}n_0^* + \mathcal{B}n_1^*. \tag{26.11}$$

26.3. *Problem description*

The problem defined by the governing equations (26.1)–(26.6) looks complicated and therefore not easy to solve. Actually, the physical mechanism it describes is relatively easy to understand: the ability of the flow in the pipeline to restart, if the initial Bingham number is larger than 1, is strongly related to the combined beneficial effects of compressibility and thixotropy. In other words, the initial compressible phase triggers shear stress in the fluid. As a result, during this compressible phase, the structure breakdown mechanism starts, that is, λ_s decreases, entailing (from relations (26.5) and (26.6)) the drop of both yield stress and viscosity. If, at the end of the early compressible transients, the yield stress has sufficiently dropped, so that the corresponding Bingham number is now less than 1, the flow restarts and eventually recovers steady flowing conditions.

An important parameter of the flow restart is $\lambda_s|_{t=0}$, assuming that one restarts (or tries to restart) the flow at $t=0$. Depending of the situation at $t=0$, three scenarios can be identified:

1. *No restart possible*, $\forall\ \lambda_s|_{t=0} \in [0, 1]$: If τ_{y0} (the constant part of the yield stress) is such that $\mathcal{B}n_0^* > 1$, that is (from (26.9))

$$\frac{2\tau_{y0}}{\varpi R} > 1, \tag{26.12}$$

the restart is impossible, even if the fluid is compressible and thixotropic.

2. *Conditional restart*: Let us assume that $\lambda_s(0)(= \lambda_s|_{t=0})$, τ_{y0} and τ_{y1} are such that

$$\mathcal{B}n_0^* = \frac{2\tau_{y0}}{\varpi R} < 1 \text{ and } \mathcal{B}n^*[\tau_{y0} + \lambda_s(0)\tau_{y1})] = \frac{2[\tau_{y0} + \lambda_s(0)\tau_{y1}]}{\varpi R} > 1. \quad (26.13)$$

If the fluid is incompressible, the restart is impossible. If, however, the fluid is compressible, the flow may restart, due to the structure breakdown mechanism described earlier. Actually, the flow and the ability to restart are driven by the compressibility and thixotropy dimensionless numbers χ^* and $\mathcal{B}$d. The higher these two numbers are, the more likely is the flow restart.

3. *Unconditional restart*: If $\lambda_s(0), \tau_{y0}$ and τ_{y1} are such that

$$\mathcal{B}n^*[\tau_{y0} + \lambda_s(0)\tau_{y1})] = \frac{2[\tau_{y0} + \lambda_s(0)\tau_{y1}]}{\varpi R} < 1, \quad (26.14)$$

the flow will restart, no matter what compressibility and thixotropy are.

Of the three above-mentioned scenarios, the most interesting is clearly the second one, explaining why, in the next section, we will focus on the situations of type 2, and investigate their restarting properties. However, it is still beyond our capabilities to give bounds for χ^* and $\mathcal{B}$d which guarantee the restart. Indeed, the number of dimensionless numbers governing the problem is fairly large. An alternative might be a computationally expensive parametric survey, but, again, this is beyond the capabilities of the serial Linux workstations used for the computations presented in this chapter. This may lead to the development of a more efficient implementation of our simulator, using parallelization, for example, and of a more sophisticated solution method (like the one advocated in VINAY, WACHS and FRIGAARD [2007]). As a consequence, our objective here is to discuss the influence of the combined effects of compressibility and thixotropy on the flow restart, via the solution of well-chosen test problems.

26.4. Results and discussion

26.4.1. Generalities

The computations have been performed with the solution methods discussed in Section 20.4; they correspond to the real-time evolution (in a physical sense) of the flow. For all the calculations, we supposed that at $t = 0$, the flow is at rest (that is $\mathbf{u} = \mathbf{0}$ and $p = 0$ in the flow region) and the material is fully gelled (that is $\lambda_s(0) \equiv 1$). The results are presented and discussed in terms of the following dimensionless dependent and independent variables: dimensionless coordinates $r^* = r/R$ and $z^* = z/L$, dimensionless time $t^* = t/t_{\text{ref}}$ (with $t_{\text{ref}} = R/u_{z\text{-max}}$), dimensionless axial velocity $u_z/u_{z\text{-max}}$, and the dimensionless numbers χ^*, $\mathcal{B}$n, $\mathcal{B}$d, and $\mathcal{R}$e.

The geometry is like the one in Section 25, implying that $L/R = 200$. As in Section 25, the space discretization of the mathematical model has been obtained from a finite volume Cartesian mesh, uniform in the r and z directions. On the basis of our numerical experiments, we can claim that including thixotropy in the governing equations does not deteriorate the robustness of our computational methodology because convergence was achieved for all the cases we investigated. Actually, this is not surprising because the class of problems under consideration can be viewed as combining the features of the problems discussed in Section 24 (space-time dependent yield stress) and Section 25 (compressibility).

26.4.2. Incompressible fully developed flow

The first test problem that we consider is the restart of the flow of an incompressible thixotropic and viscoplastic material. Our objectives here are to illustrate: (1) the structure breakdown mechanism occurring in the pipeline, and (2) the strong discontinuities of the rheological properties and of the radial derivative of the velocity (these discontinuities result from the fact that the structure breakdown takes place only in regions of nonzero shear rate). Because $\lambda_s(0) \equiv 1$ and the fluid is incompressible, the situations to be considered in this paragraph verify

$$\mathcal{B}n^*_{\max} = 2\frac{\tau_{y0} + \tau_{y1}}{\varpi R} < 1, \tag{26.15}$$

otherwise (that is if $\mathcal{B}n^*_{\max} \geq 1$), the flow cannot restart. Moreover, incompressibility and the identity $\lambda_s(0) \equiv 1$ lead to a fully developed transient flow, variable in r and uniform in z. Accordingly, we deliberately picked a mesh with a small grid size in the r-direction and a coarse one in the z-direction. In practice, we used a 50×20 finite-volume mesh. From the independence with respect to z, the results, below, are presented as r-dependent profiles. We selected $\mathcal{B}n^*_{\max} = 0.5$ and considered the three following situations:

- *Case 1*: The fluid is *fully thixotropic*, that is $\mathcal{B}n^*_0 = 0$ and $\mathcal{B}n^*_1 = 0.5$.
- *Case 2*: The fluid is *slightly thixotropic*, that is $\mathcal{B}n^*_0 = 0.25$ and $\mathcal{B}n^*_1 = 0.25$.
- *Case 3*: The fluid is *nonthixotropic*, that is $\mathcal{B}n^*_0 = 0.5$ and $\mathcal{B}n^*_1 = 0$.

The Case 1 is well-suited to illustrate how the structure breakdown mechanism affects the flow kinematics. However, we will use Cases 2 and 3 to show how the thixotropy level of the fluid modifies the steady-state velocity profile (and the corresponding flow rate). In Case 3, although the fluid is nonthixotropic (that is, the total yield stress τ_y does not depend of the structure parameter λ_s), we can still compute the time evolution of λ_s in order to compare with the results of the two other cases. In all cases considered, the velocity of reference is the one of the fully thixotropic case (Case 1) and χ^*, $\mathcal{R}e$, and $\mathcal{B}d$ are kept to 0, 2, and 0.01, respectively.

The time evolution of the flow has been reported on Fig. 26.1. Figure 26.1(a) shows the evolution of the structure parameter λ_s, from $\lambda_s \equiv 1$ at $t^* = 0$ to λ_s at $t^* = 1000$, which corresponds, essentially, to a steady state. The inspection of Fig. 26.1(a) prompts us to do the two following basic observations:

1. The Bingham number $\mathcal{B}n^*_{\max}$ has been set to 0.5. It follows then from that choice that, as the flow restarts, shear develops only in the region corresponding to $1/2 < r/R \leq 1$, the central region ($0 \leq r/R < 1/2$) being a standard plug region moving at constant velocity. The source of the structure breakdown is the local shear, implying that breakdown takes place only in the yielded (or sheared) region. This is exactly what the simulation shows: as t increases, the structure parameter λ_s drops only in the region corresponding to $1/2 < r/R \leq 1$, but $\lambda_s(r, z, t)$ stays equal to 1 if $0 \leq r/R < 1/2$. For this simple shear flow, no radial propagation of the structure breakdown takes place (assuming that the flow is stable) and the yielded (or sheared) part of the flow is determined by the initial value of τ_y, that is by the initial value of λ_s.

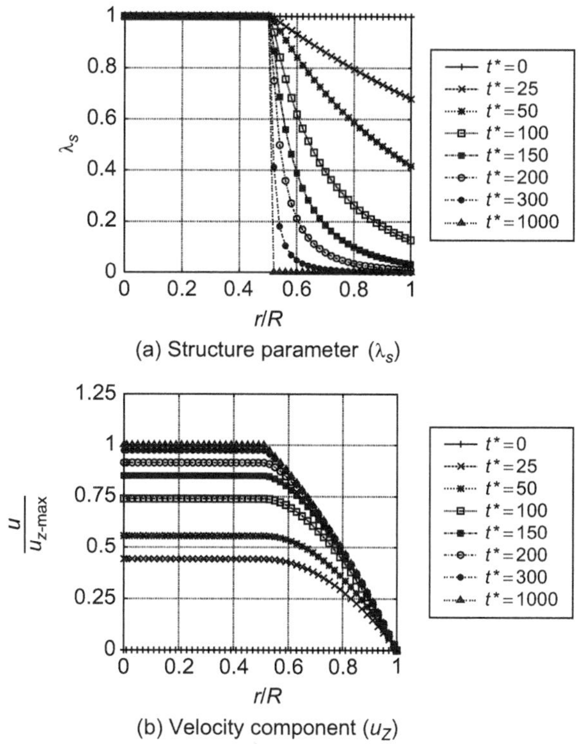

FIG. 26.1 Restart of the fully developed flow of a viscoplastic thixotropic incompressible viscous fluid: (a) time evolution of the structure parameter λ_s and (b) time evolution of u_z ($\mathcal{R}e = 2$, $\mathcal{B}n^*_{\max} = 0.5$, $\mathcal{B}n^*_0 = 0$, $\mathcal{B}n^*_1 = 0.5$, $\chi^* = 0$, and $\mathcal{B}d = 0.01$).

2. From the previous observation, it is not surprising that, at steady state, the structure parameter has a jump at $r/R = 0.5$.

The time evolution of the axial velocity is shown in Fig. 26.1(b). As the total yield stress drops in the region defined by $1/2 < r/R < 1$, the velocity increases accordingly, as expected. What is more unusual is the shape of the steady-state velocity profile: indeed, the significant discontinuity of the structure parameter (see Fig. 26.1(a)) at $r/R = 1/2$ entails a sharp discontinuity of $\frac{\partial u_z}{\partial r}$ as shown in Fig. 26.1(b) for $t^* = 1000$. This property, which does not exist for standard (that is nonthixotropic) viscoplastic flow, is a consequence of thixotropy (actually, it would be more appropriate to say that "the discontinuities of the rheological properties, yield stress in particular, due to thixotropy, are responsible for the unusual feature of the velocity field"). This particular feature of the velocity field is clearly visible on Fig. 26.2 where the velocity profiles have been visualized for the three cases considered here: Fig. 26.2(a) shows that the more thixotropic is the fluid, the lower is the total yield stress in the yielded region (and therefore, as expected, the higher is the velocity).

Figure 26.2 shows also that for this simple one-dimensional shear flow, the size of the plug region is indeed determined by $\mathcal{B}n^*_{\text{initial}} = \mathcal{B}n^*(\tau_{y0} + \lambda_s(0)\tau_{y1})$. Here, we set $\lambda_s(0) = 1$,

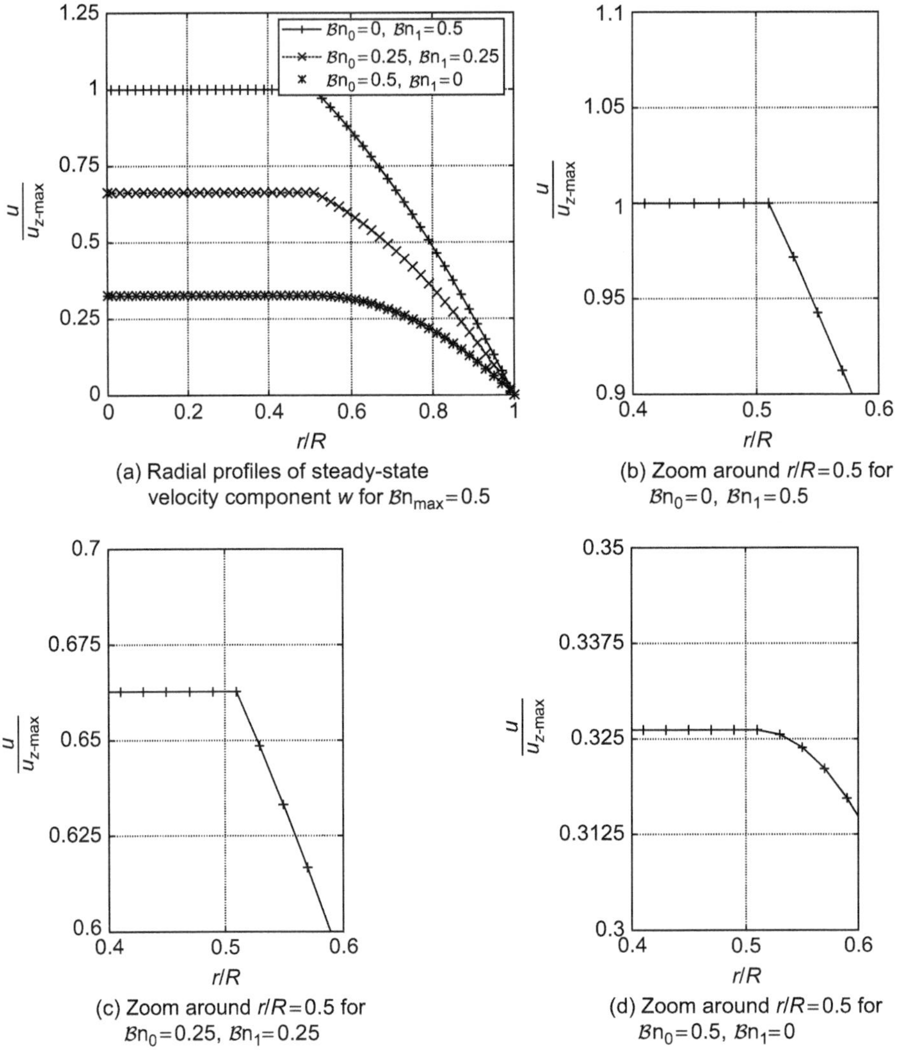

(a) Radial profiles of steady-state velocity component w for $\mathcal{B}n_{max}=0.5$

(b) Zoom around $r/R=0.5$ for $\mathcal{B}n_0=0$, $\mathcal{B}n_1=0.5$

(c) Zoom around $r/R=0.5$ for $\mathcal{B}n_0=0.25$, $\mathcal{B}n_1=0.25$

(d) Zoom around $r/R=0.5$ for $\mathcal{B}n_0=0.5$, $\mathcal{B}n_1=0$

FIG. 26.2 Restart of the fully developed flow of an incompressible viscoplastic and thixotropic fluid with $\mathcal{R}e=2$, $\mathcal{B}n^*_{max}=0.5$, $\chi^*=0$ and (a) steady-state velocity profiles. (b) Zoom of the steady-state velocity profile around $r=R/2$ for $\mathcal{B}n^*_0=0$ and $\mathcal{B}n^*_1=0.5$. (c) Zoom of the steady-state velocity profile around $r=R/2$ for $\mathcal{B}n^*_0=\mathcal{B}n^*_1=0.25$. (d) Zoom of the steady-state velocity profile around $r=R/2$ for $\mathcal{B}n^*_0=0.5$ and $\mathcal{B}n^*_1=0$.

implying that we have $\mathcal{B}n^*_{initial}=\mathcal{B}n^*_{max}=0.5$ for the three cases under consideration. As a consequence, the diameter of the plug region is $R/2$, no matter if the fluid is fully (Case 1), slightly (Case 2), or not at all (Case 3) thixotropic. To have a better look at the discontinuity of $\frac{\partial u_z}{\partial r}$ close to $r=1/2R$, we have shown in Fig. 26.2(b–d) a detailed description of the unyielded/yielded transition. It is quite striking that as soon as the fluid is thixotropic (Cases 1 and 2, Fig. 26.2(b,c)), the radial profile of the axial velocity exhibits this highly visible discontinuity, which is itself a consequence of the total yield stress discontinuity associated with the discontinuity of the structure parameter. For nonthixotropic viscoplastic materials the velocity profile is of the Poiseuille type in the yielded region and flat in the unyielded one, with no derivative discontinuity at the interface between the yielded and unyielded regions (see Fig. 26.2(d)). It is worth mentioning that this particular property of thixotropic materials has been observed experimentally and reported in ROUSSEL, LE ROY and COUSSOT [2004] for the Couette flow, another simple shear flow. Indeed, it is quite satisfactory to find in the literature evidences of experiments supporting our simulations. From these observations, we can claim that the Houska's model appears to be reliable and accurate in order to model thixotropic effects.

Finally, we have compared in Fig. 26.3 the structure breakdown rate for the three cases considered here (Case 3, where the fluid is not thixotropic, has been considered for comparison with Cases 1 and 2; we computed the time evolution of λ_s, although it does not affect the flow kinematics). We observe that the more thixotropic is the fluid, the higher is the breakdown rate. Explaining this behavior is relatively simple: if the fluid is strongly thixotropic, the total yield stress τ_y drops faster in yielded region, implying a flow rate increase, which implies in turn higher shear, and therefore in the right-hand side of equation (26.4) a negative source term whose absolute value increases; this decreases λ_s and therefore τ_y (from (26.6)), and so on; it is clearly a self-enforcing mechanism). Figure 26.3 shows that λ_s (Case 1) $<$ λ_s (Case 2) $<$ λ_s (Case 3) for two different values of t (actually, it is true for all the times t belonging to the time interval during which the flow has been simulated).

We will end this section dedicated to incompressible thixotropic viscoplastic flow by some comments on Bingham numbers. With the terminology we used, a fluid was strongly or slightly thixotropic, depending of the ratio $\dfrac{\mathcal{B}\mathrm{n}_1^*}{\mathcal{B}\mathrm{n}_0^*} = \dfrac{\tau_{y1}}{\tau_{y0}}$. Actually, this is rather simplistic because we kept $\mathcal{B}\mathrm{d}$ to a constant value (0.01, here). We remind our reader that $\mathcal{B}\mathrm{d}$ (defined by (26.7)) is used to quantify the rate of breakdown, implying that when one assesses the thixotropic properties of a viscoplastic material, one has to take into account both $\mathcal{B}\mathrm{d}$ and

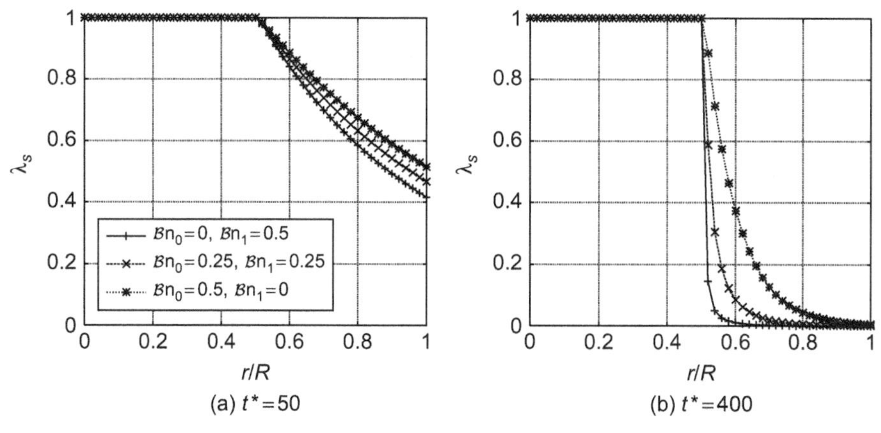

FIG. 26.3 Restart of the fully developed incompressible flow of a viscoplastic and thixotropic fluid: comparison of the radial profiles of the structure parameter at $t^* = 50$ (a) and $t^* = 400$ (b) ($\mathcal{R}\mathrm{e} = 2$, $\mathcal{B}\mathrm{n}_{\mathrm{max}}^* = 0.5$, $\mathcal{B}\mathrm{d} = 0.01$, $\chi^* = 0$, $\{\mathcal{B}\mathrm{n}_0^*, \mathcal{B}\mathrm{n}_1^*\} = \{0, 0.5\}$, $\{0.25, 0.25\}$, and $\{0.5, 0\}$).

the ratio $\dfrac{\mathcal{B}\mathrm{n}_1^*}{\mathcal{B}\mathrm{n}_0^*} = \dfrac{\tau_{y1}}{\tau_{y0}}$. We hope that our reader begins to fully understand why thixotropic flow are quite intricate; indeed, all the phenomena are fully coupled and the number of dimensionless parameters is pretty high. For this reason, in the next section, we will limit our analysis of compressible flows to an illustrative example which highlights nicely the combined effect of structure breakdown and compressibility.

26.4.3. Combined effects of thixotropy and compressibility

The last test problem to be considered includes all the physical effects described in the previous sections (with the exception of the temperature). All these effects interact in a strongly coupled way. Our goal here is to show that the shear produced during the early compressible transients may break down the gel structure of the waxy crude oil so that the flow restarts and resumes steady flowing conditions.

The situation considered here is of type 2 in the classification introduced in Section 26.3. We take $\lambda_s(0) \equiv 1$ and set $\mathcal{B}\mathrm{n}^*(\tau_{y0} + \lambda_s(0)\tau_{y1}) = \mathcal{B}\mathrm{n}^*_{\mathrm{max}} = 1.025$. We consider three particular cases:

- *Case 1*: *Incompressible and nonthixotropic fluid*
  We set $\mathcal{B}\mathrm{n}_0^* = 1.025$, $\mathcal{B}\mathrm{n}_1^* = 0$, and $\chi^* = 0$; the flow is not expected to restart.
- *Case 2*: *Compressible and nonthixotropic fluid*
  We set $\mathcal{B}\mathrm{n}_0^* = 1.025$, $\mathcal{B}\mathrm{n}_1^* = 0$, and $\chi^* = 4 \times 10^{-2}$; the flow restarts temporarily due to its compressibility, but the flow rate returns to zero eventually because the pressure drop is not large enough to maintain the flow. This case has been investigated extensively in Section 25.
- *Case 3*: *Compressible and thixotropic fluid*
  We set $\mathcal{B}\mathrm{n}_0^* = 0.1$, $\mathcal{B}\mathrm{n}_1^* = 0.925$, and $\chi^* = 4 \times 10^{-2}$; the flow restarts due to its compressibility, then, the structure breakdown entails a decrease of λ_s such that, at the end of the compression phase, the actual $\mathcal{B}\mathrm{n}^*(\tau_{y0} + \lambda_s\tau_{y1})$ is less than one, keeping thus the fluid flowing and reaching steady flow conditions.

On the basis of the numerical experiments performed in Sections 24 and 25, we used a 20×100 mesh for all the cases investigated below. Since Cases 1 and 2 have been investigated in previous sections, we will focus below on Case 3 in the particular situation where $\mathcal{B}\mathrm{d} = 0.1$ and $\mathcal{R}\mathrm{e} = 0.07$.

In Figs. 26.4 and 26.5, we have represented the time evolution of the structure parameter and of the yielded/unyielded regions, respectively. The yielded/unyielded region distribution is quite similar to the one (investigated in Section 25) associated with the restart of a compressible, nonthixotropic viscoplastic fluid. Actually, the transition from unyielded to yielded corresponds to the propagation of a compression front in the pipeline during the early transients of the compression phase. Then, once the compression phase is completed, the density variation along the pipe axis leads to a fully yielded pipe: the unyielded region has completely disappeared. Accordingly, the structure breakdown takes place everywhere in the pipeline. However because the shear rate is much larger close to the pipe wall than close to the pipe axis, λ_s decreases faster close to the pipe wall. In Fig. 26.4, at $t^* = 490$, the structure is completely broken down ($\lambda_s \approx 0$) in the half outer part of the pipe radius ($1/2 < r/R < 1$), whereas the half inner part is still fully structured ($\lambda_s \approx 1$). However, if we would have been able to perform our simulation on a much longer time interval, we

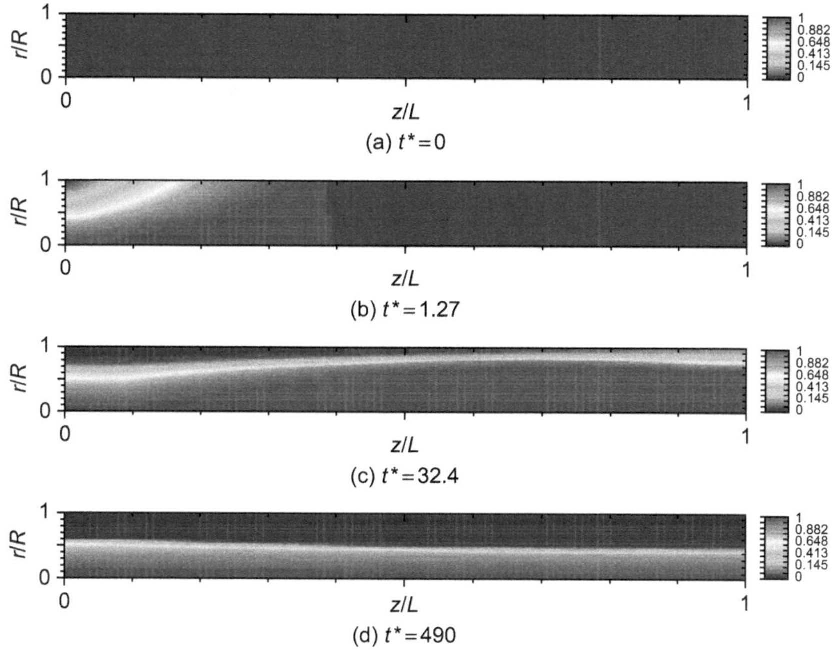

FIG. 26.4 Time evolution of the structure parameter for a successful restart ($\mathcal{B}n^*_{max} = 1.025$, $\mathcal{B}n^*_0 = 0.1$, $\mathcal{B}n = 0.925$, $\chi^* = 4 \times 10^{-2}$, $\mathcal{B}d = 0.1$, and $\mathcal{R}e = 0.07$).

would have observed the convergence to zero of the structure parameter, everywhere in the pipeline.

On Fig. 26.6, we have visualized, for $0 \leq t^* \leq 10$, the time evolution of the inlet mass flow rate for the three cases under consideration: In Case 1 (incompressibility and nonthixotropy) the flow rate sticks to zero. In Case 2 (compressibility and nonthixotropy) the flow restarts but returns "quickly" to rest. Finally, in Case 3 (compressibility and thixotropy), the flow restarts, thanks to the structure breakdown that occurs during the compression phase, and reaches a Bingham number allowing steady flowing conditions. These results are of great importance for field engineers accustomed to use the conservative relation $\mathcal{B}n^*_{max} < 1$ in order to estimate the inlet pressure necessary to resume the flow. Quite often, this relation leads to a flawed design of the pipeline installation because the restart pressure is overestimated. Indeed, we show that due to the combined effects of thixotropy and compressibility a lower inlet pressure may be sufficient to resume the flow of the crude oil.

26.5. Some remarks on transient, isothermal, and thixotropic viscoplastic flows in pipelines

In the preceding sections, we addressed the numerical simulation of the flow, in a pipeline, of both compressible and incompressible thixotropic viscoplastic fluids. Our augmented Lagrangian/finite-volume-based numerical methodology has proved to be well-suited to this task. Indeed, when applied to the above problems, our numerical methodology kept the

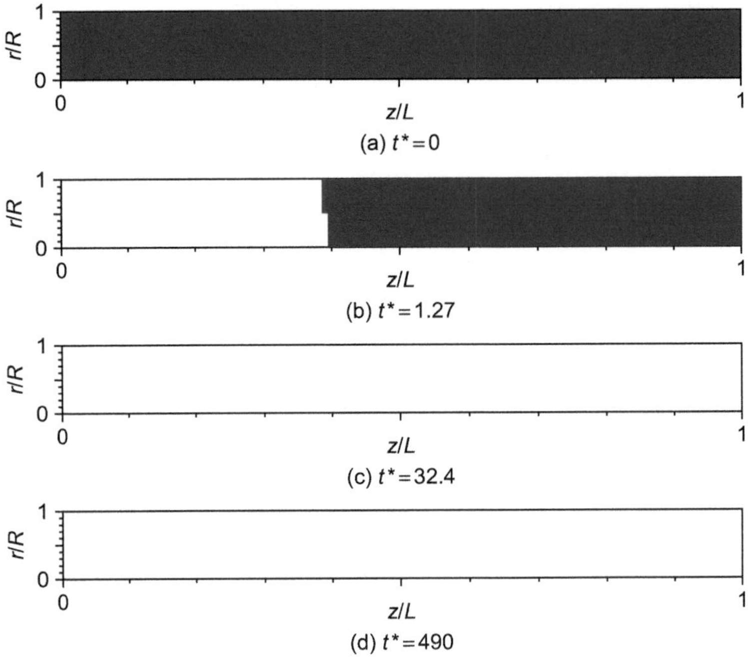

FIG. 26.5 Time evolution of the yielded/unyielded regions ($\mathcal{B}n^*_{max} = 1.025$, $\mathcal{B}n^*_0 = 0.1$, $\mathcal{B}n = 0.925$, $\chi^* = 4 \times 10^{-2}$, $\mathcal{B}d = 0.1$, and $\mathcal{R}e = 0.07$).

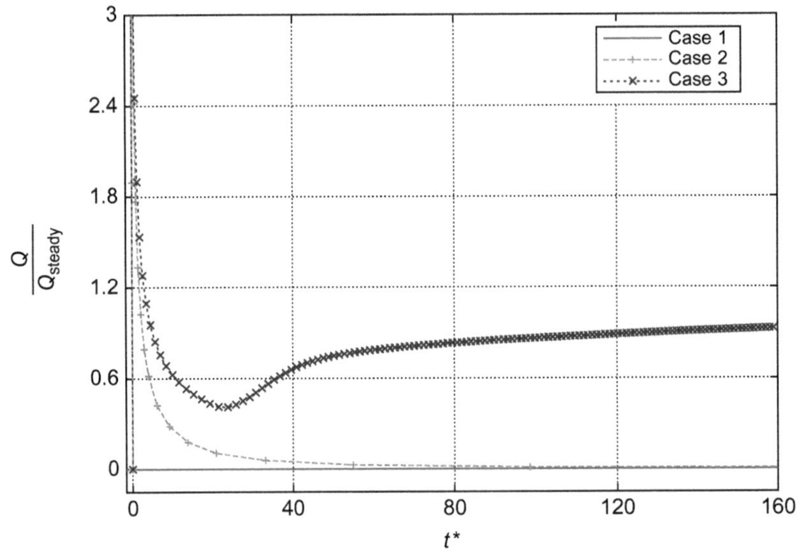

FIG. 26.6 Time evolution of the inlet mass flow rate for Cases 1, 2, and 3.

robustness and convergence properties it showed when applied to the solution of simpler problems and provided computed solutions of satisfactory accuracy. The numerical method we used had no particular difficulty in handling situations with highly discontinuous rheological properties, another evidence of its robustness.

The study of the incompressible case revealed that the related flow is quite intricate, due to the coupling of the various phenomena occurring in the fluid and to the numerous associated characteristic numbers. The fact that the structure breakdown mechanism takes place in yielded regions only leads to unusual velocity profiles, even in simple shear flow like Poiseuille's. In fact, due to the sharp discontinuity of the radial yield stress profile, the first derivative of the corresponding radial velocity profile exhibits a strong discontinuity. This observation is consistent with the experimental results reported in ROUSSEL, LE ROY and COUSSOT [2004]. Moreover, the structure breakdown rate depends on both the ratio $\mathcal{B}n_1^*/\mathcal{B}n_0^*$ and the new characteristic number $\mathcal{B}d$.

Finally, we investigated a problem closely related to real-life applications, namely, the restart of a gelled waxy crude oil contained in a pipeline, assuming that the flow is isothermal and the fluid compressible, viscoplastic, and thixotropic. Our simulations show, in particular, that the oil flow can restart even if $\mathcal{B}n^* > 1$, thanks to the combined effects of compressibility and thixotropy, the compression phase producing high levels of shear stress. However, the restarting of the oil flow occurs only if the thixotropic effects are strong enough. Indeed, the drop of μ (viscosity) and τ_y (yield stress) has to be sufficiently fast and significant to trigger the flow restart. From the simulations reported in the preceding sections, we can conclude that under favorable conditions, a compressible thixotropic viscoplastic flow can enjoy restarting, due to the combined effects of compressibility and gel breakdown mechanism; this flow restart may occur under conditions for which the restart of the corresponding incompressible flow will not occur. However, the precise determination of those favorable conditions is far from being a simple task. The main reason of this complexity has to do with the fact that the ability for the flow to restart if $\mathcal{B}n^*(\tau_{y0} + \lambda_s(0)\tau_{y1}) > 1$ is controlled by the three dimensionless numbers $\mathcal{B}n_1^*/\mathcal{B}n_0^*$, χ^*, and $\mathcal{B}d$. As today, giving accurate bounds and ranges of values, guaranteeing the flow restart, is beyond our computer capabilities, and even beyond our knowledge. This will be the subject of further investigations.

27. Additional comments on the augmented Lagrangian/finite-volume methodology: new challenges for waxy crude oil flow

In the preceding sections of this chapter, we addressed the numerical simulation of the flow of compressible and incompressible viscoplastic fluids whose yield-stress τ_y is space-time dependent. In the cases we considered, the variations of the yield stress was either a consequence of temperature variations or of the thixotropic properties of the material. To simulate these various viscoplastic flows, we advocated an augmented Lagrangian/finite-volume methodology, which proved reliable and efficient. In particular, the Uzawa's algorithm associated with the augmented Lagrangian showed robust convergence properties because convergence was achieved for all the cases that we investigated. Optimal convergence rate relies on a proper choice of the augmentation parameter r, but nonetheless the algorithm converged for all values of r, as predicted by theory. However, the finite-volume method we used was well suited to the simple geometries we considered (cavities and pipelines) and provided numerical solutions of satisfactory accuracy.

In all situations investigated, the main advantage of the augmented Lagrangian approach (compared with the regularization one, for example) is its ability to capture accurately the yielded/unyielded region interface. First, we validated our methodology using as test problem the classical lid-driven cavity flow problem, the results we obtained comparing favorably with other contributions in the literature. Next, we addressed more complicated problems where viscoplasticity was coupled with other physical properties such as temperature dependence, compressibility, and thixotropy. These attempts at including more realistic physical effects in modeling and simulation open a wide new range of potential applications that one might address. Indeed, the realm of viscoplasticity is quite large, implying that there are still many opportunities to enhance knowledge in this area of rheology, numerical simulation being one of the tools that can be used to explore it. Actually, numerical simulation is more than a predictive tool, it provides also a way to theorists and experimentalists to better understand the complex physics of viscoplastic materials.

Another area of future developments concerns the optimization of the Uzawa's algorithm. Although the algorithm is quite robust and has shown convergence for all the test problems we considered, we think that there is still room for speed of convergence improvement, by using, for example, an appropriate sequence $\{r_k\}_{k\geq 0}$ of augmentation parameters (as done in, e.g., DELBOS, GILBERT, GLOWINSKI and SINOQUET [2006] for the solution of an inverse problem from Geophysics). Another possibility is to parallelize the existing code in order to take advantage of those massively parallel Linux clusters available to IFP scientists.

Finally, going back to the transportation of waxy crude oils and to the associated restart issue, the authors hope that they helped improving the understanding of this class of flows; however, many issues still have to be addressed, such as: (1) Use a more realistic compressibility model. (2) Determine the ranges of the dimensionless parameters for which the pipe flow restarts. Because the ratio $R/L \ll 1$, the pipe flow is "almost" one-dimensional, an observation leading to a simpler (and computationally cheaper) approach based on lubrication theory (see VINAY, WACHS and FRIGAARD [2007] for details). However, multidimensional computations, like those presented in this chapter, will always be useful for those scientists looking for an accurate investigation of the local properties of the flow.

CHAPTER 4

Application of Fictitious Domain Methods to the Numerical Simulation of Viscoplastic Flow

28. Introduction. Synopsis

In the above chapters, we have discussed the numerical simulation of the flows of *single-phase viscoplastic material*. To handle such problems, we advocated a variety of *multiplier* (Lagrange's and others) based numerical methods. Although *single-phase flows* play an important role in the modeling of a large number of applications in Industry and Geophysics, situations involving *multiphase flows* are also very common, with the extra-phase(s) either solid or liquid or gas. In this chapter, we are interested by the flow of a viscoplastic material past a fixed solid obstacle and by the motion of rigid solid bodies in a viscoplastic fluid. Among the various numerical methods available to simulate such flows, those relying on *fictitious domain* approaches have shown promising possibilities. Indeed, among these fictitious domain methods, those relying on *Lagrange multipliers* fit nicely the multiplier-based simulation methods discussed in the previous chapters. It is then quite natural to attempt combining all these multiplier-based methods in order to simulate the flow of yield stress fluids in the presence of fixed or moving rigid bodies.

To the best of our knowledge, fictitious domain methods have been introduced in HYMAN [1952] (for a detailed review (including historical notes) of *fictitious domain methods*, see, e.g., GLOWINSKI [2003]). The *distributed Lagrange multiplier/fictitious domain method* (*DLM/FD*) was introduced in the mid-1990s by the first author and collaborators in order to simulate some classes of *particulate flows* and other flow around fixed or moving boundaries; it led to a substantial number of publications including GLOWINSKI, PAN, HESLA and JOSEPH [1999], GLOWINSKI, PAN, HESLA, JOSEPH and PERIAUX [2001], GLOWINSKI [2003], YU, PHAN-TIEN and TANNER [2004], YU, WACHS and PEYSSON [2006]. The principle of the DLM/FD methodology is simple and can be briefly described as follows: (1) One fills the rigid bodies located in the flow region with the surrounding fluid. (2) One uses *Lagrange multipliers* supported by these rigid bodies to impose a rigid body motion to the fluid they contain, while, typically, preserving a no-slip boundary condition at the fluid/solid interface. The main advantage of the DLM/FD methodology is related to its ability to perform computations on *fixed meshes*, thus, avoiding the necessity to update the mesh as the flow region varies with time. As in the preceding chapters, the Lagrange multipliers associated with the DLM/FD methods are computed by *Uzawa's* type algorithms. In the references given earlier, the DLM/FD methods have been combined with *operator-splitting time discretization schemes*, which enable to decouple, in a sense, the problem to be solved in a sum of simpler subproblems. As long as the time-discretization step is sufficiently small (in order to make the splitting error acceptable), operator-splitting schemes are well-suited to the simulation of *fast transient phenomena*.

In this chapter, we will consider two types of situations involving the flow of viscoplastic fluid around rigid solid bodies: in the *first case*, the rigid bodies are *fixed obstacles*, and a *fully coupled* (some practitioners say *monolithic*) solution method will be used to obtain the corresponding *steady flow*. In the *second case*, the rigid bodies are *moving freely*, the flow of the corresponding mixture being simulated via a methodology relying on a *time-discretization by operator-splitting*. In both cases, the numerical methodology benefits from the combined advantages of: (1) The *DLM/FD* method, to treat the rigid bodies. (2) The multiplier-based algorithms, discussed in the preceding chapters, to treat the *yield stress property* of the fluid.

29. Steady flow of a Bingham fluid through an eccentric annular cross-section

29.1. Generalities

As pointed out in Chapter 1, many operations in the *Oil & Gas industry* involve viscoplastic fluid flows. In *drilling operations*, in particular, the mud flows down the wellbore inside the drilling tool and flows up back to the surface in the space between the drilling tool and the wellbore (see Fig. 2.1 of Chapter 1, Section 2.2, for a visualization of the mud flow described just above). Depending on the configuration of the wellbore, it may happen that the drilling tool and the wellbore are not concentric; as a consequence, the mud flows back to the surface through an eccentric annular cross-section. For such a flow, the influence of the eccentricity is of primary importance as it modifies the flow pattern, especially in terms of *yielded* and *unyielded regions* and of the *pressure drop* required to maintain the flow at a given flow rate.

In this chapter, we will discuss the numerical simulation of the *steady flow* of a *Bingham viscoplastic fluid* through an *eccentric annular cross-section*. From a practical point of view, the consideration of several cross-sections requires the construction of as many space-discretization meshes. This operation may be time consuming; therefore, in order to avoid the construction of the meshes associated with a large number of eccentric annular geometries, we are going to treat the inner cylinder as a fictitious domain on which a zero velocity constraint will be imposed. The flow of a viscoplastic fluid in an annular cross-section cylinder has been considered by several investigators, using more conventional methods: among the related publications, let us mention the contributions of SZABO and HASSAGER [1992], WALTON and BITTLESTON [1991], NOURI, UMUR and WHITELAW [1993]. Our objective in this chapter is not to give another contribution to the physics of this type of flow. Indeed, our intention is to investigate the coupling of two types of multipliers based on computational techniques, namely the *augmented Lagrangian* methods discussed in the preceding chapters and the *DLM/FD* method; the convergence properties and the accuracy of the resulting method will be also part of our investigations. Actually, the above resulting methods will provide us with a very convenient way to investigate the influence of the eccentricity for a given flow rate, which is probably the most important parameter for this type of flow (in WACHS [2007], this type of investigation has been carried out using a *fictitious domain*-based approach very close to the one discussed in the following sections of this chapter).

29.2. Governing equations

29.2.1. Conservation and constitutive equations

In the particular case of a fully developed steady flow in a cylindrical duct of cross-section Ω, we assume that the nonaxial components of the velocity field are zero (see, e.g., WALTON and BITTLESTON [1991], and HUILGOL and YOU [2005] for details). Let us introduce the Cartesian coordinate system $\{x_1, x_2, x_3\}$, with Ox_3 parallel to the cylinder axis. From the above assumptions, the velocity field $\mathbf{u}$ verifies

$$\mathbf{u} = \{0, 0, u\}, \tag{29.1}$$

where u is a function of x_1 and x_2 only. It follows from (29.1) that

- The *velocity field* $\mathbf{u}$ is *divergence free* since $\nabla \cdot \mathbf{u} = \frac{\partial u}{\partial x_3} = 0$.
- The *stress-tensor* $\boldsymbol{\tau}$ and the *rate of strain tensor* $\mathbf{D}(\mathbf{u})$ reduce to

$$\boldsymbol{\tau} = \begin{bmatrix} 0 & 0 & \tau_{13} \\ 0 & 0 & \tau_{23} \\ \tau_{13} & \tau_{23} & 0 \end{bmatrix} \tag{29.2}$$

and

$$\mathbf{D}(\mathbf{u}) = \begin{bmatrix} 0 & 0 & D_{13}(u) \\ 0 & 0 & D_{23}(u) \\ D_{13}(u) & D_{23}(u) & 0 \end{bmatrix}, \tag{29.3}$$

respectively, with $D_{13}(u) = \frac{1}{2}\frac{\partial u}{\partial x_1}$ and $D_{23}(u) = \frac{1}{2}\frac{\partial u}{\partial x_2}$.
- The *momentum equation* reduces to

$$(\nabla \cdot \boldsymbol{\tau})_3 = \frac{\partial p}{\partial x_3} \text{ in } \Omega. \tag{29.4}$$

In (29.4), $\frac{\partial p}{\partial x_3}$ is a constant that we denote by $-\Delta P$; ΔP is nothing but the *pressure drop per unit length* (denoted by C in Chapter 2).

In order to further simplify the governing equations, we drop the useless x_3 coordinate and choose to work in the $\{x_1, x_2\}$ system. This simplification leads us to introduce:

1. The two-dimensional *stress-vector* $\boldsymbol{\tau}_3 = \{\tau_{13}, \tau_{23}\}$.
2. The two-dimensional *velocity-gradient vector* $\nabla u = \left\{\frac{\partial u}{\partial x_1}, \frac{\partial u}{\partial x_2}\right\}$ and to rewrite (29.4) and the Bingham model as follows:

$$-\nabla \cdot \boldsymbol{\tau}_3 = \Delta P, \tag{29.5}$$

$$\boldsymbol{\tau}_3 = \mu \nabla u + \tau_0 \frac{\nabla u}{|\nabla u|} \quad \text{if } |\boldsymbol{\tau}_3| > \tau_0, \tag{29.6}$$

$$\nabla u = \mathbf{0} \quad \text{if } |\boldsymbol{\tau}_3| \le \tau_0, \tag{29.7}$$

where $|\cdot|$ denotes the two-dimensional Euclidian norm defined by

$$|\mathbf{q}| = \sqrt{q_1^2 + q_2^2}, \qquad \forall \mathbf{q} = \{q_1, q_2\} \in \mathbb{R}^2.$$

We define the dimensionless *Bingham number* Bn as

$$Bn = \frac{\tau_0}{\tau_u} = \frac{\tau_0}{\tau_0 + \mu\bar{u}/L}, \tag{29.8}$$

where $\tau_u(= \tau_0 + \mu\bar{u}/L)$ denotes a characteristic shear stress, L is a characteristic length and $\bar{u}$ is the mean velocity in the cross-section.

29.2.2. *Flow geometry, boundary conditions, and problem formulation*

The *symmetry* of the problem under consideration allows us to consider only a half cross-section. In the classical formulation of the problem, the flow region is an eccentric annular cross-section, as shown in Fig. 29.1(a). If one retains the upper half of the cross-section as computational domain, the *boundary conditions* that u verifies read as follows:

- On the outer boundary Γ_1,

$$u = 0. \tag{29.9}$$

- On the inner boundary Γ_2,

$$u = 0. \tag{29.10}$$

- On the symmetry axis Γ_3,

$$\frac{\partial u}{\partial n}\left(= -\frac{\partial u}{\partial x_2}\right) = 0. \tag{29.11}$$

In order to apply the *DLM/FD* method, we fill the inner disk with the surrounding fluid, implying that this time u has to verify:

- On the outer boundary Γ_1,

$$u = 0. \tag{29.12}$$

- On the symmetry axis Γ_3,

$$\frac{\partial u}{\partial n}\left(= -\frac{\partial u}{\partial x_2}\right) = 0. \tag{29.13}$$

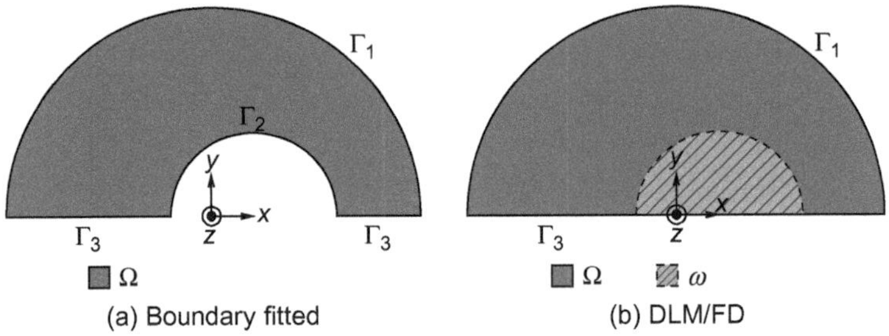

FIG. 29.1 Visualization of the half cross-section: (a) Boundary-fitted approach. (b) Fictitious domain approach.

- In the inner half-disk ω (see Fig. 29.1(b)),

$$u = 0 \text{ in } \omega. \tag{29.14}$$

From now on, the upper open half-disk will be denoted by Ω. The *fictitious domain* formulation that we will rely on reads as follows:

$$-\nabla \cdot \boldsymbol{\tau}_3 = \Delta P \text{ in } \Omega, \tag{29.15}$$

$$\boldsymbol{\tau}_3 = \mu \nabla u + \tau_0 \frac{\nabla u}{|\nabla u|} \quad \text{if } |\boldsymbol{\tau}_3| > \tau_0, \tag{29.16}$$

$$\nabla u = \mathbf{0} \qquad \text{if } |\boldsymbol{\tau}_3| \leq \tau_0, \tag{29.17}$$

$$u = 0 \text{ on } \Gamma_1, \ \frac{\partial u}{\partial n} = 0 \text{ on } \Gamma_3, \tag{29.18}$$

$$u = 0 \text{ in } \omega. \tag{29.19}$$

29.3. Numerical method

29.3.1. A pseudotransient formulation

In order to give us more flexibility, computationally speaking, we associate with (29.5)–(29.7) (in the spirit of Chapter 2) the following time-dependent model

$$\rho \frac{\partial u}{\partial t} - \nabla \cdot \boldsymbol{\tau}_3 = \Delta P \text{ in } \Omega \times (0, +\infty), \tag{29.20}$$

$$u(0) = u_0, \tag{29.21}$$

with $\boldsymbol{\tau}_3$ still verifying (29.6). The above time derivative should converge to zero as we approach steady state. Moreover, one of the advantages of the time-dependent approach is to make easy the adjustment of ΔP, in order to impose a given flow rate.

29.3.2. Variational formulations

From the earlier chapters, a conventional variational formulation of the viscoplastic flow problem under consideration reads as follows:

Find $u(t) \in H^1(\Omega \backslash \overline{\omega})$, $t \in (0, +\infty)$ such that (with $\mathrm{d}x = \mathrm{d}x_1 \mathrm{d}x_2$)

$$\rho \int_{\Omega\backslash\overline{\omega}} \frac{\partial u}{\partial t}(t)(v - u(t))\mathrm{d}x + \mu \int_{\Omega\backslash\overline{\omega}} \nabla u(t) \cdot \nabla(v - u(t))\mathrm{d}x$$

$$+ \tau_0 \left[\int_{\Omega\backslash\overline{\omega}} |\nabla v| \mathrm{d}x - \int_{\Omega\backslash\overline{\omega}} |\nabla u(t)| \mathrm{d}x \right] - \Delta P(t) \int_{\Omega\backslash\overline{\omega}} (v - u(t))\mathrm{d}x \geq 0,$$

$$\forall v \in H^1(\Omega\backslash\overline{\omega}), \ v = 0 \text{ on } \Gamma_1 \cup \Gamma_2, \tag{29.22}$$

$$u(t) = 0 \text{ on } \Gamma_1 \cup \Gamma_2, \tag{29.23}$$

$$u(0) = u_0. \tag{29.24}$$

Suppose that $u_0 \in L^2(\Omega\backslash\overline{\omega})$; it follows from, e.g., DUVAUT and LIONS [1972a, 1976] that the *parabolic variational inequality* problem (29.22)–(29.24) has a *unique solution*. This solution verifies equation (29.20) in the sense of distributions.

Following the earlier chapters, we are going to *time-discretize* problem (29.22)–(29.24) by the *backward Euler scheme*. We obtain then (with $\Delta t(> 0)$, a time-discretization step):

$$u^0 = u_0. \tag{29.25}$$

For $n \geq 1$, u^{n-1} being known, we obtain u^n from the solution of the following *elliptic variational inequality*

$$\begin{aligned}
&u^n \in H^1(\Omega\backslash\overline{\omega}),\\
&\rho \int\limits_{\Omega\backslash\overline{\omega}} \frac{u^n - u^{n-1}}{\Delta t}(v - u^n)\mathrm{d}x + \mu \int\limits_{\Omega\backslash\overline{\omega}} \nabla u^n \cdot \nabla(v - u^n)\mathrm{d}x\\
&\quad + \tau_0 \left[\int\limits_{\Omega\backslash\overline{\omega}} |\nabla v|\mathrm{d}x - \int\limits_{\Omega\backslash\overline{\omega}} |\nabla u^n|\mathrm{d}x \right] - \Delta P(n\Delta t) \int\limits_{\Omega\backslash\overline{\omega}} (v - u^n)\mathrm{d}x \geq 0,\\
&\qquad \forall v \in H^1(\Omega\backslash\overline{\omega}),\ v = 0 \text{ on } \Gamma_1 \cup \Gamma_2.
\end{aligned} \tag{29.26}$$

The iterative solution of problems such as (29.26) has been discussed in Chapter 2 (see also HE and GLOWINSKI [2000] and DEAN, GLOWINSKI and GUIDOBONI [2007]). Concerning the solution of problem (29.22)–(29.24) by a *DML/FD* method, we can either apply such a method directly on the above problem or on the time-discrete problem (29.25), (29.26). We will focus on the second approach. The fictitious domain formulation to be described below will rely on the following functional spaces, both of the *Sobolev* type:

$$\mathbf{V}_0(\Omega) = \{v \mid v \in H^1(\Omega),\ v = 0 \text{ on } \Gamma_1\}, \tag{29.27}$$

$$\mathbf{V}_0^{\omega}(\Omega) = \{v \mid v \in \mathbf{V}_0(\Omega),\ v = 0 \text{ on } \omega\}. \tag{29.28}$$

If we denote by U^n, the function defined over Ω by

$$U^n = \begin{cases} u^n & \text{in } \Omega\backslash\overline{\omega},\\ 0 & \text{in } \omega, \end{cases} \tag{29.29}$$

we clearly have *equivalence* between (29.25), (29.26) and

$$U^0 = U_0. \tag{29.30}$$

For $n \geq 1$, U^{n-1} being known, we obtain U^n from the solution of the following *elliptic variational inequality*

$$U^n \in \mathbf{V}_0^{\omega}(\Omega),$$

$$\rho \int_\Omega \frac{U^n - U^{n-1}}{\Delta t}(v - U^n)\mathrm{d}x + \mu \int_\Omega \nabla U^n \cdot \nabla(v - U^n)\mathrm{d}x$$
$$+ \tau_0 \left[\int_\Omega |\nabla v|\mathrm{d}x - \int_\Omega |\nabla U^n|\mathrm{d}x \right] - \Delta P(n\Delta t) \int_\Omega (v - U^n)\mathrm{d}x \geq 0,$$
$$\forall v \in \mathbf{V}_0^\omega(\Omega), \tag{29.31}$$

with $U_0 = \begin{cases} u_0 & \text{in } \Omega\backslash\overline{\omega}, \\ 0 & \text{in } \omega \end{cases}$ in (29.30).

Formulation (29.30), (29.31) of system (29.25), (29.26) is clearly of the *fictitious domain* type in the sense of, e.g., GLOWINSKI [2003, chapter 8]. In the following, we are going to use a *Lagrange multiplier* defined over ω to relax the condition $U^n|_\omega = 0$. This leads to the following *equivalent* formulation of (29.30), (29.31), where l^n denotes the above Lagrange multiplier:

$$U^0 = U_0. \tag{29.32}$$

For $n \geq 1$, U^{n-1} being known, we obtain $\{U^n, l^n\}$ from the solution of the following *variational inequality*

$$\{U^n, l^n\} \in \mathbf{V}_0(\Omega) \times H^1(\omega),$$
$$\rho \int_\Omega \frac{U^n - U^{n-1}}{\Delta t}(v - U^n)\mathrm{d}x + \mu \int_\Omega \nabla U^n \cdot \nabla(v - U^n)\mathrm{d}x$$
$$+ \tau_0 \left[\int_\Omega |\nabla v|\mathrm{d}x - \int_\Omega |\nabla U^n|\mathrm{d}x \right] + (l^n, v - U^n)_{1,\omega}$$
$$- \Delta P(n\Delta t) \int_\Omega (v - U^n)\mathrm{d}x \geq 0, \qquad \forall v \in \mathbf{V}_0(\Omega), \tag{29.33}$$
$$(m, U^n)_{1,\omega} = 0, \quad \forall m \in H^1(\omega); \tag{29.34}$$

in (29.33) and (29.34), $(\cdot, \cdot)_{1,\omega}$ denotes a scalar product over $H^1(\omega)$.

For $n \geq 1$, all the problems (29.33), (29.34) are of the following type:

$$\{U, l\} \in \mathbf{V}_0(\Omega) \times H^1(\omega),$$
$$\rho \int_\Omega U(v - U)\mathrm{d}x + \Delta t\mu \int_\Omega \nabla U \cdot \nabla(v - U)\mathrm{d}x$$
$$+ \Delta t\tau_0 \left[\int_\Omega |\nabla v|\mathrm{d}x - \int_\Omega |\nabla U|\mathrm{d}x \right] + (l, v - U)_{1,\omega} - \int_\Omega f(v - U)\mathrm{d}x \geq 0,$$
$$\forall v \in \mathbf{V}_0(\Omega), \tag{29.35}$$
$$(m, U)_{1,\omega} = 0, \quad \forall m \in H^1(\omega). \tag{29.36}$$

In order to solve numerically problem (29.35), (29.36), we will follow an approach which has been successful in the preceding chapters. It consists of introducing the vector-valued function $\mathbf{p} = \nabla U$, and then to treat the relation $\nabla U - \mathbf{p} = \mathbf{0}$ by a method combining (in the spirit of the *augmented Lagrangian* methods discussed in the previous chapters) *Lagrange multipliers* and *penalty*. We obtain then, with $r > 0$, the following *equivalent* formulation of problem (29.35), (29.36):

$$\{U, \mathbf{p}, \boldsymbol{\lambda}, l\} \in \mathbf{V}_0(\Omega) \times (L^2(\Omega))^2 \times (L^2(\Omega))^2 \times H^1(\omega),$$

$$\rho \int_\Omega U(v - U)\mathrm{d}x + \Delta t \mu \int_\Omega \nabla U \cdot \nabla (v - U)\mathrm{d}x$$

$$+ r \int_\Omega (\nabla U - \mathbf{p}) \cdot [\nabla (v - U) - (\mathbf{q} - \mathbf{p})]\mathrm{d}x$$

$$+ \Delta t \tau_0 \left[\int_\Omega |\mathbf{q}|\mathrm{d}x - \int_\Omega |\mathbf{p}|\mathrm{d}x \right] + (l, v - U)_{1,\omega}$$

$$+ \int_\Omega \boldsymbol{\lambda} \cdot [\nabla (v - U) - (\mathbf{q} - \mathbf{p})]\mathrm{d}x - \int_\Omega f(v - U)\mathrm{d}x \geq 0,$$

$$\forall \{v, \mathbf{q}\} \in \mathbf{V}_0(\Omega) \times (L^2(\Omega))^2, \tag{29.37}$$

$$\nabla U - \mathbf{p} = \mathbf{0}, \tag{29.38}$$

$$(m, U)_{1,\omega} = 0, \qquad \forall m \in H^1(\omega). \tag{29.39}$$

Problem (29.37)–(29.39) is *equivalent* to

$$\{U, \mathbf{p}, \boldsymbol{\lambda}, l\} \in \mathbf{V}_0(\Omega) \times (L^2(\Omega))^2 \times (L^2(\Omega))^2 \times H^1(\omega),$$

$$\rho \int_\Omega U v \,\mathrm{d}x + (\Delta t \mu + r) \int_\Omega \nabla U \cdot \nabla v \,\mathrm{d}x + (l, v)_{1,\omega} - \int_\Omega (r\mathbf{p} - \boldsymbol{\lambda}) \cdot \nabla v \,\mathrm{d}x$$

$$= \int_\Omega f v \,\mathrm{d}x, \qquad \forall v \in \mathbf{V}_0(\Omega), \tag{29.40}$$

$$(m, U)_{1,\omega} = 0, \qquad \forall m \in H^1(\omega), \tag{29.41}$$

$$r \int_\Omega \mathbf{p} \cdot (\mathbf{q} - \mathbf{p})\mathrm{d}x + \Delta t \tau_0 \left[\int_\Omega |\mathbf{q}|\mathrm{d}x - \int_\Omega |\mathbf{p}|\mathrm{d}x \right]$$

$$- \int_\Omega (r\nabla U + \boldsymbol{\lambda}) \cdot (\mathbf{q} - \mathbf{p})\mathrm{d}x \geq 0, \qquad \forall \mathbf{q} \in (L^2(\Omega))^2, \tag{29.42}$$

$$\nabla U - \mathbf{p} = \mathbf{0}. \tag{29.43}$$

The *iterative* solution of the *variational system* (29.40)–(29.43) will be discussed in the following section.

29.3.3. An iterative method for the solution of problem (29.40)–(29.43)

As expected, the rich structure of problem (29.40)–(29.43) makes it a candidate for a relatively large number of solution methods. In this article, we will focus on only one of them, namely the algorithm *ALG 2*, already encountered in Chapters 2 and 3 (see also GLOWINSKI, LIONS and TRÉMOLIÈRES [1981], FORTIN and GLOWINSKI [1982, 1983], GLOWINSKI and LE TALLEC [1989], SAMUELSSON [1993], ROQUET [2000], MOYERS-GONZALEZ and FRIGAARD [2004], HUILGOL and YOU [2005], and VINAY, WACHS and AGASSANT [2005]). Applying *ALG 2* to problem (29.40)–(29.43), one obtains:

$$\{\mathbf{p}^{-1}, \boldsymbol{\lambda}^0\} \text{ is given in } (L^2(\Omega))^2 \times (L^2(\Omega))^2. \tag{29.44}$$

For $k \geq 0$, assuming that $\{\mathbf{p}^{k-1}, \boldsymbol{\lambda}^k\}$ is known, solve

$$\begin{aligned} &\{U^k, l^k\} \in \mathbf{V}_0(\Omega) \times H^1(\omega), \\ &\rho \int_\Omega U^k v \, dx + (\Delta t \mu + r) \int_\Omega \nabla U^k \cdot \nabla v \, dx + (l^k, v)_{1,\omega} \\ &\quad - \int_\Omega (r\mathbf{p}^{k-1} - \boldsymbol{\lambda}^k) \cdot \nabla v \, dx = \int_\Omega f v \, dx, \qquad \forall v \in \mathbf{V}_0(\Omega), \end{aligned} \tag{29.45}$$

$$(m, U^k)_{1,\omega} = 0, \qquad \forall m \in H^1(\omega), \tag{29.46}$$

and then

$$\begin{aligned} &\mathbf{p}^k \in (L^2(\Omega))^2, \\ &r \int_\Omega \mathbf{p}^k \cdot (\mathbf{q} - \mathbf{p}^k) dx + \Delta t \tau_0 \left[\int_\Omega |\mathbf{q}| dx - \int_\Omega |\mathbf{p}^k| dx \right] \\ &\quad - \int_\Omega (r \nabla U^k + \boldsymbol{\lambda}^k) \cdot (\mathbf{q} - \mathbf{p}^k) dx \geq 0, \qquad \forall \mathbf{q} \in (L^2(\Omega))^2. \end{aligned} \tag{29.47}$$

Update $\boldsymbol{\lambda}^k$ by

$$\boldsymbol{\lambda}^{k+1} = \boldsymbol{\lambda}^k + r(\nabla U^k - \mathbf{p}^k). \tag{29.48}$$

Stop iterating if

$$\|U^k - U^{k-1}\|_1 \leq \text{tol}_1 \text{ and } \|\nabla U^k - \mathbf{p}^k\|_2 \leq \text{tol}_2. \tag{29.49}$$

The solution of the linearly constrained elliptic problem (29.45), (29.46) will be addressed in Section 29.3.5. Concerning problem (29.47), we clearly have

$$\mathbf{p}^k = \underset{\mathbf{q} \in (L^2(\Omega))^2}{\arg\min} \left[\frac{1}{2} \int_\Omega |\mathbf{q}|^2 dx + \Delta t \tau_0 \int_\Omega |\mathbf{q}| dx - \int_\Omega (r \nabla U^k + \boldsymbol{\lambda}^k) \cdot \mathbf{q} \, dx \right]. \tag{29.50}$$

It follows from (29.50) that

$$\mathbf{p}^k = \frac{1}{r}\left(1 - \frac{\Delta t \tau_0}{|r\nabla U^k + \boldsymbol{\lambda}^k|}\right)^+ (r\nabla U^k + \boldsymbol{\lambda}^k). \tag{29.51}$$

REMARK 29.1. The two norms in (29.49) have not been specified. A natural choice would be $\|\cdot\|_1 = \|\cdot\|_{L^2(\Omega)}$ and $\|\cdot\|_2 = \|\cdot\|_{(L^2(\Omega))^2}$. Actually, *after* an appropriate *finite element* based *space discretization* (see Section 29.3.4 for details), we have chosen for our computations $\|\cdot\|_1 = \|\cdot\|_{L^\infty(\Omega)}$ and $\|\cdot\|_2 = \|\cdot\|_{(L^\infty(\Omega))^2}$.

REMARK 29.2. Instead of algorithms (29.44)–(29.49), we could have used the close variant of it obtained by switching the roles of U and $\mathbf{p}$, namely computing $\mathbf{p}^k$ first, and then U^k.

29.3.4. Finite-element approximation

In order to achieve the *space approximation* of problem (29.40)–(29.43), we advocate the *finite-element method* we already used in Chapter 2, Sections 12 and 15. We introduce thus a *finite-element triangulation* $\mathcal{T}_h$ of Ω, and from it the following approximations of the spaces $\mathbf{V}_0(\Omega)$ and $(L^2(\Omega))^2$:

$$\mathbf{V}_{0h} = \{v \mid v \in C^0(\overline{\Omega}_h),\ v|_T \in P_1, \qquad \forall T \in \mathcal{T}_h,\ v = 0 \text{ on } \Gamma_{1h}\}, \tag{29.52}$$

and

$$\mathbf{Q}_h = \{\mathbf{q} \mid \mathbf{q} \in (L^2(\Omega_h))^2,\ \mathbf{q}|_T \in (P_0)^2, \qquad \forall T \in \mathcal{T}_h\}, \tag{29.53}$$

where in (29.52), (29.53): (1) $\overline{\Omega}_h = \bigcup_{T\in\mathcal{T}_h} T$ and $\Omega_h =$ interior of $\overline{\Omega}_h$. (2) Γ_{1h} is the approximation of Γ_1 associated with $\mathcal{T}_h$ and Ω_h. (3) P_0 (resp., P_1) is the space of the polynomials of two variables of degree 0 (resp., of degree ≤ 1).

Concerning the approximation of $H^1(\omega)$ in (29.39) and (29.41), we advocate (following, e.g., GLOWINSKI [2003, chapter 8]) the discrete space Λ_h defined by

$$\Lambda_h = \{\mu \mid \mu = \sum_{j=1}^{J} \mu_j \delta(x - x_j),\ \mu_j \in \mathbb{R}, \qquad \forall j = 1, \ldots, J\}, \tag{29.54}$$

where in (29.54): (1) $\{x_j\}_{j=1}^J (= \mathcal{P})$ is a finite set of points covering $\overline{\omega}$. (2) δ is the *Dirac measure* at $x = 0$. Among the possible sets $\mathcal{P}$, those defined as follows are particularly easy to use from a computational standpoint

$$\mathcal{P} = \mathcal{P}_1 \cup \mathcal{P}_2,$$

with $\mathcal{P}_1$ (resp., $\mathcal{P}_2$) the set of the vertices of $\mathcal{T}_h$ contained in ω and whose distance at the boundary $\partial\omega$ of ω is $\ge h$ (resp., a set of points of $\partial\omega$ such that the distance between neighboring points is of the order of h). Sets $\mathcal{P}$ of the above type have been systematically used in GLOWINSKI [2003, chapters 8 and 9] (and elsewhere) for the numerical simulation of particulate flows.

To take advantage of the point-wise nature of Λ_h, we have replaced in our computations $(\cdot,\cdot)_{1,\omega}$ by the pairing

$$\{\mu, v\} \to \langle \mu, v\rangle_h$$

with

$$\langle \mu, v \rangle_h = \sum_{j=1}^{J} \mu_j v(x_j), \qquad \forall \mu \in \Lambda_h, \forall v \in \mathbf{V}_{0h}. \tag{29.55}$$

REMARK 29.3. As pointed out in GLOWINSKI [2003, chapter 8], the *collocation* type method associated with (29.55) makes little sense for the *continuous problem* (since $H^1(\omega) \not\subset C^0(\overline{\omega})$ if $\omega \subset \mathbb{R}^2$). However, this approach is meaningful for the discrete *problem* to be described just below, since the discrete velocity fields we will encounter are all continuous over $\overline{\omega}$.

With obvious notation, the *space-discrete* analog of problem (29.40)–(29.43) reads as follows:

$$\{U_h, \mathbf{p}_h, \boldsymbol{\lambda}_h, l_h\} \in \mathbf{V}_{0h} \times \mathbf{Q}_h \times \mathbf{Q}_h \times \Lambda_h,$$

$$\rho \int_{\Omega_h} U_h v \,\mathrm{d}x + (\Delta t \mu + r) \int_{\Omega_h} \nabla U_h \cdot \nabla v \,\mathrm{d}x + \langle l_h, v \rangle_h$$
$$- \int_{\Omega_h} (r\mathbf{p}_h - \boldsymbol{\lambda}_h) \cdot \nabla v \,\mathrm{d}x = \int_{\Omega_h} f_h v \,\mathrm{d}x, \qquad \forall v \in \mathbf{V}_{0h}, \tag{29.56}$$

$$\langle m, U_h \rangle_h = 0, \qquad \forall m \in \Lambda_h, \tag{29.57}$$

$$r \int_{\Omega_h} \mathbf{p}_h \cdot (\mathbf{q} - \mathbf{p}_h) \mathrm{d}x + \Delta t \tau_0 \left[\int_{\Omega_h} |\mathbf{q}| \mathrm{d}x - \int_{\Omega_h} |\mathbf{p}_h| \mathrm{d}x \right]$$
$$- \int_{\Omega_h} (r \nabla u_h + \boldsymbol{\lambda}_h) \cdot (\mathbf{q} - \mathbf{p}_h) \mathrm{d}x \geq 0, \qquad \forall \mathbf{q} \in \mathbf{Q}_h, \tag{29.58}$$

$$\nabla U_h - \mathbf{p}_h = \mathbf{0}. \tag{29.59}$$

In order to solve the discrete variational system (29.56)–(29.59), we advocate the following discrete analog of algorithm (29.44)–(29.49) (some of the subscripts h have been dropped):

$$\{\mathbf{p}^{-1}, \boldsymbol{\lambda}^0\} \text{ is given in } \mathbf{Q}_h \times \mathbf{Q}_h \tag{29.60}$$

For $k \geq 0$, assuming that $\{\mathbf{p}^{k-1}, \boldsymbol{\lambda}^k\}$ is known, solve

$$\{U^k, l^k\} \in \mathbf{V}_{0h} \times \Lambda_h,$$

$$\rho \int_{\Omega_h} U^k v \,\mathrm{d}x + (\Delta t \mu + r) \int_{\Omega_h} \nabla U^k \cdot \nabla v \,\mathrm{d}x + \langle l^k, v \rangle_h$$
$$- \int_{\Omega_h} (r\mathbf{p}^{k-1} - \boldsymbol{\lambda}^k) \cdot \nabla v \,\mathrm{d}x = \int_{\Omega_h} f v \,\mathrm{d}x, \qquad \forall v \in \mathbf{V}_{0h}, \tag{29.61}$$

$$\langle m, U^k \rangle_h = 0, \qquad \forall m \in \Lambda_h, \tag{29.62}$$

and then

$$\mathbf{p}^k \in \mathbf{Q}_h,$$

$$r\int_{\Omega_h} \mathbf{p}^k \cdot (\mathbf{q} - \mathbf{p}^k)\mathrm{d}x + \Delta t\tau_0 \left[\int_{\Omega_h} |\mathbf{q}|\mathrm{d}x - \int_{\Omega_h} |\mathbf{p}^k|\mathrm{d}x \right]$$

$$- \int_{\Omega_h} (r\nabla U^k + \boldsymbol{\lambda}^k) \cdot (\mathbf{q} - \mathbf{p}^k)\mathrm{d}x \geq 0, \qquad \forall \mathbf{q} \in \mathbf{Q}_h. \tag{29.63}$$

Update $\boldsymbol{\lambda}^k$ by

$$\boldsymbol{\lambda}^{k+1} = \boldsymbol{\lambda}^k + r(\nabla U^k - \mathbf{p}^k). \tag{29.64}$$

Stop iterating if

$$\|U^k - U^{k-1}\|_\infty \leq \mathrm{tol}_1 \text{ and } \|\nabla U^k - \mathbf{p}^k\|_\infty \leq \mathrm{tol}_2. \tag{29.65}$$

REMARK 29.4. The fact that in (29.63) $\mathbf{p}^k$, $\mathbf{q}$, ∇U^k, and $\boldsymbol{\lambda}^k$ are *piecewise constant* over the triangles of $\mathcal{T}_h$ implies that:

(1) The solution $\mathbf{p}^k$ of problem (29.63) is given by

$$\mathbf{p}^k|_T = \frac{1}{r}\left(1 - \frac{\Delta t\tau_0}{|(r\nabla U^k + \boldsymbol{\lambda}^k)|_T|}\right)^+ (r\nabla U^k + \boldsymbol{\lambda}^k)|_T, \qquad \forall T \in \mathcal{T}_h. \tag{29.66}$$

(2) The second and third integrals in (29.61) can be computed exactly, easily. Concerning the first and fourth integrals, they can be computed exactly using the two-dimensional *Simpson rule*, or approximately using the two-dimensional *trapezoidal rule*, that is

$$\int_T \varphi\,\mathrm{d}x \approx \frac{|T|}{3}\sum_{j=1}^{3} \varphi(m_j) \qquad (\textit{Simpson rule})$$

and

$$\int_T \varphi\,\mathrm{d}x \approx \frac{|T|}{3}\sum_{j=1}^{3} \varphi(A_j) \qquad (\textit{trapezoidal rule}),$$

where $|T| =$ measure of T, A_1, A_2, A_3 are the vertices of triangle T and m_1, m_2, m_3 are the mid-points of the edges A_2A_3, A_3A_1, A_1A_2, respectively. The trapezoidal rule (resp., the Simpson rule) is exact if $\varphi \in P_1$ (resp., $\varphi \in P_2$). If one uses the trapezoidal rule to compute the first integral in (29.61), the associated *mass matrix* will be *diagonal*. It is worth observing that using either the trapezoidal rule or the Simpson rule to compute the first and fourth integrals in (29.61) produces the same steady-state solution; we advocate, thus, the first numerical integration procedure as it leads to easier computations.

The solution of problem (29.61), (29.62) will be discussed in the following section.

29.3.5. Solution of problem (29.61), (29.62)
Problem (29.61), (29.62) can be written in *matrix form* as

$$\begin{bmatrix} \mathbf{A} & \mathbf{M}^t \\ \mathbf{M} & \mathbf{0} \end{bmatrix} \begin{bmatrix} \mathbf{U} \\ \mathbf{l} \end{bmatrix} = \begin{bmatrix} \mathbf{f} \\ \mathbf{g} \end{bmatrix}, \tag{29.67}$$

where $\mathbf{A}$ is a $N \times N$ matrix symmetric and positive definite (with $N = \dim(\mathbf{V}_{0h})$), $\mathbf{M}$ is a $N \times J$ matrix (with $J = \dim(\Lambda_h)$), $\{\mathbf{U}, \mathbf{l}\} \in \mathbb{R}^N \times \mathbb{R}^J$ and $\{\mathbf{f}, \mathbf{g}\} \in \mathbb{R}^N \times \mathbb{R}^J$. Linear systems such as (29.67) are called *Kuhn–Tucker* (or *saddle-point*) systems; their solution has motivated a large number of publications. In this chapter, we follow FORTIN and GLOWINSKI [1982, 1983], GLOWINSKI and LE TALLEC [1989] by:

(1) Replacing the linear system (29.67) by the following *equivalent* one

$$\begin{bmatrix} \mathbf{A} + r'\mathbf{M}^t\mathbf{M} & \mathbf{M}^t \\ \mathbf{M} & \mathbf{0} \end{bmatrix} \begin{bmatrix} \mathbf{U} \\ \mathbf{l} \end{bmatrix} = \begin{bmatrix} \mathbf{f} + r'\mathbf{M}^t\mathbf{g} \\ \mathbf{g} \end{bmatrix}, \tag{29.68}$$

with $r' > 0$ in (29.68).

(2) Solving the augmented linear system (29.68) by an *Uzawa/conjugate gradient algorithm* like the one described just below.

Description of the Uzawa/conjugate gradient algorithm:

$$\mathbf{l}^0 \text{ is given in } \mathbb{R}^J; \tag{29.69}$$

solve

$$(\mathbf{A} + r'\mathbf{M}^t\mathbf{M})\mathbf{U}^0 = \mathbf{f} + r'\mathbf{M}^t\mathbf{g} - \mathbf{M}^t\mathbf{l}^0 \tag{29.70}$$

and set

$$\mathbf{g}^0 = \mathbf{g} - \mathbf{M}\mathbf{U}^0, \tag{29.71}$$

$$\mathbf{w}^0 = \mathbf{g}^0. \tag{29.72}$$

For $s \geq 0$, assuming that $\mathbf{l}^s$, $\mathbf{g}^s$, and $\mathbf{w}^s$ are known, the last two different from $\mathbf{0}$, we compute $\mathbf{l}^{s+1}$, $\mathbf{g}^{s+1}$, and if necessary $\mathbf{w}^{s+1}$, as follows:
Solve

$$(\mathbf{A} + r'\mathbf{M}^t\mathbf{M})\overline{\mathbf{U}}^s = -\mathbf{M}^t\mathbf{w}^s \tag{29.73}$$

and set

$$\overline{\mathbf{g}}^s = -\mathbf{M}\overline{\mathbf{U}}^s. \tag{29.74}$$

Compute

$$\rho_s = \frac{\mathbf{g}^s \cdot \mathbf{g}^s}{\bar{\mathbf{g}}^s \cdot \mathbf{w}^s} \tag{29.75}$$

and

$$\mathbf{l}^{s+1} = \mathbf{l}^s - \rho_s \mathbf{w}^s, \tag{29.76}$$

$$\mathbf{g}^{s+1} = \mathbf{g}^s - \rho_s \bar{\mathbf{g}}^s. \tag{29.77}$$

If $\frac{\|\mathbf{g}^{s+1}\|}{\|\mathbf{g}^0\|} \leq \text{tol}$ take $\mathbf{l} = \mathbf{l}^{s+1}$; else, compute

$$\gamma_s = \frac{\mathbf{g}^{s+1} \cdot \mathbf{g}^{s+1}}{\mathbf{g}^s \cdot \mathbf{g}^s} \tag{29.78}$$

and

$$\mathbf{w}^{s+1} = \mathbf{g}^{s+1} + \gamma_s \mathbf{w}^s. \tag{29.79}$$

Do $s = s + 1$ and return to (29.73).

REMARK 29.5. Because we impose $u = 0$ in the lower cylinder ω, we have, from (29.55) and (29.57), $\mathbf{g} = \mathbf{0}$ in (29.67), (29.68) and below.

REMARK 29.6. *Uzawa/conjugate gradient algorithms* such as (29.69)–(29.79) have been already encountered in Chapter 2, Section 13.5, and Chapter 3, Section 22.

REMARK 29.7. Concerning the convergence of algorithm (29.69)–(29.79), it is shown in, e.g., FORTIN and GLOWINSKI [1982, 1983] that the speed of convergence (measured in number of iterations) improves as r' increases. However, the condition number of the matrix $\mathbf{A} + r'\mathbf{M}^t\mathbf{M}$ increases because it is of the order of r'. To cope with this difficulty, we chose to solve the linear systems (29.70) and (29.73) by the *method of Cholesky* (see, e.g., CIARLET [1989]), the factorization of the matrix $\mathbf{A} + r'\mathbf{M}^t\mathbf{M}$ being done once for all, with r' chosen quite large. Indeed, when applying algorithm (29.60)–(29.65) to the solution of the discrete variational system (29.56)–(29.59), the Cholesky factorization of $\mathbf{A} + r'\mathbf{M}^t\mathbf{M}$ takes less than 10% of the whole computational time.

29.3.6. Computations with variable eccentricity

From an engineering point of view, one is interested to investigate the influence of the *eccentricity* for a *given flow rate*. Our approach to this problem can be summarized as follows: (1) We specify the lower bound $e_{\min}$ and the upper bound $e_{\max}$ of the eccentricities of interest, and then the number n_{ec} of eccentricity steps. (2) Using the methodology described in the above sections of this chapter, we compute on a fixed mesh the steady flow of the viscoplastic fluid under consideration for the eccentric positions of the inner cylinder selected in (1). (3) Because we are interested by the steady-state solutions, only, we can use a large Δt in our pseudotransient approach. (4) At every time step $t^n = n\Delta t$, we adjust the pressure drop ΔP^n so that the corresponding flow rate Q^n will match the targeted flow rate Q_0. (5) We integrate until we reach a steady-state solution whose flow rate is Q_0 (within a given accuracy).

We will start our discussion with the *adjustment of the pressure drop*:

As already mentioned, at every time step t^n the pressure drop ΔP^n is adjusted so that the corresponding flow rate Q^n will be close to Q_0. This adjustment is done through the relation

$$\Delta P^n = F(Q_0, Q^{n-1}, Q^{n-2}, \Delta P^{n-1}). \tag{29.80}$$

Assuming that $\Delta P^0 > 0$, the updating strategy associated with F is defined as follows for $n \geq 1$:

If $Q^{n-1} = 0$, take $\Delta P^n = C\Delta P^{n-1}$ (with $C > 1$); else (with ε_1 and ε_2 both positive):

If $|Q^{n-1} - Q_0| \leq \varepsilon_1$ or $(Q^{n-1} - Q_0)(Q^{n-1} - Q^{n-2}) \leq 0$ and $|Q^{n-1} - Q^{n-2}| > \varepsilon_2$ take $\Delta P^n = \Delta P^{n-1}$; otherwise, take

$$\Delta P^n = \min\left[2, \max\left(1/2, 1 + \frac{Q_0 - Q^{n-1}}{Q_0}\right)\right] \Delta P^{n-1}.$$

The steady-state solutions with flow rate Q_0, associated with the n_{ec} selected eccentricities, are obtained by the following algorithm:

(1) Compute the matrix $\mathbf{A}$ found in (29.67).

(2) Initialize with:

$$\mathbf{U}^0 = \mathbf{0}, \mathbf{l}^0 = \mathbf{0}, \mathbf{p}^0 = \mathbf{0}, \boldsymbol{\lambda}^0 = \mathbf{0} \quad \text{and} \quad \Delta P^0 = \Delta P_{\text{init}}. \tag{29.81}$$

(3) For $j = 1, \ldots, n_{ec}$

- Compute the eccentricity

$$e_j = e_{\min} + (j-1)\frac{e_{\max} - e_{\min}}{n_{ec} - 1}. \tag{29.82}$$

- Determine the set $\mathcal{P}_j$ associated with the fictitious domain treatment of the inner cylinder ω_j and denote by Λ_{jh} the corresponding Lagrange multiplier space.
- Assemble the matrix $\mathbf{M}_j$ associated with ω_j and $\mathcal{P}_j$; compute and Cholesky factorize the matrix $\mathbf{A} + r'\mathbf{M}_j^t\mathbf{M}_j$.
- Take $U_j^0 = U_{j-1}$ and for $n \geq 1$, apply algorithm (29.60)–(29.65) to the solution of the following fully discrete variational system

$$\{U_j^n, l_j^n\} \in \mathbf{V}_{0h} \times \Lambda_{jh},$$

$$\rho \int_\Omega (U_j^n - U_j^{n-1})(v - U_j^n)\mathrm{d}x + \Delta t \mu \int_\Omega \nabla U_j^n \cdot \nabla(v - U_j^n)\mathrm{d}x$$

$$+ \Delta t \tau_0 \left[\int_\Omega |\nabla v|\mathrm{d}x - \int_\Omega |\nabla U_j^n|\mathrm{d}x\right] + \langle l_j^n, v - U_j^n\rangle_h$$

$$-\Delta t \Delta P_j^n \int_\Omega (v - U_j^n)\mathrm{d}x \geq 0, \qquad \forall v \in \mathbf{V}_{0h}, \tag{29.83}$$

$$\langle m, U_j^n \rangle_h = 0, \qquad \forall m \in \Lambda_{jh}. \tag{29.84}$$

Compute the flow rate Q_j^n by

$$Q_j^n = 2\int_\Omega U_j^n \,\mathrm{d}x. \tag{29.85}$$

If

$$\|U_j^n - U_j^{n-1}\|_\infty \leq \varepsilon_3 \quad \text{and} \quad |Q_j^n - Q_0| \leq \varepsilon_1,$$

take $U_j = U_j^n$; otherwise, update the pressure drop via

$$\Delta P_j^{n+1} = F(Q_0, Q_j^n, Q_j^{n-1}, \Delta P_j^n), \tag{29.86}$$

take $n = n + 1$ and return to (29.83), (29.84).

29.4. Numerical experiments. Discussion of the numerical results

29.4.1. Generalities. Synopsis

In this section, we are going to investigate the steady flow of a Bingham viscoplastic fluid in a cylinder with an *eccentric annular cross-section.* From the symmetry of the flow, we can use a half cross-section as computational domain. The computational method to be used is the one discussed in the preceding sections; namely, it relies on the combination of a distributed Lagrange multiplier/fictitious domain method with a finite-element approximation and the augmented Lagrangian algorithm *ALG2*. The influence of the eccentricity is investigated assuming that the flow rate is a constant (denoted by Q_0 in the preceding and following sections).

The numerical results are presented and analyzed in terms of dimensionless quantities. For the *characteristic length* L mentioned in Section 29.2, we take $L = R_{\text{out}} - R_{\text{in}}$, where R_{out} (resp., R_{in}) is the radius of the *outer* (resp., *inner*) cylinder. Next, we define the dimensionless coordinates $x_1^* = \frac{x_1}{R_{\text{out}}}$ and $x_2^* = \frac{x_2}{R_{\text{out}}}$. For the *characteristic velocity*, we take $\bar{u} = \frac{Q_0}{S}$, where $S = \pi(R_{\text{out}}^2 - R_{\text{in}}^2)$ is the surface of the annular cross-section. We denote by δ the distance between the axes of the outer and inner cylinders. Further relevant dimensionless parameters and quantities are as follows:

- The *Bingham number* Bn defined by

$$B\mathrm{n} = \frac{\tau_0}{\tau_0 + \frac{\mu\bar{u}}{L}} = \frac{\tau_0}{\tau_0 + \frac{\mu Q_0}{LS}}.$$

- The *eccentricity* $e = \frac{\delta}{L}$.
- The *radius ratio* $\chi = \frac{R_{\text{in}}}{R_{\text{out}}}$.

- The *dimensionless pressure drop*

$$\Delta P^* = \frac{L\Delta P}{\tau_0 + \frac{\mu\bar{u}}{L}} = \frac{L\Delta P}{\tau_0 + \frac{\mu Q_0}{LS}}.$$

- The *dimensionless velocity* $u^* = \frac{u}{\bar{u}} = \frac{Su}{Q_0}$.

A *dimensional analysis* of the momentum equation provides a *characteristic timescale*

$$T_{\text{char}} = \frac{\rho\bar{u}L}{\tau_0 + \frac{\mu\bar{u}}{L}} = \frac{\rho Q_0 L}{S\tau_0 + \frac{\mu Q_0}{L}}, \tag{29.87}$$

which is used to estimate the time required to reach the steady state. Because one of our goals is to reach this steady state as quickly as possible, we run our computations with large time steps. This lead us to take $\Delta t = T_{\text{char}}/5$, a choice giving satisfactory results.

For comparison purposes with the results presented in SZABO and HASSAGER [1992], we provide also their definition of the Bingham number and of the eccentricity distance, that is (with obvious notation):

$$\mathcal{B}\text{n} = \frac{\tau_0}{\Delta P R_{\text{out}}} \tag{29.88}$$

and

$$\delta^k = \frac{\delta}{R_{\text{out}}} = e(l - \chi). \tag{29.89}$$

Having said that, we believe that our definition of the Bingham number is more convenient, because

If $Bn \in [0, 1)$, the fluid flows; if $Bn = 1$, there is no flow.

Our *first test problem* corresponds to the case where $e = 0$; in that particular case, the closed form of the steady-state solution is known and can be found in, e.g., SZABO and HASSAGER [1992]. Using a notation close to the one in the above reference, the dimensionless steady state velocity solution of our first test problem reads as follows in dimensionless *polar coordinates* $\left(\text{with } r^* = \sqrt{(x_1^*)^2 + (x_2^*)^2}\right)$:

$$u^* = 0 \quad \text{if} \quad \mathcal{B}\text{n} \geq \frac{1}{2}(1 - \chi). \tag{29.90}$$

If $\mathcal{B}\text{n} < \frac{1}{2}(1 - \chi)$, one has

$$u^*(x) = u_1^*(r^*) = -\frac{1}{4}(r^{*2} - \chi^2) - \mathcal{B}\text{n}(r^* - \chi) + \frac{1}{2}\beta^2 \ln\left(\frac{r^*}{\chi}\right)$$
$$\text{if } \chi \leq r^* \leq \beta_-, \tag{29.91}$$
$$u^*(x) = u_2^*(r^*) = \frac{1}{4}(1 - r^{*2}) - \mathcal{B}\text{n}(1 - r^*) + \frac{1}{2}\beta^2 \ln r^* \text{ if } \beta_+ \leq r^* \leq 1, \tag{29.92}$$
$$u^*(x) = u_2^*(\beta_+) = u_1^*(\beta_-) \text{ if } \beta_- \leq r^* \leq \beta_+, \tag{29.93}$$

where in (29.91)–(29.93):

- The quantities β, β_+, and β_- are defined by

$$\beta^2 = \beta_+(\beta_+ - 2\mathcal{B}\mathrm{n}) = \beta_-(\beta_- + 2\mathcal{B}\mathrm{n}) \tag{29.94}$$

$$\beta_+ - \beta_- = 2\mathcal{B}\mathrm{n}, \tag{29.95}$$

  β_+ being the solution of the following one variable nonlinear equation

$$2\beta_+(\beta_+ - 2\mathcal{B}\mathrm{n}) \ln\left(\frac{\beta_+ - 2\mathcal{B}\mathrm{n}}{\beta_+\chi}\right) + 4\mathcal{B}\mathrm{n}(1 - \beta_+) + (2\mathcal{B}\mathrm{n} + \chi)^2 - 1 = 0; \tag{29.96}$$

  there is no difficulty at solving equation (29.96) by the method of *Newton–Raphson*.
- One has used the following dimensionless variables:

$$u^* = \frac{u}{U_0},\ r^* = \frac{|x|}{R_{\mathrm{out}}} \quad \text{with} \quad U_0 = \frac{\Delta P R_{\mathrm{out}}^2}{\mu}. \tag{29.97}$$

- The dimensionless *flow rate* Q^* is given by

$$Q^* = \frac{\pi}{8}\left[(1-\chi^4) - 2\beta^2(1-\chi^2) - \frac{8}{3}(1+\chi^3)\mathcal{B}\mathrm{n} + \frac{16}{3}(\beta_+ - \mathcal{B}\mathrm{n})^3\mathcal{B}\mathrm{n}\right]. \tag{29.98}$$

First, we performed computations to validate our whole solution methodology, whose two main components are the *DLM/FD* method and the *updating* strategy. The test problem that we considered corresponds to $e = 0$, for which *closed form* solutions are available (see (29.90)–(29.96)) together with computational results obtained with *boundary-fitted meshes* (see, e.g., SZABO and HASSAGER [1992]). Next, we carried out a parametric survey for the values 0.2, 0.5, and 0.8 of the radius ratio χ and for $Bn = 0, 0.5, 0.75, 0.9$, and 0.98. In all the cases investigated in the parametric survey, we set $e_{\min} = 0$, $e_{\max} = 1$ and $n_{\mathrm{ec}} = 11$ (that is $\Delta e = 0.1$). Our discussion will include the description of an engineering *Response Surface Methodology* (RSM) as a tool to predict the pressure drop; the accuracy of the RSM approach will be part of our discussion.

29.4.2. On meshes and DLM/FD collocation points

Unstructured triangulations of approximately constant grid size are generated for the computations. We consider two families of such triangulations:

(1) A first family of triangulations based on a half circular geometry in order to apply the *DLM/FD*-based methodology.
(2) For comparison purposes, a second family of triangulations fitting the boundary of the half ring.

For the two families above, the most significant mesh parameter is the number of grid points located on the Ox_1-axis; let us denote this number by $2N_r$. For the *DLM/FD* meshes, $2N_r$ corresponds to the number of points located on the part of the Ox_1-axis where the fluid flows; this ensures that for a given N_r, the corresponding grid size in the region where the

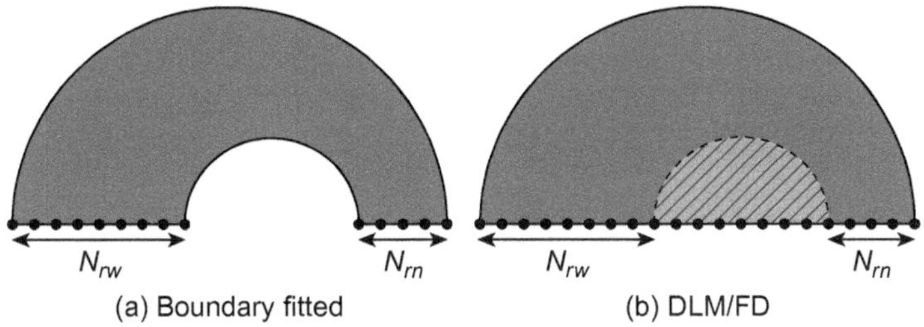

FIG. 29.2 Mesh parameter $N_r = 1/2(N_{rw} + N_{rn})$ for the boundary-fitted meshes (a) and DLM/FD (b) meshes.

TABLE 29.1
Mesh characteristics: number of triangles versus N_r for the DLM/FD and boundary-fitted meshes ($\chi = 1/2$ and $e = 1/2$)

N_r	Boundary fitted	DLM/FD
10	1120	1290
20	4683	5226
30	8880	11754
40	16520	20619
50	25228	31983

fluid flows is the same for both the DLM/FD and boundary-fitted meshes. As shown in Fig. 29.2, we have denoted by N_{rw} and N_{rn} the number of grid points located, respectively, on the wide and narrow sides. We have then

$$N_r = 1/2(N_{rw} + N_{rn}). \tag{29.99}$$

In Table 29.1, we have collected the characteristics of the meshes in the case $\chi = 1/2$ and compared, for $e = 1/2$ the total number of triangles required, respectively, by the boundary-fitted and DLM/FD approaches: for a similar grid size, the DLM/FD approach requires, approximately, 20% more triangles than the boundary-fitted one.

If the eccentric ring has a narrow gap, that is when $\chi \in [0.75, 1]$, an optimized mesh (as the one shown in Fig. 29.4) can be used. The relevant grid size is the one prevailing in the region between the inner and outer cylinders, that is where the fluid flows (if the pressure drop is large enough). Figure 29.4 shows that the grid size remains the same in this region for any value of the eccentricity. The use of a coarser mesh in the region occupied by the inner cylinder does not affect the computed solution and enables to reduce slightly the number of mesh elements (grid points and triangles). To determine the set $\mathcal{P}$ of collocation points covering the fictitious region, we applied the following strategy:

- We retained all the grid points located in the inner cylinder and whose distance at the inner cylinder boundary is greater than the grid size.
- We selected a set of points located on the inner cylinder boundary, equally spaced with a interdistance of the order of the grid size.
- We checked that every triangle contains at most one boundary point.

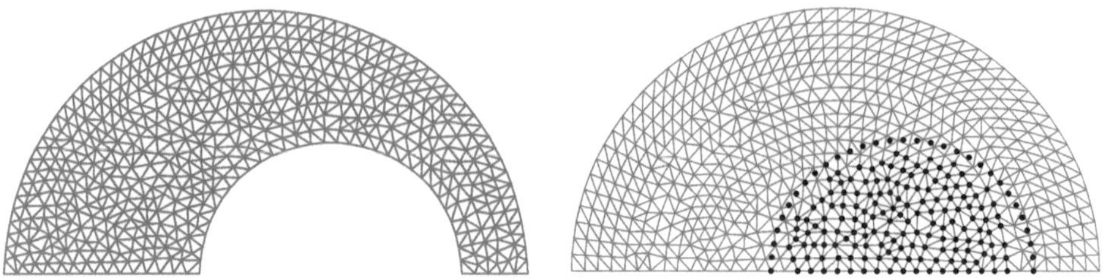

FIG. 29.3 Boundary-fitted mesh (a) and DLM/FD mesh (b). In figure (b), the set $\mathcal{P}$ covering the inner cylinder has been visualized ($N_r = 10$, $e = 1/2$, and $\chi = 1/2$).

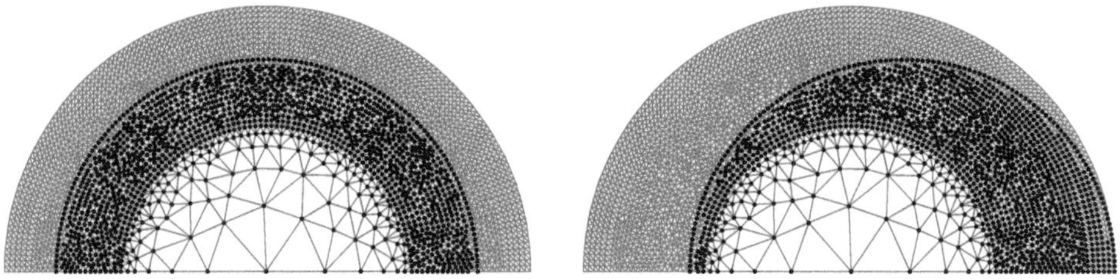

FIG. 29.4 Optimized DLM/FD mesh with $\mathcal{P}$ covering the inner cylinder ($N_r = 10$, $\chi = 0.8$, (a) $e = 0$, (b) $e = 1$).

This choice for $\mathcal{P}$ was advocated in GLOWINSKI [2003, chapter 8] and seems to give satisfactory results, as shown in the following subsections.

29.4.3. Convergence properties of the iterative methods

For all the cases considered in this chapter, the pseudotransient solution algorithm (29.81)–(29.86) was converging to a steady-state solution at the prescribed flow rate, according to the stopping criteria ε_1 and ε_3 (actually, ε_1, ε_3, and all the other stopping criteria were settled at 10^{-5}). Moreover, at every (pseudo) time step, we never encountered any problem with the convergence of ALG2; indeed, numerical simulations, done with various values of the augmentation parameter r, suggest that choosing $r = \frac{\rho}{\Delta t}$ provides good (and, in some cases, nearly optimal) convergence speed for ALG2. These unconditional convergence properties attest to the robustness of the whole solution process.

The convergence, at a prescribed constant flow rate, to a steady-state solution, for every eccentricity, relies on the updating strategy discussed in Section 29.3.6. Figure 29.5 illustrates the convergence of the dimensionless pressure drop ΔP^* to the steady-state value corresponding to a prescribed flow rate when $\chi = 1/2$, $e = 1/2$, and $Bn = 0.33$. The results reported in this figure underline the robustness of the updating strategy. In fact, whatever was the initial guess of the pressure drop, our algorithm converged to the same steady-state value of ΔP^*. However, as expected, the number of iterations required for convergence depends on the choice of the initial guess ΔP^*_{init}. For the case presented here, the converged

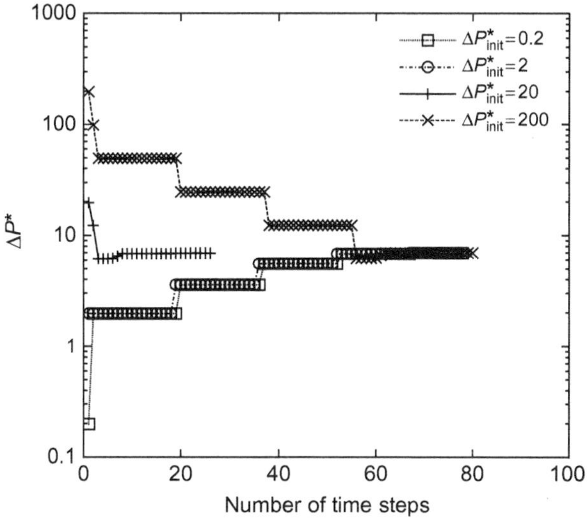

FIG. 29.5 Convergence of the pressure drop to its steady-state value at a prescribed flow rate: influence of the initial guess ($\chi = 1/2, e = 1/2$, and $Bn = 0.33$).

value of ΔP^* is 7. When we took $\Delta P^*_{\text{init}} = 20$, the convergence was obtained in 26 time steps, while we needed, approximately, 80 time steps when taking $\Delta P^*_{\text{init}} = 0.2$, 2, or 200. As mentioned already, the results reported in Fig. 29.5 correspond to $e = 1/2$; we picked this particular value of the eccentricity, in order to validate the capabilities of our methodology. After this particular test problem was solved, we ran our algorithms for e varying from $e_{\min} = 0$ to $e_{\max} = 1$, using $\Delta e = 0.1$ as increment. If one uses the notation of relation (29.82), at step j (> 2), we took for ΔP^*_{init} the value of ΔP^* computed at step $j - 1$; for $j = 2$, we took for ΔP^*_{init} the value of ΔP^* derived from relations (29.90)–(29.93), which describe the exact solution at $e = 0 (j = 1)$. With this updating strategy, the convergence of the pseudotransient algorithm was always achieved in less than 20 time steps.

29.4.4. *Accuracy of the computed solutions*

The *accuracy* of the *DLM/FD* computed solutions has been assessed in two ways:

(1) First, in the case of a *concentric annulus* ($e = 0$), we compared the computed solution with the analytical one, obtained from relations (29.90)–(29.93), in the particular case where $\chi = 1/2$ and $Bn = 0.85$ ($\mathcal{B}n = 0.1$).

(2) Then, we estimate the relative difference between approximate solutions computed on the one hand by the *DLM/FD* methodology and on the other hand by a *boundary-fitted*, finite-element method; for these comparisons, we took $\chi = 1/2$, $e = 0.75$, and $Bn = 0.9$.

The computed velocity profiles obtained for $N_r = 10$, 20, 30, 40, 50, and the analytical solution, corresponding to $e = 0$, $\chi = 1/2$, and $Bn = 0.85$, have been reported in Fig. 29.6(a) as functions of r^*. In Fig. 29.6(b) we have reported, as functions of r^*, again, the relative differences between the computed and exact solutions for the same values of N_r, e, χ, and Bn. From these figures, it is clear that for a coarse mesh ($N_r = 10$, for example),

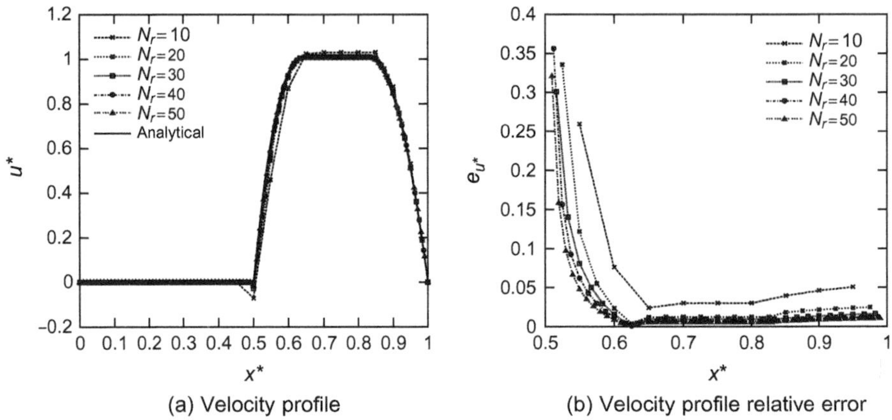

FIG. 29.6 Comparison between the computed DLM/FD solution and the exact one for $\chi = 1/2$, $e = 0$, and $Bn = 0.85$ ($\mathcal{B}n = 0.1$): (a) Velocity profiles. (b) Velocity profile relative errors.

the discrepancy between computed and exact solutions is quite large; however, for $N_r \geq 20$, the agreement improves significantly. For $N_r = 50$, the relative error drops to approximately 1% over the main part of the flowing region, except in the neighborhood of the inner cylinder. For all the meshes that we considered, the relative error on the first nodes, close to the inner cylinder boundary, was always of the order of 30% (the corresponding under-shots are quite visible in Fig. 29.6(a)); fortunately, this error decreases quickly as r^* increases, and the influence on the overall solution is quite limited. This discrepancy in the neighborhood of the internal cylinder may result from the scalar product (29.55), used to impose the no-flow condition inside the inner cylinder and from the choice of the collocation points. Another phenomenon explaining the loss of accuracy at the inner cylinder boundary is the fact that the solution of the continuous fictitious domain problem has a strong gradient discontinuity at the inner cylinder boundary; however, the discrete solution, being computed on a grid which does not match the inner cylinder boundary, will show such a discontinuity at some distance (of the order of $1/N_r$) of the inner boundary. A close inspection of our numerical results shows that mesh refinement does not reduce the relative error on the first nodes close to the boundary of the inner cylinder; however, mesh refinement shrinks the size of the large discrepancy region, implying that, if $p \in [1, +\infty)$, the L^p-norm of the error goes to zero as $N_r \to +\infty$. The additional results reported in Fig. 29.7 attest of the convergence of the DLM/FD computed solutions as $N_r \to +\infty$: in Fig. 29.7(a) (resp., Fig. 29.7(b)), we have visualized the variation, as a function of N_r, of the relative error on the maximal value of the velocity (resp., on the pressure drop); for $N_r = 50$, the relative errors on $u_{\max}$ and ΔP are approximately 0.5% and 1%, respectively; actually, the two figures above suggest that both errors are $O(1/N_r)$, approximately.

The results of comparisons between DLM/MD and boundary-fitted computed solutions have been reported in Figs. 29.8 and 29.9 in the particular case where $e = 0.75$, $\chi = 1/2$, and $Bn = 0.9$. It is clear that for $N_r \geq 20$, the solutions match "very well." For example, for $N_r = 50$, the relative difference between the values of $u_{\max}$ computed by the two methods is of the order of 0.5%; a similar result holds for ΔP.

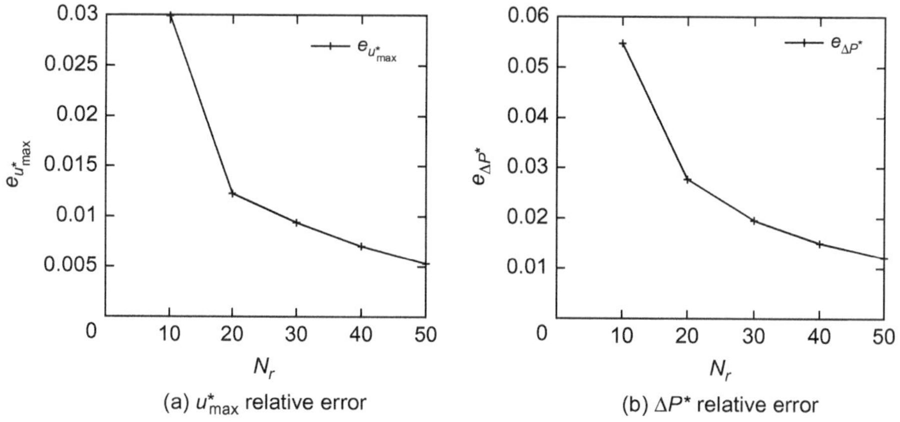

FIG. 29.7 Comparison between the computed DLM/FD solution and the exact one for $\chi = 1/2$, $e = 0$, and $Bn = 0.85$ ($\mathcal{B}n = 0.1$): (a) u_{max} relative error. (b) ΔP relative error.

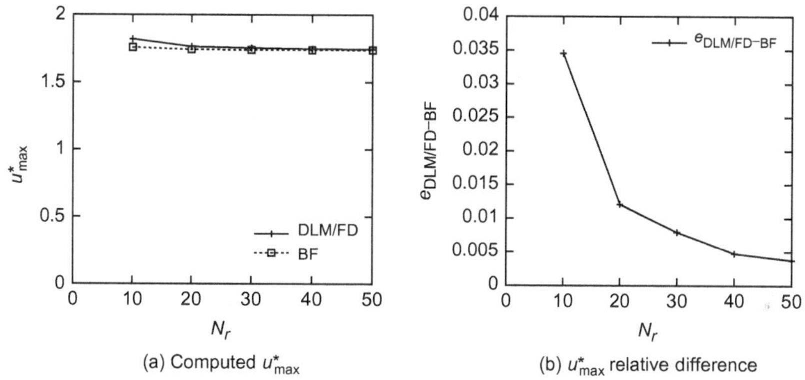

FIG. 29.8 Comparison between computed DLM/FD and boundary-fitted solutions for $\chi = 1/2$, $e = 0.75$, and $Bn = 0.9$: (a) Computed u^*_{max}. (b) u^*_{max} relative difference.

29.4.5. Analysis of the numerical results from a mechanical perspective

Based on the accuracy discussion of Section 29.4.4, we are going to comment, from a mechanical point of view, the results of numerical experiments, all obtained with $N_r = 50$. The main advantage of the DLM/FD approach used in this chapter is that the finite-element mesh we use to compute the flow velocity depends of the radius ratio χ but is independent of the eccentricity e. In this section, we are going to discuss the numerical results in terms of yielded/unyielded regions, that is in terms of flow pattern. The only cases we will consider are all associated with $\chi = 0.2$, but the following comments apply qualitatively for $\chi = 0.5$ and 0.8, for example. In Fig. 29.10, we have reported for $\chi = 0.2$, $Bn = 0.9$, and various values of the eccentricity, the pattern of the yielded (white) and unyielded (dark gray) flow regions. The unyielded region is where the strain-tensor (in fact, the gradient of the axial

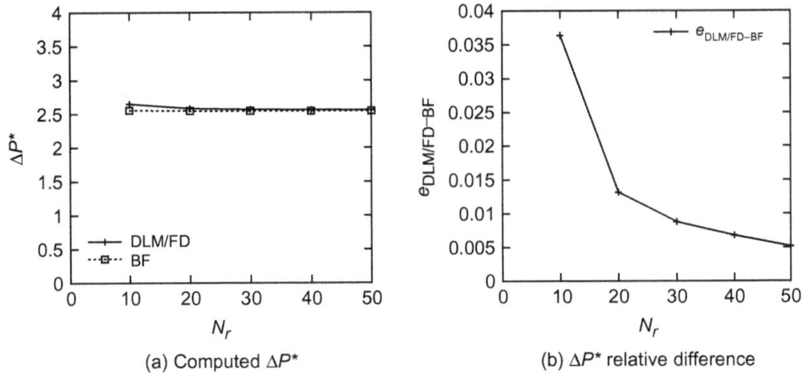

(a) Computed ΔP^* (b) ΔP^* relative difference

FIG. 29.9 Comparison between computed DLM/FD and boundary-fitted solutions for $\chi = 1/2$, $e = 0.75$, and $Bn = 0.9$: (a) Pressure drop ΔP^*. (b) ΔP^* relative difference.

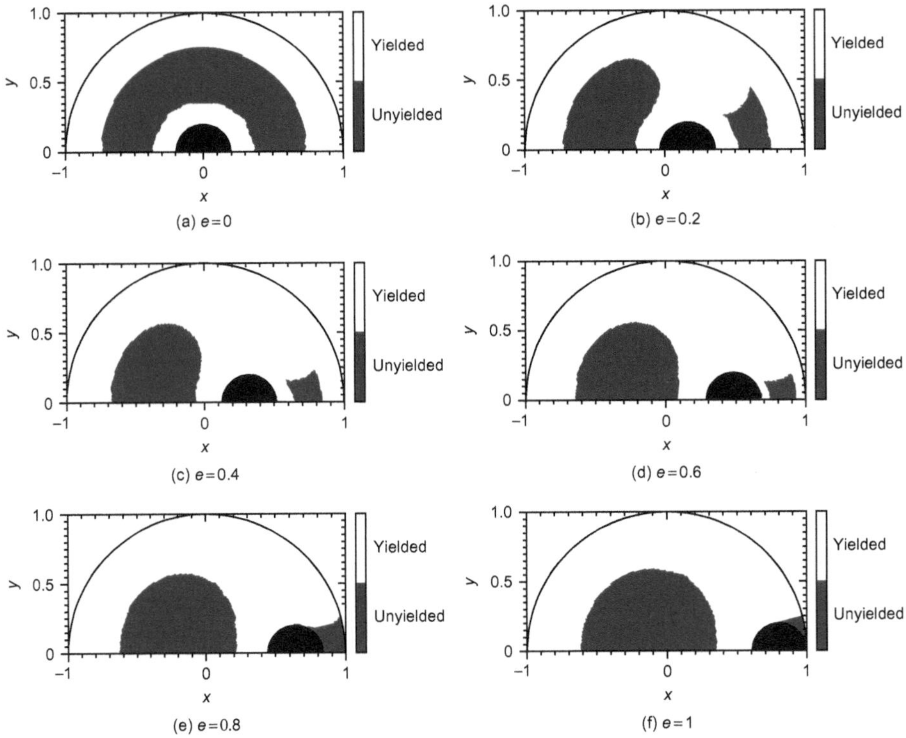

FIG. 29.10 Influence of the eccentricity on the yielded/unyielded region pattern ($\chi = 0.2$ and $Bn = 0.9$).

velocity, here) vanishes. The computed velocity being piecewise affine continuous, its gradient is piecewise constant, explaining why the interface between the yielded and unyielded regions is a polygonal line union of edges of the finite-element triangulation we use for our computations. As a consequence of the DLM/FD methodology, we use to compute the flow, the inner cylinder, where we impose $u = 0$, is part of the computed unyielded region; because this is just a numerical artifact, we have used a coal-black coloring to differentiate the inner cylinder from the physical unyielded regions.

In agreement with SZABO and HASSAGER [1992], we found *four* flow patterns:

- Pattern of type I: There is no flow, the pressure drop being too small to entail a flow. This situation did not occur here since a nonzero flow rate was prescribed.
- Pattern of type II: One moving plug region and a "dead" region. The fluid flows in the wide part of the annular cross-section but stays at rest in the narrow part. This type of situation is illustrated in Fig. 29.10(e,f); it is, usually, the results of high eccentricities and high Bingham numbers, as well.
- Pattern of type III: Two moving plug regions. The fluid flows everywhere in the cross-section, with the plug region in the wide part being larger than the one in the narrow part, as shown in Fig. 29.10(b–d).
- Pattern of type IV: One single moving plug zone. The fluid flows everywhere in the cross-section, with a single moving plug region, as shown in Fig. 29.10(a). Such a pattern occurs for small eccentricities.

In this chapter, where a nonzero flow rate has been imposed, one encounters patterns of types II, III, and IV.

In Fig. 29.11, we have visualized transitions from IV to III (Fig. 29.11(a)) and III to II (Fig. 29.11(b)). These transitions are obtained by increasing the eccentricity, while keeping Bn at 0.9.

The variation of u^*_{max} with e has been reported in Fig. 29.12. This figure provides additional information on the transitions. It shows in particular that as long as one stays in type IV, u^*_{max} stays constant. It also shows (for $Bn = 0.9$ and 0.98) that as we leave type III to go into type II, u^*_{max} is essentially a decreasing affine function of e.

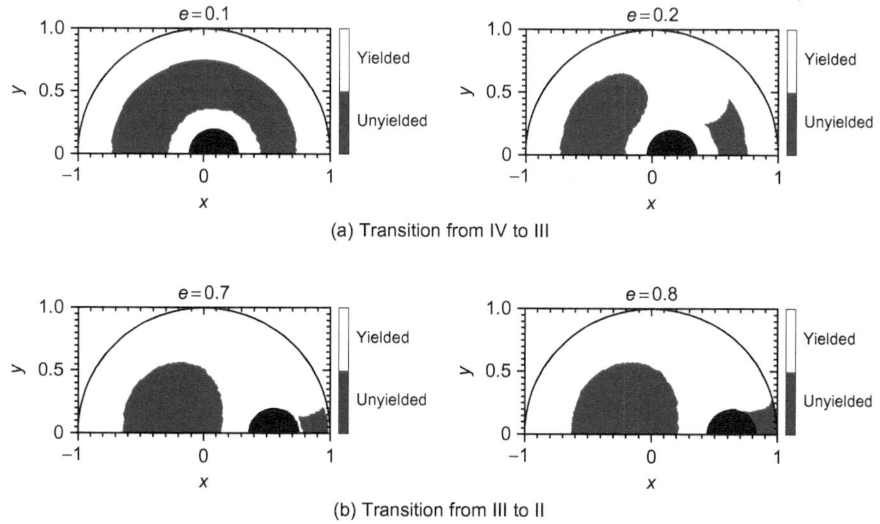

FIG. 29.11 Transitions between the different flow patterns: (a) From IV to III. (b) From III to II.

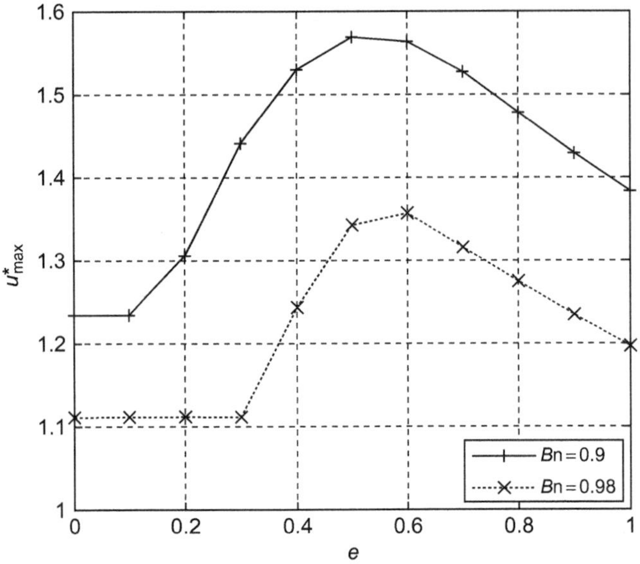

FIG. 29.12 Variation of the maximum velocity $u^*_{\max}$ as a function of the eccentricity ($\chi = 0.2$ and $Bn = 0.9$ and 0.98).

In order to compare our results with those in SZABO and HASSAGER [1992], concerning in particular the shape of the yielded and unyielded regions, we have performed additional computations, corresponding to $\chi = 0.4$, $\mathcal{B}n = 0.1$ ($Bn = 0.822$), and $\delta^* = 0.04$ ($e = 0.0667$) and 0.15 ($e = 0.25$). In Fig. 29.13, the yielded and unyielded regions obtained in SZABO and HASSAGER [1992] (using a boundary-fitted, finite-element method) are compared with those obtained in this chapter using the DLM/FD method, for $e = 0.0667$ and 0.25. We observe a very good agreement, concerning the shape of the yielded and unyielded regions.

In Fig. 29.14, we have reported the results of another comparison between our results and those in SZABO and HASSAGER [1992]: more precisely, in the particular case $\chi = 0.2$, we have visualized the regions of the plane (Bn, δ^*) associated with the patterns of types I–IV mentioned earlier. Here too, the agreement is quite good.

In Fig. 29.15, we have visualized for various values of the aspect ratio χ the graph of the function $\{e, Bn\} \to \frac{\Delta P^*}{\Delta P^*_{e=0}}(e, Bn)$. From this figure, we observe that at high Bingham numbers (at $Bn = 0.98$, for example) and small radius ratio ($\chi = 0.2$, for example), the pressure drop is almost constant for small eccentricities: in this situation, the flow pattern is of type IV, that is a single moving plug region. Accordingly, the maximum velocity does not change, as shown in Fig. 29.12, for $Bn = 0.98$ and $e \in [0, 0.3]$. For high eccentricities, the transition from type III to type II is not noticeable in Fig. 29.15; indeed, the occurrence of a dead zone in the narrow part of the annulus does not bring any particular change in the dependence to eccentricity. However, at high eccentricities, the relation between the normalized pressure drop and the eccentricity seems almost affine, a property observed already for $u^*_{\max}$ (see Fig. 29.12).

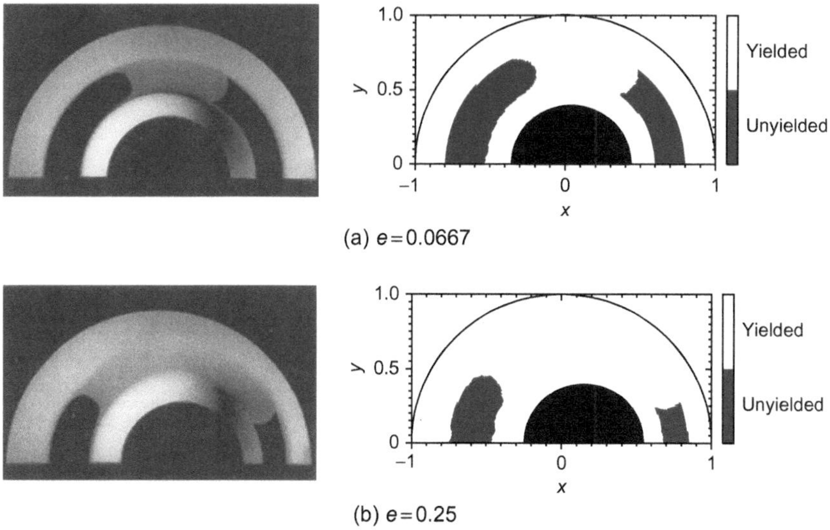

FIG. 29.13 Comparisons between the shapes of the yielded and unyielded regions obtained, respectively, in SZABO and HASSAGER [1992] using a boundary-fitted finite-element method and in this chapter using the DML/FD method [$\chi = 0.4$, $\mathcal{B}\text{n} = 0.1$ ($B\text{n} = 0.822$), $e = 0.0667$(a) and 0.25(b)].

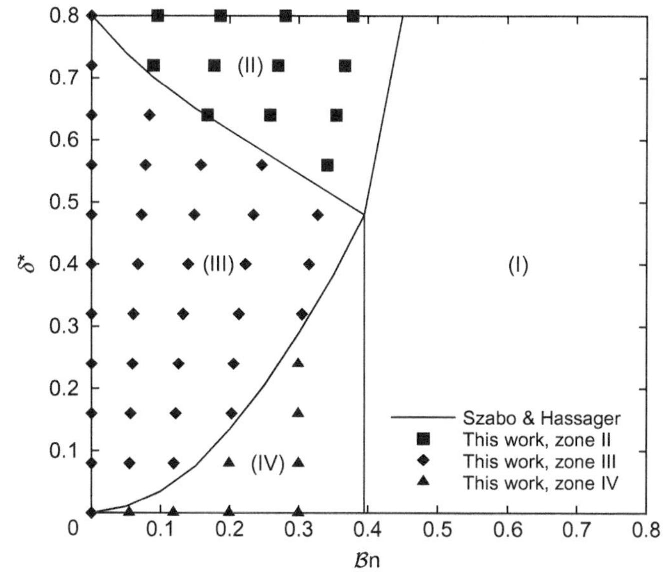

FIG. 29.14 Visualization in the plane ($B\text{n}$, δ^*) of the regions associated with the patterns of types I–IV for $\chi = 0.2$. These regions have been identified from the results obtained in SZABO and HASSAGER [1992] (—) and those obtained by the DLM/FD method discussed in this chapter (■, ◆, and ▲).

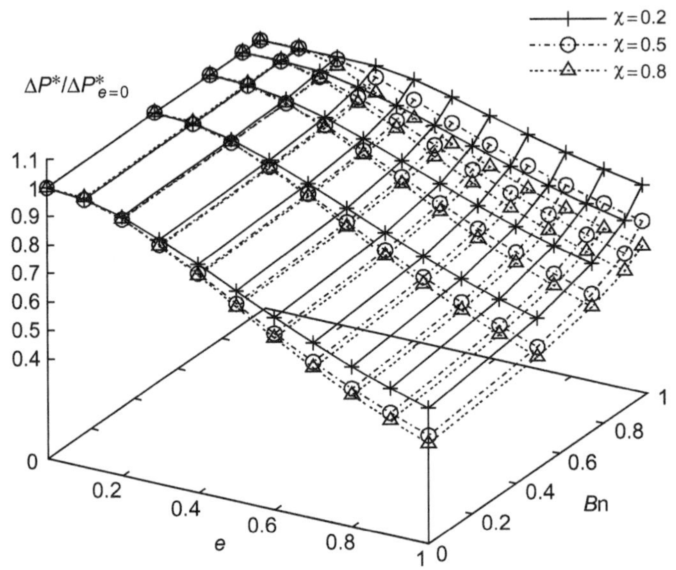

FIG. 29.15 Graph of the function $\{e, Bn\} \to \frac{\Delta P^*}{\Delta P^*_{e=0}}(e, Bn)$ computed by the DLM/FD method for $\chi = 0.2$, 0.5, and 0.8.

29.4.6. Prediction of the pressure drop by an engineering response surface methodology (RSM)

In *drilling operations*, the drilling fluid is quite commonly modeled as a *Bingham* one. On the field, oil engineers have to properly control the pumping pressure in order to drill efficiently. A good estimate of the pressure drop for a given configuration (rheological properties, geometry and flow rate) is a most valuable information. Thanks to the DLM/FD methodology discussed in this chapter, we were able to compute a large number (165, actually) of solutions parameterized by e, Bn, and χ. Because, for $e = 0$, the pressure drop can be obtained from the closed form solution given by relations (29.90)–(29.96), the value of the pressure drop for other eccentricity may be obtained using a *response surface methodology* (RSM) taking advantage of the computed solutions; a basic reference on RSM is MYERS and MONTGOMERY [2002].

Using the values computed by the DLM/FD methodology, RSM will provide a tool able to predict the pressure drop for Bingham flows in an eccentric annulus. The RSM consists in a simple multivariable *Lagrange interpolation.* To illustrate the validity of the RSM approach, we have performed two additional series of computations, namely

(1) $\chi = 0.4, Bn = 0.6, e_{\min} = 0, e_{\max} = 1, n_{\mathrm{ec}} = 7.$
(2) $\chi = 0.6, Bn = 0.93, e_{\min} = 0, e_{\max} = 1, n_{\mathrm{ec}} = 11.$

In (1) and (2), n_{ec} denotes (as in Section 29.3.6) the number of eccentricity steps.

In Fig. 29.16(a,b), we have reported the results of a comparison between the computed solutions and the ones obtained by RSM for the cases $\{\chi, Bn\} = \{0.4, 0.6\}$ and $\{0.6, 0.93\}$, respectively. From these figures, the agreement between computed solutions and RSM predictions is quite satisfactory; actually, their relative difference is less than 0.1% for the

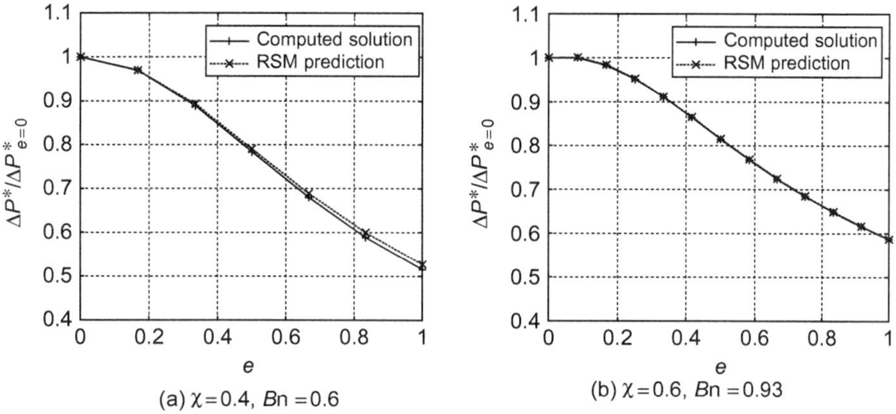

FIG. 29.16 Comparisons between computed solutions and RSM predictions for: (a) $\chi = 0.4$, $Bn = 0.6$, and (b) $\chi = 0.6$, $Bn = 0.93$.

case $\{\chi, Bn\} = \{0.6, 0.93\}$, while, for $\{\chi, Bn\} = \{0.4, 0.6\}$, the relative difference reaches its maximal value (of the order of 2%, which is still very acceptable) at $e = 1$. These results, and similar ones, show that the RSM is a valuable tool to predict at no significant extra computational cost the pressure drop in the case of the steady flow of a Bingham fluid in a pipe whose cross-section is an eccentric annulus. Of course, quality results like those above suppose that enough points have been used to construct the response surface.

29.5. Concluding remarks

From Sections 29.2–29.4, we discussed the numerical simulation of a *steady Bingham flow* in an *eccentric annular cross-section* cylinder. Assuming that the flow rate was imposed, we achieved this simulation using a methodology combining a *finite-element approximation* with a *distributed Lagrange multiplier*-based *fictitious domain method* and an *augmented Lagrangian/Uzawa algorithm*; using this methodology, we were able to compute easily and efficiently, on a fixed mesh, a large number of solutions parameterized by the eccentric position of the inner cylinder.

The results we obtained highlight different patterns for the yielded and unyielded regions of the flow. The methodology we used allowed us to identify easily and without ambiguity the unyielded regions because they correspond to a true zero strain rate tensor (reducing to a zero velocity gradient vector in this chapter). From a physical point of view, and in agreement with SZABO and HASSAGER [1992], our computational techniques were able to identify four flow patterns. We showed in particular (see Fig. 29.12) that the pattern change from type IV to type III leads to a slope discontinuity for the maximum velocity as a function of the eccentricity; on the contrary, the change of type III to type II is difficult to notice.

Thanks to the novel DLM/FD method discussed in this chapter, we could efficiently compute a large number of solutions parameterized by χ (radius ratio), Bn (Bingham number), and e (eccentricity). Using these computed solutions, we constructed a *response surface*, an engineering tool allowing the easy and accurate computation of the pressure drop for all the values of χ, Bn, and e.

Concerning the DLM/FD method that we employed, the following comments are in order:

(1) The accuracy of the computed solutions is significantly influenced by the choice of the points we use to impose, by collocation, a zero velocity inside, and at the boundary of the inner cylinder.
(2) The computed solution shows a significant error at the grid points close to the boundary of the inner cylinder. However, this error drops quickly to zero as the distance to the inner cylinder increases. Actually, it has little effect on the value of the pressure drop, as shown by the comparisons with the results in SZABO and HASSAGER [1992], obtained by a boundary-fitted finite-element method.

30. Dynamical simulation of particle sedimentation in a Bingham fluid

30.1. Introduction. Synopsis

As pointed out in Chapter 1, various *non-Newtonian* materials, such as paints, toothpastes, blood, fresh concrete, magnetite dense media in the mining industry, and drilling mud in the oil industry, exhibit a *yield stress*. Many industrial processes involve the sedimentation of particles in these viscoplastic materials. In the Oil & Gas industry, a drilling mud is used to remove the rock cuttings, resulting from the drilling at the bottom, and carry them to the surface (as shown in Fig. 2.1 of Chapter 1, Section 2.2); the fluid is designed to be viscoplastic, so that it can flow easily when circulating and, from its yield stress, prevents the settling of the rock particles when the circulation is stopped (see Section 2.2 and PEYSSON [2004]). The yield stress is also used to control the separation of mineral particles when they settle in a magnetite dense medium (as shown in HE, LASKOWSKI and KLEIN [2001]). From its importance in practical applications, the drag coefficient for a sphere settling in a viscoplastic fluid has been the object of many investigations: *theoretically* as in ANSLEY and SMITH [1967], *computationally* as in BERIS, TSAMOPOULOS, ARMSTRONG and BROWN [1985], DE BESSES, MAGNIN and JAY [2004], *experimentally* as in DEDEGIL [1987], ATAPATTU, CHHABRA and UHLHERR [1995]. So far, all the simulations we know of this phenomenon have been *static*, in the sense that the particle is *fixed* (actually, the simulators are essentially two dimensional). To fully understand the sedimentation of many particles, we need to simulate the motion of the particles resulting of their interaction with the surrounding fluid; this requires, in principle, a three-dimensional, time-dependent simulator.

From a methodological point of view, we will still use a DLM/FD methodology, but this time, we will rely on the (non-Lagrange) multiplier method with L^2-projection discussed in Chapter 2, Section 17, to overcome the difficulties associated with the nonsmoothness of the constitutive law.

The simulation of the unsteady motion of particles moving in a fluid is a nontrivial task, due to the fact that the region occupied by the fluid varies with time. Algorithms based on boundary-fitted meshes, such as *ALE/finite-element methods* are not easy to implement (see, e.g., HU, PATANKAR and ZHU [2001]), in contrast to nonboundary-fitted methods such as *lattice–Boltzmann* (see, e.g., LADD and VERBERG [2001]) and *Distributed Lagrange Multiplier-based Fictitious Domain* (DLM/FD; see, e.g., GLOWINSKI, PAN, HESLA and JOSEPH [1999], GLOWINSKI [2003, chapters 8 & 9]) methods. The DLM/FD method for the numerical simulation of particulate flow (involving, possibly, a non-Newtonian fluid) was developed by R. Glowinski, D.D. Joseph, T.W. Pan, and various collaborators, the first reviewed related publication being GLOWINSKI, PAN, HESLA and JOSEPH [1999]. The key idea with this method is to fill the interior of the particles with the surrounding fluid and to introduce, for each particle, a *Lagrange multiplier* defined over the region occupied by the particle, as a pseudo body force to enforce a rigid body motion to the fluid "inside" the particle. In addition to the above two references, the DLM/FD method is further discussed in GLOWINSKI, PAN, HESLA, JOSEPH and PERIAUX [2001]; it has been applied to a wide range of problems as shown in, e,g., GLOWINSKI [2003, chapters 8 & 9] (see also the references therein). Concerning the DLM/FD-based numerical simulation of particulate flow involving *viscoelastic* fluids (such as *Oldroyd-B*), let us mention SINGH, JOSEPH, HESLA, GLOWINSKI and PAN [2000], YU, PHAN-TIEN, FAN and TANNER [2002], GLOWINSKI [2003], HAO, PAN, GLOWINSKI and JOSEPH [2009] (a variant of the DLM/FD method is discussed in HWANG, HULSEN and MEIJER [2004], in order to investigate the rheology of a viscoelastic particle suspension in a sliding biperiodic frame). In YU, PHAN-TIEN and TANNER [2004] one investigates the settling of a sphere in a vertical tube, filled with an incompressible viscous fluid, at moderately high (based on the terminal velocity) Reynolds numbers (in the hundreds, typically).

In YU, WACHS and PEYSSON [2006], it was shown that the spatial and temporal discretizations of the Lagrange multipliers are crucial factors for the accuracy of the DLM/FD method, as are other significant differences with the original method of Glowinski, Pan et al. (we will return on these variants in Section 30.2); with this modified DLM/FD methodology, Yu, Wachs and Peysson investigated the numerical simulation of particles settling in shear-thinning fluid.

Recently, the Bingham flow-simulation method discussed in DEAN and GLOWINSKI [2002], GLOWINSKI [2003, chapter 10], DEAN, GLOWINSKI and GUIDOBONI [2007] and Chapter 2, Section 17, has been combined with the DLM/FD method described in, e.g., GLOWINSKI, PAN, HESLA and JOSEPH [1999], in order to simulate the two-dimensional motion of solid rigid particles in a Bingham fluid. Related numerical results have been reported in DEAN, GLOWINSKI and PAN [2003]; however, the method has not been validated through benchmark tests. Our objective in this chapter is to address the simulation of particle motions in Bingham fluids, using a slightly modified variant of the DLM/FD method discussed in DEAN, GLOWINSKI and PAN [2003]. Anticipating on Section 30.2, let us mention that the operator-splitting based time-discretization scheme that we use for our simulation differs in several aspects from the one in the above reference.

Our numerical method will be applied to the simulation of the sedimentation of a single and then two spheres in a *Bingham fluid* contained in a tube. We will compare the computed drag coefficients with results previously reported in the literature.

30.2. Mathematical and Numerical Modeling

For simplicity, we consider *one* particle only. We suppose that this particle is a *rigid solid* body that occupies at time t the space region $P(t) \subset \Omega \subset \mathbb{R}^3$; we denote by Γ and $\partial P(t)$ the boundaries of Ω and $P(t)$, respectively. The region occupied by the fluid is thus $\Omega \backslash \overline{P(t)}$.

30.2.1. Governing equations

Dimensional governing equations The governing equations comprise the *continuity equation*, the combined *momentum equations* and the *constitutive equations*. It follows from, e.g., GLOWINSKI, PAN, HESLA and JOSEPH [1999], GLOWINSKI [2003, chapter 8] that a *distributed Lagrange multiplier*-based *fictitious domain* formulation of the *combined fluid and particle momentum equations* reads as follows (assuming a *no-slip boundary condition at the interface fluid-particle*):

Find $\{\mathbf{u}(t), \mathbf{U}(t), \boldsymbol{\omega}(t), \boldsymbol{\lambda}_P(t)\} \in (H^1(\Omega))^3 \times \mathbb{R}^3 \times \mathbb{R}^3 \times (H^1(P(t)))^3$, such that for a.e. $t \in (0, T)$

$$\int_\Omega \left[\frac{\partial \mathbf{u}}{\partial t} + (\mathbf{u}\cdot\nabla)\mathbf{u}\right]\cdot \mathbf{v}\,\mathrm{d}x - \int_\Omega [\nabla\cdot(-p\mathbf{I} + \boldsymbol{\tau}]\cdot\mathbf{v}\,\mathrm{d}x + (\boldsymbol{\lambda}_P, \mathbf{v})_P$$

$$= \rho_f \int_\Omega \mathbf{g}\cdot\mathbf{v}\,\mathrm{d}x, \qquad \forall \mathbf{v} \in (H_0^1(\Omega))^3, \tag{30.1}$$

$$\left(1 - \frac{\rho_f}{\cdot\rho_s}\right)\left[M\left(\frac{\mathrm{d}\mathbf{U}}{\mathrm{d}t} - \mathbf{g}\right)\cdot\mathbf{V} + \frac{\mathrm{d}}{\mathrm{d}t}(\mathbf{J}\boldsymbol{\omega})\cdot\boldsymbol{\theta}\right] - (\boldsymbol{\lambda}_P, \mathbf{V} + \boldsymbol{\theta}\times\mathbf{r})_P = 0,$$

$$\forall \{\mathbf{V}, \boldsymbol{\theta}\} \in \mathbb{R}^3 \times \mathbb{R}^3, \tag{30.2}$$

$$(\mathbf{u} - (\mathbf{U} + \boldsymbol{\omega}\times\mathbf{r}), \boldsymbol{\mu})_P = 0, \qquad \forall \boldsymbol{\mu} \in (H^1(P(t)))^3. \tag{30.3}$$

In (30.1)–(30.3),

- $(0, T)$ (with $0 < T \leq +\infty$) is the time interval during which the sedimentation phenomenon is occurring.
- ρ_f and ρ_s are, respectively, the *fluid density* and the *density of the solid* material the particle is made of.
- $\mathbf{u}$ and p are the *fluid velocity* and *pressure*, respectively.
- Following Chapters 1 and 2, the *tensor-valued* function $\boldsymbol{\tau}$ is defined by

  $$\boldsymbol{\tau} = 2\mu\mathbf{D}(\mathbf{u}) + \sqrt{2}\,\tau_y\boldsymbol{\lambda}, \tag{30.4}$$

  with the tensor-valued function $\boldsymbol{\lambda}$ verifying

  $$\begin{aligned}&\boldsymbol{\lambda} \in (L^\infty(\Omega))^{3\times 3}, \boldsymbol{\lambda} = \boldsymbol{\lambda}^t,\\ &\boldsymbol{\lambda} : \mathbf{D}(\mathbf{u}) = |\mathbf{D}(\mathbf{u})| \quad \text{and} \quad |\boldsymbol{\lambda}(x)| \leq 1, \text{ a.e. in } \Omega;\end{aligned} \tag{30.5}$$

  in (30.5), $|\cdot|$ denotes the *Fröbenius-norm*, that is (with obvious notation), $\forall \mathbf{T} \in \mathbb{R}^{3\times 3}$, $|\mathbf{T}| = \sqrt{\sum_{1\leq i,j\leq 3} t_{ij}^2}$.
- $\mathbf{g}$ denotes *gravity* and $\mathbf{r} = \overrightarrow{G(t)x}$, with $G(t)$ the *center of mass* of $P(t)$.
- $M, \mathbf{J}, \mathbf{U}$, and $\boldsymbol{\omega}$ are, respectively, the *mass, inertia tensor, translational velocity*, and *angular velocity* of the particle; we have thus

  $$\frac{\mathrm{d}G}{\mathrm{d}t} = \mathbf{U}. \tag{30.6}$$

- The vector-valued function $\boldsymbol{\lambda}_P$ is a *Lagrange multiplier* vector-valued function defined over $P(t)$; the role of $\boldsymbol{\lambda}_P$ is to force the *rigid body motion* of the fluid "contained" in the particle.
- $(,\cdot,)_P$ denotes a scalar product over the space $(H^1(P(t)))^3$; several candidates for this scalar product will be shown in Remark 30.3.

- We assume that the velocity **u** verifies the following *Dirichlet boundary condition*

$$\mathbf{u}(t) = \mathbf{u}_\Gamma(t) \quad \text{on} \quad \Gamma \times (0, T), \tag{30.7}$$

with

$$\int_\Gamma \mathbf{u}_\Gamma(t) \cdot \mathbf{n} \, d\Gamma = 0,$$

n being the outward unit normal vector at Γ. There is compatibility between the Dirichlet condition (30.7) and the fact that we took $(H_0^1(\Omega))^3$ as test function space in (30.1).

Dimensionless governing equations The governing equations can be made *dimensionless* by introducing the following scales: L_c for lengths, U_c for the velocity, L_c/U_c for the time, $\rho_f U_c^2$ for the pressure, and $\rho_f U_c^2/L_c$ for the Lagrange multiplier $\boldsymbol{\lambda}_P$. We will use for convenience the same notation for the dimensionless quantities and their dimensional counterparts, unless specified otherwise. Because the viscoplastic problem is to be solved using an orthogonal projection-based algorithm, we will replace (30.5) by an equivalent fixed point type relation involving a projection operator. The complete set of dimensionless governing equations reads then as follows:

(1) **Combined momentum equations**

$$\int_\Omega \left[\frac{\partial \mathbf{u}}{\partial t} + (\mathbf{u} \cdot \nabla)\mathbf{u}\right] \cdot \mathbf{v} \, \mathrm{d}x = \int_\Omega \left[-\nabla p + \frac{1}{\mathcal{R}\mathrm{e}} \nabla^2 \mathbf{u} + \sqrt{2}\, \frac{\mathcal{B}\mathrm{n}}{\mathcal{R}\mathrm{e}} \nabla \cdot \boldsymbol{\lambda}\right] \cdot \mathbf{v} \, \mathrm{d}x$$

$$+ (\boldsymbol{\lambda}_P, \mathbf{v})_P = \mathcal{F}\mathrm{r} \int_\Omega \frac{\mathbf{g}}{g} \cdot \mathbf{v} \, \mathrm{d}x, \qquad \forall \mathbf{v} \in (H_0^1(\Omega))^3, \tag{30.8}$$

$$(\rho_r - 1)\left[V^* \left(\frac{\mathrm{d}\mathbf{U}}{\mathrm{d}t} - \mathcal{F}\mathrm{r}\frac{\mathbf{g}}{g}\right) \cdot \mathbf{V} + \frac{\mathrm{d}}{\mathrm{d}t}(\mathbf{J}^* \boldsymbol{\omega}) \cdot \boldsymbol{\theta}\right] - (\boldsymbol{\lambda}_P, \mathbf{V} + \boldsymbol{\theta} \times \mathbf{r})_P = 0,$$

$$\forall \{\mathbf{V}, \boldsymbol{\theta}\} \in \mathbb{R}^3 \times \mathbb{R}^3, \tag{30.9}$$

$$(\mathbf{u} - (\mathbf{U} + \boldsymbol{\omega} \times \mathbf{r}), \boldsymbol{\mu})_P = 0, \qquad \forall \boldsymbol{\mu} \in (H^1(P(t)))^3, \tag{30.10}$$

with $g = |\mathbf{g}|$ in (30.8), (30.9).

(2) **Weak formulation of the continuity equation**

$$\int_\Omega \nabla \cdot \mathbf{u}(t) q \, \mathrm{d}x = 0, \qquad \forall q \in L^2(\Omega), \text{ a.e. on } (0, t). \tag{30.11}$$

(3) **Constitutive equation**

$$\boldsymbol{\lambda} = P_{\boldsymbol{\Lambda}}\left[\boldsymbol{\lambda} + r\sqrt{2}\, \frac{\mathcal{B}\mathrm{n}}{\mathcal{R}\mathrm{e}} \mathbf{D}(\mathbf{u})\right], \qquad \forall r > 0. \tag{30.12}$$

The above equations have to be completed by a *boundary condition* such as

$$\mathbf{u}(t) = \mathbf{u}_\Gamma(t) \text{ on } \Gamma \times (0, T), \tag{30.13}$$

and by the following *initial conditions*

$$\mathbf{u}(0) = \mathbf{u}_0 \text{ in } \Omega\backslash\overline{P(0)}, \tag{30.14}$$

$$\mathbf{U}(0) = \mathbf{U}_0, \boldsymbol{\omega}(0) = \boldsymbol{\omega}_0, G(0) = G_0, P(0) = P_0, \tag{30.15}$$

In (30.8)–(30.12),

- The following dimensionless parameters have been introduced:

$$\text{The } \textit{density ratio } \rho_r = \frac{\rho_s}{\rho_f}. \tag{30.16}$$

$$\text{The } \textit{Reynolds number } \mathcal{R}\text{e} = \frac{\rho_f U_c L_c}{\mu}. \tag{30.17}$$

$$\text{The } \textit{Froude number } \mathcal{F}\text{r} = \frac{g L_c}{U_c^2}. \tag{30.18}$$

$$\text{The } \textit{Bingham number } \mathcal{B}\text{n} = \frac{\tau_y L_c}{\mu U_c}. \tag{30.19}$$

- $P_{\boldsymbol{\Lambda}}$ denotes the *orthogonal projection operator* from $(L^2(\Omega))^{3\times 3}$ onto the *closed convex* set $\boldsymbol{\Lambda}$ defined by

$$\boldsymbol{\Lambda} = \{\boldsymbol{\mu} | \boldsymbol{\mu} \in (L^2(\Omega))^{3\times 3}, \boldsymbol{\mu} = \boldsymbol{\mu}^t, |\boldsymbol{\mu}(x)| \leq 1, \text{ a.e. in } \Omega\}; \tag{30.20}$$

  we encountered already $\boldsymbol{\Lambda}$ and $P_{\boldsymbol{\Lambda}}$ in Chapter 2, Section 17.
- The *Froude number* measures the relative importance of gravity with respect to inertia.
- V^* and $\mathbf{J}^*$ are the dimensionless *particle volume* and *moment of inertia*. One has

$$V^* = \frac{M}{\rho_s L_c^3}, \text{ and for a spherical particle } \mathbf{J}^* = J^*\mathbf{I} \text{ with } J^* = \frac{J}{\rho_s L_c^5}.$$

  Actually, For a spherical particle,

$$V^* = \frac{4}{3}\pi (a^*)^3 \quad \text{and} \quad J^* = \frac{2}{5} V^* (a^*)^2,$$

  a^* being, here, the dimensionless radius of the particle.

From now on, only identical spherical particles will be considered in this study. For this kind of situation, it makes sense to take $L_c = d$, d being the particle diameter; we have then $a^* = 0.5$. Concerning U_c, different characteristic velocities are adopted depending on whether the *inertial effect* is strong or weak; the choice of U_c is discussed just below:

(1) *Velocity scaling in the case of small inertial effect*: In the particular case of the settling of a sphere, if the *inertial effect* is *small* one classically takes for characteristic velocity the *Stokes* velocity U_s of the same sphere settling in an unbounded domain filled with a Newtonian fluid of identical density and viscosity. Thus, we have

$$U_c = U_s = \frac{\frac{4}{3}\pi a^3 (\rho_s - \rho_f)}{6\pi a \mu} g = \frac{2a^2 (\rho_s - \rho_f)}{9\mu} g, \tag{30.21}$$

a being the sphere radius. Combining (30.21) with (30.16)–(30.18) and $L_c = 2a$, we obtain

$$\mathcal{F}\mathrm{r} = \frac{6\pi a^*}{\mathcal{R}\mathrm{e}(\rho_r - 1)V^*} = \frac{18}{\mathcal{R}\mathrm{e}(\rho_r - 1)}. \tag{30.22}$$

Back to the spherical particle settling with small inertial effect, it follows from, e.g., BLACKERY and MITSOULIS [1997], LIU, MULLER and DENN [2002] that the drag coefficient C_s is given by

$$C_\mathrm{s} = \frac{U_s}{U_T} = \frac{1}{U_T^*}, \tag{30.23}$$

where U_T (resp., U_T^*) denotes the *terminal velocity* (resp., the *dimensionless terminal velocity*).

(2) *Velocity scaling in the case of strong inertial effect*: Suppose now that the *inertial effect* is *strong*; following YU, PHAN-TIEN and TANNER [2004], it is better to define U_c by

$$U_c = U_I = \sqrt{\frac{\frac{4}{3}\pi a^3(\rho_s - \rho_f)g}{\frac{1}{2}\pi a^2 \rho_f}} = \sqrt{\frac{8a}{3}(\rho_r - 1)g}, \tag{30.24}$$

so that the standard *drag coefficient*, the so-called best number (see CLIFT, GRACE and WEBER [1978]), and the *Froude number* can be expressed, respectively, by

$$C_\mathrm{D} = \frac{\frac{4}{3}\pi a^3(\rho_s - \rho_f)g}{\frac{1}{2}\pi a^2 \rho_f U_T^2} = \frac{U_c^2}{U_T^2} = \frac{1}{|U_T^*|^2}, \tag{30.25}$$

$$N_\mathrm{D} = \frac{32 a^3 \rho_f^2(\rho_s - \rho_f)g}{3\mu^2} = \mathcal{R}\mathrm{e}^2, \tag{30.26}$$

$$\mathcal{F}\mathrm{r} = \frac{\frac{1}{2}\pi a^2}{(\rho_r - 1)V^*} = \frac{3}{4(\rho_r - 1)}; \tag{30.27}$$

hereafter, U_I will be referred to as an *inertial velocity*.

REMARK 30.1. It follows from (1) and (2), just above, that to define $\mathcal{B}\mathrm{n}$ and $\mathcal{R}\mathrm{e}$ we have chosen as *characteristic velocity*, U_c defined by either (30.21) or (30.24) (depending of the flow regime) instead of the terminal settling velocity. To avoid misunderstanding, we express the Bingham and Reynolds numbers based on the *Stokes* (resp., *inertial*) velocity as $\mathcal{B}\mathrm{n}_\mathrm{S}$ and $\mathcal{R}\mathrm{e}_\mathrm{S}$ (resp., $\mathcal{B}\mathrm{n}_\mathrm{I}$ and $\mathcal{R}\mathrm{e}_\mathrm{I}$). The Bingham and Reynolds numbers $\mathcal{B}\mathrm{n}_\mathrm{T}$ and $\mathcal{R}\mathrm{e}_\mathrm{T}$, both based on the terminal velocity U_T, can be obtained from

$$\mathcal{B}\mathrm{n}_\mathrm{T} = \mathcal{B}\mathrm{n}/U_\mathrm{T}^* \quad \text{and} \quad \mathcal{R}\mathrm{e}_\mathrm{T} = \mathcal{R}\mathrm{e}\, U_\mathrm{T}^*, \tag{30.28}$$

irrespectively of the definition of U_s. We observe that $\mathcal{B}\mathrm{n}_\mathrm{S} = \mathcal{B}\mathrm{n}_\mathrm{T}/C_\mathrm{s} = 6\tau_y^*$, where τ_y^* is a *dimensionless yield stress* (defined in BERIS, TSAMOPOULOS, ARMSTRONG and BROWN [1985]).

REMARK 30.2. Irrespectively of the choice made for the *characteristic velocity*, a main drawback with the definition (30.17) of the Reynolds number is that it does not account, explicitly, for the effect of the *yield stress*. One way to overcome this drawback is to replace μ in (30.17) by an *equivalent* (or *effective*) *viscosity* involving the effect of the *yield stress* (as done in, e.g., DEDEGIL [1987], and HE, LASKOWSKI and KLEIN [2001]). In particular, because the drag coefficient, for a steady-settling particle, depends of both $\mathcal{B}n_{\mathrm{T}}$ and $\mathcal{R}e_{\mathrm{T}}$, it was observed experimentally, in the above two references, that the drag coefficient correlates well with a *modified* (or *effective*) *Reynolds number* $\mathcal{R}e_{\mathrm{m}}$ based on the *effective viscosity*. For a *spherical* particle settling in a *Bingham fluid*, we can define, as follows, a *characteristic shear* rate $\dot{\gamma}_{\mathrm{c}}$, an *effective viscosity* η_e, and then the *modified Reynolds number* $\mathcal{R}e_{\mathrm{m}}$

$$\dot{\gamma}_{\mathrm{c}} = k\frac{U_T}{d}, \eta_{\mathrm{c}} = \mu + \frac{\tau_y}{\dot{\gamma}_{\mathrm{c}}}, \tag{30.29}$$

$$\mathcal{R}e_{\mathrm{m}} = \frac{\rho_f U_T d}{\eta_{\mathrm{e}}} = \frac{\mathcal{R}e_{\mathrm{T}}}{1 + \frac{\mathcal{B}n_{\mathrm{T}}}{k}} = \frac{\mathcal{R}e|U_T^*|^2}{U_T^* + \frac{\mathcal{B}n}{k}}, \tag{30.30}$$

where k is correction factor for the characteristic shear rate. Here, we take $k = 1$, as usually done in practice (see HE, LASKOWSKI and KLEIN [2001]).

REMARK 30.3. In the fictitious domain-based equations governing the coupled Bingham fluid flow and particle motion, we encountered the *scalar product* $(\cdot,\cdot)_P$ over the space $(H^1(P(t)))^3$. Following GLOWINSKI [2003, chapter 8], a natural choice for the above scalar product is given by

$$\{\mathbf{v},\mathbf{w}\} \to \int_{P(t)} (\mathbf{v}\cdot\mathbf{w} + \delta^2\nabla\mathbf{v} : \nabla\mathbf{w})\mathrm{d}x, \qquad \forall \mathbf{v},\mathbf{w} \in (H^1(P(t)))^3, \tag{30.31}$$

where, in (30.31), δ denotes a *characteristic distance*, such as, for example, the *diameter of the particle*. A variant of (30.31), reflecting better the physics of the phenomenon under consideration, is given by

$$\{\mathbf{v},\mathbf{w}\} \to \int_{P(t)} [\mathbf{v}\cdot\mathbf{w} + \delta^2\mathbf{D}(\mathbf{v}) : \mathbf{D}(\mathbf{w})]\mathrm{d}x, \qquad \forall \mathbf{v},\mathbf{w} \in (H^1(P(t)))^3. \tag{30.32}$$

Other candidates for $(\cdot,\cdot)_P$ are (with obvious notation):

$$\{\mathbf{v},\mathbf{w}\} \to \int_{\partial P(t)} \mathbf{v}\cdot\mathbf{w}\mathrm{d}(\partial P(t)) + \delta \int_{P(t)} \nabla\mathbf{v} : \nabla\mathbf{w}\,\mathrm{d}x, \qquad \forall \mathbf{v},\mathbf{w} \in (H^1(P(t)))^3 \tag{30.33}$$

and

$$\{\mathbf{v},\mathbf{w}\} \to \int_{\partial P(t)} \mathbf{v}\cdot\mathbf{w}\mathrm{d}(\partial P(t)) + \delta \int_{P(t)} \mathbf{D}(\mathbf{v}) : \mathbf{D}(\mathbf{w})\mathrm{d}x,$$

$$\forall \mathbf{v},\mathbf{w} \in (H^1(P(t)))^3. \tag{30.34}$$

In YU, WACHS and PEYSSON [2006], one has used as scalar product

$$\{\mathbf{v}, \mathbf{w}\} \to \int_{P(t)} \mathbf{v} \cdot \mathbf{w} \, dx, \qquad \forall \mathbf{v}, \mathbf{w} \in (H^1(P(t)))^3. \tag{30.35}$$

The choice of scalar product associated with (30.35) makes little sense, at the continuous level because the space $(H^1(P(t)))^3$ is not complete for the norm associated with the above scalar product. However, it makes sense to use (30.35) as a scalar product on the discrete analogs of the space $(H^1(P(t)))^3$ because these spaces are finite dimensional (a similar comment applies to the approximations of (30.35) obtained by *numerical integration*).

30.2.2. An operator-splitting-based computational method

Following, e.g., GLOWINSKI, PAN, HESLA and JOSEPH [1999], DEAN, GLOWINSKI and PAN [2003], YU, PHAN-TIEN and TANNER [2004], YU, WACHS and PEYSSON [2006], we use a first-order, *operator-splitting scheme* for the *time discretization* of the governing equations system (30.8)–(30.15). This will allow us to decouple the above system in a sequence of simpler subproblems of the following types (before space discretization):

(1) *Flow subproblems*

Find $\mathbf{u}^{n+1/2} \in (H^1(\Omega))^3$, $p^{n+1} \in L^2(\Omega)$, and $\boldsymbol{\lambda}^{n+1} \in \boldsymbol{\Lambda}$, such that

$$\begin{aligned}
&\int_\Omega \frac{\mathbf{u}^{n+1/2} - \mathbf{u}^n}{\Delta t} \cdot \mathbf{v} \, dx + \frac{1}{2\mathcal{R}e} \int_\Omega \nabla(\mathbf{u}^{n+1/2} + \mathbf{u}^n) : \nabla \mathbf{v} \, dx \\
&\quad - \int_\Omega p^{n+1} \nabla \cdot \mathbf{v} \, dx + \sqrt{2} \frac{\mathcal{B}n}{\mathcal{R}e} \int_\Omega \boldsymbol{\lambda}^{n+1} : \mathbf{D}(\mathbf{v}) \, dx \\
&\quad = \frac{1}{2} \int_\Omega [(\mathbf{u}^{n-1} \cdot \nabla)\mathbf{u}^{n-1} - 3(\mathbf{u}^n \cdot \nabla)\mathbf{u}^n] \cdot \mathbf{v} \, dx - \int_{P^n} \boldsymbol{\lambda}_P^n \cdot \mathbf{v} \, dx \\
&\quad + \mathcal{F}\mathrm{r} \int_\Omega \frac{\mathbf{g}}{g} \cdot \mathbf{v} \, dx, \qquad \forall \mathbf{v} \in (H_0^1(\Omega))^3,
\end{aligned} \tag{30.36}$$

$$\int_\Omega q \nabla \cdot \mathbf{u}^{n+1/2} \, dx = 0, \qquad \forall q \in L^2(\Omega), \tag{30.37}$$

$$\boldsymbol{\lambda}^{n+1} = P_{\boldsymbol{\Lambda}}[\boldsymbol{\lambda}^{n+1} + r\sqrt{2} \, \frac{\mathcal{B}n}{\mathcal{R}e} \mathbf{D}(\mathbf{u}^{n+1/2})]. \tag{30.38}$$

(2) *Particle motion and fictitious domain subproblems*

Find $\mathbf{u}^{n+1} \in (H^1(\Omega))^3$, $\omega^{n+1} \in \mathbb{R}^3$, $\mathbf{U}^{n+1} \in \mathbb{R}^3$, and $\boldsymbol{\lambda}_P^{n+1} \in (H^1(P^n))^3$, such that

$$\begin{aligned}
&\int_\Omega \frac{\mathbf{u}^{n+1} - \mathbf{u}^{n+1/2}}{\Delta t} \cdot \mathbf{v} \, dx + \int_{P^n} \boldsymbol{\lambda}_P^{n+1} \cdot \mathbf{v} \, dx = \int_{P^n} \boldsymbol{\lambda}_P^n \cdot \mathbf{v} \, dx, \\
&\quad \forall \mathbf{v} \in (H_0^1(\Omega))^3,
\end{aligned} \tag{30.39}$$

$$(\rho_r - 1)\left[V^*\left(\frac{\mathbf{U}^{n+1} - \mathbf{U}^n)}{\Delta t} - \mathcal{F}\mathrm{r}\frac{\mathbf{g}}{g}\right)\cdot \mathbf{V} + \mathbf{J}^*\frac{\boldsymbol{\omega}^{n+1} - \boldsymbol{\omega}^n}{\Delta t}\cdot \boldsymbol{\theta}\right]$$
$$- \int_{P^n} \boldsymbol{\lambda}_P^{n+1}\cdot(\mathbf{V} + \boldsymbol{\theta}\times\mathbf{r})\,\mathrm{d}x = 0, \qquad \forall\{\mathbf{V},\boldsymbol{\theta}\}\in\mathbb{R}^3\times\mathbb{R}^3, \tag{30.40}$$

$$\int_{P^n} [\mathbf{u}^{n+1} - (\mathbf{U}^{n+1} + \boldsymbol{\omega}^{n+1}\times\mathbf{r})]\cdot\boldsymbol{\mu}\,\mathrm{d}x = 0, \qquad \forall\boldsymbol{\mu}\in(H^1(P^n))^3, \tag{30.41}$$

$$\frac{G^{n+1} - G^n}{\Delta t} = \frac{1}{2}(\mathbf{U}^{n+1} + \mathbf{U}^n). \tag{30.42}$$

The above relations have to be completed by the initial and boundary conditions (resp., the initial conditions) verified by $\mathbf{u}$ (resp., G, $\mathbf{U}$, and $\boldsymbol{\omega}$); these conditions are easily obtained from relations (30.13)–(30.15).

REMARK 30.4. As suggested in YU, WACHS and PEYSSON [2006], we kept $\boldsymbol{\lambda}_P^n$ in (30.36) and (30.39) in an attempt to reduce the effect of the splitting error when computing the steady-state solution. This modification of the scheme used in, e.g., DEAN, GLOWINSKI and PAN [2003] allows us to use significantly larger time steps for simulations at low Reynolds numbers.

The flow problems (30.36)–(30.38) are solved, *iteratively*, as follows (assuming that all the quantities are known at $t^n = n\Delta t$):

$$\text{If } n \geq 1, \text{ take } \boldsymbol{\lambda}^{n+1,0} = \boldsymbol{\lambda}^n; \text{ take } \boldsymbol{\lambda}^{1,0} = \mathbf{0}. \tag{30.43}$$

For $k \geq 0$, $\boldsymbol{\lambda}^{n+1,k}$ being known, find $\mathbf{u}^{n+1/2,k+1} \in (H^1(\Omega))^3$ and $p^{n+1,k+1} \in L^2(\Omega)$, such that

$$\int_\Omega \frac{\mathbf{u}^{n+1/2,k+1} - \mathbf{u}^n}{\Delta t}\cdot\mathbf{v}\,\mathrm{d}x + \frac{1}{2\mathcal{R}\mathrm{e}}\int_\Omega \nabla(\mathbf{u}^{n+1/2,k+1} + \mathbf{u}^n):\nabla\mathbf{v}\,\mathrm{d}x$$
$$- \int_\Omega p^{n+1,k+1}\nabla\cdot\mathbf{v}\,\mathrm{d}x + \sqrt{2}\,\frac{\mathcal{B}\mathrm{n}}{\mathcal{R}\mathrm{e}}\int_\Omega \boldsymbol{\lambda}^{n+1,k}:\mathbf{D}(\mathbf{v})\,\mathrm{d}x$$
$$= \frac{1}{2}\int_\Omega [(\mathbf{u}^{n-1}\cdot\nabla)\mathbf{u}^{n-1} - 3(\mathbf{u}^n\cdot\nabla)\mathbf{u}^n]\cdot\mathbf{v}\,\mathrm{d}x - \int_{P^n}\boldsymbol{\lambda}_P^n\cdot\mathbf{v}\,\mathrm{d}x$$
$$+ \mathcal{F}\mathrm{r}\int_\Omega \frac{\mathbf{g}}{g}\cdot\mathbf{v}\,\mathrm{d}x, \qquad \forall\mathbf{v}\in(H_0^1(\Omega))^3, \tag{30.44}$$

$$\mathbf{u}^{n+1/2,k+1} = \mathbf{u}_\Gamma((n+1)\Delta t) \text{ on } \Gamma, \tag{30.45}$$

$$\int_\Omega q\nabla\cdot\mathbf{u}^{n+1/2,k+1}\,\mathrm{d}x = 0, \qquad \forall q\in L^2(\Omega), \tag{30.46}$$

and then compute $\boldsymbol{\lambda}^{n+1,k+1}$ from

$$\boldsymbol{\lambda}^{n+1,k+1} = P_\Lambda[\boldsymbol{\lambda}^{n+1,k} + r\sqrt{2}\,\frac{\mathcal{B}\mathrm{n}}{\mathcal{R}\mathrm{e}}\mathbf{D}(\mathbf{u}^{n+1/2,k+1})]; \tag{30.47}$$

as in Chapter 2, Section 16, we use $\|\boldsymbol{\lambda}^{n+1,k+1} - \boldsymbol{\lambda}^{n+1,k}\|_{(L^2(\Omega))^{3\times 3}} \leq \text{tol}$ as stopping criterion.

In order to decouple, in (30.36)–(30.38), the Navier–Stokes part from the viscoplastic one, we follow DEAN, GLOWINSKI and PAN [2003]; thus, using further splitting, we substitute to subproblem (30.36)–(30.38) the two following subproblems:

(1) Find $\mathbf{u}^{n+1/4} \in (H^1(\Omega))^3$ and $p^{n+1} \in L^2(\Omega)$ such that

$$\begin{aligned} &\int_\Omega \frac{\mathbf{u}^{n+1/4} - \mathbf{u}^n}{\Delta t} \cdot \mathbf{v}\,dx + \frac{1}{2\mathcal{R}e}\int_\Omega \nabla(\mathbf{u}^{n+1/4} + \mathbf{u}^n) : \nabla \mathbf{v}\,dx \\ &\quad - \int_\Omega p^{n+1} \nabla \cdot \mathbf{v}\,dx = -\sqrt{2}\,\frac{\mathcal{B}n}{\mathcal{R}e} \int_\Omega \boldsymbol{\lambda}^n : \mathbf{D}(\mathbf{v})\,dx \\ &\quad + \frac{1}{2}\int_\Omega [(\mathbf{u}^{n-1} \cdot \nabla)\mathbf{u}^{n-1} - 3(\mathbf{u}^n \cdot \nabla)\mathbf{u}^n] \cdot \mathbf{v}\,dx - \int_{P^n} \boldsymbol{\lambda}_P^n \cdot \mathbf{v}\,dx \\ &\quad + \mathcal{F}r \int_\Omega \frac{\mathbf{g}}{g} \cdot \mathbf{v}\,dx \qquad \forall \mathbf{v} \in (H_0^1(\Omega))^3, \end{aligned} \tag{30.48}$$

$$\mathbf{u}^{n+1/4} = \mathbf{u}_\Gamma((n+1)\Delta t) \text{ on } \Gamma, \tag{30.49}$$

$$\int_\Omega q \nabla \cdot \mathbf{u}^{n+1/4} dx = 0, \qquad \forall q \in L^2(\Omega). \tag{30.50}$$

(2) Find $\mathbf{u}^{n+1/2} \in (H^1(\Omega))^3$ and $\boldsymbol{\lambda}^{n+1} \in \Lambda$, such that

$$\begin{aligned} &\int_\Omega \frac{\mathbf{u}^{n+1/2} - \mathbf{u}^{n+1/4}}{\Delta t} \cdot \mathbf{v}\,dx + \sqrt{2}\,\frac{\mathcal{B}n}{\mathcal{R}e}\int_\Omega \boldsymbol{\lambda}^{n+1} : \mathbf{D}(\mathbf{v})\,dx \\ &\quad = \sqrt{2}\,\frac{\mathcal{B}n}{\mathcal{R}e}\int_\Omega \boldsymbol{\lambda}^n : \mathbf{D}(\mathbf{v})\,dx, \qquad \forall \mathbf{v} \in (H_0^1(\Omega))^3, \end{aligned} \tag{30.51}$$

$$\mathbf{u}^{n+1/2} = \mathbf{u}_\Gamma((n+1)\Delta t) \text{ on } \Gamma, \tag{30.52}$$

$$\boldsymbol{\lambda}^{n+1} = P_\Lambda\left[\boldsymbol{\lambda}^{n+1} + r\sqrt{2}\,\frac{\mathcal{B}n}{\mathcal{R}e}\mathbf{D}(\mathbf{u}^{n+1/2})\right]. \tag{30.53}$$

To solve the "plasticity" system (30.51)–(30.53), we use the simplified variant of algorithm (30.43)–(30.46) obtained by eliminating from (30.44) the contributions of viscosity, advection, incompressibility, and fictitious domain. The "decoupled" scheme (30.48)–(30.53) differs from the one in DEAN, GLOWINSKI and PAN [2003] in that it retains $\boldsymbol{\lambda}_P^n$ in (30.48) and $\boldsymbol{\lambda}^n$ in (30.48) and (30.51) and discards the viscous term in the "plasticity" step; the first two modifications improve the asymptotic consistence of the scheme, reducing thus the splitting error when computing steady-state solutions.

REMARK 30.5. In Chapter 2, Section 17.6, we have shown that the presence of the viscous diffusion term insures the convergence of the variant of algorithm (30.43)–(30.47) associated with (30.51)–(30.53), provided that r verifies (after space discretization) the dimensionless analog of relation (17.64). Actually, (17.64) shows also that (after space discretization) a viscous term is not necessary in the "plasticity" equation, provided that r is small enough. Indeed, our numerical experiments show that there is no difficulty with convergence when one drops the viscous term from the "plasticity" equation, at least for flows at moderate Bingham number; throughout this study, we took $r = \frac{\mathcal{R}e}{2(\mathcal{B}n)^2}$. The advantage of dropping the viscous term is that, after an appropriate space discretization, to obtain $\mathbf{u}_h^{n+1/2,k+1}$ from $\boldsymbol{\lambda}_h^{n+1,k}$ we have to solve a linear system associated with a *diagonal* matrix, making, the solution of this system pretty inexpensive. Regarding the accuracy, we will show that our "decoupled" scheme is almost as accurate as the "coupled" scheme at low-to-moderate Bingham numbers.

REMARK 30.6. The methods we use for the numerical implementation of the schemes described above have been described in detail in YU, WACHS and PEYSSON [2006]. Therefore, we will only recall the main ingredients of these methods:

(1) Concerning the *space discretization*, we use a finite-difference-based projection method on a half-staggered grid to solve the time-discrete Navier–Stokes equations (30.48)–(30.50), the plasticity multiplier being evaluated at the velocity nodes.
(2) The particle subproblems (30.39)–(30.41) is a *linear saddle-point problem*. It can be solved efficiently, using an *Uzawa/conjugate gradient algorithm*, as in, e.g., GLOWINSKI, PAN, HESLA and JOSEPH [1999], GLOWINSKI [2003, chapter 8]. To discretize the *Lagrange multiplier* λ_P^{n+1} associated with the *rigid body motion constraint* (30.41), we use here the *collocation-element* method (CE) advocated in YU, PHAN-THIEN, FAN and TANNER [2002], YU, WACHS and PEYSSON [2006]. Because, with the CE method, the elements do not fit well with the particle boundary, the computed drag coefficient is not expected to be highly accurate. The reader is referred to YU, WACHS and PEYSSON [2006] for a detailed description of the solution method for the particle subproblems.

REMARK 30.7. In this chapter, the particulate flow is always taking place in a rectangular or cuboïd domain. For more complex geometries, such as the tube considered here, an additional set of Lagrange multipliers is introduced in order to enforce a Dirichlet boundary condition on the physical boundary immersed in the extended domain. The CE method applies here also (a similar approach has been used in PAN, GLOWINSKI and HOU [2007] to investigate particle clustering phenomena in a rotating cylinder containing a fluid-particle mixture).

30.3. Sedimentation of spherical particles in a tube: Numerical results

30.3.1. Sedimentation of a single sphere at low Reynolds numbers

We start our investigations by considering the sedimentation of a sphere settling in a vertical circular tube of radius $4a$ along the tube axis. In Fig. 30.1, we have compared (as functions of time) the computed *settling velocities* obtained using both the "decoupled" and "coupled" schemes, assuming that $\mathcal{B}n_S = 0.36$ and $\rho_r = 1.1$; we suppose that at $t = 0$, the fluid and the particle are at rest, the particle being located on the axis of the tube. Figure 30.1 shows

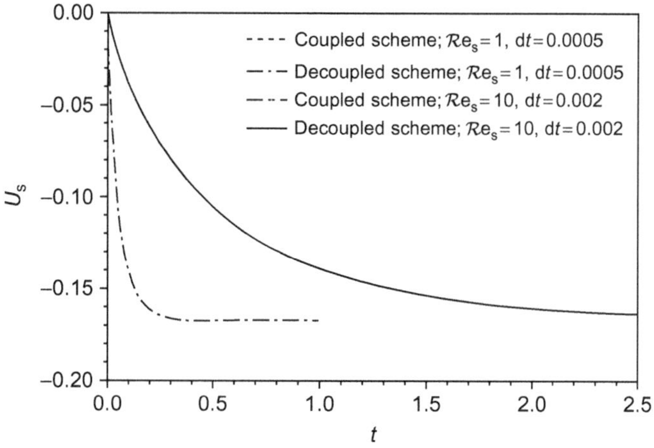

FIG. 30.1 Settling of a single sphere in a vertical circular tube of radius $4a$ at $\mathcal{B}n_S = 0.36$ and $\rho_r = 1.1$: comparison between the coupled and decoupled schemes.

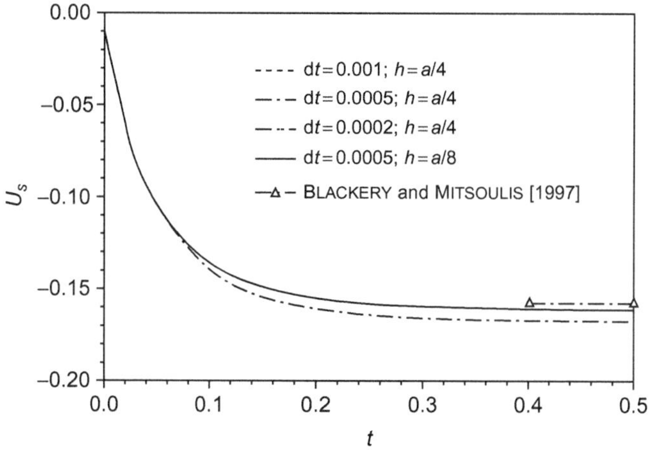

FIG. 30.2 Settling of a single sphere in a vertical circular tube of radius $4a$ at $\mathcal{B}n_S = 1$, $\mathcal{B}n_S = 0.36$, and $\rho_r = 1.1$: Influence of Δt and h, and comparison with the results in BLACKERY and MITSOULIS [1997].

that for the two values of the Reynolds number $\mathcal{R}e_S$ considered here, namely 1 and 10, the numerical results obtained through both schemes are essentially identical. We note that for $\mathcal{R}e_S = 10$, the *effective Reynolds number* $\mathcal{R}e_m$ (see (30.30)) is close to 0.5; this explains why the corresponding steady-state settling velocity (that is the inverse of the drag coefficient) is close to the one associated with $\mathcal{R}e_S = 1$ (that is $\mathcal{R}e_m = 0.05$). For a lower $\mathcal{R}e_S$, the steady-state settling velocity does not change, but the time required to reach this velocity increases, entailing the use of a smaller time step to avoid possible artificial velocity overshoot.

In Fig. 30.2, we have reported the graph of the computed *settling velocity*, as a function of time, for $\mathcal{R}e_S = 1$, $\mathcal{B}n_S = 0.36$, and $\rho_r = 1.1$; these results have been obtained for different values of the time-discretization step Δt and of the space-discretization step h. This figure shows good convergence properties as $\Delta t \to 0$; on the other hand, when compared to the results of BLACKERY and MITSOULIS [1997], the steady-state velocity for $h = 1/4a$ is overestimated by about 6%, the accuracy being significantly improved by taking $h = 1/8a$, as shown in Fig. 30.2, where some of the results in the above reference have been reported.

The *drag coefficient*, computed for different values of $\mathcal{B}n_S$, has been reported in Fig. 30.3(a). This figure shows a good agreement with the results in BLACKERY and MITSOULIS [1997] for small to moderate values of $\mathcal{B}n_S$ ($\mathcal{B}n_S < 0.6$, typically), but for $\mathcal{B}n_S >$ 0.7, our drag coefficient is overestimated. According to the results of BERIS, TSAMOPOULOS, ARMSTRONG and BROWN [1985], the sphere does not settle at $\mathcal{B}n_S \approx 0.858$. However, in our simulations the sphere velocity starts oscillating around zero at $\mathcal{B}n_S \approx 0.8$; decreasing Δt by a large factor and taking $h = 1/8a$ does not change this behavior: it is likely that decoupling, as we did, viscosity and yield stress affects the computed solutions at high $\mathcal{B}n_S$. Therefore, the method presented here is deemed to be limited to simulations at low and moderate Bingham numbers $\mathcal{B}n_S$ ($\mathcal{B}n_S < 0.7$, typically) and can not be used to decide whether the sphere settles or not. Fig. 30.3(b) shows that our results seem to be in better agreement with those in BLACKERY and MITSOULIS [1997] if one represents C_s as a function of $\mathcal{B}n_T$.

In Fig. 30.4(a,b), we have visualized, for $\mathcal{B}n_S = 0.36$ and 0.529, the *steady-state velocity field* and the corresponding *yielded* (white) and *unyielded* (black) flow regions. For both cases, the shape of the *yield surface* is in qualitative agreement with the one found in BERIS, TSAMOPOULOS, ARMSTRONG and BROWN [1985], BLACKERY and MITSOULIS [1997], LIU, MULLER and DENN [2002]. Not surprisingly, it was observed that the smoothness of the yield surface increases as h gets smaller. However, because we use a uniform mesh, the computational cost may become prohibitive for very small values of h (at least for the serial work stations that we used when these three-dimensional computations were done [around 2005/2006]). Indeed, a precise description of the yield surface would require refining the mesh, locally and dynamically, as the particle moves; this is not easy with the fictitious domain method we use. The smallest mesh size we could afford for this problem was $h = a/16$, resulting (the computational region being $(0, 8a) \times (0, 8a) \times (0, 12a)$) in

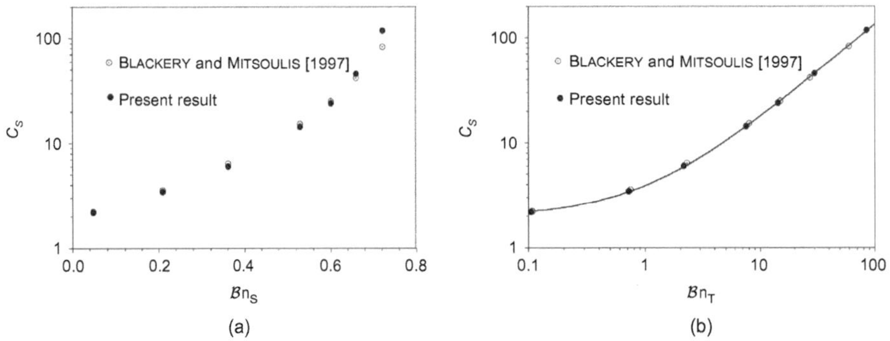

FIG. 30.3 Settling of a single sphere in a vertical circular tube of radius $4a$: (a) Drag coefficient versus $\mathcal{B}n_S$. (b) Drag coefficient versus $\mathcal{B}n_T$. In (b), the solid line represents the curve fitting the results in BLACKERY and MITSOULIS [1997].

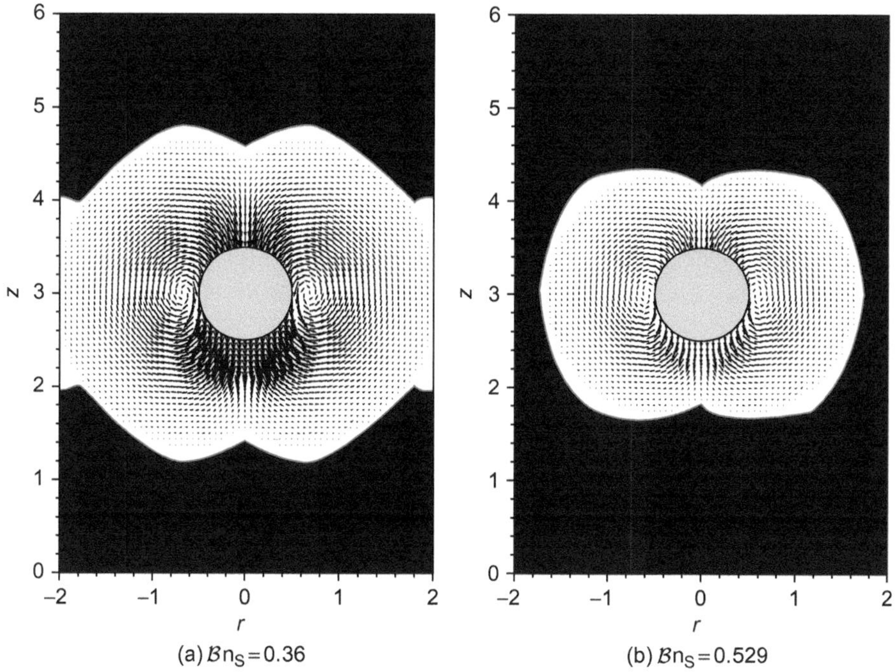

FIG. 30.4 Settling of a single sphere in a vertical circular tube of radius $4a$: Visualization of the velocity field and of the yielded (white) and unyielded (black) regions: (a) $\mathcal{B}n_S = 0.36$. (b) $\mathcal{B}n_S = 0.529$.

$128 \times 128 \times 192 (\approx 3.15 \times 10^6)$ grid points for the flow velocity. It is worth mentioning that in our computations, it was decided that a grid point X would belong to the computed yielded region at $t = n\Delta t$ if $|\boldsymbol{\lambda}_h^n(X)| \geq 1 - \varepsilon$. At $\mathcal{B}n_S = 0.529$, if one takes $\varepsilon = 5 \times 10^{-3}$, one observes a spurious yielded island attached to the boundary of the tube; this spurious yielded region disappears if one takes $\varepsilon = 10^{-3}$.

REMARK 30.8. For those readers who may be disappointed by the performances of the fictitious domain method used here, we would like to mention that one of its main advantages is the possibility of handling multiple particle situations at a cost which does not depend much on the number of particles.

30.3.2. Sedimentation of a single sphere at moderate Reynolds numbers

We consider now the sedimentation of a sphere settling in a vertical circular tube of radius $4a$ along the tube axis, at moderate Reynolds numbers. In Fig. 30.5, we have visualized, for $\mathcal{R}e_I = 100$, 200, and 400, the computed *drag coefficient* C_D as a function of the *Bingham number* $\mathcal{B}n_I$ (Fig. 30.5(a)) and of the *effective Reynolds number* $\mathcal{R}e_m$ (Fig. 30.5(b)). Figure 30.5(b) confirms the good correlation between C_D and $\mathcal{R}e_m$. Actually, our results show also that we can improve the above correlation by assuming that the shear-rate coefficient k in relation (30.30) is an increasing function of the Reynolds number $\mathcal{R}e_I$ (this is consistent with the analysis in HE, LASKOWSKI and KLEIN [2001]). The reason for this behavior is that the *thickness of the boundary layer* decreases as the *Reynolds number* increases, corresponding to a higher effective shear rate.

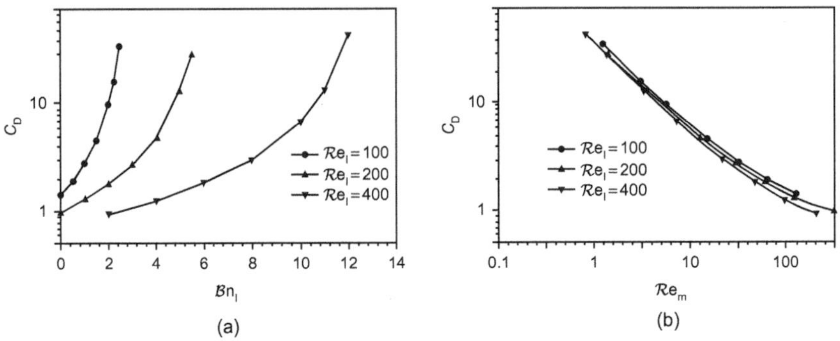

FIG. 30.5 Settling of a single sphere in a vertical circular tube of radius $4a$: (a) C_D versus $\mathcal{B}n_I$. (b) C_D versus $\mathcal{R}e_m$.

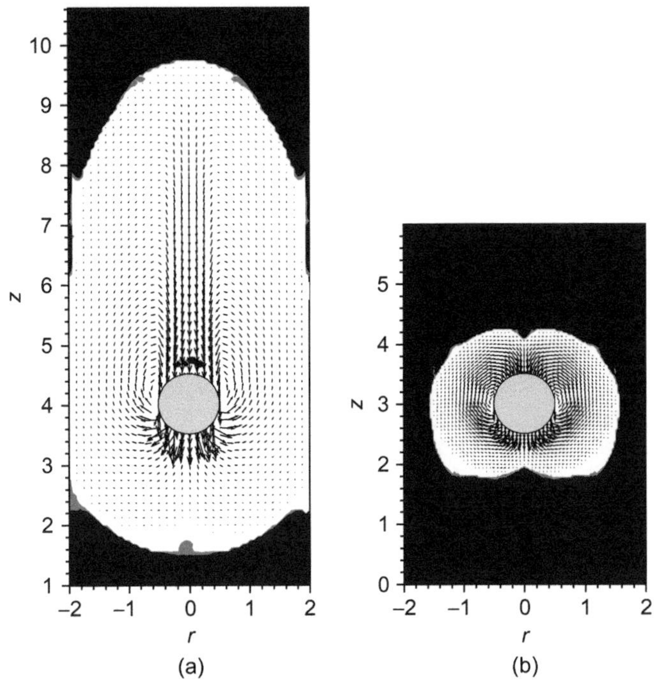

FIG. 30.6 Settling of a single sphere in a vertical circular tube of radius $4a$ at moderate Reynolds number ($\mathcal{R}e_I = 400$): Steady-state velocity field and yielded and unyielded regions at $\mathcal{B}n_I = 4$ (Fig. 30.6(a)) and $\mathcal{B}n_I = 11$ (Fig. 30.6(b)).

The steady state velocity and the yielded and unyielded regions have been visualized in Fig. 30.6, for $\mathcal{R}e_I = 400$ and $\mathcal{B}n_I = 4$ (Fig. 30.6(a)), and for $\mathcal{R}e_I = 400$ and $\mathcal{B}n_I = 11$ (Fig. 30.6(b)). For $\mathcal{B}n_I = 4$, the effective Reynolds number $\mathcal{R}e_m$ is 63.5; thus, the inertial effect is strong and the wake structure pretty large as shown in Fig. 30.6(a). For $\mathcal{B}n_I = 11$, $\mathcal{R}e_m$ decreases to 2.64, the velocity field and yield surface being reminiscent of those in Fig. 30.4(b).

30.3.3. Hydrodynamic interaction between two spheres at low Reynolds numbers

The second test problem that we consider concerns the dynamic interaction between two identical spheres translating along the axis of a vertical circular tube at low Reynolds numbers. In Fig. 30.7, we have reported the variation, as a function of the separating (scaled) distance, of the *normalized drag coefficient* $C_S/C_{S,\mathrm{single}}$ when the two spheres settle in a tube whose radius is either $4a$ or $8a$ (here, $C_{S,\mathrm{single}}$ denotes the drag coefficient of a single sphere settling in the same tube). At low Reynolds numbers, the two spheres fall faster than a single one, due to hydrodynamic interaction, but they do not approach one another. From Fig. 30.7, the relative increase in the velocity is more pronounced for the Bingham fluid than for the corresponding Newtonian one. This seems to contradict LIU, MULLER and DENN [2003], who observed larger $C_S/C_{S,\mathrm{single}}$ for the Bingham fluid than for the Newtonian one. The reason for this discrepancy is easy to explain: we fixed $\mathcal{B}n_S$ (at 0.36) in our *dynamical* simulations, whereas it is $\mathcal{B}n_T$ which is fixed in the *static* simulations reported in LIU, MULLER and DENN [2003]. Indeed, for our simulations at fixed $\mathcal{B}n_S$, as the separating distance decreases, the velocity increases, implying in turn that $\mathcal{B}n_T$ decreases (remember that $\mathcal{B}n_T = \mathcal{B}n_S/U_T^*$); thus the decrease of $C_S/C_{S,\mathrm{single}}$ that we observe is more pronounced than in LIU, MULLER and DENN [2003].

In Fig. 30.8, we have visualized the velocity field and the yielded and unyielded flow regions for two spheres settling in a tube of radius $8a$, assuming that the separating distance is $L = 5a$. The yield surfaces are in qualitative agreement with those reported in LIU, MULLER and DENN [2003].

In LIU, MULLER and DENN [2003], a slight drag reduction was observed, at $\mathcal{B}n_T = 340.7$, for two spheres getting closer, compared with a single sphere. This result is *a priori* surprising and will be investigated using our methodology. For simplicity, we take both the Reynolds number and Δt very small, so that a quasi steady-state can be reached very quickly. For validation purposes, two meshes have been used (corresponding to $h = a/8$ and $h = a/16$) to compute the normalized drag coefficient $C_S/C_{S,\mathrm{single}}$ at $\mathcal{B}n_S = 0.36$ (viscoplastic)

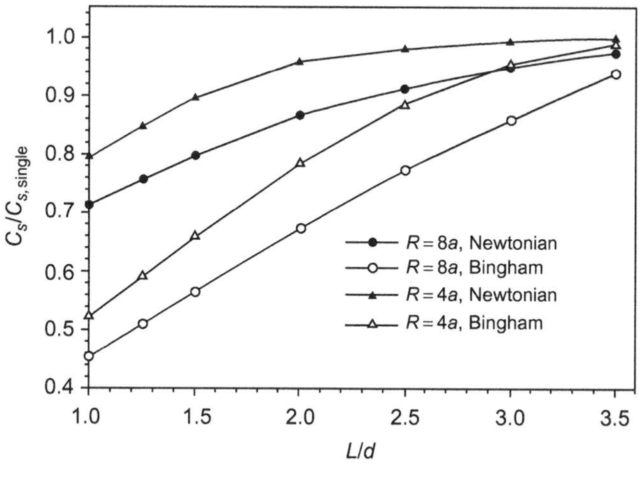

FIG. 30.7 Settling of two identical spheres in a vertical circular tube at low Reynolds number and for various radii: normalized drag coefficient versus the separating distance (for the Bingham case, $\mathcal{B}n_S = 0.36$). Here, L is the distance between the centers of the two spheres.

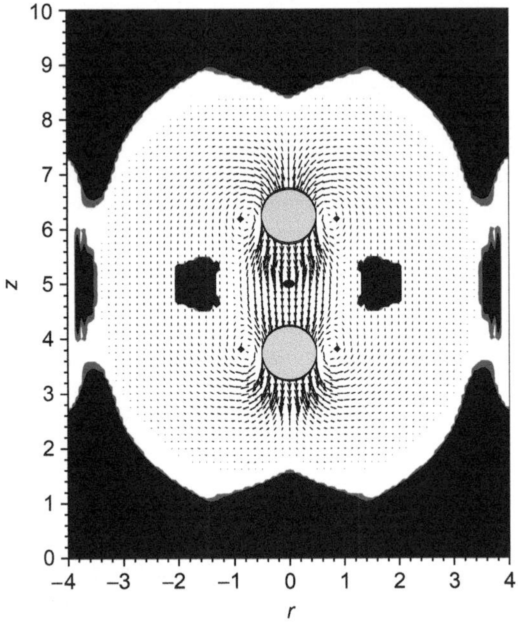

FIG. 30.8 Settling of two identical spheres in a vertical circular tube of radius $8a$ at low Reynolds number $\mathcal{R}e_S$ and for $\mathcal{B}n_S = 0.36$: steady-state velocity field and yielded (white) and unyielded (black) regions.

and $\mathcal{B}n_S = 0$ (Newtonian). For a single settling sphere, the drag coefficients computed with the two meshes agree within 1%; for each mesh, the normalized drag coefficients to be discussed below are based on the corresponding computed $C_{s,\text{single}}$. In Fig. 30.9, we have reported, as functions of the (scaled) separating distance $(L - 2a)/a$, the computed values of $C_s/C_{s,\text{single}}$, for $h = a/8$ and $a/16$ and different values of $\mathcal{B}n_S$ (including $\mathcal{B}n_S = 0$, which corresponds to a Newtonian fluid). The agreement between the results obtained with the two different meshes, when the gap is of the order of a, is quite satisfactory. Clearly, as the gap distance between the spheres decreases the agreement deteriorates, but the maximum relative difference is less than 10% when the gap is $0.2a$, a distance comparable with the mesh size; for a *fictitious domain method* using a uniform mesh, such a level of accuracy is satisfactory. From Fig. 30.9, we see that for a *Newtonian fluid*, the normalized drag coefficient increases monotonically as the gap decreases; however, for a *Bingham fluid*, the normalized drag coefficient first decreases as the gap decreases below the *first critical distance* d_{c1} (d_{c1} is related to the size of the unyielded region for a single sphere), and then start increasing and exceeds one as the gap decreases below the *second critical distance* d_{c2}). The existence of d_{c1} is easily understood, considering that there is no hydrodynamic interaction if the two spheres are located sufficiently far away from each other, so that their respective unyielded regions are disconnected. From Fig. 30.9, we observe also that if the gap distance between the particles is $2d$ then $C_s \approx C_{s,\text{single}}$ at $\mathcal{B}n_S = 0.529$, while C_s is (slightly) smaller than $C_{s,\text{single}}$ if $\mathcal{B}n_S = 0.36$; these results are consistent with the sizes of the unyielded regions shown in Fig. 30.4.

The already mentioned *drag reduction* observed in LIU, MULLER and DENN [2003] was corresponding to $C_s/C_{s,\text{single}} = 0.94$ at the gap distance $0.5a$, for $\mathcal{B}n_T = 340.7$; however,

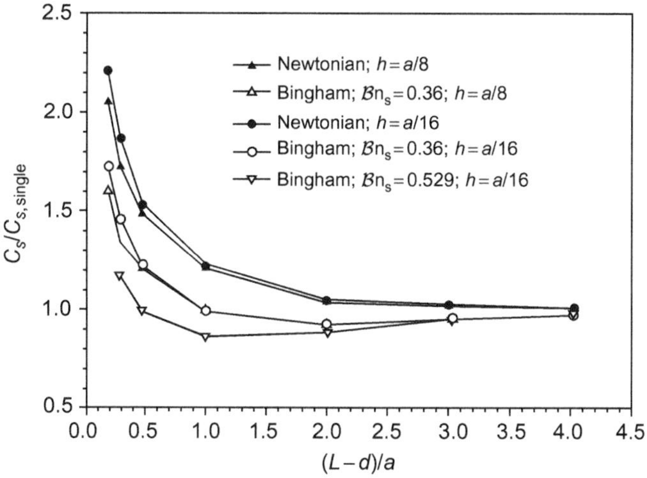

FIG. 30.9 Interaction of two identical spheres settling in a vertical circular tube of radius $4a$ at low Reynolds number: variation of the normalized drag coefficient $C_s/C_{s,\mathrm{single}}$ versus the scaled distance between the two spheres ($C_{s,\mathrm{single}}$ denotes the drag coefficient of a single sphere settling in the same tube).

the existence of the *second critical distance* d_{c2} was not reported in the above reference. Actually, Fig. 30.9 suggests that d_{c2} is a decreasing function of $\mathcal{B}n_S$, taking a value close to $0.48a$ at $\mathcal{B}n_S = 0.529$ (which corresponds to $\mathcal{B}n_T = 8.02$); thus, for $\mathcal{B}n_T = 340.7$, d_{c2} has to be much smaller than $0.48a$, explaining why it was overlooked in LIU, MULLER and DENN [2003], since in the above publication, there is no investigation of the particle interaction for gap distance less than $0.5a$. Just below, we are going to attempt giving an explanation of the type of interaction we observe for small values of the gap distance.

For a moving sphere, we can assume that the region surrounding it is *yielded*. Hence, the drag on the sphere, denoted by $\mathbf{F}_z$ here, is given by

$$\mathbf{F}_z = \int_{\partial P(t)} \left(-p\mathbf{I} + 2\mu\mathbf{D} + \tau_y \frac{\mathbf{D}}{\|\mathbf{D}\|} \right) \mathbf{n} \cdot \mathbf{e}_z \mathrm{d}(\partial P(t)), \tag{30.54}$$

where $\mathbf{e}_z$ is the unit vector of the vertical direction and $\mathbf{n}$ is the outward unit normal vector on the surface of the particle. We can split the drag force into a *lubrication* part and a *plastic* part as follows:

$$\mathbf{F}_z = \mathbf{F}_\mathrm{l} + \mathbf{F}_\mathrm{p} \tag{30.55}$$

$$\mathbf{F}_\mathrm{l} = \int_{\partial P(t)} (-p\mathbf{I} + 2\mu\mathbf{D})\mathbf{n} \cdot \mathbf{e}_z \mathrm{d}(\partial P(t)), \tag{30.56}$$

$$\mathbf{F}_\mathrm{p} = \int_{\partial P(t)} \tau_y \frac{\mathbf{D}}{\|\mathbf{D}\|} \mathbf{n} \cdot \mathbf{e}_z \mathrm{d}(\partial P(t)). \tag{30.57}$$

Here, we define $\mathbf{F}_\mathrm{l}$ as the *lubrication drag* because we consider a squeezing flow for which $\mathbf{F}_\mathrm{l}$ increases as the gap distance decreases and eventually dominates the total drag force at very small gap distance. For the *plastic drag* $\mathbf{F}_p$, we can define the *plastic viscosity* as $\mu_p = \frac{\tau_y}{2\|\mathbf{D}\|}$. Clearly, the *plastic viscosity* exhibits a *shear-thinning* property. Indeed, in LIU, MULLER and DENN [2003], one explains the drag reduction by the fact that the two sphere interaction causes a decrease of the local viscosity through the shear-thinning property of the Bingham material. However, it is our point of view that this explanation of the drag reduction in terms of the viscosity alone is not sufficient, because it may happen that the plastic stress is not shear-thinning, as is the case for a pure shear or an extensional flow. However, for a complex flow, it is possible that the term $\frac{\mathbf{D}}{\|\mathbf{D}\|}\mathbf{n} \cdot \mathbf{e}_z$ decreases as the magnitude of the shear-rate $\|\mathbf{D}\|$ increases. Indeed, our results and those in LIU, MULLER and DENN [2003] seem to show that the plastic force is shear-thinning and dominates the drag force, before the repulsive lubrication prevails when the two spheres are sufficiently close.

Actually, experimental data would be needed in order to clarify if the drag reduction observed through numerical simulation takes place for real-life viscoplastic materials, or is just a consequence of the limitations of the Bingham model. In the case of approaching spheres, it will be relatively easy to do experiments with fixed approaching velocity (that is, fixed $\mathcal{B}n_T$). Moreover, considering that a convenient experiment can also be done for a sphere approaching a horizontal bottom wall under the effect of gravity, we investigated this situation numerically considering both slip and no-slip boundary conditions at the wall. The normalized drag coefficient at $\mathcal{B}n_S = 0.36$ have been plotted on Fig. 30.10 (indeed, the problem of two approaching spheres is very close to the one of a single sphere approaching a horizontal bottom wall with free-slip boundary condition). Figure 30.10 shows that no drag reduction (resp., that drag reduction) takes place with the no-slip (resp., free-slip) boundary condition. The no-slip boundary condition related result is understandable because the repulsive lubrication force is significantly higher than for a free-slip wall. The flow fields and the yielded and unyielded regions have been visualized on Fig. 30.11; the above figure shows

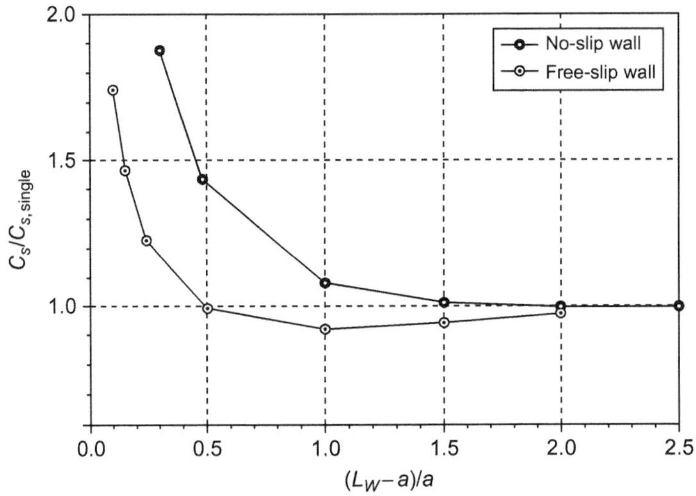

FIG. 30.10 Normalized drag coefficient versus scaled gap distance for a single sphere approaching a horizontal bottom wall in a vertical circular tube of radius $4a$ at low Reynolds number and $\mathcal{B}n_S = 0.36$: the upper (resp., lower) curve corresponds to the no-slip (resp., free-slip) wall.

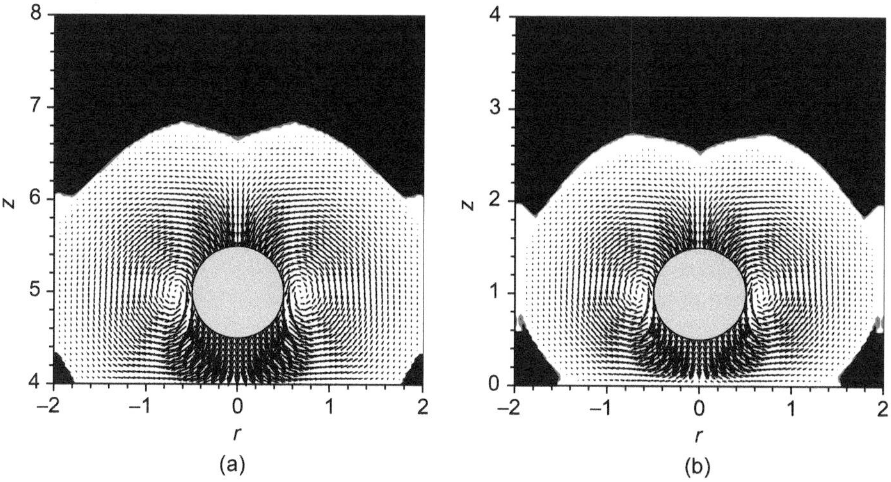

FIG. 30.11 Visualization of the velocity field and of the yielded (white) and unyielded (black) flow regions for a single sphere approaching a horizontal bottom wall in a vertical circular tube of radius $4a$ at low Reynolds number and $\mathcal{B}n_S = 0.36$: (a) Free-slip wall. (b) No-slip wall.

that the flow fields near the particle boundary are very similar for both cases we considered, suggesting that for the no-slip case, the disappearance of the drag reduction results from the predominance of the lubrication forces over the plastic ones during the entire approach. The drag reduction was not observed at $\mathcal{B}n_S = 0.529$ for either the no-slip or the free-slip boundary condition (we do not exclude that the drag reduction may occur at large Bingham numbers, but if this happens, the drag reduction being less pronounced is more difficult to observe (through numerical simulation) that in the case of two approaching spheres. Fig. 30.11 shows also that the yielded region associated with the no-slip wall is smaller than the free-slip one (this has to do with the fact that $\mathcal{B}n_T$ is larger for the no-slip wall).

30.4. Some remarks on the dynamical simulation of particle sedimentation in a Bingham fluid

In the earlier sections, we have discussed the application of a *fictitious domain method* to the numerical simulation, through an unsteady model, of the motion of particles in a Bingham viscoplastic fluid at moderate Bingham numbers. The methodology we used relies on previous work of the first author and various collaborators, particularly concerning the systematic use of operator-splitting to simplify computations. However, in the present article, we used other space-time discretization schemes to treat the governing equations. For space discretization, we use a finite-difference method and a collocation-element method to enforce the rigid body motion inside the particles. Concerning the time discretization, we used an operator-splitting method in order to decouple the Navier–Stokes, plasticity, and rigid body motion parts. A significant difference with previous splitting schemes (like the ones used in Chapter 2, Section 17, and DEAN, GLOWINSKI and PAN [2003]) is that the plasticity and rigid body motion multipliers are retained in the Navier–Stokes step, in order to reduce the splitting error (and allow, therefore, larger Δt). The present study shows also that it is not necessary to keep a viscous term in the plasticity step, although this may help for flow at large Bingham numbers.

We applied our computational methods to the sedimentation of spherical particles in a tube filled with a Bingham fluid. In the case of a single sphere, the computed drag coefficients are in good agreement with those reported in the literature. The fact that we use a fixed mesh makes it difficult for the precise determination of the yield surface. Moreover, the computed yield surface may be a very sensitive function of the small value parameters used as stopping criteria for the various iterative methods we use or used as threshold to identify the yielded and unyielded regions. Our results show that the drag coefficient for a single sphere settling in a Bingham fluid at nonzero Reynolds numbers can be well correlated with an effective Reynolds number. For two approaching spheres, there exists a critical separation distance above which a drag reduction is observed (as shown in LIU, MULLER and DENN [2003]) and below which a drag-enhancement takes place. It seems, however, that the drag reduction does not take place for a sphere falling toward a solid wall at the relatively low Bingham numbers that we considered (if we assume a no-slip boundary condition at the wall). This behavior can be explained by the competition between a shear-thinning plastic force and a repulsive lubrication force acting on the sphere.

A main drawback of our approach is that it can not be applied to the simulation of those high Bingham number flows where the primary interest is to know if the particles settle in finite time. The method is clearly not as accurate as a boundary-fitted one, but it is more efficient, making possible the direct simulation, on a standard Linux workstation, of the sedimentation of thousand of spherical particles in a Bingham fluid.

31. Further comments on distributed Lagrange multiplier/fictitious domain methods for Bingham fluid flow

In this chapter, we have combined *multipliers* (Lagrange's and others) *based methods* with *fictitious domains* to address the simulation of viscoplastic flows with fixed or solid moving boundaries. These two components of our methodology rely on *multipliers*. Indeed, in Section 29, we used an *augmented Lagrangian* approach to investigate the uniaxial flow of a Bingham fluid in a duct with eccentric annular cross-section, while in Section 30, we combined *operator-splitting* with a *multiplier-based projection method* to investigate the sedimentation of spherical particles in a vertical tube filled with a Bingham fluid.

Both approaches provide accurate computed solutions at low-to-moderate Bingham numbers. Actually, with the approach investigated in Section 29, there is no limit on the Bingham numbers which can be considered, implying that the flow cessation or a complete no-flow situation can be simulated, which is not the case with the methods discussed in Section 30. Actually, the above drawback of the methods of Section 30 stems from the use of an operator-splitting-based time discretization of the governing equations. This may be of concern for some important industrial applications.

The ability of the methods of Section 29 to handle any Bingham number (from 0 to complete no-flow situations) stems from the *fully coupled* (or *monolithic*) feature of the solution algorithm. However, this advantage has a cost because for each position of the eccentric inner cylinder, the velocity augmented matrix has to be updated and Cholesky factorized. This may become very costly if the fictitious domain changes constantly with time, as is the case for moving particles. This suggests investigating a fully coupled approach for the simulation of particle sedimentation in a Bingham fluid. If we keep a projection technique to handle the *plasticity multiplier* $\boldsymbol{\lambda}$, such a monolithic scheme (a fully coupled variant of the method discussed in Section 30) is described just below.

- Time loop $t^{n+1} = t^n + n\Delta t$, $n \geq 0$:
  - Initialization: $\mathbf{u}^{0,n+1} = \mathbf{u}^n$, $p^{0,n+1} = p^n$, $\mathbf{U}^{0,n+1} = \mathbf{U}^n$, $\boldsymbol{\omega}^{0,n+1} = \boldsymbol{\omega}^n$, $\boldsymbol{\lambda}_P^{0,n+1} = \boldsymbol{\lambda}_P^n$ and $\boldsymbol{\lambda}^{0,n+1} = \boldsymbol{\lambda}^n$.
  - For $k \geq 0$:

◆ Solve the following Stokes/fictitious domain problem: Find $\mathbf{u}^{k+1,n+1} \in (H^1(\Omega))^3$, $p^{k+1,n+1} \in L^2(\Omega)$, $\mathbf{U}^{k+1,n+1} \in \mathbb{R}^3$, $\boldsymbol{\omega}^{k+1,n+1} \in \mathbb{R}^3$ and $\boldsymbol{\lambda}_P^{k+1,n+1} \in (H^1(P^n))^3$ such that

$$\rho_f \int_\Omega \frac{\mathbf{u}^{k+1,n+1} - \mathbf{u}^n}{\Delta t} \cdot \mathbf{v}\,dx + \mu \int_\Omega \nabla \mathbf{u}^{k+1,n+1} \cdot \nabla \mathbf{v}\,dx - \int_\Omega p^{k+1,n+1} \nabla \cdot \mathbf{v}\,dx + \int_{P^n} \boldsymbol{\lambda}_P^{k+1,n+1} \cdot \mathbf{v}\,dx$$
$$= -\rho_f \int_\Omega (\mathbf{u}^n \cdot \nabla)\mathbf{u}^n \cdot \mathbf{v}\,dx - \sqrt{2}\,\tau_y \int_\Omega \boldsymbol{\lambda}^{k,n+1} : \mathbf{D}(\mathbf{v})\,dx + \rho_f \int_\Omega \mathbf{g} \cdot \mathbf{v}\,dx, \qquad \forall \mathbf{v} \in (H_0^1(\Omega))^3, \tag{31.1}$$

$$\int_\Omega \nabla \cdot \mathbf{u}^{k+1,n+1} q\,dx = 0, \qquad \forall q \in L^2(\Omega), \tag{31.2}$$

$$\mathbf{u}^{k+1,n+1} = \mathbf{u}_\Gamma(t^{n+1}) \text{ on } \Gamma, \tag{31.3}$$

$$\left(1 - \frac{\rho_f}{\rho_s}\right)\left[M\left(\frac{\mathbf{U}^{k+1,n+1} - \mathbf{U}^n}{\Delta t} - \mathbf{g}\right) \cdot \mathbf{V} + \mathbf{J}\left(\frac{\boldsymbol{\omega}^{k+1,n+1} - \boldsymbol{\omega}^n}{\Delta t}\right) \cdot \boldsymbol{\theta}\right] - \int_{P^n} \boldsymbol{\lambda}_P^{k+1,n+1} \cdot (\mathbf{V} + \boldsymbol{\theta} \times \mathbf{r})dx = 0, \quad \forall \{\mathbf{V}, \boldsymbol{\theta}\} \in \mathbb{R}^3 \times \mathbb{R}^3, \tag{31.4}$$

$$\int_{P^n} [\mathbf{u}^{k+1,n+1} - (\mathbf{U}^{k+1,n+1} + \boldsymbol{\omega}^{k+1,n+1} \times \mathbf{r})] \cdot \boldsymbol{\mu}\,dx = 0, \qquad \forall \boldsymbol{\mu} \in (H^1(P^n))^3. \tag{31.5}$$

◆ Update the tensor-valued plasticity multiplier by

$$\boldsymbol{\lambda}^{k+1,n+1} = \mathbf{P}_\Lambda[\boldsymbol{\lambda}^{k,n+1} + r\sqrt{2}\,\tau_y \mathbf{D}(\mathbf{u}^{k+1,n+1})]. \tag{31.6}$$

◆ Convergence if

$$\|\boldsymbol{\lambda}^{n+1,k+1} - \boldsymbol{\lambda}^{n+1,k}\|_{(L^2(\Omega))^{3\times 3}} \leq \text{tol}. \tag{31.7}$$

  - Set: $\mathbf{u}^{n+1} = \mathbf{u}^{k+1,n+1}$, $p^{n+1} = p^{k+1,n+1}$, $\mathbf{U}^{n+1} = \mathbf{U}^{k+1,n+1}$, $\boldsymbol{\omega}^{n+1} = \boldsymbol{\omega}^{k+1,n+1}$, $\boldsymbol{\lambda}_P^{n+1} = \boldsymbol{\lambda}_P^{k+1,n+1}$ and $\boldsymbol{\lambda}^{n+1} = \boldsymbol{\lambda}^{k+1,n+1}$.
- END

From a computational point of view, the costly part of algorithms (31.1)–(31.6) is the solution of the finite dimensional analog of system (31.1)–(31.5) obtained by finite-element or finite-difference space discretization. We are convinced that the multilevel methods discussed in, e.g., XU [2009] (see also the references therein) are well suited to the solution of such problems.

Acknowledgments

The authors would like to thank E.J. Dean, I. Frigaard, G. Guidoboni, E. Heintzé, V. Henriot, T.W. Pan, Y. Peysson, G. Vinay, D. Vola, and Z. Yu for their invaluable help, suggestions, and support. The support of IFP Energies Nouvelles, of the Institute for Advanced Studies at the Hong-Kong University of Science and Technology, and of the University of Houston is also acknowledged. Last but not least, the authors thank Robin Campbell and Yutheeka Gadhyan for their processing of the original Word files and the insertion of the figures, both nontrivial tasks indeed. The work of the first author was partially supported by the U.S. National Science Foundation (NSF) via grants DMS 9973318 and 0209066.

References

ABDALI, S.S., MITSOULIS, E., MARKATOS, N.C. (1992). Entry and exit flows of Bingham fluids. *J. Rheol.* **36** (2), 389–407.

ALLOUCHE, M., FRIGAARD, I.A., SONA, G. (2000). Static wall layers in the displacement of two viscoplastic fluids in a plane channel. *J. Fluid Mech.* **424**, 243–277.

ANSLEY, R.W., SMITH, T.N. (1967). Motion of spherical particles in a Bingham plastic. *AIChE J.* **13**, 1193–1196.

ATAPATTU, D.D., CHHABRA, R.P., UHLHERR, P.H.T. (1995). Creeping sphere motion in Herschel-Bulkley fluids: flow field and drag. *J. Non-Newton. Fluid Mech.* **59**, 245–265.

BALMFORTH, N.J., FRIGAARD, I.A. (2007a). Viscoplastic fluids: from theory to applications. *J. NonNewton. Fluid Mech.* **142** (1–3), 1–3.

BALMFORTH, N.J. FRIGAARD, I.A. (eds.) (2007b). Viscoplastic fluids: from theory to applications (a special issue of the *Journal of Non-Newtonian Fluid Mechanics*). *J. NonNewton. Fluid Mech.* **142** (1–3).

BARNES, H.A. (1999). The yield stress – a review or '*παντaρει*' – everything flows? *J. NonNewton. Fluid Mech.* **81**, 133–178.

BARNES, H.A., WALTERS, K. (1985). The yield stress myth? *Rheol. Acta.* **24**, 324–326.

BEAULNE, M., MITSOULIS, E. (1997). Creeping motion of a sphere in tubes filled with Herschel-Bulkley fluids *J. NonNewton. Fluid Mech.* **72** (1), 55–71.

BÉGIS, D., GLOWINSKI, R. (1983). Application to the numerical solution of the two-dimensional flow of incompressible viscoplastic fluids. In: Fortin, M., Glowinski, R. (eds.), *Augmented Lagrangian Methods: Application to the Numerical Solution of Boundary Value Problems* (North-Holland, Amsterdam), pp. 233–255.

BENAMOU, J.D., BRENIER, Y. (2000). A computational fluid mechanics solution to the Monge-Kantorovich mass transfer problem. *Numer. Math.* **84** (3), 375–393.

BERCOVIER, M., ENGELMAN, M. (1980). A finite-element method for incompressible non-Newtonian flows. *J. Comput. Phys.* **36**, 313–326.

BERCOVIER, M., PIRONNEAU, O. (1979). Error estimates for finite element method solution of the Stokes problem in the primitive variables. *Numer. Math.* **33** (2), 211–224.

BERIS, A.N., TSAMOPOULOS, J.A., ARMSTRONG, R.C., BROWN, R.A. (1985). Creeping motion of a sphere through a Bingham plastic. *J. Fluid Mech.* **158**, 219–244.

BETHUEL, F., BREZIS, H., HELEIN, F. (1994). *Ginzburg-Landau Vortices* (Birkhäuser, Boston).

BEVERLY, C.R., TANNER, R.I., (1992). Numerical analysis of three-dimensional Bingham plastic flow. *J. NonNewton. Fluid Mech.* **42** (1–2), 85–115.

BINGHAM, E.C. (1922). *Fluidity and Plasticity* (McGraw-Hill, New York, NY).

BIRD, R.B., DAI, G.C., YARUSSO, B.J. (1983). The rheology and flow of viscoplastic materials. *Rev. Chem. Eng.* **1** (1), 2–70.

BLACKERY, J., MITSOULIS, E. (1997). Creeping motion of a sphere in tubes filled with a Bingham plastic material. *J. NonNewton. Fluid Mech.* **70** (1–2), 59–77.

BRISTEAU, M.O. (1975). Application de la Méthode des Elements Finis à la Résolution Numérique d'Inéquations Variationnelles d'Evolution du Type Bingham, Thesis Dissertation (University Pierre & Marie Curie, Paris, France).

Bristeau, M.O., Glowinski, R. (1974). Finite element analysis of the unsteady flow of a visco-plastic fluid in a cylindrical pipe. In: Oden, J.T., Zienkiewicz, O.C., Gallagher, R.H., Taylor, C. (eds.), *Finite Element Methods in Flow Problems* (University of Alabama Press, Huntsville, AL), pp. 471–488.

Burgos, G.R., Alexandrou, A.N. (1999). Flow development of Herschel-Bulkley fluids in a sudden three-dimensional expansion. *J. Rheol.* **43** (3), 485–498.

Cawkwell, M.G., Charles, M.E. (1987). An improved model for start-up of pipelines containing gelled crude oil. *J. Pipelines* **7** (1), 41–52.

Cawkwell, M.G., Charles, M.E. (1989). Characterization of Canadian artic thixotropic pipelines gelled crude oils utilizing an eight-parameter model. *J. Pipelines* **7** (3), 251–264.

Cazaux, G. (1998). *Investigations sur la Gélification des Bruts Paraffiniques: Liens entre Structure et Rhéologie*, Technical Report 44603 (Institut Français du Pétrole, Reuil-Malmaison, France).

Cea, J., Glowinski, R. (1972). Méthodes numériques pour l'écoulement laminaire d'un fluide rigide vis-coplastique incompressible. *Int. J. Comput. Math.* **3** (1), 225–255.

Chan, T.F., Golub, G.H., Mulet, P. (1999). A nonlinear primal-dual method for total variation image restoration. *SIAM J. Sci. Comput.* **20** (6), 1964–1977.

Chang, C., Nguyen, Q.D., Ronningsen, H.P. (1999). Isothermal starting of pipeline transporting waxy crude oil, *J. NonNewton. Fluid Mech.* **87**, 127–154.

Chhabra, R.P. (2006). *Bubbles, Drops and Particles in Non-Newtonian Fluids*, Second Edition. (CRC Press, Boca Raton, FL).

Ciarlet, P.G. (1978). *The Finite Element Method for Elliptic Problems* (North-Holland, Amsterdam).

Ciarlet, P.G. (1989). *Introduction to Numerical Linear Algebra and Optimization* (Cambridge University Press, Cambridge, UK).

Ciarlet, P.G. (1991). Basic error estimates for elliptic problems. In: Ciarlet, P.G., Lions, J.L. (eds.), *Handbook of Numerical Analysis*, Volume 2 (North-Holland, Amsterdam), pp. 19–351.

Clift, R., Grace, J.R., Weber, M.E. (1978). *Bubbles, Drops, and Particles* (Academic Press, New York, NY).

Cooper, D.F., Smith, J.W., Charles, M.E., Ryan, E.J., Alexander, G. (1978). Transient temperature effects in predicting start-up characteristics of gelling-type crude oils. In: *Proceedings of the 1978 International Conference of Heat Transfer*, **Volume 4**, pp. 67–71.

Coupez, T., Zine, M.A., Agassant, J.F. (1994). Numerical simulation of Bingham fluid flow. In: Gallegos, C. (ed.), *Progress and Trends in Rheology*, Volume IV (Steinkopf, Darmstadt), pp. 341–343.

Dacorogna, B., Glowinski, R., Kuznetsov, Y.A., Pan, T.W. (2004). On a conjugate gradient/Newton/penalty method for the solution of obstacle problems. Application to the solution of an Eikonal system with Dirichlet boundary conditions. In: Glowinski, R., Korotov, S., Křìvek, M., Neittaanmäki, P. (eds.), *Conjugate Gradient Algorithms & Finite Element Methods* (Springer-Verlag, Berlin), pp. 263–283.

Davidson, M.R., Nguyen, Q.D., Chang, C., Ronningsen, H.P. (2004). A model for restart of a pipeline with compressible gelled waxy crude oil. *J. NonNewton. Fluid Mech.* **123** (2–3), 269–280.

Dean, E.J., Glowinski, R. (2002). Operator-splitting methods for the simulation of Bingham viscoplastic flow. *Chinese Ann. Math.* **23 B**, 187–204.

Dean, E.J., Glowinski, R. (2003). Numerical solution of the two-dimensional elliptic Monge-Ampère equation with Dirichlet boundary conditions: an augmented Lagrangian approach. *C. R. Acad. Sci. Paris Sér. Math.* **336** (9), 779–784.

Dean, E.J., Glowinski, R. (2006a). An augmented Lagrangian approach to the numerical solution of the Dirichlet problem for the elliptic Monge-Ampère equation in two dimensions. *Electron. Trans. Numer. Anal.* **22**, 71–96.

Dean, E.J., Glowinski, R. (2006b). Numerical methods for fully nonlinear elliptic equations of the Monge-Ampère type. *Comput. Methods Appl. Mech. Eng.* **195** (13–16), 1344–1386.

Dean, E.J., Glowinski, R., Guidoboni, G. (2007). On the numerical simulation of Bingham visco-plastic flow: old and new results. *J. NonNewton. Fluid Mech.* **142** (1–3), 36–62.

Dean, E.J., Glowinski, R., Pan, T.W. (2003). A fictitious domain method for the numerical simulation of particulate flow for Bingham visco-plastic fluids. In: Heikkola, E., Kuznetsov, Y., Neittaanmäki, P., Pironneau, O. (eds.), *Numerical Methods for Scientific Computing. Variational Problems and Applications* (CIMNE, Barcelona), pp. 11–19.

DEAN, E.J., GLOWINSKI, R., PAN, T.W. (2005). Operator-splitting methods and applications to the direct numerical simulation of particulate flow and to the solution of the elliptic Monge-Ampère equation. In: Cagnol, J., Zolésio, J.P. (eds.), *Control and Boundary Analysis* (CRC, Boca Raton, FL), pp. 1–27.

DE BESSES, B.D., MAGNIN, A., JAY, P. (2004). Sphere drag in a viscoplastic fluid. *AIChE J.*, **50**, 2627–2629.

DEDEGIL, M.Y. (1987). Drag coefficients and settling velocity of particles in non-Newtonian suspensions. *J. Fluids Eng.* **109** (3), 319–323.

DELBOS, F., GILBERT, J.C., GLOWINSKI, R., SINOQUET, D. (2006). Constrained optimization in seismic reflection tomography: a Gauss-Newton augmented Lagrangian approach. *Geophys. J. Int.* **164** (3), 670–684.

DIMAKOPOULOS, Y., TSAMOPOULOS, J. (2003). Transient displacement of a viscoplastic material by air in straight and suddenly constricted tubes. *J. NonNewton. Fluid Mech.* **112** (1), 43–75.

DUVAUT, G., LIONS, J.L. (1972a). *Les Inéquations en Mécanique et en Physique* (Dunod, Paris).

DUVAUT, G., LIONS, J.L. (1972b). Transfert de chaleur dans un fluide de Bingham dont la viscosité dépend de la température. *J. Funct. Anal.* **11** (1), 93–110.

DUVAUT, G., LIONS, J.L. (1976). *Inequalities in Mechanics and Physics* (Springer-Verlag, Berlin).

ECONOMIDES, M.J., CHANEY, G.T. (1983). The rheological properties of Prudhoe Bay oil and the effects of a prolonged flow interruption on its flow behavior. *SPE J.* **23** (3), 408–416.

EKELAND, I., TEMAM, R. (1999). *Convex Analysis and Variational Problems* (SIAM, Philadelphia, PA).

EYMARD, R., GALLOUET, T., HERBIN, R. (2000). Finite volume methods. In: Ciarlet, P.G., Lions, J.L. (eds.), *Handbook of Numerical Analysis*, Volume VII (North-Holland, Amsterdam), pp. 713–1020.

FORTIN, M. (1972). *Calcul Numérique des Ecoulements des Fluides Newtoniens Incompressibles par la Méthode des Elèments Finis*, Doctoral Dissertation (Thése d'Etat) (University Pierre & Marie Curie, Paris, France).

FORTIN, M., GLOWINSKI, R. (1982). *Méthodes de Lagrangiens Augmentés: Application à la Résolution Numérique des Problèmes aux Limites* (Dunod, Paris).

FORTIN, M., GLOWINSKI, R. (1983). *Augmented Lagrangian Methods: Application to the Numerical Solution of Boundary Value Problems* (North-Holland, Amsterdam).

FRIGAARD, I.A., HOWISON, S.D., SOBEY, I.J. (1994). On the stability of Poiseuille flow of a Bingham fluid. *J. Fluid Mech.* **263**, 133–150.

FRIGAARD, I.A., NOUAR, C. (2005). On the usage of viscosity regularization methods for viscoplastic fluid flow computation. *J. NonNewton. Fluid Mech.* **127** (1), 1–26.

FRIGAARD, I.A., SCHERZER, O. (1998). Uniaxial exchange flow of two Bingham fluids in a cylindrical duct. *IMA J. Appl. Math.* **61**, 237–266.

FRIGAARD, I.A., SCHERZER, O. (2000). The effects of yield stress variation on uniaxial exchange flow of two Bingham fluids in a pipe. *SIAM J. Appl. Math.* **60**, 1950–1976.

GERMAIN, P. (1973). *Mécanique des Milieux Continus* (Masson, Paris).

GLOWINSKI, R. (1974). Sur l'écoulement d'un fluide de Bingham dans une conduite cylindrique. *J. de Mécanique* **13** (4), 601–621.

GLOWINSKI, R. (1984). *Numerical Methods for Nonlinear Variational Problems* (Springer-Verlag, New York, NY).

GLOWINSKI, R. (2003). Finite element method for incompressible viscous flow. In: Ciarlet, P.G., Lions, J.L. (eds.), *Handbook of Numerical Analysis*, Volume IX (North-Holland, Amsterdam), pp. 3–1176.

GLOWINSKI, R. (2008). *Numerical Methods for Nonlinear Variational Problems* (Springer-Verlag, Berlin).

GLOWINSKI, R., GUIDOBONI, G., PAN, T.W. (2006). Wall-driven incompressible viscous flow in a two-dimensional semi-circular cavity. *J. Comput. Phys.* **216** (1), 76–91.

GLOWINSKI, R., HOLMSTRÖM, M. (1995). Constrained motion problems with applications by nonlinear programming methods. *Surv. Math. Ind.* **5**, 75–108.

GLOWINSKI, R., KUZNETSOV, Y.A., PAN, T.W. (2003). A penalty/Newton/conjugate gradient method for the solution of obstacle problems. *C. R. Acad. Sci. Paris Sér. I* **336** (5), 435–440.

GLOWINSKI, R., LE TALLEC, P. (1989). *Augmented Lagrangian and Operator-Splitting Methods in Nonlinear Mechanics* (SIAM, Philadelphia, PA).

GLOWINSKI, R., LIONS, J.L., TRÉMOLIÈRES, R. (1976). *Analyse Numérique des Inéquations Variationnelles* **Volumes I & II** (Dunod, Paris).

Glowinski, R., Lions, J.L., Trémolières, R. (1981). *Numerical Analysis of Variational Inequalities* (North-Holland, Amsterdam).

Glowinski, R., Marrocco, A. (1974). Sur l'approximation par éléments finis d'ordre 1 et la résolution par pénalisation-dualité d'une classe de problèmes de Dirichlet non-linéaires. *C. R. Acad. Sci. Paris* **278 A**, 1649–1652.

Glowinski, R., Marrocco, A. (1975). Sur l'approximation par éléments finis d'ordre 1 et la résolution par pénalisation-dualité d'une classe de problèmes de Dirichlet non-linéaires. *Revue Française d'Automatique, Informatique et Recherche Opérationnelle-Analyse Numérique* **R-2**, 41–76.

Glowinski, R., Pan, T.W., Hesla, T.I., Joseph, D.D. (1999). A distributed Lagrange multiplier/fictitious domain method for particulate flow. *Int. J. Multiphase Flow* **25**, 755–794.

Glowinski, R., Pan, T.W., Hesla, T.I., Joseph, D.D., Periaux, J. (2001). A fictitious domain approach to the direct numerical simulation of incompressible viscous flow past moving rigid bodies: application to particulate flow. *J. Comput. Phys.* **169**, 363–426.

Glowinski, R., Shiau, L.J., Kuo, Y.M., Nasser, G. (2006). On the numerical simulation of friction constrained motions. *Nonlinearity* **19** (1), 195–216.

Golay, F., Helluy, P. (1998). Numerical simulation of a viscous compressible fluid based on a splitting method, Technical report (Université de Toulon et du Var, La Valette, France).

Guillot, D., Hendriks, H., Callet, F., Vidick, B. (1990). Mud removal. In: Nelson, E.B. (ed.), *Well Cementing*, Chapter 5 (Schlumberger Educational Services, Sugarland, Texas).

Guyon, E., Hulin, J.P., Petit, L. (2001). *Hydrodynamique Physique*, Second Edition (Editions du CNRS, Paris).

Hao, J., Pan, T.W., Glowinski, R., Joseph, D.D. (2009). A fictitious domain/distributed Lagrange multiplier method for the particulate flow of Oldroyd-B fluids: a positive definiteness preserving approach. *J. NonNewton. Fluid Mech.* **156**, 95–111.

He, J.W., Glowinski, R. (2000). Steady Bingham fluid flow in cylindrical pipes: a time dependent approach to the iterative solution. *Numer. Lin. Algebra Appl.* **7** (6), 381–428.

He, Y.B., Laskowski, J.S., Klein, B. (2001). Particle movement in non-Newtonian slurries: the effect of yield stress on dense medium separation. *Chem. Eng. Sci.* **56**, 2991–2998.

Hénaut, I. (2002). *Etude des Bruts Paraffiniques*, Technical report (Institut Français du Pétrole, Reuil-Malmaison, France).

Hénaut, I., Brucy, F. (2001). Description rhéologique des bruts paraffiniques gélifiés. *Congrès 2001 du Groupe Français de Rhéologie*.

Hild, P., Ionescu, I.R., Lachand-Robert, T., Rosca, I. (2002). The blocking of an inhomogeneous Bingham fluid: application to landslides. *Math. Model. Num. Anal.* **36** (6), 1013–1026.

Hoppe, R.H.W., Litvinov, W.G. (2004). Problems on electro-rheological fluid flows. *Commun. Pure Appl. Anal.* **3** (4), 809–848.

Houska, M. (1981). *Engineering Aspects of the Rheology of Thixotropic Liquids*, Ph.D. Dissertation (Faculty of Mechanical Engineering, Czech Technical University of Prague, Czech Republic).

Hu, H.H., Patankar, N.A., Zhu, M.Y. (2001). Direct numerical simulation of fluid-solid systems using the arbitrary Lagrangian-Eulerian technique. *J. Comput. Phys.* **169** (2), 427–462.

Huilgol, R.R., You, Z. (2005). Application of the augmented Lagrangian method to steady pipe flow of Bingham, Casson and Herschel-Bulkley fluids. *J. NonNewton. Fluid Mech.* **128** (2–3), 126–143.

Hussain, Q.E., Sharif, M.A.R. (2000). Numerical modeling of helical flow of viscoplastic fluids in eccentric annuli. *AIChE J.* **46** (10), 1937–1946.

Hwang, W.R., Hulsen, M.A., Meijer, H.E.H. (2004). Direct simulation of particle suspensions in a viscoelastic fluid in sliding bi-periodic frames. *J. NonNewton. Fluid Mech.* **121** (1), 15–33.

Hyman, M.A. (1952). Non-iterative solution of boundary value problems. *Appl. Sci. Res. Sec. B* **2** (1), 325–351.

Ionescu, I.R., Sofonea, M. (1986). The blocking property in the study of the Bingham fluid. *Int. J. Eng. Sci.* **24**, 289–297.

Keshtiban, I.J., Belblidia, P., Webster, M.F. (2004). Numerical simulation of compressible viscoelastic liquids. *J. NonNewton. Fluid Mech.* **122**, 131–146.

KESHTIBAN, I.J., BELBLIDIA, P., WEBSTER, M.F. (2005). Computation of incompressible and weakly-compressible viscoelastic liquids: finite-element/finite volume schemes. *J. NonNewton. Fluid Mech.* **126**, 123–143.

KŘÍZEK, M., NEITTAANMÄKI, P., GLOWINSKI, R., KOROTOV, S. (eds.) (2004). *Conjugate Gradient Algorithms & Finite Element Methods* (Springer-Verlag, Berlin).

LADD, A.J.C., VERBERG, R. (2001). Lattice-Boltzmann simulations o particle-fluid suspensions. *J. Stat. Phys.* **104** (5–6), 1191–1251.

LEVEQUE, R.J. (2002). *Finite-Volume Methods for Hyperbolic Problems* (Cambridge University Press, Cambridge, UK).

LIONS, J.L., STAMPACCHIA, G. (1967). Variational inequalities. *Comm. Pure Appl. Math.* **20**, 493–519.

LITVINOV, W.G., HOPPE, R.H.W. (2005). Coupled problems on stationary non-isothermal flow of electro-rheological fluids. *Comm. Pure Appl. Anal.* **4**, 779–803.

LIU, B.T., MULLER, S.J., DENN, M.M. (2002). Convergence of a regularization method for creeping flow of a Bingham material about a rigid sphere. *J. NonNewton. Fluid Mech.* **102** (2), 179–191.

MARION, M., TEMAM, R. (1998). Navier-Stokes equations. In: Ciarlet, P.G., Lions, J.L. (eds.), *Handbook of Numerical Analysis*, Volume VI (North-Holland, Amsterdam), pp. 503–689.

MITSOULIS, E. (2004). On creeping drag flow of a viscoplastic fluid past a circular cylinder: wall effects. *Chem. Eng. Sci.* **59** (4), 789–800.

MITSOULIS, E., ABDALI, S.S., MARKATOS, N.C. (1993). Flow simulation of Herschel-Bulkley fluids through extrusion dies. *Can. J. Chem. Eng.* **71** (1), 147–160.

MITSOULIS, E., HUILGOL, R.R. (2004). Entry flows of Bingham plastic in expansions. *J. NonNewton. Fluid Mech.* **122** (1–3), 45–54.

MITSOULIS, E., ZISIS, T. (2001). Flow of Bingham plastics in a lid-driven square cavity. *J. NonNewton. Fluid Mech.* **101**, 173–180.

MODI, M.V., KISWANTO, H., MERRILL, L.S. (1994). Engineering solutions to handle waxy Kakap KRA crude. In: *Proceedings of the Annual Convention of the Indonesian Petroleum Association* (Djakarta), pp. 259–275.

MOSSOLOV, P.P., MIASNIKOV, V.P. (1965). Variational methods in the theory of the fluidity of a viscous plastic medium. *J. Mech. Appl. Math.* **29** (3), 488–492.

MOYERS-GONZALEZ, FRIGAARD, I.A. (2004). Numerical solution of duct flows of multiple visco-plastic fluids. *J. NonNewton. Fluid Mech.* **122** (1–3), 227–241.

MYERS, R.H., MONTGOMERY, D.C. (2002). *Response Surface Methodology: Process and Product Optimization Using Designed Experiments* (John Wiley & Sons, New York, NY).

NOUAR, C., BENAOUDA-ZOUAOUI, B., DESAUBRY, C. (2000). Laminar mixed convection in a horizontal annular duct. Case of thermo-dependent non-Newtonian fluid. *Eur. J. Mech.-B/Fluids* **19** (3), 423–452.

NOUAR, C., DESAUBRY, C., ZENAIDI, H. (1998). Numerical and experimental investigation of thermal convection for a thermo-dependent Herschel-Bulkley fluid in an annular duct with rotating inner cylinder. *Eur. J. Mech.-B/Fluids* **17 (6)**, 875–900.

NOURI, J.M., UMUR, H., WHITELAW, J.H. (1993). Flow of Newtonian and non-Newtonian fluids in concentric and eccentric annuli. *J. Fluid Mech.* **253**, 617–641.

NOURI, J.M., WHITELAW, J.H. (1994). Flow of Newtonian and non-Newtonian fluids in a concentric annulus with rotation of the inner cylinder. *J. Fluid Eng.* **116**, 821–827.

OLDROYD, J.G. (1947). Two-dimensional plastic flow of a Bingham fluid. *Proc. Cambridge Philos. Soc.* **43**, 383–395.

PAN, T.W., GLOWINSKI, R., HOU, S. (2007). Direct numerical simulation of pattern formation in a rotating suspension of non-Brownian settling particles in a fully filled cylinder. *Comput. Struct.* **85**, 955–969.

PAPANASTASIOU, T.C. (1987). Flow of materials with yield. *J. Rheol.* **31**, 385–404.

PERKINS, T.K., TURNER, J.B. (1971). Starting behavior of gathering lines and pipelines filled with gelled Prudhoe Bay oil. *J. Pet. Technol.* **23** (3), 301–307.

PEYSSON, Y. (2004). Solid/liquid dispersion in drilling and production. *Oil & Gas Sci. Technol. - Rev. IFP* **59** (1), 11–21.

Piar, B., Michel, B.D., Babik, F., Latché, J.C., Guillard, G., Ruggieri, J.M. (1999). CROCO: a computer code for corium spreading. In: *Proceedings of the 9th International Topical Meeting on Nuclear Reactor Thermal Hydraulics (NURETH 9), San Francisco, 1999.*

Pironneau, O. (1989). *Finite Element Methods for Fluids* (J. Wiley, Chichester).

Prager, W. (1961). *Introduction to Mechanics of Continua* (Ginn & Company, Boston, MA).

Ronningsen, H.P. (1992). Rheological behavior of gelled, waxy North Sea crude oils. *J. Pet. Sci. Eng.* **7**, 177–213.

Roquet, N. (2000). *Résolution Numérique d'Ecoulements à Effet de Seuil par Elements Finis Mixtes et Adaptation de Maillage*, Ph.D. Dissertation (University Joseph Fourier, Grenoble, France).

Roquet, N., Saramito, P. (2003). An adaptive finite element method for Bingham fluid flows around a cylinder. *Comput. Methods Appl. Mech. Eng.* **192** (31–32), 3317–3341.

Roussel, N., Le Roy, R., Coussot, P. (2004). Thixotropy modeling at local and macroscopic scales. *J. NonNewton. Fluid Mech.* **114**, 85–95.

Samuelsson (1993). *Numerical Solution of Pipe Flow Problems for Generalized Newtonian Fluids*, Thesis 367 (Linköping University, Department of Mathematics, Linköping, Sweden).

Sanchez, F. (1998). Application of a first order operator-splitting method to Bingham fluid flow simulation. *Comput. Math. Appl.* **36** (3), 71–86.

Saramito, P., Roquet, N. (2001). An adaptive finite element method for viscoplastic fluid flows in pipes. *Comput. Methods Appl. Mech. Eng.* **190** (40–41), 5391–5412.

Sestak, J., Charles, M.E., Cawkwell, M.G., Houska, M. (1987). Start-up of gelled crude oil pipelines. *J. Pipelines* **6**, 15–24.

Silva, L., Coupez, T. (2002). A unified model of the filling and post-filling stages in 3D injection moulding simulations. In: *Proceedings of the* 18th *Annual Meeting of the Polymer Processing Society, PPS-18* (Guimarães, Portugal).

Singh, P., Joseph, D.D., Hesla, T.I., Glowinski, R., Pan, T.W. (2000). A distributed Lagrange multiplier/fictitious domain method for viscoelastic particulate flows. *J. NonNewton. Fluid Mech.* **91** (2–3), 165–188.

Smith, P.B., Ramsden, R.M.J. (1978). The prediction of oil gelation in submarine pipelines and the pressure required for restarting flow. In: *Proceedings of the SPE European Petroleum Conference, 24-27 October 1978* (European Offshore Petroleum Conference and Exhibition, London, UK), pp. 283–290.

Smith, T.R., Ravi, K.M. (1991). Investigation of drilling fluid properties to maximize cement displacement efficiency. In: *Proceedings of the SPE Annual Technical Conference and Exhibition, 6-9 October 1991*, Paper 22775-MS (Society of Petroleum Engineers, Dallas, TX).

Strauss, M.J. (1973). Variations of Korn's and Sobolev's inequalities. In: Spencer, D.C. (ed.), *Partial Differential Equations* (American Mathematical Society, Providence, RI), pp. 207–214.

Szabo, P., Hassager, O. (1992). Flow of viscoplastic fluids in eccentric annular geometries. *J. NonNewton. Fluid Mech.* **45** (2), 149–169.

Talenti, G. (1976). Best constant in Sobolev inequality. *Annali di Matematica Pura ed Applicata* **116** (1), 353–372.

Tartar, L. (2007). *An Introduction to Sobolev Spaces and Interpolation Spaces* (Springer-Verlag, Berlin).

Uhde, A., Kopp, G. (1971). Pipelines problems resulting from the handling of waxy crude oils. *J. Inst. Pet.* **57**, 63–73.

Verschuur, E., Verheul, M., Den Hartog, A.P. (1971). Pilot-scale studies on restarting pipelines containing crude oil flows. *J. Inst. Pet.* **57**, 139–146.

Vinay, G. (2005). *Modélisation du Redémarrage des Ecoulements de Bruts Paraffiniques dans les Conduites Pétrolières*, Ph.D. Dissertation (Ecoles des Mines de Paris/Institut Français du Pétrole, Reuil-Malmaison, France).

Vinay, G., Wachs, A., Agassant, J.F. (2005). Numerical simulation of non-isothermal viscoplastic waxy crude oil flows. *J. NonNewton. Fluid Mech.* **128** (2–3), 144–162.

Vinay, G., Wachs, A., Agassant, J.F. (2006). Numerical simulation of weakly compressible Bingham flows: the restart of pipeline flows of waxy crude oils. *J. NonNewton. Fluid Mech.* **136** (2–3), 93–105.

Vinay, G., Wachs, A., Frigaard, I. (2007). Start-up transients and efficient computation of isothermal waxi crude oil flows. *J. NonNewton. Fluid Mech.* **143** (2–3), 141–156.

VINCENT, S. (1999). *Modélisation d'Ecoulements Incompressibles*, Ph.D. Dissertation (University of Bordeaux 1, Bordeaux, France).

VOLA, D., BABIK, F., LATCHÉ, J.C. (2004). On a numerical strategy to compute gravity currents of non-Newtonian fluids. *J. Comput. Phys.* **201**, 397–420.

VOLA, D., BOSCARDIN, L., LATCHÉ, J.C. (2003). Laminar unsteady flows of Bingham fluids: a numerical strategy and some benchmark results. *J. Comput. Phys.* **187**, 441–456.

WACHS, A. (2000). *Modèles Thermologiques et Simulations Numériques bi et tri-dimensionnelles d'écoulements non-isothermes de fluides viscoélastiques*, Ph.D. Dissertation (Institut National Polytechnique de Grenoble, Grenoble, France).

WACHS, A. (2007). Numerical simulation of steady Bingham flow through an eccentric annular cross-section by distributed Lagrange multiplier/fictitious domain and augmented Lagrangian methods. *J. NonNewton. Fluid Mech.* **142** (1–3), 183–198.

WALTON, I.C., BITTLESTON, S.H. (1991). The axial flow of a Bingham fluid in a narrow eccentric annulus. *J. Fluid Mech.* **222**, 39–60.

WANG, Y., HUTTER, K. (2001). Comparisons of numerical methods with respect to convectively dominated problems. *Int. J. Numer. Methods Fluids* **37**, 721–745.

WARDAUGH, L.T., BOGER, D.V. (1987). Measurement of the unique flow properties of waxy crude oils. *Chem. Eng. Res. Des.* **65**, 73–83.

WARDAUGH, L.T., BOGER, (1991). Flow characteristics of waxy crude oils: application to pipeline design. *AIChE* **37** (6), 871–885.

WARDAUGH, L.T., BOGER, D.V., TONNER, S.P. (1988). Rheology of waxy crude oils. In: *Proceedings of the International Meeting on Petroleum Engineering, 1-4 November 1988, (Tianjin, China)*, Paper 17625, pp. 803–810.

XU, J. (2009). Optimal algorithms for discretized partial differential equations. In: Jeltsch, R., Wanner, G. (eds.), *ICIAM 07, 6th International Congress on Industrial and Applied Mathematics, Zürich, Switzerland, 16–20 July 2007: Invited Lectures* (European Mathematical Society, Zürich), pp. 409–444.

YEE, H.C., WARMING, R.F., HARTEN, A. (1985). Implicit total variation diminishing (TVD) schemes for steady-state calculations. *J. Comput. Phys.* **57**, 327–360.

YU, Z., PHAN-TIEN, N., FAN, Y., TANNER, R.I. (2002). Viscoelastic mobility problem of a system of particles. *J. NonNewton. Fluid Mech.* **104** (2–3), 87–124.

YU, Z., PHAN-TIEN, N., TANNER, R.I. (2004). Dynamic simulation of a sphere motion in a vertical tube. *J. Fluid Mech.* **518**, 61–93.

YU, Z., WACHS, A. (2007). A fictitious domain method for dynamic simulation of particle sedimentation in Bingham fluids. *J. NonNewton. Fluid Mech.* **145** (2–3), 78–91.

YU, Z., WACHS, A., PEYSSON, Y. (2006). Numerical simulation of particle sedimentation in shear-thinning fluids by a fictitious domain method. *J. NonNewton. Fluid Mech.* **136**, 126–139.

ZHANG, J., VOLA, D., FRIGAARD, I.A. (2006). Yield stress effects on Rayleigh-Bénard convection. *J. Fluid Mech.* **566**, 389–419.

ZISIS, T., MITSOULIS, E. (2002). Viscoplastic flow around a cylinder kept between parallel plates. *J. Non-Newton. Fluid Mech.* **105** (1), 1–20.

Modeling, Simulation and Optimization of Electrorheological Fluids

Ronald H.W. Hoppe
Department of Mathematics, University of Houston, Houston, TX 77204–3008, USA
Institute of Mathematics, Universität Augsburg, D-86159 Augsburg, Germany
E-mail: rohop@math.uh.edu

William G. Litvinov
Institute of Mathematics, Universität Augsburg, D-86159 Augsburg, Germany
E-mail: litvinov@math.uni-augsburg.de

1. Introduction

Electrorheological fluids are concentrated suspensions of electrically polarizable particles of small size in the range of micrometers in nonconducting or semiconducting liquids such as silicone oils. Under the influence of an outer electric field, the particles form chains along the field lines followed by a coalescence of the chains into columns in the plane orthogonal to the field due to short-ranged potentials arising from charge-density fluctuations. The formation of the chains is a process that happens in milliseconds, whereas the aggregation to columns occurs on a timescale that is larger by an order of magnitude. On a macroscopic scale, the chainlike and columnar structures have a significant impact on the rheological properties of the suspensions. In particular, the viscosity increases rapidly with increasing electric field strength in the direction perpendicular to the field. The fluid experiences a phase transition to a viscoplastic state, and the flow shows a pronounced anisotropic behavior. Under the influence of large stresses, the columns break into continuously fragmenting and aggregating volatile structures, which tilt away from strict field alignment. As a result, the viscosity decreases and the fluid flow behaves less anisotropic. The electrorheological effect is reversible, i.e., the viscosity decreases for decreasing electric field strength such that for vanishing field strength the fluid behaves again like a Newtonian one. The fast

Numerical Methods for Non-Newtonian Fluids
Special Volume (R. Glowinski and Jinchao Xu, Guest Editors) of
HANDBOOK OF NUMERICAL ANALYSIS, VOL. XVI
P.G. Ciarlet (Editor)

ISSN 1570-8659
DOI 10.1016/B978-0-444-53047-9.00007-1

response to an outer electric field and the reversibility of the effect make electrorheological fluids particularly attractive for all technical applications, which require a controllable power transmission.

Although the discovery of the electrorheological effect is credited to WINSLO [1947] (cf. also WINSLOW [1949, 1962]), it has already been observed experimentally by PRIESTLEY [1769] during the second half of the eighteenth century and by DUFF [1896], QUINKE [1897] at the end of the nineteenth century. However, Winslow was the first scientist who conducted quantitative experiments on suspensions of silica gel particles in oils of low viscosity. He reported fibration parallel to the electric field with a solid-like behavior of the suspension at field strengths larger than 3 kV/mm. In his experiments, he also observed that the yield stress, i.e., when the shear stress is proportional to the shear rate, is proportional to the square of the electric field strength.

Winslow's work did not immediately launch tremendous research activities in the area of electrorheological fluids. In fact, it took roughly 20 to 30 more years when the availability of modern, high-resolution measurement technology on one hand and more advanced and powerful computing facilities on the other hand enabled researchers to conduct detailed experimental studies and to perform extensive numerical simulations (see BLOCK and KELLY [1988], BLOCK, KELLY, QIN and WATSON [1990], BÖSE [1998], BÖSE and TRENDLER [2001], CLERCX and BOSSIS [1993], CONRAD, SPRECHER, CHOI and CHEN [1991], DEINEGA and VINOGRADOV [1984], GAST and ZUKOSKI [1989], HANAOKA, MURAKUMO, ANZAI and SAKURAI [2002], INOUE and MANIWA [1995], KHUSID and ACRIVOS [1995], KIMURA et al. [1998], KLASS and MARTINEK [1967a,b], KLINGENBERG, VAN SWOL and ZUKOSKI [1989], KLINGENBERG and ZUKOSKI [1990], LEMAIRE, GRASSELLI and BOSSIS [1992], MARSHALL, ZUKOSKI and GOODWIN [1989], MOKEEV, KOROBKO and VEDERNIKOVA [1992], RHEE, PARK, YAMANE and OSHIMA [2003], SHULMAN and NOSOV [1985], STANGROOM [1977, 1983], STANWAY, SPROSTON and STEVENS [1987], TAO and SUN [1991b], VOROBEVA, VLODAVETS and ZUBOV [1969], WEN, HUANG, YANG, LU and SHENG [2003], WHITTLE [1990], YU and WAN [2000], and ZHAO, GAO and GAO [2002]). The experimental work focused on the creation of the chainlike and columnar structures (see KLINGENBERG and ZUKOSKI [1990], MARTIN and ANDERSON [1996], MARTIN, ANDERSON and TIGGES [1998a], and QI and WEN [2002]) (cf. Fig. 1.1 (left)) up to the formation of sheets (cf. Fig. 1.1 (right)) and body-centered tetragonal crystal lattices (see DASSANAYAKE, FRADEN and VAN BLAADEREN [2000]) (cf. Fig. 1.2) as well as on the dynamics of the process (cf., e.g., ADOLF and GARINO [1995], FOULC, ATTEN and BOSSIS [1996], KLINGENBERG [1998], KLINGENBERG and ZUKOSKI [1990], KLINGENBERG, ULICNY and SMITH [2005], MARTIN, ODINEK, HALSEY and KAMIEN [1998b], VON PFEIL, GRAHAM, KLINGENBERG and MORRIS [2002], TAM et al. [1997], UGAZ, MAJORS and MIKSAD [1994], WHITTLE, ATKIN and BULLOUGH [1999], and ZHAO and GAO [2001]). The measurements have been performed using, e.g., confocal scanning laser microscopy (DASSANAYAKE, FRADEN and VAN BLAADEREN [2000]), two-dimensional light scattering techniques (MARTIN, ODINEK, HALSEY and KAMIEN [1998b]), and nuclear magnetic resonance imaging (UGAZ, MAJORS and MIKSAD [1994]).

The potential industrial applicability of electrorheological fluids in automotive applications (BAYER and CARL SCHENCK [1998], BUTZ and STRYK [2001], COULTER, WEISS and CARLSON [1993], FILISKO [1995], GARG and ANDERSON [2003], GAVIN [2001], GAVIN, HANSON and FILISKO [1996a,b], HARTSOCK, NOVAK and CHAUNDY [1991], HOPPE, MAZUKEVICH, VON STRYK and RETTIG [2000], JANOCHA, RECH and BOELTER [1996], LORD [1996], PEEL, STANWAY and BULLOUGH [1996], SIMS, STANWAY, PEEL, BULLOUGH and JOHNSON [1999], STANWAY, SPROSTON and EL-WAHED [1996], WEYENBERG, PIALET and PETEK

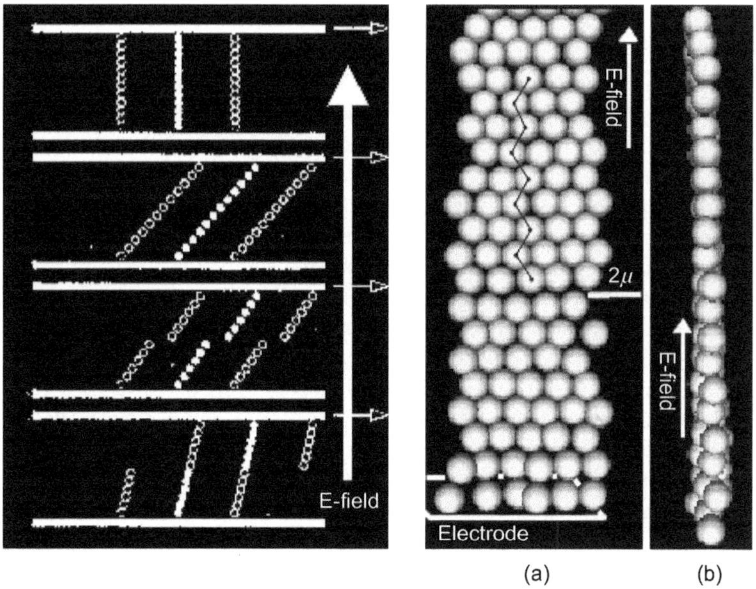

FIG. 1.1 Formation of chains aligned with the field (left) and aggregation to sheets (right).

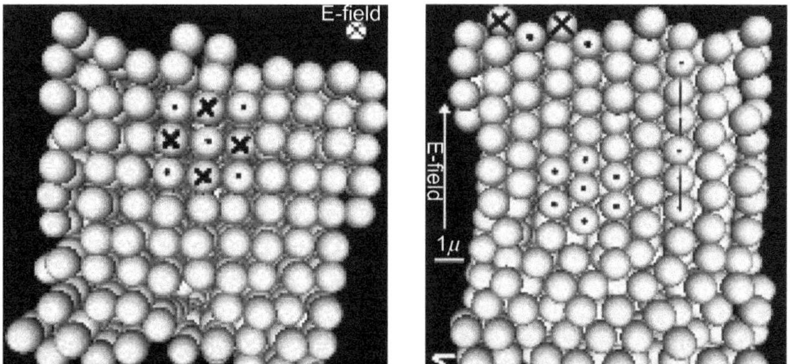

FIG. 1.2 Body-centered tetragonal crystal lattice in the xy-plane (left) and the xz-plane (right).

[1996], and ZHAO, LIU, TANG, YIN and LUO [2005]), aerospace applications (BERG and WELLSTEAD [1998], LOU, ERVIN and FILISKO [2001], and WERELEY, SNYDER, KRISHNAN and SIEG [2001]), food processing (DAUBERT, STEFFE and SRIVASTAVA [1998]), geophysics (MAKRIS [1999], and XU, QU and KO [2000]), life sciences (KLEIN et al. [2004], LIU, DAVIDSON and TAYLOR [2005], MAVROIDIS, BAR-COHEN and BOUZIT [2001], MONKMANN et al. [2003a,b], and TAKASHIMA and SCHWAN [1985]), manufacturing (KIM, LEE and KIM [2003]), military applications (DEFENSE UPDATE [2004]), and nondestructive testing (MAVROIDIS

[2002]) caused the US Department of Energy to issue a research assessment of electrorheological fluids (DOE [1993]) and popular scientific journals such as *Science and Nature* to publish overview articles (HALSEY [1992], and WHITTLE and BULLOUGH [1992]). Further references covering various aspects of experimental work, modeling efforts, and applications of electrorheological fluids can be found in BOSSIS [2002], HAO [2001], and TAO and ROY [1995].

The experimental work was always accompanied by the development of physically consistent, mathematical models, their analysis, numerical simulations, and model validations on the basis of available data from measurements and simulations. Roughly speaking, one has to distinguish between microstructural models, which combine electrostatics (see JONES [1995]), microhydrodynamics (cf. KIM and KARRILA [1991]), and liquid-state theory (see CACCANO, STEEL and HANSEN [1999]; cf. also LARSON [1999], and LUKASZEWICZ [1999]), and macroscopic models based on continuum field theories (cf., e.g., RAJAGOPAL and TRUESDELL [2000], TRUESDELL and NOLL [1965], and TRUESDELL and TOUPIN [1960]).

The simplest microscale models assume the electrorheological fluids to consist of monodisperse, neutrally buoyant hard dielectric spheres dispersed in a Newtonian continuous phase thus neglecting small conductivities in both phases, ionic impurities in the continuous phase, and triboelectric effects. Idealized electrostatic polarization methods obtain the electrostatic potential via Laplace's equation and compute the motion of the particles by Newton's equation, which requires the proper specification of the total force exerted on a particle by taking into account the interparticle forces. Since the exact solution is unavailable and the computation of all possible interparticle forces is cumbersome, the system is simplified by the point-dipole approximation (see JONES [1995], KIM and KLINGENBERG [1997], PARTHASARATHY and KLINGENBERG [1996], and VON PFEIL and KLINGENBERG [2004]) assuming that two spheres of the same size do not change their charge distributions. The resulting force equation only depends on the distance of the particles, the angle between them, the particle size, and the properties of the induced electric field. The results of the model differ by an order of magnitude from experimentally available data, since the dipole moment of the particles enhances the polarization. This has been accounted for in PARTHASARATHY and KLINGENBERG [1996] by a modified point-dipole approximation and by providing multipole models (see CONRAD, SPRECHER, CHOI and CHEN [1991], and CLERCX and BOSSIS [1993]), which are based on several electric field equations (up to four), whereas the particle interaction is performed for an N particle cluster allowing the consideration of particles in lattice structures such as body-centered tetragonal crystal lattices. The dipole-induced dipole model in YU and WAN [2000] represents a further development of the multipole models in so far as it admits spheres of different sizes. Maxwell–Wagner polarization due to accumulated charges between the interface of the particles and the continuous phase has been incorporated in PARTHASARATHY and KLINGENBERG [1996] by assuming a point dipole model for this interfacial polarization. The Maxwell–Wagner model in KHUSID and ACRIVOS [1995] further acknowledges effects of the disturbance field between particles. Microstructural models based on energy-type methods have been derived in BONNECAZE and BRADY [1992a,b]. They take into account hydrodynamic and electrostatic particle interactions using Stokesian dynamics and a model for the electrostatic energy. The latter one is determined from the capacitance matrix of the suspension. The models allow simulations of monolayers of particles for a wide range of the ratio of viscous to electrostatic forces as described by the Mason number. The macroscopic rheology can be deduced from the simulations. In accordance with experimental results, it shows that for large electric field strengths, there is a pronounced Bingham-type behavior of the suspension with a dynamic yield stress that can be related to jumps in the electrostatic energy. Numerical simulations based on microscale models are typically of molecular dynamics type (cf., e.g., HU and CHEN [1998], MELROSE [1992], MELROSE and HAYES [1993], TAO and SUN [1991a], and ZHAO and GAO [2001]) using methodologies from ALLEN and TILDESLEY [1983].

The microstructural features of electrorheological fluids have been used to derive models for a description of the macroscopic properties (cf. e.g., KLINGENBERG [1993], PARTHASARATHY, AHN, BELONGIA and KLINGENBERG [1994], PARTHASARATHY and KLINGENBERG [1995a,b, 1999], SEE [1999, 2000], VERNESCU [2002], VON PFEIL, GRAHAM, KLINGENBERG and MORRIS [2003], and WANG and XIAO [2003]). On the other hand, macroscopic models have been obtained by phenomenological approaches within the framework of mixture theory (see RAJAGOPAL [1996] and RAJAGOPAL, YALAMANCHILI and WINEMAN [1994]) and classical continuum mechanics (we refer to ATKIN, SHI and BULLOUGH [1991] as one of the first attempts in this direction (cf. also ATKIN, SHI and BULLOUGH [1999])). Since electrorheological fluids exhibit a non-Newtonian flow behavior, significant efforts have been devoted to the derivation of appropriate constitutive equations relating the stress tensor to the rate of deformation tensor by taking into account the influence of the electric field. We mention the pioneering work by RAJAGOPAL and WINEMAN [1992, 1995] (see also ENGELMANN, HIPTMAIR, HOPPE and MAZURKEVICH [2000]) and the systematic treatment by RUZICKA [2000] providing a constitutive equation of power law type (see also BUSUIOC and CIORANESCU [2003], ECKART [2000], and RAJAGOPAL and RUZICKA [2001]). Other continuum-based approaches try to incorporate microscale and mesoscale effects by using internal variables (DROUOT, NAPOLI and RACINEUX [2002]), transverse isotropy (BRUNN and ABU-JDAYIL [1998, 2004]), polar theory (ECKART and SADIKI [2001]), and more general rate-type models (SADIKI and BALAN [2003]). In this contribution, we will adopt the constitutive laws that have been suggested, analyzed, and validated in HOPPE and LITVINOV [2004] and LITVINOV and HOPPE [2005] for isothermal and nonisothermal electrorheological fluid flows, which take into account the orientation of the velocity field of the flow with respect to the outer electric field.

The content of this chapter is as follows: In Section 2, we are concerned with the balance equations and constitutive laws for isothermal and nonisothermal electrorheological fluid flows and with the existence and/or uniqueness of solutions. In Section 3, we deal with numerical methods both for steady and time-dependent fluid flows. Finally, in Section 4, we present numerical simulation results for some selected electrorheological devices and also briefly address optimal design issues.

2. Mathematical models for electrorheological fluid flows

In this section, we study balance equations and constitutive laws for isothermal and non-isothermal electrorheological fluid flows. After a general presentation in Section 2.1, in Section 2.2, we consider stationary isothermal fluid flows based on the extended Bingham-type models from HOPPE and LITVINOV [2004]. In particular, we shall be concerned with the existence and/or uniqueness results for a regularized version in 2.2.1 and for the nonregularized model in 2.2.2. In Section 2.3, we deal with time-dependent problems, whereas Section 2.4 and 2.5 are devoted to the derivation of model equations for nonisothermal fluid flows and the discussion of the existence of solutions following the approach in LITVINOV and HOPPE [2005]. We refer to DUVAUT and LIONS [1976], GALDI [1994], GLOWINSKI [2004], LADYZHENSKAYA [1969], TEMAM [1979] for general aspects related to the mathematical modeling, the analysis, and the numerical solution of fluid mechanical problems and to LITVINOV [2000] for a general treatment of optimization problems for nonlinear viscous fluids.

2.1. Balance equations and constitutive laws for isothermal fluid flows

We consider isothermal incompressible electrorheological fluid flows in $Q := \Omega \times (0, T)$, $T \in \mathbb{R}_+$, where Ω is supposed to be a bounded Lipschitz domain in $\mathbb{R}^d$, $d = 2$ or $d = 3$. We denote by $u(x,t) = (u_1(x,t), \ldots, u_d(x,t))^T$, $(x,t) \in \bar{Q}$, and $p(x,t)$, $(x,t) \in Q$, the velocity of the fluid and the pressure, whereas $E(x,t) = (E_1(x,t), \ldots, E_d(x,t))^T$, $(x,t) \in \bar{Q}$, stands for the electric field. We use the notation $u_t := \partial u/\partial t$ for the partial derivative of u with respect to time. Then, referring to $\rho \in \mathbb{R}_+$ as the density of the fluid, to $f : Q \to \mathbb{R}^d$ as a forcing term, and to $\sigma = \sigma(u, p, E)$ as the stress tensor, the balance equations (conservation of mass and momentum) are given by

$$\rho\,(u_t + (u \cdot \nabla)u) - \nabla \cdot \sigma = f \quad \text{in } Q, \tag{2.1a}$$

$$\nabla \cdot u = 0 \quad \text{in } Q, \tag{2.1b}$$

which have to be complemented by properly specified initial and boundary conditions and a constitutive law relating the stress tensor σ to the independent variables u, p, and E.

Neglecting magnetic fields, the electric field can be considered quasi-static so that for each $t \in [0, T]$, the field $E(\cdot, t)$ can be computed by $E(\cdot, t) = -\nabla\psi(\cdot, t)$ as the gradient of an electric potential $\psi(\cdot, t)$ satisfying Laplace's equation

$$\nabla \cdot (\epsilon \nabla \psi(\cdot, t)) = 0 \quad \text{in } \Omega, \tag{2.2}$$

which also has to be complemented by appropriate boundary conditions. Here, ϵ stands for the dielectric permittivity.

For the discussion of the constitutive law, we further denote by

$$\varepsilon(u) = \frac{1}{2}\left(\nabla u + (\nabla u)^T\right) \tag{2.3}$$

the rate of deformation tensor (linearized strain tensor) and by

$$I(u) = \|\varepsilon(u)\|_F^2 \tag{2.4}$$

the second invariant of the rate of deformation tensor, where $\|\cdot\|_F$ stands for the Frobenius norm. For shear flows, we refer to $\tau = \tau(u, E)$ as the shear stress, which is a field-dependent function of the shear rate

$$\gamma = (2^{-1} I(u))^{1/2}. \tag{2.5}$$

In case of flow modes such as Couette flow or Poiseuille flow, where the electric field is perpendicular to the fluid velocity, constitutive equations of the form

$$\sigma = -pI + 2\varphi(I(u), |E|)\,\varepsilon(u) \tag{2.6}$$

have been widely used. Here, $\varphi : \mathbb{R}_+ \times \mathbb{R}_+ \to \mathbb{R}$ stands for a viscosity function depending on the second invariant of the rate of deformation tensor and the electric field strength.

The most commonly used constitutive law for simple flow modes is that of a Bingham-type fluid ATKIN, SHI and BULLOUGH [1991], FILISKO [1995], PARTHASARATHY and KLINGENBERG [1996], RHEE, PARK, YAMANE and OSHIMA [2003], STANWAY, SPROSTON and EL-WAHED [1996], and WHITTLE, ATKIN and BULLOUGH [1995]. For stresses above, a field-dependent yield stress $\sigma_Y(E)$ the viscosity function φ is given by

$$\varphi(I(u), |E|) = \eta_0(E) + 2^{-1/2}\, \tau_0(E)\, I(u)^{-1/2}, \tag{2.7}$$

whereas $I(u) = 0$ for $|\sigma| \le \sigma_Y(E)$. Here, $\eta_0(E)$ is a field-dependent constant, and $\tau_0(E)$ denotes the shear stress for vanishing shear rate γ.

A related model, which can be viewed as some extensions of the Bingham fluid model, is that of CASSON [1959]. For $|\sigma| > \sigma_Y(E)$, the viscosity function

$$\begin{aligned}\varphi(I(u), |E|) &= \eta_0(E) + 2^{-1/2}\, \tau_0(E)\, I(u)^{-1/2} \\ &\quad + 2^{3/4}\, (\eta_0(E)\, \tau_0(E))^{1/2}\, I(u)^{-1/4}\end{aligned} \tag{2.8}$$

is used, whereas again $I(u) = 0$ for $|\sigma| \le \sigma_Y(E)$.

The singular character of the viscosity function φ in the Bingham and Casson fluid models requires to formulate the equations of motion (2.1a), (2.1b) as variational inequalities. A possible way to circumvent the difficulties associated with the nonmooth behavior of the viscosity function is by regularization, which in case of a Bingham model gives rise to

$$\varphi(I(u), |E|) = \eta_0(E) + 2^{-1/2}\, \tau_0(E)\, (\kappa + I(u))^{-1/2}. \tag{2.9}$$

Here, κ stands for a positive regularization parameter. For the Casson model (2.8), one may use an analogous regularization. The implications of using the classical models and the regularized models will be discussed in a more general context later in this section.

Other frequently used constitutive equations for non-Newtonian fluids assume a power law behavior (SIGNIER, DE KEE and CHHABRA [1999]). For electrorheological fluids, this leads to a viscosity function φ of the form

$$\varphi(I(u), |E|) = \begin{cases} m(E)\, \gamma_0^{n(E)-1}, & \gamma \le \gamma_0(E) \\ m(E)\, \gamma^{n(E)-1}, & \gamma > \gamma_0(E) \end{cases}, \tag{2.10}$$

where $m(E), n(E)$ are field-dependent material parameters and $\gamma_0(E)$ stands for a field-dependent shear rate. Regularizations of the power law model can be used as well. In this case, the viscosity function (2.10) is replaced by

$$\varphi(I(u), |E|) = m(E)\, (\kappa + \gamma^2)^{(n(E)-1)/2}, \kappa > 0. \tag{2.11}$$

We note that in case of steady shear flows in axially symmetric geometrical configurations, the use of the previously mentioned models in the equations of motion (2.1a), (2.1b) leads to scalar nonlinear equations that can be solved semianalytically. However, a serious drawback of the models is that the electric field strength $|E|$ occurs as a parameter in the constitutive laws thus assuming a homogeneous distribution of the electric field.

This assumption is justified for simple flows in geometrical settings, where the flow occurs between conventionally shaped electrodes at small distance from each other (cf. Sections 4.1 and 4.2), but due to experimental evidence, it does not hold true for more general configurations (cf. e.g., ABU-JDAYIL [1996], ABU-JDAYIL and BRUNN [1995, 1996, 1997, 2002], EDAMURA and OTSUBO [2004], GEORGIADES and OYADIJI [2003], OTSUBO [1997], and OTSUBO and EDAMURA [1998], and OTSUBO and EDAMURA [1999]).

One of the first systematic approaches towards a general phenomenological model based on continuum field theories has been undertaken by RAJAGOPAL and WINEMAN [1992] (cf. also RAJAGOPAL and WINEMAN [1995]), where the constitutive law is assumed to be of the form

$$\begin{aligned} \sigma =& - pI + \alpha_2 \, E \otimes E + \alpha_3 \, \varepsilon(u) + \alpha_4 \, \varepsilon^2(u) \\ &+ \alpha_5 \, (\varepsilon(u) E \otimes E + E \otimes \varepsilon(u) E) \; + \alpha_6 \, (\varepsilon^2 E \otimes E + E \otimes \varepsilon(^2 u) E). \end{aligned} \tag{2.12}$$

Here, $\otimes$ denotes the tensor product, and $\alpha_i = \alpha_i(I_1, \ldots, I_6)$, $2 \leq i \leq 6$, are scalar functions of the six invariants

$$\begin{aligned} &I_1 := \operatorname{tr}(EE^T), I_2 := \operatorname{tr}(\varepsilon(u)), I_3 := \operatorname{tr}(\varepsilon^2(u)), I_4 := \operatorname{tr}(\varepsilon^3(u)), \\ &I_5 := \operatorname{tr}(\varepsilon(u) E \otimes E), I_6 := \operatorname{tr}(\varepsilon^2(u) E \otimes E), \end{aligned}$$

where tr stands for the trace of a matrix.

Motivated by RAJAGOPAL and WINEMAN [1992, 1995], an extended Bingham-type fluid model

$$\sigma = -pI + \eta_0 \varepsilon(u) + \gamma |\varepsilon(u) E|^{-1} |E| \, (\varepsilon(u) E \otimes E + E \otimes \varepsilon(u) E) \tag{2.13}$$

has been used in ENGELMANN, HIPTMAIR, HOPPE and MAZURKEVICH [2000], HOPPE and MAZURKEVICH [2001], and HOPPE, MAZUKEVICH, VON STRYK and RETTIG [2000] in combination with a potential equation for the electric potential ψ ($E = -\nabla \psi$) to provide numerical simulations of steady electrorheological fluid flows.

In the spirit of RAJAGOPAL and WINEMAN [1992, 1995], RUZICKA [2000] has developed a model that takes into account the interaction between the electric field and the fluid flow (see also RAJAGOPAL and RUZICKA [1996, 2001]). The constitutive equation is of power law type

$$\begin{aligned} \sigma =& - pI + \gamma_1 \Big((1 + |\varepsilon(u)|^2)^{(r-1)/2} - 1 \Big) \; E \otimes E \\ &+ \Big(\gamma_2 + \gamma_3 |E|^2 \Big) \Big(1 + |\varepsilon(u)|^2 \Big)^{(r-2)/2} \varepsilon(u) \\ &+ \gamma_4 \Big(1 + |\varepsilon(u)|^2 \Big)^{(r-2)/2} (\varepsilon(u) E \otimes E + E \otimes \varepsilon E), \end{aligned} \tag{2.14}$$

where γ_i, $1 \leq i \leq 4$, are constants and $r : \mathbb{R}_+ \to \mathbb{R}_+$ is a smooth function of $|E|^2$ satisfying

$$1 \; < \; r_\infty \leq r(|E|^2) \leq r_0. \tag{2.15}$$

Here, r_0 and r_∞ are the constants

$$r_0 := \lim_{|E|^2 \to 0} r(|E|^2), \qquad r_\infty := \lim_{|E|^2 \to \infty} r(|E|^2).$$

As far as the electric field E is concerned, the quasi-static form of Maxwell's equations (ERINGEN and MAUGIN [1989], and LANDAU and LIFSHITZ [1984]) can be used such that E can be computed via the gradient of an electric potential satisfying an elliptic boundary value problem.

Due to the power law (2.14), the existence of weak solutions of the equations of motion (2.1a), (2.1b) both in the case of steady and time-dependent flows has to be studied within the framework of generalized Lebesgue and generalized Sobolev spaces (for related work see also FREHSE, MALEK and STEINHAUER [1997], LITVINOV [1982], MALEK, NECAS and RUZICKA [1996], MALEK and RAJAGOPAL [2007], and MALEK, RAJAGOPAL and RUZICKA [1995]).

A further development of Ruzicka's approach by means of an extended Casson model has been studied in ECKART [2000].

Motivated by experimental evidence (CECCIO and WINEMAN [1994], and SHULMAN and NOSOV [1985]), in HOPPE and LITVINOV [2004], a constitutive law

$$\sigma = -pI + 2\varphi(I(u), |E|, \mu(u, E))\varepsilon(u), \tag{2.16}$$

has been suggested where the viscosity function $\varphi : \mathbb{R}_+ \times \mathbb{R}_+ \times [0, 1] \to \mathbb{R}$ additionally depends on the orientation of the electric field E with respect to the velocity u of the fluid flow as described by a function $\mu : S_1^d \times S_1^d \to [0, 1]$ with S_1^d denoting the d-dimensional unit sphere. We refer to $\hat{u}$ as the velocity of the electrode. Then, for $u - \hat{u} \neq 0$ and $E \neq 0$, the function $\mu : S_1^d \times S_1^d \to [0, 1]$ is defined according to

$$\mu(u, E) := \frac{u - \hat{u}}{|u - \hat{u}|} \cdot \frac{E}{|E|}, \tag{2.17}$$

where $\cdot$ stands for the Euclidean inner product in $\mathbb{R}^d$. The function $\mu = \mu(u, E)$ is an invariant, which is independent of the choice of the reference frame and the motion of the frame with respect to the electrode. For a further discussion, we refer to HOPPE and LITVINOV [2004].

For specific electrorheological fluids, the viscosity function φ has to be determined based on experimental data for the relationship $\tau = \tau(\gamma)$ between the shear stress τ and the shear rate γ. For various electric field strengths, these data are usually available at discrete points $\gamma_i \in [\gamma_{\min}, \gamma_{\max}]$, $0 \le i \le N$, with $0 < \gamma_{\min} < \gamma_{\max} < \infty$ (cf. Table 2.1). Complete cubic spline interpolands are then used for the construction of flow curves in $[\gamma_{\min}, \gamma_{\max}]$ (cf. Fig. 2.3), and the flow curves are continuously extended to $(\gamma_{\max}, \infty)$ by straight lines $\tau(\gamma) = a_1 + a_2\gamma$ with coefficients a_i, $1 \le i \le 2$, depending on $|E|$ and $\mu(u, E)$. The extension to $[0, \gamma_{\min})$ can be done such that either $\tau(0) = \tau_0 \neq 0$ or $\tau(0) = 0$. In the former case, the viscosity function takes the form

$$\varphi(I(u), |E|, \mu(u, E)) = b(|E|, \mu(u, E))I(u)^{-1/2} + c(I(u), |E|, \mu(u, e)), \tag{2.18}$$

where $b(|E|, \mu(u, E)) = 2^{-1/2}\tau_0$ and $c : \mathbb{R}_+ \times \mathbb{R}_+ \times [0, 1] \to \mathbb{R}$ is a continuous function.

REMARK 2.1. The viscosity function φ as given by (2.18) represents an extended Bingham-type fluid model (cf. (2.7). Due to its singular behavior for $I(u) = 0$, the equations of motion (2.1a), (2.1b) have to be formulated as variational inequalities (see 2.2.2 below).

TABLE 2.1
Experimental data (shear stress – shear rate dependence) at various electric field strengths for the commercially available electrorheological fluid RHEOBAY TP AI 3565 (from BAYER [1997a]).

	Shear stress (Pa)				
Shear rate γ [per second]	0.0 V/mm	1.5 kV/mm	2.0 kV/mm	2.5 kV/mm	3.0 kV/mm
1.0×10^2	30.2	563.0	979.0	1360.0	1720.0
2.0×10^2	48.0	650.0	1070.0	1500.0	1900.0
4.0×10^2	69.3	695.0	1140.0	1600.0	2030.0
6.0×10^2	83.5	700.0	1170.0	1640.0	2070.0
8.0×10^2	100.0	712.0	1180.0	1670.0	2110.0
1.0×10^3	110.0	723.0	1200.0	1676.0	2140.0
1.2×10^3	115.0	727.0	1210.0	1686.0	2160.0
1.4×10^3	120.0	731.0	1220.0	1693.0	2180.0
1.6×10^3	225.0	735.0	1240.0	1696.0	2190.0
1.8×10^3	230.0	740.0	1250.0	1706.0	2200.0
2.0×10^3	235.0	743.0	1254.0	1710.0	2210.0

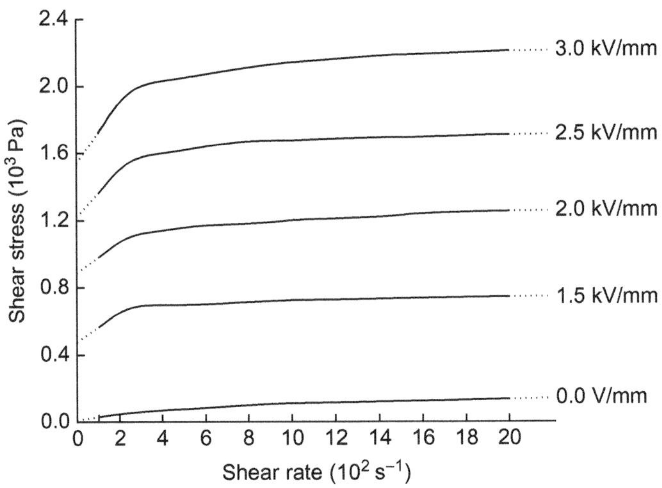

FIG. 2.3 Flow curves generated by cubic spline interpolands based on the experimental data from Table 2.1 showing the effect of the field strength (50 Hz, AC) and the shear rate γ on the shear stress τ at 40 °C.

On the other hand, if the flow curves are extended to $[0, \gamma_{\min})$ such that $\tau = 0$ for $\gamma = 0$, the viscosity function can be written as

$$\varphi(I(u), |E|, \mu(u, E)) = b(|E|, \mu(u, E))(\kappa + I(u))^{-1/2} + c(I(u), |E|, \mu(u, e)), \quad (2.19)$$

where $0 < \kappa \ll 1$ and $b : \mathbb{R}_+ \times [0, 1] \to \mathbb{R}$, $c : \mathbb{R}_+ \times \mathbb{R}_+ \times [0, 1] \to \mathbb{R}$ are continuous functions.

REMARK 2.2. The viscosity function φ of the form (2.19) can be interpreted as an extension of the regularized Bingham fluid model (2.9).

As far as the functions b, c in (2.18) and (2.19) are concerned, we assume that the following conditions are satisfied:

(A₁) c is a continuous function of its arguments, i.e., $c \in C(\mathbb{R}_+ \times \mathbb{R}_+ \times [0, 1])$, and there exist positive constants c_i, $1 \le i \le 2$ such that for all $(y_1, y_2, y_3) \in \mathbb{R}_+ \times \mathbb{R}_+ \times [0, 1]$, there holds

$$c_1 \le c(y_1, y_2, y_3) \le c_2.$$

Moreover, for fixed $(y_2, y_3) \in \mathbb{R}_+ \times [0, 1]$, the function $c(\cdot, y_2, y_3) : \mathbb{R}_+ \to \mathbb{R}$ is continuously differentiable, i.e., $c(\cdot, y_2, y_3) \in C^1(\mathbb{R}_+)$, and there exist positive constants c_i, $3 \le i \le 4$ such that for all $y_1 \in \mathbb{R}_+$, there holds

$$c(y_1, y_2, y_3) + 2\,\frac{\partial c}{\partial y_1}(y_1, y_2, y_3) \;\ge\; c_3,$$

$$\left|\frac{\partial c}{\partial y_1}(y_1, y_2, y_3)\right| y_1 \;\le\; c_4.$$

(A₂) b is a continuous function of its arguments, i.e., $b \in C(\mathbb{R}_+ \times [0, 1])$, and there exists a positive constant c_5 such that for all $(y_1, y_2) \in \mathbb{R}_+ \times [0, 1]$, there holds

$$0 \le b(y_1, y_2) \le c_5.$$

REMARK 2.3. The first condition in $(\mathbf{A_1})$ and condition $(\mathbf{A_2})$ imply that for the models (2.18) and (2.19), the viscosity function φ is bounded from below by a positive constant, and that for the regularized Bingham-type model (2.19), the viscosity function φ is bounded from above as well, whereas $\varphi(I(u), |E|, \mu(u, E)) \to +\infty$ as $I(u) \to 0$ for the extended Bingham-type model (2.18).

The second condition in $(\mathbf{A_1})$ implies that for fixed values of $|E|$ and $\mu(u, E)$, the derivative of the function $I(v) \longmapsto G(v) := 4(\varphi(I(v), |E|, \mu(u, E)))^2 I(v)$ is positive, where $G(v)$ is the second invariant of the stress deviator. The physical meaning of this condition is that in case of shear flow, the shear stress increases with increasing shear rate.

The third condition in $(\mathbf{A_1})$ imposes a restriction on the function $\partial c/\partial y_1$ for large values of y_1, which reflects the experimentally observable behavior of electrorheological fluids that their structure is destroyed at large shear rates.

On the basis of the assumptions $(\mathbf{A_1})$ and $(\mathbf{A_2})$, existence and uniqueness results for steady and time-dependent isothermal incompressible electrorheological fluid flows will be established in the subsequent Sections 2.2 and 2.3 relying on the theory of monotone operators (BREZIS [1973], BROWDER [1968], LIONS [1969], MINTY [1962], VAINBERG [1964], VISIK [1962], and ZEIDLER [1990]).

We note that under some weaker monotonicity assumptions, an existence result has been derived in DREYFUSS and HUNGERBUEHLER [2004a,b] using the theory of Young measures (see, e.g., VALADIER [1994]). We further refer to DREYFUSS and HUNGERBUEHLER [2004a,b].

Since the macroscopic behavior of electrorheological fluids is largely determined by physical processes occurring on a microscale, a natural approach to develop physically consistent macroscopic models is to use homogenization techniques within a multiscale framework. Such an approach has been undertaken in VERNESCU [2002] (cf. also BANKS et al. [1999] for a similar approach in case of magnetorheological fluids).

2.2. *Boundary value problems for steady isothermal incompressible fluid flows based on regularized Bingham-type flow models*

We adopt standard notation from Lebesgue and Sobolev space theory (cf., e.g., ADAMS [1975], GRISVARD [1985], and LIONS and MAGENES [1968]). In particular, for a bounded Lipschitz domain $\Omega \subset \mathbb{R}^d, d \in \mathbb{N}$, we refer to $L^p(\Omega)^d, 1 \leq p \leq \infty$, as the Lebesgue spaces with norms $\|\cdot\|_{p,\Omega}$ and denote by $(\cdot,\cdot)_{0,\Omega}$ the inner product in $L^2(\Omega)^d$. The spaces $W^{m,p}(\Omega)^d, m \in \mathbb{N}$, stand for the Sobolev spaces with norms $\|\cdot\|_{m,p,\Omega}$, whereas $W^{-m,q}(\Omega)^d$, $1/p + 1/q = 1, 1 \leq p < \infty$, and $W^{m-1/p,p}(\Gamma)^d, \Gamma := \partial\Omega$, refer to their dual and trace spaces, respectively. For $\Sigma \subseteq \Gamma$, the space $W^{m-p,p}_{0,\Sigma}(\Omega)^d$ denotes the space of functions $v \in W^{m,p}(\Omega)^d$ with vanishing trace on Σ, i.e., $v|_\Sigma = 0$, and $W^{m-1/p,p}_{00}(\Sigma)^d$ is the space of functions $\psi \in W^{m-1/p,p}(\Gamma)^d$ such that $\psi = v|_\Sigma$ for some $v \in W^{m,p}(\Omega)^d$ with $v|_{\Gamma\setminus\Sigma} = 0$. Furthermore, we refer to $H(\mathrm{div};\Omega) := \{v \in L^2(\Omega)^d | \nabla\cdot v \in L^2(\Omega)\}$ and $H(\mathrm{curl};\Omega) := \{v \in L^2(\Omega)^d | \nabla\times v \in L^2(\Omega)^d\}$, if $d \geq 3$, and $H(\mathrm{curl};\Omega) := \{v \in L^2(\Omega)^2 | \nabla\times v \in L^2(\Omega)\}$, if $d = 2$, as the Hilbert spaces of square integrable vector-valued functions with square integrable divergence and rotation, respectively, equipped with the standard graph norm. We denote by $H(\mathrm{div}^0;\Omega)$ and $H(\mathrm{curl}^0;\Omega)$ the subspaces $H(\mathrm{div}^0;\Omega) := \{v \in H(\mathrm{div};\Omega) | \nabla\cdot v = 0\}$ and $H(\mathrm{curl}^0;\Omega) := \{v \in H(\mathrm{curl};\Omega) | \nabla\times v = 0\}$.

Given a bounded Lipschitz domain $\Omega \subset \mathbb{R}^d$ with boundary $\Gamma = \overline{\Gamma}_D \cup \overline{\Gamma}_N, \Gamma_D \cap \Gamma_N = \emptyset$, and functions

$$f \in L^2(\Omega)^d, \quad g \in L^2(\Gamma_N)^d, \quad u^D \in W^{1/2,2}(\Gamma_D)^d, \tag{2.20}$$

we consider the following boundary value problem for steady, incompressible, isothermal electrorheological fluid flows under the Stokes approximation, i.e., we ignore inertial forces,

$$\nabla\cdot\sigma = f \quad \text{in } \Omega, \tag{2.21a}$$

$$\nabla\cdot u = 0 \quad \text{in } \Omega, \tag{2.21b}$$

$$u = u^D \quad \text{on } \Gamma_D \times (0,T), \tag{2.21c}$$

$$\nu\cdot\sigma = g \quad \text{on } \Gamma_N, \tag{2.21d}$$

where the stress tensor σ is supposed to satisfy one of the constitutive equations from the previous subsection.

As far as the electric field E is concerned, we assume that the boundary Γ features n pairs of electrodes and counter-electrodes occupying open subsets $\Gamma_i^e, \Gamma_i^c \subset \Gamma, \Gamma_i^e \cap \Gamma_i^c = \emptyset, 1 \leq i \leq n, n \in \mathbb{N}$, with voltages U_i applied to the electrodes Γ_i^e. Since we assume the electric field E to be quasi-static, it satisfies $E \in H(\text{curl}^0; \Omega)$ and $\epsilon E \in H(\text{div}^0; \Omega)$, where ϵ stands for the electric permittivity. Hence, there exists an electric potential $\psi \in W^{1,2}(\Omega)$ satisfying the elliptic boundary value problem

$$\nabla \cdot (\epsilon \nabla \psi) = 0 \quad \text{in } \Omega, \tag{2.22a}$$

$$\psi = \begin{cases} U_i & \text{on } \Gamma_i^e \\ 0 & \text{on } \Gamma_i^c \end{cases}, \quad 1 \leq i \leq n, \tag{2.22b}$$

$$\nu \cdot \epsilon \nabla \psi = 0 \quad \text{on } \Gamma \setminus \bigcup_{i=1}^{n} \overline{(\Gamma_i^e \cup \Gamma_i^c)}. \tag{2.22c}$$

Since the coupling between the electric field and the fluid is supposed to be unilateral, the boundary value problem (2.22a)–(2.22c) can be solved beforehand.

THEOREM 2.1. *Assume* $U_i \in W_{00}^{1/2,2}(\Gamma_i^e)$, $1 \leq i \leq n$, *and* $\epsilon = (\epsilon_{ij})_{i,j=1}^d, \epsilon_{ij} \in L^\infty(\Omega)$, $1 \leq i, j \leq d$, *such that for almost all* $x \in \Omega$

$$\sum_{i,j=1}^{d} \epsilon_{ij}(x)\xi_i\xi_j \geq \alpha|\xi|^2, \quad \xi \in \mathbb{R}^d, \alpha > 0.$$

Then, the boundary value problem (2.22a)–(2.22c) admits a unique weak solution $\theta \in W_{0,\Gamma^c}^{1,2}(\Omega)$, $\Gamma^c := \bigcup_{i=1}^n \Gamma_i^c$.

PROOF. Due to the assumption on the voltages U_i, there exists $\tilde{\theta} \in W^{1,2}(\Omega)$ such that $\tilde{\theta}|_{\Gamma_i^e} = U_i$ and $\tilde{\theta}|_{\Gamma_i^c} = 0$, $1 \leq i \leq n$. Defining $a(v, w) := \int_\Omega \epsilon \nabla v \cdot \nabla w dx$, $v, w \in V := W_{0,\tilde{\Gamma}}^{1,2}(\Omega)$, $\tilde{\Gamma} := \bigcup_{i=1}^n (\Gamma_i^e \cup \Gamma_i^c)$, the V-ellipticity of the bilinear form $a(\cdot, \cdot)$ implies the existence and uniqueness of $\hat{\theta} \in V$ satisfying

$$a(\hat{\theta}, v) = -a(\tilde{\theta}, v), \quad v \in V.$$

Then, $\theta = \hat{\theta} + \tilde{\theta}$ is the unique weak solution of (2.22a)–(2.22c). □

2.2.1. The regularized extended Bingham fluid model

We study the existence and uniqueness of a solution of the boundary value problem (2.21a)–(2.21d) for the electrorheological fluid model (2.19) with regularization parameter κ. We show that a weak solution of (2.21a)–(2.21d) satisfies a system of variational equations of saddle point type and establish an existence result by means of appropriate Galerkin approximations in finite dimensional subspaces of the underlying function spaces. To this end, we set

$$X := W_{0,\Gamma_D}^{1,2}(\Omega)^d, \quad V := X \cap H(\text{div}^0; \Omega) \tag{2.23}$$

and denote by $\tilde{u} \in W^{1,2}(\Omega)^d \cap H(\mathrm{div}^0; \Omega)$ the function with trace $\tilde{u}|_{\Gamma_D} = u^D$. Moreover, we introduce a functional $J_\kappa : X \times X \to \mathbb{R}$, $\kappa \in \mathbb{R}_+$, and an operator $L : X \to X^*$ according to

$$J_\kappa(v, w) := 2 \int_\Omega c(|E|, \mu(\tilde{u} + v, E))(\kappa + I(\tilde{u} + w))^{1/2}\, dx, \tag{2.24a}$$

$$\langle L(v), w\rangle := 2 \int_\Omega b(I((\tilde{u} + v), |E|, \mu(\tilde{u} + v, E))\varepsilon(\tilde{u} + v) : \varepsilon(w)\, dx, \tag{2.24b}$$

where $\langle \cdot, \cdot\rangle$ stands for the dual pairing between X^* and X.

For $\kappa > 0$, the functional J_κ is Gâteaux differentiable on X with respect to the second argument. Indeed, the partial Gâteaux derivative $\frac{\partial J_\kappa}{\partial w}(v, \cdot) \in \mathcal{L}(X, X^*)$, $v \in X$, is given by

$$\left\langle \frac{\partial J_\kappa}{\partial w}(v, w), z \right\rangle =$$

$$2 \int_\Omega c(|E|, \mu(\tilde{u} + v, E))(\kappa + I(\tilde{u} + w))^{-1/2}\varepsilon(\tilde{u} + w) : \varepsilon(z)\, dx, \quad w, z \in X. \tag{2.25}$$

We further define an operator $M_\kappa; X \times X \to X^*$, $\kappa > 0$, by

$$M_\kappa(v, v) := \frac{\partial J_\kappa}{\partial w}(v, v) + L(v), \quad v \in X. \tag{2.26}$$

We consider the problem: find $v \in V$ such that

$$\langle M_\kappa(v, v), z\rangle = \langle f + g, z\rangle, \quad z \in V, \tag{2.27}$$

where we formally view $f + g$ as an element of X^*. We will refer to $u = \tilde{u} + v$ as a weak solution of (2.21a)–(2.21d). If a pair (u, p) is a solution of (2.21a)–(2.21d), by Green's formula, it can be easily seen that $v = u - \tilde{u}$ solves (2.27). We can state (2.27) equivalently as a system of variational equations of saddle point type, if we couple the incompressibility condition by means of a Lagrange multiplier in $L^2(\Omega)$. Denoting by $B \in \mathcal{L}(X, L^2(\Omega))$ the divergence operator, i.e., $Bv = \nabla \cdot v$, $v \in X$, this leads to the following system: find $(v, p) \in X \times L^2(\Omega)$ such that

$$\langle M_\kappa(v, v), z\rangle - \langle B^* p, z\rangle = \langle f + g, z\rangle, \quad z \in X, \tag{2.28a}$$

$$(Bv, q)_{0,\Omega} = 0, \quad q \in L^2(\Omega). \tag{2.28b}$$

Lemma 2.1. *Let $v \in V$ be a solution of (2.27). Then, there exists a unique $p \in L^2(\Omega)$ such that (2.28a), (2.28b) holds true. Conversely, if $(v, p) \in X \times L^2(\Omega)$ is a solution of (2.28a), (2.28b), then the pair $(\tilde{u} + v, p)$ satisfies (2.27). Moreover, if v, p, and $\tilde{u}$ are smooth functions, then $(\tilde{u} + v, p)$ solves (2.28a), (2.28b).*

Proof. The proof follows readily from the properties of the divergence operator B. In particular, denoting by $V^\perp$ the orthogonal complement of V in X and by V^0 the polar set

$$V^0 := \{\ell \in X^* \mid \langle \ell, w\rangle = 0, w \in V\},$$

the operator B is an isomorphism from $V^\perp$ onto $L^2(\Omega)$, whereas its adjoint B^* is an isomorphism from $L^2(\Omega)$ onto V^0 (see BELONOSOV and LITVINOV [1996] and lemma 6.1.1 in LITVINOV [2000]). We note that the case $B : H_0^1(\Omega)^d \to L_0^2(\Omega)$ has been addressed, e.g., in BREZZI and FORTIN [1991], GIRAULT and RAVIART [1986], and LADYZHENSKAYA and SOLONNIKOV [1976]. □

The existence of a solution $(u, p) \in X \times L^2(\Omega)$ of (2.28a), (2.28b) will be shown by a Galerkin approximation with respect to sequences $\{X_n\}_{\mathbb{N}}$ and $\{Q_n\}_{\mathbb{N}}$ of finite dimensional subspaces that are limit dense in X and $L^2(\Omega)$, i.e.,

$$\lim_{n\to\infty} \inf_{v_n \in X_n} \|v - v_n\|_X = 0, \quad v \in X, \tag{2.29a}$$

$$\lim_{n\to\infty} \inf_{\mu_n \in Q_n} \|p - p_n\|_{0,\Omega} = 0, \quad p \in L^2(\Omega). \tag{2.29b}$$

We refer to $B_n \in \mathcal{L}(X_n, Q_n^*)$, $n \in \mathbb{N}$, as the discrete divergence operator

$$(B_n v_n, p_n)_{0,\Omega} := \int_\Omega p_n \nabla \cdot v_n \, dx, \quad v_n \in X_n, \quad p_n \in Q_n, \tag{2.30}$$

and assume that for each $n \in \mathbb{N}$, the discrete LBB condition

$$\inf_{p_n \in Q_n} \sup_{v_n \in X_n} \frac{(B_n v_n, p_n)_{0,\Omega}}{\|v_n\|_X \|p_n\|_{0,\Omega}} \geq \beta > 0 \tag{2.31}$$

is satisfied. As can be easily established, under the above assumption, the discrete divergence operators B_n, $n \in \mathbb{N}$, inherit the properties of their continuous counterpart B.

LEMMA 2.2. *Assume that $\{X_n\}_{\mathbb{N}}$ and $\{Q_n\}_{\mathbb{N}}$ are finite dimensional subspaces $X_n \subset X$, $n \in \mathbb{N}$, and $Q_n \subset L^2(\Omega)$, $n \in \mathbb{N}$. Moreover, let B_n, $n \in \mathbb{N}$, be the discrete divergence operator as given by (2.30) and suppose that the discrete LBB condition (2.31) holds true. Then, B_n is an isomorphism from $(Ker(B_n))^\perp$ onto Q_n^*, and B_n^* is an isomorphism from Q_n onto the polar set $(Ker(B_n))^0$ such that*

$$\|B_n\| \leq \beta^{-1}, \quad \left\|(B_n^*)^{-1}\right\| \leq \beta^{-1}, \quad n \in \mathbb{N}. \tag{2.32}$$

We consider the following approximating system of finite dimensional variational equations: find $(v_n, p_n) \in X_n \times Q_n$, $n \in \mathbb{N}$, such that

$$\langle M_\kappa(v_n, v_n), z_n\rangle - \langle B_n^* p_n, z_n\rangle = \langle f + g, z_n\rangle, \quad z_n \in X_n, \tag{2.33a}$$

$$(B_n v_n, q_n)_{0,\Omega} = 0, \quad q_n \in Q_n. \tag{2.33b}$$

The main result of this subsection states the solvability of the system (2.33a), (2.33b) for each $n \in \mathbb{N}$ and the existence of a subsequence $\mathbb{N}' \subset \mathbb{N}$ such that the associated sequence $\{(v_n, p_n)\}_{\mathbb{N}'}$ of solutions converges to a pair $(v, p) \in X \times L^2(\Omega)$, which solves (2.28a), (2.28b).

THEOREM 2.2. *Assume that the conditions* $(\mathbf{A_1})$, $(\mathbf{A_2})$ *are fulfilled and* f, g, *and* u^d *satisfy (2.20). Further, let* $\{X_n\}_{\mathbb{N}}$ *and* $\{Q_n\}_{\mathbb{N}}$ *be nested sequences of finite dimensional subspaces* $X_n \subset X$, $n \in \mathbb{N}$, *and* $Q_n \subset L^2(\Omega)$, $n \in \mathbb{N}$, *i.e.,*

$$X_n \subset X_{n+1}, \quad Q_n \subset Q_{n+1}, \quad n \in \mathbb{N}, \tag{2.34}$$

that are limit dense in X *and* $L^2(\Omega)$ *and suppose that the discrete LBB condition (2.31) holds true. Then, for any* $\kappa > 0$ *and* $n \in \mathbb{N}$, *there exists a solution* $(v_n, p_n) \in X_n \times Q_n$ *of the discrete saddle point problem (2.33a), (2.33b). Moreover, there exist a subsequence* $\mathbb{N}' \subset \mathbb{N}$ *and a pair* $(v, p) \in X \times L^2(\Omega)$ *such that*

$$v_n \rightharpoonup v \quad \text{in } X \quad (\mathbb{N}' \ni n \to \infty), \tag{2.35a}$$

$$p_n \to p \quad \text{in } L^2(\Omega) \quad (\mathbb{N}' \ni n \to \infty). \tag{2.35b}$$

The pair $(v, p) \in X \times L^2(\Omega)$ *is a solution of (2.28a), (2.28b).*

Theorem 2.2 will be proved by a series of Lemmas, which enable us to deduce the existence of a bounded sequence $\{(u_n, p_n)\}_{\mathbb{N}}$ of solutions of (2.33a), (2.33b) and to pass to the limit.

For $z = (\tilde{z}, z_1, z_2)$ with $\tilde{z} \in W^{1,2}(\Omega)$, $z_1 \in L^2_+(\Omega)$ and $z_2 \in L^\infty(\Omega)$, $z_2(x) \in [0, 1]$ f.a.a. $x \in \Omega$, we define $L_z : X \to X^*$ as the operator

$$\langle L_z(v), w \rangle := 2 \int_\Omega b(I(v + \tilde{z}), z_1, z_2)\varepsilon(v + \tilde{z}) : \varepsilon(w)\, \mathrm{d}x, \; v, w \in X. \tag{2.36}$$

LEMMA 2.3. *Under the assumption* $(\mathbf{A_1})$, *the operator* L_z *as given by (2.36) is a continuous, strongly monotone operator from* X *into* X^*. *In particular, for* $v, w \in X$, *there holds*

$$\|L_z(v) - L_z(w)\|_{X^*} \le C_L \, \|v - w\|_X, \tag{2.37a}$$

$$\langle L_z(v) - L_z(w), v - w \rangle \ge \gamma_L \, \|v - w\|_X^2, \tag{2.37b}$$

where $C_L := (2c_2 + 4c_4)$ *and* $\gamma_L := 2\min(c_1, c_3)$ *with* c_i, $1 \le i \le 4$, *from* $(\mathbf{A_1})$.

PROOF. For $v, w \in X$, we set $q := v - w$ and consider the function $\tau : [0, 1] \to \mathbb{R}$ which for an arbitrarily, but fixed chosen $h \in X$ is given by

$$\tau(t) := \int_\Omega b(I(\tilde{z} + w + tq), z_1, z_2)\varepsilon(\tilde{z} + w + tq) : \varepsilon(h)\, \mathrm{d}x, \; t \in [0, 1].$$

Obviously, τ satisfies

$$\tau(1) - \tau(0) = \frac{1}{2} \, \langle L_z(v) - L_z(w), h \rangle.$$

Since $\tau \in C^1([0, T])$, classical calculus tells us that for some $\xi \in (0, 1)$

$$\tau(1) = \tau(0) + \frac{\mathrm{d}\tau}{\mathrm{d}t}(\xi),$$

where $(\mathrm{d}\tau/\mathrm{d}t)(\xi)$ is given by

$$\frac{\mathrm{d}\tau}{\mathrm{d}t}(\xi) = \int_{\Omega} \Bigg(b(I(\tilde{z} + w + \xi q), z_1, z_2)\varepsilon(q) : \varepsilon(h)$$

$$+2\frac{\partial b}{\partial y_1}(I(\tilde{z} + w + \xi q), z_1, z_2)(\varepsilon(\tilde{z} + w + \xi q) : \varepsilon(q))(\varepsilon(\tilde{z} + w + \xi q) : \varepsilon(h))\Bigg) \,\mathrm{d}x. \tag{2.38}$$

In view of the inequality,

$$|\varepsilon(\tilde{z} + w + \xi q) : \varepsilon(q)| \le (I(\tilde{z} + w + \xi q))^{1/2}(I(q))^{1/2},$$

and taking $(\mathbf{A_1})(i)$ and $(\mathbf{A_1})(iii)$ into account, (2.37a) can be easily deduced.

On the other hand, we define $\eta : \mathbb{R}_+ \times \Omega \to \mathbb{R}_-$ by

$$\eta(\alpha, x) := \left(\frac{\partial b}{\partial y_1}(\alpha, z_1(x), z_2(x))^-, \quad \alpha \in \mathbb{R}_+, x \in \Omega.\right.$$

Then, if we set $\alpha := I(\tilde{z} + w + \xi q)$ and choose $h = q$ in (2.38), we obtain

$$\frac{\mathrm{d}\tau}{\mathrm{d}t}(\xi) \ge \int_{\Omega} \Big(b(I(\tilde{z} + w + \xi q), z_1, z_2)I(q)$$

$$+2g(\alpha, z_1(x), z_2(x))(\varepsilon(\tilde{z} + w + \xi q) : \varepsilon(q))^2\Big) \,\mathrm{d}x \ge \min(c_1, c_3)\, \|q\|_X^2,$$

which proves (2.37b). The continuity of the operator L_z follows from the continuity of the Nemytskii operator. □

In view of the representation of the partial Gâteaux derivative $\partial J_\kappa/\partial w$ by (2.25) and assumption $(\mathbf{A_2})$, for a given function

$$\chi \in U := \{\vartheta \in L^\infty_+(\Omega) \mid \vartheta(x) \le c_5 \text{ f.a.a. } x \in \Omega\}$$

and $\tilde{v} \in W^{1,2}(\Omega)$, we define an operator $S_\kappa : U \times X \to X^*$, $\kappa > 0$, according to

$$\langle S_\kappa(\chi, v), w\rangle := \int_{\Omega} \chi(\kappa + I(\tilde{v} + v))^{-1/2}\varepsilon(\tilde{v} + v) : \varepsilon(w) \,\mathrm{d}x, \; v, w \in X. \tag{2.39}$$

LEMMA 2.4. *Under the assumption* $(\mathbf{A_2})$, *for an arbitrarily, but fixed chosen* $\chi \in U$, *the operator* $S_\kappa(\chi, \cdot)$, $\kappa > 0$, *with* S_κ *as given by (2.39) is a continuous, monotone operator from* X *into* X^*. *In particular, there holds*

$$\|S_\kappa(\chi, v) - S_\kappa(\chi, w)\|_{X^*} \le 2c_5\kappa^{-1/2}\, \|v - w\|_X, \quad v, w \in X, \tag{2.40a}$$

$$\|S_\kappa(\chi, v)\|_{X^*} \le \left(\int_{\Omega} \chi^2 \,\mathrm{d}x\right)^{1/2}, \quad v \in X. \tag{2.40b}$$

PROOF. We set $v_1 := \tilde{v} + v$, $w_1 := \tilde{v} + w$ and define $\varphi_\kappa : \mathbb{R}_+ \to \mathbb{R}_+$, $\kappa > 0$, by

$$\varphi_\kappa(y) := \frac{1}{2}\,\chi\,(\kappa + y)^{-1/2}, \quad y \in \mathbb{R}_+. \tag{2.41}$$

Then, if we take

$$|\varepsilon(v_1) : \varepsilon(w_1)| \le (I(v_1))^{1/2}(I(w_1))^{1/2}$$

into account, it follows that

$$\begin{aligned}
\langle S_\kappa(\chi, v) - S_\kappa(\chi, w), v - w\rangle &= \langle S_\kappa(\chi, v) - S_\kappa(\chi, w), v_1 - w_1\rangle \\
= 2\int_\Omega \big(\varphi_\kappa(I(v_1))I(v_1) &+ \varphi_\kappa(I(w_1))I(w_1) \\
- (\varphi_\kappa(I(v_1)) &+ \varphi_\kappa(I(w_1)))\varepsilon(v_1) : \varepsilon(w_1)\big)\,\mathrm{d}x \\
\ge 2\int_\Omega \Big(\varphi_\kappa(I(v_1))(I(v_1))^{1/2} &- \varphi_\kappa(I(w_1))(I(w_1))^{1/2}\Big)\Big((I(v_1))^{1/2} - I(w_1)^{1/2}\Big)\,\mathrm{d}x.
\end{aligned} \tag{2.42}$$

Now, for the function φ_κ from (2.41), one easily finds

$$\varphi_\kappa(y) + 2\frac{\mathrm{d}\varphi_\kappa}{\mathrm{d}y}(y)y = \frac{1}{2}\chi(\kappa + y)^{-1/2}\Big(1 - (\kappa + y)^{-1}y\Big) > 0, \quad y \in \mathbb{R}_+. \tag{2.43}$$

Considering $\psi(z) := \varphi_\kappa(z^2)z$, we have $(\mathrm{d}\psi/\mathrm{d}z)(z) = \varphi_\kappa(z^2) + 2(\mathrm{d}\varphi_\kappa)(z^2)z^2$, which is the left-hand side in (2.43) for $z^2 = y$. It follows that ψ is a monotonously increasing function, and (2.42) implies the monotonicity of the operator $S_\kappa(\chi, \cdot)$. The boundedness (2.40b) of $S_\kappa(\chi, \cdot)$ is an immediate consequence of

$$\begin{aligned}
&|\ \langle S_\kappa(\chi, v), w\rangle\ | \\
&\le \int_\Omega \chi(\kappa + (I(\tilde{v} + v))^{-1/2}(I(\tilde{v} + v))^{1/2}(I(w))^{1/2}\,\mathrm{d}x \\
&\le \left(\int_\Omega \chi^2\,\mathrm{d}x\right)^{1/2} \|w\|_X.
\end{aligned}$$

Finally, in view of

$$\varphi_\kappa(y) \le \frac{1}{2}c_5\kappa^{-1/2}, \quad \left|\frac{\mathrm{d}\varphi_\kappa}{\mathrm{d}y}(y)\right| y \le \frac{1}{4}c_5\kappa^{-1/2}, \quad y \in \mathbb{R}_+,$$

the estimate (2.40a) can be deduced as in the proof of Lemma 2.3. □

COROLLARY 2.3. *Under the assumptions of Lemma 2.4 assume that $\{v_n\}_\mathbb{N}$ is a sequence of elements $v_n \in X$, $n \in \mathbb{N}$, and $v \in X$ such that*

$$v_n \to v \quad \textit{in } X \quad (n \to \infty),$$

$$\begin{aligned} v_n \to v \quad & \textit{a.e. in } \Omega \quad (n \to \infty), \\ \nabla v_n \to \nabla v \quad & \textit{a.e. in } \Omega \quad (n \to \infty). \end{aligned} \tag{2.44}$$

Moreover, suppose that $\{\chi_n\}_{\mathbb{N}}$ *is a sequence of elements* $\chi_n \in U, n \in \mathbb{N}$ *such that for some* $\chi \in U$ *there holds*

$$\chi_n \to \chi \quad \textit{a.e. in } \Omega \quad (n \to \infty). \tag{2.45}$$

Then, for any $\kappa > 0$, *we have*

$$S_\kappa(\chi_n, v_n) \to S_\kappa(\chi, v) \quad \textit{in } X^* \quad (n \to \infty). \tag{2.46}$$

PROOF. Straightforward estimation from above yields

$$\begin{aligned} &\|S_\kappa(\chi_n, v_n) - S_\kappa(\chi, v)\|_{X^*} \\ &\quad \le \|S_\kappa(\chi_n, v_n) - S_\kappa(\chi_n, v)\|_{X^*} + \|S_\kappa(\chi_n, v) - S_\kappa(\chi, v)\|_{X^*}. \end{aligned} \tag{2.47}$$

Due to (2.45), the second term on the right-hand side in (2.47) tends to zero as $n \to \infty$. As far as the first term is concerned, for $w \in X$, we have

$$\begin{aligned} \langle S_\kappa(\chi_n, v_n) - S_\kappa(\chi_n, v), w\rangle &= \int_\Omega \chi_n \Big((\kappa + I(\tilde{v} + v_n))^{-1/2} \varepsilon(v_n - v) : \varepsilon(w) \\ &\quad + ((\kappa + I(\tilde{v} + v_n))^{-1/2} - (\kappa + I(\tilde{v} + v))^{-1/2}) \varepsilon(\tilde{v} + v) : \varepsilon(w) \Big)\, \mathrm{d}x, \end{aligned}$$

from which we deduce

$$\begin{aligned} &\|S_\kappa(\chi_n, v_n) - S_\kappa(\chi_n, v)\|_{X^*} \\ &\quad \le \underbrace{\left(\int_\Omega \chi_n^2 (\kappa + I(\tilde{v} + v_n))^{-1} I(v_n - v)\, \mathrm{d}x \right)^{1/2}}_{=: I_1} \\ &\quad + \underbrace{\left(\int_\Omega \chi_n^2 ((\kappa + I(\tilde{v} + v_n))^{-1/2} - (\kappa + I(\tilde{v} + v))^{-1/2})^2 I(\tilde{v} + v)\, \mathrm{d}x \right)^{1/2}}_{=: I_2}. \end{aligned} \tag{2.48}$$

In view of the uniform boundedness of the sequence $\{\chi_n\}_{\mathbb{N}}$ and (2.44), obviously $I_1 \to 0$ as $n \to \infty$. On the other hand, (2.44) also implies

$$I(\tilde{v} + v_n) \to I(\tilde{v} + v) \quad (n \to \infty),$$

whence $I_2 \to 0$ as $n \to \infty$ by the Lebesgue theorem. Consequently, the first term on the right-hand side in (2.47) tends to zero as $n \to \infty$ which allows to conclude. □

We are now in a position to provide the proof of Theorem 2.2.

PROOF OF THEOREM 2.2. If $(v_n, p_n) \in X_n \times Q_n$, $n \in \mathbb{N}$, is a solution of (2.33a), (2.33b), then $v_n \in \operatorname{Ker}(B_n)$ and

$$\langle M_\kappa(v_n, v_n), w_n\rangle + \langle L(v_n), w_n\rangle = \langle f + g, w_n\rangle, \quad w_n \in \operatorname{Ker}(B_n). \tag{2.49}$$

By assumption $(\mathbf{A_2})$, for $\kappa > 0$ and $w \in X$, we have

$$\begin{aligned}
&\left|\left\langle \frac{\partial J_\kappa}{\partial w}(w, w), w\right\rangle\right| \\
&\quad = 2\left|\int_\Omega c(|E|, \mu(\tilde{u} + w, E))(\kappa + I(\tilde{u} + w))^{-1/2}\varepsilon(\tilde{u} + w) : \varepsilon(w)\, \mathrm{d}x\right| \\
&\quad \le 2\int_\Omega c(|E|, \mu(\tilde{u} + w))(I(w))^{1/2}\, \mathrm{d}x \le 2c_5|\Omega|^{1/2}\, \|w\|_X.
\end{aligned} \tag{2.50}$$

If we take assumption $(\mathbf{A_1})$ as well as (2.20) and (2.50) into account, it follows that for some $C_1 \in \mathbb{R}_+$

$$\varrho(w) := \langle M_\kappa(w, w), w\rangle - \langle f + g, w\rangle \ge \|w\|_X\, (2c_1\|w\|_X - C_1),$$

whence

$$\varrho(w) \ge 0 \quad \text{for} \quad \|w\|_X \ge r := C_1/(2c_1).$$

Then, the corollary of Brouwer's fixed point theorem in GAJEWSKI, GRÖGER and ZACHARIAS [1974] implies the existence of a solution $v_n \in \operatorname{Ker}(B_n)$ of (2.49), which satisfies

$$\|v_n\|_X \le r, \quad \|L(v_n)\|_{X^*} \le C_2, \quad n \in \mathbb{N}, \tag{2.51}$$

for some constant $C_2 > 0$. Now, for $\ell \in X^*$, let $\ell_n := \ell|_{X_n}$, $n \in \mathbb{N}$. Then, $\ell_n \in X_n^*$ and in view of (2.49), we have

$$\ell_n(M_\kappa(v_n, v_n) - (f + g)) \in \operatorname{Ker}(B_n)^0.$$

By means of Lemma 2.2, we deduce the existence of a unique $p_n \in Q_n$ such that

$$B_n^* p_n = \ell_n(M_\kappa(v_n, v_n) - (f + g)),$$

and the pair $(v_n, p_n) \in X_n \times Q_n$ solves (2.33a), (2.33b). Taking advantage of assumption $(\mathbf{A_2})$, (2.20), (2.51) and Lemmas 2.2 and 2.4, we obtain the boundedness of the sequence $\{p_n\}_\mathbb{N}$, i.e., with some $C_3 > 0$ there holds

$$\|p_n\|_{0,\Omega} \le C_3, \quad n \in \mathbb{N}. \tag{2.52}$$

Due to (2.51) and (2.52), there exist a subsequence $\mathbb{N}' \subset \mathbb{N}$ and elements $v^* \in X$, $p^* \in L^2(\Omega)$ as well as $\ell_1^*, \ell_2^* \in X^*$ such that

$$v_n \rightharpoonup v^* \quad \text{in } X \quad (\mathbb{N}' \ni n \to \infty), \tag{2.53a}$$

$$v_n \to v^* \quad \text{in } L^2(\Omega) \quad (\mathbb{N}' \ni n \to \infty), \tag{2.53b}$$

$$v_n \to v^* \quad \text{a.e. in } \Omega \quad (\mathbb{N}' \ni n \to \infty), \tag{2.53c}$$

$$p_n \to p^* \quad \text{in } L^2(\Omega) \quad (\mathbb{N}' \ni n \to \infty), \tag{2.53d}$$

$$L(v_n) \rightharpoonup \ell_1^* \quad \text{in } X^* \quad (\mathbb{N}' \ni n \to \infty), \tag{2.53e}$$

$$\frac{\partial J_\kappa}{\partial w}(v_n, v_n) \rightharpoonup \ell_2^* \quad \text{in } X^* \quad (\mathbb{N}' \ni n \to \infty). \tag{2.53f}$$

In view of (2.29a), (2.29b), and (2.53a) as well as (2.53d)–(2.53f), we pass to the limit in (2.33a), (2.33b) and obtain

$$\langle \ell_2^* + \ell_1^* - B^* p^*, w \rangle = \langle f + g, w \rangle, \quad w \in X, \tag{2.54a}$$

$$(\nabla \cdot v^*, q)_{0,\Omega} = 0, \quad q \in L^2(\Omega). \tag{2.54b}$$

We note that the action of operator L can be written as $L(v) = L(w, w)$, $w \in X$, where the mapping $(w, z) \longmapsto L(w, z)$ is from $X \times X$ into X^* according to

$$\langle L(w, z), h \rangle := 2 \int_\Omega b(I(\tilde{u} + z), |E|, \mu(\tilde{u} + w, E))\varepsilon(\tilde{u} + z) : \varepsilon(h)\, dx, \quad h \in X.$$

For $n \in \mathbb{N}'$, we define $\hat{\ell}_n \in X^*$ by

$$\begin{aligned}\hat{\ell}_n(w) := \Big\langle & \frac{\partial J_\kappa}{\partial w}(v_n, v_n) + L(v_n, v_n) \\ & - \Big(\frac{\partial J_\kappa}{\partial w}(v_n, w) + L(u_n, v) \Big), v_n - w \Big\rangle, \quad w \in X.\end{aligned} \tag{2.55}$$

The previous results show

$$\hat{\ell}_n(w) \ge 0, \quad w \in X, n \in \mathbb{N}'. \tag{2.56}$$

On the other hand, observing

$$\begin{aligned}&\Big\| \frac{\partial J_\kappa}{\partial w}(v_n, w) - \frac{\partial J_\kappa}{\partial w}(v^*, w) \Big\|_{X^*} \\ &\quad \le 2 \Big(\int_\Omega (c(|E|, \mu(\tilde{u} + v_n, E)) - c(|E|, \mu(\tilde{u} + v^*, E)))^2\, dx \Big)^{1/2},\end{aligned}$$

assumption $(\mathbf{A_2})$ in combination with (2.53b), (2.53c) and the Lebesgue theorem yield

$$\frac{\partial J_\kappa}{\partial w}(v_n, w) \to \frac{\partial J_\kappa}{\partial w}(v^*, w) \quad \text{in } X^* \quad (\mathbb{N}' \ni n \to \infty). \tag{2.57}$$

In a similar way, we obtain

$$L(v_n, w) \to L(v^*, w) \quad \text{in } X^* \quad (\mathbb{N}' \ni n \to \infty). \tag{2.58}$$

Taking $(B_n v_n, p_n)_{0,\Omega} = 0$ into account, (2.33a) and (2.53a), (2.53d) imply

$$\begin{aligned} &\langle M_\kappa(v_n, v_n), v_n\rangle = \langle f + g, v_n\rangle \to \\ &\big\langle f + g, v^*\big\rangle \quad (\mathbb{N}' \ni n \to \infty), \end{aligned} \tag{2.59a}$$

$$\begin{aligned} &\langle M_\kappa(v_n, v_n), w\rangle \to \big\langle B^*\lambda^*, w\big\rangle \\ &\quad + \langle f + g, w\rangle \quad (\mathbb{N}' \ni n \to \infty), \quad w \in X. \end{aligned} \tag{2.59b}$$

Consequently, passing to the limit in (2.55) and observing (2.54a), (2.54b) as well as (2.56)–(2.58), (2.59a), (2.59b), it follows that

$$\left(\big\langle f + g, v^* - w\big\rangle - \left\langle \frac{\partial J_\kappa}{\partial w}(v^*, w) + L(v^*, w) - B^*p^*, v^* - w\right\rangle\right) \geq 0, \quad w \in X.$$

We choose $v = u^* - \tau z$ where $\tau > 0$ and $z \in X$. The limit process $\tau \to 0$ results in

$$\big(\langle f + g, z\rangle - \big\langle M_\kappa(v^*, v^*) - B^*p^*, z\big\rangle\big) \geq 0. \tag{2.60}$$

Since this inequality holds true for all $z \in X$, we may replace z by $-z$ and deduce equality in (2.60). We have thus shown that the pair $(v^*, p^*) \in X \times L^2(\Omega)$ solves (2.28a), (2.28b). □

For further existence results in case of stationary electrorheological fluid flows and for studies of the regularity of solutions, we refer to Acerbi and Mingione [2002], Ettwein and Ruzicka [2002], Bildhauer and Fuchs [2004].

With regard to the uniqueness of a solution of (2.28a), (2.28b), we refer to Hoppe and Litvinov [2004]. We also note that electrorheological fluid flows under conditions of slip on the boundary have been studied in Hoppe, Kuzmin, Litvinov and Zvyagin [2006], Litvinov [2007].

2.2.2. *The extended Bingham-type electrorheological fluid model*

We deal now with the solution of the boundary value problem (2.21a)–(2.21d) for an extended Bingham-type electrorheological fluid model (cf. (2.18)) with viscosity function

$$\varphi(I(u), |E|, \mu(u, E)) = b(|E|, \mu(u, E))I(u)^{-1/2} + c(|E|, \mu(u, E)). \tag{2.61}$$

We assume that the function b in (2.61) satisfies $(\mathbf{A_2})$, whereas the function c is subject to the following assumption:

$(\mathbf{A_1})'$ $c : \mathbb{R}_+ \times [0, 1] \to \mathbb{R}$ is a continuous, strictly positive, and uniformly bounded function, i.e., $c \in C(\mathbb{R}_+ \times [0, 1])$, and there exist constants $c_8 > 0$ and $c_9 > 0$ such that

$$c_8 \leq c(z_1, z_2) \leq c_9, z_1, z_2 \in \mathbb{R}_+ \times [0, 1].$$

We formulate (2.21a)–(2.21d) as a variational inequality of the second kind (cf., e.g., Glowinski, Lions and Tremolieres [1981]). To this end, we denote by $\tilde{u} \in W^{1,2}(\Omega)^d \cap H(\text{div}^0; \Omega)$ the function with trace $\tilde{u}|_{\Gamma_D} = u^D$. Moreover, we introduce a functional

$J : X \times X \to \mathbb{R}$ and an operator $L : X \to X^*$ according to

$$J(v, w) := 2 \int_\Omega b(|E|, \mu(\tilde{u} + v, E)) I(\tilde{u} + w)^{1/2} \, dx, \tag{2.62a}$$

$$\langle L(v), w \rangle := 2 \int_\Omega c(|E|, \mu(\tilde{u} + v, E)) \varepsilon(\tilde{u} + v) : \varepsilon(w) \, dx, \tag{2.62b}$$

where $\langle \cdot, \cdot \rangle$ stands for the dual pairing between X^* and X.

For the constitutive equation (2.61), problem (2.27) can be written as the following variational inequality:

Find $v \in V$ such that for all $w \in V$, there holds

$$J(v, w) - J(v, v) + \langle L(v), w - v \rangle \geq \langle f + g, w - v \rangle. \tag{2.63}$$

The function $u = \tilde{u} + v$, where $v \in V$ is a solution of (2.63), is called a weak solution of (2.21a)–(2.21d) for the constitutive equation (2.61).

We will prove the existence of a solution $v \in V$ of (2.63) via an approximation of J by the functional $J_\kappa : X \times X \to \mathbb{R}$, $\kappa \in \mathbb{R}_+$, as given by (2.24a), i.e., for a sequence $\{\kappa_n\}_\mathbb{N}$ of regularization parameters $\kappa_n > 0$, $n \in \mathbb{N}$, with $\kappa_n \to 0$ as $n \to \infty$, we consider the variational problem.

Find $v_{\kappa_n} \in V$ such that for all $w \in V$, there holds

$$\left\langle \frac{\partial J_{\kappa_n}}{\partial v}(v_{\kappa_n}, v_{\kappa_n}), w \right\rangle + \left\langle L(v_{\kappa_n}), w \right\rangle = \langle f + g, w \rangle. \tag{2.64}$$

We further consider the related saddle point problem.

Find $(v_{\kappa_n}, p_{\kappa_n}) \in X \times L^2(\Omega)$ such that for all $w \in X$ and $q \in L^2(\Omega)$, there holds

$$\left\langle \frac{\partial J_{\kappa_n}}{\partial w}(v_{\kappa_n}, v_{\kappa_n}), w \right\rangle + \left\langle L(v_{\kappa_n}), w \right\rangle - \left\langle B^* p_{\kappa_n}, w \right\rangle = \langle f + g, w \rangle, \tag{2.65a}$$

$$(B v_{\kappa_n}, q)_{0,\Omega} = 0. \tag{2.65b}$$

The existence result partially relies on the following result about functionals $\Psi : U \times X \to \mathbb{R}_+$ of the form

$$\Psi(h, w) := \int_\Omega h I(w)^{1/2} \, dx, \quad h \in U, w \in X.$$

Here, $U := \{h \in L^\infty(\Omega) | 0 \leq h(x) \leq c_10 \text{ a.e. in } \Omega\}$ for some $c_10 > 0$.

LEMMA 2.5. *For an arbitrarily chosen, but fixed $h \in U$, the functional $\Psi(h, \cdot) : X \to \mathbb{R}_+$ is a continuous convex functional. Moreover, for any sequence $\{h_n\}_\mathbb{N}$ of elements $h_n \in U$, $n \in \mathbb{N}$, and any sequence $\{w_n\}_\mathbb{N}$ of elements $w_n \in X$, $n \in \mathbb{N}$, such that for $n \to \infty$*

$$h_n \to h \quad \text{a.e. in } \Omega, \quad w_n \rightharpoonup w \quad \text{in } X, \tag{2.66}$$

there holds

$$\liminf_{n\to\infty} \Psi(h_n, w_n) \;\geq\; \Psi(h, w).$$

PROOF. Assume $w_n \rightharpoonup w$ in X. In view of

$$\int_\Omega hI(w_n - w)^{1/2}\,\mathrm{d}x \leq \left(\int_\Omega h^2\,\mathrm{d}x\right)^{1/2} \left(\int_\Omega I(w_n - w)\,\mathrm{d}x\right)^{1/2},$$

for $n \to \infty$, we have

$$\int_\Omega hI(w_n - w)^{1/2}\,\mathrm{d}x \;\to\; 0\,,$$

$$\int_\Omega hI(w_n - w)^{1/2}\,\mathrm{d}x \geq \left|\int_\Omega hI(w_n)^{1/2}\,\mathrm{d}x - \int_\Omega hI(w)^{1/2}\,\mathrm{d}x\right|,$$

whence

$$\Psi(h, u_n) \;\to\; \Psi(h, w),$$

which proves the continuity of $\Psi(h,\cdot)$. For $\lambda \in [0, 1]$ and $u, v \in X$, there holds

$$\begin{aligned} I(\lambda u + (1-\lambda)v) &= I(\lambda u) + 2\lambda(1-\lambda)\,\varepsilon(u) : \varepsilon(v) + I(1-\lambda)v) \\ &\leq \left(\lambda I(u)^{1/2} + (1-\lambda)I(v)^{1/2}\right)^2, \end{aligned}$$

which implies

$$\begin{aligned} &\Psi(h, \lambda u + (1-\lambda)v) \\ &\quad = \int_\Omega hI(\lambda u + (1-\lambda)v)^{1/2}\,\mathrm{d}x \leq \lambda\Psi(h, u) + (1-\lambda)\Psi(h, v), \end{aligned}$$

and thus it proves the convexity of $\Psi(h,\cdot)$. We have

$$\Psi(h_n, w_n) = \int_\Omega \left(hI(w_n)^{1/2} + (h_n - h)I(w_n)^{1/2}\right)\mathrm{d}x, \tag{2.67a}$$

$$\left|\int_\Omega (h_n - h)I(w_n)^{1/2}\,\mathrm{d}x\right| \leq \|h_n - h\|_{0,\Omega}\,\|w_n\|_X. \tag{2.67b}$$

Due to (2.66), the right-hand side in (2.67b) goes to zero as $n \to \infty$, and hence, the convexity and the continuity of $\Psi(h,\cdot)$ as well as (2.67a), (2.67b) imply

$$\liminf_{n\to\infty} \Psi(h_n, w_n) \;=\; \liminf_{n\to\infty} \Psi(h, w_n) \;\geq\; \Psi(h, w),$$

which completes the proof of the lemma. □

THEOREM 2.3. *Assume that the conditions* $(\mathbf{A_1})'$, $(\mathbf{A_2})$ *are fulfilled and* f, g, *and* u^D *satisfy (2.20). Then, for each* $n \in \mathbb{N}$, *there exist a solution* $v_{\mathcal{K}_n} \in V$ *of (2.64) and a function* $p_{\mathcal{K}_n} \in L^2(\Omega)$ *such that the pair* $(v_{\mathcal{K}_n}, p_{\mathcal{K}_n})$ *solves the saddle point system (2.65a), (2.65b). Moreover, there exist a subsequence* $\mathbb{N}' \subset \mathbb{N}$ *and a function* $v \in V$ *such that*

$$v_{\mathcal{K}_n} \rightharpoonup v \quad \text{in } X \quad (\mathbb{N}' \ni n \to \infty), \tag{2.68a}$$

$$v_{\mathcal{K}_n} \to v \quad \text{in } L^2(\Omega)^d \quad (\mathbb{N}' \ni n \to \infty). \tag{2.68b}$$

The function v *satisfies (2.63). Further, if* $I(\tilde{u} + v) \neq 0$ *a.e. in* Ω, *the functional*

$$w \longmapsto J(v, w), \qquad w \in V,$$

is Gâteaux differentiable at the point v, *and there exists a function* $p \in L^2(\Omega)$ *such that for all* $w \in X$ *there holds*

$$\left\langle \frac{\partial J}{\partial v}(v, v), w \right\rangle + \langle L(v), w\rangle - \langle B^* p, w\rangle = \langle f + g, w\rangle.$$

PROOF. Theorem 2.2 yields both the existence of $v_{\mathcal{K}_n} \in V$ satisfying (2.64) and the existence of $p_{\mathcal{K}_n} \in L^2(\Omega)$ such that the pair $(v_{\mathcal{K}_n}, p_{\mathcal{K}_n})$ solves (2.65a), (2.65b). Moreover, it follows from the proof of Theorem 2.2 that the sequence $\{v_{\mathcal{K}_n}\}_{\mathbb{N}}$ is bounded in V. Consequently, there exist a subsequence $\mathbb{N}' \subset \mathbb{N}$ and a function $v \in V$ such that (2.68a), (2.68b) hold true. In view of Lemma 2.4, for $w \in V$, the functional $v \longmapsto J_{\mathcal{K}_n}(w, v)$ is convex, whence

$$\begin{aligned} &J_{\mathcal{K}_n}(v_{\mathcal{K}_n}, w) - J_{\mathcal{K}_n}(v_{\mathcal{K}_n}, v_{\mathcal{K}_n}) + \langle L(v_{\mathcal{K}_n}), w - v_{\mathcal{K}_n}\rangle - \langle f + g, w - v_{\mathcal{K}_n}\rangle \\ &\quad = -\left\langle \frac{\partial J_{\mathcal{K}_n}}{\partial v}(v_{\mathcal{K}_n}, v_{\mathcal{K}_n}), w - v_{\mathcal{K}_n} \right\rangle + J_{\mathcal{K}_n}(v_{\mathcal{K}_n}, w) - J_{\mathcal{K}_n}(v_{\mathcal{K}_n}, v_{\mathcal{K}_n}) \geq 0. \end{aligned} \tag{2.69}$$

Assumption $(\mathbf{A_1})'$, (2.68b) and the Lebesgue theorem imply that for $\mathbb{N}' \ni n \to \infty$

$$c(|E|, \mu(\tilde{u} + v_{\mathcal{K}_n}, E))\varepsilon(v) \to c(|E|, \mu(\tilde{u} + v, E))\varepsilon(w) \quad \text{in } L^2(\Omega), \tag{2.70}$$

whence by (2.68a)

$$\langle L(v_{\mathcal{K}_n}), w\rangle \to \langle L(v), w\rangle. \tag{2.71}$$

We define

$$M_{\mathcal{K}_n}^{(1)} := 2\int_\Omega c(|E|, \mu(\tilde{u} + v_{\mathcal{K}_n}, E))\varepsilon(\tilde{u}) : \varepsilon(v_{\mathcal{K}_n})\, \mathrm{d}x, \tag{2.72a}$$

$$M_{\mathcal{K}_n}^{(2)} := 2\int_\Omega c(|E|, \mu(\tilde{u} + v_{\mathcal{K}_n}, E))I(v_{\mathcal{K}_n})\, \mathrm{d}x, \tag{2.72b}$$

such that

$$\langle L(v_{\mathcal{K}_n}), v_{\mathcal{K}_n}\rangle = M_{\mathcal{K}_n}^{(1)} + M_{\mathcal{K}_n}^{(2)}. \tag{2.73}$$

Since (2.70) also holds true with w replaced by $\tilde{u}$, (2.68a) implies that for $\mathbb{N}' \ni n \to \infty$

$$M_{\kappa_n}^{(1)} \to 2\int_\Omega c(|E|, \mu(\tilde{u}+v, E))\varepsilon(\tilde{u}) : \varepsilon(v)\,\mathrm{d}x. \tag{2.74}$$

On the other hand, assumption $(\mathbf{A_1})'$ and (2.68b) imply that for any $w \in L^2(\Omega)$ and $\mathbb{N}' \ni n \to \infty$, there holds

$$(c(|E|, \mu(\tilde{u}+v_{\kappa_n}, E)))^{1/2} w \to (c(|E|, \mu(\tilde{u}+v, E)))^{1/2} w \quad \text{in } L^2(\Omega).$$

Consequently, (2.68a) gives

$$\int_\Omega (c(|E|, \mu(\tilde{u}+v_{\kappa_n}, E)))^{1/2} \varepsilon(v_{\kappa_n}) w\,\mathrm{d}x \to \int_\Omega (c(|E|, \mu(\tilde{u}+v, E)))^{1/2} \varepsilon(v) w\,\mathrm{d}x,$$

whence

$$(c(|E|, \mu(\tilde{u}+v_{\kappa_n}, E)))^{1/2} \varepsilon(v_{\kappa_n}) \to (c(|E|, \mu(\tilde{u}+v, E)))^{1/2} \varepsilon(v). \tag{2.75}$$

In view of (2.72b), (2.75) yields

$$\liminf_{\mathbb{N}' \ni n \to \infty} M_{\kappa_n}^{(2)} \geq 2\int_\Omega c(|E|, \mu(\tilde{u}+v, E) I(v)\,\mathrm{d}x, \tag{2.76}$$

and hence, (2.73), (2.74), and (2.76) imply

$$\liminf_{\mathbb{N}' \ni n \to \infty} \langle L(v_{\kappa_n}), v_{\kappa_n} \rangle \geq \langle L(v), v \rangle. \tag{2.77}$$

The Lebesgue theorem and (2.68b) also show that for $\mathbb{N}' \ni n \to \infty$, there holds

$$J_{\kappa_n}(v_{\kappa_n}, w) \to J(v, w). \tag{2.78}$$

We have

$$J_{\kappa_n}(v_{\kappa_n}, v_{\kappa_n}) = J_{\kappa_n}(v, v_{\kappa_n}) + 2\int_\Omega (b_{\kappa_n} - b_0)(\kappa_n + I_{\kappa_n})^{1/2}\,\mathrm{d}x, \tag{2.79}$$

where

$$b_{\kappa_n} := b(|E|, \mu(\tilde{u}+v_{\kappa_n}, E)), \quad b_0 := b(|E|, \mu(\tilde{u}+v, E)),$$
$$I_{\kappa_n} := I(\tilde{u}+v_{\kappa_n}), \quad I_0 := I(\tilde{u}+v).$$

In view of

$$\left|\int_\Omega (b_{\kappa_n} - b_0)(\kappa_n + I_{\kappa_n})^{1/2}\,\mathrm{d}x\right| \leq \left(\int_\Omega (\kappa_n + I_{\kappa_n})\,\mathrm{d}x\right)^{1/2} \left(\int_\Omega |b_{\kappa_n} - b_0|^2\,\mathrm{d}x\right)^{1/2},$$

($\mathbf{A_2}$) and (2.68a), (2.68b) imply that for $\mathbb{N}' \ni n \to \infty$

$$\int_\Omega (b_{\kappa_n} - b_0)(\kappa_n + I_{\kappa_n})^{1/2}\, dx \to 0. \tag{2.80}$$

Since $J_{\kappa_n}(v, v_{\kappa_n}) \ge J(v, v_{\kappa_n})$, we have

$$\liminf_{\mathbb{N}' \ni n\to\infty} J_{\kappa_n}(v, v_{\kappa_n}) \ge \liminf_{\mathbb{N}' \ni n\to\infty} J(v, v_{\kappa_n}). \tag{2.81}$$

Lemma 2.5 and (2.68a), (2.68b) give

$$\liminf_{\mathbb{N}' \ni n\to\infty} J(v, v_{\kappa_n}) \ge J(v, v). \tag{2.82}$$

Now, combining (2.79)–(2.82) results in

$$\liminf_{\mathbb{N}' \ni n\to\infty} J_{\kappa_n}(v_{\kappa_n}, v_{\kappa_n}) \ge J(v, v). \tag{2.83}$$

(2.65b) and (2.68a) show $v \in V$, whereas (2.69), (2.71), (2.77), (2.78), and (2.83) imply (2.63). Finally, if $I(\tilde{u} + v) \neq 0$, it is easy to verify the existence of $p \in L^2(\Omega)$ such that (2.65a), (2.65b) hold true. □

2.3. *Initial-boundary value problems for isothermal incompressible electrorheological fluid flows*

For $\bar{I} := [0, T] \subset \mathbb{R}_+$ and a closed subspace $V \subset H^1(\Omega)^d$, we refer to $L^2(I; V)$ as the space of functions $v : \bar{Q} \to \mathbb{R}^d$, $\bar{Q} := I \times \overline{\Omega}$, with $v(t, \cdot) \in V$ f.a.a. $t \in I$ with norm $\|v\|_{L^2(I;V)} := (\int_I \|v(t, \cdot)\|_{1,\Omega}^2 dt)^{1/2}$.

Given a bounded Lipschitz domain $\Omega \subset \mathbb{R}^d$ with boundary $\Gamma = \partial\Omega$, we refer to V and H as the function spaces

$$V := \{v \in H_0^1(\Omega)^d \mid \nabla \cdot v = 0\}, \quad H := \{w \in L^2(\Omega)^d \mid \nabla \cdot w = 0\}.$$

Then, given functions

$$f \in L^2(I; H^{-s}(\Omega)^d, \quad u^0 \in H, \tag{2.84}$$

where $s = 1$ for $d = 2$ and $s = 3/2$ for $d = 3$, we consider the following initial-boundary value problem for isothermal incompressible electrorheological fluid flows

$$\rho(u_t + (u \cdot \nabla)u) - \nabla \cdot \sigma = f \quad \text{in } Q, \tag{2.85a}$$
$$\nabla \cdot u = 0 \quad \text{in } Q, \tag{2.85b}$$
$$u = 0 \quad \text{on } \Gamma \times (0, T), \tag{2.85c}$$
$$u(\cdot, 0) = u^0 \quad \text{in } \Omega. \tag{2.85d}$$

Here, the stress tensor σ is supposed to satisfy either the constitutive law (2.18) or (2.19).

In case of the regularized extended Bingham fluid model (2.19), we introduce a nonlinear operator $A_\kappa : V \to V^*$ according to

$$A_\kappa(u) := (u \cdot \nabla)u + M_\kappa(u, u), \tag{2.86}$$

where $M_\kappa(\cdot, \cdot)$ is given as in (2.26) with $\tilde{u} = 0$. We are looking for a weak solution

$$u \in L^2(I; V),\ u_t \in L^2(I; H^{-s}(\Omega)^d)$$

of (2.85a)–(2.85d) such that for all $v \in L^2(I; V)$ and $w \in H$

$$\int_0^T \langle \rho u_t, v\rangle \,\mathrm{d}t + \int_0^T \langle A_\kappa(u), v\rangle \,\mathrm{d}t = \int_0^T \langle f, v\rangle \,\mathrm{d}t, \tag{2.87a}$$

$$(u(\cdot, 0), w)_{0,\Omega} = (u^0, w)_{0,\Omega}. \tag{2.87b}$$

THEOREM 2.4. *Assume that* $(\mathbf{A_1})$, $(\mathbf{A_2})$, *and (2.84) hold true. Then, the initial-boundary value (2.85a)–(2.85d) admits a weak solution.*

PROOF. We provide a constructive existence proof by means of a Galerkin approximation with respect to a sequence $\{V_n\}_{\mathbb{N}}$ of finite dimensional subspaces $V_n \subset V$, $n \in \mathbb{N}$, that are limit dense in V. We assume $V_n = \operatorname{span}\{\varphi_n^{(1)}, \ldots, \varphi_n^{(N_n)}\}$ and look for a solution

$$u_n(t) = \sum_{i=1}^{N_n} \gamma_n^{(i)}(t)\, \varphi_n^{(i)} \tag{2.88}$$

of the problem

$$\left(\rho \frac{\mathrm{d}u_n}{\mathrm{d}t}, \varphi_n^{(i)}\right)_{0,\Omega} + \left\langle A_\kappa(u_n), \varphi_n^{(i)}\right\rangle = \left\langle f, \varphi_n^{(i)}\right\rangle,\ 1 \le i \le N_n, \tag{2.89a}$$

$$u_n(0) = P_n u^0, \tag{2.89b}$$

where $P_n : H \to V_n$ is the L^2 orthogonal projection onto V_n. We note that (2.89a), (2.89b) represents an initial-value problem for a system of first-order ordinary differential equations. The assumptions $(\mathbf{A_1})$, $(\mathbf{A_2})$ guarantee the existence of a solution. Moreover, it follows that the sequences $\{u_n\}_{\mathbb{N}}$ and $\{A_\kappa(u_n)\}_{\mathbb{N}}$ are bounded in $L^p(I; V)$ and $L^2(I; H^{-s}(\Omega))$, respectively. Consequently, there exist a subsequence $\mathbb{N}' \subset \mathbb{N}$ and functions $u \in L^2(I; V)$ and $\ell^* \in L^2(I; H^{-s}(\Omega))$ such that

$$u_n \rightharpoonup u^* \quad \text{in } L^2(I; V) \quad (\mathbb{N}' \ni n \to \infty),$$
$$A_\kappa(u_n) \rightharpoonup \ell^* \quad \text{in } L^2(I; H^{-s}(\Omega)) \quad (\mathbb{N}' \ni n \to \infty).$$

Arguments from the theory of parabolic partial differential equations (cf., e.g., LIONS [1969]) show that for $\varphi \in C_0^\infty(I; V)$, there holds

$$-\int_0^T (\rho u, \varphi_t)_{0,\Omega} \,\mathrm{d}t + \int_0^T \langle A_\kappa(u), \varphi\rangle \,\mathrm{d}t = \int_0^T \langle f, \varphi\rangle \,\mathrm{d}t,$$

which gives $u \in L^2(I; V)$, $u_t \in L^2(I; H^{-s}(\Omega))$ and implies that (2.87a) holds true, since $C_0^\infty(I; V)$ is dense in $L^2(I; V)$. A similar reasoning based on an appropriate choice of a test function allows to deduce $u(\cdot, 0) = u^0$. □

We note that a generalization of Theorem 2.4 to the case of inhomogeneous Dirichlet data $u = u^D$ on $\Sigma \times I$ can be found in LITVINOV [2004].

On the other hand, if we consider the extended Bingham fluid model based on the viscosity function (2.18), we have to deal with a strongly nonlinear parabolic variational inequality. Adopting the notation from 2.2.2, we are looking for a weak solution $u \in L^2(I; V)$, $u_t \in L^2(I; H^{-s}(\Omega))$ of (2.85a)–(2.85d) in the sense that for all $v \in L^2(I; V)$ and $w \in H$, there holds

$$\int_0^T \langle \rho(u_t, v - u\rangle \, dt + \int_0^T \langle (u \cdot \nabla)u, v - u\rangle \, dt \tag{2.90a}$$

$$+ \int_0^T (J(u, v) - J(u, u)) \, dt + \int_0^T \langle L(u), v - u\rangle \, dt \geq \int_0^T \langle f, v - u\rangle \, dt,$$

$$(u(\cdot, 0), w)_{0,\Omega} = (u^0, w)_{0,\Omega}. \tag{2.90b}$$

THEOREM 2.5. *Assume that* $(\mathbf{A_1})''$, $(\mathbf{A_2})$ *and (2.84) hold true. Then, the variational inequality (2.90a), (2.90b) has a solution* $u \in L^2(I; V)$, $u_t \in L^2(I; H^{-s}(\Omega))$.

PROOF. We choose $\{\kappa_n\}_{\mathbb{N}}$ as a null sequence of positive regularization parameters. For each $n \in \mathbb{N}$, Theorem 2.4 guarantees the existence of a weak solution u_n of (2.85a)–(2.85d) with respect to the regularized extended Bingham fluid model (2.19) (with κ replaced by κ_n). The boundedness of the sequence $\{u_n\}_{\mathbb{N}}$ in $L^2(I; V)$ infers the existence of a subsequence $\mathbb{N}' \subset \mathbb{N}$ and of a function $u \in L^2(I; V)$ such that $u_n \rightharpoonup u(\mathbb{N}' \ni n \to \infty)$ in $L^2(I; V)$. Passing to the limit as in the proof of Theorem 2.3 allows to conclude. □

2.4. *Balance equations and constitutive laws for nonisothermal incompressible electrorheological fluid flows*

Nonisothermal flows of non-Newtonian fluids have been studied in a series of papers mostly in the engineering literature with respect to industrially relevant applications. Various laws of the temperature dependence of the viscosity have been assumed, e.g., a hyperbolic law for the variation of the viscosity or a Reynolds-type relation. A rigorous mathematical analysis of nonisothermal flow in a Bingham fluid can be found in DUVAUT and LIONS [1971].

As far as electrorheological fluids are concerned, it is well-known by experimental evidence that their operational behavior exhibits a dependence on the temperature (cf. BENDERSKAIA, KHUSID and SHULMAN [1980], TABATABAI [1993], and ZHIZKIN [1986]). Figure 2.4 displays the temperature dependence of the shear stress (left) and of the current density (right) for a polyurethane based electrorheological fluid under different operational conditions, i.e., electric field strengths. Mathematical models for nonisothermal electrorheological fluid flows based on a power law constitutive equation have been studied in RUZICKA [2000] (cf. also ECKART and SADIKI [2001], and SADIKI and BALAN [2003]).

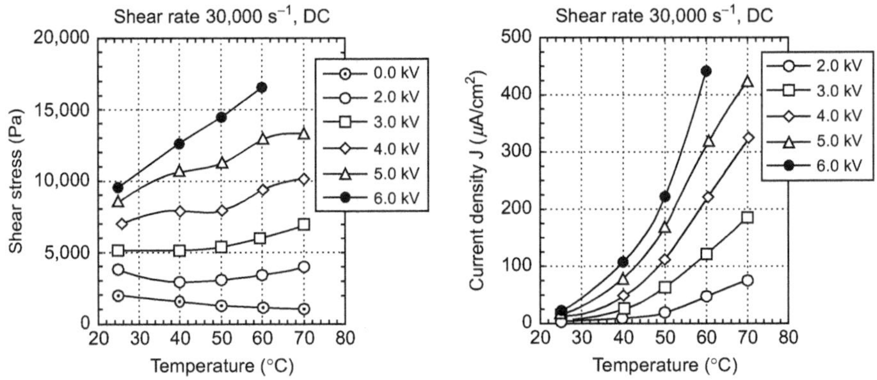

FIG. 2.4 Temperature dependence of the shear stress (left) and the current density (right) in electrorheological fluids (from BAYER [1997a]).

Here, we follow the approach in LITVINOV and HOPPE [2005]. We assume a general dependence of the viscosity function on the temperature θ and consider the following constitutive equation between the stress tensor σ and the rate of strain tensor ε

$$\sigma = -pI + 2\varphi(I(u), |E|, \mu(u, E), \theta)\varepsilon(u). \tag{2.91}$$

As in Section 2.1, u and p stand for the velocity and pressure of the fluid flow, $I(u)$ is the second invariant of the rate of strain tensor, E refers to the electric field, and $\mu(u, E)$ is the square of the cosine of the angle between the velocity and the electric field.

The equations of motion and the incompressibility condition for the fluid flow have to be completed by a thermodynamical balance equation, which can be deduced from the energy conservation law

$$e_t + u \cdot \nabla e = \sigma : \varepsilon(u) - \nabla \cdot q + f_2,$$

where e denotes the specific internal energy, q is the heat flux vector, and f_2 stands for a volumetric heat source/sink. As constitutive equations, we assume the linear Fourier law

$$e = \rho c\theta, \quad q = -k\nabla\theta,$$

where ρ, c, and k refer to the density, the specific heat, and the thermal conductivity. We are thus led to the following coupled system in $Q := \Omega \times (0, T)$

$$\rho(u_t + (u \cdot \nabla)u) - \nabla \cdot \sigma = f_1, \tag{2.92a}$$

$$\nabla \cdot u = 0, \tag{2.92b}$$

$$\rho c(\theta_t + u \cdot \nabla\theta) - k\Delta\theta - 2\varphi(I(u), |E|, \mu(u, E), \theta)I(u) = f_2, \tag{2.92c}$$

where f_1 is a volumetric force on the fluid. The equations have to be completed by appropriate initial and boundary conditions that will be discussed in detail in the subsequent subsections.

REMARK 2.4. We note that the impact of the electrical conductivity in the thermal balance equation (2.92c) has been neglected, since electrorheological fluids are electrically nonconducting.

As far as the viscosity function φ is concerned, we will assume that the following condition is satisfied:

(T$_1$) φ is a continuous function of its arguments, i.e., $\varphi \in C(\mathbb{R}_+^2 \times [0,1] \times \mathbb{R})$. For fixed $(y_2, y_3, y_4) \in \mathbb{R}_+ \times [0,1] \times \mathbb{R}$, the function $\varphi(\cdot, y_2, y_3, y_4)$ is continuously differentiable in $\mathbb{R}_+$, i.e., $\varphi(\cdot, y_2, y_3, y_4) \in C^1(\mathbb{R}_+)$. There exist positive constants c_i, $1 \le i \le 4$, such that

$$c_2 \ge \varphi(y_1, y_2, y_3, y_4) \ge c_1,$$

$$\varphi(y_1, y_2, y_3, y_4) + 2\frac{\partial \varphi}{\partial y_1}(y_1, y_2, y_3, y_4) \ge c_3,$$

$$\frac{\partial \varphi}{\partial y_1}(y_1, y_2, y_3, y_4)|\, y_1 \le c_4.$$

The first condition in **(T$_1$)** requires nonvanishing viscosity for vanishing shear rate and thus does not include Bingham-type electrorheological flow models. However, as in Section 2.1, we may consider viscosity functions of the form

$$\begin{aligned}&\varphi(I(u), |E|, \mu(u,E), \theta)\\ &\quad = b(|E|, \mu(u,E), \theta)(\kappa + I(u))^{-1/2} + c((I(u), |E|, \mu(u,E), \theta),\end{aligned} \tag{2.93}$$

where $\kappa \ge 0$, the function c is supposed to satisfy **(T$_1$)**, and the function b is subject to the assumption

(T$_2$) b is a continuous function of its arguments, i.e., $b \in C(\mathbb{R}_+ \times [0,1] \times \mathbb{R})$. There exists a positive constant c_5 such that

$$c_5 \ge b(y_1, y_2, y_3) \ge 0.$$

The case $\kappa = 0$ in (2.93) refers to a generalized Bingham-type model for nonisothermal electrorheological fluid flows, whereas $\kappa > 0$ can be interpreted as a regularization thereof. The physical relevance of these assumptions with respect to the fluid flow has been discussed in Section 2.2.

We consider the following modification of the thermal balance equation (2.92c), which gives rise to a nonlocal model:

$$\rho c(\theta_t + u \cdot \nabla\theta) - k\Delta\theta - 2\varphi(I(u)), |E|, \mu(u,E), \theta)I(P_\beta(u)) = f_2. \tag{2.94}$$

Here, $P_\beta \in \mathcal{L}(W^{1,2}(\Omega)^d, C^\infty(\overline{\Omega})^d)$, $\beta > 0$, is the regularization operator

$$(P_\beta(v))(x) := \int_{\mathbb{R}^d} \omega_\beta(|x - x'|)\,(P_E(v))(x')\,\mathrm{d}x', \quad x \in \overline{\Omega}, v \in W^{1,2}(\Omega), \tag{2.95}$$

where $P_E \in \mathcal{L}(W^{1,2}(\Omega)^d, W^{1,2}(\mathbb{R}^d))$ is an extension operator and $\omega_\beta \in C_+^\infty(\mathbb{R}_+)$ with $\text{supp}(\omega_\beta) \subset [0, \beta]$ and $\int_{\mathbb{R}^d} \omega_\beta(|x|)\,dx = 1$.

REMARK 2.5. The physical interpretation of the regularization operator P_β in the thermal balance equation (2.94) is that the dissipation of energy at a point $x \in \Omega$ only depends on the rate of strain tensor in a small vicinity of the point. We note that nonlocal models agree remarkably well with atomistic theories and experimental observations (cf., e.g., ERINGEN [2002]).

2.5. *Boundary value problems for steady nonisothermal incompressible electrorheological fluid flows*

We consider steady, nonisothermal, incompressible electrorheological fluid flow and assume $\Omega \subset \mathbb{R}^d$ to be a bounded Lipschitz domain with boundary Γ such that $\Gamma = \bar{\Gamma}_D \cup \bar{\Gamma}_N$, $\Gamma_D \cap \Gamma_N = \emptyset$. We further suppose

$$\begin{aligned} &f_1 \in L^2(\Omega)^d, \quad f_2 \in L^2(\Omega), \quad g \in L^2(\Gamma_N)^d, \\ &\quad u^D \in W^{1/2,2}(\Gamma_D)^d, \quad \theta^D \in W^{1/2,2}(\Gamma) \end{aligned} \tag{2.96}$$

to be given functions and consider the following boundary value problem

$$\nabla \cdot \sigma = f_1, \quad \nabla \cdot u = 0 \quad \text{in } \Omega, \tag{2.97a}$$

$$-\chi \Delta\theta + u \cdot \nabla\theta + 2\varrho\varphi(I(u), |E|, \mu(u, E), \theta)I(u) = f_2 \quad \text{in } \Omega, \tag{2.97b}$$

$$u = u^D \quad \text{on } \Gamma_D, \tag{2.97c}$$

$$\nu \cdot \sigma = g \quad \text{on } \Gamma_N, \tag{2.97d}$$

$$\theta = \theta^D \quad \text{on } \Gamma, \tag{2.97e}$$

where $\chi = (\rho c)^{-1}\kappa$ and $\varrho = (\rho c)^{-1}$. As in Section 2.2, we assume a unilateral coupling between the electric field E and the flow field, i.e., we suppose that E is given by means of an electrical potential ψ, which satisfies the boundary value problem (2.22a)–(2.22c).

We study the existence of a weak solution of (2.97a)–(2.97e) where the velocity is supposed to be in $W^{1,2}(\Omega)^d \cap H(\text{div}^0; \Omega)$, the pressure p in $L^2(\Omega)$, and the temperature θ in $W^{1,r}(\Omega)$ with $1 < r < 2$ for $d = 2$ and $1 < r < 3/2$ for $d = 3$. In order to accommodate the inhomogeneous Dirichlet boundary data (2.97c), (2.97e), we define $\tilde{u} \in W^{1,2}(\Omega)^d$ and $\tilde{\theta} \in W^{1,r}(\Omega)$ such that $\tilde{u}|_{\Gamma_D} = u^D$ and $\tilde{\theta}|_\Gamma = \theta^D$. We set

$$X := W^{1,2}_{0,\Gamma_D}(\Omega)^d \cap H(\text{div}^0; \Omega), \quad \|v\|_X := \left(\int_\Omega (I(v))^2\,dx\right)^{1/2}$$

and consider the operators

$$N : X \times W^{1,r}_{0,\Gamma}(\Omega) \to X^*, \quad A : X \times W^{1,r}_{0,\Gamma}(\Omega) \to W^{-1,s}(\Omega),\ s = \frac{r}{r-1},$$

which are defined according to

$$\langle N(v,\zeta),w\rangle := \tag{2.98a}$$
$$2\int_\Omega \varphi(I(\tilde{u}+v),|E|,\mu(\tilde{u}+v,E),\tilde{\theta}+\zeta)\varepsilon(\tilde{u}+v):\varepsilon(w)\,\mathrm{d}x,$$

$$\langle A(w,\zeta),\xi\rangle := \chi^{-1}\int_\Omega \Big((\tilde{\theta}+\zeta)(\tilde{u}+w)\cdot\nabla\xi \tag{2.98b}$$
$$+2\varrho\varphi(I(\tilde{u}+w),|E|,\mu(\tilde{u}+w),\tilde{\theta}+\zeta)I(\tilde{u}+w)\xi\Big)\,\mathrm{d}x.$$

Here, $\langle\cdot,\cdot\rangle$ refers to the dual product between X^* and X in (2.98a) and to the dual product between $W^{-1,s}(\Omega)$ and $W^{1,r}_{0,\Gamma}(\Omega)$ in (2.98b). For the ease of exposition, we will use the same notation. The correct meaning will always follow easily from the context.

Moreover, we refer to $B\in\mathcal{L}(X,L^2(\Omega))$ as the divergence operator $Bv=\nabla\cdot v$, $v\in X$. We consider the following system of variational equations.

Find $(v,p,\theta)\in X\times L^2(\Omega)\times W^{1,r}_{0,\Gamma}(\Omega)$ such that

$$\langle N(v,\theta),w\rangle - \langle B^*p,w\rangle = \langle f_1+g,w\rangle,\quad w\in X \tag{2.99a}$$
$$(Bv,q)_{0,\Omega}=0,\quad q\in L^2(\Omega), \tag{2.99b}$$
$$(\nabla\theta,\nabla\zeta)_{0,\Omega}-\langle A(v,\theta),\zeta\rangle = (f_3,\zeta)_{0,\Omega},\quad \zeta\in W^{1,s}_{0,\Gamma}(\Omega), \tag{2.99c}$$

where $(f_3,\zeta)_{0,\Omega}:=(f_2,\zeta)_{0,\Omega}-(\nabla\tilde{\theta},\nabla\zeta)_{0,\Omega}$. For notational convenience, we denote by θ both the solution of (2.97a)–(2.97e) and (2.99a)–(2.99c). It will be clear from the context which one is considered.

LEMMA 2.6. *Assume that (u,p,θ) is a classical solution of (2.97a)–(2.97e). Then, the triple $(u-\tilde{u},p,\theta-\tilde{\theta})$ solves (2.99a)–(2.99c). Conversely, if (v,p,θ) is a sufficiently smooth solution of (2.99a)–(2.99c), then the triple $(\tilde{u}+v,p,\tilde{\theta}+\theta)$ solves (2.97a)–(2.97c) in the classical sense.*

PROOF. The assertions are easily verified by Green's formula. □

We will prove the existence of a solution of the system (2.99a)–(2.99c) by an approximation involving the regularization operator P_β from (2.95). For that purpose, we introduce the operator

$$A_\beta : X\times W^{1,2}_{0,\Gamma}(\Omega)\to W^{-1,2}(\Omega),$$

which is given by means of

$$\langle A_\beta(w,\zeta),\xi\rangle := \chi^{-1}\int_\Omega \Big((\tilde{\theta}+\zeta)(\tilde{u}+w)\cdot\nabla\xi \tag{2.100}$$
$$+2\varrho\varphi(I(\tilde{u}+w),|E|,\mu(\tilde{u}+w),\tilde{\theta}+\zeta)I(P_\beta(\tilde{u}+w))\xi\Big)\,\mathrm{d}x.$$

Here, $\langle\cdot,\cdot\rangle$ stands for the dual product between $W^{-1,2}(\Omega)$ and $W^{1,2}_{0,\Gamma}(\Omega)$. The associated boundary value problem reads as follows.

Find $(v,p,\theta)\in X\times L^2(\Omega)\times W^{1,2}_{0,\Gamma}(\Omega)$ such that

$$\langle N(v,\theta),w\rangle-\langle B^*p,w\rangle=\langle f_1+g,w\rangle,\quad w\in X, \tag{2.101a}$$

$$(Bv,q)_{0,\Omega}=0,\quad q\in L^2(\Omega), \tag{2.101b}$$

$$(\nabla\theta,\nabla\zeta)_{0,\Omega}-\langle A_\beta(v,\theta),\zeta\rangle=(f_3,\zeta)_{0,\Omega},\quad \zeta\in W^{1,2}_{0,\Gamma}(\Omega). \tag{2.101c}$$

THEOREM 2.6. *Suppose that* $(\mathbf{T_1})$, *(2.96) are satisfied and* $E\in L^4(\Omega)$. *Then, for any* $\beta>0$, *there exists a solution* (v_β,p_β) *of (2.101a)–(2.101c), and there exist constants* $C_i>0$, $1\le i\le 2$, *such that*

$$\|v_\beta\|_X\le C_1,\quad \|p_\beta\|_{0,\Omega}\le C_2\,,\, b\in(0,a),\, a>0. \tag{2.102}$$

PROOF. We refer to LITVINOV and HOPPE [2005]. □

We will now address the existence of a solution of the system (2.99a)–(2.99c). We define an operator $\Lambda_2: V\to\mathcal{L}(W^{1,r}_0(\Omega),W^{-1,s}(\Omega))$ according to

$$\langle\Lambda_2(v)\zeta,\xi\rangle:=\chi^{-1}\int_\Omega\zeta(\tilde u+v)\cdot\nabla\xi\,dx, \tag{2.103}$$

where $v\in V$, $\zeta\in W^{1,r}_0(\Omega)$, and $\xi\in W^{1,s}_0(\Omega)$. We consider the auxiliary problem.

Find $\bar\theta\in W^{1,r}_0(\Omega)$ such that

$$\langle\nabla\bar\theta,\nabla\xi\rangle-\langle\Lambda_2(v)\bar\theta,\xi\rangle=0,\quad \xi\in W^{1,s}_0(\Omega). \tag{2.104}$$

Under these prerequisites, we now assume $\{\beta_n\}_{\mathbb{N}}$ to be a sequence of regularization parameters $\beta_n\in\mathbb{R}_+$, $n\in\mathbb{N}$, such that $\beta_n\to 0$ as $n\to\infty$ and further suppose that $\{(v_n,p_n,\theta_n)\}_{\mathbb{N}}$ is an associated sequence of solutions $(v_n,p_n,\theta_n)\in X\times L^2(\Omega)\times W^{1,2}_0(\Omega)$, $n\in\mathbb{N}$, of the system (2.101a)–(2.101c) whose existence is guaranteed under the assumptions of Theorem 2.6.

THEOREM 2.7. *Assume that* $\Omega\subset\mathbb{R}^d$, $d=2$ *or* $d=3$, *is a bounded* C^3 *domain. Further, suppose that the conditions* $(\mathbf{T_1})$ *and (2.96) hold true and the variational equation (2.104) is only trivially solvable. For a null sequence* $\{\beta_n\}_{\mathcal{N}}$ *of positive regularization parameters, let* $\{(v_n,p_n,\theta_n)\}_{\mathbb{N}}$ *be the associated sequence of solutions* $(v_n,p_n,\theta_n)\in X\times L^2(\Omega)\times W^{1,2}_0(\Omega)$, $n\in\mathbb{N}$, *of the system (2.101a)–(2.101c). Then, there exist a subsequence* $\mathbb{N}^*\subset\mathbb{N}$ *and a triple* $(v,p,\theta)\in X\times L^2(\Omega)\times W^{1,r}_0(\Omega)$ *such that for* $\mathbb{N}^*\ni n\to\infty$

$$v_n\to v\quad \text{in } X, \tag{2.105a}$$

$$p_n\to p\quad \text{in } L^2(\Omega), \tag{2.105b}$$

$$\theta_n\to\theta\quad \text{in } W^{1,r}_0(\Omega)\,. \tag{2.105c}$$

The triple (v,p,θ) *is a solution of the system (2.99a)–(2.99c).*

PROOF. We refer to LITVINOV and HOPPE [2005]. □

3. Numerical solution of electrorheological fluid flows

This section is devoted to the numerical solution of stationary and time-dependent, isothermal and nonisothermal electrorheological fluid flows. We shall begin in Section 3.1 with steady-state isothermal problems with emphasis on nonlinear Uzawa-type algorithms in 3.1.1 and augmented Lagrangian methods in 3.1.2. This includes the construction of preconditioners based on approximate inverses of the Stokes operator, which will be the subject of 3.1.3. An augmented Lagrangian approach particularly suited for nonregularized Bingham models shall be considered in 3.1.4. Time-dependent problems shall be taken care of in Section 3.2, and in Section 3.3, we shall address nonisothermal fluid flows. We refer to CROCHET [1984], ELMAN, SILVESTER and WATHEN [2005], GLOWINSKI [2004], GUNZBURGER [1989], HUANG [1998], THOMASSET [1981], and TUREK [1999] with regard to a general presentation of numerical solution techniques for Newtonian and non-Newtonian fluid flows.

3.1. Steady-state isothermal incompressible flow problems

As we have seen in Section 2.2 (cf. Theorem 2.2), steady isothermal, incompressible electrorheological fluid flows with a regularized viscosity function can be approximated by finite dimensional nonlinear saddle point problems of the form:
Find $(v_n, p_n) \in X_n \times Q_n$ such that

$$\langle S_n(v_n), w_n\rangle - \langle B_n^* p_n, w_n\rangle = \langle f + g, w_n\rangle, \quad w_n \in X_n, \tag{3.1a}$$

$$(B_n v_n, q_n)_{0,\Omega} = 0, \quad q_n \in Q_n, \tag{3.1b}$$

where $X_n \subset X := W^{1,2}_{0,\Gamma_D}(\Omega)$ and $Q_n \subset L^2(\Omega)$, $n \in \mathbb{N}$, are finite dimensional subspaces, $S_n(u_n) := M_\kappa(u_n, u_n)$ with $M_\kappa : X \times X \to X^*$ being the nonlinear operator given by (2.26), and B_n refers to the discrete divergence operator (2.30). We assume that the pairs (X_n, Q_n), $n \in \mathbb{N}$, satisfy the discrete LBB condition (2.31).

Since the nonlinear operator S_n admits an inverse S_n^{-1}, the discrete velocity field v_n can be formally eliminated from (3.1a), (3.1b), which gives rise to

$$B_n S_n^{-1}(B_n^* p_n + f_n + g_n) = 0. \tag{3.2}$$

REMARK 3.1. In the linear regime, the linear operator $B_n S_n^{-1} B_n^*$ is called the Schur complement, and (3.2) is referred to as the Schur complement system.

All numerical techniques for the solution of (3.1a), (3.1b) are nonlinear versions of methods that have been developed for linear saddle point problems, i.e., when the operator S in (3.1a) is a linear operator. The most popular numerical schemes are Uzawa-type algorithms and those based on the augmented Lagrangian approach (cf., e.g., CAO [2003], FORTIN and GLOWINSKI [1983], GLOWINSKI [1984, 2004], GLOWINSKI and LE TALLEC [1989], and LIN and CAO [2006]). In the nonlinear regime, these methods are outer–inner iterative schemes where the outer iteration takes care of the saddle point structure of the problem and the inner iteration is devoted to the nonlinear problem associated with the operator S.

3.1.1. *Nonlinear Uzawa-type algorithms*

The nonlinear Uzawa algorithm can be formally derived as a damped nonlinear Richardson iteration with damping parameter $\tau > 0$ applied to (3.2).

Given $p_n^{(0)} \in Q_n$, compute $p_n^{(\nu)} \in Q_n$, $\nu \in \mathbb{N}$, according to

$$p_n^{(\nu+1)} = p_n^{(\nu)} - \tau B_n S_n^{-1}\left(B_n^* p_n^{(\nu)} + f_n + g_n\right), \quad \nu \in \mathbb{N}_0. \tag{3.3}$$

It is used below the same notation ν for the velocity and the number of iteration. The upper index ν in brackets denotes the number of iteration. In other case ν is the velocity. Of course, we are interested in iterates $u_n^{(\nu)}$ for the discrete velocity field as well which can be obtained by means of (3.1a). Thus, we arrive at the following standard form of the nonlinear Uzawa algorithm:

Nonlinear Uzawa algorithm

Given $(v_n^{(0)}, p_n^{(0)}) \in X_n \times Q_n$ and $\tau > 0$, compute $\left(v_n^{(\nu)}, p_n^{(\nu)}\right) \in X_n \times Q_n$, $\nu \in \mathbb{N}$, as the solution of

$$\left\langle S_n v_n^{(\nu+1)}, w_n\right\rangle - \left\langle B_n^* p_n^{(\nu)}, w_n\right\rangle = \langle f + g, w_n\rangle, \quad w_n \in X_n, \tag{3.4a}$$

$$(p_n^{(\nu+1)} - p_n^{(\nu)}, q_n)_{0,\Omega} = -\tau\left(B_n v_n^{(\nu+1)}, q_n\right)_{0,\Omega}, \quad q_n \in Q_n. \tag{3.4b}$$

THEOREM 3.1. *Let $(v_n, p_n) \in X_n \times Q_n$ be the solution of (3.1a), (3.1b) and suppose that $\{(v_n^{(\nu)}, p_n^{(\nu)})\}_{\mathbb{N}}$ is the sequence of iterates generated by the nonlinear Uzawa algorithm (3.4a), (3.4b). Assume $\tau < 2\gamma_L \beta^2$ with γ_L as in Lemma 2.3 and β from Lemma 2.2. Then, for $\nu \to \infty$, there holds*

$$v_n^{(\nu)} \to v_n \quad \text{in } X, \quad p_n^{(\nu)} \to p_n \quad \text{in } L^2(\Omega).$$

PROOF. We set $e_v^{(\nu)} := v_n^{(n))} - v_n$ and $e_p^{(\nu)} := p_n^{(\nu)} - p_n$. If we subtract (3.1a) from (3.4a) and (3.1b) from (3.4b), we obtain

$$\left\langle S_n(v_n^{(\nu+1)}) - S_n(v_n), w_n\right\rangle = \left\langle B_n^* e_p^{(\nu)}, w_n\right\rangle, \quad w_n \in X_n, \tag{3.5a}$$

$$\left(e_p^{(\nu+1)} - e_p^{(\nu)}, q_n\right)_{0,\Omega} = -\tau\left(B_n e_v^{(\nu+1)}, q_n\right)_{0,\Omega}, \quad q_n \in Q_n. \tag{3.5b}$$

We choose $w_n = 2e_u^{(\nu+1)}$ in (3.5a) and $q_n = 2e_p^{(\nu+1)}$ in (3.5b). Then, multiplying (3.5a) by 2τ and adding it to (3.5b) yields

$$\begin{aligned} &\|e_p^{(\nu+1)}\|_{0,\Omega}^2 + \|e_p^{(\nu+1)} - e_p^{(\nu)}\|_{0,\Omega}^2 - \|e_p^{(\nu)}\|_{0,\Omega}^2 \\ &\quad + 2\tau\left\langle S_n\left(v_n^{(\nu+1)}\right) - S_n(v_n), e_v^{(\nu+1)}\right\rangle = 2\tau\left(B_n e_v^{(\nu+1)}, e_p^{(\nu)} - e_p^{(\nu+1)}\right)_{0,\Omega}. \end{aligned}$$

The results of Section 2.2 imply

$$\begin{aligned} &\|e_p^{(\nu+1)}\|_{0,\Omega}^2 + \|e_p^{(\nu+1)} - e_p^{(\nu)}\|_{0,\Omega}^2 - \|e_p^{(\nu)}\|_{0,\Omega}^2 \\ &\quad + 2\tau\,\gamma_L\,\|e_v^{(\nu+1)}\|_X^2 \le 2\frac{\tau}{\beta}\,\|e_v^{(\nu+1)}\|_X\,\|e_p^{(\nu+1)} - e_p^{(\nu)}\|_{0,\Omega}, \end{aligned}$$

and hence, Young's inequality gives

$$\|e_p^{(\nu+1)}\|_{0,\Omega}^2 - \|e_p^{(\nu)}\|_{0,\Omega}^2 + \tau\left(2\gamma_L - \frac{\tau}{\beta^2}\right)\|e_v^{(\nu+1)}\|_X^2 \le 0. \tag{3.6}$$

We deduce from (3.6) that the sequence $\{\|e_p^{(\nu)}\|_{0,\Omega}^2\}_{\mathbb{N}}$ is convergent which in turn gives us $e_v^{(\nu)} \to 0$ as $\nu \to \infty$. Moreover, we have

$$\left\| S_n\left(v_n^{(\nu+1)}\right) - S_n(v_n)\right\|_{X^*} \to 0 \quad \text{as} \quad \nu \to \infty. \tag{3.7}$$

On the other hand, in view of (3.5a) and Lemma 2.2, it follows that

$$\left\| S_n\left(v_n^{(\nu+1)}\right) - S_n(v_n)\right\|_{X^*} = \left\| B_n^* e_p^{(\nu)}\right\|_{X*} \geq \beta \left\| e_p^{(\nu)}\right\|_{0,\Omega}.$$

Hence, (3.7) tells us that $e_p^{(\nu)} \to 0$ as $\nu \to \infty$. □

It is well-known from the theory of linear iterative schemes that the convergence can be significantly improved by preconditioning (cf., e.g., BANK, WELFERT, and YSERENTANT [1990], BRAMBLE, PASCIAK and VASSILEV [1997], ELMAN [2002], ELMAN and GOLUB [1994], ELMAN and SILVESTER [1996], KLAWONN [1998], and RUSTEN and WINTHER [1992]). In terms of the Richardson iteration (3.3), we may use

$$p_n^{(\nu+1)} = p_n^{(\nu)} + P_n^{-1} B_n S_n^{-1}\left(B_n^* p_n^{(\nu)} + f_n + g_n\right), \quad \nu \in \mathbb{N}_0,$$

with a preconditioner $P_n : Q_n \to Q_n$, which is assumed to be a linear symmetric positive operator. This leads to the preconditioned nonlinear Uzawa algorithm.

Preconditioned nonlinear Uzawa algorithm
Let $P_n : Q_n \to Q_n$ be a linear symmetric positive operator. Then, given $(v_n^{(0)}, p_n^{(0)}) \in X_n \times Q_n$, compute $(v_n^{(\nu)}, p_n^{(\nu)}) \in X_n \times Q_n$, $\nu \in \mathbb{N}$, as the solution of

$$\left\langle S_n(v_n^{(\nu+1)}, w_n\right\rangle - \left\langle B_n^* p_n^{(\nu)}, w_n\right\rangle = \langle f + g, w_n\rangle, \quad w_n \in X_n, \tag{3.8a}$$

$$(p_n^{(\nu+1)} - p_n^{(\nu)}, q_n)_{0,\Omega} = -(P_n^{-1} B_n v_n^{(\nu+1)}, q_n)_{0,\Omega}, \quad q_n \in Q_n. \tag{3.8b}$$

REMARK 3.2. The preconditioned nonlinear Uzawa algorithm contains the standard form (3.4a), (3.4b) as a special case as can be readily seen by choosing $P_n = \tau I_n$, $\tau > 0$, with I_n denoting the identity on Q_n.

A major problem in the practical realization of the algorithm (3.8a), (3.8b) is that it requires the solution of a nonlinear problem. This issue is usually taken care of by an approximation $\tilde{S}_n$ of S_n. We will discuss feasible choices of $\tilde{S}_n$ in 3.1.3. Since in this case we do not solve (3.8a), (3.8b) exactly, the resulting scheme is referred to as a preconditioned inexact nonlinear Uzawa algorithm.

Preconditioned inexact nonlinear Uzawa algorithm
Let $\tilde{S}_n^{-1}$ be an approximate inverse of S_n^{-1} and assume that $P_n : Q_n \to Q_n$ is a linear symmetric positive operator. Then, given $(v_n^{(0)}, p_n^{(0)}) \in X_n \times Q_n$, compute $(v_n^{(\nu)}, p_n^{(\nu)}) \in X_n \times Q_n$, $\nu \in \mathbb{N}$, as the solution of

$$\left\langle \tilde{S}_n(v_n^{(\nu+1)}, w_n\right\rangle - \left\langle B_n^* p_n^{(\nu)}, w_n\right\rangle = \langle f + g, w_n\rangle, \quad w_n \in X_n, \tag{3.9a}$$

$$(p_n^{(\nu+1)} - p_n^{(\nu)}, q_n)_{0,\Omega} = -(P_n^{-1} B_n v_n^{(\nu+1)}, q_n)_{0,\Omega}, \quad q_n \in Q_n. \tag{3.9b}$$

In case of a linear symmetric positive definite operator S_n, the convergence of preconditioned inexact nonlinear Uzawa algorithms has been analyzed in BRAMBLE, PASCIAK and VASSILEV [1997], ELMAN and GOLUB [1994]. As can be expected, it requires some conditions on the approximate inverse $\tilde{S}_n^{-1}$ and on the preconditioner P_n.

3.1.2. Augmented Lagrangian methods

As we already know from 2.2.1, the nonlinear saddle point problem (3.1a), (3.1b) results from the constrained minimization problem

$$\min_{v_n \in V_n} (J_\kappa(v_n, v_n) + \langle L(v_n), v_n \rangle),$$

where $V_n := X_n \cap H(\mathrm{div}^0; \Omega)$ and $J_\kappa : X \times X \to \mathbb{R}$ and $L : X \to X^*$ are given by (2.24a) and (2.24b), if we couple the constraints $B_n v_n = 0$ by Lagrange multipliers $p_n \in Q_n$. An alternative is to use penalty methods

$$\min_{v_n \in X_n} \big(J_\kappa(v_n, v_n)) + \langle L(v_n), v_n \rangle + r(B_n v_n, B_n v_n)_{0,\Omega}\big),$$

where the constraints are taken care of by a penalty term with penalty parameter $r > 0$. The disadvantage with penalty methods is that the penalty parameter r usually has to be chosen quite large, which has a negative impact on the condition of the resulting algebraic system.

The augmented Lagrangian techniques combine the previous approaches in such a way that they work sufficiently well for a moderate choice of the penalty parameter. A convergence analysis in the symmetric case is given in FORTIN and GLOWINSKI [1983], GLOWINSKI and LE TALLEC [1989], whereas the nonsymmetric case has been addressed in AWANOU and LAI [2005].

Augmented Lagrangian algorithm

Given $\big(v_n^{(0)}, p_n^{(0)}\big) \in X_n \times Q_n$ and $r, \rho > 0$, compute $\left(v_n^{(\nu)}, p_n^{(\nu)}\right) \in X_n \times Q_n$, $\nu \in \mathbb{N}$, such that for $(w_n, q_n) \in X_n \times Q_n$, there holds

$$\left\langle S_n v_n^{(\nu+1)}, w_n \right\rangle - \left\langle B_n^* p_n^{(\nu)}, w_n \right\rangle + r\left(B_n v_n^{(\nu+1)}, B_n w_n\right)_{0,\Omega} = \langle f + g, w_n \rangle, \tag{3.10a}$$

$$\left(p_n^{(\nu+1)} - p_n^{(\nu)}, q_n\right)_{0,\Omega} + \rho\left(B_n v_n^{(\nu+1)}, q_n\right)_{0,\Omega} = 0. \tag{3.10b}$$

THEOREM 3.2. *Let $(v_n, p_n) \in X_n \times Q_n$ be the solution of (3.1a), (3.1b), and let $\{(v_n^{(\nu)}, p_n^{(\nu)})\}_{\mathbb{N}}$ be the sequence of iterates generated by the augmented Lagrangian algorithm (3.10a), (3.10b). Then, under the assumption $\rho < 2r$ for $\nu \to \infty$, there holds*

$$v_n^{(\nu)} \to v_n \quad \text{in } X, \quad p_n^{(\nu)} \to p_n \quad \text{in } L^2(\Omega).$$

PROOF. The convergence result can be verified using a similar reasoning as in the proof of Theorem 3.1. Setting $e_v^{(\nu)} := v_n^{(n)} - v_n$ and $e_p^{(\nu)} := p_n^{(\nu)} - p_n$, it follows from (3.1a), (3.1b) and (3.10a), (3.10b) that for $w_n \in X_n$ and $q_n \in Q_n$, there holds

$$\left\langle S_n(v_n^{(\nu+1)} - S_n(v_n), w_n \right\rangle + r\left(Be_v^{(\nu+1)}, Bw_n\right)_{0,\Omega} = \left\langle B_n^* e_p^{(\nu)}, w_n \right\rangle, \tag{3.11a}$$

$$\left(e_p^{(\nu+1)} - e_p^{(\nu)}, q_n\right)_{0,\Omega} = -\rho\left(B_n e_v^{(\nu+1)}, q_n\right)_{0,\Omega}. \tag{3.11b}$$

With $w_n = 2e_v^{(\nu+1)}$, $q_n = 2e_p^{(\nu+1)}$ in (3.11a), (3.11b) and the results of Section 2.2 as well as Young's inequality, we obtain

$$\|e_p^{(\nu+1)}\|_{0,\Omega}^2 - \|e_p^{(\nu)}\|_{0,\Omega}^2 + 2\rho\,\gamma\,\|e_v^{(\nu+1)}\|_X^2 + \rho\,(2r-\tau)\,\|Be_v^{(\nu+1)}\|_X^2 \le 0,$$

from which we first deduce the convergence of $\{\|e_p^{(\nu)}\|_{0,\Omega}^2\}_{\mathbb{N}}$ and then

$$e_v^{(\nu)} \to 0 \quad \text{in } X \quad (\nu \to \infty), \tag{3.12a}$$

$$Be_v^{(\nu)} \to 0 \quad \text{in } L^2(\Omega) \quad (\nu \to \infty). \tag{3.12b}$$

Now, (3.11a) and Lemma 2.2 result in

$$\|S_n(v_n^{(\nu+1)}) - S_n(v_n) + rB_n^*B_n e_v^{(\nu+1)}\|_{X^*} = \|B_n^* e_p^{(\nu)}\|_{X^*} \ge \beta\,\|e_p^{(\nu)}\|_{0,\Omega}.$$

Hence, (3.12a), (3.12b) and the continuity of S_n imply $e_p^{(\nu)} \to 0$ as $\nu \to \infty$. □

As in the case of the nonlinear Uzawa algorithm, in practical computations, we replace S_n in (3.10a) by some appropriate approximation, $\tilde{S}_n$. This leads to the inexact augmented Lagrangian algorithm.

Inexact augmented Lagrangian algorithm
Let $\tilde{S}_n$ be an approximation of S_n. Then, given $(v_n^{(0)}, p_n^{(0)}) \in X_n \times Q_n$ and $r, \rho > 0$, compute $(v_n^{(\nu)}, p_n^{(\nu)}) \in X_n \times Q_n$, $\nu \in \mathbb{N}$, such that for $(w_n, q_n) \in X_n \times Q_n$, there holds

$$\left\langle \tilde{S}_n\left(v_n^{(\nu+1)}\right), w_n\right\rangle - \left\langle B_n^* p_n^{(\nu)}, w_n\right\rangle + r\left(B_n v_n^{(\nu+1)}, B_n w_n\right)_{0,\Omega} = \langle f+g, w_n\rangle, \tag{3.13a}$$

$$\left(p_n^{(\nu+1)} - p_n^{(\nu)}, q_n\right)_{0,\Omega} + \rho\left(B_n v_n^{(\nu+1)}, q_n\right)_{0,\Omega} = 0. \tag{3.13b}$$

The convergence of the inexact augmented Lagrangian algorithm requires that $\tilde{S}_n^{-1}$ provides a sufficiently good approximation of S_n^{-1}, which also affects the choice of the parameters r and ρ.

REMARK 3.3. More efficient preconditioners can be constructed in the framework of multigrid techniques (cf. HACKBUSCH [1985]) with respect to a hierarchy of discretizations and/or domain decomposition methods (cf. QUARTERONI and VALLI [1999], and TOSELLI and WIDLUND [2005]) relying on overlapping or nonoverlapping decompositions of the computational domain. However, we are not aware of any scientific contributions where such approaches have been applied to the numerical solution of electrorheological fluid flows.

3.1.3. *Construction of approximate inverses*

There is a wide variety of possible approximate inverses $\tilde{S}_n^{-1}$ of S_n^{-1} for the realization of the inexact nonlinear Uzawa algorithm (3.9a), (3.9b) and the inexact augmented Lagrangian algorithm (3.13a), (3.13b), among them the Picard iteration, fixed point techniques, and Newton-type methods.

We recall that the operator S_n in (3.8a) and (3.11a) can be formally written as $S_n(v_n) = \hat{S}_n(v_n, v_n)$ where $\hat{S}_n : X \times X \to X^*$ is given by

$$\left\langle \hat{S}_n(v_n, w_n), z_n \right\rangle := 2 \int_\Omega \Big(b(|E|, x)(\kappa + I(\tilde{u} + v_n))^{-1/2} \varepsilon(\tilde{u} + w_n) : \varepsilon(z_n) \\ + c(I(\tilde{u} + v_n), |E|, x)\varepsilon(\tilde{u} + w_n) : \varepsilon(z_n) \Big) \, dx. \tag{3.14}$$

Then, for a given $f_n \in X_n^*$, the solution of the nonlinear variational equation

$$\langle S_n(v_n), z_n \rangle = \langle f_n, z_n \rangle, \quad z_n \in X_n, \tag{3.15}$$

can be obviously reformulated as

$$\left\langle \hat{S}_n(v_n, v_n), z_n \right\rangle = \langle f_n, z_n \rangle, \quad z_n \in X_n. \tag{3.16}$$

We first consider a Picard-type iteration (cf. MOORE and CLOUD [2007]), which in the Russian literature is also known as the Birger-Kachanov method (cf. FUCIK, KRATOCHVIL and NECAS [1973]).

Picard iteration
Given $v_n^{(0)} \in X_n$, compute $v_n^{(\nu)}$, $\nu \in \mathbb{N}$, as the solution of the linear variational equation

$$\left\langle \hat{S}_n \left(v_n^{(\nu)}, v_n^{(\nu+1)} \right), z_n \right\rangle = \langle f_n, z_n \rangle, \quad z_n \in X_n, \quad \nu \in \mathbb{N}_0. \tag{3.17}$$

THEOREM 3.3. *Let $v_n \in X_n$ be the solution of (3.15) and $\{v_n^{(\nu)}\}_\mathbb{N}$ be the sequence of iterates $v_n^{(\nu)} \in X_n, \nu \in \mathbb{N}$, generated by the Picard iteration (3.17). Then, under the assumptions* $(\mathbf{A_1})$, $(\mathbf{A_2})$ *and for $\kappa > 0$, there holds*

$$v_n^{(\nu)} \to v_n \quad \text{in } X \quad (\nu \to \infty).$$

PROOF. We refer to FUCIK, KRATOCHVIL and NECAS [1973], MOORE and CLOUD [2007]. □

We will not consider the issue how well the inverse $\tilde{S}_n^{-1}$ associated with the Picard iteration (3.17) approximates S_n^{-1} in order to access the convergence of the inexact nonlinear Uzawa algorithm or the inexact augmented Lagrangian algorithm, but instead address this question in the framework of a fixed point iteration:

We introduce $A : X \to X^*$ as a linear, continuous self-adjoint coercive operator, i.e., we assume that for $v, w \in X$

$$\langle Av, w \rangle = \langle Aw, v \rangle, \tag{3.18a}$$

$$| \langle Av, w \rangle | \le C_A \, \|v\|_X \, \|w\|_X, \tag{3.18b}$$

$$\langle Av, v \rangle \ge \gamma_A \, \|v\|_X^2. \tag{3.18c}$$

Hence, $\|\cdot\|_A := \langle A\cdot,\cdot\rangle^{1/2}$ defines a norm on X, which is equivalent to the $\|\cdot\|_X$-norm and the $\|\cdot\|_{1,2,\Omega}$-norm. We refer to $\|\cdot\|_{A^*}$ as the associated norm on the dual space X^*. Hence, the operator S_n retains its properties with respect to the $\|\cdot\|_A$- and the $\|\cdot\|_{A^*}$-norm. In particular, for $w_n, z_n \in X_n$, there holds

$$\|S_n(w_n) - S_n(z_n)\|_{A^*} \leq C_S \|w_n - z_n\|_A, \tag{3.19a}$$

$$\langle S_n(w_n) - S_n(z_n), w_n - z_n\rangle \geq \gamma_S \|w_n - z_n\|_A^2. \tag{3.19b}$$

Setting $A_n := A|_{X_n}$, for the solution of (3.15), we consider the following fixed point iteration.

Fixed point iteration
Given $v_n^{(0)} \in X_n$ and $t \in \mathbb{R}_+$, compute $v_n^{(\nu)} \in X_n$, $\nu \in \mathbb{N}$, as the solution of

$$\left\langle A_n v_n^{(\nu+1)}, z_n\right\rangle = \left\langle A_n v_n^{(\nu)}, z_n\right\rangle - t\left(\left\langle S_n\left(v_n^{(\nu)}\right), z_n\right\rangle - \langle f_n, z_n\rangle\right), \quad z_n \in X_n. \tag{3.20}$$

THEOREM 3.4. *Let $v_n \in X_n$ be the unique solution of (3.15). Assume that the operator $A \in \mathcal{L}(X, X^*)$ satisfies (3.18a)–(3.18c) and that assumptions* $(\mathbf{A_1})$, $(\mathbf{A_2})$ *hold true. Then, for $\kappa > 0$ and $t \in (0, 2\gamma_S C_S^{-2})$, the linear problem* (3.20) *has a unique solution $v_n^{(\nu+1)} \in X_n$, and there holds*

$$\|v_n^{(\nu)} - v_n\|_A \leq \frac{k(t)^\nu}{1-k(t)} \left\|S_n\left(v_n^{(0)}\right) - f_n\right\|_{A^*}, \quad \nu \in \mathbb{N}, \tag{3.21}$$

where

$$k(t) = (1 - 2\gamma_S t + C_S^2 t^2)^{1/2} < 1. \tag{3.22}$$

The optimal value is

$$k_{\mathrm{opt}} = k(t_{\mathrm{opt}}) = (1 - \gamma_S^2 C_S^{-2})^{1/2}, \quad t_{\mathrm{opt}} = \gamma_S C_S^{-2}.$$

PROOF. We denote by $J : X^* \to X$ the Riesz operator. Then, the iteration (3.20) amounts to the computation of a fixed point of the operator $T_n(t) : X_n \to X_n$ given by

$$T_n(t)(w_n) := w_n - t\,J(S_n(w_n) - f_n), \quad w_n \in X_n. \tag{3.23}$$

Taking (3.19a), (3.19b) and the isometry of J into account, from (3.23), we deduce

$$\begin{aligned}
\|T_n(t)(w_n) - T_n(t)(z_n)\|_A^2 &= \|w_n - z_n - t\,J(S_n(w_n) - S_n(z_n))\|_A^2 \\
&= \|w_n - z_n\|_A^2 - 2t\,\langle S_n(w_n) - S_n(z_n), w_n - z_n\rangle + t^2\,\|S_n(w_n) - S_n(z_n)\|_{A^*}^2 \\
&\leq \|w_n - z_n\|_A^2 - 2t\,\gamma_S\,\|w_n - z_n\|_A^2 + t^2\,C_S^2\,\|w_n - z_n\|_A^2 = k(t)^2\,\|w_n - z_n\|_A^2.
\end{aligned}$$

Hence, the assertion follows from the Banach fixed point theorem. □

REMARK 3.4. Some comments are in order with regard to an appropriate choice of the finite dimensional subspaces X_n and Q_n. In the framework of finite element approximations based on simplicial and/or quadrilateral triangulations of the computational domain, for incompressible Stokes and Navier-Stokes type fluid flow problems, various families of finite elements have been suggested. The Taylor-Hood P_k/P_{k-1}-elements, $k \in \mathbb{N}$, and its generalizations have become the most popular choice in applications. For a thorough presentation and discussion including the discrete inf-sup condition, we refer to BRAESS [2007], BREZZI and FORTIN [1991].

3.1.4. *An augmented Lagrangian approach for an extended Bingham fluid model*

In case of the extended Bingham fluid model based on the viscosity function (2.18), the fluid flow is described by the nonlinear variational inequality of the second kind (2.63). Hence, appropriate numerical methods for such variational inequalities have to be provided (cf., e.g., GLOWINSKI, LIONS and TREMOLIERES [1981]). We present here an augmented Lagrangian approach relying on a mixed formulation of the problem that has been used in ENGELMANN, HIPTMAIR, HOPPE and MAZURKEVICH [2000] for the computation of electrorheological fluid flows obeying the constitutive law (2.13). The motivation for the mixed formulation is that the nonlinearity and nonsmoothness of the problem is confined to the gradients of the components of the velocity. Hence, introducing $p = \nabla u$ as additional unknowns and using a $P1/P0$ finite element discretization of (u, p) boil down the global nonlinear problem to a sequence of local, low-dimensional nonlinear problems that can be easily solved. For simplicity, we restrict ourselves to a problem setting with full rotational symmetry where $E = E_r(r, z)e_r + E_z(r, z)e_z$ and $u = u(r, z)e_\vartheta$ with e_r, e_ϑ, and e_z denoting the unit vectors in a cylindrical coordinate system. The incompressibility condition is then automatically satisfied.

Based on the constitutive law (2.13), the steady state $u \in V := W^{1,2}_{0,\Gamma_D}(\Omega)$ of the electrorheological fluid flow corresponds to the minimizer of the global energy

$$J(u) = \inf_{v \in V} J(v). \tag{3.24}$$

Here, $J : V \to \mathbb{R}$ stands for the energy functional

$$J(v) := \gamma \int_\Omega |E||E \cdot \nabla u| r \, dr \, dz + \frac{1}{2}\eta \int_\Omega |\nabla u|^2 r \, dr \, dz - \ell(v), \quad v \in V, \tag{3.25}$$

where $\ell : V \to \mathbb{R}$ comprises volume and surface forces according to

$$\ell(v) := \langle f + g, v \rangle, \quad v \in V.$$

We introduce $p = \nabla u \in L^2(\Omega)^2$ as additional unknowns and couple the constraint $p = \nabla u$ both by a Lagrangian multiplier $\lambda \in L^2(\Omega)^2$ and by a penalty term with penalty parameter $\tau > 0$, which gives rise to the saddle point problem.

Find $(u, p, \lambda) \in V \times L^2(\Omega)^2 \times L^2(\Omega)^2$ such that

$$L^{(\tau)}(u, p, \lambda) = \inf_{v,q} \sup_{\mu} L^{(\tau)}(v, q, \mu), \tag{3.26}$$

where the augmented Lagrangian $L^{(\tau)}(\cdot, \cdot, \cdot)$ is given by

$$\begin{aligned} L^{(\tau)}(v, q, \mu) := {} & \gamma \int_\Omega |E||E \cdot p| r \, dr \, dz + \frac{1}{2}\eta \int_\Omega |p|^2 r \, dr \, dz \\ & + \int_\Omega \mu \cdot (p - \nabla u) \, dr \, dz + \frac{1}{2}\tau \int_\Omega |p - \nabla u|^2 \, dr \, dz - \ell(v) . \end{aligned}$$

For a simplicial triangulation $\mathcal{T}_h(\Omega)$ of the computational domain Ω, we use a $P1/P0$ discretization $(u_h, p_h) \in V_h \times W_h^2$ of (u, p) where V_h stands for the standard finite element space of continuous piecewise linear finite elements and W_h for the linear space of elementwise constants. If an approximation of the electric field E is obtained based on a $P1$ approximation, we define $E_h \in W_h$ locally as the elementwise integral mean of that approximation. Consequently, the discrete minimization problem amounts to the computation of $(u_h, p_h, \lambda_h) \in V_h \times W_h^2 \times W_h^2$ such that

$$L^{(\tau)}(u_h, p_h, \lambda_h) = \inf_{v_h, q_h} \sup_{\mu_h} L^{(\tau)}(v_h, q_h, \mu_h), \tag{3.27}$$

where E in the definition of $L^{(\tau)}(\cdot,\cdot,\cdot)$ has to be replaced by E_h.

The minimization problem (3.27) is solved iteratively by an operator splitting technique where each iteration step requires the solution of a global quadratic minimization problem and local, i.e., elementwise nonlinear minimization problems along with appropriate updates of the discrete Lagrangian multipliers λ_h. In particular, given sequences $\{\rho_n\}_{\mathbb{N}}$ and $\{\tau_n\}_{\mathbb{N}}$ of update parameters $\rho_n \in \mathbb{R}_+$ and penalty parameters $\tau_n \in \mathbb{R}_+$, $n \in \mathbb{N}$, as well as start vectors $(p_h^{(0)}, \lambda_h^{(1)}) \in W_h^2 \times W_h^2$, an iteration consists of the following two steps.

Step 1: Compute $u_h^{(n)} \in V_h$ as the solution of the global quadratic minimization problem

$$L^{(\tau_n)}\left(u_h^{(n)}, p_h^{(n-1)}, \lambda_h^{(n)}\right) = \inf_{v_h \in V_h} L^{(\tau)}\left(v_h, p_h^{(n-1)}, \lambda_h^{(n)}\right) \tag{3.28}$$

and update the multiplier according to

$$\lambda_h^{(n+1/2)} = \lambda_h^{(n)} + \rho_n\left(\nabla u_h^{(n)} - p_h^{(n-1)}\right). \tag{3.29}$$

Step 2: Compute $p_h^{(n)} \in W_h^2$ as the solution of

$$L^{(\tau_n)}\left(u_h^{(n)}, p_h^{(n)}, \lambda_h^{(n+1/2)}\right) = \inf_{q_h \in W_h^2} L^{(\tau)}\left(u_h^{(n)}, q_h, \lambda_h^{(n+1/2)}\right) \tag{3.30}$$

and update the multiplier according to

$$\lambda_h^{(n+1)} = \lambda_h^{(n+1/2)} + \rho_n\left(\nabla u_h^{(n)} - p_h^{(n)}\right). \tag{3.31}$$

The minimization problem (3.28) requires the solution of a linear algebraic system where the coefficient matrix corresponds to the stiffness matrix associated with the $P1$ approximation of the Laplacian $-\Delta$. On the other hand, the minimization problem (3.30) reduces to the simultaneous solution of the elementwise minimization problems: for each $T \in \mathcal{T}_h(\Omega)$, compute $p_h^{(n)}|_T \in P_0(T)^2$ such that

$$J_T^{(\tau_n)}\left(p_h^{(n)}|_T\right) = \inf_{q_h^T \in P_0(T)^2} J_T^{(\tau_n)}\left(q_h^T\right), \tag{3.32}$$

where the functional $J_T^{(\tau_n)} : P_0(T)^2 \to \mathbb{R}$ is given by

$$J_T^{(\tau_n)}\left(q_h^T\right) := L^{(\tau_n)}\left(u_h^{(n)}|_T, q_h^T, \lambda_h^{(n+1/2)}\right).$$

The local minimization problems (3.32) give rise to two-dimensional variational inequalities, which can be solved analytically.

3.2. Evolutionary isothermal incompressible flow problems

We consider the discretization of initial-boundary value problems for time-dependent incompressible isothermal electrorheological fluid problems (2.1a), (2.1b) by a difference approximation in time and by the Galerkin method in space using finite dimensional subspaces $X_n \subset X := W^{1,2}_{0,\Gamma_D}$ and $Q_n \subset L^2(\Omega)$, $n \in \mathbb{N}$ as in the previous Section 3.1. For discretization in time, we refer to

$$\bar{I}_k := \{t_m = mk \mid 0 \le m \le M, k := T/M\}, \quad M \in \mathbb{N}, \tag{3.33}$$

as a uniform partition of the time interval $[0, T]$ of step size k and approximate the time derivative $u_t(\cdot, t)$ in $t \in \bar{I}_k$ by the forward and backward difference quotients $\partial_k^{\pm} u(\cdot, t)$, which are given by

$$\partial_k^+ u(\cdot, t) := k^{-1}(u(\cdot, t+k) - u(\cdot, t)), \quad t \in \bar{I}_k \setminus \{T\}, \tag{3.34a}$$

$$\partial_k^- u(\cdot, t) := k^{-1}(u(\cdot, t) - u(\cdot, t-k)), \quad t \in \bar{I}_k \setminus \{0\}. \tag{3.34b}$$

We denote by $(u_n^{(m)}, p_n^{(m)}) \in X_n \times Q_n$ an approximation of $(u(\cdot, t_m), p(\cdot, t_m)) \in X \times L^2(\Omega)$ at time t_m. Using a convex combination of the discretizations by the forward and difference quotients in time results in the so-called Θ-scheme which at each time level amounts to the solution of the following nonlinear system of finite dimensional variational equations

$$\left\langle F_n^{(\Theta)}(u_n^{(m)}), w_n \right\rangle - \left\langle B_n^* p_n^{(m)}, w_n \right\rangle = \left\langle h_n^{(\Theta)}, w_n \right\rangle, \quad w_n \in X_n, \tag{3.35a}$$

$$\left(B_n u_n^{(m)}, q_n \right\rangle = 0, \quad q_n \in Q_n, \tag{3.35b}$$

where the nonlinear operator $F_n^{(\Theta)} : X_n \to X_n^*$ and the right-hand side $h_n^{(\Theta)} \in X_n^*$, $\Theta \in [0, 1]$, are given by

$$\left\langle F_n^{(\Theta)}(v_n), w_n \right\rangle := \rho\, k^{-1} \ \langle v_n, w_n \rangle + \Theta\ (\langle (v_n \cdot \nabla) v_n, w_n \rangle + \langle S_n(v_n), w_n \rangle), \tag{3.36a}$$

$$h_n^{(\Theta)} := f_n + g_n + k^{-1} u_n^{(m)} - (1 - \Theta) \left(\left(u_n^{(m)} \cdot \nabla \right) u_n^{(m)} + S_n \left(u_n^{(m)} \right) \right). \tag{3.36b}$$

For $\Theta = 0$ and $\Theta = 1$, we recover the standard explicit and implicit difference approximation, respectively. The difference approximation for $\Theta = 1/2$ is called the Crank–Nicolsen method. It is well-known that the Θ-scheme is consistent with the initial-boundary value problem of order $O(k)$ in time for $\Theta \neq 1/2$, whereas the Crank–Nicolsen method is consistent of order $O(k^2)$. Moreover, the Θ-scheme is only conditionally stable for $\Theta < 1/2$ and unconditionally stable for $\Theta \in [1/2, 1]$ (cf., e.g., STRIKWERDA [2004], and THOMAS [1995]). Usually, the stability condition for $\Theta \in [0, 1/2)$ imposes a severe restriction on the choice of the step size k so that the corresponding schemes are not used in practice.

The nonlinear system (3.35a), (3.35b) can be solved using the same techniques as described in Section 3.1. In particular, we may use the analog of the inexact nonlinear Uzawa algorithm (3.9a), (3.9b) and the inexact augmented Lagrangian algorithm (3.13a), (3.13b) provided we have suitable approximate inverses $(\tilde{F}_n^{(\Theta)})^{-1}$ of $(F_n^{(\Theta)})^{-1}$, $\Theta \in (1/2, 1]$, at hand. For the construction of such inverses, the Picard iteration or fixed point iterations can be used as well. The only difference is that we are faced with the additional nonlinear convective term $(v_n \cdot \nabla)v_n$, which, however, can be treated in much the same way as the nonlinearity in the operator S_n. For instance, in case of the standard implicit scheme ($\Theta = 1$), we use

$$\left\langle \tilde{F}_n^{(1)}(v_n), w_n \right\rangle := \rho\, k^{-1} \; \langle v_n, w_n \rangle + \left(\left\langle (u_n^{(m)} \cdot \nabla) v_n, w_n \right\rangle + \left\langle \hat{S}_n(u_n^{(m)}), w_n \right\rangle \right), \tag{3.37}$$

with $\hat{S}_n$ given by (3.14).

For the Crank–Nicolsen, scheme, an appropriate modification has to be used in order to retain second-order accuracy (cf., e.g., ELMAN [2002]).

3.3. Nonisothermal incompressible electrorheological flow problems

We use the notations from Section 2.5 and assume $\{X_n\}_{\mathbb{N}}$, $\{Q_n\}_{\mathbb{N}}$, and $\{Y_n\}_{\mathbb{N}}$ to be limit dense nested sequences of finite dimensional subspaces of X, $L^2(\Omega)$, and $W_{0,\Gamma}^{1,2}(\Omega)$, respectively, and we consider the following sequence of approximating systems of finite dimensional variational equations: find $(v_n, p_n, \theta_n) \in X_n \times Q_n \times Y_n$ such that

$$\langle N(v_n, \theta_n), w_n \rangle - \left\langle B_n^* p_n, w_n \right\rangle = \langle f + g, w_n \rangle, \quad w_n \in X_n, \tag{3.38a}$$

$$(B_n v_n, q_n)_{0,\Omega} = 0, \quad q_n \in Q_n, \tag{3.38b}$$

$$(\nabla \theta_n, \nabla \zeta_n)_{0,\Omega} - \left\langle A_\beta(v_n, \theta_n), \zeta_n \right\rangle = (f_3, \zeta_n)_{0,\Omega}, \quad \zeta_n \in Y_n, \tag{3.38c}$$

where $B_n \in \mathcal{L}(X_n, Q_n)$ refers to the discrete divergence operator (cf. 2.2.1).

THEOREM 3.5. *Let the assumptions of Theorem 2.6 be satisfied, and let $\{(v_n, p_n, \theta_n)\}_{\mathbb{N}}$ be a sequence of solutions of (3.38a)–(3.38c). Then, there exist a subsequence $\mathbb{N}' \subset \mathbb{N}$ and a triple $(v, p, \theta) \in X \times L^2(\Omega) \times W_{0,\Gamma}^{1,2}(\Omega)$ that solves (2.101a)–(2.101c) such that for $\mathbb{N}' \ni n \to \infty$*

$$v_n \to v \quad \text{in } X, \tag{3.39a}$$

$$p_n \to p \quad \text{in } L^2(\Omega), \tag{3.39b}$$

$$\theta_n \to \theta \quad \text{in } W_{0,\Gamma}^{1,2}(\Omega). \tag{3.39c}$$

PROOF. Setting $V_n = \mathrm{Ker}(B_n)$, (3.38a)–(3.38c) can be equivalently stated as follows: find $(v_n, \theta_n) \in V_n \times Y_n$ such that

$$\langle N(v_n, \theta_n), w_n \rangle = \langle f_1 + g, w_n \rangle, \quad w_n \in X_n, \tag{3.40a}$$

$$(\nabla \theta_n, \nabla \zeta_n)_{0,\Omega} - \left\langle A_\beta(v_n, \theta_n), \zeta_n \right\rangle = (f_3, \zeta_n)_{0,\Omega}, \quad \zeta_n \in Y_n. \tag{3.40b}$$

It follows from Theorem 2.6 that for each $n \in \mathbb{N}$ problem (3.40a), (3.40b) admits a solution $(v_n, \theta_n) \in V_n \times Y_n$. Moreover, there are constants $C_i > 0$, $1 \le i \le 2$, such that

$$\|v_n\|_X \le C_1, \quad \|\theta_n\|_{1,\Omega} \le C_2 \tag{3.41}$$

uniformly in $n \in \mathbb{N}$. We have $N(v_n, \theta_n) - (f_1 + g) \in V_n^0$, and hence, Lemma 2.2 implies that there is a unique $p_n \in Q_n$ such that

$$\langle N(v_n, \theta_n), w_n \rangle - \langle B_n^* p_n, w_n \rangle = \langle f_1 + g, w_n \rangle, \quad w_n \in X_n, \tag{3.42}$$

i.e., (v_n, p_n, θ_n) solves (3.38a)–(3.38c). Lemma 2.2 and (3.41) yield

$$\|p_n\|_{0,\Omega} \le C_3, \quad n \in \mathbb{N} \tag{3.43}$$

for some constant $C_3 > 0$. Consequently, there exist a subsequence $\mathbb{N}' \subset \mathbb{N}$ and $(v, p, \theta) \in X \times L^2(\Omega) \times W_{0,\Gamma}^{1,2}(\Omega)$ such that for $\mathbb{N}' \ni n \to \infty$

$$v_n \rightharpoonup u \quad \text{in } X, \tag{3.44a}$$

$$v_n \to v \quad \text{in } L^4(\Omega)^d, \tag{3.44b}$$

$$v_n \to v \quad \text{a.e. in } \Omega, \tag{3.44c}$$

$$p_n \rightharpoonup p \quad \text{in } L^2(\Omega), \tag{3.44d}$$

$$\theta_n \rightharpoonup \theta \quad \text{in } W_{0,\Gamma}^{1,2}(\Omega), \tag{3.44e}$$

$$\theta_n \to \theta \quad \text{in } L^4(\Omega), \tag{3.44f}$$

$$\theta_n \to \theta \quad \text{a.e. in } \Omega, \tag{3.44g}$$

$$N(v_n, \theta_n) \rightharpoonup \ell \quad \text{in } X^*. \tag{3.44h}$$

For a fixed integer $n_0 \in \mathbb{N}$, let $w_{n_0} \in X_{n_0}$ and $q_{n_0} \in Q_{n_0}$. Then, in view of (3.44a), (3.44d), and (3.44h), passing to the limit in (2.101a), (2.101b) yields

$$\langle \ell - B^* p, w \rangle = \langle f_1 + g, w \rangle, \quad w \in X_{n_0},$$
$$(Bv, q)_{0,\Omega} = 0, \quad q \in Q_{n_0}.$$

Since $n_0 \in \mathbb{N}$ was arbitrarily chosen and the sequences $\{X_n\}_{\mathbb{N}}$ and $\{Q_n\}_{\mathbb{N}}$ are limit dense in X and $L^2(\Omega)$, it follows that

$$\ell - B^* p = f_1 + g \quad \text{in } X^*, \tag{3.45a}$$

$$\nabla \cdot v = 0 \quad \text{a.e. in } \Omega. \tag{3.45b}$$

We define $L_{z_1,z_2} : X \to X^*$ according to

$$\langle L_{z_1,z_2}(w_1), w_2 \rangle$$
$$:= 2 \int_\Omega \varphi(I(\tilde{u} + w_1), |E|, \mu(\tilde{u} + z_1, E), \tilde{\theta} + z_2)\varepsilon(\tilde{u} + w_1) : \varepsilon(w_2)\, dx, w_1, w_2 \in X.$$

For $z_1 = v_n$, $z_2 = \theta_n$, Lemma 2.3 gives

$$\langle L_{(v_n,\theta_n)}(v_n) - L_{(v_n,\theta_n)}(v), v_n - w \rangle \ge 0, \quad w \in X, n \in \mathbb{N}. \tag{3.46}$$

Moreover, by (3.44b), (3.44c) and (3.44f), (3.44g) and the Lebesgue theorem,

$$L_{(v_n,\theta_n)}(w) \to L_{(v,\theta)}(w) \quad \text{in } X^*, \quad w \in X.$$

It follows that for $w \in X$, there holds

$$\lim_{\mathbb{N}' \ni n \to \infty} \langle L_{(v_n,\theta_n)}(w), v_n \rangle = \langle L_{(v,\theta)}(w), v \rangle, \tag{3.47a}$$

$$\lim_{\mathbb{N}' \ni n \to \infty} \langle L_{(v_n,\theta_n)}(w), w \rangle = \langle L_{(v,\theta)}(w), w \rangle. \tag{3.47b}$$

Observing (3.44h) and (3.45a), we obtain

$$\lim_{\mathbb{N}' \ni n \to \infty} \left(\langle L_{(v_n,\theta_n)}(v_n), w \rangle - \langle B^* p, w \rangle \right) = \langle f_1 + g, w \rangle, \quad w \in X. \tag{3.48}$$

Taking into account that

$$\langle B^* p_n, v_n \rangle = (p_n, B_n v_n)_{0,\Omega},$$

(2.101a) and (3.44a) imply that for $\mathbb{N}' \ni n \to \infty$, there holds

$$\langle L_{(v_n,\theta_n)}(v_n), v_n \rangle = \langle f_1 + g, v_n \rangle \to \langle f_1 + g, v \rangle. \tag{3.49}$$

Due to (3.47a), (3.47b) and (3.48), (3.49), we pass to the limit in (3.46) and get

$$\langle f_1 + g - L_{(v,\theta)}(w) + B^* p, v - w \rangle \geq 0, \quad w \in X. \tag{3.50}$$

If we choose $w = v - \gamma z$, $z \in X$, $\gamma > 0$, in (3.50), for $\gamma \to 0$, it follows that

$$\langle f_1 + g - N(v, \theta) + B^* p, z \rangle \geq 0, \quad z \in X.$$

Since $z \in X$ can be arbitrarily chosen, we may replace z by $-z$ and thus obtain

$$\langle N(v, \theta), z \rangle - \langle B^* p, z \rangle = \langle f_1 + g, z \rangle, \quad z \in X, \tag{3.51a}$$

$$\ell = N(v, \theta). \tag{3.51b}$$

On the other hand, (3.44a)–(3.44c) and (3.44e)–(3.44g) as well as Lebesgue's theorem imply

$$\lim_{\mathbb{N}' \ni n \to \infty} \langle A_\beta(v_n, \theta_n), \xi \rangle = \langle A_\beta(v, \theta), \xi \rangle, \quad \xi \in W^{1,2}_{0,\Gamma}(\Omega).$$

Choosing $n_0 \in \mathbb{N}$ and $\xi_{n_0} \in Y_{n_0}$ arbitrarily, but fixed, and passing to the limit in (2.101c), we get

$$(\nabla\theta, \nabla\xi_{n_0})_{0,\Omega} - \langle A_\beta(v, \theta), \xi_{n_0} \rangle = \langle f_3, \xi_{n_0} \rangle.$$

Since the sequence $\{Y_n\}_{\mathbb{N}}$ is limit dense in $W^{1,2}_{0,\Gamma}(\Omega)$, we thus have

$$(\nabla\theta, \nabla\xi)_{0,\Omega} - \langle A_\beta(v, \theta), \xi \rangle = \langle f_3, \xi \rangle, \quad \xi \in W^{1,2}_{0,\gamma}(\Omega). \tag{3.52}$$

Now, (3.45b), (3.51a) and (3.52) show that the triple (v, p, θ) is a solution of (2.101a)–(2.101c).

What remains to be shown is the strong convergence (3.39a)–(3.39c). We first note that due to (3.44a), (3.48) (with $w = v$), and (3.49)

$$\Lambda_n := \left\langle L_{(v_n,\theta_n)}(v_n) - L_{v,\theta)}(v), v_n - v\right\rangle \to 0 \quad (\mathbb{N}' \ni n \to \infty). \tag{3.53}$$

We split Λ_n according to

$$\Lambda_n = \left\langle L_{(v_n,\theta_n)}(v_n) - L_{v_n,\theta_n)}(v), v_n - v\right\rangle + \left\langle L_{(v_n,\theta_n)}(v) - L_{v,\theta)}(v), v_n - v\right\rangle. \tag{3.54}$$

In view of (3.44a) and (3.47a), (3.47b), we have

$$\left\langle L_{(v_n,\theta_n)}(v) - L_{v,\theta)}(v), v_n - v\right\rangle \to 0 \quad (\mathbb{N}' \ni n \to \infty),$$

and hence, due to (3.53), (3.54),

$$\left\langle L_{(v_n,\theta_n)}(v_n) - L_{v_n,\theta_n)}(v), v_n - v\right\rangle \to 0 \quad (\mathbb{N}' \ni n \to \infty). \tag{3.55}$$

Now, Lemma 2.3 implies

$$v_n \to v \quad \text{in } X \quad (\mathbb{N}' \ni n \to \infty), \tag{3.56}$$

whence

$$I(\tilde{u} + v_n) \to I(\tilde{u} + v) \quad \text{a.e. in } \Omega \quad (\mathbb{N}' \ni n \to \infty). \tag{3.57}$$

We choose $w = w_n \in X_n$ in (2.101a) and subtract (2.101a) from (3.38a), which shows that for $q_n \in Q_n$, there holds

$$\left\langle B^*(p_n - q_n), w_n\right\rangle = \langle N(v_n, \theta_n) - N(v, \theta), w_n\rangle + \left\langle B^*(p - q_n), w_n\right\rangle. \tag{3.58}$$

Applying Lemma 2.2 in (3.58) yields

$$\begin{aligned}\|p_n - q_n\|_{0,\Omega} &\le \sup_{w_n \in X_n} \frac{\langle B^*(p_n - q_n), w_n\rangle}{\beta\ |w_n\|_X} \\ &\le \beta^{-1}\ \|N(v_n, \theta_n) - N(v, \theta)\|_{X^*} + C\ \|p - q_n\|_{0,\Omega}, \quad q_n \in Q_n,\end{aligned}$$

where $C \in \mathbb{R}$ is a positive constant. It follows that

$$\begin{aligned}\|p - p_n\|_{0,\Omega} &\le \inf_{q_n \in Q_n} \left(\|p - q_n\|_{0,\Omega} + \|p_n - q_n\|_{0,\Omega}\right) \\ &\le \beta^{-1}\ \|N(v_n, \theta_n) - N(v, \theta)\|_{X^*} + (C + 1) \inf_{q_n \in Q_n} \|p - q_n\|_{0,\Omega}.\end{aligned} \tag{3.59}$$

Setting

$$\varphi_{nm} := \varphi(I(\tilde{u} + v_n), |E|, \mu(\tilde{u} + v_m, E), \tilde{\theta} + \theta_m), \quad n, m \in \mathbb{N}_0,$$

straightforward estimation results in

$$\frac{1}{2}\,\|N(v_n,\theta_n)-N(v,\theta)\|_{X^*}\le\left(\int_\Omega(\varphi_{nn}\varepsilon(\tilde u+v_n)-\varphi_{00}\varepsilon(\tilde u+v)^2\,\mathrm{d}x\right)^{1/2}$$

$$=\left(\int_\Omega((\varphi_{nn}(\varepsilon(\tilde u+v_n)-\varepsilon(\tilde u-v))+(\varphi_{nn}-\varphi_{00})\varepsilon(\tilde u+v))^2\,\mathrm{d}x\right)^{1/2}$$

$$\le\left(\int_\Omega\varphi_{nn}^2 I(v_n-v)\,\mathrm{d}x\right)^{1/2}+\left(\int_\Omega(\varphi_{nn}-\varphi_{00})^2 I(\tilde u+v)\,\mathrm{d}x\right)^{1/2}. \tag{3.60}$$

It follows from $(\mathbf{T_1})$, (3.44b), (3.44c), (3.44f), (3.44g), (3.56), (3.57) as well as the Lebesgue theorem that the right-hand side in (3.60) converges to zero as $\mathbb{N}'\ni n\to\infty$. Consequently,

$$\|N(v_n,\theta_n)-N(v,\theta)\|_{X^*}\to 0\quad(\mathbb{N}'\ni n\to\infty). \tag{3.61}$$

Since the sequence $\{Q_n\}_{\mathbb{N}}$ is limit dense in $L^2(\Omega)$, (3.59) and (3.61) imply

$$p_n\to p\quad\text{in } L^2(\Omega)\quad(\mathbb{N}'\ni n\to\infty). \tag{3.62}$$

Finally, from (3.44b), (3.44c), (3.44f), (3.44g), (3.56), and (3.57), we also get

$$A_\beta(v_n,\theta_n)\to A_\beta(v,\theta)\quad\text{in } W^{-1,2}(\Omega)\quad(\mathbb{N}'\ni n\to\infty). \tag{3.63}$$

Choosing $\zeta_n=\theta_n$ in (3.38c), we have

$$\|\theta_n\|_{1,2,\Omega}^2=\langle A_\beta(v_n,\theta_n),\theta_n\rangle+\langle f_3,\theta_n\rangle,$$

whence in view of (2.101c), (3.44f), and (3.63) for $\mathbb{N}'\ni n\to\infty$, we have

$$\lim_{\mathbb{N}'\ni n\to\infty}\left(\langle A_\beta(v_n,\theta_n),\theta_n\rangle+\langle f_3,\theta_n\rangle\right)=\langle A_\beta(v,\theta,\theta\rangle+\langle f_3,\theta\rangle=\|\theta\|_{1,2,\Omega}^2.$$

Consequently, $\|\theta_n\|_{1,2,\Omega}^2\to\|\theta\|_{1,2,\Omega}^2$ as $\mathbb{N}'\ni n\to\infty$, which together with (3.44f), results in

$$\theta_n\to\theta\quad\text{in } W_{0,\Gamma}^{1,2}(\Omega)\quad(\mathbb{N}'\ni n\to\infty).$$

This conclusion concludes the proof of the theorem. □

4. Numerical simulation and optimization of electrorheological devices

We shall consider the application of the algorithmic tools developed in the Section 3 to the simulation and the optimal design of electrorheological devices and systems. The most elementary devices are rheometers used for the measurement of rheological properties, which shall be discussed in Section 4.1. Examples for more advanced devices are given by electrorheological shock absorbers, which feature a much wider spectrum of damper characteristics than absorbers based on conventional fluids. The simulation of the operational behavior of such electrorheological shock absorbers, in particular their compression and rebound states, shall be treated in Section 4.2. Finally, Section 4.3 is devoted to a brief presentation of a methodology for the shape optimization of the inlet and outlet boundaries of piston ducts in electrorheological shock absorbers. For general aspects of optimization problems related to fluid mechanical processes, we refer to LITVINOV [2000], MOHAMMADI and PIRONNEAU [2001].

4.1. Electrorheological rheometers

Electrorheological rheometers are devices for the measurement of the rheological properties of electrorheological fluids. Figure 4.5 displays a simple model consisting of two coaxial cylinders of lengths l_i, l_e and radii r_r, r_e, respectively. The inner cylinder features a high-voltage lead to an external electric circuit, which supplies the lateral surface. The inner cylinder thus serves as the electrode. The lateral surface of the outer cylinder represents the counter electrode. The gap between the cylinders is filled with an electrorheological fluid.

One of the cylinders may rotate, whereas the other one remains at rest. When one of the cylinders starts revolving, the other one experiences a torque due to the viscosity of the fluid. Applying a voltage through the external electric circuit, the electrorheological effect results in an enhanced viscosity, and the strength of the torque felt by the other cylinder increases. Commercial rheometers operate within a frequency range of 10^{-7}–100 Hz and a temperature

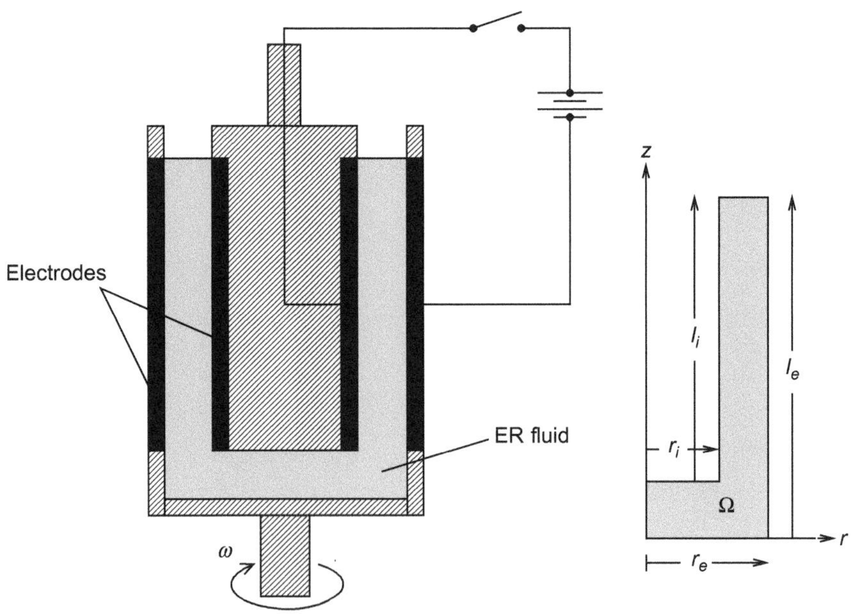

FIG. 4.5 Electrorheological clutch (left) and computational domain (right).

range of −150–1000 °C and allow angular velocities of 0–320 rad/s. The normal force range is between 10^{-3} and 50 N.

The arrangement has full rotational symmetry so that the computational domain reduces to the domain Ω as shown in Fig. 4.5 (right). Given a cylindrical coordinate system (r, α, z) with basis vectors e_r, e_α, and e_z, the velocity vector only features an angular component $u(r, z)e_\alpha$, which results in the following components of the strain tensor

$$\begin{aligned} &\varepsilon_{12}(u) = \varepsilon_{21}(u) = \frac{1}{2}\left(\frac{\partial u}{\partial r} - \frac{u}{r}\right), \varepsilon_{23}(u) = \varepsilon_{32}(u) = \frac{1}{2}\frac{\partial u}{\partial z}, \\ &\varepsilon_{11}(u) = \varepsilon_{22}(u) = \varepsilon_{33}(u) = \varepsilon_{13}(u) = \varepsilon_{31}(u) = 0. \end{aligned} \tag{4.1}$$

Hence, for the second invariant of the rate of strain tensor, we obtain

$$I(u) = \frac{1}{2}\left(\frac{\partial u}{\partial r} - \frac{u}{r}\right)^2 + \frac{1}{2}\left(\frac{\partial u}{\partial z}\right)^2. \tag{4.2}$$

In our case, $\mu(u, E) = 0$, and hence, the viscosity function φ is given by

$$\varphi(I(u), |E|, 0) := b(|E|, 0)(\kappa + I(u))^{-1/2} + c(I(u), |E|, 0), \tag{4.3}$$

where κ is the regularization parameter. Note that $\kappa = 0$ refers to the extended Bingham fluid. Assuming no volume force acting on the fluid, the steady state equations take the form

$$\begin{aligned} &\frac{\partial}{\partial r}\left(\varphi(I(u), |E|, 0)\left(\frac{\partial u}{\partial r} - \frac{u}{r}\right)\right) + \frac{\partial}{\partial z}\left(\varphi(I(u), |E|, 0)\frac{\partial u}{\partial z}\right) \\ &\quad + \frac{2}{r}\varphi(I(u), |E|, 0)\left(\frac{\partial u}{\partial r} - \frac{u}{r}\right) = 0, \end{aligned} \tag{4.4a}$$

$$\frac{\partial p}{\partial r} = \frac{\partial p}{\partial z} = 0. \tag{4.4b}$$

The incompressibility condition is automatically satisfied.

As far as the boundary conditions on $\Gamma = \partial\Omega$ are concerned, we prescribe velocities on the left boundary of Ω

$$\Gamma_\ell := \{(r, z) \mid r = 0, z \in (0, l_e - l_i)\}$$

and on the surface of the internal and external cylinder

$$\Gamma_s := \bigcup_{i=1}^{4} \Gamma_{s,i},$$

where the subsurfaces $\Gamma_{s,i}$, $1 \le i \le 4$, are given by

$$\begin{aligned} \Gamma_{s,1} &:= \{(r, z) \mid z = 0, r \in ((0, r_e)\}, \\ \Gamma_{s,2} &:= \{(r, z) \mid r = r_e, z \in (0, l_e)\}, \\ \Gamma_{s,3} &:= \{(r, z) \mid z = l_e - l_i, r \in (0, r_i)\}, \\ \Gamma_{s,4} &:= \{(r, z) \mid r = r_i, z \in ((l_e - l_i), l_e)\}. \end{aligned}$$

Moreover, surface forces are specified on

$$\Gamma_t := \Gamma \setminus (\bar{\Gamma}_\ell \cup \bar{\Gamma}_s).$$

If the inner cylinder is rotating, the boundary conditions are chosen according to

$$u(r,z) = \begin{cases} 0 & \text{on } \Gamma_\ell \cup \Gamma_{s,1} \cup \Gamma_{s,2} \\ r\omega & \text{on } \Gamma_{s,3} \\ r_i\omega & \text{on } \Gamma_{s,4} \end{cases}, \tag{4.5a}$$

$$\lim_{r\to 0} \left(\frac{\partial u}{\partial r} - \frac{u}{r} \right)(r,z) = 0, \quad z \in (0, l_e - l_i), \tag{4.5b}$$

$$\varphi(I(u), |E|, 0)\frac{\partial u}{\partial z} = 0, p = \text{const. on } \Gamma_t. \tag{4.5c}$$

On the other hand, if the outer cylinder is revolving, we have

$$u(r,z) = \begin{cases} 0 & \text{on } \Gamma_\ell \cup \Gamma_{s,3} \cup \Gamma_{s,4} \\ r\omega & \text{on } \Gamma_{s,1} \\ r_e\omega & \text{on } \Gamma_{s,2} \end{cases}, \tag{4.6a}$$

$$\lim_{r\to 0} \left(\frac{\partial u}{\partial r} - \frac{u}{r} \right)(r,z) = 0, \quad z \in (0, l_e - l_i), \tag{4.6b}$$

$$\varphi(I(u), |E|, 0)\frac{\partial u}{\partial z} = 0, p = \text{const. on } \Gamma_t. \tag{4.6c}$$

Due to the rotational symmetry, the electric field

$$E(r,z) = E_r(r,z)e_r + E_z(r,z)e_z$$

has two components E_r and E_z, which can be computed according to $E = -\nabla\psi = -(\partial\psi/\partial r, \partial\psi/\partial z)^T$ as the gradient of an electric potential $\psi = \psi(r,z)$. Denoting by

$$\Gamma_i := \{(r,z) \mid r = r_i, z \in (l_e - l_i, l_e)\},$$
$$\Gamma_e := \{(r,z) \mid r = r_e, z \in (l_e - l_i, l_e)\},$$

the lateral surfaces of the inner and outer cylinder, the electric potential ψ satisfies the boundary value problem

$$\frac{\partial}{\partial r}\left(\epsilon \frac{\partial \psi}{\partial r}\right) + \frac{\epsilon}{r}\frac{\partial \psi}{\partial r} + \frac{\partial}{\partial z}\left(\epsilon \frac{\partial \psi}{\partial z}\right) = 0 \quad \text{in } \Omega, \tag{4.7a}$$

$$\psi = U \quad \text{on } \Gamma_i, \quad \psi = 0 \quad \text{on } \Gamma_e, \tag{4.7b}$$

$$\frac{\partial \psi}{\partial r} = 0 \quad \text{on } \Gamma_0, \quad \nu_r \epsilon \frac{\partial \psi}{\partial r} + \nu_z \epsilon \frac{\partial \psi}{\partial z} = 0 \quad \text{on } \Gamma_t,$$

where U is the applied voltage, ϵ stands for the dielectric permittivity, and $\nu = (\nu_r, \nu_z)^T$ is the exterior normal unit vector.

Given a simplicial triangulation of the computational domain Ω, we have discretized (4.4a) by conforming P1 finite elements in case of a regularized viscosity function, i.e.,

$\kappa > 0$, whereas for the extended Bingham fluid model, i.e., $\kappa = 0$, we have chosen the mixed formulation from 3.1.4 and used conforming P1 elements for the primal variable and elementwise constants for the dual variables. The resulting algebraic systems have been solved by the augmented Lagrangian algorithm as described in Section 3. In both cases, the boundary value problem (4.7a), (4.7b) has been discretized by conforming P1 elements, and the resulting algebraic system has been solved by the preconditioned conjugate gradient method.

The computations have been performed for the commercially available polyurethane-based electrorheological fluid Rheobay TP AI 3565 (cf. BAYER [1997a]). Using experimental measurements for various electric field strengths, the viscosity function φ has been specified by cubic spline approximations of the $\tau(\gamma)$-flow curves (cf. Section 2).

We have considered two different geometrical configurations of the rheometer, namely a wide-gap configuration with the specifications

$$\text{Wide-gap:} \quad r_i = 35\text{ mm},\ r_e = 70\text{ mm},\quad l_i = 250\text{ mm},\ l_e = 300\text{ mm},$$
$$\omega = 125\text{ rad/s},\quad U = 0, 2, 3\text{ kV}$$

and a narrow-gap configuration with

$$\text{Narrow-gap:} \quad r_i = 24\text{ mm},\ r_e = 25\text{ mm},\quad l_i = 25\text{ mm},\ l_e = 30\text{ mm},$$
$$\omega = 5\text{ rad/s},\quad U = 0, 50, 100\text{ kV}.$$

The following results have been obtained based on the regularized viscosity function φ with $\kappa = 10^{-11}$ (for related results based on the extended Bingham fluid model, i.e., $\kappa = 0$ we refer to ENGELMANN, HIPTMAIR, HOPPE and MAZURKEVICH [2000]).

Figures 4.6 and 4.7 display the angular velocity profiles for the wide-gap configuration with revolving outer cylinder (Fig. 4.6) and revolving inner cylinder (Fig. 4.7) at applied voltages of $U = 0$ V, $U = 50$ kV, and $U = 100$ kV, respectively. In both cases, a zone with a

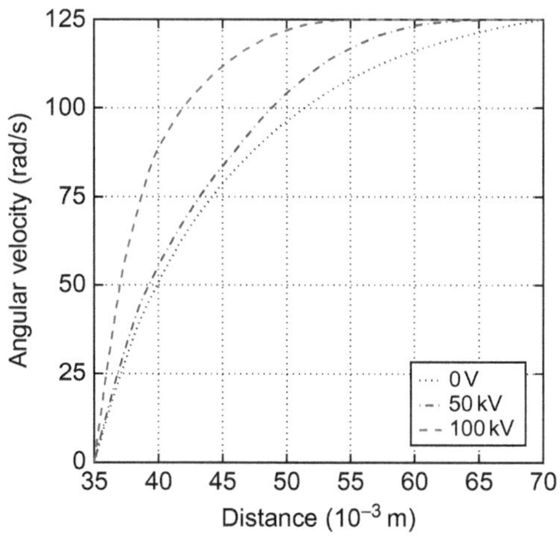

FIG. 4.6 Wide-gap configuration: angular velocity profiles (revolving outer cylinder); from HOPPE, LITVINOV and RAHMAN [2005].

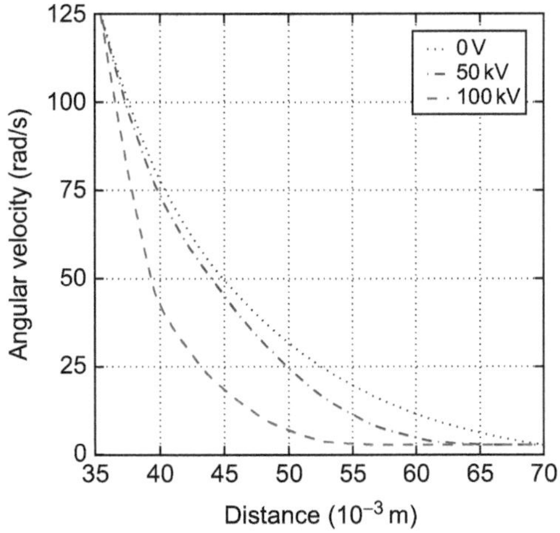

FIG. 4.7 Wide-gap configuration: angular velocity profiles (revolving inner cylinder); from HOPPE, LITVINOV and RAHMAN [2005].

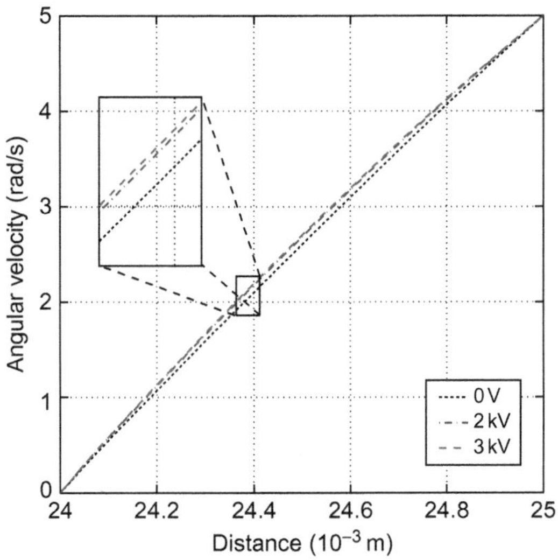

FIG. 4.8 Narrow-gap configuration: angular velocity profiles (rotating outer cylinder); from HOPPE, LITVINOV and RAHMAN [2005].

constant angular velocity occurs close to the outer cylinder, which increases for increasing voltage. This is the typical velocity profile for electrorheological Couette-type flows.

On the other hand, Figs 4.8 and 4.9 show the angular velocity profiles for the narrow-gap configuration with revolving outer cylinder (Fig. 4.8) and revolving inner cylinder (Fig. 4.9) at applied voltages of $U = 0\,\text{V}$, $U = 2\,\text{kV}$, and $U = 3\,\text{kV}$. We observe that in both cases, there

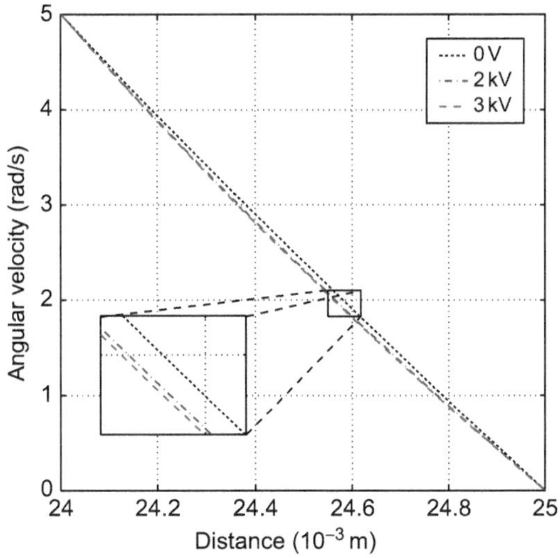

FIG. 4.9 Narrow-gap configuration: angular velocity profiles (rotating inner cylinder); from HOPPE, LITVINOV and RAHMAN [2005].

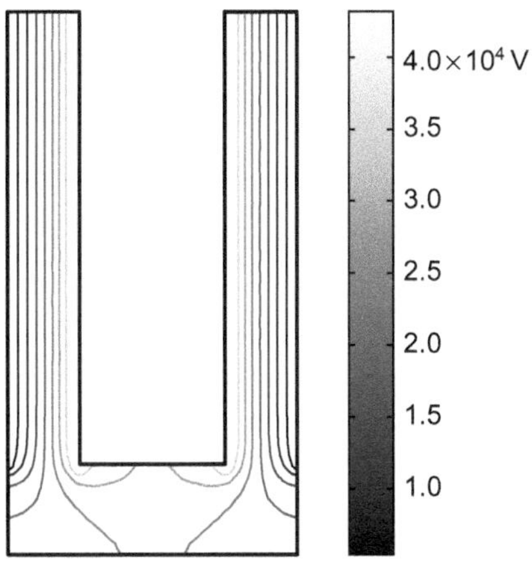

FIG. 4.10 Isolines of the electric potential (wide-gap configuration).

is no zone with a constant angular velocity. Indeed, independent of the applied voltage, the velocity profile is almost linear.

Finally, Fig. 4.10 contains the isolines of the electric potential ψ with respect to the wide-gap configuration. In fact, for both the wide-gap and the narrow-gap configuration, the electric field $E = (E_r, E_z)^T$ in the gap between the inner and outer cylinder is close to the constant vector $(U/(r_i - r_e), 0)^T$ and thus perpendicular to the velocity. The electric field decays rapidly with increasing distance to the electrodes.

4.2. Electrorheological shock absorbers

Due to their fast response to outer electrical fields, electrorheological fluids are much better suited for automotive shock absorbers than conventional oils. In fact, electrorheological shock absorbers feature a much wider characteristics than conventional ones and thus allow for an ideal adaptation to different road conditions and driving styles (cf., e.g., BAYER [1997b], BAYER and CARL SCHENCK [1998], BÖSE, HOPPE and MAZURKEVICH [2001], FILISKO [1995], GAVIN, HANSON and FILISKO [1996a,b], HOPPE, LITVINOV and RAHMAN [2003, 2007], and HOPPE, MAZUKEVICH, VON STRYK and RETTIG [2000]).

Figure 4.11 (left) displays the longitudinal section of an electrorheological shock absorber. The absorber consists of two chambers filled with an electrorheological fluid, a piston featuring two transfer ducts that connect the chambers, and a third gas-filled chamber separated from the others by a floating piston. The inner walls of the transfer ducts act as electrodes and counter electrodes. They are connected with an outer electric circuit by a high-voltage lead within the piston rod. We distinguish between the compression mode and

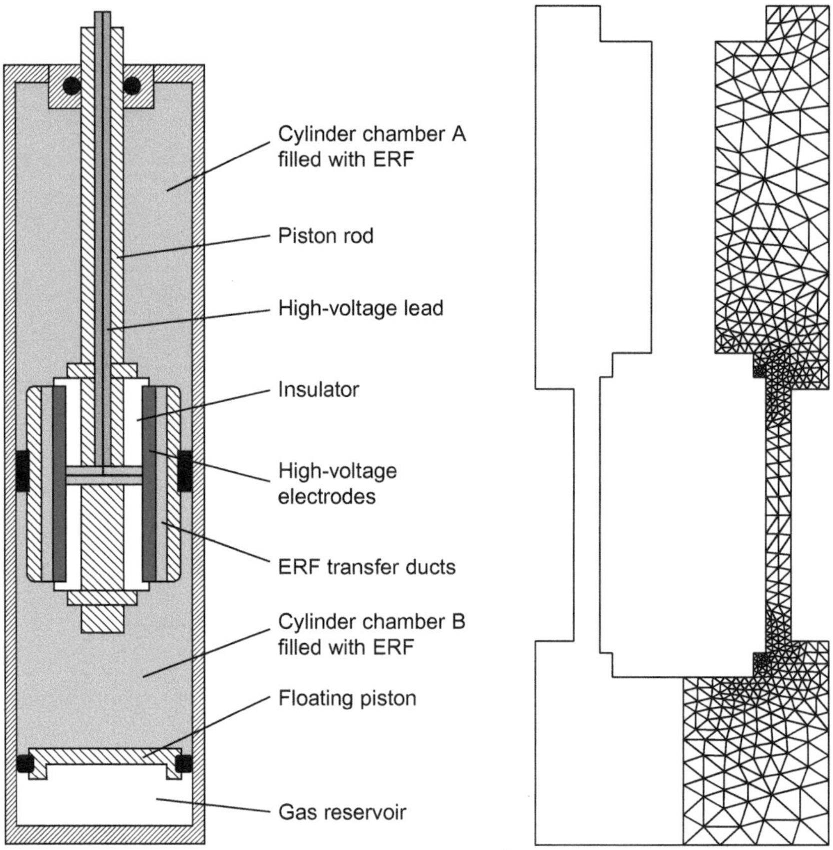

FIG. 4.11 Schematic representation of an electrorheological shock absorber (left) and simplicial triangulation of the computational domain (right).

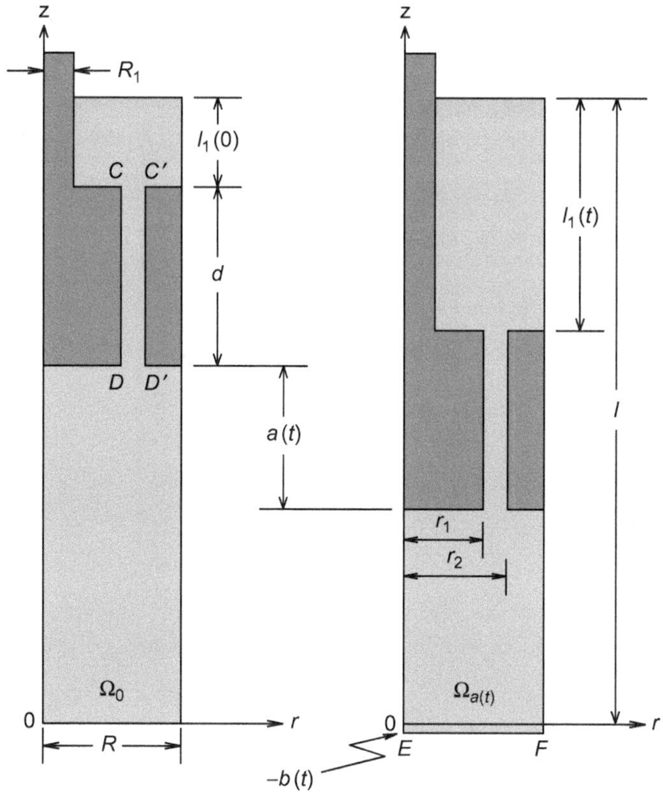

FIG. 4.12 Domain of flow of the electrorheological fluid at time instants $t = 0$ (left) and $t > 0$ (right).

the rebound mode. In the compression mode, the piston moves down and the fluid passes from the lower chamber through the ducts into the upper chamber, whereas in the rebound mode, the piston moves up and the fluid flow is in the opposite direction. The variation of the applied voltage almost instantaneously changes the viscosity of the fluid and thus allows to control the damper characteristics.

The fluid flow is assumed to be axially symmetric so that the computational domain can be restricted to the right half of the fluid chamber and displayed in cylindrical coordinates r, z. Fig. 4.12 illustrates the computational domain in the situation where the piston is at an upper position (left) and at a lower position (right). Due to the displacement $a(t)$ of the piston, the computational domain changes in time and will thus be denoted by $\Omega_{a(t)}$. If the piston is displaced by $a(t) = l_1(t) - l_1(0)$, the floating piston is displaced from its initial position by $b(t) = a(t)(R_1/R)^2$, where R and R_1 are the radii of the floating piston and the piston rod. For a proper specification of the boundary conditions, we refer to $\Gamma_{a(t)} = \partial\Omega_{a(t)}$ as the boundary of the right half of the fluid chamber. In particular, $\Gamma^{(p)}_{a(t)}$ and $\Gamma^{(f)}_{a(t)}$ stand for the boundary of the piston and the upper boundary of the floating piston. We further denote by $\Gamma^{(e)}_{a(t)}$ and $\Gamma^{(c)}_{a(t)}$ the inner wall (CD in Fig. 4.12) and the outer wall (C'D' in Fig. 4.12) of the transfer duct, which serve as the electrode and counter electrode, respectively. Finally,

$\Gamma^{(\ell)}_{a(t)} := \{(r,z) \in \overline{\Omega}_{a(t)} \mid r = 0\}$ stands for the left boundary of the computational domain, which coincides with the symmetry axis. We set $Q := \Omega_{a(t)} \times (0,T)$, $\Sigma_{a(t)} := \Gamma_{a(t)} \times (0,T)$ and use analogous notations for the other space-time domains involving the specific parts of the boundary of the computational domain.

Taking advantage of the axial symmetry, the velocity u is given by

$$u(r,z) = u_1(r,z)e_r + u_2(r,z)e_z,$$

which gives rise to the following components of the strain tensor

$$\begin{aligned}
&\varepsilon_{11}(u) = \frac{\partial u_1}{\partial r}, \quad \varepsilon_{22}(u) = \frac{u_1}{r}, \quad \varepsilon_{33}(u) = \frac{\partial u_2}{\partial z},\\
&\varepsilon_{13}(u) = \varepsilon_{31}(u) = \frac{1}{2}\left(\frac{\partial u_1}{\partial z} + \frac{\partial u_2}{\partial z}\right),\\
&\varepsilon_{12}(u) = \varepsilon_{21}(u) = \varepsilon_{23}(u) = \varepsilon_{32}(u) = 0.
\end{aligned}$$

The second invariant of the rate of strain tensor turns out to be

$$I(u) = \left(\frac{\partial u_1}{\partial r}\right)^2 + \left(\frac{u_1}{r}\right)^2 + \left(\frac{\partial u_2}{\partial z}\right)^2 + \frac{1}{2}\left(\frac{\partial u_1}{\partial z} + \frac{\partial u_2}{\partial r}\right)^2.$$

Denoting by ρ the density of the fluid, by φ the viscosity function according to (2.19), and by $f = (f_1, f_2)^T$ the volume force with the radial and axial components f_1 and f_2, the equations of motion take the form

$$\rho\left(\frac{\partial u_1}{\partial t} + u_1\frac{\partial u_1}{\partial r} + u_2\frac{\partial u_1}{\partial z}\right) + \frac{\partial p}{\partial r}$$

$$-2\frac{\partial}{\partial r}(\varphi\varepsilon_{11}(u)) - 2\frac{\partial}{\partial z}(\varphi\varepsilon_{13}(u)) - \frac{2}{r}\varphi(\varepsilon_{11}(u) - \varepsilon_{22}(u)) = f_1 \quad \text{in } Q, \tag{4.8a}$$

$$\rho\left(\frac{\partial u_2}{\partial t} + u_1\frac{\partial u_2}{\partial r} + u_2\frac{\partial u_2}{\partial z}\right) + \frac{\partial p}{\partial r}$$

$$-2\frac{\partial}{\partial r}(\varphi\varepsilon_{13}(u)) - 2\frac{\partial}{\partial z}(\varphi\varepsilon_{33}(u)) - \frac{2}{r}\varphi\varepsilon_{13}(u) = f_2 \quad \text{in } Q, \tag{4.8b}$$

$$\nabla\cdot u = \frac{\partial u_1}{\partial r} + \frac{\partial u_2}{\partial z} + \frac{u_1}{r} = 0 \quad \text{in } Q. \tag{4.8c}$$

Moreover, referring to $v^{(p)}$ as the piston velocity and to $u^{(0)}$ as some given initial velocity, the boundary conditions and the initial condition are given by

$$u_1 = 0 \quad \text{on } \Sigma_{a(t)}, \tag{4.9a}$$

$$u_2 = v^{(p)} \quad \text{on } \Sigma^{(p)}_{a(t)}, \tag{4.9b}$$

$$u_2 = v^{(p)}(R_1/R)^2 \quad \text{on } \Sigma^{(f)}_{a(t)}, \tag{4.9c}$$

$$u_2 = 0 \quad \text{on } \Sigma_{a(t)} \setminus \left(\overline{\Sigma}^{(f)}_{a(t)} \cup \overline{\Sigma}^{(\ell)}_{a(t)} \cup \overline{\Sigma}^{(p)}_{a(t)}\right), \tag{4.9d}$$

$$\frac{\partial u_2}{\partial r} = 0 \quad \text{on } \Sigma^{(\ell)}_{a(t)}, \tag{4.9e}$$

$$u(\cdot,0) = u^{(0)} \quad \text{in } \Omega_{a(t)}. \tag{4.9f}$$

The motion of the piston satisfies the initial-value problem

$$m\,\frac{\mathrm{d}v^{(p)}}{\mathrm{d}t}(t) = g(t, v^{(p)}(t), U(t)), \quad t \in (0, T), \tag{4.10a}$$

$$v^{(p)}(0) = v_0^{(p)} < 0, \tag{4.10b}$$

where m is the sum of the mass of the piston and the mass of the body that strikes the piston at $t = 0$, $U(t)$ stands for the applied voltage, and the drag force $g(t, v^{(p)}(t), U(t))$ is given by

$$g(t, v^{(p)}(t), U(t)) := -\int_{\Sigma_{a(t)}^{(p)}} (2\varphi\varepsilon_{31}(u)\nu_r + (2\varphi\varepsilon_{33}(u) - p)\nu_z)\,\mathrm{d}s. \tag{4.11}$$

The electric field E has the form

$$E(r, z) = E_1(r, z)e_r + E_2(r, z)e_z.$$

As in the previous example (cf. Section 4.1), it can be computed by means of an electric potential $\psi(t)$ which at each time instant $t \in [0, T]$ satisfies the following elliptic boundary value problem

$$\nabla \cdot (\epsilon \nabla \psi(t)) = 0 \quad \text{in } \Omega_{a(t)}, \tag{4.12a}$$

$$\psi(t) = U(t) \quad \text{on } \Gamma_{a(t)}^{(e)}, \tag{4.12b}$$

$$\psi(t) = 0 \quad \text{on } \Gamma_{a(t)}^{(c)}, \tag{4.12c}$$

$$\frac{\partial \psi}{\partial r}(t) = 0 \quad \text{on } \Gamma_{a(t)}^{(\ell)}, \tag{4.12d}$$

$$\nu_r \epsilon \frac{\partial \psi}{\partial r}(t) + \nu_z \epsilon \frac{\partial \psi}{\partial z}(t) = 0 \quad \text{elsewhere.} \tag{4.12e}$$

For the numerical simulation of the operational behavior of the electrorheological shock absorber, we have used a discretization in time with respect to a uniform partition of the time interval $[0, T]$ of step size $k := T/M$, $M \in \mathbb{N}$, using the explicit Euler scheme for the equation of motion (4.10) of the piston and the backward Euler scheme for the equations of motion (4.8a)–(4.8c) of the fluid with $\rho = 0$. Knowing the computation domain at time level t_m, $0 \le m \le M - 1$, the discretization in space has been taken care of by P_2/P_1 Taylor-Hood elements for the fluid variables and conforming P_1 elements for the electric potential with respect to a simplicial triangulation of $\Omega_{a(t_m)}$. The discretized fluid equations have been solved by the augmented Lagrangian algorithm as described in Section 3.1, whereas the preconditioned conjugate gradient method has been used for the discretized potential equation. For details, we refer to HOPPE, LITVINOV and RAHMAN [2007].

The simulations have been based on the commercial electrorheological fluid Rheobay TP AI 3565 (see BAYER [1997a]) by computing the viscosity function φ using experimentally available $\tau(\gamma)$-flow curves (cf. Section 4.1). As far as the geometry of the shock absorber is concerned, we have used the following data (cf. Fig. 4.12):

$$R := 0.023\text{ m}, \quad R_1 := 0.005\text{ m}, \quad r_1 := 0.013\text{ m}, \quad r_2 := 0.017\text{ m},$$
$$l := 0.14\text{ m}, \quad l_1(0) := 0.02\text{ m, and} \quad d := 0.04\text{ m}.$$

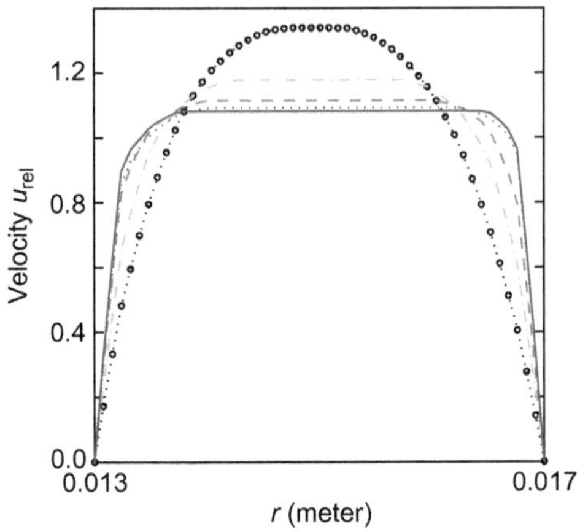

FIG. 4.13 Profiles of the relative velocity of the fluid in the piston duct for various applied voltages: $U = 0$ V (dotted-circled line), 1 kV (dashed-dotted line), 3 kV (dashed line), 6 kV (dotted line), and 9 kV (solid line).

Figure 4.13 shows the relative velocity of the fluid $u_{\rm rel} = (u - v)/\gamma$ in the piston duct for various electric field strengths, where $\gamma = (\int_{r_1}^{r_2} r\mathrm{d}r)^{-1} \int_{r_1}^{r_2} r(u - v)(r, z_1)\mathrm{d}r$ is the flow rate relative to the electrodes. In case of a vanishing electric field, we clearly observe a parabolic flow profile typical for flows of Newtonian fluids between two parallel plates. For increasing electric field strength, the profile flattens in the center of the duct with an increasing zone of constant relative velocity. This is the typical flow pattern of electrorheological fluids.

Figure 4.14 displays the isolines of the electric potential ψ for various positions of the piston assuming an applied voltage of $U = 9\,\mathrm{kV}$. Again, we see that the electric field is essentially concentrated within the transfer ducts in the direction of the r-axis and rapidly decays off the ducts.

Figures 4.15 and 4.16 contain visualizations of the velocity vector u at various stages of the compression mode (Fig. 4.15) and the rebound mode (Fig. 4.16). As has to be expected, in the transfer ducts, the direction of the velocity vector essentially coincides with the direction of the z-axis and is thus orthogonal to the electric field E.

We note that the pressure in the gas reservoir should be sufficiently large, since otherwise the fluid chamber cannot be fully filled with the fluid and cavitation may occur. For further details concerning the simulation results, we refer to HOPPE, LITVINOV and RAHMAN [2007].

4.3. *Shape optimization of electrorheological devices*

An important issue in the design of electrorheological shock absorbers is to find a suitable geometry of the inflow and outflow boundaries of the piston ducts such that both in the compression mode and in the rebound mode pressure peaks are avoided which may cause inappropriate damping profiles. This amounts to the solution of a shape optimization problem which for simplicity will be stated as a velocity and pressure tracking problem where the objective functional is given by

$$\text{minimize} \quad J(u, p, d) := \frac{\alpha_1}{2} \|u - u^d\|^2_{0,\Omega(d)} + \frac{\alpha_2}{2} \|p - p^d\|^2_{0,\Omega(d)}. \tag{4.13}$$

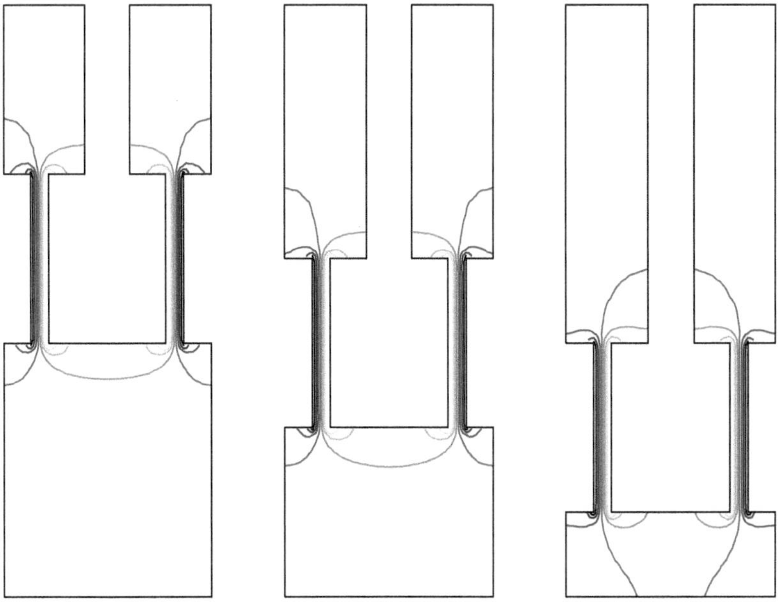

FIG. 4.14 Isolines of the electric potential at three different piston positions.

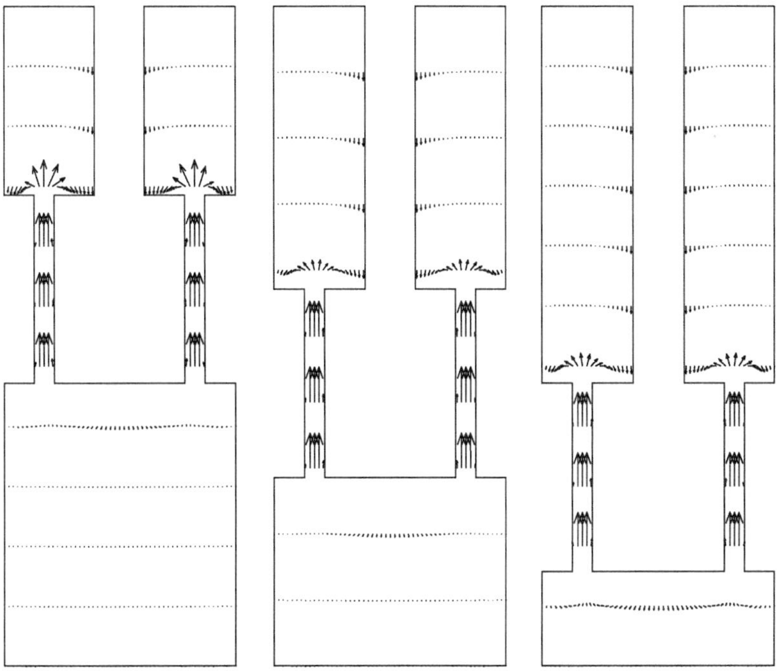

FIG. 4.15 Velocity vectors during compression.

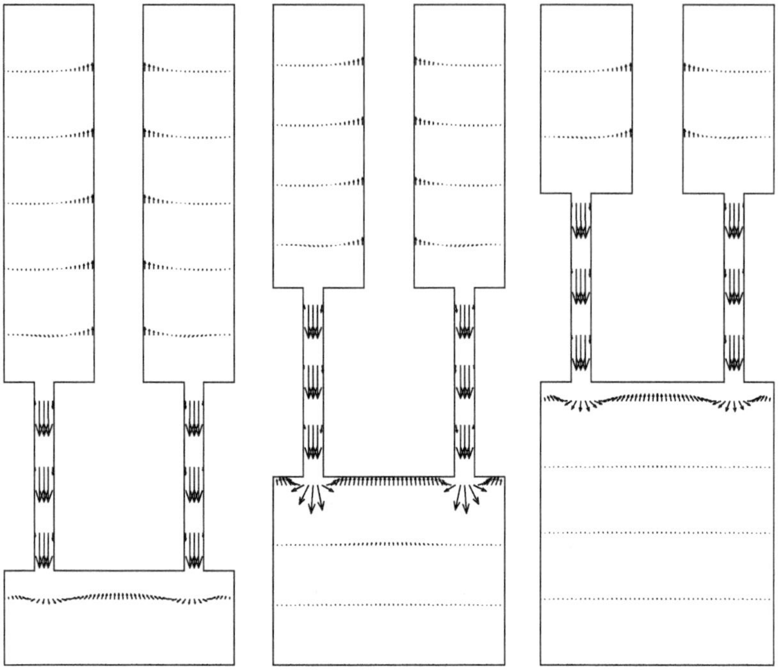

FIG. 4.16 Velocity vectors during rebound.

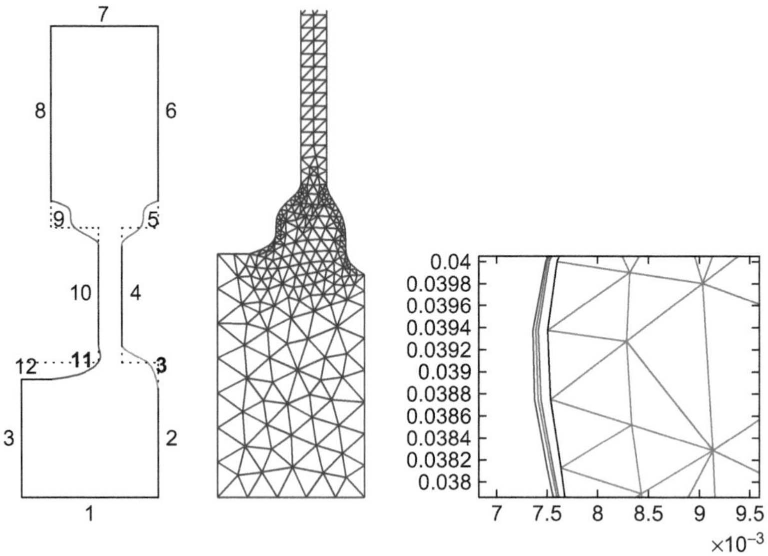

FIG. 4.17 Bézier curve representation of the inlet and outlet boundaries of a piston duct (left), optimized outlet boundary (middle), and details of the optimal design for various electric field strengths (right).

Here, $u^d \in H(\text{div}^0; \Omega(d))$ and $p^d \in L^2(\Omega(d))$ stand for a desired velocity profile and pressure distribution, respectively, $\alpha_i \in (0, 1]$, $1 \le i \le 2$, and $\Omega(d)$ is the domain occupied by the fluid which depends on the design variables $d = (d_1, \ldots, d_m)^T \in \mathbb{R}^m$. The design variables are chosen as the Bézier control points of a Bézier curve representation (cf. FARIN [2002]) of the inlet and outlet boundaries (cf. Fig. 4.17 (left)).

The PDE constraints are given by

$$-\nabla \cdot \sigma(u) = f \quad \text{in } \Omega(d), \tag{4.14a}$$
$$\nabla \cdot u = 0 \quad \text{in } \Omega(d), \tag{4.14b}$$

along with appropriate boundary conditions (cf. Section 4.2). The constitutive law is assumed to be given by

$$\sigma = -pI + 2\,\varphi(I(u), |E|, \mu(u, E))\,\varepsilon(u) \tag{4.15}$$

with a regularized viscosity function φ of the form (2.19), where the electric field E is computed via the gradient of an electric potential satisfying an elliptic boundary value problem (cf. (4.12a)–(4.12e)). We further assume bilateral constraints on the design variables according to

$$d \in K := \{d \in \mathbb{R}^m \mid d_i^{\min} \le d_i \le d_i^{\max},\ 1 \le i \le m\}. \tag{4.16}$$

Choosing $X \subset H^1(\Omega(d))^2$ and $Q := L_0^2(\Omega(d))$, we refer to $Y := X \times Q$ as the state space and denote by $S(\cdot, d)$, $d \in K$, the nonlinear Stokes operator associated with (4.14a), (4.14b). Then, the state equations can be written in operator form according to

$$S(y, d) = g, \tag{4.17}$$

where $y := (u, p)^T$ and $g := (f, 0)^T$. We choose $\hat{d} \in K$ as a reference design and refer to $\hat{\Omega} := \Omega(\hat{d})$ as the associated reference domain. Then, the actual domain $\Omega(d)$ can be obtained from the reference domain $\hat{\Omega}$ by means of an isomorphism

$$\Omega(d) = \Phi(\hat{\Omega}; d),$$
$$\Phi(\hat{x}; d) = (\Phi_1(\hat{x}; d), \Phi_2(\hat{x}; d))^T, \quad \hat{x} = (\hat{x}_1, \hat{x}_2)^T. \tag{4.18}$$

The advantage of using the reference domain $\hat{\Omega}$ is that finite element approximations of (4.17) can be performed with respect to that fixed domain without being forced to remesh for each update of the design variables.

We denote by $(\mathcal{T}_h(\hat{\Omega}))_{\mathbb{N}}$ a shape regular family of simplicial triangulations of $\hat{\Omega}$. By means of (4.18), these triangulations induce an associated family $(\mathcal{T}_h(\Omega(d)))_{\mathbb{N}}$ of simplicial triangulations of the actual physical domains $\Omega(d)$.

We use Taylor-Hood $P2/P1$ elements for the discretization of the velocity $u \in X$ and the pressure $p \in Q$ denoting the associated trial spaces by X_h and Q_h with $\dim X_h = n_1$ and $\dim Q_h = n_2$, respectively. This gives rise to an objective functional $J_h : \mathbb{R}^n \times \mathbb{R}^m$, $n := n_1 + n_2$, by means of

$$J_h(u_h, p_h, d) := \frac{\alpha_1}{2}\left(u_h - u_h^d\right)^T I_{1,h}(d)\left(u_h - u_h^d\right) + \frac{\alpha_2}{2}\, p_h^T I_{2,h}(d) p_h, \tag{4.19}$$

where $I_{\nu,h}(d)$, $1 \le \nu \le 2$, are the associated mass matrices and $u_h^d \in \mathbb{R}^{n_1}$, $p_h^d \in \mathbb{R}^{n_2}$ result from the L^2 projections of u^d, p^d onto $X_h \cap H(\mathrm{div}^0; \Omega)$ and Q_h, respectively. The discretized shape optimization problem can be stated as

$$\inf_{u_h, p_h, d} J_h(u_h, p_h, d) \tag{4.20}$$

subject to the discrete nonlinear Stokes system

$$S_h(y_h, d) = g_h \tag{4.21}$$

and the constraints

$$d \in K. \tag{4.22}$$

For notational convenience, in the sequel, we will drop the discretization index h.

Due to the dependence of the domain on the design parameters d_i, $1 \le i \le m$, the objective functional is nonconvex. Therefore, there may exist a multitude of local minima. Throughout the following, we assume that $(y^*, d^*) \in \mathbb{R}^n \times K$ is a strict local solution of (4.20)–(4.22).

We solve the discrete minimization problem by an adaptive path-following primal-dual interior-point method. To this end, we couple the inequality constraints (4.22) by logarithmic barrier functions with a barrier parameter $\beta = 1/\mu > 0$, $\mu \to \infty$, resulting in the following parameterized family of minimization subproblems

$$\inf_{y,d} B^{(\mu)}(y, d) \tag{4.23}$$

subject to (4.21), where

$$B^{(\mu)}(y, d) := J(y, d) - \frac{1}{\mu} \sum_{i=1}^{m} [\ln(d_i - d_i^{\min}) + \ln(d_i^{\max} - d_i)]. \tag{4.24}$$

The dual aspect is to couple the constraint (4.21) by a Lagrange multiplier $\lambda = (\lambda_u, \lambda_p)^T$, which leads to the saddle point problem

$$\inf_{y,d} \sup_{\lambda} L^{(\mu)}(y, \lambda, d), \tag{4.25}$$

where the Lagrangian $L^{(\mu)}$ is given by

$$L^{(\mu)}(y, \lambda, d) = B^{(\mu)}(y, d) + \lambda^T (S(y, d) - g). \tag{4.26}$$

The barrier path $\mu \longmapsto x(\mu) := (y(\mu), \lambda(\mu), d(\mu))^T$ is defined as the solution of the nonlinear system

$$F(x(\mu), \mu) = \begin{pmatrix} L_y^{(\mu)}(y, \lambda, d) \\ L_\lambda^{(\mu)}(y, \lambda, d) \\ L_d^{(\mu)}(y, \lambda, d) \end{pmatrix} = 0, \tag{4.27}$$

which represents the first-order necessary optimality conditions for (4.25).

For the solution of the parameter-dependent nonlinear system (4.27), we use an adaptive path-following predictor-corrector strategy along the lines of DEUFLHARD [2004].

Predictor Step The predictor step relies on tangent continuation along the trajectory of the Davidenko equation

$$F_x(x(\mu), \mu)\, x'(\mu) = -F_\mu(x(\mu), \mu). \tag{4.28}$$

Given some approximation $\tilde{x}(\mu_k)$ at $\mu_k > 0$, compute $\tilde{x}^{(0)}(\mu_{k+1})$, where $\mu_{k+1} = \mu_k + \Delta\mu_k^{(0)}$, according to

$$F_x(\tilde{x}(\mu_k), \mu_k)\, \delta x(\mu_k) = -F_\mu(\tilde{x}(\mu_k), \mu_k), \tag{4.29a}$$

$$\tilde{x}^{(0)}(\mu_{k+1}) = \tilde{x}(\mu_k) + \Delta\mu_k^{(0)}\, \delta x(\mu_k). \tag{4.29b}$$

We use $\Delta\mu_0^{(0)} = \Delta\mu_0$ for some given initial step size $\Delta\mu_0$, whereas for $k \geq 1$, the predicted step size $\Delta\mu_k^{(0)}$ is chosen by

$$\Delta\mu_k^{(0)} := \left(\frac{\|\Delta x^{(0)}(\mu_k)\|}{\|\tilde{x}(\mu_k) - \tilde{x}^{(0)}(\mu_k)\|} \frac{\sqrt{2}-1}{2\Theta(\mu_k)} \right)^{1/2} \Delta\mu_{k-1}, \tag{4.30}$$

where $\Delta\mu_{k-1}$ is the computed continuation step size, $\Delta x^{(0)}(\mu_k)$ is the first Newton correction (see below), and $\Theta(\mu_k) < 1$ is the contraction factor associated with a successful previous continuation step.

Corrector step As a corrector, we use Newton's method applied to $F(x(\mu_{k+1}), \mu_{k+1}) = 0$ with $\tilde{x}^{(0)}(\mu_{k+1})$ as a start vector. In particular, for $\ell \geq 0$ and $j_\ell \geq 0$, we compute $\Delta x^{(j_\ell)}(\mu_{k+1})$ according to

$$F'(\tilde{x}^{(j_\ell)}(\mu_{k+1}), \mu_{k+1})\, \Delta x^{(j_\ell)}(\mu_{k+1}) = -F(\tilde{x}^{(j_\ell)}(\mu_{k+1}), \mu_{k+1}) \tag{4.31}$$

and $\overline{\Delta x}^{(j_\ell)}(\mu_{k+1})$ as the associated simplified Newton correction

$$F'(\tilde{x}^{(j_\ell)}(\mu_{k+1}), \mu_{k+1})\, \overline{\Delta x}^{(j_\ell)}(\mu_{k+1}) = -F(\tilde{x}^{(j_\ell)}(\mu_{k+1}) + \Delta x^{(j_\ell)}(\mu_{k+1}), \mu_{k+1}). \tag{4.32}$$

We monitor convergence of Newton's method by means of

$$\Theta^{(j_\ell)}(\mu_{k+1}) := \|\overline{\Delta x}^{(j_\ell)}(\mu_{k+1})\| / \|\Delta x^{(j_\ell)}(\mu_{k+1})\|.$$

In case of successful convergence, we accept the current step size and proceed with the next continuation step. However, if the monotonicity test

$$\Theta^{(j_\ell)}(\mu_{k+1}) < 1 \tag{4.33}$$

fails for some $j_\ell \geq 0$, the continuation step has to be repeated with the reduced step size

$$\Delta\mu_k^{(\ell+1)} := \left(\frac{\sqrt{2}-1}{g(\Theta^{(j_\ell)})} \right)^{1/2} \Delta\mu_k^{(\ell)}, \quad g(\Theta) := \sqrt{\Theta+1} - 1 \tag{4.34}$$

until we either achieve convergence or for some prespecified lower bound $\Delta\mu_{\min}$ observe

$$\Delta\mu_k^{(\ell+1)} < \Delta\mu_{\min}.$$

In the latter case, we stop the algorithm and report convergence failure.

Actually, we perform the correction step by an inexact Newton method featuring right-transforming iterations. The derivatives have been computed by automatic differentiation. For details, we refer to ANTIL, HOPPE and LINSENMANN [2007], HOPPE, PETROVA and SCHULZ [2002], HOPPE and PETROVA [2004], HOPPE, LINSENMANN and PETROVA [2006], WITTUM [1989].

Figure 4.17 (middle) shows the optimized design of the outlet boundary of a piston duct in the rebound stage (cf. Section 4.2) and details of the optimized outlet boundary for various electric field strengths (the lines show the different designs for increasing electric field strengths from right to left). Although the designs do not differ that much, the specification of a best compromise is the subject of a further optimization routine.

Acknowledgments

The work presented in this contribution has been supported by the German National Science Foundation DFG within the Collaborative Research Center DFB 438 and by the US National Science Foundation NSF under Grant No. NSF-DMS0511611. The authors would also like to express their sincere thanks to H. Antil, C. Linsenmann, and T. Rahman for their most valuable assistance in performing the numerical simulations.

Bibliography

ABU-JDAYIL, B. (1996). *Electrorheological Fluids in Rotational Couette Flow, Slit Flow and Torsional Flow (Clutch)* (Shaker, Aachen).

ABU-JDAYIL, B., Brunn, P.O. (1995). Effects of nonuniform electric field on slit flow of an electrorheological fluid. *J. Rheol.* **39**, 1327–1341.

ABU-JDAYIL, B., BRUNN, P.O. (1996). Effects of electrode morphology on the slit flow of an electrorheological fluid. *J. Non-Newton. Fluid Mech.* **63**, 45–61.

ABU-JDAYIL, B., BRUNN, P.O. (1997). Study of the flow behavior of electrorehological fluids at shear- and flow-mode. *Chem. Eng. Process.* **36**, 281–289.

ABU-JDAYIL, B., BRUNN, P.O. (2002). Effect of electrode morphology on the behavior of electrorheological fluids in torsional flow. *J. Intell. Mater. Syst. Struct.* **13**, 3–11.

ACERBI, E., MINGIONE, G. (2002). Regularity results for stationary electrorheological fluids. *Arch. Ration. Mech. Anal.* **164**, 213–259.

ADAMS, R.A. (1975). *Sobolev Spaces* (Academic Press, New York).

ADOLF, D., GARINO, T. (1995). Time-dependent dielectric response of quiescent electrorheological fluids. *Langmuir* **11**, 307–312.

ALLEN, M.P., TILDESLEY, D.J. (1983). *Computer Simulation of Liquids* (Oxford University Press, Oxford).

ANTIL, H., HOPPE, R.H.W., LINSENMANN, C. (2007). Path-following primal-dual interior-point methods for shape optimization of stationary flow problems. *J. Numer. Math.* **15**, 81–100.

ATKIN, R., SHI, X., BULLOUGH, W. (1991). Solution of the constitutive equations for the flow of an electrorheological fluid. *J. Rheol.* **35**, 1441–1461.

ATKIN, R., SHI, X., BULLOUGH, W. (1999). Effect of non-uniform field distribution on steady flows of an electro-rheological fluid. *J. Non-Newton. Fluid Mech.* **86**, 119–132.

AWANOU, G.M., LAI, M.J. (2005). On convergence rate of the augmented Lagrangian algorithm for nonsymmetric saddle point problems. *Appl. Numer. Math.* **54**, 122–134.

BANK, R., WELFERT, B.D., YSERENTANT, H. (1990). A class of iterative methods for solving saddle point problems. *Numer. Math.* **56**, 645–666.

BANKS, H.T., ITO, K., JOLLY, M.R., LY, H.V., REITICH, F.L., SIMON, T.M. (1999). Dynamic simulation and nonlinear homogenization study for magnetorheological fluids. In: Varadan, V.V. (ed.), *Proceedings of SPIE, Smart Structures and Materials 1999: Mathematics and Control in Smart Structures* (Int. Soc. for Optical Engrg., Bellingham,WA), pp. 92–100.

BAYER, A.G. (1997a). Provisional product information. Rheobay TP AI 3565 and Rheobay TP AI 3566. Report No. AI 12601e, Bayer AG, Silicones Business Unit, Leverkusen.

BAYER, A.G. (1997b). Technology based on ERF. Rheobay for applications in fluid mechatronics (in cooperation with IFAS, RWTH Aachen, and Carl Schenck AG, Darmstadt). Report No. AI 12666d+e, Bayer AG, Silicones Business Unit, Leverkusen.

BAYER, A.G., CARL SCHENCK, A.G. (1998). Active ERF-vibration damper (a joint development by Carl Schenck AG and Bayer AG). Report No. AI 12668d+e, Bayer AG, Silicones Business Unit, Leverkusen.

BELONOSOV, M.S., LITVINOV, W.G. (1996). Finite element methods for nonlinearly viscous fluids. *Z. Angew. Math. Mech.* **76**, 307–320.

BENDERSKAIA, S.L., KHUSID, B.M., SHULMAN, Z.P. (May–June 1980). Nonisothermal channel flows of non-Newtonian fluids. *Akad. Nauk SSSR, Izvestiia Mekhanika Zhidkosti i Gaza* 3–10 (in Russian).

BERG, C.D., WELLSTEAD, P.E. (1998). The application of a smart landing gear oleo incorporating electrorheological fluid. *J. Intell. Mater. Struct.* **9**, 592–600.

BILDHAUER, M., FUCHS, M. (2004). A regularity result for stationary electrorheological fluids in two dimensions. *Math. Methods Appl. Sci.* **27**, 1607–1617.

BLOCK, H., KELLY, J.P. (1988). Electro-rheology. *J. Phys. D* **21**, 1661–1677.

BLOCK, H., KELLY, J.P., QIN, A., WATSON, T. (1990). Materials and mechanisms in electrorheology. *Langmuir* **6**, 6–14.

BONNECAZE, R.T., BRADY, J.F. (1992a). Dynamic simulation of an electrorheological fluid. *J. Chem. Phys.* **96**, 2183–2202.

BONNECAZE, R.T., BRADY, J.F. (1992b). Yield stresses in electrorheological fluids. *J. Rheol.* **38**, 73–115.

BÖSE, H. (1998). Investigations on zeolite-based ER fluids supported by experimental design. In: Nakano, M., Koyama, K. (eds.), *Proc. 6th Int. Conf. on Electrorheological Fluids, Magnetorheological Suspensions and Their Applications 1998* (World Scientific, Singapore), pp. 240–247.

BÖSE, H., HOPPE, R.H.W., MAZURKEVICH, G. (2001). Mathematical modelling and numerical simulation of electrorheological devices. In: Hoffmann, K.H. (ed.), *Smart Materials* (Springer, Berlin-Heidelberg-New York), pp. 39–50.

BÖSE, H., TRENDLER, A. (2001). Comparison of rheological and electric properties of ER fluids based on different materials. *Int. J. Mod. Phys. B* **15**, 626–633.

BOSSIS, G. (ed.) (2002). Electrorheological fluids and magnetorheological suspensions. Proc. Eighth Int. Conf., Nice, France, July 9–13, 2001, World Scientific, Singapore.

BRAESS, D. (2007). *Finite Elements: Theory, Fast Solvers and Applications in Solid Mechanics*, Third Edition (Cambridge University Press, Cambridge).

BRAMBLE, J.H., PASCIAK, J.E., VASSILEV, A.T. (1997). Analysis of the inexact Uzawa algorithm for saddle point problems. *SIAM J. Numer. Anal.* **34**, 1072–1092.

BREZIS, H. (1973). *Opérateurs Maximaux Monotones Et Semi-Groupes De Contractions Dans Les Espaces De Hilbert* (North-Holland, Amsterdam).

BREZZI, F., FORTIN, M. (1991). *Mixed and Hybrid Finite Element Methods* (Springer, Berlin-Heidelberg-New York).

BROWDER, F.E. (1968). Nonlinear maximal monotone operators in Banach space. *Math. Ann.* **175**, 89–113.

BRUNN, P.O., ABU-JDAYIL, B. (1998). Fluids with transverse isotropy as models for electrorheological fluids. *Z. Angew. Math. Mech.* **78**, 97–107.

BRUNN, P.O., ABU-JDAYIL, B. (2004). A phenomenological model of electrorheological fluids. *Rheol. Acta* **43**, 62–67.

BUSUIOC, V., CIORANESCU, D. (2003). On the flow of a Bingham fluid passing through an electric field. *Int. J. Non Linear Mech.* **38**, 287–304.

BUTZ, T., VON STRYK, O. (2001). Modelling and simulation of electro- and magnetorheological fluid dampers. *Z. Angew. Math. Mech.* **82**, 3–20.

CACCANO, C., STEEL, G., HANSEN, J.P. (1999). *New Approaches to Problems in Liquid State Theory: Inhomogeneities and Phase Separation in Simple, Complex and Quantum Fields* (Springer, Berlin-Heidelberg-New York).

CAO, Z. (2003). Fast Uzawa algorithm for generalized saddle point problems. *Appl. Numer. Math.* **43**, 157–171.

CASSON, N.A. (1959). Flow equation of pigment oil suspension of printing ink type. In: Mill, C.C. (ed.), *Rheology of Disperse Systems* (Pergamon Press, London), pp. 84–120.

CECCIO, S., WINEMAN, A. (1994). Influence of orientation of electric field on shear flow of electrorheological fluids. *J. Rheol.* **38**, 453–463.

CLERCX, H.J.H., BOSSIS, G. (1993). Many-body electrostatic interactions in electrorheological fluids. *Phys. Rev. E* **48**, 2721–2738.

CONRAD, H., SPRECHER, A.F., CHOI, Y., CHEN, Y. (1991). The temperature dependence of the electrical properties and strength of electrorheological fluids. *Soc. Rheol.* **35**, 1393–1410.

COULTER, J.P., WEISS, K.D., CARLSON, J.D. (1993). Engineering applications of electrorheological fluids. *J. Intell. Mater. Syst. Struct.* **4**, 248–259.

CROCHET, M.J. (1984). *Numerical Simulation of Non-Newtonian Flow* (Elsevier, Amsterdam).

DASSANAYAKE, U., FRADEN, S., VAN BLAADEREN, A. (2000). Structure of electrorheological fluids. *J. Chem. Phys.* **112**, 3851–3858.

DAUBERT, C.R., STEFFE, J.F., SRIVASTAVA, A.K. (1998). Predicting the electrorheological behavior of milk chocolate. *J. Food Process Eng.* **21**, 249–261.

DEFENSE UPDATE. (2004). Future force warrior project. In: *International Online Defense Magazine*, Issue **4**, 2004, http://www.defense-update.com/features/du-4-04/FFW.htm

DEINEGA, Y.F., VINOGRADOV, G.V. (1984). Electric fields in the rheology of disperse systems. *Rheol. Acta* **23**, 636–651.

DEUFLHARD, P. (2004). *Newton Methods for Nonlinear Problems. Affine Invariance and Adaptive Algorithms* (Springer, Berlin-Heidelberg-New York).

DOE - US DEPARTMENT OF ENERGY. (1993). *Electrorheological (ER) Fluids - A Research Needs Assessment, DOE/ER/30172* (US Department of Energy,Washington DC).

DREYFUSS, P., HUNGERBUEHLER, N. (2004a). Navier-Stokes systems with quasimonotone viscosity tensor. *Int. J. Differ. Equ.* **9**, 59–79.

DREYFUSS, P., HUNGERBUEHLER, N. (2004b). Results on a Navier-Stokes system with application to electrorheological fluid flow. *Int. J. Pure Appl. Math.* **14**, 241–271.

DROUOT, R., NAPOLI, G., RACINEUX, G. (2002). Continuum modeling of electrorheological fluids. *Int. J. Mod. Phys. B* **16** (17 & 18), 2649–2654.

DUFF, A.W. (1896). The viscosity of polarized dielectrics. *Phys. Rev.* **4**, 23–38.

DUVAUT, G., LIONS, J.L. (1971). Transfert de chaleur dans un fluide de Bingham dont la viscosité dépend de la température. *J. Funct. Anal.* **11**, 93–110.

DUVAUT, G., LIONS, J.L. (1976). *Inequalities in Mechanics and Physics* (Springer, Berlin-Heidelberg-New York).

ECKART, W. (2000). Phenomenological modeling of electrorheological fluids with an extended Casson-model. *Continuum Mech. Thermodyn.* **12**, 341–362.

ECKART, W., SADIKI, A. (2001). Polar theory for electrorheological fluids based on extended thermodynamics. *Int. J. Appl. Mech. Eng.* **6**, 969–998.

EDAMURA, K., OTSUBO, Y. (2004). Electrorheology of dielectric liquids. *Rheol. Acta* **43**, 180–183.

ELMAN, H.C. (2002). Preconditioners for saddle point problems arising in computational fluid dynamics. *Appl. Numer. Math.* **43**, 75–89.

ELMAN, H.C., GOLUB, G.H. (1994). Ineaxact and preconditioned Uzawa algorithms for saddle point problems. *SIAM J. Numer. Anal.* **30**, 1645–1661.

ELMAN, H.C., SILVESTER, D. (1996). Fast nonsymmetric iterations and preconditioning for Navier-Stokes equations. *SIAM J. Sci. Comp.* **17**, 33–46.

ELMAN, H.C., SILVESTER, D., WATHEN, A.J. (2005). *Finite Elements and Fast Iterative Solvers with Applications in Incompressible Fluid Dynamics* (Oxford University Press, Oxford).

ENGELMANN, B., HIPTMAIR, R., HOPPE, R.H.W., MAZURKEVICH, G. (2000). Numerical simulation of electrorheological fluids based on an extended Bingham model. *Comput. Visual. Sci.* **2**, 211–219.

ERINGEN, A.C. (2002). *Nonlocal Continuum Field Theories* (Springer, Berlin-Heidelberg-New York).

ERINGEN, A.C., MAUGIN, G. (1989). *Electrodynamics of Continua* Volume I and II. (Springer, Berlin-Heidelberg-New York).

ETTWEIN, F., RUZICKA, M. (2002). Existence of strong solutions for electrorheological fluids in two dimensions: steady Dirichlet problem. In: Hildebrandt, S., Karcher, H. (eds.), *Geometric Analysis and Nonlinear Partial Differential Equations* (Springer, Berlin-Heidelberg-New York), pp. 591–602.

FARIN, G. (2002). *Curves and Surfaces for CAGD. A Practical Guide*, Fifth Edition (Morgan Kaufman Publishers, San Francisco).

FILISKO, F. (1995). Overview of ER technology. In: Hawelka, K. (ed.), *Progress in ER Technology* (Plenum Press, New York), pp. 3–18.

FORTIN, M., GLOWINSKI, R. (1983). *Augmented Lagrangian Methods: Applications to the Numerical Solution of Boundary-Value Problems* (North-Holland, Amsterdam).

FOULC, J.N., ATTEN, P., BOSSIS, C. (1996). Correlation between electrical and rheological properties of electrorheological fluids. *J. Intell. Mater. Syst. Struct.* **7**, 579–582.

FREHSE, J., MALEK, J., STEINHAUER, M. (1997). An existence result for fluids with shear dependent viscosity - steady flows. *Nonlinear Anal. Theory Meth. Appl.* **30**, 3041–3049.

FUCIK, S., KRATOCHVIL, A., NECAS, J. (1973). Kachanov-Galerkin method. *Comments Math. Univ. Carol.* **14**, 651–659.

GAJEWSKI, H., GRÖGER, K., ZACHARIAS, K. (1974). *Nichtlineare Operatorgleichungen und Operatordifferentialgleichungen* (Akademie-Verlag, Berlin).

GALDI, G.P. (1994). *An Introduction to the Mathematical Theory of the Navier-Stokes Equations* (Springer, Berlin-Heidelberg-New York).

GARG, D., ANDERSON, G. (2003). Structural damping and vibration control via smart sensors and actuators. *J. Vib. Control* **9**, 1421–1452.

GAST, A.P., ZUKOSKI, C.F. (1989). Electrorheological fluids as colloidal suspensions. *Adv. Colloid Interface Sci.* **30**, 153–202.

GAVIN, H.P. (2001). Multi-duct ER dampers. *J. Intell. Mater. Syst. Struct.* **12**, 353–366.

GAVIN, H.P., HANSON, R.D., FILISKO, F. (1996a). Electrorheological dampers, Part I: analysis and design. *J. Appl. Mech.* **63**, 669–675.

GAVIN, H.P., HANSON, R.D., FILISKO, F. (1996b). Electrorheological dampers, Part II: testing and modeling. *J. Appl. Mech.* **63**, 676–682.

GEORGIADES, G., OYADIJI, S.O. (2003). Effects of electrode geometry on the performance of electrorheological fluid valves. *J. Intell. Mater. Syst. Struct.* **14**, 105–111.

GIRAULT, V., RAVIART, P. (1986). *Finite Element Approximation of the Navier-Stokes Equations* (Springer, Berlin-Heidelberg-New York).

GLOWINSKI, R. (1984). *Numerical Methods for Nonlinear Variational Problems* (Springer, Berlin-Heidelberg-New York).

GLOWINSKI, R. (2004). *Handbook of Numerical Analysis: Numerical Methods for Fluids* (Elsevier, Amsterdam).

GLOWINSKI, R., LETALLEC, P. (1989). Augmented Lagrangian and operator-splitting methods in nonlinear mechanics. *SIAM Studies in Applied Mathematics* Volume 9 (SIAM, Philadelphia).

GLOWINSKI, R., LIONS, J.L., TREMOLIERES, R. (1981). *Numerical Analysis of Variational Inequalities* (North-Holland, Amsterdam).

GRISVARD, P. (1985). *Elliptic Problems in Nonsmooth Domains* (Pitman, Boston).

GUNZBURGER, M. (1989). *Finite Element Methods for Viscous Incompressible Fluids* (Academic Press, New York).

HACKBUSCH, W. (1985). *Multi-Grid Methods and Applications* (Springer, Berlin-Heidelberg-New York).

HALSEY, T.C. (1992). Electrorheological fluids. *Science* **258**, 761–766.

HANAOKA, R., MURAKUMO, M., ANZAI, H., SAKURAI, K. (2002). Effects of electrode surface morphology on electrical response of electrorheological fluids. *IEEE Trans. Electron Devices* **9**, 10–16.

HAO, T. (2001). Electrorheological fluids. *Adv. Mater.* **13**, 1847–1857.

HARTSOCK, D., NOVAK, R., CHAUNDY, G. (1991). ER fluid requirements for automotive devices. *J. Rheol.* **35**, 1305–1326.

HOPPE, R.H.W., KUZMIN, M.Y., LITVINOV, W.G., ZVYAGIN, V.G. (2006). Flow of electrorheological fluids under conditions of slip on the boundary. *Abstract and Applied Analysis*, Volume 2006, Special issue 'Topological and variational methods of nonlinear analysis and their applications,' 1–14.

HOPPE, R.H.W., LINSENMANN, C., PETROVA, S.I. (2006). Primal-dual Newton methods in structural optimization. *Comput. Visual. Sci.* **9**, 71–87.

HOPPE, R.H.W., LITVINOV, W.G. (2004). Problems on electrorheological fluid flow. *Commun. Pure Appl. Anal.* **3**, 809–848.

HOPPE, R.H.W., LITVINOV, W.G., RAHMAN, T. (2003). Mathematical modeling and numerical simulation of electrorheological devices and systems. In: Heikkola, E., Kuznetsov, Y., Neittaanmäki, P., Pironneau, O. (eds.), *Numerical Methods for Scientific Computing: Variational Problems and Applications* (CIMNE, Barcelona), pp. 81–93.

HOPPE, R.H.W., LITVINOV, W.G., RAHMAN, T. (2005). Problems of stationary flow of electrorheological fluids in a cylindrical coordinate system. *SIAM J. Appl. Math.* **65**, 1633–1656.

Hoppe, R.H.W., Litvinov, W.G., Rahman, T. (2007). Model of an electroreheological shock absorber and coupled problem for partial and ordinary differential equations with variable unknown domain. *Eur. J. Appl. Math.* **8**, 513–536.

Hoppe, R.H.W., Mazukevich, G. (2001). Modeling and simulation of electrorheological devices. In: Feistauer, M., Kozel, K., Rannacher, R. (eds.), *Numerical Modelling in Continuum Mechanics* (Matfyzpress, Prague), pp. 153–162.

Hoppe, R.H.W., Mazukevich, G., Von Stryk, O., Rettig, U. (2000). Modeling, simulation, and control of electrorheological automobile devices. In: Bungartz, H.J., Hoppe, R.H.W., Zenger, C. (eds.), *Lectures on Applied Mathematics* (Springer, Berlin-Heidelberg-New York), pp. 251–276.

Hoppe, R.H.W., Petrova, S.I. (2004). Primal-dual Newton interior-point methods in shape and topology optimization. *Numer. Linear Algebra Appl.* **11**, 413–429.

Hoppe, R.H.W., Petrova, S.I., Schulz, V. (2002). A primal-dual Newton-type interior-point method for topology optimization. *J. Optim. Theory Appl.* **114**, 545–571.

Hu, Y., Chen, M.W. (1998). Computer simulation of polarization of mobile charges on the surface of a dielectric sphere in transient electric fields. *J. Electrostat.* **43**, 19–38.

Huang, H.C. (1998). *Finite Element Analysis of Non-Newtonian Flow: Theory and Software* (Springer, Berlin-Heidelberg-New York).

Inoue, A., Maniwa, S. (1995). Electrorheological effect of liquid crystalline polymers. *J. Appl. Polym. Sci.* **55**, 113–118.

Janocha, H., Rech, B., Boelter, R. (1996). Practice-relevant aspects of constructing ER fluid actuators. *Int. J. Mod. Phys. B* **10**, 3243–3255.

Jones, T.B. (1995). *Electromechanics of Particles* (Cambridge University Press, New York).

Khusid, B., Acrivos, A. (1995). Effects of conductivity in electric-field-induced aggregation in electrorheological fluids. *Phys. Rev. E* **52**, 1669–1693.

Kim, S., Karrila, S.J. (1991). *Microhydrodynamics* (Butterworth-Heinemann, Boston).

Kim, Y.D., Klingenberg, D.J. (1997). An interfacial polarization model for activated electrorheological suspensions. *Korean J. Chem. Eng.* **14**, 30–36.

Kim, W.B., Lee, S.J., Kim, Y.J. (2003). The electromechanical principle of electrorheological fluid-assisted polishing. *Int. J. Mach. Tools Manuf.* **43**, 81–88.

Kimura, H., Aikawa, K., Masubuchi, Y., Takimoto, J., Koyama, K., Uemura, T. (1998). 'Positive' and 'negative' electro-rheological effect of liquid blends. *J. Non-Newton. Fluid Mech.* **76**, 199–211.

Klass, D., Martinek, T. (1967a). Electroviscous fluids. I. Rheological properties. *J. Appl. Phys.* **38**, 67–74.

Klass, D., Martinek, T. (1967b). Electroviscous fluids. II. Electrical properties. *J. Appl. Phys.* **38**, 75–80.

Klawonn, A. (1998). An optimal preconditioner for a class of saddle point problems with a penalty term. *SIAM J. Sci. Comput.* **19**, 540–552.

Klein, D., Rensink, D., Freimuth, H., Monkmann, G.J., Egersdörfer, S., Böse, H., Baumann, M. (2004). Modeling the response of a tactile array using electrorheological fluids. *J. Phys. D Appl. Phys.* **37**, 794–803.

Klingenberg, D.J. (1993). Simulation of the dynamic response of electrorheological suspensions: demonstration of a relaxation mechanism. *J. Rheol.* **37**, 199.

Klingenberg, D.J. (1998). Polarization in electrorheological suspensions. *Mat. Res. Soc. Bull.* **23**, 30–34.

Klingenberg, D.J., Ulicny, J.C., Smith, A. (2005). Effects of body forces on electro- and magnetorheological fluids. *Appl. Phys. Lett.* **86**, 104101.

Klingenberg, D.J., Van Swol, F., Zukoski, C.F. (1989). Dynamic simulation of electrorheological suspensions. *J. Chem. Phys.* **91**, 7888.

Klingenberg, D.J., Zukoski, C.F. (1990). Studies of the steady shear behavior of electrorheological suspensions. *Langmuir* **6**, 15–26.

Ladyzhenskaya, O. (1969). *The Mathematical Theory of Viscous Incompressible Flow* (Gordon and Breach, New York).

Ladyzhenskaya, O., Solonnikov, V. (1976). Some problems of vector analysis and generalized formulation of boundary value problems for the Navier-Stokes equations. *Zap. Nauchn. Semin. Leningrad. Otdel. Mat. Inst. Steklov (LOMI)* **59**, 81–116 (in Russian).

LANDAU, L.D., LIFSHITZ, E.M. (1984). *Electrodynamics of Continuous Media* (Pergamon Press, Oxford).

LARSON, R.G. (1999). *The Structure and Rheology of Complex Fluids* (Oxford University Press, Oxford).

LEMAIRE, E., GRASSELLI, Y., BOSSIS, G. (1992). Field induced structure in magnetoand electro-rheological fluids. *J. Phys. II France* **2**, 359–369.

LIN, Y., CAO, Y. (2006). A new nonlinear Uzawa algorithm for generalized saddle point problems. *Appl. Math. Comput.* **175**, 1432–1454.

LIONS, J.L. (1969). *Quelques Méthodes De Résolution Des Problémes Aux Limites Non Linéaires* (Dunod, Paris).

LIONS, J.L., MAGENES, E. (1968). *Problèmes Aux Limites Non Homogènes* Volume I-III (Dunod, Paris).

LITVINOV, W.G. (1982). *Motion of Nonlinear Viscous Fluid* (Nauka, Moscow).

LITVINOV, W.G. (2000). *Optimization in Elliptic Problems with Applications to Mechanics of Deformable Bodies and Fluid Mechanics* (Birkhäuser, Basel).

LITVINOV, W.G. (2004). Model and problem on flow of a magnetorheological fluid. *Nonlinear Phenom. Complex Syst.* **7**, 332–346.

LITVINOV, W.G. (2007). Problem on stationary flow of electrorheological fluids at the generalized conditions of slip on the boundary. *Comm. Pure Appl. Anal.* **6**, 247–277.

LITVINOV, W.G., HOPPE, R.H.W. (2005). Coupled problems on stationary non-isothermal flow of electrorheological fluids. *Comm. Pure Appl. Anal.* **4**, 779–803.

LIU, Y., DAVIDSON, R., TAYLOR, P. (2005). Investigation of the touch sensitivity of ER fluid based tactile display. In: *Proc. of SPIE, Smart Structures and Materials: Smart Structures and Integrated Systems* **5764**, 92–99.

LORD COOPERATION. (1996). *VersaFlo Fluids Product Information. ER-100 Fluid Form PI101-ER100A* (Lord Cooperation, Cary, NC).

LOU, Z., ERVIN, R.D., FILISKO, F.E. (2001). Electrorheological fluids for aircraft flight control: feasibility of a position servomechanism. Report UMTRI-91–30, Transportation Research Institute University of Michigan, Ann Arbor.

LUKASZEWICZ, G. (1999). *Micropolar Fluids. Modeling and Simulation in Science, Engineering & Technology* (Birkhäuser, Basel).

MAKRIS, N. (1999). Rigidity-plasticity-viscosity: can electrorheological dampers protect base-isolated structures from near-source ground motions? *Earthquake Eng. Struct. Dyn.* **26**, 571–591.

MALEK, J., NECAS, J., RUZICKA, M. (1996). On the non-Newtonian incompressible fluids. *M3AS* **3**, 35–63.

MALEK, J., RAJAGOPAL, K. (2007). Incompressible rate type fluids with pressure and shear-rate dependent material moduli. *Nonlinear Anal. Real World Appl.* **8**, 156–164.

MALEK, J., RAJAGOPAL, K., RUZICKA, M. (1995). Existence and regularity of solutions and the stability of the rest state for fluids with shear dependent viscosity. *M3AS* **5**, 789–812.

MARSHALL, L., ZUKOSKI, C.F., GOODWIN, J.W. (1989). Effects of electric fields on the rheology of non-aqueous concentrated suspensions. *J. Chem. Soc. Faraday Trans.* **185**, 2785–2795.

MARTIN, J.E., ANDERSON, R.A. (1996). Chain model of electrorheology. *J. Chem. Phys.* **104**, 4814–4827.

MARTIN, J.E., ANDERSON, R.A., TIGGES, C.P. (1998a). Simulation of the athermal coarsening of composites structured by a uniaxial field. *J. Chem. Phys.* **108**, 3765–3787.

MARTIN, J.E., ODINEK, J., HALSEY, T.C., KAMIEN, R. (1998b). Structure and dynamics of electrorheological fluids. *Phys. Rev. E* **57**, 756.

MAVROIDIS, C. (2002). Development of advanced actuators using shape memory alloys and electrorheological fluids. *Res. Nondestr. Eval.* **14**, 1–32.

MAVROIDIS, C., BAR-COHEN, Y., BOUZIT, M. (2001). Haptic interfaces using electrorheological fluids. In: *Electroactive Polymer Actuators as Artificial Muscles: Reality, Potentials, and Challenges* (SPIE Opt. Engrg. Press, Bellingham, WA), pp. 567–594.

MELROSE, J.R. (1992). Brownian dynamics simulation of dipole suspensions under shear: the phase diagram. *Mol. Phys.* **76**, 635–660.

MELROSE, J.R., HAYES, D.M. (1993). Simulations of electrorheological and particle mixture suspensions: agglomerate and layer structures. *J. Chem. Phys.* **98**, 5873–5886.

MINTY, G.J. (1962). Monotone (nonlinear) operators in Hilbert space. *Duke Math. J.* **29**, 341–346.

MOHAMMADI, B., PIRONNEAU, O. (2001). *Applied Shape Optimization for Fluids* (Oxford University Press, Oxford).

MOKEEV, A., KOROBKO, E., VEDERNIKOVA, L. (1992). Structural viscosity of electrorheological fluids. *J. Non-Newton. Fluid Mech.* **42**, 213–230.

MONKMANN, G., BÖSE, H., ERMERT, H., BAUMANN, M., FREIMUTH, H., MEIER, A., EGERSDÖRFER, S., BRUHNS, O.T., RAJA, K. (2003a). Technologies for haptic systems in telemedicine. In: Nerlich, M., Schächinger, U. (eds.), *Integration of Health Telematics into Medical Practice* Volume 97. (IOS Press, Amsterdam), pp. 83–94.

MONKMANN, G., EGERSDÖRFER, S., MEIER, A., BÖSE, H., BAUMANN, M., ERMERT, H., KHALED, W., FREIMUTH, H. (2003b). Technologies for haptic displays in teleoperation. *Ind. Rob.* **6**, 525–530.

MOORE, R.E., CLOUD, M.J. (2007). *Computational Functional Analysis* (Horwood Publishing, Chichester).

OTSUBO, Y. (1997). Effect of electrode pattern on the column structure and yield stress of electrorheological fluids. *J. Colloid Interface Sci.* **190**, 466–471.

OTSUBO, Y., EDAMURA, K. (1998). Viscoeleasticity of a dielectric fluid in nonuniform electric fields generated by electrodes with flocked fabrics. *Rheol. Acta* **37**, 500–507.

OTSUBO, Y., EDAMURA, K. (1999). Electric effect on the rheology of insulating oils in electrodes with flocked fabrics. *Rheol. Acta* **38**, 137–144.

PARTHASARATHY, M., AHN, K.H., BELONGIA, B.M., KLINGENBERG, D.J. (1994). The role of suspension structure in the dynamic response of electrorheological suspensions. *Int. J. Mod. Phys. B* **8**, 2789.

PARTHASARATHY, M., KLINGENBERG, D.J. (1995a). A microstructural investigation of the nonlinear response of electrorheological suspensions I. Start-up of steady shear flow. *Rheol. Acta* **34**, 417–429.

PARTHASARATHY, M., KLINGENBERG, D.J. (1995b). A microstructural investigation of the nonlinear response of electrorheological suspensions II. Oscillatory shear flow. *Rheol. Acta* **34**, 430–439.

PARTHASARATHY, M., KLINGENBERG, D.J. (1996). Electrorheology: mechanisms and models. *Mat. Sci. Eng.* **R17**, 57–103.

PARTHASARATHY, M., KLINGENBERG, D.J. (1999). Large amplitude oscillatory shear of electrorheological suspensions. *J. Non-Newton. Fluid Mech.* **81**, 83–104.

PEEL, D.J., STANWAY, R., BULLOUGH, W.A. (1996). A generalised presentation of valve and clutch data for an ER fluid, and practical performance predictionmethodology. *Int. J. Mod. Phys.* **10**, 3103–3114.

PRIESTLEY, J. (1769). *The History and Present State of Electricity with Original Experiments* (Dodsley, London).

QI, Y., WEN, W. (2002). Influence of geometry of particles on electrorheological fluids. *J. Phys. D Appl. Phys.* **35**, 2231–2235.

QUARTERONI, A., VALLI, A. (1999). *Domain Decomposition Methods for Partial Differential Equations* (Oxford Science Publications, Oxford).

QUINKE, G. (1897). Die Klebrigkeit isolierender Flüssigkeiten im konstanten elektrischen Felde. *Ann. Phys. Chem.* **62**, 1–13.

RAJAGOPAL, K. (1996). An introduction to mixture theory. In: Galdi, G.P., Malek, J., Necas, J. (eds.), *Mathematical Theory in Fluid Mechanics* Pitman Research Notes in Mathematics, Volume 354 (Longman, Boston), pp. 86–113.

RAJAGOPAL, K., RUZICKA, M. (1996). On the modeling of electrorheological materials. *Mech. Res. Commun.* **23**, 401–407.

RAJAGOPAL, K., RUZICKA, M. (2001). Mathematical modeling of electrorheological materials. *Continuum Mech. Thermodyn.* **13**, 59–78.

RAJAGOPAL, K., TRUESDELL, C. (2000). *An Introduction to the Mechanics of Fluids* (Birkhüser, Basel).

RAJAGOPAL, K., WINEMAN, A. (1992). Flow of electrorheological materials. *Acta Mech.* **91**, 57–75.

RAJAGOPAL, K., WINEMAN, A. (1995). On constitutive equations for electrorheological materials. *Continuum Mech. Thermodyn* **7**, 1–22.

RAJAGOPAL, K., YALAMANCHILI, R.C., WINEMAN, A. (1994). Modeling electrorheological materials through mixture theory. *Int. J. Eng. Sci.* **32**, 481–500.

RHEE, E.J., PARK, M.K., YAMANE, R., OSHIMA, S. (2003). A study on the relation between flow characteristics and cluster formation of electrorheological fluids using visualization. *Exp. Fluids* **34**, 316–323.

RUSTEN, T., WINTHER, R. (1992). A preconditioned iterative method for saddlepoint problems. *SIAM J. Matrix Anal. Appl.* **13**, 887–904.

RUZICKA, M. (2000). *Electrorheological Fluids: Modeling and Mathematical Theory, Lecture Notes in Mathematics 1748* (Springer, Berlin-Heidelberg-New York).

SADIKI, A., BALAN, C. (2003). Rate-type model for electro-rheological material behavior cocsistent with extended thermodynamics: application to a steady viscometric flow. *Proc. Appl. Math. Mech.* **2**, 174–175.

SEE, H. (1999). Advances in modeling the mechanisms and rheology of electrorheological fluids. *Korea Aust. Rheol. J.* **11**, 169–195.

SEE, H. (2000). Constitutive equation for electrorheological fluids based on the chain model. *J. Phys. D Appl. Phys.* **33**, 1625–1633.

SHULMAN, Z.P., NOSOV, B.M. (1985). Rotation of nonconducting bodies in electrorheological suspensions. *Nauka i Technika, Minsk* (in Russian).

SIGNIER, D.A., DE KEE, D., CHHABRA, R.P. (1999). *Advances in the Flow and Reheology of Non-Newtonian Fluids* (Elsevier, Amsterdam).

SIMS, N.D., STANWAY, R., PEEL, D.J., BULLOUGH, W.A., JOHNSON, A.R. (1999). Controllable viscous damping: an experimental study of an electrorheological longstroke damper under proportional feedback control. *Smart Mater. Struct.* **8**, 601–615.

STANGROOM, J.E. (1977). Electric field responsive fluids, U.S. Patent 4,033,892.

STANGROOM, J.E. (1983). Electrorheological fluids. *Phys. Technol.* **14**, 290–296.

STANWAY, R., SPROSTON, J.L., EL-WAHED, A.K. (1996). Application of electrorheological fluids in vibration control: a survey. *Smart Mater. Struct.* **5**, 464–482.

STANWAY, R., SPROSTON, J.L., STEVENS, N.G. (1987). Non-linear modelling of an electro-rheological vibration-damper. *J. Electrostat.* **20**, 167–184.

STRIKWERDA, J.C. (2004). *Finite Difference Schemes and Partial Differential Equations*, Second Edition (SIAM, Philadelphia).

TABATABAI, S. (1993). Radiative heat transfer in electrorheological fluids, Report Michigan State University, East Lansing.

TAKASHIMA, S., SCHWAN, H.P. (1985). Alignment of microscopic particles in electric fields and its biological implications. *Biophys. J.* **47**, 513–518.

TAM, W.Y., YI, G.H., WEN, W., MA, H., LOY, M.M.T., SHENG, P. (1997). New electrorheological fluid: theory and experiment. *Phys. Rev. Lett.* **78**, 2987–2990.

TAO, R., ROY, G.D. (eds.) (1995). *Electrorheological Fluids: Mechanisms, Properties, Technology, and Applications* (World Scientific, Singapore).

TAO, R., SUN, J.M. (1991a). Ground state of electrorheological fluids from Monte Carlo simulations. *Phys. Rev. Lett.* **44**, 44–47.

TAO, R., SUN, J.M. (1991b). Three-dimensional structure of induced electrorheological solid. *Phys. Rev. Lett.* **67**, 398–401.

TEMAM, R. (1979). *Navier-Stokes Equations* (North-Holland, Amsterdam).

THOMAS, J.W. (1995). *Numerical Partial Differential Equations. Finite Difference Methods* (Springer, Berlin-Heidelberg-New York).

THOMASSET, F. (1981). *Implementation of Finite Element Methods for Navier-Stokes Equations* (Springer, Berlin-Heidelberg-New York).

TOSELLI, A., WIDLUND, O. (2005). *Domain Decomposition Methods - Algorithms and Theory* (Springer, Berlin-Heidelberg-New York).

TRUESDELL, C., NOLL, W. (1965). *The Non-Linear Field Theories of Mechanics. Handbuch Der Physik* Volume III/3 (Springer, Berlin-Heidelberg-New York).

TRUESDELL, C., TOUPIN, R.A. (1960). *The Classical Field Theories*. Handbuch Der Physik Volume III/1 (Springer, Berlin-Heidelberg-New York).

TUREK, S. (1999). *Efficient Solvers for Incompressible Flow Problems* (Springer, Berlin-Heidelberg-New York).

UGAZ, V.M., MAJORS, P.D., MIKSAD, R.W. (1994). Measurements of electrorheological fluid flow through a rectangular channel using nuclear magnetic resonance imaging, American Society for Mechanical Engineers (ASME), pp. 15–27, FED205/AMD 190.

VAINBERG, M.M. (1964). *Variational Methods for the Study of Nonlinear Operators* (Holden Day, San Francisco).

VALADIER, M. (1994). A course on Young measures. *Rend. Istit. Mat. Univ. Trieste* **26**, 349–394.

VERNESCU, B. (2002). Multiscale analysis of electrorheological fluids. *Int. J. Mod. Phys. B* **16** (17 & 18), 2643–2648.

VISIK, I.M. (1962). Solubility of boundary-value problems for quasi-linear parabolic equations of higher orders. *Mat. Sb. (N.S.)* **59**, 289–325.

VON PFEIL, K., GRAHAM, M.D., KLINGENBERG, D.J., MORRIS, J.F. (2002). Pattern formation in flowing electrorheological fluids. *Phys. Rev. Lett.* **88**, 188301.

VON PFEIL, K., GRAHAM, M.D., KLINGENBERG, D.J., MORRIS, J.F. (2003). Structure evolution in electrorheological and magnetorheological suspensions from a continuum perspective. *J. Appl. Phys.* **93**, 5769–5779.

VON PFEIL, K., KLINGENBERG, D.J. (2004). Nonlocal electrostatics in heterogeneous suspensions using a point-dipole model. *J. Appl. Phys.* **96**, 5341–5348.

VOROBEVA, T.A., VLODAVETS, I.N., ZUBOV, P.I. (1969). The size distribution of oriented aggregates formed in suspension with the application of an alternating electric field. *Colloid J. USSR* **31**, 533–537.

WANG, B., XIAO, Z. (2003). A general constitutive equation of an ER suspension based on the internal variable theory. *Acta Mech.* **163**, 99–120.

WEN, W., HUANG, X., YANG, S., LU, K., SHENG, P. (2003). The giant electrorheological effect in suspensions of nanoparticles. *Nat. Mater.* **2**, 727–730.

WERELEY, N.M., SNYDER, R., KRISHNAN, R., SIEG, T. (2001). Helicopter damping. In: Braun, S.G., Ewins, D.J., Rao, S.S. (eds.), *Encyclopedia of Vibration* (Academic Press, London), pp. 629–642.

WEYENBERG, T.R., PIALET, J.W., PETEK, N.K. (1996). The development of electrorheological fluids for an automotive semi-active suspension system. *Int. J. Mod. Phys.* **10**, 3201–3209.

WHITTLE, M. (1990). Computer simulation of an electrorheological fluid. *J. Non-Newton. Fluid Mech.* **37**, 233–263.

WHITTLE, M., ATKIN, R.J., BULLOUGH, W.A. (1995). Fluid dynamic limitations on the performance of an electrorheological clutch. *J. Non-Newton. Fluid Mech.* **57**, 61–81.

WHITTLE, M., ATKIN, R.J., BULLOUGH, W.A. (1999). Dynamics of a radial electrorheological clutch. *Int. J. Mod. Phys. B* **13**, 2119–1226.

WHITTLE, M., BULLOUGH, W.A. (1992). The structure of smart fluids. *Nature* **358**, 373.

WHITTLE, M., FIROOZIAN, R., PEEL, D.J., BULLOUGH, W.A. (1995). Electrorheological relaxation times derived from pressure response experiments in the flow mode. *J. Non-Newton. Fluid Mech.* **57**, 1–25.

WINSLOW, W.M. (1947). Translating electrical impulses into mechanical force, U.S. Patent 2,417,850.

WINSLOW, W.M. (1949). Induced fibration of suspensions. *J. Appl. Phys.* **20**, 1137–1140.

WINSLOW, W.M. (1962). Field responsive force transmitting compositions, U.S. Patent 3,047,507.

WITTUM, G. (1989). Multigrid methods for the Stokes and the Navier-Stokes equations. *Numer. Math.* **54**, 543–564.

XU, Y.L., QU, W.L., KO, J.M. (2000). Seismic response control of frame structures using magnetorheological/electrorheological dampers. *Earthquake Eng. Struct. Dyn.* **29**, 557–575.

YU, K.W., WAN, J.T.K. (2000). Interparticle force in polydisperse electrorheological fluids. *Comput. Phys. Commun.* **129**, 177–184.

ZEIDLER, E. (1990). *Nonlinear Monotone Operators* (Springer, Berlin-Heidelberg-New York).

ZHAO, X., GAO, D. (2001). Structure evolution in Poiseuille flow of electrorheological fluids. *J. Phys. D Appl. Phys.* **34**, 2926–2931.

ZHAO, X., GAO, X.Y., GAO, D.J. (2002). Evolution of chain structure of electrorheological fluids in flow model. *Int. J. Mod. Phys. B* **16** (17 & 18), 2697–2703.

ZHAO, X., LIU, S., TANG, H., YIN, J.B., LUO, C.R. (2005). A new kind of self-coupled electrorheological damper and its vibration character. *J. Intell. Mater. Syst. Struct.* **16**, 57–65.

ZHIZKIN, G.V. (1986). Nonisothermal Couette flow of a non-Newtonian fluid under a pressure gradient. *J. Appl. Mech. Tech. Phys.* **27**, 218–220.

Index

C

D

E

FIG. 2.1 Rock cutting removal.

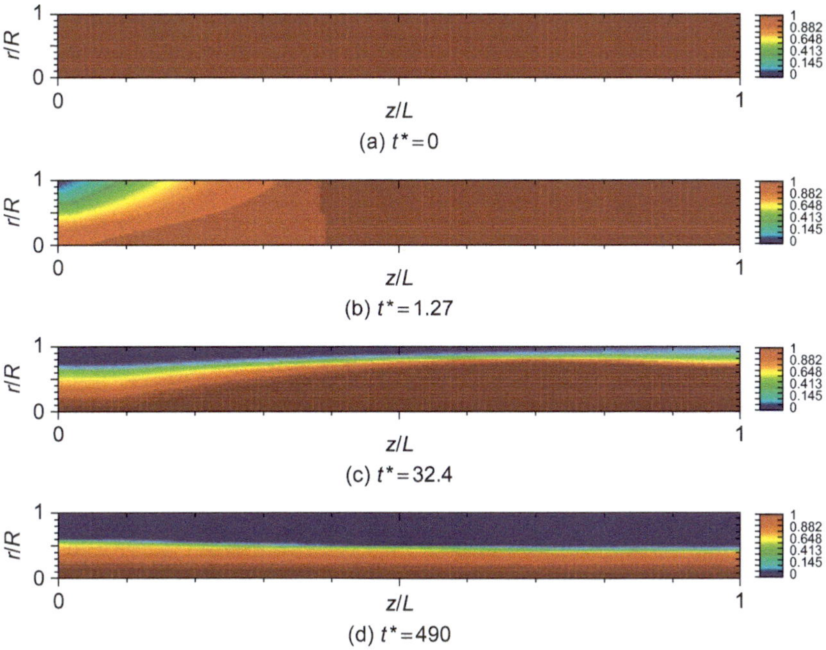

FIG. 26.4 Time evolution of the structure parameter for a successful restart ($\mathcal{B}n^*_{max} = 1.025$, $\mathcal{B}n^*_0 = 0.1$, $\mathcal{B}n = 0.925$, $\chi^* = 4 \times 10^{-2}$, $\mathcal{B}d = 0.1$, and $\mathcal{R}e = 0.07$).

Kolibri 360 GmbH
Ferdinand-Jühlke-Straße 7
99095 Erfurt, Deutschland
produktsicherheit@kolibri360.de